# Stephenson Conference Research for Railways 2015

London, United Kingdom
21 - 23 April 2015

ISBN: 978-1-5108-5583-0

**Printed from e-media with permission by:**

Curran Associates, Inc.
57 Morehouse Lane
Red Hook, NY 12571

CURRAN ASSOCIATES INC. proceedings.com

**Some format issues inherent in the e-media version may also appear in this print version.**



Printed by Curran Associates, Inc. (2018)

For permission requests, please contact Institution of Mechanical Engineers (IMechE) at the address below.

Institution of Mechanical Engineers (IMechE)
One Birdcage Walk
London, SW1H 9JJ
United Kingdom

library@Imeche.org
www.imeche.org

**Additional copies of this publication are available from:**

Curran Associates, Inc.
57 Morehouse Lane
Red Hook, NY 12571 USA
Phone: 845-758-0400
Fax: 845-758-2633
Email: curran@proceedings.com
Web: www.proceedings.com

# CONTENTS

## CASE STUDIES



## AUTHOR INDEX

**Stephenson Conference Scientific Committee**

| | |
|---|---|
| Stefano Beretta | POLIMI, Italy |
| Mats Berg | KTH, Sweden |
| Andrew Broadbent | RSSB, UK |
| Mark Burstow | Network Rail, UK |
| Daniel Cadet | Alstom, France |
| Sylvia Churba | GHD, UK |
| Andy Doherty | Network Rail, UK |
| Bridget Eickhoff | RSSB, UK |
| Mike Forde | University of Edinburgh, UK |
| Pierre-Etienne Gautier | SYSTRA, France |
| Roger Goodall | Loughborough University, UK |
| Richard Gostling | Past Railway Division Chair, IMechE, UK |
| Andreas Haigermoser | Siemens, Austria |
| Stuart Hillmansen | University of Birmingham, UK |
| Francis How | RIA, UK |
| Simon Iwnicki | University of Huddersfield, UK |
| David Johnson | DGauge, UK |
| Terry Johnson | Independent, UK |
| Rick Jones | Acoustics Consultant, UK |
| Andrew King | Alstom, UK |
| David Knights | RSSB, UK |
| Alan Lawton | ALEC Ltd, UK |
| Jon Leigh | Faiveley, UK |
| Brian Marquis | Volpe, USA |
| Manuel Pereira | IST Lisbon, Portugal |
| William Powrie | University of Southampton, UK |
| Clive Roberts | University of Birmingham, UK |
| Rebeka Sellick | Interfleet, UK |
| Helen Simpson | Porterbrook, UK |
| John Tunna | TTCI, USA |
| Alan Wickens | Loughborough University, UK |
| Peter Woodward | Herriot Watt University, UK |
| Tom Worsley | ITS Leeds, UK |

**Organising Committee**

| | |
|---|---|
| Francis How | RIA, UK |
| Simon Iwnicki | University of Huddersfield, UK |
| Andrew Broadbent | RSSB, UK |
| Richard Gostling | Past Chairman, Railway Division, UK |

# From research to rulemaking in the Federal Railroad Administration

**J Tunna**
Federal Railroad Administration, Office of Research and Development,
US Department of Transportation, USA

**ABSTRACT**

The Federal Railroad Administration's (FRA) research and development program provides the engineering and scientific foundation for rail rulemaking and enforcement in the United States. This paper describes how the program is funded and executed. Examples of projects in the track, rolling stock, train control and human factors disciplines are used to illustrate strategic direction, project selection, stakeholder engagement, technology transfer, performance measurement and evaluation. Changes in regulations from being prescriptive to performance-based are discussed, as is the shift to risk reduction and system safety management approaches.

This paper gives the reader a general understanding of rail research and development in the U.S. It is intended to encourage contacts between rail R&D practitioners for the exchange of information and to identify opportunities for future collaboration.

## 1 BACKGROUND

FRA's R&D program began in the 1970s with the High Speed Ground Transportation Act. This saw the creation of the Transportation Technology Center in Pueblo, Colorado and the development and testing of novel, high-speed trains.

Today's focus for the program is to improve railroad safety. Figure 1 shows the recent trends in safety performance for the U.S. railroad industry. The improvements in safety are due, in part, to outputs from FRA's R&D program. This paper provides some examples of those contributions.

The safety information presented in Figure 1 and elsewhere in this paper comes from FRA's accident and incident database. This database is available on-line at http://safetydata.fra.dot.gov/officeofsafety.

FRA's R&D program is funded by the U.S. Congress with annual appropriations of typically $35 million. The funding is broken down into five program areas: Track, Rolling Stock, Train Control and Communications, Human Factors and Railroad System Issues.

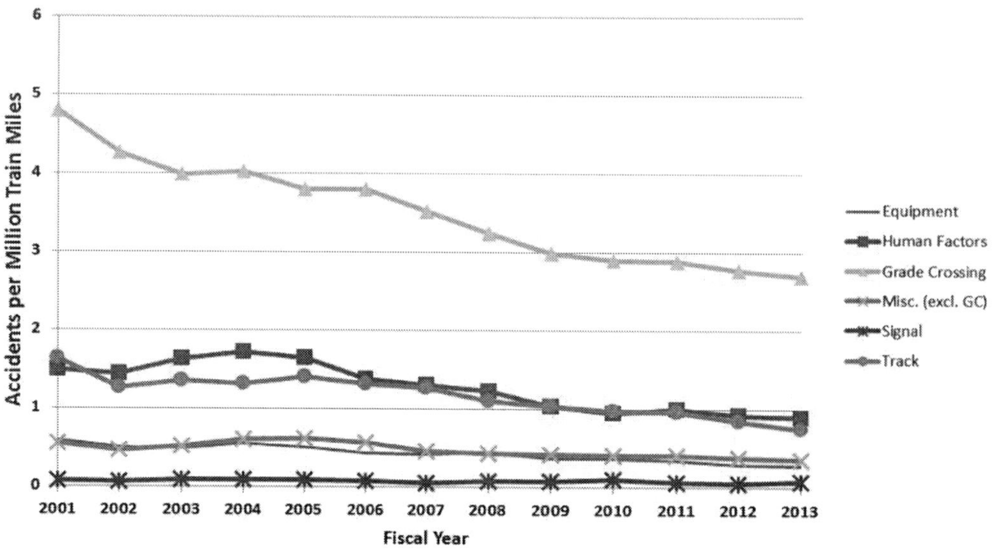


**Figure 1. Safety Trends in the U.S. Railroad Industry**

Figure 2 shows the level of funding plotted against accident rate for four accident causes. The size of the circles is proportional to the harm in terms of fatalities and injuries for each accident cause.

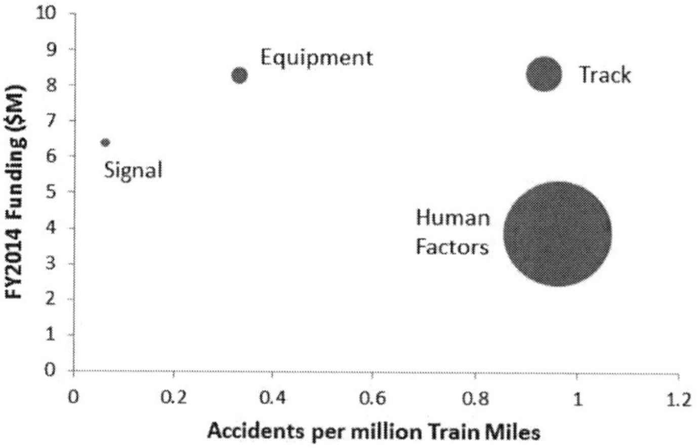


**Figure 2. R&D Funding Levels and Accident Rates**

Grade crossing and trespass accidents result in 90 percent of railroad fatalities in the U.S. The R&D budget includes $1.6 million of annual funding to improve safety in this area.

The budget also includes funding for Railroad Systems Issues, which cover safety risk analysis and R&D project prioritization and evaluation. It also funds projects aimed at improving environmental sustainability and reducing energy usage.

FRA conducts its R&D through other government agencies, railroad engineering companies, and universities. The Volpe Center, which is part of the U.S. Department of Transportation, is a major supplier of technical support to FRA's R&D program. There is good synergy between the FRA's R&D program and the Strategic Research Initiative program funded by the Association American Railroads (AAR). AAR's subsidiary, the Transportation Technology Center Inc., manages FRA's Transportation Technology Center and performs R&D at the facility for both the AAR and FRA. Ensco has the contract for maintenance and operation of FRA's fleet of track inspection vehicles, which includes two cars and a road-rail vehicle used mainly for R&D.

FRA publishes 4-page research result briefs and full length technical reports. These are available at http://www.fra.dot.gov/eLib/find and from the National Transportation Library at http://ntl.bts.gov/.

## 2 RULEMAKING AND ENFORCEMENT

FRA's Office of Railroad Safety creates and enforces safety regulations for the U.S. railroad industry. It has 14 divisions that serve as technical experts on matters of railroad safety, provide technical assistance to field personnel, and aid in the development of regulations. Staff includes 400 Federal safety inspectors who operate out of eight regional offices. Each regional administrator is supported by two deputy regional administrators, chief inspectors, supervisory specialists, grade crossing safety managers and safety inspectors for five safety disciplines focusing on compliance and enforcement in: hazardous materials, motive power and equipment, operating practices, signal and train control, and track.

The goal of FRA regulations, as with other U.S. Federal regulations, is to improve public health, welfare and safety, and the environment while promoting economic growth, innovation, competitiveness and job creation. Regulations are often initiated by Acts of Congress (e.g. the Rail Safety and Improvement Act of 2008 (RSIA)). They can also be driven by compelling safety data, and they can be created in response to public petitions.

Regulations are supported by analysis of the costs and benefits of implementation. Safety benefits can be converted to financial values using the Value of a Statistical Life and an Abbreviated Injury Scale published by the Department of Transportation. Costs of travel delays are also calculated using published guidance (http://www.dot.gov/office-policy/transportation-policy/).

In 1996, FRA established the Railroad Safety Advisory Committee (RSAC) to develop new regulatory standards, through a collaborative process, with all segments of the rail community working together to fashion mutually satisfactory solutions to safety regulatory issues. RSAC working groups are important forums for presenting and discussing FRA's research results. This is the principal means by which those results inform regulations.

In recent years there has been a growing emphasis on safety risk reduction. RSIA created a risk reduction program in which each railroad carrier will systematically evaluate and manage safety risks on its system. For passenger rail operators, FRA is proposing a similar system safety approach. This represents a shift from prescriptive regulations to a mode in which the regulated body says how it plans to achieve safety and the regulator approves and audits the plans.

Research outcomes can be achieved through regulation or voluntary adoption of new technologies or operating practices to reduce safety risk. Several examples follow.

## 3 PROGRAM EXAMPLES

### 3.1 Tank Car Structural Integrity

There was an intensive period of research and rulemaking in the 1970s aimed at improving tank car safety. Puncture of the head of tanks cars in accidents was a common failure mode. Research was conducted into the appropriate thickness and material properties for head shields to increase the tank car's structural integrity. Figure 3 shows the result of an impact test performed at FRA's Transportation Technology Center as part of this research. The impactor simulates a coupler on an adjacent car in an accident.

**Figure 3. Tank Car Head Shield Test in the 1970s**

The results of the research were published in 36 reports between 1970 and 1980 (e.g. (1) and (2)). The results were used to inform three new regulations published in 1974, 1977 and 1981. The regulations required head shields and a new design of coupler to be fitted to certain types of tank cars. The new coupler has shelves that prevent the coupler from disengaging in accidents. As the industry complied with the new regulations, the numbers of accidents with release of hazardous materials through the heads of tank cars reduced dramatically as shown in Figure 4.

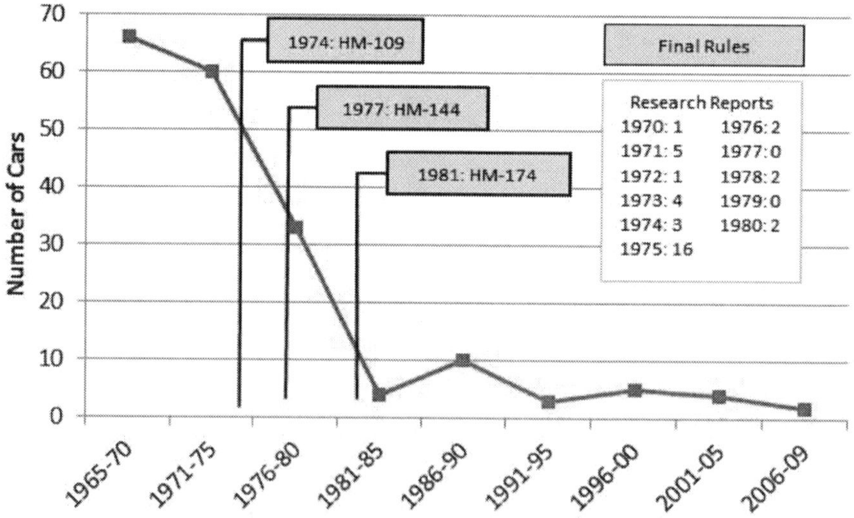


**Figure 4. DOT-105,112 and 114 cars with lading loss through heads in accidents – 5 year periods**

A more recent example of these safety benefits occurred at a derailment on January 14, 2008 near Springfield, IL. Figure 5 shows the aftermath of the accident. The tank car was carrying ethylene oxide, which is classified as a poisonous by inhalation hazardous (PIH) material. The head of the tank car was protected by a

17.46 mm (11/16 inch) thick shield as required by the regulations. The indentation caused by the coupler on the adjacent car (very similar to that in the earlier testing shown in Figure 3) did not result in rupture and release of hazardous material.

**Figure 5. Post-derailment at Springfield, IL in January 2008**

### 3.2 Safety Culture

In 1999, the FRA reported (3) on the results of a focus group meeting with North American Railroads to discuss the general issue of operating rule compliance. The discussions revealed the prevalence of a military style, command and control culture. There was a punitive approach to enforcing compliance to rules and a blame-based approach to accident investigations. Labor-management relations were negative and impeded safety communications. This created a desire to improve safety culture in the industry.

Table 1 lists several safety culture improvement pilot programs that were tried and evaluated. At the time, these programs were not required by regulation. The recent regulatory shift to risk reduction and system safety means programs such as those shown in Table 1 are possible steps towards compliance.

**Table 1. Safety culture pilot programs**

<table>
<tr><th colspan="2">Approach</th><th>Carriers</th><th>Start</th></tr>
<tr><td colspan="2">Participative Safety Rules Revision</td><td>ACBL, CSXT, KCS, CN-IC</td><td>1999</td></tr>
<tr><td colspan="2">ISROP: Investigation of Safety Related Occurrences Protocol</td><td>Canadian Pacific</td><td>2003</td></tr>
<tr><td rowspan="3">Clear Signal for Action (CSA)</td><td>EAGLES: Employee Alliance for Greater Levels of Excellence in Safety</td><td>Amtrak</td><td>2001</td></tr>
<tr><td>CAB: Changing At-risk Behavior</td><td>Union Pacific</td><td>2005</td></tr>
<tr><td>STEEL: Safety Through Employees Exercising Leadership</td><td>Union Pacific</td><td>2006</td></tr>
<tr><td colspan="2">C3RS: Confidential Close Call Reporting System</td><td>Union Pacific<br>Canadian Pacific<br>New Jersey Transit<br>Amtrak</td><td>2007<br>2008<br>2009<br>2010</td></tr>
</table>

The most recent approach to improving safety culture, the Confidential Close Call Reporting System (C3RS), is founded on the notion that a close call is an opportunity to improve safety practices in a situation or incident that has a potential for more serious consequences. C3RS allows railroad workers who observe noteworthy events to submit a confidential report. Peer review teams, consisting of labor, railroad management and FRA, then analyze groups of events to identify safety hazards and develop solutions to the threats.

Safety improvements from these pilot projects have been evaluated by FRA's Office of R&D. At one pilot site a safety culture survey was performed at the outset and after two years. Interviews were also held with management and labor. Table 2 summarizes the results and shows significant improvements (4).

**Table 2. Safety culture improvements at mid-term from a C3RS pilot**

| | Significant Improvement | |
|---|---|---|
| | Managers | Labor |
| Labor-Management Relations | * | * |
| Organizational Fairness During Change | * | |
| Supervisor Fairness | * | * |
| Supervisor-Employee Relationship | | * |
| Management Safety | * | * |
| Raising Concerns with Supervisors | | * |
| Work Safety Priorities | | * |
| Helping Behavior | | * |
| Coworker Safety | * | |

The survey also showed that employees became more engaged in the C3RS program. At midterm, 74 percent of them were willing to submit a C3RS report versus 58 percent at the beginning. Both labor and management were more convinced at midterm that C3RS was effective and that the identities of those who reported close-call events remained confidential. A very positive finding was that disciplinary cases had reduced by 90 percent over the first two years of the C3RS pilot.

At the same pilot site the effect of C3RS on the number of cars moved between derailment incidents was studied. A 31 percent improvement was found after two years into the C3RS pilot. In comparison, at two other similar sites where C3RS had not been implemented, there was no statistically significant change in derailment rates.

A similar effect was seen at a second C3RS pilot site. In this case derailments due to running through switches set in the wrong position were reduced by 50 percent (5). Prior to C3RS a switch run-through was a rule violation subject to disciplinary action. After C3RS, analysis of run-through switch incidents identified multiple factors and corrective actions to address the problems.

### 3.3 Rail Joint Bar Inspection

On Saturday, May 27th, 2000 a Union Pacific freight train derailed in Eunice, Louisiana. This derailment resulted in a release of hazardous materials with explosions and fires that required the evacuation of more than 3,500 people from the surrounding affected area. There was $35 million in estimated equipment and infrastructure damage. The National Transportation Safety Board (NTSB) determined that the probable cause of the derailment was the failure of a set of defective joint bars that were not found during required track inspections.

A failure of both joint bars can result in the rail ends moving relative to each other and provide a discontinuous running surface that may cause train wheels to derail. From the years 2000 to 2011, 152 derailments were attributed to joint bar failures.

Traditionally, cracks and breaks in joint bars are detected by inspectors walking along the track. The cracks being sought after can barely be seen with the naked eye, and some are easily missed. The manual, visual inspection is difficult, slow, subjective, labor intensive and inconsistent. As a result, the FRA decided to investigate other methods to locate cracked joint bars using advanced machine-vision-based technologies.

The chosen technology has cameras on-board a rail vehicle to capture images that are then evaluated to locate cracks and other defects (6). The main advantage of this approach is that the cameras can be set up to have the ideal viewing position and magnification, proper lighting, and consistent quality control evaluation processes. The approach has some disadvantages including the inability to find internal cracks that are not visible on the surface and the potential for false positive notifications.

Figure 6 shows an example of the production system that records at speeds up to 60 mph. Figure 7 shows an image taken by the system with the crack highlighted using the automated detection algorithms.

**Figure 6. Production joint-bar inspection system fitted to a rail car**

**Figure 7. Image of broken joint bar detected by the joint-bar inspection system**

During the many field tests of the system, several railroads showed interest in using the system before its final completion. FRA's contractor built several systems and provided an inspection service for the railroads. Table 3 shows the number of joint bars inspected between January 2008 and July 2011. The system found many defects, including fifty double (both joint bars) center breaks, which was same type of defect that caused the freight train derailment in Eunice, Louisiana.

**Table 3. Automated joint bar inspections from January 2008 to July 2011**

| |
|---|
| 19,105 miles surveyed<br>• 7,205 CWR<br>• 11,900 Jointed<br><br>~2,600,000 joints evaluated (5,200,000 joint bars)<br><br>17,070 defective joints found including:<br>• 5,640 center cracks<br>• 1,600 center breaks<br>• 7,400 quarter cracks and breaks<br>• 150 double center cracks<br>• 50 double center breaks<br>• 2,230 FRA classified bolt defects |

Figure 8 shows the history of accidents related to joint bars between 2000 and 2011. The number of accidents increased between the years 2000 and 2006, and suddenly decreased from 2007 to 2011. Several activities contributed to this decrease including a Railroad Safety Advisory Committee (RSAC) rule development and publication, the railroad's increased awareness of the problem and their incorporation of change in their processes, and the implementation of the joint bar inspection system technology. It is difficult to isolate which activity had the most significant impact on the reduction in accidents, but the turning point occurred when the new inspection technology was implemented by the railroads.

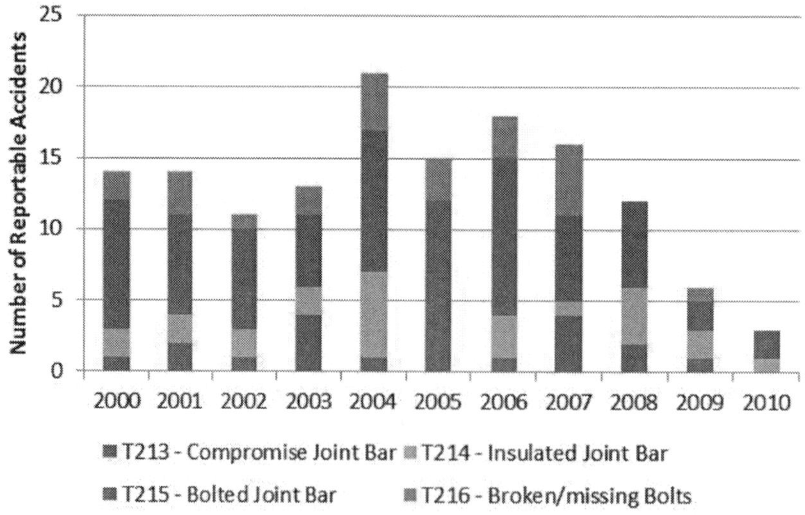


**Figure 8. History of accidents related to joint bars**

## 3.4 Employee-in-Charge Portable Remote Terminal

In response to several fatal accidents between 2002 and 2008, the U.S. Congress passed the Rail Safety and Improvement Act of 2008 (RSIA). Among the requirements was the implementation of positive train control (PTC) on certain routes by the end of 2015. In addition to preventing train-to-train collisions and over-speed derailments, PTC is required to prevent unauthorized train incursions into established work zones. FRA, in partnership with a railroad, has developed a portable remote terminal (PRT), used by the employee-in-charge of the work zone, that meets this requirement.

Current operating practice is for the engineer to stop the train before a track work zone and contact the employee-in-charge of the work zone for instructions. The employee-in-charge gives permission for the train to enter the work zone with instructions on speed and any stopping locations.

With PTC the train will be forced to stop before the work zone. The employee in charge will then use the PRT to send instructions on speed restrictions and work limits to the control center. The control center automatically sends the instructions to the PTC computer on-board the train for the engineer to follow.

Figure 9 shows an example of the PRT user interface after the employee-in-charge has entered instructions to proceed at a reduced speed and then stop. The text in the box on the right is generated from selections made by the employee-in-charge in the center of the screen. The format of the text is compatible with that currently used on manual forms.

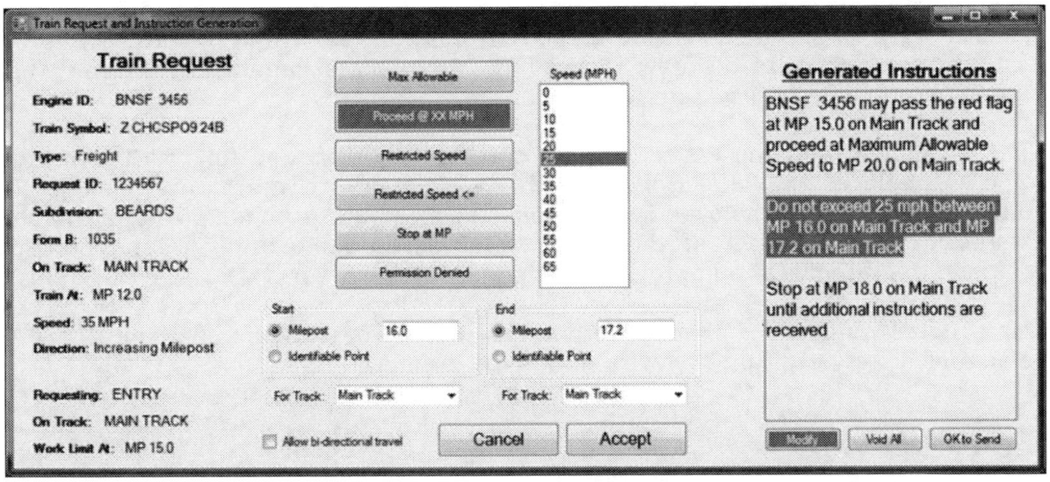


**Figure 9. Employee-in-charge portable remote terminal screenshot**

The PRT clearly has safety benefits over the current procedures. It avoids the possibility of error through verbal communications. PTC enforces the work limits and speed restrictions. The employee-in-charge can alter the instructions as the situation in the work zone changes. The PRT must also meet demanding standards for safety assurance.

The safety assurance of the PRT must follow the requirements in 49 Code of Federal Regulations (CFR) Part 238 Subpart I. This regulation invites the application of the IEEE-1483 specification for the "Verification of Vital Functions in Processor-Based Systems Used in Rail Transit Control." The PRT was analyzed according to these requirements and the analysis was integrated with that conducted by one of the railroads for approval of its PTC implementation plan.

## 4 CONCLUSIONS

Safety is the number one goal of the U.S. Department of Transportation and the main driver of the Federal Railroad Administration's R&D program. Outputs from the R&D program help create and enforce regulations, which lead to improved safety. They can also produce direct safety benefits when new technology is developed and implemented, often in partnership with the railroad industry. This paper includes examples of these different approaches.

Decades of research and testing have led to improvements in the performance of tank cars in accidents. Research results have been used to inform new regulations that the industry has followed. The benefits take several years to appear because the new regulations often apply exclusively to new tank cars, and there is a period during which older designs are phased out.

Modern regulations tend to be based on system safety and risk reduction. They rely on the railroads having a sound safety culture. The Confidential Close Call Reporting System is being adopted by several railroads as a way to improve safety culture. It began as an R&D project with a small number of pilot sites. Evaluations of these pilots showed convincing safety benefits.

The joint bar inspection system that was developed under FRA's R&D program was implemented by the railroads without the need for regulation. Public funds were used to develop and demonstrate a successful prototype. Then private funds were used to commercialize the new technology. Immediate benefits were seen when the joint bar inspection system was used.

The legislation requiring Positive Train Control drove the need for new technology. One example of the FRA working with the industry to meet this need is the development of the employee in charge portable remote terminal. The partnership started from the outset of the project and ensured a compliant and practical solution that promises to improve the safety of track workers.

More detail on these and many other FRA R&D projects can be found in FRA's electronic library at http://www.fra.dot.gov/eLib/find.

## REFERENCE LIST


1) Hohenemser, K. H., et al (1975), "Computer Simulation of Tank Car Head Puncture Mechanisms", Federal Railroad Administration, DOT/FRA/ORD/75-23
2) Peters, D. A., et al (1980), "Tank Car Head Puncture Mechanisms", Federal Railroad Administration, DOT/FRA/ORD/76-269
3) Coplen, M. (1999), "Compliance with Railroad Operating Rules and Corporate Culture Influences - Results of a Focus Group and Structured Interviews", Federal Railroad Administration, DOT/FRA/ORD/99-09
4) Ranney, J. & Raslear, T. (2012), "Derailment Decrease at a C3RS Site at Midterm", Federal Railroad Administration, RR 12-04
5) Ranney, J. & Raslear, T. (2013), "Another C3RS Site Improves Safety at Midterm", Federal Railroad Administration, RR 13-49
6) Gilbert-Serra, X. et al (2007), "A Machine Vision System for Automated Joint Bar Inspection from a Moving Rail Vehicle", JRCICE2007-40039, Proceedings of ASME/IEEE Joint Rail Conference & Internal Combustion Engine Spring Technical Conference

# Video measurement techniques for understanding wheel-induced lateral forces and track component deflections

**P Waterfall[1,2], B Temple[3], C Hardwick[3], J Sharam[1], A Plumb[1]**
[1] Imetrum Ltd, UK
[2] University of Bristol, UK
[3] LB Foster Rail Technologies, UK

## 1. INTRODUCTION

Digital Image Correlation (DIC) and Video Extensometry (VE) are growing in acceptance as techniques for structural monitoring and provide a form of Video Gauge. The potential of these techniques for understanding train / track interaction was recognised by Darren Bowness [1] although there has not been much further work on the technique published by Southampton University or others.

This author's doctoral research has developed and tested VE outdoors, including benchmarking of a system for monitoring track displacement under train loading. A snapshot of this work was presented in 2011 [2], and in other published work with a focus on bridges [3] & [4].

A video gauge has the potential to be a useful tool for assessing a number of important parameters for the rail and trackbed engineer including:

- Dynamic deflection for ride quality (and subsequent speed limits) - see section 4.
- Detailed investigation of track behaviour and voiding at switches and crossings [5].
- Track monitoring during nearby construction works -case study produced by Railassist BV.
- Track stiffness determination - see section 3.
- Determination of lateral forces under load - see section 6.

All of these are possible by measuring deflection, rotation and strain from a point of safety, several metres from track. Building on the concepts proposed by Bowness, and work at Bristol University on VE from the late 1980s [6] & [7], the innovation presented is to (1) measure multiple points simultaneously (2) outside in (3) 3-dimensions, using cameras running at over 100Hz from a position of safety.

We outline the academic background and current technical advances and challenges inherent in using a video gauge on track, using a selection of the case studies generated over the past four years. The implementation for the modelling of calculating 3D forces is still a work in progress and will be presented in future papers, but the basic resolution of the system, and therefore the potential for being able to generate these measurements, is presented here.

## 1.1. Background

### *1.1.1. Rail Monitoring*

There are many aspects of railway track design and maintenance that are important to get right, which historically have been checked on a regular basis by track walkers. Since budget realities and the requirement for centralised information capture and analysis removed the opportunity to employ such locally specialised people, track monitoring has been attempted in other ways. A wide range of techniques have been used to monitor track, including those outlined below, all of which offer some element of dynamic displacement measurement, but none of which quite offer the same measurement possibilities of video systems.

- The Flying Banana and other track measurement trains are fitted with accelerometers and lasers to monitor dynamic track geometry e.g. [8] and [9].
- WILD systems [10] & [11], use dynamic response to assess many train parameters including train weight and wheel flat-spots, as was modelled in the 1970s (e.g. by [12]), but are unable to give the full track system deflection.
- Geophones have been used for understanding dynamic response through the depth of a trackbed, and were used along with video cameras by Southampton University [1].
- Train mounted systems have been mooted for around a decade, including Ground Penetrating Radar (GPR) [13] and Laser Doppler [14], but to date these have not been adopted.
- Conventional electro-mechanical sensors such as strain gauges, accelerometers and tilt meters are also used, but have been found to be problematic in ballasted track.

The ability to monitor dynamic deflection from a point of safety was taken up by this author, and first reported in 2011 (Malcolm, Yu, & Waterfall, 2011), where a similar measurement quality to Bowness et al was gained, but using a much wider field of view (approximately 8m, as opposed to 200mm). The narrow-view approach has also recently been adopted by researchers in Canada (Murray & Take, 2014), who were however able to obtain excellent resolutions and use multiple cameras simultaneously to characterise a larger length of track.

Using cameras to measure was not a new idea however - the first known example of the technique is the Tacoma Narrows Bridge (Washington, University of, 1954), which collapsed in 1940. In recent years, the number of researchers investing time and effort in this field are growing considerably, with over 20 papers published in 2013 alone. Most of these refer back to the work of either (Stephen, Brownjohn, & Taylor, 1993) or (Olaszek, 1999).

The Bristol system developed by Stephens et al remained in use, particularly by Macdonald, initially for his work on the second Severn crossing [7], and then for ongoing investigations into dynamic structures, particularly cables, both on site and in the laboratory. It was developed further to increase its resolution by [15] and is the basis of the system being used for the work reported here. Many authors talk about the need to verify the camera system, and so extensive benchmarking has been conducted by researchers, such that by 2003 Wahbeh et al could state:

> *"However, vision-based methods have demonstrated promising results in both laboratory and field experiments. Vision-based approaches offer the potential of both high spatial and temporal resolution."* [16]

With this in mind, further development of the technique needs to focus on tailoring it to specific applications, and working on practical field implementations of hardware and software.

## 2. DEFLECTION OF RAIL TRACK AND COMPONENTS

### 2.1. Introduction

Conceptually, the option of being able to take detailed, dynamic measurements without getting too close to a train had potential to be a useful tool. With that in mind, a demonstration was set up at a local historic railway to check the initial concept, and produce a case study. This initial case study has been followed by interest from rail survey companies, UK, US and Australian based resellers of rail surveying equipment and also Prorail, the operator of rail track infrastructure in the Netherlands. Recent uses include investigations of switches and crossings, voiding, track stiffness, monitoring during construction of under-track pipe crossings, and on-track component testing. There is also ongoing development of the technique, working with LB Foster Rail Technologies to demonstrate the capability of the system for 3 dimensional displacement measurements (and therefore calculation of lateral forces).

### 2.2. Initial Case Study – Washford Station

#### *2.2.1. Experimental Setup*

A video camera was set up on a wooden surveyor's tripod, approximately 8m away from the running line of the West Somerset Railway, at Washford station, as indicated in Figure 1. The camera was connected to a laptop running the Imetrum Video Gauge software, which was used to calculate vertical displacements of the chairs used to connect the rail to the sleepers, as shown in Figure 2. The red boxes indicate the areas used by the software as targets.

**Figure 1: Monitoring setup at Washford Station**

A second setup was also tried to examine the possibility of collecting data from a longer section of track (Figure 3). This allowed approximately triple the length of track to be viewed, albeit at lower resolution.

**Figure 2: Camera view 1 and monitoring points at Washford Station**

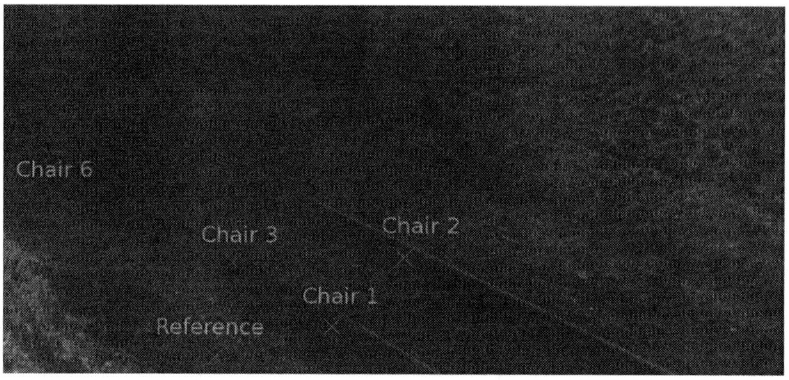


**Figure 3: Camera view 2 looking down a length of track in a cutting**

### *2.2.2. Measurement Results*

Both views were able to successfully record displacements, as illustrated by the sample displacement traces (Figure 4, Figure 5). The video system was able to pick up the detailed impact of individual axles, and the bogies on the axles. Additionally, the system showed robustness at being able to respond well to a temporary loss of targets. Measurement resolution was better for view 1 than view 2, but both were <0.5 mm.

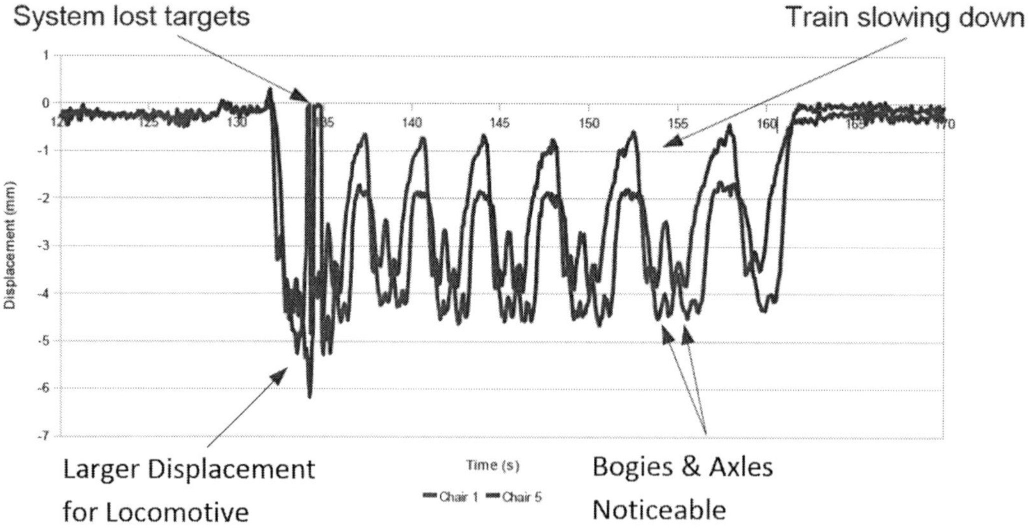


**Figure 4: Camera View 1: Vertical Displacements**

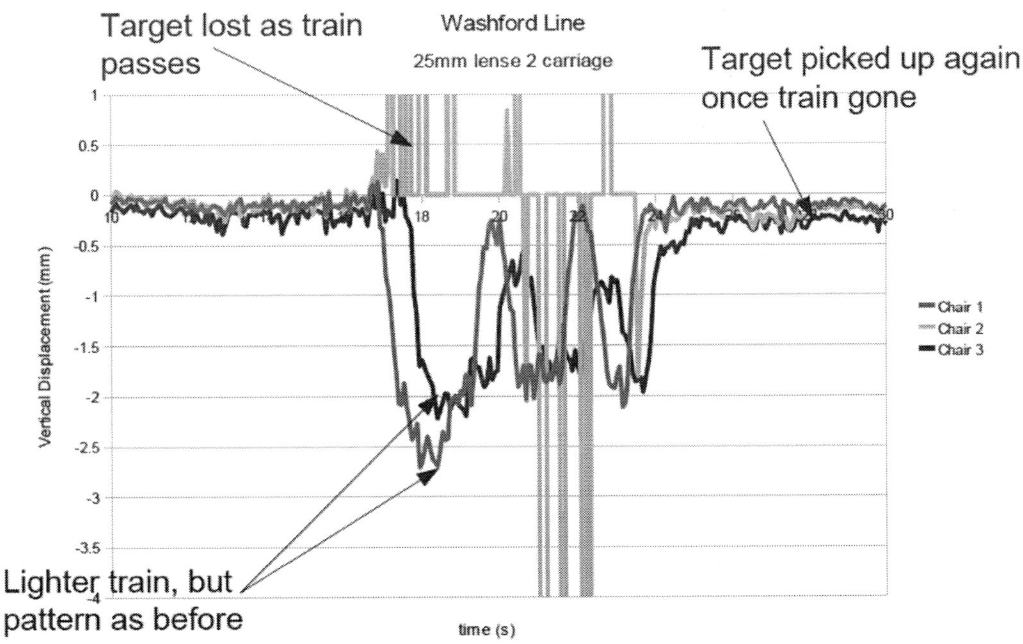


**Figure 5: Camera View 2: Vertical Displacements**

### *2.2.3. Conclusions and Further Work*

These results were a sufficiently good proof of concept to be able to start to look for opportunities to talk about the system to potential users within the railway industry, in order to understand measurement requirements, potential benefits and market opportunities. The results of some of this ongoing research are presented below.

## 3. BENCHMARKING – TRACK STIFFNESS

### 3.1. Background to the Benchmarking Exercise

Matthew Lock (Korec), and Nick Matthews (Network Rail), proposed that using a video gauge could provide an alternative methodology for calculating track stiffness. A video system requires minimal access to site meaning a safer, quicker and cheaper job, whilst also providing a high quality of data. At present, a falling weight deflectometer (FWD) is often used as an indirect means of calculating stiffness (along with other characteristics of the track system) [17]. However, there are limitations to its use for this purpose. A good treatment of ideal user requirements, as well as potential methods for determining track stiffness is given by De Bold [13]. Whilst his work focused on train mounted GPR, he also highlights the benefits of an easy to use, portable system over the FWD (in his case choosing a lightweight impact technique) for stiffness calculation. To verify measurements and demonstrate ease-of-use, a test was set up at a siding on the West Coast Mainline (WCML) at Forge Lane, Stoke.

### 3.2. Experimental Setup

A single camera video gauge system was set up to capture deflection of the main line during passage of trains under a normal timetable. Setup was done in less than 30 minutes, with a tripod positioned approximately 8m back from the line, and videos recorded at 30 frames/second. Deflection values for marked locations (Figure 7) were generated in real time, displayed on screen and then discussed on-site, to see if the deflections obtained were in keeping with those that would be expected for this particular section of track, which was well known to the engineers in attendance.

Simply, track stiffness (k) can be calculated from deflection (d) using the equation:

$$k(t) = F(t)/d(t)$$

Where F is the force applied. As noted by [18] the track stiffness is also a function of the frequency of the loading, and so for an accurate assessment, this also needs to be taken into account. Thus, if axle loading is known, track stiffness can be calculated, although with dynamic variation of the load this will be an approximation. Track stiffness can also be derived using data from strain gauges such as may be found on a WILD system. Given that a video system is able to calculate deflections for each axle, it also has an opportunity to understand some of the dynamic factors involved in the stiffness calculation. However, it may also be used as a suitable alternative method to determining performance of the whole track system, assisting the engineering team with research into track system behaviour, and future guidance on design and maintenance strategies.

**Figure 6: Monitoring location for Track Stiffness Benchmarking**

### 3.3. Results

Sample results are shown in Figure 8. The deflection recorded by these traces, when combined with axle loads, was then used on site by Matthews to calculate track stiffness. The value calculated corresponded with that which could be expected for a well maintained section of track, providing assurance as to the validity of the technique.

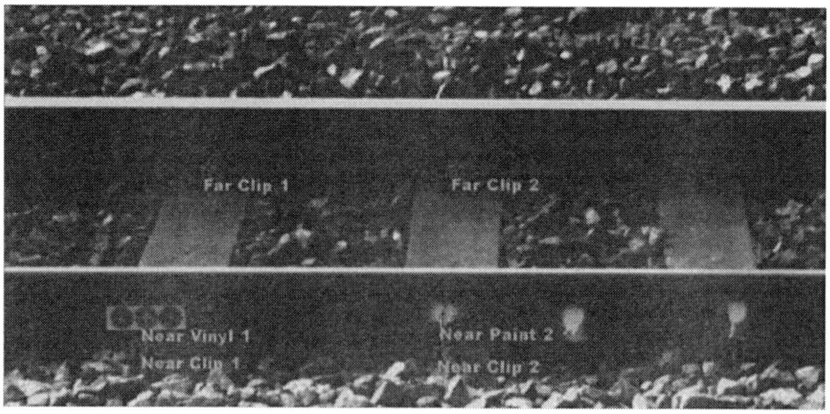


**Figure 7: View of track on WCML showing sleepers, clips, spray-painted targets and self-adhesive vinyl targets**

This trial was also useful test for the tracking algorithms for a range of different targets. So, the software was used to capture 4 different types of target:

- The natural pattern of the near rail clip
- The natural pattern of the far rail clip
- A self-adhesive vinyl target, computer generated and printed on
- A blob of spray-paint applied on site

Figure 8 and Table 1 below show the results of this investigation, during the passing of a train. As can be seen from the traces, when a pattern has been applied to the rail, the software tracks it well. Where the natural pattern of the clip is used, then in some cases the target is not tracked, whilst in other cases, it is tracked, but displays an erroneous displacement magnitude – this is because the changing appearance of the natural pattern as the train passes over (in particular the different shadow) is erroneously reported as a movement.

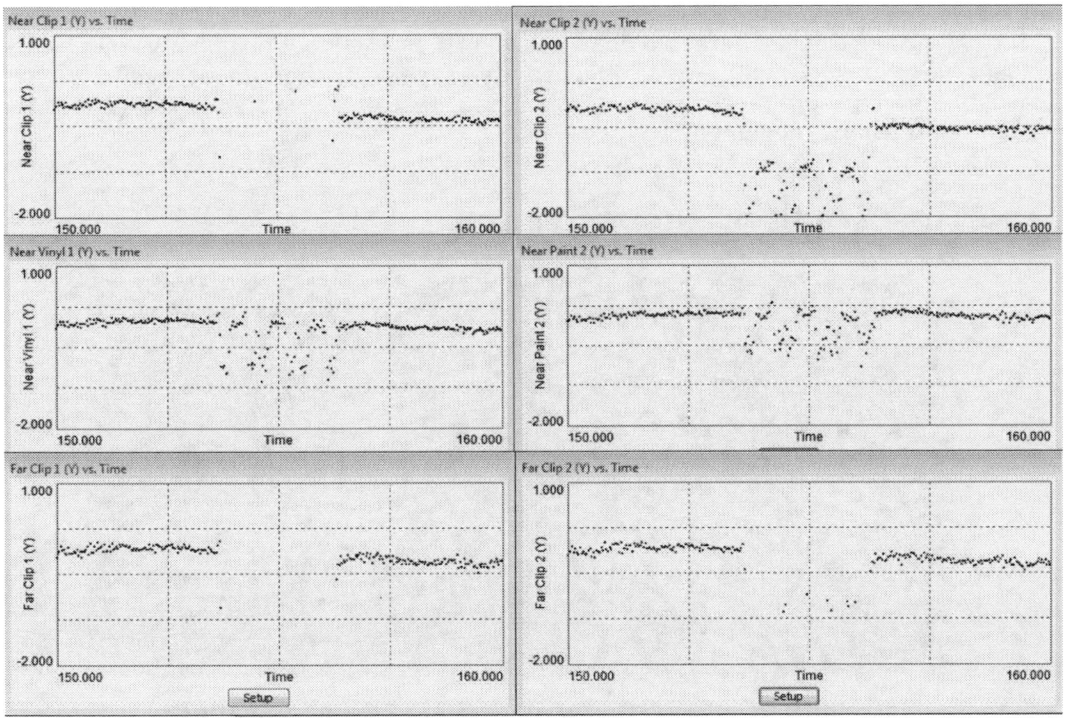


**Figure 8: Vertical Deflection under load using vinyl targets, paint, and clips for tracking by the software**

**Table 1: Resolution and Tracking consistency for WCML trial**

| | Near Clip 1X | Near Clip 1Y | Near Clip 2X | Near Clip 2Y | Near Vinyl 1X | Near Vinyl 1Y | Near Paint 2X | Near Paint 2Y | Far Clip 1X | Far Clip 1Y | Far Clip 2X | Far Clip 2Y |
|---|---|---|---|---|---|---|---|---|---|---|---|---|
| % measurements captured | 90.3% | 90.3% | 100.0% | 100.0% | 100.0% | 100.0% | 100.0% | 100.0% | 89.7% | 89.7% | 90.8% | 90.8% |
| Unloaded Resolution (mm) | 0.028 | 0.028 | 0.028 | 0.032 | 0.024 | 0.030 | 0.027 | 0.026 | 0.037 | 0.040 | 0.032 | 0.045 |
| % difference from vinyl target resolution | 4% | 4% | 2% | 19% | -12% | 12% | 0% | -5% | 35% | 48% | 19% | 64% |

Note that for the near rail, there is very little difference in unloaded measurement resolution, indicating that the natural and man-made patterns are equally as good when there is little lighting change. For the far clips, the noticeably worse resolution is because the natural features are too small for the tracking algorithms.

### 3.4. Conclusions and Further Work

This setup started to demonstrate some of the capability of the system for track deflection, and consequently deriving track stiffness. As can be seen from the traces where a variety of targets are applied, vertical deflection measurements are possible using a modern railway, not just the preserved railway with large chairs as seen in section 2.2. It also shows that there is difficulty in using the natural pattern of the clips to track deflection where there is a strong shadow from a train.

The difficulty in undertaking a comparative benchmarking exercise at this time with an FWD illustrated one of the benefits of using a video based system – there was no need to close the line for track access. FWD is a very useful device for understanding a range of track conditions by measuring wave propagation through the track system and it can also provide track stiffness indirectly. A video gauge is quicker and simpler to use, will give you deflection and enable you to infer stiffness if you understand load, however will not be able to characterise the trackbed in the same way as FWD.

The promising results here did lead to a further testing opportunity with Peter Musgrave (Network Rail), which allowed the system to be run at over 100 fps, investigate vibration effects and also enabled a test of the system's robustness for monitoring ballast movement.

## 4. EXAMPLE PROJECT – HIGH SPEED LINE

**Figure 9: Site Setup for High Speed Demonstration**

### 4.1. Experimental Setup

Figure 9 shows the site setup during this test. The camera was positioned on an embankment, just over 7m from the track, with a lens zoomed in on a couple of sleepers, as can be seen in Figure 10. Site conditions on this occasion were wet, with rain falling throughout the test. The site was chosen as it was shortly to undergo ground stabilisation works, and was therefore known to have relatively large amounts of train-induced movement.

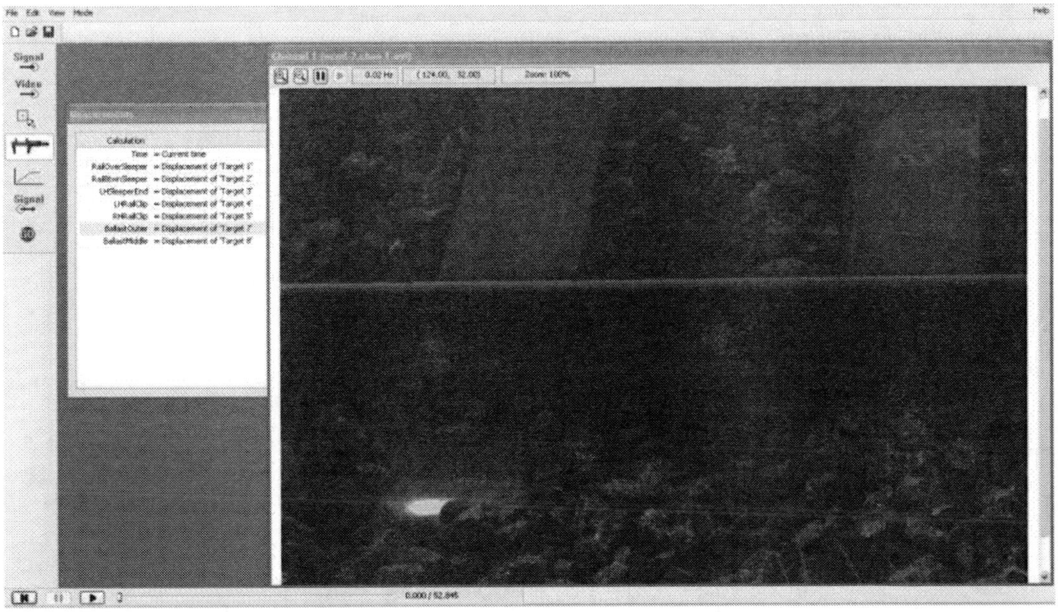

**Figure 10: Screen shot showing field of view and monitoring locations**

## 4.2. Results – Displacement

As can be seen in Figure 11 and Figure 12, vertical and horizontal displacements were successfully achieved on track, sleeper, clips and ballast, during passage of a High Speed Train (HST) at 90 mph. It was even possible to track displacement of an individual stone forming part of the ballast between the rails - zero values were reported when a wheel was blocking the view, but it was tracked immediately afterwards.

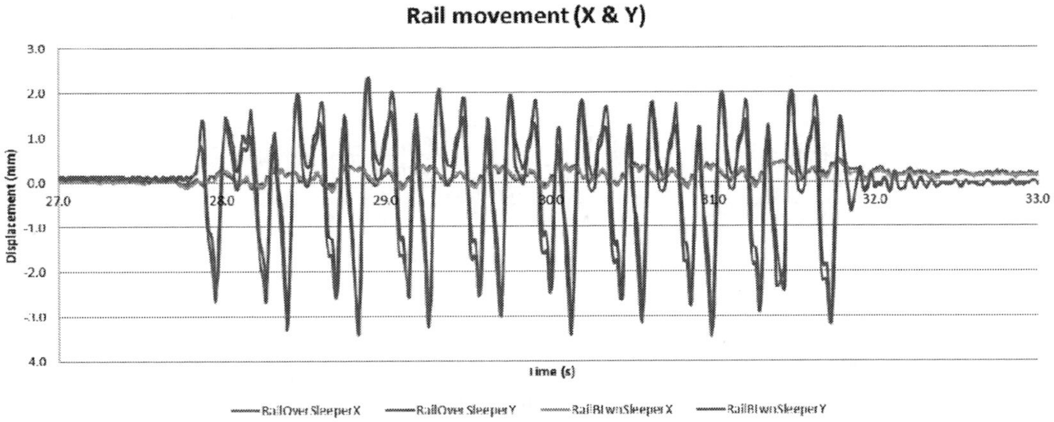


**Figure 11: Vertical (Y) and Horizontal (X) deflection of rail using spray-painted targets. Note that ability to track horizontal deflection could be used to characterise longitudinal rail movement and rail creep**

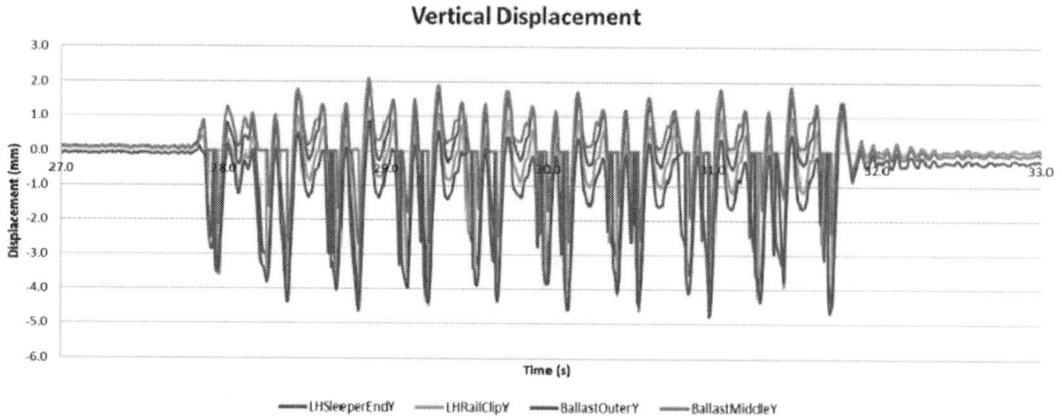


**Figure 12: Vertical Deflection of Ballast, Clips and Sleeper. Note that the light blue line is the ballast between the rails, which was tracked when the train is on top, as the train cast a limited shadow on the overcast day**

Four load events were recorded - two on the far track and two on the near track. Table 2 gives maximum and minimum displacement values for the targets on the near rail for each of these. In order to minimise time on site, real time displacement was displayed for a single point in real time, and videos recorded for more detailed analysis. Subsequent benchmarking has shown that more than 10 points can be tracked in real time with 100 fps camera speed.

**Table 2: Maximum and Minimum Horizontal (X) and Vertical (Y) Displacements (mm) during high speed demonstration**

| Train Event: | RailOver SleeperX | RailOver SleeperY | RailBtwn SleeperX | RailBtwn SleeperY | LHSleeper EndX | LHSleeper EndY | LHRail ClipX | LHRail ClipY | RHRail ClipX | RHRail ClipY | Ballast OuterX | Ballast OuterY | Ballast MiddleX | Ballast MiddleY |
|---|---|---|---|---|---|---|---|---|---|---|---|---|---|---|
| 1 - Far Rail | | | | | | | | | | | | | | |
| MAX | 0.14 | 0.53 | 0.14 | 0.52 | 0.28 | 0.51 | 0.15 | 0.53 | 0.11 | 0.55 | 0.31 | 0.36 | 0.29 | 0.55 |
| MIN | -0.24 | -0.36 | -0.19 | -0.40 | -0.21 | -0.33 | -0.26 | -0.35 | -0.25 | -0.38 | -0.17 | -0.63 | -0.18 | -0.71 |
| 2 - Near Rail | | | | | | | | | | | | | | |
| MAX | 0.53 | 2.33 | 0.47 | 2.31 | 3.96 | 1.91 | 1.41 | 1.33 | 1.12 | 1.78 | 2.62 | 1.40 | 1.15 | 2.11 |
| MIN | -0.15 | -3.23 | -0.23 | -3.46 | -0.61 | -4.35 | -0.82 | -4.67 | -0.54 | -4.55 | -0.52 | -4.77 | -0.96 | -2.70 |
| 3 - Near Rail | | | | | | | | | | | | | | |
| MAX | 0.43 | 2.42 | 0.45 | 2.35 | 1.20 | 1.95 | 0.76 | 1.60 | 0.72 | 2.00 | 1.01 | 1.88 | 1.18 | 2.07 |
| MIN | -0.21 | -3.55 | -0.26 | -3.59 | -0.63 | -4.29 | -0.71 | -5.06 | -0.46 | -4.69 | -0.72 | -4.77 | -1.06 | -2.78 |
| 4 - Far Rail | | | | | | | | | | | | | | |
| MAX | 0.11 | 0.40 | 0.11 | 0.42 | 0.23 | 0.46 | 0.11 | 0.44 | 0.16 | 0.47 | 0.14 | 0.33 | 0.21 | 0.41 |
| MIN | -0.21 | -0.45 | -0.16 | -0.46 | -0.26 | -0.35 | -0.24 | -0.38 | -0.18 | -0.39 | -0.25 | -0.50 | -0.18 | -0.65 |

### 4.3. Results – Vibration

Table 3 gives a summary of measurement resolution obtained. This has been calculated as the standard deviation of 100 measurement points when there is no load being applied. Variations in measurement resolution within a single image are principally down to the quality of the target as the software sees it. Imetrum quotes a typical measurement resolution of 1/100$^{th}$ pixel. Here a 1024 x 768 pixel camera observed an 1100 mm track length, giving an anticipated resolution of 0.011 mm would be expected. Observed average resolutions are broadly similar with the applied pattern giving a slightly worse average resolution (0.014 mm) than those using natural features (0.011 mm).

**Table 3: Displacement resolution during high speed demonstration**

| Resolution (mm) | Rail | | | | Sleeper | | Clips | | | | Ballast | | | |
|---|---|---|---|---|---|---|---|---|---|---|---|---|---|---|
| 1 Far | 0.013 | 0.011 | 0.012 | 0.015 | 0.008 | 0.008 | 0.008 | 0.009 | 0.008 | 0.012 | 0.008 | 0.009 | 0.006 | 0.010 |
| 2 Near | 0.014 | 0.021 | 0.017 | 0.023 | 0.013 | 0.019 | 0.011 | 0.020 | 0.009 | 0.021 | 0.010 | 0.019 | 0.009 | 0.019 |
| 3 Near | 0.007 | 0.010 | 0.007 | 0.010 | 0.008 | 0.008 | 0.007 | 0.008 | 0.008 | 0.007 | 0.006 | 0.006 | 0.006 | 0.008 |
| 4 Far | 0.014 | 0.017 | 0.011 | 0.018 | 0.012 | 0.015 | 0.011 | 0.016 | 0.010 | 0.017 | 0.009 | 0.016 | 0.009 | 0.016 |
| Average Res. (mm) | 0.014 | | | | 0.011 | | 0.011 | | | | 0.010 | | | |

# 5. EXAMPLE PROJECT – 3D DEFLECTION OF A TRACK COMPONENT

## 5.1. Background to the Project

Barnaby Temple and Paul Insley (LB Foster Rail Technologies) saw the potential for using the system for a range of applications including:

- Testing and verification of track components and equipment
- Obtaining a detailed understanding of the response of the whole track system under live loading
- Measuring lateral forces in rails.

At present, a new design or device may be instrumented with strain gauges, LVDTs, geophones and accelerometers, and measurements generated this way. In a track environment, this is a time consuming and costly process, meaning that this is rarely done. As it is important to validate any new system prior to use on a network, then the potential for quick and simple measurements could help overcome one of the key barriers to innovation in the rail industry.

This project was designed to see if there was an easier way of getting measurement values of a similar quality to what could already be achieved, without the problems inherent in attaching strain gauges to track. As part of ongoing technical developments, the software team at Imetrum (Chris Setchell and James Sharam), had developed a method for using synchronised stereo cameras to generate measurements live in 3 dimensions, an essential element to being able to monitor lateral movement (perpendicular to camera sensor).

## 5.2. Site Setup

**Figure 13: Twin camera system mounted on a bar with EDM to assist with calibration. Area of rail under test is on left hand edge of photo**

The Rail Alliance test facility at Long Marston was used for this project, allowing use of a dedicated diesel vehicle and driver. The video cameras were mounted on a 2m bar and set up 6m from the near rail (Figure 13). The synchronised cameras were set to record video at 130 fps, and focused on the component (Figure 14). X Y Z displacements of 11 points were gathered, along with 3 strains. Self-adhesive targets were applied to the areas of interest, as there were not many places where a suitable natural pattern was visible for the software to track. It was raining for most of the duration of the test, which will have reduced measurement resolution slightly [19].

**Figure 14: Twin camera views in Video Gauge software, with measurement locations indicated**

### 5.3. Sample Results

Approximately 100 train loading events were recorded, to assist the manufacturers of the track component verify their performance models and gain a rapid understanding of any likely failure mechanisms. The details of the particular component are company confidential however it is possible to give an indication of the measurements generated (Figure 16, Table 4).

Running two synchronised 'full HD' cameras at 130 fps worked smoothly, and it was possible to track movement of a few points in real time. Benchmarking prior to arriving on site had indicated that it would not be possible to take all 15 measurements in 3 dimensions at this speed, so videos were recorded, and re-processed afterwards The displacement values were benchmarked by measuring the areas where paint had been scraped off moving parts (Figure 15).

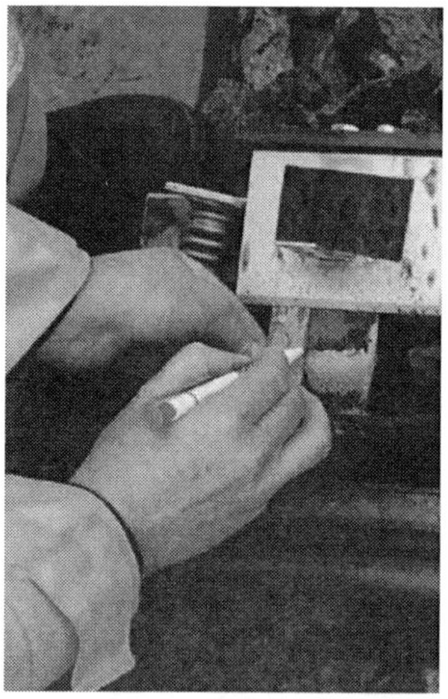

**Figure 15: Comparative measurements using a ruler and paint**

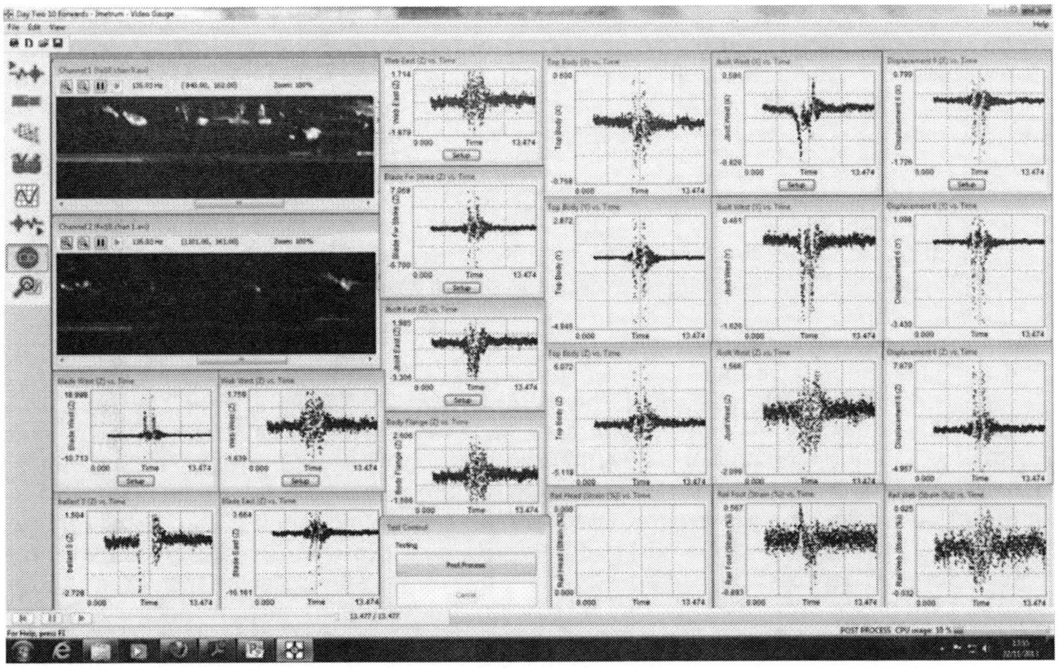

**Figure 16: Screen-shot of software demonstrating deflection measurements**

Four random samples of data were chosen to investigate displacement resolution for this setup (Table 4), to see if the system has potential for use in a range of track applications, including the calculation of lateral forces. They indicate an out-of-plane measurement resolution (Z) four times worse than the in-plane measurement resolution (X & Y). Given the geometry of the setup then this order of difference is not unexpected.

**Table 4: X, Y & Z (out-of-plane) measurement resolution for 3D displacement**

| Displacement Points | 1X | 2X | 3X | 1Y | 2Y | 3Y | 1Z | 2Z | 3Z |
|---|---|---|---|---|---|---|---|---|---|
| Train event | Displacement Resolutions (mm) - 1/100th pixel = 0.008 mm | | | | | | | | |
| 5 forwards | 0.006 | 0.010 | 0.010 | 0.010 | 0.009 | 0.010 | 0.037 | 0.039 | 0.037 |
| 12 backwards | 0.009 | 0.005 | 0.014 | 0.011 | 0.009 | 0.012 | 0.055 | 0.051 | 0.045 |
| 27 backwards | 0.012 | 0.012 | 0.012 | 0.013 | 0.013 | 0.013 | 0.047 | 0.051 | 0.050 |
| 36 forwards | 0.005 | 0.006 | 0.005 | 0.012 | 0.011 | 0.013 | 0.043 | 0.038 | 0.042 |
| Average resolution: | 0.009 | | | 0.011 | | | 0.045 | | |

## 6. DEVELOPMENT – 3D DEFLECTION FOR CALCULATION OF LATERAL EFFECTS INCLUDING DEFLECTIONS, STRAINS AND FORCES

### 6.1. Background to the Project

Wheel/rail interaction is a complex subject that is influenced by many parameters including contact profiles of wheel and rail, track geometry, vehicle suspension characteristics and friction conditions. Determining the influence of changes on curving behaviour is a challenge on site, with effects such as wear and contact fatigue taking time to become apparent. Measuring the forces imparted at the interface can be a complex and costly exercise, but can be demonstrated by attaching an array of strain gauges along the rails. Analysing the outputs of these, combined with knowledge of the properties of the rail can give a measure of lateral force and other desired parameters. Doing this however involves substantial time and expense to organise, which can be hard to justify, so such installations are usually confined to long term sites for monitoring train loading.

A method of measuring lateral effects quickly and simply would be a useful tool, but also has wider benefits for verifying track designs or components therein and investigating specific areas of particular concern such as sharp radius curves, switches, bridge transition zones and areas of hard or soft trackbed.

### 6.2. Proof of Concept Proposal

Video based measurement is a well-established technique for monitoring strain on a laboratory scale, in the plane of the camera sensor. What is required here is a means of replacing an out-of-plane strain measurement, quickly, from a safe distance, to a high resolution (20 microstrain equivalent or better). This resolution is not easily achievable with a video system in the out-of-plane direction, and so an alternative methodology is proposed, of using deflection to calculate relative bending in the out-of-plane direction, which can then be used to calculate the lateral force. Theoretically, this method should be equally valid to the strain gauge method currently employed, but as it is using a different input parameter, a project has been developed to represent bending parameters of the rail under load and test the capability of the 3D video method to capture it effectively.

Stage 1 of this involves work in a laboratory using a test frame to replicate wheel loading, and initial measurement results of this are presented below, with modelling work ongoing. Step 2 will involve site based measurements, ideally at a location that is also instrumented with strain gauges. Step 3 will involve rail professionals developing wider experience to test the robustness and practicality of the method through using it in a variety of circumstances.

### 6.3. Initial Results

A bar-mounted 2 camera system was set up to mimic on-site fields of view (3m distance from rail), and static and cyclic load cycles performed. Results plotted below are load signal and displacements for a cyclic test (0 to 250kN), applied to a vertical rail at 5Hz. Tests on an inclined rail, to generate lateral flexure in the rail to an approximately equivalent level to that which can be induced by a train, were then also performed following establishment of the measurement accuracy in the vertical setup. Shown are vertical and lateral and bending displacements as cyclic loading commences, along with a resolution investigation to calculate standard deviations of displacement and bending traces prior to loading.

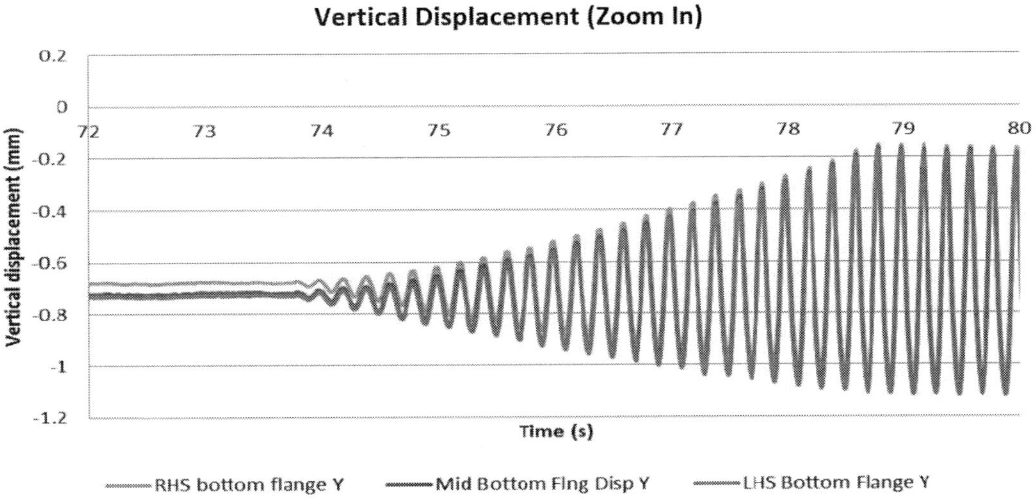


**Figure 17: Vertical displacement caused by nominal vertical cyclical loading of 0 to 250kN on rail**

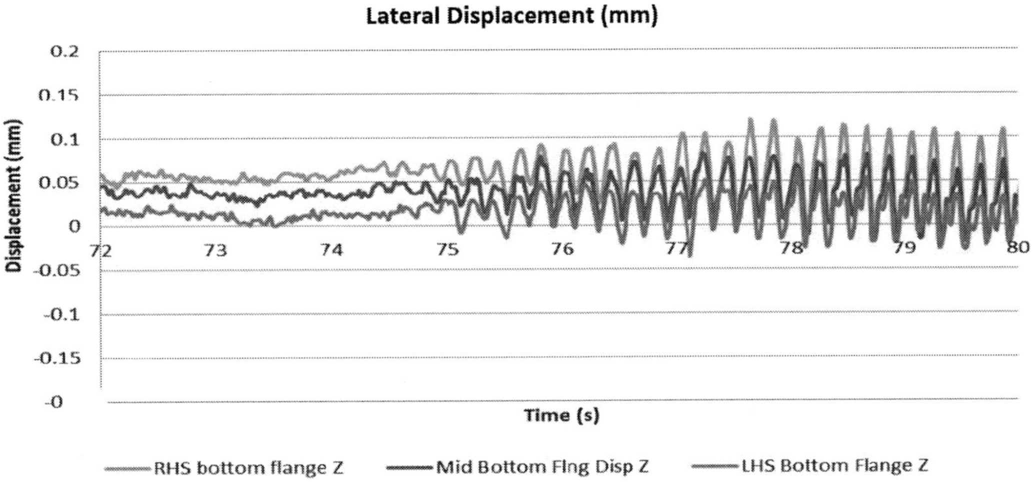


**Figure 18: Lateral displacement caused by nominal vertical cyclical loading of 0 to 250kN on rail**

When calculating bending, it is possible to mitigate for camera movement, and also take out any translation of the rail, by performing a relative calculation. The displacement for the bend profile can be calculated as:

$$\text{Bending Displacement} = \Delta_{mid} - (\Delta_{left} + \Delta_{right})/2$$

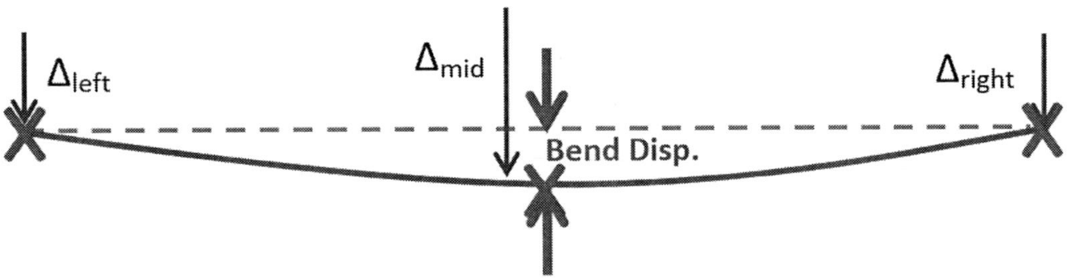


**Figure 19: Diagram of Bending Displacement Calculation**

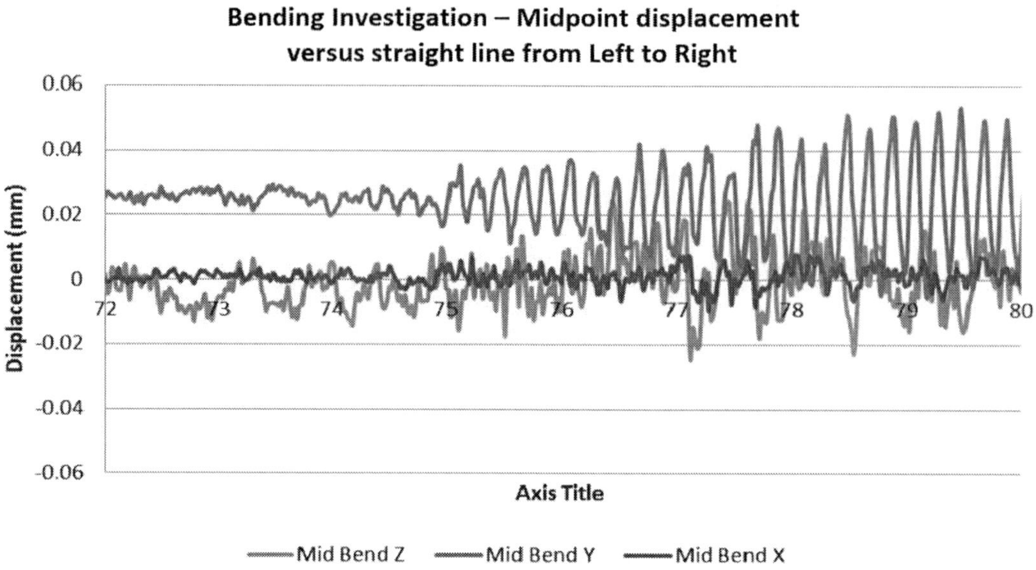


**Figure 20: Bend Displacements under nominal vertical 0 to 250 kN cyclical loading**

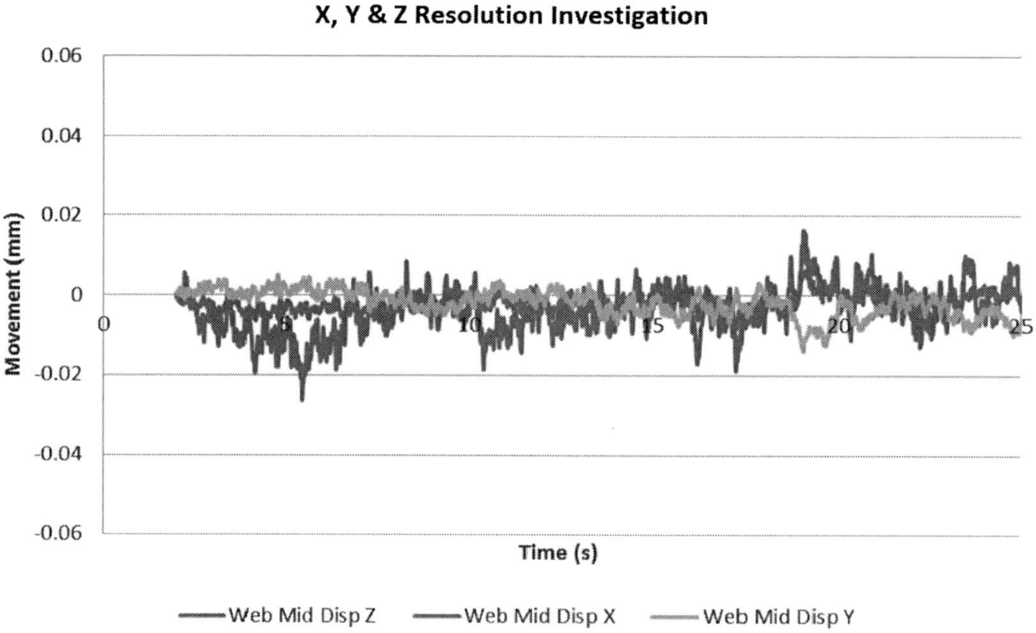


**Figure 21: Sample (unloaded) 3D displacement plot of lab setup (1 pixel = 0.2mm approximately)**

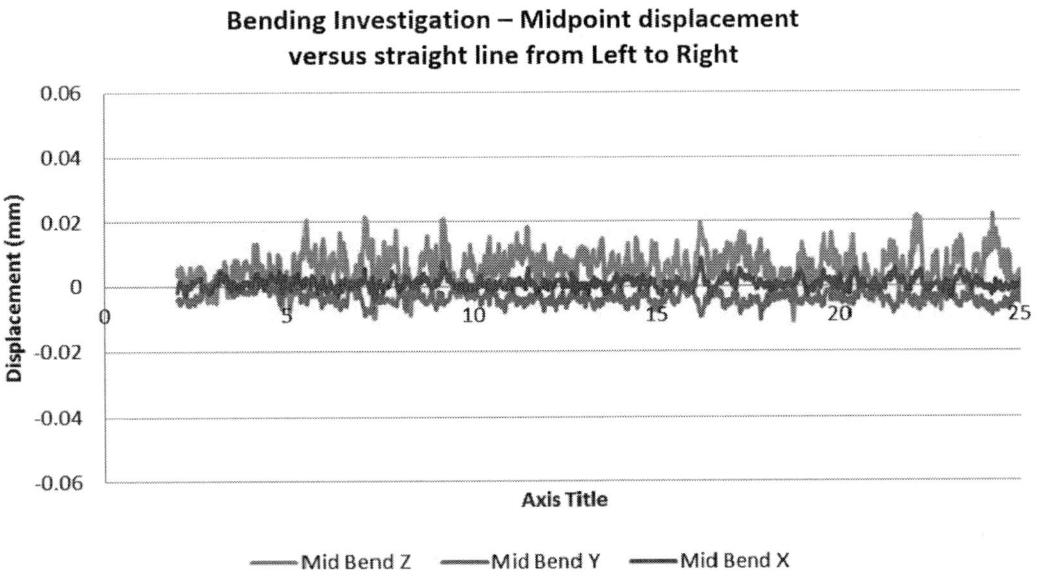


**Figure 22: Sample (unloaded) Bend Displacement plot**

**Table 5: Resolutions in mm for 0 to 250 kN cyclical loading test (standard deviations of traces over 10 seconds)**

| Resolution (mm) | X (horizontal) | Y (vertical) | Z (lateral) |
|---|---|---|---|
| Displacement | 0.002 | 0.003 | 0.006 |
| Bending | 0.002 | 0.002 | 0.005 |

### 6.4. Conclusions and Further Work

Having proved an out-of-plane displacement resolution of 5 microns at a 500mm field of view, with cameras that are running at 100frames/second, the measurement system proposed has been shown to have sufficient resolution for the task in hand. It is now important to complete the modelling of lateral effects in a rail and to build use-experience, gaining confidence that the system can be used regularly by field engineers for the measurement of wheel induced lateral forces on rails outdoors.

## 7. SUMMARY

Sections 2.2 to 6 above have shown the development, and some of the projects, which have explored the use of video measurement as a tool for use on rail track. Time and effort has been spent developing an understanding of end user requirements, in order to identify and test the opportunities for the technology, and what barriers there are to be overcome. The technique has been developed to be suitable for use in rail applications, including hardware and software specification and selection, and developing a clear understanding of the operational parameters of video measurement for a track monitoring scenario.

There is no doubt that the speed requirements, vibration considerations, and optical variation during the passing of a train have led to new understandings and developments, meaning that video measurement is now able to address some real-world challenges and problems efficiently and effectively. There is further testing and trialling work to be done, along with work by engineering teams to develop suitable deployment strategies prior to widespread adoption as an investigative tool, but initial technical results are compelling, and mean that with the off-track advantages for deployment, market interest is picking up.

## 8. WORKS CITED


[1] D. Bowness, A. C. Lock, W. Powrie, J. A. Priest and J. D. Richards, "Monitoring the dynamic displacements of railway track," *Proc. IMechE Vol 221 part F: Journal of Rail and Rapid Transit,* pp. F1: 13 - 22, 2007.

[2] K. Malcolm, H.-T. Yu and P. Waterfall, "Recent Advances in Geotechnical and Structural Instrumentation - from a manufacturers point of view," in *Proceedings of the International Conference on Advances in Geotechnical Engineering*, Perth, 2011.

[3] P. W. Waterfall, J. H. G. Macdonald and N. J. McCormick, "Targetless Precision Monitoring of Road and Rail Bridges using Video Cameras," in *Proceedings of the 6th International Conference on Bridge Maintenance, Safety & Management*, Stresa, 2012.

[4] N. McCormick and P. W. &. O. A. Waterfall, "Optical imaging for low cost structural measurements," *Proceedings of the ICE - Bridge Engineering, Volume 167, Issue 1,* pp. 33-42, 2013.

[5] X. Liu, V. L. Markine and I. Shevtsov, "Dynamic experimental tools for condition monitoring of railway turnout crossing," in *The Second International Conference on Railway Technology: Research, Development and Maintenance,* Ajaccio, 2014.

[6] G. A. Stephen and J. M. W. &. T. C. A. Brownjohn, "Measurements of static and dynamic displacement from visual monitoring of the Humber Bridge," *Engineering Structures, Volume 15, Number 3,* pp. 197 - 207, 1993.

[7] J. H. G. Macdonald, E. L. Dagless and B. T. &. T. C. A. Thomas, "Dynamic Measurements of the Second Severn Crossing," *Proceedings of the ICE: Transport, Volume 123, Number 4, pp 241 - 248,* 1997.

[8] P. W. Lindgren, "Project HISTEP," in *Proceedings of the 1968 Annual Convention*, Chicago, 1968.

[9] B. Solomon, Railway Maintenance: The Men and Machines That Keep the Railroads Running, St Paul: MBI Publishing Company, 2001.

[10] Strainstall, "WILD - Wheel Impact Load Detection," 15 September 2014. [Online]. Available: http://www.strainstall.com/what-we-do/our-systems/load-measurement-weighing-systems/wheel-impact-load-detection/. [Accessed 15 September 2014].

[11] Salient Systems, "Wheel Impact Load Detector," 01 September 2014. [Online]. Available: http://www.lbfoster-salientsystems.com/Wheel_Impact_Load_Detector.asp#mstto=. [Accessed 21 September 2014].

[12] S. G. Newton and R. A. Clark, "An Investigatino into the Dynamic Effects on the Track of Wheelflats on Railway Vehicles," *Journal of Mechanical Engineering Science, vol 21 no.4,* pp. 287 - 297, 1979.

[13] R. P. De Bold, *Non-destructive evalation of railway trackbed ballast,* Edinburgh: The University of Edinburgh, 2011.

[14] G. Hildebrand and S. Rasmussen, "Development of a High Speed Deflectograph," Road Directorate, Danish Road Institute, Roskilde, 2002.

[15] A. Towse, C. Setchell, D. Potter, A. B. Clarke, J. H. G. Macdonald, M. R. Wisnom and R. D. Adams, *Use Experience With a Developmental General Purpose Non-Contacting Extensometer With High Resolution,* West Conshohocken, PA: ASTM, 2001.

[16] A. M. Wahbeh, J. P. Caffrey and S. F. Masri, "A vision-based approach for the direct measurement of displacements in vibrating systems," *Smart Materials & Structures,* pp. 785-794, 2003.

[17] P. Sharpe, "Trackbed Investigation," *Journal and report of Proceedigns - Permanent Way Institution, 118 (3),* pp. 238 - 254, 2000.

[18] A. Robinet, M. Hosseingholian and A. Quibel, *Track Stiffness Assessment,* UIC - International Union of Railways, 2008.

[19] N. McCormick, G. Klimaytys and J. Lord, "MAT49 - A review of techniques suitable for measuring position or strain in Civil Engineering structures," NPL, London, 2011.

[20] P. W. Waterfall and C. Setchell, "Real-time dynamic monitoring using video cameraa," in *Proceedings of the 12th KTA International Symposium on Mechanized Tunnelling Technology*, Seoul, 2011.

[21] D. White, W. Take, M. Bolton and S. Munachen, "A deformation measurement system for geotechnical testing based on digital imaging, close-range photogrammetry and PIV image analysis," in *Proceedings of the International Conference on Soil Mechanics and Geotechnical Engineering*, Istanbul, 2001.

# A 2D DEM analysis of railway ballast vibration under cyclic loads

**X Zhang, C Zhao, W Zhai**
State Key Laboratory of Traction Power, Southwest Jiaotong University, China

**ABSTRACT**

This paper presents a two dimensional (2D) discrete element simulation of railway ballast under cyclic loads. Each ballast particle was simulated as a clump with irregular shape using numerical image technology. The distribution characteristic of the contact forces between ballast particles was analyzed and the force chains in this model were identified. Further, the distribution characteristics of velocities and accelerations of ballast particles were obtained. The results show that only a few ballast particles under sleepers form some strong force chains to bear most of the load, and the vibration of ballast particles attenuates with the increase of depth.

**Keywords:** railway ballast, vibration, discrete element model, force chain

## 1 INTRODUCTION

The ballast is used to bear, transmit and disperse train loads in the conventional railway track. As loose granular mixtures, the ballast can mitigate track vibration, reduce stress at the subgrade surface and facilitate free drainage condition. However, due to the non-uniformity and poor stability of granular mixtures, the ballast would occur degradation under the long-term repeated loads, such as the abrasion and breakage of ballast particles, the fouling and hardening of ballast bed, et al. Ballast degradation would increase the rail irregularity, and then affect the train running safety and ride comfort of passengers.

In order to reveal the deterioration mechanism and settlement laws of railway ballast, a large number of research works have been carried out about this topic around the world. Early researches mainly depended on laboratory tests and field surveys. From these researches, only some qualitative understanding and empirical conclusions were derived to guide the design, construction and maintenance of the ballasted track. Then, many numerical simulations have been carried out based on continuum mechanics methods, for example, using the finite element method or developing different theoretical models including the viscoelastic bar model[1], the continuous elastic layer model[2], and the separated spring-mass model[3], et al. However, the meso and micro-scale mechanical behavior of railway ballast can't be reflected with these classic methods because the discrete nature of granular material is ignored in these models. Consequently, the degradation mechanism of railway ballast is still not clear. To this end, new methods based on discrete mechanics should be developed to simulate the movement of a single ballast particle, the interaction between particles, and even the breakage of particles.

In the last decade, the discrete element method (DEM) has made great progress, and it was applied successfully to pharmaceutical engineering, geotechnical engineering, and chemical engineering et al. Meanwhile, some researchers have studied the mechanical behavior of railway ballast using the discrete element method. Sebastian and Vallejo[4] studied the influence of particle breakage on the settlement of ballast under cyclic loading. He pointed out that the permanent deformation strongly increase even though only a few particles were broken, but only simple circle elements were used in this model. Tutumluer et al[5] developed a discrete element model using polyhedron elements, and simulated the mechanical behavior of ballast. In his work, the predicted settlement of ballast from the discrete element model under repeated train loads fits well with the field test result. However, the irregular shape of ballast particles was not adequately considered in these previous researches, and it might lead to inaccurate numerical results on the meso and micro scale mechanical behavior of ballast particles. Moreover, most of previous researches pay little attention to the distribution of force chains and vibration characteristics of railway ballast.

This paper focuses on dynamic behavior of railway ballast under cyclic loads, and tries to provide some new insights into mechanical behavior of railway ballast. The remainder of this paper is organized as follows: in Section 2, a 2D discrete element model of railway ballast is established considering the geometry shape of ballast in detail; then, the space distribution of contact forces and force chains in the ballast are analyzed in Section 3; the vibration characteristics of ballast particles are analyzed in Section 4; some conclusions are drawn in Section 5.

## 2 DISCRETE ELEMENT MODEL OF RAILWAY BALLAST

### 2.1 2D clump model for irregular ballast particle

Ballast particles can be regarded as coarse particles with the particle size of about 10~60mm. Generally, the high-grade ballast materials use crushed stones, and the ballast particles always have irregular shape with sharp corners and rough surfaces. Lu and McDowell[6] pointed out that the irregular shape can significantly affect mechanical behavior of coarse granular materials. In fact, there are multi contact points around a certain ballast particle and the irregular shape of particles may increase the number of contacts, i.e. the interlocking between irregular particles may be more strong relative to circle ones. Therefore, the complex geometry shape and sharp corners of the ballast should be taken into consideration in the discrete element model.

In order to capture the irregular shape of a ballast particle, a photo of the ballast sample was taken firstly using a digital camera. Then, binary image processing was made, and the 2D edge of the ballast particle can be extracted. Taking this edge line as a planar boundary, small circles with lattice orientation arrangement can be generated on the planar and those circles within the boundary can be regarded as a 2D clump model for this ballast particle. Finally, taking notes of these circles centers together with their radiuses, a clump model of the ballast particle was created by the $PFC^{2D}$ software. As shown in Fig.1, the clump models generated in this way can meet the 2D irregular shape of ballast particles well.

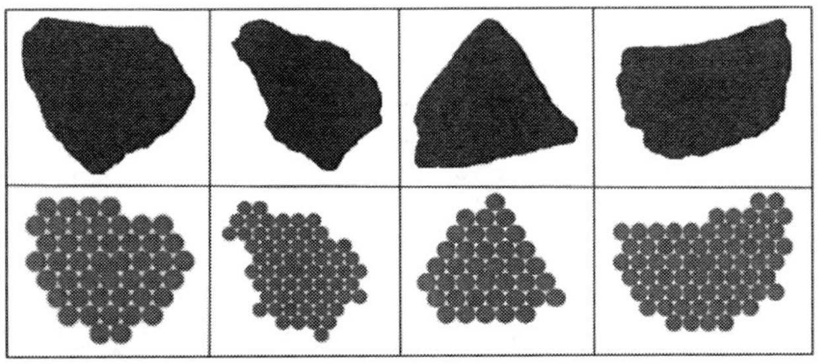

**Fig.1 Binary images and 2D clump models of ballast particles**

## 2.2 Discrete element model of the sleeper-ballast structure

A number of photos of ballast samples which had been sieved in advance were taken and the corresponding 2D clump models were created as described in Section 2.1. Thus, a database composed of many clump models of ballast particles has been established. In this database, the information of ballast particles including the particle size, coordinates of centroids, sequence numbers were stored and used to create clump models. Then, sample particles together with their property files can be randomly chosen from this database to construct discrete element models of ballast conveniently.

The discrete element model of sleeper-ballast structure was built by the $PFC^{2D}$ software. Firstly, straight boundary walls were generated. Secondly, the numbers of necessary ballast particles in different particle size range were calculated according to the grading requirement of Chinese super-grade ballast (TB/T2140-2008), and the information of the corresponding particles was randomly chosen from the clump model database. Thirdly, using these chosen files, clumps were generated and packed in the region enclosed by boundary walls. The packed model was also compacted by an extra wall to obtain a dense model. The model was built and compacted into three layers. Finally, the clumps located in the sleeper regions were deleted and three clumps for sleepers were created in these blank spaces (see Fig.2). The whole model was balanced until the ratio of the maximum unbalanced force to the sum of applied force and contact force was cycled to be less than $1\times10^{-4}$.

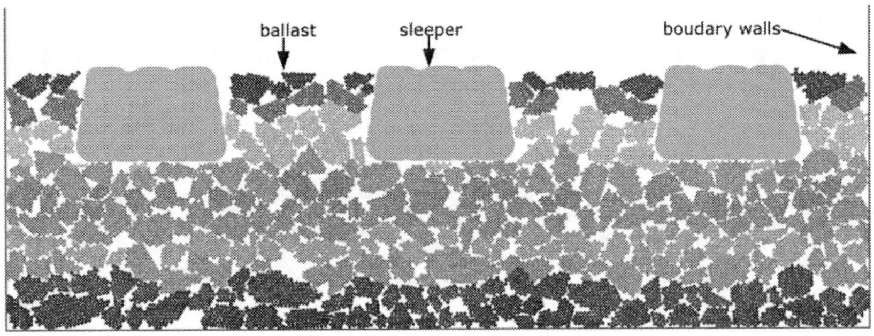


**Fig.2 The 2D discrete element model for sleeper-ballast structure**

The linear contact model was adopted to calculate the interaction forces between contacting clumps and walls in this model. The normal contact stiffness was set to $2.88\times10^{8}$ N/m, and the ratio of shear contact stiffness to normal contact stiffness was set to 0.8. The friction coefficient was set to 0.5. The length of sleepers-ballast model in the longitudinal direction is 1.8m and the sleeper pitch is 0.6m. The thickness of ballast below the bottom of sleepers is 0.35m which meets the related requirement of Chinese high speed railway.

### 2.3 Cyclic loads

Considering the real weight of rail, each sleeper was applied a constant force in the gravity direction firstly, which is equivalent to the weight of 0.6m long rail. The whole model was balanced under the constant load. Then assume that there is a dynamical wheel rail interaction force *P* above the middle sleeper, and it can be described as:

$$P = P_0(1 + \sin(2\pi ft) / 2) \tag{1}$$

where $P_0$ is the static axle load, the symbol *f* represents the load frequency, and the symbol *t* is loading time. According to the actual axle loads of Chinese high speed trains with an operation speed range of 200~250km/h, the value of $P_0$ and *f* was set to 140kN and 4Hz, respectively. However, the wheel rail interaction force wasn't directly applied to this model because there is no rail. Hence, in order to simulate the railway ballast vibration under cyclic loads, the approximate rail-sleeper force acted on each sleeper induced by the wheel rail interaction force was calculated and applied to this model. Ultimately, the cyclic loads applied to three sleepers were 0.25*P*, *0.5P and 0.25P* from the left one to the right one, respectively, due to the fact that the rail-sleeper forces on adjacent sleepers are not uniform.

## 3 CONTACT FORCE OF BALLAST PARTICLES UNDER CYCLIC LOADS

As it is known, the train loads are transmitted into ballast through the contact interactions between ballast particles. The distribution of contact forces during cyclic loading can be obtained from this model which also shows the advantage of discrete element method comparing with the methods based on continuum mechanics. Moreover, the particles carrying large contact forces and composing strong force chains in the ballast will also be identified in this section. These will contribute to analyzing the vibration characteristics of ballast in the next section, because the vibration level of particles depends on the amplitude of contact forces carried by the particles.

### 3.1 Distribution of contact forces

Fig.3 shows the distribution of the contact forces in this model when the cyclic loads reach the maximum value. It can be seen that only forces at a few contacts under sleepers are very large, while most other contacts carry low forces. It means that the load acted on each sleeper is directly transmitted downward, and a few ballast particles beneath the sleeper bear most of the load.

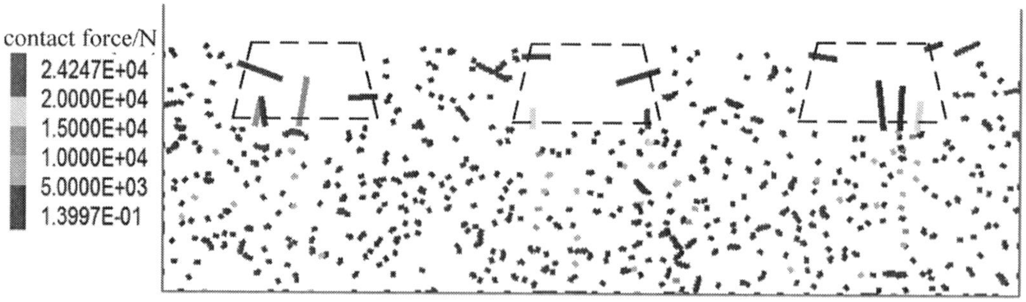


**Fig.3 Distribution of contact forces between ballast particles (straight lines connecting the centers of two contacting circles represent contact forces, the color of lines represents the magnitude of contact force and the black dotted lines represents the locations of sleepers)**

Fig.4 presents the probability density distribution of the ratio of contact force to the mean contact force in this model when the loads reach different values. The mean contact force here is defined as the sum of total contact forces divided by the number of contacts. Fig.4 shows that the probability density under different load conditions is coincidence with each other, which decreases exponentially with the increase of the ratio of contact force to the mean force. Considering that the number of contacts in the model changes little during the cyclic loading, this suggests that the variance of the magnitude of contact force at each contact is approximately proportional to the variance of the load.

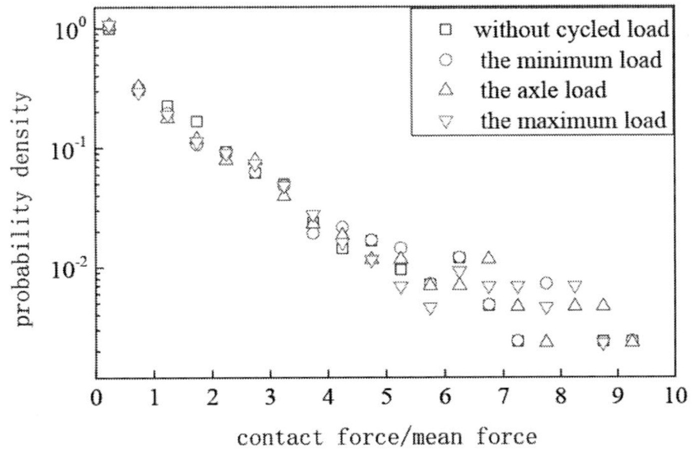


**Fig.4 The probability density distribution of the ratio of contact force to mean contact force under different load conditions**

### 3.2 The phenomenon and identification of force chains

It is shown in Section 3.1 that only a few ballast particles under the sleepers carry large contact forces while the others resist very small forces. These ballast particles carrying large contact forces can form some stable granular structures and provide paths to transmit the load in the ballast. These stable granular structures can be referred as force chains[7] according to the mechanics of granular material.

J.F. Peters et al[8] presented a detailed description about how to identify the force chains in granular material. In this paper, a brief explanation on identification of force chains in our model was given here. The stress tensor $\sigma_{ij}$ of a ballast particle is defined as:

$$\sigma_{ij} = \frac{1}{V}\sum_{N_c} f_i r_j \tag{2}$$

where $V$ is the volume of the clump, $N_c$ is the number of contacts around the clump, $r_i$ is the $i$th component of the branch vector from the center of the clump to the contact point, $f_j$ is the $j$th component of contact force at this contact. Then, the first principal stress of each clump, which is also the maximum compressive principle stress, was calculated according to its stress tensor. Thirdly, those clumps whose first principal stresses are greater than the average principal stress were selected to form the strong force chains. Finally, an angle of 60 degrees was introduced here to identify whether a highly stressed clump belongs to a certain force chain or not. The strong force chains found in this model is shown in Fig.5.

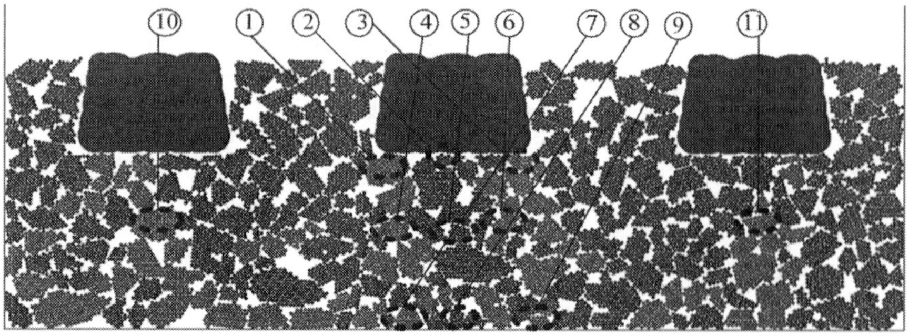


**Fig.5 Force chains found in the DEM model (red clumps are parts of force chains, while blue clumps are not; ballast particles located at the black dotted lines are measuring points)**

Comparing the above figure with Fig.3, it can be seen that only the ballast particles carrying large contact forces under the sleepers form force chains in the model. These force chains are the main and stable structures to bear the load and clearly show us the main propagation paths of contact forces in this model. From this figure, it can be concluded that the loads are transmitted directly downward under the sleeper, and a small angle of diffusion may occur during the propagation of contact forces. This validates the hypothesis of the load being transmitted within a cone region in the ballast by Ahlbeck et al[9] in some way.

## 4 VIBRATION OF BALLAST PARTICLES UNDER CYCLIC LOADS

In order to analyze the vibration characteristics of ballast, time histories of some ballast particles were saved during the loading. The positions of these particles, namely, the measuring points are shown in Fig.5. Measuring points from No.1 to No.9 are located at three different depths of 0m 0.15m and 0.35m under the middle sleeper, so that the attenuation of vibration along the depth in the ballast can be analyzed. Among these nine particles, six are part of a force chain while the other three are not part of a force chain. This contributes to analyzing the difference between the vibration of a ballast particle belonging to a force chain and that not belonging to a force chain. Point 10 under the left sleeper and point 11 under the right sleeper are located almost at the same depth of points No.4 to No.6. This contributes to analyzing the variation of ballast vibration along the longitudinal direction. Only the vertical vibration characteristics of ballast particles are analyzed in this paper.

### 4.1 The velocity of ballast particles

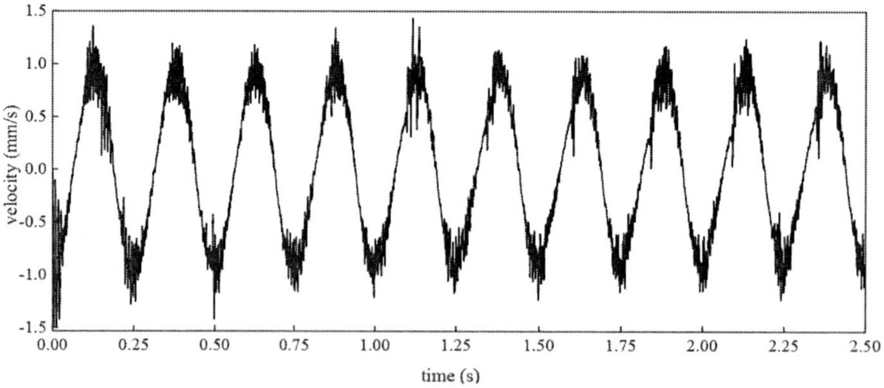


**Fig.6 Time history of vertical velocity of the ballast particle at point 4**

Fig.6 shows the time history of vertical velocity of the ballast particle at measuring point 4 during the cyclic loading. An obvious periodical variation with the frequency of 4Hz which conforms to the load frequency can be seen in this figure. On this time history curve, there are also lots of small peaks which correspond to high frequency up to 300Hz. These high frequency peaks may occur due to the contact vibration of ballast particles. Similar time histories of vertical velocities were obtained for those particles located at other measuring points.

Fig.7 shows the histogram of the amplitude of vertical velocity of ballast particles at measuring points. The velocity amplitudes are close to each other for the three clumps at the same depth under the middle sleeper. That is to say, under the sleeper, no obvious difference between the velocity amplitudes of clumps belonging to force chains and that not belonging to a force chain was found in this model. A remarkable attenuation trend along the depth under the middle sleeper can be seen in this figure. Specifically, average amplitude of velocity for the three clumps at each depth decreases from 1.73mm/s at 0 m to 1.02mm/s at 0.15m and 0.89mm/s at 0.35m, which corresponds to a reduction ratio of 40.5% and 48.5%, respectively. The velocity amplitudes of particles at 0.15m under the left and right sleeper are smaller than that under the middle one mainly due to the lower load exerted on these two sleepers.

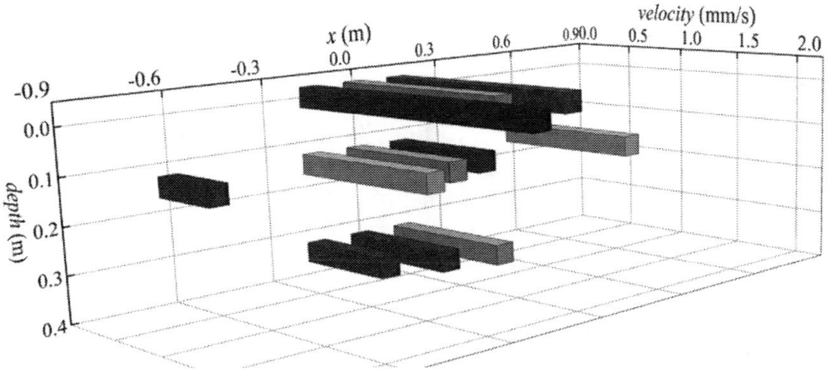


**Fig.7 The amplitudes of velocities of ballast particles at measuring points**

### 4.2 The acceleration of ballast particles

Fig.8 shows the time history of vertical acceleration of the ballast particle at measuring point 4 during the cyclic loading. An obvious periodical variation with the frequency of 4Hz can also be seen in this figure. The high frequency of the acceleration induced by the contact vibration of ballast particles reaches up to 3000Hz. Similar time histories of vertical accelerations were also obtained for those ballast particles located at other measuring points.

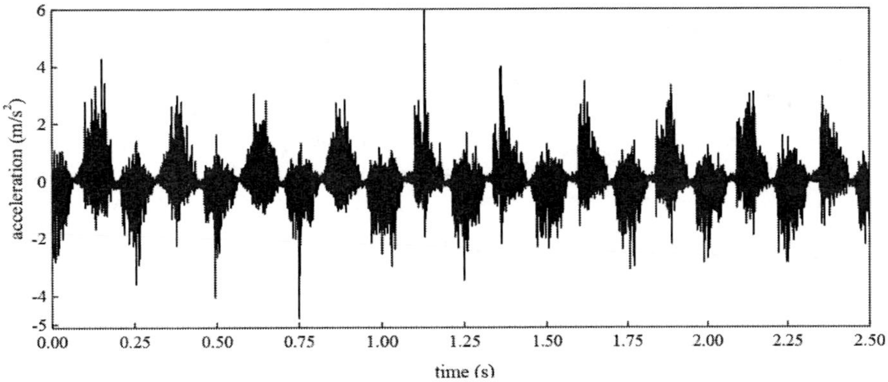


**Fig.8 Time history of vertical acceleration of the ballast particle at point 4**

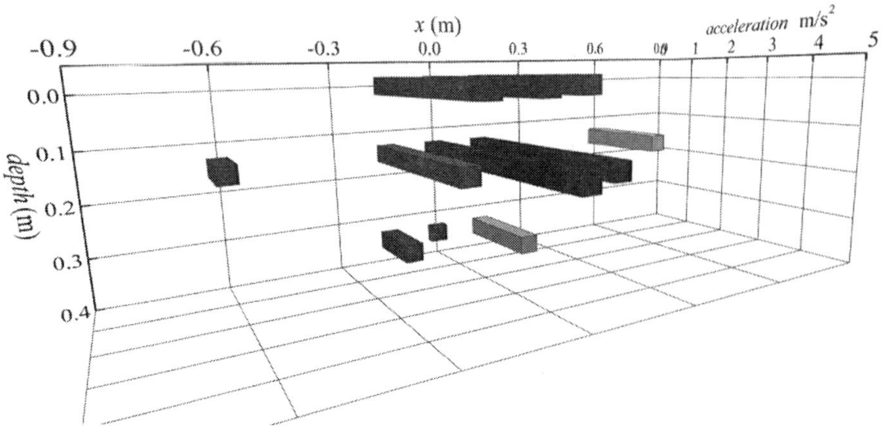


**Fig.9 The amplitudes of accelerations of ballast particles at measuring points**

Fig.9 shows the histogram of the amplitude of vertical acceleration of ballast particles at measuring points. It is worth noting that the acceleration amplitude of ballast particle at point 8 is much smaller than that at point 7 and point 9. The acceleration amplitudes of six particles at 0m and 0.15m under the middle sleeper are close to each other. The average acceleration amplitude is 4.0m/s$^2$ at 0m and 4.3m/s$^2$ at 0.15m, respectively. The average acceleration amplitude of particles at point 7 and point 9 is 1.6m/s$^2$, which means a reduction ratio of 62.8% comparing with the particles at 0.15m under the middle sleeper. The acceleration amplitudes of particles at 0.15m under the left and right sleeper are also much smaller than that under the middle one mainly due to the lower load.

## 5 CONCLUSIONS AND DISCUSSIONS

A 2D discrete element model of railway ballast has been established in this paper. In this model, each ballast particle was regarded as a clump consisting dozens of circle pebbles to catch its irregular shape. By applying cyclic loads on sleepers, the dynamic response of ballast was calculated. The distribution characteristics of contact forces between ballast particles and the ballast vibration in the model were analyzed. Some conclusions can be drawn as follows:

(1) Only a few ballast particles under the sleepers carry large contact forces and form some strong force chains to mainly bear the load. The variance of the magnitude of contact force at each contact is approximately proportional to the variance of the load.

(2) The load applied to the sleepers is transmitted directly downward along the direction of force chains in the ballast and a small angle of diffusion may occur during the propagation of contact forces.

(3) The attenuation of velocity of ballast particles along the depth under a sleeper is remarkable. However, no continuous attenuation of acceleration along the depth was seen in this model. The amplitude of acceleration of ballast particles at 0.15m is close to that at 0 m under the sleeper, and the reduction ratio at 0.35m is 62.8% comparing with the particle at 0.15m.

## ACKNOWLEDGEMENTS

This research was supported by the National Natural Science Foundation of China (NSFC) under grant No.U1234209 and the National Basic Research Program of China (973 Program) under grant 2013CB036200.

## REFERENCES

[1] Knothe, K. and Wu, Y., 1998. Receptance behavior of railway track and subgrade. *Archive of applied mechanics*, 68(7-8): 457-470.

[2] Bode, C., Hirschauer, R., and Savidis, S. A., 2000. Three-dimensional time domain analysis of moving loads on railway tracks on layered soils. *In: Chouw & Schmid: wave*, 3-12.

[3] Zhai, W. M., Wang, K. Y., and Lin, J. H., 2004. Modelling and experiment of railway ballast vibrations. *Journal of sound and vibration*, 270:673-683.

[4] Sebastian, L. G. and Vallejo, L. E., 2006. Discrete element method analysis of railtrack ballast degradation during cyclic loading. *Granular Matter*, 8(3-4): 195-204.

[5] Tutumluer, E., Qian, Y., Hashash Y M A, Ghaboussia, J. and Davis, D. D., 2013. Discrete element modelling of ballasted track deformation behavior. *International Journal of Rail Transportation*, 1(1-2): 57-73.

[6] Lu, M. and McDowell, G. R., 2007. The importance of modelling ballast particle shape in the discrete element method. *Granular matter*, 9(1-2): 69-80.

[7] Bouchaud, J. P., Cates, M. E. and Claudin, P., 1995. Stress distribution in granular media and nonlinear wave equation. *Journal de Physique I*, 5(6): 639-656.

[8] Peters, J. F., Muthuswamy, M., Wibowo, J., Tordesillas, A., 2005. Characterization of force chains in granular material. *Physical review E*, 72(4): 041307.

[9] Ahlbeck, D. R., Meacham, H. C. and Prause, R. H., 1978. The development of analytical models for railroad track dynamics. *In: railroad track mechanics & technology*, Pergamon Press, Oxford.

# Railway bridge maintenance modelling

**B Le, J Andrews**
Nottingham Transportation Engineering Centre, University of Nottingham, UK

## ABSTRACT

Railway bridges are inspected regularly and the condition of their elements recorded in the form of scores. The majority of asset management models have used these scores to predict the condition of the bridge/ structure, given a maintenance strategy. However defining the model states based on such scores has many deficiencies which limit the usefulness of the models when used to establish the maintenance strategy. This paper develops a more subtle and sophisticated bridge model which employs the Petri net approach. The model states are defined based on actual degraded conditions which can be related to the specific maintenance actions required to restore the condition, creating a significantly more accurate and useful output for railway asset managers.

**Keywords:** bridge, asset management, maintenance modelling, whole life costs, Petri net.

## 1 INTRODUCTION

Difficulties in managing railway bridges in the UK are presented by the large number of bridges in operation coupled with the increasing age and utilisation of these structures. To support asset management decision-making, several bridge models have been developed and they can be categorised as Markov (1-6), semi-Markov (7-10) or lifetime analysis based models (11-14). All of these models are stochastic and state-based. They can be used to evaluate the effect of a specified maintenance strategy. The model prediction is often the probability of an asset being in each of the potential condition states. Most of the models developed are based on the Markov approach, since Markov models are relatively easy to construct with a fast solution time. However, these models assume a constant deterioration rate and the effect of an increasing need for maintenance as the failure rate increases with age is not captured (15). Semi-Markov bridge models are capable of modelling the commonly experienced increasing deterioration rates by employing the Weibull distribution to model the time of bridge elements residing in the different states. However, the number of model states increases exponentially, which can present difficulties when it comes to modelling complex problems (16). In the lifetime analysis-based models, the degradation process of bridges or bridge elements is modelled using the lifetime analysis method. Studies have demonstrated that the bridge elements follow a non-constant deterioration process and that the Weibull distribution provides the best fit to the life time data of bridge elements (15).

Bridge assets in the US are rated in discrete values between 0-9 (17, 18), whereas in the UK, railway bridge condition is described by values from 0-100 (19). Most of the bridge history condition databases contain these condition scores/ratings and hence the majority of bridge models developed define model states based on these values. The model states correspond to either a single score or a range of these scores. Bridge models developed based on these scores rarely indicate the maintenance action required to fix the problem (20). This is particularly true where the same scoring system is used for different bridge elements with different deterioration processes. The scores/ratings are also subjective, with the reliability of the ratings depending on the experience and subjective evaluation of bridge inspectors (21). Despite all these limitations for accurate model prediction, the condition scores are commonly used as there is little other data available to enable alternative model formulations.

This paper proposes a bridge model based on the Petri net (PN) approach. The bridge model states are defined based on the main deterioration processes that relate to the maintenance action required to restore the element condition. For each deterioration process, the level of deterioration is expressed in terms of the percentage of the affected area of the bridge element. Sub-models are developed for major bridge elements and combined to form a complete bridge model. The PN method employed in this paper allows the modelling of dynamic processes using Weibull distributions for the state residence times. The flexibility of this method enables the model developed to consider, in detail: the deterioration processes; the interaction between the conditions of different elements of the bridge system; and the inspection and maintenance processes; whilst maintaining a manageable analytical problem. The methodology is demonstrated through an application to a typical metal railway bridge structure.

## 2 BRIDGE ELEMENTS AND THEIR DETERIORATION MODES THAT TRIGGER MAINTENANCE

The bridge type used to demonstrate the asset condition modelling process has reinforced concrete decks, metal girders and masonry abutments as illustrated in Figure 1. Waterproofing, whilst not a major bridge element, is important to consider, as it directly affects the degradation processes of the bridge deck and abutments, thus it is included in the analysis in this paper. Over time, each bridge element deteriorates and structural defects start to appear. According to the observation scheme currently employed by several bridge authorities in the UK (20), the severity and extent of these structural degradation defects can be used to describe the condition of each bridge element. Adopting this approach, this paper identifies the condition of the structure in terms of the main degradation processes and their extent (% coverage) as illustrated in Table 1. There can be several deterioration processes for a single element and they can either be independent, dependent, or represent the progression from another process. Levels of degradation are used to define the condition state of an element and at these states, the appropriate type of intervention that is necessary (Table 1).

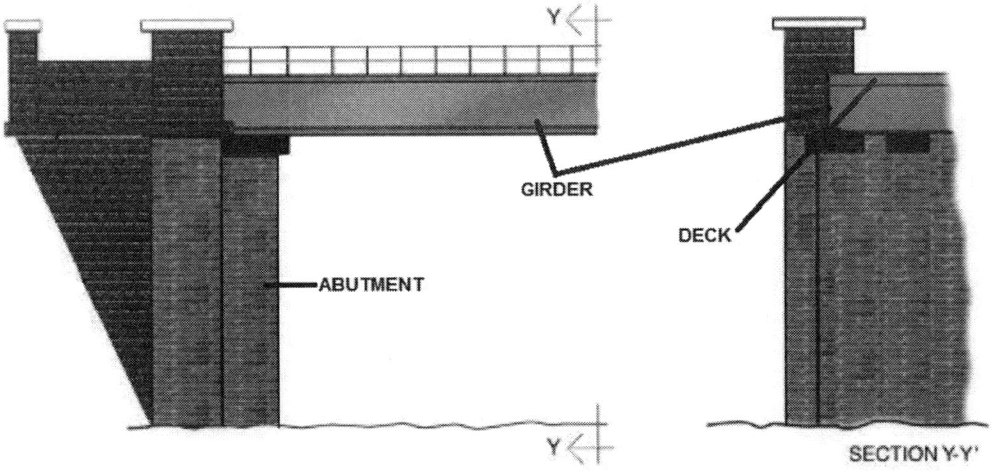


**Figure 1: Typical bridge structure**

**Table 1: Degradation levels and corresponding interventions**

| Element | Material | Degradation process | Levels of degradation (% coverage of defects) | Maintenance actions |
|---|---|---|---|---|
| Abutment | Masonry | Vegetation infestation | 5%, 10%, 15%, 20%, ≥25% | Vegetation remove |
| | | Pointing degradation | 5%, 10%, 15%, 20%, ≥25% | Re-pointing work |
| | | Brick erosion | 5%, 10%, 15%, 20%, 25%, ≥30% | Brickwork repair |
| Girder | Metal | Paint flaking | 5%, 10%, 15%, 20%, 25%, ≥30% | Patch painting/full painting |
| | | Minor corrosion | 5%, 10%, 15%, 20%, 25%, ≥30% | Minor corrosion repair (corrosion clean and re-paint) |
| | | Major corrosion | 5%, 10%, ≥20% | Major corrosion repair (welding/plating and re-paint) |
| | | | | Renew when major corrosion ≥20% |
| Deck | Reinforced concrete | Surface spalling | 5%, 10%, 20%, 30%, 40%, ≥50% | Concrete patching |
| | | Deep spalling | 5%, 10%, 20%, ≥30% | Filling/concrete patching |
| | | Steel corrosion | 5%, 10%, 20%, ≥30% | Structure repair |

## 3 PETRI NET METHOD

The Petri-Net (PN) method (22) is increasingly being used to model dynamic systems in many engineering, science and business fields due to its flexibility (23). The concept of a PN (24) is a directed graph with two types of nodes called places and transitions, they are linked by arcs. A place represents a particular state or condition of the system. Tokens are located in the places to indicate the current state of the system. A transition enables a token to move from one place to another. The firing of tokens changes the state of the system, thus the dynamic behaviour of the system is modelled. Firing happens when a transition is enabled: this is when there are tokens in all the input places of a transition. Once a transition has fired, tokens in input places are cleared and deposited in the output places according to the arc (arrows). A simple PN and the firing process is illustrated in Figure 2, which includes inhibitor arcs. An inhibitor arc is denoted by an arc with a round head, instead of an arrow. It inhibits the firing process of a transition, once there is a token in its input place.

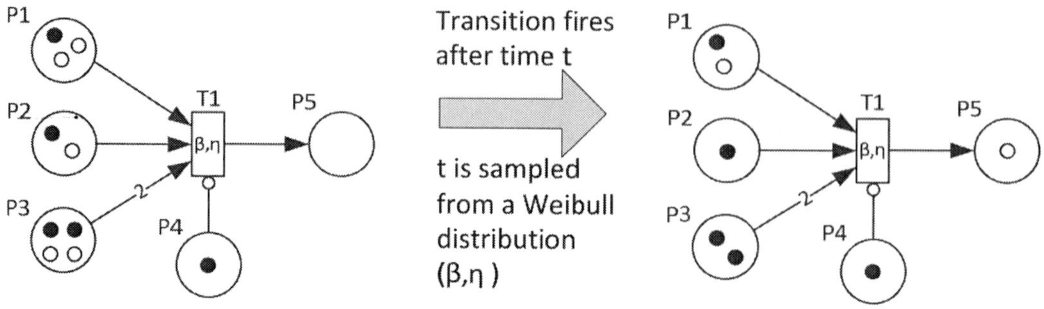


**Figure 2: Features of a PN and the firing process**

It was found that in order to produce an efficient model, new types of transitions were needed to accommodate certain tasks. They are labelled as illustrated in Figure 3. A *reset transition(16, 25)*, when it fires, resets the marking of specified places in the PN to the desired number of tokens. This feature is useful for when maintenance action has happened and bridge element condition is restored. A *place conditional transition* (16, 25) is a transition in which the delay time is sampled from different distributions depending on the number of tokens residing in a specific place in the network to which they are linked by a dashed arc. This transition allows easy modelling of dependent deterioration processes. An *opportunistic transition* is used in the modelling of an opportunistic maintenance action (as defined in 4.3g below) on bridge elements where each element is modelled by different types/colours of token.

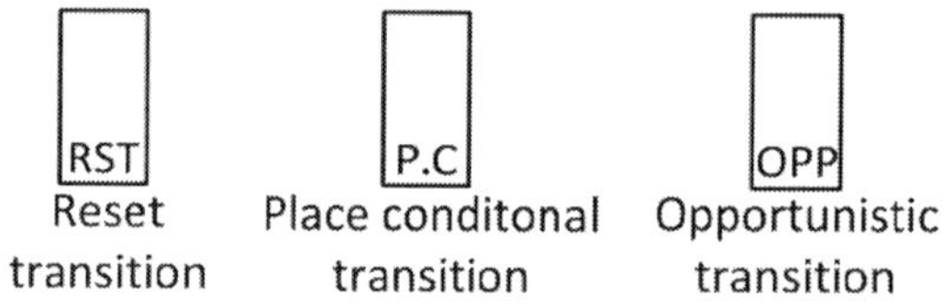


**Figure 3: Representation of the new types of transitions**

## 4 DEVELOPMENT OF PN BRIDGE MODEL

The complete bridge model is constructed in modules (sub-models) which correspond to each individual bridge element. These individual section models are linked together using common places. Tokens are added to the net where each token represents a single bridge element. The dynamic process of the token moving around the complete net models the associated degradation, inspection, maintenance and renewal processes of each bridge element over the bridge life time. The state of the bridge is represented by the combined marking of the places in the modelled elements. The model complexity can be increased by adding more rules and relationships if required, this can easily be accommodated within the approach presented. The following sub-sections demonstrate some parts of the PN that were developed to model different processes for the bridge elements considered (deck, girders, and abutments). Note that, only some parts of the net are presented to save repetition because some of the nets modelling the same process for different bridge elements would have similar structure.

### 4.1 Deterioration process

For a concrete bridge deck, the section of the net modelling the degradation process is illustrated in Figure 4. The degradation process exhibits three progressive modes: surface spalling, deep spalling and steel corrosion (Table 1) represented by places P1-17. The levels of degradation are described in terms of the percentage areas of the defects over the structure. Between these degraded states are transitions T1-14 which specify the transition times of moving between different levels of deterioration of the bridge deck. In this paper, the transition times are sampled from the appropriate Weibull distributions with parameters ($\beta_1,\eta_1,$ $\beta_2,\eta_2$ ... $\beta_{14},\eta_{14}$). It can also be seen that these transitions are place conditional transitions (denoted P.C), because the deterioration rate of the bridge deck is dependent on the condition of the waterproofing, represented by places P(WP.1-3). With worse waterproofing conditions, the degradation rate of the bridge deck increases slightly and the mean time to move to a more degraded states decreases. There are also dependencies between the three degradation mechanisms such that: deep spalling cannot start until the concrete surface is in the spalled condition and the reinforcing steel corrosion can only start when deep spalling has occurred.

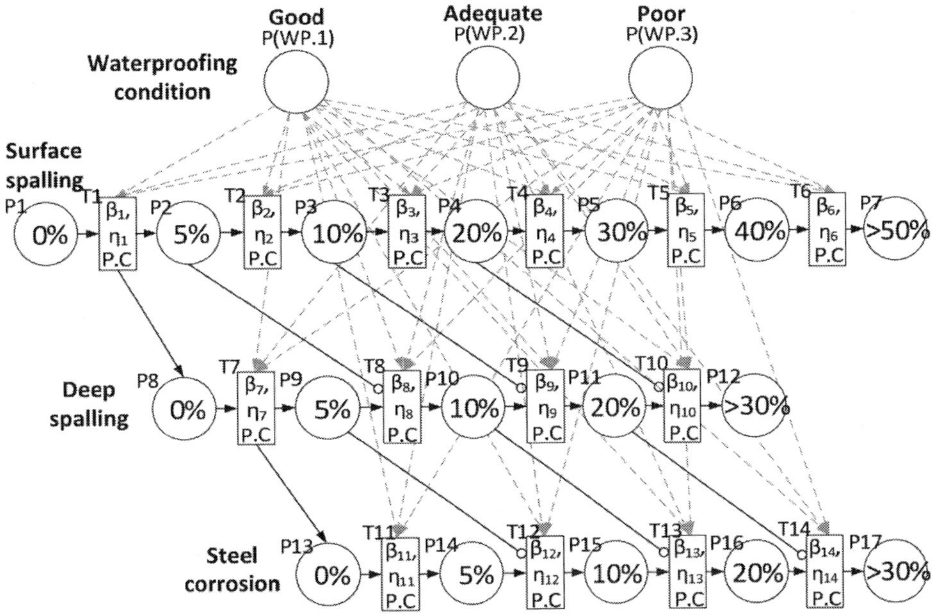


**Figure 4: PN of the degradation process of the bridge deck**

Similarly, for the bridge abutments and girders, the deterioration processes are modelled using the PNs as illustrated in Figure 5 and Figure 6 respectively. For the bridge abutments, there are three separate processes that drive the degradation: vegetation infestation, pointing degradation and brick degradation. These processes are also dependent on the condition of the waterproofing given by places P(WP.1-3) and illustrated in Figure 5. The main deterioration process for bridge metal girders is corrosion, and the corrosion process is enabled by the degradation of the protective paint condition. It is recognised that patch painting only repaints the area of the flaked paint, other areas that have already started the degradation process would have reduced times of reaching the degraded state. This effect is captured in the model.

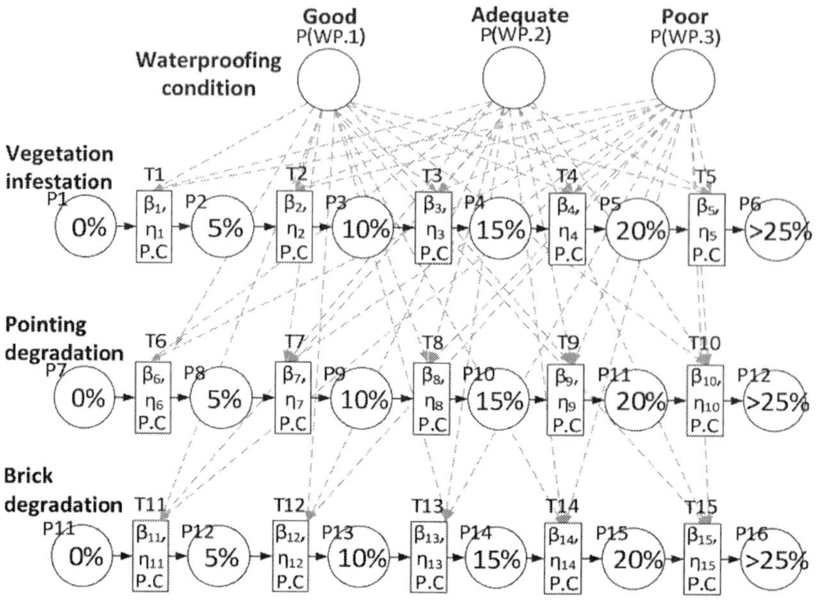


**Figure 5: PN of the degradation process of the bridge abutments**

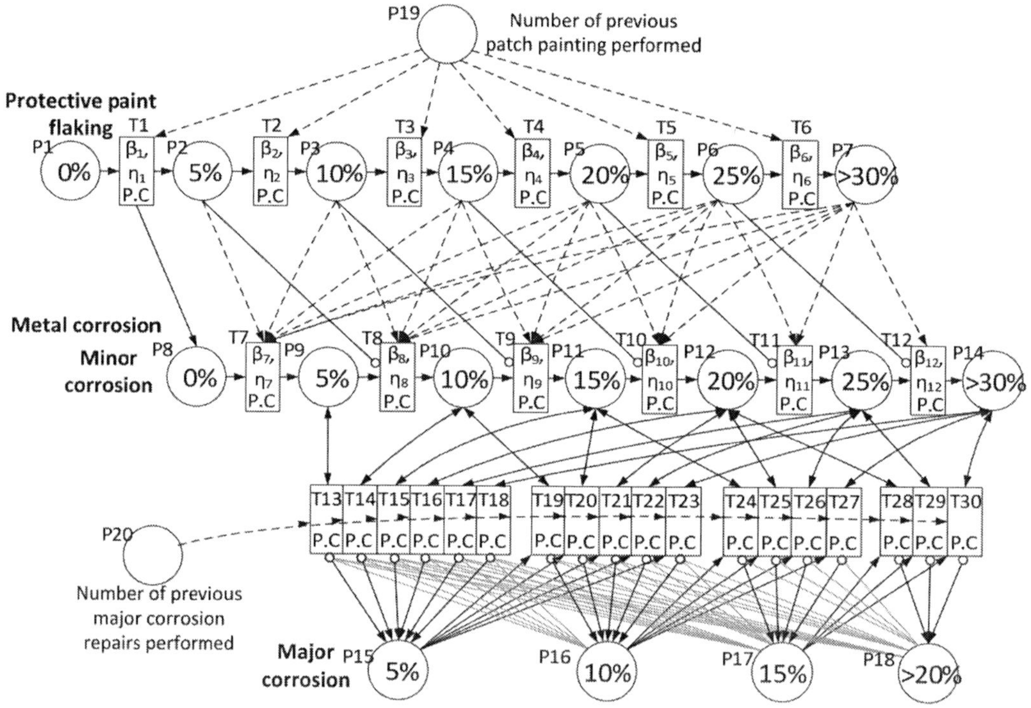


**Figure 6: PN of the degradation process of the bridge girders**

### 4.2 Inspection process

The state of the element will only be revealed following an inspection. Inspection of a bridge deck is modelled using the inspection transitions (marked with INSP) in Figure 7. It should be noted that the deterioration process of the bridge deck continues after the inspection. This allows the model to effectively capture the scenario where the bridge element condition degrades to a poorer state during the repair scheduling period (the time between when the inspection happens and the repair takes place). There are two types of inspections, with different inspection periods. Minor inspection takes place every year but can only detect surface spalling. A major inspection happens every 4 years and reveals any deep spalling or reinforcing steel corrosion. Places P32-34 ensure that an inspection will not take place again whilst scheduled for repair. The sections of the PN modelling the inspection process for other bridge elements follow the same structure.

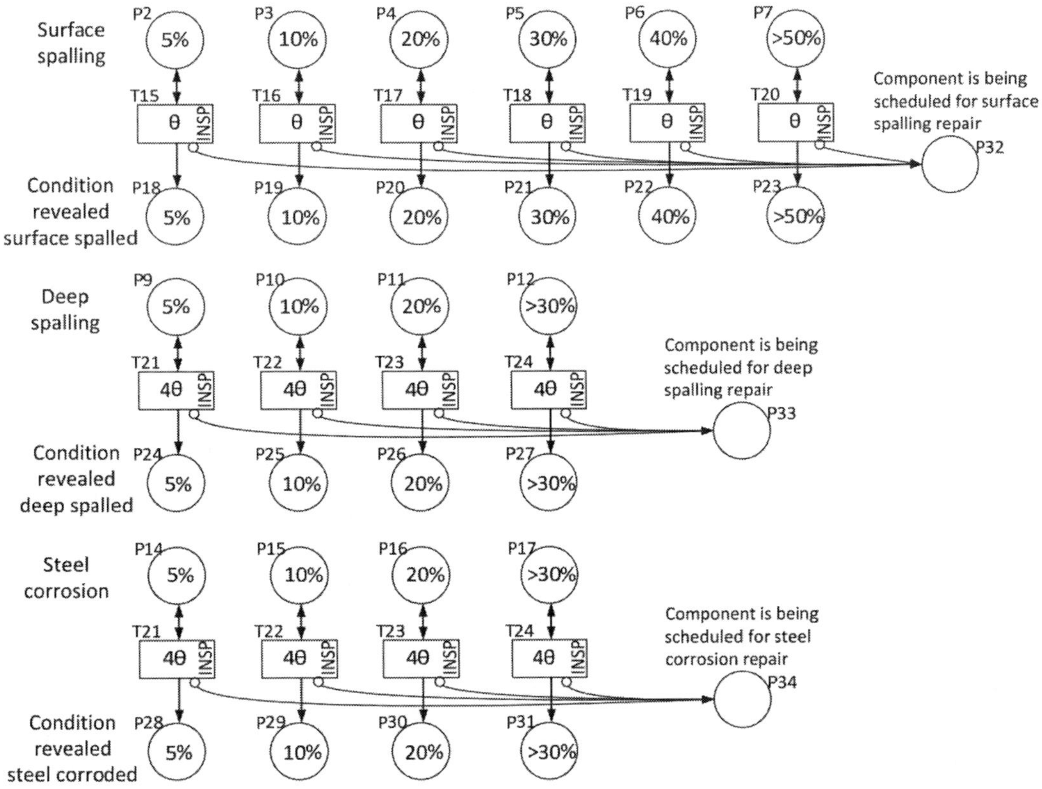


**Figure 7: PN of the deck inspection process**

### 4.3 Maintenance and scheduling process

When the condition of the bridge element is revealed after an inspection, then the appropriate type of maintenance work can be applied. Figure 8 illustrates part of the net that models the complete maintenance process of the bridge girders. For other bridge elements, the nets follow a similar structure. In this paper, the following bridge asset maintenance issues are considered:

1. Intervention actions can have scheduling delay times or can be carried out immediately.
2. The repair times are different depending on the type of repair and the severity of the defects, i.e. it would take a longer repair time for a larger percentage of degraded area.
3. The element continues to degrade after the inspection while it is being scheduled for repair, thus the appropriate maintenance must be based on the true condition of the bridge element at the point when repair actually begins.
4. For the bridge deck, deep spalling and reinforced steel corrosion are not revealed through minor inspection, however they will be revealed as soon as the repair process for the surface spalling begins and the appropriate repair is triggered to rectify these undetected defects.
5. For the bridge girders, major corrosion is revealed through the repair process for minor corrosion even if they are undetected during a minor inspection.

6. Flaked paint is patch painted, however patch painting is only allowed up to 5 times, after this, the whole girder is fully repainted.
7. When repairing corrosion, the corroded area will also be repainted.
8. The girder is renewed once major corrosion covers more than 20%.
9. Opportunistic maintenance is considered on adjacent girders if their condition, while not currently triggering maintenance, is expected to do so in the near future.
10. For the abutments, any repointing and brickwork repair would also clear any vegetation on the structure.

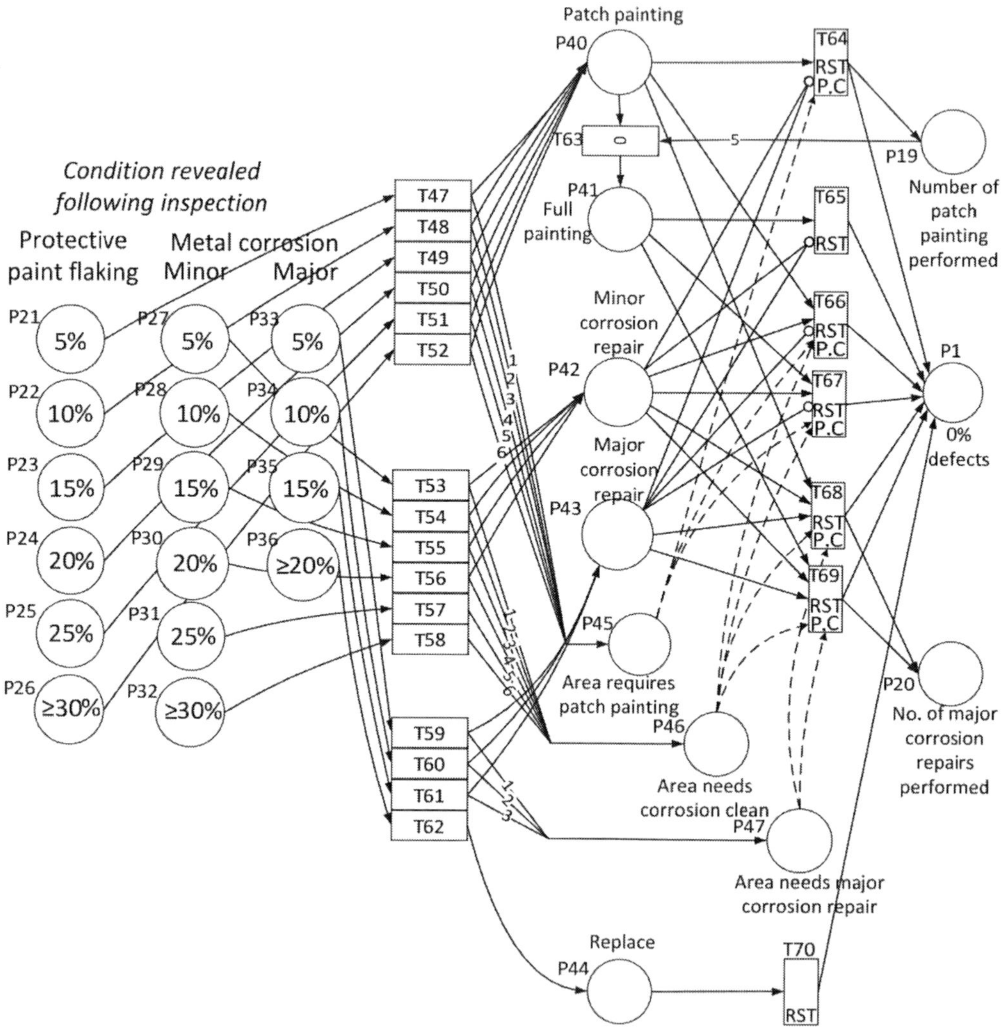

Figure 8: PN for the maintenance, scheduling and renewal process of the bridge main girders

## 4.4 Renewal process

At some time the bridge element will reach a point where it is cheaper to renew it rather than continue to maintain it. There are many possibilities as to how this time can be established. It can be after a predetermined lifetime; after the condition is deemed to be too poor; or after a set number of interventions. Any of these strategies can be implemented within the PN. For the bridge deck, it is renewed after a certain number of specific interventions and Figure 9 illustrates the section of the PN that implements this renewal strategy. For the girder, the girder is renewed when major corrosion covers more than 20%. The renewal of the waterproofing is after every 15 years.

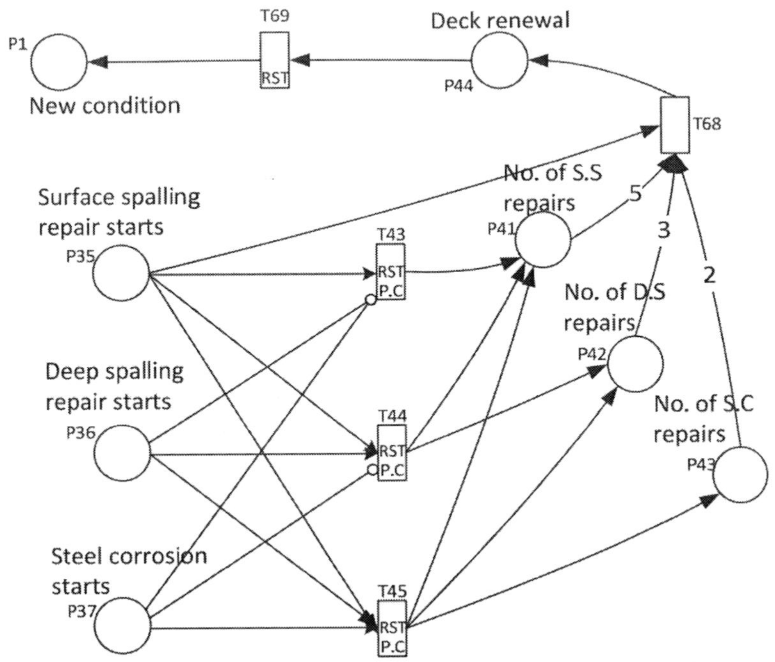


**Figure 9: PN net for the renewal process**

### 4.5 Maintenance policy

Different maintenance policies can be set by choosing the level of degradation which will trigger maintenance. Figure 10 and Table 2 illustrate the maintenance options available for the bridge deck. The bridge deck repair strategy is set by placing the appropriate number of tokens into places P45-47. Similarly, different maintenance policies can be set for other bridge elements.

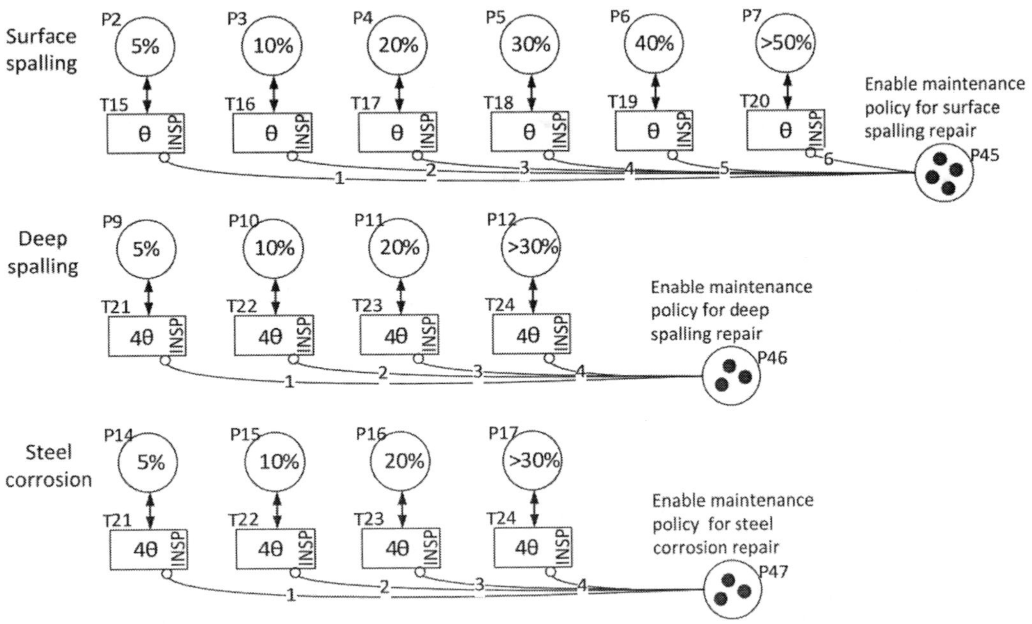


**Figure 10: PNs for setting maintenance policy for the bridge deck**

**Table 2: Maintenance policies implemented in the PN for bridge deck**

| No of token in P45 | Maintenance action is triggered when surface spalling is | No of token in P46 | Maintenance action is triggered when deep spalling is | No of token in P47 | Maintenance action is triggered when steel corrosion is |
|---|---|---|---|---|---|
| 0 | ≥5% | 0 | ≥5% | 0 | ≥5% |
| 1 | ≥10% | 1 | ≥10% | 1 | ≥10% |
| 2 | ≥20% | 2 | ≥20% | 2 | ≥20% |
| 3 | ≥30% | 3 | ≥30% | 3 | ≥30% |
| 4 | ≥40% | ≥4 | No intervention | ≥4 | No intervention |
| 5 | ≥50% | | | | |
| ≥6 | No intervention | | | | |

### 4.6 Opportunistic maintenance

Opportunistic maintenance considers carrying out maintenance on elements which have a deteriorated condition but would not normally trigger maintenance. It is possible to implement opportunistic maintenance in the developed model. Figure 11 illustrates part of the net that performs opportunistic patch painting on a group of girders represented by the black and white tokens.

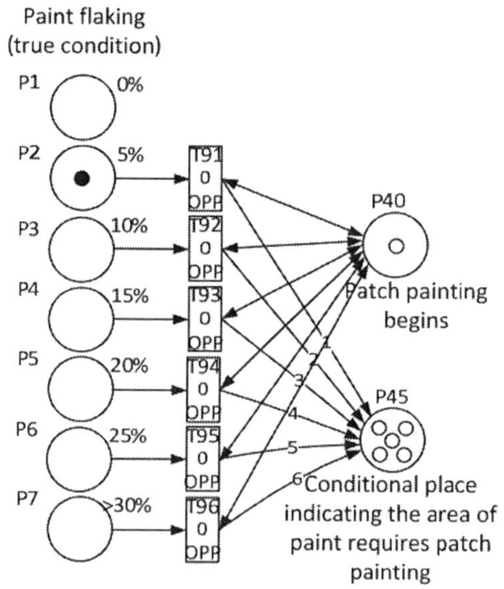


**Figure 11: PNs for opportunistic maintenance of different girders modelled using different tokens**

## 5 MODEL SOLUTION AND RESULTS

### 5.1 Model construction and inputs

A computer program was written to generate the solution of the PN model. There are two main types of model inputs: information regarding the degradation characteristics of bridge elements (26) which can be obtained from studying historical data or from expert opinions; and parameters that are used to specify a selected maintenance and renewal strategy which is set by maintenance decision makers.

### 5.2 Monte Carlo simulation

The solution of the model was achieved using a Monte Carlo simulation (27). The lifetime of the bridge is simulated using the PN for a number of simulations and once the results are checked for convergence, the performance statistics are collected.

## 5.3 Model results and discussion

An example analysis of all the element models is presented in this paper. All elements are assumed to have the initial condition of as good as new and the repair strategy is to repair as soon as any defect is revealed. Each simulation was performed over a lifetime of 60 years. With random sampling, the results must be checked for convergence, and this was found to be achieved in all analyses after 200 simulations, with a solution time for the model of under 10 minutes.

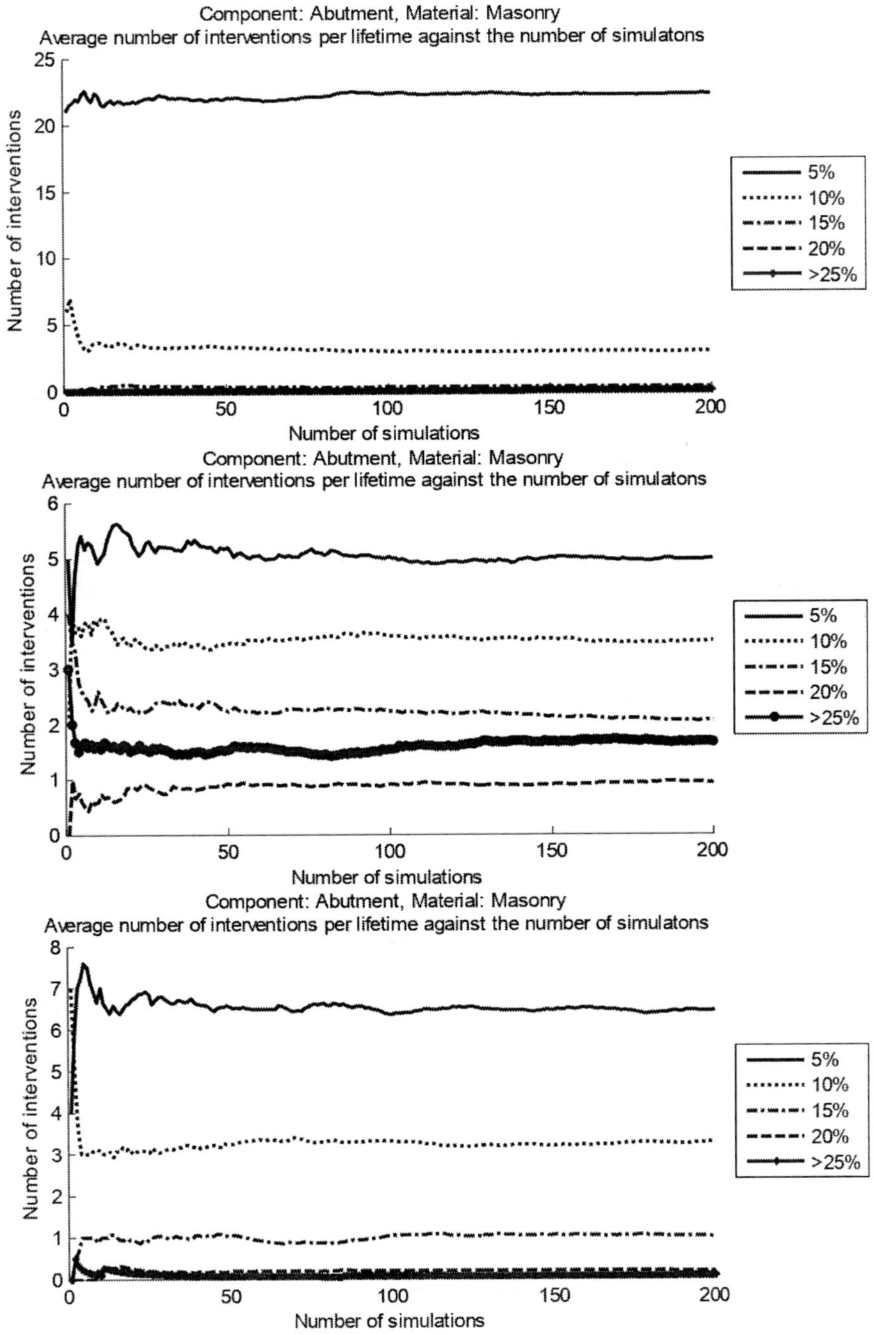


**Figure 12: Average number of interventions on the bridge abutments (a) Vegetation clearing (b) Repointing works (c) Brickwork repair**

Figure 12 shows the average number of interventions performed on the abutment over its lifetime. Plots 12(a), (b) and (c) show the number of maintenance actions for vegetation clearing, repointing and brickwork repair respectively. Furthermore, each of these plots shows the expected number of specific actions, for example, the third plot shows that there are 6.5 repairs for when there is 5% of brick defects; about 3 brickwork repairs at the point where 10% of the brickwork has degraded, about 1 repair at the 15% degraded brick condition and almost no repairs at the more degraded conditions.

For other bridge elements, this information can also be easily obtained, Table 3 shows the statistics obtained for a single bridge girder. It can be seen that patch painting is required about 12 times and on average the girder will be fully repainted 1.68 times. The number of interventions required for corrosion is generally low, being under 2 times over the 60-year lifetime prediction and no girder renewal is expected.

**Table 3: Statistics of the number of interventions on a single bridge girder**

| **Type of interventions** | | **Level of degradation** | | | | | |
|---|---|---|---|---|---|---|---|
| | | **5%** | **10%** | **15%** | **20%** | **25%** | **>30%** |
| **Patch painting** | Min | 2 | 0 | 0 | 0 | 0 | 0 |
| | Max | 20 | 5 | 1 | 1 | 0 | 0 |
| | Avg | 11.30 | 1.03 | 0.06 | 0.01 | 0 | 0 |
| | S.D | 2.89 | 1.02 | 0.23 | 0.07 | 0 | 0 |
| **Full painting** | Min | 0 | | | | | |
| | Max | 3 | | | | | |
| | Avg | 1.68 | | | | | |
| | S.D | 0.57 | | | | | |
| **Minor corrosion repair (corrosion clean and repaint)** | Min | 0 | 0 | 0 | 0 | 0 | 0 |
| | Max | 7 | 1 | 0 | 0 | 0 | 0 |
| | Avg | 1.42 | 0.02 | 0 | 0 | 0 | 0 |
| | S.D | 1.34 | 0.14 | 0 | 0 | 0 | 0 |
| **Major corrosion repair (welding/ plating)** | Min | 0 | 0 | 0 | - | - | - |
| | Max | 2 | 0 | 0 | - | - | - |
| | Avg | 0.10 | 0 | 0 | - | - | - |
| | S.D | 0.32 | 0 | 0 | - | - | - |
| **Renewal** | Min | 0 | | | | | |
| | Max | 0 | | | | | |
| | Avg | 0 | | | | | |
| | S.D | 0 | | | | | |

Figure 13(a) shows the probabilities of the bridge deck residing in states with surface spalling for different degrees of coverage, over the 60 year lifetime. Figure 13(b) shows similar results for the bridge deck residing in deep spalling states.

Figure 14 shows the distributions of times the waterproofing resides in different conditions. The distributions clearly show that, over the pre-determined lifetime of 15 years until the waterproofing is re-applied, it spends more time in poor condition than in good or adequate conditions.

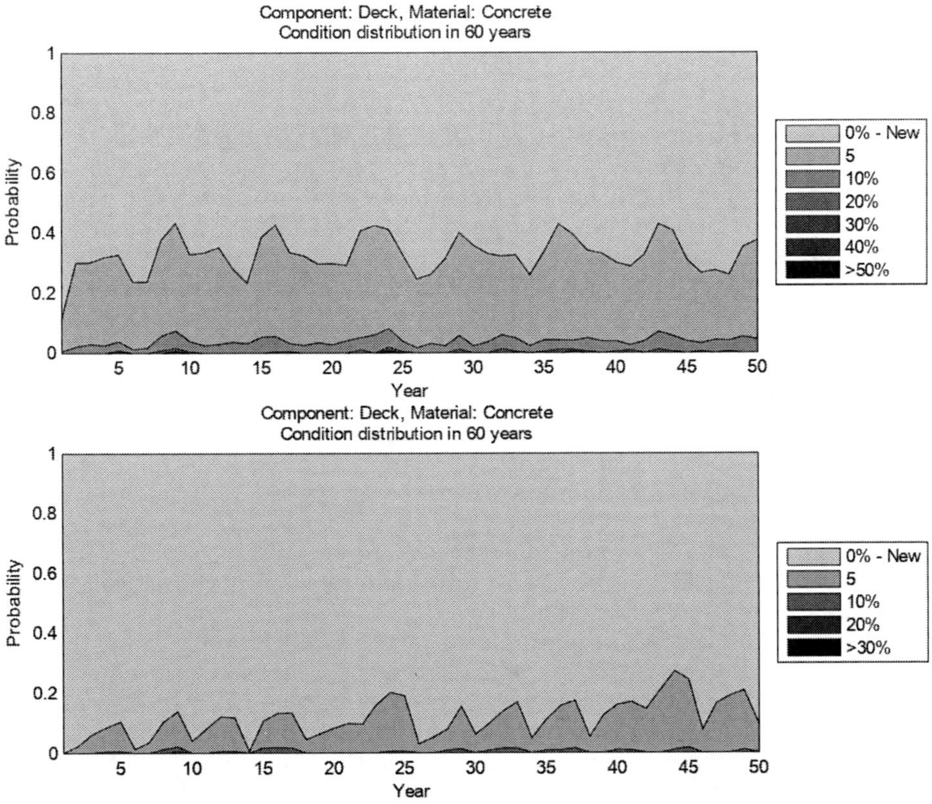


**Figure 13: Probabilities of the bridge deck being in different condition states: (a) states represent level of degradation of surface spalling (b) states represent level of degradation of deep spalling**

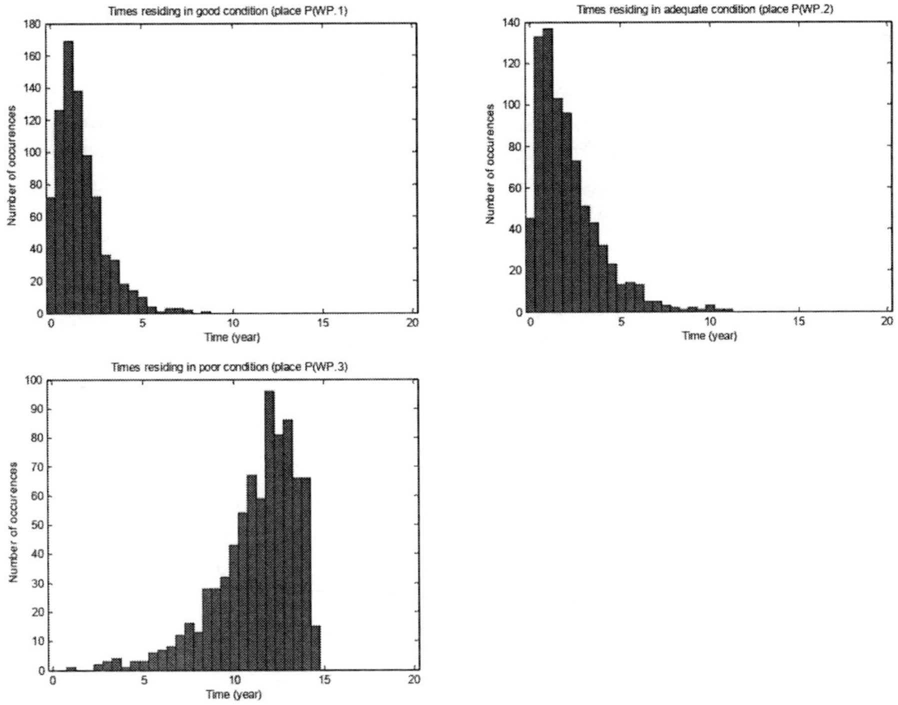


**Figure 14: Distributions of times of waterproofing residing in different conditions**

## 6 CONCLUSION

In conclusion, the benefits of the bridge model developed based on the PN modelling method are:

- Model states are based on actual bridge element deterioration conditions that relate to required maintenance actions for their restoration.
- Bridge elements can have several deterioration processes, which can be independent or dependent of each other, and which may result in different deteriorated conditions.
- The model has the capability of modelling bridge elements with changing deterioration rates.
- The model considers the inspection processes through which the asset condition is revealed. Depending on the maintenance strategy selected, appropriate maintenance actions are triggered.
- The model can account for maintenance scheduling times during which the asset may continue to deteriorate further. The types of maintenance action are adjusted at the point of execution to match the required restoration work.
- Using the model, some defects which would be undetected at minor inspections, will be revealed and acted upon opportunistically when the maintenance work is executed.
- The model records maintenance actions performed, and their future effectiveness to control the state of the asset is also captured.
- The model incorporates the potential to include several routine maintenance, opportunistic maintenance and renewal strategies, each set by placing tokens in controlling places in the PN bridge model.
- The same PN bridge model can be used to model bridge elements of the same type, thus the model states and size remain the same even if the number of similar modelled elements increases in a more complex structure.

## ACKNOWLEDGEMENT

John Andrews is the Royal Academy of Engineering and Network Rail Professor of Infrastructure Asset Management. He is also Director of Lloyd's Register Foundation (LRF)[1] for Risk and Reliability Engineering at the University of Nottingham. Bryant Le is conducting a research project funded by Network Rail. They gratefully acknowledge the support of these organisations.

## REFERENCE LIST

(1) Jiang, Y. and K.C. Sinha, *Bridge service life prediction model using the Markov chain.* Transportation Research Record, 1989(1223): p. 24-30.

(2) Robelin, C.-A. and S.M. Madanat, *History-dependent bridge deck maintenance and replacement optimization with Markov decision processes.* Journal of Infrastructure Systems, 2007. **13**(3): p. 195-201.

(3) Cesare, M.A., et al., *Modeling bridge deterioration with Markov chains.* Journal of Transportation Engineering, 1992. **118**(6): p. 820-833.

(4) Ortiz-García, J.J., S.B. Costello, and M.S. Snaith, *Derivation of transition probability matrices for pavement deterioration modeling.* Journal of Transportation Engineering, 2006. **132**(2): p. 141-161.

---

[1] Lloyd's Register Foundation supports the advancement of engineering-related education, and funds research and development that enhances safety of life at sea, on land and in the air.

(5) Chase, S.B. and L. Gáspár, *Modeling the reduction in load capacity of highway bridges with age.* Journal of Bridge Engineering, 2000. **5**(4): p. 331-336.
(6) Morcous, G., *Performance Prediction of Bridge Deck Systems Using Markov Chains.* Journal of Performance of Constructed Facilities, 2006. **Vol. 20**(No. 2): p. 146–155.
(7) Ng, S. and F. Moses, *Prediction of bridge service life using time-dependent reliability analysis.* Bridge management, 1996. **3**: p. 26-32.
(8) Kleiner, Y., *Scheduling inspection and renewal of large infrastructure assets.* Journal of Infrastructure Systems, 2001. **7**(4): p. 136-143.
(9) Mishalani, R.G. and S.M. Madanat, *Computation of infrastructure transition probabilities using stochastic duration models.* Journal of Infrastructure Systems, 2002. **8**(4): p. 139-148.
(10) Yang, Y.N., H.J. Pam, and M.M. Kumaraswamy, *Framework Development of Performance Prediction Models for Concrete Bridges.* Journal of Transportation Engineering, 2009. **135**(8): p. 545-554.
(11) DeStefano, P.D. and D.A. Grivas, *Method for estimating transition probability in bridge deterioration models.* Journal of Infrastructure Systems, 1998. **4**(2): p. 56-62.
(12) Frangopol, D.M., S.J. Kong, and S.E. Gharaibeh, *Reliability-based life-cycle management of Highway bridges.* Journal of Computing in Civil Engineering, 2001. **15**(1): p. 27-34.
(13) Noortwijk, J.M.v. and H.E. Klatter, *The use of lifetime distribution in bridge maintenance and replacement modelling.* Computers and Structures, 2004. **82**: p. 1091-1099.
(14) Sobanjo, J., P. Mtenga, and M. Rambo-Roddenberry, *Reliability-based modeling of bridge deterioration hazards.* Journal of Bridge Engineering, 2010. **15**(6): p. 671-683.
(15) Agrawal, A.K., A. Kawaguchi, and Z. Chen, *Deterioration rates of typical bridge elements in New York.* Journal of Bridge Engineering, 2010. **15**: p. 419-429.
(16) Andrews, J., *A modelling approach to railway track asset management.* Proceedings of the Institution of Mechanical Engineers, Part F: Journal of Rail and Rapid Transit, 2013. **227**(1): p. 56-73.
(17) American Association of State Highway Transportation Officials. Subcommittee on Bridges Structures, *The Manual for Bridge Evaluation*2011: American Association of State Highway and Transportation Officials.
(18) Federal Highway Administration, *Specifications for the National Bridge Inventory*, 2006.
(19) Network Rail, *Structures Condition Marking Index Handbook for Bridges (formerly RT/CE/C/041)*, 2004.
(20) Madanat, S., R. Mishalani, and W.H.W. Ibrahim, *Estimation of infrastructure transition probabilities from condition rating data.* Journal of Infrastructure Systems, 1995. **1**: p. 120.
(21) Office of Rail Regulation, *Annual assessment of Network Rail 2006-07*, 2007.
(22) Petri, C.A., *Fundamentals of a Theory of Asynchronous Information Flow.* Proc. of IFIP Congress 62. --- Amsterdam: North Holland Publ. Comp., 1963: p. 386-390.
(23) Girault, C. and R. Valk, *Petri nets for systems engineering: a guide to modeling, verification, and applications* 2002: Springer.
(24) Schneeweiss, W.G., *Petri nets for reliability modeling:(in the fields of engineering safety and dependability)* 1999: LiLoLe-Verlag.
(25) Prescott DR, A.J., *A Railway Track Ballast Maintenance and Inspection Model for Multiple Track Sections*, in *In Proceedings of PSAM 11 (Probablistic Safety Assessment and Management) / ESREL 2012 (European Safety and Reliabilty Conference)* 2012: Helsinki, Finland.

(26) Le, B. and J. Andrews, *Modelling railway bridge asset management.* Proceedings of the Institution of Mechanical Engineers, Part F: Journal of Rail and Rapid Transit, 2013. **227**(6): p. 644-656.
(27) Rubinstein, R.Y. and D.P. Kroese, *Simulation and the Monte Carlo Method* 2011: Wiley.

# Multi-domain simulation of hybrid diesel-driven railway vehicles and comparison of electrical energy storage systems

**S Schmid [1], K Ebrahimi [2], W Commerell [3]**
[1] Voith Turbo GmbH & Co. KG, atr – Research and Development, Germany
[2] University of Bradford, School of Engineering, Design and Technology, UK
[3] UAS Ulm, Institute of Energy and Drive Technologies, Germany

1D simulation models of hybrid diesel-driven railway vehicles are evaluated with the help of a Python based simulation tool. Two diesel-hydromechanic hybridized propulsion systems in combination with three different electrical energy storage systems based on batteries, double-layer capacitors and flywheel technology are regarded. The fuel saving potentials of the two hybrid diesel multiple units on a regional and a suburban service profile are investigated for different energy management strategies. Detailed life cycle cost analyses, taking into account the operation mode of a diesel multiple unit, show the cost savings which can be achieved with each type of storage system from a vehicle operators' point of view.

## 1 INTRODUCTION

Due to their flexibility and independence from catenaries, diesel-driven railway vehicles are an important tool for passenger transportation all over the world. In order to reduce life cycle costs (LCC) for their operators, railway vehicle manufacturers and suppliers are looking for ways to improve driveline efficiency of diesel multiple units (DMU). One way of achieving this goal is by hybridizing DMU propulsion systems with the help of energy storage systems. In the past decade several examples of hybrid DMUs being tested or already in service have appeared in literature [1, 2, 3, 4]. In addition, a study conducted by the Department of Transport under the direction of Dr S. Hillmansen also showed the high potential of railway vehicle hybridization for the British railway system [5, 6]. In a joint research project, Voith Turbo, the University of Bradford and the University of Applied Sciences Ulm investigate on the potentials of hybrid railway vehicles with the help of model-based design in order to meet future demands.

In the present paper, the main results of the first year of the project are highlighted and a conclusion for further research areas based on preliminary simulation results is given. Section 2 describes the analyzed hybrid railway vehicles and their main characteristics. Through the use of a Python-based simulation tool, described in section 3, the potentials in terms of fuel reduction are analyzed for both vehicles for the UNIFE TecRec 100 suburban and regional service profiles [7]. The simulation results and LCC analysis for one of the regarded DMUs are discussed in section 4. The paper concludes with a discussion of the simulation results and a recommendation for further research in terms of use of electrical energy storage systems for railway vehicles.

## 2 ANALYSED HYBRID DMU CONFIGURATIONS

### 2.1 Main vehicle and track parameters

For the simulative study conducted in this work, two common DMUs will be analysed in terms of their hybridization potential with the help of model-based design. The first one is a Mobile 2 by Bom Sinal, which is commonly used for commuter transport in South America. The vehicle is propelled by two 6-cylinder inline diesel engines by MAN with a nominal output of 338 kW each. A hydromechanic 4-speed DIWA transmission by Voith [8] is used to convert the engines speed. Other main characteristics can be found in Table 1. It is common for DMUs, that the diesel engine, gearbox and auxiliary systems are connected to a supporting frame and referred to as railpack.

**Table 1. Vehicle parameters Bom Sinal Mobile 2**

| Parameter | Unit | Value |
|---|---|---|
| Train mass (wet incl. passengers) | [t] | 98.0 |
| Adhesion mass | [t] | 40.0 |
| Number of railpacks | [-] | 2 |
| Max. speed | [km/h] | 90 |
| Nom. engine power | [kW] | 338 kW |

For regional drive cycles in Germany, a Siemens Desiro Classic is a DMU mostly operated by the DB as class 642. It is equipped with two six cylinder diesel engines with nominal outputs of 390 kW connected to a six speed automatic gearbox. Table 2 shows the main data of the vehicle used for the simulation.

**Table 2. Siemens Desiro Classic vehicle parameters**

| Parameter | Unit | Value |
|---|---|---|
| Train mass (wet incl. passengers) | [t] | 86.0 |
| Adhesion mass | [t] | 57.0 |
| Number of railpacks | [-] | 2 |
| Max. speed | [km/h] | 125 |
| Nom. engine power | [kW] | 390 kW |

In order to analyse the fuel saving potential of the Bom Sinal Mobile 2 the TecRec 100 suburban service profile is used for the simulation studies. For the Siemens Desiro Classic DMU, the regional service profile is used. For both tracks, timetables with section wise driving and station dwell times as defined in [9] are stored in the simulation tool (see section 3.2). No gradients are used for both tracks in this study and their speed profile and station stops are depicted in Figure 1.

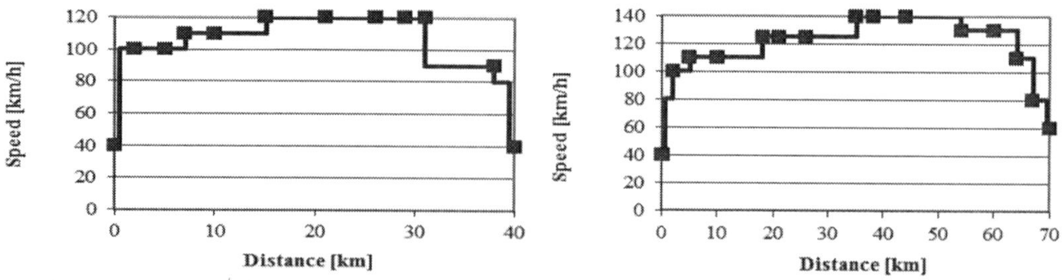


**Figure 1. Speed limits and station stops for CleanER-D suburban (left) and regional (right) service profile [7]**

To determine fuel consumptions of the DMUs, an outward journey and a return journey connected by a 20 minute break with idling engine is simulated.

## 2.2 Energy storage technologies

Driven by past and current developments of hybrid and full electric vehicles in the automotive sector, different energy storage technologies have emerged with the aim of reducing vehicle fuel consumption. With respect to the required power and energy contents needed for railway vehicles, three different electrical energy storage technologies have been chosen for further investigation in this work: double-layer capacitors (DLCs), flywheels and li-ion batteries. Further information on the examined storage technologies can be found in [10]. Each railpack of the DMUs is equipped with an individual energy storage system (ESS), made up of the energy storage, converter and energy management system.

Referring to the battery ESS, a 200s1p configuration consisting of Nano NMC (lithium **n**ickel **m**anganese **c**obalt oxide) [11] cells is used for the simulative studies. Table 3 highlights the most important parameters of the li-ion ESS used in the simulation models.

**Table 3. Simulation parameters of Li-Ion battery ESS [11]**

| | Unit | Value |
|---|---|---|
| State of charge at start of simulation | [%] | 50 |
| Number of parallel branches | [-] | 1 |
| Number of cells in series | [-] | 200 |
| Capacity of one cell | [Ah] | 46 |
| Cell maximum continuous charge/discharge current | [A] | 230/265 |
| Cell maximum pulse charge/discharge current (10 sec) | [A] | 367/561 |
| Cell minimum/maximum voltage | [V] | 3.493/4.110 |
| Efficiency of converter | [%] | 0.95 |
| Specific energy of ESS | [Wh/kg] | 100 |
| Weight of ESS incl. E-Motor(s) and converter per railpack | [kg] | 878 |

The second type of ESS considered is based on double-layer capacitors (DLC). Six BMOD0063 DLC modules by Maxwell Technologies® with a rated capacitance of 63F [12] are connected in series to provide power to the E-Motor of the hybrid propulsion systems. This type of module has already been successfully used in a hybrid bus application [13]. Table 4 shows the parameters of the DLC ESS.

**Table 4. Simulation parameters of DLC ESS [12]**

| | Unit | Value |
|---|---|---|
| State of charge at start of simulation | [%] | 1 |
| Number of parallel branches | [-] | 1 |
| Number of cells in series | [-] | 6 |
| Rated capacitance of one cell | [F] | 63 |
| Cell maximum continuous charge/discharge current | [A] | 240/240 |
| Cell minimum/maximum voltage | [V] | 12.5/125.0 |
| Efficiency of converter | [%] | 0.95 |
| Specific energy of ESS | [Wh/kg] | 2.3 |
| Weight of ESS incl. E-Motor and converter per railpack | [kg] | 856 |

Flywheels have received increased attention over the past years in various hybrid vehicle use cases and a lot of effort is put into their further development [14]. A scaled down version of a flywheel prototype developed for a project by DB and Alstom [15] is used for the simulation (see Table 5 for parameters).

**Table 5. Simulation parameters of Flywheel ESS [cf. 14]**

| | Unit | Value |
|---|---|---|
| State of charge at start of simulation | [%] | 0 |
| Maximum power output of flywheel E-Motor | [kW] | 140 |
| Efficiency of flywheel E-Motor | [%] | 85 |
| Maximum speed of flywheel | [1/min] | 25000 |
| Minimum speed of flywheel | [1/min] | 12500 |
| Energy Content of Flywheel | [Wh] | 1500 |
| Specific Energy of ESS | [Wh/kg] | 4.615 |
| Efficiency of converter | [%] | 95 |
| Weight of ESS incl. E-Motor and converter per railpack | [kg] | 825 |

### 2.3 Parallel hybrid powertrains

Two different parallel hybrid powertrains are considered for the presented work. The railpack of the Siemens Desiro Classic DMU is equipped with an inline crankshaft starter generator placed between diesel engine and automatic gearbox. For the Bom Sinal Mobile 2 DMU, a summation gearbox is used to connect the e-motor to the output shaft of the DIWA transmission. The general outline of both concepts is depicted in Figure 2. For simplification purposes, converters, cooling systems and auxiliary systems are not shown. As already mentioned before, both vehicles use two railpacks for propulsion.

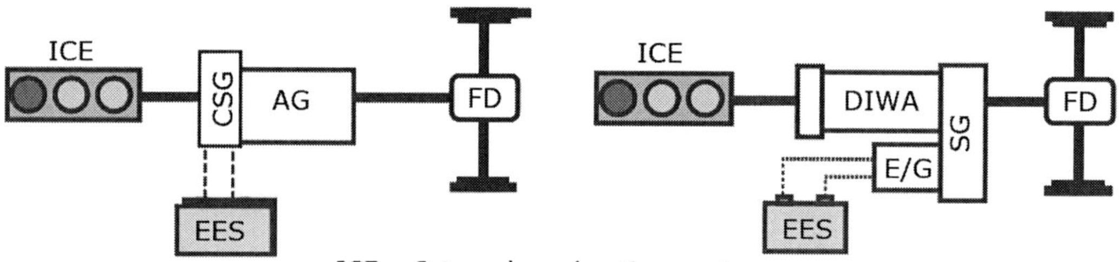


ICE = Internal combustion engine
AG = Automatic gearbox
CSG = Crankshaft starter generator
EES = Electrical energy storage
FD = Final drive
SG = Summation gearbox
E/G = E-Motor / generator unit

**Figure 2. Hybrid powertrain architectures of Siemens Desiro (left) and Bom Sinal Mobile 2 (right) used in the simulation models**

## 3 SIMULATION PROCEDURE AND POST PROCESSING

### 3.1 Automated parameterization

A Python-based software tool (see Figure 3 for main GUI) is programmed to evaluate the hybridization potential of the DMUs by means of model-based design. A combination of vehicle, diesel engine, power transmission, energy storage system and track is chosen according to the data of the railway vehicles as described in section 0. After selection, the tool automatically opens a LMS Imagine.Lab Amesim [16] simulation model and imports all necessary parameters of the chosen configuration.

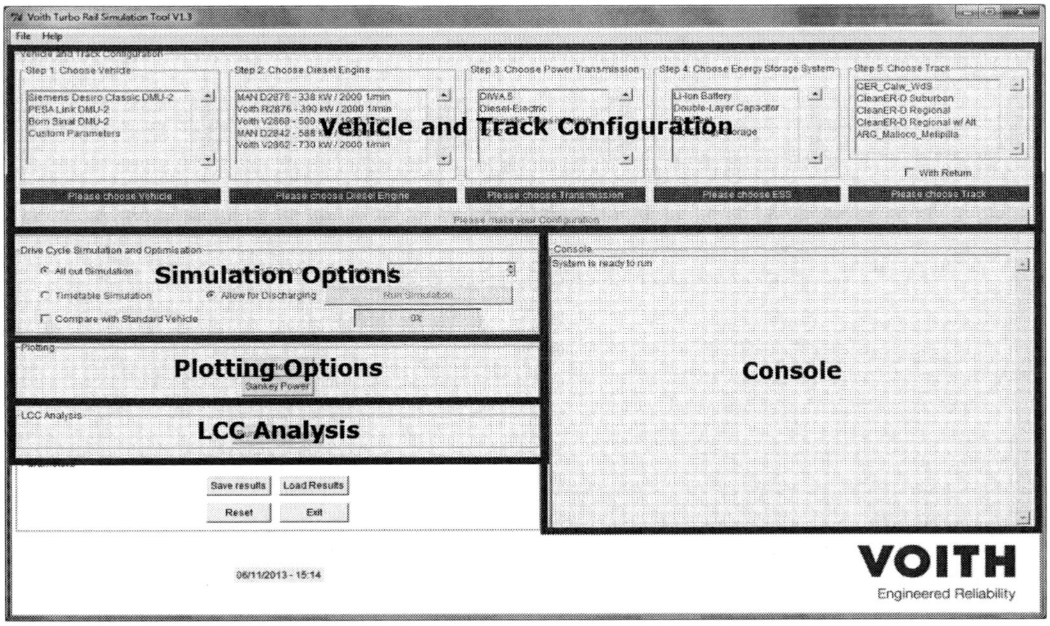


**Figure 3. Main GUI of software tool**

Before starting a simulation run, different settings for the operation mode have to be made. The user chooses if a fast driving style is applied (e.g. when a DMU is running past schedule) or a fuel saving "sawtooth" driving style (see section 3.2) according to the timetable of the chosen track is used. For the analyses presented here the DMUs stick to the timetable as defined in [9].

For both vehicles, a standard configuration without ESS is simulated and used as reference to calculate the fuel savings of the hybrid configurations under consideration of the occurring vehicle weight differences. With these values and further information on the costs of the hybrid components and operation of the vehicle, a LCC analysis is carried out as described in section 4.1.2.

### 3.2 Section wise simulation and timetable optimisation

Figure 4 shows a typical speed trajectory of a railway vehicle driving between two stations. Based on Voith experience and analysis of real driver behaviour, a "sawtooth" driving style is used for the simulation to further reduce fuel consumption in addition to the benefits of hybridization. After accelerating up to the desired speed, the driver model lets the vehicle coast until the speed falls below 95 % of the current speed limit on the track section. After that, the vehicle accelerates again. Before braking to a lower speed limit on the track or into a station, coasting phases with no tractive effort at the wheels are used to save fuel and adjust the driving time to the specified timetable time on the section. To define the coasting distances, two parameters *coasting_1* [-] and *coasting_2* [-] are optimized according to the time deviation of the railway vehicle. They define the ratio of coasting distance to braking distance. Once the optimization reaches a time deviation of 1 second compared to the timetable time, the next section is simulated.

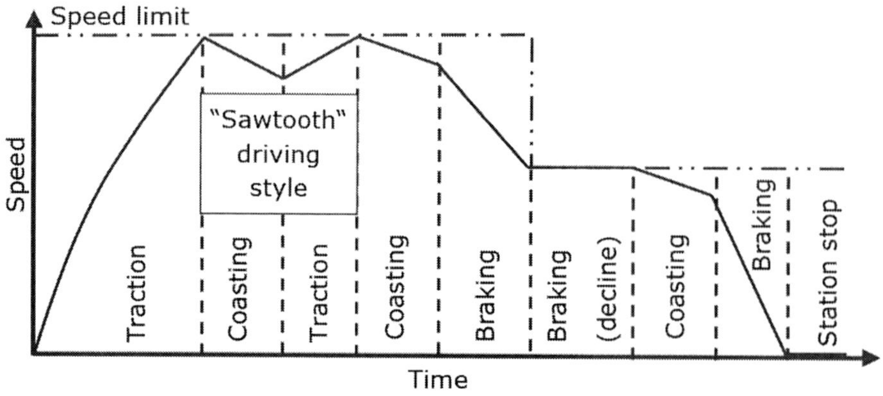


**Figure 4. Representative speed trajectory of a railway vehicle**

### 3.3 Energy management strategies

Two different energy management strategies can be chosen in the software tool and applied to the simulation models of the hybrid railway vehicles: boost mode and load point shifting (LPS) mode. In boost mode the electric motor of the hybrid powertrains (cf. Figure 2) increases the maximum tractive effort at the wheels in comparison to the non-hybrid vehicle. A parameter *boost [-]* is used to define the ratio of power provided by the e-motor to power of the diesel engine. When LPS mode is chosen, for the current rotational speed the diesel engine is run at its best operating mode. As soon as the ESS reaches its minimum SOC, LPS is shut off and the engine operates normally in the whole possible torque range. The same applies when the vehicle reaches a certain speed defined by the parameter *speed_lps [km/h]*.

When a battery ESS is chosen, it is ensured that the final SOC of the battery after the return journey is the same as at the start of the outward journey. This is achieved by optimizing the parameters *boost [-]* and *speed_lps [-]*.

### 3.4 ESS aging

#### *3.4.1 DLC ESS lifetime*

One of the main advantages of a DLC is its high cyclic lifetime. For the ESS lifetime, a total number of 1,000,000 cycles is used according to [12]. The cycles in the simulation are summarized independent on the depth of discharge (DOD) of the cycles. With the journey time [sec] for one simulation $t_{dc}$, the number of cycles during this journey $N_{C\_used}$ and the vehicle operation hours per year $t_{Op\_h}$, the DLC ESS lifetime $T_{ESS\_DLC}$ in [a] is calculated with Formula 1.

$$T_{ESS_{DLC}} = \frac{1000000}{N_{C_{used}}} \cdot \frac{t_{dc}}{3600} \cdot \frac{1}{t_{Op_h}} \quad (1)$$

#### *3.4.2 Flywheel ESS lifetime*

The lifetime of a flywheel $T_{ESS\_FW}$ [a] is very hard to estimate, since it mainly consists of mechanical parts and depends on the material, bearings and other components used in each individual case. For a first estimation, the same formula as for the DLC is used, with 1,000,000 possible cycles (see Formula 2).

$$T_{ESS_{FW}} = \frac{1000000}{N_{C_{used}}} \cdot \frac{t_{dc}}{3600} \cdot \frac{1}{t_{Op_h}} \quad (2)$$

### *3.4.3 Battery ESS lifetime*

The chemical deterioration processes of a battery have many influences such as temperature, charge and discharge currents and calendaric ageing. Extensive research has been carried out in the last years to develop lifetime prediction models for li-ion cells under various use scenarios. The lifetime value presented in section 4.1.2 for the regarded NMC cells are based on research by J. Schmalstieg et al (ISEA Aachen, Germany) [17] and M. Masih-Tehrani et. al [18]. The presented formulae in their work are adapted to the cells described in section 2.2.

# 4 SIMULATION RESULTS

## 4.1 Bom Sinal Mobile 2 on suburban service profile

### *4.1.1 Fuel savings*

When being equipped with an ESS, the hybrid Bom Sinal Mobile 2 achieves fuel savings for all types of storage technologies in the simulation runs on the suburban service profile (cf. Table 6). The highest saving of 12.4 % is reached with a flywheel ESS in boost mode, followed by the battery ESS in boost mode with 10.5 % and the DLC ESS in boost mode with 10.0 %.

**Table 6. Fuel savings of two-coach hybrid Bom Sinal Mobile 2 on suburban service profile**

| | | boost mode | LPS mode |
|---|---|---|---|
| Hybrid Vehicle with battery ESS | [%] | 10.5 | 10.1 |
| Hybrid Vehicle with DLC ESS | [%] | 10.0 | 5.1 |
| Hybrid Vehicle with flywheel ESS | [%] | 12.4 | 7.7 |

For all three technologies, the results show that for the suburban service profile a boost mode energy management strategy achieves higher savings than applying load point shifting (LPS mode) to the diesel engine. To explain this phenomena, Figure 5 shows the speed trajectories of the hybrid Bom Sinal Mobile 2 on the second section of the suburban service profile with an explanation of the different phases. This track section has a length of 3 kilometres and a travel time of 186 seconds according to the timetable. The standard vehicle and the hybrid DMU with LPS mode have nearly the same run of the curves, since the same tractive effort is applied to the wheels.

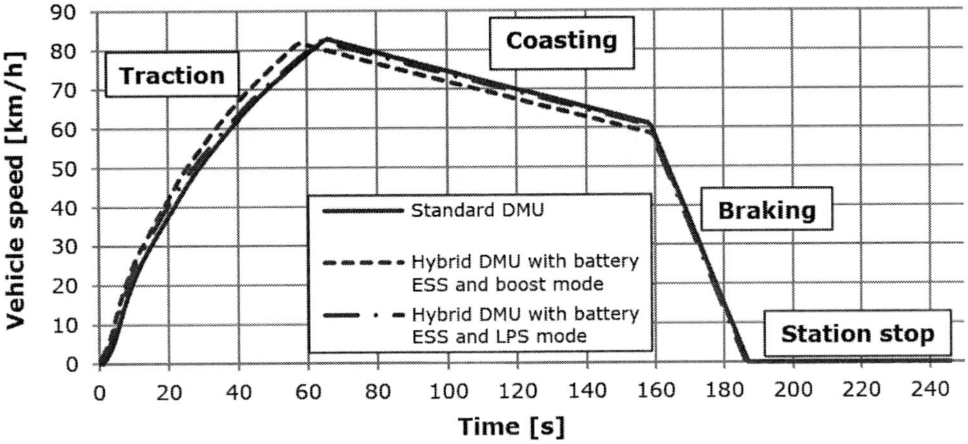


**Figure 5. Speed trajectory of Bom Sinal Mobile 2 on second section of suburban service profile**

For the hybrid DMU with battery ESS in boost mode, a higher acceleration is achieved in the initial traction phase (cf. Figure 5). Therefore, the virtual driver can sooner go into coasting mode before finally braking into the station. In terms of both regarded energy management strategies, a trade-off between higher acceleration and longer coasting phases (boost mode) or the same speed trajectory with lower diesel engine power output (LPS mode) can be seen. Both strategies lead to fuel savings on this section as shown in Figure 6.

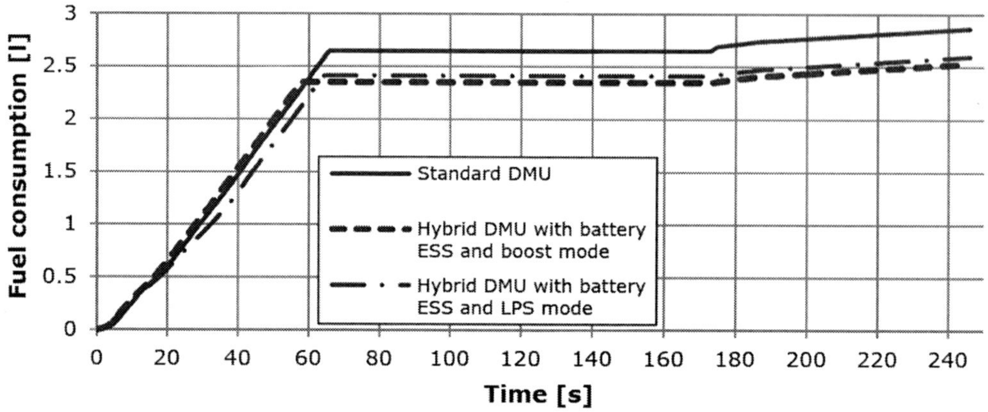


**Figure 6. Fuel consumption of Bom Sinal Mobile 2 on second section of regional service profile**

Figure 7 shows the time slice in terms of engine speed and torque for the regarded second section of the suburban service profile. In boost mode the engine operates longer at low torques (0-400 Nm) compared to the standard vehicle, because of making use of extended coasting phases as explained before (cf. Figure 8). When a LPS mode is used for the hybrid Bom Sinal Mobile 2 with battery ESS, a general shift towards lower torque outputs of the diesel engine can be seen as depicted in Figure 9.

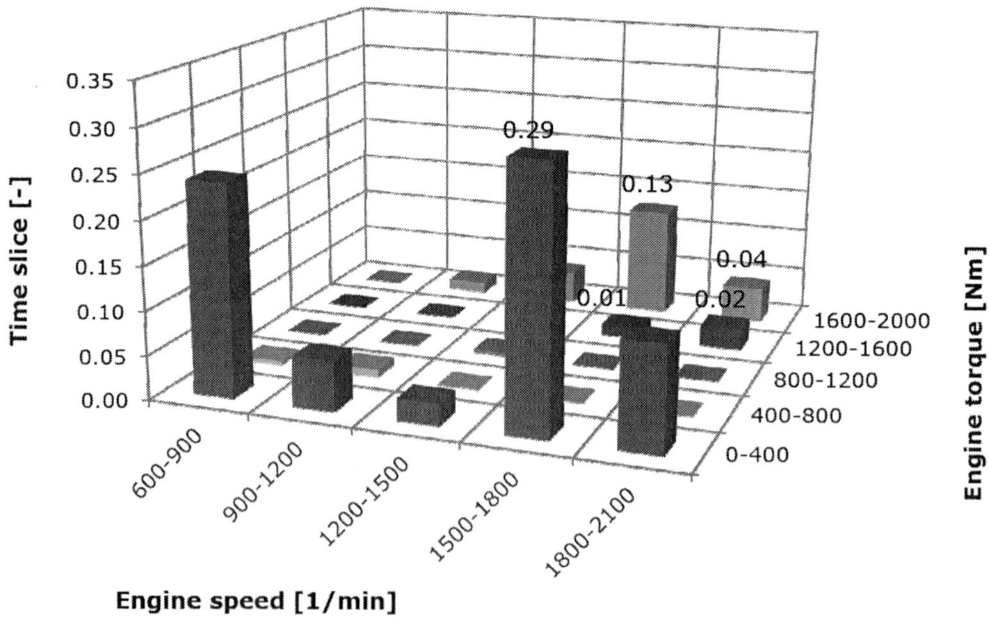


**Figure 7. Engine load of standard Bom Sinal Mobile 2 on second section of suburban service profile**

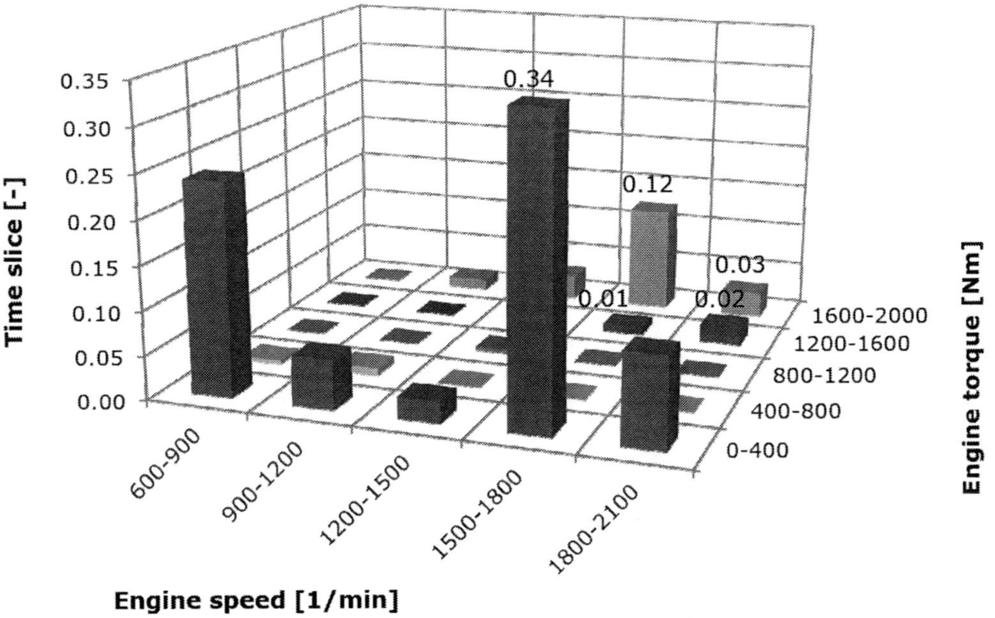


**Figure 8. Engine load of hybrid Bom Sinal Mobile 2 with battery ESS in boost mode on second section of suburban service profile**

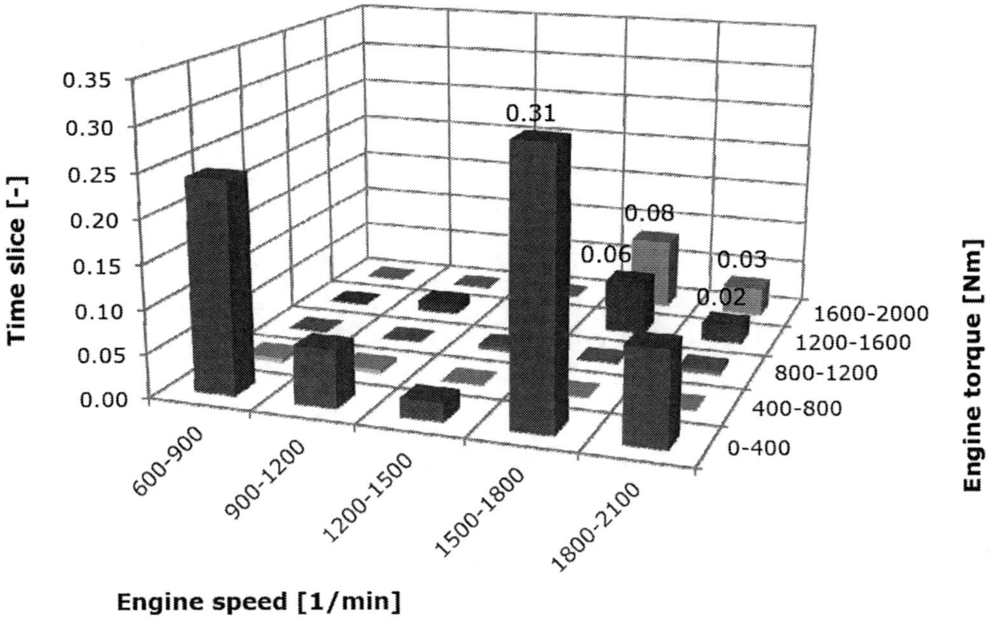


**Figure 9. Engine load of hybrid Bom Sinal Mobile 2 with battery ESS in LPS mode on second section of suburban service profile**

Therefore, both energy management strategies not only lead to fuel consumption reductions, but also reduced engine load. In a later project stage it has to be evaluated if this could also lead to extended service intervals of the engine and hence reduced life cycle costs for the vehicle operator.

### 4.1.2 *Life cycle cost analysis*

This section aims at highlighting the importance of considering the operating costs of a hybrid vehicle in addition to the fuel savings. Therefore, a LCC analysis has been incorporated as a post processing feature in the software tool. For the LCC analysis presented here, a time period of 20 years is evaluated with a yearly operating time of 3500 hours for the Bom Sinal Mobile 2 DMU. The current diesel fuel price in Germany is at 1.43 €/L (http://www.fuel-prices-europe.info/, 04-09-14). This price may vary depending on the country of investigation. An annual fuel price increase of 2.3 % is used for the investigation according to [19]. Due to non-disclosure agreements, ESS prices cannot be published in this work and no actual values for the return on investment of the hybrid vehicles can be given.

The following Figure 10, Figure 11 and Figure 12 show the return on investment from a vehicle operator's point of view for the three different ESS technologies (battery, flywheel and DLC) for both energy management strategies. The height of the bars give an indication of the best ESS solution for the regarded use case. The initial negative bar shows the additional investment which has to be made when purchasing a hybrid DMU compared to a standard DMU. Positive bars equal a return on the investment.

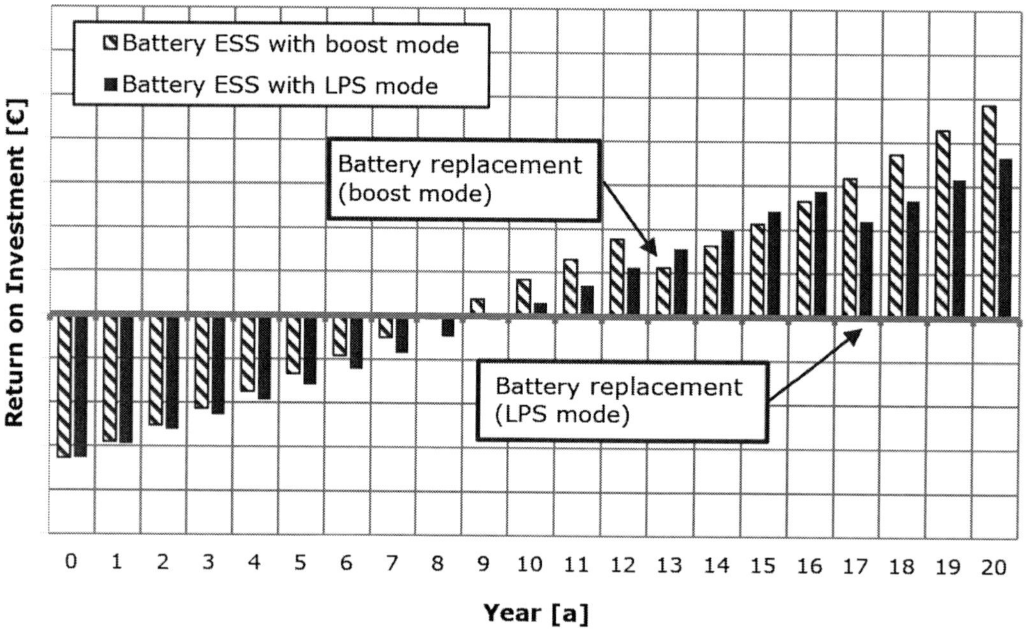


**Figure 10. LCC analysis of hybrid Bom Sinal Mobile 2 with battery ESS on second section of regional service profile**

When using a battery ESS, the li-ion cells have a lifetime of 13 years (boost mode) and 17 years (LPS mode) before they have to be replaced. The replacement of the cells results in additional costs and reduced money savings in the given year as shown in the bar representation of the LCC analysis (cf. Figure 10). The highest return on investment is achieved with a DLC ESS in boost mode as shown in Figure 11. For the given use case, the DLCs do not have to be replaced during the 20 year period.

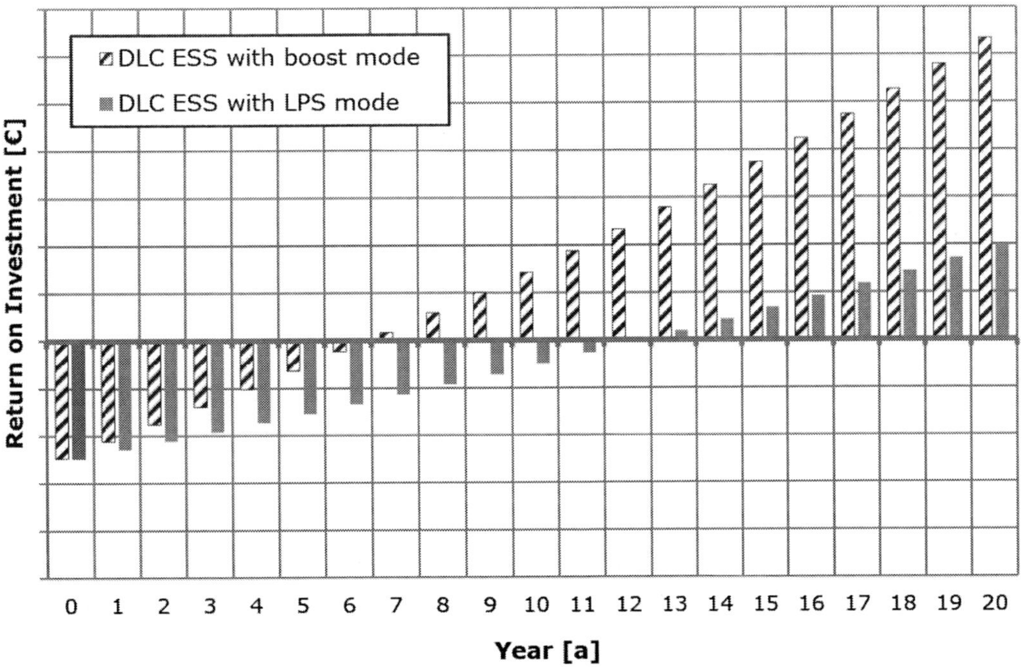


**Figure 11. LCC analysis of hybrid Bom Sinal Mobile 2 with DLC ESS on second section of regional service profile**

The flywheel ESS in boost mode ends up with nearly the same profit as the DLC ESS in boost mode after the 20 year service life of the DMU (see Figure 11 and Figure 12). No replacement of the flywheel ESS is necessary.

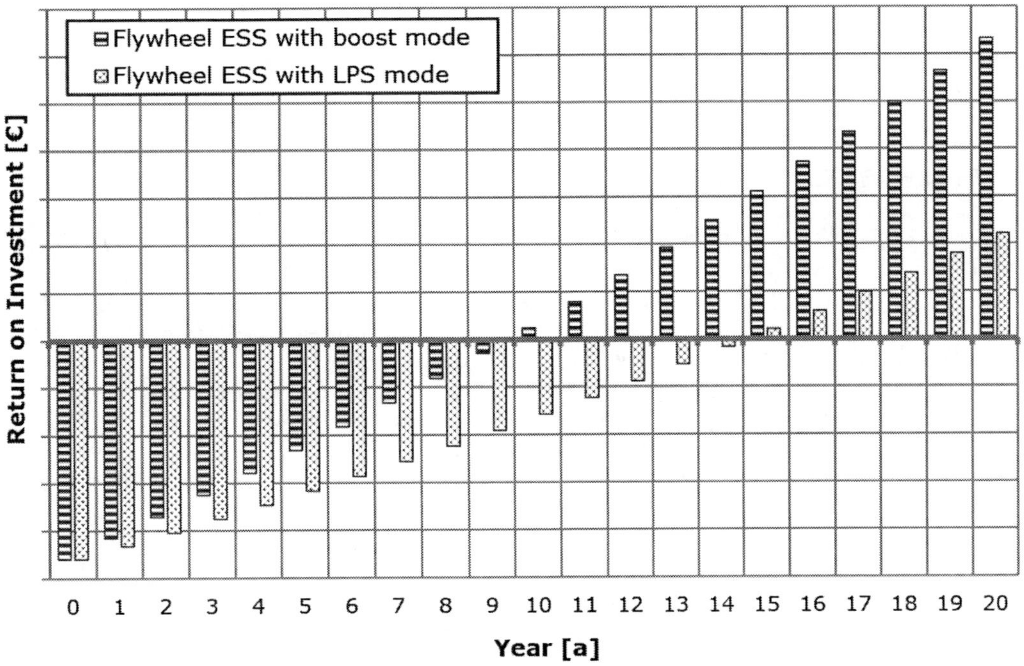


**Figure 12. LCC analysis of hybrid Bom Sinal Mobile 2 with flywheel ESS on second section of regional service profile**

## 4.2 Siemens Desiro Classic on regional service profile

According to the simulation results, a hybridisation of the Siemens Desiro Classic DMU on the regional service profile would lead to a fuel saving in the range of 12.8 % to 14.9 % (cf. Table 7). The highest reduction in fuel consumption of 14.9 % is achieved when the DMU is equipped with a battery ESS or a flywheel ESS.

**Table 7. Fuel savings of two-coach hybrid Siemens Desiro Classic on regional service profile**

| | | boost mode | LPS mode |
|---|---|---|---|
| Hybrid Vehicle with battery ESS | [%] | 14.9 | 14.0 |
| Hybrid Vehicle with DLC ESS | [%] | 12.8 | 13.5 |
| Hybrid Vehicle with flywheel ESS | [%] | 14.9 | 14.1 |

# 5 CONCLUSION AND FURTHER RESEARCH AREAS

Studies with 1D simulation models of two DMUs (Bom Sinal Mobile 2 and Siemens Desiro Classic) show, that the hybridisation of diesel-driven railway vehicles could achieve significant fuel savings. Depending on the type of service profile, ESS and energy management, fuel savings of up to 14.9 % compared to the non-hybrid vehicles are achieved. LCC analyses demonstrate that fuel savings alone are insufficient to evaluate if hybridization is economically reasonable. The prices of additional components (e.g. storage systems, converters and e-motors) have to be taken into consideration.

With the help of model-based design, railway vehicle manufacturers and suppliers have a powerful tool at hand to develop the best type of energy storage system for a given use case in cooperation with the vehicle operator. Further project research will focus on energy management functions to achieve higher fuel savings and the combination of energy storage technologies (e.g. DLC and li-ion) in order to obtain longer ESS lifetimes.

# REFERENCE LIST


(1) T. Fujii, N. Teraya and M. Osawa, 2004. *Development of a NE Train*. JR East Technical Review - No. 4, pp. 62-70.

(2) N. Shiraki, H. Satou and S. Arai, 2010. *A Hybrid System for Diesel Railcar Series Ki-ha E200*. The 2010 International Power Electronics Conference, Sapporo, Japan, 21-24 June 2010, pp. 2853-2858.

(3) Grantham, A., 2007. *Hybrid High Speed Train unveiled*. URL: http://www.railwaygazette.com/nc/news/single-view/view/hybrid-high-speed-train-unveiled.html (Accessed: 26 August 2014).

(4) I. Lehmann et al., 2011. *Hybrid-Powerpack für nachhaltigen und umweltfreundlichen Triebwagenantrieb*. Eisenbahntechnische Rundschau No. 9, September 2011, pp. 18-22.

(5) S. Hillmansen et al., 2008. *Final Report: Concept Validation for Hybrid Trains DfTRG/0078/2007*. Birmingham Research and Development Ltd., Birmingham, UK.

(6) S. Hillmansen et al., *2009. Draft Final Report: DMU Hybrid Concept Evaluation -Follow on Work DfTRG/0078/2007*. Birmingham Research and Development Ltd., Birmingham, UK.

(7) UNIFE / UIC, 2010. *Technical Recommendation: Specification and verification of energy consumption for railway rolling stock*. URL: http://www.unife.org/page.asp?pid=164 (accessed 26-08-2014).
(8) Voith Publication, 2013. *Combining Ride Comfort with Economy*. DIWA.5. URL: http://resource.voith.com/vt/publications/downloads/483_e_483_e_g_1747_e_diwa5_2012-08_screen.pdf (accessed 14-08-2014).
(9) European Commission, 2013. Clean European Rail-Diesel, *D7.2.1: Detailed Specification: Parameters definition*. Brussels (FP7 – 234338).
(10) S. Schmid, *Simulation of hybrid Railway Vehicles and Comparison of electrical Energy Storage Systems*, ASIM STS/GMMS Workshop 2014, Reutlingen, 21-02-2014.
(11) AKASOL GmbH Publication, 2012. *AKAMODULE. HIGHLY FUNCTIONAL. VALIDATED. UNIQUELY POWERFUL*. URL: http://www.akasol.com/fileadmin/Kundendaten/pdf/datenblatt/update_dezember_2012/AKASOL_Datenblatt_AKAMODULE_E_12_2012.pdf (accessed 14-08-2014).
(12) Maxwell Technologies, 2013. *125 V Heavy Transportation Module BMOD0063 P125 B04/B08 Datasheet*. URL: http://www.maxwell.com/products/ultracapacitors/docs/datasheet_bmod0063_1014696.pdf (accessed 14-08-2014).
(13) Green Autoblog, 2014. *Meet the articulating Solaris Urbino 18 DIWA hybrid city bus* . URL: http://green.autoblog.com/2011/03/30/meet-the-articulating-solaris-urbino-18-diwa-hybrid-city-bus/ (accessed 04-09-2014).
(14) Ricardo Press Release, 2014. *Ricardo's 'TorqStor' receives SAE 2014 World Congress Tech Award*. URL: http://www.ricardo.com/en-GB/News--Media/Press-releases/News-releases1/2014/Ricardos-TorqStor-receives-SAE-2014-World-Congress-Tech-Award/ (accessed 27-08-2014).
(15) Deutsche Bahn AG, 2001. *Applications for energy storage flywheels in vehicles of Deutsche Bahn AG*. World Congress on Railway Research 2001, Cologne, Germany, 25 – 29 November 2001. URL: http://www.uic.org/cdrom/2001/wcrr2001/start.htm (accessed 14-08-2014).
(16) LMS Imagine.Lab Amesim: Siemens PLM Software. Information available at: http://www.plm.automation.siemens.com/de_de/products/lms/imagine-lab/amesim/index.shtml (accessed 01-09-2014).
(17) J. Schmalstieg et al., *A holistic aging model for Li(NiMnCo)O2 based 18650 lithium-ion batteries*. Journal of Power Sources 257 (2014) pp. 325-334.
(18) M. Masih-Tehrani et. al, 2013. *Optimum sizing and optimum energy management of a hybrid energy storage system for lithium battery life improvement*. Journal of Power Sources 244 (2013) pp. 2-10.
(19) U.S. Energy Information Administration, 2014. *AEO2014 Early Release Overview*. URL: http://www.eia.gov/forecasts/aeo/er/early_prices.cfm (accessed 04-09-2014).

# Prediction of wheel-rail contact forces based on a heavy haul locomotive dynamic inverse modelling

**Y Q Sun, M Spiryagin, C Cole**
Central Queensland University, Centre for Railway Engineering, Australia

**ABSTRACT**

A prediction model based on the acceleration measurement on a locomotive has been developed to evaluate wheel-rail contact dynamic forces. In order to verify the prediction model, a comprehensive wagon model has been generated using Gensys software. The accelerations on the car body and the bogies from the Gensys simulations due to a typical type of track geometry irregularity were used as the inputs of the prediction model. The reasonable agreement has been achieved through the comparison between the predicted and simulated primary and secondary suspension forces and wheel-rail contact dynamic forces.

## 1. INTRODUCTION

The rail vehicles (locomotives and wagons) generate and are subjected to dynamic forces in the vertical and lateral directions when they run on tangential or curved tracks with geometry irregularities. Depending on track geometry irregularities, significant dynamic forces occur in the lateral and vertical directions. Simultaneous occurrence of the lateral and vertical dynamic loading has the potential to increase damage and degradation of rails and wheels, and/or cause derailments.

Understanding the heavy haul locomotive dynamic behaviour, e.g., wheel-rail contact dynamic forces, is very important for railway operations. Currently, the accelerations on locomotive components can be easily obtained due to the excellent range of transducers available. The problem is how to predict wheel-rail contact dynamic forces based on the locomotive responses. In fact, such a problem belongs to one of the inverse dynamics identification problems. The inverse identification problem can be defined as finding system inputs based on given responses, boundary conditions and system modelling. Inverse identification problems are of paramount importance for monitoring and quality control purposes arising in critical applications in engineering [1] ~ [5]. Mathematical modelling and the numerical study of these problems require high competence in computational mechanics and applied optimisation. To solve the inverse identification problem, an inverse model of the system has to be known.

The limited applications of inverse identification problems in vehicle-track systems are reviewed as follows for both prediction of wheel-rail contact dynamic forces and track geometry irregularities. The method based on Tikhonov's regularisation [6] was applied to the identification of wheel-rail contact forces based on the axle box accelerations. An inverse wagon model [7] - [9] was developed to estimate the wheel-rail contact forces using the measurements of wagon body acceleration responses as inputs. The model combines integration and partial modal matrix techniques. In order to estimate the high-frequency components of the forces at the wheel-rail interface, a grey box-based inverse wagon model [10] was developed. A recurrent neural network (RNN) architecture [11] has been proposed to predict the wheel-rail contact dynamic forces due to short wavelength defects. A method based on an Auto-Regression model with eXtra inputs (ARX) [12] was used to estimate track geometry irregularities by measuring accelerations of a passenger car instead of direct measurement by using a conventional track inspection car. A method for track geometry assessment considered the vehicle-track interaction using 'representative' transfer functions was put forward [13]. A Structural Health Monitoring system was developed based on vibration measurements during the rail vehicle operation simultaneously with GPS position and velocity estimation. Track irregularities are identified using this procedure [14].

The possibility of establishing a relationship between the wheel-rail dynamic forces and the accelerations on locomotive components has been theoretically worked out in this paper. The research has been, at first, carried out on the case study due to a cross level irregularity using a very detailed locomotive model generated using Gensys package. Then, an inverse locomotive model has been developed to predict the dynamic forces between the locomotive component connections, the wheel-rail contact forces based on the acceleration inputs. The accelerations from the Gensys locomotive model have been as the inputs to this model. The prediction calculations were carried out by using a numerical integration modified Newmark - $\beta$ method. The predicted outputs - dynamic forces have been compared with those from the detailed Gensys model. In the paper, the inverse locomotive model will be described in detail. The comparison between the simulations and the predictions will be illustrated and discussed.

## 2. LOCOMOTIVE DYNAMICS INVERSE MODELLING

The lateral and vertical locomotive dynamic inverse modelling is shown in Fig. 1 (a) and (b).

The inverse modelling can be described as follows:

- The determination of secondary suspension forces. In this paper, through the lateral and vertical accelerations of the locomotive car body and two bogie frames at the secondary suspension positions respectively, the double integrations are carried out to obtain the velocities and displacements. Therefore, the secondary suspension forces can be determined by multiplying the damping and stiffness coefficients of the suspension springs and dampers.
- The dynamic equivalent equations are established for the bogie frames and wheelsets. Such a dynamic system is subjected to the secondary suspension forces.
- A numerical method is applied to solve the equations. Finally, the primary suspension and wheel-rail contact forces can be determined.

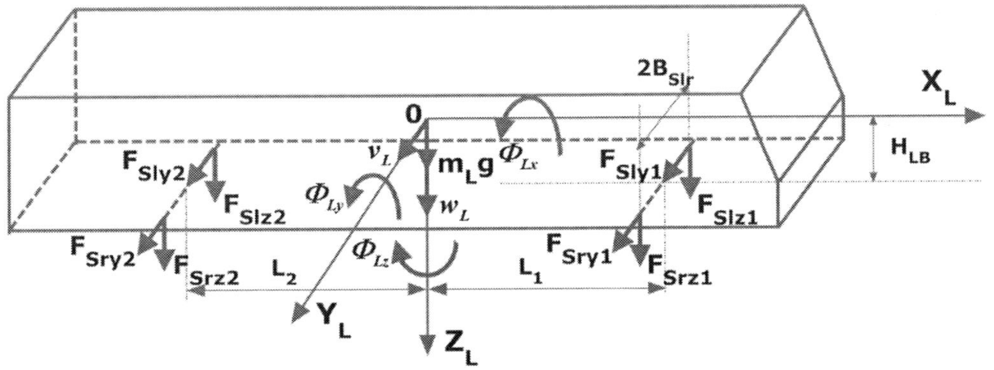


(a) Locomotive Car Body

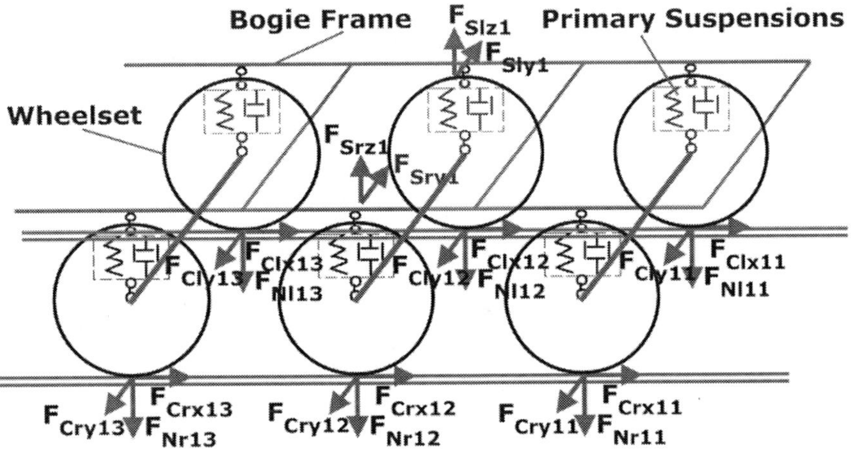


(b) Bogie

**Fig. 1 Inverse Locomotive Modelling**

### 2.1 Bogie Frame

Five degrees of freedom (DoFs) describing the motions of the bogie frame, namely, the lateral and vertical displacements ($v_B$ and $w_B$) and the roll, the pitch and the yaw rotations ($\emptyset_{Bx}$, $\emptyset_{By}$ and $\emptyset_{Bz}$) about the $X_B, Y_B$ and $Z_B$ axes are used. In Fig. 1 (b), the dynamic equivalent equations of each bogie frame ($i$ = 1, 2):

$$\begin{bmatrix} m_B & 0 & 0 & 0 & 0 \\ 0 & m_B & 0 & 0 & 0 \\ 0 & 0 & J_{Bx} & 0 & 0 \\ 0 & 0 & 0 & J_{By} & 0 \\ 0 & 0 & 0 & 0 & J_{Bz} \end{bmatrix} \begin{Bmatrix} \ddot{v}_{Bi} \\ \ddot{w}_{Bi} \\ \ddot{\emptyset}_{Bxi} \\ \ddot{\emptyset}_{Byi} \\ \ddot{\emptyset}_{Bzi} \end{Bmatrix} = \begin{Bmatrix} -F_{Sryi} - F_{Slyi} + \sum_{j=1}^{3}\left(F_{Pryij} + F_{Plyij}\right) \\ -F_{Srzi} - F_{Slzi} + \sum_{j=1}^{3}\left(F_{Przij} + F_{Plzij}\right) \\ (F_{Slzi} - F_{Srzi})B_{Slr} + \left[\sum_{j=1}^{3}\left(F_{Przij} - F_{Plzij}\right)\right]B_{Plr} \\ \left[-(F_{Przi1} + F_{Plzi1}) + (F_{Przi3} + F_{Plzi3})\right]L_W \\ \left[\left(F_{Pryi1} + F_{Plyi1}\right) - \left(F_{Pryi3} + F_{Plyi3}\right)\right]L_W \end{Bmatrix}$$

$$[M_B]\{\ddot{q}_B\} = \{F_{SPi}\} \qquad (i = 1, 2) \qquad (1)$$

In Eq. (1), $m_B$ and $J_B$ represent the mass and the mass inertia moment of bogie frame. $F_S$ and $F_P$ represent the secondary and primary suspension forces. $B_{Slr}$ and $B_{Plr}$ are the lateral distances between secondary and primary suspensions respectively. $L_W$ is the longitudinal distance between two adjacent wheelsets. It is assumed that the mass centre of bogie frame is in the same vertical plane as the secondary suspension forces and the primary suspension forces on the second wheelset.

## 2.2 Wheelset

Four degrees of freedom (DoFs) describing the motions of a wheelset, namely, the lateral and vertical displacements ($v_W$ and $w_W$) and the roll and the yaw rotations ($\emptyset_{Bx}$ and $\emptyset_{Bz}$) about the $X_W$ and $Z_W$ axes. In Fig. 1 (b), the dynamic equivalent equations of each wheelset (*i* = 1, 2 and *j* =1, 2, and 3) can be written as:

$$\begin{bmatrix} m_W & 0 & 0 & 0 \\ 0 & m_W & 0 & 0 \\ 0 & 0 & J_{Wx} & 0 \\ 0 & 0 & 0 & J_{Wz} \end{bmatrix} \begin{Bmatrix} \ddot{v}_{Wij} \\ \ddot{w}_{Wij} \\ \ddot{\emptyset}_{Wxij} \\ \ddot{\emptyset}_{Wzij} \end{Bmatrix} = \begin{Bmatrix} F_{PWyij} \\ F_{PWzij} \\ M_{PWxij} \\ M_{PWzij} \end{Bmatrix}$$

$$[M_W]\{\ddot{q}_{Wij}\} = \{F_{PWij}\} \qquad (i = 1, 2 \text{ and } j = 1, 2, \text{ and } 3) \qquad (2)$$

$$\begin{Bmatrix} F_{PWyij} \\ F_{PWzij} \\ M_{PWxij} \\ M_{PWzij} \end{Bmatrix} =$$

$$\begin{Bmatrix} -F_{Plyij}-F_{Pryij} + F_{Clyij}+F_{Cryij}+F_{Nlyij}+F_{Nryij} \\ -F_{Plzij}-F_{Przij} + F_{Nlzij}+F_{Nrzij} \\ (J_{Wy}-J_{Wz})\frac{V}{r_0}\dot{\emptyset}_{Wzij}-r_0(F_{Clyij}+F_{Cryij}+F_{Nlyij}+F_{Nryij}) + B_{Wlr}(F_{Nrzij}-F_{Nlzij}) + B_{Plr}(F_{Plzij}-F_{Przij}) \\ (J_{Wx}-J_{Wy})\frac{V}{r_0}\dot{\emptyset}_{Wxij}+B_{Wlr}(F_{Clxij}-F_{Crxij})+B_{Wlr}\emptyset_{Wzij}(F_{Clyij}-F_{Cryij}+F_{Nlyij}-F_{Nryij}) \end{Bmatrix}$$

In Eq. (2), $m_W$ and $J_W$ represent the mass and the mass inertia moment of wheelset. $F_C$ and $F_N$ represent the tangent creep and the normal forces between wheel and rail.

## 2.3 Wheel-Rail Contact Forces

The wheel-rail normal force can be determined by the following equation:

$$\begin{cases} F_{Nlij} = C_H[(w_{Wij} - B_{Wlr}\emptyset_{Wxij}) - \mu_l(x)]^{\frac{3}{2}}, & if\ [(w_{Wij} - B_{Wlr}\emptyset_{Wxij}) - \mu_l(x)] \le 0, \quad F_{Nlij} = 0 \\ F_{Nrij} = C_H[(w_{Wij} + B_{Wlr}\emptyset_{Wxij}) - \mu_r(x)]^{\frac{3}{2}}, & if\ [(w_{Wij} + B_{Wlr}\emptyset_{Wxij}) - \mu_r(x)] \le 0, \quad F_{Nrij} = 0 \end{cases} \qquad (3)$$

in which $C_H$ is Hertz contact coefficient, $B_{Wlr}$ is the lateral distance between the wheel-rail contact point and the track centreline, $\mu(x)$ is the track geometry irregularity in the normal direction of wheel-rail contact, which includes the data measured by using a track inspection car - cross level, lateral alignment, top surface and gauge variations, and the short-wavelength track defects - joint dip, squat, etc.

The longitudinal and lateral creep forces are determined based on the Polach's theory:

$$F_C = \frac{2F_N\mu}{\pi}\left(\frac{\varepsilon}{1+\varepsilon^2} + arctan\varepsilon\right) \text{ with } \varepsilon = \frac{1}{4}\frac{G\pi abc_{mm}}{F_N\mu}s \qquad (4)$$

In which $G$ is the shear module; $a, b$ are the half-axes of the contact ellipse; $\mu$ is the coefficient of friction; $s$ is the total creep and $c_{mm}$ is the coefficient from Kalker's linear theory, which are obtained as follows:

$$s = \sqrt{{s_x}^2 + {s_y}^2} \text{ and } c_{mm} = \sqrt{\left(c_{11}\frac{s_x}{s}\right)^2 + \left(c_{22}\frac{s_y}{s}\right)^2}$$

$c_{11}, c_{22}$ are the coefficients from Kalker's linear theory; $s_x, s_y$ are the creep in longitudinal ($X$) and lateral ($Y$) directions, which are obtained as follows:

$$\begin{cases} s_{lxij} = +B_{Wlr}\frac{\dot{\emptyset}_{Wzij}}{V} + \frac{\lambda y}{r_0} \\ s_{rxij} = -B_{Wlr}\frac{\dot{\emptyset}_{Wzij}}{V} - \frac{\lambda y}{r_0} \end{cases}, \quad \begin{cases} s_{lyij} = \frac{\dot{v}_{Wij}}{V} - \emptyset_{Wzij} \\ s_{ryij} = \frac{\dot{v}_{Wij}}{V} - \emptyset_{Wzij} \end{cases}$$

Based on Eq. (1) ~ (4), the dynamic equivalent equations of an inverse locomotive model can be written as the matrix form as:

$$[M_w]\{\ddot{q}_w\} + [C_w]\{\dot{q}_w\} + [K_w]\{q_w\} = \{F_w\} \tag{5}$$

In which, $[M_w]$, $[C_w]$ and $[K_w]$ are the mass, damping and stiffness matrixes of inverse Locomotive modelling. $\{q_w\}$ and $\{F_w\}$ are the displacement vector and the force vector. The total of degrees of freedom is 34. In Eq. (5), the secondary suspension forces are known. Therefore, through the Eq. (5), the wheel-rail contact forces can be predicted through calculations using the numerical integration Newmark - β method.

## 3. LOCOMOTIVE DYNAMICS MODELLING USING GENSYS

Fig. 2 shows the locomotive model. This model is used for the case study. In the case study, the model passes through a particular track short-wave length defect. The lateral and vertical accelerations of the locomotive car body and two bogie frames at the secondary suspension positions are measured respectively. Then, the double integrations are conducted to determine the velocities and displacements. Finally, the secondary suspension forces can be determined.

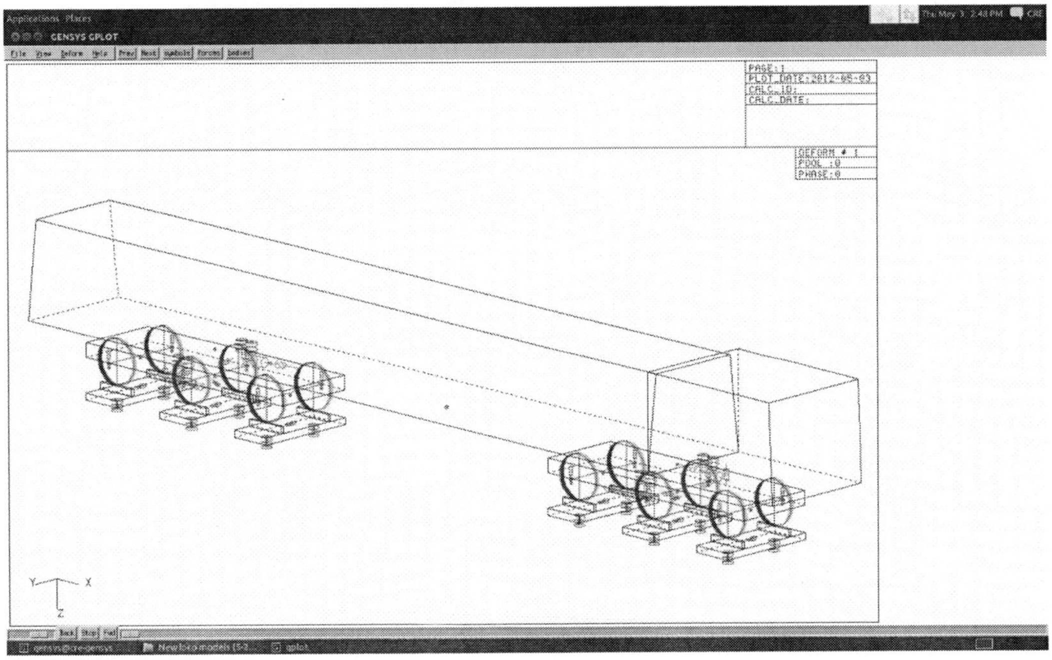

(a) Locomotive

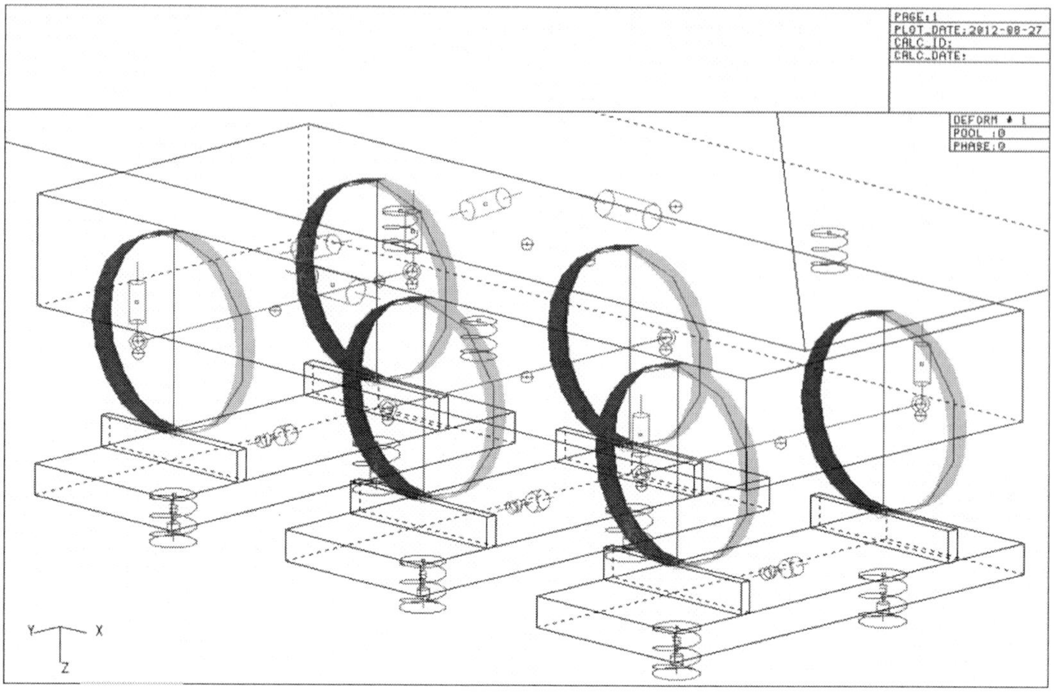


(b) Bogie

**Fig. 2 Gensys Locomotive Model**

The model connections - secondary suspensions and primary suspensions are described below:

Secondary suspensions to connect the locomotive body and one bogie frame:

- Four coil springs;
- Two stiffness elements representing traction rods at the direction specified by the coupling's attachment points;
- One lateral bumpstop element;
- Two vertical bumpstop elements;

Primary suspensions to connect the bogie frame and two axle boxes of one wheelset:

- Four coil springs;
- Two stiffness elements representing traction rods at the direction specified by the coupling's attachment points;
- Two vertical damping elements representing hydraulic shock absorbers at the direction specified by the coupling's attachment points;
- Two lateral bumpstop elements;
- Two vertical bumpstop elements.

Primary suspensions to connect the bogie frame and one traction motor:

- One vertical stiffness element and one vertical damping element representing traction motor hydraulic shock absorbers at the direction specified by the coupling's attachment points;
- Two lateral bumpstop elements;
- Two vertical bumpstop elements.

The parameters used for the simulations are listed in Table 1.

**Table 1 Locomotive Parameters**

| Parameter | Value | Unit |
|---|---|---|
| Locomotive car body mass | 87180 | kg |
| Mass moment of inertia of locomotive car body about X axis (Roll) | 168550 | $kgm^2$ |
| Mass moment of inertia of locomotive car body about Y axis (Pitch) | 3610410 | $kgm^2$ |
| Mass moment of inertia of locomotive car body about Z axis (Yaw) | 3590650 | $kgm^2$ |
| Mass of bogie frame | 14860 | kg |
| The mass moment of inertia of bogie frame about X axis | 6520 | $kgm^2$ |
| The mass moment of inertia of bogie frame about Y axis | 45370 | $kgm^2$ |
| The mass moment of inertia of bogie frame about Z axis | 50300 | $kgm^2$ |
| Wheelset mass | 2850 | kg |
| Mass moment of inertia of wheelset about X, Z axis | 1789 | $kgm^2$ |
| Mass moment of inertia of wheelset about Y axis | 1200 | $kgm^2$ |
| Stiffness coefficient of secondary suspension along Z axis | $2.138\times10^6$ | N/m |
| Stiffness coefficient of secondary suspension along X and Y axis | $1.2\times10^6$ | N/m |
| Longitudinal distance from the mass centre of locomotive car body to the mass centre of the front and the rear bolsters | 7.095 | m |
| Semi lateral distance between the left and the right secondary suspensions in a bogie | 1.012 | m |
| Height between the mass centres of the locomotive car body and the bogie frame | 1.93 – 0.733 | m |
| Height between the mass centres of bogie frame and wheelset | 0.733 – 1.067/2 | m |
| Height of car body floor | 1.5 | m |
| Stiffness coefficient of primary suspension along Z axis | $0.73\times10^6$ | N/m |
| Stiffness coefficient of primary suspension along X axis | $24\times10^6$ | N/m |
| Stiffness coefficient of primary suspension along Y axis | $3.5\times10^6$ | N/m |
| Damping coefficient of primary suspension along Z axis | $5\times10^3$ | Ns/m |
| Damping coefficient of primary suspension along X axis | $25\times10^3$ | Ns/m |
| Damping coefficient of primary suspension along Y axis | $15\times10^3$ | Ns/m |
| Semi lateral distance between the left and the right primary suspensions | 1.012 | m |
| Longitudinal distance between two adjacent wheelsets | 1.85 | m |
| Wheel diameter | 1.067 | m |

## 4. CASE STUDY

A track geometry irregularity of cross level with 20mm dip on high rail and 20mm rise on low rail and semi-span of 6m is selected for the case study, as shown in Fig. 3.

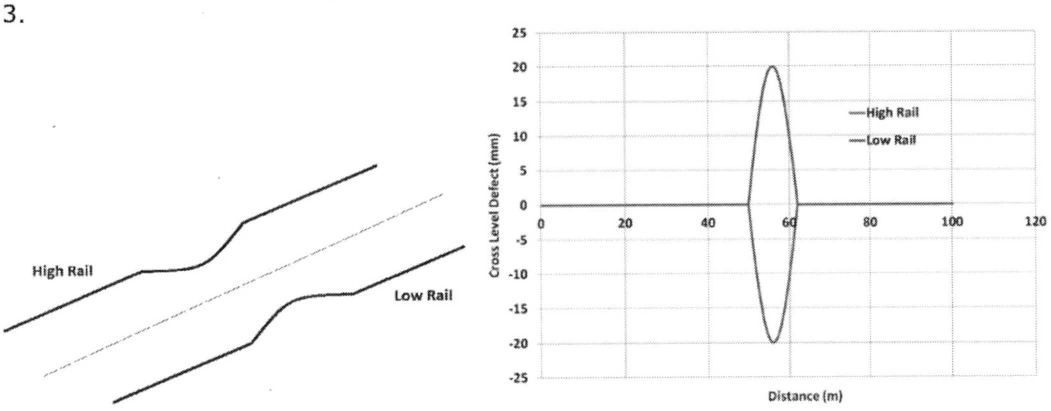


**Fig. 3 Track cross level defect shape and size**

At this stage, the vertical wheel-rail forces are only concerned. Therefore, the vertical accelerations of the locomotive car body and two bogie frames at the secondary suspension positions are measured respectively and considered as the inputs of inverse locomotive model. Fig. 4 shows the vertical accelerations of the locomotive car body and the bogie frame at the front secondary suspension positions.

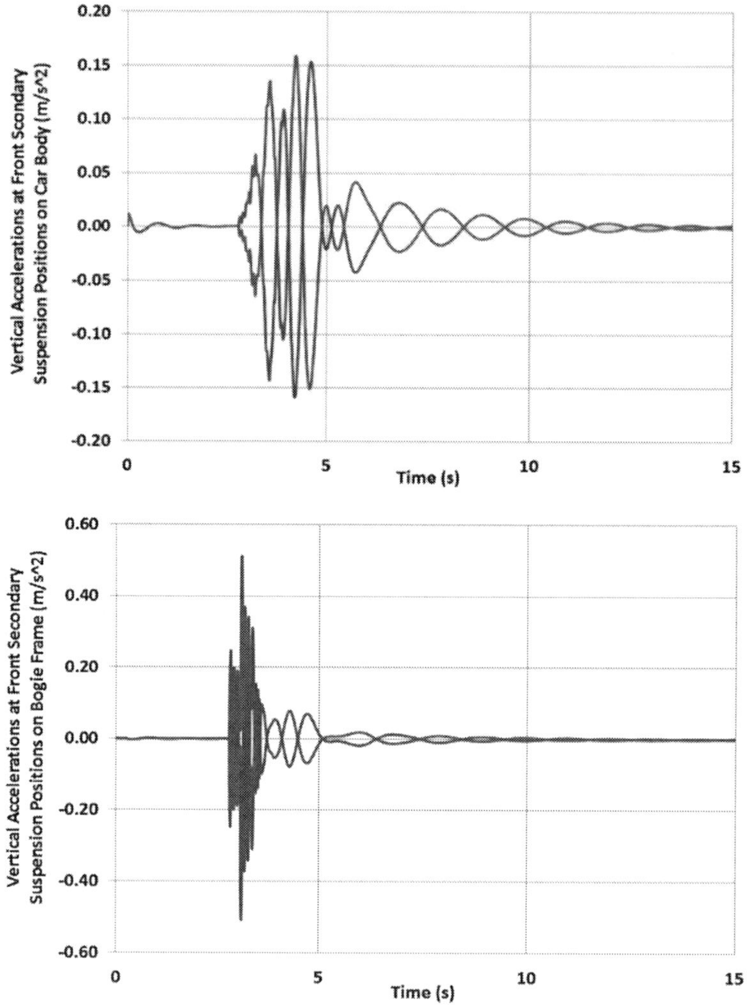


**Fig. 4 Vertical accelerations**

The secondary suspension forces can be determined based on the double integrations of these accelerations. The vertical secondary suspension forces from the Gensys simulation and from the inverse modelling are shown in Fig. 5.

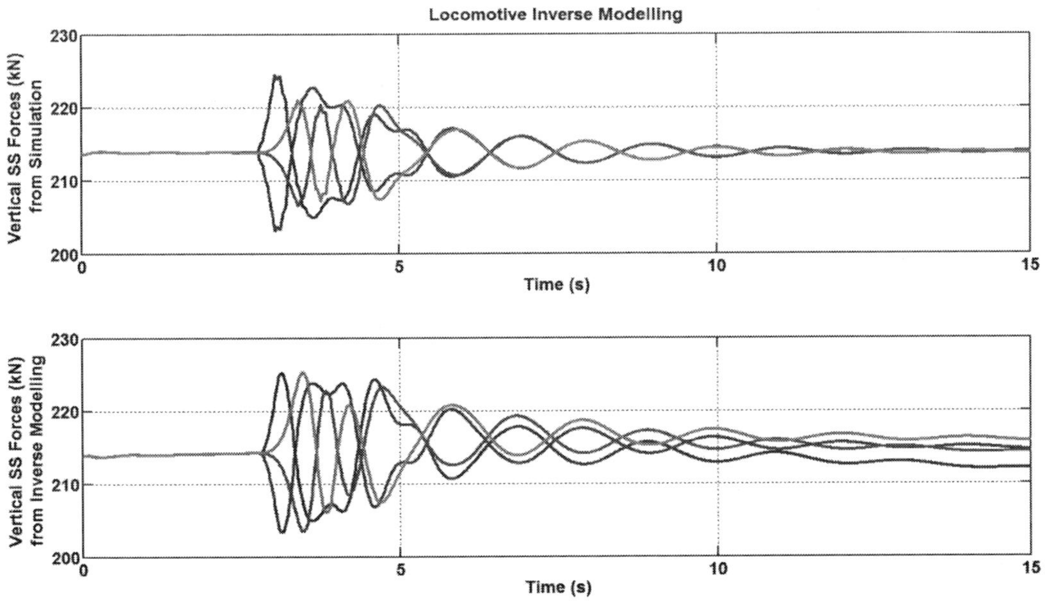


**Fig. 5 Vertical Secondary Suspension Forces from Simulation and from Inverse Modelling**

The forces shown in Fig. 5 are considered as the external force, which are acting on the two bogies. Fig. 6 and Fig. 7 show the vertical primary suspension forces and the vertical wheel-rail contact forces.

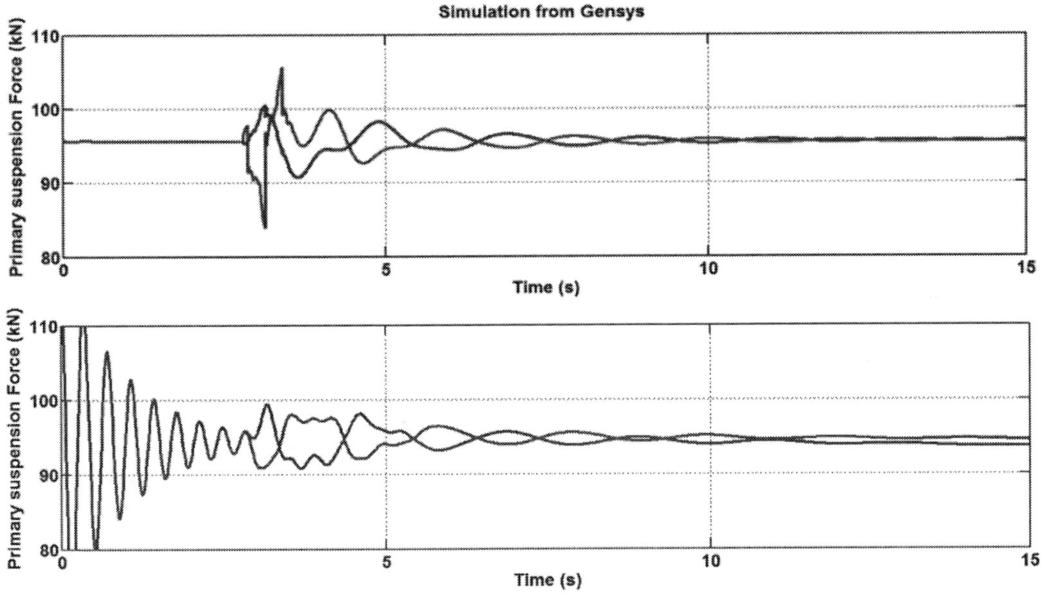


**Fig. 6 Vertical Primary Suspension Forces from Simulation and from Inverse Modelling**

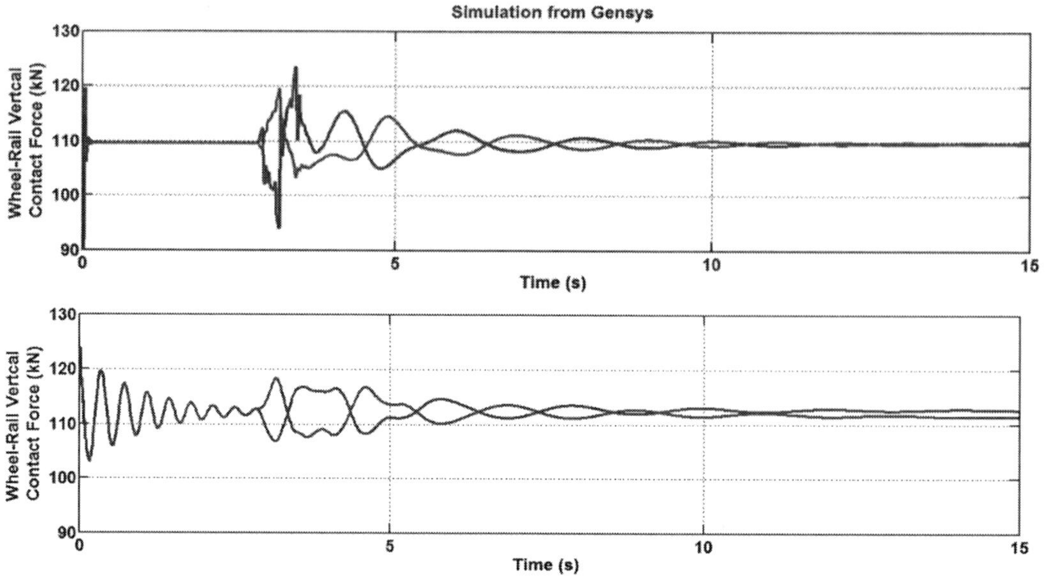


**Fig. 7 Vertical Wheel-Rail Forces from Simulation and from Inverse Modelling**

## 5. CONCLUSIONS

An inverse locomotive modelling has been theoretically developed. The simple inverse locomotive model can give the reasonable and useful predictions of locomotive component connection forces and the wheel-rail contact forces based on the acceleration measurements on locomotive components due to track geometry irregularities. It has been noticed that the results from the inverse model have some minor differences from those using a detailed Gensys model. It is also noticed that this is our preliminary stage on the investigation on the inverse locomotive dynamics modelling. The further improvements for the inverse model and the better selection of the integration and/or prediction methods and a lot of simulations on the different cases are necessary.

## REFERENCES


[1] FRISWELL, M. I., 2007. Damage identification using inverse methods. Phil. Trans. R. Soc. A (2007) 365, 393-410.

[2] H. U. VOSS, J. TIMMER, and J. U. KURTHS, 2004. NONLINEAR DYNAMICAL SYSTEM IDENTIFICATION FROM UNCERTAIN AND INDIRECT MEASUREMENTS. International Journal of Bifurcation and Chaos, Vol. 14, No. 6 (2004) 1905-1933.

[3] G. Rus, R. Gallego, 2007. Solution of identification inverse problems in elastodynamics using semi-analytical sensitivity computation. Engineering Analysis with Boundary Elements 31 (2007) 343-360.

[4] M. G. ZHANG, 2007. INVERSE MODEL IDENTIFICATION OF NONLINEAR DYNAMIC SYSTEM USING NEURAL NETWORK Proceedings of the Sixth International Conference on Machine Learning and Cybernetics, Hong Kong, 19-22 August 2007.

[5] Y. M., Mao and X. L. Guo, 2009. Experiment study on dynamic force identification by a parameter estimation method Proceedings of the SEM Annual Conference June 1-4, 2009 Albuquerque New Mexico USA.

[6] Tadeusz Uhl, 2007. The inverse identification problem and its technical application Arch. Appl Mech (2007) 77: 325-337.

[7] Xia, F, Cole, C. and Wolfs, P., 2008. WHEEL RAIL CONTACT FORCES PREDICTION AND VALIDATION WITH FIELD TESTS, Conference On Railway Engineering, Perth, Australia, p73-82.

[8] Xia, F, Cole, C. and Wolfs, P., 2007. An inverse railway wagon model and its applications. Vehicle System Dynamics, Vol. 45, No. 6, June 2007, 583-605.

[9] T. Zhu, S. Xiao, G. Yang, W. Ma & Z. Zhang (2013): An inverse dynamics method for railway vehicle systems, Transport, DOI:10.3846/16484142.2013.789979.

[10] Xia, F, Cole, C. and Wolfs, P., 2008. Grey box-based inverse wagon model to predict wheel-rail contact forces from measured wagon body responses Vehicle System Dynamics Vol. 46, Supplement, 2008, 469-479.

[11] Hossein-Zadeh, N., Dhanasekar, M. and Sun, Y.Q., 2004. ANN Application to Predict the Wheel-Rail Impact Forces due to the Short Wavelength Defects in Rail. Proc. of The Second International Conference on Artificial Intelligence in Science And Technology (AISAT 2004), Hobart, Australia, 21 - 25 November, 212-217.

[12] Junji Kawasaki and Kamal Youcef-Toumi, 2002. Estimation of Rail Irregularities. Proceedings of the American Control Conference, Anchorage, AK May 8-10.

[13] B. Luber: Railway track quality assessment method based on vehicle system identification. e&i Elektrotechnik und Informationstechnik, 126, 180-185, 2009.

[14] T. Uhl, K. Mendrok, and A. Chudzikiewicz. 2010. Rail Track and Rail Vehicle Intelligent Monitoring System. THE ARCHIVES OF TRANSPORT, VOL. XXII, NO 4, 495-510.

# Plasticity in wheel-rail contact and its implications on vehicle-track interaction

**K Six[a], A Meierhofer[a], G Trummer[a], C Marte[a], G Müller[a], B Luber[a], P Dietmaier[b], M Rosenberger[a]**
[a] VIRTUAL VEHICLE Research Center, Austria
[b] Institute for Applied Mechanics, Graz University of Technology, Austria

## ABSTRACT

Physical processes within the wheel-rail contact highly influence the vehicle-track interaction in railway operation. Thus, prediction models are of high importance. Within the contact zone two plasticity effects occur: "global" plastification in the order of millimetres due to the general loading situation and "tribological" plastification in the order of microns due to surface effects (e.g. roughness) which is always accompanied by the wear process. State of the art models do not take into account these effects sufficiently. Hence, a new physical based overall wheel-rail contact and damage model has been developed.

## 1 INTRODUCTION

Physical processes within the wheel-rail contact have a high influence on the vehicle-track interaction in railway operation. Thus, accurate models describing these processes are of high importance e.g. for reliable investigations on vehicle dynamics, for optimising the transmission of traction and braking forces or for predictions on wheel-rail wear and rolling contact fatigue (RCF). Generally, the wheel-rail contact problem can be divided into a normal and a tangential contact problem. The aim is to calculate the shape of the contact patch as well as the normal stress and shear stress distributions within the area of contact. These quantities depend on the geometry of the contacting bodies, the normal loading and the relative motion between the two bodies in tangential direction prescribed by the rigid body creepages (longitudinal, lateral, spin). The solution of the contact problem is highly influenced by the material behaviour of the contacting bodies and by tribological effects within the contact. Figure 1 gives an overview of the classical contact problem and its implications. The wheel-rail material behaviour, which has to be considered within the contact problem directly, also influences the damage behaviour of wheel and rail.

If there is friction between wheel and rail, which is generally the case for the wheel-rail contact, the normal and tangential contact problem can be solved one after another (tangential contact has no influence on normal contact) as long as wheel and rail behave purely elastic and as long as they have the same material properties (Young's modulus and Poisson's ratio). Because wheel and rail consist of steel, the Young's modulus and the Poisson's ratio can be assumed to be the same for both. The remaining question is if wheel and rail behave purely elastic. State of the art contact models for calculating the contact patch geometry and the contact stress distributions are often based on elastic half space theory. Thus, purely elastic behaviour of wheel and rail is a central assumption. For example, Kalker's FASTSIM algorithm [1] represents the state of the art to calculate creep forces in multi body system (MBS) simulations of railway vehicles. This algorithm, which is a simplification of Kalker's exact theory [2], assumes purely elastic material behaviour of wheel and rail, smooth surfaces and elliptic contact patches (Hertzian contact geometries). Therefore, the normal contact problem has to be solved first e.g. by Hertzian theory, where also pure elastic material behaviour is assumed. For general (non-elliptic) contact conditions Kalker's exact theory, implemented in CONTACT [2], can be applied where again pure elastic material behaviour is assumed for solving the normal and tangential contact problem.

For consideration of tribological effects within the contact both mentioned state of the art models as well as other models from the literature assume Coulomb's law of friction. This law limits the shear stresses within the contact locally. The limits are calculated from the normal stresses by multiplying it with the coefficient of friction (COF). If the local shear stress is below the limit, wheel and rail stick on each other. If the shear stress exceeds the limit, sliding occurs. Most of the models assume constant COF's, some models take into account for example sliding velocity dependent COF's [3, 4].

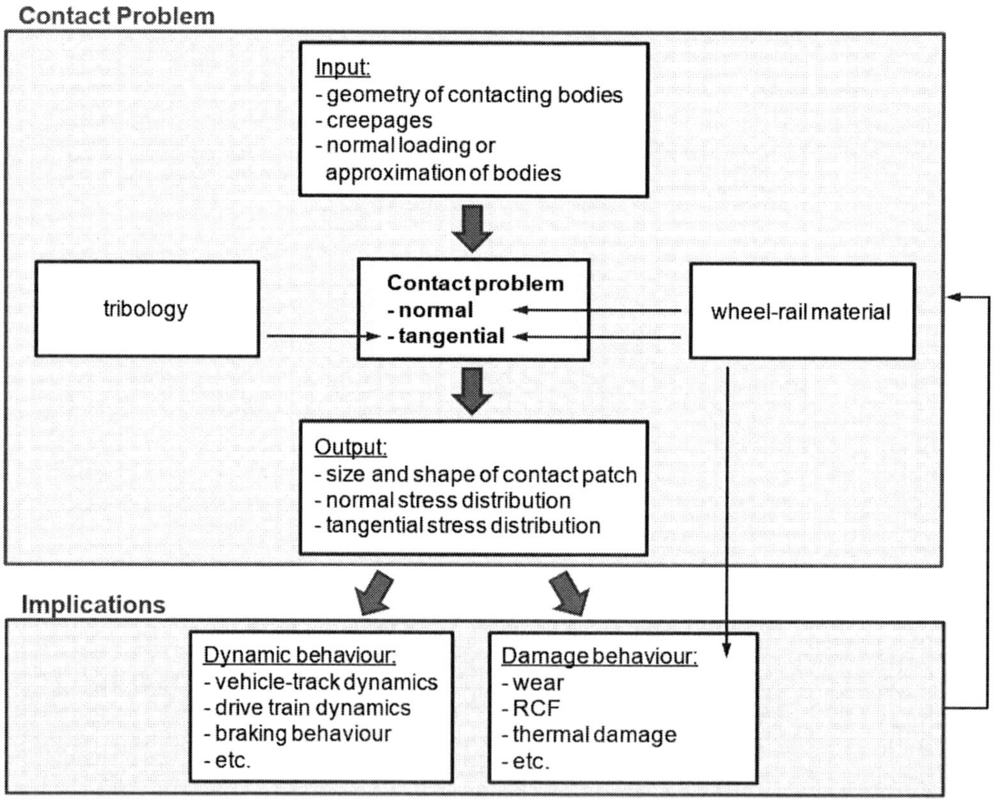


**Figure 1: Structure of the classical contact problem and its implications.**

To summarize, all mentioned models assume pure elastic material behaviour and do not take into account any plasticity effects, even though the Coulomb law of friction can be interpreted as ideal plastic material law describing the shear behaviour of the contacting asperities [5]. The question is, whether this assumption of pure elastic behaviour of wheel and rail is valid in real railway operation or not?

## 2 PLASTICITY IN WHEEL-RAIL CONTACT

Taking a closer look on wheels and rails in real railway operation, a plastically deformed material layer on the surfaces can be observed frequently, as shown in Figure 2. The extent of this plastic shear deformation usually increases towards the surface. Hence, plastic deformations occur in regular railway operation. Such plastic deformations cause hardening of the wheel-rail material, introduce residual stresses and can change also the geometry of wheels and rails (in the sense of wheel-rail profiles).

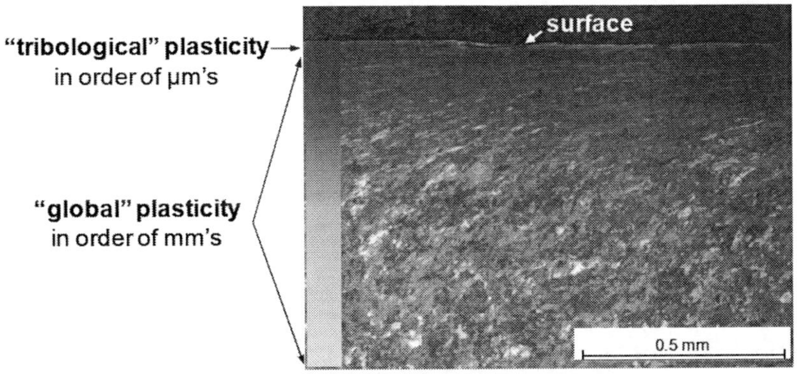


**Figure 2: Longitudinal metallographic section through a rail showing a typical plastically deformed near-surface layer.**

The plastically deformed near-surface layer plays a key role in the initiation and propagation of RCF cracks [6, 7, 8] (e.g. Head Checks on rails). Thus, it is of high importance to understand how plastic deformation occurs, how it interacts with the load situation in the contact and how it contributes to the development of damage.

As a first step it is important to understand, whether plasticity within the contact occurs all the time in railway operation or only during a limited number of cycles starting with new or newly maintained wheels and rails. Therefore, two processes regarding plasticity of wheels and rails in two different orders should be distinguished. Hence, the following classification has been introduced: "global" plasticity and "tribological" plasticity (see Figure 2).

### 2.1 "Global" plasticity

This kind of plasticity is caused by the "global" loading situation within the contact in normal and tangential direction. It occurs in the order of millimetres below the surface (up to the size of the contact patch) depending on the loading and the size of the contact patch. The "global" plasticity is independent from the surface roughness. In regular railway operation, this kind of plasticity occurs generally very quickly (low number of cycles) at the beginning of the lifetime of new or newly maintained wheels and rails. Wheels and rails deform plastically, they change their profiles, residual stresses develop and hardening effects occur, but after a certain number of cycles more or less stable conditions exist. The material gets into elastic shakedown, further plasticity effects occur only in time scale as wheel and rail profiles vary due to wear. Of course, for example at heavy haul operation this kind of plasticity can occur over a longer period of time but at the end the system has to get into shakedown not to be destroyed.

In this case of plastification, where the system reaches the state of elastic shakedown after a certain number of cycles, the assumption of elastic material behaviour is feasible. Hence, state of the art models as mentioned in section 1 for calculation of the contact patch geometry and contact stress distributions are sufficient.

Due to the high importance of the plastic shear deformation according to "global" plasticity on the initiation and propagation of RCF cracks, appropriate prediction models for plastification are important. Therefore, generally full 3D investigations based on state of the art finite element analysis (FEA) software packages are carried out. Elastic-plastic material behaviour of wheel and rail are taken into account and realistic shear deformations according to the "global" plasticity effect are calculated. This approach allows for investigations even on the first cycles where generally the "global" plasticity occurs and the system gets into elastic shakedown. The influence of the plasticity on the contact stresses during the first cycles can be investigated in detail as well as the development of residual stresses and changes of the wheel-rail profiles. The main disadvantage of this approach is the still very high calculation time of FEA. Thus, fast approximate models are an alternative. The Approximate Wear and Damage Model (AWD-Model) presented in [6] approximates the "global" plastic shear deformations below the wheel and rail surfaces from the stress situation calculated with state of the art contact models assuming elastic material behaviour (section 1). This approach reduces the calculation times significantly. The accuracy of the model is sufficient as long as the plastic shear deformation is dominated by the contact shear stresses and as long as the yield stress of the material is not exceeded significantly. It has to be mentioned that both approaches, FEA and the AWD Model, still assume Coulomb's law of friction not taking into account any hardening effects from plastification on asperity (roughness) scale.

## 2.2 "Tribological" plasticity

This kind of plasticity occurs in the order of microns (order of asperities / roughness) below the surface. The roughness of wheel and rail, together with solid third body layers (3BL) [9, 10, 11] in between the contacting surfaces, cause local stress peaks within the contact patch which lead to plastic deformations within a very limited surface-near layer. These plastic deformations are again associated with the development of residual stresses and material hardening. Additionally, again the geometry is influenced by this process, but in the sense of changes in the surface roughness instead of variations in the wheel-rail profiles. This kind of plastic deformation is accompanied by the wear process. Asperities get into contact, slide and stick against each other, deform plastically, harden and get sheared off after reaching a certain limit depending on the wear resistance of the material (wear debris). At the same time new asperities get into contact undergoing the same process. This wear process is accompanied by the generation of micro-cracks which have to be distinguished from the RCF cracks mentioned before in section 2.1.

In contrast to "global" plasticity, "tribological" plasticity is a continuous process which occurs all the time. On asperity level the material can never reach an elastic shakedown state due to the continuous wear process. Now it is important to recognize, that this continuous "tribological" plasticity determines the general shear stress distribution within the contact. Looking at the state of the art models, always Coulomb's law of friction (with constant or variable COF) is assumed. As mentioned before, this law can be interpreted as ideal plastic material law as described in [5]. For this, it is assumed that the wheel and the rail surfaces (asperities) adhere to each other all the time (no sliding). Due to normal loading the rough surfaces (asperities) get into contact and form a metallic contact area. Lateral relative motions introduced by the creepages cause shear stresses. If these shear stresses reach the yield stress of the asperities (normal stress multiplied by COF), ideal plastic deformation occurs. But this ideal plastic deformation is still in conflict with the continuous "tribological" plasticity and its related hardening effects. Hence, it was necessary to develop a more physical Extended Creep Force Model (ECF-Model) taking into account effects from "tribological" plasticity and 3BL's. The model is described in detail in [12, 13].

Because the "tribological" plasticity influences the general shear stress distribution in the contact, the "global" and the "tribological" plasticity are not independent of each other as expected at the first glance. Thus, for reliable predictions regarding all aspects of vehicle-track interaction (vehicle-, drive train-, braking-dynamics, wear, RCF, etc.) an overall consideration of the wheel-rail contact is necessary. This was the starting point for a research focus on that issue in collaboration between VIRTUAL VEHICLE Research Center, Graz University of Technology and industrial partners (DB Systemtechnik GmbH, L.B.Foster Rail Technologies, Corp., ÖBB Infrastruktur AG, SBB AG, Siemens AG Deutschland, Siemens AG Österreich, voestalpine Schienen GmbH, Wiener Linien GmbH & Co KG).

## 3 OVERALL WHEEL-RAIL CONTACT AND DAMAGE MODEL (OCD-MODEL)

The main idea of the presented Overall Wheel-Rail Contact and Damage Model (OCD-Model) is to combine several physical sub-models in a way to allow for reliable physical based predictions on contact stresses (contact forces) as well as on wear and RCF phenomena. These sub-models are mostly approximate but still fully physical (taking into account the main effects) to keep the calculation time sufficiently low. The OCD-Model consists of two main modules, the ECF-Model which takes into account the main tribological effects ("tribological" plasticity) in the contact between wheel and rail, and the AWD-Model for calculation of the "global" plasticity and the initiation of RCF cracks.

### 3.1 Extended Creep Force Model (ECF-MODEL)

The frictional behaviour of the wheel-rail contact highly influences all aspects of vehicle-track dynamics. Reliable prediction models taking into account the main tribological effects are of high interest. For predictions on wear and RCF not only the resulting creep forces are of interest, but as realistic as possible contact stress distributions are needed as well. State of the art models are often not sufficient. Hence, a new extended creep force model called ECF-Model has been developed [12, 13]. Figure 3 gives an overview of the model. The core of the model is a 3BL-Model taking into account "tribological" plasticity effects.

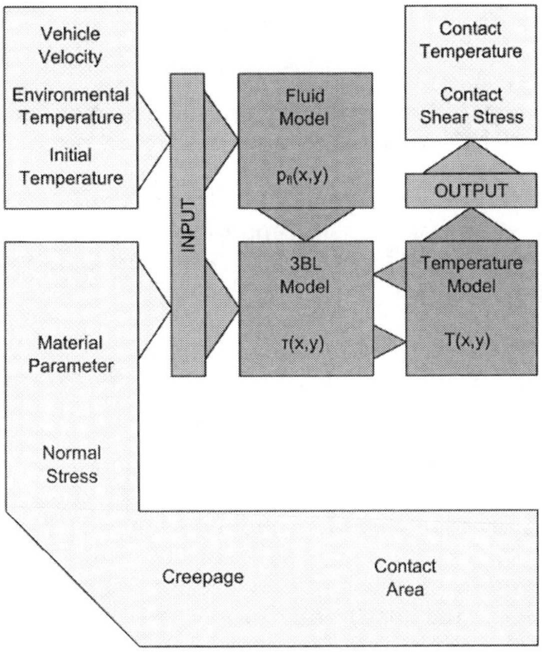


**Figure 3: Overview of ECF-Model.**

This 3BL is defined as follows: Additionally to real solid third bodies (e.g. sand, wear debris), the 3BL comprises also all kind of irregularities of the surface-near layers of wheel and rail (e.g. roughness, micro-cracks). Wheel and rail are considered as purely elastic, the 3BL behaves elastic-plastic to account for "tribological" plasticity.

For parameterisation and validation of the 3BL Model, tests have been carried out on a so called high pressure torsion (HPT) test rig. At this test rig, two discs are pressed against each other on their face side and rotated relative to each other. The normal load and the required torque are recorded. Figure 4 shows results for the measured mean normal contact stress $p_{Test}$ plotted over the mean displacement $u_{mean}$ calculated for the mean radius of the discs. The normal contact stress has been increased up to 1000 MPa. Furthermore, the measured mean shear stress $\tau_{Test}$ is plotted as well as the mean shear stress $\tau_{3BL\text{-}Model}$ calculated with the 3BL-Model. The test has been carried out with dry contact conditions. The results show that for a constant normal load the shear stress still increases which confirms the hardening due to "tribological" plasticity. Additionally, it has been shown that the ratio between the shear stress and the normal stress decreases with increasing normal load. Both effects are in conflict with the commonly used Coulomb's law of friction.

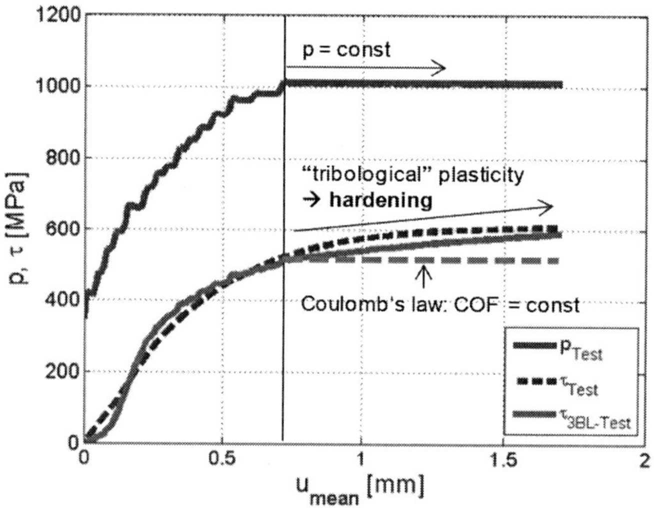


**Figure 4: Measured normal and shear stresses from HPT tests for dry contact conditions and predicted shear stresses.**

This normal load dependency has been considered in the 3BL-Model. Furthermore, the 3BL-Model parameters are chosen temperature dependent to take into account changes of the 3BL material behaviour due to the dissipated energy in the contact caused by high sliding velocities.

The normal contact stress distribution for the calculation of the pressure dependency of the 3BL-Model parameters is calculated externally e.g. by Hertzian theory or CONTACT (non-elliptic contact patches). The contact temperature distribution T(x,y) for the calculation of the temperature dependency of the 3BL-Model parameters is calculated by the Temperature Model iteratively. The ECF-Model also includes an Interfacial Fluid Model from the approach presented in [5] taking into account interactions between the fluid pressure $p_{fl}(x,y)$ and 3BL's.

Figure 5 shows a comparison between the traction coefficient f (longitudinal creep force divided by the normal load) measured at vehicle tests with a locomotive (circles) and calculated with the ECF-Model (smooth curves). The results are plotted over the longitudinal creepage $c_x$. Three different vehicle speeds have been investigated for dry contact conditions. The ECF-Model is able to reproduce the measured characteristic sufficiently accurate. It is important to mention that the predicted results for the three different speeds have been calculated with one set of ECF-Model parameters.

As a consequence of these results, the following important conclusion can be derived: The friction between wheel and rail at a certain position on rail cannot be described by only one number for the COF measured e.g. by a tribometer. Friction is strongly influenced by vehicle speed, normal loading, contact geometry, etc.

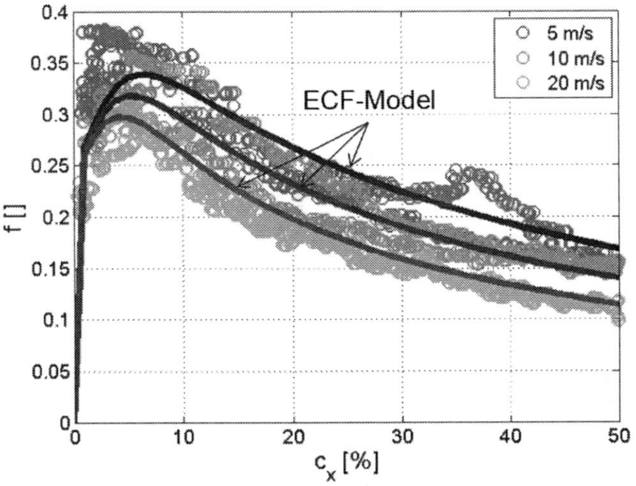


**Figure 5: Comparison between results from vehicle tests and results calculated with the ECF-Model, three different vehicle speeds, dry contact conditions.**

### 3.2 Approximate Wear and Damage Model (AWD-Model)

Wear and damage on wheels and rails cause high maintenance and renewal costs and can have a negative impact on the system safety as a final consequence. Head checks are a very common RCF damage pattern on rails as shown in Figure 6. Reliable prediction models are therefore of high interest during the development process of vehicles as well as during their operation for planning of maintenance activities.

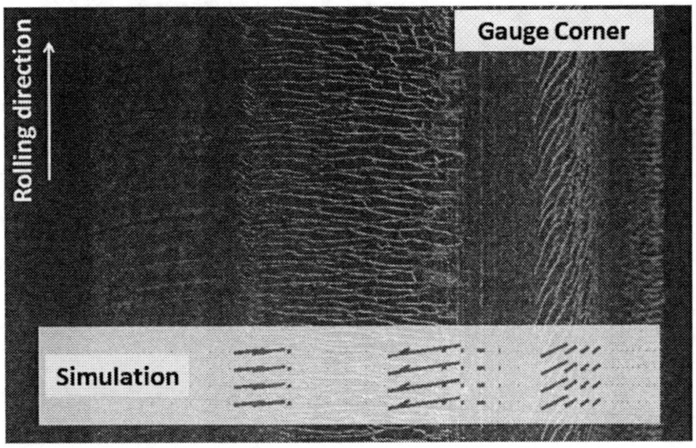


**Figure 6: RCF on the high rail in metro operation, lines in the insert are simulation results, indicating the expected orientations of the fatigue cracks at the surface, which are oriented normal to the direction of plastic deformation [6].**

There are several challenges developing such models. Due to high loads in the contact patch severe "global" plastic deformation (shearing) on and below the surface of wheels and rails occurs. This causes anisotropy in the material and results in the alignment of microstructural features, such as grain boundaries, pearlite structures or pro-eutectoid ferrite at grain boundaries. Generally, RCF cracks follow these aligned structures as long as the cracks reach the undeformed material. Hence, models describing the plastic shear situation below the surface are an important basis for reliable RCF prediction models. Parallel to the growth of RCF cracks material from the wheel-rail surfaces gets removed by wear. Hence, a reliable RCF prediction model must be able to account for that.

The AWD-Model shown in Figure 7 represents an approximate approach. In this way the calculation time can be reduced significantly compared to FEA. The core of this approach is the Plasticity Model presented in [6]. Based on elastic stresses ($\sigma(x,y,z)$, $\tau(x,y,z)$) calculated with the elastic half space theory (Bulk Stress Model), the plastic shear deformation ($\gamma_{xz}(y,z)$, $\gamma_{yz}(y,z)$) in the near-surface material layer is estimated. The model has been parameterized and validated by twin disc experiments at different normal loading (maximum normal contact stress) and creepage conditions, see Figure 8. A good agreement between the measured and predicted microstructure orientations $\alpha$ over the whole depth z can be seen. It is again important to mention that the predicted results for the four loading conditions have been calculated with one set of model parameters.

Depending on the calculated plastic shear deformation, the probability of RCF crack initiation is calculated from the elastic stress situation within the Fatigue Model (damage D). Additionally, the influence of the wear process is taken into account by continuous removal of material from the wheel-rail surfaces ($\Delta z(y)$) as calculated by the Wear Model.

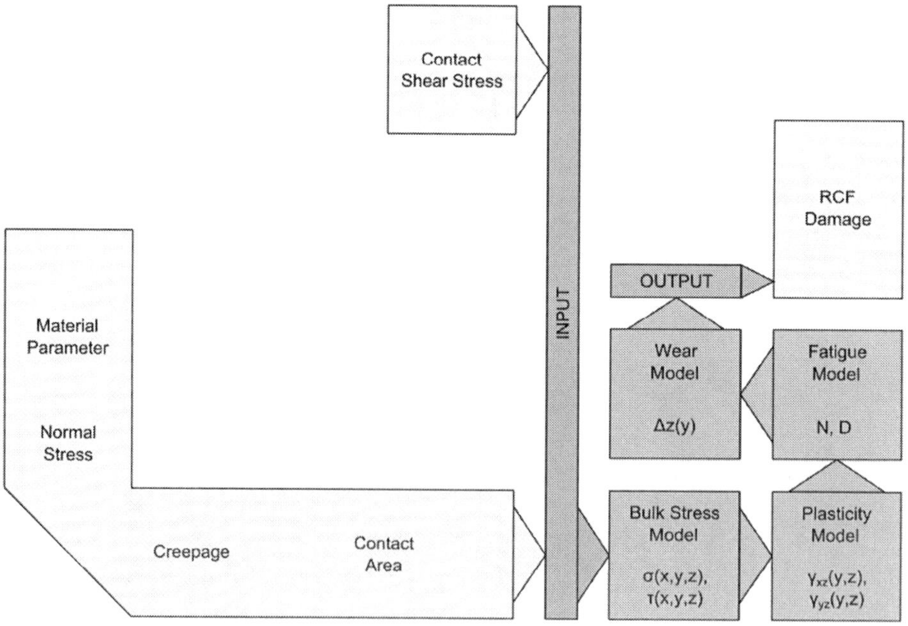


**Figure 7: Overview of AWD-Model.**

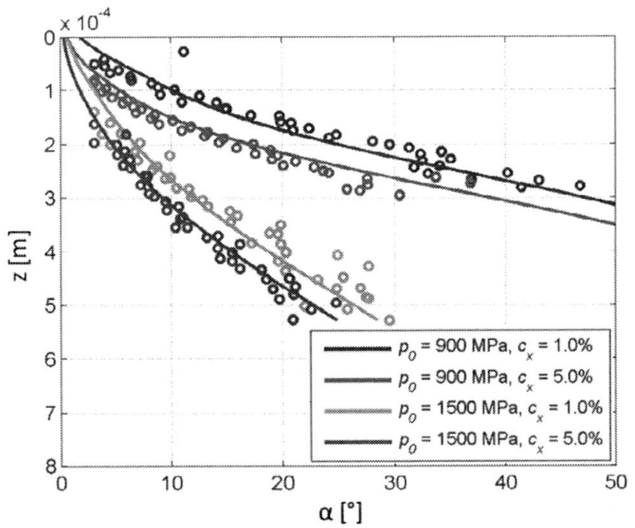


**Figure 8: Results of the plasticity model parameterization based on twin disc experiments. Dots: Measured local microstructure orientation α depending on the depth z, Solid lines: α calculated with AWD-Model [6].**

The AWD-Model represents a good compromise between calculation efficiency and accuracy. The red lines in the insert of Figure 6 show results from the AWD-Model based on MBS simulations of vehicles in metro operation. The red lines indicate the expected orientations of the fatigue cracks at the rail surface, which are oriented perpendicular to the direction of the predicted plastic deformation. A good agreement with the observed crack patterns could be reached.

The COF according to Coulomb's law of friction is still an uncertain input parameter within the AWD-Model. Tribological effects, like "tribological" plasticity, its local influence of the contact stress and contact temperature are not taken into account sufficiently. To this end, the AWD-Model has been combined with the ECF-Model in a next step. The structure of this Overall Wheel-Rail Contact and Damage Model (OCD-Model) is shown in Figure 9.

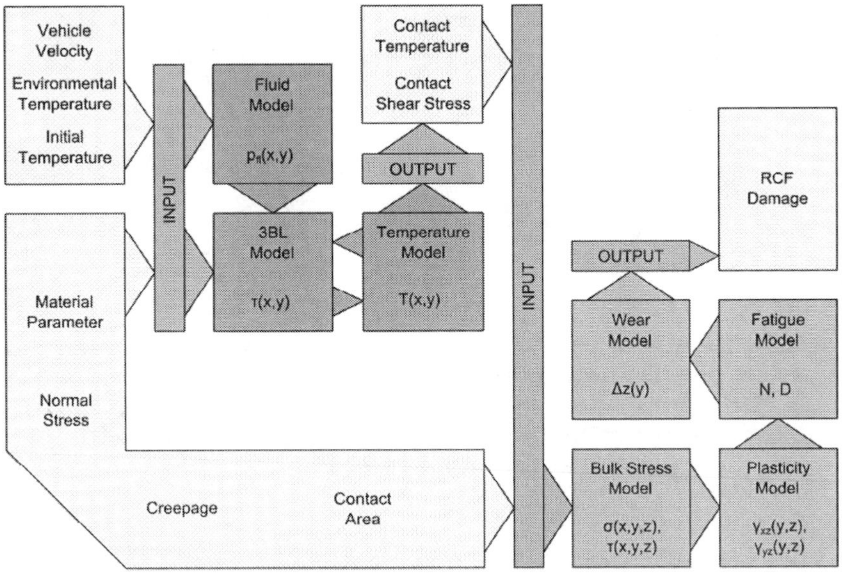


**Figure 9: Overview of OCD-Model consisting of ECF-Model (blue) and AWD-Model (yellow).**

Within the OCD-Model the shear stress distribution needed within the AWD-Model is calculated directly by the ECF-Model taking into account the main tribological effects. Hence, the COF according to the Coulomb's law of friction is no longer an uncertain model input parameter for the AWD-Model. Variations in the contact shear stress distribution caused e.g. by variations in the vehicle speed, normal loading or contact geometry and its influence on the "global" plasticity and initiation of RCF cracks are taken into account implicitly in the OCD-Model. At the top of Figure 10 the influence of the wheel-rail contact geometry on the resulting traction coefficient is shown by variation of the rail head radius R. Reduction of R yields smaller contact patches with increased normal contact stresses resulting in significant reduced traction coefficients. These reduced traction coefficients are accompanied by reduced contact shear stresses (compared to FASTIM assuming a constant COF in both cases) influencing the "global" plasticity and RCF behaviour. At the bottom of Figure 10 the according contact shear stress distributions for a longitudinal creepage of 5 % are depicted calculated with the OCD-Model (centre of the contact patch, longitudinal direction). The red shear stress distributions show the results calculated with FASTSIM assuming a constant COF of 0.4, which is a good estimate for the case of R = 300 mm. For the case of R = 50 mm the contact shear stress according to a COF of 0.4 is much too high. Only by an empirical adaption of the COF, this physical effect can be taken into account appropriately. Within the new OCD-Model, this effect is considered implicitly.

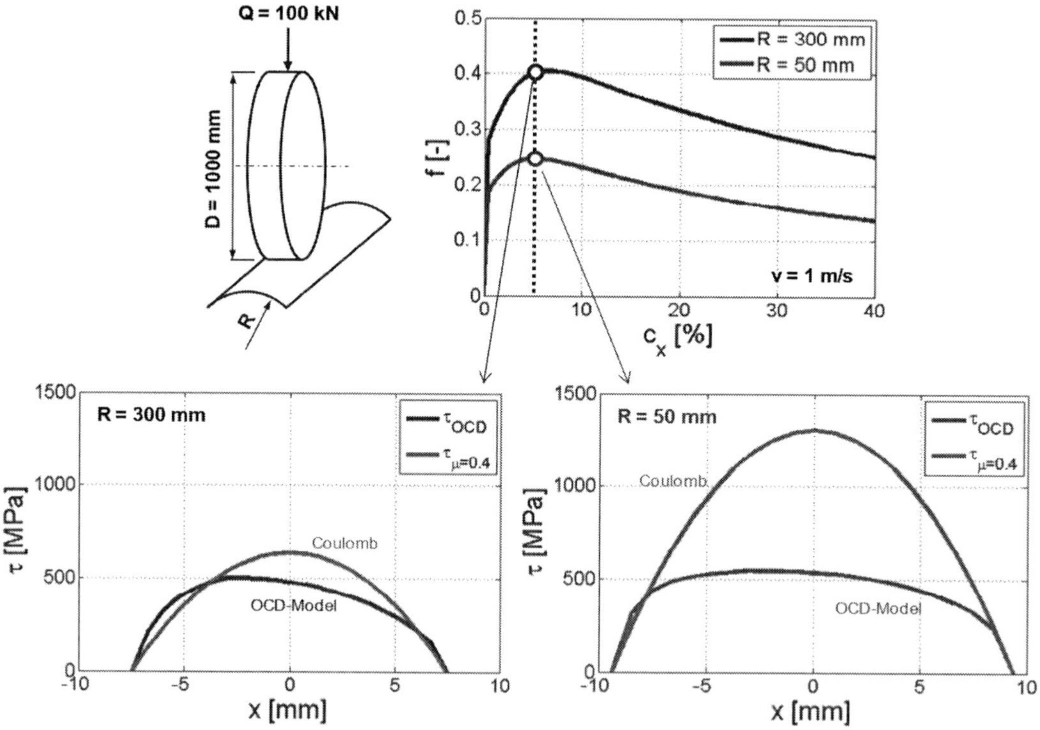


**Figure 10: Influence of contact geometry on traction coefficient (top) and shear stress distribution (bottom), cylindrical wheel, pure longitudinal creepage, 1000 mm wheel diameter, 100 kN normal load, 1 m/s vehicle speed. Red curves in the bottom figures are calculated with FASTSIM assuming a COF of 0.4.**

## 4 CONCLUSIONS

The wheel-rail contact represents the interface between railway vehicles and the track system. Physical processes within this contact have a high influence on all aspects of vehicle-track dynamics as well as on the development of wear and damage on wheels and rails. Thus, prediction models are of high importance. Such models can be divided in contact models describing the shape of the contact patch as well as the normal and tangential contact stress distributions and models describing the wear and damage behaviour resulting from this stress situation.

In this context, plastic deformations of wheel and rail play a key role. Two plasticity effects in different orders should be distinguished and the following classification has been introduced: "global" plasticity in the order of millimetres and "tribological" plasticity in the order of microns (asperities). "Global" plasticity occurs generally only during the first load cycles and stays approximately constant afterwards. The resulting shear deformation highly influences RCF crack initiation and propagation. "Tribological" plastic deformations occur all the time due to the accompanying wear process. Furthermore, this "tribological" plasticity influences the general contact shear stress distribution and therewith interacts with the "global" plasticity. State of the art models do not take into account these plasticity effects sufficiently.

Thus, a new overall wheel-rail contact and damage model has been developed taking into account both of these plasticity effects on a physical basis. In this way the influence of e.g. the vehicle speed, normal load, contact geometry on the traction coefficient and therewith on the shear stress distribution within the contact can be described more accurately. This increases the prediction quality regarding all aspects of vehicle-track dynamics. Together with the embedded "global" plasticity module also the prediction quality regarding RCF crack initiation is increased.

## ACKNOWLEDGEMENT

The authors would like to acknowledge the financial support of the "COMET - Competence Centers for Excellent Technologies Programme" of the Austrian Federal Ministry for Transport, Innovation and Technology (bmvit), the Austrian Federal Ministry of Science, Research and Economy (bmwfw), the Austrian Research Promotion Agency (FFG), the Province of Styria and the Styrian Business Promotion Agency (SFG).

We would furthermore like to express our thanks to our supporting industrial and scientific project partners, namely DB Systemtechnik GmbH, L.B. Foster Rail Technologies, Corp., ÖBB Infrastruktur AG, SBB AG, Siemens AG Deutschland, Siemens AG Österreich, voestalpine Schienen GmbH, Wiener Linien GmbH & Co KG and to the Graz University of Technology.

## REFERENCE LIST

[1] J.J. Kalker, A fast algorithm for the simplified theory of rolling contact, Vehicle System Dynamics, 11 (1982), pp. 1-13.

[2] J.J. Kalker, Three-dimensional elastic bodies in rolling contact, Kluwer Academic Press, 1990.

[3] O. Polach, Creep forces in simulations of traction vehicles running on adhesion limit, Wear, 258 (2005), pp. 992-1000.

[4] E.A.H. Vollebregt, Numerical modeling of measured railway creep versus creep-force curves with CONTACT, Wear, 314 (2014), pp. 87-95.

[5] C. Tomberger, P. Dietmaier, W. Sextro and K. Six, Friction in wheel-rail contact: a model comprising interfacial fluids, surface roughness and temperature, Wear, 271 (2011), pp. 2-12.
[6] G. Trummer, K. Six, C. Marte, P. Dietmaier, C. Sommitsch, An approximate model to predict near-surface ratcheting of rails under high traction coefficients, Wear, 314 (2014), pp. 28-35.
[7] R. Stock, R. Pippan, RCF and wear in theory and practice – The influence of rail grade on wear and RCF, Wear, 271 (2011), pp. 125-133.
[8] N. Larijani, J. Brouzoulis, M. Schilke, M. Ekh, The effect of anisotropy on crack propagation in pearlitic rail steel, Wear, 314 (2014), pp. 57-68.
[9] S. Descartes, C. Desrayaud, E. Niccolini, Y. Berthier, Presence and role of the third body in a wheel-rail contact, Wear, 258 (2005), pp. 1081-1090.
[10] E. Niccolini, Y. Berthier, Wheel-rail adhesion: laboratory study of natural third body role on locomotives wheels and rails, Wear, 258 (2005), pp. 1172-1178.
[11] A. Meierhofer, C. Hardwick, R. Lewis, K. Six and P. Dietmaier, Third body layer – experimental results and a model describing its influence on the traction coefficient, Wear, 314 (2014), pp. 148-154.
[12] A. Meierhofer, A new physical creepforce model to investigate the influence of a third body layer on the steady and unsteady traction characteristic, Ph.D. diss., Graz University of Technology, 2014, in review.
[13] K. Six, A. Meierhofer, G. Müller, P. Dietmaier, Physical processes in wheel-rail contact and its implications on vehicle-track interaction, Vehicle System Dynamics, Special Issue on "Modelling of Traction in Railway Vehicles". (2015), in review.

# Development of a Tram-Train wheel profile for dual-operation running

**D Crosbee, P D Allen**
University of Huddersfield, UK
**R Carroll**
Sheffield Supertram, UK

## ABSTRACT

This paper explores the problematic interface between a Tram-Train vehicle and two very different railway infrastructures, detailing the analysis and design process required to develop an optimised wheel profile for dual operation running.

One of the key issues in developing a dual-operation wheel profile is managing the contact conditions within the wheel/rail interface. The interface is critical not only to the safe running of the vehicle but also to maximise wheelset life and to minimise wheel-rail damage. A combination of vehicle dynamic simulations and bespoke software were used to allow the development of a new wheel profile for Tram-Train operations.

## 1 INTRODUCTION

The Tram-Train concept provides a dual-mode vehicle that operates as a tram on light rail infrastructure and a conventional train on heavy rail infrastructure, offering a seamless journey to the passenger into the heart of city centres and relieving capacity from mainline stations. The first Tram-Train scheme was in Karlsruhe, Germany, in the early 1990s and has spread successfully to several other European cities – but not yet to the UK.

To demonstrate that the benefits of Tram-Train can be realised in the UK, the Tram-Train pilot project was set up by the Department for Transport (DfT) with Network Rail, Northern Rail, Stagecoach Supertram and South Yorkshire Passenger Transport Executive (SYPTE) as partners. The chosen route runs for 6.5km from the centre of Sheffield on the Sheffield Supertram (SST) light-rail network before connecting with Network Rail (NR) heavy-rail infrastructure at Tinsley. From here the Tram-Train vehicle will run for a further 5.7km on the NR track to Rotherham. In order to run on both networks there are many interface issues to deal with such as signalling, platform heights, overhead line equipment and wheel/rail interaction.

Network Rail (NR) requested the support of the Institute of Railway Research (IRR) at the University of Huddersfield (UoH) to develop a wheel profile design suitable for Tram-Train operation over NR heavy-rail infrastructure and the Sheffield Supetram (SST) light-rail network. This paper discusses the challenges and the design process followed during the evolution of the wheel profile and the rationale behind the design choices that were made. The paper also describes the required design assurance which has been carried out to ensure the new profile has sufficient resistance to derailment, is compatible with switches and crossings (S&C) and has acceptable performance in terms of wheel-rail rolling contact fatigue and wear.

## 2 PROFILES

There is a combination of wheel and rail profiles used on the SST and NR infrastructure, all of which needed to be considered. The following sub-sections present the various wheel and rail profiles currently in use on these two systems.

### 2.1 Comparison of existing wheel profiles

Wheel profiles are normally designed or selected to be compatible with the rail profiles used on a particular system. As a result of this, the wheel profiles used on SST and NR are very different when compared to one another; Figure 1 shows the two profiles and key dimensions. The SST vehicles use a wheel profile (1) that is typical of tramway applications; low conicity tread compatible with the flat crown (rail head) of rail profiles such as 55G2 and 39E1 used on the SST system, a steep 76° flange for derailment resistance on tight curves and a flat flange tip for flange tip running through embedded (S&C). On NR infrastructure there are a range of approved profiles which are defined in the Railway Wheelsets Railway Group Standard, GM/RT2466 (2). The most common profile on the NR section of the Tram-Train route is the British Rail (BR) P8 profile which is commonly used throughout the UK and is compatible with all NR infrastructure. The P8 profile has a shallower flange angle of 68° and its tread is a 'pre-worn' shape suited to the smaller crown radii of NR rail sections, and results in higher levels of wheel-rail conicity when compared to the SST profile. The SST profile has a narrower flange than the P8 profile required for running in grooved rail sections, this means that in common with the majority of tramway vehicles, the SST profile is installed at a greater effective wheelset Back-to-Back spacing, thereby ensuring a standard rail flangeway clearance on standard gauge track is maintained. This can be seen in Figure 1 which shows the two profiles at the correct lateral position relative to one another.

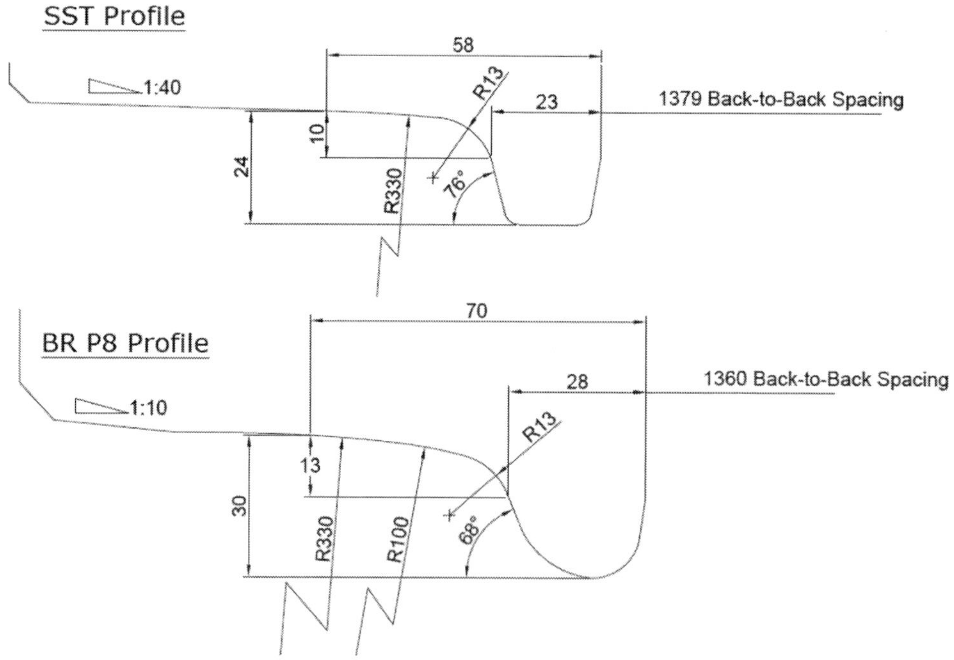


**Figure 1: Comparison of existing SST and NR wheel profiles**

### 2.2 Comparison of existing rail profiles

The Tram-Train route will have a combination of new and worn rail profiles of both grooved and vignole (flat bottom) type. In order to develop an optimised wheel profile for dual-operation on both systems it was necessary to consider all the profiles that Tram-Train will encounter. Table 1 lists the installed rail profiles on the Tram-Train route.

**Table 1: Rail Profiles used on Tram-Train route**

| Profile | Standard | System | Rail Type | Inclination | Track Type |
|---|---|---|---|---|---|
| BS80A (39E1) | EN 13674-4 (3) | SST | Vignole | Vertical | S&C |
| | | | | 1:40 | Ballasted Track |
| 55G2(41GP) | EN 14811 (4) | SST | Grooved | Vertical | Embedded Track |
| 54E1 (UIC 54) | EN 13674-1 (5) | SST | Vignole | Vertical | Viaduct |
| BR113A (56E1) | EN 13674-1 (5) | SST | Vignole | Vertical | Ballasted Track |
| | | NR | Vignole | 1:20 | Ballasted Track |
| SST Worn | - | SST | - | - | - |
| NR Worn | - | NR | - | - | - |

When considering the worn rail profile shapes, the former Rail Technology Unit (RTU) at Manchester Metropolitan University compiled a report (6) on the suitability of different wheel profiles for Tram-Train. The report demonstrated that measured rail profiles of SST BS80A and 55G2 had the same worn shape and therefore only one worn rail profile for SST is considered. This is typical of 'closed' rail systems whereby the infrastructure sees only one vehicle and wheel profile type.

The IRR holds a significant number of measured worn NR profiles and a representative worn rail profile was selected for use in this study.

Figure 2 shows the most prevalent new rail profiles and the corresponding worn profiles found on the Tram-Train route. The new BS80A and 55G2 profiles which are installed on SST have virtually the same gauge corner and head profile giving similar contact conditions, the key difference is the keeper rail on the 55G2. When comparing the new and worn SST and NR profiles it can be seen that there is a large variation in shapes, the key differences being with the gauge corner and crown radii.

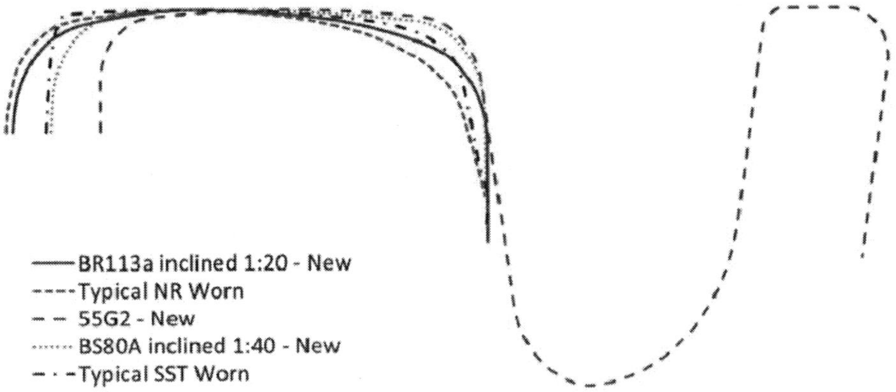


**Figure 2: Prevalent rail profiles on Tram-Train route**

Typically as a rail wears it tends to adopt the shape of the wheels running on it, such that the NR worn rail shape is similar to a BR P8 wheel and the SST worn rail profile is similar to the SST wheel profile. This conformality is demonstrated in Figure 3.

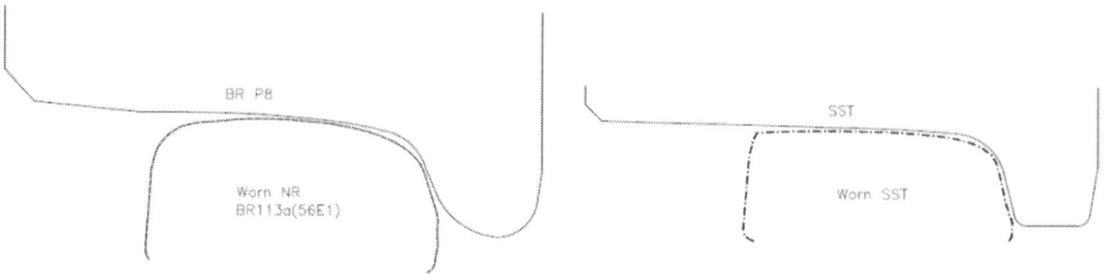


**Figure 3: Comparison of wheel profiles and worn rail profiles**

## 3 PROFILE DEVELOPMENT

The Tram-Train vehicle will be the minority vehicle running on both NR and SST infrastructure. As a result, following introduction of the Tram-Train service, the existing worn rail shapes on both NR and SST will remain similar to their current shape. The design of the Tram-Train profile has therefore been optimised for the worn rail shapes on NR and SST, with the best achievable compatibility with new 55G2 (41GP), BR113a (56E1) and BS80A (39E1).

In addition, to assist in the approvals process, the profile design has also been focussed on meeting Railway Group Standard for wheelsets, GM/RT2466 (2).

The initial study conducted by the RTU considered a range of existing wheel profiles for use on Tram-Train (6). The study concluded that none of the current UK tramway wheel profiles including the current SST wheel were suitable for use on NR infrastructure due to severe two-point contact at relatively large curve radii, resulting in accelerated wheel and rail wear. However the work did find that an BR P8 wheel profile generated similar wheel-rail wear rates on SST infrastructure as the current tramway wheel profile and therefore the P8 wheel could form the basis of a possible Tram-Train profile. This is demonstrated in Figure 4 which shows the Tγ (contact patch energy – the product of creepage and creep force) generated on the tread and flange for a range of curve radii. Higher Tγ values equate to higher wear rates as shown by work carried out by British Rail Research (BRR) (7). The plots in Figure 4 for NR infrastructure show that the generated Tγ is slightly higher on the tread for the SST wheel when compared to the NR P8 and considerably higher on the flange. Plots on SST infrastructure show that the P8 and SST profiles perform in a very similar manner.

It was therefore decided that a hybrid profile was required, based upon a combination of the SST and the P8 profile to ensure interoperability on the two systems. The profile required the following features:

- P8 type geometry from flange face to tread, this maintains geometry that is compatible with NR rail profiles and avoids severe two-point contact generated using conventional tram profiles.
- 68° flange angle for compatibility with NR switch toes and facilitate the required 3mm of residual switch opening (RSO).
- SST flat flange tip for compatibility with flange tip running through crossings in grooved rail on SST.
- SST flangeback angle and position to create similar keeper/check rail contact conditions.
- Cut-out in back of wheel to provide two checking faces – one for NR and one for SST.

Figure 5 shows the Tram-Train profile which combines all of the required features listed above.

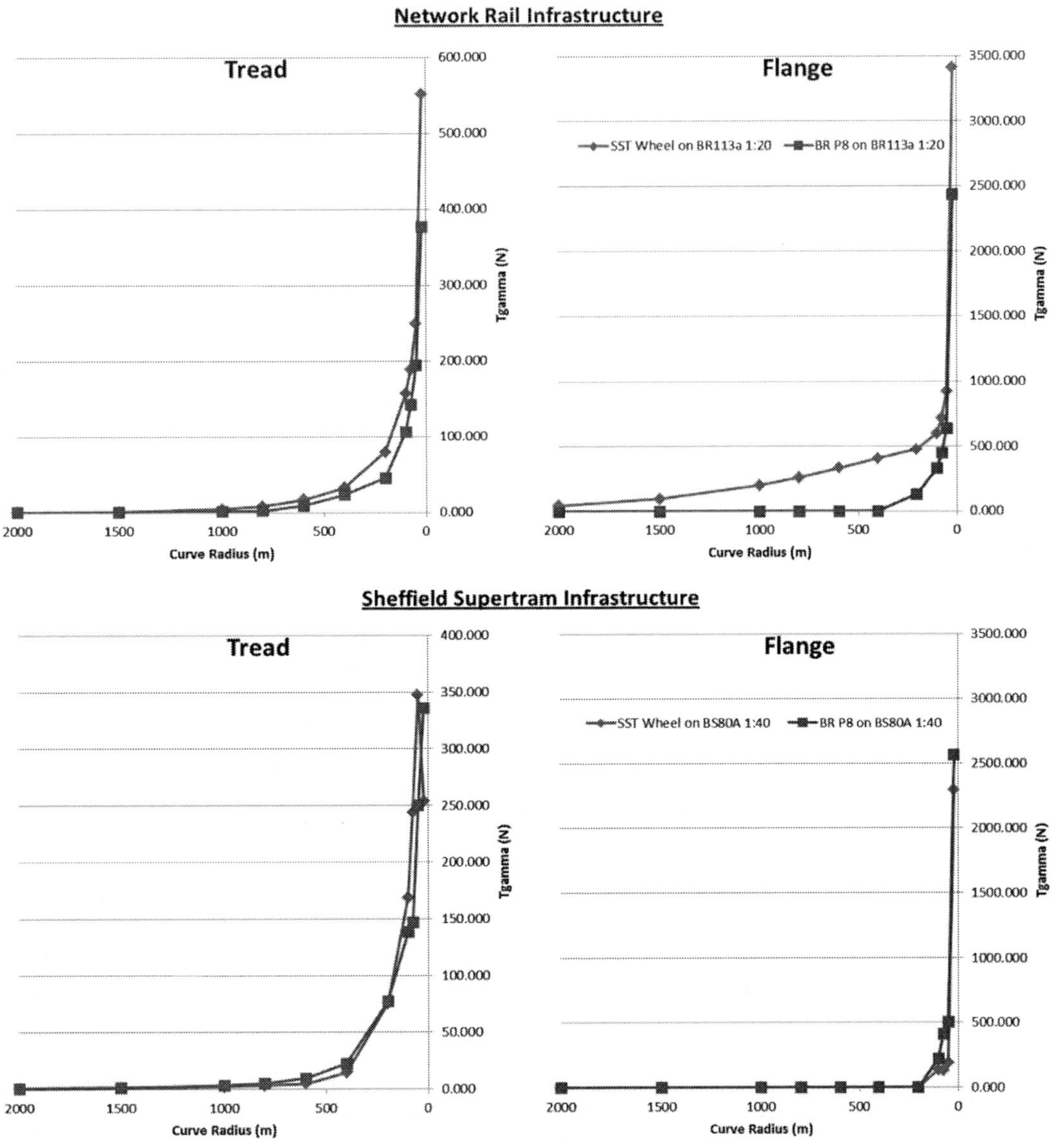


**Figure 4: Tγ vs Curve Radius for SST and NR P8 wheel profiles on NR and SST infrastructure**

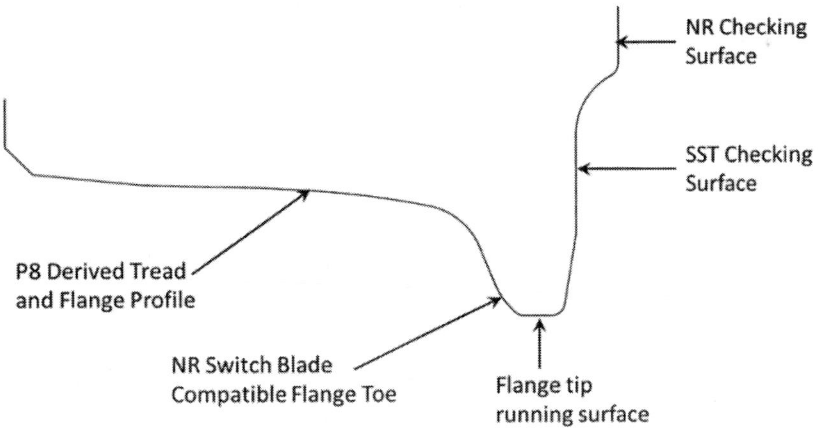


**Figure 5: Tram-Train profile**

Beginning with the first design iteration of the Tram-Train profile, a series of other profiles with stepwise developments were made. The methodology used for profile development was:

1. Determine new and worn rail section crown and gauge corner radii using relevant standards and 2D CAD.
2. Modify profiles to improve conformality between wheel and rail profiles to avoid distinctly separated two-point contact.
3. Develop flange and flangeback geometry to ensure checking faces are in the correct location.
4. Assess contact conditions for suitability:
    - Rolling Radius Difference
    - Equivalent Conicity
    - Contact Patch Area and Contact Stress
    - Tγ and Contact Stress (Curving Performance)
5. Review design and repeat as necessary

## 4 CONTACT CONDITIONS

For each wheel/rail profile combination on the Tram-Train route, contact data was generated. The contact data was created using the Contact Data Generation program within the VAMPIRE vehicle dynamics simulation software. The contact data contains information describing the geometric contact conditions between wheel and rail for a series of lateral positions. The values of interest were rolling radius difference, contact angle and contact patch area. The Contact Data Generation program also calculates the equivalent conicity of the wheel rail combination.

### 4.1 Rolling radius difference

The plot of rolling radius difference (RRD) between left and right rails, provides an indication of the level of conicity that a wheel profile generates as well as helping identify if, or when, two-point contact occurs. Severe two-point contact is avoided to prevent excessive Tγ levels and high wheel-rail wear rates. The desired RRD plot should, if possible, have a smooth transition from tread to flange contact without significant jumps which would indicate two-point contact. In addition, the gradient of the rolling radius difference plot does not want to be too great, as this would result in a high conicity wheel/rail profile combination. The conicity of the profile is important as a high conicity will provide good curving performance but will make the vehicle more susceptible to lateral instability. Although good curving performance is desirable, poor lateral stability should be avoided as it is detrimental to passenger ride comfort and also results in increased wheel and rail wear.

Figure 6 shows the RRD plots for new and worn SST wheel profiles and the Tram-Train wheel profile on worn SST rail. The new and worn SST wheel profiles provide little RRD up to the point of flange contact, at which point the RRD rises rapidly. The jump in the RRD for the worn SST profile between 6 and 7mm lateral shift is due to the way that the side worn rail profile interacts with the wheel flange resulting in the contact patch jumping up and then back down the flange. The Tram-Train wheel profile provides a higher amount of RRD on the tread indicating higher conicity and therefore better curving performance. The transition on to the flange is less abrupt than for the SST wheel profiles and does not exhibit two point contact which will help minimise rail side wear.

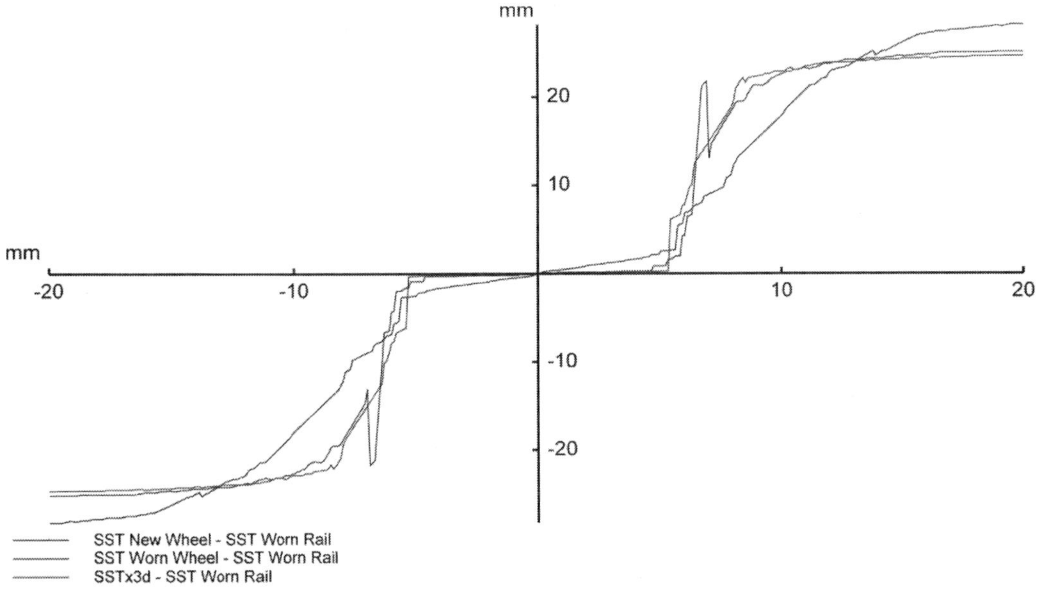


**Figure 6: Rolling Radius Difference plot on SST infrastructure**

## 4.2 Contact angle

The contact angle is the angle of the contact patch relative to the horizontal plane. The contact angle has been considered as a measure of flange climb derailment resistance. A steeper flange contact angle maintains a lower risk of flange climbing. Figure 7 presents a plot of contact angle for new and worn SST wheel profiles and the Tram-Train profile on SST infrastructure

The plots show that as the profiles enter flange contact, the maximum contact angle reached is equal to the wheel profile flange angle. For the current SST profile that angle is 76° and for the developed Tram-Train profiles, the angle is 68° - the same as a BR P8. The plots also show the distance over which the angle is maintained in terms of wheelset lateral shift, with a larger lateral shift being more favourable as this offers the greatest protection against flange climb.

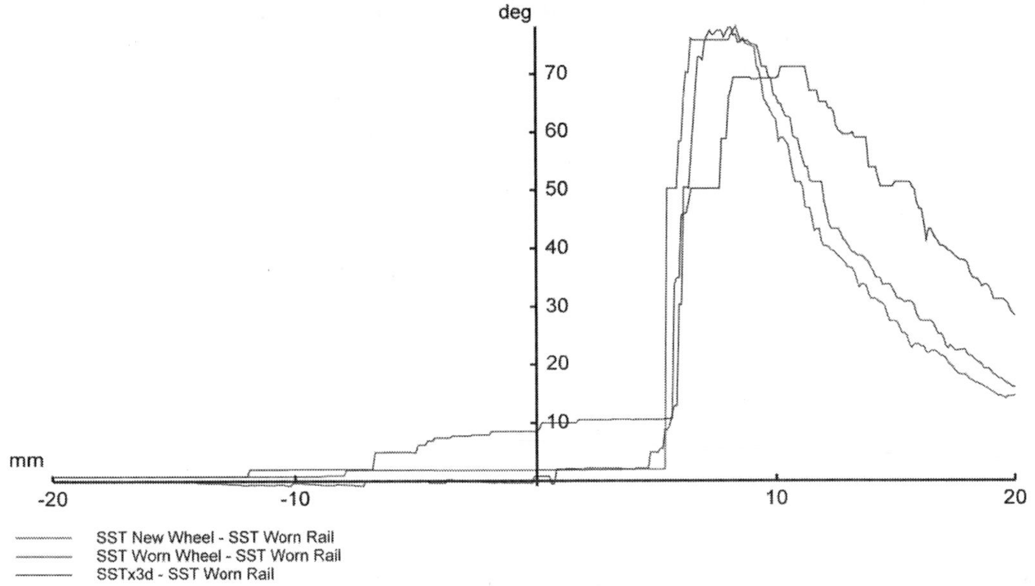


**Figure 7: Contact Angle plot for right wheel on SST infrastructure**

For all of the Tram-Train profiles developed, the flange angle is reached and sustained without a significant rate of reduction in flange angle as the contact patch moves up the flange. This characteristic provides sustained flange climb protection under more demanding contact conditions.

### 4.3 Contact patch area

It is desirable to maximise the contact patch area in order to reduce the contact stress. Increased levels of contact stress can result in greater damage within the wheel/rail interface in the form of wear, Rolling Contact Fatigue (9,10) and Rail Squats (11). Higher contact stresses will cause increased levels of wear and could possibly cause plastic flow of material within the wheel and rail. The aim of this aspect of the study was to develop a new Tram-Train profile with similar or greater contact patch areas than the existing profiles on the SST and NR systems. Figure 8 shows that although the contact patch area is slightly smaller on the flange for the Tram-Train profile, the contact patch area is similar overall to the current SST profiles.

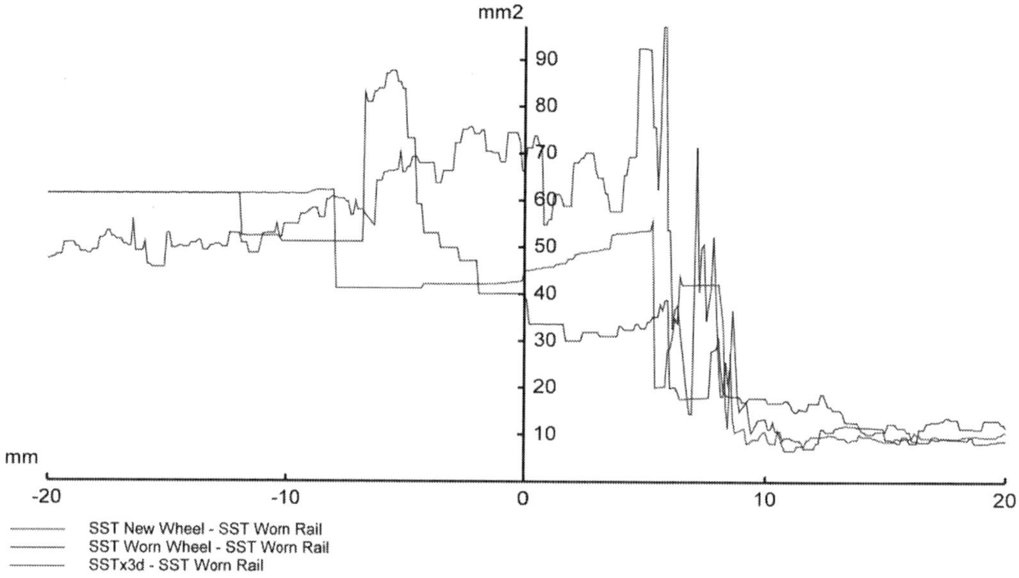


**Figure 8: Contact Patch Area plot for right wheel on SST infrastructure**

### 4.4 Equivalent conicity

Equivalent conicity provides an indication of the vehicle's curving performance and lateral stability. The conicity values of existing SST and NR wheel/rail combinations have been calculated and are used as a benchmark for the levels of conicity that the Tram-Train vehicle should be able to accommodate without increased risk of stability issues. Table 2 contains the equivalent conicity values for the different wheel/rail combinations on the Tram-Train route. It can be seen that the conicities generated by the Tram-Train profile do not exceed the maximum conicities generated by the SST and NR wheel/rail profiles.

**Table 2: Equivalent Conicity values for different profile combinations**

| Wheel | Rail | | | | |
|---|---|---|---|---|---|
| | **55G2** | **39E1 1:40** | **SST Worn** | **56E1 1:20** | **NR Worn** |
| P8 New | -- | -- | -- | 0.174 | 0.103 |
| P8 Worn | -- | -- | -- | 0.264 | 0.163 |
| SST New | 0.338 | 0.083 | 0.024 | -- | -- |
| SST Worn | 0.264 | 0.352 | 0.066 | -- | -- |
| Tram-Train | 0.23 | 0.201 | 0.22 | 0.185 | 0.13 |

## 5 DYNAMIC SIMULATIONS

The VAMPIRE Curving Analysis program was used to assess the curving performance of all profile combinations. The simulations were carried out using a coefficient of friction of 0.45 on the tread and flange and were run at balancing speed – the speed at which the lateral forces from curving are cancelled out by the cant of the track.

The outputs from the simulations were the Tγ and the Contact Stress in the tread and flange contact patches.

### 5.1 Tγ

Tγ is the work done or energy dissipated in the contact patch (as defined in Section 3) and provides an accepted method of quantifying the wear at the wheel/rail interface as shown by work carried out by British Rail Research (BRR) (7). Figure 9 presents the wear function developed by BRR which relates Tγ to a rail wear rate.

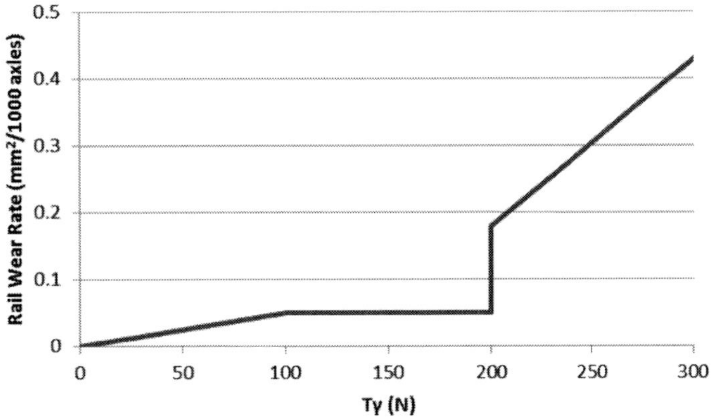


**Figure 9: British Rail wear function**

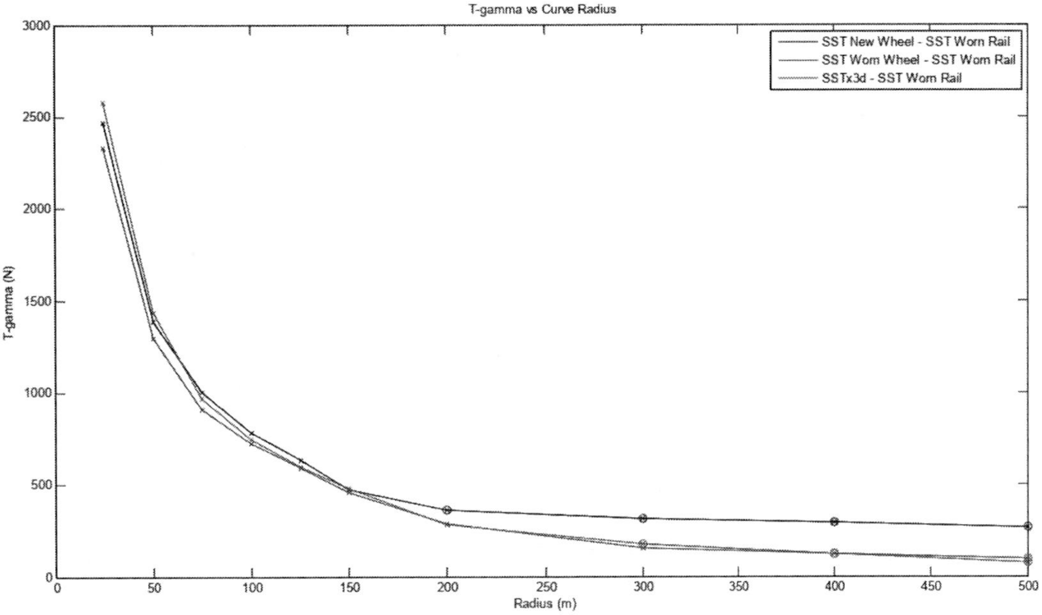


**Figure 10: Tγ vs Curve Radius for Supertram Infrastrucutre**

The wear function shows that higher Tγ values equate to higher levels of wear so from this perspective it is desirable to keep Tγ values as low as possible. Figure 10 shows a typical plot of Tγ curving results for a range of curve radii. This plot shows the current new and worn SST and new Tram-Train wheel profiles on worn SST rails. It can be seen that the worn SST wheel generates much lower Tγ values than the new SST wheel on curve radii greater than 150m. The Tram-Train profile performs in a similar way to the worn SST wheel profile which demonstrates good performance on worn SST rails.

### 5.2 Contact stress

The contact stress calculation is linked to the contact area calculated in Section 4.3 but the calculation of contact stress takes into account the effect of the dynamic forces generated by curving and the distribution of loads between tread and flange contact patches. Figure 11 shows how the contact stresses vary with curve radius for the Tram-Train vehicle on SST infrastructure. On the right tread the Tram-Train profile generates lower contact stresses than the new SST profile for all curve radii. On the left wheel tread the Tram-Train profile generates the highest stress on the tread out of all the profiles at large curve radii but the corresponding stress on the flange is below that of the current SST until the curve radii drops below 200m. This indicates that flange and rail side wear will be reduced when using the Tram-Train profile.

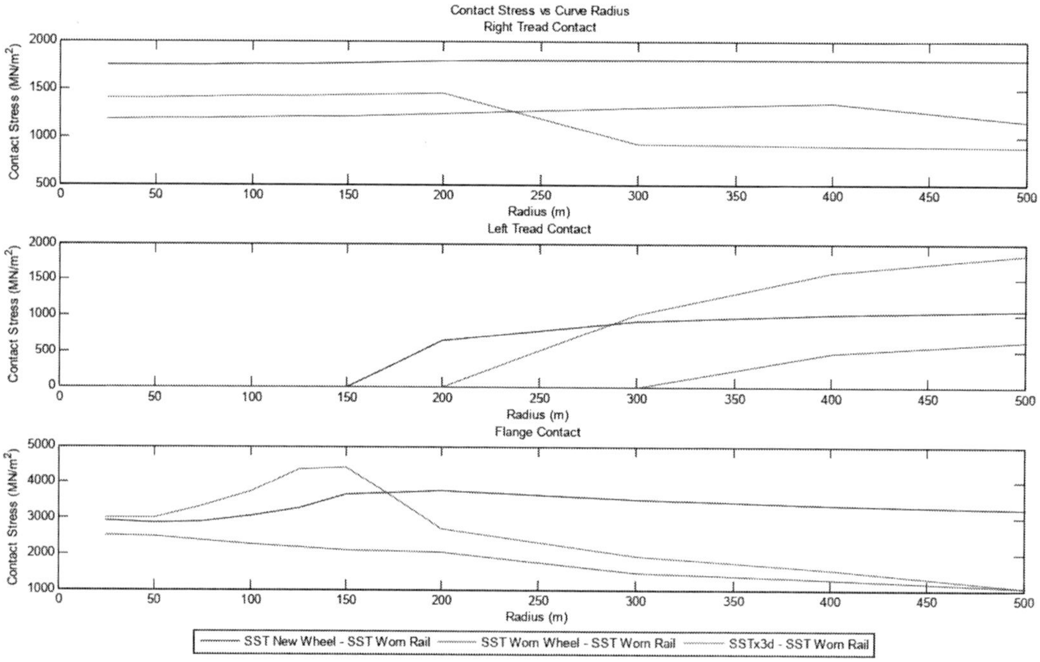


**Figure 11: Contact Stress vs Curve radius for Supertram Infrastructure**

## 6 GEOMETRIC ASSESSMENT

A series of geometric assessments were undertaken to ensure safe passage of the Tram-Train wheelset through all trackforms and S&C.

### 6.1 Grooved rail

When a vehicle negotiates a curve the wheelsets assume an Angle of Attack (AoA) producing a geometric effect which increases the effective wheel flange width. It is important that the effective flange width does not exceed the rail groove width and thereby ensures free wheel passage in all curves on the system. If the effective flange width is too great then both flange face and flangeback can come into contact with either side of the groove simultaneously; known as forcing of the wheel flanges in the rail groove. This scenario can pose an increased derailment risk and will also result in excessive wheel and rail wear.

The maximum effective flange width, or minimum groove width, was calculated for the maximum AoA which occurs at the smallest curve radii, 25m for SST. The calculation was performed using IRR in-house software which uses the Filkins-Wharton method to determine the effective flange width and generates a Nytram plot (12) showing the locations of the flange and flangeback contacts. Figure 12 shows the Nytram plot for the worst case - new wheels, minimum Back-to-Back spacing and maximum track gauge. The green points illustrate the flange contacts on the high and low rails and the red point the flangeback contact which would cause flange forcing. The purple point highlights where keeper rail contact would occur and hence defines the minimum groove width required to prevent keeper contact. The minimum calculated groove width for the SST system was found to be 26.7mm which is less than the 40.7mm groove width of new 55G2 groove rail. Therefore forcing of the flanges in grooved rail sections will not occur.

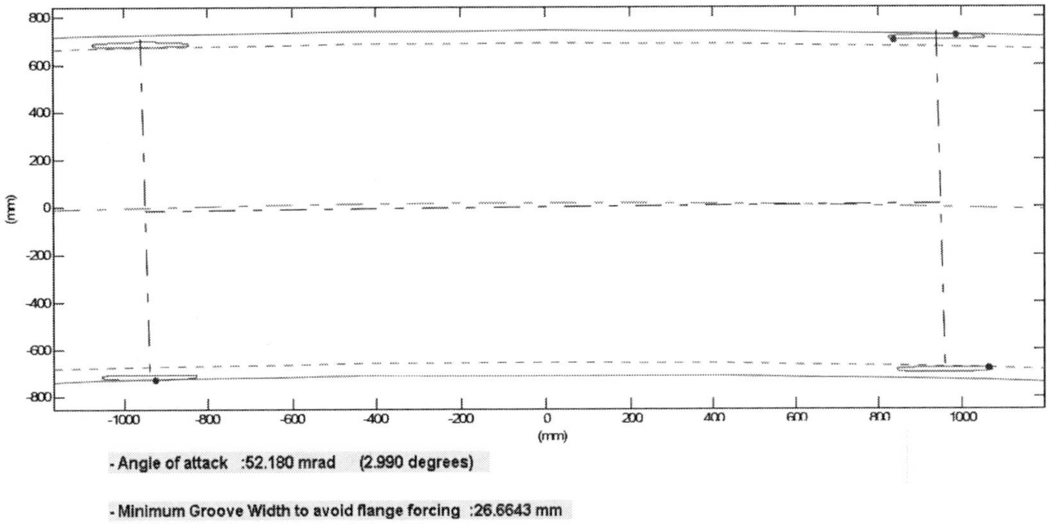


**Figure 12: Nytram Plot**

## 6.2 Switch and Crossing Interaction

Several methods have been applied to ensure that the Tram-Train wheelsets safely negotiate S&C. These methods have been taken from BOStrab guidelines, Network Rail standards and EuroNorms.

### *6.2.1 Blade vertical overlap*

BOStrab clause 3.10.2 (13) looks at the switch toe and requires that the wheel flange overlaps the switch toe by a minimum of 4mm. Figure 13a shows the dimensions that must overlap, H and h. Figure 13b shows the Tram-Train wheel profile located at an NR full depth and shallow switch toe with 3mm of residual opening and the dimensions H and h giving an acceptable overlap of 6.8mm.

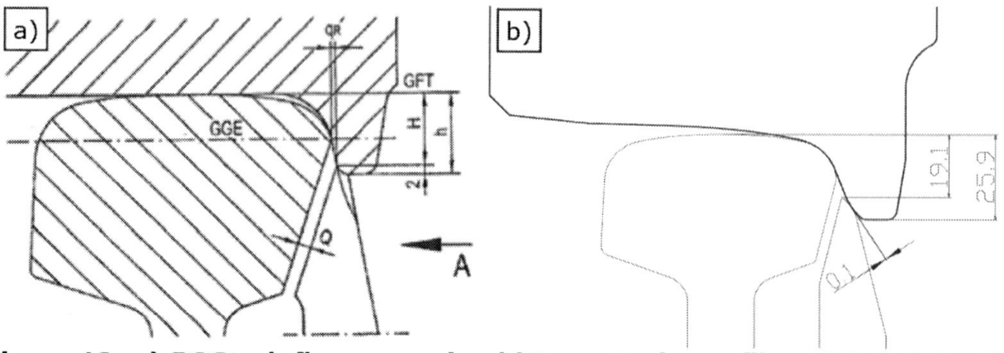


**Figure 13: a) BOStrab flange overlap b)Tram-Train profile at NR full depth switch toe with 3mm residual switch opening showing flange overlap**

### *6.2.2 Minimum contact angle*

This NR Standard for 'Inspection and repair to reduce the risk of derailment at switches' (14), states that as a general rule, the point at which the wheel-rail contact angle reduces to 60° should typically be no less than 20-25mm below the rail head (See Figure 14a). It also mandates that contact should not occur with the switch blade at an angle lower than 60°. This is to protect against flange climb derailment at the switch toes. Figure 14b shows the Tram-Train profile on an NR switch blade. The distance to the 60° is greater than the 20mm minimum distance but the minimum contact angle does drop below 60° to 56.8°. This is however not considered a problem in this application, as the standard BR P8 profile, which has an excellent operational safety record, also fails to meet this requirement with a minimum contact angle of 50.8° under the same conditions.

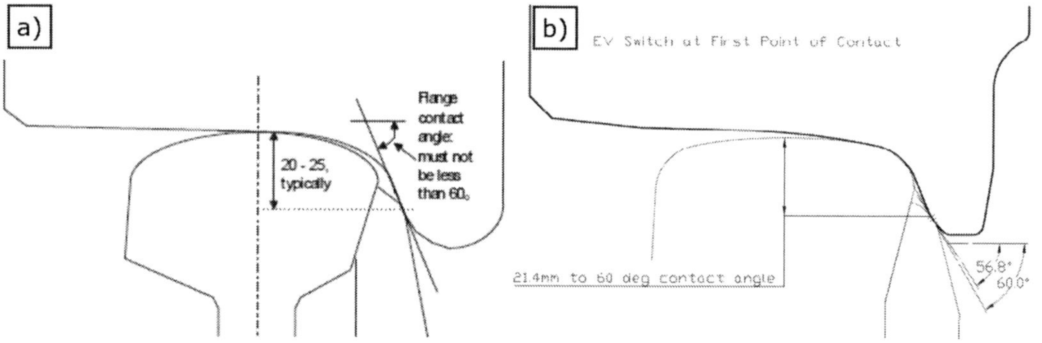


**Figure 14: a) Minimum contact angle b) Tram-Train profile at first point of contact with NR Full Depth switch blade showing contact angles and positions**

### *6.2.3 Secant contact angle*

Secant contact occurs when the wheel encounters an object on its route – in this case the end of the switch toe. EN 13232-9 (15) states that contact with the switch toe should not occur in the contact danger zone which is an area around the flange tip where the contact angle falls below a certain limit. This assessment protects against flange climb and switch splitting derailment.

The standard defines that the danger zone extends from the point at which the flange angle drops to 40° and extends around the flange tip, Figure 15a shows an example of a safe contact condition. Figure 15b shows the Tram-Train wheel profile located at the proposed NR switch toes with 3mm of residual opening. The 'danger zone' is highlighted in red and extends through an angle of 80° around the flange tip. It can be seen that the wheel flange does not contact the switch toes at any point within the defined sector and therefore the Tram-Train profile meets the requirements in this assessment.

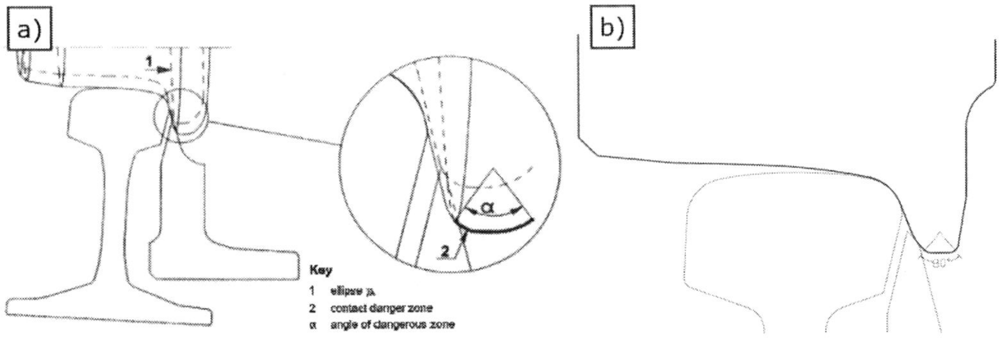


**Figure 15: a) Secant contact angle b) Tram-Train profile at NR full depth switch toe with 3mm residual switch opening showing 'Danger zone'**

### 6.3 Check Rail Interaction

The Tram-Train profile design required a cut-out in the flangeback to provide a checking face at 1379mm Back-to-Back spacing for compatibility with SST grooved rail whilst retaining a checking surface further up the flangeback with a spacing of 1360mm for compatibility with NR S&C. The cut-out extends up the flangeback to a height which was selected to ensure that the wheel profile can operate safely on SST grooved rail even when the rail head reaches its vertical wear limit.

The provision of a flangeback cut-out raises the checking face for NR infrastructure further up the flangeback, therefore NR check rails must be raised to maintain correct and safe contact conditions with the Tram-Train profile flangeback. This process requires that route gauging clearance is required for all vehicles running on the NR infrastructure as the lift takes the check rail beyond the standard NR structure gauge.

CAD software was used to determine the required check rail lift, Figure 16 shows the check rail in the nominal and raised position. The minimum amount that the check rail should be raised is 40mm to bring the vertical checking surface of the check rail in line with the NR checking surface on the flangeback.

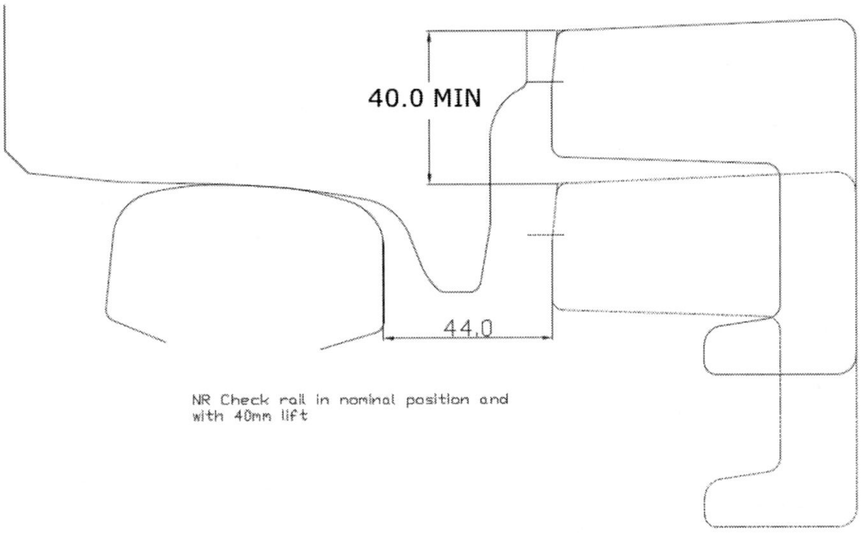


**Figure 16: NR Check rail in nominal and raised position**

## 7 CONCLUSIONS

A dual operation wheel profile has been designed to run on Network Rail and Sheffield Supertram infrastructure. The design incorporates several features to meet the requirements of the two rail systems such as

- Cut-out in the flange back to provide two checking surfaces for NR and SST infrastructure
- 68° flange angle with bespoke flange toe profile to provide required clearance on NR switch toes
- Flat flange tip to facilitate flange tip running through SST embedded switches and crossings
- Tread geometry derived from the BR P8 profile that avoids hard two-point contact, reduces wear and improves curving

The final wheel profile design is illustrated in Figure 17.

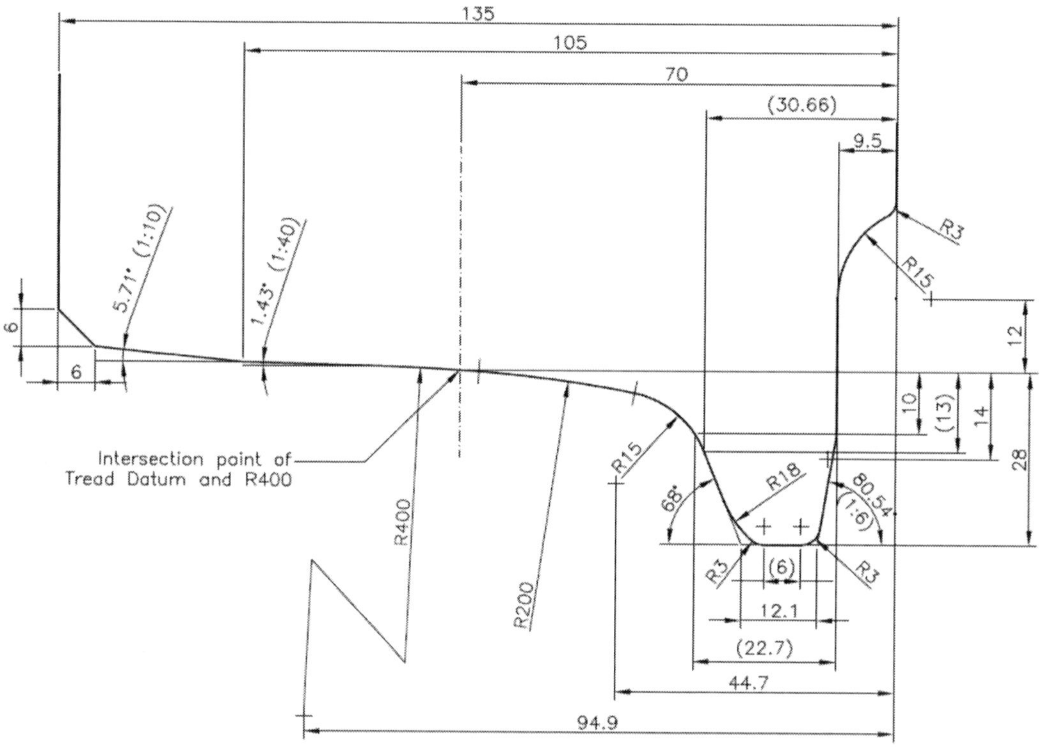


**Figure 17: Final Tram-Train wheel profile**

## REFERENCE LIST

1. Duewag Drawing No. 3-D-02-Y-09006320, Sheffield Supertram Wheel Tyre
2. Rail Safety and Standards Board (2012), "Railway Wheelsets", GM/RT2466
3. "Railway applications - Track - Rail - Part 1: Vignole railway rails 46 kg/m and above", BS EN 13674-1:2003+A1:2007
4. "Railway applications – Track - Special purpose rail - Grooved and associated construction", BS EN 14811:2006+A1:2009
5. "Railway applications – Track - Rail - Part 4: Vignole railway rails from 27kg/m to, but excluding 46 kg/m", EN 13674-4:2006+A1:2009
6. Allen, P. (2008) "Support for the Introduction of Tram-Train Vehicles on Network Rail Infrastructure", RTU Report 81/13 Issue 2

7. McEwan I. J. and Harvey R. F. (1986), "Technical Memorandum – Interpretation of Wheel Wear Numbers", British Rail Research Report TM-VDY-004
8. Rail Safety and Standards Board (2012), "Railway Wheelsets", GM/RT2466
9. Ringsberg, J.W. (2000), *Cyclic Ratchetting and Failure of a Pearlitic Rail Steel.* Fatigue & Fracture of Engineering Materials & Structures. **23**(9): p. 747-758
10. Beynon, J.H., J.E. Garnham, and K.J. Sawley (1996), Rolling Contact Fatigue of Three Pearlitic Rail Steels. Wear. 192(1-2): p. 94-111
11. Sinclair, J.C. and Allery M.B.P. (1991), "Development of a Failure Parameter for Rail Squats", British Rail Research Report TM MF 211
12. Transportation Research Board (2012),"Track Design Handbook for Light Rail Transit, Second Edition", ISBN 978-0-309-25824-1
13. "Regulations on the Guidance of Rail Vehicles in accordance with the German Federal Regulations on the Construction and Operation of Light Rail Transit Systems (BOStrab) - Guidance Regulations (SpR)", March 2004, English translation by J Snowden Sept. 2008
14. Network Rail (2008),"Inspection and repair to reduce the risk of derailment at switches", NR/L2/TRK/0053 Issue 5
15. "Railway applications - Track - Switches and crossings - Part 9: Layouts", BS EN 13232-9:2006+A1:2011

# An experimental study of dynamic performance of slab track using a full scale test rig

**M Wang, X Yuan, C Cai, S Zhu**
State Key Laboratory of Traction Power, Southwest Jiaotong University, P.R. China

**ABSTRACT**

The dynamic stiffness and damping of railway track are significant parameters for the dynamic simulation. This paper investigated the track dynamic performance by carried out wheel-dropping experiments on a full scale China Railway Track System (CRTS) II slab track model. The vertical vibration transmitting characteristics of the CRTS II slab track under impact load were analyzed and summarized. Meanwhile, the dynamic response of rail fastening tension clamp under impact load was obtained. Finally, the whole dynamic stiffness and damping of the slab track can be determined by the wheel-dropping test. It is found that impact load imposes great excitation on the tension clamp and it is more liable to induce damage of the tension clamp. The dynamic stiffness and damping of CRTS II slab track are 128.9 kN/mm and 108.4 kN·s/m, respectively.

**Keywords:** CRTS II slab track; dynamic stiffness; damping; tension clamp vibration.

## 1 INTRODUCTION

In the past decade, China high-speed railway developed rapidly and has developed CRTS II slab track, which is based on the introduced German slab track technology. The Beijing-Tianjin and Beijing-Shanghai high-speed railways have constructed in the form of CRTS II slab track, and acquired good operation effect. However, high-speed railway operation occurs more and more significant railway wheel/rail system dynamic problems and the wheel/rail dynamic interaction become more and more intensive. On one hand, the damage of track structure induced by high speed train is more and more serious, especially lacking of systematic study on non-ballast track damage induced by high speed trains. On the other hand, track dynamic stability and track geometry condition have more significant influence on the train operation safety and ride comfort. The track dynamic performance, especially track structure parameter identification, is the key step to carry out theoretical and numerical simulation of wheel/rail dynamic interaction under high speed condition.

At present, studies on high speed railway track dynamic performance mainly carried out by theoretical analysis, numerical simulation and experiment. Since the high-speed railway track-subgrade system dynamic interaction problem is quite complex, including the nonlinear dynamic properties of some components, such as mortar layer and fastener, the theoretical research is weak. It is hard to choose dynamic parameters accurately, and calculation results will be inevitably without certain error. Therefore, high speed railway track dynamic performance can be studied with the help of scientific instruments and test conditions can be controlled artificially by built up full scale high speed railway track model which has similar principle with the origin one. Comparing with field test which influenced greatly by environment factors, indoor test can control test parameters strictly and test repeatedly. There are already several researchers have studied track and subgrade dynamic performance by experiment with models[1-3].

Reference [4] indicates that the dynamic performance of non-ballast track can be evaluated by wheel/rail vertical force, wheel/rail horizontal force, derailment coefficient, wheel load reduction ratio, track horizontal stability coefficient, rail support point pressure, CA mortar dynamic stress, rail vertical displacement, dynamical gauge expansion, track slab displacement, subgrade surface displacement, subgrade surface dynamic stress and other indexes. However, all the indexes mentioned above need measure by field test, and the measure results are closely related to test vehicle conditions (such as wheel eccentricity and wheel flat), external surrounding influence (such as temperature, wind and rain), and track conditions (such as track irregularity and rubber pad under rail aging).

Therefore, the interference factors mentioned above can be avoided by measuring track dynamic stiffness and damping directly inside the laboratory, which will provide reference for theoretical and numerical simulation in large system environment when study wheel/rail dynamic interaction under high speed operation condition.

## 2 FULL-SCALE CRTS II SLAB TRACK MODEL

The full-scale model is constructed according to the CTRS2 slab track system, which including three parts, the subgrade model, track model and dynamic response test system. The full-scale model is on the straight section, its total length is 15.6m and it is set in a stable cement tank. The tank's width is 6m, height is 3.87m and the thickness of the wall is 2m.

### 2.1 Subgrade model

The length, width and height of the subgrade model are 15.6m, 6m 3.1m, respectively, and the model is put in the cement tank. The bottom layer of subgrade is 2.7m thick, which constructed with A, B fillings; the surface layer of subgrade is 0.4m thick and filled with graded broken stone. China railway classifies subgrade fillings into 4 groups according to the performance, which by performance from high to low grouping as A, B, C, and D. The performance indexes of each filling see reference [5]. This model can weaken the influence of boundary conditions to the experiment since the two end sides of subgrade model in longitudinal direction connecting with other type full-scale track models, whose components are exactly the same with this model.

**Table 1 Subgrade filler compaction standard**

| Subgrade | Filler | Compaction coefficient $K$ | Foundation coefficient $K_{30}$(MPa/m) | Dynamic modulus of deformation $E_{vd}$ (MPa) |
|---|---|---|---|---|
| Bottom layer of subgrade | Sandy soil and gravel soil | ≥0.95 | ≥130 | ≥40 |
| | Broken stone and cobbly soil | ≥0.95 | ≥150 | ≥40 |
| Surface layer of subgrade | Graded broken stone | ≥0.97 | ≥190 | ≥55 |

Subgrade model is constructed strictly in accordance with standards and the filling compaction criterion see table 1 [6], where the $K_{30}$ represents the stiffness of the foundation, and the $E_{vd}$ denotes the dynamic modulus of the foundation. The cross section of subgrade model is shown as figure 1.

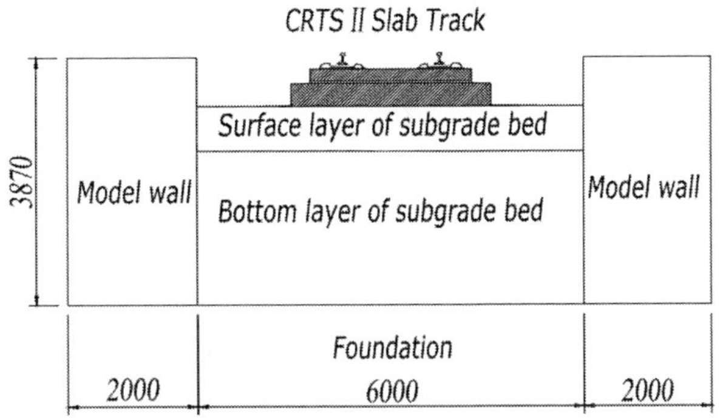


**Figure 1 Subgrade cross section map**

## 2.2 Track model

The full-scale track model is constructed according to the CRTS II slab track construction criteria. C20 pure concrete is paving on the subgrade surface layer as supporting layer. Mortar layer is above supporting layer, whose elasticity modulus is 7000-10000MPa and able to weaken supporting layer vibration efficiently. Slabs are above the mortar layer, including a complete CRTS II slab and two 70% length CRTS II slabs, which are connected together by connectors and bearing longitudinal force. Slab connects CHN60 rails, which length is 12.5m, by WJ-8 fasteners and other rails at both ends by clamping plates. This model can eliminate boundary conditions well during experiment since the length of model reaches 15.6m and both ends of track model are free ends except the rails. Track model is shown as figure 2.

**Figure 2 Track model photo**

## 2.3 Track dynamic response test system

The dynamic response test system is installed to test the dynamic response of track and subgrade. Subgrade dynamic response test system won't introduce in detail since this article aims at track dynamic response. Track dynamic response test system composes sensors, data acquisition devices and workstation. Sensors are installed at 2 test sections, accelerometers, displacement meters and strain rosettes, which testing wheel/rail force, are installed at the main test section. Track model test sensors layout see as figure 3.

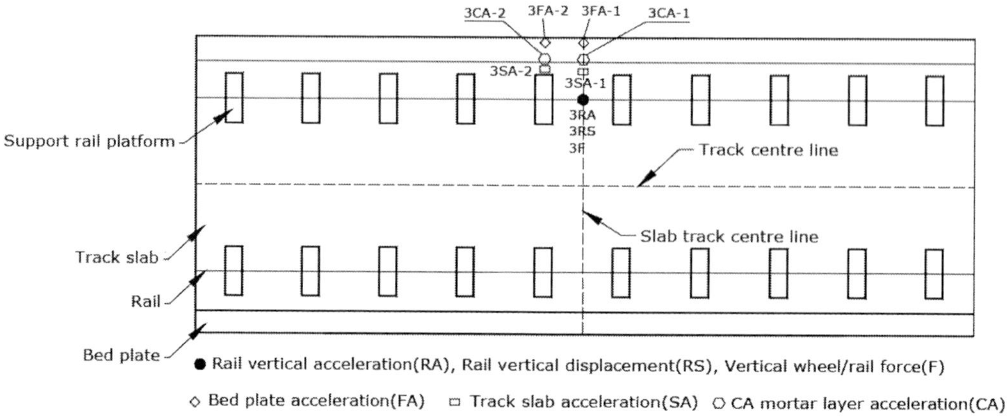


**Figure 3 Sensors layout on track**

## 2.4 Loading device

The experiment applies load by wheel-droppingset to impact the track, which carried out by a wheel-dropping car, shown as figure 4. This wheel-dropping car composes a movable chassis, a wheelset and a lifting structure which using for lifting wheelset. At the experiment, firstly drives the car to the loading position, and then raising the loading car until car wheels separate from rails. Lift the wheelset to the experiment height, finally dropping the wheelset suddenly to make it impact the track to finish loading. The weight of the wheelset is 1020kg, and the lifting height can change continuously at the range of 0~ 200mm, which usually within 50mm during actual uses. Simulated wheel drop test by FEM, the impact force can reach 120kN when wheelset is lifted to 20mm high, which is able to stimulate track structure vibration sufficiently and reach the goal of testing its dynamic performance.

**Figure 4 Wheel drop loading car**

## 3 TEST SCHEME

The dynamic performance of track is significant to operate high speed train steadily, safely for track irregularity, wheel eccentricity, wheel flat and other factors will lead to vehicle- track coupling vibration when train operating at high speed.

There are already several researchers have carried out tests on various types of track since Y. Sato from Japan proposed the theory to measure track stiffness and damping by wheel drop test. This theory simplifies the track at wheel drop impact point as a spring element and a damping element, see as figure 5.

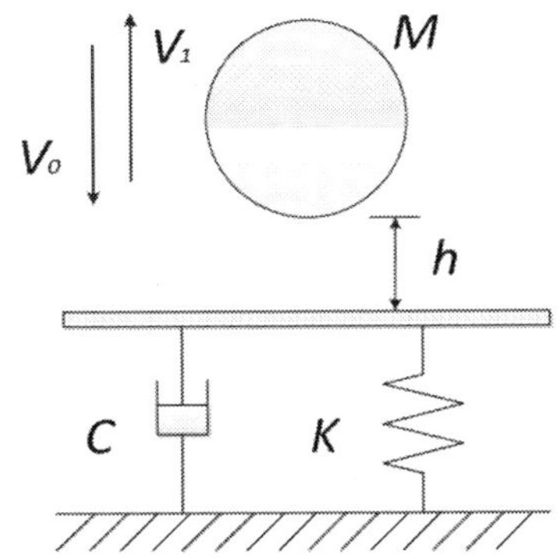


**Figure 5 Wheel drop impact model**

Track stiffness $K$ and damping $C$ can be obtained according to (1) (2) (3).

$$K = \frac{W\left(\pi - \arctan\frac{2\zeta\sqrt{1-\zeta^2}}{1-2\zeta^2}\right)^2}{\left(1-\zeta^2\right)^2 t_0^2} \tag{1}$$

$$C = 2\zeta\sqrt{W \cdot K / g} \tag{2}$$

$$\frac{V_1}{V_0} = e^{\frac{-\zeta\left(\pi - \arctan\frac{2\zeta\sqrt{1-\zeta^2}}{1-2\zeta^2}\right)}{\sqrt{1-\zeta^2}}} \tag{3}$$

In the formula, $K$ is dynamic stiffness, kN/mm; $W$ is wheel load, N; $g$ is gravitational acceleration, m/s$^2$; $\xi = C/(2\sqrt{m \cdot k})$ is critical damping coefficient ratio; $C$ is damping coefficient, kN·s/m; $t_0$ is impact time, ms; $V_0$ is wheel/rail impact speed; $V_1 = gT_1$ is bounce speed. Duration time $t_0$ and interval time $2T_1$ should obtain according to wheel/rail force waveform graph. Wheel/rail force measured by 90°strain rosettes which stuck to rail neutral axis on both sides according to shear force method.

This article measures track entirety dynamic stiffness and damping on CRTS II slab track full-scale model, as well as testing dynamic response of every track component, and then analyzes track dynamic performance. Test scheme is shown as table 2. For the convenience of comparing with other type of tracks, wheel-dropping test was carried out 8 times at the height of 20mm.

**Table 2 Test scheme**

| Wheel drop height(mm) | Test times | Test items |
|---|---|---|
| 10 | 3 | Wheel/rail vertical force, rail acceleration, elastic strip acceleration, track slab acceleration, mortar layer acceleration, support layer acceleration, rail vertical displacement |
| 15 | 3 | |
| 20 | 8 | |
| 25 | 3 | |
| 30 | 3 | |

## 4 TEST RESULTS AND ANALYSIS

### 4.1 Track structure dynamic response

Statistic of test results for track structure dynamic response at different wheel-dropping height is shown in table 3. It can be seen from table 3 that the vibration acceleration of track slab and other components below it are decreased greatly for the damping effect of fastener system. The mortar elastic modulus is only 7000~10000MPa, mortar vibration acceleration is larger than track slab's which is above it. This reflects that mortar absorbs energy and alleviates impact for the substructure by its vibration; the acceleration of base is only about 1/3 of track slab. Acceleration of tension clamp was tested when wheel-dropping height is 20mm, as figure 6 shows. In actual operation conditions, the impact effect may occur which is similar to 20mm high wheel-dropping test when operation speed is high enough as well as wheel flat damage reaches a certain extent. The average value of tension clamp acceleration in this test is 123g, while tension clamp acceleration maximum value is about 4g which tested on the same type of track when EMU passing according to reference [7]. Therefore, impact load has great excitation on tension clamp and it is more liable to result in damage of tension clamp.

**Table 3 Track structure vertical acceleration peak value**

| Test point | | Rail | Track slab | | Mortar layer | | Base | | Tension Clamp |
|---|---|---|---|---|---|---|---|---|---|
| | | 3RA-1 | 3SA-1 | 3SA-2 | 3CA-1 | 3CA-2 | 3FA-1 | 3FA-2 | 3TCA-1 |
| acc (g) | 10mm | 153 | 4.6 | 4.5 | 5.6 | 5.2 | 1.2 | 1.2 | - |
| | 15mm | 165 | 5.1 | 5.3 | 5.6 | 5.9 | 1.5 | 1.4 | - |
| | 20mm | 188 | 6.0 | 6.0 | 7.6 | 6.6 | 1.9 | 1.8 | 123 |
| | 25mm | 278 | 6.8 | 6.2 | 9.9 | 7.8 | 2.9 | - | - |
| | 30mm | 320 | 8.7 | 7.9 | 11.7 | 13.0 | 3.4 | - | - |

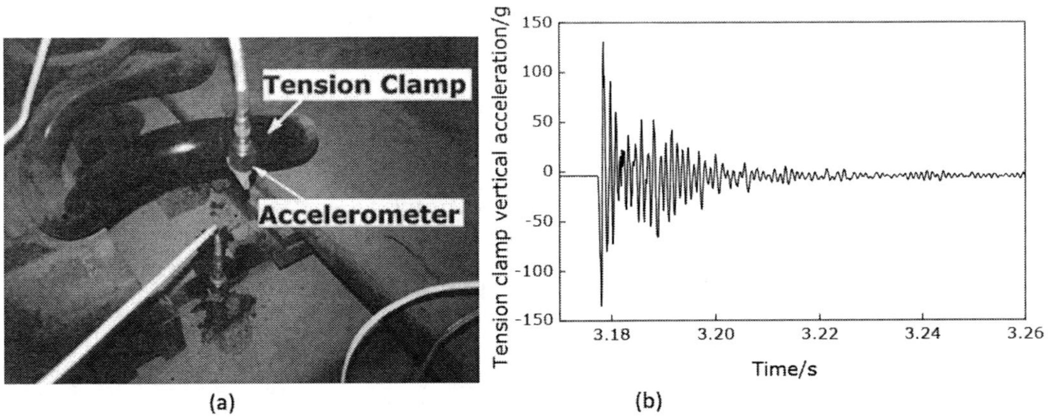


**Figure 6 Tension clamp acceleration test and response waveform: a. Tension clamp acceleration test photo; b. Tension clamp response waveform**

### 4.2 Wheel/rail vertical force waveform

The wheel /rail contact time and wheelset bounce height can be measured by the test for wheel/rail vertical force waveform, and the track structural stiffness and damping can be calculated according it. Wheel/rail vertical force waveform sees as figure 7, which shows distinct high frequency P1 force and low frequency P2 force. The values of P1 force and P2 force are related to the dynamic performances of track and vehicle. P1 force and P2 force will be different when impact on different type of tracks in the same wheel-dropping test. High frequency P1 force influences vehicle dynamic performance greatly while low frequency P2 force mainly influences track structure.

In figure 7, wheel/rail impact time is $t_0$=8.1ms, then detects a succession of negative wheel/rail force, which is due to the rail bounces back upwards after impact and wheelset has already separated from rails at that time. The time interval between wheelset separating from rails and second impact is $2T_1$=83ms, which can be used to calculate wheelset bounce speed and then calculate track structure stiffness and damping.

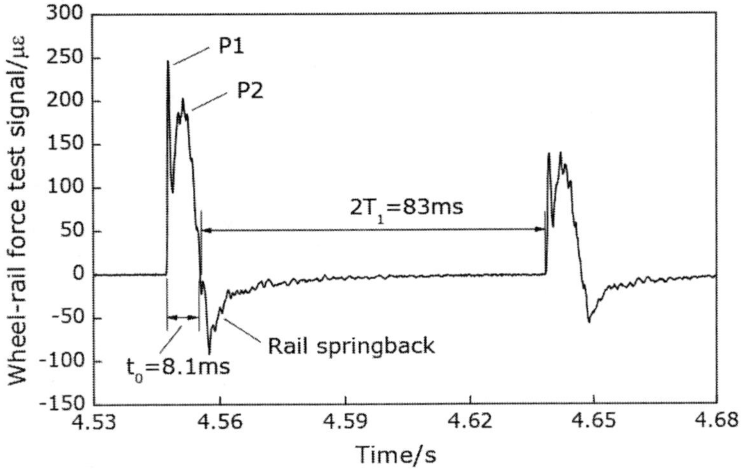


**Figure 7 Wheel/rail vertical force waveform**

### 4.3 Track structure dynamic stiffness and damping

The track structure dynamic stiffness and damping can be calculated according to formula (1) (2) (3) and the tested wheel/rail vertical force waveform. It is tested that the dynamic stiffness and damping of CRTS II slab track are respectively 128.9 kN•mm$^{-1}$ and 108.4 kN•s•m$^{-1}$.

**Table 4 Track structure dynamic stiffness and damping**

| Wheel-dropping height $h$/mm | Test No. | Impact time $t_0$/ms | Time interval $2T_1$/ms | Bounce speed $V_1$/m·s$^{-1}$ | Impact speed $V_0$/m·s$^{-1}$ | Speed ratio $V_1/V_0$ | Damping ratio $\xi$ | Dynamic stiffness $K$/kN·mm$^{-1}$ | Damping $C$/kN·s·m$^{-1}$ |
|---|---|---|---|---|---|---|---|---|---|
| 10 | 1 | 8.2 | 55.6 | 0.272 | 0.443 | 0.615 | 0.171 | 126.0 | 124.0 |
| | 2 | 8.2 | 53.4 | 0.262 | 0.443 | 0.591 | 0.187 | 124.6 | 134.5 |
| | 3 | 8.1 | 59.4 | 0.291 | 0.443 | 0.657 | 0.146 | 131.7 | 108.0 |
| 15 | 1 | 8.2 | 70.6 | 0.346 | 0.542 | 0.638 | 0.157 | 127.4 | 114.4 |
| | 2 | 8.2 | 71.2 | 0.349 | 0.542 | 0.643 | 0.154 | 127.7 | 112.3 |
| | 3 | 8.2 | 71.7 | 0.351 | 0.542 | 0.648 | 0.151 | 128.0 | 110.3 |
| 20 | 1 | 8.5 | 84.6 | 0.415 | 0.626 | 0.662 | 0.143 | 119.9 | 100.7 |
| | 2 | 8.1 | 82.3 | 0.403 | 0.626 | 0.644 | 0.153 | 130.9 | 113.3 |
| | 3 | 8.1 | 83.8 | 0.411 | 0.626 | 0.656 | 0.146 | 131.7 | 108.4 |
| | 4 | 8.1 | 84.9 | 0.416 | 0.626 | 0.664 | 0.141 | 132.2 | 104.9 |
| | 5 | 8.2 | 82.8 | 0.406 | 0.626 | 0.648 | 0.151 | 128.0 | 110.3 |
| | 6 | 8.1 | 85.3 | 0.418 | 0.626 | 0.668 | 0.139 | 132.5 | 103.4 |
| | 7 | 8.4 | 84.1 | 0.412 | 0.626 | 0.658 | 0.145 | 122.6 | 103.4 |
| | 8 | 8.1 | 83.3 | 0.408 | 0.626 | 0.652 | 0.149 | 131.4 | 110.1 |
| 25 | 1 | 8.3 | 96.4 | 0.472 | 0.700 | 0.675 | 0.135 | 126.6 | 98.3 |
| | 2 | 8.2 | 94.1 | 0.461 | 0.700 | 0.659 | 0.145 | 128.6 | 106.1 |
| | 3 | 8.1 | 94.3 | 0.462 | 0.700 | 0.660 | 0.144 | 132.0 | 106.4 |
| 30 | 1 | 8.2 | 104.7 | 0.513 | 0.767 | 0.669 | 0.139 | 129.3 | 101.7 |
| | 2 | 8.1 | 106.3 | 0.521 | 0.767 | 0.679 | 0.133 | 133.2 | 99.0 |
| | 3 | 8.1 | 106.4 | 0.521 | 0.767 | 0.680 | 0.133 | 133.2 | 98.8 |
| Average value | - | - | - | - | - | - | - | 128.9 | 108.4 |
| Stander deviation | - | - | - | - | - | - | - | 3.6 | 8.6 |

## 5 CONCLUSIONS

A full scale test model has been built up according to real CRTS II slab track structure. Further, the dynamic response of track structure components under impact load has been tested, and the track dynamic stiffness and damping are identified with the wheel-dropping method. Some conclusions can be drawn from the test results as below.

(1) The vibration acceleration level of track structure components shows a decrease tendency from top to bottom. The fasteners and the mortar layer have a certain contribution to the vibration attenuation.
(2) The vibration acceleration of the tension clamp under the wheel-dropping impact load is much larger than that obtained from the field tests when high speed train passing on the same type of slab track. It shows that impact load has great excitations on the tension clamp and it is most likely to aggravate the tension clamp damage.
(3) The dynamic stiffness and damping of the CRTS II slab track are 128.9 kN • $mm^{-1}$ and 108.4 kN·s/m, respectively.

Further research will focus on the dynamic performance tests of the full scale model of CRTS I slab track, CRTS III slab track, and CRTS I double-block track. And then the difference of dynamic performances of different type of tracks will be investigated. In the process of the experimental test, much attention should be paid to the influence of some factors on test results in order to improve the test precision.

## REFERENCES


[1] Wang Qiyun, Zhang Jiasheng, Meng Fei, et al. Study on Track-subgrade Model of High-speed Railway and Dynamical Loading. Journal of The China Railway Society, 2012, 34(12): 90-95
[2] Jiang Hongguang, Bian Xuecheng, Xu Xiang, et al. Full-scale model tests on dynamic performances of ballastless high-speed railways under moving train loads. Chinese Journal of Geotechnical Engineering, 2014, 36(2): 354-362
[3] S.J. Cox, A. Wang, C. Morison, et al. A test rig to investigate slab track structures for controlling ground vibration. Journal of Sound and Vibration, 2006(293) : 901–909
[4] Cai Chengbiao, Zhai Wanming, Wang Kaiyun. Calculation and Assessment Analysis of the Dynamic Performance for Slab Track on Sui-Yu Railway. China Railway Science, 2006, 27(4): 17-21
[5] Code for Design on Subgraed of Railway (TB10001-2005). Beijing. China Railway Publishing House. 2005, 12-17
[6] Code for Design of High Speed Railway (TB10621-2009). Beijing. China Railway Publishing House. 2009
[7] Zhu Shengyang, Cai Chengbiao, Yin Qiang, et al. Dynamic Analysis of Rail Fastening Clip in High-Speed Railway. Engineering Mechanics, 2013, 6:254-258

# The influence of support conditions on short- and long-term track behaviour

**I Grossoni, Y Bezin**
Institute of Railway Research, University of Huddersfield, UK

**ABSTRACT**

Railway track support conditions are known to deeply affect the dynamic performance of vehicle-track interaction, influencing the state of the track system both in the short and in the long term. Exactly how much and how is not precisely understood and the notion of track stiffness, although thought to be a key parameter of the track quality, is currently not being monitored systematically. This paper seeks to analyse the influence on the ballast behaviour of track vertical stiffness and especially its spatial non-uniformity, using available experimental data measured at different sites. Mathematical models are developed and the effectiveness of applying under sleeper pads is also investigated. Finally, an iterative procedure based on Guerin's settlement law is used to take into account the long-term behaviour of the ballast. Such models can help to understand mitigation solutions as well as predicting track quality evolution over time.

**NOTATION**

| Symbol | Meaning |
|---|---|
| $\alpha, \beta$ | Soil parameters for the Guerin's settlement law |
| $\Delta N$ | Incremental traffic |
| $\Delta y$ | Incremental settlement |
| $\delta y_{ball,max}$ | Maximum elastic ballast deformation |
| **Abbreviation** | **Meaning** |
| FWD | Falling Weight Deflectometer |
| KS | Kolmogorov-Smirnov test |
| SD | Standard Deviation |
| USPs | Under Sleeper Pads |

## 1. INTRODUCTION

The performance of the railway system in terms of dynamic loading strongly depends on the track support conditions. Usually, the track stiffness is used as the main parameter to describe the support conditions and is defined as the ratio of the load applied to the rail over the vertical rail deflection [1]. Ideally that parameter is constant, but in reality this condition is very unlikely to happen, for example due to a non-uniformly compacted ballast layer, local drainage problems or presence of voids. Therefore, there is a non-uniform amplification of track forces which leads to further non-uniform track loading and, thus, to further non-uniform track degradation.

Several studies have been carried out in order to understand the influence of non-constant support stiffness on the vehicle and track dynamics, such as those in [1-4], and on the track settlement, such as in [2, 5]. Nevertheless, the influence of the actual support statistical distribution has not been fully covered.

A vehicle-track model that takes into account the vertical dynamic interaction is described in Section 2. The track irregularities and vertical support stiffness used in this study were measured at four UK sites. The influence of the statistical distribution of the non-uniform support conditions on the ballast performance is mathematically analysed in Section 3. Also the effectiveness of applying under sleeper pads is considered in Section 3. Finally, an iterative model using Guerin's settlement law is presented in Section 4 and the long-term effects of non-uniform support conditions are investigated.

## 2. MODELLING THE VEHICLE-TRACK SYSTEM

The vehicle-track model used in the present study is shown in Figure 1.

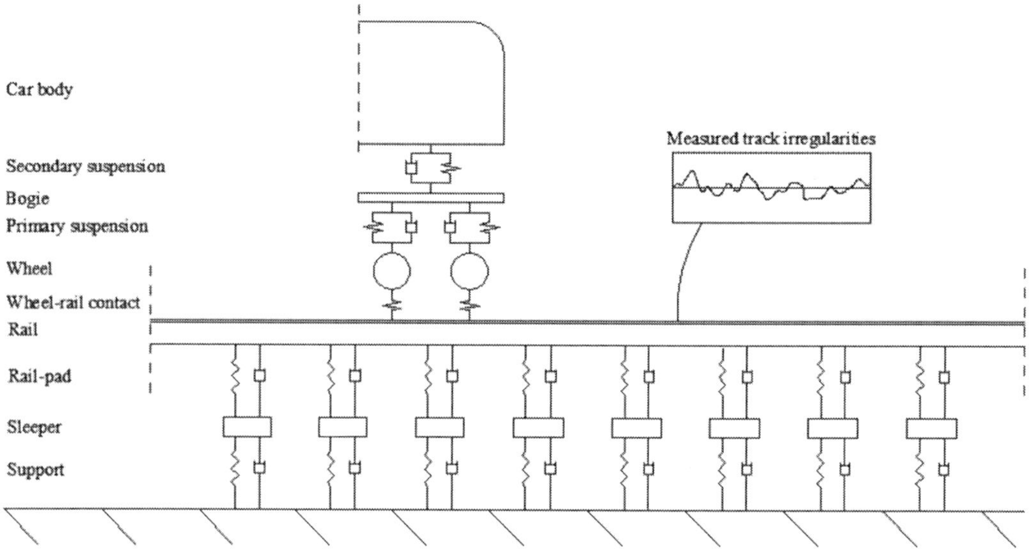


**Figure 1 – Vertical vehicle-track model including measured track irregularities and measured support stiffness.**

The track consists of two-layer ballasted track, including the rail-pad layer, the sleeper mass and the support layer. The rail is modelled as a Timoshenko beam. Four beam elements are considered within each sleeper-spacing in order to achieve a good resolution of results [6].

The main track parameters used in the model are:

- *Rail section*: 60E1;
- *Rail pad vertical dynamic stiffness*: 270 MN/m (medium-hard rail pad);
- *Sleeper mass*: 308 kg (typical concrete sleeper);
- *Sleeper spacing*: 0.65 m.

The vehicle is a typical freight vehicle characterised by:

- *Axle load*: 22.5 t, corresponding to circa 110 kN per wheel;
- *Suspensions*: primary and secondary suspensions, including linearized stiffness (respectively 13 and 6.2 MN/m) and linearized damping (respectively 90 and 100 kNs/m).

## 2.1. Experimental measurements of the support conditions

Four sets of sleeper support stiffness data have been analysed, whose main characteristics are reported in Table 1. The data was measured using Falling Weight Deflectometer (FWD) equipment.

**Table 1 – Main characteristics of the measured sites.**

| SITE | N. of sleepers | Support stiffness mean value [kN/mm/sleeper end] | Support stiffness SD [kN/mm/sleeper end] | KS test *p*-value |
|---|---|---|---|---|
| **1** | 155 | 84.6 | 14.4 | 0.32 |
| **2** | 70 | 68.0 | 18.1 | 0.95 |
| **3** | 80 | 110.4 | 16.2 | 0.90 |
| **4** | 81 | 71.0 | 8.6 | 0.76 |

The curve fitting is shown in Figure 2, assuming a normal distribution of the support stiffness. This hypothesis has been validated performing the Kolmogorov-Smirnov (KS) test and checking that the *p*-value of each set of data (Table 1) is not less than the 10% significance level.

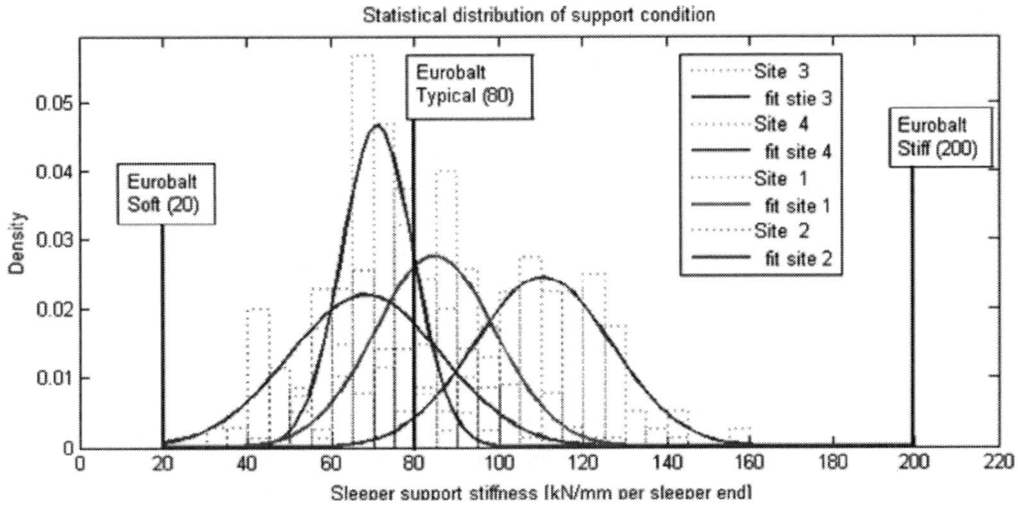


**Figure 2 – Curve fitting for the four sites considered.**

In Figure 2 the values reported in the EUROBALT project for 'soft', 'typical' and 'stiff' are also shown for comparison. This highlights the fact that the mean values for all sites are representative of typical values, i.e. situated around 80 kN/mm. Some of the extreme low values are just above 'soft' while the extreme high values are close to 160 kN/mm.

## 3. SHORT-TERM BEHAVIOUR

In this section, the influence of the support conditions on the track short-term behaviour is assessed. In particular, the response in terms of ballast forces without (Section 3.1) and with (Section 3.2) USPs are analysed.

The speeds considered in the present study are 80, 120 and 140 km/h, which are speeds considered by the authors as part of the European project SUSTRAIL investigating the possibility to raise maximum speed for freight wagon from 120 to 140 km/h on the basis of track and vehicle running gears innovation.

It is worth underlining that the ballast forces quoted hereafter are for one rail as the model is that of a half track, assuming symmetrical vehicle-track configuration and loading.

### 3.1. Without under sleeper pads

Figure 3(a) shows, for the three speeds considered in this study, the maximum ballast forces at each site. The relative differences between these values and the mean value of the four sites are depicted in Figure 3(b).

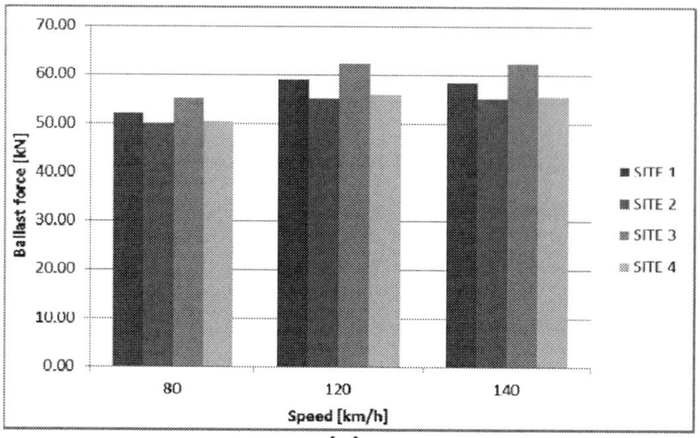


(a)

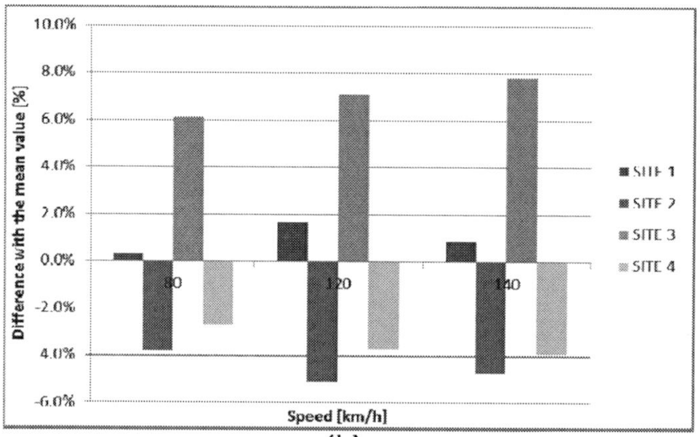


(b)

**Figure 3 – (a) Maximum ballast forces at each site for speeds 80, 120 and 140 km/h; (b) Relative differences between these values and the mean value of the four sites.**

From Figure 3(a) and Table 2 it can be observed that there is an increase in ballast force from speed 80 to 120km/h, while the forces seem to level out between 120 and 140km/h. Force fluctuations (difference with mean values across the four sites) are quoted with the higher values observed at site 3 (highest support stiffness mean value) and the lowest values observed at site 2 (lowest support stiffness mean value).

**Table 2 – Mean values of ballast force as a function of speed.**

| Speed [km/h] | Ballast force mean value [kN] |
|---|---|
| **80** | 52 |
| **120** | 58 |
| **140** | 58 |

The standard deviation of the ballast force as a function of the mean value of the support stiffness is shown in Figure 4(a), and in Figure 4(b) as a function of the standard deviation of the support stiffness.

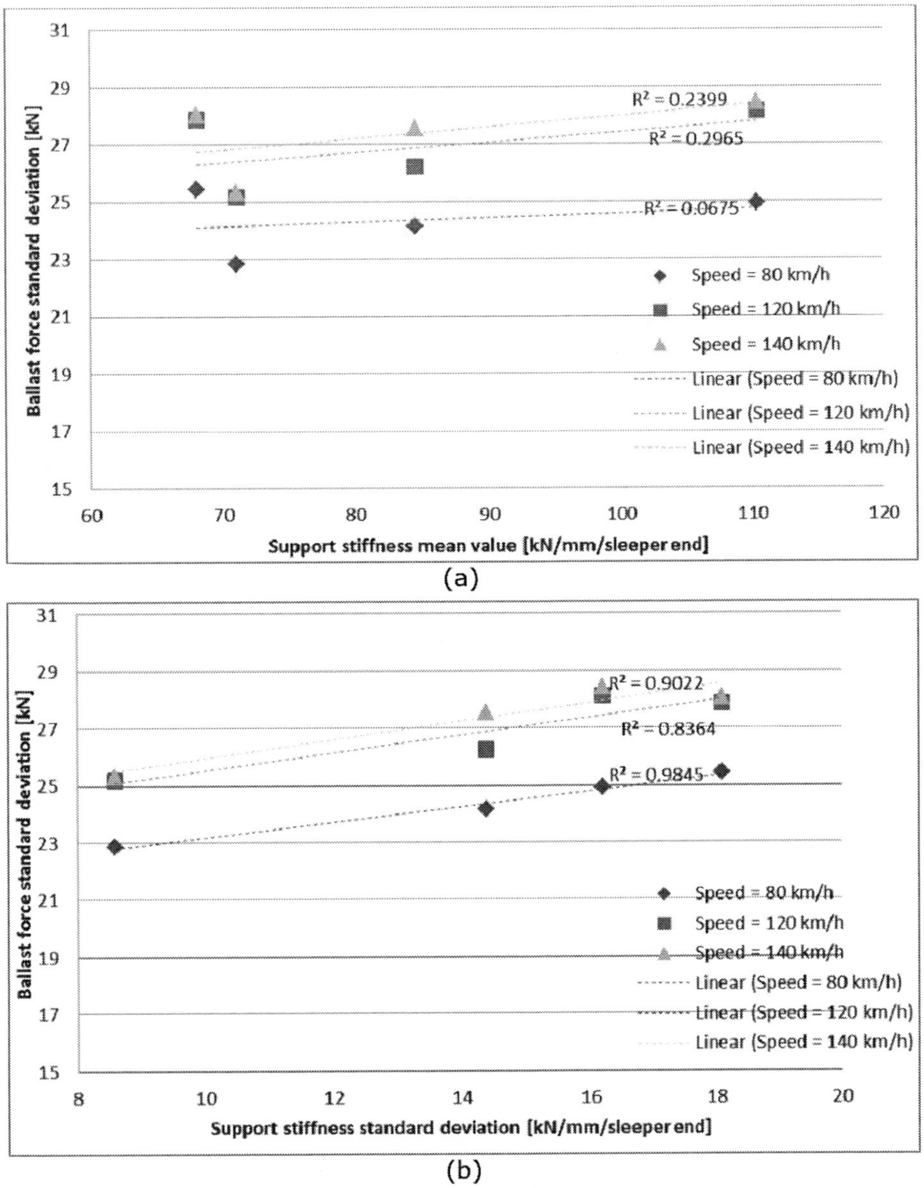


**Figure 4 – Ballast force SD as function of (a) support stiffness mean value and (b) support stiffness SD.**

Regarding the plots in Figure 4 and analysing the $R^2$ values, it is possible to notice that the support stiffness SD influences the variation of the ballast forces, more than the support stiffness mean value. This finding is in line with the previous studies on the vertical dynamics, according to which the main driver of the dynamic amplification and differential settlement is the distribution of support conditions rather than the mean value [4, 7]. In fact, assuming a linear trend the $R^2$ values are very high in all the speed cases when analysing the influence of support SD and quite low when analysing the influence of support mean values. Moreover, increasing the speed leads to an average increase of the ballast force SD (Figure 4), as expected.

A detailed analysis of the results, which is not presented in the present article, allows verifying that the variation of ballast force attributed to support stiffness can lead to an increase of up to 30% (25% on average) with respect to the case where stiffness is considered homogeneous. At the same time certain sleepers can see a reduction of force by up to 26% (-23% on average). Therefore, it can be concluded that a variability in the support conditions leads to a strong variability in the short-term behaviour of ballast forces and, thus, in the long-term behaviour.

### 3.2. With under sleeper pads

In Figure 5 the maximum ballast forces versus USP stiffness for different speed are presented per each site.

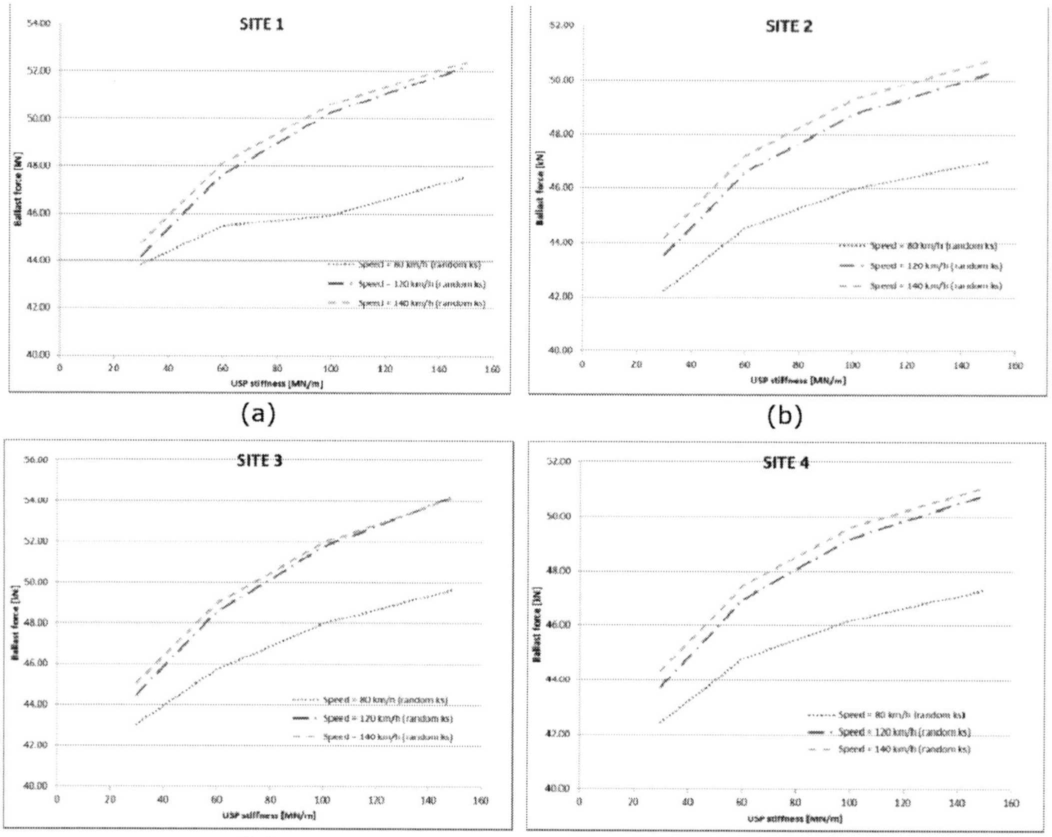


(a) (b)

(c) (d)

**Figure 5 – Maximum ballast forces versus USP stiffness for different speeds at (a) site 1, (b) site 2, (c) site 3 and (d) site 4.**

From Figure 5 it can be deduced that there is a parabolic increase of ballast force with increasing USP stiffness in case of all the speed and sites considered. For example, in case of site 1 there is an average decrease of 14% with respect the situation where no USP is present for speed equal to 80 km/h and an average decrease of 19% for speed equal to 140 km/h.

From a more detailed analysis of all the results not presented in this article, it is possible to notice that the greatest decrements in maximum forces, up to 30-40%, are in case of highest speed values (120-140 km/h), the highest mean support stiffness (site 1 and site 3) and the lowest USP stiffness (30-60 MN/m). Instead, when the mean support stiffness is already low (site 2 and site 4), the advantage of using USPs is greatly reduced, becoming almost negligible for the highest USP stiffness values.

Figure 6 shows the ballast force SD as function of the support stiffness SD for different USP stiffness values, as well as the related trend lines.

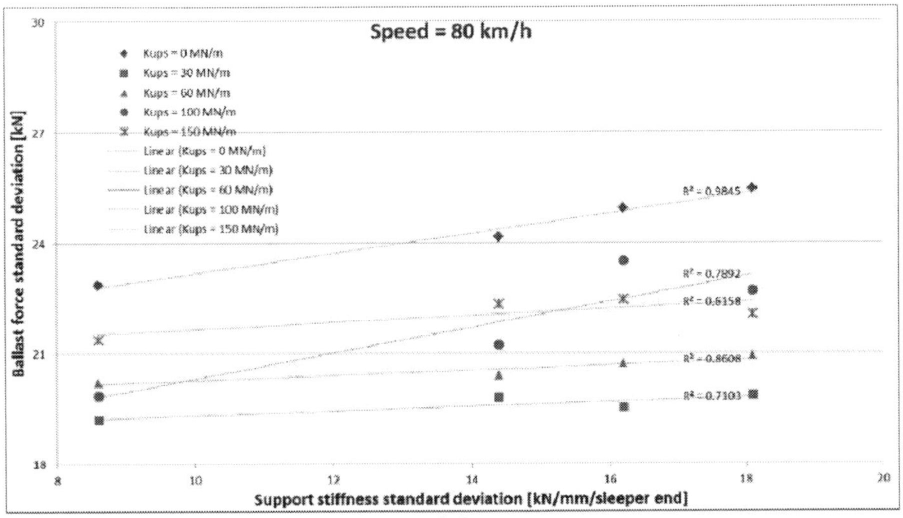


(a)

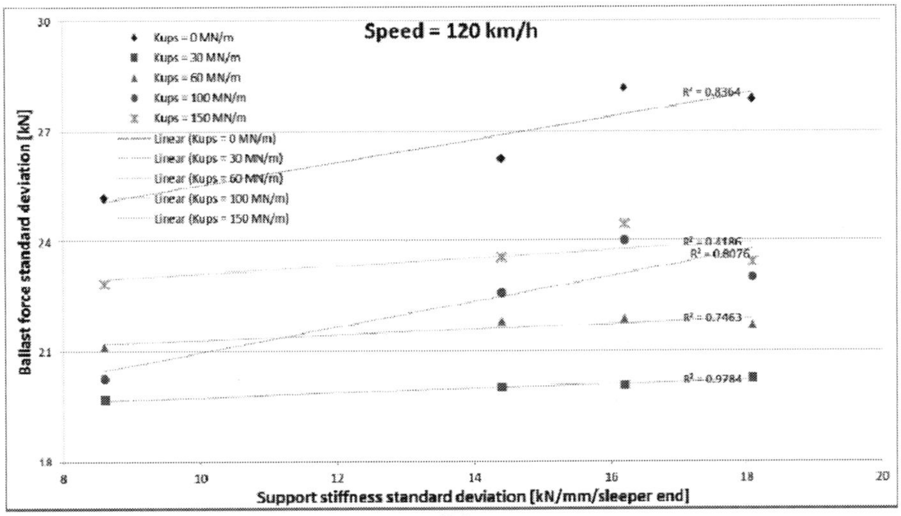


(b)

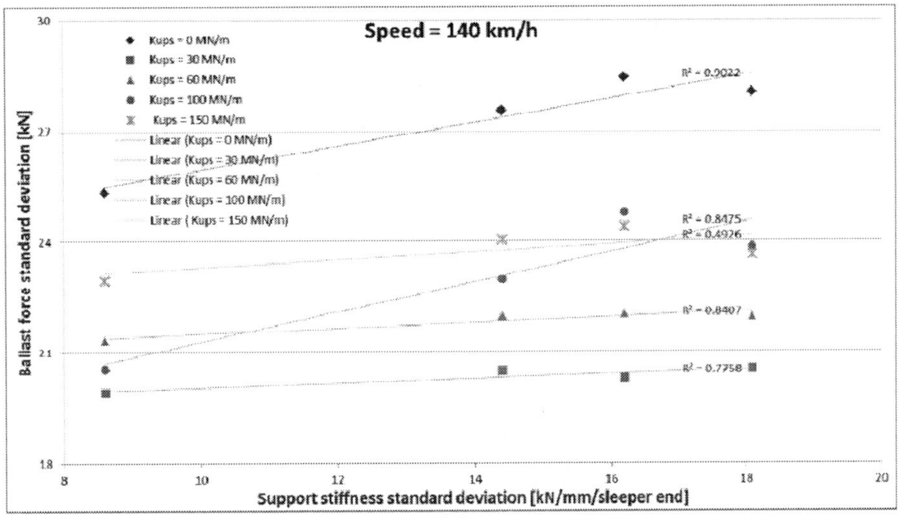


(c)

**Figure 6 – Ballast force SD as function of support stiffness SD for different USP stiffness values for speed (a) 80 km/h; (b) 120 km/h; (c) 140 km/h.**

Figure 6 confirms that the presence of USPs also helps to reduce the dispersion in the ballast forces in all speed cases analysed. This reduction follows the same trend of reduction in the ballast forces, i.e. the greatest reductions are in case of highest speed values, the highest mean support stiffness and the lowest USP stiffness. Moreover, it is possible to notice that the $R^2$ value is very high in case of low speed (80 km/h) and in case of low USP stiffness (30-60 MN/m) because of the lower dynamic dispersion.

To conclude, all the graphs show how the presence of the USPs can effectively help the reduction of force peak value and dispersion in the ballast forces and how this reduction depends on both the actual support stiffness distribution and the chosen USP stiffness. This reduction in force would in practice lead to a further reduction of pressure in the ballast and therefore a significant reduction in differential settlement as shown in Section 4.

## 4. LONG-TERM BEHAVIOUR

The track long-term behaviour in terms of settlement has been calculated following an iterative procedure (Figure 7): after the initialization with the vehicle, track and vertical rail profile data, the dynamic response of the coupled system is calculated in terms of contact forces and displacements. Later, the track settlement law is applied and the incremental settlement Δy due to the incremental traffic ΔN is calculated. Finally, a check in terms of maximum settlement and maximum traffic is performed to decide on the continuation of the iterative process.

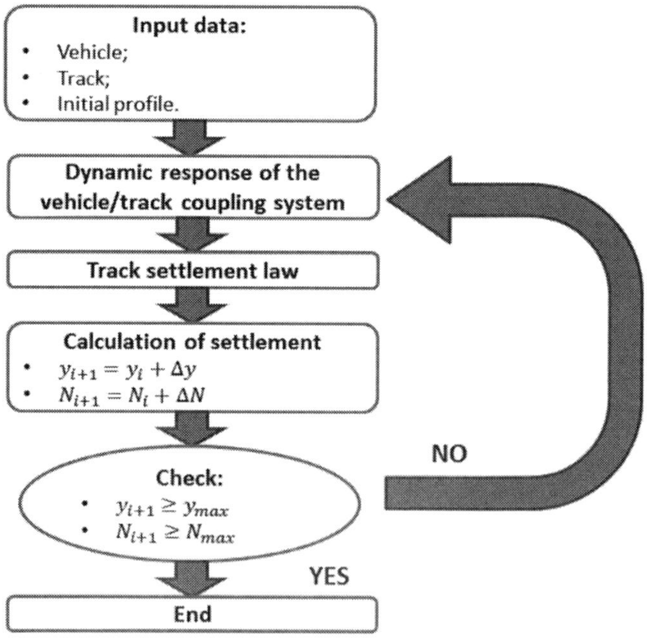


**Figure 7 – Iterative process to calculate the track long-term behaviour.**

In the present study, the Guerin's law [8] adopted and the incremental settlement is calculated as:

$$\Delta y = \Delta N \cdot \alpha \cdot \delta y_{ball,max}{}^{\beta}$$

Where $\delta y_{ball,max}$ is the maximum elastic ballast deformation and α and β two coefficients.

An example of the evolution of vertical track geometry is shown in Figure 8.

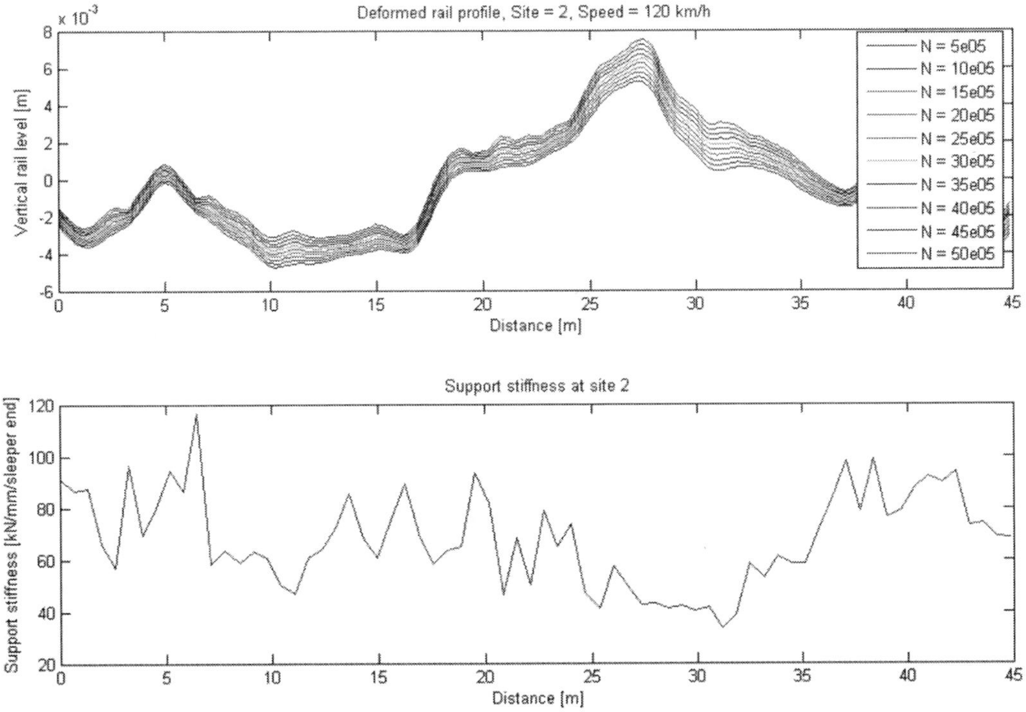


**Figure 8 – Evolution of the vertical track geometry for various cumulative traffic (top), and support stiffness along distance at site 2 (bottom).**

From that figure, it is possible to deduce how the two properties are strictly linked. In fact, the locations characterized by the highest support stiffness values (e.g. at circa 6.5 m the total settlement is 0.085 mm) experience lower long-term deformation then the locations characterized by the lowest support stiffness values (e.g. at circa 31.5 m the total settlement is 0.29 mm). This result is in line with what is expected.

The track data obtained in the simulations have been treated with a band-pass filter between 3 m and 25 m, as suggested in [9] for speeds lower than 200 km/h. Usually, track quality is estimated as the standard deviation of top level and alignment over 200-meter long sections (or approximately "an eight of a mile"). In this study, it is not possible to apply that rule because of the limited available support stiffness data (Table 1). In further work this aspect will be taken into account developing a methodology in order to extrapolate from the real data and produce a new set of support conditions with the same statistical characteristics.

The speed plays an important role: increasing the speed leads to an increase of the settlement, although increasing from 120 to 140 km/h shows very little difference. An example for site 3 is shown in Figure 9.

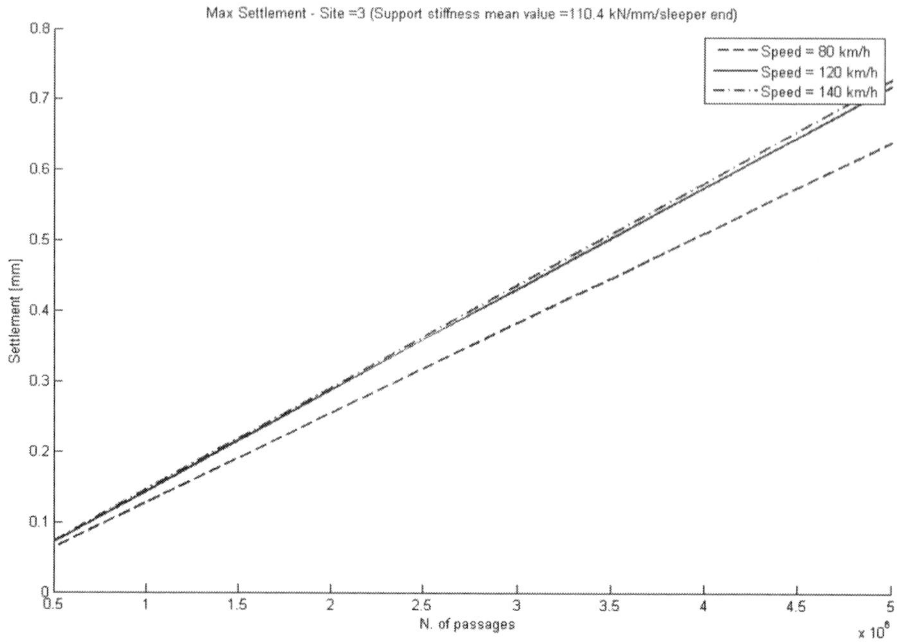


**Figure 9 – Mean settlement in site 3 versus number of passages for different speeds.**

In order to study the influence of the stiffness SD, it is necessary to isolate the effect of the track stiffness mean value fixing a common mean value (e.g. 80 kN/mm/sleeper end) and artificially varying the SD (e.g. 2/7/12/17/22 kN/mm). In Figure 10 the mean settlement values versus number of passages varying the support stiffness SD is presented.

As expected, Figure 10 shows that as the variation in support stiffness increases, the mean settlement values increases accordingly. This is the case for all speeds.

To conclude, the track long-term behaviour is greatly influenced by both the stiffness mean value and standard deviation, as well as by the vehicle speed.

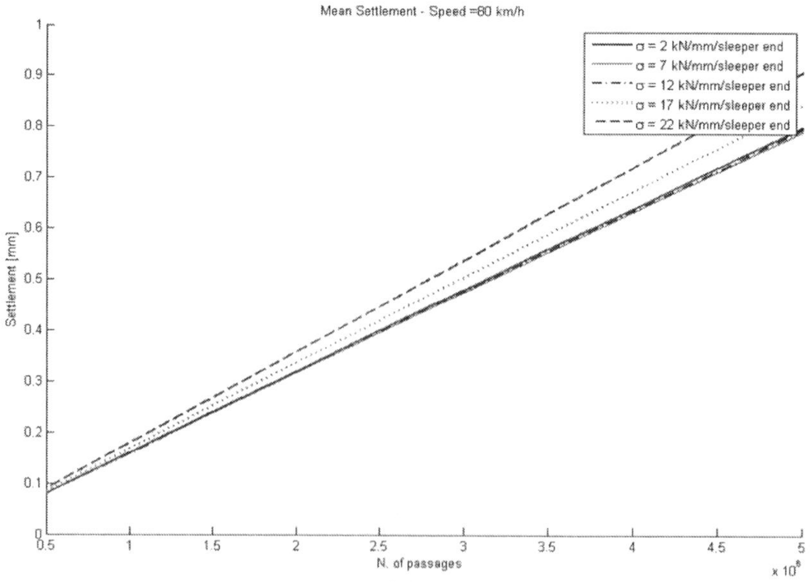


(a)

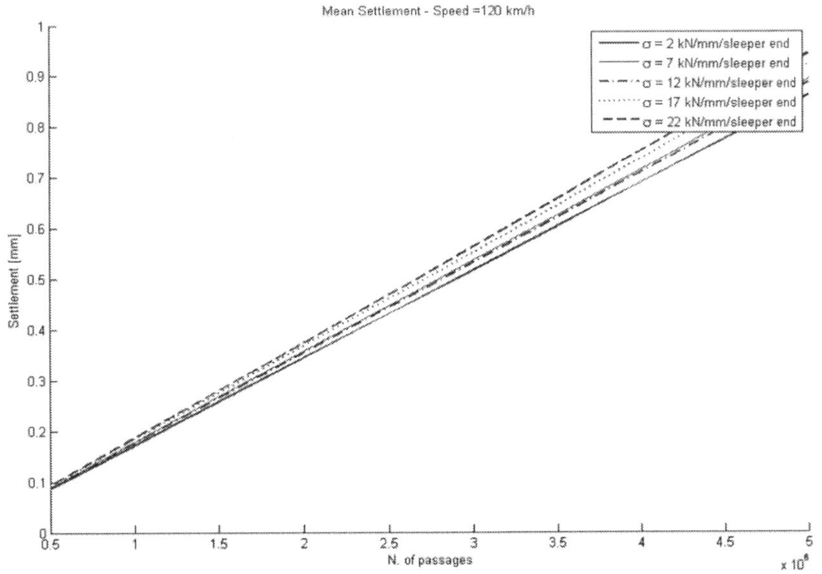


(b)

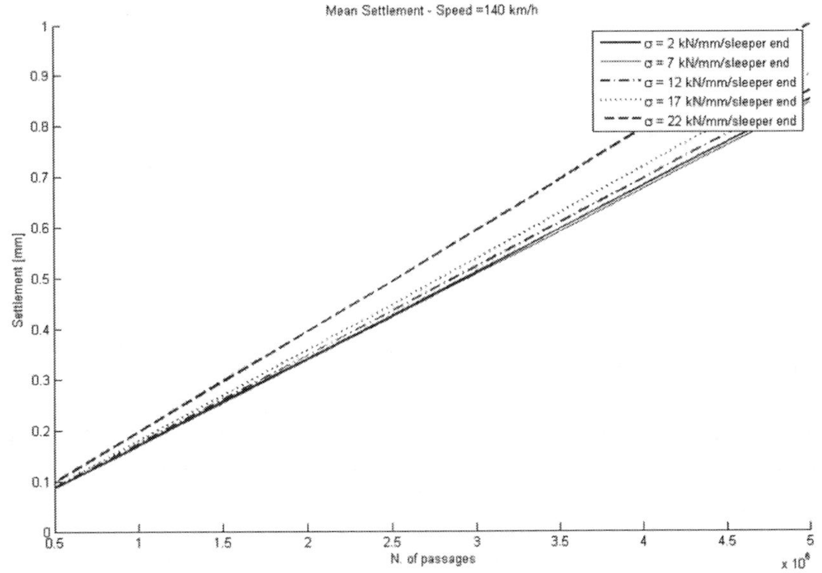


(c)

**Figure 10 – Mean settlement versus number of vehicle passage for different support stiffness SD and for speed (a) 80 km/h; (b) 120 km/h; (c) 140 km/h).**

## 5. CONCLUSIONS

The primary objective of this study was to investigate the influence of the support condition, most specifically the global stiffness, on both track short- and long-term behaviour. A vertical model of vehicle-track interaction, including measured stiffness data from four UK sites, has been used for this purpose. For these sets of data the hypothesis that vertical support stiffness follows a normal distribution is not rejected through the KS goodness-of-fit statistical tests.

Regarding the short-term behaviour, it has been demonstrated how the variability of the support stiffness, described through the standard deviation, leads to high variation in imposed ballast forces. The numerical model used also shows that the presence of USPs can also effectively help reducing both peak forces and the variation in ballast forces. The actual reduction depends on both the support stiffness distribution and the chosen USP stiffness.

Regarding the long-term behaviour, the results show that track settlement is strongly depending on the support condition. Not only the support stiffness mean values affects the track settlement but also the standard deviation.

Finally, it is worth underlining that this paper describes a methodology which can mathematically evaluate the dynamic reaction of vehicle and track including varying support stiffness. This therefore represents a robust platform for further work to understand long-term track settlement and evaluate remediation actions (e.g. here with USPs). The work will be further developed to include a statistical approach which will allow realistic extrapolation of the site specific measured stiffness data, to produce larger sets of data with identical characteristics (mean value and spatial variability).

## ACKNOWLEDGMENTS

The present work has been supported by the European Commission within the FP7 Sustrail project [grant n. 265740].

The authors would also like to thank Dr Antonio Andrade (University of Huddersfield) for his help in the statistical analysis.

## REFERENCES

1. Dahlberg, T., *Railway track stiffness variations – consequences and countermeasures.* International Journal of Civil Engineering, 2010. **8**(1): p. 12.
2. Frohling, R.D., *Deterioration of railway track due to dynamic vehicle loading and spacially varying track stiffness* in *Faculty of Engineering*1997, University of Pretoria: Pretoria.
3. Li, M.X.D. and E.G. Berggren, *A study of the effect of global track stiffness and its variations on track performance: Simulation and measurements.* Proceedings of the Institution of Mechanical Engineers, Part F: Journal of Rail and Rapid Transit, 2010. **224**: p. 8.
4. Lopez Pita, A., P.F. Teixeira, and F. Robuste, *High speed and track deterioration: The role of vertical stiffness of the track.* Proceedings of the Institution of Mechanical Engineers, Part F: Journal of Rail and Rapid Transit, 2004. **218**: p. 10.
5. Frohling, R.D., H. Scheffel, and W. Ebersohn, *The vertical dynamic response of a rail vehicle caused by track stiffness variations along the track.* Vehicle System Dynamics, 1996. **25 Suppl.1**: p. 13.
6. Grossoni, I., et al., *Dynamics of a vehicle–track coupling system at a rail joint* Proceedings of the Institution of Mechanical Engineers, Part F: Journal of Rail and Rapid Transit, To be published.
7. Birmann, F. and J. Eisenmann, *Serial measurements on rails to determine wheel lateral forces.* Eisenbahntechnische Rundschau, 1966. **15**(5): p. 10.
8. Guerin, N., *Approche experimentale et numerique du comportement du ballast des voies ferrees*, in *Structures et materiaux*1996, Ecole nationale des ponts et chausses: Paris.
9. European Committee for Standardization, *Railway applications. Testing for the acceptance of running characteristics of railway vehicles. Testing of running behavior and stationary tests*, 2005.

# Scoping prediction of railway ground-borne vibration in the presence of variable soils

**D P Connolly [1], P K Woodward [1], O Laghrouche [1], B Suryanto [1], G Medero [1], G Kouroussis [2], M C Forde [3]**
[1] Heriot-Watt University, UK
[2] Department of Theoretical Mechanics, Dynamics and Vibrations, University of Mons, Belgium
[3] University of Edinburgh, UK

## ABSTRACT

This paper describes the development of a railway ground-borne noise and vibration prediction approach for environmental impact assessments. The software is called 'ScopeRail', and is capable of predicting in-door noise and vibration, inside buildings close to high speed railway lines. The software is empirically based and can quickly predict vibration levels over large lengths of track. To validate the software, experimental data is collected on European high speed lines and combined with existing data sets from across several countries. Using this data, predicted results are benchmarked against an alternative model and it is found that ScopeRail may give elevated accuracy.

## 1. INTRODUCTION

Growing high speed rail (HSR) infrastructure has resulted in an elevated number of properties residing near high speed rail lines [1]. The resulting ground vibrations generated from such lines may impact negatively on the local environment, causing wall/floor shaking and indoor noise. This can make residential properties uninhabitable, and can result in the loss of building functionality (e.g. buildings with specialised machinery such as hospitals and manufacturing centres [2]).

To minimise these negative effects, a vibration assessment is undertaken before line construction. Typically, a scoping study is first performed to determine which sections of track are most likely to give rise to elevated vibration levels. Then, a detailed study is undertaken to more accurately quantify vibration levels at these specific locations.

A challenge with ground vibration prediction is that detailed modelling is expensive. Therefore its use should be minimised, which is achieved by only performing detailed analysis at locations where vibration will prove problematic. To prevent the unnecessary deployment of detailed models (i.e. 'false positives'), it is vital that the initial scoping study is as accurate as possible.

In an attempt to improve scoping model accuracy, [3] proposed a mathematical tool to predict absolute vibration velocity levels in decibels and root mean squared values. Predicted values were benchmarked against data collected in [4] and it was found to provide reasonably accurate approximations.

Alternatively, [5] proposed a model for high speed lines that could account for train speed, track offset, and used empirical coefficients to model vibration transfer into nearby structures. In addition, [6] detailed a model that could perform similar analysis, for a variety of vibration metrics.

This paper builds upon the approach undertaken by [6], and describes a model that can predict vibration decibel levels in the presence of layered soils, different track types and for various buildings. It uses a combination of machine learning and empirical relationships to increase accuracy and run times, in comparison to traditional approaches.

## 2. MODELLING APPROACH

ScopeRail was constructed in two stages. Initially, a finite element (FE) railway track model was developed to predict ground vibration time histories. This model was executed a large number of times, for a wide range of soil material permutations, to construct a database of velocity time histories. When the database had been populated a statistical machine learning approach was used to map the input parameters (e.g. train speed, distance from track and soil layering) to the desired outputs (i.e. velocity levels). This resulted in a model that could rapidly predict vibration levels for arbitrary soil layering conditions.

### 2.1 FE model development

A train, track and soil model was developed to simulate the passage of high speed trains (Figure 1). The model was executed using a time domain explicit analysis and only half of all model components required modelling due to symmetry. The soil was linear elastic and was modelled using three dimensional brick elements, as in [7]. It was terminated on 4 sides using an absorbing boundary condition to prevent reflections contaminating the modelling space.

The railway track was also modelled in 3 dimensions thus overcoming the 2D assumptions associated with the track models presented by [8]. The track conformed to the layout commonly found on high speed rail lines in mainland Europe and was constructed from subgrade, subballast and ballast layers, supporting evenly spaced sleepers at 0.6m spacings. The rail was connected to the sleepers and modelled using 0.1m long beam elements. The track material properties are show in Table 1.

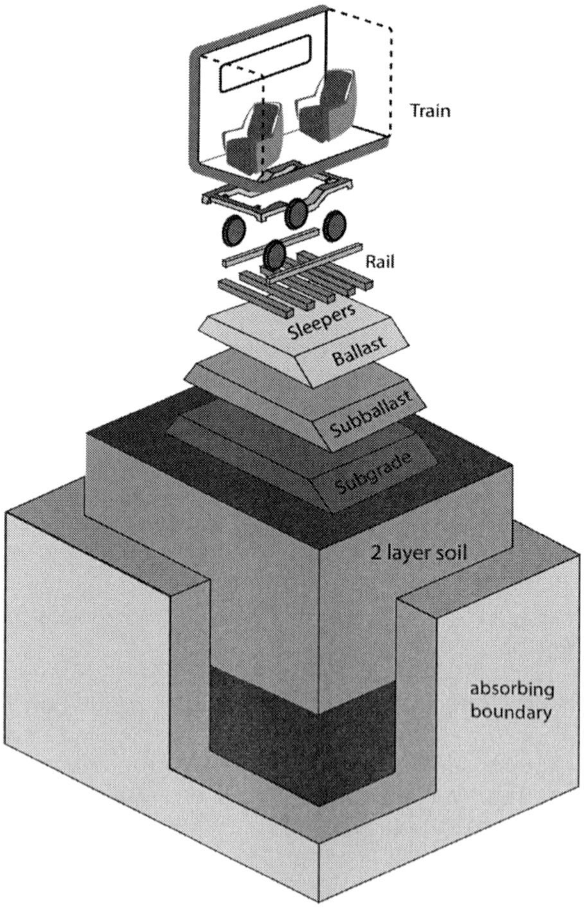


**Figure 1 – FE model schematic**

**Table 1 – Track material properties**

| | Young's modulus (MPa) | Poisson's ratio | Density ($kg/m^3$) | Layer thickness (m) |
|---|---|---|---|---|
| Rail | 210,000 | 0.28 | 7,900 | |
| Sleepers | 20,000 | 0.25 | 2,400 | |
| Ballast | 200 | 0.3 | 1600 | 0.3 |
| Subballast | 130 | 0.3 | 2000 | 0.2 |
| Subgrade | 90 | 0.3 | 2000 | 0.5 |

The vehicle was modelled using a lumped mass, multi-body dynamics approach. Three masses were used to simulate the car, bogie and wheel respectively, and were connected using spring/damper elements. Although the soil and track equations of motion were developed directly using the commercial FE software ABAQUS, the equations of motion for the vehicle were written manually and then coupled with the ABAQUS solver. This allowed for an efficient method to compute the force input to the track ([9], [10]). The vehicle was assumed to be a Thalys high speed train with properties as defined in [11]

## 2.2 ScopeRail model development

Although the FE model could model railway tracks in detail and was able to predict vibration time histories, its run times were too long for it to be used for railway vibration scoping assessments. Therefore it was used as the basis to develop another model, with much lower computational demands, known as ScopeRail. To do so, a machine learning approach was used in an attempt to map train, track and soil characteristics to resultant absolute ground vibration levels.

Firstly, a sensitivity analysis was undertaken to determine the most influential parameters that contributed to the generation of ground borne vibration from rail lines. The least influential parameters and those with large standard deviations (e.g. wheel/rail defects) were excluded from model development. A more detailed explanation of these tests can be found in [6], [12].

One of the most important parameters affecting ground vibration propagation is soil characteristics [13]. To include soil properties within the scoping model, two alternative approaches were undertaken. The first approach was to consider the soil as a homogenous half space (i.e. a single layer) and the second was to consider the soil as a two layer medium. It should be noted that although an infinite number of soil configurations exist in practice, for the purpose of a scoping model, a limited number of input parameters was desirable. This was because it is difficult to obtain detailed soil information for large geographical areas. Therefore, extending the model to three or more layers was considered undesirable. The key input/output model parameters for the two layer soil model are shown in Figure 2.

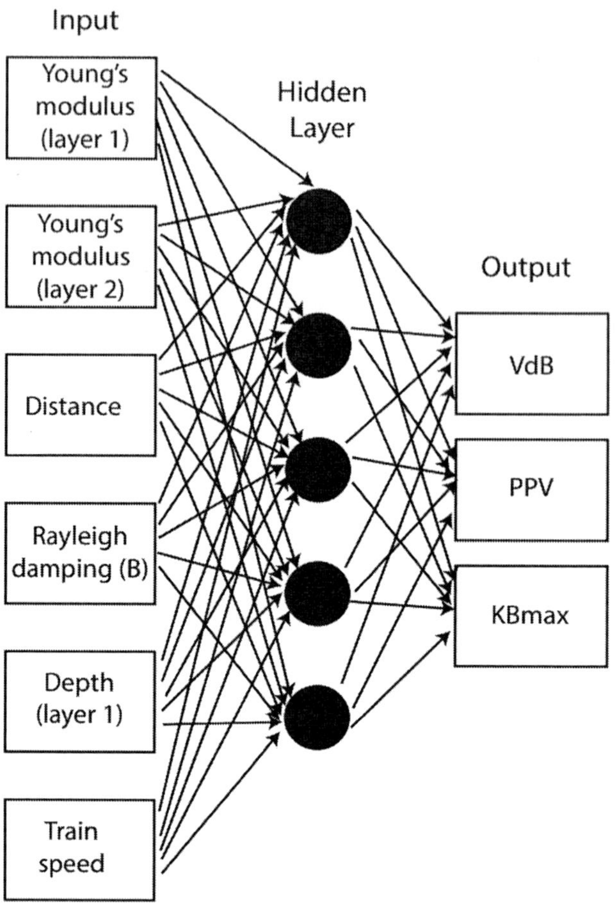


**Figure 2 – Two layer neural network schematic**

## 2.3 Vibration metrics

Rather than attempt to predict velocity time histories or frequency bands, several of the most commonly used absolute vibration metrics were preferred. This was because for scoping studies, it is uncommon to investigate frequency bands, and also machine learning approaches are less well suited for frequency prediction. These were calculated using raw vibration time histories (found via numerical modelling) and then used as the targets for neural network construction.

ScopeRail was designed to predict vibration levels for 3 different metrics: PPV, VdB, $KB_{Fmax}$. VdB is focused on in this report – further details of the PPV and $KB_{Fmax}$ investigations can be found in. VdB is defined in decibels ($VdB_{max}$ - hereafter denoted simply VdB) as outlined in [5]. VdB was calculated using Equation 1, where $V_{rms}$ was the running average of the velocity trace, measured over a one second time interval ('slow' setting). $V_0$ was the reference level of background vibration, and although commonly varies between countries, a constant value of $2.54x10^{-6}$ m/s was chosen.

$$VdB = 20\log_{10}\frac{v_{rms}}{v_0} \qquad \text{Equation 1}$$

## 2.4 Soil layer mapping

In-situ soils are typically heterogeneous and can consist of an arbitrary number of layers. Despite this, due to modelling constraints, ScopeRail was only capable of computing a discrete number of input soil layers. Therefore, to allow the model to be used at a wider variety of test sites, a soil layer mapping technique was used to convert arbitrary soil layering configurations into a 2 layer soil:

$$E_{eq} = \frac{\sum H_i E_i}{\sum H_i} \qquad \text{Equation 2}$$

where $E_{eq}$ = equivalent Young's modulus, $H_i$ = each layer thickness and $E_i$ = Young's modulus of each layer.

## 3. STRUCTURAL VIBRATION, MITIGATION AND EXCITATION FACTORS

The machine learning approach was based upon the prediction of vibration levels from a train-track-soil model. Therefore it could only predict absolute vibration levels on a soil surface, and only for one type of track structure. This was undesirable because building vibration is typically of primary concern, and there are also a variety of track structures in operational use.

To increase ScopeRail versatility and overcome these challenges it was combined with empirical modification factors. These allowed for special train/track features to be modelled including: wheelflats, corrugated track, switches/crossings, ballast mats, floating slabs, resilient fasteners/ties and earthwork profiles. Similar modification factors were used to convert the ground vibration levels into building vibration. A list of these factors can be found in [5].

## 4. MODEL VALIDATION

### 4.1 Field work

An experimental campaign was undertaken to collect vibration records for model comparison. These were recorded at high speed rail test sites in both Belgium (cutting, at-grade and embankment profiles) and England (at-grade and embankment). The Belgian data was recorded on the Paris to Brussels high speed line [14], [15] and denoted 'Mons 2012'. The English data was recorded on the London-Paris high speed line (HS1) and denoted 'HS1 2012'. Vibrations were measured up to 100m from the track using one and three component geophones. The dynamic characteristics of the soil were calculated using a multi-channel analysis of surface waves (MASW) method, with an instrumented hammer used as an excitation source (Figure 4). Rather than using a lazer/camera to record train speed, it was computed at the post-processing stage by comparing experimental and theoretical frequency spectrum, using cepstral analysis, dominant frequency analysis and regression techniques [16]. The results were combined with existing results as collected by additional independent European researchers ([17], [18], [19], [4], [20]). A selection of these, plotted together, are shown in Figure 5.

**Figure 3 – Test sites, left: 'Mons 2012' test site, Right: 'HS1 2012' test site**

**Figure 4 – MASW testing and recording station**

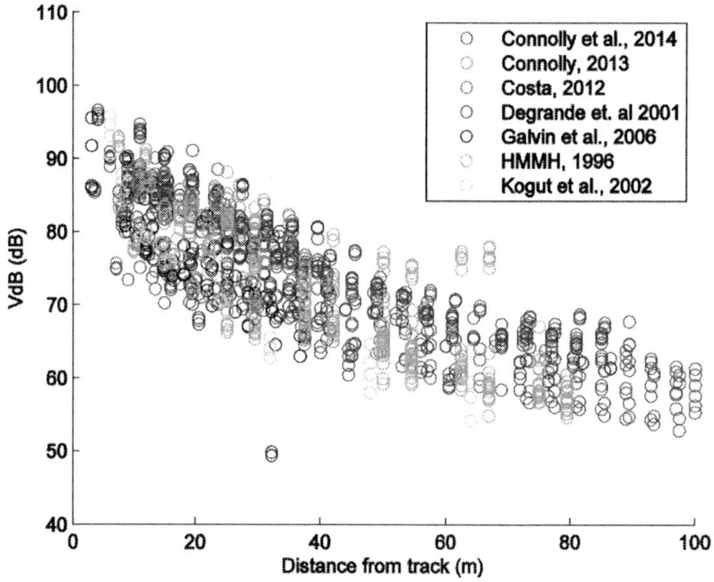


**Figure 5 – Vibration attenuation with distance for several European test sites**

### 4.2 Model validation

The ScopeRail model was benchmarked against the experimental data from Belgium and other countries [21], as well as against an alternative scoping model proposed by [5]. Figure 6 shows that both models were conservative in their predictions. Despite this, ScopeRail was found to perform better for all test cases and for some datasets such as [17] the increase in accuracy was high. For the HS1 2012 site the vibrations exhibited a low attenuation with distance and thus the accuracy achieved using both models was reduced.

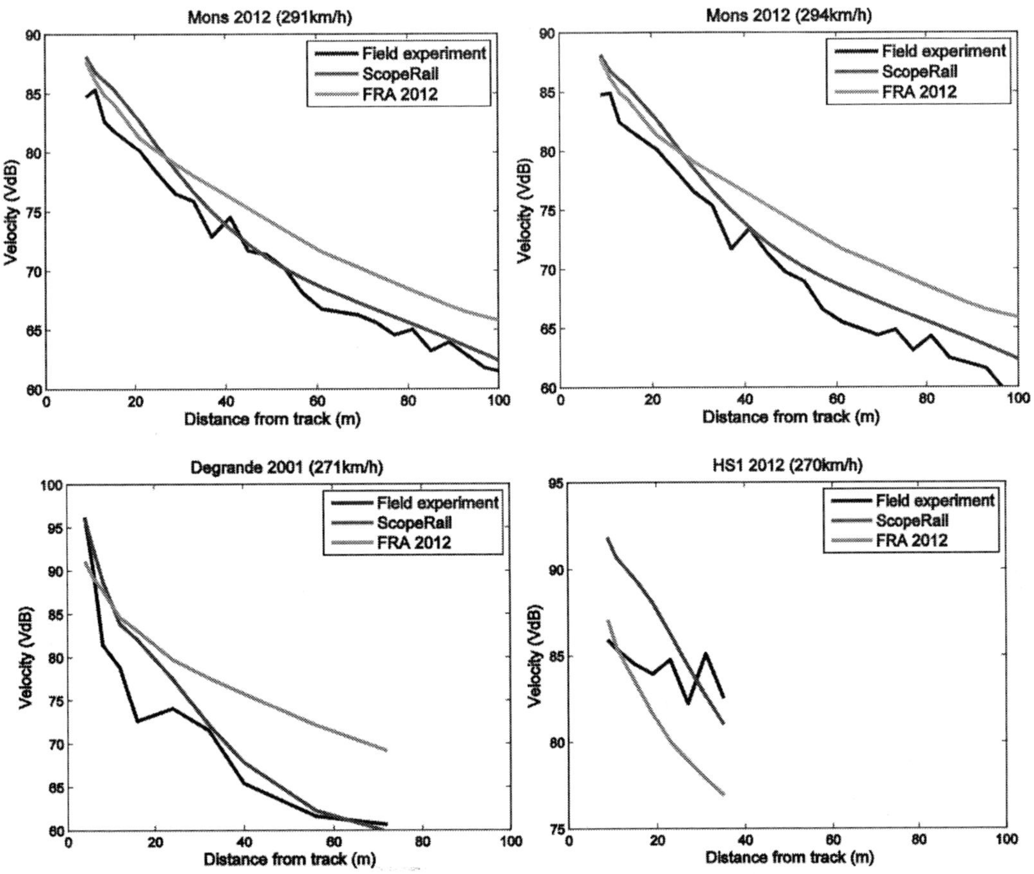


**Figure 6 – Two layer model performance, Top left: Mons 2012 (291 km/h), Top right: Mons 2012 (294 km/h), Bottom left: Degrande 2001 (271 km/h), Bottom right: HS1 2012 (270km/h)**

## 5. CONCLUSIONS

Ground-borne vibration from railways is a growing environmental problem, particularly in urban environments. Therefore a new prediction model (ScopeRail) was developed to predict in-door noise and vibration in buildings close to railway lines. Firstly, a 3D finite element model was developed to simulate train passages on high speed rail lines. It was run hundreds of times, each with varying input parameters to build a database of vibration time histories. These results were statistically analysed to determine the relationship between input variables and vibration levels. This resulted in a model that could instantly predict ground vibration levels in the presence of different train speeds and soil profiles. To facilitate the prediction of in-door noise and vibration in buildings, ScopeRail was coupled with empirically derived structural response metrics. It was found that the new model could predict vibration levels with elevated accuracy in comparison to an alternative approach.

## 6. ACKNOWLEDGEMENTS

The authors thank Heriot Watt University, the University of Edinburgh and the University of Mons for the resources provided for the undertaking of this research. Additionally, the funding provided by Engineering and Physical Sciences Research

Council (EP/H029397/1) and the Natural Environment Research Council is also greatly appreciated. A selection of the experimental field results mentioned in this work can be found at www.davidpconnolly.com.

## REFERENCE LIST


[1] D. P. Connolly, G. Kouroussis, O. Laghrouche, C. L. Ho, and M. C. Forde, "Benchmarking Railway Vibrations - Track, Vehicle, Ground and Building Effects," *Constr. Build. Mater.*, *10.1016/j.conbuildmat.2014.07.042*, 2014.

[2] K. Vogiatzis, "Noise & Vibration theoretical evaluation & monitoring program for the protection of the ancient 'Kapnikarea Church' from Athens Metro Operation," *Int. Rev. Civ. Eng.*, vol. 1, no. 5, pp. 328–333, 2010.

[3] F. Rossi and A. Nicolini, "A simple model to predict train-induced vibration: theoretical formulation and experimental validation," *Environ. Impact Assess. Rev.*, vol. 23, no. 3, pp. 305–322, *10.1016/S0195-9255(03)00005-2*, May 2003.

[4] Harris, Miller, and Hanson, "Summary of European high speed rail noise and vibration measurements (HMMH Report No. 293630-2)," 1996.

[5] Federal Railroad Administration, "High-Speed Ground Transportation Noise and Vibration Impact Assessment. U.S. Department of Transportation," 2012.

[6] D.P. Connolly, G. Kouroussis, A. Giannopoulos, O. Verlinden, P.K. Woodward, and M. C. Forde, "Assessment of railway vibrations using an efficient scoping model," *Soil Dyn. Earthq. Eng.*, vol. 58, pp. 37–47, *10.1016/j.soildyn.2013.12.003*, Mar. 2014.

[7] A. El Kacimi, P. K. Woodward, O. Laghrouche, and G. Medero, "Time domain 3D finite element modelling of train-induced vibration at high speed," *Comput. Struct.*, vol. 118, pp. 66–73, *10.1016/j.compstruc.2012.07.011*, Mar. 2013.

[8] G. Kouroussis, O. Verlinden, and C. Conti, "Free field vibrations caused by high-speed lines: Measurement and time domain simulation," *Soil Dyn. Earthq. Eng.*, vol. 31, no. 4, pp. 692–707, *10.1016/j.soildyn.2010.11.012*, Apr. 2011.

[9] K. Johnson, *Contact mechanics*. Cambridge, UK: Cambridge University Press, 1985, pp. 1–468.

[10] D. Connolly, A. Giannopoulos, W. Fan, P. K. Woodward, and M. C. Forde, "Optimising low acoustic impedance back-fill material wave barrier dimensions to shield structures from ground borne high speed rail vibrations," *Constr. Build. Mater.*, vol. 44, pp. 557–564, *10.1016/j.conbuildmat.2013.03.034*, Jul. 2013.

[11] G. Kouroussis, D. P. Connolly, and O. Verlinden, "Railway induced ground vibrations - a review of vehicle effects," *Int. J. Rail Transp.*, vol. 2, no. 2, pp. 69–110, *10.1080/23248378.2014.897791*, 2014.

[12] D. P. Connolly, G. Kouroussis, P. K. Woodward, A. Giannopoulos, O. Verlinden, and M. C. Forde, "Scoping prediction of re-radiated ground-borne noise and vibration near high speed rail lines with variable soils," *Soil Dyn. Earthq. Eng.*, vol. 66, no. 2014, pp. 78–88, *10.1016/j.soildyn.2014.06.021*, Nov. 2014.

[13] L. Auersch, "Train induced ground vibrations: different amplitude-speed relations for two layered soils," *Proc. Inst. Mech. Eng. Part F J. Rail Rapid Transit*, vol. 226, no. 5, pp. 469–488, *10.1177/0954409712437305*, Feb. 2012.

[14] G. Kouroussis, D. P. Connolly, M. C. Forde, and O. Verlinden, "An experimental study of embankment conditions on high-speed railway ground vibrations," in *Proceedings of the 20th International congress on Sound and Vibration (ICSV20)*, 2013, pp. 1–8.

[15] D. P. Connolly, G. Kouroussis, P. K. Woodward, O. Verlinden, P. Alves Costa, and M. C. Forde, "Field testing and analysis of high speed rail vibrations," *Soil Dyn. Earthq. Eng.*, vol. 67, pp. 102–118, *10.1016/j.soildyn.2014.08.013*, 2014.
[16] G. Kouroussis, D. P. Connolly, M. C. Forde, and O. Verlinden, "Train speed calculation using ground vibrations," *Proc. Inst. Mech. Eng. Part F J. Rail Rapid Transit*, vol. 0, no. 0, pp. 1–18, *10.1177/0954409713515649*, Dec. 2014.
[17] G. Degrande and L. Schillemans, "Free Field Vibrations During the Passage of a Thalys High-Speed Train At Variable Speed," *J. Sound Vib.*, vol. 247, no. 1, pp. 131–144, *10.1006/jsvi.2001.3718*, Oct. 2001.
[18] P. Alves Costa, R. Calçada, and a. Silva Cardoso, "Track–ground vibrations induced by railway traffic: In-situ measurements and validation of a 2.5D FEM-BEM model," *Soil Dyn. Earthq. Eng.*, vol. 32, no. 1, pp. 111–128, *10.1016/j.soildyn.2011.09.002*, Jan. 2012.
[19] P. Galvin and J. Domínguez, "Experimental and numerical analyses of vibrations induced by high-speed trains on the Córdoba–Málaga line," *Soil Dyn. Earthq. Eng.*, vol. 29, no. 4, pp. 641–657, *10.1016/j.soildyn.2008.07.001*, Apr. 2009.
[20] J. Kogut, G. Degrande, W. Haegeman, and L. Karl, "Free field vibration due to the passage of an IC train and a Thalys HST on the high speed track L2 Brussels-Koln," *IABSE Symp. Rep.*, *10.2749/222137803796330176*, 2002.
[21] D. P. Connolly, P. Alves Costa, G. Kouroussis, P. Galvin, P. K. Woodward, and O. Laghrouche, "Large Scale International Testing Of Railway Ground Vibrations Across Europe," *Soil Dyn. Earthq. Eng.*, *10.1016/j.soildyn.2015.01.001*, 2015.

# Study on radiation contribution of wheel/rail system noise

**Q Yin, C Cai, S Zhu, Z Chen**
State Key Laboratory of Traction Power, Southwest Jiaotong University, China

**ABSTRACT**

This paper presents a framework to investigate the sound radiation characteristics of wheel/rail noise on the basis of the hybrid finite element-boundary element numerical method (FEM-BEM). Firstly, a finite element model is built to calculate the vibration of wheel and rail. Then, a boundary element model is adopted to obtain the frequency-dependent acoustic power of each component in wheel/rail system, including the rim and the web of wheel and the head, the web and the foot of rail. Finally, the primary acoustic sources of the wheel and the rail are determined, respectively.

## 1 INTRODUCTION

Analyzing the component of the noise source and finding out the biggest noise source is the precondition to manage the vibration noise. Wheel/rail rolling noise is usually the most important source of environmental noise, even at high speeds. The noise is usually radiated by the vibrations of the wheel and the rail, being excited by the roughness of the wheel and rail interface [1-3].

There are a large number of studies on the most important source of wheel/rail noise in the past years. Most researchers in Europe regarded the noise of wheel radiation as the predominant, while noise of rail radiation plays a primary role for the researchers in the USA [4]. However, different results would appear with many factors, such as different vehicle track system and different route. For example, soft rail-pads are commonly equipped in the Chinese high-speed railways, while stiffer rail-pads are widely used in European and American railways. A softer pad increases the proportion of rail noise and vibration, as a straight and a curved route have different wheel/rail noise radiation characteristics.

In this paper, for the Chinese high-speed railways, the structure of wheel/rail systems is refined by adopting the hybrid finite element method. The wheel/rail system is divided into five parts, describing the wheel rim, the wheel web, the rail head, the rail waist and the rail foot. The acoustic power of each part is analyzed in the effect of the wheel/rail vertical and lateral forces, and the octave analysis of the acoustic power is implemented. Through the above analysis, the most important acoustic radiation source of wheel and rail in each case could be obtained.

## 2 FEM-BEM

### 2.1 Finite element formulation

The vibration equation of a structure with harmonic excitation is given as

$$\left(-\omega^2[M]+i\omega[C]+[K]\right)\{x\}=\{f\} \tag{1}$$

where, $[M]$, $[C]$ and $[K]$ are the mass, damping and stiffness matrices, respectively, $\{x\}$ is the displacement vector for the node, $\{f\}$ is the external excitation load vector.

### 2.2 Sound field boundary element formulation

The pressure $p(r,\omega)$ generated in the external fluid medium with the harmonic excitation should meet the Helmholtz differential equation (Eq. 2), fluid-solid interface condition (Eq. 3) and the Sommerfeld radiation condition (Eq. 4)

$$\nabla^2 p + k^2 p = 0 \tag{2}$$

$$\frac{\partial p}{\partial n} = i\omega\rho v_n \tag{3}$$

$$\lim_{r\to\infty}\left[r\left(\frac{\partial p}{\partial r}-ikp\right)\right]=0 \tag{4}$$

where, $\omega$ is the circular frequency, $r = |Q - P|$, Q is any point on the surface of the structure S, P is an arbitrary point in space, $k$ is the wave number, C is the speed of sound in the fluid medium, n is S outside normal unit vector, ρ is the density of the fluid medium, $v_n$ for the normal on the S velocity.

### 2.3 Analysis method

The basis of the theoretical model contained in this paper is summarized in Fig. 1. In the FEM section, unit harmonic forces are acted on the wheel or rail finite element model. The forces acting on the wheel conclude vertical force at tape circle, lateral force at tape circle and lateral force at flange. The forces acting on rail conclude vertical force at the middle point of rail head, lateral force the middle point of rail head and lateral force at the side of rail head. The finite element model and the result of FEM calculation provide the inputs to the BEM section. Based on the vibration data, the acoustic radiation of wheel rim, wheel web, rail head, rail waist and rail foot could be obtained by using boundary element predictions.

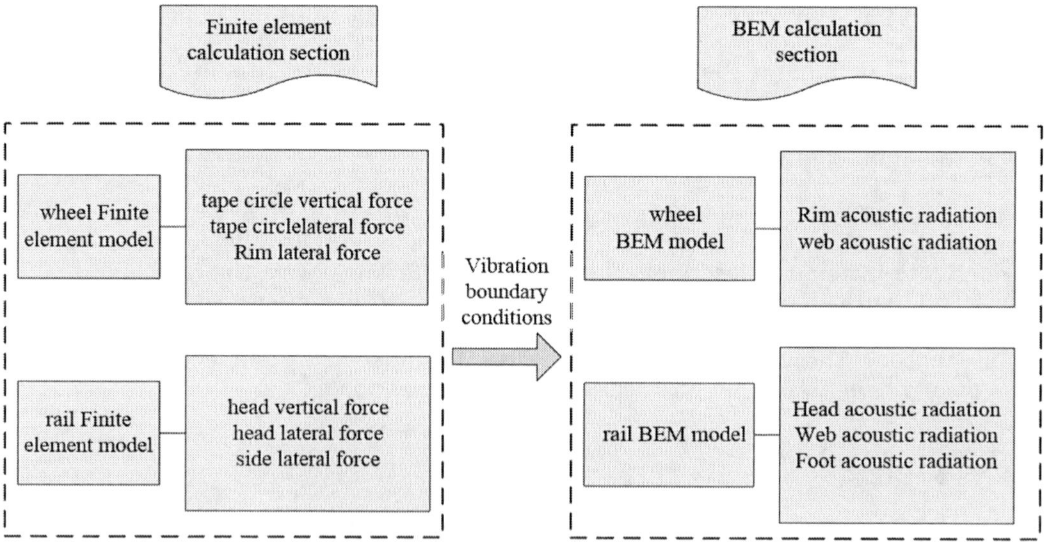


**Fig. 1 Research route map**

## 3 FINITE- BEM CALCULATION MODEL

### 3.1 FEM

The web shape of the railway wheel serviced in China high-speed railways can be broadly categorized into a curve type, a straight type and a corrugated type. In this study, the most widely used type of wheel is modeled, in which the diameter are 915mm, and the thickness of the web is 20 mm to 25 mm. The wheel is modeled with 3D solid element, and the finite element model can be determined.

Fig. 2(a) shows the wheel/rail contact point on the wheel and Fig. 2(b) is the wheel/rail contact point on the rail. The point A and B represent the point on tape circle and flange, respectively, while points C and D preform the point on rim and web, respectively. In the Fig. 2(b), the points A' and B' represent the middle point and the side of rail head, respectively. Point C' and D' represent a point on rail waist and rail foot respectively. The points would be used later as the positions of forces.

Wheel/rail interaction will produce vertical compression at the contact point, also vertical and lateral slip. The corresponding wheel/rail interaction forces are wheel/rail normal positive pressure, wheel and rail longitudinal creep force and wheel/rail lateral creep force, respectively.

When the vehicle runs smoothly in a straight line, the longitudinal creep forces and lateral creep forces are small. But if there is hunting of the vehicle, with the small radius curves and curves with off-balance superelevation and so on, wheel/rail lateral stick slip will become violent, generating a larger wheel/rail noise.

Due to the complexity of the problem, this paper only takes the vertical force and lateral force at point A (A') and the lateral force at point B (B') into consideration. In the process of wheel vibration analysis, applying fixed constraints within the boundaries of the hub hole, without considering the influence of the axle. There are some errors in this simulation for the low frequency modes (0 diameter and 1 diameter), as there is a strong coupling vibration between wheels and axles. But there is a higher modal damping with these modes, so the effect on the vibration and noise is not great. Moreover, existing research [3] shows that, this simulation has a good approximation with the radiation predicting of two diameters and larger diameters modes which have much more effects on vibration and acoustic radiation.

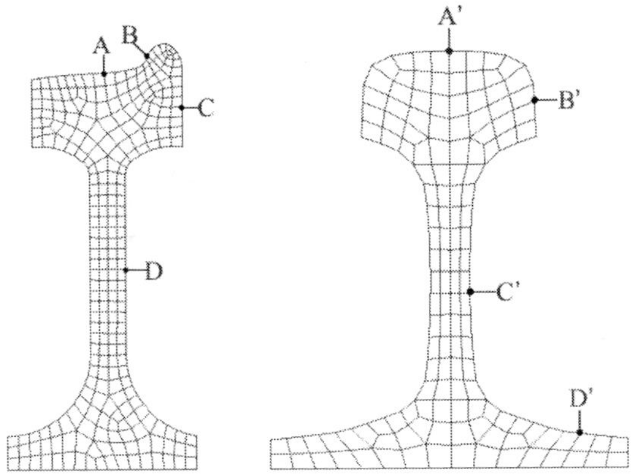


**a) contact point on the wheel** **b) contact point on the rail**

**Fig. 2 Wheel/rail force point**

The track system is modeled as rail – pads system, as this work is focusing on analyzing the rail acoustic radiate characteristics, and the Chinese high-speed railway rail-pad stiffness is very small (i.e. a weak coupling relationship between the rail and the understructure). The pads are considered as discrete spring – damper elements.

In the calculation, the CHN 60 rail is used, where, the young's modulus is $2.1 \times 10^{11}$N / $m^2$, the density is 7850 kg / $m^3$, the poisson ratio is 0.3, the rail loss factor is 0.01. . The vertical and lateral stiffness of the pad are set to be 60MN/m and 30 MN/m, respectively, and the loss factor is 0.25.

### 3.2 BEM

The finite element model and the result of FEM calculation provide an input data to the BEM calculation section. The wheel/rail radiated noise is calculated for each condition by adopting the acoustic BEM software SYSNOISE. The direct boundary element method is used to calculate the acoustic radiation. In the calculation, the air density is 1.21kg / $m^3$, the air speed of sound is 344 m / s, the frequency is in the range of 100Hz to 4 000Hz, and the step size is 10Hz. Boundary element mesh is one of the key factors affecting the accuracy of the analysis. For the requirements of the SYSNOISE acoustic boundary element mesh, there must be at least 6 units in the smallest analysis wavelength.

In acoustic radiation calculation process, the division of the wheel and rail surface area has been shown in Fig. 3. The wheel surface is divided into three parts, including the rim, the web and the hub. While the rail is divided into the rail head surface, rail waist and the rail foot. The acoustic radiation power of each component is calculated with different forces conditions.

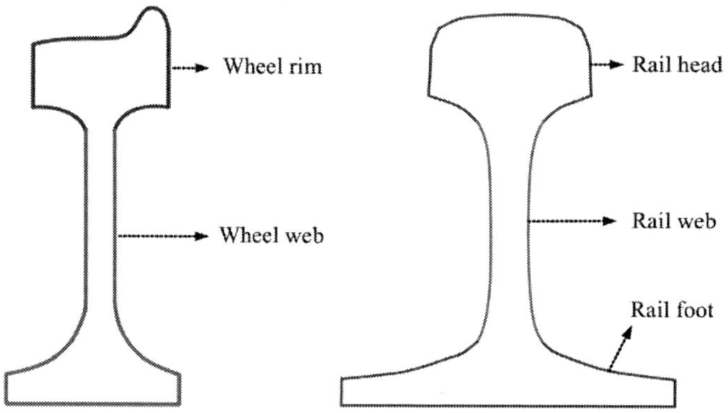


**Fig. 3 Boundary Element zoning**

## 4 RESULTS AND DISCUSSION

### 4.1 Frequency response function

In the acoustic process, the normal vibration on the wheel/rail interface is closely related to the acoustic power. The vibration amplitudes of the surface of the wheel/rail system are extracted by the results of the finite element model, and the radiating capacity can be judged.

#### *4.1.1 Wheel acoustic*

A unit harmonic force is acted on the wheel, and the figures 4-6 show the wheel vibrations.

Due to the asymmetry of the wheel section, vertical force could cause a significant lateral vibration, as shown in Fig. 4. Peak effect is evident in the response of wheel vibration at the resonant frequency. As a whole, there is a bigger vibration of the tread in the most frequency bands, followed by the vibration of the web, and that of the rim is the least one. It can be seen from Fig. 5 that vibration level of the wheel rim and wheel web are equivalent when a lateral force acting at point A. Fig. 6 shows the vibration of point C and point D with a lateral force acting at point B. From the Fig. 6, it can be seen that the wheel rim and wheel web show an equivalent vibration too.

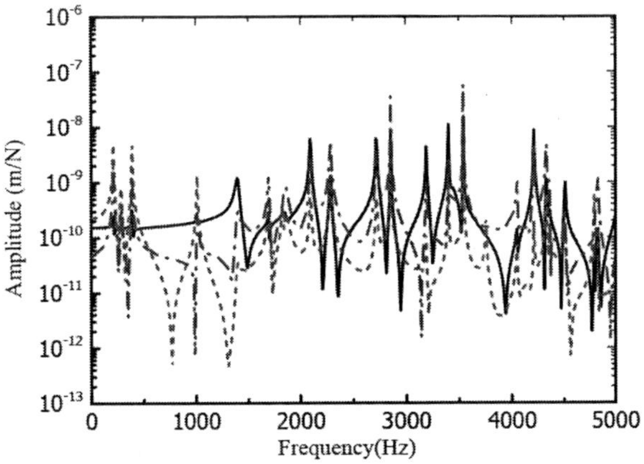


**Fig. 4 The displacements of the wheel surface along the normal direction with the vertical force at point A**
**——at point A, ----- at point C, -.-.- at point D**

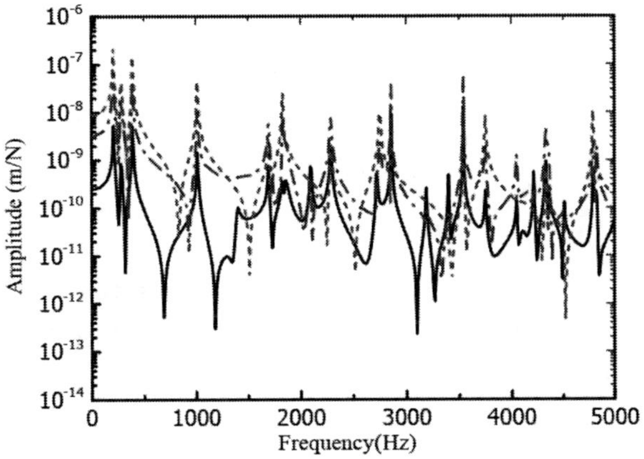


**Fig. 5 The displacements of the wheel surface along the normal direction with the lateral force at point A**
**——at point A, ----- at point C, -.-.- at point D**

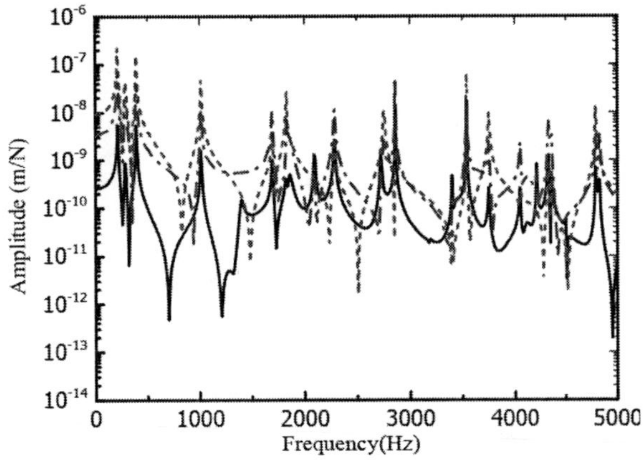


**Fig. 6 The displacements of the wheel surface along the normal direction with the lateral force at point B**
**——at point A, ----- at point C, -.-.- at point D**

#### *4.1.2 Rail acoustic*

Applying a unit harmonic force on the rail, the rail vibration is shown in figures 7-9.

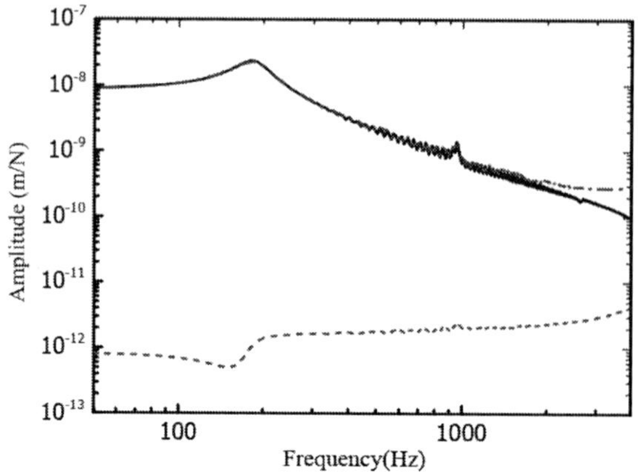


**Fig. 7 The displacements of the rail surface along the normal direction with the vertical force at point A'**
**——at point A', ----- at point C', -.-.- at point D'**

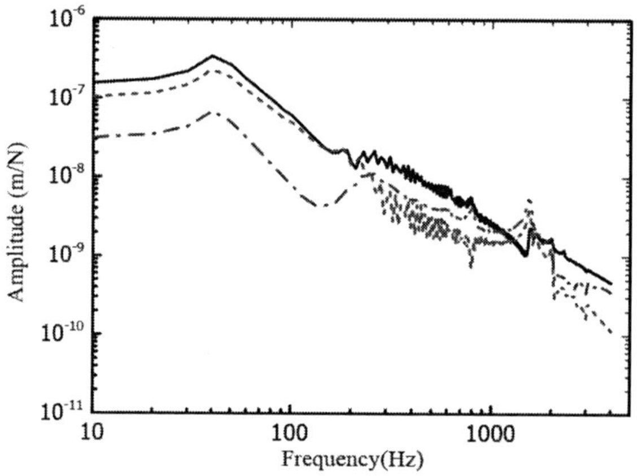


**Fig. 8 The displacements of the rail surface along the normal direction with the lateral force at point A'**
**——at point A', ----- at point C', -.-.- at point D'**

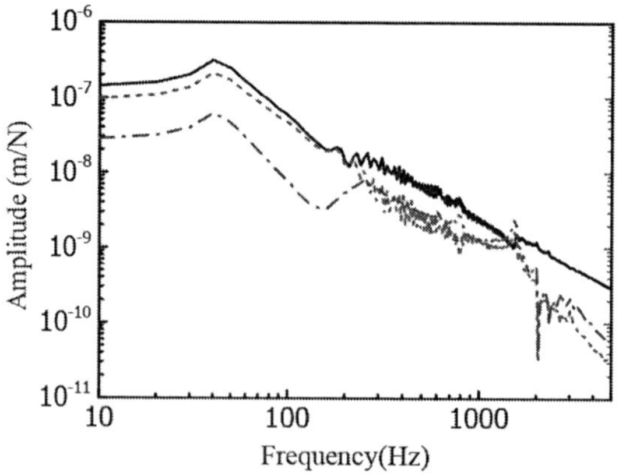


**Fig. 9 The displacements of the rail surface along the normal direction with the lateral force at point B'**
**——at point B', ----- at point C', -.-.- at point D'**

As can be seen from Fig. 7, due to a large damping of the track system, the vibration of the rail is different from the wheel. The vibration response does not occur a spike effect as the wheel. It can be seen from Fig. 7 that the vibration level of rail head and rail foot is much higher than that of the rail waists obviously. Due to the rail section is symmetry, its lateral vibration is smaller. The rail head vibration is coincident with that of the rail foot in a wide range of low frequency. However, the demarcation point appears between 1000Hz to 2000Hz. At low frequencies, up to 1000Hz, the rail vibrate in an overall form, while with higher frequencies, higher than 1500Hz, local vibration of the rail appears. It implied that the response of the rail can't be represented by a simple beam at high frequency.

It can be seen from the response spectrum in Fig. 8 that the rail head has a more severe vibration than other components in all frequency bands with lateral force applying on the point A'. And the rail waist has a more severe vibration than the rail foot below 200Hz. In addition, the web vibrate more severely than the foot between frequencies from 200Hz to 1500Hz, while a reverse happened at a higher frequency bands.

Fig. 9 shows the vibration of the rail with the lateral force of tape circle point. The vibrations in the frequency range above 200Hz all shows that the vibration of rail head is the biggest, and the rail feet is the second one, while the web is the least one.

## 4.2 Acoustic radiation power

According to the results of the finite element model, the frequency-dependent acoustic radiation power of the wheel and rail is calculated with the boundary element method (BEM). The acoustic power of each component of the wheel and rail is calculated with a frequency interval of 10Hz, and it is plotted in a narrowband curve. In order to show the magnitude difference of the various parts more clearly, the amplitude of octave frequency band (The RMS. value of the amplitude of the band) is also given.

### *4.2.1 Acoustic radiation power of the rail*

The acoustic radiation power of rail components and the sum under different load conditions are shown in Fig. 10 to Fig. 12.

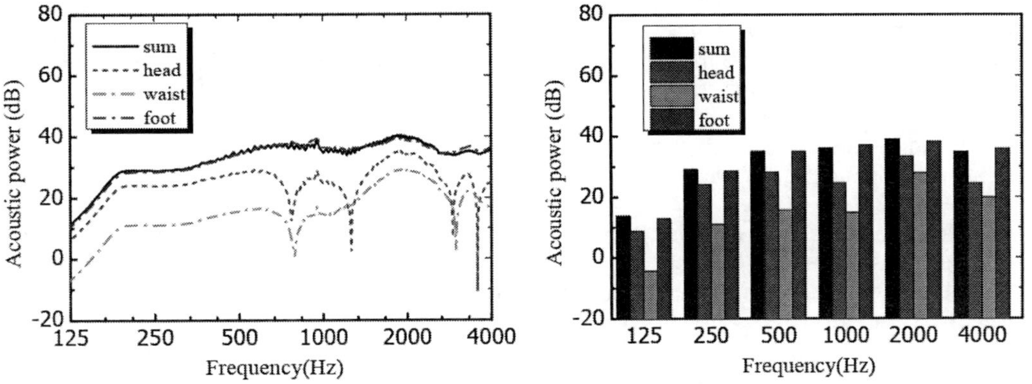


**Fig. 10 Acoustic power of the rail when a vertical force is applied at the top of the rail head**

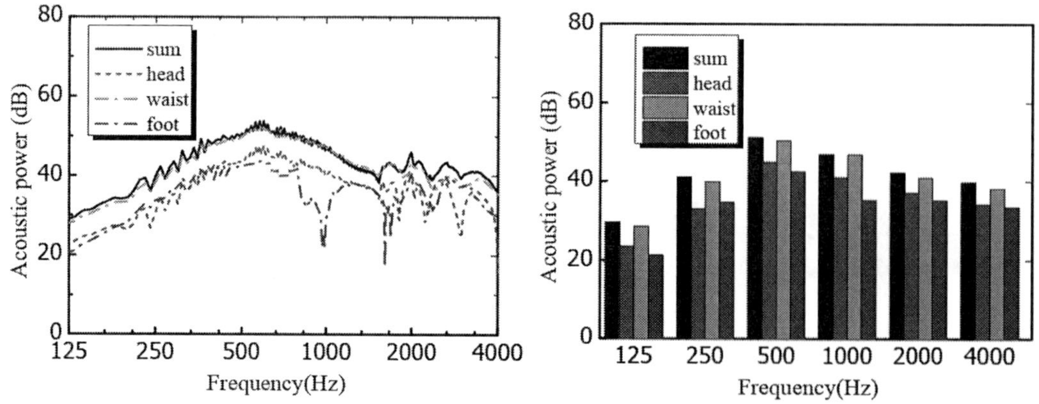


**Fig. 11 Acoustic power of the rail when a lateral force is applied at the top of the rail head**

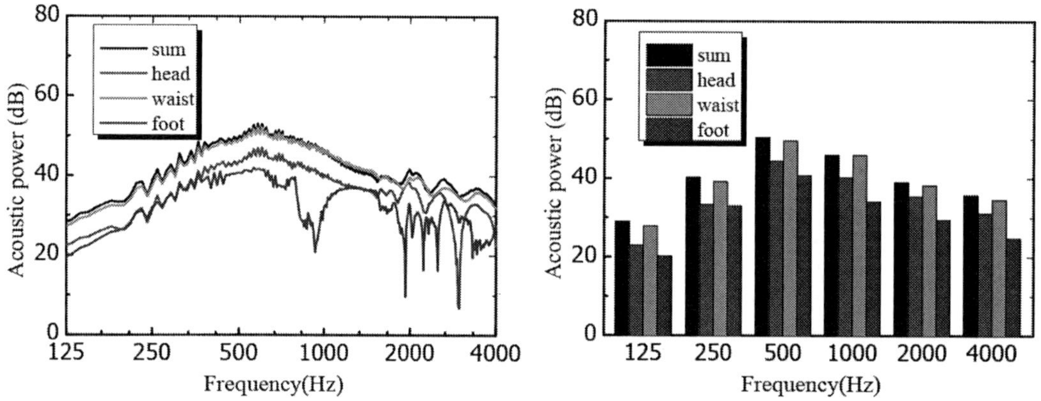


**Fig. 12 Acoustic power of the rail when a lateral force is applied at the side of the rail head**

As shown in Fig. 10, the rail foot dominants the rail acoustic in the whole frequency bands. In the octave analysis, the acoustic power of the rail head in each octave bandwidth is at least 5dB lower than that of the rail foot. It suggests that if the rail is simplified as the rail foot, the prediction error of rail acoustic radiation power will be less than 1.2dB. The result that the rail head shows a lower acoustic power than the rail foot is consistent with the prior result of FEM. The reason is that the acoustic area of the rail foot is bigger than rail head.

According to the above analysis, the rail noise is mainly radiated from the rail foots if only the wheel/rail vertical force is applied. To this end, reducing the vibration of the rail bottom can remarkably reduce the rail noise in the case of straight route. The measures include the use of narrow foot rail, etc.

Fig. 11 shows that when a lateral force is applied on the top of rail head, the rail waist radiates the most power, and the acoustic power of the rail foot is the lowest. So, when the vehicle runs on a curve route, the most important objective is the rail waist to reduce the rail acoustic power. For this purpose, using the short rail or pasting damping on the rail waist can reap the benefits[5].

The acoustic powers of rail components when a lateral force is applied at the side of the rail head are plotted in Fig. 12. This case corresponding to a severe curves routes, the rail waist is the dominant source of rail acoustic radiation. Therefore, the noise reduction measures could be the same with the former situation.

### *4.2.2 Acoustic radiation power of the wheel*

The acoustic radiation power of wheel components and the sum are shown in Fig. 13 to Fig. 15.

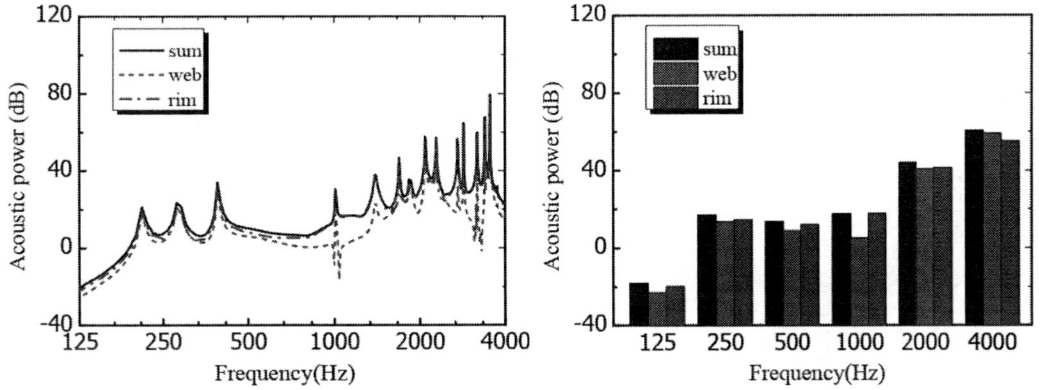


**Fig. 13 Acoustic power of the wheel when a vertical force is applied at the wheel tape circle**

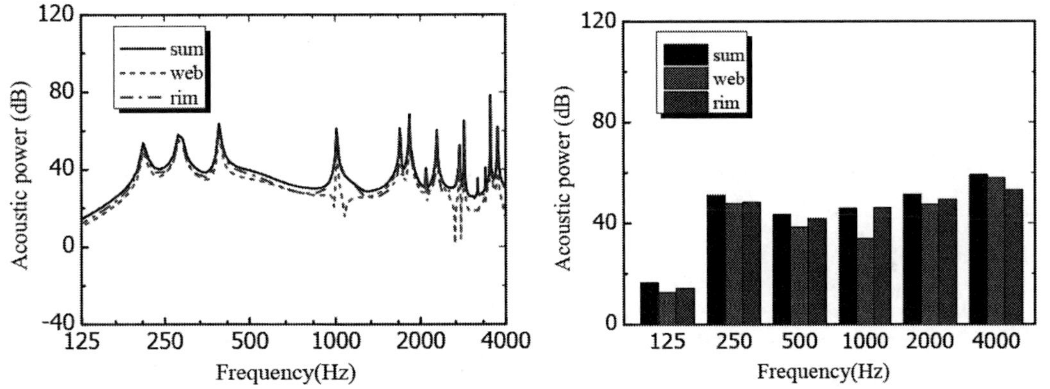


**Fig. 14 Acoustic power of the wheel when a lateral force is applied at the wheel tape circle**

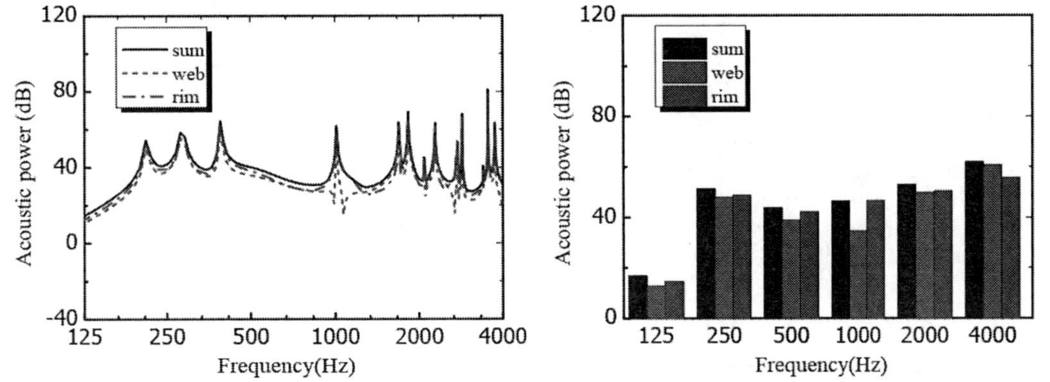


**Fig. 15 Acoustic power of the wheel when a lateral force is applied at the wheel rim**

As shown in Fig. 13, the wheel rim dominants the wheel acoustic in the whole frequency band. It can be seen from the octave analysis that the rail head acoustic more power than wheel web in each octave bandwidth except 4000Hz. However, the difference is not significant. This implies that under the tape vertical force, the radiation contribution of wheel rim and wheel web are almost equivalent. So the simplifications that representing the wheel by a love's ring and taking the wheel web acoustic power equal to the wheel rim in the Ref. [6] are acceptable. The same rule can be seen in Fig. 13 and Fig. 14. So the wheel rim and wheel web should be paid the same attention at different situations to control the acoustic radiation power of the wheel.

## 5 CONCLUSION

The sound radiation characteristics of wheel/rail noise have been investigated with the finite element-boundary element method. The results show that, under the wheel/rail vertical force, corresponding to the case of straight route, the rail acoustic power is mostly radiated from the rail foot, and the rail waist dominates the rail acoustic radiation under wheel/rail horizontal force, corresponding to a severe curves routes. However, in all cases the wheel's performance is consistent the rim dominants the wheel acoustic radiation below 2000Hz while the web radiates more noise at higher frequencies.

## ACKNOWLEDGEMENT

This work was supported by the National High-tech R&D Program (863 Program) (Grant No. 2011AA11A03), the 2015 Doctors' Innovation Fund of Southwest Jiaotong University and the Fundamental Research Funds for the Central Universities.

## REFERENCES

[1] Wakabayashi Y, Kurita T, Yamada H, et al. Noise measurement results of Shinkansen high-speed test train (FASTECH360S, Z)[M]//Noise and Vibration Mitigation for Rail Transportation Systems. Springer Berlin Heidelberg, 2008: 63-70.

[2] Okada T. A study on predicting Shinkansen noise levels using the sound intensity method[J]. JSME International Journal Series C, 2004, 47(2): 508-511.

[3] Poisson F, Gautier P E, Letourneaux F. Noise sources for high speed trains: A review of results in the TGV case[M]//Noise and Vibration Mitigation for Rail Transportation Systems. Springer Berlin Heidelberg, 2008: 71-77.

[4] Remington P J. Wheel/rail rolling noise: What do we know? What don't we know? Where do we go from here?[J]. Journal of Sound and Vibration, 1988, 120(2): 203-226.

[5] Thompson D J, Gautier P E. Review of research into wheel/rail rolling noise reduction[J]. Proceedings of the Institution of Mechanical Engineers, Part F: Journal of Rail and Rapid Transit, 2006, 220(4): 385-408.

[6] Remington P J. Wheel/rail rolling noise, I: Theoretical analysis[J]. The journal of the Acoustical Society of America, 1987, 81(6): 1805-1823.

# Modelling of Train Induced Vibration

**E Ntotsios[1], S G Koroma[1], W I Hamad[2], D J Thompson[1], H E M Hunt[2], J P Talbot[2], M F M Hussein[3]**
[1] ISVR, University of Southampton, UK
[2] Engineering Department, University of Cambridge, UK
[3] Department of Civil and Architectural Engineering, Qatar University, Qatar

**ABSTRACT**

This paper reports on recent developments in techniques for modelling ground vibration from railways. The modelling considers both surface and underground railways, and accounts for the main dynamic systems involved, i.e. tracks (both ballasted and slab), tunnels and multi-layered ground. Results are presented to illustrate the modelling capabilities and the efficiency of computations for the models proposed. The work presented is part of the MOTIV project (Modelling of Train Induced Vibration), which is a collaboration between the Universities of Southampton and Cambridge. Future development of models and plans within the project are also addressed.

## 1. INTRODUCTION

The MOTIV project aims to develop a better understanding of key issues relating to the generation and propagation of railway-induced ground vibration. The project primarily addresses the efficient modelling of the main sub-systems involved in railway infrastructure and the dynamic interaction between them. These systems are the resilient elements of the track (ballasted or slab), the tunnel structure (for the case of underground railways), the surrounding soil and the nearby buildings, including those with pile foundations.

Predictions of ground-borne vibration from railways using numerical models are important as they help significantly in developing a better understanding of the physics of the generation and propagation and provide insight into results of field measurements and vibration problems encountered in practice. Numerical models are particularly useful in situations where a new railway line is to be constructed close to existing buildings or where a new building is to be built close to an existing track or tunnel. Moreover, modelling is also an essential pre-requisite to the development of prediction tools that can be used by engineers when designing vibration mitigation measures aimed at reducing vibration to an acceptable level. A large number of numerical models for predicting vibration from surface and underground railways have been presented in the literature. An overview of the state of the art on railway induced ground vibration models can be found in (1). These prediction models range from simple multi-degree-of-freedom models to two-dimensional and more comprehensive three-dimensional models. Most of these models are formed by coupling sub-models for the train, the track, the tunnel, the soil and the nearby buildings.

Within the MOTIV project, the modelling of both surface and underground railways is considered. From the available numerical models proposed for the prediction of ground-borne vibration from surface and underground railways, two were selected as the starting point for the project. The first model is the 'TGV: Train-induced Ground Vibration' model, developed within the ISVR at the University of Southampton for the prediction of ground vibration due to surface trains, based on the work of Sheng et al. (2). The second model, for the case of underground trains, is the 'PiP: Pipe-in-Pipe' model developed by Forrest and Hunt (3), and Hussein and Hunt (4) at Cambridge University. Both models are considered as semi-analytical models, where invariancy in the direction of the track is assumed. This allows the use of efficient solution procedures, based on a Fourier transform with respect to the coordinate along the track, for calculating vibration from railways in a three-dimensional field from a two-dimensional geometry. The results are transformed to the space domain using the inverse Fourier transform. Within the objectives of MOTIV, these models are being revisited and improved adding a number of developments needed to increase their efficiency, capabilities and applications.

This paper reports current and future development of the numerical models and plans within the MOTIV project. In the first part, the TGV and PiP models for simulating ground vibration generated by surface and underground trains are introduced and their capabilities in predicting the vibration in the free-field are addressed in two representative numerical examples. The last part of the paper addresses the forthcoming plans and objectives of the project, which involve the theoretical and experimental investigations on the effect of pre-load and the non-linear behaviour of the resilient elements of tracks, and the effect of coupling the railway and the surrounding piled buildings.

## 2. TGV – A NUMERICAL PREDICTION MODEL FOR SURFACE RAILWAYS

### 2.1 Model description

The TGV model is a semi-analytical model developed for the prediction of ground vibration generated by surface trains. The model can be used to analyze three components of vibration generated by the wheels of a train, namely (a) dynamic loads generated at fixed points on the track, (b) moving constant axle loads, and (c) moving dynamic loads applied through the wheels. It should be noted that the solution for the cases of moving axle loads is achieved by working in a frame moving with the velocity of the train.

In the TGV model, the train vehicles are described as linear 2D multiple rigid-body systems and only the vertical dynamics are considered (Figure 1a). The track is modelled as multiple beams supported by vertical springs with consistent mass (Figure 1b). The rails are represented as a single Euler-Bernoulli beam and the rail pads are modelled as a distributed vertical stiffness. The sleepers are modelled as a continuous mass per unit length of the track and the ballast is modelled as a continuous distributed vertical spring stiffness with consistent mass. An embankment, if present, can be modelled in the same way as the ballast. The vertical profile of the rail may be decomposed into a spectrum of discrete harmonic components. Furthermore, a Hertzian contact spring can be introduced between each wheelset and the rail.

The ground is represented by horizontal layers on a homogenous half-space or rigid bedrock. The railway track is aligned in the $x$ direction and has an invariant contact width $2b$ with the ground (Figure 1c). An efficient semi-analytical model developed by Sheng et al. in (5) is used for the prediction of the ground response excited by moving constant or harmonic loads acting directly on the ground, or for loads acting via a coupled track structure. The model uses a two-dimensional Fourier transform in the wavenumber domain with $\beta$, $\gamma$ the wavenumbers corresponding to the coordinates $x$, $y$ along and normal to the track. The coupling of the ground with the railway track is carried out by taking into account the continuity of the displacements and the equilibrium of the stresses in the plane of contact between them, rendering it possible to calculate the Fourier transformed response of the ground surface and the track elements. Results in the frequency-space domain are calculated using the inverse Fourier transform.

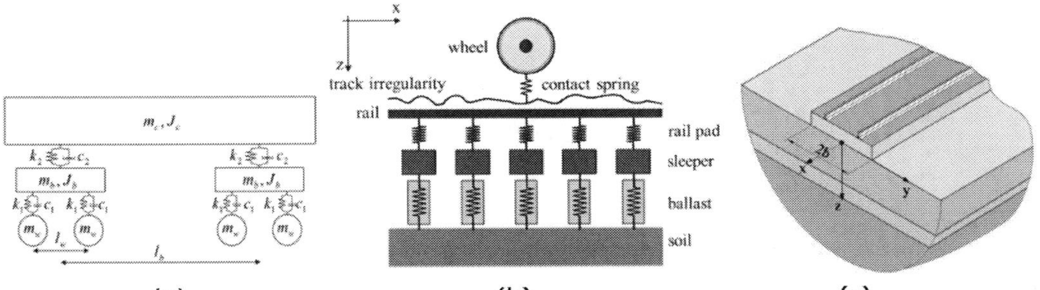


(a) (b) (c)

**Figure 1. (a) 10 DOF multi-body model of vehicle, (b) coupling of the wheelset with the track (rail, pad, sleeper, ballast) and the ground and (c) the geometry of the coupled track-soil system.**

The TGV model uses the moving axle loads and vertical rail irregularities as its inputs. Outputs include the dynamic forces at the wheel-rail interface and the response (displacement/velocity/acceleration amplitudes or spectra) of the track and the ground. Railway track unevenness is described as a stationary random process by means of the PSD of the irregularity in terms of the wavenumber along the railway track. In this concept, the axle loads of the vehicles represent a stationary random process as well; nevertheless, this is not the case for the vibrations observed in the free field. This is due to the fact that, for a receiver at a fixed position in the free field, the vibration level depends on the position $x$ of the axle load along the rail, or on the time $t$, and is therefore a non-stationary random process. The calculation of a non-stationary response PSD when the correlation of axle loads is taken into account can prove highly computationally demanding (6). However, in the TGV model, the calculation of the vibration PSD is treated efficiently as a stationary process using two approaches. The first approach is based on modelling the train at fixed positions on the track which are excited by a "moving roughness"; the roughness is 'pulled through' between the wheels and track with the velocity of the train, assuming that each wheel is excited by the same roughness apart from a time lag. In the second approach, the vibration PSD is calculated at a fixed point for the moving train. First the response is calculated at a point that is moving with the speed of the train (moving frame) and a Fourier transform with respect of time is applied (5) leading to PSD formulation that does not vary in time (2).

## 2.2 Numerical example

In order to highlight some of the output predictions of the TGV model, a numerical example is presented for a range of practical train parameters and velocities. The results are compared with a coupled 2.5D finite element/boundary element (FE/BE) model developed in (7). The coupled FE/BE model uses the moving roughness excitation approach. For the TGV model, results for both approaches are presented and it should be noted that in the results below the 'moving roughness' approach is denoted as TGV-MR.

In the example, two different types of soil are considered. The first soil has properties similar to a measurement site at Horstwalde in Germany, which is a sandy soil that can be represented as a homogeneous half-space of moderately soft soil (8). The second is a layered half-space with a soft ground corresponding to a measurement site at Greby in Sweden (9). For both soil cases the same track is used, which is modelled as a continuously supported track with the properties reported in (8). The rail roughness profile (Figure 2) for all simulations was chosen according to FRA class 3 (10). The train parameters used for the simulations are chosen to correspond to those used in (9). These are based on a modified Bombardier Regina EMU ('Gröna Tåget'). Unlike (9) a four-car train is used with a total length of 106.4 m. The train speed is 150 km/h.

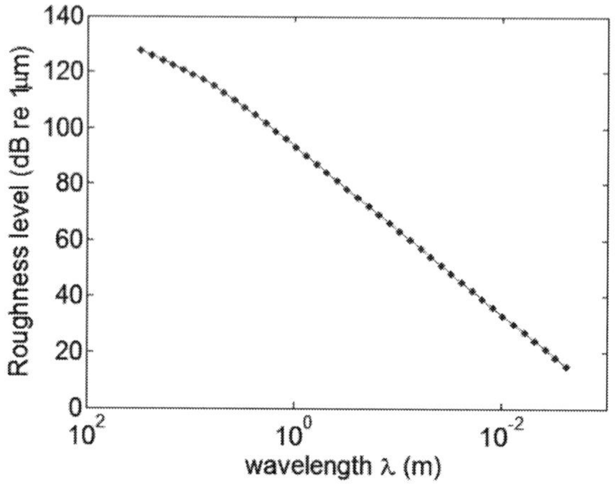


**Figure 2. Assumed roughness spectrum in one-third octave.**

The predicted dynamic response of the rail for both soil types is shown in Figure 3 in one-third octave. The results from the FE/BE model are also presented for the Horstwalde site. The motion of the train is included only in the TGV case. Figure 4 shows the one-third octave dynamic response in the free field for the Horstwalde site at distances of 0 m (under the track), 8 m, 32 m and 100 m from the track. For the distances 0, 8 and 32 m the velocity predicted using the FE/BE model is also shown. For all cases the TGV-MR and the FE/BE models show very good agreement. They show the same trends as the TGV model, although lacking some of the spectral detail which is caused by the motion of the dynamic loads past the receiver point. Note that this detail is only significant at the rail, it being essentially lost at distance within the free-field.

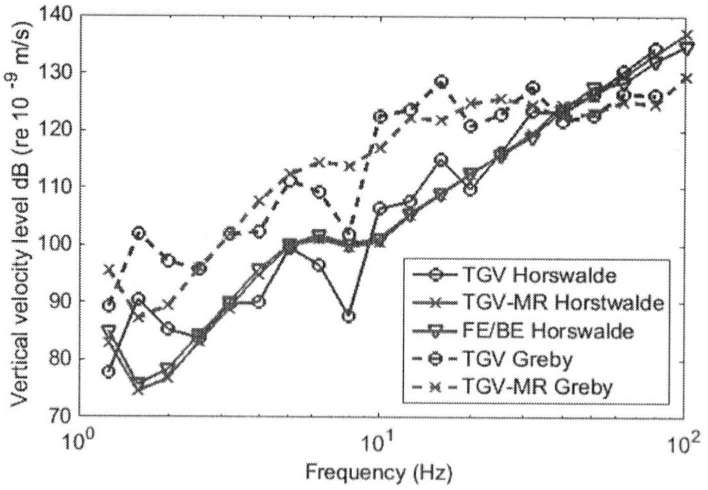


**Figure 3. One-third octave comparison for the dynamic velocity rail response level for both sites.**

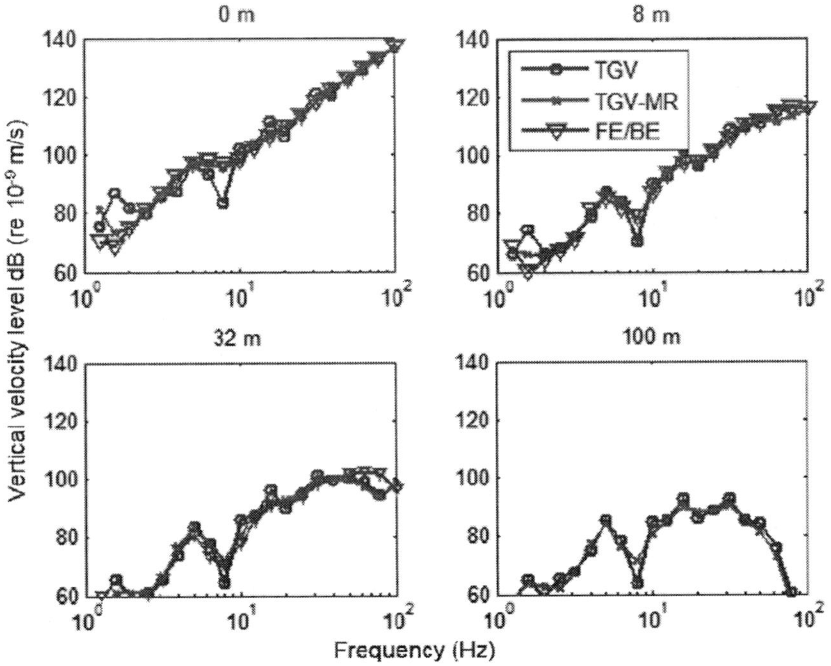


**Figure 4. One-third octave dynamic velocity response level in the Horstwalde free field at 0, 8, 32 and 100 m from the track.**

## 3. PIPE-IN-PIPE MODEL OF UNDERGROUND RAILWAYS

### 3.1 Model description

The PiP model is a semi-analytical model that predicts the vibration from a tunnel embedded in a multi-layered half-space (Figure 5). The tunnel and the surrounding soil are modelled as concentric thick cylindrical shells. A floating slab track is modelled as an Euler–Bernoulli beam and is coupled to the tunnel wall. The main assumption used in the model is that the near-field displacement of the tunnel is not influenced by the existence of a free surface or ground layers. The accuracy of this assumption increases with the increasing depth of tunnel. Similar to the TGV model, the tunnel wall and the surrounding soil are assumed to be invariant in the longitudinal direction. Hence, all calculations are performed in the frequency-wavenumber domain and results in the frequency-space domain are calculated using the inverse Fourier transform. It has been shown that the efficiency of computations of PiP is significantly higher than discretization models such as the coupled FE-BE model (11, 12, 13).

In PiP a train of infinite length is represented by an infinite number of wheelset masses, with a constant spacing, moving on a track as shown in Figure 6. Due to the low stiffness of primary suspensions of modern trains, and as the interest for trains in tunnels is usually limited to frequencies above 10 Hz, it is reasonable to ignore sprung masses in such a model. As shown in Figure 6, a model of double-beams supported on an elastic foundation is used to calculate forces at the wheel-rail interface. The source of excitation in the model is the track irregularity which causes relative displacements between the axles and the rail. The relative displacements are defined as uncorrelated random inputs and the outputs are calculated at points in the soil in the same cross-section as one of the wheelsets. These calculations are based on the assumption that vibration does not vary along a line parallel to the tunnel. The assumption is reasonable at distances from the tunnel that are large compared with the axle spacing (3).

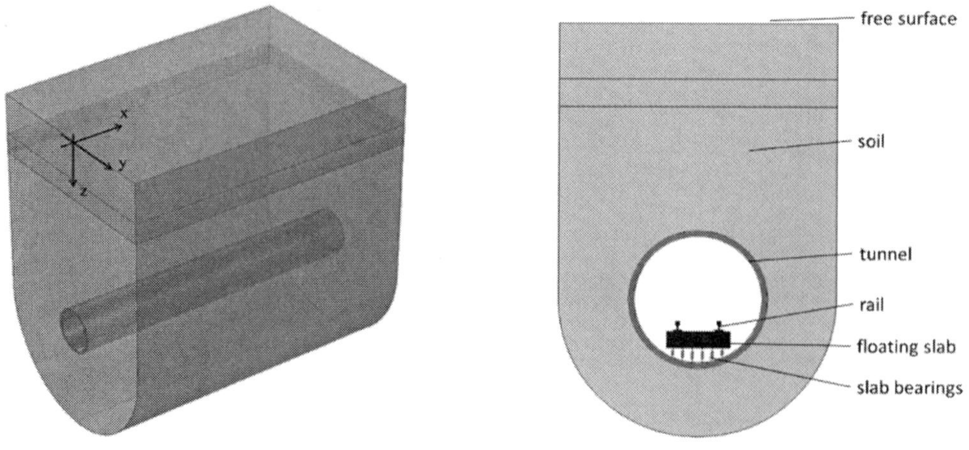


**Figure 5. Layout of the model; a tunnel embedded in a multi-layered half-space. The figure shows two soil layers on a half-space.**

To calculate displacements in the soil due to the forces applied at the rails, a model of a double-beam coupled to a tunnel embedded in a layered half-space is used. The model is shown in Figure 5b and it is used to calculate transfer functions between the rails and soil. The track is coupled to the tunnel-soil system in the wavenumber-frequency domain by using frequency response functions of the double-beam system and the tunnel-soil system. To calculate the transfer functions for the tunnel-soil system, it is assumed that displacements at the tunnel-soil interface due to a source inside the tunnel are the same whether or not there is a free surface. More specifically, first, the displacements at the tunnel-soil interface are calculated using a model of a tunnel embedded in a full space. Next, the internal source that produces the same displacements is calculated using a full-space model based on two concentric pipes. Finally, the vibration in the far field is calculated using these internal sources together with the Green's functions for an elastic layered half-space. The Green's functions calculations are performed using the elastodynamic toolbox developed at KU Leuven (14).

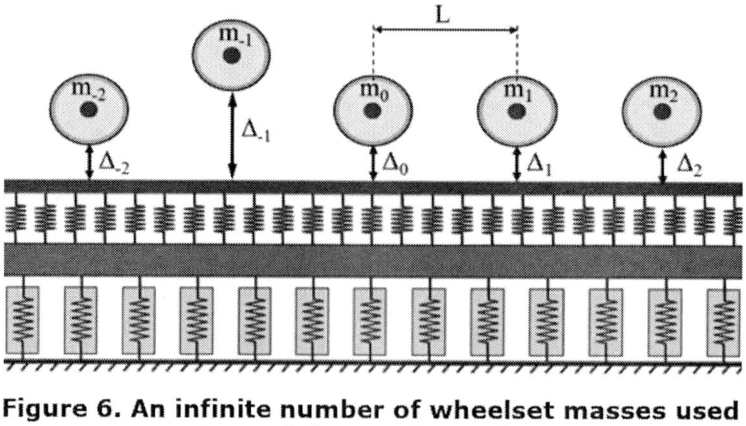


**Figure 6. An infinite number of wheelset masses used to model a moving train.**

### 3.2 Numerical example

In this section, the near-field and far-field response of a tunnel embedded in a layered half-space is calculated using the PiP model for a harmonic load applied at the invert of the tunnel, at ($x=y=0$ m, $z=16.75$ m). The response is calculated by the PiP model and then compared with that calculated by a coupled FE/BE model (12). The cross-section of the tunnel and the ground at $x=0$ is shown in Figure 7.

The tunnel has an external radius of $r_o$=3 m and thickness $t$=0.25 m. The tunnel is made of concrete with a density $\rho_t$=2500 kg/m$^3$, longitudinal wave velocity $c_p$=5189 m/s, shear wave velocity $c_s$=2774 m/s and hysteretic damping loss factor $\eta$=0.03. The soil comprises a 6 m surface layer on top of a half-space. The surface layer (type 1, Figure 7) has longitudinal wave velocity $C_{p1}$=1964 m/s and shear wave velocity $C_{s1}$=275 m/s. For the half-space (type 2, Figure 7), the longitudinal wave velocity is $C_{p2}$=1571 m/s and the shear wave velocity is $C_{s2}$=220 m/s. The density for both layers was set to $\rho$=1980 kg/m$^3$ and the hysteretic material damping ratio $\zeta_s$= $\zeta_p$=0.03 in shear and volumetric deformation. The response is calculated at 4 points: A ($x$=0, $y$=0, $z$=17); B ($x$=0, $y$=0, $z$=11); C ($x$=10, $y$=10, $z$=6); and D ($x$=10, $y$=10, $z$=0).

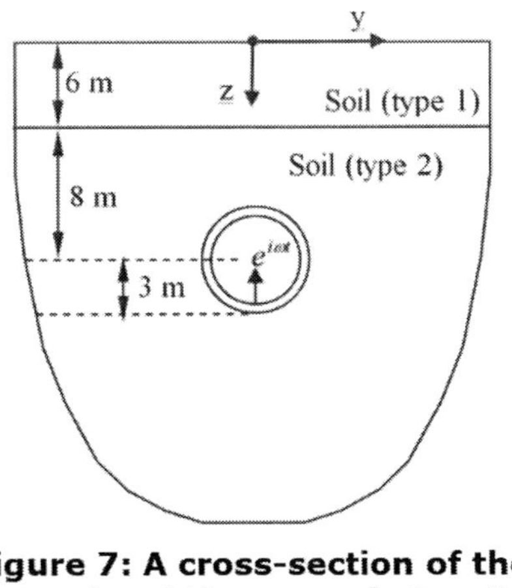


**Figure 7: A cross-section of the tunnel and the ground at x=0. A harmonic load is applied at the tunnel invert at ($x$,$y$=0 m, $z$=16.75 m).**

The results are presented for the frequency range of 1 to 80 Hz. Higher frequencies are not attempted due to the high computational cost of the FE/BE model as a result of the finer mesh required at high frequencies. The current frequency range is the predominant perceptible range for ground-borne vibration in buildings and it is sufficient for the purposes of validation.

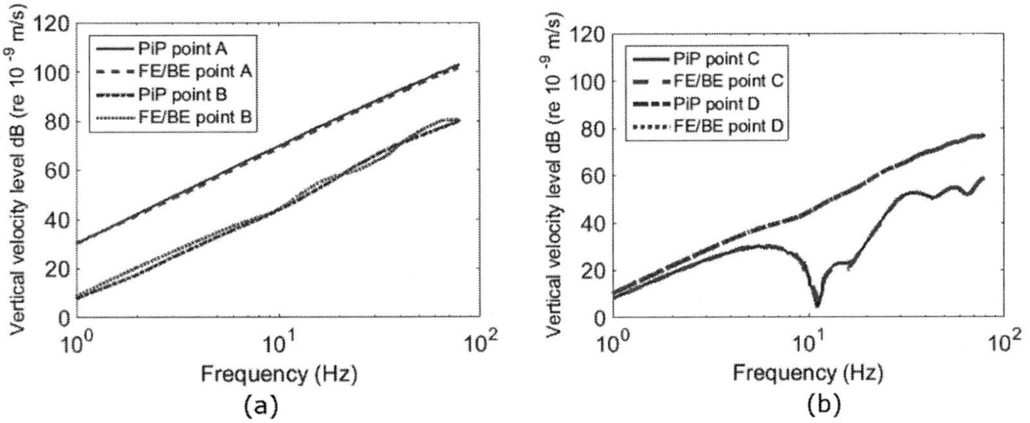


**Figure 8: The velocity response level at (a) near-field (point A and point B) and (b) at far-field (point C and point D).**

Figure 8a shows the vibration level at the tunnel-soil interface, at the tunnel invert (A) and the tunnel crown (B). A good agreement is observed between the PiP and FE/BE models, which confirms that the near-field vibration is not influenced by layers away from the tunnel. Figure 8b show the far-field displacements at point C, at the interface of the two types of soil, and point D at the free surface. These results confirm that the PiP model calculates the far-field vibration for a tunnel embedded in a layered ground with the same accuracy as significantly more computationally expensive FE/BE models. This example confirms that the near-field vibration is controlled by the dynamics of the tunnel and the soil layer surrounding the tunnel.

## 4. OBJECTIVES AND IMPACT OF MOTIV PROJECT

As well as the work improving the TGV and PiP models, MOTIV aims to address the non-linear behaviour of track elements and the dynamic tunnel-soil-pile interaction for single as well as twin tunnels. The main hypothesis of the proposed work is that the level of accuracy of predictions will be significantly improved by accounting for these mechanisms when modelling ground-borne vibration from railways. The hypothesis will be tested by a series of comparisons of the models against measurements. The models will be incorporated in end-user software that will be released at the end of the project.

The effect of pre-load dependence and non-linear behaviour of resilient elements of tracks is investigated using excitation models that account for non-linearity not only of railpads but also of all other elements within conventional tracks, i.e. ballast, slab and trough tracks. The modelling effort uses FE methods and the Dynamic Stiffness approach that accounts for both discretely and continuously supported tracks. The track behaviour and the reaction force at the trackbed are investigated for tracks due to the movement of trains with roughness excitation; the models are developed in the time domain to allow the non-linear track stiffness and damping values to be taken into account. The effect of non-linear track stiffness is quantified by comparing the results from this model with that resulting from the same model when ignoring the non-linear behaviour (15).

Regarding the interaction of the railway with the surrounding soil, a novel BE model is under development for a multi-layered ground with one and two cylindrical cavities in the horizontal direction to allow for the coupling of single and twin tunnels respectively, and multiple cavities in the vertical direction to allow for coupling of piles. The model uses periodic structure theory to decrease the computational cost as only two slices of the problem need to be discretized: i) a generic slice with one or two cylindrical cavities; and ii) a central slice containing cavities for the piles beside the tunnel cavities. The piles are modelled using a bar formulation for axial deformation and a beam formulation for bending. The tunnels are modelled using thin-shell theory. The slices are joined, before coupling the tunnels and piles, to generate a solution for a full problem with horizontal and vertical cylindrical cavity(ies). Coupling is achieved with tunnels with infinite lengths by using the transfer functions for the two systems, following the method developed by Hussein and Hunt (16) in modelling floating slab tracks with discontinuous slabs in underground railway tunnels. The model is used to understand and validate the effect of weak coupling between tunnels and between tunnels and pile foundations. Moreover, the tunnel model is compared with the PiP model while the piled-foundation model is validated against previous FE/BE models (17).

## 5. CONCLUSIONS

The current paper reports recent developments in techniques for modelling vibration from surface and underground railways within the MOTIV project. The results presented illustrate the modelling capabilities and the efficiency of computations for the proposed numerical models. The models presented show good accuracy with reduced computational requirements compared with coupled FE/BE models. Further advancements within the work of the MOTIV project aim to address the non-linear behaviour of the track elements and the dynamic tunnel-soil-pile interaction for single and twin tunnels.

## ACKNOWLEDGEMENTS

The work described has been funded by EPSRC through grants EP/K006665/1, EP/K005847/1 and EP/K006002/1, Modelling of Train Induced Vibration. The project is a collaboration between the Universities of Southampton and Cambridge. The stakeholders group includes KU Leuven, London Underground Limited, Centro (West Midlands Passenger Transport Executive), Network Rail, Pandrol, Arup, and URS.

## REFERENCES

1. Lombaert G., Degrande G., Francois S., Thompson D.J. (2013) Ground-borne vibration due to railway traffic. *Proceedings of the 11th International Workshop on Railway Noise, IWRN11*, J. Nielsen et al. (Eds), Uddevalla, Sweden, 266-301, September 2013.
2. Sheng X., Jones C.J.C., Thompson D.J. (2004) A theoretical model for ground vibration from trains generated by vertical track irregularities. *Journal of Sound and Vibration*, **272**(3-5), 937–965.
3. Forrest J.A. and Hunt H.E.M. (2006) A three-dimensional tunnel model for calculation of train-induced ground vibration. *Journal of Sound and Vibration,* **294**(4–5), 678–705.
4. Hussein M.F.M. and Hunt H.E.M. (2007) A numerical model for calculating vibration from a railway tunnel embedded in a full-space. *Journal of Sound and Vibration*, **297**(1–2), 37–54.
5. Sheng X., Jones C.J.C., Petyt M. (1999) Ground vibration generated by a load moving along a railway track. *Journal of Sound and Vibration*, **228**(1), 129–156.
6. Lombaert G., Degrande G., Clouteau D. (2004) The non-stationary freefield response for a moving load with a random amplitude. *Journal of Sound and Vibration*, **278**, 611–635.
7. Sheng X., Jones C.J.C., Thompson D.J. (2006) Prediction of ground vibration from trains using the wavenumber finite and boundary element methods. *Journal of Sound and Vibration*, **293**, 575 – 586.
8. Toward M.G.R, Jiang J., Dijckmans A., Coulier P., Thompson D.J., Degrande G., Lombaert G., Hussein M.F.M. (2014) Mitigation of railway induced vibrations by using subgrade stiffening and wave impeding blocks. *Proceedings of the 9th International Conference on Structural Dynamics (EURODYN 2014)*, A. Cunha, E. Caetano, P. Ribeiro, G. Müller (eds.), 767-774, Porto, Portugal, 30 June - 2 July 2014.
9. Mirza A.A., Frid A., Nielsen J.C.O., Jones C.J.C. (2010) Ground vibrations induced by railway traffic - the influence of vehicle parameters. *Proceedings of the 10th International Workshop on Railway Noise, IWRN10*, Nagahama, Japan, October 2010.
10. Garg V. and Dukkipati R. (1984) Dynamics of railway vehicle systems. Academic Press, Canada.
11. Hussein M.F.M, Hunt H.E.M., Rikse L., Gupta S., Degrande G., Talbot J.P, Francois S., and Schevenels M. (2007) Using the PiP Model for Fast Calculation of Vibration from a Railway Tunnel in a Multi-layered Half-Space. *Proceedings of the 9th International Workshop on Railway Noise, IWRN9*, München, Germany, 136-142, September 2007.
12. Gupta S., Hussein M.F.M., Degrande G., Hunt H.E.M., Clouteau D. (2007) A comparison of two numerical models for the prediction of vibrations from underground railway traffic. *Soil Dynamics and Earthquake Engineering*, **27**, 608–624.
13. Hussein M.F.M., Francois J.P., Schevenels S., Hunt H.E.M., Talbot J.P., Degrande G. (2014) The fictitious force method for efficient calculation of

vibration from a tunnel embedded in a multi-layered half-space. *Journal of Sound and Vibration*, **333**(25), 6996–7018.
14. Schevenels M., Francois S., Degrande G. (2009) EDT: An ElastoDynamics Toolbox for MATLAB. *Computers & Geosciences*, **35**(8), 1752–1754.
15. Koroma S.G., Hussein M.F.M., Thompson D.J., Ntotsios E. (2014) The effect of non-linearity of railway tracks on ground-borne vibration. *Proceedings of the 21st International Congress on Sound and Vibration, ICSV21*, Beijing, China, July 2014.
16. Hussein M.F.M. and Hunt H.E.M. (2009) A numerical model for calculating vibration due to a harmonic moving load on a floating-slab track with discontinuous slabs in an underground railway tunnel. *Journal of Sound and Vibration*, **321**(1–2), 363–374.
17. Hamad W.I., Hunt H.E.M., Hussein M.F.M., Talbot J.P. (2014) Tunnel-soil-pile interaction in the prediction of vibration from underground railways: validation of the sub-models. *Proceedings of the 9th International Conference on Structural Dynamics (EURODYN 2014)*, A. Cunha, E. Caetano, P. Ribeiro, G. Müller (eds.), 767-774, Porto, Portugal, 30 June - 2 July 2014.

# Determining system-wide energy use in an established metro network

**J P Powell, A González-Gil, R Palacín, P Batty**
Newcastle University, UK

## ABSTRACT

Energy optimisation is becoming increasingly important in all transport systems. However, when evaluating alternative methods for energy optimisation, it is vital to obtain a thorough understanding of the energy flows within the system. While new systems generally have comprehensive automatic energy metering and monitoring of their components, there is much legacy infrastructure without this capability. This paper develops a methodology to determine a breakdown of energy use for the Tyne and Wear Metro, the results of which are compared to other railway systems. This is complemented by the summary of an investigation into the energy consumption of stabled rolling stock.

## NOMENCLATURE

| | | |
|---|---|---|
| AWS | - | Automatic Warning System |
| DB | - | Deutsche Bahn |
| DBTW | - | DB Regio Tyne & Wear Ltd., the Tyne and Wear Metro operator |
| ICE | - | Intercity-Express |
| Nexus | - | Tyne and Wear Metro infrastructure owner/maintainer (a public body) |
| RSSB | - | Rail Safety and Standards Board |
| TPWS | - | Train Protection and Warning System |

## 1. INTRODUCTION

RSSB research (1) has identified energy as one of five major 'game changer' issues facing the railway in Great Britain in the next few decades, with prices rising significantly and potential availability risks, due to small generating capacity margins and dwindling oil supplies. The conclusions from the report are that energy is likely to form a significantly greater proportion of the costs in the railway industry, rising from 4-5% to around 19% of the total in the next few decades. The strategy to mitigate this is threefold: increasing self-generation, protecting against the peak/off-peak energy price/availability disparity and implementing energy efficiency measures. However, to effectively target energy efficiency measures requires a good understanding of where energy is consumed within the railway system.

The energy use of both vehicles and infrastructure should be considered over their entire life cycles, encompassing the energy consumed directly by the construction, operation, maintenance and disposal phases, and indirectly embodied in the materials and processes used. Several methods are available, depending on the objectives of the work in question. The relative merits of different approaches were considered in a paper by van Wee *et al.* (2).

Life cycle assessments that compare different transport modes for both passengers and freight have been the subject of previous detailed studies, for example in Europe (3) and the USA (4, 5). The broad trends are all reasonably similar; rail outperforms both road and air for energy efficiency (the energy consumed per passenger-km or freight.tonne-km), and has a similar performance to water transport. The investigation of individual railway systems is required to draw more specific conclusions, given their design, characteristics and operating environment can be significantly different.

The majority of the railway system life cycle energy assessments in recent years have focussed on high speed rail, in response to the need for policy guidance. A study of the ecology profile of the Deutsche Bahn-operated ICE services was carried out by von Rozycki *et al.* (6). The majority of the life cycle energy use was accounted for by traction energy and passenger trips to or from the station, and to a lesser extent the embodied energy from infrastructure. The most important factors that influence energy use are therefore the number of trains running, their duty cycles and load factors, and a smaller (but still notable) contribution from the number and size of major civil works such as bridges and tunnels. Such findings have been echoed in other high-speed rail infrastructure studies in Sweden (7) and Great Britain (8). There are plenty of examples of built environment literature now available that detail the methodologies available for calculating the life cycle energy demand of different elements of the railway system (8-15).

It can be concluded from the above literature that approximately two-thirds of the energy consumed by railway systems is accounted for by direct operational energy use, itself dominated by traction energy. Differences in the exact breakdown of energy use vary according to the individual characteristics of the railway system considered, the extent of the analysis and the inherent uncertainties in the data used. Nevertheless, the broad trends are similar and as such, this paper focusses on understanding how direct operational energy is used in rail systems. The Tyne and Wear Metro in the north-east of England is used as a case study, with comparisons made to other rail systems. The energy consumption of stabled rolling stock – a hitherto relatively uninvestigated area in literature – is also considered to aid attempts to optimise energy consumption.

## 2. ESTABLISHING THE DIRECT OPERATIONAL ENERGY CONSUMPTION OF LEGACY INFRASTRUCTURE

### 2.1. The Tyne and Wear Metro

Energy metering for different subsystems within the railway provides the data required for analysis of energy use, and the general introduction of computer control and monitoring systems means that this practice has become standard for new build infrastructure. However, this level of information is not necessarily available for legacy infrastructure, and further investigations are required to derive sufficient data to effectively target measures to improve energy efficiency. The Tyne and Wear Metro, a light rail system centred on Newcastle upon Tyne in the north-east of England, is used as a case study for this paper.

The first 54 km of the network were opened progressively from 1980, using a combination of existing heavy-rail alignments converted to Metro use, new tunnels through the centre of Newcastle and Gateshead, and a new bridge across the river Tyne. The network was subsequently extended to Newcastle International Airport in 1991 and to South Hylton in 2002, the latter extension sharing tracks with heavy rail services between Pelaw and Sunderland (16, 17).

The original fleet of 90 twin-section Metrocars remain in service today, and operate services in pairs. They were refurbished between 1995 and 2000, and are currently undergoing life extension work to run until the mid-2020s. Traction is provided by two 185 kW series-wound DC motors per Metrocar, with air/oil camshaft resistance control, and a combination of friction and rheostatic braking. Waste heat is recovered from the resistors for saloon heating, supplemented by an auxiliary heater when required; no air conditioning is fitted.

The trains are manually driven, with a mix of 2, 3 and 4-aspect lineside colour light signalling controlled by track circuits. The Indusi train protection system is fitted across the network, as well as AWS and TPWS on the section between Pelaw and Sunderland.

The system is electrified using overhead wires at 1500 V DC, fed from 10 substations connected to the National Grid. There is an internal 11 kV network that covers the central underground section in Newcastle; this feeds two further traction substations, as well as stations and signalling supplies. The remaining stations across the system are connected directly to National Grid local distribution networks, except those adjacent to a traction substation, which are connected to the substation. The original electrical network was described in a paper by Prickett (18); subsequently a second transformer/rectifier was added at Kenton Bank for the airport extension, and for the Sunderland extension new substations were built at Fellgate, Seaburn and Pallion. As well as the electrical supplies, a small amount of gas is used in Gosforth depot and in the control centre at South Gosforth station.

## 2.2. Methodology

There are some energy meters in place on the Tyne and Wear Metro system, but not all of the subsystems are metered individually. Therefore, the overarching methodology in this paper is to use the existing sub-metered data, together with specifications and estimated duty cycles for subsystems with no specific metered data; the totals (and inherent uncertainties) can then be compared with the high level metered data used for billing purposes. The period from 1$^{st}$ April 2012 to 31$^{st}$ March 2013 was analysed in detail, with the analysis broken down and classified into subsystems that allow a comparison with other railway systems, discussed in later sections. The total energy consumption predicted from the estimates for each subsystem matched the actual overall metered value of 69 GWh closely, comfortably inside the margin of uncertainty calculated for all of subsystems.

### *2.2.1 Traction*

Seven Metrocars were fitted with different designs of energy meter during 2012. The energy consumed by Metrocar number 4067 for the entire year was used as a basis for estimating the fleet's consumption, by comparing the distance travelled (as recorded by the energy meter) against the corresponding distance for both the unit and the overall fleet from the maintenance records.

This gave a total energy use of 53 GWh for the energy consumed by the Metrocars for traction, with an uncertainty of approximately 5% from the energy meter measurement accuracy for energy and distance. The potential variation in mileage between different Metrocars in the fleet, as well as the transmission losses in the overhead wires, increase this uncertainty to around 10%.

### *2.2.2 Depot*

The 11 kV/415 V transformer that supplies the depot was replaced in 2011 and an energy meter fitted. Readings were taken at four week intervals from July 2012. To estimate the missing data, the average daily consumption for each four-week period was first calculated. This was used to obtain the minimum, average and maximum four-weekly energy consumption figures for July 2012 to March 2013, which were in turn used to derive minimum, central and maximum estimates for April 2012 – July 2012. This resulted in an estimated energy consumption of approximately 1.8 GWh ± 4% for the full year.

#### *2.2.3 Stations*

The stations fed from the National Grid directly are individually metered, and the total energy consumption of 4.4 GWh is therefore known. A survey of all of the stations in the network was carried out over several months in 2012, using portable meters to measure the load on the main intake position at each station for a seven day period. The metered data for the stations fed by the National Grid was then used to provide two further pieces of information: the accuracy of the survey (by comparison with the relevant four-week period of data), and also the variation in energy consumption throughout the year in response to seasonal variations in temperature and hours of daylight.

This allowed an estimate of 8.1 GWh to be made for the total annual energy consumption of the stations fed directly from traction substations. The cumulative uncertainties resulting from a number of steps throughout the process are rather large, and the potential range of this value is between 5.5 and 11.0 GWh.

#### *2.2.4 Infrastructure*

The principal infrastructure components on the Tyne and Wear Metro are signalling and electrical switchgear/auxiliary equipment in the traction substations. The signalling energy consumption was determined from typical equipment specifications and the number of signals on the network. The starting point was a power consumption of 30 W for signal lamps. Track circuit power supplies vary between a few Watts to a couple of tens of Watts, depending on design and block length. The Indusi balises are passive, energised by the aerials on the Metrocars, and therefore draw no power from the lineside. On the shared section to Sunderland, AWS magnets consume around 10 W in operation, and the TPWS system typically tens of Watts. Therefore, a nominal estimate of 100 W per signal was made, with a possible range of 50 to 200 W. Point motors draw more power than signalling equipment, at approximately 150 W, but the duty is very intermittent. Level crossing equipment consumes a similar amount of power in operation but the duty is also intermittent, and as a result the power consumed by point motors and level crossings may be considered to be negligible by comparison with the range of uncertainty in the signalling estimates.

The energy consumed by the equipment in the traction substations was estimated from information provided by Network Rail, taking an average for substation energy consumption and multiplying this by the number of substations on the Metro network. This gives a total estimate of 1.2 GWh for infrastructure, although the approximations result in a large range of uncertainty of 0.5 to 1.7 GWh. It should nevertheless be noted that this range is small by comparison with the energy consumed in the whole system.

#### *2.2.5 Offices*

The offices category includes the Gosforth control centre and the administrative offices in Newcastle city centre. The metered energy consumption at South Gosforth station was approximately ten times that of other stations of a similar design, therefore the difference was taken to be the power consumed by the control centre. The energy consumption of the city centre offices was estimated from the floor area and benchmarks for a typical UK office building of similar design. This gave an estimate of 0.9 GWh, with a range of uncertainty of 0.8 to 1.1 GWh.

### 2.3 Calculating energy breakdown for other rail systems

Having estimated the breakdown of energy use between different subsystems in the Tyne and Wear Metro, it was considered beneficial to compare these characteristics with other rail systems in Britain. Hence, data were derived from analysis of contemporary carbon footprint reports: a Transport for London summary of London Underground (19) and an RSSB research report into the carbon footprint of the national rail network in Great Britain (20).

Although carbon emissions do not map directly to energy consumption, the direct operational energy use identified as contributing to the carbon footprint can be extracted from the reports. An approximate overall breakdown for each system was therefore inferred, and is shown in Figure 1 in Section 3.

In total for 2008/2009, the mainline railway used 4.4 TWh of electricity (of which 2.9 TWh was for traction), 692 million litres of diesel fuel and 417 million kWh of gas. By contrast, London Underground consumed approximately 1 TWh of electricity, and a relatively small amount of gas. To obtain the overall energy consumption on a like-for-like basis between the different sources, the Greenhouse Gas Conversion Factors for Company Reporting (21) provided by the UK Government were used.

## 3. RESULTS AND ANALYSIS

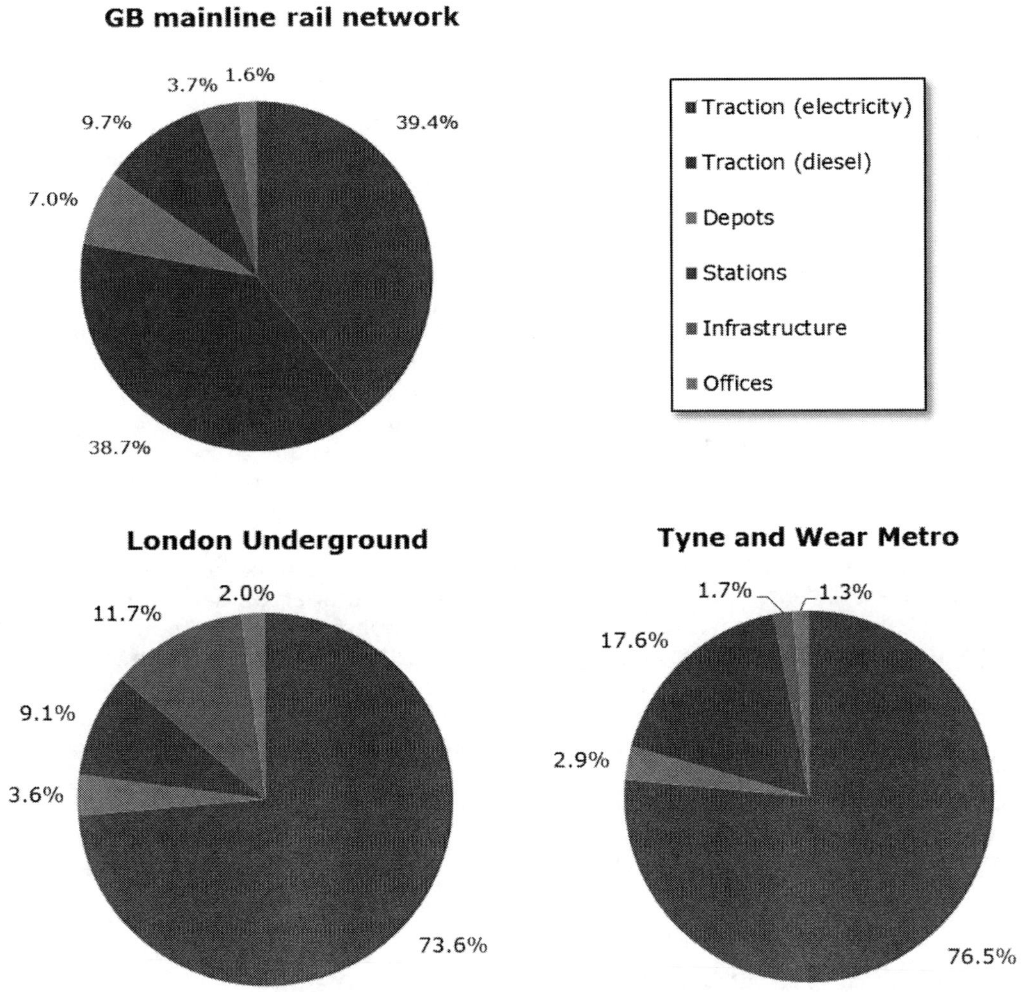


**Figure 1: Direct operational energy use in three GB railway systems**

The traction energy consumptions for each of the three systems shown in Figure 1 are all very similar, with 78.1%, 73.6% and 76.5% for the Great Britain mainline network, the London Underground network and the Tyne and Wear Metro respectively. This is remarkably consistent, given the differences between the systems. As such, data from JR East in Japan (22) is also included here as a point of comparison: traction energy accounts for 70.5% of the total energy used for direct operation of the railway. It should be noted that JR East has achieved a certain level of notoriety through its strong business diversification strategy that has developed a strong foothold in the real-estate field (in terms of housing, retail and entertainment), both at stations, and also alongside tracks (23). When the energy consumed by these extra station developments is taken into account, traction energy only accounts for 61.1% of the total energy consumption. By contrast, around 93% of the energy consumed by Indian Railways is accounted for by traction energy (24).

London Underground runs a very high density service over its 402 km network, with many stop/start cycles, approximately 45% of which is in tunnels with a high resistance to motion. Infrastructure also consumes more energy on the London Underground than other networks, on account of the necessity for tunnel ventilation fans and groundwater pumps for the extensive tunnel network. The GB mainline network has a lower train service density than London Underground, with proportionally very few underground stations or tunnels; much of the infrastructure is also older than the Tyne and Wear Metro.

The energy consumed by London Underground stations varies significantly over the network, but in total comprises 9.1% of the total consumption. Within the London Underground system, there are 408 escalators and 112 lifts, with an average of 1.48 escalators and 0.41 lifts per station. By contrast, the Tyne and Wear Metro has 0.58 escalators and 0.58 lifts per station, yet 17.6% of the total energy consumption is taken up by stations. The main contributors to this figure are the 11 underground stations within the network, in which all 35 escalators and 24 of the 35 lifts in the network operate. As such, the implementation of energy efficiency measures within these underground stations is an area worth future consideration, and is an ongoing area of work for the Tyne & Wear Metro (slowing escalators for example). It is also likely that the Tyne and Wear Metro stations consume proportionally more of the total energy consumption than the London Underground due to the less intensive train service; note also the mainline network has a lower train service and station density than the London Underground to offset the reduced consumption of above-ground stations.

The energy consumption of depots is highest on the mainline, on account of the relatively less intense train service and a greater number of heavy maintenance sites and remote sidings. Problematically, it is not fully understood how many systems categorise the energy supplied to stabled trains within depots, since this can be considered as either traction energy or depot energy. Given the level of variance between the energy consumptions of depots in this study, this merits further investigation into the effects of the energy consumed by stabled trains on the overall energy consumption. As such, to further understand the energy flows within such systems, Section 4 summarises the findings of a separate study by the authors which analyses the stabled energy consumption of the Metrocars in the Tyne and Wear Metro system; an area which is commonly overlooked in scientific literature.

## 4. ENERGY CONSUMPTION DURING STABLING HOURS

Due to the fleet size of the Tyne and Wear Metro, many vehicles are stabled outdoors at Gosforth depot, with others stabled in the inspection shop (heated) and running shed (effectively unheated). In principle, all of the Metrocar's auxiliary systems remain in operation when stabled, unless isolated from the overhead power supply for certain maintenance tasks. This means that the Metrocar heating maintains the target interior temperature to the same level as it is in service throughout stabling. The investigation involved using the energy meter data collected from on board the aforementioned Metrocar number 4067, and analysing the overhead line voltage, current and power drawn from the overhead line, and the output voltage from the motor-alternator set that powers some of the auxiliaries. These data were then compared with the historical climate data to understand the relationship between the ambient temperature and the vehicle's energy consumption while stabled, and the factors that affect it.

It was found that, of the annual energy consumption of a Tyne and Wear Metrocar, approximately 11% can be attributed to on-board auxiliary systems while the vehicle is stabled. The greatest contributor to this amount was found to be the heating; shown to be responsible for approximately 45% of this consumption, with the lighting and compressed air systems contributing 10% and 4% respectively to the overall figure. A combination of other energy consumers, such as fans and control circuits, represent the remaining 41% of the total stabled energy consumption. Further details of the study can be found in Powell *et al.* (25).

In the context of this paper, the energy consumed by stabled Metrocars therefore accounts for around 8% of the total system direct energy consumption. This is more than all of the above-ground stations together, or alternatively, more than the combined consumption of the depot, infrastructure and offices.

## 5. CONCLUSIONS

Understanding the breakdown of energy consumption within rail systems is key when attempting to focus investment on energy efficiency improvements to a system. This investigation developed a solution to enable rail systems without automated energy metering capabilities to achieve a greater understanding of this breakdown, through the use of a combination of measurements and modelling to provide an indication of the energy usage by the various sub-systems. This methodology has shown that underground stations are particularly important in the case of the Tyne and Wear Metro, and has highlighted the energy consumption of stabled vehicles as an area worthy of additional research.

The comparison between different railway systems has also provided insight into how the relative energy consumption of different subsystems differs, depending on the characteristics of a particular system under investigation. The methods developed can be applied to other systems to help ensure rail systems retain their environmental superiority over other mass transit modes.

## ACKNOWELDGEMENTS

The authors would like to thank DBTW for their support during this investigation.

## REFERENCE LIST

1. Rail Safety and Standards Board, *Energy Game Changer - Macro energy risks affecting the railway in Great Britain*. 2012.
2. van Wee, B., Janse, P. and van den Brink, R., *Comparing energy use and environmental performance of land transport modes.* Transport Reviews, 2005. 25(1): 3-24.
3. Bos, A.J.M., *Direction indirect - The indirect energy requirements and emissions from freight transport*. 1998, Rijksuniversiteit Groningen: Groningen.
4. Facanha, C., *Life-cycle air emissions inventory of freight transportation in the United States*. 2006, University of California at Berkeley: Berkeley.

5. Chester, M.V., *Life-cycle Environmental Inventory of Passenger Transportation in the United States*. 2008, University of California at Berkeley: Berkeley.
6. von Rozycki, C., Koeser, H. and Schwarz, H., *Ecology profile of the German high-speed rail passenger transport system ICE*. The International Journal of Life Cycle Assessment, 2003. 8(2): 83-91.
7. Åkerman, J., *The role of high-speed rail in mitigating climate change – The Swedish case Europabanan from a life cycle perspective*. Transportation Research Part D: Transport and Environment, 2011. 16(3): 208-217.
8. Chau, C., Soga, K., O'Riordan, N. and Nicholson, D., *Embodied energy evaluation for sections of the UK Channel Tunnel rail link*. Proceedings of the ICE-Geotechnical Engineering, 2012. 165(2): 65-81.
9. Miyauchi, T., Nagatomo, T., Tsujimura, T., and Tsuchiya, H., *Fundamental Investigations of LCA of Shinkansen Vehicles*. Quarterly Report of RTRI, 1999. 40(4): 204-209.
10. Struckl, W., Stribersky, A. and Gunselmann, W., *Life cycle analysis of the energy consumption of a rail vehicle*. Workshop of the Allianz pro Schiene, 2006, Berlin.
11. Hammond, G.P. and Jones, C.I., *Embodied energy and carbon in construction materials*. Proceedings of the ICE - Energy, 2008. 161(2): 87-98.
12. Kiani, M., Parry, T. and Ceney, H., *Environmental life-cycle assessment of railway track beds*. Proceedings of the ICE-Engineering Sustainability, 2008. 161(2): 135-142.
13. Jorge, R.S., Hawkins, T.R. and Hertwich, E.G., *Life cycle assessment of electricity transmission and distribution—part 1: power lines and cables*. The International Journal of Life Cycle Assessment, 2012. 17(1): 9-15.
14. Jorge, R.S., Hawkins, T.R. and Hertwich, E.G., *Life cycle assessment of electricity transmission and distribution—part 2: transformers and substation equipment*. The International Journal of Life Cycle Assessment, 2012. 17(2): 184-191.
15. Wang, X.C. and Sanders, L., *Energy consumption and carbon footprint of high-speed rail projects: Using CAHSR and FHSR as examples*. Proceedings of the Institution of Mechanical Engineers, Part F: Journal of Rail and Rapid Transit, 2012. 226(1): 26-35.
16. Howard, D.F., *Tyne and Wear Metro — A modern rapid transit system*. Proceedings of the Institution of Mechanical Engineers, 1976. 190: 121-136.
17. Mackay, K.R., *Sunderland Metro - Challenge And Opportunity*. Proceedings of the ICE - Municipal Engineer, 1999. 133(2): 53-63.
18. Prickett, B.R., *Electrification of the Tyne and Wear Metro*. Electric Power Applications, IEE Proceedings B, 1981. 128(2): 81-91.
19. Transport for London, *LU Carbon Footprint Report 2008*. 2009.
20. Rail Safety and Standards Board, *Whole life carbon footprint of the rail industry*. 2010.
21. Department for Environment, Food and Rural Affairs, *Guidelines to Defra's Greenhouse Gas Conversion Factors for Company Reporting*. 2008.
22. Shinbo, M., *Railway System Optimization Seen from an Energy Management Perspective*. JR East Technical Review 2012. 23: 5-8.
23. van de Velde, D.M., *Learning from the Japanese railways: Experience in the Netherlands*. Policy and Society, 2013. 32(2): 143-161.
24. Kumar, S. *The Voice of Members – Indian Railways*. UIC Energy Efficiency Days, 2014, Antwerp.
25. Powell, J.P., González-Gil, A., and Palacin, R., *Experimental assessment of the energy consumption of urban rail vehicles during stabling hours: Influence of ambient temperature*. Applied Thermal Engineering, 2014. 66(1–2): 541-547.

# Viable flywheel system for rail

**J C Wheals, W To, J Dalby, M Vigar, J Hodgson, J Buchanan**
Ricardo UK Ltd, Shoreham Technical Centre, UK

**A Robertson, J Macpherson, J Taylor**
Artemis Intelligent Power Ltd, UK

**W Lanoe, M Heaton**
Bombardier Transportation, UK

## ABSTRACT

The paper reports live development activity between Bombardier, Artemis Intelligent Power and Ricardo at the stage of rig measurement of system efficiency and functional validation ahead of final product design. The specific configuration is focused upon storing retardation energy in a flywheel system to avoid high losses in the transmission during pull-away in certain classes of Diesel Multiple Unit (DMU). Arising from development of an existing 1MJ flywheel, two 4.5MJ flywheels are required to capture half of the kinetic energy of the DMU when decelerating from 100kph. Pull-away to 40kph is achieved with the engine idling so reducing local particulate emissions and noise. System hardware is described in conjunction with simulation and test rig results. A payback model is presented quantifying the commercial advantages and challenges that sustain the business case: 10.2% annual fuel saving valued at £13k pa/DMU with 4.6 year payback.

## 1. INTRODUCTION

### 1.1 Objectives

The project will measure the three elements of wheel-to-wheel efficiency for a rail hybrid, namely: the efficiency with which a hydraulic system can transfer energy to a flywheel on approach to a station, the ability of the flywheel to maintain its state-of-charge whilst the train dwells in a station, and the efficiency of converting stored energy to vehicle motion during pull-away from the station. This data will validate the assumptions within a simulation of fuel usage on a selected route. The fuel savings are the basis of a payback model, as seen from an operator's perspective that also includes amortisation of engineering effort and components costs across a likely DMU fleet.

It is anticipated that a positive outcome will result in the development of a rail demonstrator for use in traffic. The benefits of such a system include aspects readily accepted on balance sheets, but also aspects that are currently less tangible, but none-the-less worthwhile, that might be emphasised by regulators. Potential rail operator savings are identified in Table 1, the benefit of those in bold type are quantified in this project; broadly, many aspects may also be relevant to Electric Multiple Unit (EMU) application.

Further attributes of the system that impact the passengers' experience also support adoption of the technology, namely:

- Cleaner air in stations will have a benefit for passengers, but the use of zero emission shunting in maintenance yards reduces risk to rail workers in ill-ventilated facilities who suffer disproportionate adverse effects. [1,US] indicates that rail workers are 2.5 times as likely to suffer lung disorders as unexposed workers. Increasing awareness of the injurious effect of airborne Particulate Matter ($PM_{10}/PM_{2.5}$) from road vehicles would argue that greater scrutiny is inevitable in all sectors [2,UK]. Table 2, derived from [2], provides the attributable fraction of deaths and life-years lost, with Westminster having the highest concentration of $PM_{2.5}$ in the UK with a mean of 14.9 $\mu g\ m^{-3}$. In Waverley Rail Station [3, UK] the following pollutant levels were recorded:
    - NOx averaged 205 to 304 micrograms per cubic metre vs annual average air quality standard of 40 required by European law. (associated with respiratory illnesses and heart attack)
    - $PM_{10}$ was twice as high as the air quality standard, and up to 10 times higher than in nearby streets. (associated with cancer, stroke, heart attack)
    - Toxic diesel pollutants known as polycyclic aromatic hydrocarbons, or PAHs, were 4 times higher than the relevant air quality standard. (associated with cancers)
- Reduced noise during launch from stations. Depending upon the requirement to run ancillary loads whilst static, the diesel engines may be shut down and only restarted once the flywheels have accelerated the train beyond the confines of the station.
- Increased performance may offer modest reduction in journey times, but enable more trains to be run improving the prospect of seated travel.

**Table 1: Rail operator savings**

| Operating Modes | Benefits | Comments |
|---|---|---|
| ***Launch from station*** | ***Less fuel used***<br>Less local emissions<br><br>Increased performance | Disproportionately large benefit due to poor efficiency of certain transmission types during launch<br><br>Approach EMU performance, to increase traffic density |
| ***Brake on approach to station*** | Less brake wear | Small reduction in maintenance costs |
| Retardation (store) down a slope | Less fuel used on flat/ascent | Significant storage capacity required. n.b. 1m elevation = 0.44MJ per DMU/EMU |
| Autonomous power | Bridge short catenary gap | For electric trains, reduced restrictions during maintenance work |

**Table 2: UK Mortality rates due to $PM_{2.5}$ particulate**

UK baseline population, modelled population-weighted mean concentrations (μg $m^{-3}$) and estimated effects on annual mortality in 2010 of anthropogenic $PM_{2.5}$ air pollution

| Area | Population age 25+ ($x10^3$) | Deaths age 25+ | Mean anthropogenic $PM_{2.5}$ (μg $m^{-3}$) | Attributable fraction(%) | Attributable deaths age 25+ | Associated life-years lost |
|---|---|---|---|---|---|---|
| ENGLAND | 35878 | 458743 | 9.9 | 5.6 | 25002 | 264749 |
| SCOTLAND | 3660 | 53800 | 6.8 | 3.9 | 2094 | 22474 |
| WALES | 2075 | 31041 | 7.5 | 4.3 | 1320 | 13549 |

## 1.2 Energy Losses

In 2010, Artemis conducted a feasibility study with ScotRail. A Class 158 was fitted with sensors and data-logging equipment during operation on two Edinburgh routes.

- Edinburgh-Dunblane: 134 km, 21 stops, 2.5 hours, avg. 5.7 km between stations
- Fife Circle: 97 km, 22 stops, 1.8 hours, avg. 4.6 km between stations

The data was analysed using a simulation model of the existing engine, transmission and accessories. Of the total energy at the engine output:

- >50% is lost as heat in the brakes
- 15-20% is lost as heat in the transmission
- 5% is consumed by losses in hydraulic fan & alternator drives

The individual energy sinks are shown in Figure 1 from which it is evident that 50% of fuel is wasted in braking. This project assesses the commercial viability of addressing this inefficiency using a flywheel-type battery.

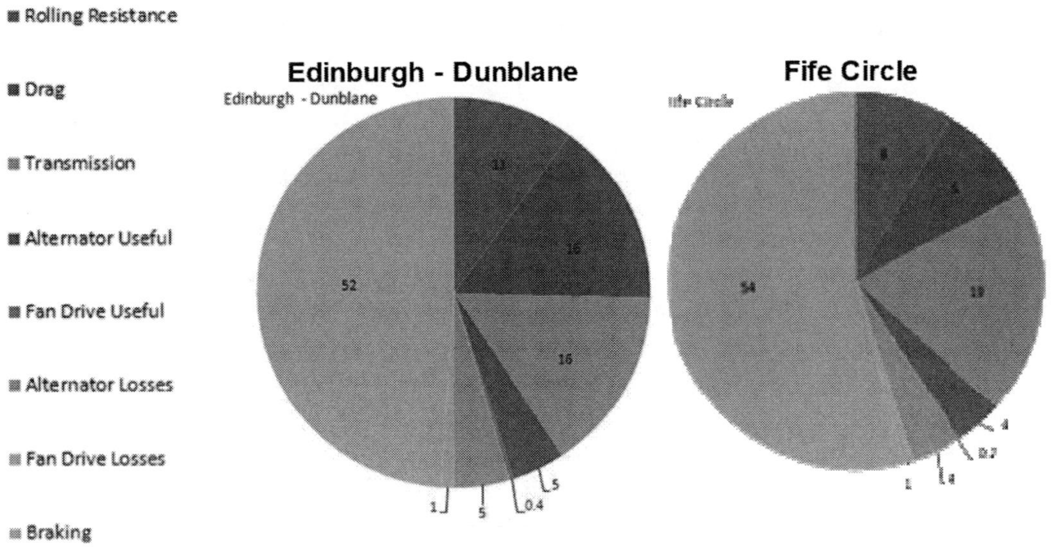


**Figure 1. Energy sinks for Class 158**

## 2. APPLICATIONS

The UK fleet of Turbostar DMUs are fitted with either Voith T211(2-stage hydrodynamic transmission) on the Class 170 or 171, or ZF Ecomat (rail version of planetary bus transmission) on the Class 172 drivelines. This investigation applied the flywheel system to overcome the low efficiency of the hydrodynamic transmission during launch. The transmission and its full load efficiency characteristics are shown in Figure 2. Illustrated is the mean mechanical efficiency during launch to 20 mph: c. 30%.

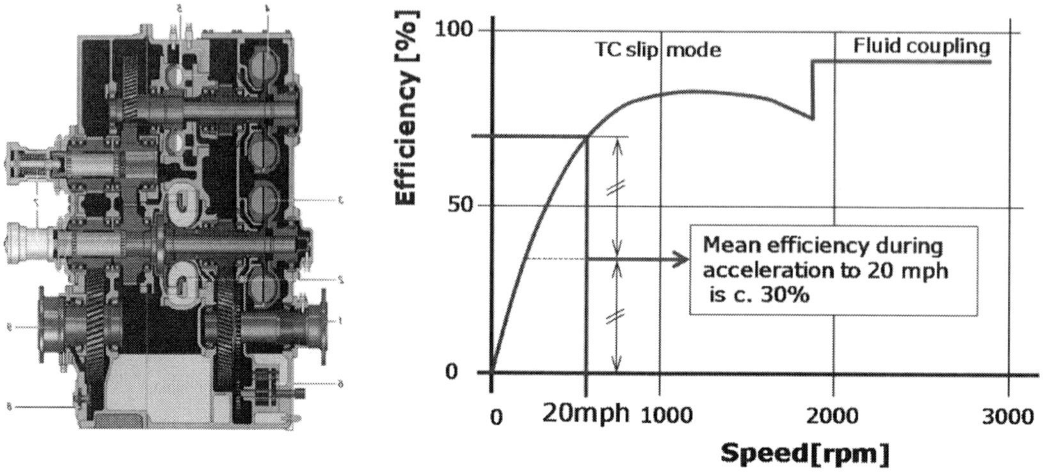


**Figure 2. Voith T211 Transmission and Efficiency map**

When considered together, the loss of 50% fuel energy to braking and further significant losses during launch, a great potential for improvement exists.

## 3. DESCRIPTION OF SYSTEM

### 3.1 Overview

The core technology employs a filament wound flywheel supported on a unique bearing solution to permit high speed in conjunction with high shock loads (21Kg spinning mass, > 45,000 rpm). Torque is transferred to the flywheel in a vacuum chamber by a form of magnetic coupling through the chamber wall eliminating a critical seal that might otherwise compromise vacuum integrity. Speed matching is achieved by a high efficiency Digital Displacement® hydraulic pump/motor in which individual cylinders are 'enabled' by digitally controlled solenoid valves.

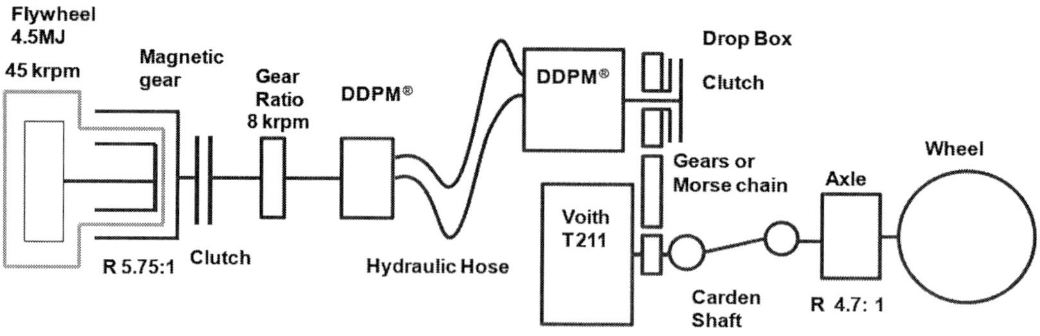


**Figure 3. System Schematic**

Figure 3 illustrates the main elements and their application to a DMU driveline. As will be later discussed, the use of a hydraulic connection to the flywheel module enables this element to be mounted resiliently to the body and so avoid the high levels of vibration associated with the axles or bogeys; further, installation is facilitated by flexible hoses rather than shafts or immediate local attachment to allow designers to use existing space with minimised interruption to existing equipment.

The hydraulic pump/motor is connected to the driveline with a Morse chain engaging a sprocket accommodated on a modified PTO flange, allowing ready retrofit.

## 4. DESCRIPTION OF MAIN COMPONENTS

### 4.1 Flywheel Module (Flywheel, Bearings and Containment)

This section of the paper describes aspects of a 1MJ flywheel module implemented on other Ricardo projects. Proprietary knowledge and design techniques would be re-applied to the particular case of larger flywheels. It is worth noting that Ricardo have been unable to cause a flywheel failure as part of other validation programmes despite attempts with over-sped components modified by gross synthetic faults.

The safe engineering of an energy storage device is paramount for applications in which the consequence of energy release is high, arguing ultimate implementation according to Level 3-4 ISO System Integrity. Fundamentally, the severity relates to the rate of energy release, rather than the quantity, since 4.5MJ equates approximately to just 0.1 litre of diesel fuel. Currently, there is insufficient empirical data, public or otherwise, associated with the failure of filament wound flywheels; to address this, Ricardo is leading a research activity (Flysafe project [4]) to observe intended failure of a variety of flywheels to advise the following:

- Nature of flywheel decomposition: dust, large fragments or combination. This will identify the requirement to contain a pressure blast and or piercing missiles.
- Release of intact rotor due to shaft or bearing failure.

#### *4.1.1 Flywheel assembly*

The flywheel assembly comprises a modular arrangement in which the flywheel and magnetic gear are each supported on separate shafts, these shafts being selectably connected by a clutch as shown in Figure 4. This arrangement allows the flywheel to be isolated during dwell periods in stations and is a feature introduced specifically for rail applications. The details of a clutch system preferred for this in-vacuum application are described later.

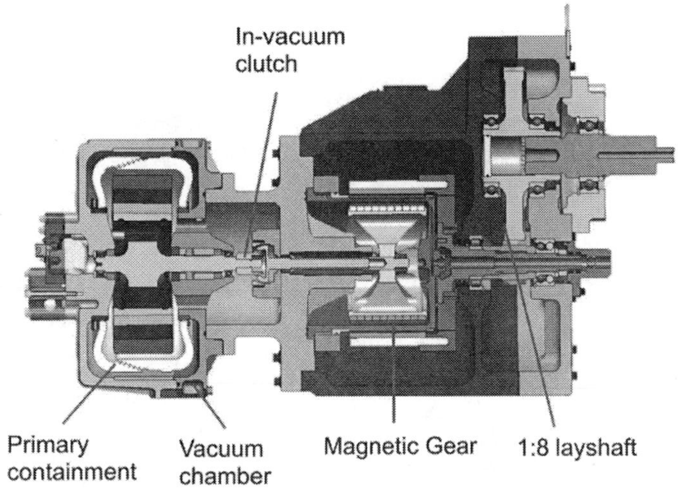


**Figure 4. Section of flywheel assembly (for test rig)**

The flywheel illustrated is smaller than that intended for 4.5MJ which has a mass of 22kg and a nominal diameter 280mm, but the fundamentals pertain. A display unit replicating the system fitted to Ricardo's excavator demonstrator ("HFX", 2012) is shown in Figure 5 along with the unit from an earlier bus application.

**Figure 5. Flywheel module – Ricardo HFX excavator demonstrator 2012 (left) and bus unit (right)**

### *4.1.2 Flywheel*

The flywheel shown in Figure 6 comprises T4 carbon-fibre filament wound rings upon a steel hub. The composite is wound wet under controlled tension and is subject to a balancing process following baking. First sample prototypes designed for 60krpm have been proof testing to 90krpm without incident. Whilst reassuring, it did identify unduly conservative design factors. Appropriate safety factors will be applied to the component elements according to the degree of variability anticipated in first series production and the susceptibility of the component to excitation of the failure modes arising within an assumed life duty cycle. Beyond dynamic modal analysis, the following have been considered within the design:

- Cyclic fatigue
- Composite creep during sustained loading
- Ageing
- Thermal cycling and temperature gradient from extreme cold start
- Peripheral heating due to vacuum degradation

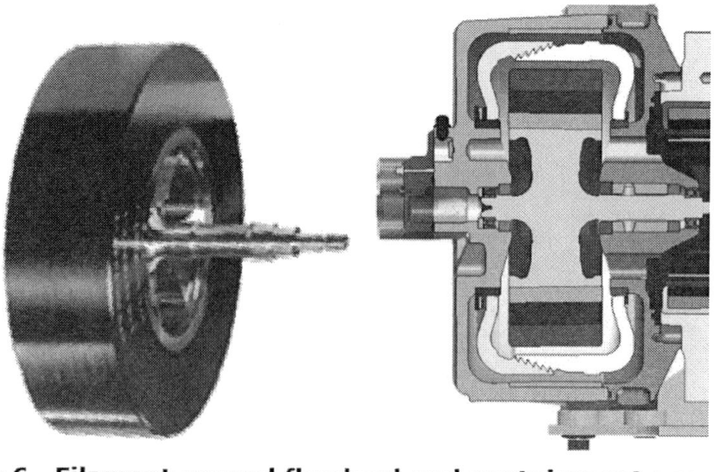

**Figure 6. Filament-wound flywheel and containment assembly**

#### *4.1.3 Containment*

Until sufficient test data exists, and to address anxiety over the introduction of novel technology, the philosophy of absolute containment has been adopted. Figure 6 also shows a steel "horseshoe" designed to contain both a fragmented flywheel and also a released intact rotor since the horseshoe can spin on its own "saver" bearings to dissipate energy and to prevent localised perforation.

Shown in Figure 7 is the maximum distortion of the two-piece horseshoe under the action of a shock-wave arising from 100% composite fragmentation. The condition arises at a time of 340μs after first crack development. The scale of the distortion is amplified many times to aid observation. Consideration has been applied to the plastic deformation of the horseshoe during the normal design process as have deviations from steady-state properties of steels under rapid rates of strain.

**Figure 7. Horseshoe under pressure wave (scaled)**

Determination of appropriate safety factors necessarily defers to close parallels from other disciplines. [5] relates to the design of blast containment vessels for development of explosives and reports significant asymmetry (a factor of 3) in the measured pressure distribution arising from a central explosive source.

#### *4.1.4 Magnetic gear*

The means of applying torque to the flywheel is achieved by a form of magnetic coupling *through* the vacuum chamber wall using a three element magnetic gear. This confers the following advantages over a design using a shaft with a seal:

- Integrity of vacuum chamber not compromised so the device is maintenance-free without requirement for a vacuum pump.
- speed increasing ratio of c. 1:5
- defined torque limit provides a re-sealing "fuse"

Proprietary techniques arising from extensive development has proven necessary to reduce losses at high speed. Figure 8 shows the device in schematic form and an illustration of the magnetic analysis undertaken. The device comprises magnets attached to an inner and an outer rotor between which are located a circular array of ferrous pins. Figure 9 shows early prototype parts with a binding layer over the magnets on the inner (high speed rotor) to overcome centripetal loading, and also shown are the ferrous pins located in a plastic section of the vacuum chamber. [6] provides an introduction to the principles behind magnetic gears for the interested reader.

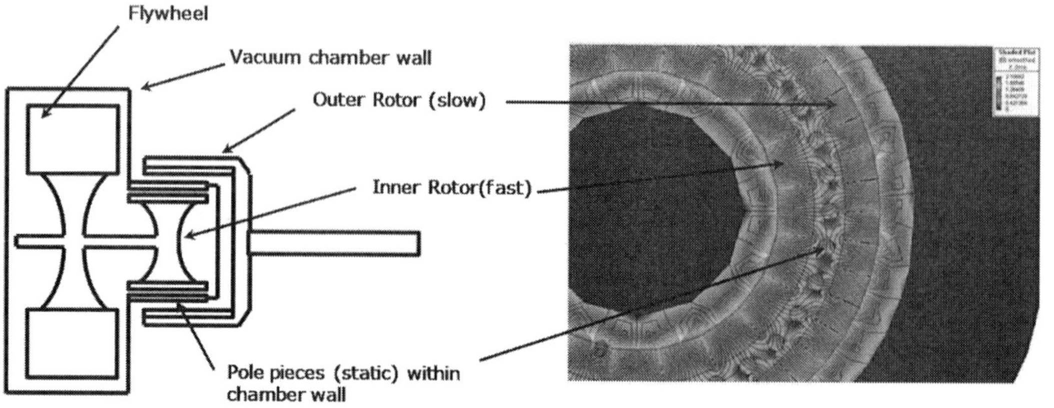


**Figure 8. Magnetic gear: device and flux analysis**

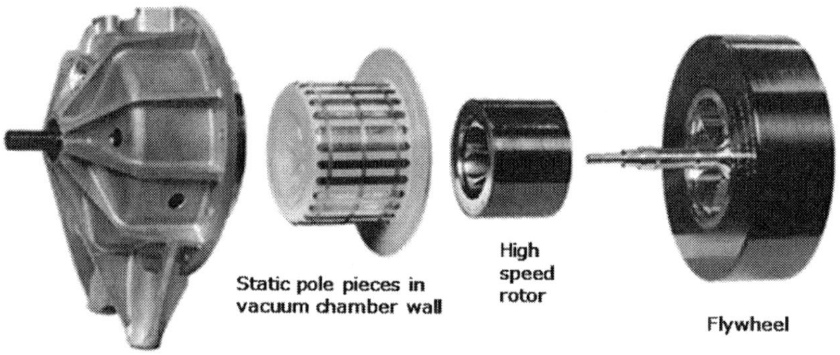


**Figure 9. Early prototype elements of magnetic gear trans-vacuum torque coupling**

### *4.1.5 De-coupling clutch between flywheel and magnetic gear*

Figure 10 shows the clutching system to de-couple the magnetic gear from the flywheel during station dwell, thus the only drain on the flywheel is its own bearing losses. Conventional friction clutches emit non-ferrous friction material which cannot be easily localised, and they require sustained axial clamp loads, making them unsuited for this unusual application.

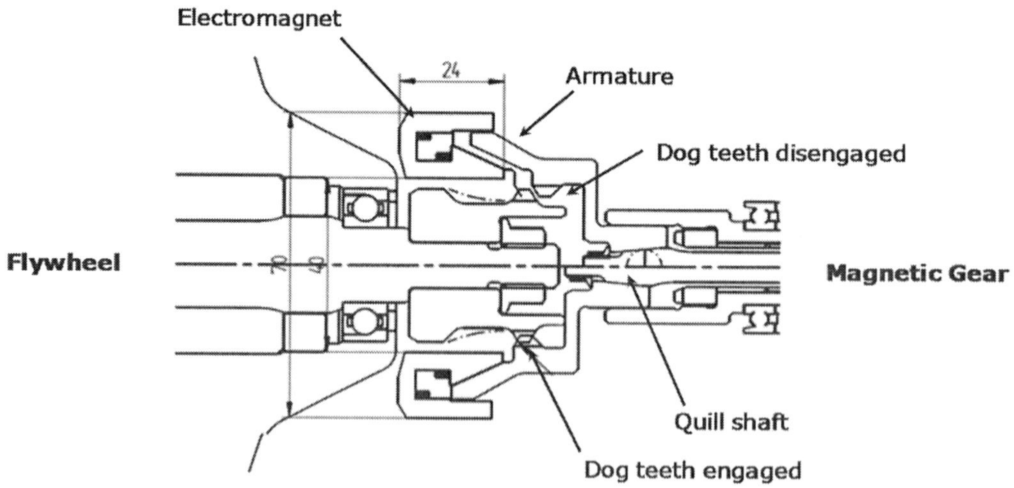


**Figure 10. In-vacuum decoupling clutch between flywheel and magnetic gear**

The selected approach comprises a dog clutch engaged by a specifically designed electromagnet. The quill shaft attached to the armature transmits torque via a sliding spline to the magnetic gear and also accommodates misalignment tolerance between the flywheel axis and the axis of the magnetic gear, each having its own pair of bearings. Attention to precise speed matching is required to avoid damage to the dog teeth.

Around the assembly (not shown) sits a local shroud with rings of magnets to trap ferrous particle debris arising from engagement of the dogs which would otherwise contaminate the vacuum chamber.

#### 4.1.6 Bearings

A significant challenge arises in defining durable bearings for a device with a 20 year life, that operates for 12 hours per day at speeds of up to 45,000 rpm, that is subject to the rigours of rail vibration, that operates within a vacuum, for which efficiency is critical. At the time of writing, Ricardo solutions are currently subject to ongoing patent application. What may be usefully discussed is the approach taken using novel sensor technologies both for development purposes and for condition monitoring of the bearings when in service, since bearing faults are progressive enabling viable intervention, unlike potential modes of the composite which are effectively instantaneous.

Of the modes of fault that allows an intact rotor to escape, bearing failure is the most likely first stage ahead of shaft and casing fracture, and early detection has value. Inference of condition arising from vibration identifies late stage failure only. Knowledge of the loads may inform an accumulated damage model. The loads would include gyroscopic loads arising from external shock loads acting upon the flywheel module. Such a model would necessarily include validated models of the prevailing physics-of-failure, rather than assumptions of failure through high-cycle fatigue.

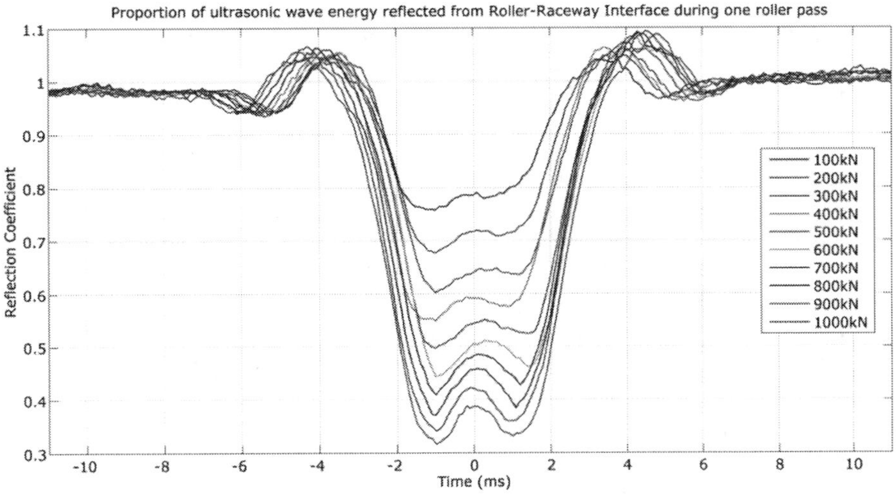


**Figure 11. Load inference from ultrasonic reflection coefficient**

Oil-film measurement derived from the reflection of ultrasonic signals has been successfully applied to assess the tribology state within bearings for more than ten years. A recent advance in this area is the accurate inference of inter-component load derived from additional analysis of metrics associated with the oil-film thickness. Early stage results (courtesy of Sheffield University) are presented in Figure 11 in which a metric associated with the reflection coefficient varies in close linearity with applied load for a single roller in a rolling element bearing. This degree of insight with readily applied instrumentation makes practical improved open-loop component life models. The degree to which additional sensors are added for closed-loop confirmation of progression rates depends upon the fault type and the degree of confidence required to sanction preventative intervention. Taking the case of sub-surface crack propagation as a likely fault mode for raceways, the rate of progress may be established through the use of acoustic emission sensors which detect the release of strain energy arising from crack propagation. Having applied both technologies within a wind turbine gearbox under the DECC OWDIn project, both sensor technologies will be applied to the first track-bound application of the flywheel.

### 4.2 Hydraulics

The hydraulic system comprises one machine attached to the driveline and one coupled to the flywheel module; the variable speed ratio being achieved by variation in the relative displacement of the two machines. Traditionally, hydraulics are regarded as flexible, torque/power dense and providing favourable control characteristics. Efficiency, particularly at part load operation, has been a disincentive to application.

Artemis have developed a variant of radial piston pump known as a Digital Displacement Pump Motor (DDPM®) which differs from foregoing radial piston pumps in that whilst the fixed displacement of each cylinder is similar, the valves in each cylinder are controlled electronically. The valves are large poppet-types which reduces pumping losses and are less susceptible to contamination. Within a bank of cylinders, the valves are selectively enabled to deliver the required flow rate and achieve improved part-load efficiency as shown in Figure 12 [7]. A modest improvement is observed at full load, but a more dramatic improvement is clear at part load (20%).

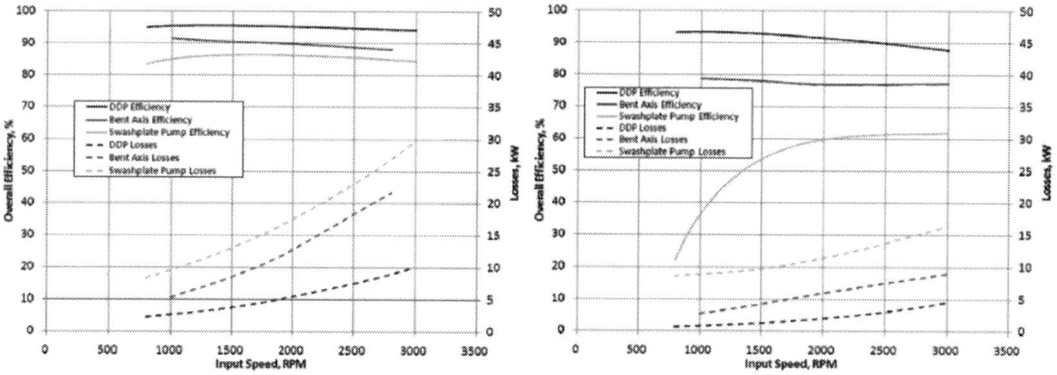


**Figure 12. Relative efficiency of different pump types, 100% (left) and 20% load (right)**

DDPM® devices have been applied with hydraulic accumulators as the energy storage medium in parallel hybrids for city busses, for which savings in excess of 30% have been predicted. The low energy density of hydraulic accumulators compared to chemical or kinetic batteries makes them unsuited for rail application, hence, in part, consideration of flywheels. An Artemis engineer provides a helpful guide as to the scale of the hydraulic machine for the rail application in Figure 13.

**Figure 13. 480 cc/rev DD Pump/ Motor® on test stand (Artemis)**

## 5. SIMULATION RESULTS

A fuel-usage simulation was developed with detailed efficiency maps for the transmission and driveline elements in addition to the BSFC map for the engine. In conjunction with gradient data, this was then exercised across velocity/distance data from an On Train Monitoring Recorder (OTMR) for a route with the following characteristics:

- Length = 198.7km
- Stops = 15
- Maximum distance between stops = 23.3 km
- Minimum distance between stops 1.85 km
- Average station dwell = 45.4 s

Figure 14a shows the operation of the system on approach to a station, dwell and launch. Figure 14b shows the torque applied via the magnetic gear to the flywheel. Figure 14c shows the difference between the torque requested by the simulation driver and that which is actually delivered by the hydraulic system. It is evident from Figure 14c during the braking phase (2.1 – 2.13 hours) that both the magnetic coupling and the pump saturate and the braking torque is augmented by the dissipative system.

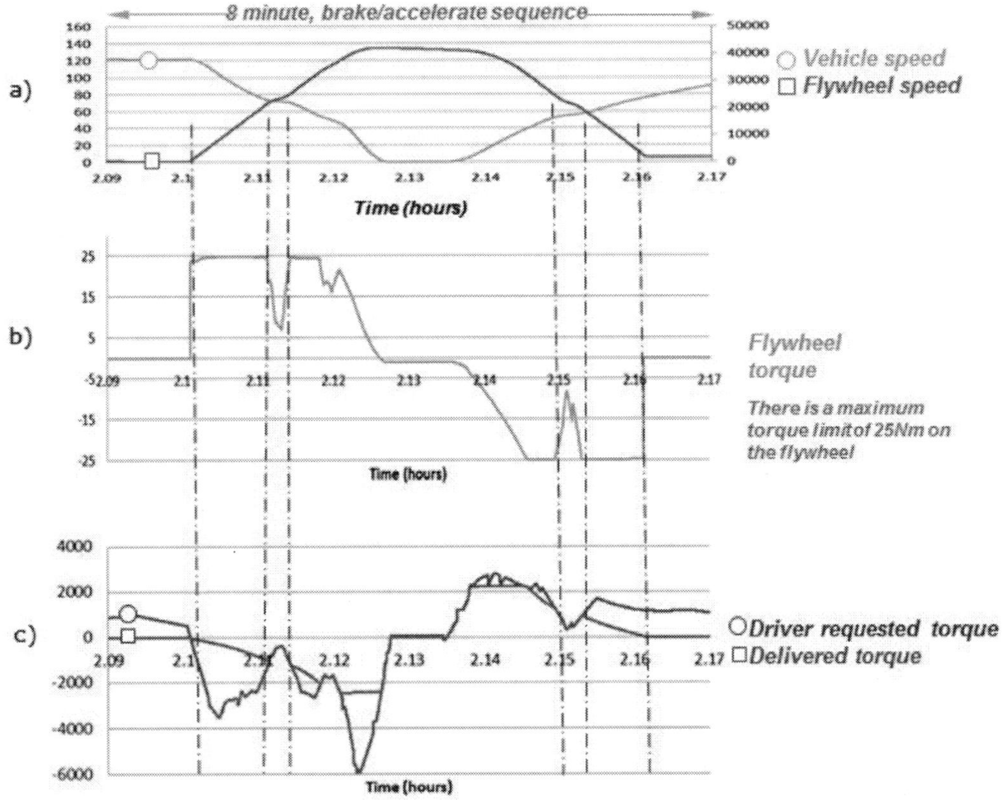


**Figure 14. System Simulation**

Early in the programme, sensitivity studies were undertaken to define the capacity of the system elements and to quantify any penalty arising from application of standard components. Figure 15 shows the impact of torque capacity of the magnetic coupling upon fuel saving for three levels of flywheel pre-charge applied to a fixed flywheel capacity. The outcome with a pre-charge of 20,000 rpm is insensitive to coupling torque capacity, but the fuel saving is limited by the flywheel capacity; whereas the outcome with a pre-charge of 1000 rpm is entirely limited by the capacity of the coupling.

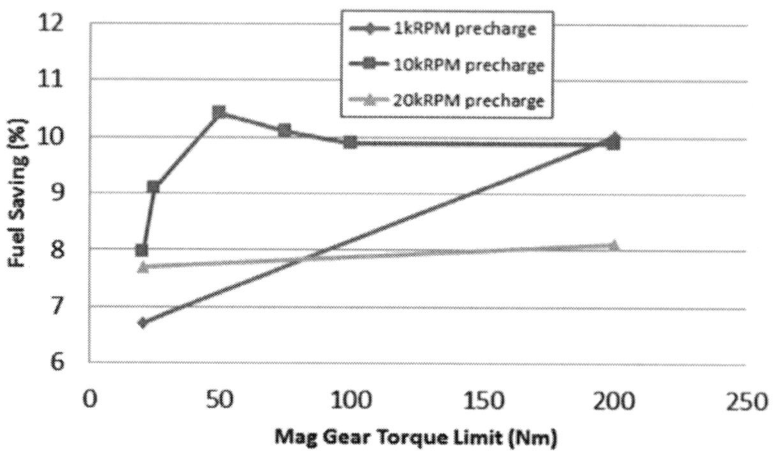


**Figure 15. Effect of Magnetic Gear Torque Limit on Fuel Saving**

The rate at which energy can be accepted by a flywheel varies with the square of its speed, so a sensitivity study was undertaken to identify the optimum pre-charge (minimum speed) of the flywheel. Increasing the pre-charge speed then reduces the energy that can be ultimately accepted, and it is this that causes the decline in fuel saving shown in Figure 16. The optimum was a fuel saving of 10.2% with a pre-charge speed of 14,600rpm. This was achieved without decoupling the flywheel from the rest of the hybrid driveline including the magnetic gear.

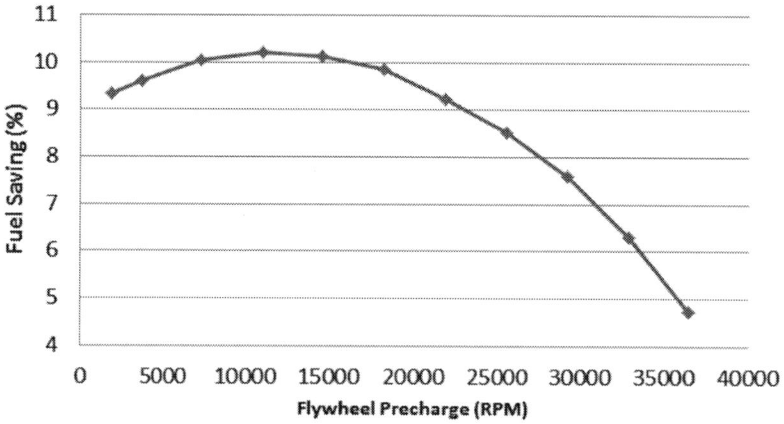


**Figure 16. Effect of flywheel pre-charge on fuel saving**

Without flywheel decoupling, the estimated wheel to wheel efficiency was 59.6%. If decoupled this would be improved to 64.7%. It is anticipated that further bearing development will allow this to exceed 70%.

## 6. BUSINESS CASES

A detailed payback model has been developed that addresses fuel savings arising during launch of DMUs fitted with the Voith T211 transmission and a connection node for the flywheel module downstream of the transmission. The main operating and commercial assumptions are listed below:

- 22 stops from 100kph per route
- routes per day, 6 days per week
- 8% discount rate applied to CAPEX
- £2k pa OPEX
- 5%pa fuel cost escalator (65p/litre 2013)

- Installed cost included
- Amortisation of non-recurrent engineering costs over a fleet of 150 units

The calculation for flywheel energy usage is described below:

- All flywheel energy is expended to accelerate the train before using the engine, thus avoiding the lowest efficiency region of the gearbox operation.
- The resulting train speed is calculated accounting for system efficiencies and discharge losses.
- The average transmission efficiency up to this speed is then calculated.
- The fuel saving is then calculated according to the overall powertrain efficiency

A sensitivity study is presented in Table 3 according to flywheel capacity and installed cost. It is seen that a payback of 4.5 years can be achieved for an installed cost of £50k for 9MJ capacity (2x 4.5MJ units), based upon detailed production cost estimation. Volume effects arising for bus applications, different technology and yet further improved efficiency make 3.5 years payback a viable target with continued annual fuel savings then exceeding £13k/DMU.

**Table 3: Payback projection according to Flywheel Capacity**

| Flywheel Capacity [MJ] | Hybrid Launch Speed [mph] | Mean Trans Efficiency [] | Cost Installed [£k] | Payback [years] |
|---|---|---|---|---|
| **1** | 13.02 | 0.2 | 25 | 13.9 |
| **2** | 18.42 | 0.325 | 27 | 11.1 |
| **3** | 22.56 | 0.375 | 30 | 8.7 |
| **4.5** | 27.6 | 0.4 | 35 | 6.6 |
| **9** | **37.07** | **0.425** | **40** | **3.5** |
| **9** | **37.07** | **0.425** | **50** | **4.6** |
| **9** | 37.07 | 0.425 | 60 | 5.53 |
| **13.5** | 47.8 | 0.5 | 72 | 5.1 |

## 7. TEST PROGRAMME TO CONFIRM FUNCTION AND VALIDATE EFFICIENCY

The test programme has the following objectives:

- Confirmation and development of system function – this being the first occasion upon which these components have been operated in concert.
- Validation of efficiency assumptions within the payback model. The efficiency will be measured using existing Ricardo flywheel modules with some rail-specific adaptations, such as the de-coupling clutch previously described.
- Opportunity for observation of a functional system by interested parties.

Functionality confirmed, accurate projections can be made for larger 4.5MJ flywheels, using the measured data. Figure 17 identifies the system and associated instrumentation to be tested. The test facilities are to be located at the Artemis site in Edinburgh, UK. Previously, the flywheel module was run in a specialist Ricardo test cell with a certain energy containment capability and guaranteed isolation from staff, as shown in Figure 18. Figure 19 shows the design of a precautionary containment vessel to be used elsewhere comprising two skins: the inner to prevent perforation by fragments, and the outer to contain a pressure wave, having higher ductility. The gap will be filled with a damping medium, ideally a self-coalescing ballistic powder. A sliding door is provided to enable access and a second hemispherical end cap is bolted to complete the containment vessel.

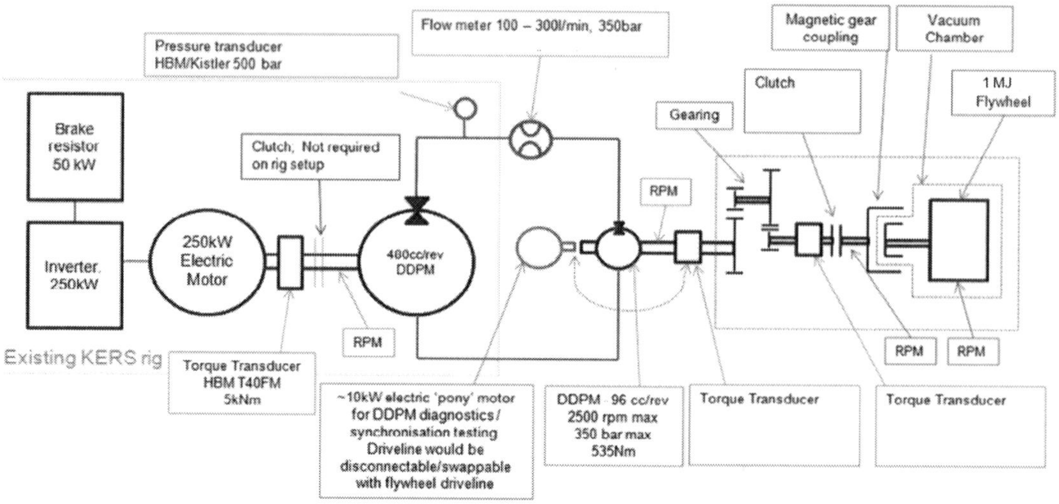


**Figure 17. Test rig layout with 1MJ flywheel**

**Figure 18. Flywheel module installed at Ricardo test facility**

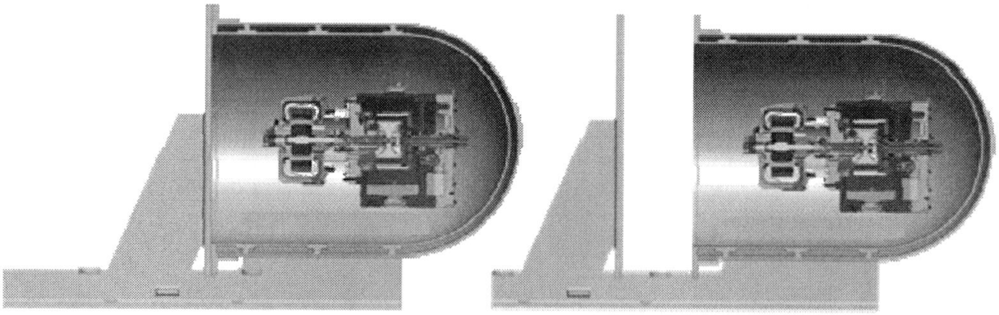

**Figure 19. Duplex-skinned containment vessel for flywheel module**

## 8. INSTALLATION

### 8.1 System to be installed

Figure 20 presents a single flywheel module fitted with a DDPM® unit. Figure 21 shows the 9MJ system to be fitted in schematic form inclusive of sensors and de-coupling elements necessary for validation of efficiency during first vehicle trials. The flexibility of hydraulic coupling is manifest, at least at a superficial level:

additional energy capacity is located and simply "plumbed in". Beyond the general requirements of fixing equipment to trains and matters such as service access and immunity to impact from rail debris, the following must be addressed

- Vibration of the flywheel module
- Interface to main transmission
- Crash integrity (addressed later)

**Figure 20. CAD model of flywheel module with hydraulic machine fitted**

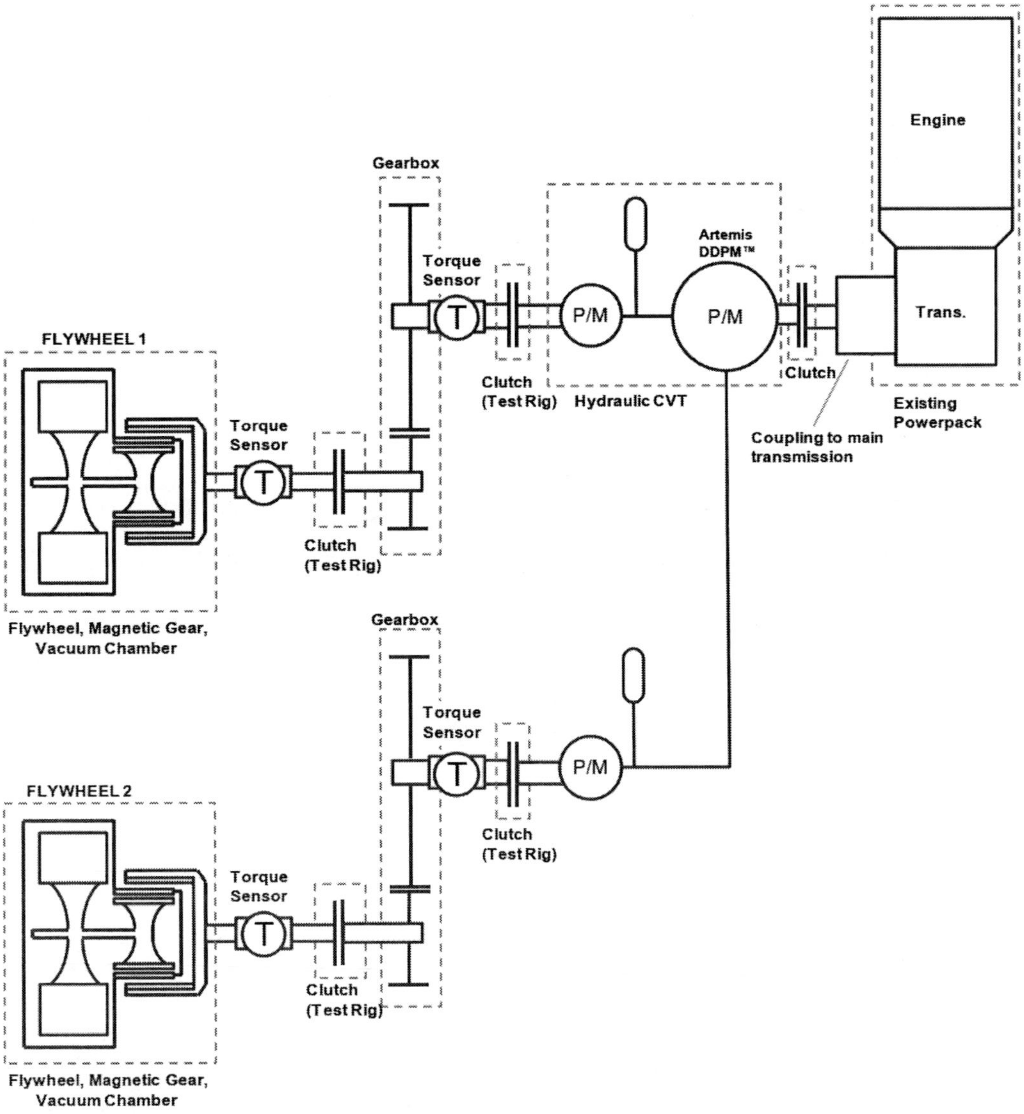


**Figure 21. Schematic of equipment to be fitted for 9MJ (2x4.5MJ)**

## 8.2 Interface to transmission

Figure 22 shows a preferred approach to interface the wheel-side hydraulic machine (DDPM®) to the main transmission in which the PTO flange is modified to accommodate a sprocket for a Morse chain housed in a chain box. The chain coupling is clutched at one end to disconnect the hydraulic machine to avoid an over-speed limit of 2600rpm at high rail speeds, and also to reduce losses when not active.

This topology of connection does not change the duty of the gearbox, and the configuration of chain box is able to accommodate existing projections from the transmission, in combination simplifying retrofit.

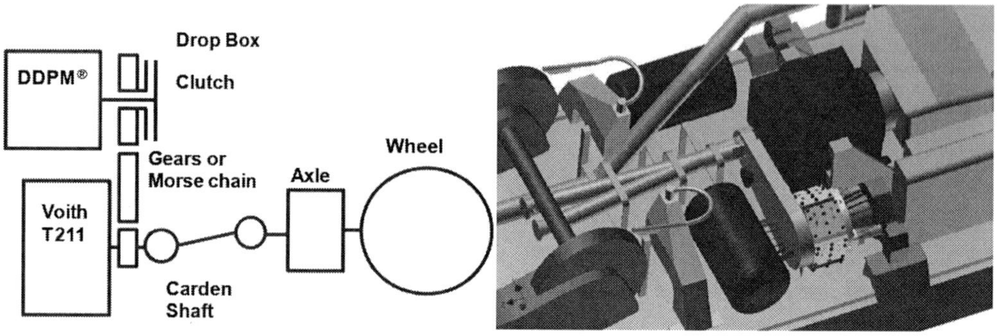


**Figure 22. Interface of hydraulic machine with main transmission**

The final installation of the system will be defined by Bombardier engineers as the Design Authority for the classes of vehicles being modified.

## 8.3 Vibration

Table 4 provides extracts from GM/RT2100 [8] for both fatigue and proof loads. This provides valuable guidance in selecting the orientation of the spin axis of the flywheel to reduce gyroscopic loads upon the flywheel shaft and bearings, and also the importance of mounting the flywheel assembly on the vehicle body. The hydraulic system provides a critical degree of isolation.

**Table 4: Vibration specifications GM/RT2100**

| | Longitudinal | | Lateral | | Vertical | |
|---|---|---|---|---|---|---|
| | Proposed | Current | Proposed | Current | Proposed | Current |
| AXLEBOX MOUNTED EQUIPMENT | 12 | N/A | 10 | 3 | 70* | 25 |
| Bogie Mounted Equipment | 7 | 0.5 | 7 | 1.5 | 10 | 10 |
| Body Mounted Equipment (Passenger Vehicle) | 0.2 | 0.2 | 0.4 | 0.15 | 0.45 | 0.15 |
| Body Mounted Equipment (Non-passenger vehicle excl. freight) | 0.2 | 0.2 | 0.4 | 0.3 | 0.45 | 0.3 |

Fatigue Acceleration for Vehicle Mounted Equipment (+/- g)

| | Longitudinal | | Lateral | | Vertical | |
|---|---|---|---|---|---|---|
| | Proposed | Current | Proposed | Current | Proposed | Current |
| AXLEBOX MOUNTED EQUIPMENT | 18 | 5 | 20 | 5 | 155* | 50 |
| Bogie Mounted Equipment | 8 | 5 | 11 | 3 | 20 | 20 |
| Body Mounted Equipment | Unchanged from GM/RT2100 | | | | | |

Proof Acceleration for Vehicle Mounted Equipment (+/- g)

## 9. INTEGRATION

The project raises a number of integration challenges that are being assessed by the project partners. Bombardier's experience of rail system upgrades and new technology integration partnered with the system specific integration expertise of Ricardo and Artemis has been supported by ongoing discussions with the Office of Rail Regulation (ORR) and the Rail Safety Strategy Board (RSSB) to address the following:

- Mechanical interface for energy transfer
- Installation of the equipment within the existing space frame
- Impact on the brake system control – specifically brake blending and Wheel Slide Protection
- Impact on vehicle power control (effect of additional power input)
- Potential interface with train control system
- Mechanical limitations on the driveline
- Effect of shock and vibration on the equipment
- Mechanical fixing and mounting of equipment (GM/RT 2100)
- Impact protection of equipment
- Effect on gauging from installation of new equipment (GE/RT8073)
- Compliance with EMC standards (EN50121)
- Functional integration software will be guided by ISO 26262 aligned to due FMEA studies
- Compliance to applicable Group Standards and EuroNorms

The system should be transparent to the driver, but trials will be undertaken to assess whether maximum advantage will be derived from current or sympathetic driving styles.

## 10. SAFETY

The philosophy of applying absolute containment until statistical proof exists from rig and field components from stable series manufacture techniques has been described and will safeguard passengers and track-side rail staff.

The safety of those in the vicinity has also been addressed for the cases of train crash and general servicing, for which the following precautions have been considered:

- The flywheel module must be immediately identifiable as a hazard with the casing to carry appropriate markings that cannot be obscured by rail contamination after many years of service.
- The module should provide an indication of its state of charge by at least two independent means, with one being apparent without specialist equipment such as a handheld monitor.
- The flywheel module shall have self-contained function to allow isolation from the hydraulic system in case of crash/derailment.
- Disassembly shall be prevented by interlocks whilst the flywheel is rotating
- Means of discharge to a state of minimal risk shall be provided: one for use by maintenance technicians, and the other for use in case of emergency by rescue staff. Proprietary techniques have been developed for meet these two requirements.

## 11. FUTURE ACTIVITY AND APPLICATIONS

This project has focused upon deriving a fuel saving from the capture and re-use of energy on approach to stations. At the completion of the project, the technical basis and a significant aspect of the business case necessary to justify the investment in a rail demonstration vehicle will have been provided.

The other areas highlighted in Table 1, beyond fuel saving, may also be investigated potentially justifying developments to the core technologies of the flywheel module and hydraulic system. The following are anticipated:

- Increased power – higher torque and higher speed within flywheel module
- Increased energy – higher speed flywheels to capture energy on hill descent
- Low mass containment – a benefit per se, but also to avoid elevation to a higher rail access charge band.

It is likely that application to EMUs would argue a significant increase in energy capacity to allow autonomous operation, in addition to the performance enhancement possible and reduction in disruption to local grids.

## 12. CONCLUSIONS

Significant advantages have been identified for DMUs and potentially EMUs arising from the fitment of high capacity flywheels. The investigation has quantified the fuel saving arising from capture of energy on approach to a station and use to accelerate the train from the station for DMUs fitted with one type of hydrodynamic transmission that has low efficiency during launch.

The current project has confirmed technical and practical viability of a high durability flywheel system for retrofit, initially to DMUs fitted with Voith T211 unit for which simulation using OTMR velocity data for the Waverley route showed a 10.2% fuel reduction with 9MJ storage capacity.

A detailed payback model supported sensitivity studies have shown that a total of 9MJ comprising two 4.5MJ units provides payback of 4.6 years with ongoing savings of c. £13kpa/DMU for stop-start commuter routes. This may reduce to c. 3.5 years if less conservative assumptions are validated by rig results.

A thorough business model has been developed which allows assessment of different routes and determination of tailored storage capacity for specific routes. This could include assessment of capture of gradient energy given that the incremental cost of flywheel storage is low compared to alternatives such a chemical batteries.

The technologies presented have the potential for high reliability through the avoidance of a vacuum seal through the flywheel chamber by use of a magnetic gear coupling, unique bearing solutions, and a CVT technology applied in large offshore wind turbine transmission for which a service life of 20 years is anticipated. Despite this, contingency for maintenance has been included within the payback model.

The hydraulic technology applied (DDPM®) has uniquely high part load efficiency and has applications beyond use as a component of a parallel hybrid system. The technology is currently applied in 7MW wind turbines and could be scaled for use as a main rail transmission, or as a 15kW ancillary drive.

Installation considerations have been addressed. The flexibility afforded by use of a hydraulic system to couple multiple flywheel units to the driveline facilitates retrofit.

Safety considerations associated with the high energy flywheel module have been addressed for passengers but also for rail workers and emergency rescue crews.

Evidence has been presented relating to the health impacts of diesel-derived air pollution both in the UK and US. The flywheel system presented would reduce ambient levels of $PM_{2.5}$ in stations to the advantage of both passengers and rail workers.

It is understood that franchise bids for operation of rail services in the UK are required to seek and apply technologies that bring substantial improvements to both commercial operation, but also to passengers' experience. The fuel saving and clean/quiet station launch provided by a hydraulically coupled flywheel system meets these objectives, with further advantages relating to increased performance according to operational imperatives.


## ACKNOWLEDGEMENTS

The authors wish to offer the thanks to the following:

- F. McKinney (Project Monitoring Officer, Innovate UK) and A. Broadbent (Head of Engineering Research, RSSB) for financial support of the project.
- Directors of Ricardo for permission to publish this paper and financial support of the project and the research from which it emanates.
- Bill Reeves (Commercial Director at Transport Scotland) for review of project and guidance.
- Prof. A. Atkins, J. Dalby and S. Shepherd and innumerable colleagues (Ricardo UK) for development of flywheel and magnetic gear for light applications.
- S. McCauley (Ricardo UK) for technical guidance relating to locomotive transmissions.
- Prof. Rob Dwyer Joyce (Head of Mechanical Engineering, University of Sheffield) for access to development and application guidance of advanced sensor systems for measurement of oil-film thickness, interface loads and crack propagation.
- Prof. Geraint Jewell (Head of Electrical Engineering, University of Sheffield) for generous technical support in electromagnetic design.


## REFERENCES


1. J.E. Ginda (advisor to US Senate). "Adverse Health Effects of Diesel Particle Air Pollution". EPA, USA, 2010.
2. Estimating Local Mortality Burdens associated with Particulate Air Pollution. A.M. Gowers[1], B.G. Miller[1] and J.R. Stedman[2]. [1]Institute of Occupational Medicine, [2]Ricardo-AEA. Public Health England, 2014.
3. R. Edwards, Herald Scotland, 16-12-2012. "Air pollution at Waverley Station seven times higher than safety limits".
4. Flysafe_http://gtr.rcuk.ac.uk/project/2525C213-8835-4CFA-996B-A24818AE3AC7
5. Q. Dong, B.Y.Hu, S.Y. Chen, Y. Gu. Institute of Fluid Physics, China Academy of Engineering Physics. "Engineering design of a multiple-Use spherical explosion containment vessel subject to internal blast loading from 25Kg TNT high explosive". Journal of pressure vessel technology. ASME April 2012. Vol. 134.
6. X. Li, K.-T. Chau, M. Cheng, W. Hua. "Comparison of magnetic-geared permanent magnet machines". Progress in Electromagnetics Research, Vol. 133, 177-198, 2013.
7. L. Wadsley, Sauer-Danfoss (US) Company. "Optimal System Solutions Enabled by Digital Pumps". IFPE (International Exposition for Power Transmission) 2011.
8. RSSB Engineering Standard. "Vibration environment for rail vehicle mounted equipment".

# Metro traction and power system energy optimisation

**N Zhao[1], Z Tian[1], S Hillmansen[1], C Roberts[1], M Yuan[2], J Li[2], H Shi[3], K Li[3]**

[1] The Birmingham Centre for Railway Research and Education, School of Electronic, Electrical and Computer Engineering, University of Birmingham, UK

[2] Guangzhou Metro Corporation, P.R. China

[3] Guangzhou Metro Design & Research Institute Co., Ltd., P.R. China

**ABSTRACT**

This paper presents a train trajectory optimisation for a Guangzhou Metro Line in China. A train simulator was developed specifically for this study. It is able to simulate the movement of railway vehicles and calculate the detailed power system energy consumption with different train trajectories when implemented on a typical AC or DC powered metro line. An Enhanced Brute Force algorithm was developed in order to achieve the optimisation quickly and efficiently. Analysis of the results shows that by implementing an optimal train trajectory, the energy usage in network can be reduced with both human driving system and Automatic Train Operation systems.

## 1 INTRODUCTION

With increasing energy prices and environmental concerns, train energy-efficient operation techniques have been the focus of more attention in recent years. Transportation is facing increasing pressure to optimise power management strategies. The driving style, which is conventionally controlled by drivers or by modern ATO (Automatic Train Operation) control system, becomes one of the most popular areas for research because of the possibility of significant energy saving [1-3].

Researchers have studied a number of different methods to optimise train operation as described in recent literature. A genetic algorithm (GA) has been applied to optimise train movements using appropriate coasting control [4]. Bocharnikov proposed a novel approach which is a combination of physical principles and driving strategies with a modern searching method including a GA and a fuzzy logic technique [5]. In [6], a distance-based train movement model is established, and three optimisation algorithms are implemented to search for the optimal train speed trajectory, namely ant colony optimisation (ACO), dynamic programming (DP) and genetic algorithm (GA). After comparing the performance of the three methods, it was found that DP is better than either GA or ACO; nevertheless it is sensible to use multiple methods in optimizing a trajectory to ensure robustness of the results. An enhanced brute force algorithm was applied to minimize energy cost and impact of train delays has been considered for differing railway train control systems [7]. In order to speed the computing response and develop an online optimisation system, Howlett utilised the Pontryagin principle to find necessary conditions on an optimal strategy for a continuous control problem [8-11]. The main achievement is formulating and solving a generalised train control problem using a general motion equation, and both continuous control and discrete control are taken into account. From a theoretical point of view, Rongfang Liu developed a calculation algorithm and computer program for an energy-efficient train control [12]. An analytical solution which gives the sequence of optimal controls is presented. In order to achieve a global optimality of driving strategy and optimal timetable, Shuai Su analysed a hierarchy of energy-efficient train operation and proposed an integrated algorithm to generate a globally optimal operation schedule [13-15].

In this paper, a train traction simulator, which can simulate both the details of the traction infrastructure and the train movement, is developed for computing the energy consumption and journey time. An Enhanced Brute Force Algorithm is illustrated to optimise the train trajectory based on both human driving and Automatic Train Operation systems. In the end of the paper, some case results based on the infrastructure and operation data of Guangzhou Metro Line are presented, which illustrate that the proposed approach can obtain a good energy-saving performance for driving systems and ATO systems.

## 2 SIMULATOR DEVELOPMENT

In order to consider the fundamental physics of train motion, Lomonossoff's Equations are used in the simulator development to solve the dynamic movement equations, which have previously been described in earlier works [1, 16, 17].

$$M_{tr}\frac{d^2s}{dt^2} = F_{tr} - R - F_{grad} \quad (1)$$

where $s$ is the train location; $dt$ is the time derivative; $F_{grad}$ is the force due to the gradient; $F_{tr}$ is the traction force effort; $M_{tr}$ is the effective mass; $R$ is the vehicle resistance.

The resistance is also referred as the Davis equation [18], which can be calculated by the following equation with constants $a$, $b$, $c$. The constants are relative to the aero-dynamic and track resistance.

$$R(v) = a + b|v| + cv^2 \quad (2)$$

Figure 1 and Table 1 show the four typical movement modes for train motion obtained based on Equation 1 and Equation 2.

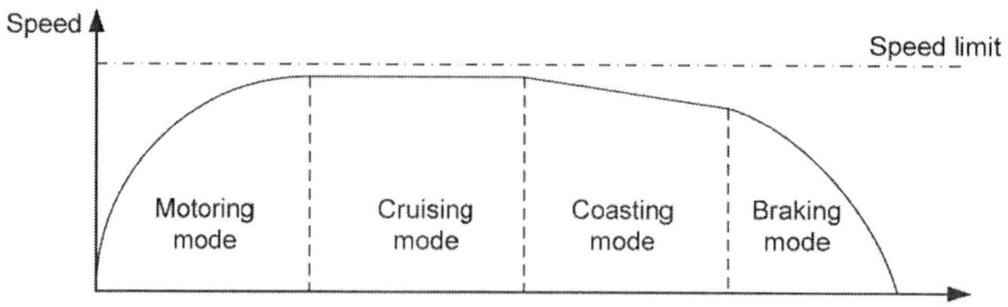


**Figure 1. Train movement modes.**

**Table 1. Four modes of train movement.**

| Mode | Equations |
|---|---|
| Motoring | $F_{total} = F_{tr} - R - F_{grad} > 0$ |
| | $a = F_{total} / M_{tr} > 0$ |
| Cruising | $F_{total} = F_{tr} - R - F_{grad} = 0$ |
| | $a = F_{total} / M_{tr} = 0$ |
| Coasting | $F_{total} = -R - F_{grad}$ |
| | $a = F_{total} / M_{tr}$ |
| Braking | $F_{total} = F_{tr} - R - F_{grad} < 0$ |
| | $a = F_{total} / M_{tr} < 0$ |

The system flowchart is shown on Figure 2. The parameters that are not changed throughout the simulation process are defined as fixed parameters, such as train traction data, gradient, line speed limits. The parameters that are changed are associated with vehicle speed, location, timetable, and gradient and are considered as dynamic parameters.

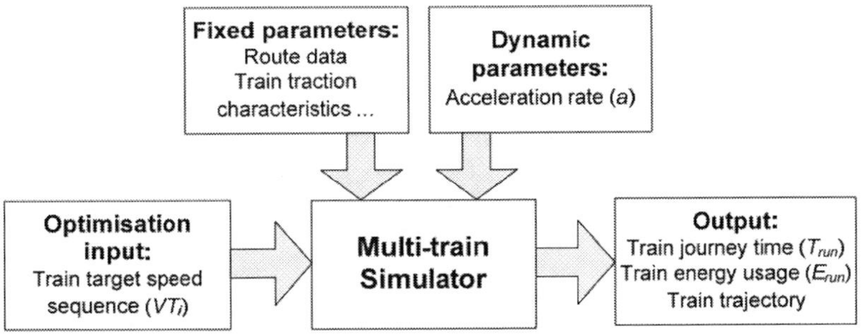


**Figure 2. Flowchart of the developed simulator.**

## 3 TRAIN TRAJECTORY OPTIMISATION

In this research, the route is divided into twelve sections depending on the gradient changes and line speed limits changes, by vertical dotted lines shown in Figure 3. Using a different target speed in each section will result in different train running performance. By using the developed simulator, the train trajectory can be controlled by implementing different train target speed sequences, as shown in Figure 2.

The aim of the train trajectory optimisation in this research is to calculate the most appropriate train target speed sequence ($VT_1$ to $VT_i$) to minimise the train energy usage within a given journey time.

$$E_{run} = \int_0^{T_{run}} f[v(t)]v(t)dt \qquad if \;\; T_{min} \leq T_{run} \leq T_{max} \tag{3}$$

$$T_{run}, E_{run} = g(VT) \tag{4}$$

where $T_{max}$ and $T_{min}$ are the permitted scheduled journey times; $T_{run}$ is the simulation journey time; $E_{run}$ is the train energy usage; where $f[v(t)]$ is the maximum tractive effort at the current vehicle speed $v(t)$.

Both human driving systems and Automatic Train Operation systems are considered in the research as two case studies. For human driving systems, only three different speed target sections are considered, considering the difficulty of human driver's operation ability.

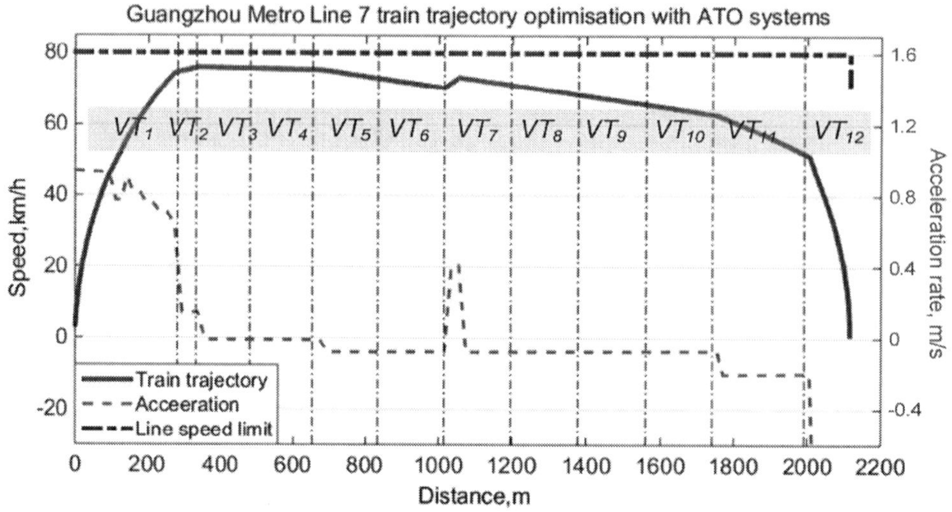


**Figure 3. An example of the train trajectory optimisation for Automatic Train Operation systems (without cruising mode).**

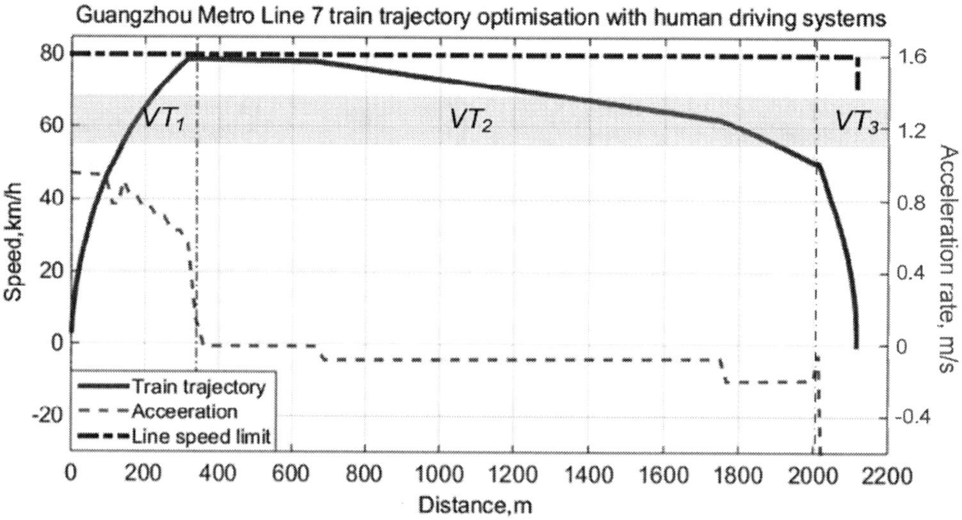


**Figure 4. An example of the train trajectory optimisation for human driving systems (with cruising mode).**

In this research, each target speed sequence is assumed as a candidate solution. The solution domain depends on the assumed search boundary and interval. A conventional brute force algorithm produces all possible candidates in the solution domain to find the optimum [19]. For complex problems, this algorithm rapidly becomes impractical due to enormous computation time [20].

In order to overcome this problem, an enhanced brute force algorithm is developed [7, 21]. The method firstly produces an estimated sequence based on the given journey time by comparing the train service interval with the minimum line headway. Then only the candidate solutions close to the estimated value are considered by the algorithm. The solution domain therefore can be reduced significantly, thereby decreasing the computation time. Finally, the method enumerates the candidate solutions and evaluates them using Equation (3). The top ranking solution will be selected as the most appropriate.

## 4 TRAIN TRAJECTORY OPTIMISATION

The case study is designed for a Guangzhou Metro Line. The route and station configurations of the line are shown in Figure 5 and Table 2 respectively. Figure 6 shows the traction system characteristics the vehicle.

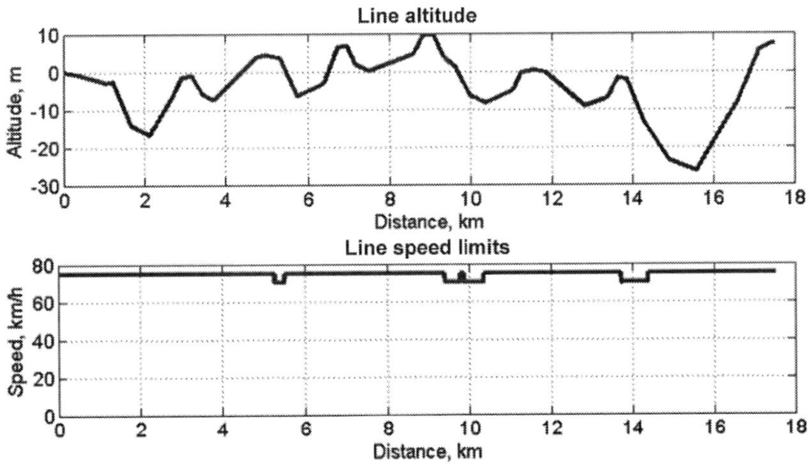


**Figure 5. Guangzhou Metro Line gradient and speed limits.**

**Table 2. Scheduled timetable.**

| Station sequence | Station location, m | Dwell time, s | Scheduled journey time, s |
|---|---|---|---|
| 1 | 0 | 25 | N/A |
| 2 | 1100 | 35 | 82 |
| 3 | 3048 | 30 | 119 |
| 4 | 5190 | 30 | 134 |
| 5 | 6852 | 35 | 107 |
| 6 | 8929 | 35 | 130 |
| 7 | 11354 | 30 | 144 |
| 8 | 13700 | 40 | 145 |
| 9 | 17508 | 30 | 216 |

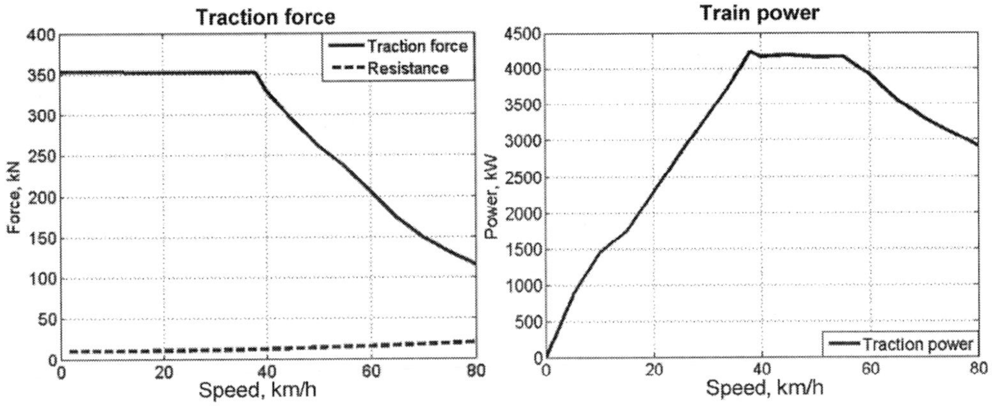


**Figure 6. Train traction system characteristic.**

The train trajectory optimisation results for Automatic Train Operation systems and human driving systems are shown in Figure 7 and Figure 8. ATO systems are usually operated with two levels. The higher level aims to calculate the train target speed. And the lower one is responsible for supervising the train actual speed. In practice, it is difficult for a train operating at a constant speed (cruising mode). If a train trajectory with such a mode is implemented to the ATP system, the lower level will control the train switching between acceleration and braking frequently, which will cause extra energy usage. Therefore, as shown in Figure 7, cruising mode has not been considered for the ATO system case study.

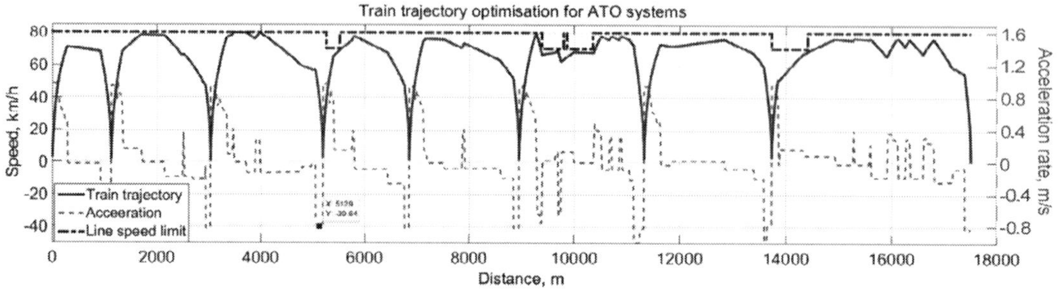


**Figure 7. The optimisation results for Automatic Train Operation systems.**

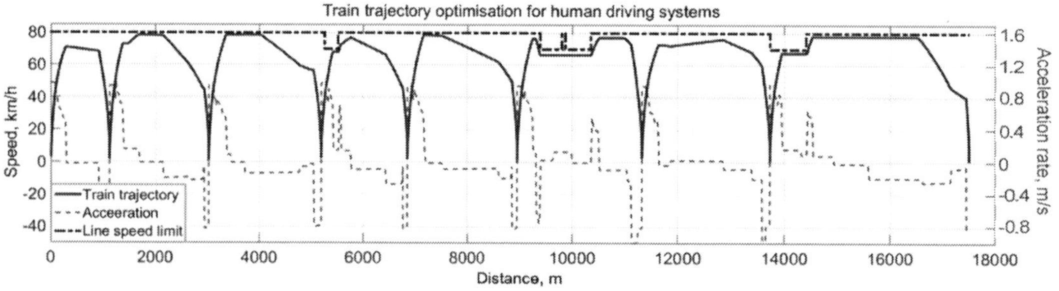


**Figure 8. The optimisation results for human driving systems.**

Table 3 shows the optimisation comparison for different driving systems. The simulation journey times for both two case studies are the same as the scheduled journey time. Therefore the optimisation results are able to be applied in practice without modifying the operation timetable. The ATP driven train achieves 3.5% lower energy usage than the human driven train. It is important to note that, similar to the ATO systems, most of the trains operating by human drivers are unable to operate at a constant speed. Therefore, the human driving trains will achieve a higher energy usage in practice.

**Table 3. Optimisation comparison for different driving systems.**

| Station sequence | Dwell time, s | Scheduled journey time, s | For human driving systems | | For Automatic Train Operation systems | |
|---|---|---|---|---|---|---|
| | | | Simulation journey time, s | Energy usage, kWh | Simulation journey time, s | Energy usage, kWh |
| 1 | 25 | | | | | |
| | | 82 | 82 | 20.52 | 82 | 20.52 |
| 2 | 35 | | | | | |
| | | 119 | 119 | 19.37 | 119 | 18.34 |
| 3 | 30 | | | | | |
| | | 134 | 134 | 27.52 | 134 | 26.84 |
| 4 | 30 | | | | | |
| | | 107 | 107 | 18.28 | 107 | 18.28 |
| 5 | 35 | | | | | |
| | | 130 | 130 | 22.43 | 130 | 22.41 |
| 6 | 35 | | | | | |
| | | 144 | 144 | 32.9 | 144 | 32.82 |
| 7 | 30 | | | | | |
| | | 145 | 145 | 22.48 | 145 | 22.48 |
| 8 | 40 | | | | | |
| | | 216 | 216 | 41.61 | 216 | 36.31 |
| 9 | 30 | | | | | |
| In total | | 1302 (Including dwell time) | 1302 (Including dwell time) | 205.11 | 1302 (Including dwell time) | 197.98 (-3.5%) |

## 5 CONCLUSION

A Brute Force algorithm application for train trajectory optimisation on a Guangzhou Metro Line has been presented in this paper. The optimisation aims to minimise energy usage within a given journey time for different driving systems. A train movement simulator was developed specifically for this research.

The simulation results identify that the developed simulator is able to provide accurate train movement performance. And the enhanced brute force algorithm is shown to be able to achieve optimal results with different constraints. Furthermore, the comparison results show that the trains equipped with ATO systems are able to achieve better energy usage when compared the ones with human driving systems.

## 6 ACKNOWLEDGMENT

This research is jointly supported by Guangzhou Metro Corporation and Guangzhou Metro Design & Research Institute Co., Ltd. This research is funded by Guangzhou Municipal Science and Technology Project. Project reference: 2013J4500068.

## REFERENCES

[1] S. Hillmansen, and C. Roberts, "Energy storage devices in hybrid railway vehicles: A kinematic analysis," *Proceedings of the Institution of Mechanical Engineers, Part F: Journal of Rail and Rapid Transit,* vol. 221, no. 1, pp. 135-143, 2007.

[2] K. K. Wong, and T. K. Ho, "Coast control of train movement with genetic algorithm." pp. 1280-1287 Vol.2.

[3] K. K. Wong, and T. K. Ho, "Coast control for mass rapid transit railways with searching methods," *Electric Power Applications, IEE Proceedings -,* vol. 151, no. 3, pp. 365-376, 2004.

[4] C. S. Chang, and S. S. Sim, "Optimising train movements through coast control using genetic algorithms," *Electric Power Applications, IEE Proceedings -,* vol. 144, no. 1, pp. 65-73, 1997.

[5] Y. V. Bocharnikov, A. M. Tobias, C. Roberts, S. Hillmansen, and C. J. Goodman, "Optimal driving strategy for traction energy saving on DC suburban railways," *Electric Power Applications, IET,* vol. 1, no. 5, pp. 675-682, 2007.

[6] S. Lu, S. Hillmansen, T. K. Ho, and C. Roberts, "Single-Train Trajectory Optimization," *Intelligent Transportation Systems, IEEE Transactions on,* vol. 14, no. 2, pp. 743-750, 2013.

[7] N. Zhao, C. Roberts, and S. Hillmansen, "The application of an enhanced Brute Force Algorithm to minimise energy costs and train delays for differing railway train control systems," *Proceedings of the Institution of Mechanical Engineers, Part F: Journal of Rail and Rapid Transit,* vol. 228, no. 2, pp. 158-168, 2014.

[8] P. G. Howlett, I. P. Milroy, and P. J. Pudney, "Energy-efficient train control," *Control Engineering Practice,* vol. 2, no. 2, pp. 193-200, 4//, 1994.

[9] P. G. Howlett, "The Optimal Control of a Train," *Annals of Operations Research,* vol. 98, no. 1-4, pp. 65-87, 2000/12/01, 2000.

[10] P. G. Howlett, P. J. Pudney, and V. Xuan, "Brief paper: Local energy minimization in optimal train control," *Automatica,* vol. 45, no. 11, pp. 2692-2698, 2009.

[11] A. R. Albrecht, P. G. Howlett, P. J. Pudney, and X. Vu, "Energy-efficient train control: From local convexity to global optimization and uniqueness," *Automatica,* vol. 49, no. 10, pp. 3072-3078, 10//, 2013.

[12] R. Liu, and I. M. Golovitcher, "Energy-efficient operation of rail vehicles," *Transportation Research Part A: Policy and Practice,* vol. 37, no. 10, pp. 917-932, 2003.

[13] S. Su, T. Tang, X. Li, and Z. Gao, "Optimization of Multitrain Operations in a Subway System," *Intelligent Transportation Systems, IEEE Transactions on,* vol. 15, no. 2, pp. 673-684, 2014.

[14] S. Su, T. Tang, X. Li, and Z. Gao, "A Subway Train Timetable Optimization Approach Based on Energy-Efficient Operation Strategy," *Intelligent Transportation Systems, IEEE Transactions on,* vol. 14, no. 2, pp. 883-893, 2013.

[15] S. Su, T. Tao, L. chen, and B. Liu, "Energy-efficient train control in urban rail transit systems," *Proceedings of the Institution of Mechanical Engineers, Part F: Journal of Rail and Rapid Transit*, 2014.

[16] S. Lu, S. Hillmansen, and C. Roberts, "A power-management strategy for multiple-unit railroad vehicles," *Vehicular Technology, IEEE Transactions on,* vol. 60, no. 2, pp. 406-420, 2011.

[17] H. Chen, "The traction characteristics on high-speed motor train," *Urban mass transit,* vol. 11, no. 7, pp. 20-24, 2008.

[18] J. R. Loumiet, and W. G. Jungbauer, *Train Accident Reconstruction and FELA and Railroad Litigation*, Fourth ed., p.^pp. 126-127: Lawyers & Judges Pub Co, 2005.

[19] B. Preneel, C. Paar, and J. Pelzl, *Understanding Cryptography: A Textbook for Students and Practitioners*, 1st ed.: Springer, 2009.

[20] A. Levitin, *Introduction to The Design & Analysis of Algorithm*, p.^pp. 497: Addison Wesley, 2003.

[21] H. M. Faheem, "Accelerating motif finding problem using grid computing with enhanced Brute Force." pp. 197-202.

# Modelling critical velocities on soft soils using finite element methods

**S Mezher, P K Woodward, O Laghrouche, G Kouroussis, G Medero, D P Connolly**
Institute for Infrastructure and Environment, Heriot-Watt University, UK

**ABSTRACT**

This paper investigates critical velocity effects and predicts the response of high speed rail lines constructed over soft soils. It is of benefit to designers for the prevention of passengers discomfort and the improvement of safety criteria. As train speeds approach the critical velocity of the underlying soil, ground Mach cones can develop. This phenomenon is associated with the generation of surface waves and ground vibrations beneath the track, and can increase the risk of track damage. Critical velocity is lower for a site with soft soil, thus making these the primary areas of concern. To study the problem, a three-dimensional model of Ledsgaard, a high speed rail site with soft soil in Sweden, is developed. The model follows the Finite Element Method and operates in the time domain. The outcomes of the simulations are presented in terms of vertical deflections and the dynamic behaviour of the ground. The critical velocity effects are observed and the relationship between train speed and track vibrations is presented.

**Keywords:** High-speed, Soft Soils, Ground Mach Cones, Finite Element

## 1. INTRODUCTION

As a result of increasing speeds in the past few years, critical velocity effects have been given major attention. These effects include high vibrations that lead to formation of surface waves, track uplift, and resonance at soil layers beneath the ground. Critical velocity was defined as the speed when the coincidence of surface wave velocity (Rayleigh waves speed) and speed of the train occurs. The arising problems of the critical velocities are of a major concern to passengers' safety and comfort as they result in the damage of the track and surrounding structures. The track vibrations were noticeable mostly on weak soft soil (2). When the train is in motion, the weight creates a pattern of stress on the ground that moves as the train is moving causing differential induced stresses into the soil layers surrounding the train, and these stresses are higher in value for weak soft soil. Many sources of vibration were introduced by Hall (3): structural response of the track, discontinuity on the track, and wheel-rail interactions. The track structural response depends on the wheel axle spacing, the weight of the axle, and the speed of the train (1).

The main focus of the paper is based on the X-2000 train in Ledsgaard, Sweden that connects Gothenburgh and Malmo and runs at a maximum speed of approximately 200 km/h. The site is characterized by weak clay soil that has a depth of approximately 50 m. It was chosen for study purposes as seismic tests were run for the site to obtain the dynamic behaviour of the soil by Banverket. The critical velocity effects were observed on the site and investigations were performed before and after ground reinforcement (3). Being able to predict a similar behaviour to the measured response using 3-dimensional finite element analysis techniques will be beneficial for future projects. Monitoring the dynamic problems that are associated with train-induced ground vibrations will help in finding appropriate and cost-effective mitigation strategies, thus, resulting in maintenance reduction.

Much research was carried out on the X-2000 train to confirm the critical velocity effects. The studies clearly indicated the increase in dynamic amplification as the speed of the train approaches critical conditions. It is important to create realistic models that take into account the interaction forces between the train and the track considering the dynamic excitation at the contact points. Madshus (4) was able to produce results of vertical displacements of the track that are with good agreement to the measured response. As the speed increases, an over prediction of results was noticeable in the simulated displacements of Madshus compared to the actual displacements. The research also included the simulations of a higher speed of 252 km/h. In this paper, the simulations are performed using DART3D, Dynamic Analysis of Railway Track 3D software, which is a finite element software developed at Heriot-Watt University. The results are to be compared to Madshus results in order to come up with a better prediction of soil behaviour. The governing equations and properties that are used are presented in this paper. The vertical displacements of the track are plotted against the distance along the track. The 3-dimensional observations of surface waves formations are presented.

## 2. FINITE ELEMENT ANALYSIS

Dynamic analysis of railway track 3D (DART3D) is a finite element analysis software that is used to model the X-2000 train. The software uses a full coupled finite element model making it easier to solve dynamic-related problems. Controlling the wave reflection is performed by using viscous boundaries, while dealing with complex geometry is done easily by the finite element method. The model allows the prediction of ground vibrations and dynamic response that are initiated by track faults or rail surface irregularities. The software is also capable of modelling material non-linearity, train-track interactions, multi-layered soil, and damping (5). 8-core P7 Processor computer was used due to long 3D computations.

### 2.1 Train and track models

The rail is modelled as a 3D Euler-Bernoulli beam. A total of 424 beam elements are used. The rail is connected directly to the sleepers, and the geometric irregularities are considered. Only half of the car body is considered in the analysis of using symmetry of loading on the rail, and a quarter model is used for analysing the dynamic vehicle. The components of the model are car body, bogies, and suspension systems. Two suspension systems are defined: the primary suspension system connects the bogie and the wheel, while the secondary suspension system connects the bogie and the car body. A Hertzian contact spring is modelled to represent the interactions between wheel and rail. Figure 1 shows the components of the train model.

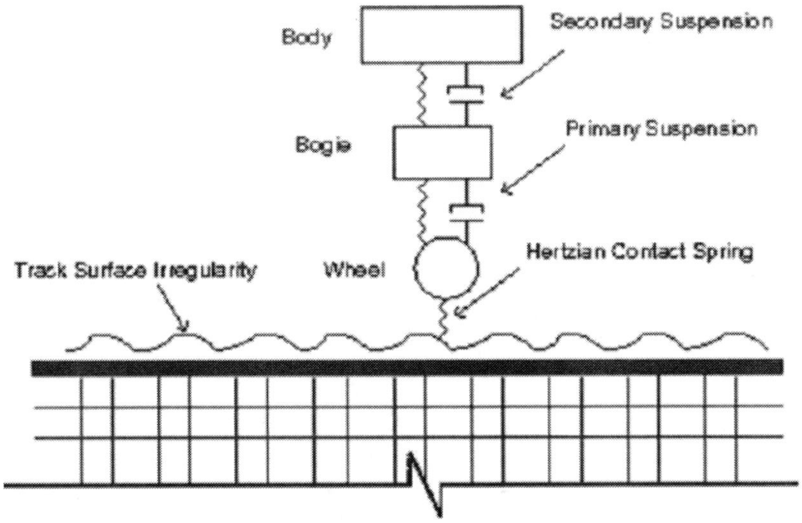


**Figure 1. Diagram layout of the quarter train model (5)**

The ordinary differential equation and wheel-rail interaction force are given below (5):

$$\begin{pmatrix} k_c & -k_c & 0 \\ -k_c & k_c + k_b & -k_b \\ 0 & -k_b & k_b \end{pmatrix}\begin{pmatrix} w_c \\ w_b \\ w_w \end{pmatrix} + \begin{pmatrix} c_c & -c_c & 0 \\ -c_c & c_c + c_b & -c_b \\ 0 & -c_b & c_b \end{pmatrix}\begin{pmatrix} \dot{w}_c \\ \dot{w}_b \\ \dot{w}_w \end{pmatrix}$$

$$+\begin{pmatrix} \overline{m}_c & 0 & 0 \\ 0 & \overline{m}_b & 0 \\ 0 & 0 & \overline{m}_w \end{pmatrix}\begin{pmatrix} \ddot{w}_c \\ \ddot{w}_b \\ \ddot{w}_w \end{pmatrix} = \begin{pmatrix} \overline{m}_c g \\ \overline{m}_b g \\ m_w g + F_{wr} \end{pmatrix} \quad \text{..... (1)}$$

$$F_{wr} = k_H (w_w - w_r - r)^{3/2}, \quad \text{..... (2)}$$
(if $w_w > w_r + r$, otherwise, $F_{wr} = 0$)

$k_c, k_b$ are the secondary and primary stiffness constants, respectively (N/m). $c_c, c_b$ are the secondary and primary damping coefficients, respectively (Ns/m). $\overline{m}_c$ is the mass of the car body and is equal to $\overline{m}_c = m_c/8$ (kg). $\overline{m}_b$ is the mass of the car body and is equal to $\overline{m}_b = m_b/4$ (kg). $\overline{m}_w$ is the mass of the wheel (kg). $w_c, w_b$ and $w_w$ are the vertical displacements of the car body, bogie and wheel, respectively (m). $g$ is the gravitational acceleration (m/s2). $k_H$ is the Hertzian constant (kN/m3/2). $w_r$ is the vertical displacement of the rail (m). $r$ is the rail surface irregularity (m). The train parameters that are used in DART3D are given in table 1 (10).

**Table 1. Train suspension parameters used in the linear analysis**

| **Axle Load (kN)** | 181 |
|---|---|
| **Primary Suspension** | |
| Stiffness (MN/m) | 3.28 |
| Damping (kNs/m) | 90 |
| **Secondary Suspension** | |
| Stiffness (MN/m) | 1.31 |
| Damping (kNs/m) | 30 |

Five carriages were considered for the train model, along with four rail elements between sleepers. Twenty axle loads were applied. Figure 2 shows the spacing used for the first two carriages.

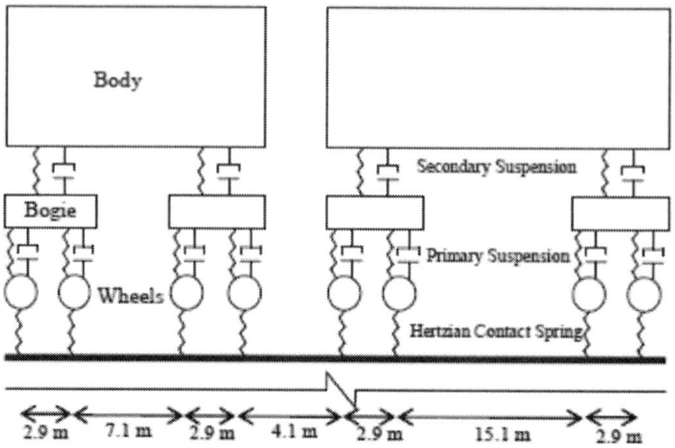


**Figure 2. Components and used spacing**

Moving to track properties, 20-noded brick elements are used for the substructure. The total depth of the mesh is 15 m. Each node has 6 degrees of freedom: 3 translational and 3 rotational. The mesh is presented in figure 3. The width of the ballast is 2.8 m. The subgrade is assumed to be homogeneous and consists of clay soil. Table 2 shows the parameters used for the ballast and subgrade; depth, Young's modulus, Poisson's ratio and density (10).

**Table 2. Track parameters used in the linear analysis**

| Layer | D (m) | E (MPa) | $\gamma$ | $\rho(kg/m^3)$ |
|---|---|---|---|---|
| Top Ballast | 1 | 10 | 0.4 | 1600 |
| Sub Ballast | 3 | 100 | 0.35 | 1800 |
| Crust | 2 | 13 | 0.49 | 1500 |
| Organic Clay | 2 | 4 | 0.496 | 1260 |
| Upper Marine Clay | 2 | 16 | 0.499 | 1475 |
| Lower Marine Clay | 3 | 32 | 0.498 | 1475 |
| Half Space Marine Clay | 2 | 45 | 0.498 | 1475 |

### 2.2 Explicit time integration

For DART3D simulations, explicit time integration is carried out. The dynamic equation of motion, equation 1, is performed for a finite number of discrete time instants. Different approximation schemes can be used; some require behaviour information at both ends of the interval and others require information from more time instants. For the K, C, and M matrices, no previous transformation is needed; however, initial conditions are applied for displacement, $w(0)$, and velocity, $\ddot{w}(0)$. Starting with a known response $t_i$, the ground acceleration is determined by a set of discrete values, then the time $t_{i+1} = t_i + \Delta t$ is applied (6). The chosen time step, $\Delta t = 8 \times 10^{-6}\ s$, is small to ensure accuracy.

### 2.3 Material damping

The Rayleigh damping is described as a linear combination of mass and stiffness matrices (7), and it is given by the following equation, where $\alpha$ and $\beta$ are constants, and D is the damping ratio:

$$\mathbb{C} = \alpha\mathbb{M} + \beta\mathbb{K} \qquad \text{...... (3)}$$

$$\alpha = \frac{2D\omega_1\omega_2}{\omega_1+\omega_2} \qquad \text{...... (4)}$$

$$\beta = \frac{2D}{\omega_1+\omega_2} \qquad \text{...... (5)}$$

**Figure 3. 3D mesh used in ground dynamics simulations**

$\omega_1$ and $\omega_2$ are the first natural frequency, and the highest natural frequency that the vibration modes of the response produce. The damping ratio is taken as 5%. The mass damping is more important at lower frequencies. Nevertheless the stiffness damping becomes the controlling factor at higher frequencies. The damping ratio increases as the strain amplitude increases, which are affected by the soil plasticity.

### 2.4 Boundary conditions

The reflection of stress waves that expand into the surrounding media should be prevented as they cause errors in the analysis. To ensure that the radiation damping is approaching infinity, Lysmer and Kuhlemeyer (8) viscous boundary conditions are used. The wave energy is absorbed at the defined boundaries. This type of boundary does not absorb the Rayleigh waves entirely; however, it provides a suitable way to validate the model in infinite media. The period of interest must not be exceeded as this will require a larger mesh and more computation time (9). Table 3 shows the mathematical computations behind the viscous boundary definition.

**Table 3. Mathematical computations of viscous damping boundary**

| | |
|---|---|
| Wave Stress (Pa) | $\{\sigma\} = \mathbb{C}_v \left\{\frac{\partial u}{\partial t}\right\}$ |
| Damping (Ns/m) | $\mathbb{C}_v = \alpha_v c_p n n^T + \beta_v c_s (t_1 t_1^T + t_2 \bar{t}_2)$ |
| Vector Force (N) | $f = \int_A [N]^T [\mathbb{C}_v][N]\{\frac{\partial u}{\partial t}\}_n dA$ |
| Viscous Damping Matrix (Ns/m) | $[C^b] = \int_A [N]^T [\mathbb{C}_v][N] dA$ |

Where N and A are the boundary element shape function and area, respectively. $t_1$ and $t_2$ are orthogonal tangent unit vectors to the boundary. $\alpha_v$ and $\beta_v$ are positive constant coefficients. $\{\frac{\partial u}{\partial t}\}_n$ is the nodal velocity of the boundary element. In order to form the global damping matrix, the viscous damping matrix is added to the provided material damping matrix.

## 3. SIMULATIONS AND COMPARISON

The analysis is carried out for 70 m/s speed and vertical deflection time history is obtained. Only linear analysis is considered in this paper. The presented run is part of an initial calibration study, and is not considered a final analysis. Figure 4 shows the 3D plot of deflection. The formation of the ground Mach cone is clearly noted. It is generated when the ground wave velocity is exceeded, causing a transmission of surface waves. The transmission depends on the soil density, Poisson's ratio, and stiffness of the ground. It is important to use numerical methods to be able to read the complicated profiles that are generated due to the changing Rayleigh wave velocities in the multi-layered soil. The ground Mach cone ratio, for a homogenous soil, can be determined using the following relationship:

$$GM = \frac{V_T}{V_R} \quad ..... (6)$$

Where $V_T$ is the train's speed and $V_R$ is the Rayleigh wave velocity. A ratio of 1 is an indication of critical conditions.

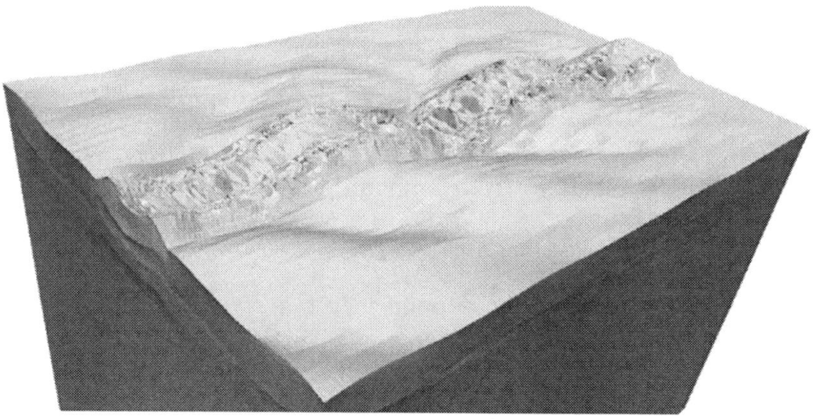

**Figure 4. 3D plot of track deflection**

Figure 5 shows the obtained displacement response from an early calibration study that is compared to the approximated results of Madshus (4). His simulated results are also presented in Figure 6 for the speed of 51 m/s, which are compared to the measured response of the actual field specified with a thicker line. The track uplift, which is caused due the stress waves transmission, is clearly identifiable with positive values. It is an indication of track failure. DART3D and Madshus's results are showing similar response: however, the difference between peaks values is high.

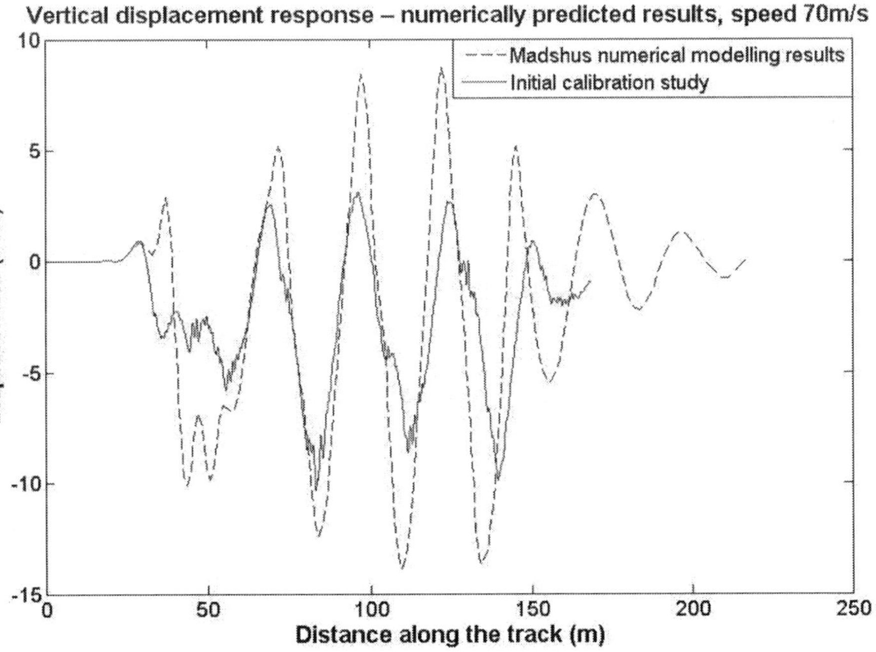


**Figure 5. Vertical displacement response, speed 70 m/s**

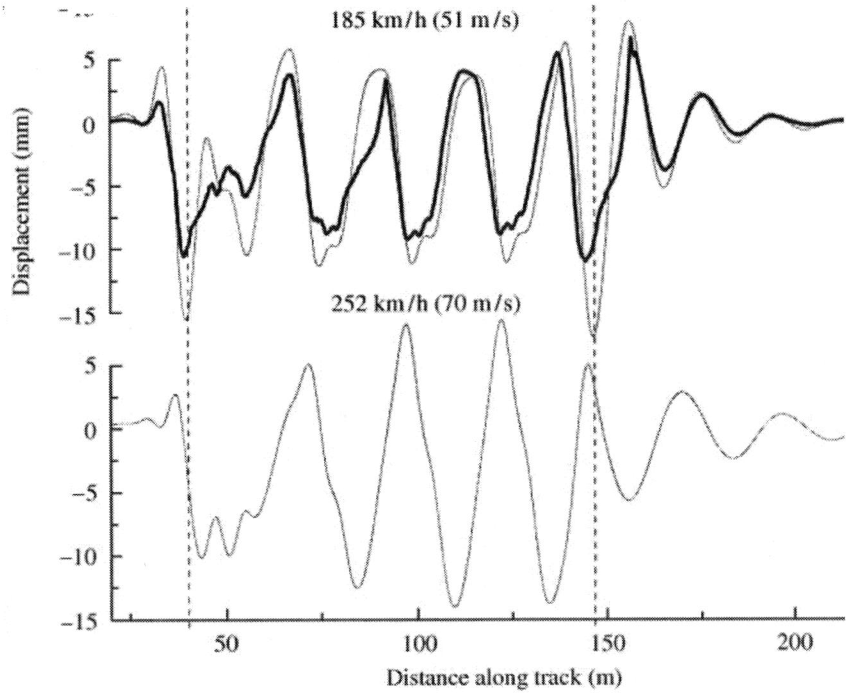


**Figure 6. Vertical displacement response, speeds 51 and 70 m/s (4)**

Referring to the peaks of Figure 6 (a comparison of an existing model from the literature), Madshus's prediction results of 51 m/s speed are slightly higher than the actual response. This is highly observed at 40 m and 142 m of the distance along the track. Error percentages exist between the measured and simulated results. Larger peak values are also detected for uplift. In Madshus's approximation of the 70 m/s response, the uplift value could reach 9 mm in deflection, which is considerably higher. The percentage of error between DART3D and Madshus's results (70 m/s speed) is relatively similar to what is found between Madshus's results compared to the actual site measurements (51 m/s speed). Thus, DART3D is able to produce realistic results of the ground behaviour. The overall shape of the calibration study reflects good agreement with the measured behaviour of X-2000. The DART3D simulations are ongoing with future analyses being run and compared for different speeds of X-2000.

## 4. CONCLUSION

The ground response of the X-2000 train in Ledsgaard is confirmed using DART3D finite element analysis software. The model configuration proposed train and track properties, time integration, material damping, and boundary conditions that were used. The model was validated for having similar ground behaviour and wave propagation as the actual field tests. The development of ground Mach cone and uplift effects were noted. The peaks of vertical displacements proved to be less than a previously simulated analysis run by Madshus, which resulted in an improved representation of displacement values. This is an ongoing work, and it is important to note that the results can be improved by taking the non-linearity of the ground into account.

## REFERENCES


(1) Krylov, V. V., Dawson, A. R., Heelis, M. E., & Collop, A. C. (2000). Rail movement and ground waves caused by high-speed trains approaching track-soil critical velocities. Proceedings of the Institution of Mechanical Engineers, Part F: Journal of Rail and Rapid Transit, 214(2), 107-116.

(2) Banimahd, M., Woodward, P., Kennedy, J., & Medero, G. (2012). Three-dimensional modelling of high speed ballasted railway tracks. Proceedings of the ICE-Transport, 166(2), 113-123. http://dx.doi.org/10.1680/tran.9.00048

(3) Hall. L. (2000). Simulations and analyses of train-induced ground vibrations, a comparative study of two and three-dimensional calculations with actual measurements. PhD thesis, Royal Institute of Technology, Sweden.

(4) Madshus, C., & Kaynia, A. M. (2000). High-speed railway lines on soft ground: dynamic behaviour at critical train speed. Journal of Sound and Vibration, 231(3), 689-701. doi:10.1006/jsvi.1999.2647

(5) El Kacimi, A., Woodward, P. K., Laghrouche, O., & Medero, G. (2013). Time domain 3D finite element modelling of train-induced vibration at high speed. Computers & Structures, 118, 66-73. doi.org/10.1016/j.compstruc.2012.07.011

(6) Chopra, A.K. (2001). Dynamics of structures: theory and applications to earthquake engineering. New Jersey: Prentice Hall.

(7) Caughey, T. K. (1960). Classical normal modes in damped linear dynamic systems. Journal of Applied Mechanics, 27(2), 269-271.

(8) Lysmer, J., and Kuhlemeyer, R.L. (1969). Finite dynamic model for infinite media. Journal of the Engineering Mechanics, ASCE, pp. 859-877.

(9) Hung, H.H., and Yang, B.Y. (2001). A review of researches in ground-borne vibrations with emphasis on those induced by trains. Proceedings of National Research Council, part A: Phys. Sci. & Eng., ol. 25, no. 1, pp. 1-16.

(10) Sheng, X., Jones, C. J. C., & Thompson, D. J. (2003). A comparison of a theoretical model for quasi-statically and dynamically induced environmental vibration from trains with measurements. Journal of Sound and Vibration, 267, 621–635. doi:10.1016/S0022-460X(03)00728-4

(11) Takemiya, H. (2003). Simulation of track-ground vibrations due to high-speed train: the case of X-2000 at Ledsgard. Journal of Sound and Vibration 261, 503-526.

(12) Paolucci, R., Maffeis, A., Scandella, L., Stupazzini, M., and Vanini, M. (2003). Numerical prediction of low-frequency ground vibrations induced by high-speed trains at Ledsgard, Sweden. Soil Dynamics and Earthquake Engineering 23, 425-433. doi:10.1016/S0267-7261(03)00061-7
(13) Costa, P. A., Calcada, R., Cardoso, A. S., & Bodare, A. (2010). Influence of soil non-linearity on the dynamic response of high-speed railway tracks. Soil Dynamics and Earthquake Engineering, 30, 221–235. doi:10.1016/j.soildyn.2009.11.002
(14) Shih, J.-J., Thompson, D., & Zervos, A. (2014). Assessment of track-ground coupled vibration induced by high speed trains. In 21st International Congress on Sound and Vibration (pp. 13–17).
(15) Connolly, D., Giannopoulos, A., & Forde, M. C. (2013). Numerical modelling of ground borne vibrations from high speed rail lines on embankments. Soil Dynamics and Earthquake Engineering, 46, 13–19. doi:10.1016/j.soildyn.2012.12.003
(16) Connolly, D. P., Kouroussis, G., Woodward, P. K., Verlinden, O., Alves Costa, P., & Forde, M. C. (2014). Field testing and analysis of high speed rail vibrations. Soil Dynamics and Earthquake Engineering, 67, 102–118. doi:10.1016/j.soildyn.2014.08.013

# Bayesian reconstruction of three dimensional railway track geometry: improvements and developments in the system toward practical use

**A Yoshimura**
Tokyo University of Technology, School of Computer Science, Japan

**ABSTRACT**

A new method of reconstructing a three dimensional (3D) railway track geometry from measurements of "3-point asymmetrical chord offset for level and alignment" is described. In the new method, 3D track geometry is reconstructed without distinction of components of design track and track irregularity, as far as a signal to noise ratio permits and also for any kind of railway sections such as curve, transition curve, etc. Reconstructed data include not only the geometrical shape of the track but also the variations of angles and curvatures in the vertical and horizontal directions along the track. These data will be very useful for the analysis of the vehicle-track dynamic interaction.

## 1 INTRODUCTION[(1),(2)]

In most railways a measurement of railway track irregularities for level and alignment is practiced by the 3-point asymmetrical chord offset methods in the track inspection car. In this paper, hereafter we use abbreviations of "offset measurements", "offset for level" and "offset for alignment" in order to avoid the use of this lengthy name for the measurement method. Unfortunately, it does not give a real track geometry. A track geometry is a crucial element in most aspects of vehicle dynamics. Recently, we have developed a new method of reconstructing a 3D railway track geometry from offset measurements for level and alignment. A new method uses the technique of the recursive nonlinear filtering based on the Bayesian approach which has been developed in the statistical and engineering fields for last few decades. The implementation is realized by the sequential Monte Carlo particle filter which has become possible by the advent of the high-speed and large mass storage computer.

Such remarkable advance of the technology has made possible the solution of various complex and difficult inverse problems. A reconstruction of the railway track geometry from its measurements is one of the typical inverse problems in the railway engineering. A new method described in this paper can not only reconstruct a geometric shape of the track involving the design track component but also disclose variations of angles, curvatures in the vertical and horizontal directions along the track. In this paper, firstly various methods adopted in order to apply the nonlinear filtering to reconstruction problem are described. They include a parametric representation of 3D track geometry driven by local curvatures which are given by random walk models, a measurement model for 3-point asymmetrical chord offsets for level and alignment, a self-organizing state space model enabling an estimation of unknown curvatures. In our inverse problem, it is necessary to take into account a compromise between fitting the measurement data and eliminating unwanted features in the solution, just like the Tikhonov regularization in the linear inverse problem. In the Bayesian framework, the prior density plays an important role on the regularization from the statistical interpretation's point of view. We also mention about recent studies concerning obtaining more accurate and smoother solutions.

## 2 MEASUREMENT METHOD OF LEVEL AND ALIGNMENT(3),(4),(5)

Figure 1 illustrates the chord offset method for level and alignment.

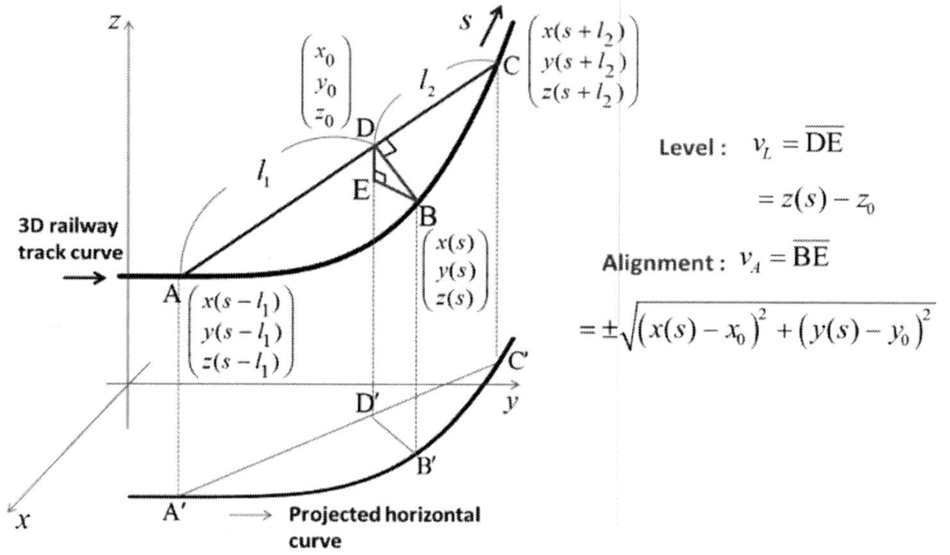


**Figure 1 3-point asymmetrical chord offset for level and alignment**

In this study, it is assumed that the component in the vertical direction of the offset $\overline{BD}$ is the offset for level $v_L$ and the component in the horizontal one is the offset for alignment $v_A$, respectively. In this case, following relations between 3D track geometry and measurements for level and alignment are simply obtained.

$$v_L = z(s) - z_0 = z(s) - \left(\alpha z(s-l_1) + \beta z(s+l_2)\right)$$

$$v_A = \pm\sqrt{\{x(s)-x_0\}^2 + \{y(s)-y_0\}^2}$$

*where* $x_0 = \alpha x(s-l_1) + \beta x(s+l_2), \quad y_0 = \alpha y(s-l_1) + \beta y(s+l_2),$

$$z_0 = \alpha z(s-l_1) + \beta z(s+l_2), \quad \alpha = l_2/L, \quad \beta = l_1/L$$

Measurements of level and alignment are denoted by $d_L$ and $d_A$, respectively. In this study, assuming additive measurement noises $w_L$ and $w_A$ having known probability densities $p_{w,L}(\cdot)$ and $p_{w,A}(\cdot)$, respectively, the measurement model is expressed as

$$\mathbf{v}_n = \left[v_{L,n}, v_{A,n}\right], \quad \mathbf{w}_n = \left[w_{L,n}, w_{A,n}\right], \quad \mathbf{d}_n = \left[d_{L,n}, d_{A,n}\right] = \mathbf{v}_n + \mathbf{w}_n.$$

## 3 RECONSTRUCTION BASED ON THE BAYESIAN APPROACH[(1),(5),(6)]

We solve the inverse problem of reconstructing the railway track geometry using the Bayesian nonlinear filtering. In the Bayesian framework, two models are required: first, a system model and second, a measurement model. A system model is given as

$$\mathbf{u}_n = \mathrm{F}(\mathbf{u}_{n-1}, \mathbf{r}_{n-1}).$$

F is a known, nonlinear function. $\mathbf{u}_n$ is a state vector representing a track geometry. $\mathbf{r}_{n-1}$ is a vector of white noises causing the evolution of the system. A system model must satisfy the condition of a Markov process of order one. A measurement model is given as already described, assuming additive measurement noise $\mathbf{w}_n$.

$$\mathbf{d}_n = \mathbf{v}_n + \mathbf{w}_n, \quad where \quad \mathbf{v}_n = \mathrm{H}(\mathbf{u}_n).$$

H is a nonlinear function expressing a measurement method. A measurement model is required to be defined so that its output depends only on the current state of the system. Bayesian filtering consists of two stages: prediction and filtering. At the prediction stage, supposing that the posterior density $p(\mathbf{u}_{n-1}|\mathbf{D}_{n-1})$ at distance $n-1$ is available, the prediction density $p(\mathbf{u}_n|\mathbf{D}_{n-1})$ of the state at distance $n$ is calculated via the Chapman-Kolmogorov equation:

$$p(\mathbf{u}_n|\mathbf{D}_{n-1}) = \int p(\mathbf{u}_n|\mathbf{u}_{n-1})\, p(\mathbf{u}_{n-1}|\mathbf{D}_{n-1}) d\mathbf{u}_{n-1},$$

where $\mathbf{D}_{n-1} \triangleq \{\mathbf{d}_i, i = 1, \ldots, n-1\}$ is a sequence of all available measurements up to distance $n-1$. A prior probability distribution $p(\mathbf{u}_n|\mathbf{u}_{n-1})$ is determined by the known statistics of $\mathbf{r}_{n-1}$. At the filtering stage, at distance step $n$ when a measurement $\mathbf{d}_n$ becomes available, the prediction density is updated via the Bayes' rule:

$$p(\mathbf{u}_n|\mathbf{D}_n) = \frac{p(\mathbf{d}_n|\mathbf{u}_n) p(\mathbf{u}_n|\mathbf{D}_{n-1})}{\int p(\mathbf{d}_n|\mathbf{u}_n)\, p(\mathbf{u}_n|\mathbf{D}_{n-1}) d\mathbf{u}_n}.$$

These processes are executed recursively and finally a sequence of posterior probability distribution for the state is obtained. An optimal estimate for the track geometry can be calculated from the posterior density with respect to any criterion. For example, the minimum mean-square error (MMSE) estimate is the conditional mean of:

$$\hat{\mathbf{u}}_{n|n}^{MMSE} \triangleq E\{\mathbf{u}_n|\mathbf{D}_n\} = \int \mathbf{u}_n \cdot p(\mathbf{u}_n|\mathbf{D}_n) d\mathbf{u}_n.$$

## 4 CONSTRUCTION OF THE SYSTEM MODEL

### 4.1 Parametric representation of three dimensional curve[(5)(7),(8)]

In order to apply a recursive nonlinear filtering, it is necessary to describe 3D railway track geometry as a dynamic system. Taking the Cartesian coordinates $(x, y, z)$ in the 3D space, as shown in Figure 2, a point $\mathrm{P}(x, y, z)$ on the three dimensional curve is represented by the vector $\mathbf{r}(s) = \overrightarrow{OP}$. A unit vector tangent to a curve at the point P is expressed as

$$\frac{d\mathbf{r}}{ds} = \lim_{\Delta s \to 0} \frac{\Delta \mathbf{r}}{\Delta s} = \left(\frac{dx}{ds}, \frac{dy}{ds}, \frac{dz}{ds}\right), \quad where \quad \left|\frac{d\mathbf{r}}{ds}\right| = 1$$

In addition, by defining a spherical coordinate $(r, \varphi, \theta)$ , the tangent vector $d\mathbf{r}/ds$ on the 3D curve is expressed by a set of differential equations:

$$\frac{dx}{ds} = \cos\theta\cos\varphi, \quad \frac{dy}{ds} = \cos\theta\sin\varphi, \quad \frac{dz}{ds} = \sin\theta, \quad \frac{d\theta}{ds} = \kappa(s), \quad \frac{d\varphi}{ds} = \tau(s),$$

where $\theta(s)$ and $\varphi(s)$ denote an angle of inclination in the vertical direction and an angle of direction in the horizontal direction, respectively. By differentiating the angles, two local curvatures $\kappa(s)$ and $\tau(s)$ are introduced. Obviously, a 3D curve is completely determined if we can know local curvatures $\kappa(s)$ and $\tau(s)$, except for its global positioning and orientation. Thus in the 3D space, we can consider a railway track geometry as if it were a dynamic system driven by local curvatures.

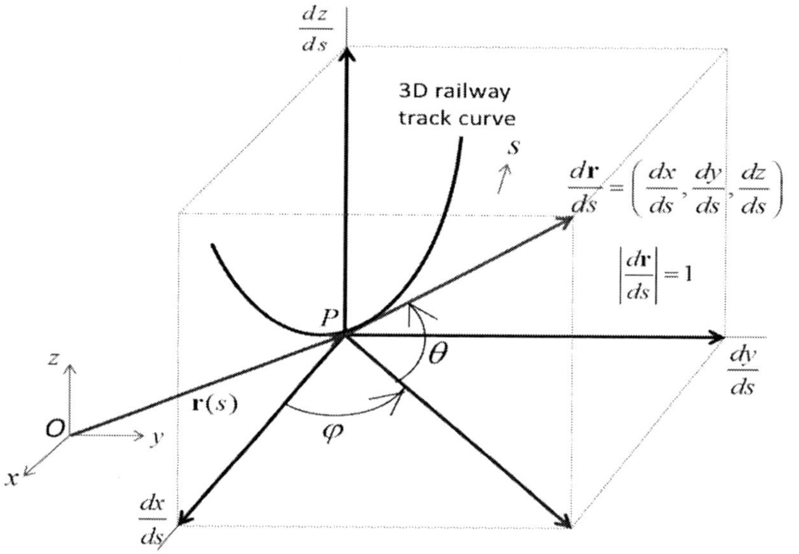


**Figure 2 Parametric representation of 3D curve**

To get a discrete model, the above differential equations are discretised applying the Backward differentiation formulae (BDF), where $\alpha_1, \ldots, \alpha_6, \beta_0$ are coefficients of BDF method of order 6 and $h \equiv \Delta s$ is a sampling interval for the distance s along the track.

$$x_n = -\sum_{i=1}^{6} \alpha_i x_{n-i} + h\beta_0 \cos\theta_n \cos\varphi_n, \qquad y_n = -\sum_{i=1}^{6} \alpha_i y_{n-i} + h\beta_0 \cos\theta_n \sin\varphi_n$$

$$z_n = -\sum_{i=1}^{6} \alpha_i z_{n-i} + h\beta_0 \sin\theta_n, \quad \theta_n = -\sum_{i=1}^{6} \alpha_i \theta_{n-i} + h\beta_0 \kappa_n, \quad \varphi_n = -\sum_{i=1}^{6} \alpha_i \varphi_{n-i} + h\beta_0 \tau_n$$

### 4.2 Definition of the state vector for the railway track geometry[5],[6]

In order for the output of the measurement model to depend only on the current state of the system, it must be possible to calculate a chord offset from the current state of the railway track geometry. Therefore, each state vector must hold consecutive railway track geometry along the track, at least, corresponding to the length of the 3-point chord. We define a state vector for the railway track geometry as a set of multi-dimensional vectors, as follows. A state vector $\mathbf{u}_n$ denoting a current state of a railway track geometry has a consecutive railway track geometry data ranging from $s = s_{n-a}$ to $s = s_{n+b}$, where $a$ and $b$ are taken to be $ah = l_1$, $bh = l_2$, respectively.

$$\mathbf{u}_n = \left[\mathbf{x}_n, \mathbf{y}_n, \mathbf{z}_n, \boldsymbol{\theta}_n, \boldsymbol{\varphi}_n, \boldsymbol{\kappa}_n, \boldsymbol{\tau}_n\right], \quad where \quad \mathbf{x}_n = \left[x_{n+b}, \cdots, x_n, \cdots, x_{n-a}\right],$$

$$\mathbf{y}_n = \left[y_{n+b}, \cdots, y_n, \cdots, y_{n-a}\right], \quad \mathbf{z}_n = \left[z_{n+b}, \cdots, z_n, \cdots, z_{n-a}\right],$$

$$\boldsymbol{\theta}_n = \left[\theta_{n+b}, \cdots, \theta_n, \cdots, \theta_{n-a}\right], \quad \boldsymbol{\varphi}_n = \left[\varphi_{n+b}, \cdots, \varphi_n, \cdots, \varphi_{n-a}\right],$$

$$\boldsymbol{\kappa}_n = \left[\kappa_{n+b}, \cdots, \kappa_n, \cdots, \tau_{n-a}\right], \quad \boldsymbol{\tau}_n = \left[\tau_{n+b}, \cdots, \tau_n, \cdots, \tau_{n-a}\right].$$

### 4.3 Evolution of the railway track geometry by random walk models[5],[6]

In order to realize the evolution of the railway track geometry and at the same time estimate its real values of curvatures, unknown curvatures κ(s) and τ(s) have been included in the definition of the state vector. Specifically, unknown curvatures are defined as random walk models, as follows.

$$\begin{cases} \kappa_{n+b} = \kappa_{n-1+b} + r_{\kappa,n-1} \\ \tau_{n+b} = \tau_{n-1+b} + r_{\tau,n-1} \end{cases}, \qquad \mathbf{r}_{n-1} = \left[r_{\kappa,n-1}, r_{\tau,n-1}\right]$$

where $r_{\kappa,n-1}$ and $r_{\tau,n-1}$ are white noises and their probability distributions are specified in advance. Though curvatures are unknown, as the recursive filtering makes progress, random walk models work well and good estimates for curvatures are obtained. This outstanding feature in the Bayesian estimation is referred to as a self-organizing state-space model.

## 5 IMPLEMENTATION BY THE MONTE CARLO PARTICLE FILTER[1],[5],[6]

In order to implement the recursive nonlinear filtering in practice, the Monte Carlo particle filter is used. In the particle filter, each probability distribution is approximated by many independent realizations from that distribution. They are called particles. Two kinds of particles approximating a prediction distribution and a filter distribution, respectively, are defined, as follows.

$$\mathbf{p}_n^{(j)} = \left[\mathbf{x}_n^{(j)}, \mathbf{y}_n^{(j)}, \mathbf{z}_n^{(j)}, \boldsymbol{\theta}_n^{(j)}, \boldsymbol{\varphi}_n^{(j)}, \boldsymbol{\kappa}_n^{(j)}, \boldsymbol{\tau}_n^{(j)}\right] \sim p(\mathbf{u}_n \mid \mathbf{D}_{n-1}) \quad : prediction$$

$$\mathbf{f}_n^{(j)} = \left[\mathbf{x}_n^{(j)}, \mathbf{y}_n^{(j)}, \mathbf{z}_n^{(j)}, \boldsymbol{\theta}_n^{(j)}, \boldsymbol{\varphi}_n^{(j)}, \boldsymbol{\kappa}_n^{(j)}, \boldsymbol{\tau}_n^{(j)}\right] \sim p(\mathbf{u}_n \mid \mathbf{D}_n) \quad : filtering$$

$$j = 1, \cdots, M \qquad M : \text{the number of particles}$$

The number of particles in the set has a great influence on the accuracy of the reconstructed track geometry. Prediction and filtering operation practiced in each particle is illustrated in Figure 3.

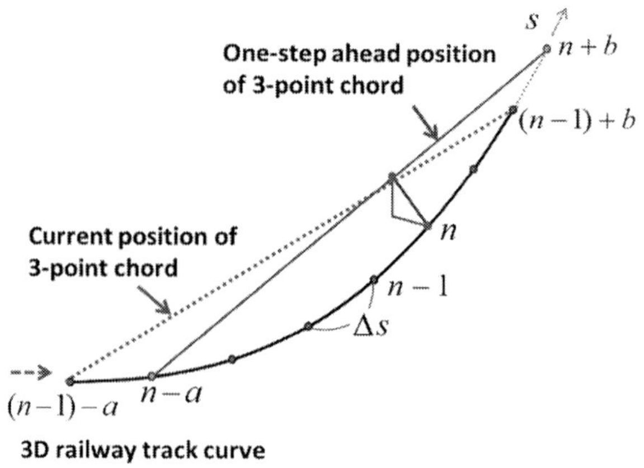


**Figure 3 Practice of prediction and filtering at one-step evolution**

At each step of the reconstruction of the railway track geometry, local curvatures are generated by random walk models and substituting them into the differential equations a new railway track position is calculated.

$$\begin{cases} \kappa_{n+b}^{(j)} = \kappa_{n-1+b}^{(j)} + r_{\kappa,n-1}^{(j)} \\ \tau_{n+b}^{(j)} = \tau_{n-1+b}^{(j)} + r_{\tau,n-1}^{(j)} \end{cases}$$

Then, taking a new position as a leading point, offsets for level and alignment are calculated. Then, they are compared with the measured data at the corresponding position. Using the probability densities $p_{w,L}(\cdot)$, $p_{w,A}(\cdot)$ of the measurement noises for level and alignment, respectively, a probability for new position can be calculated.

$$\underbrace{\begin{cases} w_{L,n}^{(j)} = d_{L,n} - v_{L,n}^{(j)} \\ w_{A,n}^{(j)} = d_{A,n} - v_{A,n}^{(j)} \end{cases}}_{\textit{measurement noise}} \Rightarrow \underbrace{\begin{cases} L_{L,n}^{(j)} = p_{w,L}(w_{L,n}^{(j)}) \\ L_{A,n}^{(j)} = p_{w,A}(w_{A,n}^{(j)}) \end{cases}}_{\textit{probability}} \Rightarrow \underbrace{\begin{cases} \bar{L}_{L,n}^{(j)} = L_{L,n}^{(j)} \Big/ \sum_{k=1}^{M} L_{L,n}^{(k)} \\ \bar{L}_{A,n}^{(j)} = L_{A,n}^{(j)} \Big/ \sum_{k=1}^{M} L_{A,n}^{(k)} \end{cases}}_{\textit{normalization}} \Rightarrow \underbrace{L_n^{(j)} = \bar{L}_{L,n}^{(j)} \cdot \bar{L}_{A,n}^{(j)}}_{\textit{likelifood of particle}}$$

Each particle is weighted in proportion to its likelihood. For the set of weighted particles, resampling is carried out in order to get a new set of particles with uniform weight. When a recursive filtering has completed, we have a sequence of a set of particles. Using this sequence we can calculate an optimal estimate of curvature at every point of the track and substituting it into the differential equation of the system once again, optimal estimates of all other system variables are obtained along the track.

## 6 EXAMPLE OF THE BAYESIAN RECONSTRUCTION

### 6.1 Input measured data and parameter specifications

Figure 4 and Figure 5 show a design track geometry of the railway track section taken as an example in the Bayesian reconstruction. It is characterized by the elevation in the vertical direction and the curvature in the horizontal one. A design track geometry is only a target one. A real track geometry cannot avoid deteriorations due to frequent passages of trains and has a track irregularity.

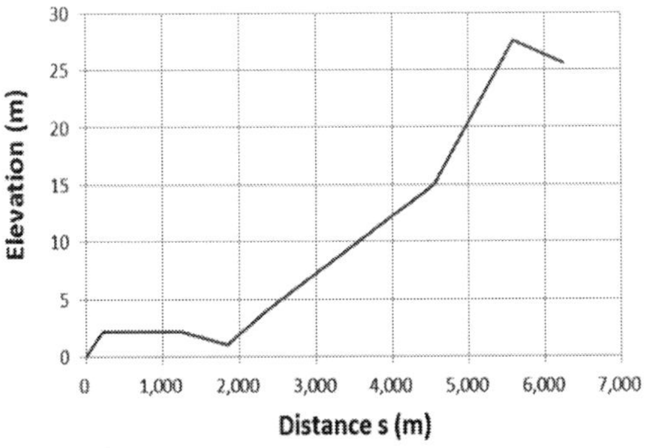


**Figure 4 Elevation in the vertical direction**

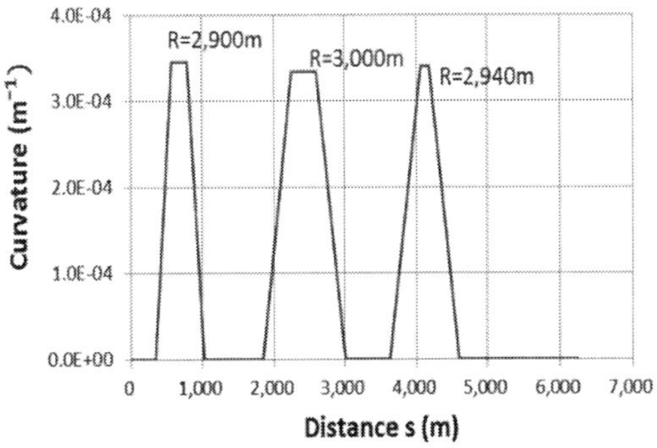


**Figure 5 Curvature in the horizontal direction**

Figure 6 and Figure 7 show measured 3-point chord offsets for level and alignment. Parameter specifications in the reconstruction are summarized in Table 1.

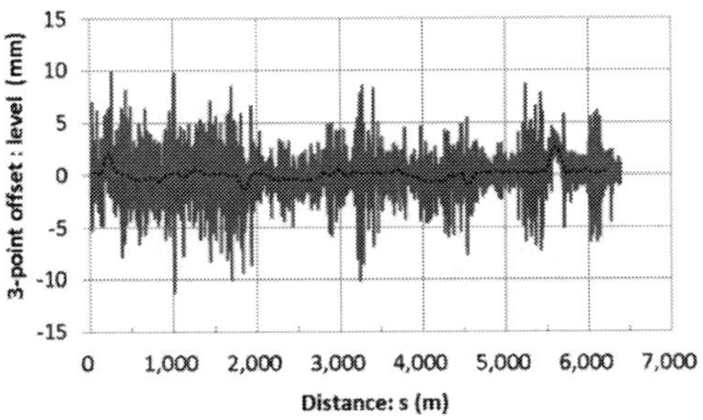


**Figure 6 Measured 3-point chord offset for level**

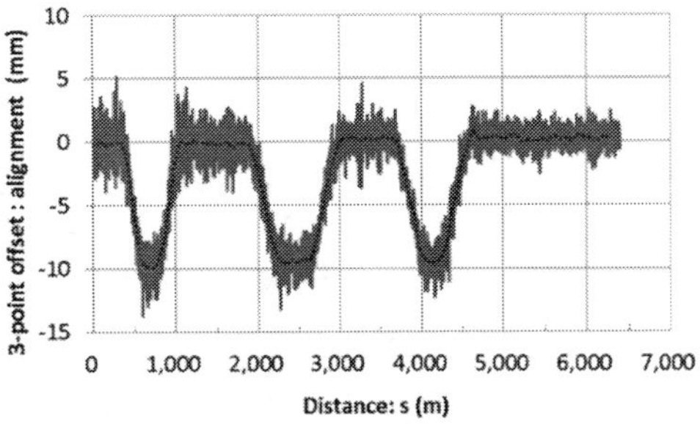


**Figure 7 Measured 3-point chord offset for alignment**

**Table 1 Parameter specifications in the reconstruction**

| *Item* | *Value* |
|---|---|
| Length of track section | 6,250 m |
| Number of measurement data | 25,002 |
| Sampling interval | $h = \Delta s = 0.25$ m |
| 3-point chord : level | $L = 20$ m ,$l_1 = 17.5$ m, $l_2 = 2.5$ m |
| 3-point chord : alignment | $L = 20.8$ m ,$l_1 = 17.5$ m, $l_2 = 3.3$ m |
| The number of particles | 20,000 |
| Dimension of particle | 422 |
| Random walk model 1 : κ(s) | Gaussian $\sigma_\kappa = 0.0021$ |
| Random walk model 2 : τ(s) | Gaussian $\sigma_\tau = 0.00067$ |
| Measurement noise 1 : level | Cauchy $\sigma_L = 0.3$ mm |
| Measurement noise 2 : alignment | Cauchy $\sigma_A = 0.4$ mm |

By the way, in addition to the Bayesian reconstruction, there is a simple method of reconstructing only a design track geometry from measured 3-point chord offsets for level and alignment. As shown in Figure 8, an offset $e$ of the 3-point chord subtending a circle with a constant radius $R$ is determined easily from the geometrical relation, as follows.

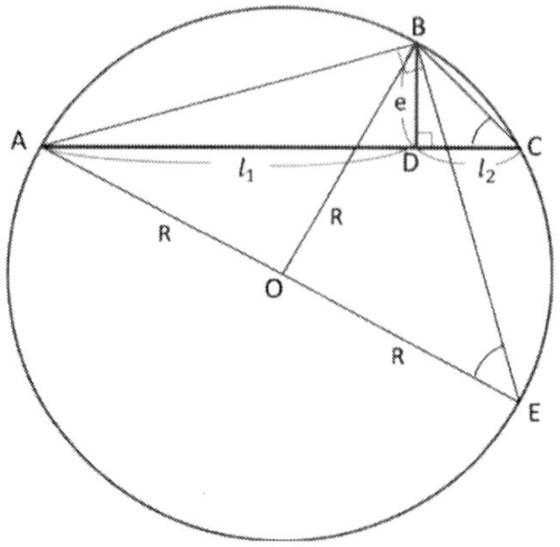


$$\frac{1}{R} = \frac{2e}{\sqrt{l_1^2 + e^2}\sqrt{l_2^2 + e^2}}$$

$$\approx 2e \Big/ l_1 l_2 \left[ 1 + \frac{1}{2} \left\{ \left( \frac{e}{l_1} \right)^2 + \left( \frac{e}{l_2} \right)^2 \right\} \right],$$

where $l_1$, $l_2$ are the respective chord lengths of the 3-point chord. $1/R$ is a curvature of a circle.

**Figure 8 3-point chord subtending a circle with a constant radius**

Using this equation, all measured 3-point chord offsets for level and alignment are transformed into curvatures. Eventually it makes up sequences of instantaneous curvatures along the track. Then for this sequence of instantaneous curvatures, smoothing operations are carried out, for example, by taking a moving average.

As a result, two sequences of smoothed curvatures $\kappa(s)$ and $\tau(s)$ are obtained corresponding to level and alignment measurement data. Smoothed curvatures are considered to be "track irregularity-free" curvatures, in other words, curvatures of a design track geometry. So substituting them into the differential equation of a 3D space curve (section 4.1), a design track geometry is reconstructed. Hereafter, we call it a "track irregularity-free" design track geometry. While a nominal design track geometry often does not reflect a current state of the track, a "track irregularity-free" design track geometry reflects it faithfully because it is reconstructed based on the latest measurement.

In Figure 6 and Figure 7, 3-point chord offsets calculated for the "track irregularity -free" design track geometry are plotted with dotted line.

### 6.2 Results of the Bayesian reconstruction

Figure 9 to Figure 14 show results of the Bayesian reconstruction. In these figures, corresponding parameters resulting from the "track irregularity-free" design track geometry are also plotted with dotted line. Every data pertaining to the real track geometry is developing along close to the corresponding data of the "track irregularity-free" design track geometry. This indicates the usefulness of the "track irregularity-free" design track geometry.

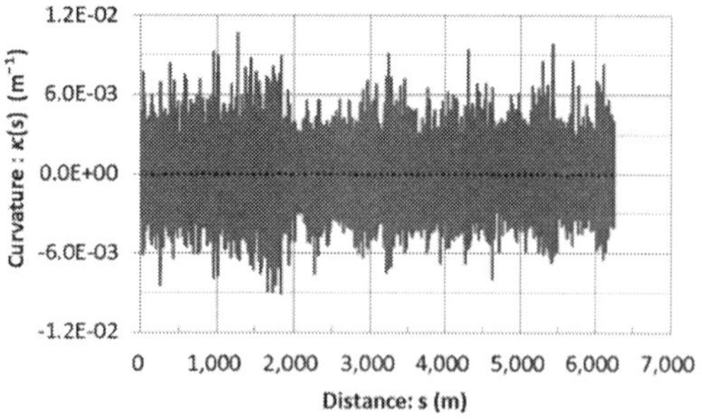


**Figure 9 Vertical curvature** $\kappa(s)$

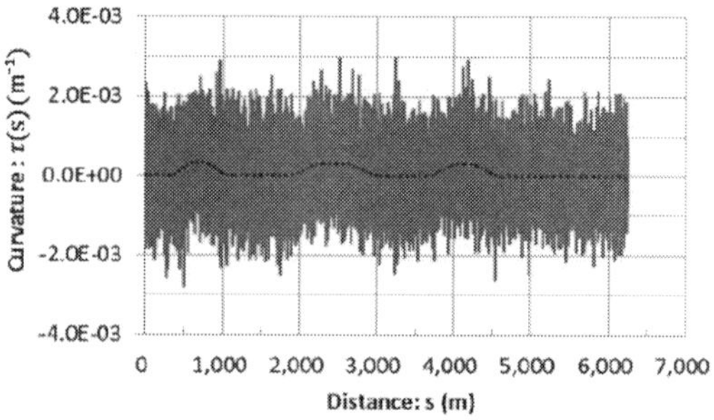


**Figure 10 Horizontal curvature** $\tau(s)$

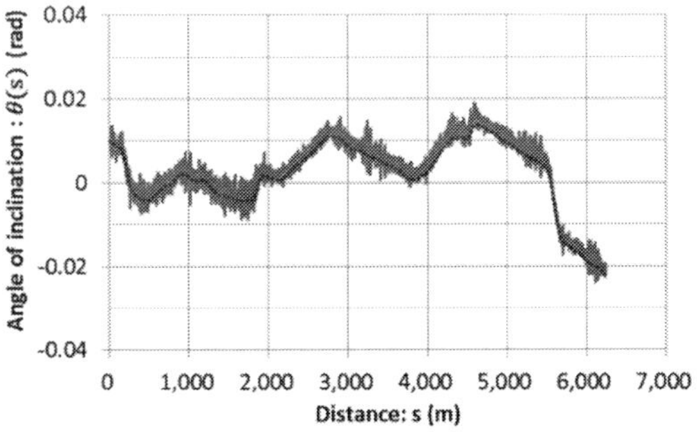


**Figure 11 Angle of inclination** $\theta(s)$

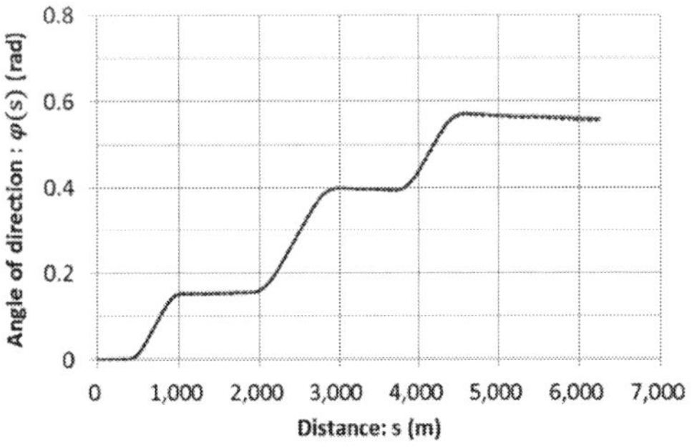


**Figure 12 Angle of direction** $\varphi(s)$

Figure 13 and Figure 14 show a vertical curve and planar curve in the reconstructed track geometry, respectively. A real track geometry consists of both components of design track and track irregularity. But a track irregularity is very small and invisible in the scale of these figures.

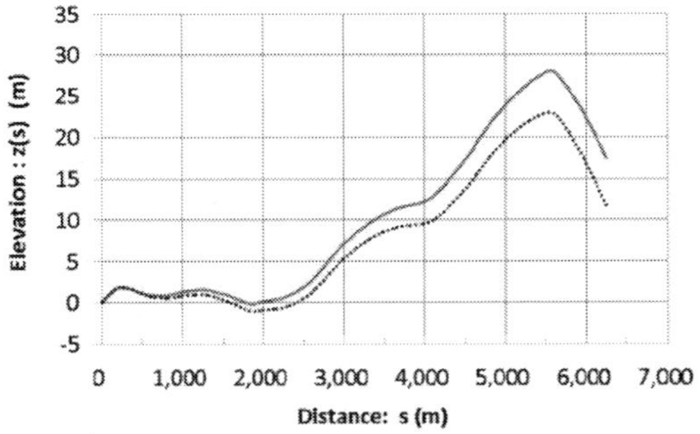


**Figure 13 Vertical curve** $z(s)$

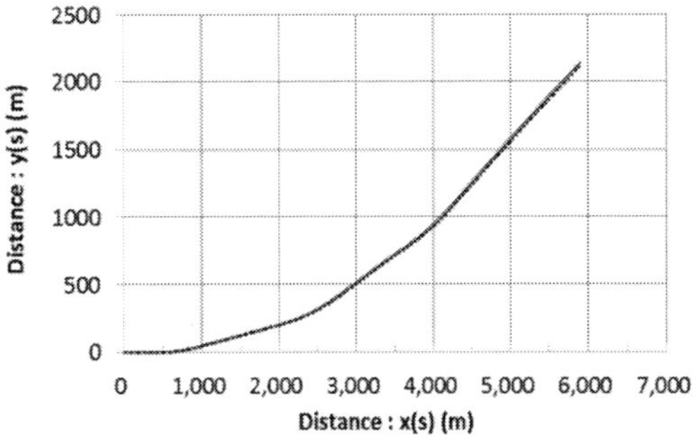


**Figure 14 Planar curve** $(x(s), y(s))$

To see a track irregularity is exactly reconstructed, 3-point chord offsets are calculated for the reconstructed track geometry. Then they are compared with the measured ones. Figure 15 and Figure 16 show graphs plotted with them superposed. A good agreement can be seen. To know more details about this agreement, Figure 17 and Figure 18 show results of the power spectral analysis of these data. As far as a signal to noise ratio permits, power spectra of the measured 3-point chord offset and the calculated one develop a good correspondence in the wide frequency range. A real track geometry is reconstructed very accurately by the Bayesian estimation.

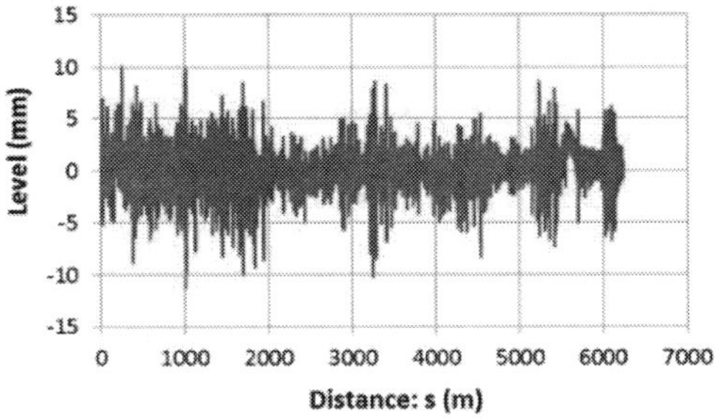


**Figure 15 Measured and calculated 3-point chord offsets for level**

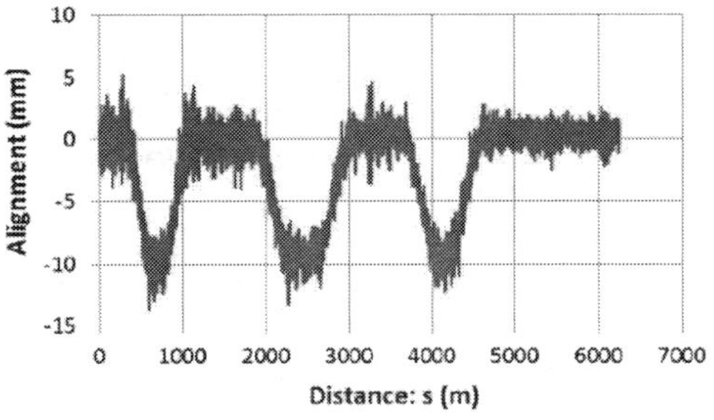


**Figure 16 Measured and calculated 3-point chord offsets for alignment**

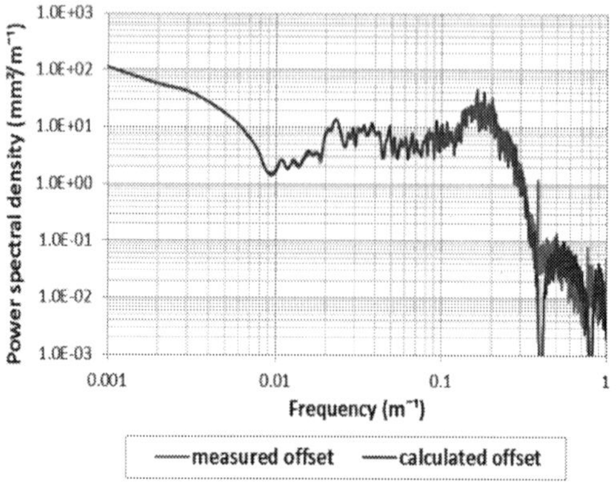


**Figure 17 Power spectra of measured and calculated 3-point chord offsets for level**

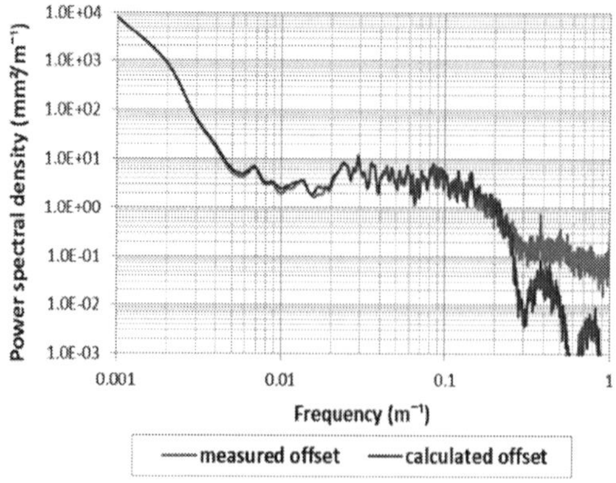


**Figure 18 Power spectra of measured and calculated 3-point chord offsets for alignment**

By the way, in Figure 13 the vertical curve of the reconstructed real track geometry is a little bit out of it of the "track irregularity-free" design track geometry. To know this cause, it is necessary to investigate influences of the measurement noise on the reconstruction. By taking a difference between a measured 3-point chord offset and a calculated one, a measurement noise can be estimated. Figure 19 and Figure 20 show measurement noises of 3-point chord offsets for level and alignment $\widehat{w}_L$, $\widehat{w}_A$, respectively. Their standard deviations are $\hat{\sigma}_L = 0.31\,mm$, $\hat{\sigma}_A = 0.45\,mm$. These are a little bit larger than those specified in advance in Table 1. This means that there is yet room for improvement in the Bayesian reconstruction.

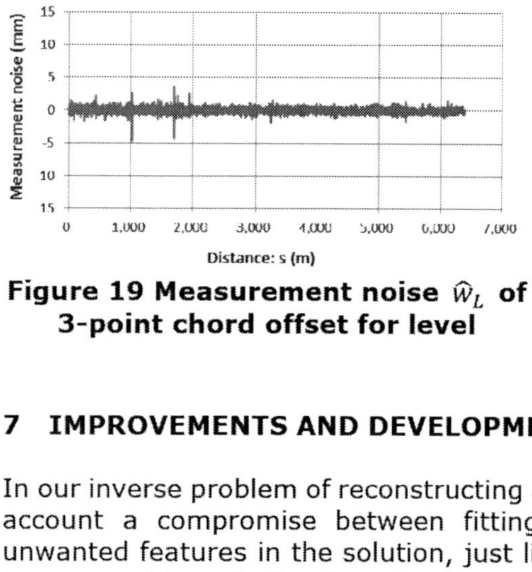


**Figure 19 Measurement noise $\widehat{w}_L$ of 3-point chord offset for level**

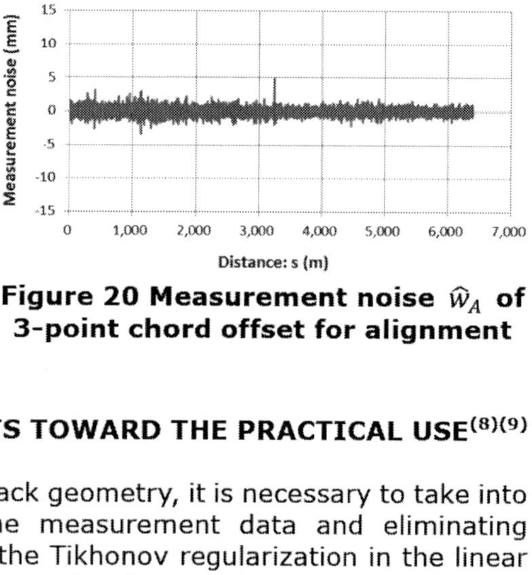


**Figure 20 Measurement noise $\widehat{w}_A$ of 3-point chord offset for alignment**

## 7 IMPROVEMENTS AND DEVELOPMENTS TOWARD THE PRACTICAL USE[8][9]

In our inverse problem of reconstructing a track geometry, it is necessary to take into account a compromise between fitting the measurement data and eliminating unwanted features in the solution, just like the Tikhonov regularization in the linear inverse problem. In the Bayesian framework, a prior density plays an important role on the regularization from a statistical interpretation's point of view. Specifically, a variance of a random input in the random walk model for a curvature has to be carefully adjusted concerning a smoothness of the solution. Recently, in order to obtain physically smoother curvatures, we fitted the Auto-Regressive time series model to the curvature obtained using random walk models and applied it again to the evolution of the system, instead of the random walk model. As a result, we got to smaller standard deviations of the estimated measurement noises of $\hat{\sigma}_L = 0.21\,mm$, $\hat{\sigma}_A = 0.37\,mm$, compared with the previous results. This means that the great fidelity of the solution to the measured data has been attained while keeping the smoothness of the solution.

Figure 21 to Figure 24 show results concerning the elevation in the vertical direction among the reconstruction using the AR model. The vertical curve of the reconstructed real track geometry corresponds very well to it of the "track irregularity-free" design track geometry, as shown in Figure 23.

Figure 25 shows a histogram from a set of 25,000 data points of the measurement noise $\hat{w}_L$. The characteristic bell-shaped curve in the histogram might make it appear at first that the measurement noise is normally distributed. However, as shown in Figure 26, a normal Q-Q plot (Q: quantile) for measurement noise $\hat{w}_L$ makes it apparent that $\hat{w}_L$ contains substantially more extreme values than the normal distribution would predict. This indicates that the assumption of the Cauchy distribution for the measurement noise specified in Table 1 works well. The Cauchy distribution is equivalent to the t distribution with one degree of freedom. The Student's t distribution has a very wide wings in the distribution than the normal distribution. According to the experiments practiced so far, it is perceived the t distribution is superior to the Gaussian distribution concerning the removal of unexpected impulsive noise.

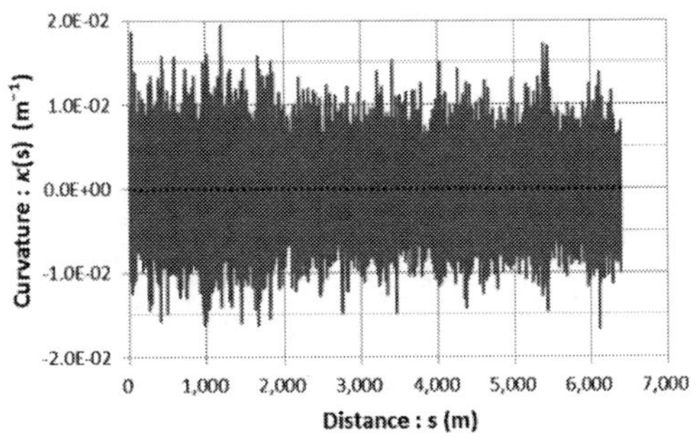


**Figure 21 Vertical curvature $\kappa(s)$**

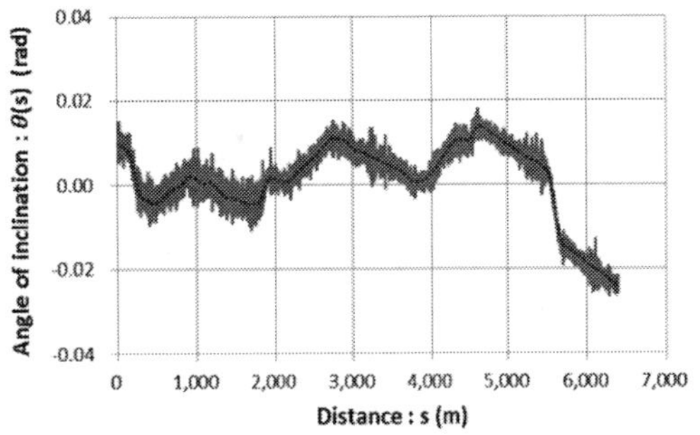


**Figure 22 Angle of inclination $\theta(s)$**

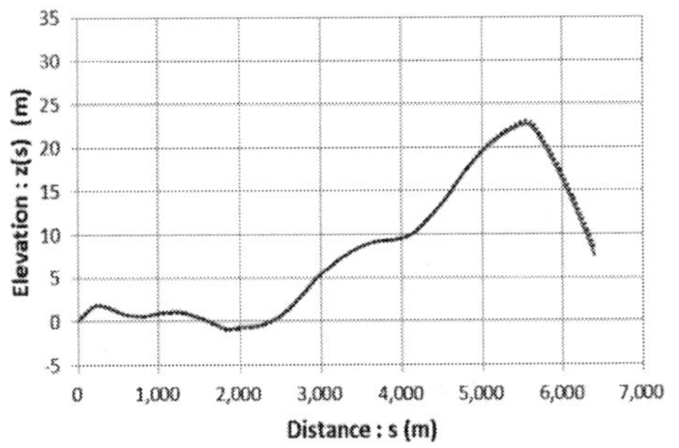


**Figure 23 Vertical curve $z(s)$**

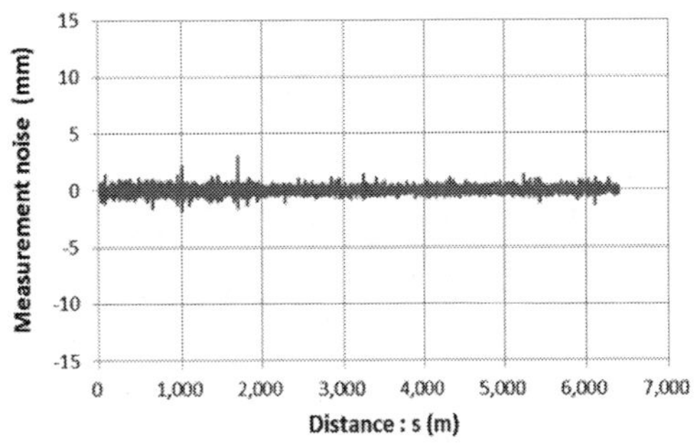


**Figure 24 Measurement noise $\hat{w}_L$ of 3-point chord offset for level**

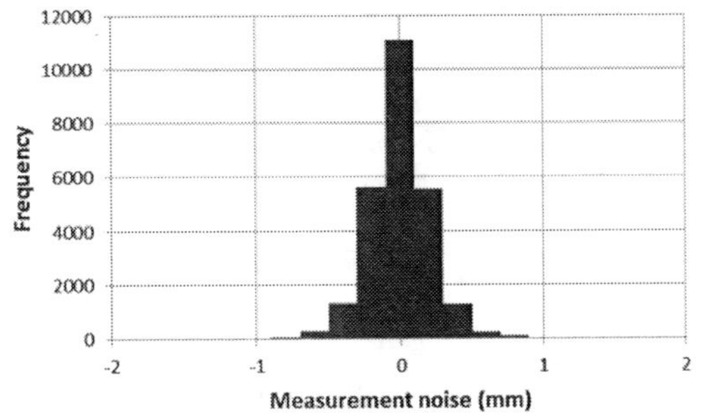


**Figure 25 Histogram of noise $\hat{w}_L$**

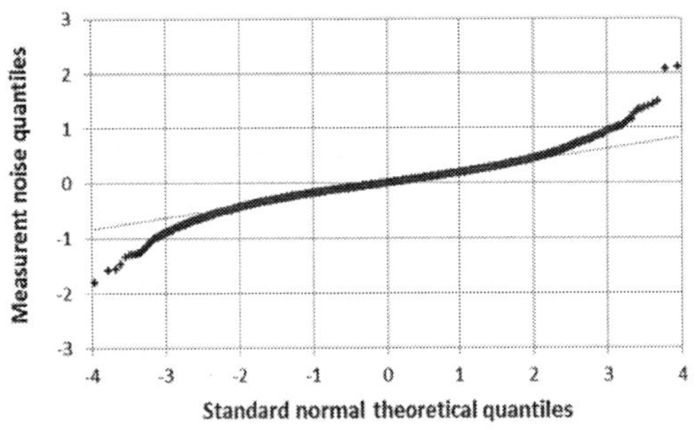


**Figure 26 Q-Q plot for noise $\hat{w}_L$**

We have attained a great improvement on the feature of the solution in both effectiveness and efficiency. This is referred to as a smoothness prior. It goes without saying that a variance of the measurement noise has a large influence on the fidelity of the solution to the measured data.

## 8 CONCLUSION

In this paper a new method reconstructing many useful information such as design track, track irregularities, curvatures, angles of inclination and direction, about the railway track geometry in 3D space from 3-point chord offset measurements is described. They have been considered to be impossible so far. Especially local curvatures along the track in the vertical and horizontal directions will be very useful not only for railway track maintenance but also for railway vehicle dynamic simulations. Originally the primary objective of the chord offset measurement was to know the variation of curvatures along the track in order to secure the running safety and maintain a comfortable riding quality of trains. From this point of view, a new method is expected to bring a new innovation in the research field of the railway track maintenance and vehicle track dynamic interaction. I am going on with my research toward further improvement.


## ACKNOWLEDGEMENTS

The author would like to thank Y. Kawasaki and Y. Naganuma in JR Central for useful discussions.


## REFERENCES


(1) B. Ristic, S. Arulampalam and N. Gordon, "Beyond the Kalman Filter, Particle Filters for Tracking Applications", Artech House, 2004.

(2) J. Evans and M. Berg. "Challenges in simulation of rail vehicle dynamics", Vehicle System Dynamics, Vol. 47, No. 8, August 2009, 1023-1048.

(3) A. Yoshimura and Y. Naganuma, "A New Method to Reconstruct the Track Geometry from Versine Data Measured in the Curved Track Using the Monte Carlo Particle Filter", Railway Engineering 2013, 12th Int. Conf., Imperial College, London, UK, 10th-11th July 2013.

(4) A. Yoshimura and Y. Naganuma, "A Novel Method of Reconstructing the Railway Track Geometry Using the Particle Filter and the Possibility of the Reconstruction of the 3D Track Geometry", 10th World Congress on Railway Research, WCRR2013, Sydney, Australia, 25th-28th November 2013.

(5) A. Yoshimura and Y. Naganuma, "Bayesian reconstruction of 3D railway track geometry by particle filter", in J. Pombo, (Editor), Proc. of the 2nd Int. Conf. on Railway Technology: Research, Development and Maintenance, Civil-Comp Press, Stirlingshire, UK, Paper 88, 2014, doi:10.4203/ccp.104.88.

(6) G. Kitagawa, "A Self-Organizing State-Space Model", Journal of the American Statistical Association, 93(443), 1203-1215, 1998.

(7) U. M. Ascher and L. R. Petzold, "Computer Methods for Ordinary Differential Equations and Differential-Algebraic Equations", chap.5 129-131, SIAM, Philadelphia, 1998.

(8) G. Kitagawa and W. Gersch, "Smoothness Priors Analysis of Time Series", Lecture Notes in Statistics 116, Springer, 1996.

(9) D. Cavetti and E. Somersalo, "Introduction to Bayesian Scientific Computing, Ten Lectures on Subjective Computing", Surveys and Tutorial in the Applied Mathematical Sciences Vol.2, Springer, 2007.

# Design, construction and operation of a REPOINT laboratory demonstrator

**S D Bemment, R Dixon, R Goodall, C P Ward, N Wright**
School of Electronic, Electrical and Systems Engineering,
Loughborough University, UK

## ABSTRACT

The REPOINT project, led by Loughborough University, has been active since March 2011. It seeks to improve the reliability, safety and maintainability of track switching technology, with the aim of increasing network capacity and lowering operating costs. To do this, the project is exploring combining mature concepts from other industries such as fault tolerance, line-replaceable units and passively safe design, with novel mechanical arrangements, in order to bring about a step change in performance. One design, based around a stub-switch arrangement, has showed particular promise and is the currently the subject of three patent applications covering the novel mode of operation. A laboratory-scale demonstrator of all key subsystems is currently under construction, under funding from the FutureRailway.org team. This is integrated with test and monitoring equipment, alongside a rapid-prototyping control system. This first-generation design will be used to prove the concept of operation and to develop the associated control and monitoring technology.

The goal of this paper is to provide an overview of the REPOINT project to date, and the design and operation of the proposed novel REPOINT design. This paper firstly introduces the REPOINT project and highlights of the proposed novel design. It then discusses the simulation, modelling and design of the demonstrator rig, and the associated test and development equipment. The conclusions highlight the progress so far – on the REPOINT project and the Demonstrator rig - and comment upon potential next steps towards network deployment.

## 1 INTRODUCTION

Railway track switching provides necessary flexibility to a rail network, allowing vehicles to take a multitude of routes. However, track switches represent single points of failure, and even when operational can introduce capacity constraints due to the design of physical track components and the associated signalling systems for control and operation. Switches are expensive and complex designs when compared to equivalent plain line (1, 2). Their population is therefore generally optimised at design time - alongside a known timetable - in order to minimise initial outlay and substantial ongoing maintenance costs. This can compound the negative effects upon network delay performance during timetable perturbation. With the anticipated move towards cab-signalling, the only remaining 'active' line-side assets will be switches and level crossings, and these will thus contribute an ever increasing portion of network delay totals without significant further work to improve performance.

In the United Kingdom, open-access statistics (3) show passenger counts are at their highest level since re-privatisation, with some lines now running at or near operational vehicular capacity. This fact, when coupled with cross-industry initiatives such as the '24/7 Railway' (4), 'On-time Railway', and increasing overnight freight utilisation as suggested in the recent IMechE 'Rail Freight Report' (5), is much reducing the portion of time available to take maintenance possessions of infrastructure. Importantly, it is often not the physical maintenance act itself which is expensive in monetary terms, but instead the time the asset is out of use - whether this be for a planned maintenance intervention or unanticipated failure. This monetary cost is therefore associated to a capacity cost.

It is commonplace in some industries to replicate critical components in order to increase whole-system reliability and/or as a step to enable improved maintainability (6). Commercial aircraft, for instance, would typically have triplex or quadruplex redundancy built in to flight control systems as repair or replacement - due to an incipient fault in a single subsystem - must necessarily wait until after the aircraft has landed. This approach has made some inroads into certain rail industry assets – signalling bulbs and SSI (Solid State Interlocking) being prime examples. Indeed, SSI was state of the art in safety critical computing when under development in the 1970's, and other industries adopted the principles therein soon afterwards (7). However, multi-channel redundancy and/or extensive fault tolerance have not yet been adopted for track switches.

Recognising the restrictions switches place upon network performance, and the opportunities to exploit technologies, techniques and approaches from other industries, the REPOINT project was initiated in 2011. The goal was to investigate what could be changed in existing switch designs, and to propose novel concepts, in both physical design and operating rules. The project was also to quantify what effect these changes would have. The project background is covered in more detail in section two. As part of the investigation, a novel design of track switch was devised, which is introduced in section three. This paper covers the modelling, simulation, and construction for a 384mm gauge laboratory demonstrator of this novel arrangement. This work is presented in sections four and five. The demonstrator is now complete and operational, and the next steps in the research and development work are discussed in section six.

## 2 BACKGROUND

The REPOINT project (now referred to as REPOINT1) was initiated in 2011 by the United Kingdom's RSSB (Rail Safety and Standards Board) as part of a call for proposals titled: 'Railway Capacity: overcoming the constraints caused by nodes (stations and junctions) on the rail network'. REPOINT was one response to that call, to examine whether the redundancy and fault-tolerant concepts which are readily accepted in other safety critical environments could be used beneficially for track switching on railway networks. The project set out to answer the research question:

*'Could a fundamental re-think of railway track switching ease some of the current route-setting constraints to provide higher capacity, and provide a significant reduction in operational unreliability arising from points failures?'*

REPOINT proposed that due to the critical nature of operations at nodes, they require some built in redundancy to faults - fundamentally changing the nature of a junction to result in higher reliability of the individual assets and improved capacity. The capacity improvement comes partly from the reliability improvement and partly due to changes in the operating rules that become possible when the switches are redesigned to be intrinsically safe. The capacity benefits of re-engineering switches have been previously published by the authors, for instance in (8, 9).

Another significant output of this project was concept design for a novel arrangement of track switch, which enables this multi-channel approach. This switch uses multi-channel actuation elements, in a similar manner to aircraft flight control surfaces, and has the potential to provide the improved switch performance discussed in (8, 9). 'Performance', in this instance, refers to reduced lifecycle cost, increased availability, and improved maintainability. A detailed design of this proposal is available in more detail within the associated patent disclosures (10, 11). The concept design utilises four principles novel to track switching design, some inspired by the aerospace and nuclear industries: the LRU (Line-replaceable unit, as formally defined in standard MIL-PRF-49506 (12), mission-critical subsystem redundancy, fault tolerant control, and model-based condition monitoring. These have been combined with an idea fundamental to the rail sphere: Design-safe operation.

There currently exists a knowledge-gap regarding the practical implementation and operational aspects of this novel design. To bridge this gap, and move towards a prototype design, REPOINT 2 Phase 1 was initiated in May 2013 with the goal of constructing a functioning laboratory scale demonstrator. This demonstrator is now complete and fully operational in the control systems group laboratory at Loughborough University.

## 3 NOVEL TRACK SWITCHING ARRANGEMENT

The general layout of the 'traditional' design of track switch is well discussed in literature (See for example 1, 2) and is not included here for brevity. Generally, upon request from the signalling system, a single actuator moves two switch blades via a linkage, before locking the blades in place and communicating the detected position of the blades and lock back to the interlocking. In some installations there are multiple actuation units, or power take-offs known as 'back-drives' to ensure the entire movable length of blade moves correctly, however this is not multi-channel redundancy. Trains can be issued a movement authority (either by radio in communication-based signalling systems, or else by a line-side signal aspect) to pass the switch only once the correct detection signal is received by the interlocking. Movement of the switch blades normally takes several seconds. Around 8 seconds is allowed in British signalling practice, see, for example (13) for a more detailed discussion of switch control). As a switch represents a derailment danger when between positions and/or unlocked, the interlocking prevents signals being cleared in this state, and thus a service-affecting failure may ensue after a subsystem failure in any one of the elements listed above.

Adding multi-channel actuation to a track switch is non-trivial, as unlike the example of flight control surfaces, there is a requirement for the switch to be locked in a particular position for passing traffic (14). This essentially means each actuator must act through a common locking mechanism (and therefore a single point of failure), or else be able to unlock all other elements in the bank (perhaps necessitating an increasingly complex trackside mechanism as the number of actuation elements grows - with all associated maintenance issues). REPOINT has devised - and has patent pending upon - a simple yet novel arrangement referred to as 'passive locking', which overcomes these issues and for the first time may enable parallel multi-channel, actuation of track switches.

The actuation arrangement and passive locking could be applied to a traditional design of track switch, or more radically, a stub switch - the general layout of which is shown in figure 1. Use of a stub switch eliminates several of the more common failure points and modes of traditional track switches, including blockages between the switch and stock rails. However, practical engineering knowledge of stub switches upon a modern railway is limited, so additional development work would be required to bring this design to a deployable state. The stub switch reverses the traditional arrangement of 'heels' and 'toes', as described in (1), and utilises full-section rails throughout. It also allows arrangements with more than two routes from a single switch, impossible with the traditional design. All work described here on in in is related to the stub switch.

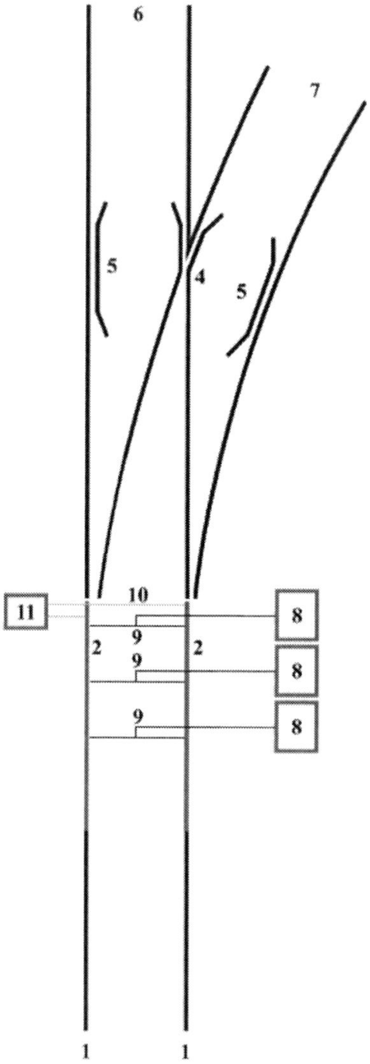


**Figure 1: Proposed high-availability stub-switch concept with redundant actuation paths. 1 (Black, Bold) Stock Rails; 2 (Grey, Bold) Moveable Switch Rails; 4 Common Crossing; 5 Check Rails; 6 Straight Route; 7 Turnout Route; 8 Multi-Channel Actuators; 9 (Black) Drive Rod and Linkages; 10 Detection Rods; 11 Blade Position Detection and Feed-back Unit. Reproduced from (9)**

Actuation elements and passive locking elements are located in bearers along the movable length of rails, termed 'Actuator-bearers'. All elements are connected to a trackside cabinet, with a processor in a layer abstracted from the signalling system, capable of isolating (but crucially not commanding movement from) individual actuation elements should the in-built condition monitoring suspect an incipient fault. Individual elements can also be isolated by visiting repair teams, such that minor repairs can be effected whilst the switch is still allowing traffic to pass.

Each actuation element consists of a hollow bearer containing line-replaceable motor/gearbox unit, and a minimal set of five moving parts. The general arrangement of these moving parts is shown in figure 2. The motor/gearbox transmits power to a sliding, toothed actuation rod through a gearhead arrangement. This rod has further teeth meshing with two cams, the lobes of which engage with the underside of a component referred to as the 'hopper'. When the rack moves, the cams are forced to rotate. A 180-degree movement of the cam effects a change of route by firstly lifting the rails before traversing them, and lowering them in a second position, in a semi-circular arc. When in the lowered position, the hopper sits upon a set of 'locking blocks' – essentially machined and matched mating protrusions on the underside of the hopper, and base of the hollow bearer, preventing any lateral or longitudinal movement of the hopper. The general arrangement of these blocks is shown in figure 3. This figure shows a 3-position REPOINT bearer, as per the laboratory demonstrator. The mass of a passing vehicle is also transmitted to the bearer casing, and thus ballast, through these blocks. When lifted, the hopper is free to move laterally, but not longitudinally. In order for the design to offer full passive safety, the force required to back-drive the motor must be less than that provided by the bending of the rails in the vertical plane. This will ensure that if there is a power loss during motion, the switch will fall back down to a safe, locked state. The lifting motion is also critical for the parallel-channel actuation:- each actuator bearer is capable of moving the switch alone, as each is capable of lifting the rails alone, the action of which will unlock all other actuator bearers.

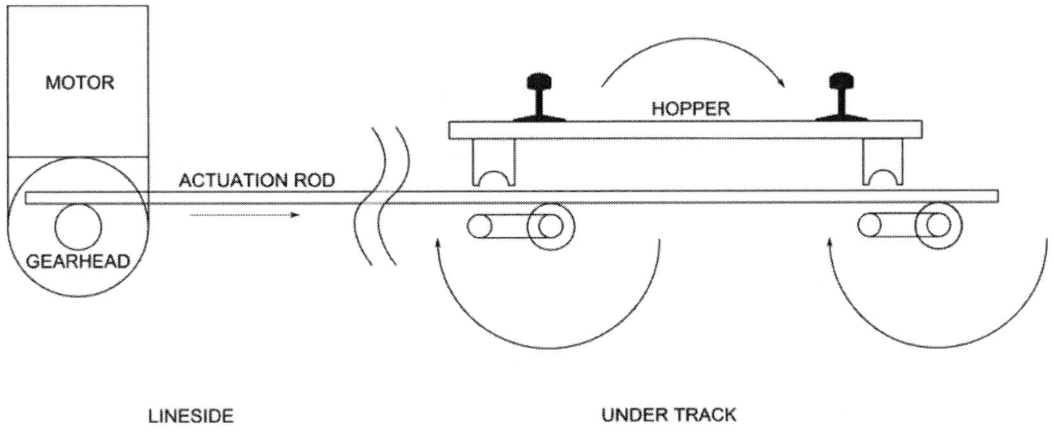


**Figure 2: Mechanical transmission elements of each actuator-bearer unit**

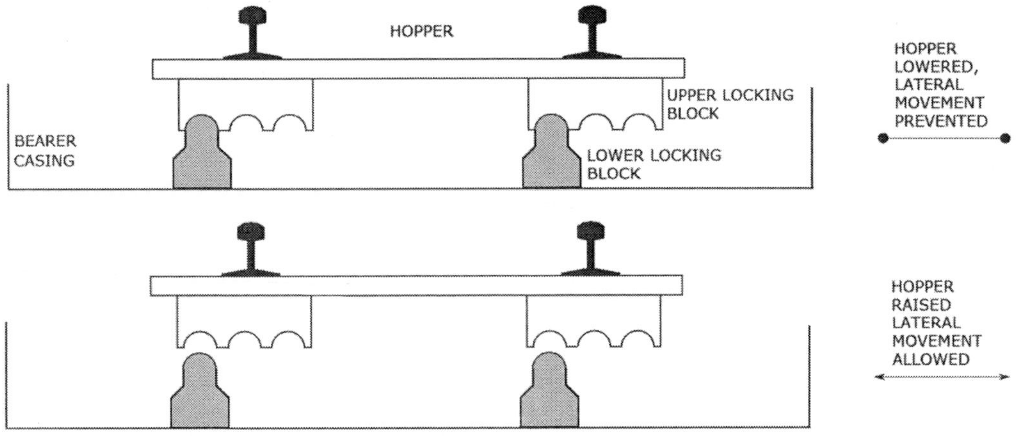


**Figure 3: Passive locking elements of each actuator-bearer unit**

## 4 MODELLING AND SIMULATION

For reasons of available space, it was proposed that the laboratory demonstrator be constructed at 384mm gauge. This is the gauge of the RHDR (Romney, Hythe and Dymchurch railway) in Kent, which has been identified as a possible technology demonstrator site for the next phase of development. This meant that the system modelling tasks were twofold. Firstly, to ensure a switch at this gauge was plausible, and to size the actuation units accordingly. Secondly, to model a standard gauge switch to ensure that the design would scale, and that bending full-section rail over the length of a typical NR (Network Rail) installation is plausible. NR-type switches, of lengths 'C' and 'F', were identified by NR as being examples of common short, station-throat type and long, high-speed types, respectively.

Initial appraisal and static calculations showed the dominant force through the actuation phase was bending of the full rail section in the vertical plane. Using Macaulay's technique, the rail pair are treated as a cantilever beam from a reaction anchor point ($R_A$), as shown in Figure 4. For a triple-redundant system, three actuation forces ($P_{1-3}$) were then applied, resisted by the uniformly distributed weight of the rail (q) and hopper/rail mounting masses ($B_{A1-3}$). The goal was to create a particular vertical rail stub toe deflection, which for a semi-circular tip actuation path equates to half the rail spacing between adjacent routes. Each actuator-bearer needs to be able to operate the switch alone for true triplicate redundancy. Therefore, the force requirement for each actuator bearer is simply the maximum of ($P_{1-3}$), plus a margin. Given the peak force requirements, the switching time of a given arrangement can then be calculated for any given power supply.

The resolved forces for both RHDR and NR type switches were not only remarkably close to each other, but also to those developed by existing switch designs. The effect of larger section - and therefore stiffer - CEN60 rail of the NR example was balanced by the much longer turnout radius and therefore effective beam length. This offered the opportunity to construct a 384mm gauge bearer, but with a motor and drive correctly sized for a NR example. The results of this modelling are shown in Table 1.

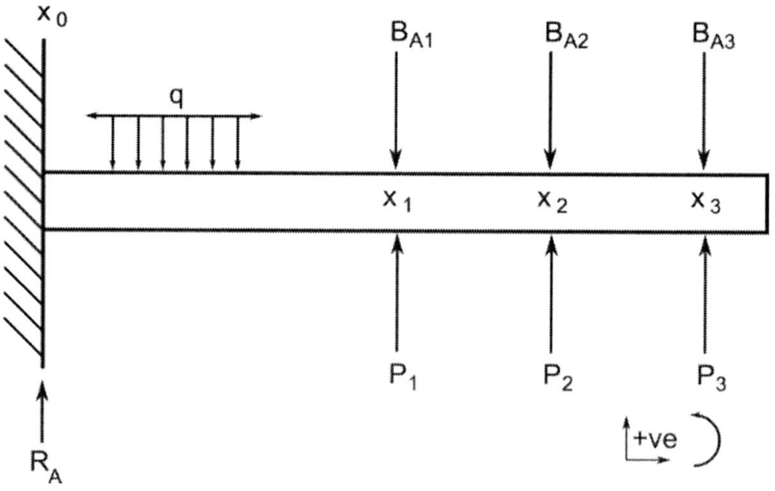


**Figure 4: Simple beam equation for switch simulation and bearer interaction**

**Table 1: Results of rail pair bending simulation (*defined here as toe to heel)**

| | RHDR (384mm) | NR (C - type) | NR (F – type) |
|---|---|---|---|
| **Natural switch length(mm)* / angle** | 4500 / 10 | 11900 / 9.25 | 20800 / 18.5 |
| **Movable rail length (Repoint, mm)** | 5000 | 7800 | 15800 |
| **Tip deflection (Horizontal) (mm)** | 120 | 100 | 100 |
| **Tip deflection (Vertical) (mm)** | 60 | 50 | 50 |
| **Resultant rod load, peak (N)** | 950 | 1380 | 920 |

Once feasibility was established, a more complex mathematical system model was created using the MATLAB/Simulink environment. This model was derived from a first principles physical analysis of the component parts of a REPOINT actuator-bearer. The objective was to provide an easily accessible, transparent description of the physical behaviour of a switch. Different switch installations may have differing parameters, but the model and interaction of the subcomponents will remain the same. As such, the structure of the model mirrors the physical structure of the plant. The structure of the model is shown in Figure 5. It has five main components: the brushless DC (BLDC) motor model, the gearhead model, the mechanical linkage model, the cam and hopper model, and a rail pair model linking each actuator bearer model, as described above. Figure 6 shows an example plot of load force vs cam angle for an RHDR switch example, as listed in table 1 and shown in Figure 7. The vertical load force is clearly dominant over the horizontal load, as would be expected when bending rail – an I-beam- in this manner. The resultant load curve represents the force required on the actuation rod in order to drive the cam. Due to the nature of the semi-circular actuation path providing a variable force vector, peak load on the rod is generally only around half of that required to move the switch. Note also that the total power requirement, which is proportional to the area under the resultant load curve, sums closer to zero than the total power transfer, which can be deduced from the sum of areas under the vertical and horizontal load curves. This is due to the motor optionally acting as a brake in the second half of actuation, when the force requirement turns negative as the spring and weight of the rails tends to force the switch back into a locked (lowered) position.

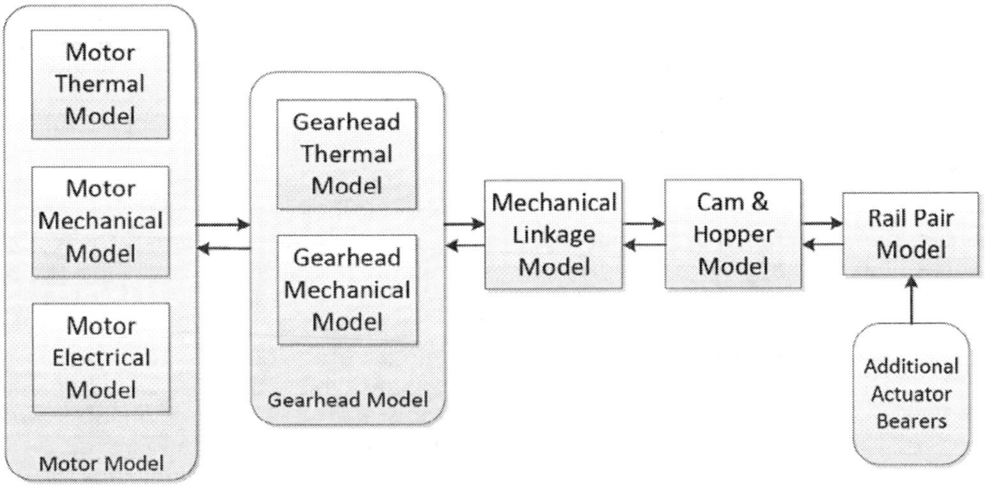


**Figure 5: Whole-switch simulation schematic**

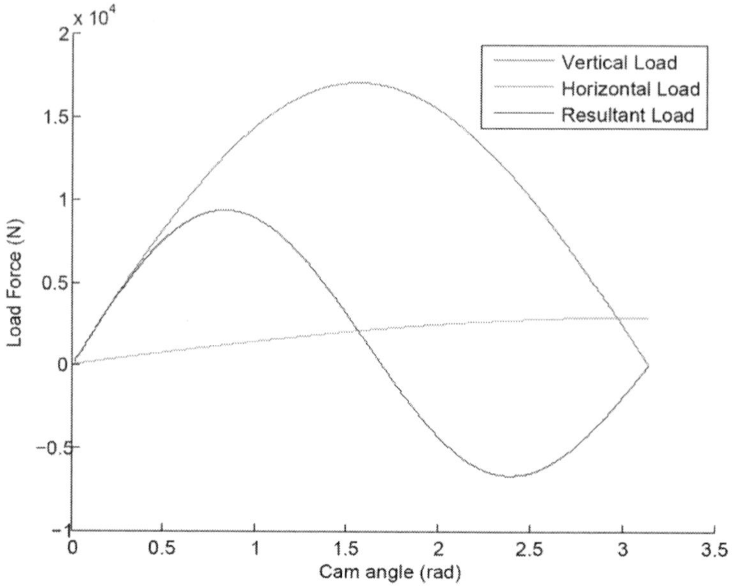


**Figure 6: Example switch loading during actuation cycle (RHDR example)**

The model was utilised in the model-based design process in order to correctly size components. However, it would also provide the basis for a co-simulation of actuator-bearers in the laboratory demonstrator, as well as a fundamental part of the model-based condition monitoring scheme of the functioning bearer. Extensive mathematical detail of each modelled element is described in (15) alongside sample outputs demonstrating load forces for particular switch arrangements.

## 5 LABORATORY DEMONSTRATOR

Using the results of this modelling exercise, a laboratory scale demonstrator has been designed and constructed. As of September 2014, the demonstrator is complete and undergoing a period of shakedown testing. The demonstrator consists of a control panel and mock signalling interface, a real-time processor, mock trackside power supply, single actuator bearer unit, and single bearer with machined stub rail ends for a 3-route switch. Figure 7 shows a view of the arrangement of demonstrator components in the laboratory. Figure 8 shows a close-up view of the first practical implementation of the passive locking arrangements described in figure 3.

For the first demonstrator iteration, Dexion bracing was used to connect the individual mechanical parts of the bearer. These parts were designed to fit inside a typical RHDR bearer envelope. Currently, only a single actuator-bearer unit has been constructed, which forms bearer number nine in the demonstrator layout shown in Figure 6. Actuator-bearer units seven and eight are software simulations using the model derived in section 4, and parameters established through testing the real unit, which run in conjunction with the physical mock-up when the switch is commanded to change position. These co-simulations run on the real-time D-Space processor unit.

As intended for a mainline installation, the demonstrator utilises COTS (Commercial off-the-shelf) components and technology wherever possible to keep costs relatively low. As such, the motor and gearbox are sealed units on an industry standard mounting pattern. The drive rod and all gears are standard sizes. The machined rail ends are held to the hopper and static bearer using Pandrol e-clip type clips, though the type of rail mounting is not critical to the function of the unit.

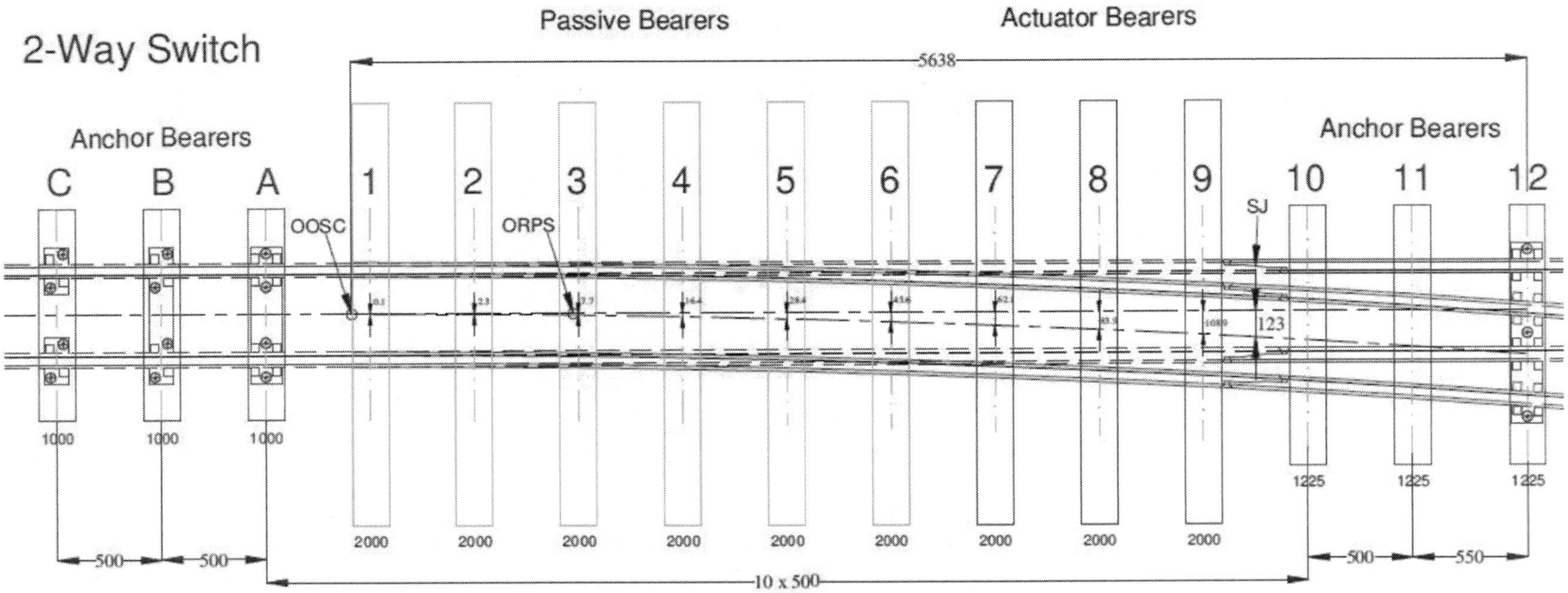


**Figure 7: 384mm gauge switch layout as in laboratory demonstrator**

**ORPS: Original Real Point of Switch**
**SJ: Switch Joint**
**OOSC: Original Origin of Switch Curvature**

**Figure 8: 384mm scale actuator bearer in laboratory**

**Figure 9: Close up photograph of actuation cam and locking block protrusions**

The switch is actuated through a computer mock-up of an NX-type signalling panel. The computer has various other displays to demonstrate what information could be presented to different stakeholders – for instance local maintenance teams, or central asset managers.

## 6 CONCLUSIONS AND NEXT STEPS

This paper has introduced a novel mechanical arrangement of railway track switch which is under development as part of the REPOINT project at Loughborough University. The background to the project, the theory behind the novel design and the design features have been covered in some detail. The modelling and simulation of the design has been described, and the layout of the laboratory demonstrator shown. The demonstrator is now functional in a laboratory at Loughborough University.

The next phase of REPOINT is to take the general demonstrator arrangement, and complete a prototype installation with triplicate redundancy at a suitable test site. This test site may be at scale or standard gauge.


## ACKNOWLEDGEMENTS

The authors wish to acknowledge the financial support provided by the United Kingdom EPSRC (Engineering and Physical Sciences Research Council) and the UK RSSB (Railway Safety and Standards Board) in grant number EP/I010823/1, for the project titled 'REPOINT: Redundantly engineered points for enhanced reliability and capacity of railway track switching'. The authors also wish to acknowledge the support of the UK's EIT (Enabling Innovation Team), for providing follow-on research funding towards concept demonstrator design and construction (http://www.futurerailway.org/).


## REFERENCE LIST


1. Morgan, J. C. *British Railway Track 7th Edition, Volume 1, Design Part 2: Switches and Crossings.* Derby, UK: The Permanent Way Institution; 2009.
2. Cope, D. D. L., Ellis, J. B. *British Railway Track 7th Edition, Volume 5: Switch and Crossing Maintenance.* Derby, UK: The Permanent Way Institution; 2002.
3. ORR (Office of Rail Regulation) Online Data Portal - Total journey count reporting (Accessed 2013-10-12) http://dataportal.orr.gov.uk/displayreport/report/html
4. Network Rail. *"Network Rail October 2007 Strategic Business Plan Supporting document:- Seven Day Railway"*, Technical report, October 2007.
5. Institution of Mechanical Engineers, *"Rail Freight: Getting on the Right Track"*, Technical report, Institution of Mechanical Engineers, February 2009.
6. Isermann, R., *Fault-Diagnosis Systems: An Introduction from Fault Detection to Fault Tolerance*, Springer, Berlin, 2006.
7. Cribbens, A. H., *Solid-state interlocking (SSI): An integrated electronic signalling system for mainline railways.* P102-112, Electric Power Applications, IEE Proceedings B, 1987.
8. Bemment, S., Dixon, R., Goodall, R., *"REPOINT Redundantly Engineered Points for Enhanced Reliability and Capacity of Railway Track Switching"*, in Proc.10th World Congress on Railway Research (WCRR). Sydney, Australia, November 2013.
9. Bemment, S., Dixon, R., Goodall, R., Brown, S., *"Redundantly Engineered Track Switching for Enhanced Railway Nodal Capacity"*, in Proc. 1st IFAC ACATTA Workshop. Istanbul, Turkey, September 2013.
10. (Patent) Loughborough University, "Railway points operating apparatus (GB 1322660)", 20 December, 2013.
11. (Patent) Loughborough University, "Railway points (GB 1322641)", 20 December, 2013.
12. US Army. *MIL-PRF-49506 US Military Specification: Performance Specification:- Logistics Management Information*, November 1996.
13. Hadaway, H. W., *IRSE 'Green Booklet' no.5:- Principles of power point control and detection (British Practice).* The Institution of Railway Signal Engineers, London, 1950.
14. Genner, R., *Railway Group Standard GK/RT0062: Control Of Points.* London, 1997.
15. Wright, N., Bemment, S., Ward, C., Dixon, R., *A Model of a Repoint Track Switch for Control,* Proceedings of UKACC International Conference on Control, Loughborough, UK, June 2014.

# Identification of bridge bearing damage based on Radial Basis Function Neural Network

**Z Chen, Q Yin, D Zhang**
State Key Laboratory of Traction Power, Southwest Jiaotong University, China

**ABSTRACT**

This paper presents a method to identify bridge bearing damage based on Artificial Neural Network theory. On the basis of analyzing the relationship between the beam vibration mode and the bearing damage, a Radial Basis Function Neural Network (RBFNN) is built. The beam vibration mode and damage information are the input and output of the neural network, respectively. The network is trained by a large number of data samples formed by numerical method. Finally, numerical tests are carried out to verify the reliability of the RBFNN. Results show that the RBFNN is effective to identify the bearing damage.

**Keywords:** Radial Basis Function Neural Network, damage identification, bridge bearing, vibration mode

## 1 INTRODUCTION

The bridge bearing is an important connection element between the bridge superstructure and the substructure. As the bridge accounts for a large proportion in the Chinese high-speed railways, the bearing is also widely adopted [1-2]. Being influenced by the reciprocal action of the wheel load and the change of the environment, the performance of the bearing could be changed. And even crack and aging appear in the bearing. With the damaged bearings, the running safety and riding comfort of the high speed trains passing through the bridges would be affected. Therefore, the exact identification of bridge bearing damage is an important precondition to ensure the running safety and ride comfort of the trains.

At present, there are many studies focusing on the bridge bearing [3-7]. Ali et al. [3] built a finite element model of a bridge, considering the non-linear characteristic of the bearing. Adopting the model, the vibration of bridge system under seismic action was analyzed. Mutobe et al. [4] also built a beam-bearing finite element model, and studied the reasonable value of bearing parameters (stiffness, damping). Further, Yang and Kim [5-6] studied the influence of elastic bearing on the dynamic response of the bridge system. By now, there are only a few studies on the bearing damage. Kim et al. [7] considered the damaged bearing as a friction element, and analyzed the effect of the bearing damage on the bridge vibration under seismic action.

Meanwhile, as the neural network theory has a rapid development in recent years, it has been widely used in the area of damage identification [8-15], especially in the respect of the bridge damage identification [11-15]. When the neural network is adopted to identify the damage of a bridge, the modal information of the beam body is widely used as the inputs of the networks. The reason is that the structure modal is more sensitive to the damage of structure, and the modal information (especially the low-order modals) is easy to be obtained by field test in practical engineering [16-20]. However, the current researches all focus on the damage identification of bridge beam body, and few studies pay attention to the bearing damage identification.

Based on the analysis of the relationship between the vibration mode of beam body and the bearing damage, the Radial Basis Function Neural Network (RBFNN) is built. The vibration mode of beam body and the bearing damage information are taken as the input and the output of the neural network, respectively. Meanwhile, the finite element model of bridge structure is built, in which the bearings are modeled as springs. The bearing damage is considered as the change of the bearing spring stiffness, and the vibration modes of the beam body with different spring stiffness are analyzed adopting the finite element model. A large number of data samples are formed by the correspondence between different bearing stiffness and different beam vibration modes, which are used to train the network. Finally, a series of numerical tests are carried out to verify the reliability of the RBFNN.

## 2 BASIC THEORY OF THE RBFNN

The RBFNN is a kind of forward network. There is no feedback in the network, and the structure of the network is simple. Comparing with other kinds of forward neural networks, RBFNN takes less training time and has higher accuracy, so it is widely used. In this work, the RBFNN is used to identify the damage of the bridge bearings.

The RBFNN is made up of a hidden layer (RBF layer) and a linear output layer, as shown in Fig.1.

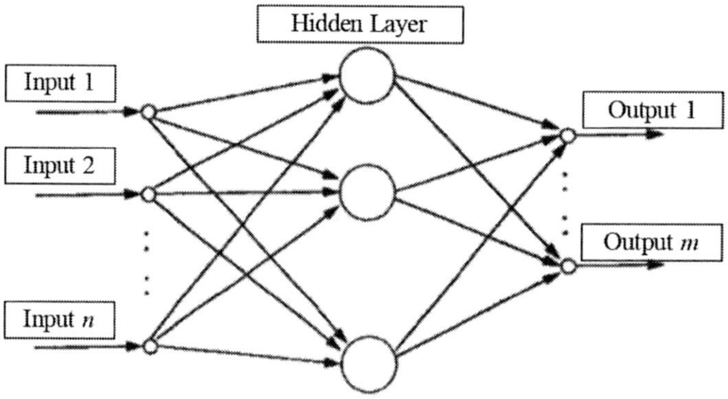


**Fig.1 Structure of the RBFNN**

A simple linear function is widely used in the output layer of the RBFNN, and Gauss Basis Function (Eq. (1)) is usually chosen as the basis function of the hidden layer.

$$u_j = \exp\left[-\frac{(X-C_j)^{\mathrm{T}}(X-C_j)}{2\sigma_j^2}\right] \quad j=1,2,\ldots,N_{\mathrm{h}} \tag{1}$$

where $u_j$ is the output of the $j$th node in the hidden layer; $X$ is the input of the sample; $C_j$ is the central value of the Gauss Function; $\sigma_j$ is a constant; $N_{\mathrm{h}}$ is the number of nodes in the hidden layer.

The output of the RBFNN can be expressed as the linear combination of the output of nodes in the hidden layer

$$y_i = \sum_{j=1}^{N_h} w_{ij} u_j - \theta \quad i = 1,2,\dots,m \tag{2}$$

where $w_{ij}$ is the weight value and $\theta$ is the threshold value.

Learning and training can make a neural network more accurate. The learning and training process of the RBFNN can be divided into two stages. The first stage is to decide the central value of the Gauss Function $C_j$ and the constant $\sigma_j$ by the inputs. The second stage is to determine the weight value of the output layer on the basis of the least square method.

## 3 ANALYSIS OF THE RATIONALITY ABOUT REFLECTING THE BEARING DAMAGE BY THE CHANGE OF BEAM VIBRATION MODE

In this work, the change of the beam vibration mode is taken as the input of the neural network, and the bearing damage can be identified. Therefore, it is the precondition of the bearing damage identification that whether the changes of the beam vibration modes can reflect the different bearing damages. In order to ascertain the relationship between the vibration modes and the bearing damages (i.e. the change of the bearing stiffness), a simplified beam body–bearing model is built, in which the beam is modeled as an Euler-beam and the bearings are described as support springs, as shown in Fig.2. The cross section of the beam is adopted as the standard section of the 32.6m simply supported girder bridge widely used in China, shown in Fig.2 (b). In the figure, $k_1$ and $k_2$ are the support stiffness of the two ends and $l$ is the length of the beam.

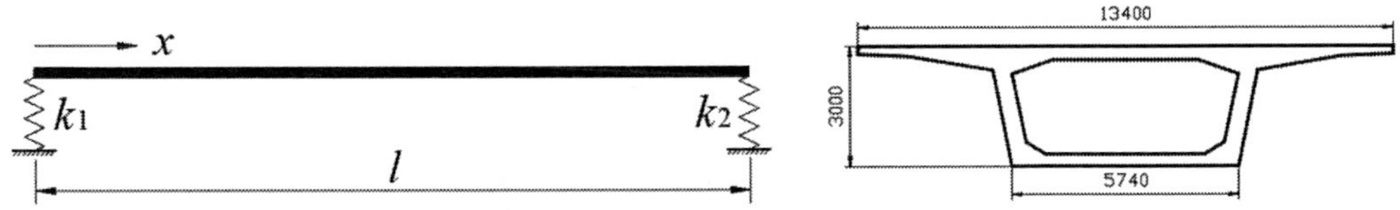


**a) Model of the beam** **b) Cross section of the beam**

**Fig.2 Spring-supported Euler-beam in Both Ends**

According to the related knowledge of the vibration mechanics [21], the vibration mode function $Y(x)$ of the Euler-beam can be expressed as

$$Y(x) = A\sin kx + B\cos kx + C\sinh kx + D\cosh kx \tag{3}$$

in which

$$k^4 = \frac{\rho A}{EI}\omega^2 \tag{4}$$

where $x$ is the coordinate of any point on the beam; $A\sim D$ is the coefficients; $\omega$ is the natural vibration frequency of the beam; $\rho A$ is the mass per unit length of the beam; $E$ and $I$ are the elastic modulus and the section inertia of the beam, respectively.

The boundary conditions of the beam above could be written as

$$\begin{cases} Y''(0) = 0 \\ Y''(l) = 0 \\ Y(0)\cdot k_1 = Y'''(0) \\ Y(l)\cdot k_2 = Y'''(l) \end{cases} \tag{5}$$

Solving the Eq. (3) and Eq. (5) together, the following equation can be obtained.

$$\begin{cases} C = A + 2k_1 B \\ D = B \end{cases} \tag{6}$$

Further, the frequency equation of the beam can be described as

$$\frac{\cos kl - 2k_1 \sinh kl - \cosh kl}{\sinh kl - \sin kl} = \frac{k^3 \sin kl - k_2 \cos kl + (2k^3 k_1 - k_2)\cosh kl + (k^3 - 2k_1 k_2)\sinh kl}{k_2 \sin kl + k^3 \cos kl + k_2 \sinh kl - k^3 \cosh kl} \tag{7}$$

The natural frequencies of the beam can be obtained by solving Eq. (7) with the numerical methods. It can be seen from Eq. (7) that $k$ is only related to the bearing support stiffness $k_1$ and $k_2$ in a given structure (i.e. the section and the material property of the beam are constant). For $k$ is a univalent function of the beam natural frequencies, the natural frequencies of the beam are also only related to $k_1$ and $k_2$. Meanwhile, the coefficients $A \sim D$ of the vibration mode function are the expression of the bearing support stiffness, too. That is to say, the change of the bearing support stiffness $k_1$ and $k_2$ would lead to the changes of the beam's natural frequencies and the vibration modes. It can be concluded that, for a given structure, the change of the bearing stiffness results in the change of the beam vibration modes directly, and the change of the beam vibration modes are only affected by the bearing support stiffness.

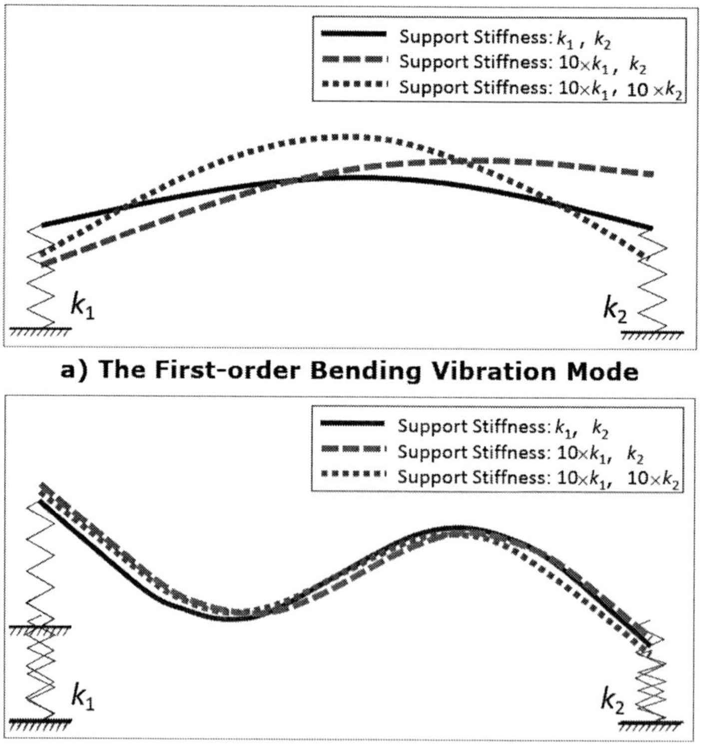


**a) The First-order Bending Vibration Mode**



**b) The Second-order Bending Vibration Mode**

**Fig.3 The Influence of the Bearing Support Stiffness on the Beam Vibration Modes**

It indicates that the bearing damage can be reflected by the change of the beam vibration modes. Fig.3 shows the influence of the change of the bearing support stiffness on the beam vibration modes. It can be known from the figure that the beam vibration modes change with the change of the support stiffness, which also indicates the rationality about adopting the change of the beam vibration modes to reflect the different bearing damages.

## 4 BEARING DAMAGE IDENTIFICATION ON THE BASIS OF RBFNN

Based on the analysis of the relationship between the beam vibration mode and the bearing damage, the RBFNN is adopted to identify the damages of the bearings in the simply supported girder bridge system.

### 4.1 Model of the RBFNN

In order to obtain the RBFNN used for bearing damage identification accurately, reasonable inputs and outputs of the neural network should be determined, and also a large number of learning and training are needed.

#### *4.1.1 Input parameters*

The efficiency and the accuracy of the bearing damage identification increase by selecting the reasonable input parameters. It can be known from the Section 3 that the change of the beam vibration mode can describe the bearing damage, so the input of the neural network should be the displacements of specific points on the beam body in the specific vibration modes. Nevertheless, the three problems below should be solved firstly before finally determining the input parameters.

① Choose the points (on the beam) whose displacements need to be extracted in different vibration modes.
② Determine the order of vibration modes and obtain the specific vibration modes.
③ Whether the displacements along all directions (the longitudinal direction, the lateral direction and the vertical direction) of the chosen points should be obtained or not.

For the problem ①, the change of the beam vibration modes would result in the displacements of almost all the points on the beam in different vibration modes. However, it's obviously unbefitting to take the displacements of all the points on the beam as the inputs of the neural network, so the specific points should be chosen. Thus, the workload is reduced by picking the reasonable points whose displacements should be extracted in different vibration modes. It can be seen from Fig.3 that the displacements at the bearings locations are more sensitive and regular with the change of the bearing support stiffness. Thus, the displacements at the four bearing locations are extracted the displacements in different vibration modes.

As to the problem ②, in the engineering practice, the displacements of different points in each vibration mode can be obtained by experimental methods [16-20]. While in the lab, the different vibration modes can be determined by numerical methods, such as finite element method and so on. Meanwhile, the first-order vertical vibration mode and the first-order lateral vibration mode of the beam can be tested and obtained accurately and easily. So, the displacements of the bearing points on the beam in the two beam vibration modes are chosen to identify the bearing damage.

The problem ③ can be solved by analyzing the sensibility of the displacements in different directions to different bearing damages, and a finite element model of the simply supported girder bridge is built to do the work, as seen in Fig.4. In the finite element model, the beam body is modeled as beam elements (the total length of the bridge is 32.6m and the cross section of the beam is shown in Fig.2 (b)) and the bearings are modeled as springs. The bearing node is connected to the beam with a rigid arm of a big stiffness and a small mass. Fig.5 shows the arrangement plan of the bearings.

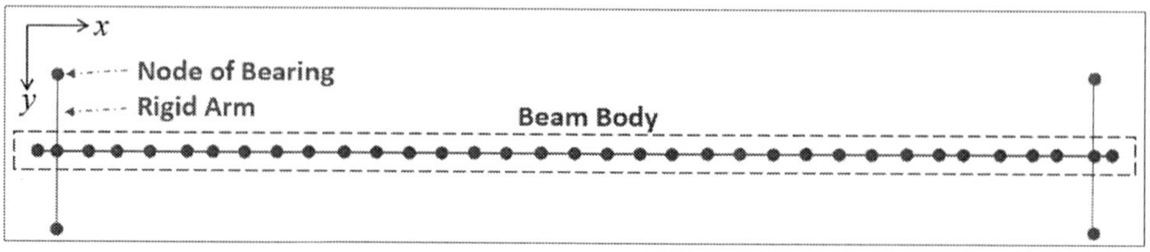


**Fig.4 Finite Element Model of the Simply Supported Girder Bridge**

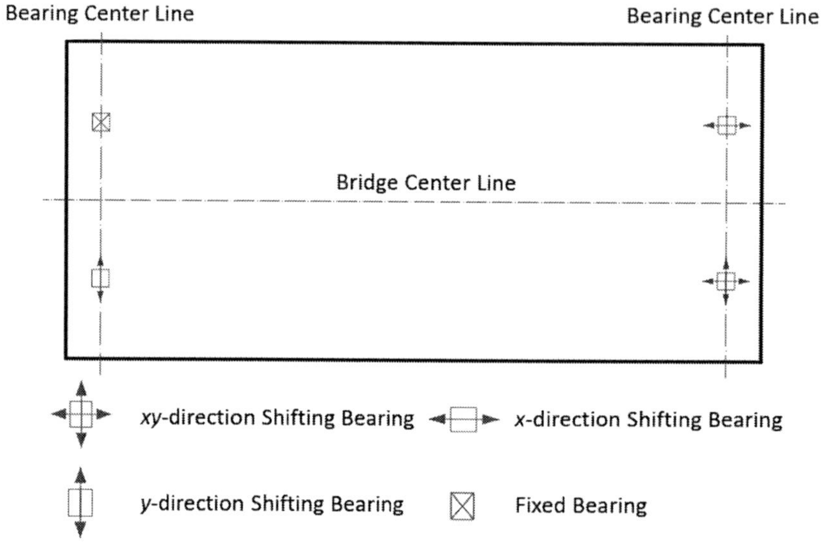


**Fig.5 Arrangement Plan of the Bearings**

Any kind of bearing damage can be broken up into the change of bearing stiffness in the three directions of *x*, *y* and *z*, and the beam vibrations (in the range of low-frequency) in the orthogonal directions of *x*, *y* and *z* are uncoupled [21]. Thus, the stiffness changes of the bearing in different directions would lead to different changes of the beam vibration modes. Via the modal analysis for the finite element model, the influences of the stiffness changes in different directions on different beam vibration modes can be obtained. Fig.6~ Fig.8 show the effects of the stiffness changes of the fixed bearing in different directions on the beam vibration modes.

As shown in Fig.6, for the longitudinal stiffness change of the fixed bearing, there is a deflection angle in the horizontal plane for the first order vertical vibration mode of the beam body, leading to the longitudinal displacements of the bearing nodes. The deflection angle is related to the change of the bearing stiffness. Nevertheless, the change of the longitudinal stiffness has almost no effects on the lateral vibration mode, so the longitudinal displacements of the four bearing points in the first-order vertical vibration mode can be used for identifying the bearing damage in the longitudinal direction. It can be known from Fig.7 that the lateral stiffness change of the fixed bearing has more effect on the first-order lateral vibration mode than the vertical vibration mode. Thus, the lateral displacements at the four bearing locations in the first-order lateral vibration mode can be used for identifying the bearing damage in the lateral direction. Similarly, the damage in the vertical direction can be identified by the vertical displacements of the bearing points, as seen in Fig.8.

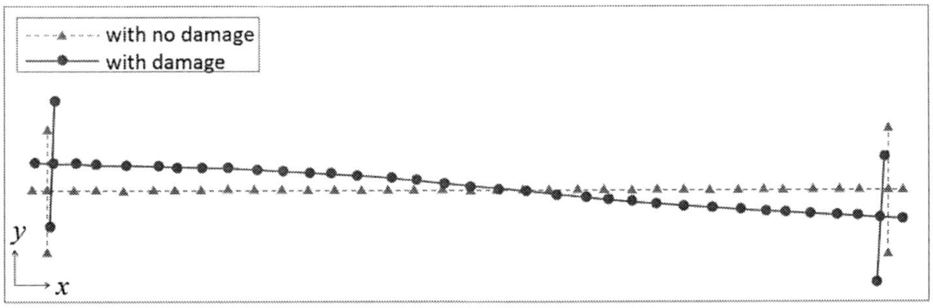


**a) First-order Vertical Vibration Mode**

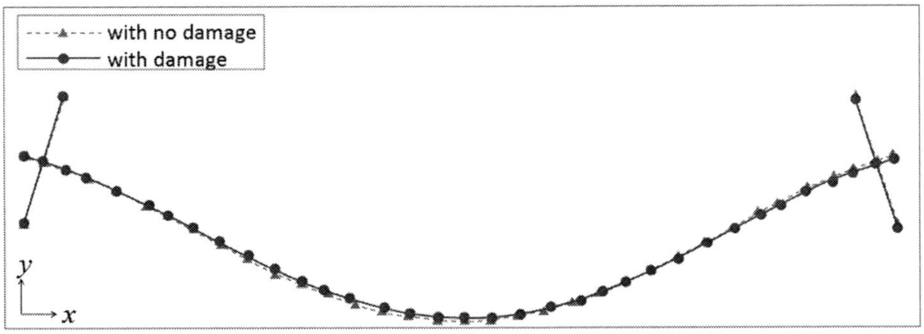


**b) First-order Lateral Vibration Mode**

**Fig.6 Effects of the Longitudinal Stiffness Changes of the Fixed Bearing on Different Beam Vibration Modes**

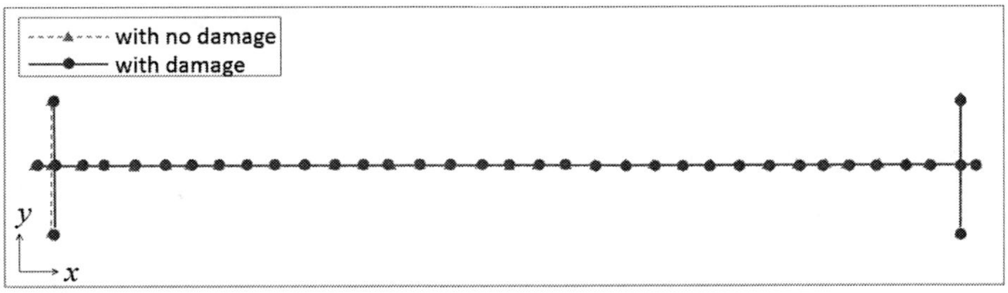


**a) First-order Vertical Vibration Mode**

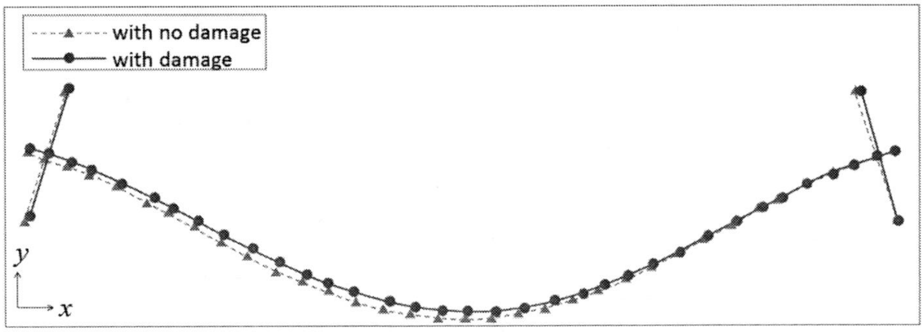


**b) First-order Lateral Vibration Mode**

**Fig.7 Effects of the Lateral Stiffness Changes of the Fixed Bearing on Different Beam Vibration Modes**

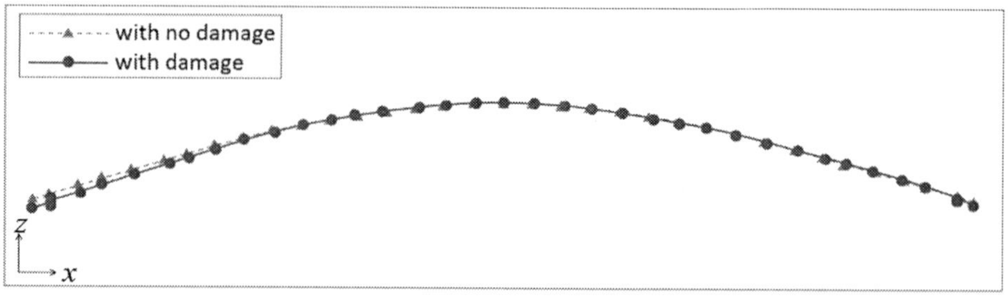


**a) First-order Vertical Vibration Mode**

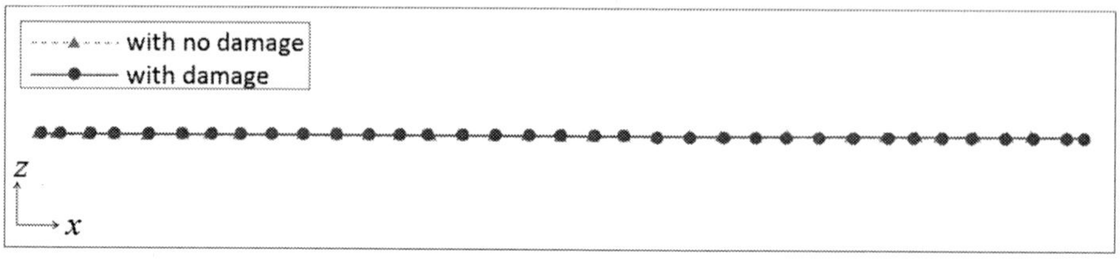


**b) First-order Lateral Vibration Mode**

**Fig.8 Effects of the Vertical Stiffness Changes of the Fixed Bearing on Different Beam Vibration Modes**

In conclusion, for the simply supported girder bridge, 12 displacements are needed to identify the bearing damage in different directions, describing the longitudinal and vertical displacements at the 4 bearing locations in the first-order vertical vibration mode and the lateral displacements of the 4 bearing points in the first-order lateral vibration mode. After putting the displacements in the same direction into normalization processing (divide the displacements by the maximum displacement in the same direction), 12 dimensionless values are obtained, which are the final input parameters of the neural network.

#### *4.1.2 Output parameters*

For the simply supported girder bridge, the damage can appear in 8 locations and directions, including the *x*, *y*, *z* directions of the fixed bearing, the *y*, *z* directions of the *x*-direction shifting bearing, the *x*, *z* directions of the *y*-direction shifting bearing and the *z*-direction of the *xy*-direction shifting bearing. Therefore, there are 8 output parameters for the neural network (Tab.1), corresponding to the health information of the 8 places, respectively. A damage of one place appears when the health information value of the output channel is in the range of 1±20% (i.e. 0.8~1.2, considering an error of 20%).

**Tab.1 The Output Parameters of the Neural Network**

| Output channel | Health information |
|---|---|
| 1 | *x*-direction of the fixed bearing |
| 2 | *y*-direction of the fixed bearing |
| 3 | *z*-direction of the fixed bearing |
| 4 | *y*-direction of the *x*-direction shifting bearing |
| 5 | *z*-direction of the *x*-direction shifting bearing |
| 6 | *x*-direction of the *y*-direction shifting bearing |
| 7 | *z*-direction of the *y*-direction shifting bearing |
| 8 | *z*-direction of the *xy*-direction shifting bearing |

Note: assuming there is no damage of the shifting bearing in its shifting direction.

#### *4.1.3 Learning and training*

Learning and training are important to a neural network, which make the network more accurate. Adopting the finite element model aforementioned, a large number of modal analysis are carried out by taking the different damage conditions of different bearings in different directions into consideration. A large number of data samples are formed by the correspondence between different bearing stiffness and different normalized-displacements at the bearing locations in different beam vibration modes, and the complete data samples are used for training the network.

### 4.2 Bearing damage identification

Adopting the RBFNN, two examples of bearing damage identification are carried out, which can also verify the method of identifying the bearing damage by the RBFNN. The two damage conditions are listed as: ①The lateral stiffness of the fixed bearing increases by 10 times; ② The vertical stiffness of the *x*-direction shifting bearing increases by 15 times, while the longitudinal stiffness of the *y*-direction shifting bearing reduces to 1/5 of the original stiffness. For the two work conditions, the normalized-displacements at the bearing locations are listed in Tab.2~Tab.3.

**Tab.2 Normalized-displacements in Different Vibration Modes (Work Condition 1)**

| | Fixed bearing | *x*-direction shifting bearing | *y*-direction shifting bearing | *xy*-direction shifting bearing |
|---|---|---|---|---|
| Longitudinal displacements in the first-order vertical vibration mode | 1 | -0.989 | 1 | -0.989 |
| Lateral displacements in the first-order lateral vibration mode | 1.000 | 0.995 | 1.000 | 0.995 |
| Vertical displacements in the first-order vertical vibration mode | 1 | 1 | 1 | 1 |

**Tab.3 Normalized-displacements in Different Vibration Modes (Work Condition 2)**

| | Fixed bearing | *x*-direction shifting bearing | *y*-direction shifting bearing | *xy*-direction shifting bearing |
|---|---|---|---|---|
| Longitudinal displacements in the first-order vertical vibration mode | 0.993 | 0.993 | 1 | 0.985 |
| Lateral displacements in the first-order lateral vibration mode | 1 | 1 | 1 | 1 |
| Vertical displacements in the first-order vertical vibration mode | 0.864 | 0.068 | 1 | 0.847 |

Putting the normalized-displacements into the neural network as inputs, the outputs are shown in Fig.9.

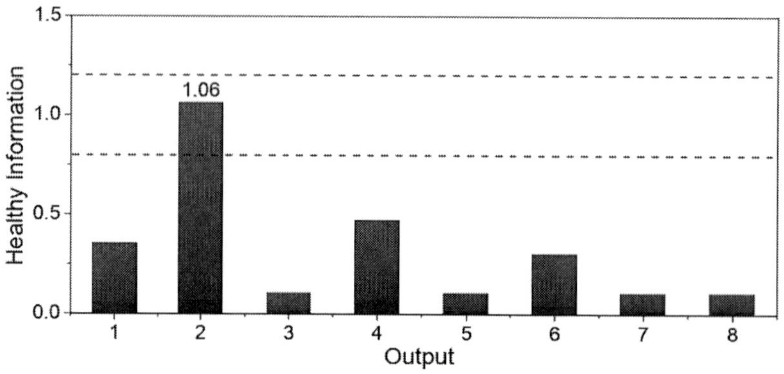


**a) Identification Result of Work Condition 1**

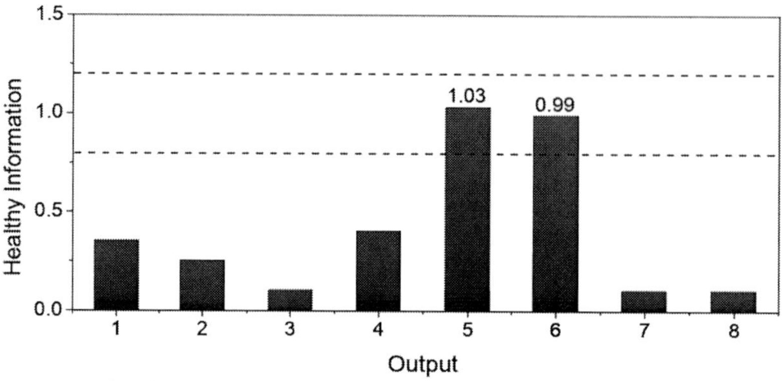


**a) Identification Result of Work Condition 2**

**Fig.9 Bearing Damage Identification Results**

It can be seen from Fig.9a) that the healthy information of the second output channel is 1.06, which is between 0.8 and 1.2, indicating that there is a damage in the place of output channel 2. Compared with the output parameters in Tab.1, it can be known that there is a damage in the lateral direction of the fixed bearing, which is consistent with the damage condition 1. Similarly, as shown in Fig.9b), there are damages in the $5^{th}$ and the $6^{th}$ output locations, which is also consistent with the damage condition 2. From the two damage identifications, it can be concluded that the RBFNN works well in bearing damage identification, indicating that the RBFNN can be used for identifying the damages of the bearings.

## 5 CONCLUSION

Based on the analysis of the relationship between the beam vibration modes and the different bearing damages, a Radial Basis Function Neural Network has been built in this paper for the identification of the bearing damage, and numerical tests have been carried out to verify the neural network. Results show that the RBFNN shows a good performance in the bearing damage identification, indicating that the RBFNN can be used for identifying the damages of the bearings. It should be noted that, the bearing damage identification of the simply supported girder bridge is carried out in this paper, bearing damage identification in other kinds of bridge systems can also be conducted by adopting this method.

## ACKNOWLEDGEMENTS

This work was supported by the National Basic Research Program of China ("973" Program) (Grant No. 2013CB036206, No. 2013CB036205), the 2015 Doctors' Innovation Fund of Southwest Jiaotong University and the Fundamental Research Funds for the Central Universities.

## REFERENCE LIST


(1) Zhai W, Xia H, Cai C, et al. High-speed train–track–bridge dynamic interactions –part I: theoretical model and numerical simulation. International Journal of Rail Transportation, 2013, 1(1-2): 3-24.

(2) Zhai W, Wang S, Zhang N, et al. High-speed train–track–bridge dynamic interactions –part II: experimental validation and engineering application. International Journal of Rail Transportation, 2013, 1(1-2): 25-41.

(3) Ali H M, Abdel A M. Modeling the nonlinear seismic behavior of cable-stayed bridges with passive control bearings. Computers and Structures, 1995, 54(3): 461-492.

(4) Mutobe R M, Cooper T R. Nonlinear analysis of a large bridge with isolation bearings. Computers and Structures, 1999, 72:279-292.

(5) Yang Y B, Lin C L, Yau J D, Chang D W. Mechanism of resonance and cancellation for train-induced vibrations on bridges with elastic bearings. Journal of Sound and Vibration, 2004, 269: 345-360.

(6) Kim C W, Kawatani M, Hwang W S. Reduction of traffic –induced vibration of two-girder steel bridge seated on elastomeric bearings. Engineering Structures, 2004, 26: 2185-2195.

(7) Kim S H, Mha H S, Lee S W. Effects of bearing damage upon seismic behaviors of a multi-span girder bridge. Engineering Structures, 2006, 28: 1071-1080.

(8) Min J, Park S, Yun C B, et al. Impedance-based structural health monitoring incorporating neural network technique for identification of damage type and severity. Engineering Structures, 2012, 39: 210-220.

(9) Shu J, Zhang Z, Gonzalez I, et al. The application of a damage detection method using Artificial Neural Network and train-induced vibrations on a simplified railway bridge model. Engineering Structures, 2013, 52: 408-421.

(10) Abbassi F, Belhadj T, Mistou S, et al. Parameter identification of a mechanical ductile damage using Artificial Neural Networks in sheet metal forming. Materials and Design, 2013, 45: 605–615.

(11) Lee J J, Lee J W, Yi J H, et al. Neural networks-based damage detection for bridges considering errors in baseline finite element models. Journal of Sound and Vibration, 2005, 280: 555–578.

(12) Tadesse Z, Patel K A, Chaudhary S, et al. Neural networks for prediction of deflection in composite bridges. Journal of Constructional Steel Research, 2012, 68: 138–149.

(13) Yeung W T, Smith J W. Damage detection in bridges using neural networks for pattern recognition of vibration signatures. Engineering Structures, 2005, 27:685–698.

(14) Bakhary N, Hao H, Deeks A J. Damage detection using artificial neural network with consideration of uncertainties. Engineering Structures, 2007, 29: 2806–2815.

(15) Mehrjoo M, Khaji N, Moharrami H, et al. Damage detection of truss bridge joints using Artificial Neural Networks. Expert Systems with Applications, 2008, 35: 1122–1131.

(16) Ubertini F, Gentile C, Materazzi A L. Automated modal identification in operational conditions and its application to bridges. Engineering Structures, 2013, 46:264–278.

(17) Fan J, Zhang Z, Hua H. Data processing in subspace identification and modal parameter identification of an arch bridge. Mechanical Systems and Signal Processing, 2007, 21:1674–1689.
(18) Hea X H, Huab X G, Chenb Z Q, et al. EMD-based random decrement technique for modal parameter identification of an existing railway bridge. Engineering Structures, 2011, 33: 1348–1356.
(19) Chena J, Xua Y L, Zhangb R C. Modal parameter identification of Tsing Ma suspension bridge under Typhoon. Victor: EMD-HT method. Journal of Wind Engineering and Industrial Aerodynamics, 2004, 92: 805–827.
(20) Ataei S, Mohammadzade S. Modal shape identification of the vibration data of bridge dynamic test using fuzzy clustering. Expert Systems with Applications, 2010, 37: 5813–5817.
(21) Clough R W, Penzien J. Dynamic of structures (3rd edition). Berkeley: Computers & Structures Inc., 2003.

# Dynamic simulations in support of installation of light rail tracks on the Homer H. Hadley Memorial floating bridge

**C Ketchum [1], T Cooper [2], A Foan [3], R Joy [1], H Sederat [4], J Sleavin [5]**
[1] Transportation Technology Center, Inc., USA
[2] Parsons Brinckerhoff, Inc., USA
[3] Andy Foan Ltd, UK
[4] SC Solutions, USA
[5] Sound Transit, USA

## ABSTRACT

A unique application of NUCARS®[1] vehicle dynamics software and ADINA finite element analysis software was used to predict the performance of a transit vehicle over a new track structure being developed to support light rail traffic on Interstate 90's Homer H. Hadley Memorial Floating Bridge east of Seattle, Washington, USA.

Led by Parsons Brinckerhoff, Inc. (PB), the designing team of Transportation Technology Center, Inc. (TTCI), SC Solutions (SCS), and Andy Foan, Ltd. provided services to Sound Transit (ST) in support of its East Link Extension project, which will take light rail vehicles across the Lake Washington I-90 floating bridge from Seattle to Mercer Island, Washington, continuing to Bellevue and Redmond, Washington.

A novel piece of special trackwork, named a "track bridge", is required for light rail to operate across the joints between the fixed approach spans and onto the floating bridge. The track bridge must accommodate multidimensional movement at the transition zone while providing a smooth running surface for the vehicle. To achieve this, the PB team developed a track-bridge design employing a pair of innovative rail joints called the Curved Element Supported Rail (CESuRa).

TTCI used NUCARS and ADINA simulation results from SCS to predict vehicle dynamic response and wheel-rail forces over the CESuRa. Rail deflections from ADINA were fed into the NUCARS model, and the resulting wheel-rail forces were fed back to ADINA. Results from both models were compared, and the models were modified until an acceptable level of correlation was reached.

The models were validated by suspension characterization testing of a light rail vehicle in the ST shops and a full-scale test at the Federal Railroad Administration's Transportation Technology Center in Pueblo, Colorado.

## 1 INTRODUCTION

Sound Transit (ST), as part of its East Link Extension project, will operate light rail vehicles (LRV) across the Homer M. Hadley Memorial Bridge (that currently routes Interstate 90 across Lake Washington) from Seattle to Mercer Island, Washington,

[1] NUCARS® is a registered trademark of Transportation Technology Center, Inc.

continuing to Bellevue and Redmond, Washington. The Homer M. Hadley Memorial Bridge and adjacent Lacey V. Murrow Memorial Bridge (Figure 1) are among the longest floating bridges in the world (1).

**Figure 1: Homer M. Hadley Memorial Bridge (left) and the Lacey V. Murrow Memorial Bridge (right) looking east toward Mercer Island**

A novel piece of special trackwork, a track bridge, will be required for light rail to operate across the joints between the fixed approach spans and onto the floating bridge. Each track bridge must accommodate multidimensional movement at the transition zone while providing a smooth running surface for the vehicle. To achieve this, a design team led by Parsons Brinckerhoff, Inc. (PB), developed an innovative track-bridge design called the Curved Element Supported Rail (CESuRa). Eight track bridges will be required to accommodate the planned double-track at both ends of the floating bridge.

The complex nature of the design required a program of dynamic modeling and testing to evaluate its performance. Numerical modeling and analysis of the CESuRa system were performed using three primary software programs: ADINA, version 8.8.4, nonlinear finite-element program developed by ADINA R&D, Inc., Watertown, Massachusetts, LARSA, from LARSA, Inc., Melville, New York, and NUCARS®, a simulation program developed and owned by Transportation Technology Center, Inc. (TTCI) that is used to model vehicle and track interaction.

ADINA and LARSA were used to perform structural analysis, whereas ADINA and NUCARS were used to analyze light-rail vehicle response and rider comfort. Results of the ADINA and NUCARS models were then reconciled using an iterative process to take advantage of the nonlinear structural capabilities of ADINA and the detailed wheel and rail interaction effects available from NUCARS. ADINA modeling was conducted by SC Solutions (SCS), Sunnyvale California.

A program of laboratory and field tests was also completed to validate the models and to evaluate the dynamic response of the track bridge under a moving vehicle.

## 2 CESURA DESIGN

The track-bridge system needs to provide a smooth, load-bearing curved track across a hinged bridge joint. The bridge joint is constantly moving because of seasonal changes in the lake level, temperature, wind and wave action, roadway vehicle loading, and LRV loading. The track bridge also needs to accommodate extreme weather conditions, which can cause movement in the bridge structure.

To accommodate these movements, CESuRa, a variable radius transition joint, was designed. It is based on the interaction of a pair of curved-track supports in two planes that adjust automatically to angular movements in the transition spans of the bridge. The innovative approach accommodates pitch and roll and yaw resulting in a uniform profile under all bridge movement combinations. Figure 2 illustrates how the CESuRa transitions between the fixed span and the floating span.

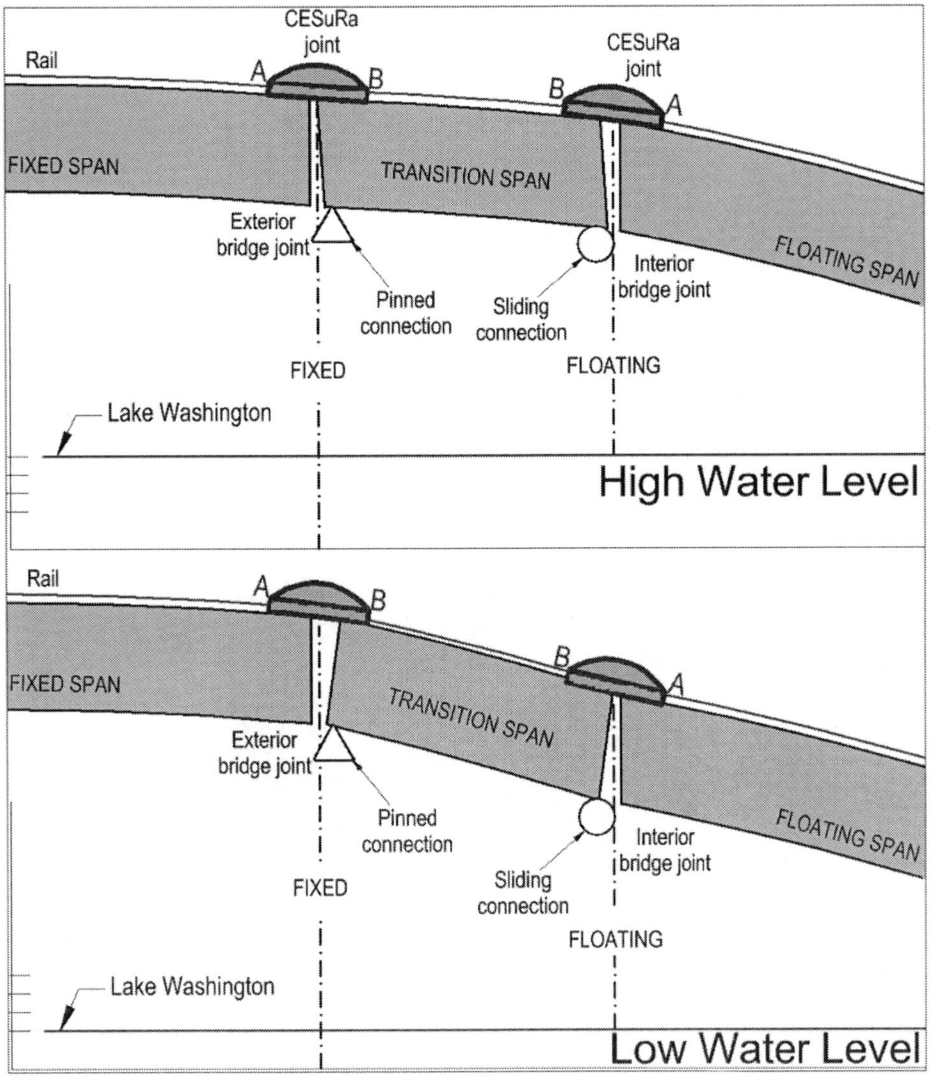


**Figure 2: CESuRa track bridge transition for high water and low water level conditions**

Figure 3 illustrates the CESuRa concept. Two bridge decks join at a hinge. The wings are comprised of two triangular secondary planes with one long edge, shown as AB, perpendicular to the hinge and vertex, denoted as C, on or near the hinge axis. Bearings permitting rotation and some sliding are located at A, B, and C between the wings and the bridge decks. On the surface of each wing are a series of bearer bars, which perform similar function as sleepers or bearers. These bearings are aligned along a curve, called the yoke. Rails are mounted using a typical elastic fastening system onto the bearer bars. Guardrails are affixed, if necessary.

With this configuration, when the hinge angle is zero and the wings are lying flat, an observer looking from the side in the direction of the hinge axis will see the wings (ACB) and yokes as straight lines. As the hinge angle increases, the wings will naturally incline inward as the long edges (AB) are forced upward above the hinge axis (C). The observer will now see a developing curvature as the yokes rise on the sloping wings, and the bearer bars will now appear to the observer to lie on a smooth yet continuously variable curve.

By selecting the right geometric parameters, the curves will always be tangential at either end to the incoming and outgoing tracks. When the hinge angle is positive upward, the road or track has a smooth, vertical curve or sag. When the hinge angle is negative downwards, the road or track has a smooth vertical curve or crest.

As the wings rotate, the bearer bars move vertically through various distances and the rail bends in the vertical plane (illustrated in Figure 3). The vertical curvature of the rails can be designed to provide a smooth curved progression from the transition span to the adjoining fixed and floating spans, while the rails are fully supported along the length of the CESuRa joint by choosing a suitable horizontal curve on which the yoke and bearings lie, and a suitable location for the supports for the wings by choosing a suitable horizontal curve on which the bearing lie, and the location of the supports for the wings.

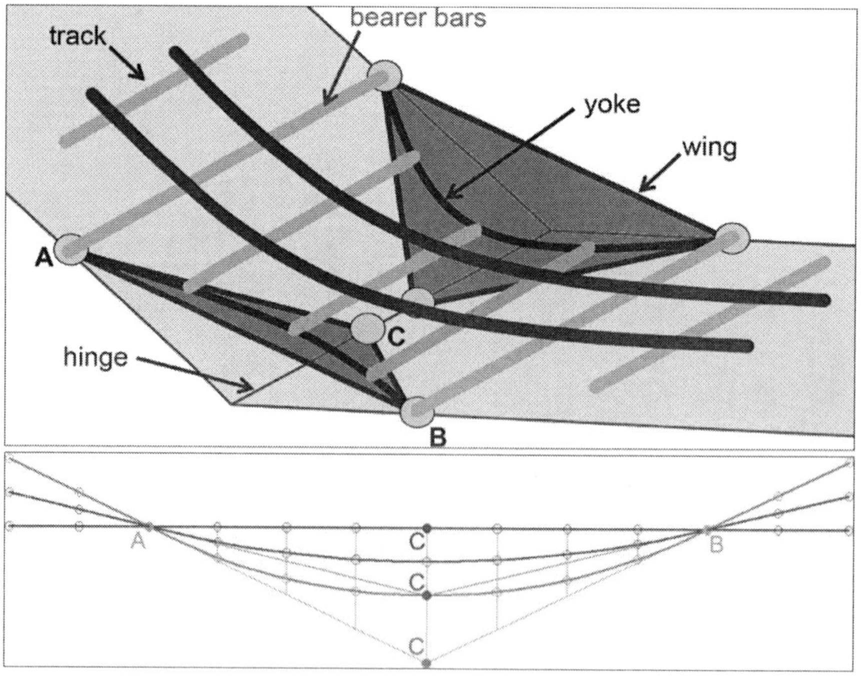


**Figure 3: CESuRa track-bridge concept**

These relationships were studied in more detail using a three-dimensional geometric computer program in which the sensitivity of the rail geometry to variations in structure positions was considered. Three-dimensional models were made using MicroStation software, developed by Bentley Systems, Inc., Exton, Pennsylvania, to evaluate the physical interfaces of the track bridge elements with one another and with the bridge. Ensuring that the components of the proposed project do not interfere with the existing structure is a key concern in developing a robust long term solution.

The refined CESuRa system, as Figure 4 illustrates, acts like a pin-ended beam that spans the hinge point between the transition and fixed span or between the transition and floating span. The main elements of the system are as follows:

- Two longitudinal box girders or wings
- Transverse bearer bars that serve a purpose similar to that of sleepers or bearers in a conventional track supported on the wings by unidirectional friction pendulum bearings (FPB) that allow rotation and limited translation
- Rails that rest on and are secured to the bearer bars

**Figure 4: Refined CESuRa system design**

Because the bearings on each wing are placed in the shape of a horizontal circular curve, the bearer bars are all of different lengths, with those at the ends of the wings being the longest. The wings are supported from below on the transition and the fixed or floating spans and thus move vertically with the transition span in response to lake level changes. However, the wings are supported eccentrically with respect to their longitudinal axes so that, in addition to vertical movement, they undergo rigid body rotation about their longitudinal axes. As the wings rotate, the bearer bars move vertically through different distances, and the rails bend in the vertical plane. To ensure a smooth curved transition of the rails between the transition and the fixed or floating span, the horizontal curve radius on which the bearings lie, as well as the location of the supports for the wings, had to satisfy particular geometric requirements. The length of the track bridge is designed to ensure that the rail stresses resulting from this bending are tolerable. Despite the movements of the various components, the rails are fully supported along the length of the CESuRa track bridge.

There are 17 bearer bars in the full scale joint, and the overall length is 13 meters long (42.5 feet). Figure 5 shows a full-scale prototype installed for testing at the Federal Railroad Administration's (FRA) Transportation Technology Center (TTC), Pueblo, Colorado, USA.

**Figure 5: Full-scale prototype of CESuRa track bridge**

Both the structural elements of the track bridge and the rails themselves experience substantially lower stresses with the CESuRa concept than the stress levels in previous design options. These differences in performance could improve ST operations, lower life-cycle costs, and enhance system reliability. Besides traditional loads for bridges, the design loading for the track bridge included an evaluation of the system, when it is subjected to movements of the floating bridge and transition spans. These conditions were defined as follows:

- Pitch caused by seasonal lake level changes of ± 0.46 meters (1.5 feet)
- Yaw caused by sway; i.e., horizontal movement, resulting from wind
- Roll resulting from wind and "plunging" of the pontoon when light-rail vehicles are present
- Expansion and contraction of the bridge and track by as much as ±0.61 meter (2 feet) in response to changes in temperature and lake level.

## 3 MODELING WORK WITH ADINA AND NUCARS

In order to understand the vehicle-structure interaction, a new approach was used. A combination of ADINA and NUCARS modeling output was used to evaluate the performance of the track-bridge and LRV as a system. The ADINA model was used to simulate bridge response to LRVs passing across the fixed spans, track bridges, and transition span and then onto the floating span. Results from these analyses provided predictions of wing and rail stresses, and displacement input for the NUCARS models, which were used to evaluate rider comfort and vehicle performance.

### 3.1 NUCARS® model

TTCI developed a vehicle model of the three-section passenger ST LRV (Figure 6) from information provided by PB, the vehicle manufacturer, and ST. In addition, TTCI (with assistance from ST staff) conducted resonance tests and quasi-static suspension stiffness measurements on the LRV to confirm air suspension characteristics, primary suspension stiffness, primary suspension damping, carbody moments of inertia, and center of gravity locations (Figure 7). Results were applied to NUCARS and ADINA modeling.

**Figure 6: Sound Transit light rail vehicle and naming convention**

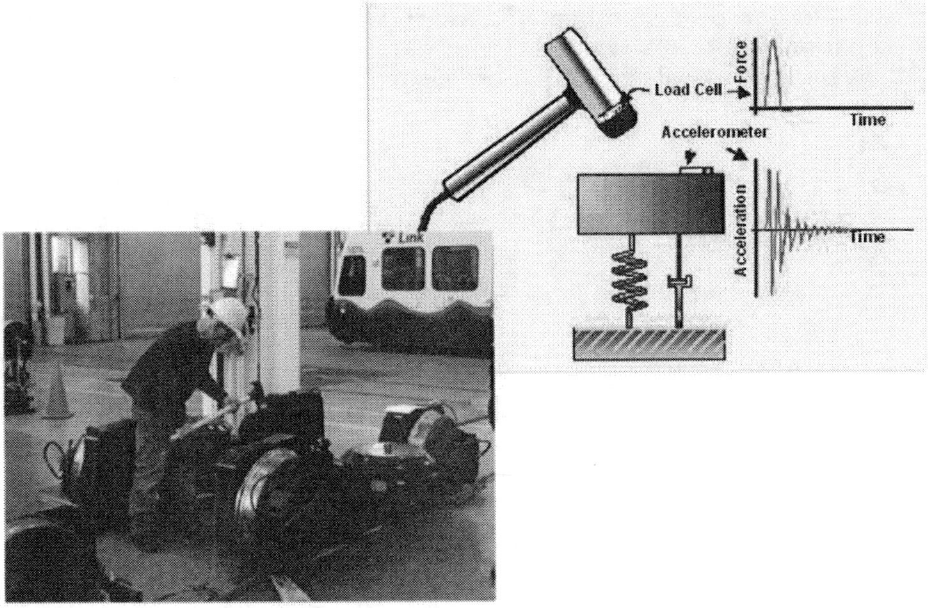


**Figure 7: Truck resonance test conducted at ST shops**

Figure 8 shows a representation of the single-layer track model was developed to represent the CESuRa track bridge in NUCARS. The track model consists of a left and right rail flexible rail suspended on the support structure with multiple parallel springs and dampers along the track. The stiffness and damping of the support structure are represented by the multiple spring and damper characteristics. The rails are modeled as Euler-Bernoulli beams. The flexible rails have roll, lateral, and vertical rigid displacement degrees of freedom. The rail flexible modes include torsion, lateral bending, and vertical bending.

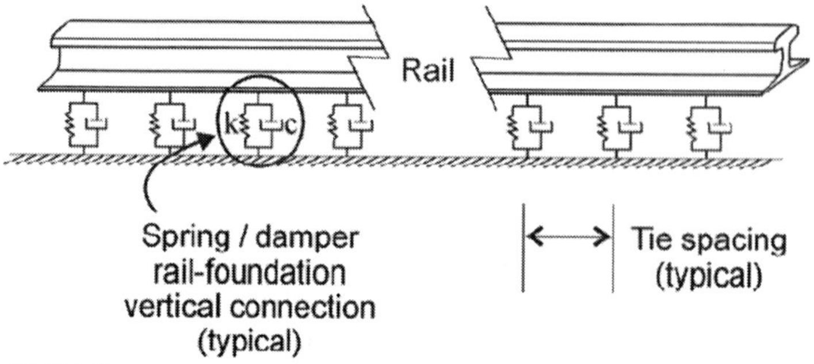


**Figure 8: Representation of NUCARS track model**

### 3.2 ADINA model

The bridge, including floating pontoon, anchor cables, piers, concrete slabs, rails, fasteners, FPBs and CESuRa track bridges were modeled in detail using results obtained from an earlier study commissioned by ST and validated by laboratory and in-track testing.

The concrete pontoons were modeled explicitly using elastic 8-noded shell elements. The floatation was provided by nonlinear springs. The anchor cables were modeled explicitly using beam elements. All piers were modeled with linear elastic beam elements. The slabs were modeled using 8-noded shell elements with linear elastic material properties.

A detailed model of the CESuRa track bridges was developed and inserted in the global model at the interior and exterior joints (Figures 9 and 10). Each CESuRa includes 34 friction pendulum bearings, which were modeled in detail using three-dimensional solid elements (Figure 11). In the iterative analysis between TTCI and SCS, the vehicle was modeled with moving wheel loads.

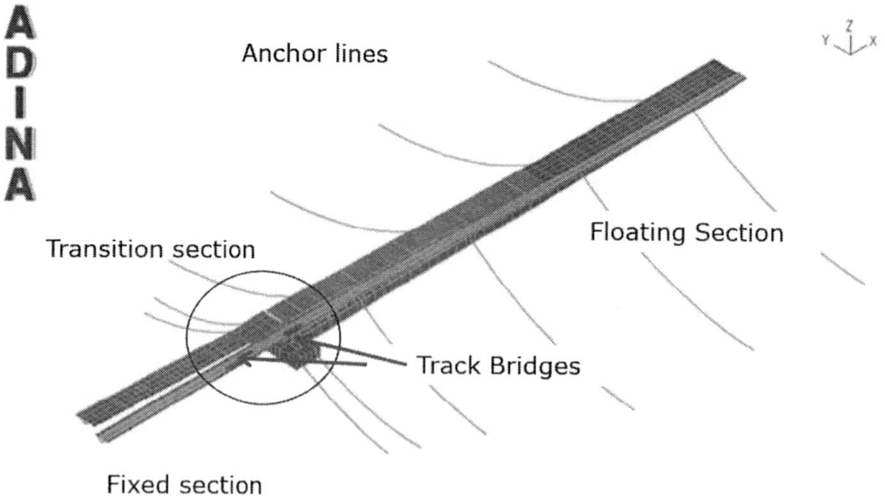


**Figure 9: Isometric view of the ADINA FE model of the floating bridge**

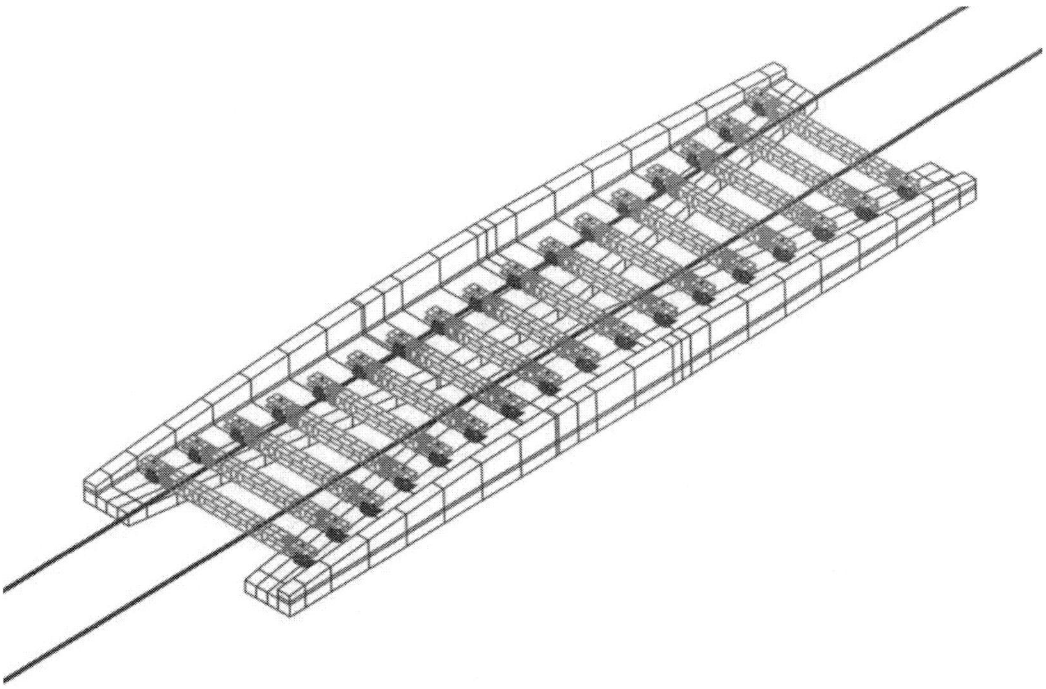

**Figure 10: Isometric view of the track bridge with bearer bars, FPB, and rails**

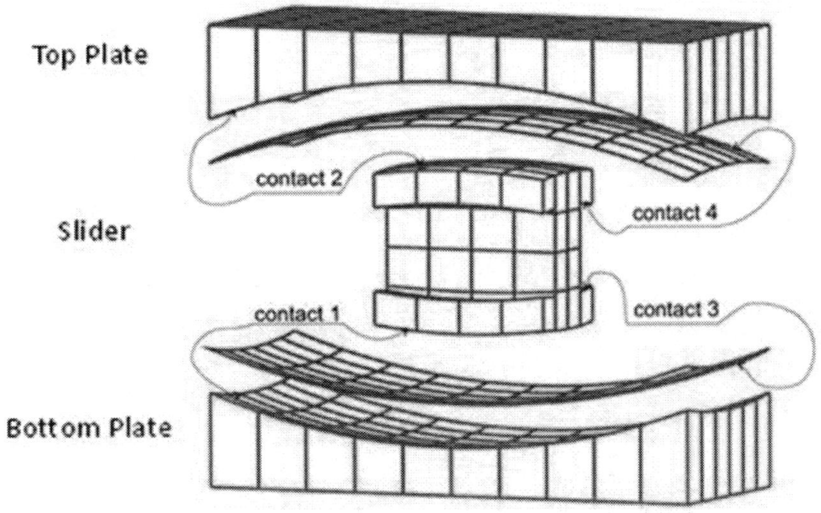


**Figure 11: Finite element model of the double FPB**

### 3.3 ADINA and NUCARS model comparison

To ensure the accuracy of the behaviors reported by these two models, an iterative convergence method was established, in which the deformations predicted from the dynamic analysis conducted in ADINA were fed into the NUCARS model and the wheel-rail reaction force predictions from NUCARS were returned to ADINA.

When a train crosses a track bridge, the rails will move as a result of deflection and wing rotation. SCS produced a 3-dimensional deflected rail "profile time history" ($D_{initial}$) using preliminary wheel-rail force predictions from the ADINA model. $D_{initial}$ was then used as a dynamic input to the NUCARS model as the LRV crossed the bridge. Time histories of vertical, lateral, and longitudinal forces between each wheel and rail were then output from NUCARS. These "force time histories" ($F_2$) were applied to the ADINA model of the bridge and a second three-dimensional profile time history developed ($D_2$). While $D_2$ results were fairly close to $D_{initial}$, an additional iteration was made to close the loop ($F_3$ and $D_3$). Figure 12 illustrates the process.

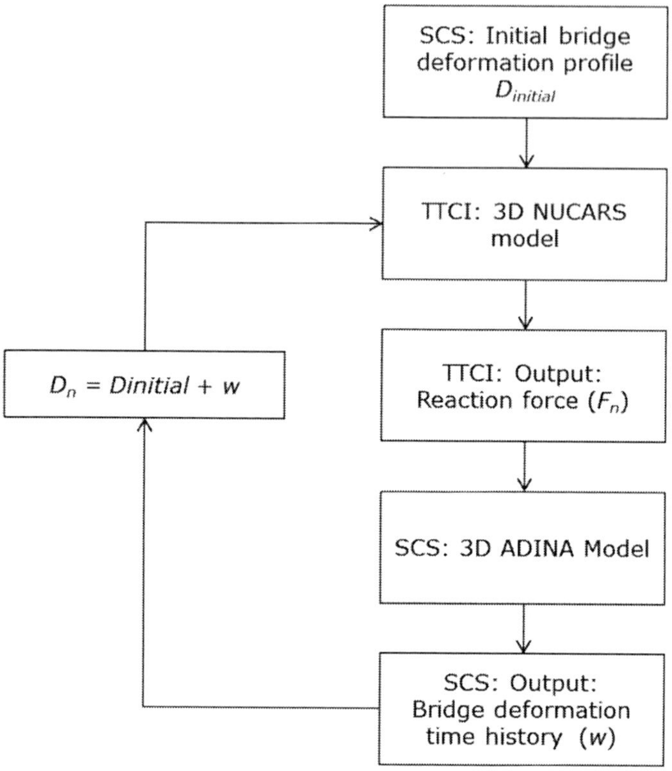


**Figure 12: ADINA and NUCARS iterative convergence method**

The following describes example comparisons for an 88.5 km/hr (55 mph) run with 0.5 degrees of pitch, 0.1 degrees of yaw, and 0.7 degrees of roll angle in the track bridge track geometry. Figure 13 compares the vertical component of $D_{initial}$, $D_2$ and $D_3$ for the left rail at Axle 1. Figure 14 compares the lateral component for the same position. Figure 11 shows NUCARS vertical wheel-rail force outputs from the first and second iterations. Figure 14 and 15 show vertical and lateral wheel-rail forces $F_2$ and $F_3$.

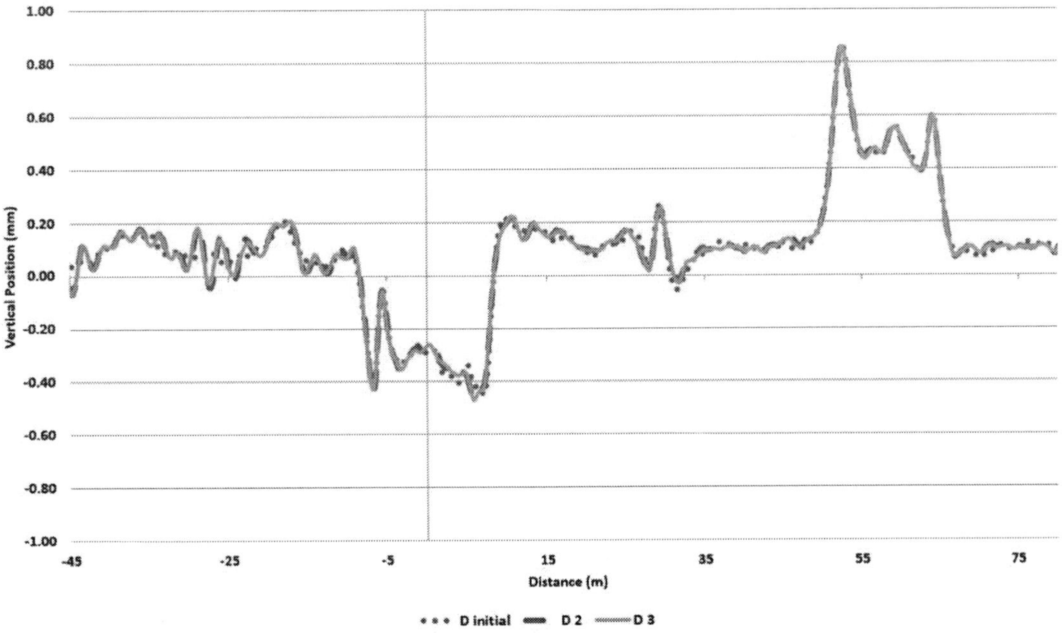


**Figure 13: Vertical deflections for left rail at axle 1**

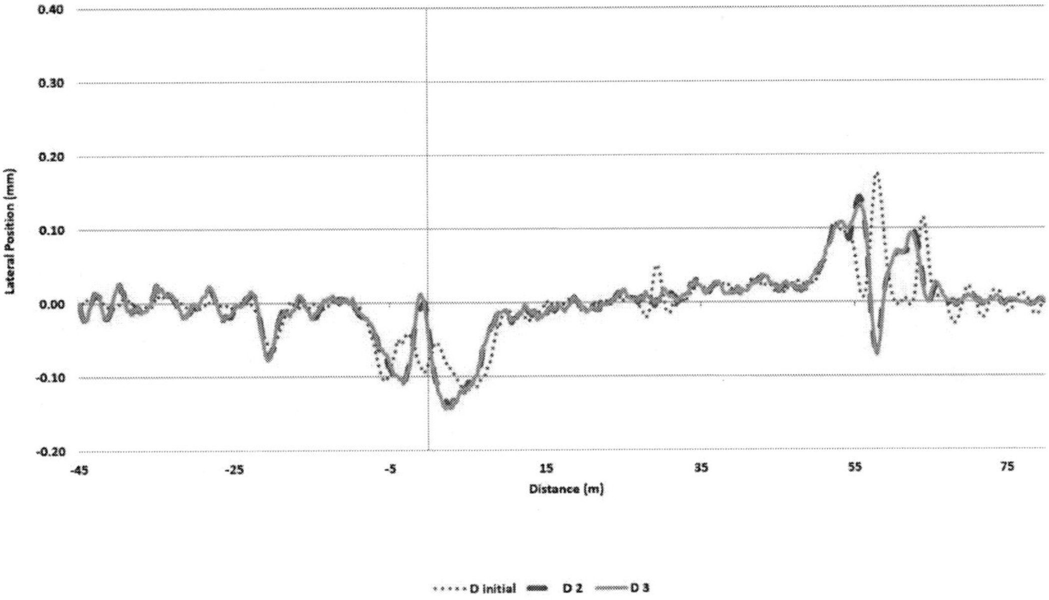


**Figure 14: Lateral deflections for left rail at axle 5**

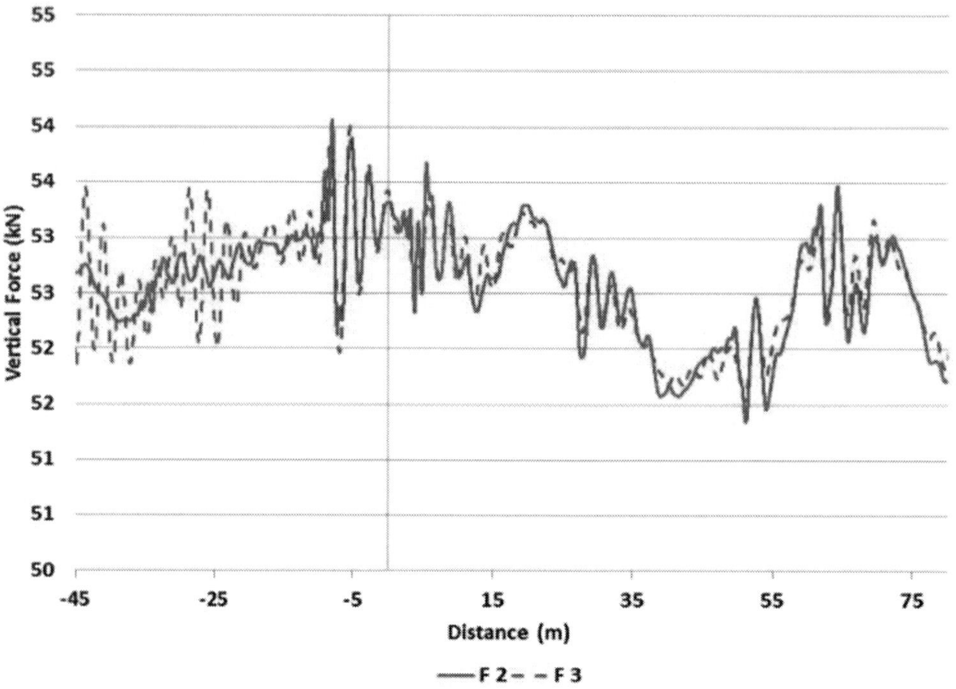


**Figure 15: Vertical forces for left wheel at axle 1**

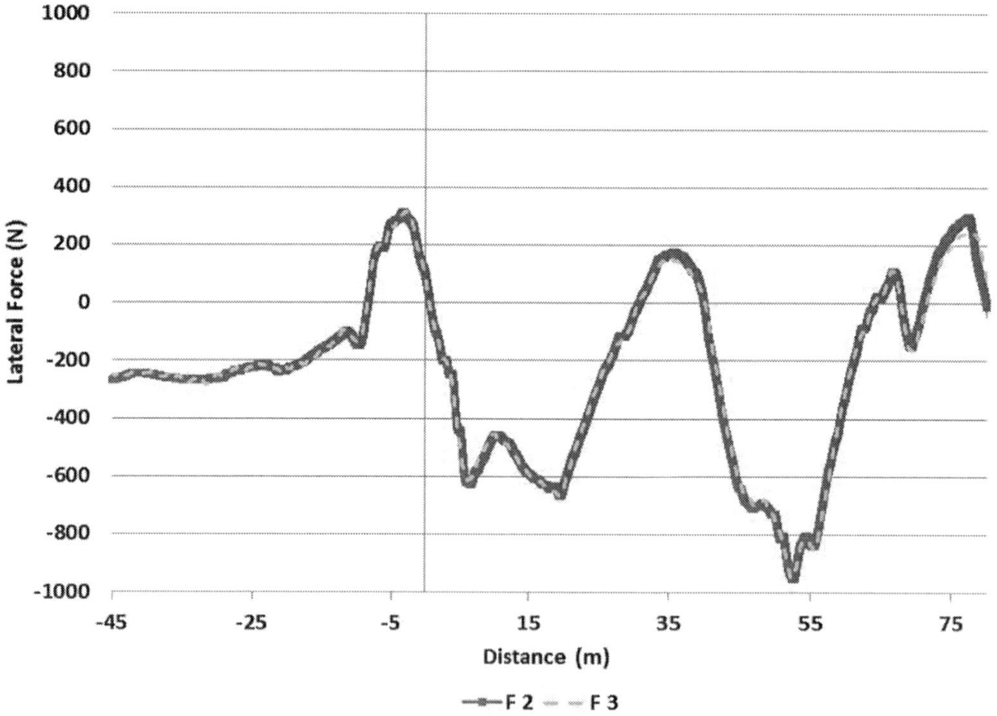


**Figure 16: Lateral forces for left wheel at axle 1**

### 3.4 Model results

Figure 17 shows predicted ride quality from NUCARS using the $D_3$ profile time history from ADINA. Vehicle performance was well within International Organization for Standardization ride quality criteria (2). Results were also well within FRA safety criteria (3) that were used for this project. Other simulations were conducted to evaluate the effects of traction and braking and forces and vehicle lateral stability. All results met criteria established by ST.

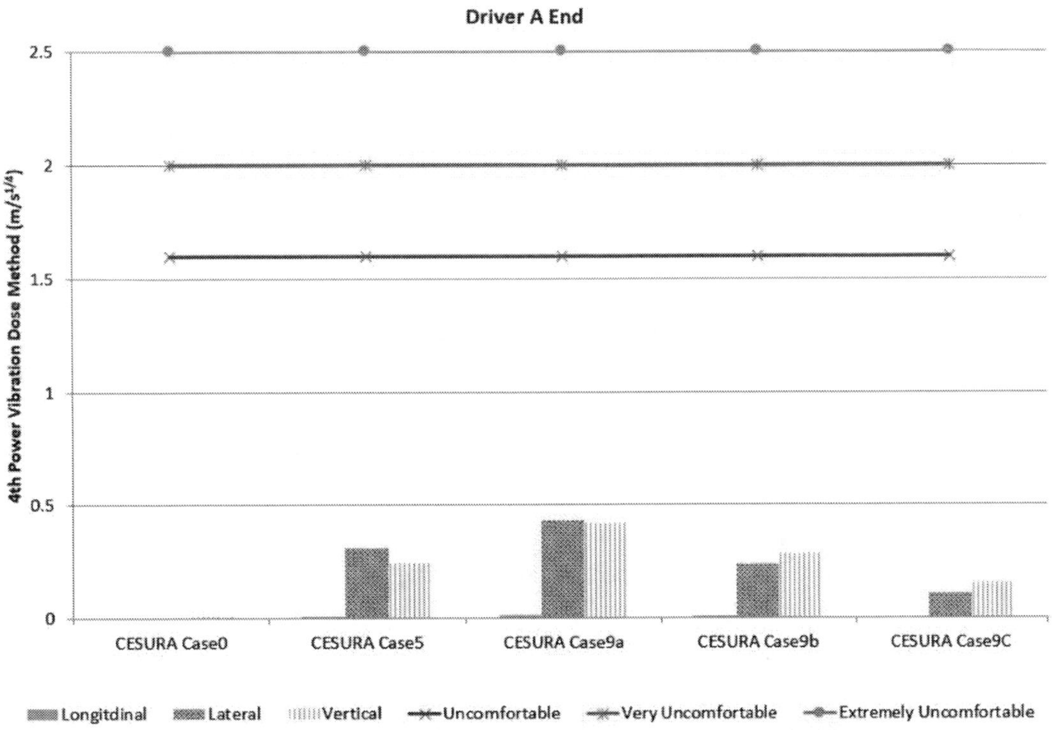


**Figure 17: Predicted ride quality from NUCARS**

This unique approach to modeling provided a good understanding of bridge and vehicle interaction and proved to be a useful tool in the design process providing valuable data to enhance the design for optimal structural and vehicle performance.

## 4 FULL-SCALE COMPONENT TESTING

Jesse Engineering Company, Tacoma, Washington, manufactured two full-scale prototype track bridges. During the design phase of prototype development, the project team conducted full-scale component testing through the University of Washington's Structural Research Laboratory and reported the results to PB. The goal was to verify, before full-scale prototype testing, that the wings, bearer bars, bearings, and rails would undergo the required movements in the manner anticipated in the design. Jesse Engineering Company also conducted functional testing after the prototypes were constructed. Resulting recommendations of the testing were included in the prototype design that was ultimately tested (1).

## 5 FULL-SCALE TRACK TESTING

Full-scale in-track testing of two prototype track bridges was conducted between August 25, 2013, and October 14, 2013. The testing objectives included validation of the computer analysis results and confirmation that the design met the performance criteria for safety and rider comfort, reliability, and constructability.

Other test objectives included measuring rail-to-earth electrical resistance of the prototype track bridge, measuring resonant frequencies of the prototype track bridge, determining wayside noise levels generated by LRVs crossing the track bridges (Wilson, Ihrig and Associates, Inc.), and verifying performance of a secondary data acquisition system, which included components to be used in a permanent monitoring system to be installed on the track bridges in service (Direct Measurements, Inc.).

A new 1,524-meter-long (5,000-foot-long) double-ended siding with overhead power, called the ST Test Track (STTT), was constructed at TTC. The STTT was constructed by Balfour Beatty Rail Inc. under contract to PB expressly for testing the prototype track bridges. Figures 17 and 18 show the STTT.

Two prototype track bridges were installed at the crest of a 61-meter-long (200-foot-long) transition zone on a 3-percent grade and in a 13-meter (42-foot) sag curve at the downhill end of the transition to replicate the geometry of the west end of the I-90 bridge at low lake level.

While not all of the floating bridge structural systems were incorporated, considerable effort was made to replicate the track geometry and displacements of the transition and floating spans, reflecting the service condition limits.

The STTT provided sufficient length for a loaded two-car train to accelerate up to 88.5 km/hr (55 mph), cross the track bridge testing area, and safely decelerate.

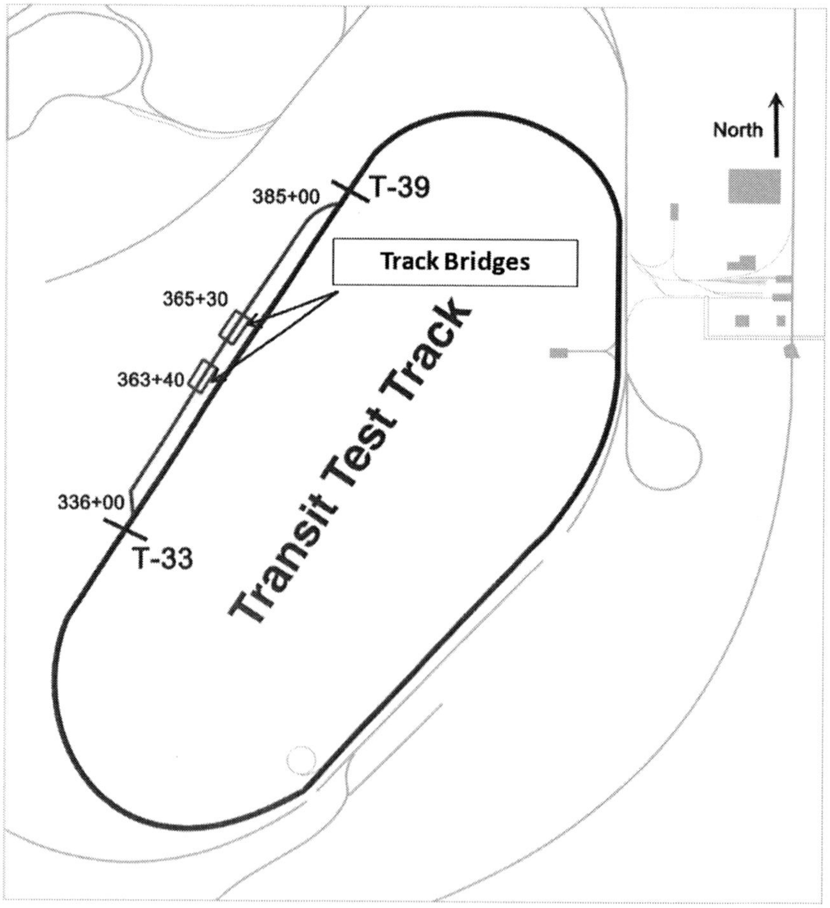


**Figure 18: STTT test track for full-scale testing at TTC**

**Figure 19: Prototype track bridges in the STTT**

ST provided two LRVs for testing. One of the vehicles was instrumented with accelerometers to measure ride quality and to validate modeling results, displacement transducers to measure yaw rotations at articulation joints and displacement transducers across primary and secondary suspension systems, gyroscopes to measure vehicle roll, and various transducers for speed and position.

Track bridge performance was monitored throughout the test with over 200 channels of instrumentation data collected on each track bridge. Instrumentation included the following:

- Strain gages for structural stress and rail forces
- Displacement transducers for displacements and rotations
- Thermocouples for monitoring rail, structure, and ambient temperatures
- Optical sensors to monitor vehicle position in the test zone

Testing was conducted in three phases: (Phase 1) tangent track with simulated low-lake level geometry, (Phase 2) a lateral shift of the interior track bridge by 203.2 mm (8 inches) to create 0.24 degrees of yaw, and (Phase 3) the same configuration as Phase 2, but with 19.05 mm (0.75 inch) of superelevation built into the interior track bridge to represent 0.7 degrees of roll.

The test data was compared to the simulation results for each test phase. The simulation data shows the same frequency and similar content to test data. Figures 20–24 show comparisons between simulation and test data for an unloaded vehicle at 88 km/hr during Phase 3.

The model matches the overall dynamic response of the test fairly well; however, the simulations consistently calculate higher acceleration levels than measured during the tests, a conservative result. A spike in the predicted response that does not match the test results can be seen at approximately 213 meters (700 feet) in each of the figures. This location is approximately where an expansion joint is located. The spike is likely related to an anomaly in track geometry measurements.

Other differences between model and test results could be due to small differences in weight distribution of the carbodies and differences in track geometry measurements as recorded on unloaded track compared to actual loaded conditions. In addition, the STTT bridge approaches were of ballasted construction, while the profile time histories from ADINA assumed direct fixation track throughout. It is also important to note that while the stiffness change between the ballasted track and track bridges was evident in the data, people riding the train during test operations did not perceive the "bumps" due to the transition.

No safety criteria were exceeded for either test or simulation results (3).

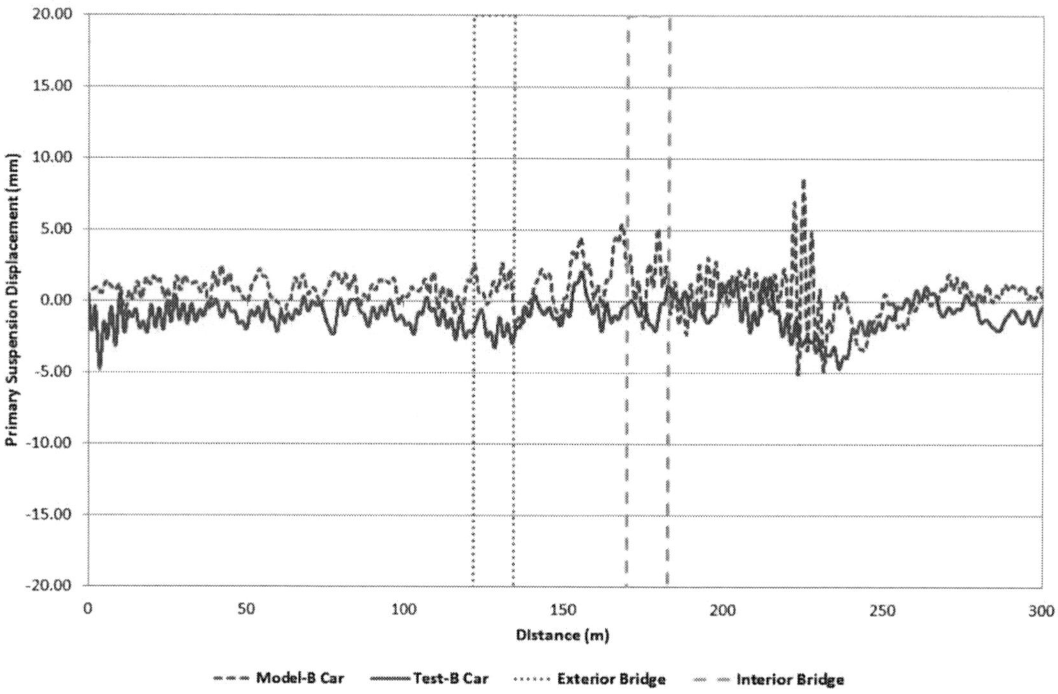


**Figure 20: Displacement measured across the primary suspension B-car**

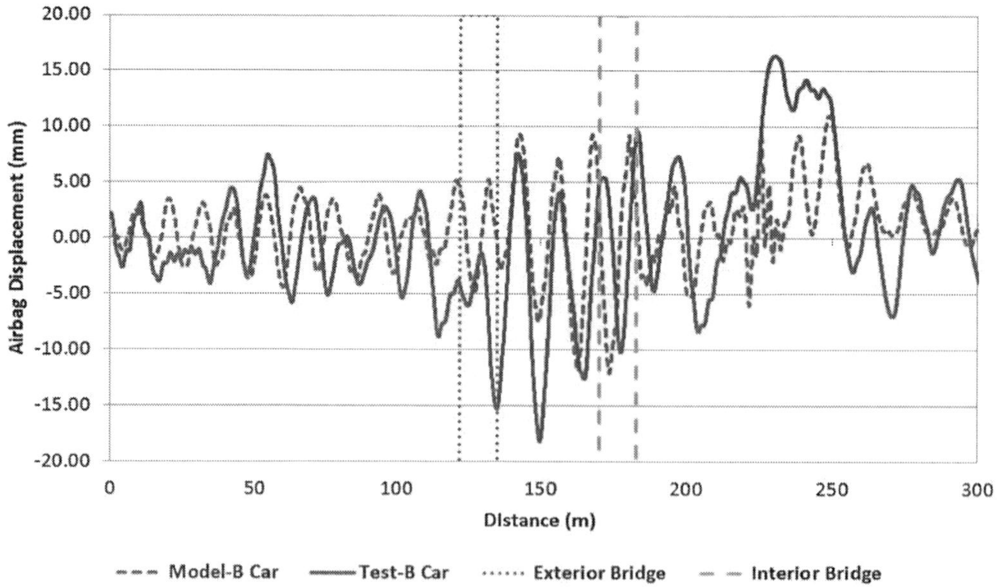


**Figure 21: Displacement measured across the secondary suspensions B-car**

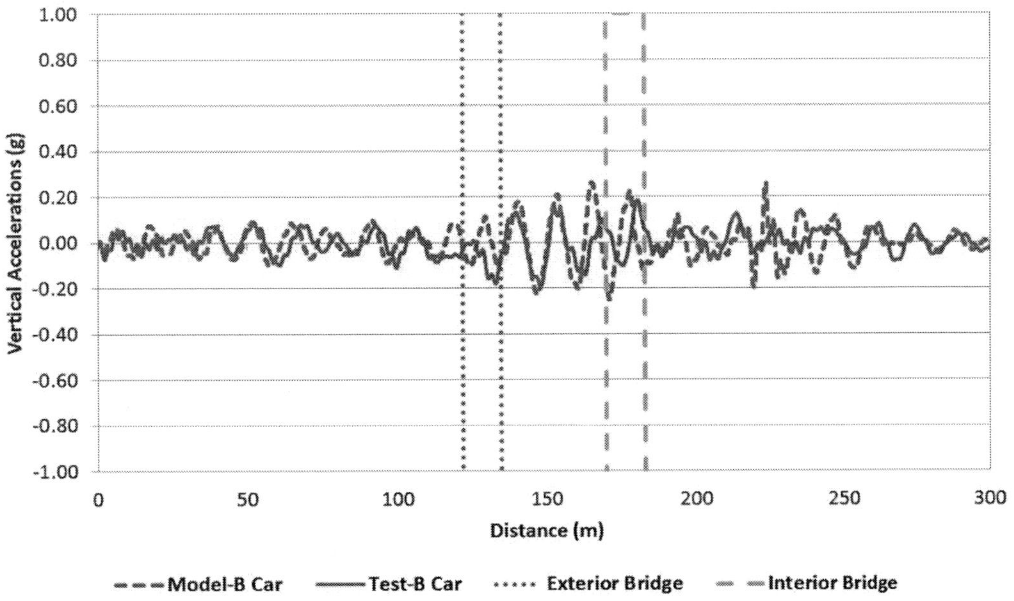


**Figure 22: Vertical accelerations measured in the B-car**

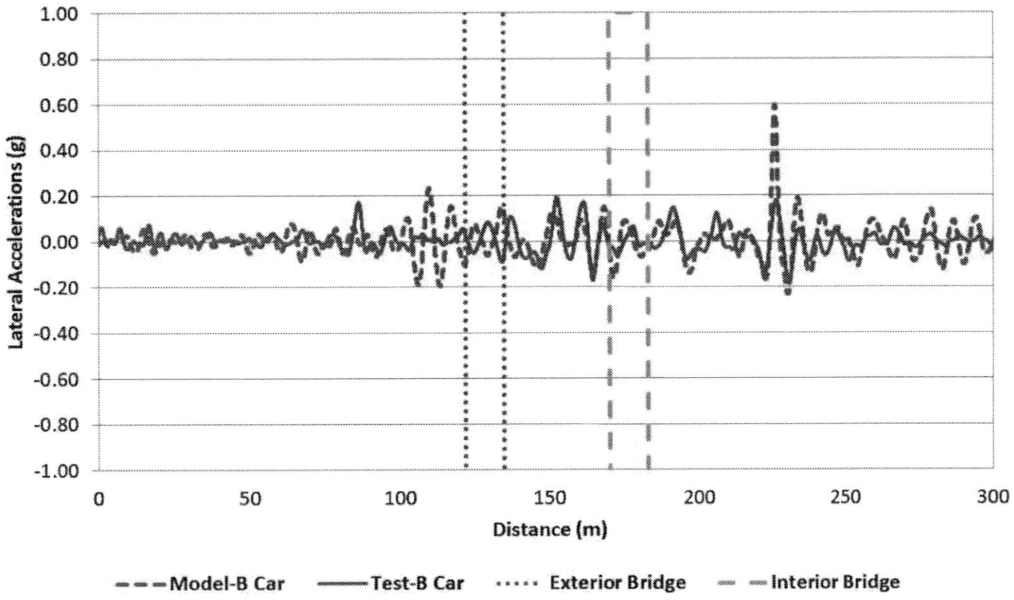


**Figure 23: Lateral accelerations measured in the B-car**

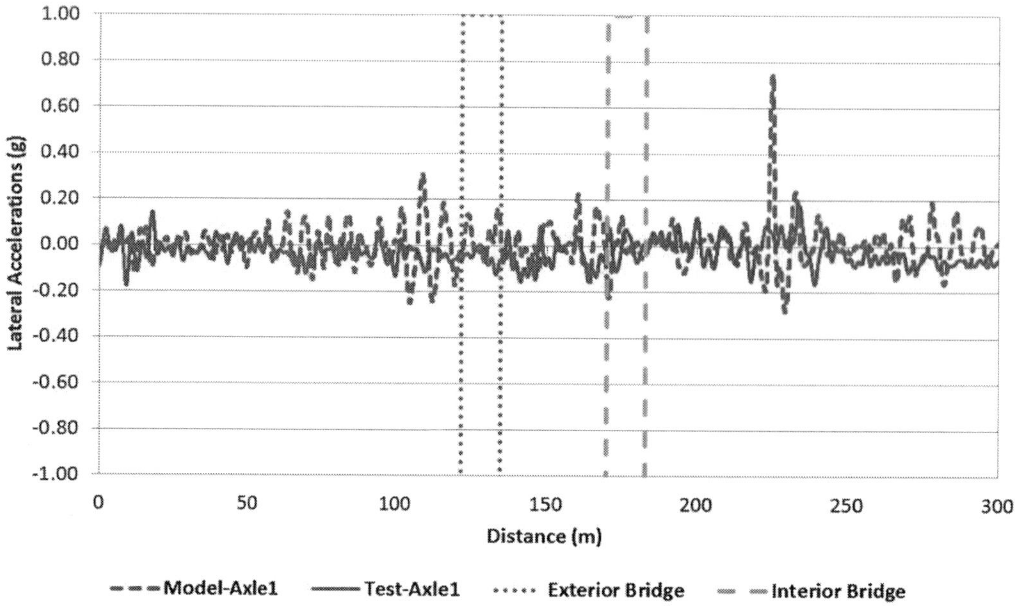


**Figure 24: Lateral accelerations measured at axle 1**

Figures 25 and 26 compare simulated ride quality with test results. The simulation results tended to under predict the passenger ride quality index for the baseline test configurations prior to ballast settlement. For later configurations (Phases 2 and 3), the simulations tended to over predict the passenger ride quality with results generally within 10 percent of the test data. The model results are therefore a conservative estimate of ST LRV performance.

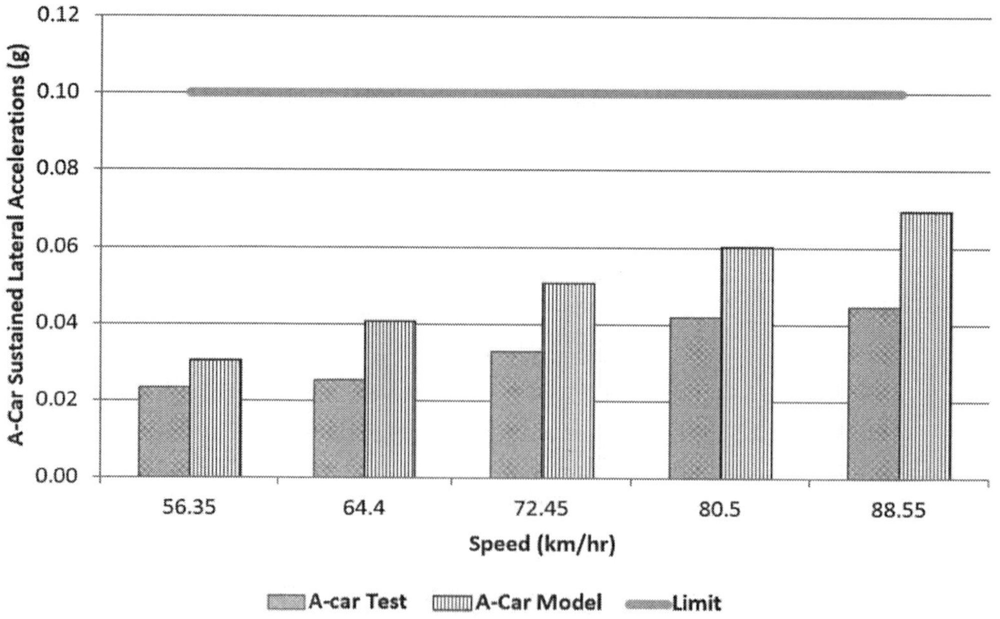


**Figure 25: Ride quality — comparison of simulated lateral accelerations with test results**

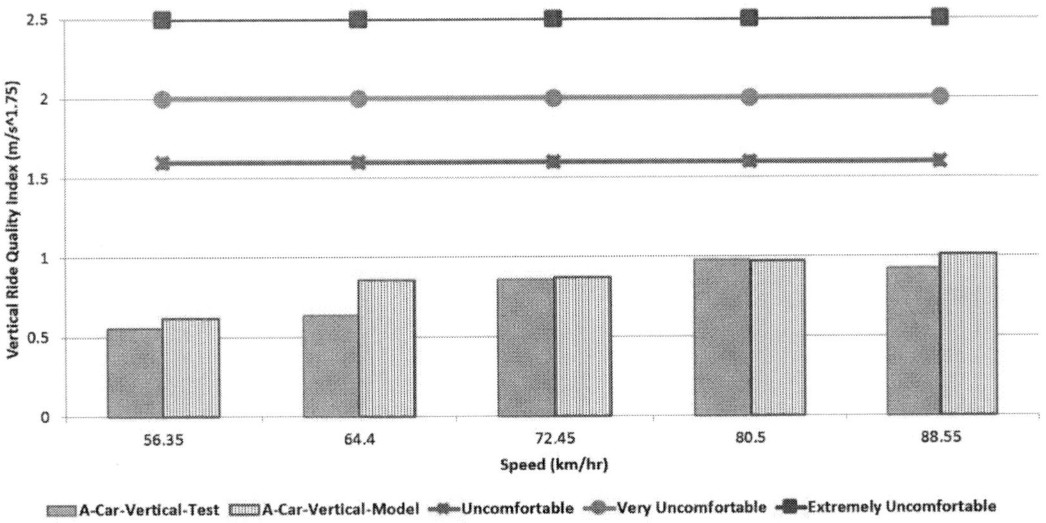


**Figure 26: Ride quality — comparison of simulated vertical accelerations with test results**

## 6 SUMMARY AND CONCLUSIONS

The complex nature of the track bridge design required a unique design, dynamic modeling, and test program to ensure safe operation. The CESuRa track bridge was shown to accommodate multidimensional movement while providing a smooth running surface for the rail vehicle to traverse from fixed bridge span to movable bridge span or vice versa. The dynamic modeling and test program showed that CESuRa not only provided a smooth transition but met all safety and passenger ride quality design criteria.

The system approach to modeling, utilizing the advanced features of NUCARS and ADINA provided valuable data to enhance the track bridge design for optimal vehicle and structural performance. Full-scale testing allowed the models to be updated and fully validated to provide the most accurate information possible. The models can now be used to investigate operation conditions that may be unsafe or too costly for a full-scale test.

This type of system modeling allows for structures and vehicle response to be modeled simultaneously and adverse reaction of the vehicle or structure to be mitigated in the design process. There is potential for this type of modeling to be used for other track structure and vehicle response designs.

## REFERENCE LIST

1. Cooper, T., P.E., P.Eng., J. Sleavin, P.E., A. Foan, and T. Thonstad. "Unprecedented connections," *Civil Engineering—ASCE*, Vol. 84, No. 3, March 2014, pp. 56-63, 76.
2. International Organization for Standardization (ISO).1997. Mechanical vibration and shock — Evaluation of human exposure to whole-body vibration Part 1: General requirements. ISO 2631-1:1997 (E), Second edition corrected and reprinted 1997-07-15, Switzerland.
3. Track Safety Standards Part 213, Subpart G, Class of Track 6 and Higher, Department of Transportation, Federal Railroad Administration, Office of Safety, Oct 5, 2004, Washington, DC.

# Estimating the damage and marginal cost of different vehicle types on rail infrastructure: combining economic and engineering approaches

**A S J Smith[a], A Kaushal[b], K Odolinski[c], S Iwnicki[b], P Wheat[a]**
[a] University of Leeds, UK
[b] University of Huddersfield, UK
[c] Swedish National Road and Transport Research Institute, Department of Transport Economics, Sweden

## ABSTRACT

EU legislation requires that European infrastructure managers set access charges based on the marginal cost of running trains on their networks. Two methods have been used in the literature for this purpose. Top-down methods relate actual costs to traffic volumes. Bottom-up methods use engineering models to simulate damage and then translate damage into costs based on assumptions about interventions and their unit costs. Whilst top down methods produce sensible results for marginal cost overall, they have struggled to differentiate between traffic types. The challenge for bottom-up approaches is how to translate damage into cost, with numerous assumptions being required which may be invalid.

This paper proposes a new, two stage approach to estimating the marginal cost of rail infrastructure usage. The first stage uses engineering models to simulate damage caused by vehicles on the network. The second stage seeks to establish a statistical relationship between actual costs and damage. It is thus possible to convert damage estimates into costs using actual cost data, rather than through a set of potentially invalid assumptions as in previous approaches.

Only the first stage is implemented in this paper. We show that it possible to produce total (annualised) damage measures for three damage mechanisms on five actual track sections in Sweden. Once extended, it will be possible to model the relationship between damage and actual costs for the first time; and thus better understand the relative costs of the different damage mechanisms and in turn inform the level and structure of track access charges.

## 1. INTRODUCTION

European policy since the mid-1990s has emphasised the promotion of within-mode competition as a way of revitalising the fortunes of Europe's railways. Progressively freight and international passenger services have been opened up to competition. The proposals contained in the European Commission's Fourth Railway package (European Commission, 2013) for further reforms of Europe's railways, mean that, once enacted, it will be compulsory to introduce competitive tendering for passenger services run under public service contracts and open access for commercial services across the whole of Europe. Vertical separation of infrastructure and operations or, at least, fair access to infrastructure and transparent prices for access, is seen by the Commission as a key enabler of competition in the sector.

The above developments mean that understanding the cost, and in particular, the marginal (infrastructure) cost, of running an extra service on the network has become more important than ever. Existing legislation requires that charges for access to the infrastructure must be based on "costs directly incurred as a result of operating the train service". This can be interpreted as what economists would call the short-run marginal (or incremental) cost imposed on the infrastructure by the service running on the network. This paper focuses on one element of short-run marginal cost, namely the additional maintenance and renewal cost required to rectify the incremental damage caused a train service (the marginal wear and tear cost).

Of course, the need to estimate marginal cost of infrastructure use is not merely for the purpose of meeting EU legislation. It is important from the purpose of economic efficiency (in terms of making best use of the existing network) that train operators pay at least the short-run marginal cost of running trains on the network. Further, track access charges that vary according to the different damage and cost imposed by different vehicles, should ensure that the "right" vehicles are run on the network and potentially that new rolling stock designs are developed that reduce whole system costs (operator and infrastructure managers costs).

The previous literature, and practice by bodies responsible for charging, contains three approaches for estimating marginal costs of traffic and also different types of traffic: cost allocation methods; statistical approaches; bottom-up engineering methods. In practice the approaches used by charging setting bodies can be a hybrid of approaches, such as that used by the British Office of Rail Regulation (ORR). However, each of the approaches has significant drawbacks. The contribution of this paper is to propose an alternative, two-stage approach that combines engineering and statistical approaches in a way that seeks to combine the best features of both; and thus overcome some of the weaknesses of previous approaches.

The remainder of the paper is structured as follows. In section 2 we explain the previous approaches used to estimate marginal rail infrastructure costs and their weaknesses. The proposed methodology is set out in section 3 and the data used for the empirical work outlined in section 4. Section 5 presents the results and section 6 concludes.

It should be noted that at this stage of the research we have only been able to implement stage 1 of the methodology. The second stage results are currently being generated and will be presented at the conference. This conference paper therefore reports and discusses the methodology and the first stage results.

## 2. PREVIOUS APPROACHES

To summarise briefly, there are two methods for producing estimates of marginal costs (Wheat and Smith, 2008). Top-down methods relate actual costs to traffic volumes, controlling for characteristics of the infrastructure. Bottom-up methods can be characterised as using some form of engineering model to estimate the damage inflicted by different types of vehicle on the network. Then assumptions can be made about the intervention / remediation required to deal with that damage, combined with estimates of unit costs of that remediation activity, to give the marginal cost estimates. These approaches are summarised in Figure 1. A "third" method, the so-called cost allocation method, can be thought of as a hybrid that utilises engineering judgement and econometric evidence and other rules of thumb to establish the variability of different cost categories. This approach is therefore not discussed further.

- Method 1: engineering approach
  - Simulate damage done by traffic (engineering model)
  - Determine action need to remedy damage (e.g. tamping)
  - Activity volume * Unit cost of activity = (marginal) Cost
- Method 2: top down statistical approach
  - Relate actual costs to passenger and freight tonne-km (regression)
  - E.g. Log Cost = a + b* Log Passtonne + c * Freight tonne
  - Compute marginal costs from the parameter estimates (the a and b) from that model

**Figure 1: Alternative approaches for estimating marginal costs**

Both methods have strengths and weaknesses. The advantage of top-down methods is that they use actual cost data. Their weakness lies in the fact that it is likely to be very hard to capture the complexity of factors that will affect the relationship between traffic and cost, and in particular, it has proved difficult to get any sensible estimates of the relative cost of passenger and freight vehicles. The bottom-up method is very good at capturing complexity and it is possible to model and gain estimates of the relative damage of different vehicle types. The problem is how then to translate these damage estimates into cost.

It is worth noting, more precisely, why there is a difficulty in getting from damage to cost in bottom-up approaches. First, traffic results in different types of damage. In practice, one vehicle may cause more of one type of damage and less of another, thus meaning that information is needed on the relative cost of the different types of damage to obtain estimates of relative marginal cost. This leads to a second problem, namely that assumptions are needed on what type of activity and how much of it, are needed to rectify the damage done. This potentially requires a very detailed model or alternatively simplifying assumptions are needed which might be wrong.

Added complexities include the fact that the mix of damage types will affect what activity should be undertaken (e.g. some traffic types might cause damage but at the same time alleviate the need for other forms of remediation, such as rail grinding) and the fact that some damage mechanisms lead to more maintenance activity, whilst others result in more renewal (and the costs of these can be very different). Finally it is hard to estimate unit costs of activities as these will depend on, inter alia, the location, the nature of the job, the length of possession and the scale of the activity.

The research question therefore is as follows: how can we obtain better estimates of the relative cost of different damage mechanisms (which in turn can then help estimate the relative marginal cost of different vehicle types).

In terms of its relation to other approaches, in the past ORR has used a top-down cost allocation approach, based on engineering judgement, to determine the general level of cost variability, and then used an engineering formulae to allocate costs to vehicles based on their relative damage (vertical forces only). Since then Network Rail has developed its bottom-up cost modelling approaches (its infrastructure cost model (ICM) model, which in turn is based on the Vehicle Track Interaction Strategic Model (VTISM)) to estimate marginal costs from the bottom up). This approach measures the overall variability of costs with respect to traffic. Engineering approaches, based on both vertical and horizontal forces, are then used to allocate that element of costs that is deemed variable down to individual vehicles (see, ORR, 2013). This approach means that Britain, unlike other European countries, has highly differentiated charges, by vehicle, which should incentivise the use of more track friendly vehicles.

At the same time, top-down econometric methods relating actual costs to traffic volumes, controlling for other factors have been extensively used and the results used by the European Commission (see, for example, Johansson and Nilsson (2004), Wheat et. al. 2009 for a summary, and Andersson et. al., 2012 for subsequent developments in modelling of renewals costs). These studies have covered a range of European countries, and suggest that the marginal cost of rail infrastructure maintenance is in the region of 20-35% of maintenance costs (or up to 45% for heavily used sections). Wheat et. al. (2009) found that the available evidence was much less strong for renewals, though suggested an indicative overall cost variability proportion of around 35% of renewal costs. More recent evidence has put this at a higher level; at approximately 55% (Andersson et. al., 2009). As noted above, these methods have been useful in determining the extent of cost variability with traffic in general, but less effective at allocating to types of traffic or vehicle. It is worth further noting that the engineering based bottom-up approach used by Network Rail puts the variability proportion at less than 10% which is out of line with the top-down econometric evidence from across Europe.

Our proposed approach is therefore positioned within the existing literature and would particularly enhance ORR's approach to allocating variable costs to vehicle by providing new evidence on the relative cost of different damage mechanisms, in turn leading to better estimates of the relative marginal cost of different vehicle types. The approach could also be used to determine absolute marginal cost and cost variability levels to compare against the results of top-down and other engineering models.

## 3. METHODOLOGY

Our proposed method essentially consists of two stages. The first stage involves an engineering simulation exercise in which traffic (of certain vehicles and mixes of vehicles) is run down a network of known characteristics, to produce estimates of the resulting damage (denote these D1, D2, and D3, to represent the three main damage mechanisms: settlement, wear and rolling contact fatigue). For this exercise we choose actual track sections from Sweden where we have data on the maintenance and renewal costs, the traffic volumes and the infrastructure characteristics. The second stage involves establishing a statistical relationship between actual costs (maintenance and or renewal) for actual track sections on a network (in our case the Swedish network) and damage. The approach is summarised in Figure 2 below.

The three damage mechanisms are defined as follows. Wear of the rail is a natural process in which material is removed from the head and/or the gauge corner of the rail when railway vehicles run. The rate of removal is affected by the forces and contact conditions. Severe wear can change the cross sectional profile of the rail, resulting in a change of the running surface seen by the wheel. Rolling contact fatigue (RCF) occurs if the rail surface is subjected to repeated plastic deformation as is often caused by repeated wheel passages.

The 'Tgamma' value provides a measure of the wear and RCF performance of the vehicle. Tgamma is the product of the tangential or creep forces and the slippage or creepage in the contact patch between wheel and rail. Tgamma combined with a non-linear damage function produces a RCF damage index as shown in 3. This index is then used to interpret whether the vehicle is damaging the track due to wear, RCF or a combination of both.

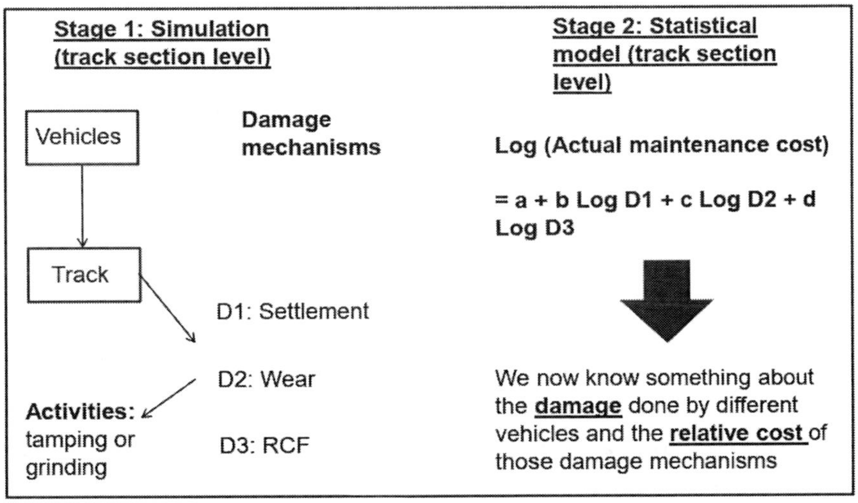


**Figure 2: Overview of the methodology**

With reference to 3, as Tgamma increase from 0 to 15 N, no RCF damage is generated as there is insufficient energy to initiate RCF cracks. As Tgamma increase from 15 to 65 N, the probability of RCF incitation increase, to a maximum of 1 at a Tgamma value of 65 N. As Tgamma increase further from 65 to 175 N, the level of energy is such that the dominant form of surface damage is wear (rather than crack initiation), therefore the probability of RCF damage decreases as wear increases. Negative values of RCF damage indicate values of Tgamma greater than 175 N, resulting in wear and no RCF initiation. The units of the RCF damage index are 10-5 per axle. This indicates that for a damage index of 1, 100000 (One hundred thousand) axle passes would result in RCF initiation.

Finally, track settlement may be defined as the sinking of the track (in the vertical plane) into the ballast under a variety of conditions. A number of models have historically been used to predict track settlement. Initially the Technical University of Munich (TUM) model and the Sato model were considered. It is assumed that the settlement is going to be proportional to the vertical force on the wheelset.

Estimates of the relative cost of different damage mechanisms can be estimated from the parameters in this second stage regression (the b, c and d parameters in Box 2), which in turn allows us to estimate the relative cost of different vehicles. Of course, a more complex relationship could be assumed in the second stage statistical model, for example to include interactions between damage types (e.g. D1*D2). The detailed assumptions for stage 1 of the approach, as they apply in practice, are set out in section 5, after the description of the data. As noted in the introduction, stage 2 of the approach is not implemented in this paper due to lack of data at this time.

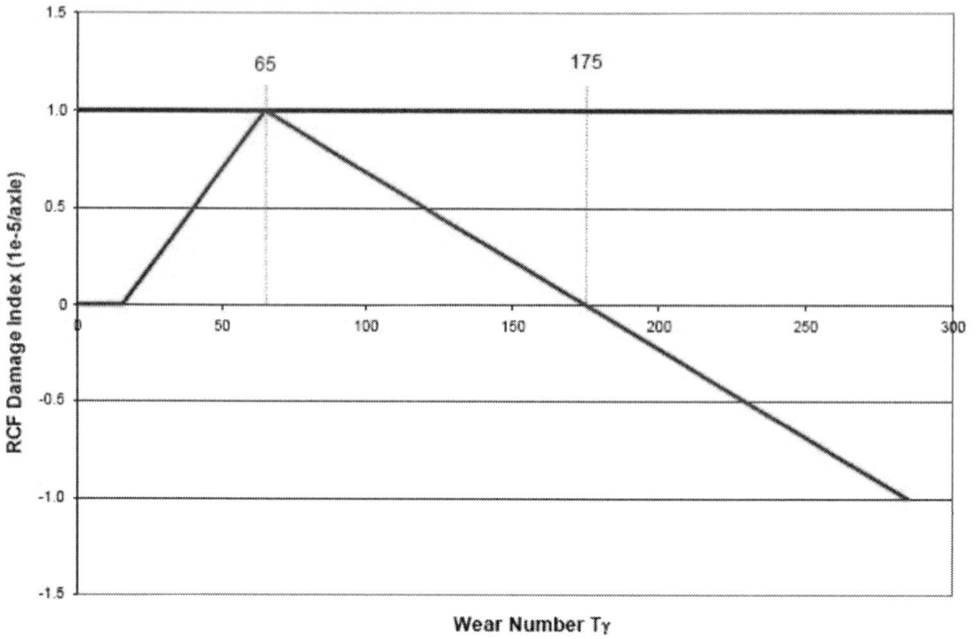


**Figure 3: Relationship between wear number (Tγ) and RCF damage index**

## 4. DATA

There are four main aspects to the dataset, which has been provided by the Swedish Transport Administration (*Trafikverket*). These are, for the damage simulations, data on track alignment and vehicle characteristics. Secondly, for the second stage modelling, data on costs (maintenance and renewals), traffic volumes and infrastructure characteristics.

So far the analysis has focussed on five track sections. Data for another one hundred sections has been requested but has not been received in time for inclusion in this conference paper. The additional data points will be used in subsequent analysis to implement the second stage.

### 4.1 Vehicle data

Three vehicle types have been modelled running down the selected five sections (see Table 1). These include two types of freight locomotives (with associated wagons) and one EMU. The data provided by Trafikverket revealed the generic type of vehicles and their usage on the network. However, the information was not sufficiently detailed to identify the exact vehicle. Hence generic freight and EMU vehicle types from the EU funded research project, INNOTRACK[1], were used. These types are considered to represent typical freight and EMU vehicles running on the EU network. In future work it is hoped to obtain more detailed information on the exact vehicles running on the Swedish network.

The vehicle models were described in terms of mass properties, geometry, axle load, unsprung mass, wheel radius and suspension characteristics. The VAMPIRE simulation package uses the model description to generate equations of motion for the vehicle which it then solves with the track input. Once the simulations are completed, outputs are available for wheel-rail forces vehicle motions and can be used to estimate damage.

---

[1] http://www.innotrack.net/article/presentation

## 4.2 Track data

Track alignment data were obtained using a track geometry coach which measures the curvature, cross level, vertical irregularity, lateral irregularity and gauge at fixed intervals of 25cm. This data was then used to create VAMPIRE track input files for the chosen track sections. An example of the track data for section 629 is shown in Figure 4. The five graphs represent:

1. Cross level versus Distance along the track
2. Curvature versus Distance along the track
3. Lateral Irregularity versus Distance along the track
4. Vertical Irregularity versus Distance along the track
5. Gauge Variation versus Distance along the track

**Table 1: The vehicles running on each track section**

| Track Sections | Vehicle ID | Type |
|---|---|---|
| 326 | RC2 | Loco |
| | RC4 | Loco |
| | 4-axle Wagon | Wagon |
| | 2-axle Wagon | Wagon |
| 629 | RC2 | Loco |
| | RC4 | Loco |
| | 4-axle Wagon | Wagon |
| | 2-axle Wagon | Wagon |
| 652 | X12 | EMU |
| | X14 | EMU |
| 654 | X12 | EMU |
| | X14 | EMU |
| 821 | X11 | EMU |
| | X31 | EMU |

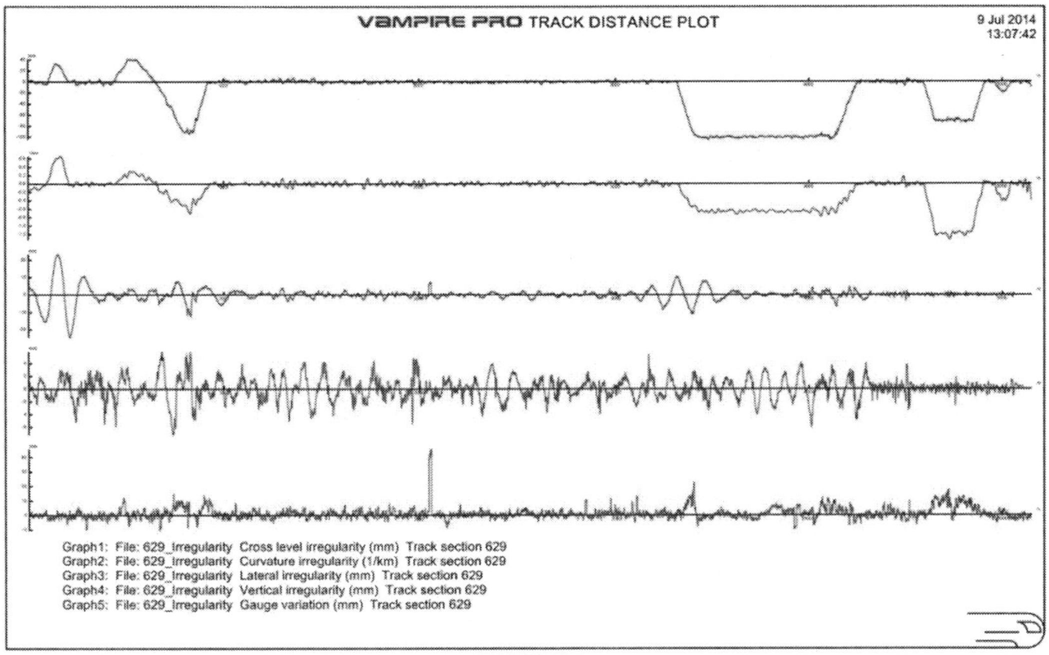


VAMPIRE Plot

**Figure 4: Track alignment data for track section 629**

### 4.3 Second stage data: costs, traffic and track characteristics

Costs for maintenance and renewals are collected in Sweden at the track section level. Included in these costs are activities conducted to maintain the railway assets as well as replacement and refurbishments to restore the assets to original condition. Snow removal is defined as maintenance in Sweden, though these costs are excluded in our data set. Renewal activities are not as frequent as maintenance, and a renewal cost is only observed for one of the track sections (see Table 2).

**Table 2 – Costs, SEK**

| Track section no. | Maintenance costs | Renewal costs |
|---|---|---|
| 326 | 5 757 077 | 132 707 |
| 629 | 323 503 | 0 |
| 652 | 15 609 488 | 0 |
| 654 | 8 735 644 | 0 |
| 821 | 3 566 033 | 0 |

The simulation models produce an overall estimate of damage per tonne-km for each vehicle type. It is therefore necessary to scale up these measures based on the volume and mix of traffic that is actually running on the network. Traffic data include the declared train kilometres, weight (gross tonnes), number of wagons, and train length for different types of locomotives or EMUs/DMUs. The data is summarised in Table 3.

In the second stage it is also important to recognise that the cost of addressing different levels of damage will depend on the characteristics of the infrastructure. Additional data has therefore also been collected on measures such as rail weight and linespeed (these are not shown to keep the discussion tractable).

**Table 3 – Traffic data**

| | *Train-km* | | | *Tonne-km* | | |
|---|---|---|---|---|---|---|
| Track section no. | Passenger trains | Freight trains | Empty trains and service trains | Passenger trains | Freight trains | Empty trains and service trains |
| 326 | 8 150 | 212 521 | 15 247 | 1 676 878 | 192 837 314 | 1 208 129 |
| 629 | 18 | 2 983 | 95 | 2 810 | 684 166 | 5 502 |
| 652 | 285 707 | 15 728 | 22 586 | 38 554 466 | 15 262 308 | 3 442 515 |
| 654 | 312 434 | 5 368 | 3 404 | 33 160 000 | 5 327 958 | 338 685 |
| 821 | 325 744 | 16 927 | 53 390 | 53 677 213 | 14 306 152 | 8 709 709 |

## 5. RESULTS AND DISCUSSION

The output from the simulation runs are the vertical track force and 'Tgamma', which has been described earlier. The vertical track force was used to calculate a damage index, which is the vertical force raised to the power 1.21. This damage index is then used to calculate track settlement. Tgamma was used to calculate wear and RCF damage. All damages for each track section were measured per gross tonne-km.

It should be noted that damage is estimated at a number of points along the section (that is, each section is split up into segments covering 40 or 200 metres for wear/RCF and settlement respectively). Thus it is necessary to produce a measure of overall damage for each of the three damage types over the whole section. In the case of wear and RCF these are the summations of the damage over each of the segments. In the case of settlement, an average measure is taken, as a summation would not make sense from an engineering perspective.

The summary damage measures per gross-tonne km are then scaled up based on the total annual gross tonne-km run by each of the vehicle types in Table 1 in order to produce an annual total estimated damage. It is this measure that can then be related to annual costs in the second stage of the modelling. However, a mix of other vehicle types has also run on these track sections accounting for a small proportion of the total gross tonne-km. The damage, wear and RCF from these vehicle types are assumed to be proportionate to the damage measures caused by the vehicle types in Table 1. We have therefore scaled up the measures with respect to the total tonne-km run on the sections. The summary, total damage measures are shown in Table 4 below.

**Table 4 – Total damage, wear and RCF**

| Track section no. | Vehicle | Loading condition | Damage | Wear | RCF |
|---|---|---|---|---|---|
| 629 | RC2 | Laden | 9 235 914 | 274 163 841 | 886 093 |
| | RC4 | Laden | 13 895 739 | 408 431 752 | 1 285 242 |
| | Wagon 1 | Tare | 96 482 | 2 001 759 475 | 783 707 |
| | | Laden | 58 945 934 | 916 350 922 | 7 014 644 |
| | Wagon 2 | Tare | 7 371 | 6 614 805 | 70 194 |
| | | Laden | 673 991 | 5 098 989 | 57 969 |
| 326 | RC2 | Laden | 1 175 899 756 | 47 387 842 433 | 68 569 622 |
| | RC4 | Laden | 1 200 996 700 | 48 100 608 862 | 68 487 534 |
| | Wagon 1 | Tare | 375 216 324 | 3 136 983 770 | 33 453 667 |
| | | Laden | 23 018 827 311 | 366 396 576 247 | 2 788 072 442 |
| | Wagon 2 | Tare | 1 217 851 938 | 4 364 299 435 | 14 699 661 |
| | | Laden | 4 079 811 306 | 9 770 716 020 | 67 225 065 |
| 821 | X11 | Tare | 9 496 796 | 16 774 577 118 | 47 132 486 |
| | X11 | Laden | 42 756 976 | 59 680 115 333 | 150 847 091 |
| | X31 | Tare | 56 019 027 | 100 016 274 153 | 279 023 469 |
| | X31 | Laden | 334 761 323 | 471 596 050 478 | 1 187 786 786 |
| 654 | X12/14 | Tare | 0 | 0 | 0 |
| | X12/14 | Laden | 1 165 046 538 | 14 559 357 588 564 | 544 327 218 |
| 652 | X12/14 | Tare | 27 638 191 | 2 740 211 270 | 20 480 736 |
| | X12/14 | Laden | 2 277 950 252 | 132 724 585 837 | 1 371 949 647 |

The damage measures in Table 4 are thus summary, total simulated damage measures per section, per year. It should therefore now be possible, once the same modelling has been carried out for more track sections, to estimate a relationship between these damage measures and actual costs on those sections. This second stage will enable us to understand the relative cost of different damage mechanisms and also how the different damage types and the cost of correction interact with each other to affect overall cost. This new information on the relative cost of damage, based on actual reported cost measures, will be an important contribution to the literature, where currently the link between damage and costs relies on a set of assumptions that may not be valid.

## 6. CONCLUSIONS

This paper proposes and implements the first stage of a new approach to estimating the marginal cost of rail infrastructure usage. The approach fills an important gap between the modelling of damage – which is well understood – to obtaining measures of marginal cost, where currently the relative cost of different damage mechanisms is not well understood. The approach combines engineering and statistical approaches in a way that seeks to combine the best features of both; and thus overcome some of the weaknesses of previous approaches.

A new dataset is utilised, comprising detailed information on individual track sections of the Swedish network. This data includes information, at the section level, on track alignment and actual maintenance and renewal costs, as well as information on the vehicle volumes and types running on those sections.

Through testing this approach out on five sections, we conclude that, as expected, it is possible and indeed relatively straight forward to model the damage resulting from running vehicles on the network. We have further shown, however, that is possible to produce summary, total (annualised) damage measures for each of the three damage mechanisms. Once the approach is extended to an increased number of track sections, as planned, it will then be possible to explore the relationship between these damage measures and actual costs for the first time; and thus better understand the relative costs of the different damage mechanisms. Such information will be highly valuable to academics, the industry and policy makers as it will allow the cost implications of different technologies to be more clearly assessed. Further, from a track access charge perspective, the approach will permit improved estimation of the relative cost of different vehicle types, allowing more cost reflective charges for different vehicles on the network. In turn, more cost reflective access charges should incentivise the development and use of more track friendly vehicles.

There are a number of aspects to the research where assumptions have been used, for example, concerning the precise nature of the vehicles running on the network (we could identify passenger versus freight vehicles, and locomotive versus EMU, but had to use generic types for these rather than the actual vehicle characteristics). It has also been assumed that the damage caused by one vehicle is independent of the damage caused by other vehicles, and that the damage measures from one vehicle run can be scaled up in a simply manner. These are limiting but pragmatic and sensible assumptions; however, relaxing them would be interesting and useful for future research. Validation of the damage estimates, as compared to actual measurements would also be a useful addition if the necessary information can be obtained.

## REFERENCES


Andersson, M., Smith, A.S.J., Wikberg, A., and Wheat, P.E. (2012), 'Estimating the marginal cost of railway track renewals using corner solution models', Transportation Research Part A, 46 (6), 954–964.

European Commission (2013): The Fourth Railway Package – Completing the Single European Railway Area to Foster European Competitiveness and Growth.

Johansson, P., and J-E. Nilsson (2004): 'An Economic Analysis of Track Maintenance Costs', Transport Policy, 11(3), 277-286.

ORR (2013), Final determination of Network Rail's outputs and funding for 2014-19.

Wheat, P.E. and Smith, A.S.J. (2008), 'Assessing the marginal infrastructure maintenance wear and tear costs for Britain's railway network', *Journal of Transport Economics and Policy,* vol. 42 (2), pp. 189-224.

Wheat, P., A.S.J. Smith, and C. Nash (2009): 'Deliverable 8 - Rail Cost Allocations for Europe', CATRIN (Cost Allocation of TRransport INfrastructure cost).

# Studying variations of skyhook method for comfort improvement

**A Qazizadeh[a], S Stichel[a], R Persson[b]**
[a] Division of Rail Vehicles, KTH Royal Institute of Technology, Sweden
[b] Rolling Stock Central & Northern Europe and Asia, Bombardier Transportation, Sweden

**ABSTRACT**

To improve ride comfort on high speed passenger trains, active suspension technology can be used. An active secondary suspension is already developed for a Regina 250 EMU within a collaboration between KTH and Bombardier. Vertical dampers were replaced by four vertical actuators and the skyhook method was used to calculate reference forces for actuators. The basic idea of the skyhook method is that the force is proportional to the absolute velocity. The idea with this study is to explore variations of the skyhook method where the actuator force is proportional to other kinematic variables than just the absolute velocity of the carbody. These variables could be e.g. carbody-bogie relative position, carbody-bogie relative velocity, carbody acceleration square etc. The aim of this study is to investigate possible advantages of these variations in improving ride comfort and carbody positioning compared to the original skyhook concept. Simulation results show that there are variations which perform better than the original skyhook method and improve either ride comfort or the relative displacement between carbody and bogie.

## 1 INTRODUCTION

Controlling active suspension can be done by different control methods like $H_\infty$, model predictive control (MPC), skyhook, etc. [1] [2] [3]. The skyhook method was first proposed by D. Karnopp [4] where the actuator reference force is calculated based on the absolute velocity of the degree of freedom desired to be controlled. This solution has been widely used and is known to be effective and easy to implement.

The performance of an active suspension based on skyhook control was studied on a passenger train within a collaboration between KTH and Bombardier. This was tested successfully in May 2013 and the results are published in [5]. However, variations of the skyhook method could be used in which the actuator reference force is proportional to other kinematic variables measured on the train. These variables can be set in different ways as inspired by [6]. Table 1 shows the original skyhook method and a selection of different variations studied here.

**Table 1: Different variations of skyhook divided into three types**

| Type 1 | Type 2 | Type 3 |
|---|---|---|
| $\dot{x}_c$ | $\dot{x}_c(x_{ac} - x_{ab})$ | $\dot{x}_c[1 + (x_{ac} - x_{ab})]$ |
| $\dot{x}_c{}^{1.5}$* | $\dot{x}_c(\dot{x}_{ac} - \dot{x}_{ab})$ | $\dot{x}_c[1 + (\dot{x}_{ac} - \dot{x}_{ab})]$ |
| $\dot{x}_c{}^{2}$† | $\dot{x}_c{}^{1.5}(x_{ac} - x_{ab})$ | $\dot{x}_c{}^{1.5}[1 + (x_{ac} - x_{ab})]$ |
| $x_c$ | $\dot{x}_c{}^{1.5}(\dot{x}_{ac} - \dot{x}_{ab})$ | $\dot{x}_c{}^{1.5}[1 + (\dot{x}_{ac} - \dot{x}_{ab})]$ |
| $x_c^{1.5}$ | $x_c(x_{ac} - x_{ab})$ | $x_c[1 + (x_{ac} - x_{ab})]$ |
| $x_c^2$ | $x_c(\dot{x}_{ac} - \dot{x}_{ab})$ | $x_c[1 + (\dot{x}_{ac} - \dot{x}_{ab})]$ |
| $\ddot{x}_c$ | $x_c^{1.5}(x_{ac} - x_{ab})$ | $x_c^{1.5}[1 + (x_{ac} - x_{ab})]$ |
| | $x_c^{1.5}(\dot{x}_{ac} - \dot{x}_{ab})$ | $x_c^{1.5}[1 + (\dot{x}_{ac} - \dot{x}_{ab})]$ |

The Kinematic variables used in the table can be explained as:

1- $\ddot{x}_c$:
is the modal acceleration on the carbody. This can be bounce, pitch or roll and is calculated with help of four accelerometers on the carbody floor above the air springs. The different modes are achieved by combining the four accelerometer signals, cf. Section 2.

2- $\dot{x}_{ac} - \dot{x}_{ab}$:
is the modal relative velocity. This can be again bounce, pitch and roll and is calculated in the same way as modal acceleration but instead of the four acceleration signals, four relative velocities are used. These relative velocities are each measured at the two ends of each vertical actuator in the line of action of the actuator. Figure 1 shows the location of the actuators and the accelerometers on the carbody.

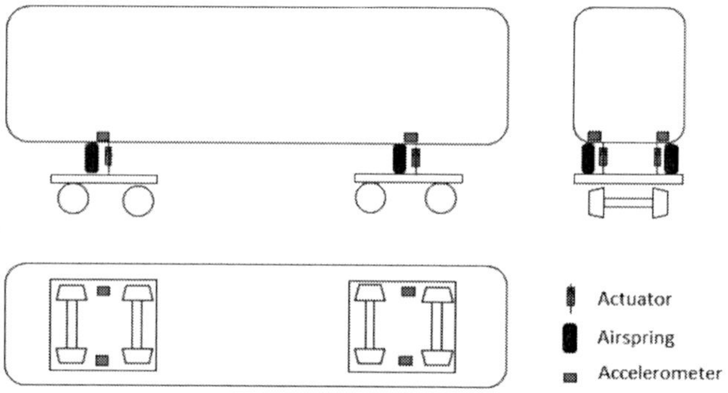


**Figure 1: Schematic view of the carbody equipped with actuators and accelerometers. From left to right: side view, rear view and top view**

The different variations in Table 1 are divided into three different types. The first type is based on the modal acceleration of the carbody and its integrations ($\dot{x}_c$ or $x_c$) and also powers of these signals. Using $\ddot{x}_c$, $\dot{x}_c$ and $x_c$ will lead to creation of a force which resists acceleration, velocity (original skyhook) and displacement respectively and powering signals is introduced to compare linear and non-linear solutions. Powering signals by 1.5 and 2 without losing the signal sign is shown in Section 3. The second type is achieved by multiplying the first type by the modal relative velocity or its integration. The aim is to achieve a better control over the

---

* Powering the signal by 1.5 is performed in a way that the signal sign is maintained, see Figure 4

† Powering the signal by 2 is performed in a way that the signal sign is maintained, see Figure 5

suspension deflection while improving comfort. Type 3 is the summation of those from type 1 and 2 which have a common factor e.g. $\ddot{x}_c^{1.5}[1 + (\dot{x}_{ac} - \dot{x}_{ab})]$ is a summation of the second item from Type 1 and the fourth item from Type 2 where the common factor is $\ddot{x}_c^{1.5}$. One can extend the list of variations even further, but in this study it is decided to limit the list to the variations in Table 1.

The aim of this study is to investigate possible advantages of these variations in improving ride comfort and carbody positioning compared to the original skyhook concept. This study is performed on a model of the Regina 250 which was developed by Bombardier in Simpack environment. Satisfactory solutions may be considered for testing on a test train in future experiments.

## 2 METHODOLOGY

To explain how different variations of the skyhook method are implemented, they are first described on a two degrees of freedom system. Consider the two degrees of freedom system shown in Figure 2 in which the secondary damper is replaced by an actuator. The force of the actuator is calculated according to plant G which is fed by acceleration of the secondary mass and also the relative velocity of the two ends of the actuator. The plant G is where different control variations mentioned in Table 1 are implemented. Then the damping coefficient $c_s$ is used to scale the force to the required level.

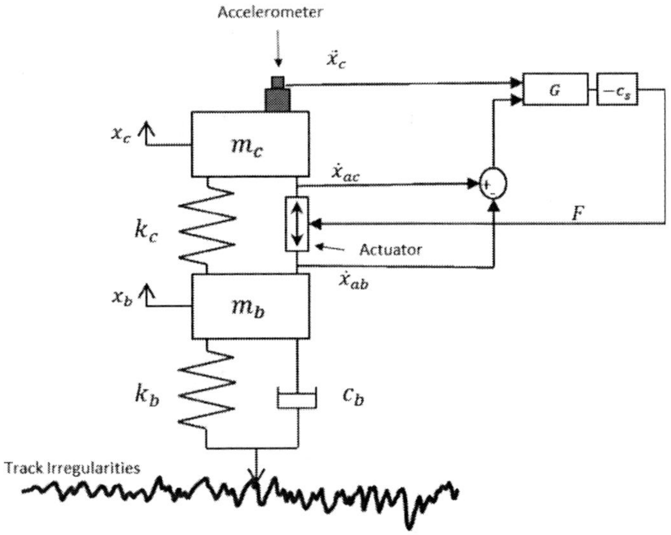


**Figure 2: Two degrees of freedom system with active suspension**

The same concept is used for controlling carbody vibration. However, in case of a carbody, the system is flexible, continuous and has infinite degrees of freedom. As it is impossible to control infinite degrees of freedom with just four actuators, one should consider controlling mode shapes instead. A 3 dimensional flexible structure has 6 rigid body modes and infinite number of flexible modes. In this study just three of the rigid modes are considered for controlling which are bounce, roll and pitch modes. These modes are controllable with vertical actuators and their natural frequencies are low enough compared to cut-off frequency of most of the actuators.

The complete controller for the carbody is shown in Figure 3. The figure shows how the vehicle model in Simpack and the controller in Simulink interact. The vehicle model from Simpack has eight outputs which are carbody floor acceleration on four locations above the air-springs and relative velocity between two ends of the vertical secondary dampers. Having these signals, average bounce, roll and pitch modes are calculated in the yellow blocks (No. 1 and 2) for the acceleration and relative velocity signals. Each of the modes is then treated in the same way as shown for the two degrees of freedom system in Figure 2 with plant G. This is done for the vehicle model in the three orange blocks (No. 3 to 5), each for a specific mode. Each orange block also includes a damping coefficient which is called 'modal damping coefficient'. In this study these coefficients are selected so that the coefficient for roll is three times larger compared to the other two coefficients to compensate for the smaller lever arm compared to pitch. The output of each orange block is a modal force which should be distributed between the four actuators with the right sign. Therefore the force of each actuator is a superposition of the three modal forces and this is performed in the green blocks (No. 6 to 9). The last step is to scale the amplitude of the force to the right level by using a constant gain. This gain works in the same way as $c_s$ in Figure 2 and is called 'global damping coefficient'. Each calculated force is then sent to the vehicle model with positive and negative sign to be applied at the two opposite ends of each actuator. Finally vehicle model interaction with the controller in Simulink is shown by the cyan block (No. 10).

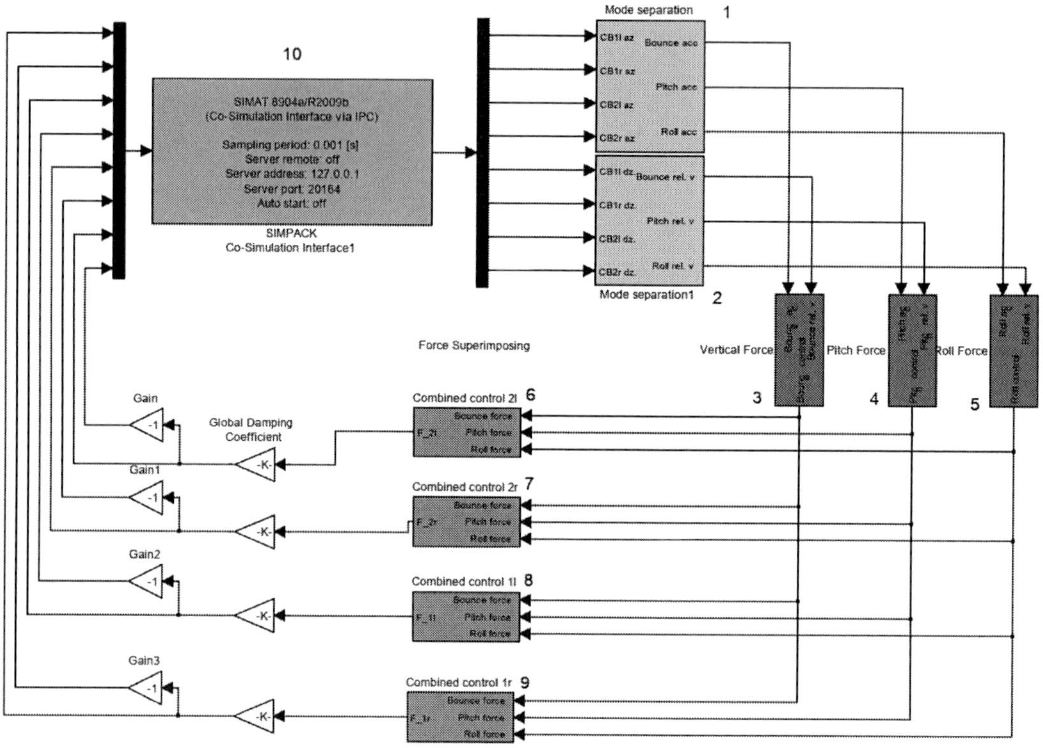


**Figure 3: Controller layout in Simulink**

This study is divided into two parts where in the first part forces are applied through an ideal actuator and in the second part actuator dynamics are also considered. The actuator is modelled as a first order low pass filter with cut-off frequency at $f_{cutoff} = 8$ Hz and a time delay of 0.01 s. The actuator transfer function can be written as

$$\frac{e^{-0.01s}}{s/(2\pi f_{cutoff}) + 1}$$

Though the model is very simple, it allows approximating those limitations of a hydraulic actuator that are interesting to this study. These limitations are the time delay and the limited bandwidth of a hydraulic actuator.

# 3 SIMULATIONS

Simulations of different variations of the skyhook method are performed in Simulink and Simpack environment. The model of the Regina 250 which is a Bombardier test train is modelled in Simpack and the controller is modelled in Simulink.

The Regina 250 is modelled with all parts as rigid, except for the carbody which is modelled as flexible with the first 17 modes. Simulations are performed on straight track with measured track irregularity. Eleven seconds of running is simulated for each test at the speed of 250 km/h, i.e. a distance of 764 m is covered. The Simpack solver used is called 'SODASRT' and the time step is selected to be 0.001 seconds.

The structure of the controller was illustrated and described in the previous section; however, there are other details which require further explanation. These details are discussed in the following paragraphs.

Calculating the reference force signal based on different variations leads to different force signals. Force signals can be quite different both in level and frequency content. Large force level difference makes it impossible to compare different variations with each other in a fair way. To solve this issue it was decided to keep the root mean square (RMS) of the force signal in a specified range. The center value of this range is decided to be RMS of force applied by the original skyhook case as this case is the reference solution against which other solutions are compared. Then the RMS value of the force for all other variations should remain inside a ±20% tolerance of the force RMS applied by the skyhook case. To fulfill this requirement for each of the variations, the global damping coefficient should be tuned which may require repeating the simulation of each variation several times.

Another issue is integration of signals which can lead to large drifts due to presence of low frequency content. This can be avoided by using high pass filters. In the current simulations whenever an integrator is used it is followed by a first order high pass filter with the following transfer function:

$$\frac{s}{s + 2\pi f_l}$$

where $f_l$ is cut-off frequency and is selected to be 0.3 Hz.

The other issue is powering signals which are required for several skyhook variations mentioned in Table 1. Here it is described how this is done without losing signal sign and also some other considerations are explained. Figure 4 shows how the signal is powered by 1.5. Once acceleration is integrated, the signal is split into two lines. The main signal is multiplied by the square root of the signal to achieve 1.5 power. To avoid imaginary numbers, the square root block is fed with absolute values of the signal. The final result would be a signal with the same sign as the original signal.

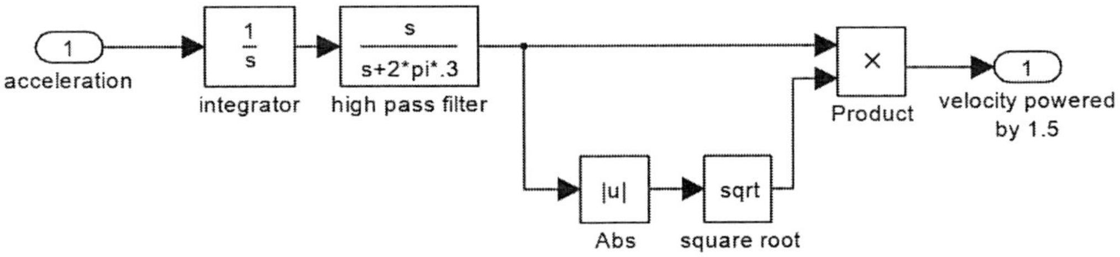


**Figure 4: Powering signal by 1.5**

Normalization of the signal and squaring it is shown in Figure 5. When squaring values below one, the result would be smaller and it is the other way around for values larger than one. This fact can be used in treating low and high levels of vibrations in different ways. So before squaring the signal it can be normalized by an arbitrary value which is believed to be the boundary of low and high vibration level. This value is selected to be 0.025 in Figure 5 for case 4 (cases are introduced in Table 2). The same technique is used for case 7 with 0.04 as normalization factor. This will give the active suspension a more aggressive characteristic towards high levels of vibration compared to low levels. Finally to keep the sign of the original signal, it is multiplied with its absolute value as shown in the figure.

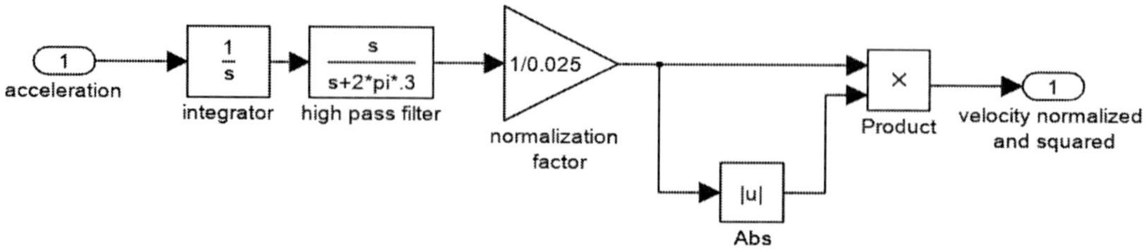


**Figure 5: Scaling and powering signal by two**

## 4 RESULTS AND DISCUSSIONS

In this section the performance of different variations is studied by comparing ride comfort and RMS of relative displacement. The reference cases are the passive and skyhook cases. The results are divided into two parts. First all cases are studied with ideal actuator thereafter the effect of actuator dynamics is added.

### 4.1 Ideal actuator

To keep the results of different cases comparable, the global damping coefficient was set for each simulation in a way that the RMS values of force during the 11 seconds simulation fall into a ±20% tolerance of 1381 N which is the force RMS for the reference case no. 2, the original skyhook case. To make sure that the tolerance range is good enough, case 25 is introduced. This case is the same as case 2 but with a force level that is higher than the other cases. Comparing results for cases 25 and 2 in Table 2, it can be concluded that increasing the force has had just a minor effect on performance. Therefore it can be concluded that the tolerance is good enough to allow comparing different cases with case 2.

The results in vertical direction are shown in Table 2. Three properties are tabulated for each case. The first two properties are ride comfort on the centerline of the carbody floor at the front and in the middle. Ride comfort values are calculated according to EN12299 [7]. The third property is the RMS of relative displacement measured on the rear actuator on the left side. Case 1 is the result for a Regina 250 with conventional passive dampers. Cases 2 and 25 are the original implementation of skyhook control and the remaining cases are variations of the skyhook method as explained earlier.

**Table 2: Results of different variations in vertical direction with ideal actuator**

| Case No. | Case | Comfort front $[m/s^2]$ | Comfort middle $[m/s^2]$ | Relative displacement $[mm]$ |
|---|---|---|---|---|
| 1 | Passive | 0.29 | 0.31 | 3.4 |
| 2 | $\dot{x}_c$ | 0.084 | 0.156 | 3.16 |
| 3 | $\dot{x}_c^{\ 1.5}$ | 0.08 | 0.165 | 3.3 |
| 4 | $\dot{x}_c^{\ 2}$ | 0.08 | 0.16 | 3.2 |
| 5 | $x_c$ | 0.104 | 0.202 | 2.3 |
| 6 | $x_c^{1.5}$ | 0.111 | 0.197 | 3.5 |
| 7 | $x_c^2$ | 0.106 | 0.21 | 2.2* |
| 8 | $\ddot{x}_c$ | 0.109 | 0.245 | 9.1 |
| 9 | $\dot{x}_c(x_{ac} - x_{ab})$ | 0.075 | 0.181 | 4.4 |
| 10 | $\dot{x}_c(\dot{x}_{ac} - \dot{x}_{ab})$ | 0.084 | 0.128 | 4.5 |
| 11 | $\dot{x}_c^{\ 1.5}(x_{ac} - x_{ab})$ | 0.077 | 0.144 | 4.6 |
| 12 | $\dot{x}_c^{\ 1.5}(\dot{x}_{ac} - \dot{x}_{ab})$ | 0.086 | 0.129 | 4.5 |
| 13 | $x_c(x_{ac} - x_{ab})$ | 0.12 | 0.186 | 4.8 |
| 14 | $x_c(\dot{x}_{ac} - \dot{x}_{ab})$ | 0.183 | 0.36 | 6.6 |
| 15 | $x_c^{1.5}(x_{ac} - x_{ab})$ | 0.114 | 0.183 | 4.8 |
| 16 | $x_c^{1.5}(\dot{x}_{ac} - \dot{x}_{ab})$ | 0.11 | 0.2 | 5.8 |
| 17 | $\dot{x}_c[1 + (x_{ac} - x_{ab})]$ | 0.063* | 0.128* | 3.5 |
| 18 | $\dot{x}_c[1 + (\dot{x}_{ac} - \dot{x}_{ab})]$ | 0.07 | 0.144 | 3.6 |
| 19 | $\dot{x}_c^{\ 1.5}[1 + (x_{ac} - x_{ab})]$ | 0.069 | 0.14 | 3.5 |
| 20 | $\dot{x}_c^{\ 1.5}[1 + (\dot{x}_{ac} - \dot{x}_{ab})]$ | 0.073 | 0.136 | 3.7 |
| 21 | $x_c[1 + (x_{ac} - x_{ab})]$ | 0.108 | 0.193 | 2.9 |
| 22 | $x_c[1 + (\dot{x}_{ac} - \dot{x}_{ab})]$ | 0.104 | 0.195 | 2.3 |
| 23 | $x_c^{1.5}[1 + (x_{ac} - x_{ab})]$ | 0.107 | 0.178 | 3.5 |
| 24 | $x_c^{1.5}[1 + (\dot{x}_{ac} - \dot{x}_{ab})]$ | 0.107 | 0.194 | 2.7 |
| 25 | $\dot{x}_c$ | 0.08 | 0.156 | 3.2 |

The original skyhook solution shows large ride comfort improvement compared to the passive conventional suspension, see Figure 6. In this figure it can be observed that the comfort improvement is more significant above the front bogie than in the middle. The worse comfort in the middle could be the effect of uncontrolled bending mode.

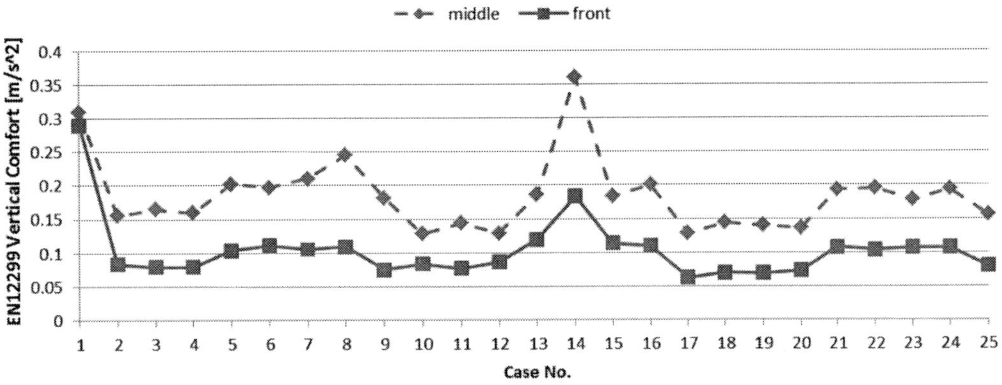


**Figure 6: Comparing ride comfort between different cases in the middle and at the front of the carbody floor**

Cases better than the original skyhook case (case 2) are marked by yellow in Table 2. The best cases regarding ride comfort and relative displacement are cases 17 and 7 respectively which are marked by *.

Case 17 is built up of cases 2 and 9. The signal achieved by case 2 is about 6 times larger compared to case 9. Therefore case 17 can be considered to be the same as the original skyhook case (case 2) with a correction part (case 9).

Looking at the cases which improve ride comfort and those which improve relative bogie-carbody displacement, it can be observed that there is no single case that improves both characteristics at the same time compared to the original skyhook case. This indicates that these two characteristics are contradictory which can be observed in Figure 7. So the solutions can be categorized into two groups where the first group is helpful in improving ride comfort and the second group is effective in reducing relative displacement. The cases which have been successful in reducing relative displacement are almost all based on displacement signals and cases successful in controlling and enhancing ride comfort are mostly based on velocity signals.

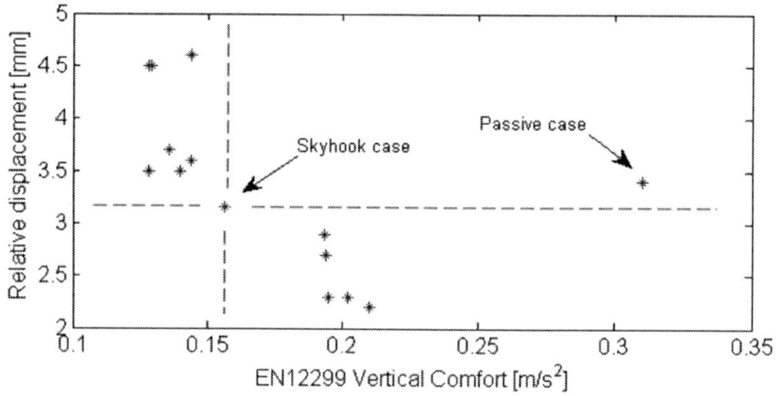


**Figure 7: Relative displacement as function of vertical comfort in the middle. There is no single case which improves both comfort and relative displacement compared to the original skyhook case. (Only good cases, marked in Table 2, are included)**

### 4.2 Modelled actuator

In this section the same study as in the previous section is repeated but this time actuator dynamics is also considered through a simple model introduced in Section 2.

The results are shown in Table 3. For some cases the behavior got unstable. This is due to the time delay and phase shift imposed by the actuator dynamics, leading to exertion of force at the wrong time. The original skyhook case does not deteriorate much after adding actuator dynamics. Other cases better than skyhook are marked in yellow and the best solution in each column is marked by a *.

**Table 3: Results of different variations in vertical direction with a modelled actuator**

| Case No. | Case | Comfort front $[m/s^2]$ | Comfort middle $[m/s^2]$ | Relative displacement $[mm]$ |
|---|---|---|---|---|
| 1 | Passive | 0.29 | 0.31 | 3.4 |
| 2 | $\dot{x}_c$ | 0.095 | 0.176 | 3 |
| 3 | $\dot{x}_c^{1.5}$ | 0.094 | 0.173 | 3 |
| 4 | $\dot{x}_c^{2}$ | 0.097 | 0.174 | 3 |
| 5 | $x_c$ | 0.109 | 0.208 | 2.3* |
| 6 | $x_c^{1.5}$ | 0.114 | 0.212 | 3.6 |
| 7 | $x_c^2$ | unstable | | |
| 8 | $\ddot{x}_c$ | 0.088 | 0.165 | 8.2 |
| 9 | $\dot{x}_c(x_{ac} - x_{ab})$ | unstable | | |
| 10 | $\dot{x}_c(\dot{x}_{ac} - \dot{x}_{ab})$ | 0.134 | 0.157 | 4.4 |
| 11 | $\dot{x}_c^{1.5}(x_{ac} - x_{ab})$ | unstable | | |
| 12 | $\dot{x}_c^{1.5}(\dot{x}_{ac} - \dot{x}_{ab})$ | 0.105 | 0.156 | 4.5 |
| 13 | $x_c(x_{ac} - x_{ab})$ | unstable | | |
| 14 | $x_c(\dot{x}_{ac} - \dot{x}_{ab})$ | unstable | | |
| 15 | $x_c^{1.5}(x_{ac} - x_{ab})$ | unstable | | |
| 16 | $x_c^{1.5}(\dot{x}_{ac} - \dot{x}_{ab})$ | unstable | | |
| 17 | $\dot{x}_c[1 + (x_{ac} - x_{ab})]$ | 0.09 | 0.144 | 3.4 |
| 18 | $\dot{x}_c[1 + (\dot{x}_{ac} - \dot{x}_{ab})]$ | 0.087 | 0.139* | 3.5 |
| 19 | $\dot{x}_c^{1.5}[1 + (x_{ac} - x_{ab})]$ | 0.084* | 0.154 | 3.4 |
| 20 | $\dot{x}_c^{1.5}[1 + (\dot{x}_{ac} - \dot{x}_{ab})]$ | 0.093 | 0.171 | 3.6 |
| 21 | $x_c[1 + (x_{ac} - x_{ab})]$ | 0.125 | 0.23 | 2.9 |
| 22 | $x_c[1 + (\dot{x}_{ac} - \dot{x}_{ab})]$ | 0.127 | 0.246 | 2.4 |
| 23 | $x_c^{1.5}[1 + (x_{ac} - x_{ab})]$ | unstable | | |
| 24 | $x_c^{1.5}[1 + (\dot{x}_{ac} - \dot{x}_{ab})]$ | 0.135 | 0.25 | 2.6 |
| 25 | $\dot{x}_c$ | 0.095 | 0.172 | 3 |

Cases 17 to 20 show good comfort improvement at the front and in the middle regardless of using an ideal actuator or the modelled one. This shows robustness of these solutions to actuator dynamics added in this study. On the other hand case 11 was among the best solutions in the previous subsection but adding actuator dynamics has made this case unstable. On the other hand, cases 3 and 8 are among the best solutions just after adding actuator dynamics. Case 3 shows much less performance deterioration compared to the original skyhook case once actuator dynamics is added and this is why case 3 is among the best solutions. Case 8 is the only solution that shows performance improvement after addition of actuator dynamics. This is discussed in more details in the following paragraph.

In case 8, the calculation of the reference force is based on the acceleration signal. Adding the actuator dynamics has a surprisingly positive effect on ride comfort. It is believed that the behavior of case 8 in Table 3 is due to the performance similarity of this case to the successful skyhook case with an ideal actuator in Table 2. To explain how this similarity is possible one should note that on one hand the acceleration signal is ahead of the velocity signal in time and predicts it and on the other hand a delay is imposed by the actuator in case 8. If the time gained by predicting velocity compensates the actuator time delay, working on acceleration with the modelled actuator can be equivalent to working on velocity with an ideal actuator. However, despite good performance of case 8 in improving ride comfort, its poor behavior regarding relative displacement is a major drawback.

Figure 8 and Figure 9 compare ride comfort in the middle and relative displacement respectively for the ideal and modelled actuators. Once actuator dynamics is added, comfort is deteriorated for all cases except for case 8 which was discussed earlier. Relative displacement is measured on the left rear actuator and one can observe that there is not much difference between including actuator dynamics or not.

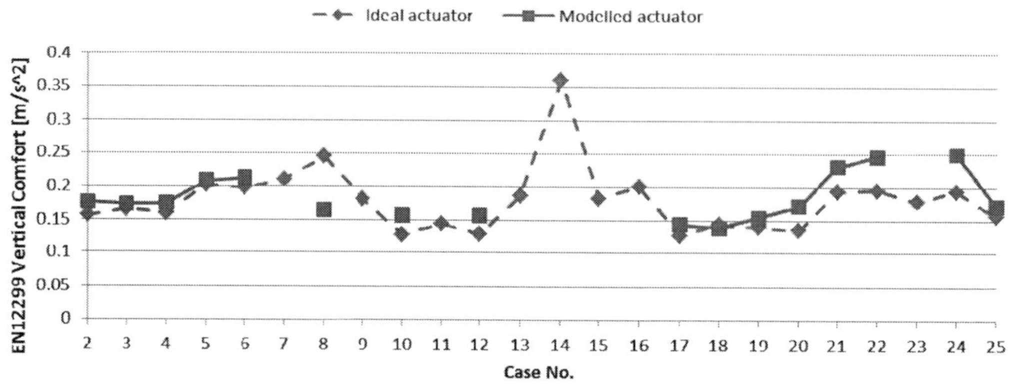


**Figure 8: Comparison of ride comfort results with ideal and the modelled actuator at the front**

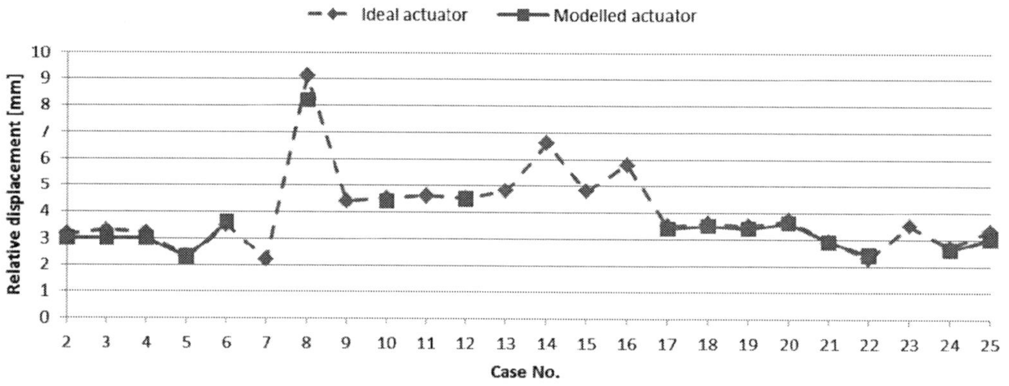


**Figure 9: Comparison of relative displacement results with ideal and the modelled actuator**

Finally, difficulty levels for implementation of different successful cases (3, 8 and 17 to 20) should be taken into account. Cases 8 and 3 are the easiest to implement as they just need measurement of acceleration on the carbody floor. Cases 17 to 20 are one step harder to implement, as not only acceleration on the carbody floor has to be measured but also relative velocity between carbody and bogie is needed which possibly could be achieved by derivation of the actuators position.

## 5 CONCLUSION

In this paper the use of different variations of the skyhook method controlling the active suspension is studied and the results are compared with the original skyhook method under the condition that the RMS of the force is about the same. It is found that better solutions exist but it is not possible to improve both ride comfort and relative displacement at the same time compared to the original skyhook case. It was also concluded that cases successful in controlling and enhancing ride comfort are mostly based on velocity signals and cases which have been successful in reducing relative displacement are mostly based on displacement signals. Some of these variations may bring further difficulty due to extra measurements needed.

Investigating the effect of adding actuator dynamics on relative displacement and ride comfort shows that the former is just slightly affected but the latter deteriorates for all cases except for case 8. The different behavior of this case was concluded to be due to compensation of actuator time delay by predicting the velocity signal which makes this case similar to the skyhook case with an ideal actuator.

Finally, some cases proved to show significant comfort improvement compared to the original skyhook solution. Among those are case 17 with ideal actuators and case 18 with the modelled actuators. To our opinion these solutions would be worth to be tested on a real train. As discussed earlier all solutions with better comfort performance, increase relative displacements. A summary of the results for these two cases is shown in Table 4.

As a future step simulations on curves/gradients should be considered to evaluate performance of the different skyhook variations regarding comfort and suspension deflection.

**Table 4: Good solutions for improving ride comfort with ideal and modelled actuator**

| Case No. and equation | Modelled actuator | Comfort improvement Front | Comfort improvement Middle | Relative displacement deterioration |
|---|---|---|---|---|
| 17 $\dot{x}_c[1 + (x_{ac} - x_{ab})]$ | No | 25% | 18% | -6% |
| 18 $\dot{x}_c[1 + (\dot{x}_{ac} - \dot{x}_{ab})]$ | Yes | 8.4% | 21% | -14% |

## REFERENCES


[1] T. Hirata, S. Koizumi and R. Takahashi, "H∞ Control of Railroad Vehicle Active Suspension," *Automatica,* vol. 31, no. 1, pp. 13-24, 1995.

[2] P. E. Orukpe, X. Zheng, I. M. Jaimoukha, A. C. Zolotas and R. M. Goodal, "Model Predictive Control Based on Mixed H2/H∞ Control Approach for Active Vibration Control of Railway Vehicles," vol. 46, no. sup1, pp. 151-160, 2008.

[3] A. Orvnäs, On Active Secondary Suspension in Rail Vehicles to Improve Ride Comfort, Stockholm: KTH Royal Institute of Technology, 2011.

[4] D. Karnopp, "Active Damping in Road Vehicle Suspension Systems," *Vehicle System Dynamics,* vol. 12, no. 6, pp. 291-316, 1983.

[5] A. Qazizadeh, R. Persson and S. Stichel, "On-Track Tests and Simulation of Active Secondary Suspension on a Rail Vehicle," in *Proceedings of the Second International Conference on Railway Technology: Research, Development and Maintenance*, Stirlingshire, UK, 2014.

[6] S. Mousavi Bideleh and V. Berbyuk, "Application of Semi-Active Control Strategies in Bogie Primary Suspension System," in *Proceedings of the Second International Conference on Railway Technology: Research, Development and Maintenance*, Stirlingshire, UK, 2014.

[7] EN12299, "Railway Applications —Ride Comfort for Passengers — Measurement and Evaluation — English version of DIN EN 12299:2009-08," 2009.

# Material considerations on the formation of Vertical Split Rim wheel failures

**D H Stone**
ESI, USA

**S L Dedmon**
Standard Steel, LLC, USA

## ABSTRACT

Vertical Split Rim (VSR) failures have become the leading cause of wheel failures in North American freight car service. VSRs are known to initiate from rolling contact tread surface shelling. Previous research suggests this failure mode is associated with high impact wheels and that wheel rim residual stress state may play a role in this type of failure.

This paper covers the material aspects of the conditions considered necessary to promote the formation of VSRs. Specifically, mechanical property changes induced by cyclic deformation of the wheel tread surface, coupled with changes in the manufactured residual stress patterns result in stress states favorable for the formation of a Vertical Split Rim crack.

## 1 INTRODUCTION

A previously rarely seen failure mode in railway wheels has appeared in North American rail service. The failures cause a large portion of the rim to sudden fractures that occur vertically from rolling contact fatigue tread shells on the wheel tread surface. Many, but not all, of these vertical split rim failures have been associated with high wheel/rail impacts. Other studies have suggested that residual tensile stress may be a significant factor in crack propagation. While tread shells and spalls are present at initiation sites, the mechanisms for producing the vertical fracture mode remain unknown. The data presented herein will show that continued plastic deformation decreases the elastic modulus independent of stress ratio.

If Melan's[1] theory of shakedown is correct, then the yield condition is the linear elastic limit. Because the linear elastic limit is unaffected by prior plastic deformation, and if the cyclic load is uniform, then the changes in residual stresses will move the system to shakedown. Further, the directions of the residual stresses will become orientated such that cracking will be possible in the axial direction below the wheel tread. According to the Association of American Railroads (AAR) the Why Made Code 68 from the MD-115 Broken Wheel database accounts for the majority of recent broken wheel failures in North American freight car service[2]. VSR cracks account for almost all of the broken wheels of this cause code. Because VSR failure rate is greater than other broken wheel causes, and because of the potential for derailments much recent research has been conducted on describing the defect and its causes[3]. The TTCI sponsored research through Vanderbilt University with the assistance of wheel manufacturers to perform complex simulations which incorporated known defects (cracks) at various positions, sizes and orientations on the tread surface to determine if any of these factors had a significant effect on VSR formation[4]. Even incorporating manufacturing residual stresses failed to isolate a predominant risk factor. Concurrent with the Finite Element Analyses (FEA) was an extensive review of over 50 VSR failures accumulated from recent failures in both US and Canadian service. A team of wheel experts reviewed these failed wheels to determine wheel type, geometry, crack initiation site, microcleanliness and other potential variables which could be related to VSR formation. Factors were ruled out as significant contributors, but no clear failure mechanism was determined. C. Lonsdale, et.al. attempted to cause a VSR by impact loading a wheel repeatedly but without success[5]. This research was followed by significant and well researched analyses of axial residual stresses on cast, forged, rim treated and untreated wheels. These results indicated no significant risk factors for cast versus forged wheels, but did show a higher risk factor for heat treated wheels over untreated wheels[6]. D. Stone researched using a neural network to isolate potential contributors to VSR formation[7]. Neural networks are used when many potential variables exist, but the number of defects is small, seemingly an ideal condition for evaluating VSRs. Once again, no strong correlations were found.

Based on these investigations factors determined not to be critical to forming VSR cracks include geometry of the rim (rim thickness, hollow rim, false flange), whether cast or forged. Factors determined to play a significant role were whether rim spray heat treated or untreated, residual stresses, a history of impacts and the presence of a tread or subsurface crack to act as an initiator. The last two may be somewhat related in that a tread cracked (shelled) wheel is more likely to also have a history of impacts as measured by wayside impact detectors. Two other observations: almost 90% of VSRs break out to the field side of the rim, the balance involve the flange side of the wheel. Most VSRs initiate at tread cracks and a smaller percentage appear to initiate at a subsurface location. See Figures 1 and 2.

**Figure 1: A still intact VSR at the point of breaking out. Note the spreading of the rim toward the field side. The picture to the right shows a typical VSR fracture surface.**

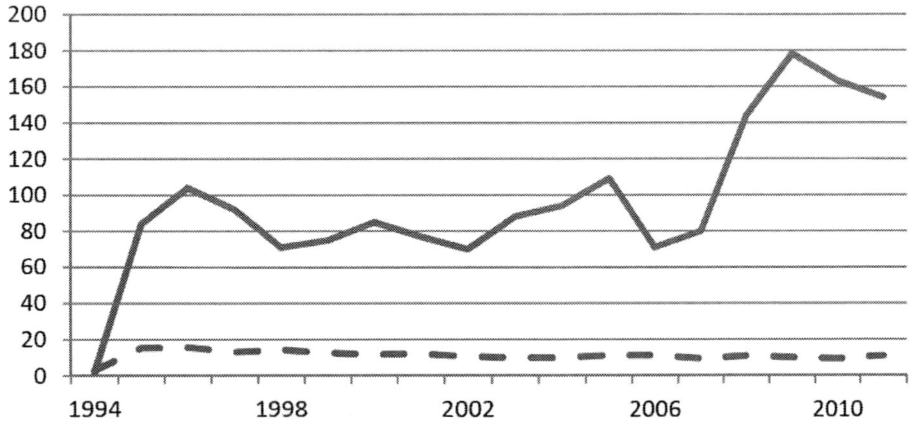


**Figure 2: History of Why Made Code 68 as reported through the AARs MD-115 reporting system. Data for wheels reported as broken from 1994 and before was not available. Number of failures reported annually is shown as a solid line; average age of the broken rim wheels is a dashed line.**

## 2 INVESTIGATION

### 2.1 Evaluation of manufactured residual stresses

Manufactured residual stresses are generated from stresses due to time dependent (creep) and time independent plastic deformations occurring from high thermal gradients during rim spray heat treatment. Elevated temperature tensile properties and creep properties were determined. See Figure 3 and 4. These properties were used to create deformation models which were incorporated into an ANSYS finite element model. The residual stress patterns calculated by the Finite Element Analysis agreed reasonably well to surface stresses measured on heat treated wheels which have not been machined.

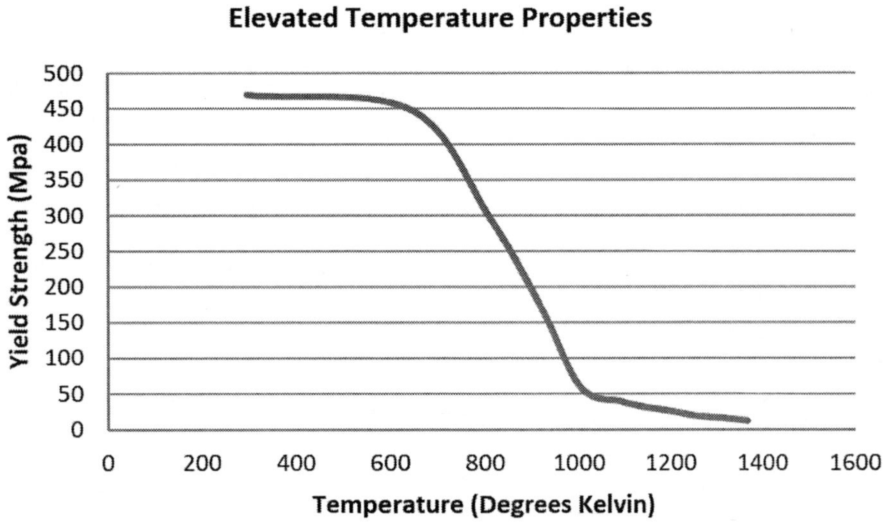


**Figure 3: Elevated temperature tensile results of a Class C wheel used in this study.**

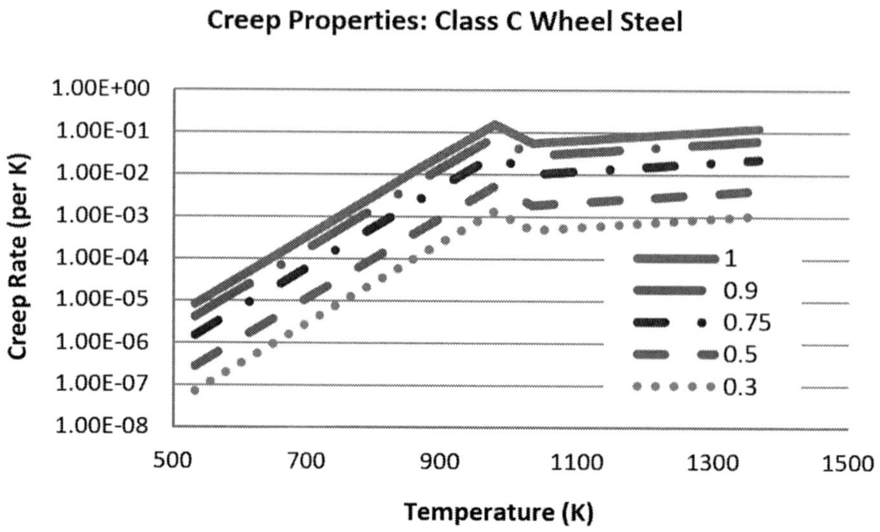


**Figure 4: High temperature creep properties of a Class C wheel steel. Lines in the chart represent different stress to yield strength ratios from 0.3 to 1.0.**

The second component to manufactured residual stresses was the prediction of stresses after removing stock from the outside diameter and end faces of the rim as would occur during machining of a forged wheel. This was accomplished by changing the material properties (modulus of elasticity) of those elements we desired to remove subsequent to the heat treating simulation. Again, the resulting residual stress pattern agreed reasonably well to surface residual stress results from actual tests. The axial and hoop direction results are shown in Figures 5 to 8.

## 2.2 Evaluation of mechanical properties

To calculate the effect of cyclic loading (rolling contact between the rail and the wheel tread surface) the effect of deformation on tensile properties had to be determined. Tensile properties were evaluated using a modification of the normal tensile test. Full sized tensile (12 mm diameter with 50 mm gage) specimens were prepared in accordance with ASTM A 370 and were taken from the rim at 1 inch (25 mm) below the tread surface, at mid-width and in the longitudinal (hoop) direction. The specimens were stress relieved at 100C prior to testing to remove any residual stresses generated by sample preparation. Each specimen had an extensometer attached at the initial gage points. The test specimen was loaded to a total strain of 1%, after which the specimen was unloaded, and the extensometer was reset for the next cycle. The cycles were repeated, until clear evidence of necking of the specimen occurred. A stress versus strain curve was generated for each cycle, and for the last cycle, which was loaded to breaking. The plastic portion of the strain was calculated for each cycle, and correlated against changes to the Elastic Modulus, the 0.002%, 0.2% offset yield strengths, the stress at 1% total strain and the slope of the straight portion of the stress-strain curve between the 0.2% offset yield strength and 1% total strain, sometimes referred to as the plastic modulus. The stress at rupture, percent elongation and percent reduction of cross-sectional area was also recorded for the last cycle. See Figures 9 and 10.

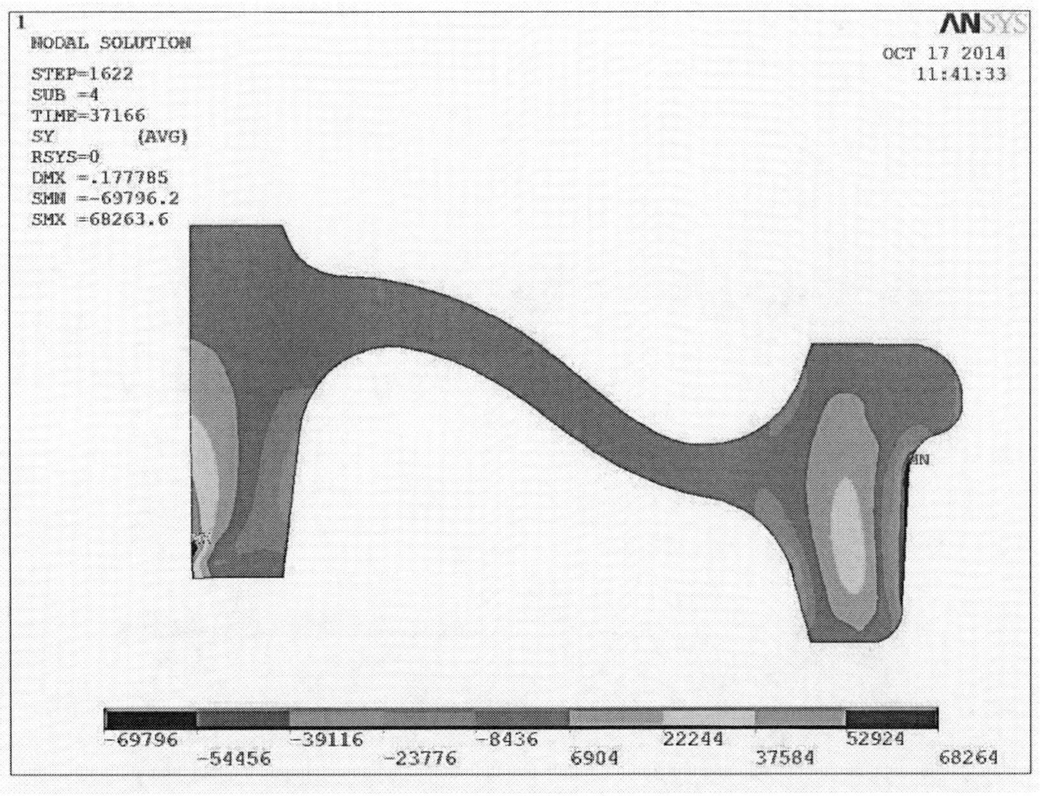


**Figure 5: Axial residual stresses of Finite Element Modeling of the Rim Spray heat treatment an H36 freight car wheel. Compare to Figure 6.**

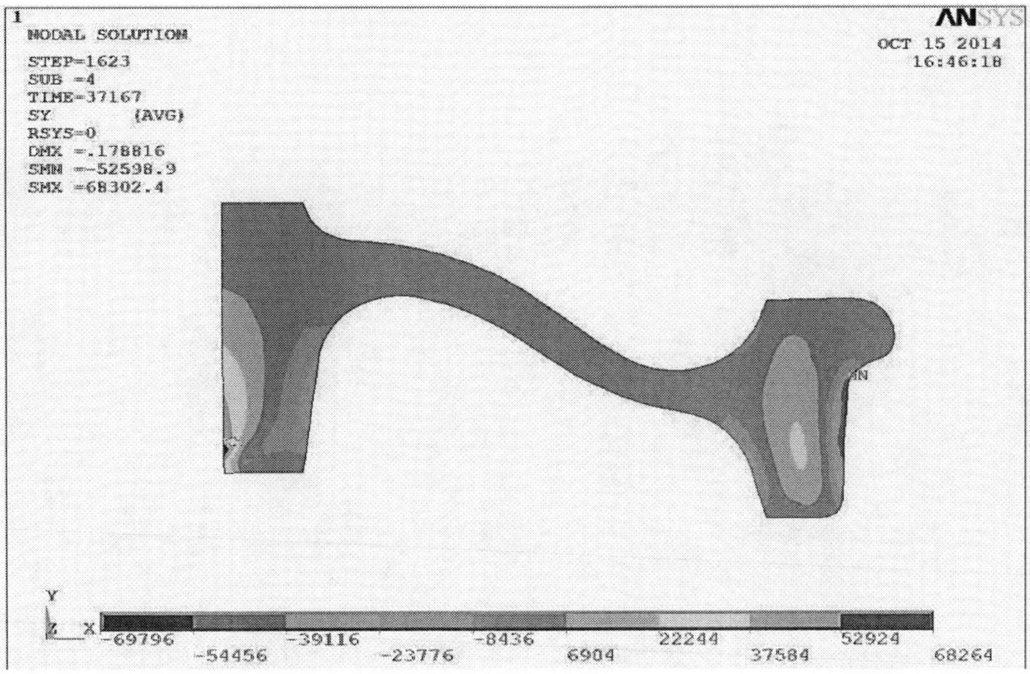


**Figure 6: Axial residual stresses of an FEA of the heat treatment of an H36 freight car wheel after removing stock to simulate the machining of the flange and tread.**

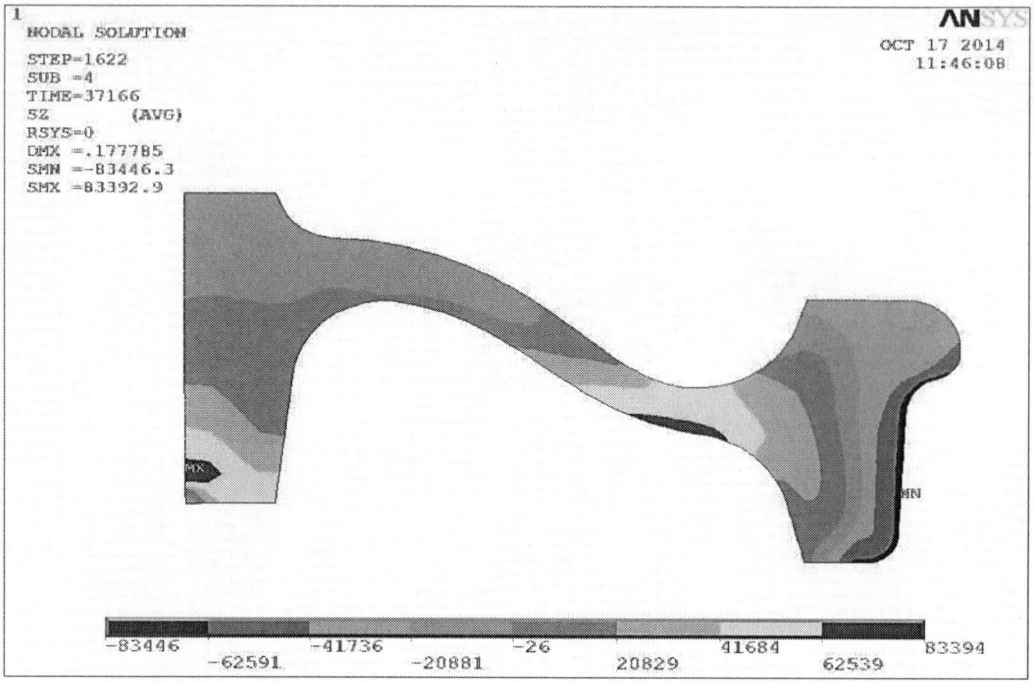


**Figure 7: Hoop residual stresses of Finite Element Modeling of the Rim Spray heat treatment an H36 freight car wheel. Compare to Figure 8.**

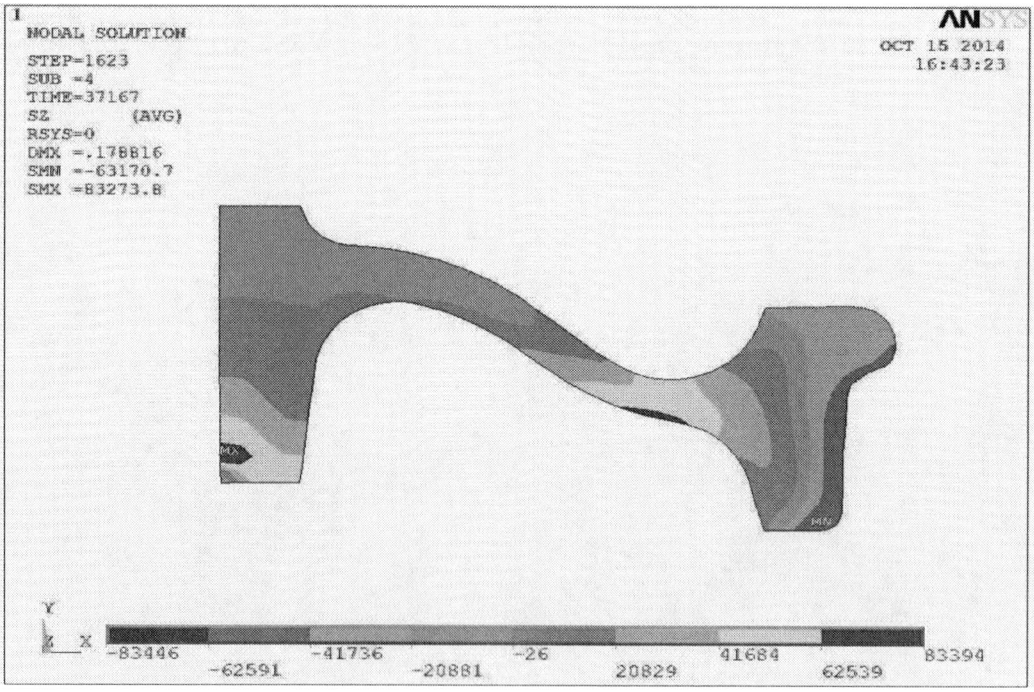


**Figure 8: Hoop residual stresses of an FEA of the heat treatment of an H36 freight car wheel after removing stock to simulate the machining of the flange and tread. Stresses are displayed in psi.**

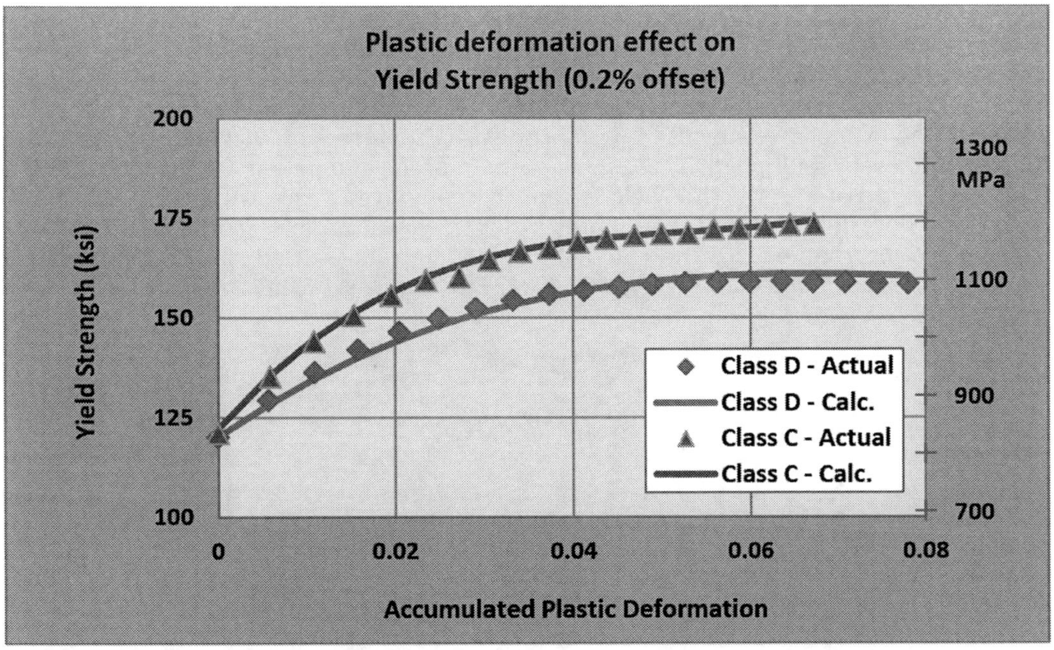


**Figure 9: The effect of plastic deformation on the 0.2% offset yield strength under strain hardening conditions. The material for testing was taken from a fully heat treated Class C freight car wheel.**

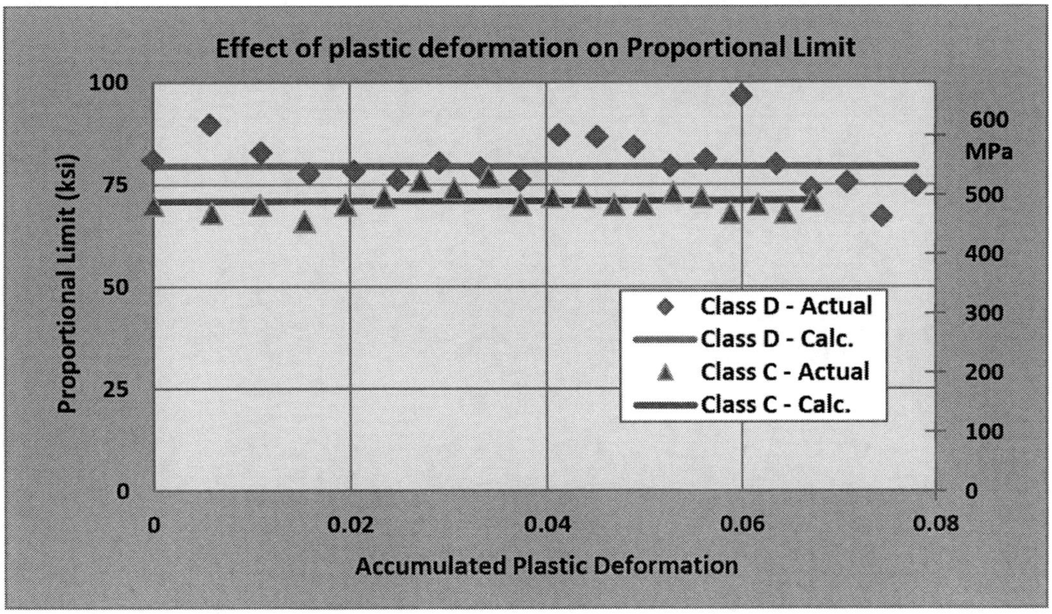


**Figure 10: The effect of plastic deformation on the proportional limit under strain hardening conditions. The material for testing was taken from a fully heat treated Class C freight car wheel.**

## 2.3 Evaluation of used wheel hardness profile

The next step was to evaluate the effect of cold work on the hardness of the tread. The wheel was sectioned into radial/axial slices from which metallographic specimens were prepared for micro-hardness evaluations. See Figures 11 and 12. During shakedown, cyclic loads resulting in stresses which exceed the proportional limit on the rolling contact surface will continue to result in plastic deformation until residual stresses increase enough to result in only elastic behavior. The resulting hardness is a picture of the state of shakedown of the tread. The tread hardness was used to estimate the amount of plastic deformation that had occurred up to the time of the wheel's removal. The hardness (converted to Vickers) of plastically deformed steels multiplied by 2.3 is approximately equivalent to the 0.2% offset yield strength (in MPa). The yield strength (converted from tread hardness) was used to determine the accumulated plastic deformation from Figure 9. An equivalent thermal load to the calculated plastic strain was applied to simulate the residual stresses induced by rolling contact. See Figures 13 and 14.

**Figure 11: Photomicrograph showing cross-section of wheel tread. Note the highly deformed pearlite near tread surface where strain hardening conditions exist.**

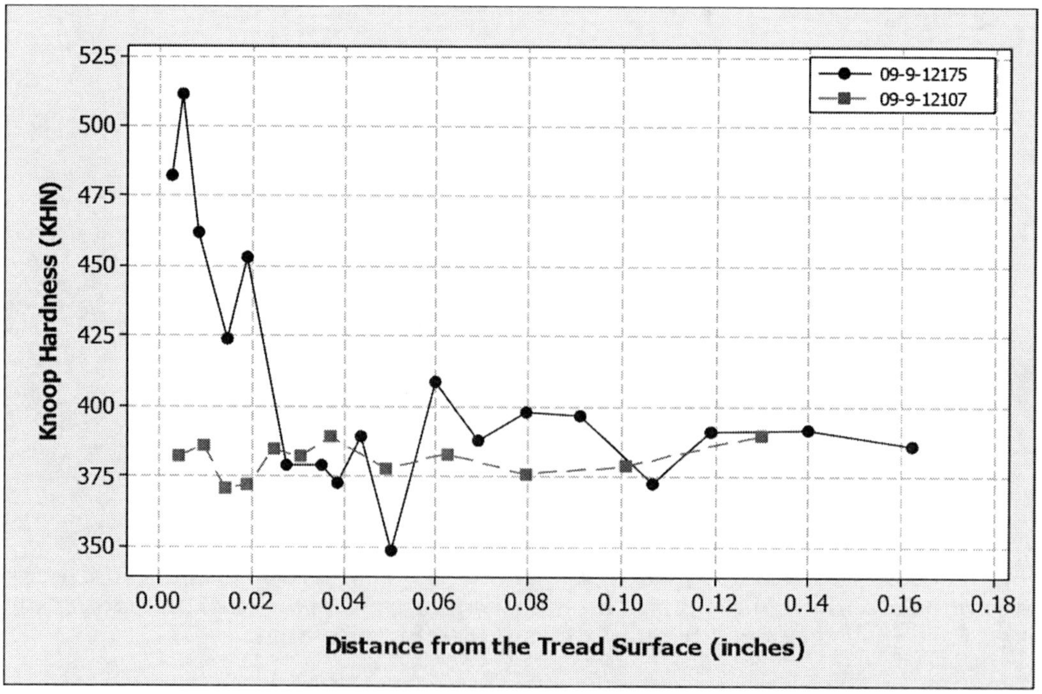


**Figure 12: Hardness of the cold worked tread surface and subsurface areas of a Class C freight car wheel having been in service for several years. These results were converted to plastic strains, which were then applied to the FEA model.**

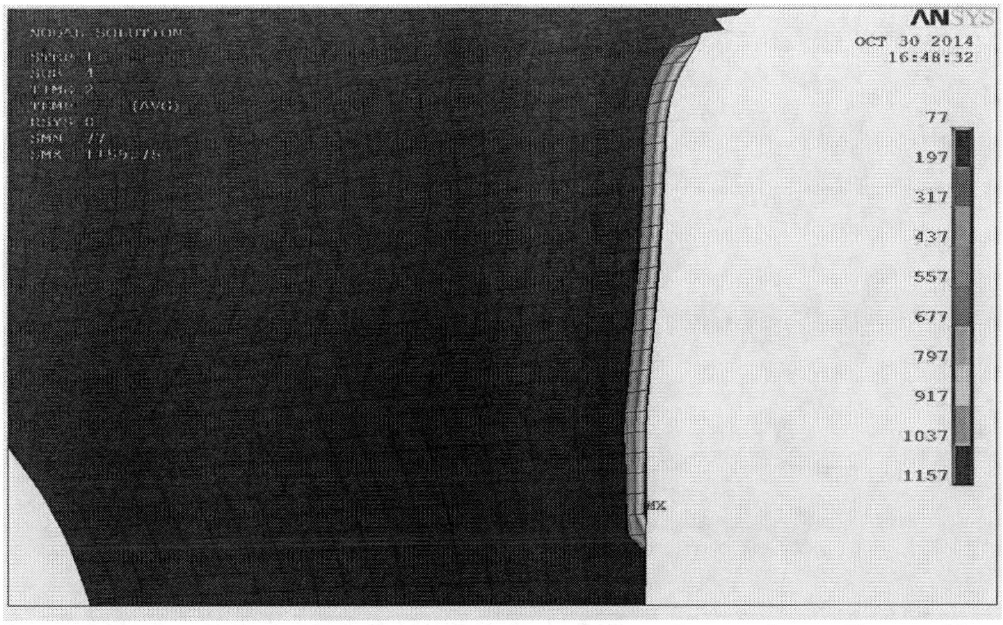


**Figure 13: Simulation of deformation to tread surface using a thermal load across the contact surface. Average thermal load was 300C (540F) in one element thickness of 2 mm. This thermal load was calculated to be equivalent to a deformation of 0.66%. Temperature scale of figure is displayed in degrees Fahrenheit.**

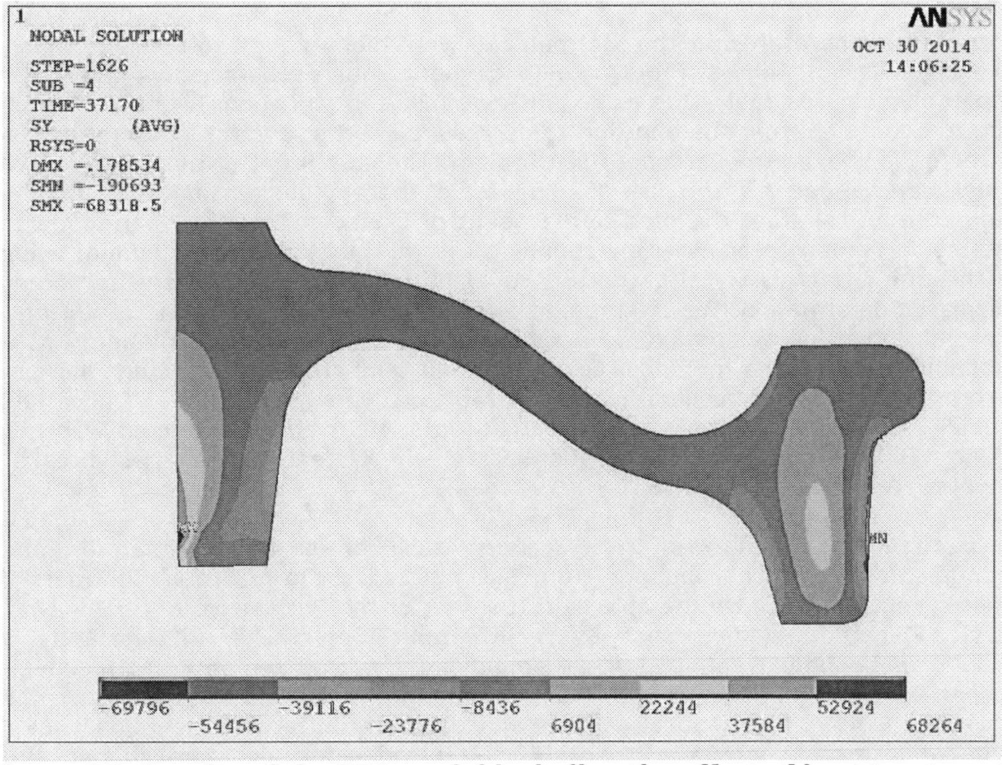


**Figure 14: Results of the FEA model including the effect of heat treatment, machining and work from rolling contact on axial residual stresses. The tensile axial residual stress pool has increased in size and has moved slightly closer to the tread surface.**

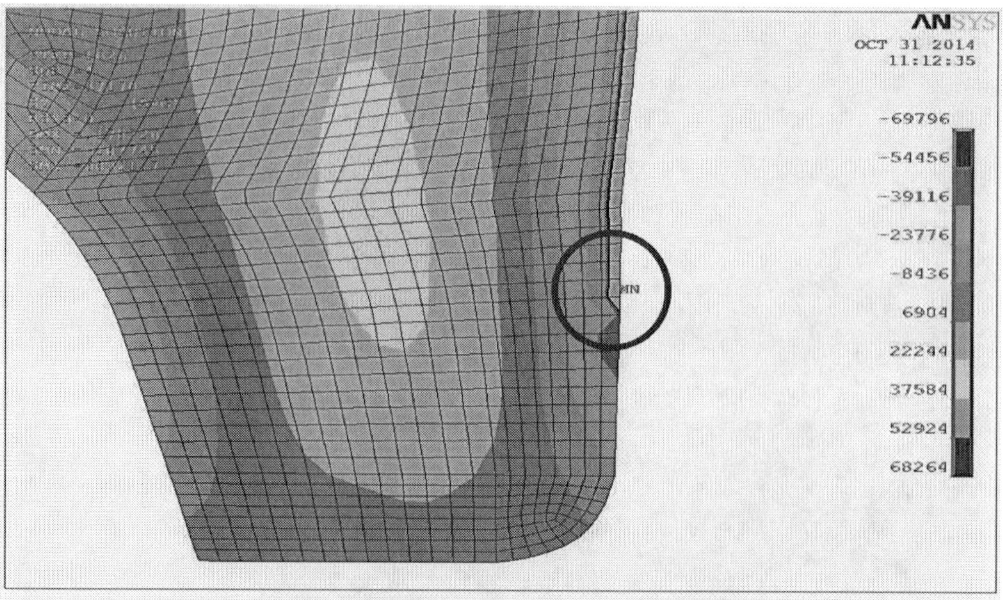


**Figure 15: Axial Residual stresses with approximately 0.66 percent plastic strain on tread surface. Note the circled area of significantly reduced compressive stresses at the edge of the rolling contact zone. At greater amounts of plastic strain, this zone should become tensile.**

## 3 SUMMARY

Predicted temperatures for the thermal load were high enough to result in changes to mechanical properties. From a finite element analysis perspective this creates two problems: residual stresses extant at subsurface locations were understated due to creep deformations and due to decreased yield strength of the hotter steel. Three options exist to overcome this problem. The coefficient of thermal expansion could be changed for the last loading cycle thereby decreasing the maximum temperature to induce the predicted amount of plastic strain; the heat conduction rate could be decreased, thereby confining the thermal load to a minimum volume of the rim; the mechanical properties could be changed to decrease the potential for plastic deformation. We attempted the first two, and changing the coefficient of thermal expansion seemed to work best. However, in future simulations we will probably use all three methods. Given that the temperature restrictions increased error, the amount of residual stresses predicted were understated. Only 0.0066 (0.66%) plastic strain was used when the amount of strain predicted was much greater (6% or more). Note that for Class C wheel steel a tensile test breaks at about 13 to 20% elongation.

A finite element analysis that includes residual stresses imparted during manufacturing plus contributions from a simulated cyclic environment treats the wheel in such a way as to be able to better predict the occurrence of relatively rare service generated defects such as Vertical Split Rim cracks. In this study, we simulated the heat treating process employing elevated temperature mechanical properties. Attenuation of the heat treatment residual stress patterns was accomplished by mathematically removing tread and rim face elements as would be done when machining of the tread surface. Introducing mechanical damage from rolling contact between the wheel tread and the head of the rail was somewhat more difficult. A typical used wheel was hardness tested to determine the degree of work hardening which was generated during rolling contact. By determining the relationship between the increase in yield strength during strain hardening and the amount of accumulated plastic deformation, a correlation was made possible between plastic deformation, yield strength and hardness. The deformation map was applied to the finite element model thereby creating a stress model.

The model also explains quite well why VSR crack formation is so rare. Since the location of the residual stress field favorable to VSR formation is confined to such small areas (field side, flange side and subsurface) pre-existing cracks are less likely to be found in these areas. Also, since rolling contact deformations require a significant number of loaded cycles, the residual stress field would rarely juxtapose with a tread defect outside the normal rolling contact area. The conditions necessary to generate a crack outside the normal rolling contact zone are similarly rare (unusual rail cant during curving, etc.). Subsurface VSR crack initiations are even rarer. A subsurface crack existing at the proper depth is unusual, but subsurface cracks usually break out to the surface creating a shell. Plastic deformations occur nearer to the surface than the crack, which creates hoop distortions (bubbling) which eventually break out to the tread surface. A subsurface crack would have to be oriented at an angle to the tread for a breakout not to occur. Since most subsurface cracks are parallel to the tread surface, most subsurface cracks result in shells and not VSRs.

The mechanism of VSR crack formation and propagation appears to follow a four step process. Firstly, a crack must be present in an area favorable for propagation. Secondly, residual stresses from rolling contact must overcome the compressive residual axial stresses induced by manufacturing to permit rolling contact stresses to propagate the crack. Once the crack has propagated to a critical depth, then further propagation in the roughly radial direction must occur. Since the residual tensile axial stresses exist at a depth of about 12 mm, once a crack propagates to this depth, then rapid fracture or low cycle fatigue will be accelerated. Fourthly, hoop stress differentials from the rim face (which is highly compressive) to deeper axial positions (which are less compressive, or even tensile) results in circumferential propagation of the crack. Since the protective residual compressive hoop stresses prevent the crack from breaking out to the surface, propagation of the crack continues (most VSR cracks extend 30 cm to 50 cm before finally breaking out). Eventually, the rim bows out from the rest of the wheel and gets mechanically removed in a switch or frog or some other obstruction. See Figure 16.

**Figure 16: An excellent example of an as-yet intact VSR showing the last stage of hoop stress assisted crack propagation, and bowing of the field side rim face just prior to failure.**

This model can be used to evaluate the impact of artificial defects, to predict how and if a given crack will grow and predict its path upon applying a mechanical and thermal load.

## 4 CONCLUSIONS

The combination of manufacturing and service induced residual stresses creates conditions likely to result in a vertical split rim crack.

Residual stresses resulting from rolling contacts are favorable for propagating a surface crack deeper into the rim, especially at the field side of the tread, and to a lesser extent to the flange side. The model also predicts the fairly rare subsurface initiation. Axial residual stresses are predicted to drive the crack deeper into the rim, and hoop direction residual stresses are predicted to drive the crack circumferentially. Final fracture is not predicted by the model, but is probably due to out-pouching of the VSR crack which eventually would be knocked off by switch, frog or other interference.

## REFERENCES


1. Melan, E., Zur Plastizität des Räumlichen-kontinuums, (1938), Ing. Arch.,9, 116-126.
2. Association of American Railroads, Manual of Standards and Recommended Practices, Section G and GII, (2013), Washington D.C.
3. D. Stone, S. Dedmon, J. Pilch and S. Cummings, Vertical Split Rim Wheels (2010) Proceedings of the 16h International Wheelset Congress, Cape Town, South Africa.
4. V. Sura and S. Mahadevan, Vertical split Rim Failure Analysis in Railroad Wheels, (2010) Proc. Of the 2010 ASME Joint Rail Conf. Paper JRC2010-36024.
5. C. Lonsdale, T. Rusin and T. Hay, (2009) Proc. Of the 2009 ASME Joint Rail Conf. Paper JRC2009-63026.
6. C. Lonsdale and J. Oliver, (2012) Proc. Of the 2012 ASME Joint Rail Conf. Paper JRC2012-74010.
7. Meddah, A., and Stone D., A Neural Network Analysis of the Causes of Vertical Split Rim Wheel Rim Failure, (2013), Proceedings of the 17th International Wheelset Congress, Kiev, Ukraine.

# Design of the running gear for the SUSTRAIL freight vehicle

**S D Iwnicki**[1], **A Orlova**[2], **P-A Jonsson**[3], **M Fartan**[4]
[1] Institute of Railway Research, University of Huddersfield, UK
[2] Petersburg State Transport University, Russia
[3] Division of Railway Vehicles, KTH Royal Institute of Technology, Sweden
[4] REMARUL Engineering, Romania

## ABSTRACT

The authors are working together in the 'SUSTRAIL' project to design an innovative, sustainable freight vehicle for use in new European markets. Market research has determined that the SUSTRAIL freight vehicle needs to run at up to 140km/h with an axle load of up to 25T. In order to reduce the impact of the increased freight traffic a requirement for a 50% reduction in lateral forces on the track has also been set.

These very demanding requirements can not be met without a corresponding step change in suspension design. After a thorough review it was therefore decided that the conventional Y25 type suspension was not appropriate for the SUSTRAIL freight vehicle. The Y25 suspension is very common in European railway operations and this brings advantages for operation and maintenance however it does have a high level of longitudinal stiffness after the small amount of longitudinal clearance at the axle box has been exceeded and this can result in relatively poor curving and high lateral wheel-rail forces. As an alternative a modified version of the Y25 suspension with a double Lenoir link primary suspension has been investigated and shown to produce much lower longitudinal primary stiffness while still utilizing standard components and methods which are well established within the railway industry.

Computer simulation tools have been used by the project team to optimize the precise geometric design and the component parameters. In order to improve stability of the vehicle and allow higher speed operation it was decided to assess the benefit of linkages providing longitudinal stiffness between the axleboxes using a radial arm. Various previous designs were considered and the final proposal involves a novel cross bracing arrangement. A prototype SUSTRAIL freight vehicle is now being constructed and will be tested in 2015.

## 1. INTRODUCTION

The aim of the SUSTRAIL project is to promote modal shift of freight in Europe from road to rail. A key objective of the policy is the shift of 30% of road freight over 300km to other modes such as rail or waterborne transport by 2030 (and 50% by 2050) as targeted by the European Commission. The SUSTRAIL project intends to provide the approach, structure, and technical content to support this modal shift through improvements in the railway freight system including innovations in rolling stock and in track components that aim to reduce general maintenance cost. The project includes work packages focused on market research, vehicles, infrastructure and assessment of cost benefits. The work described here is part of work package 3: 'The freight vehicle of the future'.

## 2. THE SUSTRAIL FREIGHT VEHICLE

The main scientific and technological innovations being considered for the SUSTRAIL freight vehicle are:

- The development of advanced vehicle dynamics concepts based on new wheel profiles and improvements in suspension design responding to the needs of a mixed traffic railway;
- Developments in the traction and braking systems for high speed low impact freight operation;
- Novel designs and materials for lightweight high performance freight wagon body vehicles and bogie structures;
- Advanced condition based predictive maintenance tools for critical components of both railway vehicles and the track;
- Identification of performance based design principles to move towards the zero maintenance ideal for the vehicle/track system.

Partners in the project have carried out a technology review to identify the potential innovative technologies to meet the above requirements and the results have been ranked and two concept vehicles are being designed. The '*Conventional*' vehicle will use optimised existing technology and a demonstrator for this is being built as part of the project. The '*Futuristic*' vehicle will utilise technology which has not yet been proven in the railway field but has potential to make greater improvements. The results of the concept design were assessed within the project against a benchmark vehicle with Y25 bogies and using Life Cycle Cost (LCC) and Reliability, Availability, Maintainability and Safety (RAMS) methods.

## 3. RUNNING GEAR SIMULATIONS

The authors are working together to carry out simulations of the dynamic behaviour of the concept design vehicles running on typical track in tare, part laden and fully laden cases. In line with the target of a 50% reduction in lateral forces on the track and stable running at 140km/h a suspension using double Lenoir linkages, longitudinal linkages between axle boxes. An improved centre pivot arrangement was also considered. Computer simulation has been used to optimise the suspension and to select suitable parameters for the various components. Assessment of the results is based on:

- Stability: stable running on typical European track at the design speed of 140km/h must be ensured and ride quality (vertical lateral and longitudinal accelerations experienced by the goods transported) will be assessed.
- Reduced track forces: track geometrical deterioration (ballast settlement and horizontal level, alignment and buckling), rail surface damage (wear, rolling contact fatigue – RCF) and track components damage (sleeper cracking, rail pad deterioration, rail fatigue, fastening deterioration) will all be assessed.

A benchmark vehicle has been selected based on a Y25 bogie and flat bed wagon and has been used to allow quantification of the benefits of the new design.

## 4. DOUBLE LENOIR LINK

Freight vehicle suspensions are challenging to design due to the very large difference between tare and laden vehicle mass and consequent challenge of controlling the dynamic behaviour of the vehicle in both conditions. The Y25 type suspension tackles this by using a 'Lenoir link' to transfer some of the vertical load onto a friction surface and thus providing progressive damping. A disadvantage of this arrangement is that it results in a very high longitudinal stiffness after the small amount of longitudinal clearance has been exceeded. A number of radical innovations were considered during the technology review stage of the project but it was decided that the use of double Lenoir link primary suspension (Figure 1), would be investigated. The double Lenoir link suspension provides much lower longitudinal primary stiffness while still utilising standard components and methods which are well established within the railway industry.

**Figure 1: A suspension with double Lenoir links**

As part of the optimisation of the primary suspension the following parameters were varied:

- The vertical coil spring stiffness
- The angle of the Lenoir link
- The length of the Lenoir link
- The friction coefficient at the sliding surfaces (through changing material)
- The vertical clearance to the bump stop.

These parameters are illustrated in figure 2.

**Figure 2: The Lenoir link showing the parameters varied in this work**

A model of the SUSTRAIL vehicle was set up with double Lenoir links using the computer simulation tool Gensys and the influence of variations in the suspension parameters on the critical speed of the wagon was simulated. Straight track was used for this simulation and an initial lateral disturbance was introduced followed by ideal track with no irregularities. Axle load is 22.5T, wheel profile is S1002 and rail profile UIC60 inclined at 1:40. The wheel rail coefficient of friction is set at 0.35.

The wagon speed was reduced from an initial 170km/h and critical speed assumed to have been reached when $\sum Y$ drops below 2.5kN. An example is shown in Figure 3.

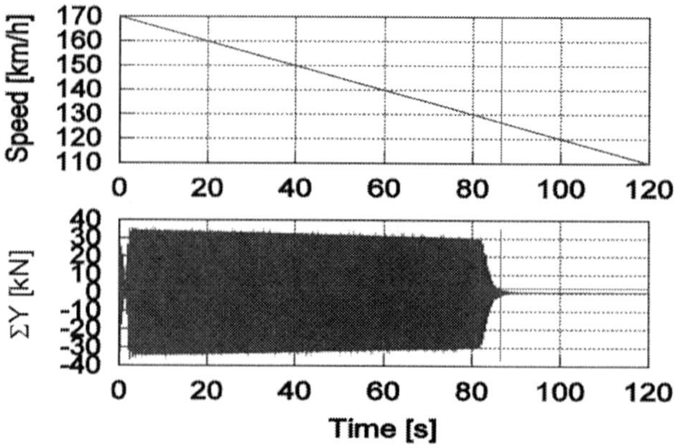


**Figure 3: A sample simulation results showing the establishment of the critical speed for the SUSTRAIL vehicle with double Lenoir links**

The effects of the various suspension parameters on the critical speed are summarised in Figure 4.

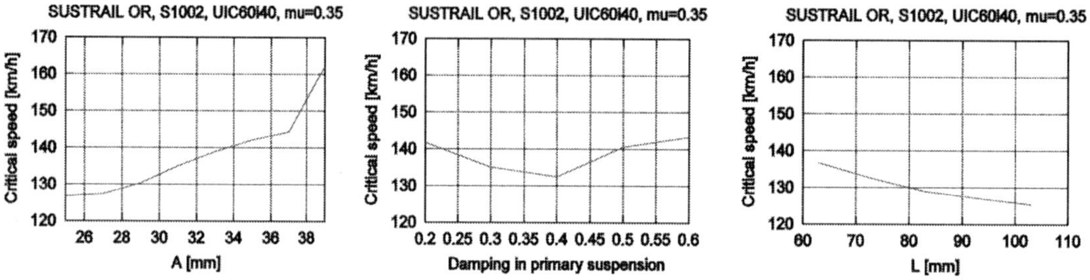


**Figure 4: The effect of Lenoir link angle, friction coefficient and length on the critical speed of the SUSTRAIL vehicle**

The simulations were repeated with a speed of 120km/h on straight track with measured irregularities and the maximum vertical track force was established for each track section as shown in Figure 5.

Further variations were carried out and the effect of the friction coefficient and stiffness within the suspension on the maximum contact force is shown in Figure 6.

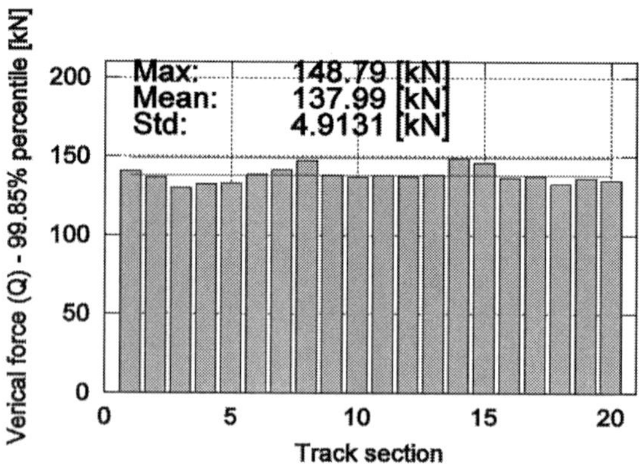


**Figure 5: Maximum vertical force on the rail for the SUSTRAIL vehicle running at 120km/h**

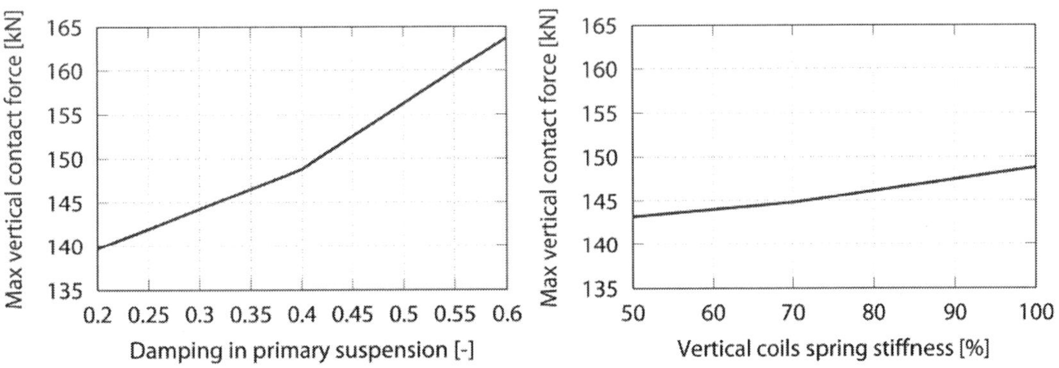


**Figure 6: The effect of friction coefficient and spring stiffness on the contact force**

It can be seen that the maximum vertical contact forces tends to increase with the coefficient of friction and with the spring stiffness.

## 5. LONGITUDINAL LINKAGES

In order to improve the running behavior of the SUSTRAIL vehicle it was decided to assess the benefit of linkages providing longitudinal stiffness between the axle boxes using a radial arm. The radial arm was first designed by Dr. Scheffel and is shown in Figure 7. This has been studied previously in the Infra-Radial project [1] which aimed to develop a bogie for heavy haul vehicles (axle loads over 25T) with reduced life cycle costs.

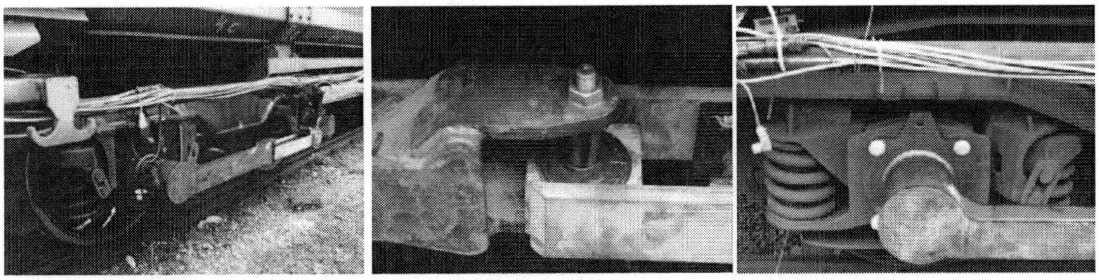

**Figure 7: The radial arm as used in the Infra-radial project**

Tests using the radial arm with four different primary suspension types showed good results with stable running and radially aligned wheelsets in curves. Wear of the wheels was seen to reduce significantly [2].

In the work reported here simulation was carried out using MEDYNA for the SUSTRAIL vehicle with double Lenoir links and radial arms. Wagon movement was simulated on a straight track with irregularities positioned at the distance of 40 m from the start with velocity reducing from 160 km/h to 40 km/h.

Diagrams of the dependence of critical speed for laden and empty wagons on longitudinal ($K_x$) and lateral ($K_y$) stiffness of radial arm are presented in figure 8. Stiffness is given per one Radial Arm on one side of the bogie.

Analysis of the results showed:

1) Critical speed of laden wagon on bogies with two Lenoir dampers on each axle box and without radial arms is 107 km/h and it is 80 km/h for empty wagon.

2) The highest critical speed (not less than 140 km/h) can be archived by the following stiffness of radial arm:
   - for laden wagon $K_y$ more than 750 kN/m (critical speed of laden wagon doesn't almost depend on longitudinal stiffness $K_x$);
   - for empty wagon $K_y$ more than 40 kN/m and $K_x$ not more than 250 kN/m or $K_y$ and $K_x$ more than 250 kN/m.
3) According to simulation of the empty and laden wagon, to achieve the critical speed not less than 140 km/h for the wagon on bogies with two Lenoir dampers on each axle box, the Radial Arm should provide 750 kN/m lateral stiffness. No introduction of the longitudinal stiffness in Radial Arm is necessary.

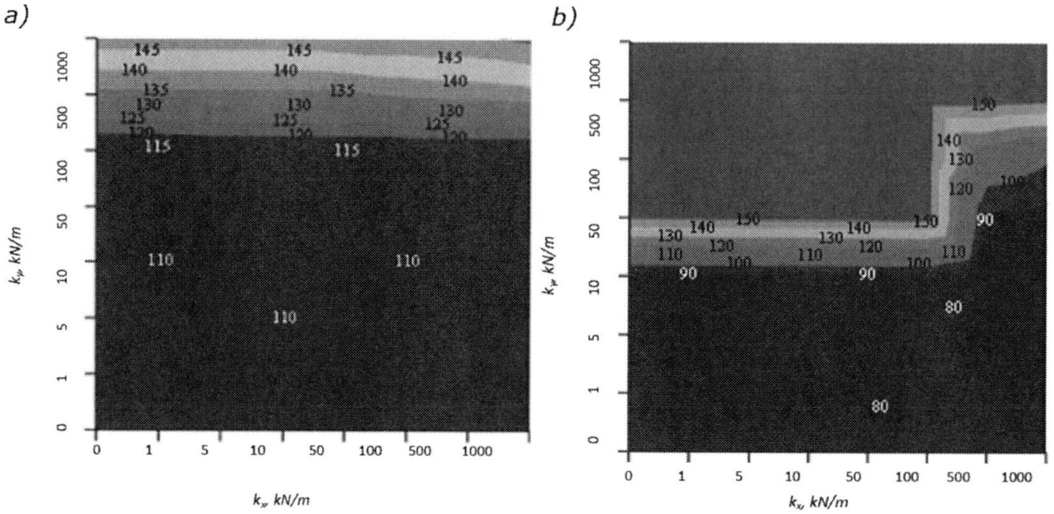


**Figure 8: Diagram of critical speed for laden wagon with 22.5 t axleload (a), for empty wagon with 4.69 t axleload (b)**

### 6. THE PROTOTYPE SUSTRAIL VEHICLE

A prototype of the SUSTRAIL freight vehicle is being constructed by REMARUL engineering. In addition to the innovative suspension described in this paper the SUSTRAIL vehicle will have disk brakes with an electronic control system. The bogie design is shown in figure 9.

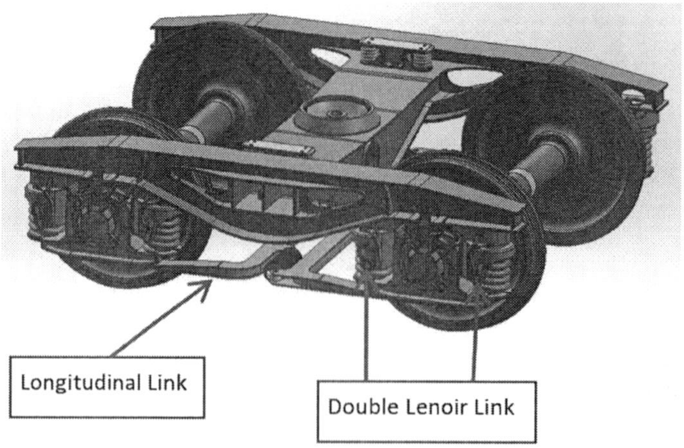


**Figure 9: The prototype SUSTRAIL freight bogie**

## 7. CONCLUSIONS AND NEXT STEPS

A novel high speed low impact freight vehicle has been proposed based on existing innovations which have the potential to improve the dynamic behaviour and reduce track forces even at relatively high speeds. This vehicle has been modelled and the results of the simulations compared against an existing freight vehicle. Double Lenoir links and longitudinal arms linking the axle boxes are included in the primary suspension and a resilient secondary suspension added below the standard centre bowl at the secondary suspension. These innovations have been modelled and conclusions drawn regarding the stability and modal behaviour of the vehicle on straight track. Construction of the vehicle is now under way and testing is planned during 2015.

## REFERENCES


[1] Kik W., Scholdan D., Stephanides J; 'Project INFRA-RADIAL – bogies for axle loads of 25 t – test and simulation'; XXI Century Rolling Stock: Ideas, Requirements, Projects Conference, St. Petersburg, 2007.

[2] Scheffel H. A new design approach for railway vehicle suspension; Rail International. – 1974. - -№10.-P. 638-651.

[3] International Union of Railways: Wagons - Running gear – Normalisation. UIC 510-1, International Union of Railways. 1978.

[4] Hecht, Markus; Keudel, Johannes: LeiLa-DG. Bericht Nr. 30/2004, Technische Universität Berlin, 2004.

# Effect of disabled cement asphalt mortar layer on dynamic behavior of slab track

**J Han[1], G T Zhao[1,2], L Ling[1], X B Xiao[1], Z F Wen[1], X S Jin[1]**
[1.] State Key Laboratory of Traction Power, Southwest Jiaotong University, China
[2.] China Railway Corporation, China

## ABSTRACT

One of the problems about high-speed tracks in service is the damage of the cement asphalt mortar (CAM) layer. The development of CAM damage leads to the gradual failure of the slab supporting. The CAM damage seriously affects the dynamic behavior of the track. A three-dimensional coupling dynamic model of a vehicle and a CRTS-I slab track is developed. By using the proposed model, the wheel-rail contact forces, and the displacements and the accelerations of the track are calculated to study the influences of the size and location of the damaged CAM zone on the dynamic characteristics of the track system.

## 1 INTRODUCTION

A new type of slab track, so called China Railway Track System I (CRTS-I), is developed and has been applied on the high-speed lines of Chongqing-Suining, Shijiazhuang-Taiyuan, Shanghai-Nanjing and Haerbin-Dalian in China (1). In their operation, different degrees of damages occurred on the cement asphalt mortar (CAM) filling layers of the tracks. The damages include cracks, shelling and deterioration due to the effect of water in the tracks, as shown in Fig. 1 (2, 3). Few researches are conducted on the vehicle-track coupling system considering CAM damage. Xiang (4) studied the effect of the voided slab induced by the deterioration of the CAM layer on vibration responses of the slab track with speed variation. Based on the perspective of the system energy, he used the Wilson-$\theta$ numerical integral method to solve the track vibration equations. Wang (5) studied the effect of the CAM debonding on the dynamic properties of the CRTS II slab track. He used the LS-DYNA to solve the dynamic equations. Both of Xiang and Wang treated the rail as a continuous Euler beam, and their models only considered the vibration in vertical direction. And the influence of the size and location of the damaged zone of CAM on the dynamic characteristics of the slab track was not studied.

**Fig. 1 Damages of the CAM ((2), (3))**

This paper develops a three-dimensional coupling dynamic model of a high-speed vehicle and a CRTS-I slab track. Using the proposed model, the wheel-rail contact forces, the displacements and the accelerations of the track are calculated in the case of the CAM damage at high speed. The influences of sizes and locations of the damaged CAM zone on the dynamic characteristics of the track system are analyzed.

## 2 COUPLING DYNAMIC MODEL OF VEHICLE AND CRTS-I SLAB TRACK

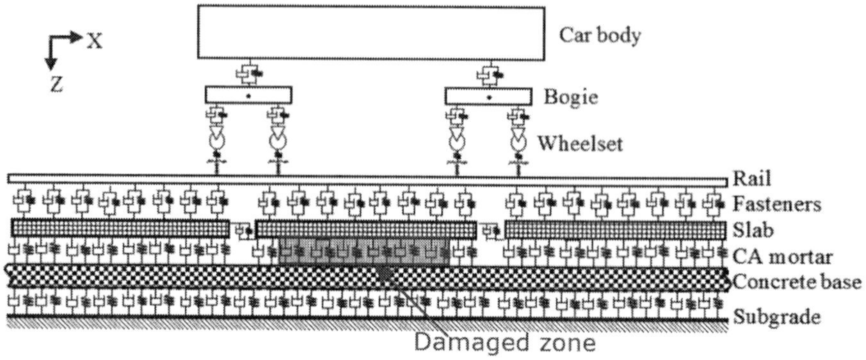


(a) Elevation

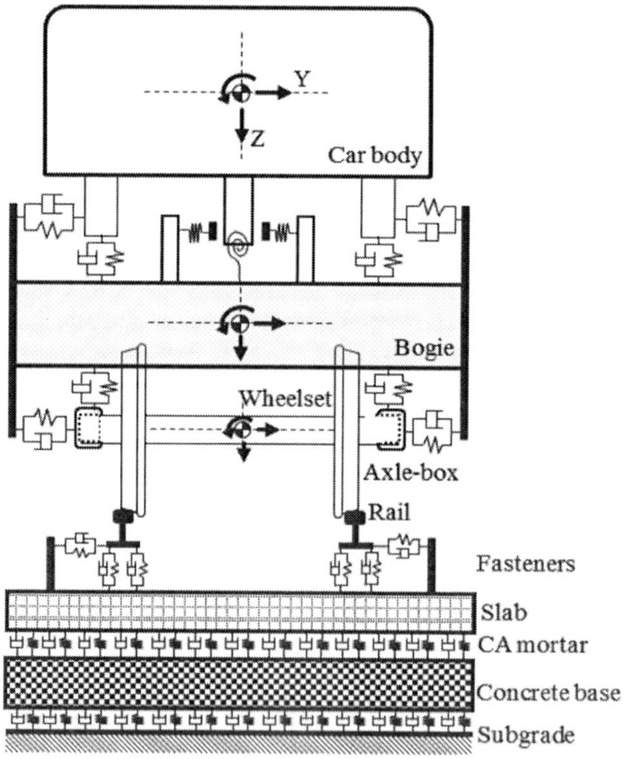


(b) End view

**Fig. 2 Coupling dynamic model of a vehicle and a CRTS-I slab track**

Fig. 2 illustrates the coupling dynamic model of a high-speed vehicle and the CRTS-I slab track to study the effect of disabled CAM layer on the dynamic behavior of the track. In the numerical simulation, the slab supporting stiffness is considered as being zero to approximately model the CAM damage. A moving structure under the rail is adopted as a new vehicle-track coupling interface excitation model (called as "Tracking Window") (6, 7). This excitation model is closer to the real moving vehicle under the excitation of the discrete sleepers, and reduces computation time. The vehicle-track coupling system equations are solved by means of a new explicit integration method (8).

## 2.1 Dynamic model of vehicle subsystem

According to the vehicle dynamic theory (9), a high-speed railway vehicle is considered as a rigid multi-body model, in which the car body is supported on two double-axle bogies with the primary and the secondary suspension systems. For the connecting parts (primary vertical damper, secondary lateral damper, secondary yaw damper and lateral stopping block) with nonlinear characteristics, a piecewise linear simulation is used. As shown in Fig. 2, each component of the vehicle has five degrees of freedom (DOFs): the lateral displacement, the vertical displacement, the roll angle, the yaw angle and the pitch angle. The total DOFs of the vehicle are 35. Based on the coordinate system moving along the track at a constant speed of the vehicle, the equations of the vehicle subsystem motion are the second order differential equations in time domain:

$$\boldsymbol{M}_v\ddot{\boldsymbol{u}}_v + \boldsymbol{C}_v\dot{\boldsymbol{u}}_v + \boldsymbol{K}_v\boldsymbol{u}_v = \boldsymbol{F}_v \tag{1}$$

where $\boldsymbol{M_v}$ is the mass matrix of the vehicle, $\boldsymbol{C_v}$ and $\boldsymbol{K_v}$ are the damping and the stiffness matrices, respectively. $\ddot{\boldsymbol{u}}_v$, $\dot{\boldsymbol{u}}_v$ and $\boldsymbol{u}_v$ are the vectors of displacements, velocities and accelerations of the vehicle subsystem, respectively, and $\boldsymbol{F_v}$ is the vector of generalized loads acting on the vehicle subsystem

## 2.2 Dynamic model of slab track subsystem

The dynamic model of the slab track subsystem includes rails, fastener systems, slabs, CAM layers and concrete base, as shown in Fig. 5. The rail is treated as a continuous Timoshenko beam resting on rail pads, and the lateral, vertical and torsion motions of rails are simultaneously taken into account (10). The slabs and the concrete base are modeled by using three-dimensional finite element method. The rail fastener systems and the CAM layer are modeled by using periodic discrete viscoelastic units.

The finite element model of the slab has 20600 solid elements and 26520 degrees of freedom. The length of the slab is 4.962 m. The geometric dimensioning of its cross section is 0.19 m × 2.4 m. The vibration of the slab can be described in the second order differential equations in terms of generalized coordinates, as expressed by equation (2). The modal analysis on the slab is carried out by means of ANSYS to obtain the first 16 order elastic modes. By using the mode shapes of the elastic modes and 6 rigid motions, equation (2) is decoupled and solved according to the modal superposition principle.

$$\boldsymbol{M}_{si}\ddot{\boldsymbol{u}}_{si} + \boldsymbol{C}_{si}\dot{\boldsymbol{u}}_{si} + \boldsymbol{K}_{si}\boldsymbol{u}_{si} = \boldsymbol{F}_{rsi} + \boldsymbol{F}_{sci} \tag{2}$$

where $\boldsymbol{M}_{si}$, $\boldsymbol{C}_{si}$ and $\boldsymbol{K}_{si}$ are, respectively, the mass, damping and stiffness matrices of the *i*th slab. $\ddot{\boldsymbol{u}}_{si}$, $\dot{\boldsymbol{u}}_{si}$ and $\boldsymbol{u}_{si}$ are, respectively, the acceleration vector, velocity vector and displacement vector. $\boldsymbol{F}_{rsi}$ is the load vector between the rail and the *i*th slab. $\boldsymbol{F}_{sci}$ is the load vector between the slab and the concrete base.

The model of the concrete base is similar to the slab. The concrete base model has 433956 solid elements and 515424 degrees of freedom. The length of the concrete base is 60 m. The geometric dimensioning of its cross section is 0.3 m × 3 m.

### 2.3 Model of wheel-rail in rolling contact

Wheel/rail dynamic interaction modeling is the key of vehicle-track coupling dynamic model. In this paper, when calculating the dynamic response of the vehicle-track, the tracing-curve-method (11) is adopted to locate the wheel-rail spatial contact geometry. Therefore, the method of two-dimensional scanning can be replaced by one-dimensional scanning through a tracing curve. It can greatly reduce the computational time. The non-linear Hertzian elastic contact theory is used to calculate the wheel-rail normal contact forces. The tangential wheel-rail creep forces are calculated by using the Shen-Hedrick-Elkins non-linear theory (12).

## 3 RESULTS AND ANALYSIS

By using the vehicle-track coupling dynamic model, the dynamic responses of the track system influenced by the sizes and locations of the damaged CAM zone are analyzed, as shown in Fig. 3. Fig. 3(a) illustrates the size change of the damaged CAM zone. A1-A3 represent different sizes, the length of which are 2 sleeper pitches, 4 sleeper pitches and 6 sleeper pitches, respectively. And A0 represents the case without the CAM damage. Fig. 3(b) illustrates the location change of the damaged CAM zone without size change. P1-P3 represent different locations, the length of which are 2 sleeper pitches. And P0 represents the case without the CAM damage. Fig. 3(c) illustrates the position of the damaged CAM zone occurred at the two ends of the slab. Fig. 3(d) illustrates the position of damaged CAM zone occurred at the two sides of the slab.

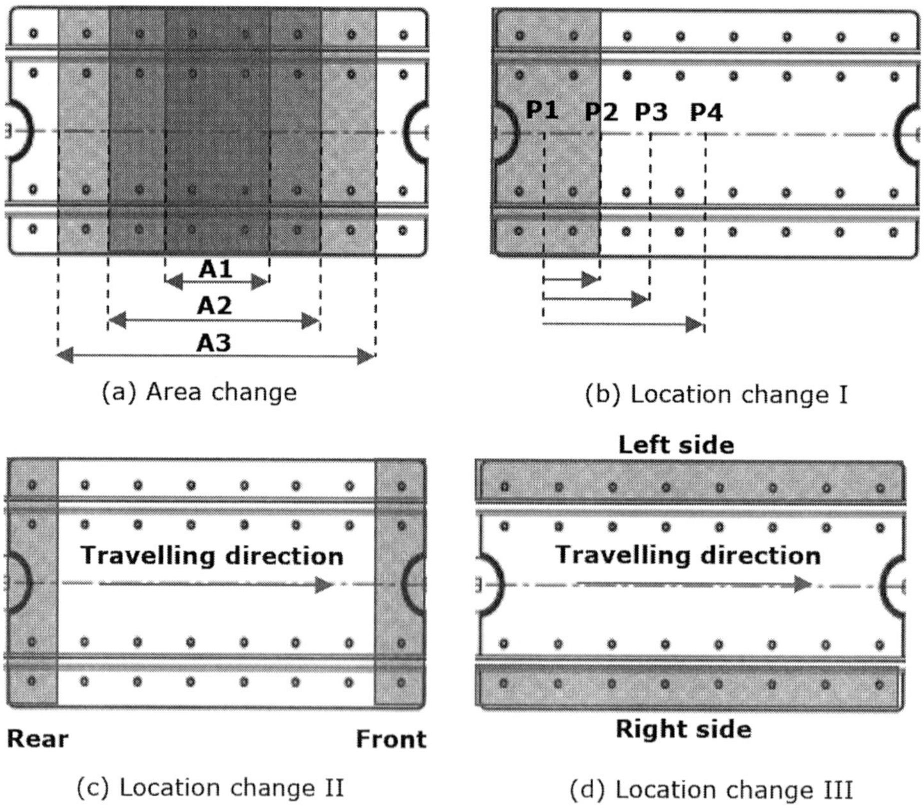


**Fig. 3 Changes of size and location of damaged CAM zone**

### 3.1 Effect of damaged CAM area

In this analysis, a high-speed train runs on a tangent slab track at 250km/h. The track is symmetrical with respect to its center line, so here the dynamic responses of the left side of the track are presented. Taking the leading wheelset as the example, Fig. 4 shows the wheel/rail normal forces with the travelling distance in the cases without the CAM damage (A0) and with the different damaged areas (A1 to A3). As shown in Fig. 4, when the wheelset is about to pass through the damaged area, the wheel/rail normal force starts deloading. When the wheelset is passing through the area, wheel/rail normal force increases significantly and its maximum value occurs in the middle of the damaged area. With the increase of damaged CAM area, wheel/rail normal force increases.

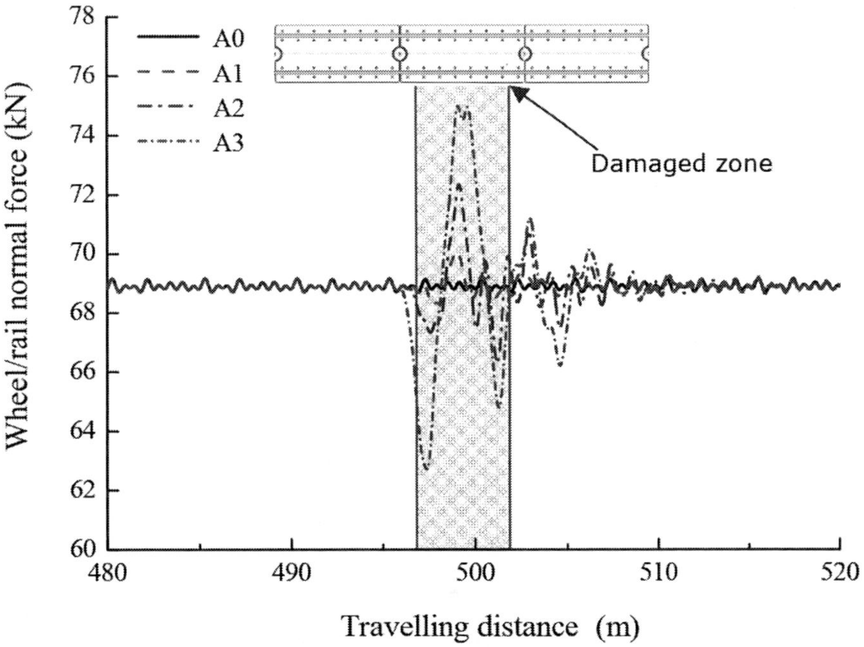


**Fig. 4 Wheel/rail normal forces of the leading wheelset in the case of the damaged CAM area**

Fig. 5 shows the maximum wheel/rail normal forces of the four wheelsets in the cases of the different damaged CAM areas. W1 (W3) and W2 (W4) indicate the leading wheelset and trailing wheelset of Bogie 1 (2), respectively. When the damaged area reaches A3 (longitudinal length 3.72 m > wheelbase 2.5 m), the leading wheelset has not yet left the damaged area, however, trailing wheelset has entered the damaged area. In such a situation, the maximum wheel/rail normal force of the trailing wheelset is greater than that of the leading wheel.

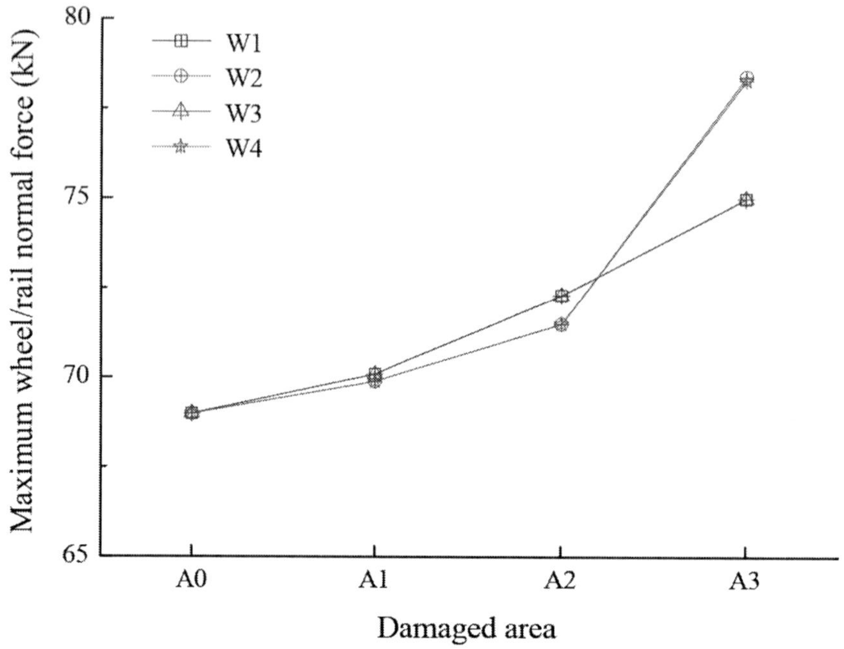


**Fig. 5 Maximum wheel/rail normal force with the damaged CAM area increase**

Fig. 6(a) shows the vertical displacement of the rail at the first wheelset with the travelling distance in the cases of CAM damage (A0, A1, A2, A3 and A3). As shown in Fig. 6(a), with the increase of the damaged CAM area, the vertical displacement of the rail increases. The vertical displacement of the rail in the case of A3 increases most obviously.

Fig. 6(b) shows the vertical displacement of the rail in the middle of the damaged CAM zone. In Fig. 6(b), there are four evident displacement peaks, which denote the largest displacements of the rail at the four wheels W1,W2,W3 and W4. The displacement of the rail in the damaged CAM zone is much greater than that in the case without the CAM damage. With the increase of the damaged CAM area, the vertical displacement of the rail increases. In the case of A3, the vertical displacement of the rail increases obviously. The rail displacement under W2 is greater than that under W1, and the displacement under W4 is greater than that under W3.

The vertical displacement of rail is usually less than the benchmark 1.5mm, shall not be greater than the maximum limit 2mm (13). In the case of A3, the vertical displacement of the rail exceeds the maximum limit.

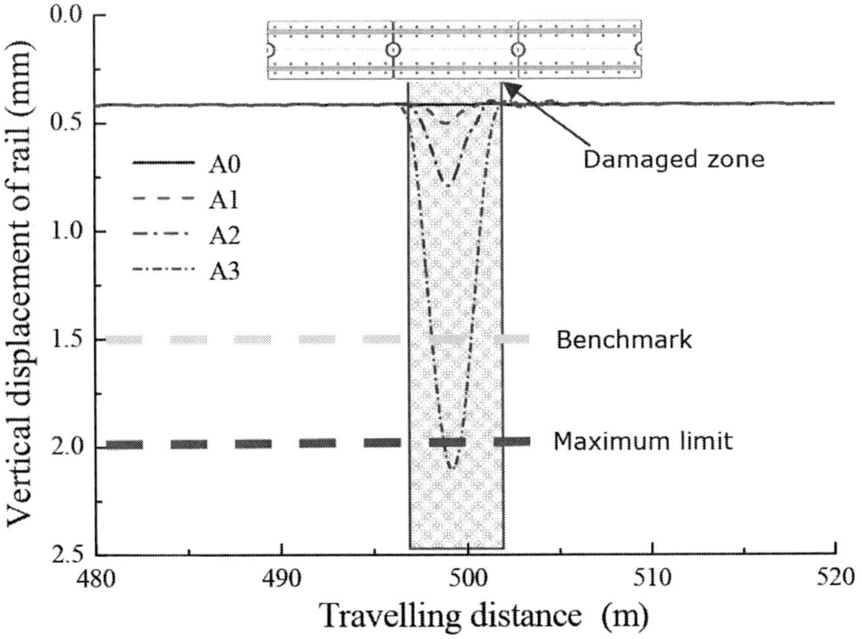

(a) Vertical displacement of the rail under the leading wheelset

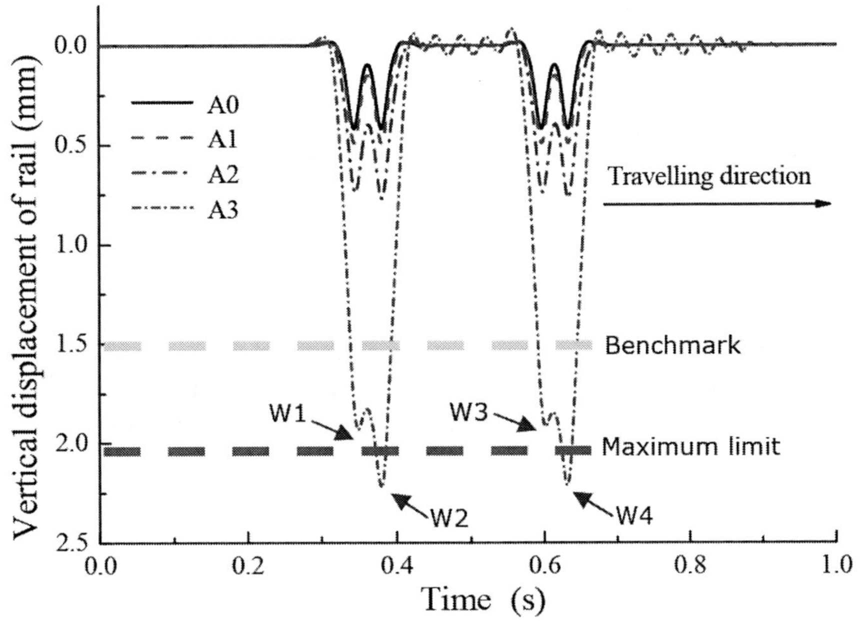

(b) Vertical displacement of the rail in the damaged CAM zone

**Fig. 6 Vertical displacement change of the rail influenced by the damaged CAM area**

Fig. 7(a) shows the vertical displacement of the slab at the fasteners in the cases of A0-A3. As shown in Fig. 7(a), when the vehicle passes through the damaged CAM zone, the vertical displacement of the slab increases obviously. With the increase of the damaged CAM area, the vertical displacement of the rail increases. The vertical displacement of the rail in the case of A3 increases most obviously.

Fig. 7(b) shows the vertical displacement of the slab in the middle of the damaged CAM zone. As shown in Fig. 7(b), there are four evident displacement peaks when the vehicle passes through the monitoring points. The displacement of the slab in the damaged CAM zone is significantly greater than that in the case without the CAM damage. With the increase of the damaged CAM area, the vertical displacement of the slab increases. In the case of A3, the vertical displacement of the rail increases obviously. At the same time, the displacement peaks contain the combined contributions by the passing wheels W1 (W3) and W2 (W4).

The vertical displacement in the middle of the slab is usually less than the benchmark 0.2mm, shall not be greater than the maximum limit 0.3mm (13). In the case of A2 and A3, the vertical displacement in the middle of the slab exceeds the maximum limit.

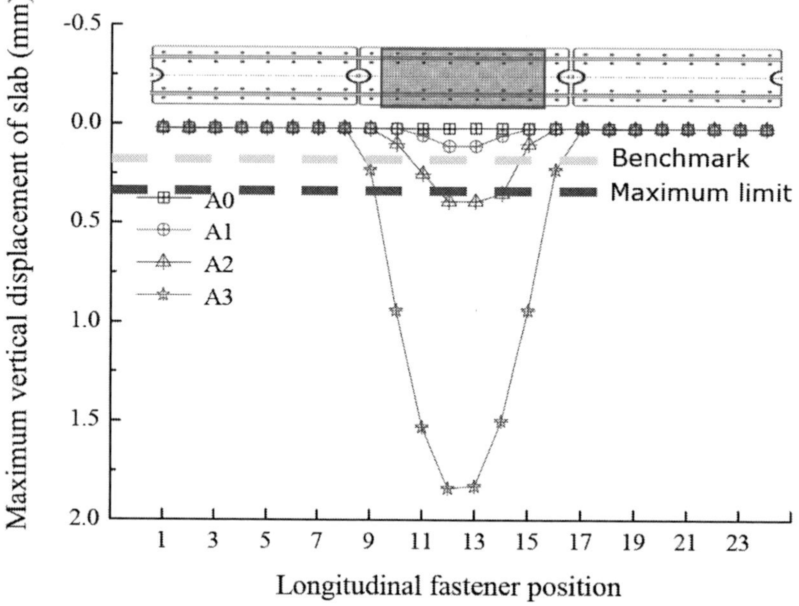


(a) Vertical displacement of the slab at the fasteners

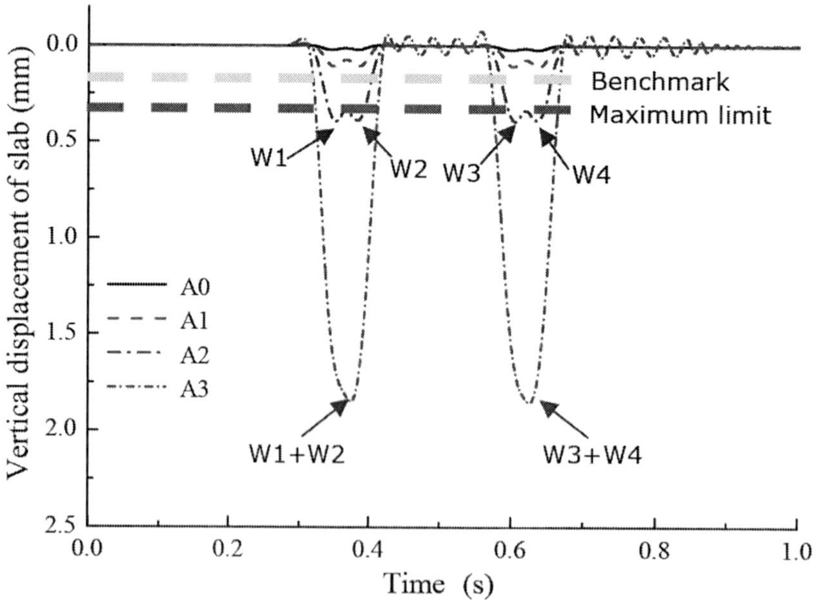


(b) Vertical displacement of the slab in the damaged CAM zone

**Fig. 7 Vertical displacement of the rail influenced by the damaged CAM area**

### 3.2 Effect of location of damaged CAM zone

Fig. 8 shows the wheel/rail normal forces with the travelling distance in the case that considers the normal CAM (A0) and the different positions (P1 - P4) of the damaged CAM zones. For simplicity, the case is indicated by (I). The damaged CAM areas occur from the end to the middle of a slab along the longitudinal direction. As is shown in Fig. 8, when the wheelset is about to pass through the damaged zone, the wheel/rail normal force appears deloading in the first place. Then when the wheelset passes through the CAM damaged zone, the wheel/rail normal force increases significantly. The damage CAM occurring at the end of the slab (P1) causes the most significant wheel/rail normal force.

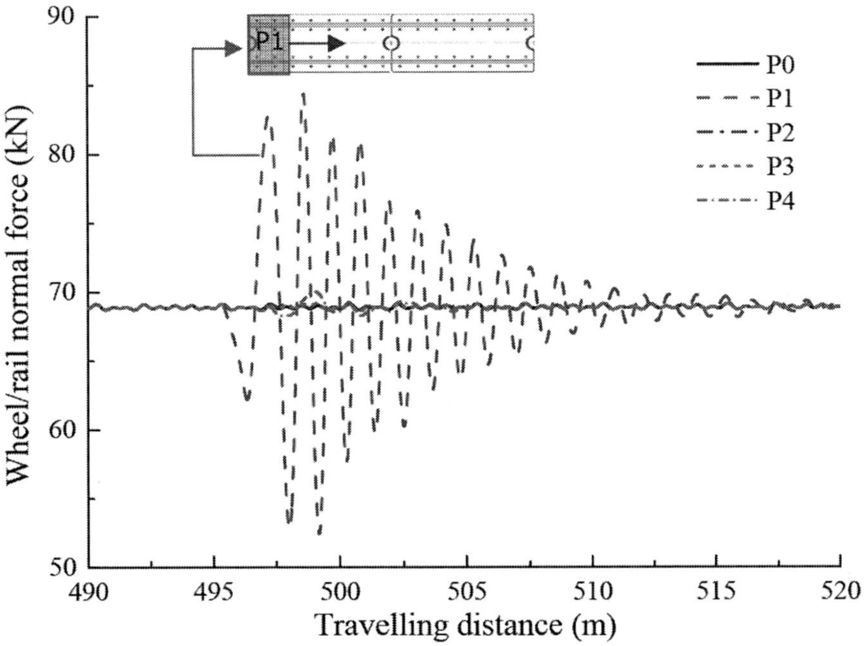


**Fig. 8 Wheel/rail normal force change of the leading wheelset in (I)**

Fig. 9 shows the vertical displacements of the rail with the travelling distance in (I). When the damaged CAM zone occurs at the end of the slab (P1), the vertical displacement of the rail is the most significant, and when the damage occurs in the middle of the slab (P4), the vertical displacement of the rail is the second significant, which do not exceed the Benchmark.

Fig. 10 shows the vertical displacements of the slab at the fasteners in the longitudinal direction positions in (I). The damaged CAM zone (P1) causes the most significant vertical displacement of the slab, which exceeds the Benchmark 0.4 and maximum limit 0.5 (standard in the end of the slab) (13).

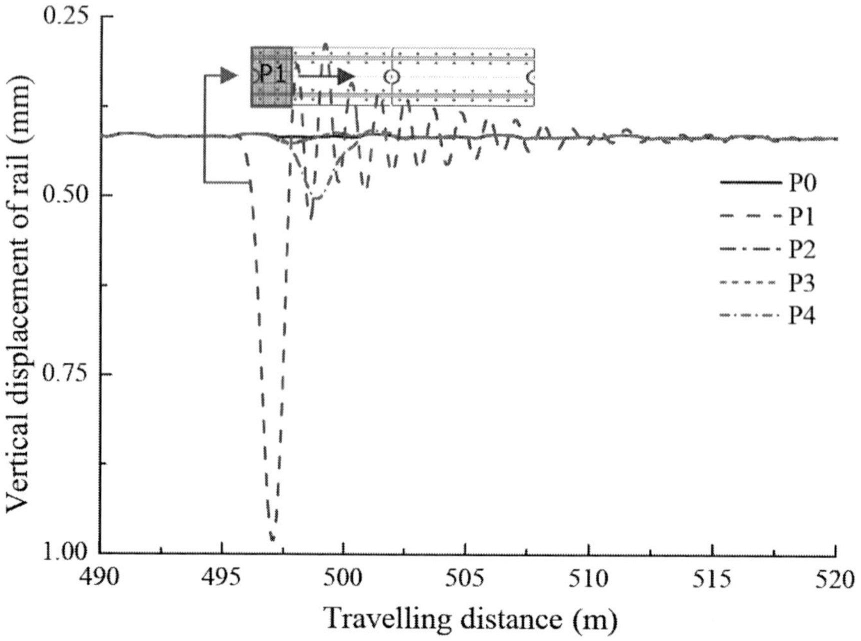


**Fig. 9 Vertical displacement of the rail under the leading wheelset in (I)**

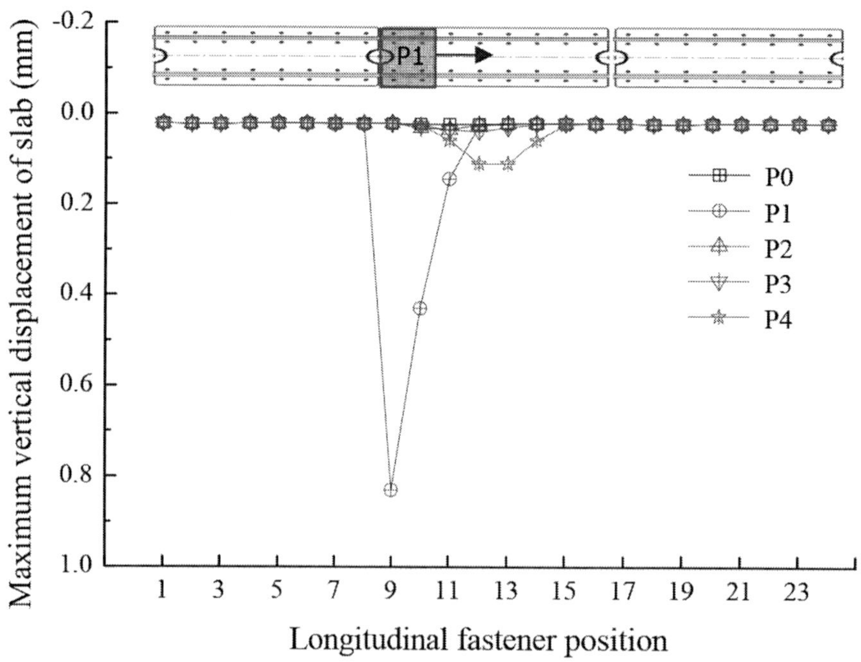


**Fig. 10 Vertical displacement of the slab at the fasteners in (I)**

Fig. 11 shows the wheel/rail normal force with the travelling distance in the case which considers the normal CAM, the damaged CAM area occurring at the rear of the slab, and at both the front and rear of the slab. For simplicity, the case is denoted by using (II). In Fig. 11, the black line indicates the result of the normal CAM, the dash line represents the result of the damaged CAM zone at the rear, and the dash dot line represents the results at the front and the rear. When the vehicle passes through the damaged CAM zone at the rear, it is found that the wheel/rail normal forces of the two kinds damaged CAM zones are basically identical. However, in the damaged CAM zone in the front of the slab with the vehicle passing through it, the fluctuating amplitudes of the wheel/rail normal force is enlarged again.

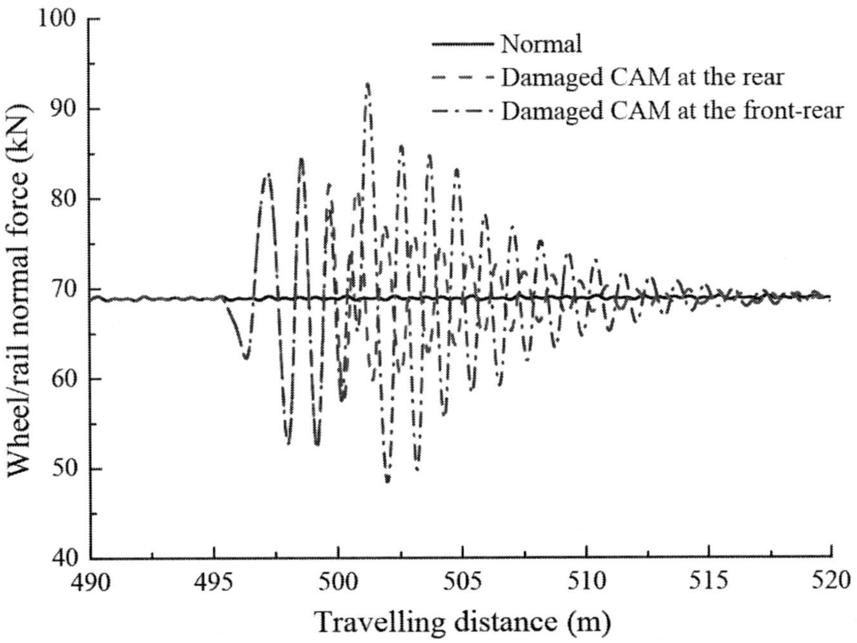


**Fig. 11 Wheel/rail normal force of the leading wheelset in (II)**

Fig. 12 shows the vertical displacement of the rail with the travelling distance in the same case as shown in Fig. 11. As is shown in Fig. 12, when the vehicle passes through the damaged CAM zone at the rear, the vertical displacements of the rail in the two cases of the damaged CAM zones are quite close and have a great peak (Peak1). In the damaged CAM zone at the front of the slab with the vehicle passing through it, the vertical displacement of the rail presents the second significant peak (Peak2), and the displacement fluctuates more fiercely after Peak2.

Fig. 13 shows the vertical displacement of the slab at the fasteners. As shown in Fig. 13, when the vehicle passes through the damaged CAM zone in the rear, the vertical displacements of rail in the two damaged CAM zones are close, and have a large peak (Peak1). However, in the damaged CAM zone in the front of the slab, the vertical displacement of the rail has a second high peak (Peak2).

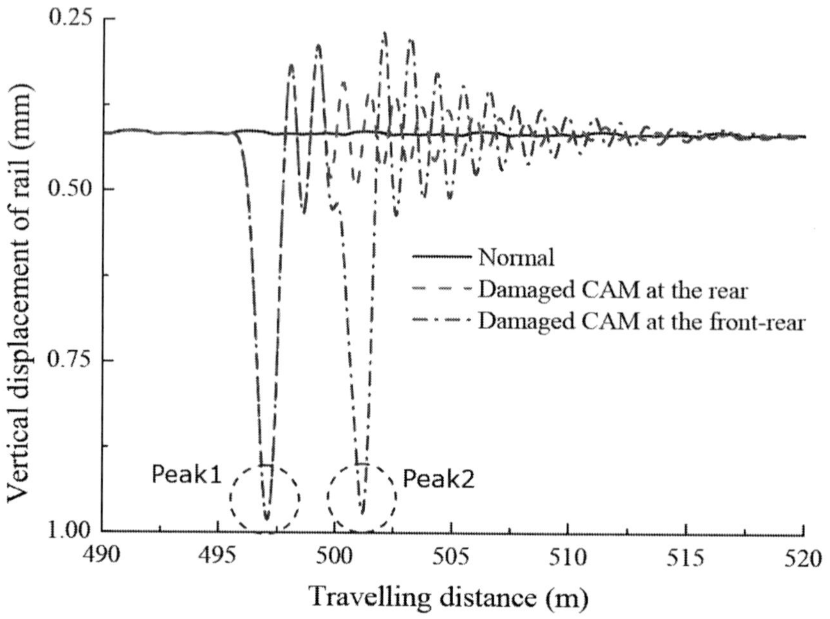


**Fig. 12 Vertical displacement of rail under leading wheelset in (II)**

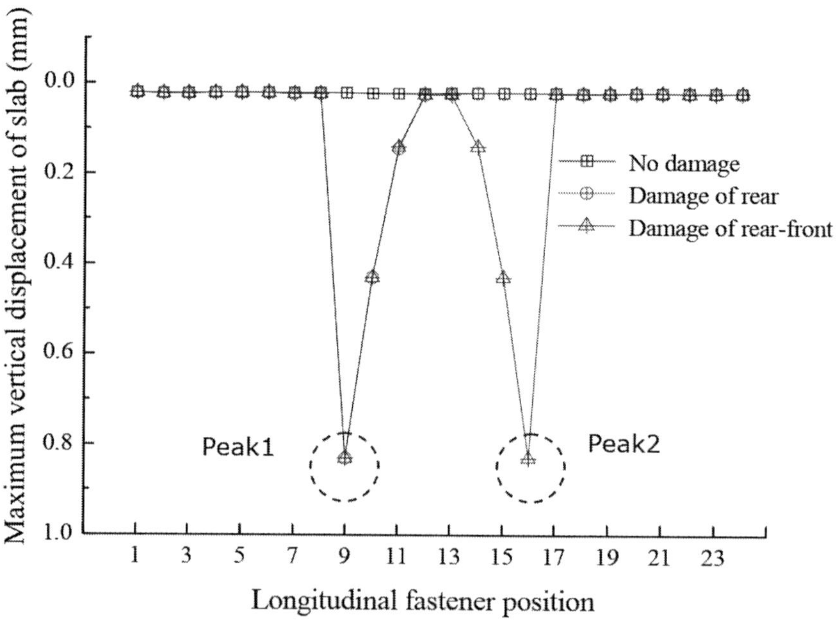


**Fig. 13 Vertical displacement of the slab at the fasteners in (II)**

Fig. 14 shows the wheel/rail normal force acting on the left rail with the travelling distance in the case that includes the normal CAM and the damaged CAM on the left/right sides of the slab. For simplicity, the case is denoted by using (III). As shown in Fig. 14, the dash line represents the CAM damaged zone on the left side, and the dash dot line represents the damaged CAM zone on the left and right sides. The influence of case (III) on the wheel/rail normal force is not obvious. Enlarging the damaged zone, when the vehicle passes through it, the fluctuating amplitude of the wheel/rail normal force increases, and the fluctuating amplitude of the wheel/rail normal force in the case of the damaged CAM areas on the left and right sides is larger than that on the left side.

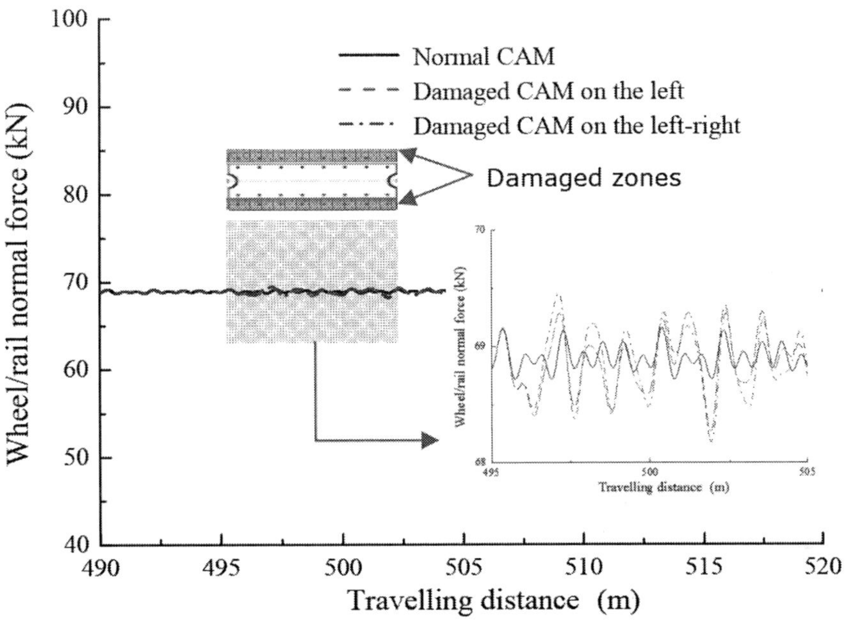


**Fig. 14 Wheel/rail normal force of leading wheelset in (III)**

Fig. 15 shows the vertical displacement of the left rail with the travelling distance in (III). As shown in Fig. 15, when the vehicle passes the damaged CAM zone, the rail displacement increases, and after a stable period, it begins to decrease. The rail displacement in the case of the damaged CAM zone on the left side is slightly larger than that on the left and right sides at the same time. The main reason is that the damage only on the left side causes the unbalance supporting between left and right sides of the track.

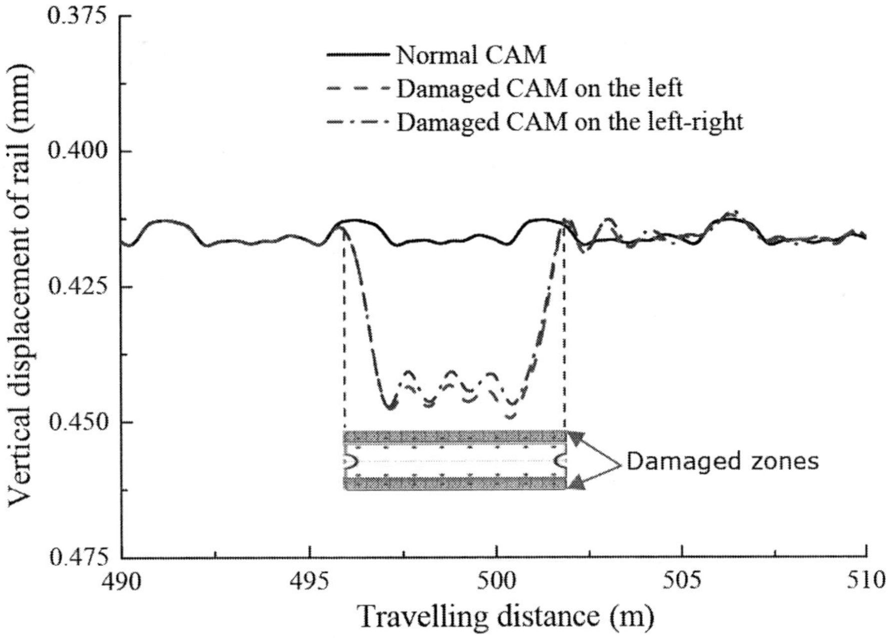


**Fig. 15 Vertical displacement of rail under leading wheelset in (III)**

Fig. 16 shows the slab vertical displacement under the left rail in the fasteners in (III). As shown in Fig. 16, when the vehicle passes through the damaged CAM zone, the slab vertical displacements at the two ends of it are larger than that in the middle of it. The slab displacement in the damaged CAM zone on the left side of the slab is slightly larger than that on the left and right sides of the slab at the same time. The main reason is that the damage only on the left side causes the unbalance supporting between the left and right sides of the slab.

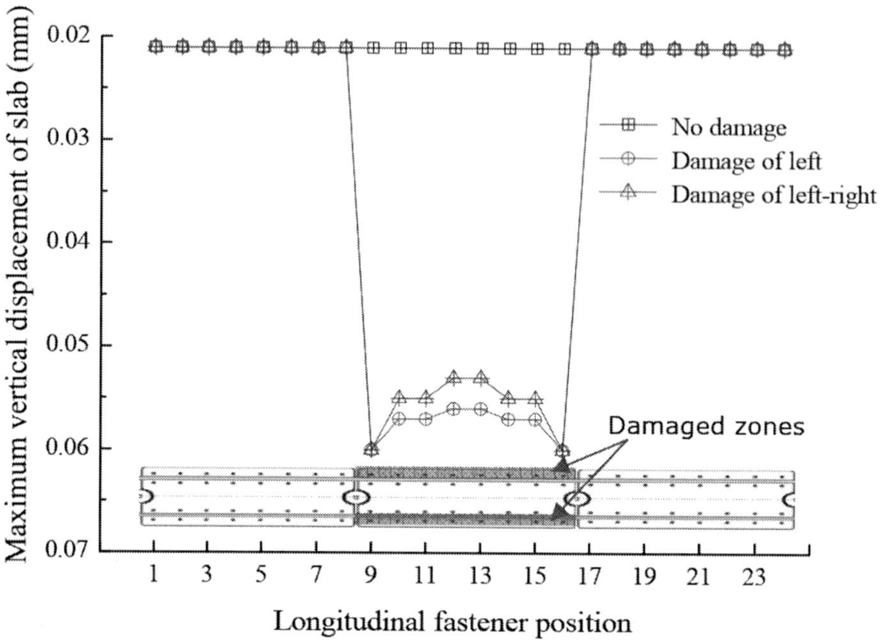


**Fig. 16 Vertical displacement of the slab at the fasteners in (III)**

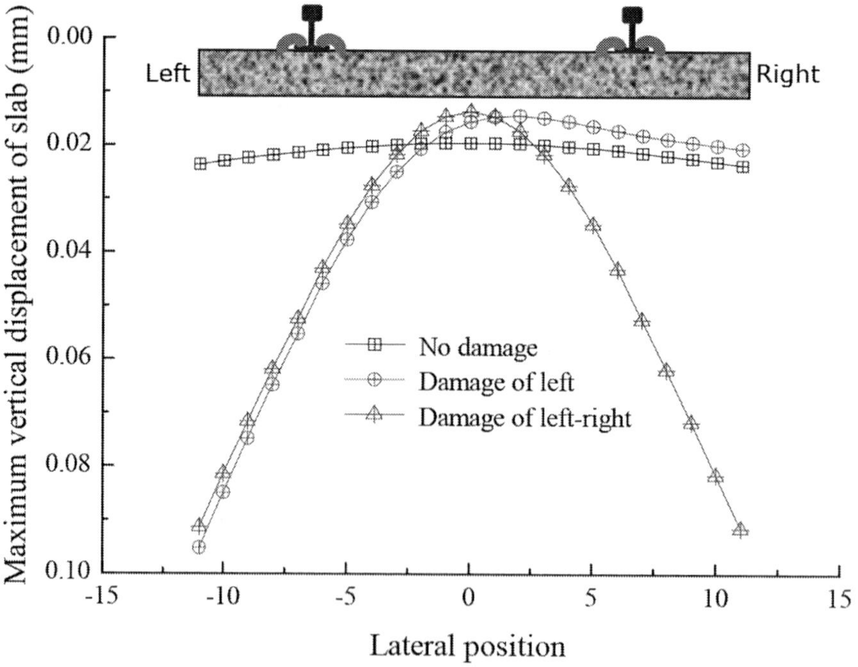


**Fig. 17 Vertical displacement of the slab in the lateral direction in (III)**

Fig. 17 shows the maximum vertical displacement of the slab surface in the lateral direction in (III). As is shown in Fig. 17, when the vehicle passes through the cross section of the slab, the part of the slab with the damaged CAM areas bands down significantly, and the rest part of the slab bends upwards compared to the slab without the CAM damage.

It can be seen from Figs. 15 – 17 that the displacement of rail and slab do not exceed the benchmark and maximum limit. However, this kind of damage (case III) is common in practice, and it is easy to develop into a comprehensive damage. So it is of interest to pay attention on this kind of damage.

## 4 CONCLUSIONS

A three-dimensional coupling dynamic model of a vehicle and a CRTS-I slab track is developed to study the influences of the size and location of the damaged CAM zone under the slab on the dynamic characteristics of the track system. The following conclusions can be drawn.

(1) When the vehicle passes through the damaged CAM zone, the dynamic responses of the track increase significantly and their maximum values appear in the middle of the damaged zone. With the increase of the damaged CAM area, the dynamic responses of the track increases. And it is recommended that the longitudinal length of the damaged CAM zone is not larger than the wheelbase of the vehicle. Because vertical displacement of rail and slab with the CAM damage in this case is much larger than the maximum limit.

(2) The CAM damage occurred at one end of the slab causes the most significant dynamic responses of the track, compared to the damaged CAM zones occurring in the other places. The vertical displacement of slab with the CAM damage occurred at one end is beyond the maximum limit.

(3) When the CAM damage occurs at the left or right side, the part of the slab with CAM damage bends down significantly, and the rest part of the slab bends upwards compared to the slab without CAM damage. The dynamic responses of the track with the CAM damage on the one side of the slab are slightly larger than those on the left and right sides of the slab at the same time.

Further research can focus on the effect of disabled CAM layer on vehicle running safety at a curved track.

## ACKNOWLEDGEMENTS

The present work is supported by the National Natural Science Foundation of China (U1134202), the National Basic Research Program of China (2011CB711103), the Program for Changjiang Scholars and Innovative Research Team in University (IRT1178, SWJTU12ZT01), and 2015 Doctoral Innovation Funds of Southwest Jiaotong University.

## REFERENCES


(1) Zhao GT. High-speed railway ballastless track. Beijing: China Railway Press, 2005.

(2) Liu Y. Study on characteristics and influences of CRTS II slab track early temperature field. Ph.D. thesis, Southwest Jiaotong University, China, 2013.

(3) Lin HS. Research on the static and dynamic property of ballastless track based on fracture and damage mechanics. Ph.D. thesis, Southwest Jiaotong University, China, 2009.

(4) Xiang J, He D and Zeng QY. Effect of cement asphalt mortar disease on dynamic performance of slab track (in Chinese). *Journal of Central South University* 2009; 40(3): 791-796.

(5) Wang P, Xu H and Chen R. Effect of cement asphalt mortar debonding on dynamic properties of CRTS II slab ballastless track. *Advances in Materials Science and Engineering* 2014; 2014: 1-8.

(6) Xiao XB, Jin XS and Wen ZF. Effect of disabled fastening systems and ballast on vehicle derailment. *Journal of Vibration and Acoustics-Transactions of the ASME* 2007; 129: 217-229.

(7) Jin XS, Wen ZF. Effect of discrete track support by sleepers on rail corrugation at a curved track. *Journal of Sound and Vibration* 2007; 315: 279-300.

(8) Zhai WM. Two simple fast integration methods for large-scale dynamic problems in engineering. *Internal Journal for Numerical Methods in Engineering* 1996; 39(24): 4199-4214.

(9) Zhai WM. Vehicle-track coupling dynamics (Third edition). Beijing: Science Press, 2007.

(10) Xiao XB, Jin XS, Deng YQ et al. Effect of curved track support failure on vehicle derailment. *Vehicle System Dynamics* 2008; 46(11): 1029-1059.

(11) Chen G and Zhai WM. A new wheel/rail spatially dynamic coupling model and its verification. *Vehicle System Dynamics* 2004; 41: 301-322.

(12) Shen ZY, Hedrick JK, Elkins JA. A comparison of alternative creep force models for rail vehicle dynamic analysis. *Vehicle System Dynamics* 1983; 12(1-3): 79-83.

(13) The Ministry of Railways of the People's Republic of China. TB 10716-2013, Technical regulations for dynamic acceptance for high-speed railways construction. Beijing: China Railway Publishing House, 2013.

# The effects of railway traffic on trackbed stability and deformation due to clay fill performance

**G R Taylor, R Tan, A S O'Brien**
Mott MacDonald Limited, UK

**J S Lane**
Rail Safety and Standards Board Ltd, UK

**E Evans**
Network Rail, UK

**ABSTRACT**

Most railway embankments in the UK were built in the Victorian era and are of end-tipped construction using materials (usually cohesive) excavated from adjacent cuttings, resulting in a clod-and-matrix structure. Historically there has been a lack in understanding of the mechanical behaviour of such railway embankments. In the next decade railway traffic in the UK, particularly freight, is forecast to grow considerably. Consequently there is a need to improve the fundamental understanding of how increases in rail traffic loading may influence the mechanical behaviour of railway embankments and thus track performance. The Rail Safety and Standards Board (RSSB) in conjunction with Network Rail (NR) is currently undertaking a programme of applied research into this topic.

The objective of the research programme is to measure the deformation of embankment fills, particularly those of clay core construction, which is attributable to railway traffic loading, and subsequently to develop analytical models which can be used at both strategic and tactical levels for assessment of embankment response due to changes in train axle loads and/or frequency of passing trains. This paper summarises the work done to date, the conclusions drawn and proposals for future work, including a programme of instrumentation and monitoring at a number of sites on the UK rail network to validate and inform further development of the preliminary analytical models.

## 1 INTRODUCTION

In Britain earthwork engineering knowledge developed during the canal construction era where empiricism and trial-and-error were utilized in the construction of earthworks. Construction of the subsequent earthworks for the railway network had their basis in this era. The construction techniques for these early earthworks were crude and labour intensive. Soil and rock excavated from hillsides to form cuttings was transported to valleys by hand barrow or horse drawn cart and tipped on the ground as fill to form embankments – a process known as 'end-tipping'. Common practice on railways where two tracks were to be built was to form one line of track and then form the embankment to support the other line by tipping excavated fill from the side of wagons – 'side-tipping'.

As excavation of the cutting proceeded, so the embankments could be extended until either the materials won from the cuttings were exhausted, or it became uneconomic to continue to place the fill, and viaduct or underbridge structures were then built to continue the track alignment across the valley. Because of this construction technique, embankment composition may reasonably be assumed to be derived from the local soils and rocks. However, where additional fill materials were required, they were obtained from 'borrow areas' excavated into the adjacent land, or from whatever source happened to be available. The composition of embankments can therefore be unpredictable and highly variable over very short distances.

The opening of the Manchester to Liverpool railway in 1830 marked the start of the main phase of railway construction in the UK. In the majority of cases, the age of a railway earthwork will date back to the original construction date of the particular line; earthwork reconstruction is rare. The typical age of railway earthworks is therefore of the order of 150 years.

Deterioration of railway embankments/trackbed occurs naturally over time. However, there has been a marked increase in embankment/trackbed deformation on the railway network in Great Britain due to railway traffic over the past ten years or so, sometimes resulting in failure, for example the failure of the embankment at Mottingham in Kent during 2001. The increase in damage has been accompanied by an increase in railway traffic loading due to higher axle loads and increased train frequency. Currently, there is no means of assessing the potential impact on embankments/trackbed where an increase in railway traffic loading is anticipated.

In the next decade railway traffic in the UK, particularly freight, is forecast to grow considerably. Consequently there is a need to improve the understanding of how increases in railway traffic loading may influence the mechanical behaviour of railway embankments and thus track performance. In response to this situation, the Rail Safety and Standards Board in conjunction with Network Rail has commissioned a programme of applied research. The objective of the research programme is to measure the deformation of embankment fills, particularly those of clay core construction, which is attributable to railway traffic loading, and subsequently to develop analytical models which can be used at both strategic and tactical levels for assessment of the embankment response due to changes in train axle loads and/or vehicle frequency.

## 2 DESK STUDY

Desk study revealed that the failure/deformation mechanisms associated with railway traffic loading are different to classical geotechnical slope stability failures. Railway traffic induces fatigue-type failure within medium to high plasticity clay supporting the trackbed, which becomes evident as excessive deformation of the track. Over time the speed-related limits set for correction of vertical track geometry faults can be exceeded resulting in an increase in track maintenance. The deformation is likely to be concentrated in the upper layers of the embankment below the subgrade.

Such behaviour may be classified as a 'fatigue-type' serviceability failure which develops slowly with an increasing number of load cycles under service loading. Therefore, dependent upon the freight traffic intensity on a particular line, it will usually take several years for failures to develop and then for remediation of the track alignment to become necessary.

## 3 LABORATORY TESTING PROGRAMME

As a result of the complex construction history described above, railway embankments were usually uncompacted at the time of their construction and thus potentially have a clod and matrix structure (O'Brien et al, 2004, O'Brien, 2007) as shown in Figure 1. Hence old railway embankments often differ significantly from natural materials or more recent well-compacted materials. Undisturbed testing is required to realistically investigate potential deformation behaviour and this is rare.

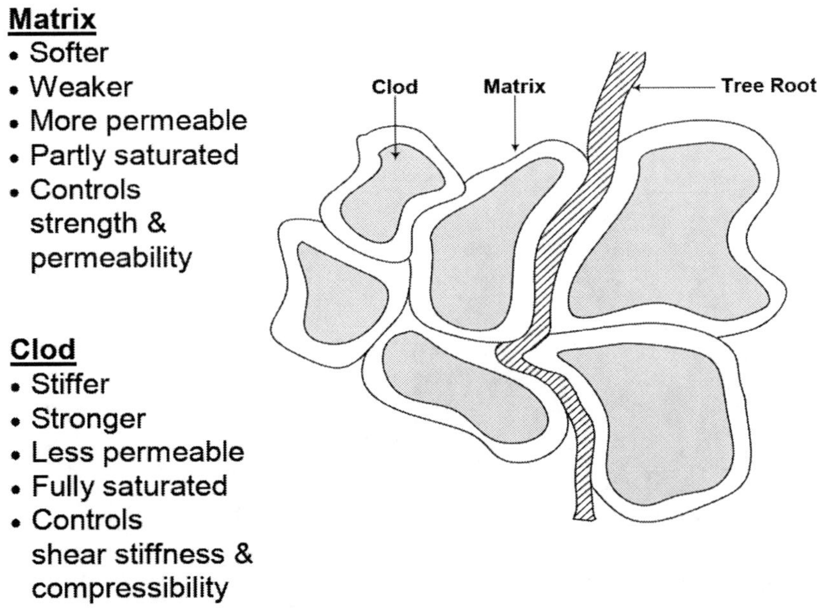


**Figure 1: Conceptual model of railway embankment fill**

To facilitate the programme of laboratory testing undertaken as part of the applied research project clay fill samples were sourced from an existing railway embankment of Victorian age at Tumpy Green to the north of Bristol on the Bristol to Gloucester railway line (ELR-BGL). The published geological map covering the Tumpy Green Embankment site, British Geological Survey Map Sheet 234 (1975), shows the site to be underlain by Lower Jurassic strata of Lower Lias Clay (i.e. a high plasticity clay), bounded by outcrops of Pleistocene and Recent drift deposits.

Undisturbed samples of the embankment clay fill (i.e. block samples and driven open tube U100 samples) were obtained to facilitate the programme of laboratory element testing. A large bulk disturbed sample of the clay fill was also obtained to facilitate the complementary programme of laboratory based physical model testing.

Two fundamentally different types of testing were undertaken as part of the laboratory testing programme:

a) Conventional "element" testing which was used to assess soil behaviour under a wide variety of cyclic loading regimes (both remote from and close to failure). Cyclic load testing of in-situ railway embankment fill, in the 'as-constructed' condition, in order to assess behaviour under railway traffic loading is not known to have been undertaken previously. The main objective of the laboratory element testing programme was to measure the accumulation of plastic strain and the fatigue strength under cyclic loading, of three different types of clay fill specimen: intact samples obtained from an existing railway embankment, 'reconstituted' samples prepared in the laboratory, and 'remoulded' samples obtained on completion of one of the laboratory physical model tests. The tests were carried out in the triaxial apparatus.

b) Physical model tests – these tests are of a much larger scale than those described in (a), and enabled 1-g load tests to be carried out on models of the whole system performance (i.e. track/ballast/embankment fill). The main objective of the 1-g laboratory physical model testing was to measure the accumulation of plastic strain in a clay fill model embankment when the embankment is subject to repeated loading. The physical model tests were not undertaken at full scale; based on the results of numerical studies performed to investigate potential test set-ups a scale factor, $\lambda$=2.0, was adopted. Additionally, they were constructed in a way that simulated 'Victorian' embankment construction techniques.

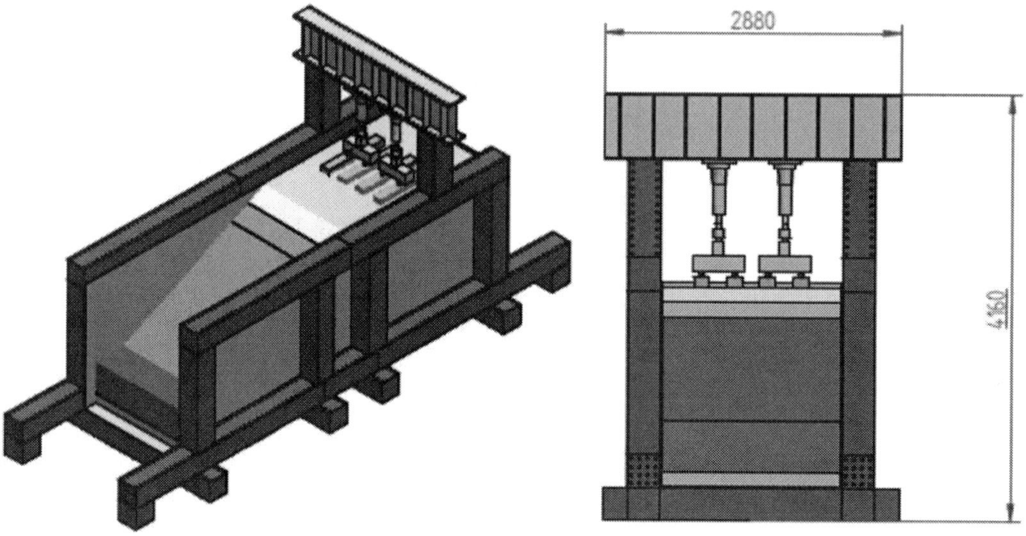


**Figure 2: Schematic of physical model test apparatus**

The physical model test embankments were subjected to repeated loading in a series of four tests, with loading applied in stages, as summarised in Table 1.

**Table 1: Physical Model Tests/Stages**

| Test | Equivalent Axle Load (t) per Loading Stage | | | | |
|---|---|---|---|---|---|
| | 1 | 2 | 3 | 4 | 5 |
| 2a | 15 | 25 | 37.5 | 25 | 15 |
| 2b[1] | 15 | 25 | 37.5 | 25 | 15 |
| 2c | 37.5 | 25 | 15 | 25 | 37.5 |
| 2d[2] | 15 | 25 | 37.5 | 25 | 15 |

Notes
1. Out-of-phase loading applied during Test 2b; otherwise loading applied in-phase.
2. In Test 2d loading jacks inclined at 14.5° to simulate the application of a coincident horizontal load.

Prediction of the deviatoric stress level that causes instability of a soil under cyclic loading (i.e. fatigue strength) was investigated in the laboratory element tests. This is defined by Frost et al. (2004) as the threshold stress level "under which the rate of accumulation of deformation increases exponentially". The threshold stress level is a key property of fine grained materials such as clay fill and an important parameter in determining the effect of cyclic loading on the stability of the railway trackbed. Values for the threshold stress were determined from the laboratory element testing and may be considered to be the limiting stress above which the accumulation of plastic deformation increases exponentially (Table 2 and Figure 3).

**Table 2: Threshold Stresses**

| Sample Type | Threshold Stress (kPa) |
|---|---|
| Intact | 40-50 |
| Reconstituted | 30-35 |
| Remoulded | <35 |

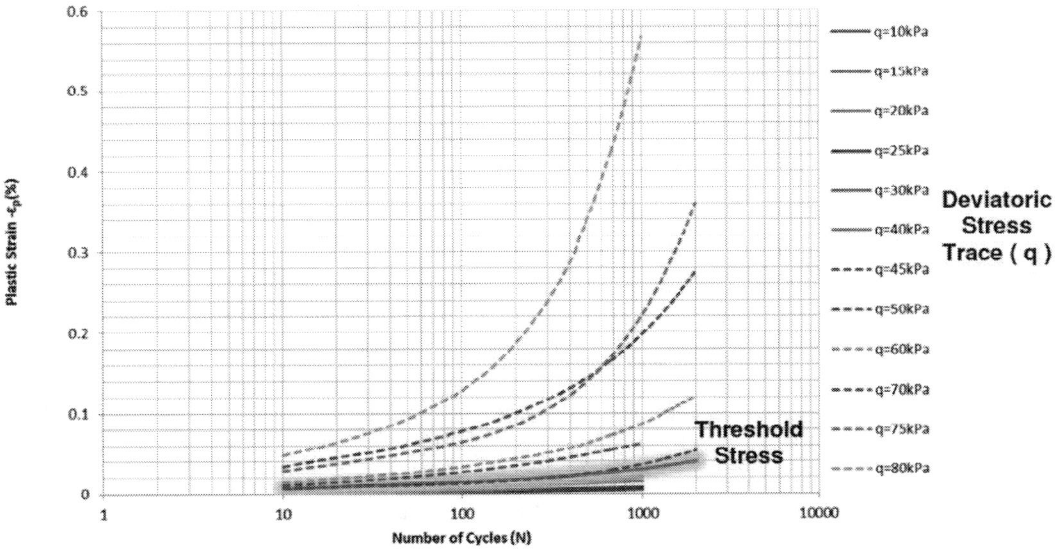


**Figure 3: Accumulation of plastic strain with number of applied stress cycles for TA CY UD1**

The key findings from the physical model testing are:

- The response of the embankment clay fill is independent of previous load history. As shown in Figure 4 the effect of the initial load increments (or primary loading) is to impose a 'pre-stress' within the soil.
- The influence of principal stress rotation as a train passes a point within the clay fill embankment, is not significant for the clay material tested.
- For each load stage the rate of accumulation of plastic deformation of the embankment clay fill is virtually constant after about 1000 cycles.
- Lateral displacements are generally small in the absence of an applied lateral load.

The laboratory element and physical model testing are reported in detail in Taylor et al 2015a and Taylor et al 2015b (both in preparation).

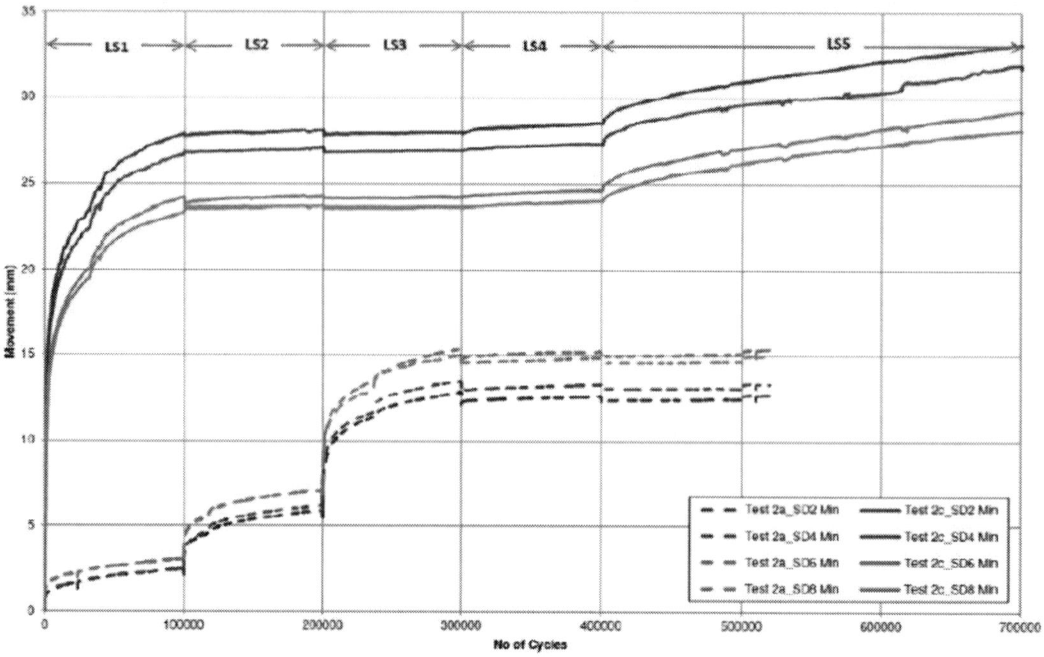


**Figure 4: Comparison of Tests 2a and 2c – Cumulative Surface Displacement – all load stages (SD2, 4, 6 & 8)**

## 4 ANALYTICAL MODEL DEVELOPMENT

The passage of an individual train over a railway embankment will induce both elastic and plastic deformations within the earthwork. Although the embankment deformation is largely elastic and fully recoverable, irrecoverable plastic strains will accumulate gradually over a large number of cycles; the magnitude of permanent deformation increases with train axle load. This phenomenon was observed by Selig & Sluz (1978) and Yoo & Selig (1979) following a comprehensive full scale testing programme.

The accumulation of permanent plastic deformations for fine grained soils under repeated loading has been summarised by Li & Selig (1996). While loading under any given cycle may be largely recoverable, after a large number of cycles the permanent plastic component of the deformation can become significant. One of the most commonly used models is empirical and based on a power law relationship; Li and Selig (1996) proposed the following predictive model for determining the cumulative plastic strain in fine grained soils given a number of applied stress cycles:

$$\varepsilon_p = AN^b \text{............}(1)$$

where

$\varepsilon_p$ is the plastic strain in the soil (%),
N is the number of stress cycles applied, and
A and b are constants to be determined.

Parameter A can be expressed in the form given below:

$$A = a(\frac{\sigma_d}{\sigma_s})^m \text{............}(2)$$

where

$\sigma_d$ is the applied deviatoric stress (kPa),

$\sigma_s$ is the static soil strength (which is related to the 'threshold' strength, $\sigma_t$, of the soil, in kPa), and

a and m are constants to be determined.

Values of the parameters – a, b and m – for this power law relation were derived from the laboratory test data which are specifically relevant for medium to high plasticity NR railway embankment clay fills.

As part of the research programme practical analytical models at both strategic and tactical levels are to be developed. The models should be capable of producing output varying from strategic level prioritisation data, through to site-specific tactical level calculations and be developed in the context of technical rigour and practicality for asset management. At this stage in the research programme, only a general strategic level model can be defined, based on the plasticity of the embankment fill and anticipated train axle loading (Tan et al, 2011).

Analytical model development included the preparation of three-dimensional numerical models which simulate the train loading, the track and the underlying response of the ballast/fill layers. Firstly, the physical model test embankment set-up was established as a numerical model and this then calibrated against the results from the testing. This model was then extrapolated to full-size and parametric studies undertaken. Numerical analysis was also used to determine scale factors for use in back analysis of the physical model tests. The scale factors reflect the non-linear stress-strain behaviour exhibited by clod and matrix embankment clay fills and are different to the scaling factors commonly used for linear elastic systems.

Once scaling factors had been derived and strains representative of a 'full-size' embankment determined, the physical model test set-up was back-analysed. The resulting 'full-size physical model test embankment was divided into a number of layers and the representative change in deviatoric stress for each layer was determined for each load stage of each of the physical model tests. Knowing the respective 'full-size' plastic strains, representative change in deviatoric stress (q) and undrained shear strength ($s_u$), values of the ratio ($q/2s_u$) were determined and plotted on graphs of plastic strain versus ($q/2s_u$).

Two load scenarios were subsequently identified:

(a) Reload - the embankment clay fill has previously been subject to the level of deviatoric stress being imposed, and

(b) Virgin - the embankment clay fill is yet to experience such a level of deviatoric stress.

For each load scenario two assessment lines were defined – median and upper bound. It was concluded that in most cases the reload load scenario is most representative of historic embankments and that it should be considered in conjunction with the Median Assessment Line in impact assessment. The Upper Bound Assessment Line will usually provide a worst case estimate of the deformation of the embankment, particularly where there is a history of embankment/track defects.

The plot of plastic strain versus the change in deviatoric stress normalised by undrained shear strength ($q/2s_u$) for the back-analysed physical model tests (Reload Scenario) for N=10,000 cycles, is presented in Figure 5. For this plot the

($q/2s_u$) ratio is limited to a maximum of 0.5 which covers the range of values (typically between 0.3 and 0.5) that are likely to be obtained in impact assessment for freight trains. For passenger trains the ratio will be lower. The lines derived from the application of the 'Li and Selig' parameters for high plasticity clays, and the median and upper bound lines derived from back analysis of the physical model tests, are shown on this plot. The power law parameters a, b and m for each of these cases are summarised in Table 3.

For the reload loading scenario the plastic strains indicated by the median line range up to 1% for a ($q/2s_u$) ratio up to 0.5, depending upon the number of load applications considered; for the upper bound line plastic strains of up to almost 12% are suggested. In comparison, the plastic strains obtained for the virgin load scenario and median line range up to 3% for a ($q/2s_u$) ratio up to 0.5; for the upper bound line plastic strains of up to almost 14% are obtained. It is considered likely that for the clay fill investigated, failure will occur at plastic strains of the order of 6% and therefore beyond this point the strains indicated by the upper bound lines are not applicable.

As shown in Figure 5 the median line derived from back analysis of the physical model tests for the reload loading scenario, is in relatively good agreement with the individual data points obtained from back analysis of the physical model tests. However, the agreement is poorer for lower numbers of load applications.

The 'Li and Selig' derived average line for high plasticity clays provides a higher estimate of plastic than the median line determined from the back analysis of the physical model tests (reload loading scenario). The greatest estimate of plastic strain is obtained for the upper bound line derived from the back analysis of the physical model tests (reload loading scenario).

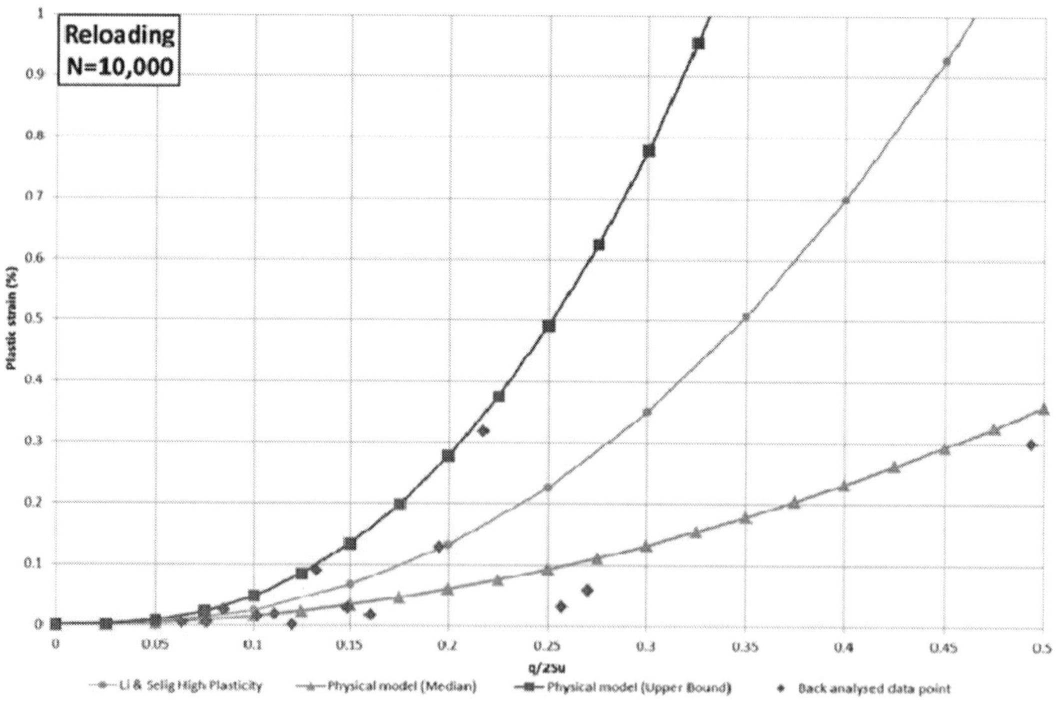


**Figure 5: Plastic Strain versus (q/2su) from Physical Model Tests, N=10,000 cycles, Reload Scenario, for (q/2su) up to 0.5**

**Table 3: Power Law Parameters for UK Medium to High Plasticity Embankment Clay Fill**

| Scenario | Power Law Parameters | | |
|---|---|---|---|
| | a | b | m |
| Reload – Median | 0.21 | 0.2 | 2.0 |
| Reload – Upper Bound | 0.38 | 0.4 | 2.5 |
| Virgin – Median | 3.4 | 0.04 | 1.3 |
| Virgin – Upper Bound | 1.0 | 0.3 | 1.7 |
| Li & Selig Average and Range for High Plasticity Clay | 1.2 (0.82-1.5) | 0.18 (0.12-0.27) | 2.4 (1.3-3.9) |

From this work a framework to assess the impact of an increase in railway traffic loading on embankments of clay core construction, taking account of considerations for both the Ultimate and Serviceability Limit States, has been developed. In common with the design of other geotechnical structures, potential failure of a railway embankment can be classified as exceedance of the Ultimate or Serviceability Limit States. Failure of the embankment or track subgrade at the Ultimate Limit State (ULS) would involve collapse resulting in loss of support to the track, whereas the Serviceability Limit State (SLS) involves excessive deformation involving exceedance of track geometry limits. Within these limit states there are a range of failure mechanisms which would have different impacts on the railway, from ULS failure of the embankment to SLS failure involving the need for the correction of track geometry defects. The infrastructure supported by the embankment (the railway) is potentially particularly sensitive to small changes in line and level caused by embankment movements that would not in other situations be significant. Serviceability limit states are likely to be particularly important as a measure of trackbed performance.

The Ultimate Limit State (ULS, or cyclic failure) check compares the predicted deviatoric stress change within the clay fill under a particular freight train axle load, against an allowable (or 'threshold') stress for the clay fill under consideration. The Serviceability Limit State (SLS, or cyclic deformation) check compares the predicted deformation which will accumulate under repeated axle loading against an allowable limit for variation of the vertical alignment of the track due to ground movement.

## 5 DISCUSSION

Initially embankment damage due to railway traffic loading becomes evident through the need for increased maintenance of the vertical track alignment. In the investigations to date, four principal contributory factors have been identified as the cause of traffic induced deformation, namely:

- Train axle load;
- Embankment clay fill plasticity;
- Trackbed configuration (i.e. ballast thickness), and
- Effectiveness of the track drainage.

Track alignment features such as horizontal curvature, and local structural features ('hard spots') are regarded as secondary contributory factors to track deformation. Embankment height, shoulder width and slope angle will have an impact on track deformation, but only where certain critical thresholds are exceeded such as a two metre embankment height, one metre shoulder width and 1 in 2 slope inclination, for example.

The potential benefits of this research once it is complete are:

(1) The development of analytical models (at both strategic and tactical levels) that can be used as a means of assessing the deformation of medium/high plasticity clay fill embankments/trackbed where a change in railway traffic loading and/or vehicle frequency is anticipated,

(2) An initial assessment framework has been developed that can be used to support the development of business cases for embankment/trackbed renewal works where excessive deformation is anticipated for increased railway traffic load, and

(3) Enable usage of the rail network to be maximised.

The outputs from the research will enable assessments to be made of the potential impact on embankments/trackbed due to changes in train axle loads and/or vehicle frequency.

It is common practice to express railway traffic loading in terms of the overall tonnage applied to the railway line, i.e. Million Gross Tonnes (MGT). The research project has demonstrated that the key variables affecting earthwork performance are the axle load and the number of load applications (N), which should be considered separately. Axle load is the primary parameter, with load frequency N having a relatively smaller impact. Hence, for a given MGT, a line subject to a larger axle load will be far more vulnerable to damage than a line exposed to a smaller axle load and a proportionately increased number of load cycles.

## 6 CONCLUSIONS

Most railway embankments in the UK were built in the Victorian era and are of end tipped construction using materials excavated from adjacent cuttings, resulting in a clod-and-matrix structure. Historically there has been a lack in understanding of the mechanical behaviour of such railway embankments when subjected to load. However, over the past ten years there has been an increase in embankment damage on the railway network in Great Britain, due to increased railway traffic loading as a result of higher axle loads and/or increased train speed, sometimes resulting in failure. Over the next decade railway traffic in the UK is forecast to grow considerably, particularly freight. Consequently, there is a need to improve our understanding of how increases in railway traffic loading may influence the mechanical behaviour of railway embankments/trackbed and thus the performance of the track.

This paper has presented a summary of the investigations carried out to date into the effects of railway traffic loading on embankment stability, particularly in relation to deformation of the embankment clay fill supporting the track. Strategic and tactical level models have been developed for assessment of clay fill embankments; these models produce different types of output, varying from strategic level prioritisation data, through to site-specific tactical level calculations. The models have been developed in the context of technical rigour and practicality for the purposes of asset management. Programmes of laboratory element and physical model testing were undertaken to support the development of the analytical models. The key findings of these laboratory testing programmes have been presented in this paper. A preliminary framework for assessment of the response of clay fill to repeated axle loading has thus been developed, incorporating both Ultimate and Serviceability Limit State Considerations, and has been summarised herein.

Although significant steps have been taken to provide an improved understanding of the complex relationship between the train loading (axle load and speed of travel/frequency) and the embankment/trackbed response, work remains to be undertaken to confirm the validity of the analytical models developed. The models and assessment procedure developed within this research programme have been passed to NR for development, implementation and verification. The following recommendations are made for further work.

- Validation and testing of the analytical models through the instrumentation and monitoring of actual railway embankment sites;
- 'Back analysis' of problem sites, where relevant monitoring data is available;
- Further numerical and analytical modelling to support validation;
- Further development of the models to take account of ballast thickness and drainage conditions (which is likely to be dependent upon weather conditions and train drainage effectiveness);
- Additional laboratory testing (both small-scale element and large scale 1-g physical model) as necessary to support further model development.

## ACKNOWLEDGEMENTS

The support of RSSB in funding this research and the support of Network Rail in providing data and other information are acknowledged.

## REFERENCES

Li, D. and Selig, E. T. (1996). *Cumulative plastic deformation for fine-grained subgrade soils*. Journal of Geotechnical Engineering, 122(12): 1006-1013.

O'Brien, A. S., Ellis, E.A. and Russell, D. (2004). *Old railway embankment clay fill - laboratory experiments, numerical modelling and field behaviour*. Proceedings of Advances in Geotechnical Engineering: The Skempton Conference, London 2004, London: Thomas Telford. Vol. 2, 911-921.

O'Brien, A. S. (2007). *Rehabilitation of Urban Railway Embankments-Investigation,* Analysis and Stabilisation. Proceedings of the 14th International Conference Soil Mechanics and Geotechnical Engineering, Madrid, 1, 125–143.

Selig, E. T. and Sluz, A. (1978). *Ballast and subgrade response to train loads*. Transport Research Record, 694, 53-60.

Yoo, T-S. and Selig, E. T (1979). *Field observations of ballast and subgrade deformations in track*. Transport Research Record, 733, 6-12.

Tan, R, Taylor, G. R., O'Brien, A. S., Lane, J. S. and Evans, E. R. (2011). *The effects of railway traffic on embankment stability*. Proceedings of Railway Engineering 2011, London, June 2011.

Taylor, G. R., Tan, R, O'Brien, A. S., Smethurst, J., Watson, G. V. R., Lane, J. S., and Evans, E. R. (2015). *The use of laboratory testing in the characterisation of embankment clay fill from the UK rail network (IN PREPARATION)*. ECSMGE 2015, The Fifteenth International Conference on Soil Mechanics and Geotechnical Engineering, Edinburgh, Scotland, September 2015.

Taylor, G. R., Tan, R, O'Brien, A. S., Baeßler, M., Cuéllar, P., Lane, J. S., and Evans, E. R. (2015). *The use of physical model testing in the development of models for potential impact assessment on the UK rail network (IN PREPARATION)*. ECSMGE 2015, The Fifteenth International Conference on Soil Mechanics and Geotechnical Engineering, Edinburgh, Scotland, September 2015.

# Formalised finite element analysis and design of cast manganese crossings for switch and crossing layouts

**H Persson**[1][2], **J M Ritchie**[1], **M Westoby**[2], **Y Chen**[1]

[1] Institute of Mechanical, Process and Energy Engineering,
School of Engineering & Physical Sciences, Heriot-Watt University, UK

[2] Progress Rail Services UK Limited, UK

## ABSTRACT

By developing and implementing a framework for the design development of cast crossings, the design process and finite element analysis (FEA) can become standardised and constructive, leading to effective and rigorous analyses of crossings. It will serve to create benchmarks and enable true and thorough analyses to be performed on proposed design changes prior to any in-track trials being undertaken. The framework will determine and validate the FEA model, utilising in-track data and historic failures, thus giving the confidence of proposed design changes. Through industry-wide use the framework would ensure that each crossing manufacturer would be analysing their unique crossing designs to the same standard, thus ensuring the customer has the best possible product.

## 1. INTRODUCTION

A switch and crossing (S&C) layout allows a train to move from one track to another safely and efficiently in mainline and light rail systems. Crossings and other components of the S&C layout are highly safety critical components and their quality, design and maintenance is of the utmost importance to the travelling general public. The end of life failure mode for these components must be fail-safe, in all of the approximate 21,000 S&C layouts within the UK. (1)

However, when seeking guidance on the design, analysis and development of a crossing, the information available is limited. Studies have been performed to understand how such parameters, amongst others, impact on wheel/rail interaction, with the majority of the literature focusing on the wheel transfer area (WTA) (see figure 1) of the crossing, as opposed to the whole crossing.

### 1.1 Wheel transfer area

Within the WTA various aspects have been investigated, ranging from assessing support conditions, through to optimising geometry. Andersson and Dahlberg (2) investigated the vertical dynamics in an S&C layout, paying particular interest in the WTA of the crossing. The authors noted that as the transition became more irregular the contact forces increased.

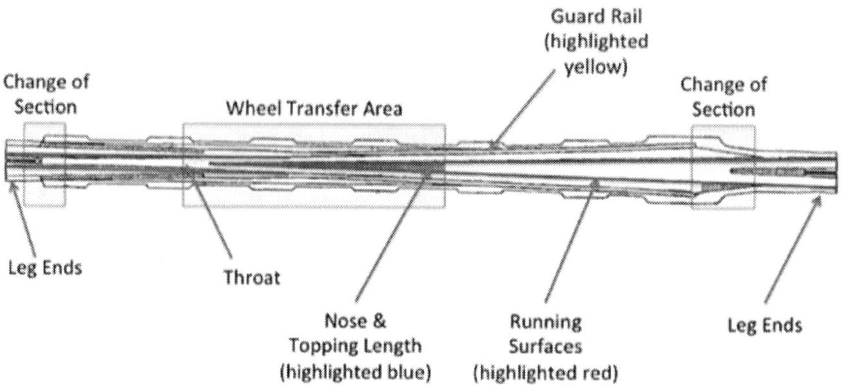


**Figure 1: A common crossing with key features highlighted**

Crossing geometry can promote good wheel/rail interaction and a study by Nicklisch et al. (3) investigated the WTA geometry to protect the crossing nose. The options were to encourage wheels to impact far from the nose tip and to promote a good wheel transfer between the wing rail and nose, thus reducing the impact forces experienced. Batter damage and shelling often found in-service would be reduced. It was also found reducing track stiffness would reduce the contact forces experienced, but could in turn shorten the fatigue life of the cast manganese crossing (CMX) due to excessive bending. This study, whilst successfully minimising contact forces and stresses, did not take into account what is happening in the body of the crossing.

Wiest et al. (4, 5) addressed wheel impact occurring on the nose, firstly with one wheel only and then with multiple wheels, using a simplified model investigating wear and plastic deformation. The study found the contact patch size decreased in size from the first impact to the tenth impact in the AMS crossing. It was also noted in a material comparison between AMS and composite steel that the composite steel experienced a maximum vertical reaction force of approximately 310kN over 5 cycles, compared to 210kN for the AMS, which decreased as the cycles progressed. The authors noted that due to the work-hardening ability of AMS, it would alter its geometry to withstand the loading unlike composite steel, thus explaining the reduction in forces.

Pletz et al. (6, 7) investigated two materials and their suitability for crossings: non-pre-hardened manganese steel and maraging steel. The austenitic manganese steel (AMS) crossing was found to deform whilst experiencing varying speeds and axle loads, whereas the study indicated that maraging steel, when applied in crossings might have a benefit in instances where high loads would induce plastic deformation. However, maraging steel is very expensive and does not lend itself to maintenance. No benefit was seen in low axle cases, therefore making the use of maraging steel over AMS for crossings unjustifiable.

Later, the effect of a 1mm crack in the crossing nose was analysed for each material (8), using FEA to understand the development of rolling contact fatigue (RCF). However, it can be argued that cracks in the nose and running surfaces would be identified in maintenance checks at appropriate timescales and monitored or immediately subjected to remedial grinding to remove the cracks. Cracks occurring internally within the crossing are more serious as their detection can on occasion be extremely difficult, if impossible, to detect until the damage is so great. At that stage, replacement would be the only viable option. In the studies true geometry from the wing rail to the nose was modelled. However, only the upper section of the crossing was analysed. The study did adopt a realistic approach by sharing the load on both the wing rail and the nose as the wheel transitioned from one to the other.

Wiest et al. (9) considered one wheel on a crossing nose comparing the Hertzian contact theory (10); J.J. Kalker's theory in the program, CONTACT (11) and two FEA models: one elastic only and elastic-plastic, to predict the contact patch. The contact area and the maximum contact pressure produced extremely similar values and contact patch shapes in all, except for the latter which experienced an increased contact area of 83% and a decrease of 42% in maximum contact pressure.

However, models require validation. Kassa and Nielsen (12) recorded track data obtaining both vertical and lateral wheel/rail contact forces and real-life geometries to input the data into their model. The results from the layout were then used as a means of validating their predicted results.

Table 1 summarises the various FEA analyses performed on the WTA.

**Table 1: Summary of FEA and areas of interest in the WTA**

| Author | Year | Package | Area of Interest |
|---|---|---|---|
| Wiest et al. (5) | 2005 | ABAQUS/ Explicit | Nose – Dynamic Analyses |
| Wiest et al. (4) | 2008 | ABAQUS/ Standard | Nose – Stress, Damage and Material Comparison |
| Pletz et al. (6) | 2012 | ABAQUS/ Standard | Nose and Wing Rail – Dynamic Analyses and Material Comparison |
| Pletz et al. (7) | 2012 | ABAQUS/ Explicit | Nose and Wing Rail – Dynamic Analyses |
| Wiest et al. (9) | 2008 | ABAQUS/ Standard | Nose – Contact Patch |
| Pletz et al. (8) | 2014 | ABAQUS/ Explicit | Nose and Wing Rail – Crack Behaviour & Propagation/Material Comparison |

### 1.2 Crossing remainder

Guo et al. (13) analysed cross-sectional areas of a CMX out with the WTA, using MSC Marc, focussing on the contact patches during wheel/rail interaction. When the wheel was on plain rail the contact patch shape was oval and large, with the minor axis positioned in the running direction. Whereas, once the wheel was on the crossing, the contact patch was a narrow and long oval shape. This time, the major axis was located in the running direction. It was seen that whilst the contact patches exceeded the yield stress, the remainder of the crossing did not.

### 1.3 Whole crossing

There are only two known studies, conducted by Xiao et al. (14) and Guo et al. (15), which consider a CMX in its entirety. However, the authors focussed on a localised region within the WTA as opposed to the overall behaviour, with the majority of the loads applied to the nose of the CMX.

Xiao et al. assessed the contact patch by applying loads at four points on the crossing nose using the FEA package ANSYS. However, only the nose and contact patch were of true interest. The contact patch and its associated stress were analysed with the maximum stress being experienced at the point of contact.

Guo et al. assessed contact patch stresses at seven locations throughout the CMX, five of which were in the WTA. In the investigation the authors used an extremely high yield stress for the entire CMX model, 1000MPa. This is due to the work hardening of AMS, which creates an extremely hard and brittle layer overlying a soft tough substrate. This hard layer is only a layer and no consideration has been given to the transition from this 1000MPa yield stress to a 'normal' value. The high yield stress would only be applicable once the CMX had been in-service and undergone work hardening. It is unrealistic to assume such as high value throughout the whole CMX. As expected the greatest stresses occurred near the CMX nose tip due to its narrow nature, which gradually dissipated as the nose grew wider.

## 1.4 Component degradation

Today's rail networks around the world want track components that can withstand increased axle loads and traffic, but with that increase there is the risk of exacerbating and quickening damage formation and reduced component life. Support conditions, impact forces and the general wear of wheels all have their part to play in its degradation. The majority of crossings replaced are due to reaching an irreparable state as deemed by track maintainers, via inspections such as outlined in the inspection, maintenance and repair for cast, welded and fabricated crossings standard (16). However, it is vital to understand the failure mechanisms that exist within the rail network so as to have an awareness of limitations that may exist when prolonging the CMX lifespan.

### *1.4.1 Wheel transfer area*

Studies have been performed to establish a relationship between wheel/rail interaction and failure mechanism propagation, including wear and plastic deformation and RCF.

When a crossing is installed, the nose starts to suffer from plastic deformation when the first wheel lands on it. This behaviour was investigated by Wiest et al. (4, 5) and each wheel had elastic-only properties, implying no permanent damage or deformation occurred. This suggests rolling stock wheels are perfectly maintained and wheel flats do not exist, when it is known they do and create higher impact forces. However, adopting this approach reduced the solution computation time and as a result the method will not yield the worst-case scenario.

The study by Pletz et al. (8) considered small cracks within the surface of a crossing nose, over three materials, AMS, Hardox and a maraging steel, which could further develop into RCF. Whilst such small cracks can be removed using grinding, it is worthwhile to understand how such cracks can behave, and if left unattended, deteriorate. As expected with AMS being the softest material and able to plastically deforming to cope with the induced loads, it produced the lowest stress in the crack tip alongside a reduced crack growth angle.

Studies have been conducted to introduce control measures so as to reduce or eliminate the risk of failure propagation before an S&C component is manufactured. The study by Nicklisch et al. (3) evaluated the design of an S&C layout to optimise the WTA geometry and reduce the forces experienced. The authors only considered S&C optimisation with one rolling stock design and in the UK, optimisation of S&C may prove difficult as much of the rail network is mixed-mode, consisting of high-speed and freight traffic. It would be difficult to achieve the optimum in design and performance when it has to carry both. Therefore a compromise design would be required and would not reduce the damage as well if fully optimised. In this study only the nose and wing rails were subjected to optimisation and no consideration was given to the impact of the running surface changes on the body of the crossing.

### *1.4.2 Whole crossing*

Another fault that can occur, either during manufacture or in service, is the development of cracks. Hazra and Mandal (17) sought to reduce the cracking frequency of their in-service crossings. Numerous design changes were implemented immediately, whilst altering the chemical composition. No time was given to monitor each alteration and assess its merit. If significant benefits had been evident in FEA modelling, then multiple design changes could have been introduced with confidence. The authors concluded that the cracking frequency dropped, but it is not known how long after the improvements had been implemented that this conclusion was drawn.

In an ideal situation, the previously discussed failure mechanisms would not occur. However, they do and it is the responsibility of the S&C Design Engineer and Track Maintainer to ensure components have a prolonged life and are eventually fail-safe and such control measures can be introduced.

Grinding, and the risk of overgrinding, can be avoided through the implementation of explosive depth hardening (EDH) on crossings, a technique discussed by Westoby (18) and Rommelaere and Maujean (19). They highlighted how by utilising EDH on CMX the maintenance regime can be reduced (table 2). Both studies presented comparisons of EDH and non-EDH CMX, with the EDH CMX experiencing less material loss over the lifespans.

**Table 2: Comparison of maintenance and repair regime of EDH and non-EDH crossings based on Rommelaere and Maujean's (19) discussions**

| Timescale tonnage (MGT) | Non-EDH crossing | EDH Crossing |
|---|---|---|
| 60 | 1st Maintenance Repair | - |
| 90 | Further Inspection | - |
| 110 | 2nd Maintenance Repair | 1st Maintenance Repair |
| 150 | 3rd Maintenance Repair | - |
| 180 | Replacement | 2nd Maintenance Repair |
| 250 | - | Replacement |

An INNOTRACK document (20) presented key parameters in S&C layouts highlighting the consequences of potential failure modes ranging from reduced line speeds (which incur heavy penalties for the track maintainer giving rolling stock companies compensation) through to the potentially fatal derailments.

### 1.5 Conclusions

Through conducting a literature review gaps have been identified within the existing knowledge, predominantly in the FEA of crossings: analysis of a full crossing as opposed to a section. The level of the validation of FEA results presented in the literature is relatively low, especially given the easy nature of inadvertently producing false results with FEA. Table 3 considers the analysis techniques employed by each study highlighting the lack of validation on desktop studies. Validation techniques have varied between using historical data to measuring track components and wheel profiles. However, there are no clear examples of comparing a whole crossing using bespoke experiments, FEA modelling and in-track data.

The Rail Technical Strategy (21) has identified the need for S&C components to undergo "...in-depth theoretical modelling" and to also have reduced costs and the ability to sustain increased maintenance regimes. They must be 'future-proof' in preparation for increased rolling stock.

Tables 4 and 5 show the investigated parameters pertaining to CMX and rolling stock, emphasising the majority of the investigative studies centre on the WTA with interest in specific factors such as material and supports conditions. No thought has been given to the effect of the rest of the crossing, which is a key gap in the knowledge which must be rectified.

As such the literature and in-sector studies have pointed to the need for some form of overall crossing design and analysis framework to guide and support future CMX development and, possibly, drive industry standards. This should incorporate modern numerical methods within its operation for more accurate and experimentally validated guidance.

**Table 3: Analysis techniques comparison**

| Author | FEA | MBS | Experimental/ In-Track Data |
|---|---|---|---|
| Andersson & Dahlberg (2) | | | Y |
| Nicklisch et al. (3) | | Y | Y |
| Wiest et al. (4) | Y | | |
| Wiest et al. (5) | Y | | |
| Pletz et al. (6) | Y | | |
| Pletz et al. (7) | Y | | |
| Pletz et al. (8) | Y | | |
| Wiest et al. (9) | Y | Y | |
| Kassa & Nielsen (12) | | Y | Y |
| Guo et al. (13) | Y | | |
| Xiao et al. (14) | Y | | |
| Guo et al. (15) | Y | | Y |
| Hazra & Mandal (17) | | Y | |
| Westoby (18) | | Y | |
| Rommelaere & Maujean (19) | | Y | |

**Table 4: Summary of CMX features investigated**

| Authors | Wheel Transfer Area | Crossing Remainder | Whole Crossing |
|---|---|---|---|
| Andersson & Dahlberg (2) | Y | | |
| Nicklisch et al. (3) | Y | | |
| Wiest et al. (4) | Y | | |
| Wiest et al. (5) | Y | | |
| Pletz et al. (6) | Y | | |
| Pletz et al. (7) | Y | | |
| Pletz et al. (8) | Y | | |
| Wiest et al. (9) | Y | | |
| Kassa & Nielsen (12) | Y | | |
| Guo et al. (13) | | Y | |
| Xiao et al. (14) | | | Y |
| Guo et al. (15) | | | Y |
| Hazra & Mandal (17) | | | Y |

**Table 5: Rolling stock wheel/rail parameters investigated for wheel/rail interaction**

| Authors | Speed | Axle Load | Direction (Facing/ Trailing) | Direction (Through/ Diverging) |
|---|---|---|---|---|
| Andersson & Dahlberg (2) | Y | | | |
| Pletz et al. (6) | Y | Y | Y | |
| Pletz et al. (7) | Y | | Y | |
| Kassa & Nielsen (12) | | | | Y |
| Xiao et al. (14) | Y | Y | | |
| Guo et al. (15) | Y | | | |

## 2. FRAMEWORK DEVELOPMENT

It has been ascertained that there is currently no means of robustly analysing a full crossing in its entirety or a structured methodology for doing so. This has been verified through the extensive literature review and through experience gained within the S&C industry. Today's customers are demanding the use of modern analysis techniques and digital tools such as FEA, to improve and validate these safety critical components prior to supply.

A CMX can be analysed to a degree using traditional bending beam calculations but it neglects the finer features. It is impossible to understand the loading and resultant stress behaviour of both external and internal features. Up to now this has been acceptable; with CMX failures below the neutral axis seemingly ignored. So much focus has rightly been given to the study of wheel/rail interaction in CMX and the WTA. If contact forces are reduced, then so too are bending stresses. We have reached a stage where it is vital to delve deeper and look beneath the CMX running surface. If the industry is to achieve the production and installation of components that have extended intervals between maintenance periods, and ultimately lead to fewer personnel on track as envisaged by the Rail Technical Strategy (21), then the focus must be broadened in a rigorous and conscientious manner. Through the development and implementation of an industry CMX analysis framework this can be achieved.

The intention is to develop a framework for applying FEA, with supporting and validating empirical measures, to the design of cast crossings. Design features will be analysed whilst considering the effects of external factors, such as the support conditions and wheel/rail interaction. It is envisaged this will naturally develop into an analysis process for the overall S&C layout.

The framework may employ the use of traditional bending beam theory; FEA; an experimental programme and the use of trackside measurement data from strain gauges and historical data. The proposed or current design would serve as a benchmark for future design alterations and as such any benefits would be evident.

By employing the use of an experimental programme and trackside measurement data, the FEA results will be more readily validated, thus giving greater confidence in taking a proposed design change from paper to reality. This would allow the approval time for such design alterations to be shortened and increase the service life of each cast crossing.

The framework will be based on the principle that three fundamental primary components (figure 2) related to crossing design and performance need to be taken into consideration: rolling stock, track, and support conditions, which when working together create three key inter-relationships (figure 3): wheel/rail interaction, track and support condition interaction, which in turn cause eventual component degradation. Through this basis, the framework will look to bind together existing research regarding wheel/rail interaction and component degradation, whilst opening up the examination of the overall crossing behaviour within such a system, rather than focussing on localised geometry.

Through the implementation of a framework, S&C manufacturers can confidently develop and manufacture crossings that are initially robust as per international and local standards and then further develop them to be able to withstand poor maintenance regimes, both in track and in rolling stock. There would still be a strong focus on wheel/rail interaction but consideration has to be given to the rest of the crossing as its failure is not as easily remedied and maintained as the running surface profile. By considering the system as a whole the failures of S&C will begin to decrease. If track conditions are to deteriorate, new improved designs may only last as long as their predecessor but perform under greater duress. However, it must be known 'what lies beneath' the crossing, and the overall layout, and it is vital there is a true understanding of the support conditions.

**Figure 2: Framework primary components**

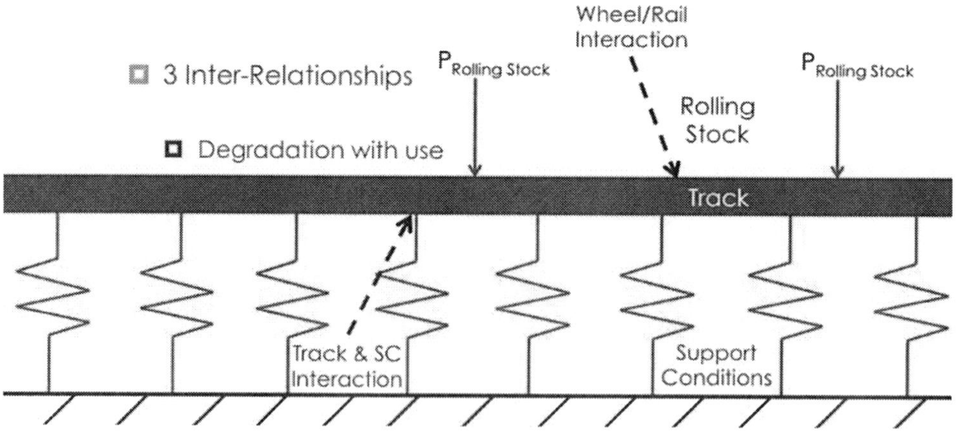


**Figure 3: Diagram demonstrating the three primary components and the resulting inter-relationships**

## 2.1 Influencing factors

There are numerous factors to consider when analysing a crossing. The worst-case scenario would be developed initially with the appropriate refinement to develop a realistic case. Just some of the factors to be considered for each component and the corresponding inter-relationships are highlighted in tables 6 and 7, highlighting common factors and demonstrating the complex relationships present in an S&C

layout. Some factors are currently addressed in the literature, whilst others have been raised by industry. The framework will seek to address some of the gaps and factors raised by industry.

**Table 6: Primary component factors to be considered when analysing a crossing in the framework**

| Rolling Stock | Track (Crossing) | | Support Conditions (S/C) | |
|---|---|---|---|---|
| Axle Speed | Layout Type | Dip Angles | Ballast | Layout Geometry |
| Axle Weight | Crossing Type | Traffic Frequency | Overall Track Stiffness | Track Curvature |
| Wheel Conditions | Geometry | Layout Location | Loss of Support | Subgrade |
| Wheel Flats | Joint Types | Ease of Maintenance | Voiding | S/C Maintenance |
| Traffic Frequency | Component Conditions | Ease of Installation | Fixation Methods | Component Replacement |
| Number of Wheels | Fatigue Life | Ease of Inspection | S/C Restoration | Repairability |
| Wheel Material | Support Conditions | Ease of Assembly | S/C Components | Local Landscape |
| Wheel Profiles | Topping | Ease of Manufacture | Tamping | Ballast Stiffness |
| Maximum Bogie Weights (particularly Freight) | Trailing/ Facing | Track Alignment | Pads (Materials/ Thickness/ Stiffness) | |
| | Material | Tolerances | Ballast Reinforcement Polymer | |
| Vehicle Speed | Product Quality | Failure Mechanisms | Bearer Types (Materials/Lengths) i.e. shrouds, jointed, | |
| Suspension Type | Weldability | Repairability | | |
| Unsprung Mass | Through-route/Diverging | | | |
| | EDH | CMX/Rail Profiles | | |
| | Track Tension | Metallurgy | **Key**<br>Framework Factor = Literature Review<br>Framework Factor = Industrial (CMX Expert/Meetings/On-track Experience) | |
| | Local Landscape (Inclined/Flat) | | | |
| | Fail-safe | | | |
| | Clips/Bolts present & in position | | | |

**Table 7: Key inter-relationships factors to be considered when analysing a crossing in the framework**

| Wheel/Rail Interaction | | Track and Support Condition Interaction | Component Degradation with Use | |
|---|---|---|---|---|
| Loss of Support | Axle Weight | Loss of Support | Inspection Frequency | Tamping |
| Voiding | Ballast Stiffness | Voiding | Material | Failure Modes |
| Track Stiffness | Voiding | Maintenance | Track Conditions | Fail-Safe |
| Component Material | Tamping | | Component Condition | Repair Methods |
| Component Condition | Topping | | Severity | Repairability |
| Dip Angles | Wheel Profiles | | Maintenance | Fatigue Life |
| Trailing/ Facing | Wheel Flats | | Loss of Support | Grinding |
| Grinding | Axle Speed | | Voiding | Track Re-alignment |
| Fixation | Wheel Condition | | Topping Restoration | |
| Gauging (Track Alignment) | | | Consequences | |
| Pads (Materials/ Thickness/Stiffness) | | | Knowledge of Materials/ Specifications/Care | |
| Bearers (Materials/ Lengths) | | | | |
| Mixed-mode/Dedicated Railway | | | | |
| Contact Patch Size & Shape | | | | |
| Vertical & Lateral Forces | | **Key**<br>Framework Factor = Literature Review<br>Framework Factor = Industrial<br>(CMX Expert/Meetings/On-track Experience) | | |
| Crossing Running Profiles | | | | |
| Through-route/ Diverging | | | | |

## 3. CONCLUSIONS

The framework will help to improve and develop improved designs whilst utilising real-life and experimental data to back up computational predictions. There is an urgent requirement to have this tool in place since crossing degradation and life expectancies are significant costs to the rail industry, addressing a need identified by the Rail Technical Strategy.

"New designs of switches and crossings reduce failures to negligible levels and reduce the costs and disruptions associated with maintenance interventions" (21)

By identifying where problems are within a crossing, the design can be altered so as to remove stress concentrations and increase the crossing life. This would satisfy rail infrastructure owners but also give manufacturers comfort that their product should satisfy the market demand and track conditions. New crossing products and variants should also be able to be developed and analysed confidently.

As noted with the current development of the framework there are numerous factors to consider in the analysis of just one crossing, but by developing this work such important analysis can adopt a structured format in assessing the true impact of factors such as lateral forces, curved geometry, etc. As identified in the literature and the Rail Technical Strategy (21) support conditions need to be fully understood and this can be further explored and supported by the use of existing theory, bespoke experiments and in-track trials.

The framework will look to highlight cast crossing design issues in a series of worst case scenarios, whilst allowing the framework user to develop the optimum design and meet industry requirements. It will also serve as a tool to introduce and assess the suitability of design changes. Using the mix of techniques and validation to assess the original crossing design, design changes will be assessed in a timely manner, helping to reduce the overall product qualification time and associated costs. Whilst the initial focus will be on the development of a cast crossing design framework, it may develop into a design framework considering the whole S&C layout.

## 4. REFERENCES


(1) Coleman, I. & Cornish, A. Predicting the future for switches & crossings; European Railway Review, Issue 5, 2010

(2) Andersson, C. & Dahlberg, T. Wheel/rail impacts at a railway turnout crossing; Proceedings of the Institution of Mechanical Engineers, Part F, Vol. 212, 1998, pages 123-134

(3) Nicklisch, D., Kassa, E., Nielsen, J., Ekh, M. & Iwnicki, S. Geometry and Stiffness optimization for switches and crossings, and simulation of material degradation; Proceedings of the Institution of Mechanical Engineers, Part F: Journal of Rail and Rapid Transit, Vol. 224, 2010, pages 279-292

(4) Wiest, M., Daves, W., Fischer, F.D. & Ossberger, H. Deformation and damage of a crossing nose due to wheel passages, Wear 265, 2008, pages 1431-1438

(5) Wiest, M., Daves, W., Fischer, F.D. & Ossberger, H. Plastification and damage in wheel-rail rolling contact - case study on a crossing, Proceedings of Applied Mathematical Mechanics 5, 2005, pages 67-70

(6) Pletz, M., Daves, W. & Ossberger, H. A wheel passing a crossing nose: dynamic analysis under high axle loads using finite element modelling, Proceedings of the Institution of Mechanical Engineers, Part F: Journal of Rail and Rapid Transit Journal Vol. 226, 2012, pages 603-611

(7) Pletz, M., Daves, W. & Ossberger, H. A wheel set/crossing model regarding impact, sliding and deformation – explicit finite element approach, Wear 294-295, 2012, pages 446-456
(8) Pletz, M., Daves, W., Yao, W. and Ossberger, H. Rolling contact fatigue of three crossing nose materials – Multiscale FE approach, Wear 314, 2014, pages 69-77
(9) Wiest, M., Kassa, E., Daves, W., Nielsen, J.C.O. & Ossberger, H. Assessment of methods for calculating contact pressure in wheel-rail/switch contact, Wear 265, 2008, pages 1439-1445
(10) Iwnicki, S. Handbook of Railway Dynamics, CRC Press, 2006, Chapter 4
(11) Kalker, J.J., Computational contact mechanics of the wheel-rail system, Rail Quality & Maintenance for Modern Railway Operation, 1993, pages 151-164
(12) Kassa, E. & Nielsen, J.C.O., Dynamic interaction between train and railway turnout: full-scale field test and validation of simulation models; Vehicle Systems Dynamics: International of Vehicle Mechanics and Mobility, Vol. 46, Supplement, 2008
(13) Guo, X., Zhang, J., Sun, C. & Zhang, J. Study on irregularity under railway vehicle passing through a frog, WASE Conference – Information Engineering, 2010, pages 303-306
(14) Xiao, J., Zhang, F. & Qian, L. Numerical simulation of stress and deformation in a railway crossing; Engineering Failure Analysis 18, 2011, pages 2296-2304
(15) Guo, S.L., Sun, D.Y., Zhang, F.C., Feng, X.Y. and Qian, L.H. Damage of a Hadfield steel crossing due to wheel rolling impact passages, Wear 305, 2013, pages 267-273
(16) Network Rail Standard: NR/L2/TRK/1054 – Inspection, maintenance and repair procedures for cast, welded and fabricated crossings in the track, Issue 4, June 2012
(17) Hazra, M.K. & Mandal, A.K., Investigations into cracking problem in austenitic manganese steel crossings, Indian Foundry Journal, 1996
(18) Westoby, M. Explosive depth hardening (EDH) of cast manganese crossings or is it destructive testing of cast manganese crossings? Institute of Rail Welding, 2004, pages 3-4
(19) Rommelaere, M. & Maujean, J.-P. Pre-hardened crossings can cut life-cycle costs, Railway Gazette, Vol. 168, Issue No. 2, 2013
(20) INNOTRACK. List of key parameters for switch and crossing monitoring, Project No. TIP5-CT-2006-031415, Deliverable Report D3.3.1, 2008
(21) Technical Strategy Leadership Group. Railway Technical Strategy: The Future Railway, 2012

# A frequency-domain finite element approach for modelling railway induced vibration in soils

**N Hamdan, O Laghrouche, P K Woodward, D Connolly**
Institute for Infrastructure & Environment, Heriot-Watt University, UK

**ABSTRACT**

A finite element model combining the paraxial and consistent transmitting boundary conditions is used in this paper to model ground wave propagation under plane strain conditions in the frequency domain. These boundary conditions allow waves to transmit laterally and vertically so they simulate the semi-infinite extent of the medium. The model incorporates unstructured triangular mesh grids so that complicated geometries can be studied. The efficiency of this modelling criterion is first assessed against given analytical solutions for layered soil over bedrock or homogenous half-space media. The model is used then to simulate ground borne vibration from a conventional railway soil embankment and a highly compacted geo-materials embankment with vertical retaining walls. Last, examples of ground vibration reduction are presented.

**Keywords**: Finite element, consistent transmitting boundary, paraxial boundary condition, ground borne vibration, embankment

## 1. INTRODUCTION

There has been recently a high demand to construct high speed railway lines. Vibration level is one of the challenges encountered with such lines. This is due to the high speed of the trains as well as the heavier applied loads. This has become a real problem due to the fact that it is sometimes difficult to change the route of the line where it may pass near buildings such as hospitals with sensitive equipment or any other sensitive machines. As a consequence, the transmitted vibrations may affect the functionality of these equipment. Also, the level of nuisance to populated areas in the vicinity of the line is a major concern. Thus, it is crucial to understand the dynamic behaviour of a proposed site for a train passage. If the site has the potential to amplify the induced vibration levels, then some measures should be taken into account.

In fact, the issue of ground borne vibration induced by a moving load or oscillatory machines has been extensively examined over the past four decades. Various approaches have been used, for example, Woods [1] and Haupt [2] used, respectively, on site full scale tests and laboratory model tests to examine the efficiency of using empty trenches and various types of wave barriers in mitigation of ground vibrations. Segol *et al.* [3] investigated the efficiency of open and bentonite-slurry-filled trenches on vibration reduction using the Finite Element Method (FEM) with special non-reflecting conditions applied at the boundaries of the domain. Beskos [4] examined the efficiency of open trenches and in-filled trenches using the boundary element method. Ahmad & Al-Hussaini [5] used the boundary element method and developed empirical design expressions for vibration isolation using both empty and in-filled trenches. A combined finite element-infinite element model was developed by Yang *et al.* [6] to simulate ground borne vibrations from underground tunnels. Laghrouche & Le Houédec [7] also investigated the use of empty trenches on the reduction of ground surface vibration. Adam & Schmid [8] employed the boundary element method to model a practical example where a railway embankment is supported by a vertical wall from one side due to a lack of space. More recently, Alzawi [9] used full scale field tests to investigate the efficiency of using empty trenches and in-filled trenches with GeoFoam on the reduction of dynamic loads induced ground surface vibration. On the other hand, numerical studies using several techniques have been developed. A three dimensional finite element model was presented and used by Connolly *et al.* [10] to investigate the effect of embankment materials on the ground borne vibrations at various distances from the track. Kouroussis *et al.* [11] presented a three dimensional numerical model where they employed Lysmer and Kuhlemeyer absorbing boundary conditions and infinite elements to simulate the semi-infinite extent of the domain. They used their model to study ground borne vibrations induced by the passage of a tram and its effect on a nearby structure. The efficiency of various shapes of wave barriers has been investigated by Hamdan *et al.* [12] where a finite element model coupled to the thin layer method is used. Moreover, the authors also investigated the use of inclusions as ground surface reduction measures. Both rigid and soft mats were used in their analyses.

This paper presents a frequency domain finite element model under plane strain conditions capable of dealing with ground borne vibration problems. A brief background about the model is first presented. Then, the numerical model is validated against known analytical results of the natural frequencies of a soil layer over rigid base. It is also validated against an approximate solution in order to evaluate the fundamental frequency of a multi-layered soil profile over rigid bedrock. Then, a test example of wave propagation in homogenous half-space is considered. The numerical model is employed to deal with ground borne vibrations induced by stationary harmonic loads on railway embankments. The use of Highly Compacted Geo-materials with Retaining Walls (HCG-RW) is compared with the use of conventional railway embankment. An example of vibration reduction from surface harmonic load induced wave propagation on half-space media is presented and compared with published results. Concluding remarks based on the numerical results are presented in the last section.

## 2. THE FINITE ELEMENT MODEL

The FEM has proved efficient in modelling complex geometry domains. However, when dealing with dynamic loading special transmitting boundary conditions should be imposed at the boundaries of the model in order to allow wave propagation through these boundaries. In this work, the FEM is used to model a finite size domain and the consistent transmitting boundary conditions, developed by Waas [13], are used to simulate the semi-infinite extent of the domain in the lateral directions. In such a way, the domain is assumed to be resting over rigid bedrock. The domain of interest is modelled by dividing the semi-infinite domain into three regions, a central region and two lateral domains. The FEM is used in the central region as it is possible to include discontinuities and consider complex geometries within this region as it is shown in Figure 1. The two lateral domains are modelled to simulate the semi-infinite extent of the medium in the horizontal direction. An automatic mesh generator is incorporated in the numerical model. It is used to mesh the central region into unstructured 3-node triangular elements with two degrees of freedom per node, the vertical and horizontal displacements.

Unstructured mesh grids are more practical as they allow the inclusion of complex geometries in the domain and modelling practical engineering problems where structured mesh grids could not be easily obtained. This meshing criterion has been also employed within the context of the Thin Layer Method (TLM) by Hamdan *et al.* [12] and Hamdan [14].

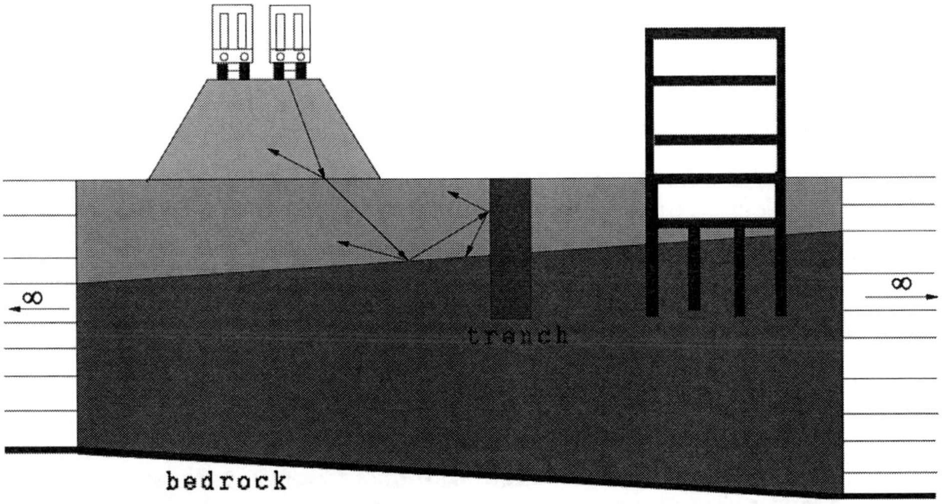


**Figure 1: Schematic diagram of a soil-structure interaction problem.**

The consistent transmitting boundary conditions are very powerful for simulating the semi-infinite extent of the finite element model in the horizontal direction. They are based on discretising the domain in the direction of layering and applying analytical solution in the horizontal direction. These regions are divided into sub-layers. The displacement is assumed to vary linearly with respect to depth within each sub-layer and exponentially with the wave number $k$ in the horizontal direction. Therefore, the thickness of each sub-layer should not exceed one-tenth of the Rayleigh wavelength of the sub-layer in order to satisfy the linear variation. Using the governing differential equation of motion for each individual layer and considering the free vibration of the entire system with ensuring compatibility of stresses and displacements at the layer interfaces and applying the boundary conditions, free boundary at the surface, a global second order eigenvalue problem is obtained by Waas [13]. It is of the form

$$[k^2\mathbf{A} + \mathrm{i}k\mathbf{B} + \mathbf{G} - \omega^2\mathbf{M}]\mathbf{U} = 0 \tag{1}$$

In expression (1), $k$ refers to the wave number, **U** is the displacement of the lateral region and **A, B, G, M**, are obtained from the geometry and material properties of the horizontal sub-layers. They are of dimensions (2n+1) x (2n+1), where n is the number of the sub-layers. They are reduced to 2n x 2n in the case of a rigid base. As both the rigid base case in addition to the half-space media are considered in this work, the exact dynamic stiffness matrix of Kausel & Roësset [15] is used to simulate the effect of half-space. It is first expanded using Taylor series, retaining only the first three terms in order to obtain a second order approximation of the impedance matrix in the horizontal wave number. In such a manner, waves are assumed to impinge onto the half-space at nearly normal incidence and therefore their direction of propagation is close to the vertical axis. Hence, this type of boundary condition is called paraxial. More details about the coupling the Paraxial Boundary Condition (PBC) and the consistent transmitting boundary conditions can be found, for example, in Hamdan [14]. Employing this technique allows modelling of the half-space by considering the contribution of the half-space elements and the corner nodes of the domain.

Solving the eigenvalue problem (1), after coupling the PBC, leads to finding (2n+1) eigenvalues and eigenvectors for each lateral boundary. These eigenvalues and eigenvectors are combined to form the displacement field in the lateral regions. Hence, the dynamic stiffness matrices of the coupled boundary conditions of the lateral forces applied at these boundaries, **R** and **L**, can be evaluated. It is worth noting, that Hamdan *et al.* [16] proposed a technique to select a few eigenvalues and their corresponding eigenvector in the analysis, instead of using all modes. This has led to significant reduction in the computational cost of the analysis. The system equation to be solved has the form

$$[(1+2i\xi)\mathbf{K}-\omega^2\mathbf{M}+\mathbf{R}+\mathbf{L}]\mathbf{U}=\mathbf{P} \tag{2}$$

In expression (2), ξ refers to the damping coefficient and **P** is the external forces vector. A direct solver is used to solve the linear system (2) where the results are obtained as vertical and horizontal displacements.

## 3. VALIDATION OF THE NUMERICAL MODEL

The numerical model is validated in the next sections against known analytical and approximate solutions of some problems. First, the natural frequencies of a soil layer over a rigid base are computed using available theoretical solutions and using the current model. Then, the fundamental frequency of a multi-layered soil profile is computed numerically and compared with approximate solutions. The case of wave propagation in homogenous half-space is validated in the last example.

### 3.1 Homogenous soil layer over bedrock

The dynamic behaviour of a soil layer over rigid bedrock is a very important aspect to consider when analysing dynamic soil-structure interaction problems such as machine foundations and passing trains over a soil stratum over bedrock. This is due to the fact that the response of the soil layer will be amplified if the frequency of the excitation load approaches the natural frequencies of the soil layer and most importantly the fundamental frequency. This leads to excessive ground borne vibrations, and as a consequence causing nuisance to nearby buildings, in addition to uncomfortable ride. Therefore, it is important that designers avoid these frequencies by altering the applied frequency or by changing the properties of the soil profile.

Next, an example of a soil layer over rigid bedrock subjected to a vertical harmonic load is presented. A 1D wave propagation theory has shown analytically that a soil layer over a rigid base has an infinite number of natural frequencies. These frequencies are obtained from the amplification factors of the vertical and horizontal directions. The theoretical solution presented in the textbook of Kramer [17] is used to compute the amplification factor of the vertical response. The amplification factor is given in this case as

$$F^v(\omega)=\frac{1}{\cos(\omega H/c_p)} \tag{3}$$

The natural frequencies of the vertical response are derived from the previous expression and expressed as

$$f_n^v=(2n-1)\frac{c_p}{4H} \tag{4}$$

with $c_p$ being the pressure wave velocity of the soil layer and $H$ is its thickness. Let us consider a soil layer of 10m in depth overlaying a rigid base subjected to a vertical harmonic load. The soil layer has a density of 1700kg/m$^3$, elasticity modulus of 100MPa and 0.3 Poisson's ratio. The undamped case is considered here. The amplification factor of the soil layer is computed and presented in Figure 2. As it is shown in the figure the amplification occurs at certain frequencies, natural frequencies, around 7Hz and 21Hz, for the first and second respectively. Next, the numerical model is used to simulate the dynamic response of the soil layer where the length of the model is taken as 30m. The model is meshed into finite elements with size not exceeding one-tenth of the associated Rayleigh wavelength. The displacements at the surface of the domain are computed for each applied frequency and the area, *A*, underneath each vertical displacement curve is computed using a numerical integration scheme [14]. The area is normalised by a unit area $A_0$. The results are depicted in Figure 3. As it is shown in the figure, the numerical model predicted the first and second natural frequencies of the soil model to be around 7.05Hz and 21.1Hz, respectively. Good agreement with the analytical solution is obtained employing the adopted numerical model.

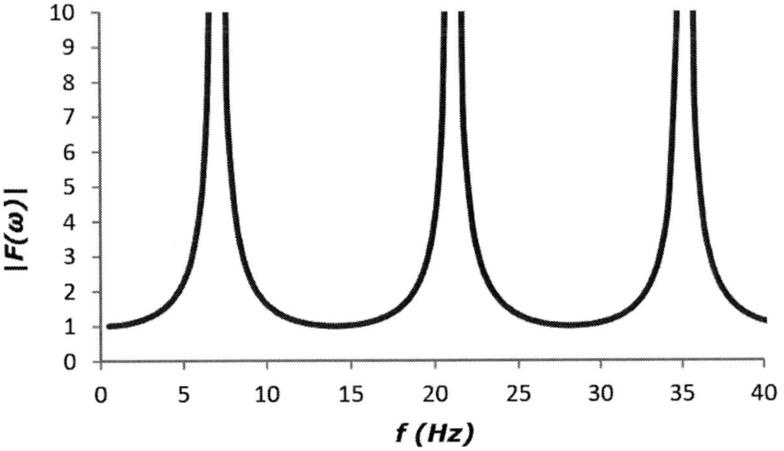


**Figure 2: Theoretical amplification factor for vertical excitation of undamped soil layer over rigid bedrock.**

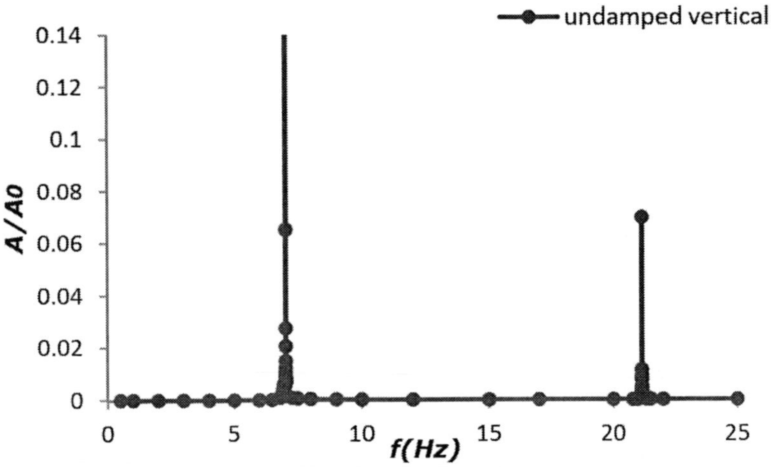


**Figure 3: Numerical amplification for vertical displacement of undamped soil layer over rigid bedrock.**

## 3.2 Multi-layered soil profile

It is difficult to guarantee the presence of a soil layer over a rigid base all times. Therefore, it is more practical to consider a multi-layered soil profile and simulate its dynamic behaviour. Hadjian [18] developed approximate solutions to estimate the fundamental period of a two-layer soil profile. The equivalent fundamental frequencies of Hadjian expressions are given as

$$f = \frac{1}{\left[1+\beta\left(\frac{f_1}{f_2}\right)^n\left(1+\frac{\rho_1 H_1}{\rho_2 H_2}\right)^n\right]^{1/n}} \quad \text{for } \frac{H_1}{H_2} \leq 1 \tag{5}$$

$$f = \frac{1}{\sqrt{\frac{\pi^2}{8}\left[0.75+\left(\frac{f_1}{f_2}\right)^2\left(1+2\frac{\rho_1 H_1}{\rho_2 H_2}\right)\right]}} \quad \text{for } \frac{H_1}{H_2} > 1 \tag{6}$$

With $\beta = 1 - 0.2\left(\frac{\rho_1 H_1}{\rho_2 H_2}\right)^2$ and $n = 4 - 1.8\frac{\rho_1 H_1}{\rho_2 H_2}$

In the previous expressions, $H_1$ and $H_2$ denote, respectively, the thickness of the upper and bottom soil layers, $\rho_1$ and $\rho_2$ refer to the density of the soil layers and $f_1$ and $f_2$ are their fundamental frequencies. The previous approximate solutions were used by Hamdan *et al.* [12] to validate the model for a two-layer soil profile. Here, a multi-layered soil profile is considered.

A soil domain of 30m in length and 10m depth is used in this example. For simplicity, all layers are assumed to have density of 1550kg/m$^3$, Poisson's ratio of 0.3 and a damping ratio of 2.5%; as it is shown in Table 1. The Successive-Use of Two-Layer solution is used here to compute the fundamental frequency of the multi-layered medium. First, the fundamental frequency of the first two layers is computed, then the two layers are replaced by one equivalent layer and the fundamental frequency of the equivalent layer and the underlain layer is computed. The process is repeated until reaching the last soil layer. This approach is not very accurate as it uses successive approximations, however, numerical solutions can tackle these problems more efficiently.

**Table 1: Soil properties used in the multi-layered soil profile.**

| Layer | Density[kg/m$^3$] | E[MPa] | υ[-] | β[%] |
|---|---|---|---|---|
| 1 | 1550 | 26 | 0.3 | 2.5 |
| 2 | 1550 | 39 | 0.3 | 2.5 |
| 3 | 1550 | 52 | 0.3 | 2.5 |
| 4 | 1550 | 65 | 0.3 | 2.5 |
| 5 | 1550 | 78 | 0.3 | 2.5 |
| 6 | 1550 | 91 | 0.3 | 2.5 |
| 7 | 1550 | 104 | 0.3 | 2.5 |
| 8 | 1550 | 117 | 0.3 | 2.5 |
| 9 | 1550 | 130 | 0.3 | 2.5 |
| 10 | 1550 | 143 | 0.3 | 2.5 |

The domain is meshed into finite elements with size less than one-tenth of the lowest Rayleigh wavelength. The vertical displacement curves are obtained and an integration scheme is used to compute the area under the vertical displacement curves. The fundamental frequency of the layered soil profile is estimated from expressions (5) and (6) to be around 7.9Hz while a fundamental frequency of 7.3Hz was computed using the presented numerical model. The two results are close enough, despite the very approximate nature of the analytical method, and hence the two approaches are in good agreement.

### 3.3 Homogenous half-space

The coupled paraxial-consistent transmitting boundary condition model is used in the current section to simulate wave propagation in a homogenous half-space soil media. It is validated against the analytical solution of coupling the Green's function of a homogenous layer and a homogenous half-space. In this example, a homogenous half-space domain with density of $1700kg/m^3$, elasticity modulus of 100MPa, Poisson's ratio of 0.3 and damping ratio of 5% is used. A vertical harmonic load of 1N is applied at the surface of the domain with a frequency of 25Hz. The depth of the domain is taken as $3\lambda_R$ and its length is $10\lambda_R$, with $\lambda_R$ being around 5.58m. It is worth noting that the depth of the half-space model is recommended to be at least $1\lambda_P$, Hamdan [14] and Hamdan *et al.* [19]. The vertical displacement at the surface of the half-space is plotted in Figure 4. Good agreement is obtained by the coupled finite element model and hence the current model is capable of dealing with wave propagation in half-space soil media.

## 4. APPLICATIONS

The numerical model is employed in the following analysis to deal with some problems of practical interest related to ground vibration reduction. First, ground borne vibrations induced by stationary harmonic loads operating on railway embankment overlaying rigid bedrock is considered. Then, a surface ground borne vibration reduction of harmonic load on homogenous half-space is presented followed by an example of train induced vibration and isolation of a nearby structure.

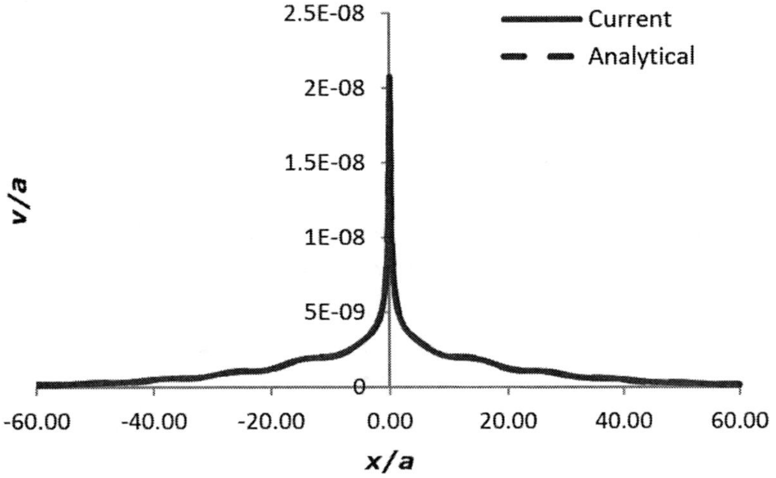


**Figure 4: Surface vertical displacement of homogenous half-space subjected to surface harmonic load.**

### 4.1 Trains induced vibrations on railway embankment structures

Two examples are presented in this section to show the ability of the numerical model in simulating ground borne vibrations from railway embankments. In the first example, a conventional railway embankment is used where four stationary harmonic loads of magnitude 83kN each are applied at the surface of the embankment as illustrated in Figure 5. The embankment is assumed to be overlaying soft soil underlain by rigid base. The railway embankment is replaced by HCG-RW.

Both the embankment and HCG-RW structure have a height of 5m. Table 2 shows material properties used in this example. The embankment has 7m width at the top and is inclined by an angle of 35° each side leading to a base width of 21m. Using the HCG-RW leads to 14m saving in land as it is depicted in Figure 5. The examined domain has a thickness of $4\lambda_R$ and a distance of $0.5\lambda_R$ is taken from the left side edge of the embankment while $3\lambda_R$ distance is considered beyond the right side edge of the embankment. The domain is modelled by adopting unstructured mesh grids with three nodes per elements.

**Table 2: Material properties used in the embankment and HCG-RW structures.**

| | Density[kg/m$^3$] | μ[MPa] | υ[-] | β[%] |
|---|---|---|---|---|
| Embankment | 1700 | 18.8 | 0.33 | 2.5 |
| Soil layer | 1500 | 7.4 | 0.35 | 2.5 |
| HCG | 2000 | 83.3 | 0.2 | 1 |
| RW | 2500 | 6521 x 10$^6$ | 0.15 | 1 |

The surface of the embankment is subjected to four harmonic loads. The frequency of the loads is varied from 0 to 26Hz and the vertical displacement curves over the $3\lambda_R$ distance are computed for the considered range of frequency. The area under the vertical displacement curves is computed as explained in the previous sections. It is difficult sometimes to construct railway embankments due, for example, to the land space restrictions in urban areas and crowded cities in addition to the high cost of land. Therefore, the embankment is replaced by highly compacted geo-materials with retaining walls. In order to show the efficiency of using HCG-RW structure on the vibration transmission, the same procedure employed for the embankment is used again. The frequency of the applied loads is varied and the vertical displacement curves are computed. The area under each curve is computed. It is then normalised by the area of the corresponding area for the embankment case.

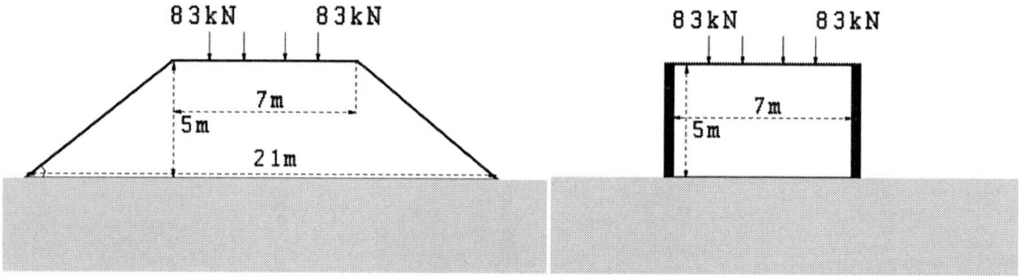


**Figure 5: Configuration of railway embankment (left) and its equivalent HCG-RW (right).**

As it is shown in Figure 6, the use of HCG-RW leads to a reduction of the soil surface displacements in most cases of loading frequency and that better reduction is achieved at the highest considered frequencies. At very low frequency, however, amplification occurs and resonance takes place at around 4Hz. It is worth indicating that a similar practical example was presented by Adam and Schmid [8]. Their example consisted of a railway embankment supported by a vertical retaining wall only from one side due to a physical space restriction and the problem was examined using the BEM where the response at the surface of the embankment was computed. The response after some distance beyond the vertical wall was analysed and they are also reported amplification due to resonance for frequencies close to the fundamental frequency of the embankment.

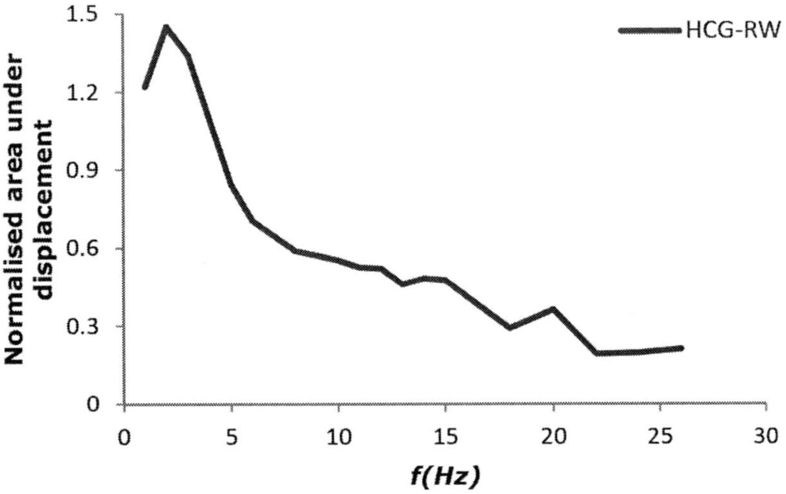


**Figure 6: Vibration reduction when using conventional railway embankment and HCG-RW.**

## 4.2 Ground borne vibration reduction

The coupled paraxial-consistent transmitting boundary condition finite element model is used in this section to deal with ground surface vibration reduction. It is not the scope of this paper to examine the efficiency of wave barriers in vibration isolation, as this has been already investigated extensively, for example by Segol *et al.* [3], Laghrouche & Le Houédec [7], and Ahmad and Al-Hussaini [5]. In this example, a half-space soil domain of $20\lambda_R$ in length and $4\lambda_R$ in depth is subjected to a surface vertical harmonic load, at the middle of the domain, and with a frequency of 31Hz. The depth of the model as $4\lambda_R$ is very close to $0.5\lambda_P$ as was recommended by Hamdan [14] in order to avoid the reflected wave from the location of the paraxial boundary. An empty trench is placed at a distance of $5\lambda_R$ from the harmonic load. The vertical displacements at the surface of the domain are computed with and without the presence of the trench. The amplitude reduction factor, $A_r$, which is the ratio between the displacements after installing the trench to the displacements without the trench is computed for an assumed distance of $5\lambda_R$ after the trench location. Figure 7 shows the amplitude reduction factor using the current model and also other published results are shown for comparison. As it is shown in the figure, the presented model predicted the vibration reduction well when comparing to other solutions.

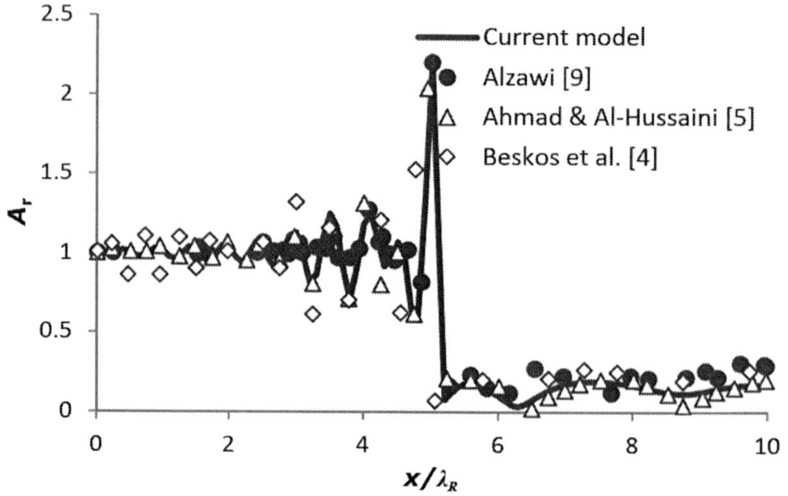


**Figure 7: Amplitude reduction factor at the surface of a half-space using an empty trench.**

Last, an example of a railway embankment subjected to harmonic loading induced ground surface vibrations transmitted to nearby structures is supplied. Figure 8 illustrates the configuration of the model and the used mesh grids. It is clear that using such unstructured mesh grids is very useful to deal with problems containing discontinuities and other man-made structures. An empty trench is installed before the structure where it is aimed to reduce the vibration level. Figure 9 depicts the contour plots of the vertical displacement without the trench while Figure 10 shows the contour plots of the vertical displacements after installing the empty trench. It is clear when comparing both figures that the vertical displacements have significantly reduced.

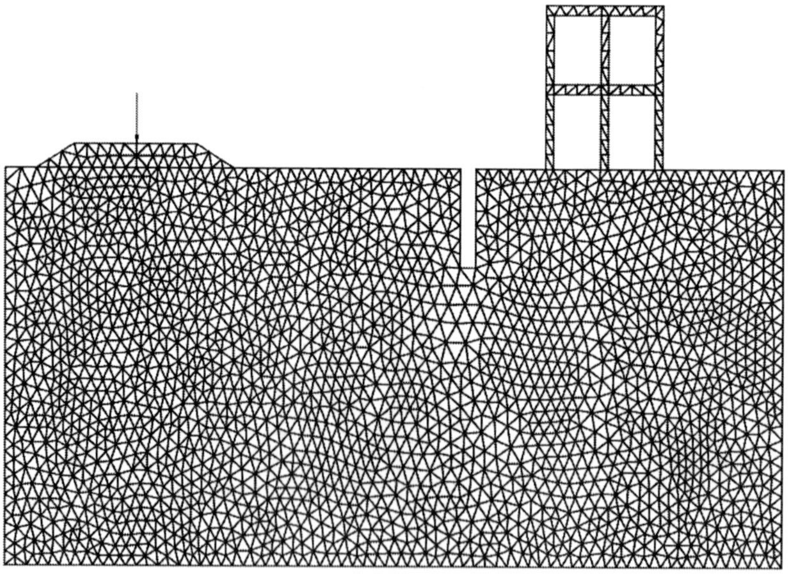

**Figure 8: Finite element representation of vibration isolation using wave barrier.**

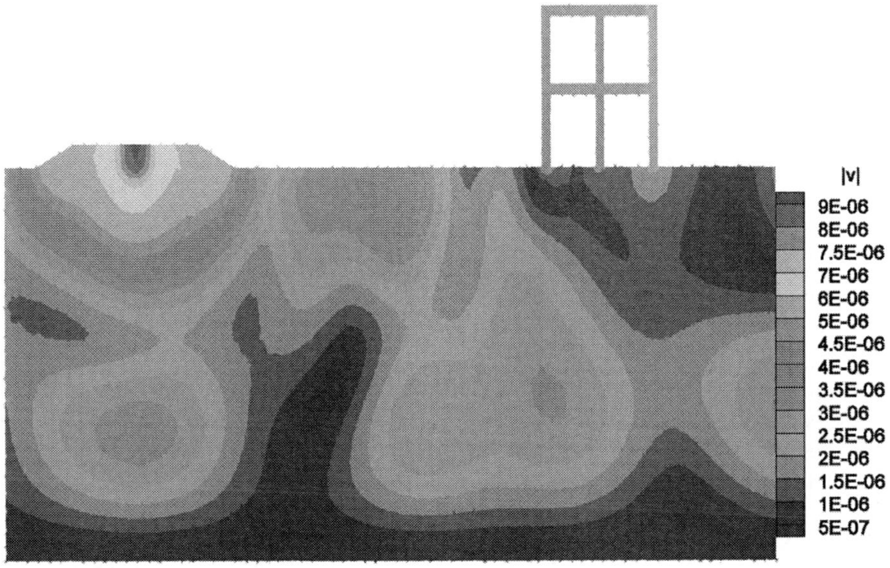


**Figure 9: Contour plots of the vertical displacement without wave barrier.**

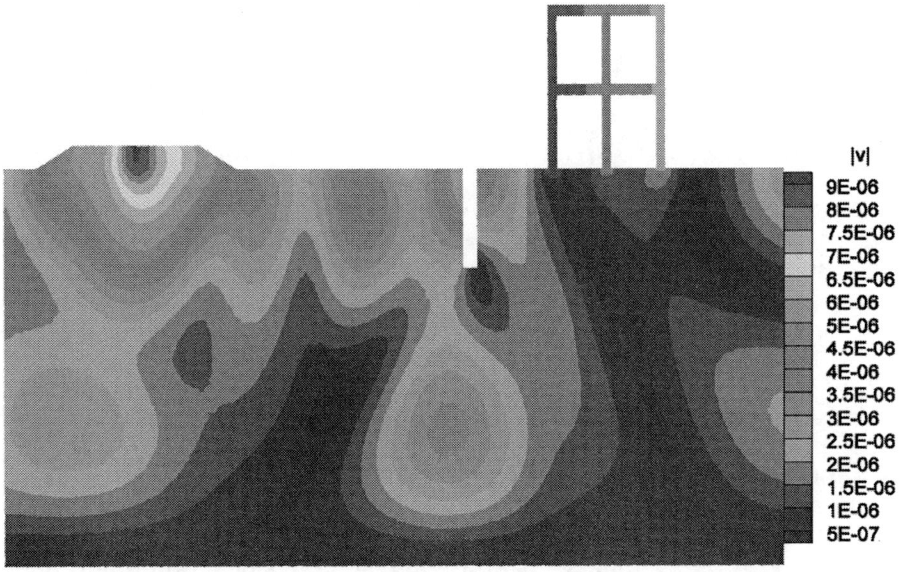


**Figure 10: Contour plots of the vertical displacement when using wave barrier.**

## 5. CONCLUSIONS

A frequency domain numerical model under plane strain conditions based on coupling the FEM with the consistent transmitting boundary and the paraxial boundary condition is employed in the current paper to simulate wave propagation in soil media. Structured and unstructured mesh grids are used in the numerical examples. The model is first validated by simulating the natural frequencies of a soil layer over rigid bedrock where the analytical solution is known. Then, it is used to estimate the fundamental frequency of a multi-layered soil profile where an approximate analytical solution is used for comparison. The coupled model, for simulating wave propagation in half-space soil media, is thereafter validated against available analytical solution. The numerical results showed that the model is capable of simulating ground borne vibration in a soil layer over a rigid base, multi-layered soil profile over bedrock as well as homogenous half-space. The use of a highly compacted geo-material embankment with retaining walls is compared to the use of a conventional railway embankment. The comparison of the induced vertical surface displacements over a certain distance showed that, overall, the former leads to a reduction of the ground surface transmitted vibrations except at very low frequencies. In fact, for frequencies below around 5Hz the displacements are amplified with resonance occurring at around 4Hz. The adopted unstructured mesh grids showed the flexibility of the model in dealing with complex geometry models of practical interest.

As it is shown in the numerical examples, it is difficult to deal with more complicated problems when using analytical solutions and, hence, using numerical solutions such as the presented model can be employed to deal with such problems. For example, it is of practical interest to investigate the effect of soil stiffening techniques with constant modulus columns, for example, to further reduce the transmitted surface vibration.

## 6. REFERENCES


[1] Woods, R. D. (1968). Screening of surface waves in soils. J Soil Mech Found Div ASCE 94(4):951–79.

[2] Haupt, W. A. (1981). Model tests on screening of surface waves. In: Proceedings of the 10th International Conference on Soil Mech. Found. Eng., Vol. 3, pp. 215–222, Stockholm, Sweden.

[3] Segol, G. Lee, C. Y. & Abel, J. F. (1978). Amplitudes reduction of surface waves by trenches. J Eng Mech Div ASCE; 104(EM3):621–41.

[4] Beskos, D. E. Dasgupta, G. & Vardoulakis, I. G. (1986). Vibration isolation using open or filled trenches, part1: 2-D homogeneous soil. Computational Mechanics, 1:43-63.

[5] Ahmad, S. & Al-Hussaini, T. M. (1991). Simplified design for vibration screening by open and in-filled trenches. Journal of Geotechnical Engineering. Vol. 117 (1), pp. 67-88.

[6] Yang, Y. B. Kuo, S. R. & Hung, H. H. (1996). Frequency-independent infinite elements for analysis semi-infinite problems. Int J Numer Methods; 39:3553–69.

[7] Laghrouche, O. & Le Houédec, D. (1994). Etude comparative de quelques moyens de réduction de vibrations dans le sol. Méc Ind Matér 1994;47(2):35–7.

[8] Adam, M. A. & Schmid, G. (2002). Investigation of ground vibrations in the vicinity of a train track embankment. In: Wave 2002, Okayama, Japan.

[9] Alzawi, A. (2011). Vibration isolation using in-filled Geofoam trench barriers. PhD thesis, The University of Western Ontario, London, Ontario, Canada.

[10] Connolly, D. Giannopoulos, A. & Forde, M. C. (2013). Numerical modelling of ground borne vibrations from high speed rail lines on embankments. Soil Dynamics and Earthquake Engineering 46: 13-19

[11] Kouroussis, G. Van Parys, L. Conti, C. & Verlinden, O. (2014). Using three-dimensional finite element analysis in time domain to model railway-induced ground vibrations. Adv Eng Softw;70:63–76.

[12] Hamdan, N. Laghrouche, O. Woodward, P. K. & Mahmood, M. S. (2014). Ground vibration reduction analysis using a frequency-domain finite element approach. Construction and Building Materials, doi.org/10.1016/j.conbuildmat.2014.04.140

[13] Waas, G. (1972). Linear two-dimensional analysis of soil dynamics prolems in semi-infinite layered media. PhD Thesis, University of California, Berkeley.

[14] Hamdan, N. (2013). Numerical modelling of wave propagation in soil media. PhD Thesis, Heriot-Watt University, UK.

[15] Kausel, E. & Roësset, J. M. (1981). Stiffness matrices for layered soils. Bulletin the Seismological Society of America 71 (6):1743-1761.

[16] Hamdan, N. Laghrouche, O. El Kacimi, A. & Woodward, P. K. (2013). Consistent transmitting boundary conditions with a reduced number of eigenmodes for wave propagation in elastic media. Computers and Geotechnics 53:9-16.

[17] Kramer, S. L. (1996). Geotechnical Earthquake Engineering. Prentice Hall Upper Saddle River, New Jersey.

[18] Hadjian, A. H. (2002). 1D Fundamental period and mode shape of layered soil profiles. Soil Dynamics and Earthquake Engineering 22:885-891.

[19] Hamdan, N. Laghrouche, O. Woodward, P. K. & El Kacimi, A. (2015). Combined paraxial-consistent boundary conditions finite element model for simulating wave propagation in elastic half-space media. Soil Dynamics and Earthquake Engineering 70:80-92.

# Application of dynamic vibration double-absorbers in low-frequency vibration control of floating slab tracks under train–track interaction

**S Zhu[1], J Yang[2], H Yan[2], C Cai[1], L Zhang[1]**
[1] State Key Laboratory of Traction Power, Southwest Jiaotong University, P.R. China
[2] China Railway Eryuan Engineering Group Co. Ltd., P.R. China

## ABSTRACT

This paper aims to effectively mitigate the low-frequency vibrations of floating-slab tracks (FSTs) due to train–track interaction using double-dynamic vibration absorbers (DDVA). First, the optimal locations where the DDVAs are attached are determined by modal analysis with a finite element (FE) model of the FST. Further, by identifying the equivalent mass of the concerned modes, the optimal stiffness and damping coefficient of each DDVA are obtained to minimize the resonant vibration amplitudes. Finally, a three-dimensional coupled dynamic model of a metro vehicle and the FST with the DDVAs is developed to investigate the effect of the DDVAs on the vibration absorption of the FST. The sensitivity of the mass ratio of DDVAs is studied with the help of the insertion loss in one-third octave frequency bands. The numerical results show that the proposed DDVAs could effectively suppress low-frequency vibrations of the FST. The insertion loss due to the attachment of DDVAs increases as the mass ratio increases.

**Keywords:** floating-slab track; vibration control; dynamic vibration absorber; vehicle–track coupled dynamics; insertion loss.

## 1. INTRODUCTION

Ground-borne vibrations due to underground railways have become important environmental issues with the rapid development of the urban rail transit. The vibrations of underground railway tracks transmit through the tunnel and the surrounding soil into nearby buildings, which cause disturbance to people. This problem can be more serious in some circumstances such as where an underground tunnel is passing below sensitive buildings, e.g. concert halls, hospitals or precise instruments. The occurring of these undesired vibrations to the nearby buildings is mainly attributed to the fact that vibration at higher frequencies is generally attenuated rapidly with distance along the transmission path through the ground, while the mitigation effect is negligible for the vibration at lower frequencies, especially for the frequencies below 20 Hz which is quite sensitive to buildings and precise instruments. One of the effective means for reducing vibrations from underground railways is the use of floating slab tracks (FSTs) [1, 2]. The FSTs basically consists of concrete slabs supported on resilient elements such as rubber bearings, glass fiber, or steel springs. They have been used to control vibration from underground trains for years [3]. The slabs can be cast in situ, leading to a continuous length of concrete, or can be constructed in discrete pre-cast sections connected by construction joints. Although a large number of literature reported its effectiveness, it remains controversial for underground railways due to the interactions with the tunnel and surrounding soil at lower frequencies[4-7]. The analysis of the isolation performance using simple mass-spring models reveals that vibrations can be isolated only for the frequency above $\sqrt{2}$ times the natural frequency of the system. Therefore, the FST is undesirable in attenuating vibrations at the sensitive frequency below 20 Hz due to the fact that the first natural frequency of the FST is generally around 13 Hz. For mitigating the vibrations of nearby buildings induced by underground railways, the traditional method is to further decrease the natural frequency of the FST by decreasing the stiffness or increasing the mass of the system. Nevertheless, the variation of the natural frequency is not significant attributed to the fact that increasing the mass of the system is limited for the consideration of the construction cost and the tunnel space. On the other hand, the rail displacement will be notably increased by decreasing the stiffness of the system. Consequently, it will influence the running behavior of subway trains in terms of the riding comfort and derailment safety. Therefore, modern railway designers face challengeable requirements for finding an effective solution to control low-frequency vibrations of underground railways.

One of the well developed approaches for mitigating vibration levels is to add a secondary linear oscillator to the existing system. This secondary oscillator, commonly called dynamic vibration absorbers, which have been proven to be an efficient and reliable way to attenuate undesirable vibration of structures at certain frequencies, either globally for rigid structures or locally for flexible structures[8, 9]. These absorbers have been widely applied in different fields of engineering such as mechanical, civil and aerospace structures due to their unique specifications such as simple maintenance, uncomplicated design, high reliability and excellent performance[10-12].

The aim of the present paper is to use the DDVAs to suppress the low-frequency vibrations of a FST with discontinuous slabs. This paper is organized as follows: Section 2 briefly introduces the design methodology of DDVAs attached to the floating slab. Section 3 shows the derivation of the three-dimensional dynamic model of a metro vehicle and the FST attached with DDVAs. Section 4 discusses the optimal locations where the DDVAs are attached, and the optimal parameters of the DDVAs to minimize resonance vibrations. Further, sensitivity analysis of the mass ratio of DDVAs is conducted in detail.

## 2. DESIGN METHODOLOGY OF DDVAS FOR FSTS

In this work, the low-vibration energy of the FST is suppressed and absorbed by attaching the DDVA to the FST, as shown in Figure 1. The optimal stiffness and damping coefficient of this DDVA as a function of the mass ratio ($\mu_a$ and $\mu_b$) can be given as follows[13]

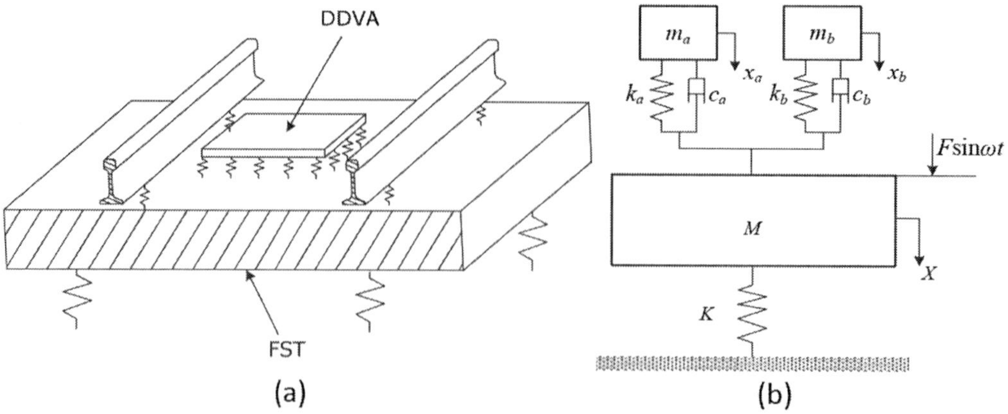


**Figure 1. Floating slab track attached with the DDVA; (a) full view, and (b) DDVA system.**

$$k_a = m_a \frac{K}{M}(0.403(\mu_a + 0.131)^{-0.437})^2 \tag{1}$$

$$c_a = 2m_a\sqrt{\frac{K}{M}}\left((\mu_a / 17.66)^{0.285} - 0.06\right) \tag{2}$$

$$k_b = m_b \frac{K}{M}(-0.72\mu_b + 1.03)^2 \tag{3}$$

$$c_b = 2m_b\sqrt{\frac{K}{M}}\left((\mu_b / 3.06)^{0.377} - 0.05\right) \tag{4}$$

where

$$\mu_a = m_a/M\ ,\quad \mu_b = m_b/M \tag{5}$$

where $k_a$ and $k_b$ are the optimal stiffness (tuning ratio) of each dynamic vibration absorber, respectively; $c_a$ and $c_b$ are the optimal damping coefficients (damping ratio) of each dynamic vibration absorber, respectively; $K$ and $M$ are the equivalent stiffness and mass of the FST. Clearly, the expression of the optimal stiffness in Equations (1) and (3) is independent of the DDVA's damping. It is worth pointing out that the maximum amplitude magnification factor will change significantly when the tuning ratio changes slightly, whereas it changes very little even when the damping ratio changes considerably[14]. Therefore, the optimal damping coefficient of DDVAs for damped primary structures can be determined by using the Equations (2) and (4) for undamped primary structures.

The above described approach to determine optimal parameters of DDVAs attached to a single degree of freedom (DOF) cannot be applied directly to the FST, due to the fact that the FST is a continuous system. Ozer and Royston[15] extended Den Hartog's method to multi-DOF structure and derived the optimal dampers and mounting locations. They further employed Sherman-Morrison matrix inversion formula to calculate the optimal parameters for a damped multi-DOF absorber system[16]. In this paper, the FST is first discretized as a multi-DOF structure. An uncoupled system composed of a collection of single DOFs is then derived based on the modal analysis and equivalent mass identification method. Consequently, the optimal parameters of DDVAs for the FST can be obtained using the design method of DDVAs for a single DOF. In this regard, the starting point is the determination of the equivalent mass of the FST at one single frequency.

According to the principle of modal analysis, a system with $N$ DOFs (shown in Figure 2) can be treated as a collection of $N$ single DOFs uncoupled each other. The kinetic energy of the system at the $i$th modal vibration can be cast as

$$T_{total} = \frac{1}{2}\omega_i^2\left(m_1x_1^2 + m_2x_2^2 + \cdots + m_jx_j^2 + \cdots + m_Nx_N^2\right) \tag{6}$$

where $\omega_i$ denotes the $i$th circle frequency; $\{x_1 \cdots x_j \cdots x_N\}^T$ represents the eigenvector of the system.

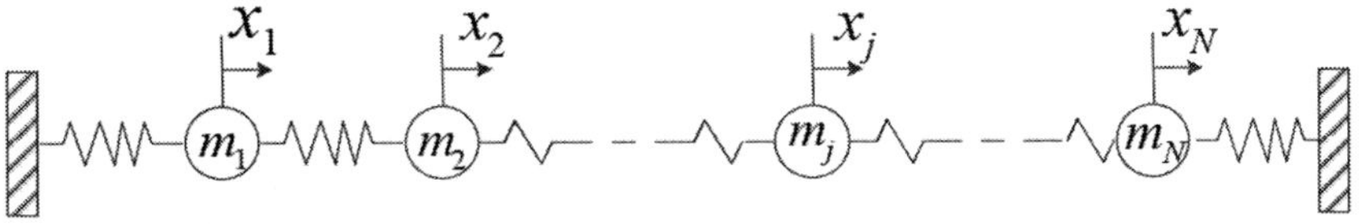


**Figure 2. System with *N* DOFs at the *i*th mode.**

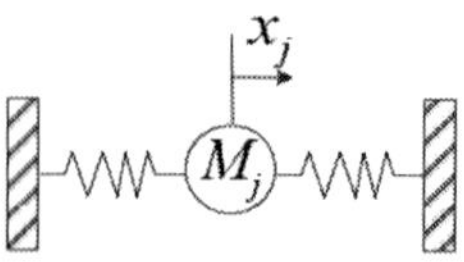


**Figure 3. A single DOF model for mass point *j* at the *i*th mode.**

Focus on the movement of the mass point *j*, which is described by a single DOF system (shown in Figure 3), its movement is identical with that of the mass point *j* in Figure 2. The kinetic energy of the single DOF system can be expressed as

$$T_j = \frac{1}{2} M_{ij} \left( \omega_i x_j \right)^2 \tag{7}$$

Setting $T_{total} = T_j$, the equivalent mass of the *N* DOFs system at *i*th mode can be determined as

$$M_{ij} = 2T_{total} / \left( \omega_i x_j \right)^2 = m_1 \left( \frac{x_1}{x_j} \right)^2 + m_2 \left( \frac{x_2}{x_j} \right)^2 + \cdots + m_j + \cdots m_N \left( \frac{x_N}{x_j} \right)^2 \tag{8}$$

Clearly, the equivalent mass for the *i*th mode is minimum when the displacement of point *j* undergoes maximum displacement during each vibration cycle. This point with maximum displacement is referred to as modal antinodes location. Whereas the equivalent mass is infinite when the displacement of point *j* is zero. This point with no displacement is referred to as modal nodes location, and it is not influenced by excitation forces. Therefore, the DDVA should be placed at the modal antinodes location of the *i*th mode, meanwhile, this location should best be the modal nodes position of other modes. This arrangement can minimize the mutual effect of DDVAs for suppressing vibrations.

Using the equivalent mass, the equivalent stiffness of the *N* DOFs system at *i*th mode can be obtained as

$$K_{ij} = M_{ij} \cdot \omega_i^2 \tag{9}$$

Making the use of Eqs. (8) and (9), the optimal parameters of DDVAs for the FST can be identified through the same procedure for DDVAs design of a single DOF system.

## 3. COUPLED DYNAMIC MODEL OF A VEHICLE AND THE FST WITH DDVAS

The vehicle-track coupled dynamics has been a fundamental method to investigate the interactions between train and track. It has wide application in different research topics such as dynamic characteristics of railway structures[17, 18], damage and fracture of track structures[19, 20], evaluation of riding safety and comfort of high-speed vehicles[21]. In this paper, a three-dimensional coupled dynamic model of a vehicle and the FST with DDVAs is established to study the effect of the DDVAs on vibration attenuations of the FST at low frequencies, as shown in Figure 4.

The vehicle model is treated as a rigid multi-body model consisting of the car body, the bogies, and the wheelsets. The car body is supported on two double-axle bogies with the primary and the secondary suspensions. The primary suspensions, connecting the wheels and the bogie frames, are represented by spring-damper elements. The secondary suspensions, connecting the bogie frames and the car body, are also modeled as spring-damper elements. Each component of the vehicle has five DOFs: the vertical displacement *Z*, the lateral displacement *Y*, the roll angle *Φ*, the yaw angle

$\psi$, and the pitch angle $\beta$. As a result, the total DOFs of the vehicle are 35. The rails are modeled as Bernoulli-Euler beams supported on a viscoelastic foundations. The vertical, lateral, and torsion motions of the rails are simultaneously taken into account in the model.

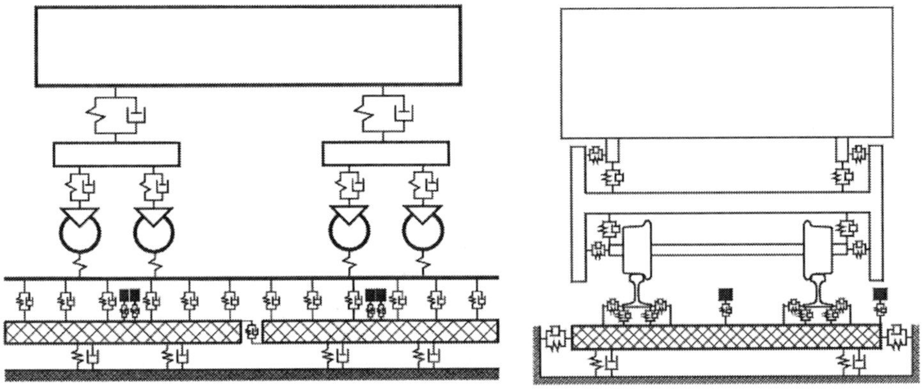

(a) (b)

**Figure 4. Three-dimensional coupled dynamic model of a vehicle and the FST with DDVAs; (a) elevation, and (b) end view.**

The motion equation of the floating slab in the global coordinate system can be cast as

$$[M]\{A_f\}+[C]\{V_f\}+[K]\{X_f\}=\{F_{rf}\}+\{F_{fs}\}+\{F_{DDVA}\} \tag{10}$$

where $[M]$, $[C]$ and $[K]$ are the mass matrix, the damping matrix, and the stiffness matrix, respectively; $\{X_f\}$, $\{V_f\}$ and $\{A_f\}$ are the vectors of displacements, velocities and accelerations of the floating slab, respectively; and $\{F_{rf}\}$, $\{F_{fs}\}$ and $\{F_{DDVA}\}$ are, respectively, the interaction forces between the rail and the floating slab, the floating slab and the subgrade, and the DDVA and the floating slab. They are given as

$$\{F_{rf}\}=\{K_{rf}\}[\{X_r\}-\{X_f\}]+\{C_{rf}\}[\{V_r\}-\{V_f\}] \tag{11}$$

$$\{F_{fs}\}=\{K_{fs}\}\{X_f\}+\{C_{fs}\}\{V_f\} \tag{12}$$

$$\{F_{DDVA}\}=\{K_{DDVA}\}\{X_{DDVA}\}+\{C_{DDVA}\}\{V_{DDVA}\} \tag{13}$$

where $\{K_{rs}\}$, $\{C_{rs}\}$, $\{K_{fs}\}$, $\{C_{fs}\}$, $\{K_{DDVA}\}$ and $\{C_{DDVA}\}$ are the matrices of the stiffness and damping coefficients of the rail fasteners, the floating slabs, and the DDVAs, respectively; and $\{X_r\}$, $\{V_r\}$, $\{X_f\}$, $\{V_f\}$, $\{X_{DDVA}\}$ and $\{V_{DDVA}\}$ are the vectors of displacements and velocities of the rails, the floating slab and the DDVAs, respectively.

Using the modal superposition principle and the normalized shape functions of the concrete slab, the vibrations of the floating slab governed by Equation (10) can be expressed as a series of uncoupled equations in terms of the generalized coordinates, as presented by Equation (14). These equations can be solved with the time-stepping integration method.

$$M_i A_{fi}+C_i V_{fi}+K_i X_{fi}=P_i \qquad (i=1\sim N_{\text{mode}}) \tag{14}$$

where $N_{mode}$ is the total number of the mode shape used in the modal superposition calculation; $X_{fi}$, $V_{fi}$ and $A_{fi}$ are the generalized vectors of displacements, velocities and accelerations of the floating slab, respectively; and $M_i$, $C_i$, $K_i$ and $P_i$ are the generalized mass, damping, stiffness, and load for the $i$th mode, respectively. They are defined as follows

$$M_i = \{\psi\}_i^T [M] \{\psi\}_i \qquad (i=1 \sim N_{mode}) \tag{15}$$

$$C_i = \{\psi\}_i^T [C] \{\psi\}_i \qquad (i=1 \sim N_{mode}) \tag{16}$$

$$K_i = \{\psi\}_i^T [K] \{\psi\}_i \qquad (i=1 \sim N_{mode}) \tag{17}$$

$$P_i = \{\psi\}_i^T [F_{rf}] + \{\psi\}_i^T [F_{fs}] + \{\psi\}_i^T [F_{DDVA}] \qquad (i=1 \sim N_{mode}) \tag{18}$$

where $\{\psi\}_i$ is the $i$th mode shape vector, which is obtained through the modal analysis using the FE model of the concrete slab (shown in Figure 5). Then, the displacement vector of the concrete slab can be obtained as

$$\{X_f\} = \sum_{i=1}^{N_{mode}} \{\psi\}_i X_{fi} = \{\Psi\}\{X_f\} \tag{19}$$

As shown in Figure 4, the vehicle and the FST are spatially coupled by the wheel-rail interface. In this paper, the trace curve method[22] is adopted to locate the wheel-rail spatial contact points which are only on a curve named trace curve. Therefore, 2D scanning can be replaced by 1D scanning through the trace curve. The nonlinear Hertzian contact theory is used to calculate the wheel-rail normal contact forces which depends on the displacements of the wheels and rails. The tangential wheel-rail creep forces are calculated first by using the Kalker's linear creep theory and then modified by the Shen-Hedrick-Elkins nonlinear model[23]. If track irregularities are known, the dynamic responses can be obtained with the help of the coupled dynamic model solved by means of a new explicit integration method[24]. Detailed descriptions of the equations of the vehicle-track coupled dynamic system can be found in Ref.[17, 18].

## 4. RESULT AND DISCUSSION

### 4.1 Optimal locations of DDVAs attached to floating slabs

To identify the optimal locations and parameters of DDVAs attached to the floating slab, a three-dimensional FE model of the FST with discontinuous slabs is created based on the software ANSYS, as shown in Figure 5. In the model, linear line elements of Beam188 are used for the rails, and hinged boundaries are adopted for the rails to prevent free body motions. Linear hexahedral elements of Solid45 are applied for the floating slab, and symmetrical constraints are applied on the two ends of the floating slab to simulate the constraints of construction joints between the slabs. Spring-damper elements of Combin14 are employed to represent rail fasteners and steel-springs. The nodes of these spring-damper elements can move in the vertical, lateral and longitudinal directions with the corresponding stiffness and damping coefficient. All the nodes at the bottom of the steel-spring are subjected to fixed constraints. The FST system is represented by 6740 elements and 5204 degrees of freedom, and all the elements in the present study are assumed to behave purely elastically.

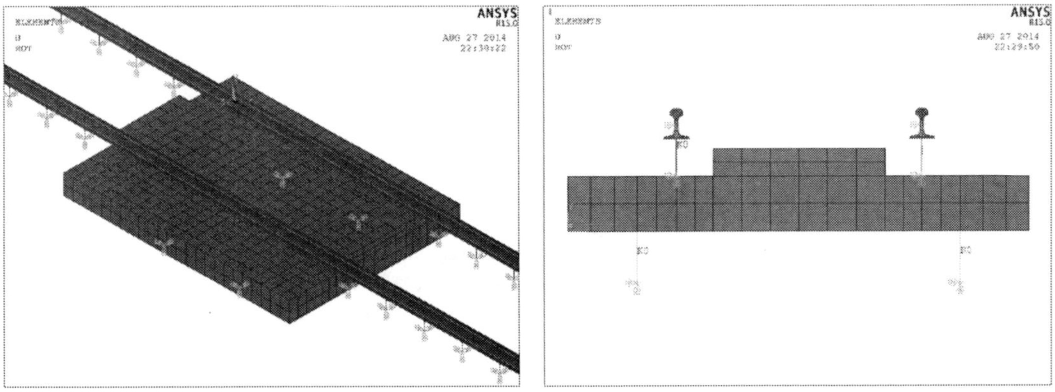


(a) (b)

**Figure 5. Three-dimensional FE model of the FST; (a) the entire FE mesh, and (b) end view of the FE model.**

Due to the fact that vibrations can be isolated by the FST system only for the frequency above $\sqrt{2}$ times the natural frequency of the system, the FST is undesirable in attenuating vibrations at low frequencies below 20 Hz which are sensitive to the nearby buildings. Therefore, the natural frequencies of the FST system below 20 Hz are considered in the determination of optimal locations and optimal parameters of the DDVAs.

Using the FE model of the FST system, it can be known from modal analysis that the first natural frequencies of the FST system is 13.09 Hz with the up-and-down translation mode shape, and the second natural frequencies is 14.91 Hz with the lateral rotation mode with respect to longitudinal center line of the floating slab, as presented in Figure 6. Obviously, every node of the floating slab is at the mode antinodes position of the first mode, while the mode nodes locations of the second mode are at the longitudinal center line of the floating slab. Therefore, the DDVA for suppressing the first mode vibration can be placed at the center of the floating slab (Point A) which is the mode antinodes position of the first mode and the mode nodes position of the second mode, as shown in Figure 7(a). And the DDVA for mitigating the second mode vibration can be attached to the lateral end of the floating slab (Point B), as shown in Figure 7(b). The mass of the DDVAs are modeled with elements Mass21, and the stiffness and damping are modeled with elements Combin14.

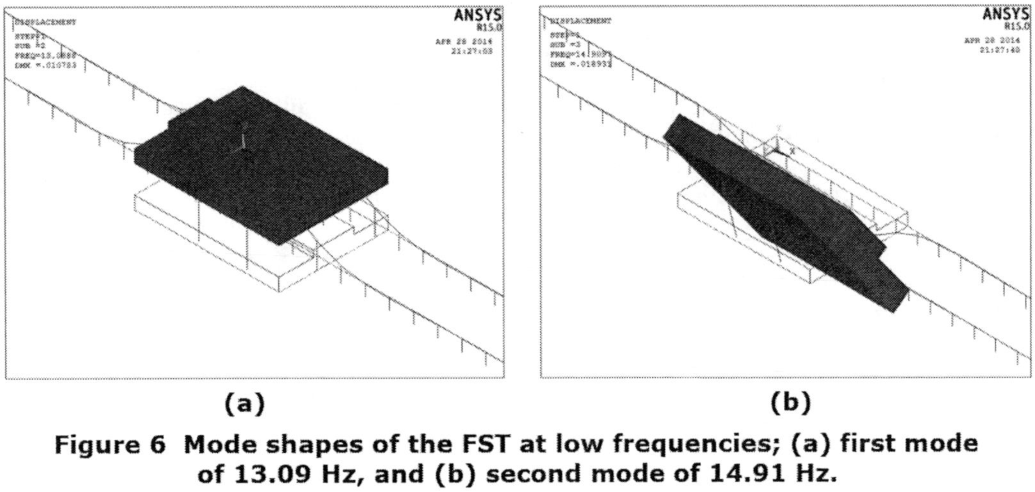


(a) (b)

**Figure 6 Mode shapes of the FST at low frequencies; (a) first mode of 13.09 Hz, and (b) second mode of 14.91 Hz.**

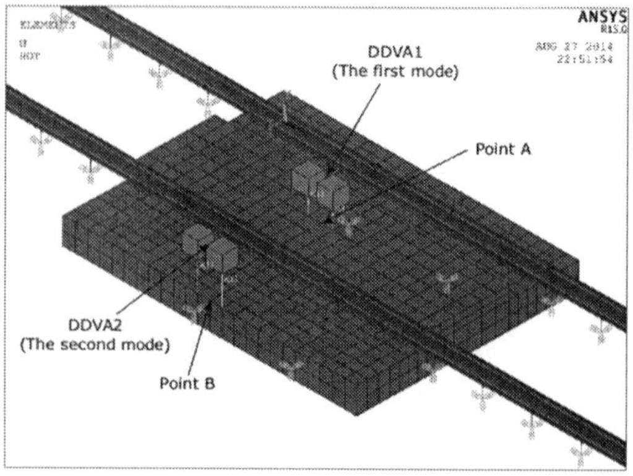


**Figure 7 Optimal locations of DDVAs attached to the floating slab.**

### 4.2 Optimal parameters of DDVAs attached to floating slabs

As can be seen in Figure 5, the FST is a damped system. In this regard, Equations (1) and (3) are first employed to obtain initial stiffness of the DDVAs, then the optimal stiffness can be identified by appropriately reducing this value according to the fact that the optimal stiffness for a damped system is smaller than that provided by Equations (1) and (3). This optimal stiffness together with the optimal DVA's damping obtained with Equations (2) and (5) will minimize the maximum vibration of the primary structure. For this purpose, using Equations (8) and (9), the optimal stiffness and damping coefficient as functions of the mass ratio are identified based on the modal analysis. Table 1 list these parameters of DDVAs for suppressing the first and second mode vibrations of the FST when the mass ratios of DDVA1 and DDVA2 are all set to be 0.1.

**Table 1 Optimal parameters of the DDVAs**

| DDVA | Mass ratio $\mu_a$ and $\mu_b$ | Optimal stiffness $k_a$ and $k_b$ (kN/mm) | Optimal damping $c_a$ and $c_b$ (kN·s/m) | Mass $M_a$ and $M_b$(kg) |
|---|---|---|---|---|
| DDVA1 (The first mode) | 0.1/0.1 | 3.81/5.98 | 26.8/35.8 | 965.1/965.1 |
| DDVA2 (The second mode) | 0.1/0.1 | 1.42/2.23 | 8.78/11.7 | 277.2/277.2 |

### 4.3 Vibration-absorbing performance of the DDVAs to the vehicle and FST systems

The track irregularity routinely appears at the wheel-rail interface and directly influences the wheel-rail interactions. According to the operation characteristics of the FST, the track irregularity generated with the fifth grade track spectra of US railways is adopted in the investigation of vibration-absorbing performance of the DDVAs to the FST and vehicle system. This track irregularity generated by means of a time–frequency transformation technique[22, 25], is characterized by wavelengths from 1 m to 30 m. The metro vehicle is assumed to move along the FST at a constant speed of 80 km/h in this paper.

Figures 8-9 show the dynamic response of the vehicle and the FST attached with DDVAs in time and frequency domains, juxtaposed with results obtained by applying the FST without DDVAs.

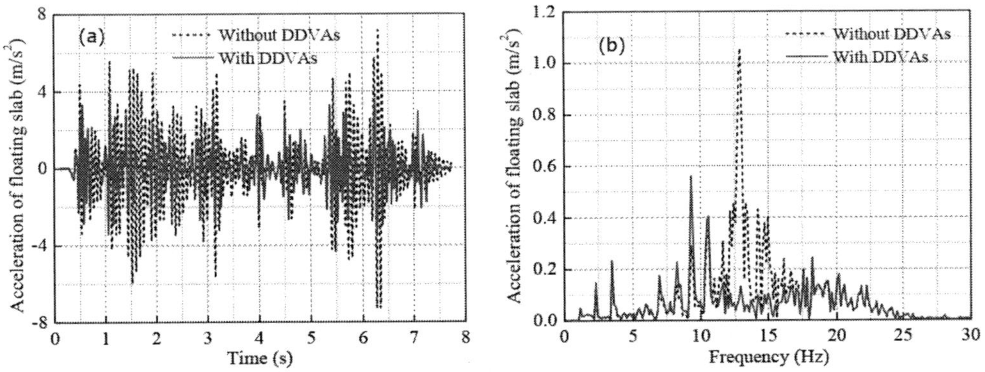


**Figure 8 Floating slab acceleration; (a) time histories, (b) frequency contents.**

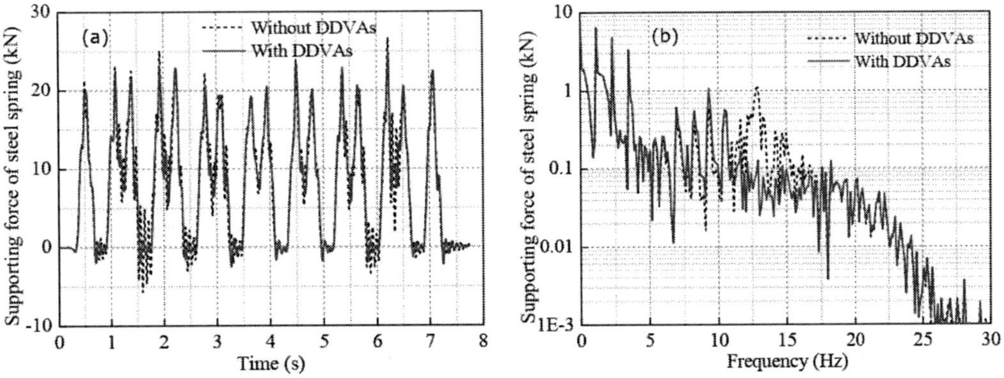


**Figure 9 Steel-spring supporting force; (a) time histories, (b) frequency contents.**

Figure 8(a) and 8(b) show the time history and frequency content of the floating slab acceleration at Point B when the vehicle passes through the FST, respectively. As it can be seen, the acceleration of the floating slab with DDVAs attenuates faster than that without DDVAs when the vehicle passes, and the acceleration amplitudes decrease because of vibration absorption of the DDVAs at low frequencies. Obviously, peak accelerations in Figure 8(b) can be distinguished at the first and second natural frequencies of around 13 and 15 Hz when the vehicle passes through the FST without DDVAs. Whereas when the DDVAs are attached on the floating slab, vibration energy of the floating slab is transferred to the attached DDVAs through the linked springs and dampers. As a result, the acceleration amplitudes of the floating slab are greatly reduced to negligible proportions and the vibrations of the DDVAs are significant. The above phenomena also can be found in the dynamic response of the steel-spring supporting force (shown in Figure 9). These results demonstrate that the DDVAs have a great effect on the vibration reduction of the FST by effectively suppressing the low-frequency vibrations of the first and second modes.

### 4.4 Effect of the mass ratio of the DDVA on its vibration-absorbing performance

It can be clearly known from Equations (1)-(4) that the optimal parameters of DDVAs are depended on the mass ratio, thus the DDVA with different mass ratios might exhibit various effects on the vibration reduction. Therefore, it is important to investigate the effect of the mass ratio of DDVA on its vibration-absorbing performance. Due to the fact that the contribution of the second mode to the FST vibration is smaller than that of the first mode, the mass ratio of DDVA1 selected here are not bigger than that of DDVA2. Moreover, this configuration can, to some extent, avoid the interference of DDVA2 on the vibration-absorbing performance of DDVA1. In this regard, ten cases with different mass ratios for DDVA1 and DDVA2 are adopted to conduct the present analysis to capture the effect of the damping ratio on the vibration-absorbing performance of DDVAs, based on the three-dimensional coupled dynamic model of the vehicle and the FST attached with DDVAs. The optimal parameters for each case are provided by using the above described method, as shown in Table 2.

**Table 2 Optimal parameters of DDVAs for different mass ratios**

| | DDVA1 (The first mode) | | | | DDVA2 (The second mode) | | | |
|---|---|---|---|---|---|---|---|---|
| Case | $\mu_a$ and $\mu_b$ | $k_a$ and $k_b$ (kN/mm) | $c_a$ and $c_b$ (kN·s/m) | $M_a$ and $M_b$ (kg) | $\mu_a$ and $\mu_b$ | $k_a$ and $k_b$ (kN/mm) | $c_a$ and $c_b$ (kN·s/m) | $M_a$ and $M_b$ (kg) |
| 1 | 0.1/0.1 | 3.81/5.98 | 26.8/35.8 | 965.1/965.1 | 0.1/0.1 | 1.42/2.23 | 8.78/11.7 | 277.2/277.2 |
| 2 | | | | | 0.05/0.05 | 0.88/1.2 | 3.32/4.2 | 138.6/138.6 |
| 3 | | | | | 0.025/0.025 | 0.5/0.62 | 1.22/1.47 | 69.3/69.3 |
| 4 | | | | | 0.01/0.01 | 0.22/0.25 | 0.31/0.34 | 27.7/27.7 |
| 5 | 0.05/0.05 | 2.36/3.22 | 10.2/12.9 | 482.5/482.5 | 0.05/0.05 | 0.88/1.2 | 3.32/4.2 | 138.6/138.6 |
| 6 | | | | | 0.025/0.025 | 0.5/0.62 | 1.22/1.47 | 69.3/69.3 |
| 7 | | | | | 0.01/0.01 | 0.22/0.25 | 0.31/0.34 | 27.7/27.7 |
| 8 | 0.025/0.025 | 1.34/1.67 | 3.74/4.49 | 241.3/241.3 | 0.025/0.025 | 0.5/0.62 | 1.22/1.47 | 69.3/69.3 |
| 9 | | | | | 0.01/0.01 | 0.22/0.25 | 0.31/0.34 | 27.7/27.7 |
| 10 | 0.01/0.01 | 0.29/0.68 | 0.93/1.04 | 96.5/96.5 | 0.01/0.01 | 0.22/0.25 | 0.31/0.34 | 27.7/27.7 |

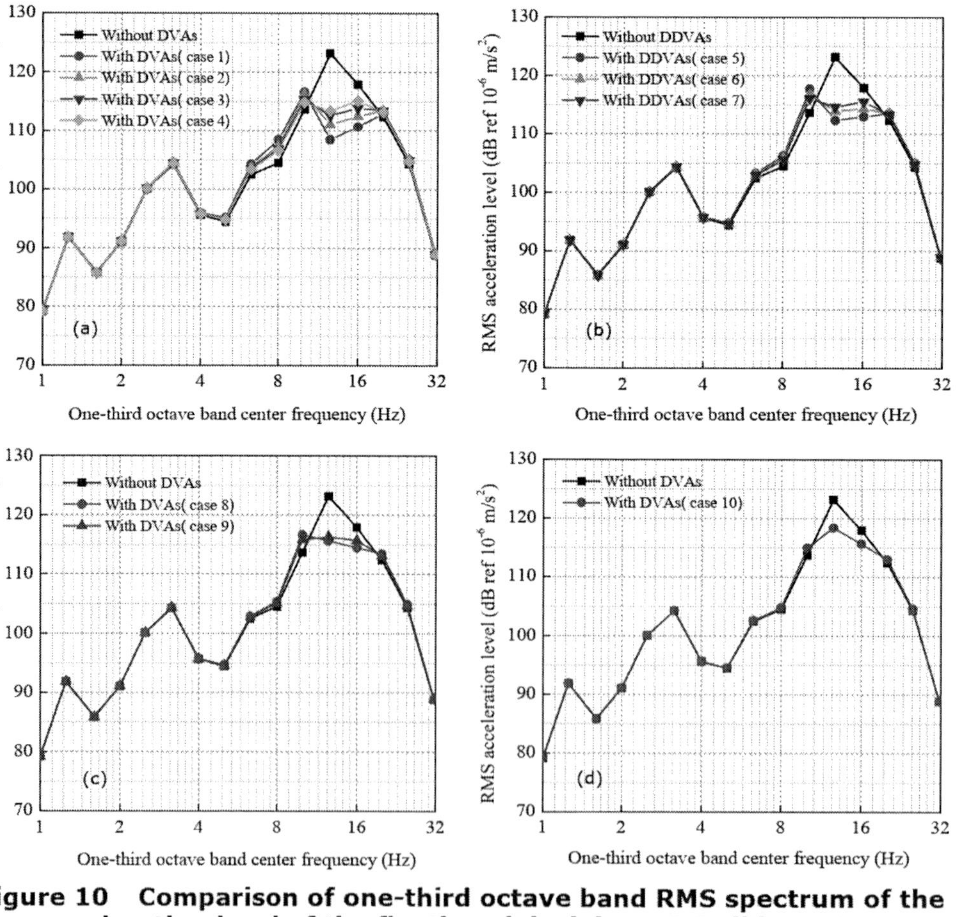


**Figure 10 Comparison of one-third octave band RMS spectrum of the acceleration level of the floating slab. (a) $\mu_1$=0.2, (b) $\mu_1$=0.1, (c) $\mu_1$=0.05, (d) $\mu_1$=0.02.**

The RMS acceleration is an important indicator to evaluate the vibration level of structures. Figure 10 displays the comparison of the one-third octave band RMS spectrum of the acceleration level of the floating slab at Point B for the ten cases, together with the result without DDVAs.

Obviously, as the mass ratios of the DDVAs increase, a considerable reduction of the RMS acceleration level can be observed at the frequencies of 10-20 Hz. The maximum insertion loss for the floating slab given in Figure 11 is found to be 15 dB in case 1 which is approximately 3 times that in case 10. Therefore, it can be argued that the DDVA with a larger mass ratio can control the low-frequency vibration of the FST more effectively, which is beneficial for the running comfort and stability of metro trains as well as structural safety. Whereas the construction cost of DDVAs and space limitation of the FST should be considered in selection of the mass ratios of DDVAs.

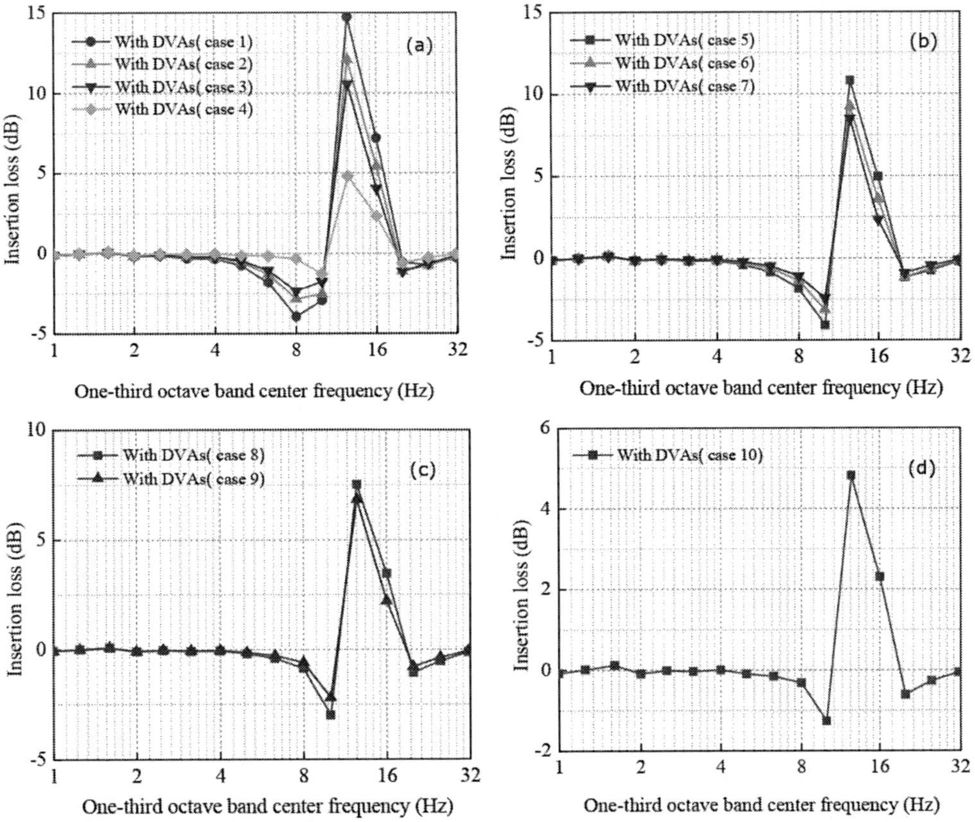


**Figure 11 Comparison of the insertion loss due to the DDVAs with different mass ratios. (a) $\mu_1$=0.2, (b) $\mu_1$=0.1, (c) $\mu_1$=0.05, (d) $\mu_1$=0.02.**

## 5. CONCLUSIONS

In this paper, a three-dimensional coupled dynamic model of a metro vehicle and the FST attached with DDVAs has been developed to investigate the effect of the DDVAs on the low-frequency vibration reduction of the FST. For this purpose, the optimal locations where the DDVAs are attached have been first determined by modal analysis with the FE model of the FST. Further, by identifying the equivalent mass of the concerned modes, the optimal stiffness and damping coefficient of each DDVA have been obtained to minimize the resonant vibration amplitudes. Finally, the performance of the DDVAs attached to the FST subjected to the vehicle dynamic loads has been evaluated considering a variation of the mass ratio of DDVAs. The numerical results can be summarized as follows.

(1) The DDVAs have a significant effect on the low-frequency vibration reduction of the FST by effectively suppressing the vibrations of the first and second mode. The application of DDVAs are quite useful in mitigating low-frequency vibrations of the FST which is sensitive to the nearby buildings.
(2) The peak values of the floating slab acceleration and steel-spring supporting force at the first and second natural frequencies of around 13 and 15 Hz are substantially decreased to negligible proportions when the vehicle passes through the FST attached with the DDVAs.
(3) Compared to the result obtained by using the FST without DDVAs, the RMS acceleration of the floating slab exhibits a notable reduction at the frequencies of 10-20 Hz, where positive insertion loss can be indentified due to the attachment of the DDVAs.
(4) The insertion loss increases with the increasing of the mass ratio of the DDVA, illustrating that DDVAs with a larger mass ratio can control the low-frequency vibration of the FST more effectively.

## ACKNOWLEDGEMENTS

This work was supported by the National key Basic Research Program of China ("973" Program) (Grant No. 2013CB036202, No. 2013CB036206), and the Innovative Practice Program for Graduate Students of Southwest Jiaotong University (Grant No. YC201421139).

## REFERENCES

1. Lombaert G, Degrande G, Vanhauwere B, Vandeborght B, François S. The control of ground-borne vibration from railway traffic by means of continuous floating slabs. *J Sound Vibr*. 2006; 297 946–61.
2. Nelson JT. Recent developments in ground-borne noise and vibration control. *J Sound Vibr*. 1996; 193: 367–76.
3. Wilson GP, Saurenman HJ, Nelson JT. Control of ground-borne noise and vibration. *J Sound Vibr*. 1981; 87: 339–50.
4. Krylov V, Ferguson C. Calculation of low-frequency ground vibrations from railway trains *Appl Acoust*. 1994; 42: 199–213.
5. Krylov VV. Low-frequeney ground vibrations from underground train. *Journal of Low Frequeney Noise and Vibration*. 1995; 14: 55–60.
6. Schillemans L. Impact of sound and vibration of the North–South high-speed railway connection through the city of Antwerp Belgium. *J Sound Vibr*. 2003; 267: 637-49.
7. Wolf S. Potential low frequency ground vibration (<6.3Hz) impacts from underground LRT operations. *J Sound Vibr*. 2003; 267: 651-61.
8. Ibrahim RA. Recent advances in nonlinear passive vibration isolators. *J Sound Vibr*. 2008; 314: 371–452.
9. Kela L, Vähäoja P. Recent Studies of Adaptive Tuned Vibration Absorbers/ Neutralizers. *Applied Mechanics Reviews*. 2009; 62: 060801-1--9.
10. Gong D, Zhou J-s, Sun W-j. On the resonant vibration of a flexible railway car body and its suppression with a dynamic vibration absorber. *J Vib Control*. 2012; 0: 1–9.
11. Thompson DJ, Jones CJC, Waters TP. A tuned damping device for reducing noise from railway track. *Appl Acoust*. 2007; 68: 43-57.
12. Wu TX. On the railway track dynamics with rail vibration absorber for noise reduction. *J Sound Vibr*. 2008; 309: 739-55.
13. Den Hartog JP. *Mechanical vibrations*. New York: McGraw-Hill, 1956.

14. Anh ND, Nguyen NX. Design of non-traditional dynamic vibration absorber for damped linear structures. *Proceedings of the Institution of Mechanical Engineers Part C-Journal of Mechanical Engineering Science*. 2013; 0: 1–11.
15. Ozer MB, Royston TJ. Extending den Hartog's vibration absorber technique to multi-degree-of-freedom systems. *Journal of Vibration and Acoustics*. 2005; 127: 341-50.
16. Ozer MB, Royston TJ. ApplicationofSherman-Morrison matrix inversion formula to damped vibration absorbers attached to multi-degree of freedom systems. *J Sound Vibr*. 2005; 283: 1235-49.
17. Zhai W, Wang K, Cai C. Fundamentals of vehicle-track coupled dynamics. *Veh Syst Dyn*. 2009; 47: 1349-76.
18. Zhai W, Xia H, Cai C, et al. High-speed train–track–bridge dynamic interactions – Part I: theoretical model and numerical simulation. *International Journal of Rail Transportation*. 2013; 1: 3-24.
19. Zhu SY, Fu Q, Cai CB, Spanos PD. Damage evolution and dynamic response of cement asphalt mortar layer of slab track under vehicle dynamic load. *Sci China-Technol Sci*. 2014.
20. Zhu SY, Cai CB. Interface damage and its effect on vibrations of slab track under temperature and vehicle dynamic loads. *International Journal of Non-Linear Mechanics*. 2014; 58: 222-32.
21. Zhai W, Wang K. Lateral Hunting Stability of Railway Vehicles Running on Elastic Track Structures. *Journal of Computational and Nonlinear Dynamics*. 2010; 5: 520-30.
22. Chen G, Zhai WM. A new wheel/rail spatially dynamic coupling model and its verification. *Veh Syst Dyn*. 2004; 41: 301–22.
23. Shen Z, Hedrick J, Elkins J. A comparison of alternative creep-force models for rail vehicle dynamic analysis. *Proceedings of the 8th IAVSD symposium*. 1984: 591-605.
24. Zhai W. Two simple fast integration methods for large-scale dynamic problems in engineering *Int J Numer Methods Eng*. 1996; 39: 4199-214.
25. Spanos PD, Zeldin BA. Monte Carlo treatment of random fields: A broad perspective. *Applied Mechanics Reviews*. 1998; 51: 219–36.

# Investigation of ground vibration from circular tunnels using a 2.5D FE/BE model of tunnel and ground

**Q Jin, D J Thompson**
Institute of Sound and Vibration Research, University of Southampton, UK

## ABSTRACT

A numerical model of a circular tunnel using the 2.5D finite element and boundary element method is described. This makes use of the constant geometry in the axial direction so that a two-dimensional model is solved for a series of wavenumbers. The full 3D solution can be recovered by a Fourier transformation. The response of the tunnel structure is analysed first. Then, the vibration at certain points on the ground surface is predicted using the tunnel-soil model. Finally, a parametric study is carried out to show the influence of different aspects of tunnel design to the vibration in the ground.

## 1. INTRODUCTION

Train-induced vibration and noise from underground railways has gradually attracted public concern, as the subways are rapidly expanding under densely populated cities in many countries. Therefore, recently, many prediction models have been developed in response to the requirement of reducing the effect from ground vibration and ground-borne noise. Two types of prediction models are popularly used: analytical models and numerical models. The analytical method provides a rapid solution and is useful for engineering practice. However, to cater for more detailed design studies a numerical approach is required.

An example of an analytical model is the PiP model, developed by Hussein et al. to calculate the far-field displacement due to a train running in a tunnel (1). The tunnel is represented by a circular cylinder embedded in a whole space ground and the track is represented by a separate analytical model. The model has recently been extended to include the free ground surface and ground layering. However, the efficiency of this analytical approach is achieved at the cost of model simplification and application limitations. In order to study the tunnel and track design in more detail, to meet the requirements for vibration and noise reduction, numerical models which can provide an authentic modelling of tunnel and track structure are required.

One such numerical approach is a coupled finite element and boundary element model in which the tunnel structure is modelled using finite elements and the ground using boundary elements which avoid problems of reflections at artificial boundaries (2). In (3), the FE/BE models in two and three dimensions are compared; 2D models are found to be lacking in precision whereas 3D models are computationally expensive. The so-called 2.5D approach is a compromise between these two models. A series of 2D models are used with the third dimension represented by wavenumber. This approach has been employed for example by

Sheng et al. (4) and Francois et al. (5) to predict the ground vibration from trains with good results.

In this paper the 2.5D FE/BE approach is used to model a circular tunnel. The response of the tunnel structure (without soil) to a harmonic excitation is analysed first to illustrate its dynamic behaviour. Then, the vibration at certain points on the ground surface is predicted using the tunnel-soil model. Finally, a parametric study is carried out to show the influence of different aspects of tunnel design, such as the tunnel lining and invert, on the vibration of the ground.

## 2. MODEL DESCRIPTION

The research object is a straight circular tunnel embedded in a homogeneous half space. In the present work the track is omitted but this can readily be included in the model.

The tunnel and ground are modelled using the 2.5D coupled finite element / boundary element software WANDS (Wave-Number-Domain FE-BE Software) (6). Based on the assumption that the soil and tunnel structure properties are invariant in the longitudinal direction, the 2.5D model is formed from a sequence of 2D FE/BE models of the cross-section of tunnel and soil, in terms of the wavenumber in the axial direction (*x*- direction). The 3D solution can be recovered by using an inverse Fourier transform over wavenumber. As shown in Figure 1, the tunnel structure is modelled using 8-noded solid elements; the ground is modelled using 3-noded boundary elements around the tunnel and at the ground surface (not shown). A harmonic force is applied directly on the tunnel lining or invert. Ground layers can also be included as necessary. The parameters chosen in the wavenumber domain depend mainly on the soil properties. Details of the model will be given in the following section.

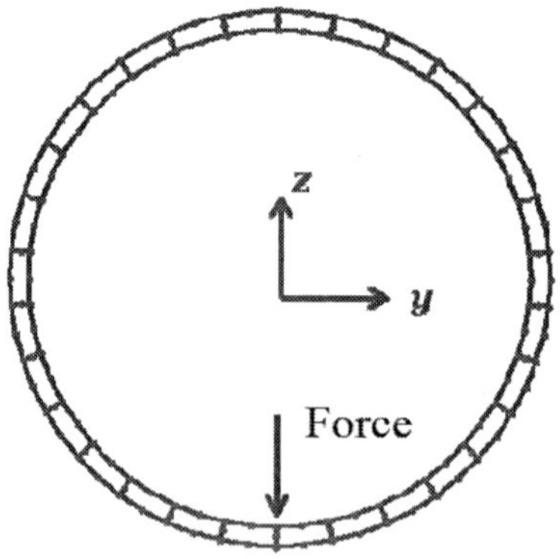


**Figure 1 Model of tunnel lining in WANDS**

## 3. MODE ANALYSIS OF FE TUNNEL MODEL

### 3.1 Model description

Before considering the coupled tunnel-ground problem, the circular tunnel is first modelled alone, without the consideration of the surrounding soil. The tunnel lining is modelled with a single layer of 8-noded solid elements, as shown in Figure 1. The parameters used in the model are listed in Table 1. The frequency range used in the calculation is from 1~100 Hz with a logarithmic spacing. A discussion of the parameters used in the wavenumber domain can be found in (7). Compared to the model with soil in Section 4, a relatively fine wavenumber resolution ($\Delta\beta$) is chosen here to give precise results in the axial (*x*) direction. The total number of wavenumbers is given by $\beta_{max} = 0.5 \cdot \Delta\beta \cdot N_\beta$. A vertical harmonic force is applied at the bottom of the tunnel. The parameters of the tunnel are based on a typical concrete tunnel. In addition two types of tunnel invert are considered, described in Section 3.3 below. They are also modelled with the 8-noded solid elements and the same material properties.

**Table 1 Parameters and properties used in tunnel model**

| **Parameters in Wavenumber Domain** | **Tunnel properties** |
|---|---|
| Frequency: 1~100 Hz<br>Number of wavenumber [$N_\beta$]: 2048<br>Wavenumber resolution [$\Delta\beta$]: $6.28\times10^{-4}$ rad/m<br>Maximum wavenumber [$\beta_{max}$]: 0.64 rad/m | Inner diameter: 3.81 m<br>Outer diameter: 4.11 m<br>Young's Modulus: 50 MPa<br>Density: 2500 kg/m$^3$<br>Poisson's ratio: 0.3<br>Loss factor: 0.03 |

### 3.2 Mode analysis

The 2.5D FE model of the tunnel structure provides a comprehensive insight into the deformation modes of the tunnel. The magnitude of the response at each frequency and wavenumber, when the tunnel is excited by a unit force at the tunnel bottom, is plotted in Figure 2. The responses at the forcing point in each direction are shown separately, although it should be noted that the force is in the vertical (*z*) direction in each case. Also plotted are the dispersion curves (free wavenumbers) of a beam which has the same geometry and properties as the tunnel structure.

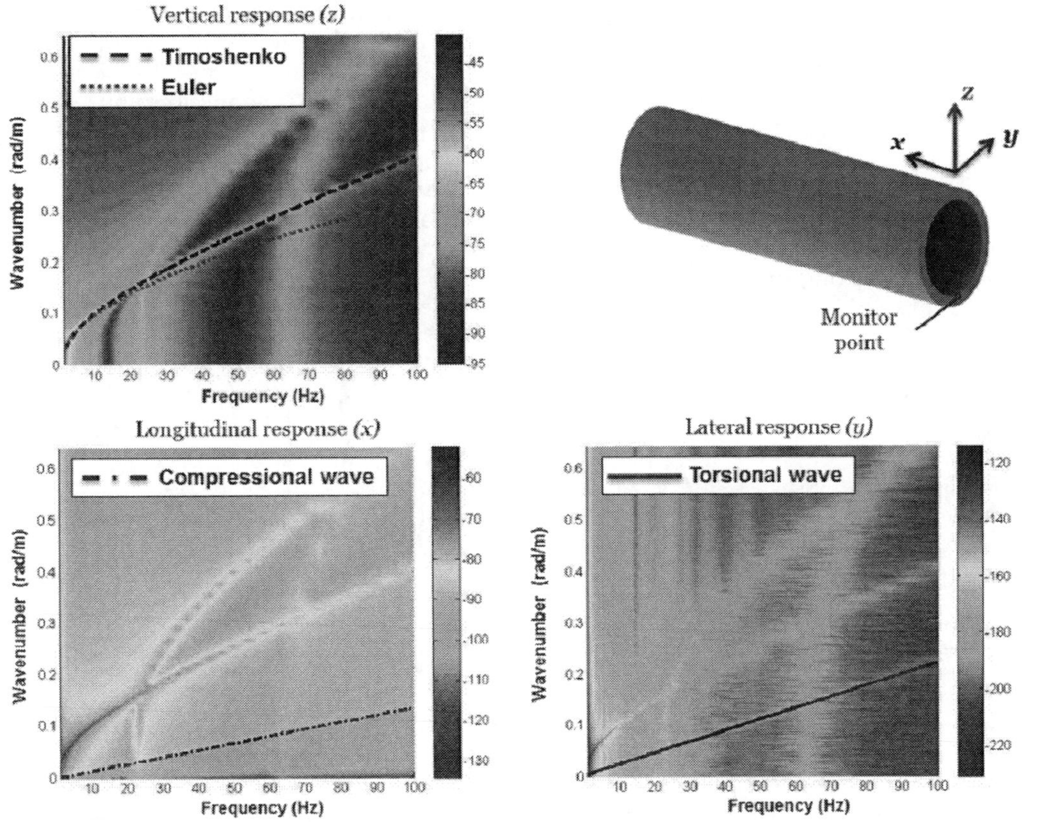


**Figure 2 Dispersion diagrams at tunnel bottom for a vertical force applied at the same location**

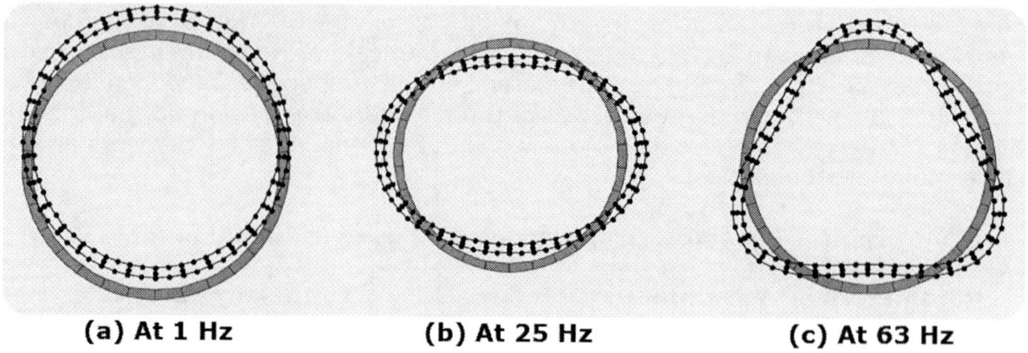


**Figure 3 In-plane deformation of the tunnel lining under a vertical force at tunnel bottom**

The maximum response at the tunnel bottom is in the vertical ($z$) direction, the results for which are shown at the top left. Three bright lines can be seen, each representing a deformation mode of the tunnel ring. The cut-on frequencies for each mode (the frequencies at which the mode starts to propagate) can also be found in the dispersion diagram as 0, 23 and 63 Hz. Figure 3 shows examples of the cross-section deformation close to the cut-on frequency of each of these modes. The first wave to occur corresponds to a rigid body movement of the cross-section, which shows a very good agreement with the bending wave of the corresponding Timoshenko beam (see Figure 2). The other two modes are higher order shell modes.

The response in the other two directions is much smaller than in the vertical direction; theoretically it should be zero but due to numerical issues a small non-zero response is found. Apart from the three modes found in the vertical response, two other curves can be observed in the dispersion diagrams for these two directions. They reflect the compressional wave in longitudinal direction and the torsional wave in the lateral direction.

### 3.3 Influence of lining thickness and invert

In this section, the response to a harmonic excitation at the tunnel bottom is analysed. The influence is studied of changing the thickness of the tunnel lining, as well as introducing an invert with various shapes and height.

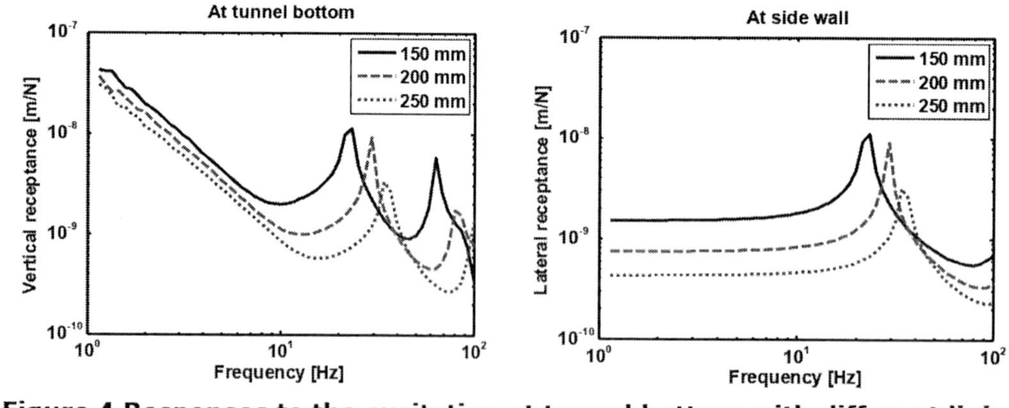


**Figure 4 Responses to the excitation at tunnel bottom with different lining thickness**

To study the influence of the thickness of the tunnel lining, the responses at the tunnel bottom and side wall are compared in Figure 4 from FE tunnel models with different lining thicknesses. These are shown as receptance (displacement for a unit force). All the tunnels have the same inner radius of 3.81 m. It can be seen from the figure that the cut-on frequencies (visible as peaks in the response) increase with the thickness of the tunnel lining, while the average amplitude of the response decreases. The responses at the crown of the tunnel lining (not shown) are almost identical to those at the bottom when there is no soil. The lateral response at the side wall is smaller at low frequency as the first mode has no response here. At the cut-on frequency of the second mode the response is similar to that at the tunnel bottom. The third mode also does not appear in the lateral deformation of the side wall.

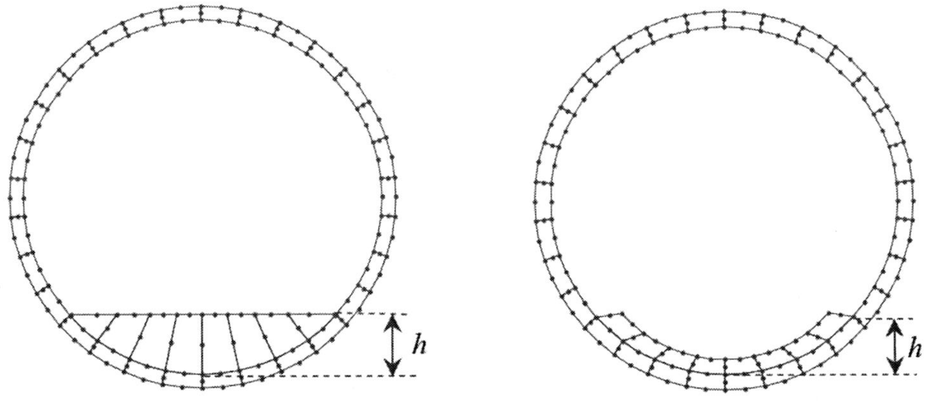


**(a) Full invert** **(b) Thickened base**

**Figure 5 FE tunnel model with two types of invert**

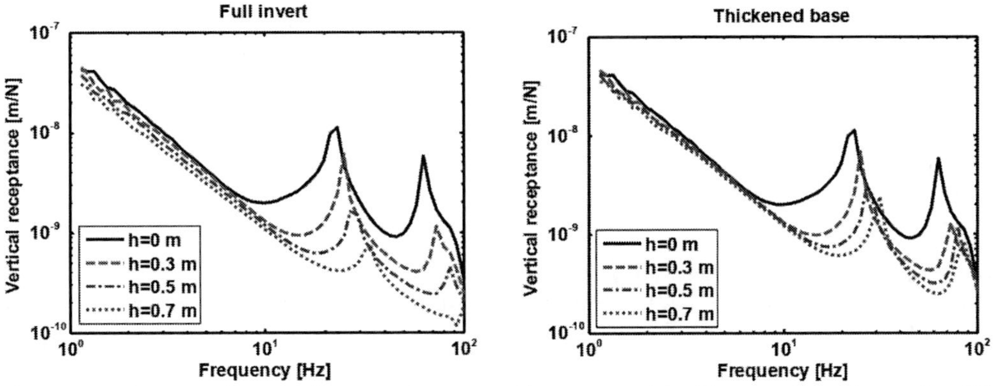


**Figure 6 Comparison of responses at the tunnel bottom when invert with different shape and height is used**

The tunnel invert is an important element connecting the track with the tunnel lining. Although the tunnel invert may have various shapes and geometry, it is often ignored or simplified in the modelling. Here, it will be studied how the invert influences the dynamic behaviour of the tunnel structure.

Two different shapes of tunnel invert are considered based on those which are commonly used in practice, as shown in Figure 5. The one on the left can be used with ballasted track, booted-sleeper track and so on, while the one on the right is used particularly with floating slab track.

Figure 6 compares the responses at the tunnel bottom when different shapes and heights of invert are used. For the full invert, $h$ is the height of the invert, as shown in Figure 5(a). For the thickened base, the thickness is kept constant and $h$ defines the extent of the invert, as shown in Figure 5(b). In each case $h$=0 means the tunnel without the invert. Compared with the tunnel model without invert, the responses of those with an invert are lower. The cut-on frequencies of the deformation modes also increase with the height of the invert.

## 4. RESULTS OF TUNNEL-GROUND MODEL

### 4.1 Model description

The tunnel is now considered embedded in soil. Some characteristics of the modal behaviour of tunnel structure discussed above will assist in interpreting the results obtained here.

The tunnel modelled here is the same as the one in the previous section, apart from the tunnel thickness, which is varied from 150~250 mm; unless otherwise stated, it is 150 mm as before. The tunnel is buried at a depth of 15 m or 25 m below the ground surface, the distance in each case being to the tunnel centre. The ground is formed by 3-noded boundary elements. The free ground surface is modelled to a distance of 30 m on one side and 40 m on the other side of the tunnel. At the edge of the ground surface mesh, special edge elements are used to minimise reflections (3). The element size on the ground surface is 0.25 m. The tunnel lining is modelled using 30 elements, which have a length of about 0.3 m. Two types of tunnel invert are considered as previously. The soil properties are based on London clay, as listed in Table 2. The parameters used in the wavenumber domain for the tunnel-ground model are different from those used in tunnel model, and are mainly determined by the soil properties. In order to get a precise prediction of vibration, a large maximum wavenumber is used, while the wavenumber resolution is less fine compared with that used in the tunnel model.

**Table 2 Parameters and properties used in tunnel-soil model**

| **Parameters in Wavenumber Domain** | **Soil properties** (8) |
|---|---|
| Frequency: 1~100 Hz | S-wave speed: 220 m/s |
| Number of wavenumber [$N_\beta$]: 1024 | P-wave speed: 1751 m/s |
| Wavenumber resolution [$\Delta\beta$]: $6.28\times10^{-3}$ rad/m | Density: 1980 kg/m$^3$ |
| Maximum wavenumber [$\beta_{max}$]: 3.2 rad/m | Loss factor: 0.078 |

### 4.2 Results overview and verification

Using the tunnel-ground model built in WANDS, the vibration produced by the vertical force acting on the tunnel bottom can be predicted.

Figure 7 shows the response on the tunnel at different positions. Compared with Figure 4, the responses are much flatter due to the radiation damping of the soil. The cut-on frequencies are not evident at all at the tunnel bottom, although the response is clearly affected by the lining thickness. The second and third cut-on frequencies appear at the side wall and tunnel crown respectively, but their amplitudes are much smaller than in Figure 4.

Figure 8 shows the waves propagating on the ground surface for different excitation frequencies. The $x$-axis is the longitudinal direction of tunnel. The force is applied on the tunnel bottom at $x=y=0$. It can be seen in Figure 8 that, as the frequency increases, the strength of the waves in the longitudinal direction is gradually moderated due to the existence of the tunnel, and the wavelength is also lengthened relative to that in the transverse direction.

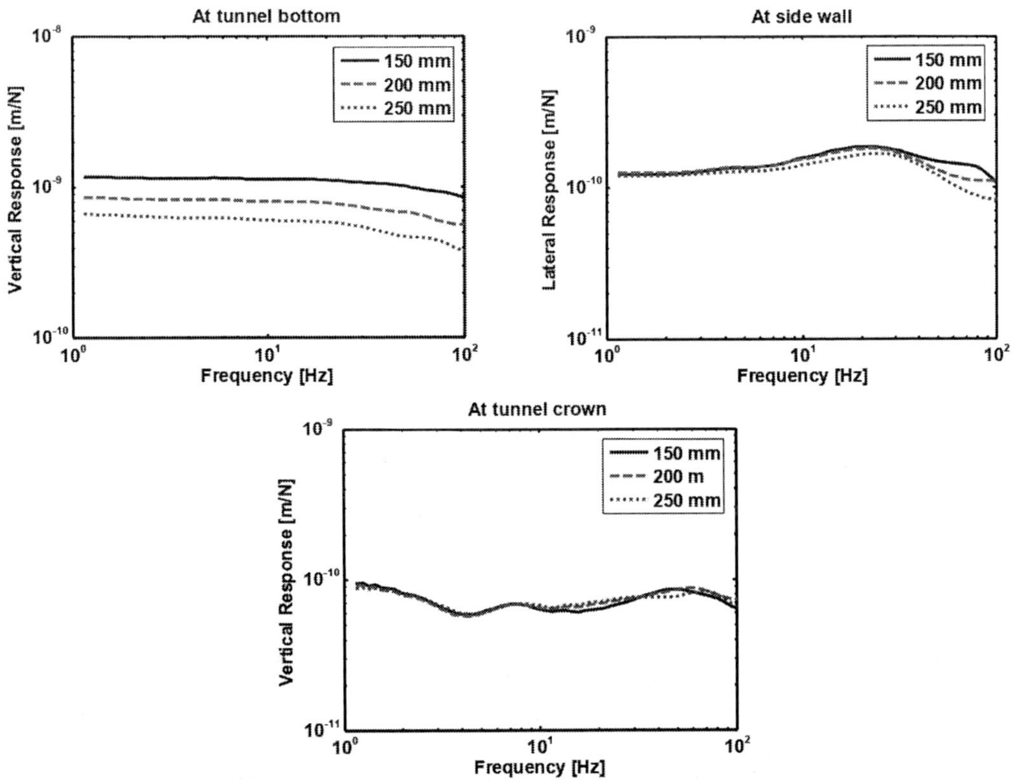


**Figure 7 Responses to the excitation at tunnel bottom with different lining thickness for tunnel in soil at 25 m depth**

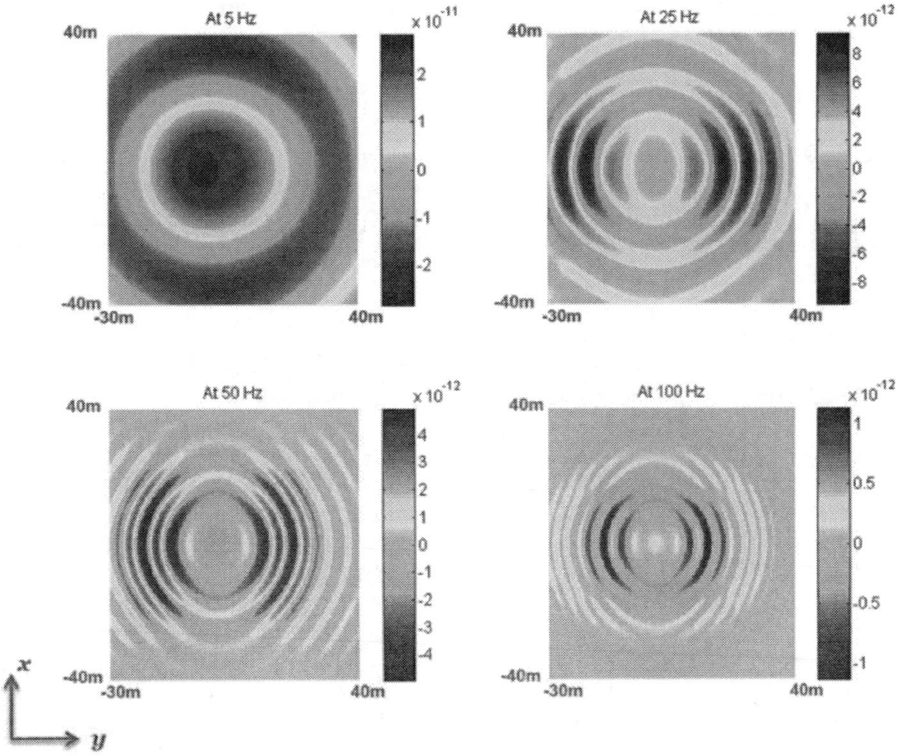


**Figure 8 Surface plot of instantaneous vertical displacement (m/N) on the ground surface on the ground surface excited at different frequencies**

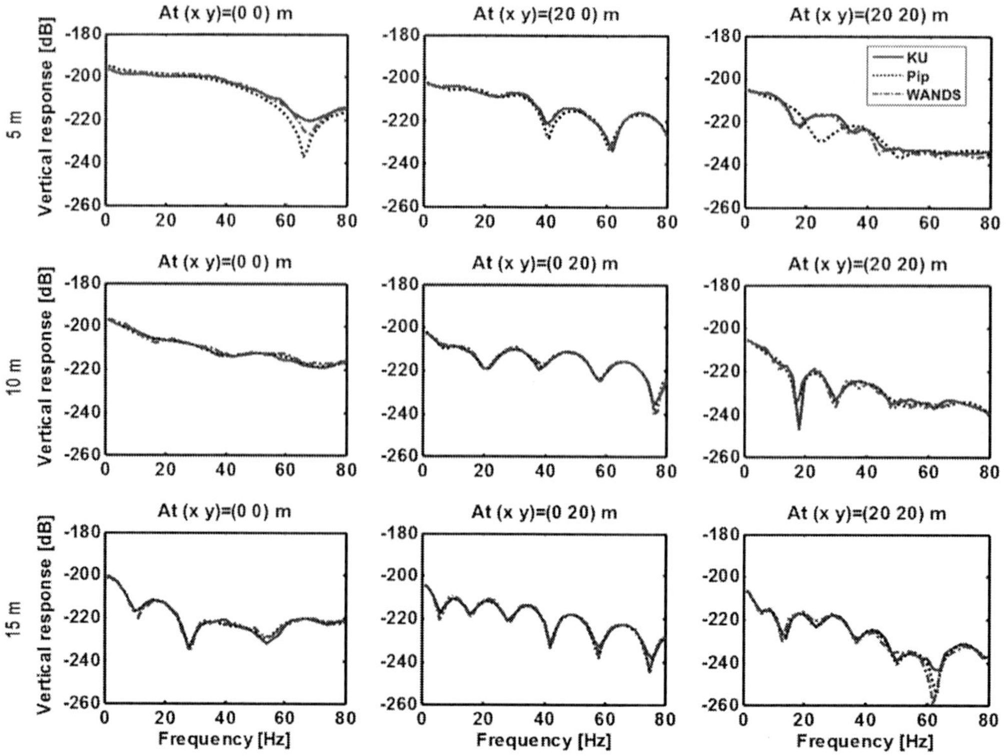


**Figure 9 Comparison of vertical responses on the the ground surface between WANDS, PiP and KU Leuven model (9)**

To verify the correctness of the 2.5D tunnel-ground model, a comparison is made between the results from WANDS, PiP (1) and a model from KU Leuven (5). PiP (1) is an analytical model while the KU Leuven model is a numerical model similar to WANDS. The parameters used in the comparison model are different from those introduced previously. For the tunnel, the inner radius is 2.75 m and the lining thickness is 0.25 m. For the soil, the shear and compressional wave speeds are 200 m/s and 400 m/s, with density of 1800 kg/m$^3$ and loss factor of 0.04. The vibration predicted by the three models at three different points on the ground surface and for three different tunnel depths is compared in Figure 9. The comparison shows a good agreement, especially for the tunnels with buried depths of 10 m and 15 m.

### 4.3 Parametric analysis

Results are presented for a series of models with different tunnel size and invert depth to study the impact on the responses at the ground surface. They are all converted to one-third octave bands for ease of comparison.

Figure 10 and Figure 11 show the responses in different directions at various positions on the ground surface from the models with the same buried depth of 25 m. Above the tunnel axis, in Figure 10, there is no response in the lateral direction, while in the cross-section of the forcing point, Figure 11, the longitudinal response is zero. At 10 m from the tunnel, the lateral response exceeds the vertical response above 6 Hz, but further away the vertical response increases. This is due to the shadow effect (2) of the tunnel structure, see Figure 8.

These results illustrate that the vibration on the ground surface is not only dominated by the vertical component; the longitudinal and lateral components are also important at the points away from $(x, y)=(0, 0)$. Therefore, a way of evaluating the total vibration is adopted from (2) known as the pseudo-resultant amplitude:

$$u = \sqrt{|u_z|^2 + |u_y|^2 + |u_x|^2}$$

where $u_z$, $u_y$ and $u_x$ are the complex amplitudes of vertical, lateral and longitudinal responses, respectively.

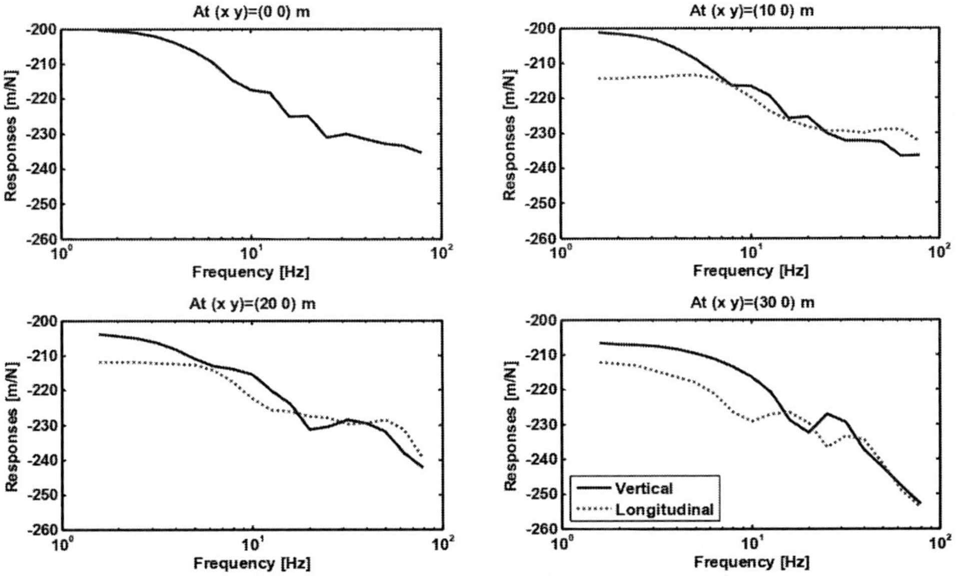


**Figure 10 Responses at various locations along x direction on the ground surface predicted for a tunnel at a depth of 25 m**

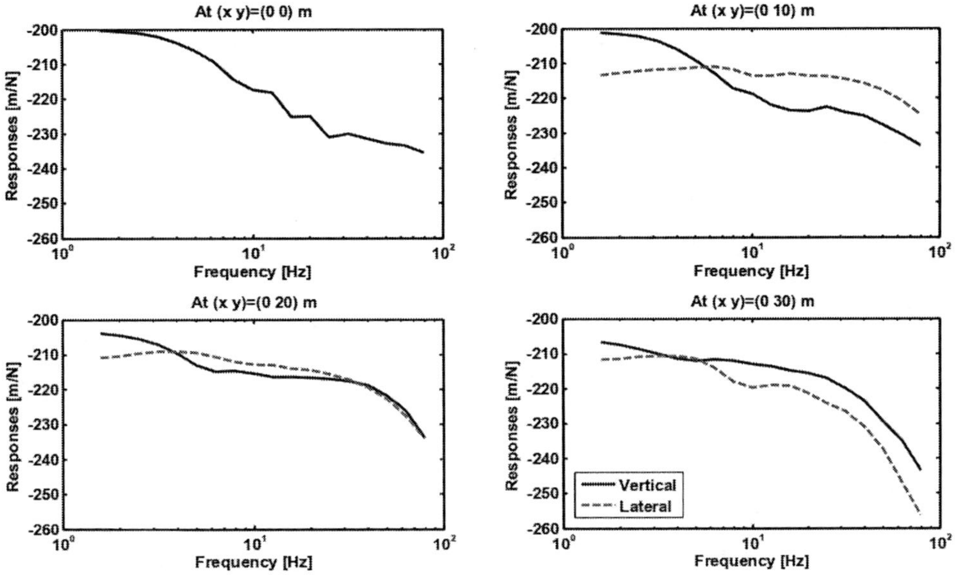


**Figure 11 Responses at various locations along y direction on the ground surface predicted for a tunnel at a depth of 25 m**

The pseudo-resultant vibration from models with different tunnel depth and lining thickness is compared in Figure 12. The upper lines from the shallow tunnel model are higher than the lower lines from the deep tunnel model. This gap increases slightly with frequency. The resultant amplitude is higher at the points further away from the tunnel (figures on the right when $y$=10 m) than that directly above the tunnel (figures on the left when $y$=0 m). The differences due to the tunnel thickness are small and mainly limited to high frequencies. Furthermore, the difference caused by the tunnel thickness is more evident in the lower figures for $x$=10 m.

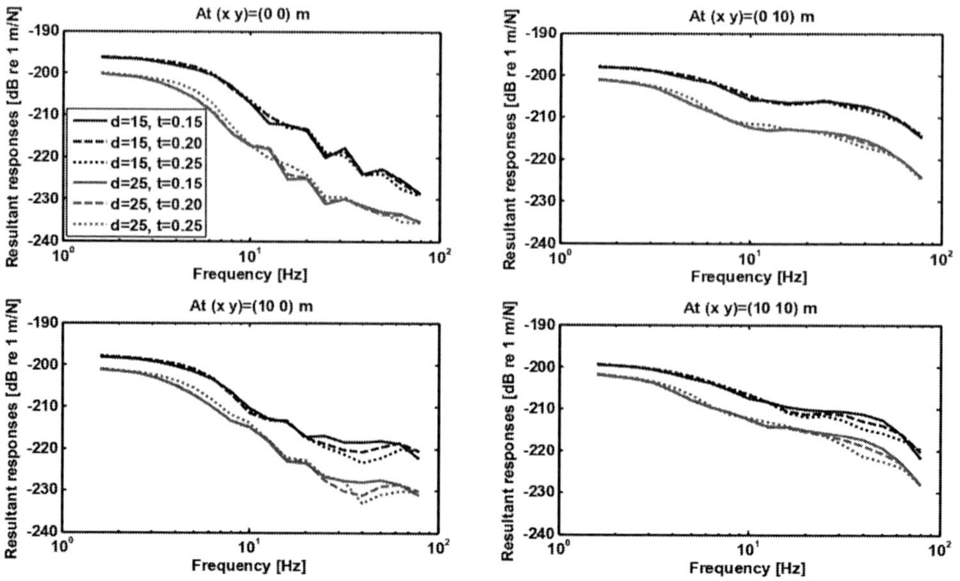


**Figure 12 Pseudo-resultant response predicted for tunnels with different depth (d) and lining thickness (t) (unit: m)**

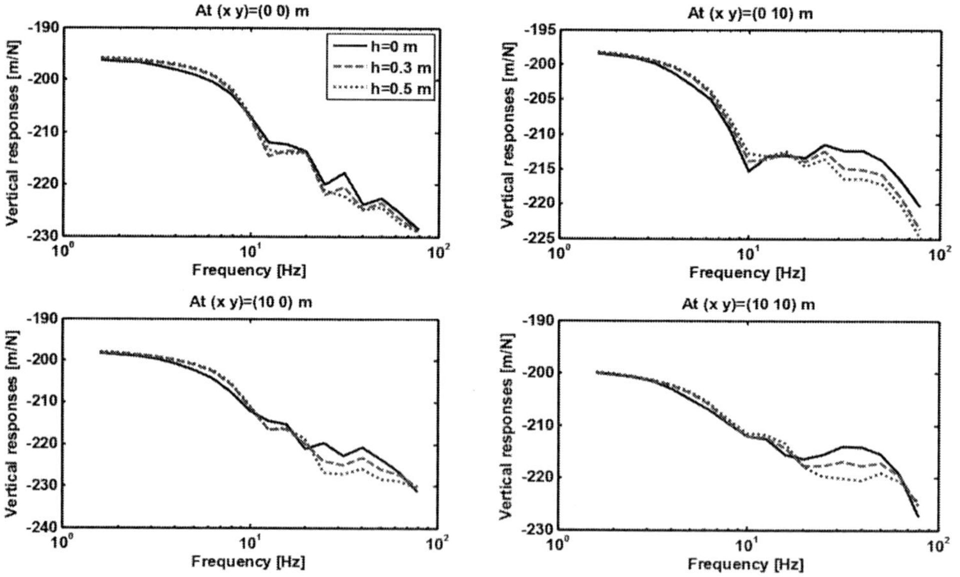


**Figure 13 Vertical responses predicted with and without tunnel invert**

Figure 13 compares the ground responses between models with and without invert. $h$ is the height of the invert, as indicated in Figure 5. The tunnel geometry and soil properties of the model are based on those introduced previously, while the tunnel depth is 15 m. The invert adopted in the model is the full invert with the same material properties of the tunnel lining, see Figure 5(a). The black solid lines give the results from the model without invert. Generally speaking, the existence of an invert slightly increases the response on the ground surface at low frequencies, while decreasing the responses at high frequencies (above about 10 Hz).

## 5. CONCLUSIONS

In this paper, a tunnel with and without ground is modelled with the 2.5D finite element and boundary element method. With these models, the modes of the tunnel structure are investigated and a parametric study is carried out to investigate the influence of the depth, tunnel lining and invert structure.

The behaviour of the tunnel structure can be interpreted in terms of the deformation modes of a cylindrical shell. An increase in the lining thickness will increase the stiffness of the tunnel structure; therefore, the overall response amplitude is reduced but the cut-on frequencies of the deformation modes increase. In the same way, the tunnel invert also plays an important role in the deformation. The influence of the invert may differ for different types, but the tendency of the changes with the increase of the height of invert is the same as that of the tunnel thickness.

The responses at different points on the ground surface to a vertical harmonic force at the tunnel bottom are compared with two existing models and show good agreement. From a parametric study, the following conclusions can be made:

1) The decay rate against distance is slightly higher in the axial direction than that in the lateral direction. Above 6 Hz, the lateral displacements at 10 m away from the tunnel exceed the vertical displacements, but the vertical displacements are higher at further distances.

2) For a deeper tunnel, the vibration on the ground surface is predicted to be smaller for all excitation frequencies.

3) The influence of lining thickness and tunnel invert on the ground vibration is limited to high frequencies. However, there is a greater difference between the decay rates against distance in the tunnel direction and the direction perpendicular to the tunnel in the presence of a tunnel invert.

It is clear that the geometry of the tunnel structure has a significant influence on the distribution and radiation of vibration in the ground. Therefore, it is important to include such details in the modelling of the tunnel-ground system.

## REFERENCES

(1) Hussein, M. F. M., Hunt, H. E. M., Rikse, L., Gupta, S., Degrande, G., Talbot, J. P., François, S. & Schevenels, M. (2008) *Using the PiP model for fast calculation of vibration from a railway tunnel in a multi-layered half-space.* Noise and Vibration Mitigation for Rail transportation systems. Springer Berlin Heidelberg, 136-142.
(2) Jones, C. J. C., Thompson, D. J., & M. Petyt. (2002) *A model for ground vibration from railway tunnels.* Proceedings of the ICE-Transport 153.2: 121-129.
(3) Andersen, Lars, & Jones, C. J. C. (2006) *Coupled boundary and finite element analysis of vibration from railway tunnels—a comparison of two-and three-dimensional models.* Journal of Sound and Vibration 293.3: 611-625.
(4) Sheng, X., Jones, C. J. C. & Thompson. D. J., (2005) *Modelling ground vibration from railways using wavenumber finite-and boundary-element methods.* Proceedings of the Royal Society A: Mathematical, Physical and Engineering Science 461.2059: 2043-2070.
(5) Francois, S., Schevenels, M., Lombaert, G., Galvın, P. & Degrande, G. (2010) *A 2.5D coupled FE-BE methodology for the dynamic interaction between*

*longitudinally invariant structures and a layered halfspace*. Computer Methods in Applied Mechanics and Engineering, 199(23-24):1536–1548.
(6) Nilsson, C. M., Jones, C. J. C. (2007) *Theory manual for WANDS 2.1*, ISVR Technical Memorandum N. **975**. University of Southampton.
(7) Jin, Q. & Thompson, D. J. (2014) *Predictions of ground vibration from trains using a multi-body vehicle model coupled to a 2.5D FE/BE model of the track and ground*. The 21st International Congress on Sound and Vibration, Beijing, July.
(8) Clouteau, D., Othman, R., Arnst, M. Chebli, H., Degrande, G., Klein, R., Chatterjee, P. & Janssens, B. (2004) *A numerical model for ground-borne vibrations from underground railway traffic based on a periodic FE-BE formulation*. In Proceedings of the Eighth International Workshop on Railway Noise.
(9) Hussein, M. F. M., Francois, S., Schevenels, M., Hunt, H. E. M., Talbot, J. P. & Degrande, G. (2014) *The fictitious force method for efficient calculation of vibration from a tunnel embedded in a multi-layered half-space*. Journal of Sound and Vibration. 333.25 (2014): 6996-7018.

# Addressing nodal constraints on railway capacity

**J Armstrong[1], J Preston[2], C Potts[3], T Bektas[4], D Paraskevopoulos[5]**

[1] University of Southampton, Civil, Maritime and Environmental Engineering and Science, and Arup, UK

[2] University of Southampton, Civil, Maritime and Environmental Engineering and Science, UK

[3] University of Southampton, School of Mathematics, UK

[4] University of Southampton, Southampton Management School, UK

[5] University of Bath, School of Management, UK

## ABSTRACT

As demand for passenger and freight transport on Britain's railways increases, providing additional capacity and making the best use of the existing infrastructure are priorities for the industry. Since the stations and junctions forming the nodes of the railway network tend to form the constraints on route and network capacity, improved understanding of their operation and capacity characteristics is particularly important.

This paper describes research undertaken to improve the understanding of nodal capacity and capacity utilisation, and to route and schedule trains more efficiently through nodes, thus improving service quality and/or releasing capacity for additional train services.

## 1 INTRODUCTION

Railway capacity is an increasingly scarce and valuable commodity. The UK Department for Transport (DfT) summarised the situation in its 2011 High-Speed Rail Consultation (1) as follows:

> *Britain's rail network is seeing a continuing pattern of steeply rising demand. As a result, rail capacity is under increasing strain and services are growing more crowded.*

Capacity constraints at nodes (i.e. junctions and stations) in the railway network are a particular concern, since they tend to form the bottlenecks which limit the capacity of the overall network. As observed by the Engineering and Physical Sciences Research Council (EPSRC), the Rail Safety and Standards Board (RSSB) and DfT in their 2010 request for proposals for research into overcoming these nodal capacity constraints (2),

> *nodes in rail networks (stations and junctions) are vital elements of a railway infrastructure, but they are frequently also a major constraint on capacity.*

This paper describes work undertaken for the OCCASION (Optimising Capacity Constraints: A Simulation Integrated with Optimisation of Nodes) project, one of five research projects undertaken in response to the above request, to develop means of increasing nodal capacity without major investments in infrastructure enhancements. It also describes subsequent and ongoing work in this area, undertaken partly in collaboration with European railway Infrastructure Managers (IMs).

Following this introduction, measures of and issues relating to railway capacity and capacity utilisation are described and discussed, and the extension of an existing capacity utilisation measure from network links to nodes is described. Next, approaches to the re-routeing and re-scheduling of trains at individual network nodes to reduce capacity utilisation, and thus to increase potential capacity and/or service reliability, are described. Methods for extending the approach from individual nodes to a wider network are then considered. Finally, some conclusions are drawn and subsequent and ongoing work is described, followed by a list of cited references.

## 2 RAILWAY CAPACITY AND CAPACITY UTILISATION

Railway capacity is a somewhat abstract and elusive measure: as noted by UIC (*Union International des Chemins de Fer*, or International Union of Railways), it is difficult, if not impossible, to define a specific capacity value for any given railway track, route or network: "capacity as such does not exist [and] railway infrastructure capacity depends on the way it is utilised" (3). Achievable capacity therefore depends not only upon the characteristics of the infrastructure, including the signalling system used, but also upon the performance characteristics and mix of trains using the route, the timetable in operation, and also the target levels of reliability and punctuality to be achieved by the timetabled trains.

Two important railway capacity-related objectives are: (i) making the best use of the capacity that is potentially available, i.e. aiming to maximise the number of trains that can be operated, subject to other constraints (service mix, stopping patterns, etc.), and (ii) the related issue of maintaining quality (i.e. reliability and punctuality) of service by not operating so many trains that service quality is reduced below an acceptable level. Capacity utilisation (or capacity consumption) is a measure of the extent to which the available capacity under a given set of circumstances is being used or consumed, and thus of the potential to operate additional services. Capacity Utilisation assessments inform both of the above objectives, in that (i) the greater the number of trains being run for a given level of capacity utilisation, the greater the level of capacity being attained; and (ii) the greater the level of capacity utilisation, the less likely the service is to be robust and reliable, with particularly severe levels of service quality deterioration being seen above capacity utilisation levels in excess of 60%-70% over extended periods, and in excess of 75%-85% over shorter, peak periods (3).

As implied in the preceding paragraph, a wide range of capacity values may thus be provided for any given level of capacity utilisation; conversely, different levels of capacity utilisation may be achieved for a given level of capacity provision, depending upon how that capacity is provided. The situation is summarised as follows by Armstrong et al. (4):

> *where capacity is scarce, the objective will typically be to maximise the capacity for the maximum level of capacity utilisation that is consistent with the provision of a stable, reliable quality of service. In order to achieve this, it is necessary to have measures of both*

*provided capacity and capacity utilisation for all parts of the route or network in question, i.e. at nodes (junctions and stations) as well as links (sections of plain line between nodes). Well-defined capacity utilisation measures are currently available for links, but not generally for nodes.*

The first element of the development work undertaken for the OCCASION project was therefore the extension of capacity utilisation measurement techniques to railway network nodes.

## 2.1 Extending the CUI approach

Two widely-used approaches to the measurement of capacity utilisation are the international UIC 406 approach (3) and, in Britain, the Capacity Utilisation Index (CUI) method, as described by Gibson et al. (5). The two methods are similar in approach, both employing the technique of 'timetable compression', as illustrated in Figure 1, reproduced from (5), summarising the CUI approach.

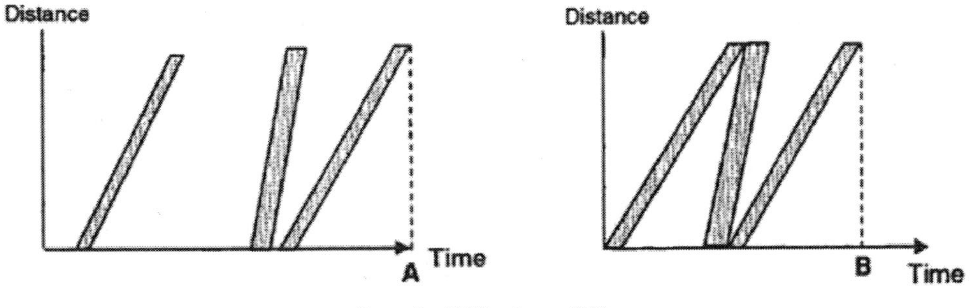


**Figure 1: Definition of the Capacity Utilisation Index**

As can be seen in Figure 1, the timetable compression approach entails the 'squeezing together', or compressing, of train paths, while maintaining their planned sequence, so that successive trains are separated by the minimum headway applicable to the route section in question. The CUI is then calculated as the ratio of the compressed time, B, to the time interval under consideration, A (typically one or more hours), and usually expressed as a percentage. (Capacity utilisation measures on railways are thus similar to the Ratio of Flow to Capacity (RFC) measure applied to highway traffic.) The UIC 406 and CUI methods vary in the level of detail at which they are applied: the UIC 406 approach is applied at the signal block level, whereas the CUI approach is applied to longer route sections, and does not consider individual block sections.

Initial work conducted on very simple node examples (4) was 'scaled up' for application to the case study area chosen for the OCCASION project: the East Coast Main Line (ECML) between Huntingdon and Grantham (59 and 106 miles respectively north of London King's Cross). A detailed node- and link-based network model was developed for the area, and the detailed routeings of trains through the modelled area were determined from electronic timetable data for the route, using a Perl script developed for this purpose. Having obtained the route, and thus the timings, of each train through the individual nodes (point ends, crossings and platforms) and intermediate links comprising the study area, the timings for each element of the network were aggregated, sorted and compressed to produce individual CUIs, from which overall minimum, maximum and average values were calculated, again using the script referred to above. The CUI calculation process was repeated for the optimised timetable with additional trains (as described in the following section), and the results for both timetables are summarised in Table 1 for the weekday period between 07:00 and 09:00.

**Table 1: CUI summary results for existing and optimised ECML timetables**

| Timetable | Network Element | CUIs | | |
|---|---|---|---|---|
| | | Minimum | Maximum | Average |
| May 2011 | Nodes | 0% | 51% | 23% |
| | Links | 0% | 64% | 25% |
| Optimised | Nodes | 0% | 73% | 24% |
| | Links | 0% | 81% | 26% |

The 0% values occur on nodes and links that are unused during normal timetabled operation; it can be seen that the maximum and average vales for both links and nodes increase when additional trains are included in the optimised timetable.

### 2.2 Ongoing development work

Except in the cases of the simplest junctions and stations, overall average values are unlikely to provide a representative measure of capacity utilisation; maximum values may be more helpful (see Table 2 and the accompanying text, below), but are again unlikely to be sufficiently representative. Work is therefore ongoing to develop the methodology further. This work is being undertaken as part of the UIC-sponsored ACCVA (Assessment of Capacity Calculation Values) project, whose aim is to extend the UIC 406 method to calculate capacity utilisation values for nodes and to identify guideline maximum values, similar to those already available for links.

The project is being undertaken in collaboration with IMs from the Austrian, German, French, Italian, Czech, Slovak and Hungarian national railway organisations, and the proposed nodal updates to UIC 406 will be tested against existing methods employed by the German and Czech IMs, and compared with the results obtained from OCCASION. To facilitate the comparison, it is intended that the ECML study area will be used for the comparative study.

## 3 RE-ROUTEING AND RE-SCHEDULING TRAINS

Having established a measure of capacity utilisation for network nodes, the second stage of the OCCASION project entailed the re-routeing and re-sequencing of trains through the junctions and stations in the case study area, with the objective of minimising nodal capacity utilisation, and thus maximising reliability and potentially releasing capacity for additional services. This element of the project made use of well-established workshop scheduling techniques, whereby trains are treated as 'jobs' to be processed by 'machines' comprising the links and nodes of the railway network, train movements are treated as the 'operations' performed on jobs, and the sectional running times of trains through track sections are treated as the corresponding jobs' processing times. The relationship between train scheduling and production scheduling has been known for some time, since the pioneering work conducted by Szpigel (6); however, its use for solving practical problems is comparatively recent, including, for example, its application by Liu and Kozan (7), in the form of a blocking parallel machine Job Shop scheduling problem, to the Train Timetabling Problem (TTP) on Australian railways.

'Time windows' were used in conjunction with the Job Shop model, to reflect the varying degrees of 'flex' available for application to different train services when adjusting the timetable. 'Hard' time windows were applied to long-distance passenger services, preventing their alteration; 'regular' time windows were applied to shorter-distance/local passenger services, enabling them to be modified, albeit at the expense of a penalty being applied to the objective function; and soft or no time windows were applied to freight services and 'empty coaching stock' moves, enabling them to be adjusted significantly, and reflecting their relative flexibility.

When adding trains to the timetable, replicates of existing services (in terms of running times and stopping patterns) were used, both to limit the 'search space' and to reflect existing commercial service patterns – while a more comprehensive approach to re-scheduling would almost certainly reduce capacity utilisation further, and enable the introduction of a greater number of additional services, this would entail a more-or-less complete recast of the existing timetable, and would require significant liaison with operators and other stakeholders to ensure a commercially attractive and realistic outcome.

A construction heuristic was used to generate initial solutions, reducing/eliminating scheduled waiting time (i.e. 'pathng' and other allowances included in the timetable), and thus reducing capacity utilisation. Since this instance of the Job Shop scheduling problem is NP-hard, and exact solutions are not available, a two-phase metaheuristic was developed to provide a solution to the next two steps: in the first phase, a serial insertion heuristic was used to add replicate trains to the existing timetable, thus providing a set of initial solutions; in the second phase, a Tabu Search (8) algorithm was used to improve the initial solutions, enabling the optimisation process to escape from local optima. Finally, a CPLEX solver was used to minimise variations in service frequency, thus enhancing the regularity of the timetable pattern. The results of the overall process are summarised in Table 2.

**Table 2: ECML Re-Routeing and Re-Scheduling Results**

| Period | Initial Services | Total Services | Additional Services | % Increase in Services | CUI-type measure |
|---|---|---|---|---|---|
| 07:00 – 09:00 | 53 | 67 | 14 | 26 | 79% |

24.5 minutes of scheduled waiting time were removed from the timetable by means of the re-routeing and re-scheduling process, and, as can be seen from Table 2, 14 additional services were accommodated, a 26% increase in service numbers, indicating a 'CUI-equivalent' measure for the existing timetable of 53/67 = 79%. This value is significantly higher than the average values shown in Table 1, but similar to the maximum values shown for the optimised timetable. Although the values are not strictly comparable, the similarity with the maximum values for the optimised timetable seems intuitively correct, in that the Job Shop analysis takes account of multiple constraints across the modelled network, and its results are therefore likely to reflect maximum, rather than average, CUI values on the network.

It should be noted that none of the 14 additional trains accommodated in the optimised timetable is a Main Line service to or from London: all are regional/cross-country passenger services operating on regional lines. However, this is a realistic outcome, reflecting capacity constraints between Peterborough and London, and high levels of capacity utilisation in the morning peak period. While there is unlikely to be much passenger demand for the additional identified train paths, they are of potential use for freight traffic.

## 4 CONSIDERATION OF ADJACENT NODES AND THE WIDER NETWORK

While optimising the routeing and scheduling of trains through a single junction or station may provide local capacity benefits, it does not guarantee an overall optimal outcome, since an optimal solution for one node may not be feasible for one or more adjacent nodes, which may be subject to additional traffic demands or capacity constraints. The third and final major element of the OCCASION project was therefore to extend the application of the techniques already developed from single to multiple, adjacent nodes, to ensure that maximising the throughput of one node does not cause problems elsewhere in the network.

For the consideration of multiple network nodes, a Multi-commodity Capacitated Network Design Problem (MCNDP) approach (9) was used. The MCNDP can be defined on a graph of nodes and links (10), as shown in Figure 2.

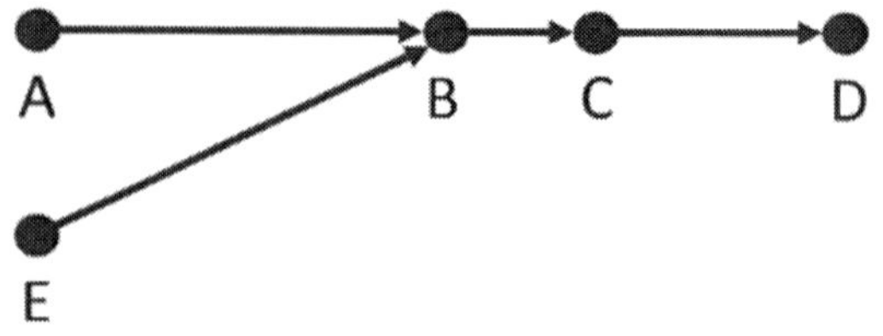


**Figure 2: Node-link graph**

The links are typically directed and represent the arcs of the network. The nodes represent the origins and/or destinations of a particular transport demand of single or multiple commodities (e.g. freight, passengers). Each link has a fixed cost, length and capacity. The fixed cost is considered when a particular link is used, while there are variable costs, coupled with each used link, related to the commodity flow. The capacity depicts the volume of the traffic that a link can accommodate. The objective is to select a subset of links of the network to satisfy all the transportation requests of the nodes and to determine how much and which commodity will travel via each link such that the total fixed and variable costs are minimised.

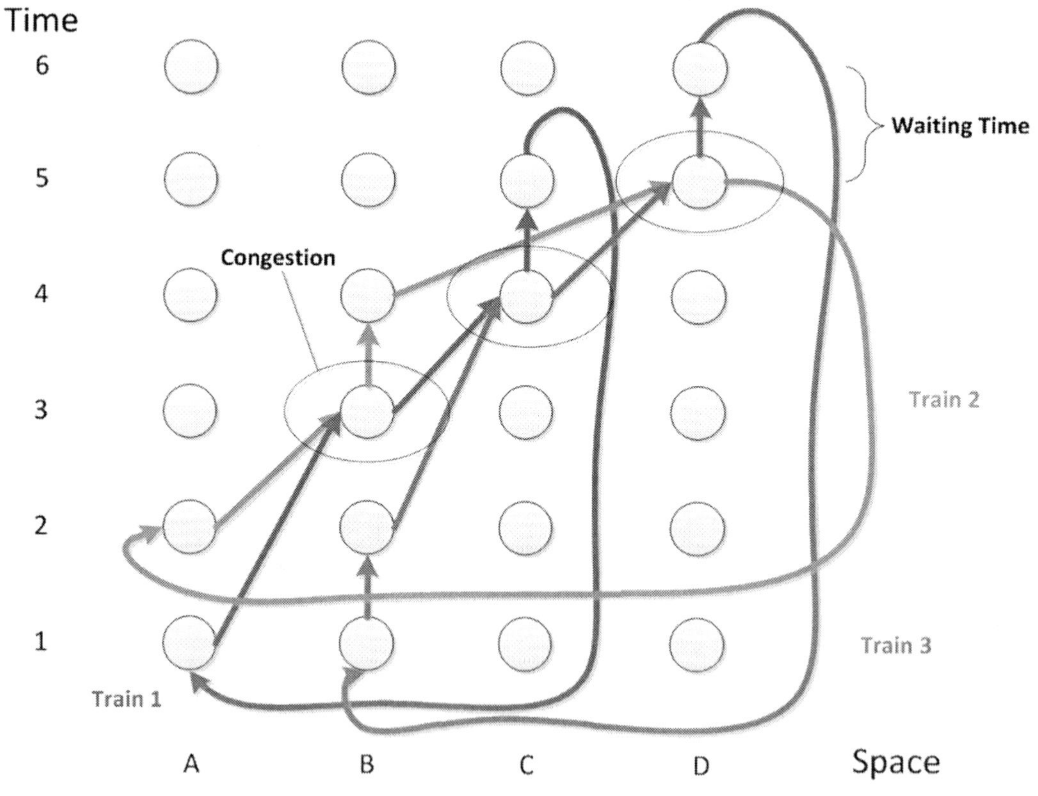


**Figure 3: Cyclic time-space network**

To model the MCNDP, a two-dimensional time-space representation was used, as shown in Figure 3, with the horizontal axis depicting the stations and/or junctions, and the vertical axis the time horizon of 'time buckets'. Each node of the network thus reflects the origin and/or destination at each time bucket of the time horizon.

The capacity of each node is considered to be one, i.e. only one train can arrive at the station/junction and only one can depart at a specific time bucket. Solving the MCNDP for such a network is equivalent to selecting the appropriate flows to satisfy the passenger and freight demands, such that the total cost is minimised. In the context of the OCCASION project, the objective was to minimise the capacity utilisation of each station, which in turn minimises the potential for congestion and provides an opportunity to add capacity by introducing additional services. An efficient metaheuristic algorithm was developed, which produces high-quality solutions for large-scale problem instances; its performance was tested on a benchmark set of problem instances (9), and the computational results indicated that the performance of the algorithm compared favourably with the current state of the art. The MCNDP model was integrated with the Job Shop scheduling model to provide a unified solution framework.

## 5 CHANGES TO TIMETABLE PLANNING RULES AND INFRASTRUCTURE TECHNOLOGY

Investigations were undertaken of the potential capacity benefits arising from changes to the Timetable Planning Rules (specifically, the reduction of planning headways) and infrastructure technology (the introduction of higher-speed turnouts at critical nodes, i.e. those nodes with the highest levels of capacity utilisation).

However, it was found that no additional Main Line services to/from London could be accommodated without reducing the headway to a value of 90s, due primarily to the two-track section of line between Peterborough and Huntingdon. In respect of the improved turnouts, it was found that, where they could provide significant capacity benefits, there was insufficient space within the constraints of the existing infrastructure to accommodate them, and, where they could be accommodated, they would not provide significant capacity benefits.

## 6 CONCLUSIONS AND FURTHER WORK

This paper describes novel approaches to (i) the assessment of railway station and junction capacity utilisation and (ii) the re-routeing and re-scheduling of train services to reduce capacity utilisation and thus provide improved service reliability and/or additional capacity. Capacity utilisation (CUI) analysis of the individual nodes (point ends, crossings and station platforms) and intermediate links comprising the OCCASION study area indicated relatively low (23%-26%) average values, but maximum values in the range of 51%-81%. The subsequent optimisation process suggests that these maxima constrain the scope for the introduction of additional services. It was nonetheless found that the re-scheduling of services could both eliminate scheduled waiting time for trains, and enable the accommodation of additional timetabled services, although not on the ECML between the study area and London, where the potential benefits are likely to be greatest. Modest changes to operational rules (in the form of reduced headways) or to infrastructure technology (through the use of higher-speed turnouts) were not found to substantively change this conclusion.

The work undertaken provides a basis for improvements to operational resilience and/or capacity at network nodes, and thus across the railway network. It will be further developed in collaboration with the Universities of Leeds and Swansea in the course of the DITTO Railway Systems (Developing Integrated Tools to Optimise Railway Systems) project, funded by RSSB under the FuTRO (Future Traffic Regulation Optimisation) research programme. DITTO Railway Systems aims to (i) enable increased levels of capacity utilisation while maintaining standards of operational reliability, (ii) consider real-time movements of individual trains in the context of system-wide traffic management to produce optimal outcomes, (iii) enable the production of optimal timetables that can be amended in real time in response to operational perturbations, and (iv) to underpin the preceding three objectives by ensuring operational safety. These objectives are consistent with the Operational Philosophy for the national railway, whose development has been facilitated by Arup with support from the University of Southampton, on behalf of RSSB and the wider railway industry.

The methods and tools already developed in the course of the OCCASION project have been further developed and applied in commercial projects, notably in the recalibration of Network Rail's Capacity Charge by Arup and the University of Southampton. For the purposes of the recalibration, hourly and three-hourly CUIs were calculated across the national network by assigning weekday, Saturday and Sunday timetables, totalling in excess of 60,000 trains, to a ~6,000-link representation of the national network, and calculating the resulting levels of capacity utilisation.

Work is ongoing, in collaboration with Arup and the ACCVA project partners, to further refine the CUI and UIC 406 calculation processes, and to calibrate both measures in order to establish suitable upper limits for overall station and junction nodes. For the CUI development work, it is planned to use a combination of simulation techniques and historic timetable and performance data to test the robustness and sensitivity of the models developed, including their application to a larger set of case studies.

## REFERENCE LIST


(1) DfT (2011), *High Speed Rail: Investing in Britain's Future* [online], Crown copyright. Available from http://webarchive.nationalarchives.gov.uk/20110405154200/http:/highspeedrail.dft.gov.uk/sites/highspeedrail.dft.gov.uk/files/hsr-consultation.pdf [Accessed 8th September 2014].

(2) RSSB (2013), *EPSRC/RSSB/DfT Strategic Partnership – Capacity at nodes* [online], RSSB. Available from http://www.rssb.co.uk/Library/about-rssb/2013-05-Newsletter-InformationBulletin.pdf [Accessed 8th September 2014].

(3) UIC (2004), *Leaflet 406: Capacity*, Union Internationale Des Chemins De Fer (UIC), Paris.

(4) Armstrong, J., Blainey, S. Preston, J., Hood, I. (2011), *Developing a CUI-based Approach to Network Capacity Assessment*. Proceedings of the 4th International Seminar on Railway Operations Modelling and Analysis, RailRome2011.

(5) Gibson, S., Cooper, G., Ball, B. (2002), *Developments in Transport Policy: The Evolution of Capacity Charges on the UK Rail Network*, Journal of Transport Economics and Policy, Vol. 36, Part 2, pp. 341-354, May 2002.

(6) Szpigel, B. (1973), *Optimal train scheduling on a single line railway*, Operational Research 72, M. Ross (ed.) North Holland, Amsterdam, 344–351.

(7) Liu, S.Q., Kozan, E. (2009), *Scheduling trains as a blocking parallel-machine job shop scheduling problem*. Computers & Operations Research 36 2840 – 2852.

(8) Glover, F. (1986), *Future paths for integer programming and links to artificial intelligence*, Computers and Operations Research 13, 533–549.

(9) Crainic, T.G. (2000), *Service network design in freight transportation*, European Journal of Operational Research 122, 272–288.

(10) Ahuja R., Magnanti, T., Orlin, J. (1993), Network Flows: Theory, Algorithms and Applications, Prentice Hall: Englewood Cliffs, NJ.

# Investigation into the effect of moisture on rail adhesion

**J Fisk[1], T Armitage[1], R Lewis[2], D Fletcher[2], M Evans[2], A Arnall[2]**
[1] Arup, UK
[2] The University of Sheffield, Department of Mechanical Engineering, UK

**ABSTRACT**

This paper presents the results of a study, funded by RSSB on behalf of the Adhesion Research Group and the Vehicle Track System Interface Committee, into the effect of the presence of moisture at the wheel / rail interface during dew and damp conditions. Arup and the University of Sheffield have collaborated to perform a comprehensive and detailed review of the current body of knowledge in this area and an analysis of the latest relevant operational performance data.

For the purposes of the assessment of data the study has utilised a proven methodology (based upon the approach of the National Intelligence Model) to assess the quality and relevance of the information based on profiling each piece of 'evidence' related to wet-rail syndrome. The evidence was categorised and a knowledge map produced which shows the position of each piece of evidence (based upon its topic).

In addition extensive performance data was analysed from a variety of sources.

The available written evidence is well supported by information derived from performance data records and has led to the development of the following evidence based definition:

**Wet Rail Phenomenon:**
*Poor adhesion conditions caused when low levels of moisture are present at the wheel / rail interface. These conditions are associated with dew on the rail head, very light rain, misty conditions and the transition between dry and wet rails at the onset of rain. They are not necessarily associated with the additional presence of other (non-water) rail head contaminants.*

## 1. INTRODUCTION

The quantitative effect of the presence of moisture at the wheel / rail interface during dew and damp conditions is not widely understood. This is however an area which is subject to much debate and supposition. It is also an area where some academic and industrial research has been undertaken, though the results of these studies have not been collated into a single, readily useable body of knowledge on the subject.

On behalf of the Adhesion Research Group (ARG) this RSSB project aims to review and collate the available knowledge base on the subject, to determine what robust information is available and thereby:

- Deliver an overall definition of the "Wet Rail" phenomena
- Document the tried and tested mitigation that exists
- Facilitate targeted and prioritised future activities
- Provide a better understanding of the subject to allow industry to target and prioritise further work and investigation

## 2. BACKGROUND

Understanding adhesion levels between the wheel and the rail is an important aspect of ensuring safe and reliable railway operations. Considerable work has been undertaken by the industry over many years to build a body of knowledge dedicated to adhesion at the wheel / rail interface. Much of this research has been directed at understanding and mitigating low adhesion conditions during the autumn period which is defined in NR/L2/OCS/021 'Weather – Managing the Operational Risks' as 1st October to 13th December. Considerable progress has been achieved in this area, evidenced by reducing the critical KPIs of signals passed at danger (SPADs), station overruns, wrong-side track circuit failures (WSTCF) and the corresponding delay minute attribution to adhesion related issues during autumn periods up to 2012. Low adhesion challenges remain during autumn, but the industry is better equipped to recognise and manage them.

Given the progress up to 2012 of managing adhesion during autumn, some of the industry's focus is now moving to low adhesion events occurring outside of the autumn (leaf-fall season), or rather to events occurring at any time of the year (including autumn) but perhaps unrelated to leaf contamination. These events may now be more prominent as the mitigation of low adhesion caused by contamination is much improved and these events used to mask the 'secondary' causes. One phenomenon that has been identified is the occurrence of low adhesion throughout the year, when no rail head contamination is visible and there are reports of the presence of moisture on the rail head. Presence of low adhesion effects in these circumstances has come to be referred to as the 'wet rail' phenomenon.

## 3. METHODOLOGY

### Evidence Review

The project considered over 280 sources of evidence pertaining to wet rail syndrome (1). Each piece of evidence is assigned a primary category: General Railway Adhesion, Wet Rail Adhesion, Wheel / Rail Contact, Low Adhesion Mitigation, Railway Operational Performance, Braking Systems. A sub-category has also been assigned to each piece of evidence: Academic Research, Laboratory Testing, Quantitative On-track Testing, Qualitative In-service Observations, OTMR data, Train Performance Data.

Finally, the quality of the document has been assessed using a proven methodology (based on the National Intelligence Model). The evidence has been assessed for clarity, completeness, consistency, correctness, integrity and uniqueness.

### Knowledge Map

The Evidence Review has been used as the basis for a knowledge map (Figure 1) – the primary and sub categorisation have provided nodes on the knowledge map around which the documents have been positioned, whilst the consistencies form links between areas of the knowledge map. The map also graphically represents the quality of each piece of evidence using a traffic light system.

**Full Knowledge Map**

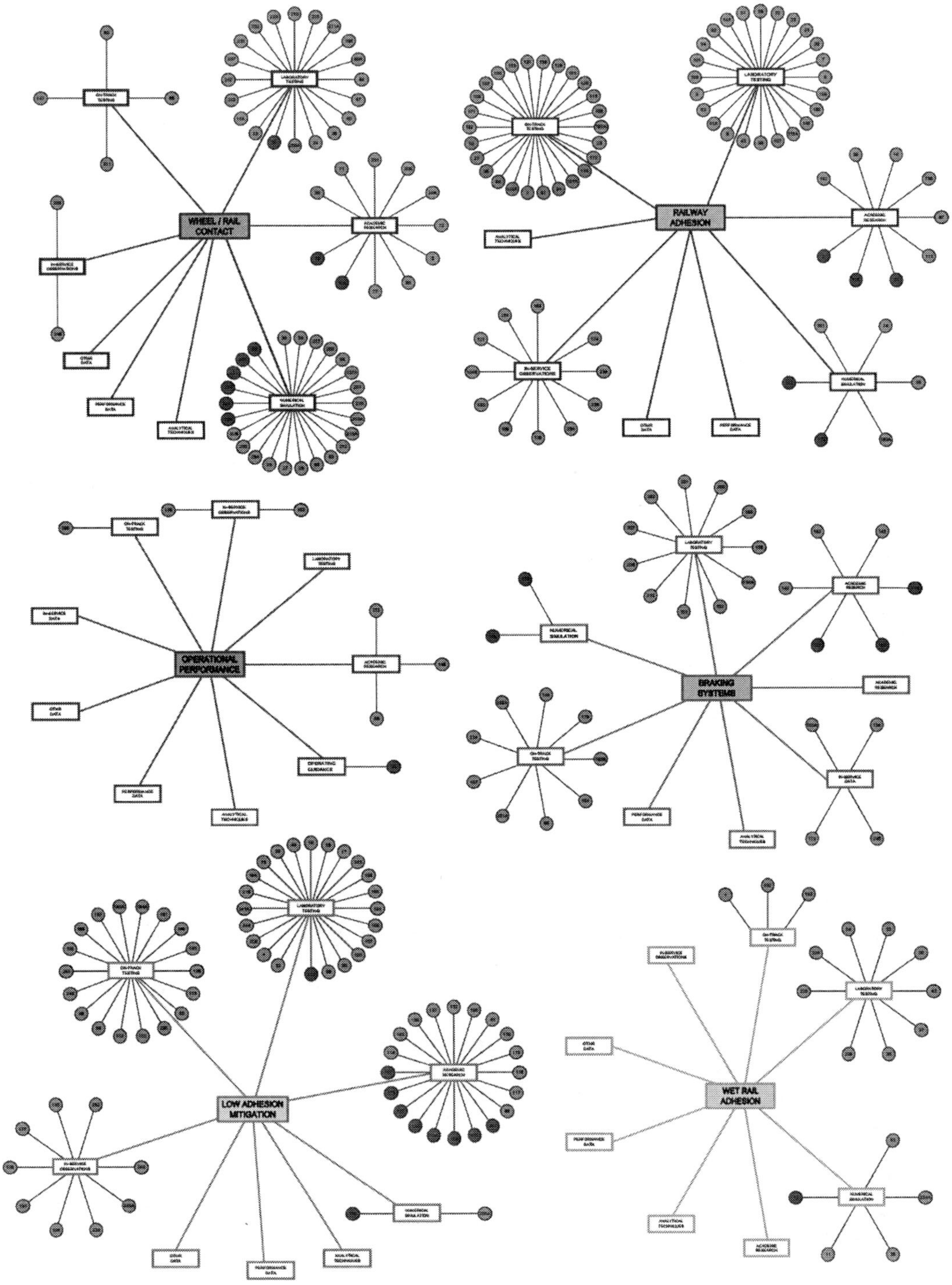


**Figure 1 - Mapping the full body of knowledge analysed in this review.**

**Operational Performance Data Review**
A range of recent performance related data has also been made available to improve understanding of the underlying causes of low adhesion at the wheel-rail interface. The key data set is a record of station overruns by hour of day during which they occurred, as a key performance indicator pertaining to rail adhesion. The data are normalised and plotted alongside national average weather conditions in order to make some suggestions on the effect of atmospheric condition.

## 4. EVIDENCE REVIEW

The results of the evidence review are presented in Figure 1. Each evidence ID was scored out of a maximum of six, based on the assessment described above. The knowledge map follows system analogous to a traffic light system, where those scoring 5 or 6 out of 6 are represented as white, those scoring 3 or 4 out of 6 are represented as black and those scoring 2 or below are represented as grey. (Or green, amber and red respectively on the high resolution colour version of the knowledge map, available in the conference proceedings.)

It is important to highlight that every piece of evidence presented in Figure 1 was ranked on one consistent system with no weighting given to a particular category or competency. For example, a peer reviewed paper focussing on small scale laboratory experimentation and containing no operationally validated testing data would not score a full six out of six, despite the fact that it may be of excellent rigour and very high standard, and despite the fact that the paper itself, and the authors behind it never set out to do anything other than that which they achieved in the publication. Careful consideration was given to this system of ranking, and it was decided that the most robust way to compare each piece of evidence, and the clearest way to present the results (knowledge map) as a whole, was to ensure that the system was consistent.

## 5. OPERATIONAL PERFORMANCE DATA REVIEW

### Environmental Factors

#### *Autumn Incident Data*
The large number of incidents in the Autumn Incident data-sets permitted the daily incident analysis for the whole autumn season and the hourly incident analysis throughout an average autumn day. For the 2012 Autumn Incident data-set the daily incident analysis was combined with national average weather conditions throughout the same period, which is publicly available from smtweather.co.uk. For the purposes of this report, weather data was averaged out over the whole of the UK for the autumn season. More localised analysis on particular days and areas of interest is recommended for future work.

Station overrun data by time of day was provided in categories (wet/dry/ unrecorded). The total number of station overruns normalised by station stops attempted is plotted in Figure 2.

Figure 2 shows that there is a peak in terms of number of incidents between the hours of 06:00 and 09:00 which is not simply a function of increased traffic. Furthermore, there appears not to be a significant peak at around 17:00 hours (when evening traffic is high), further implying that low adhesion events are not synonymous with high levels of traffic. There is another peak in the data at around 22:00 hours, and these two peaks (early morning and late evening) might be attributable to environmental factors such as dew forming or drying on the rail head. [1]

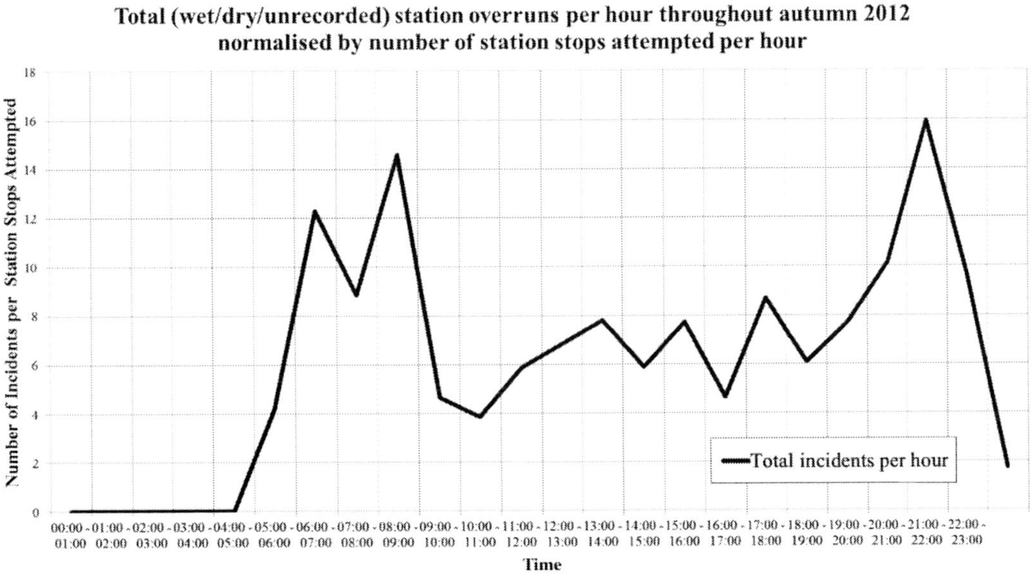


**Figure 2 - Plotting total number of station overruns per hour normalised by hourly traffic**

It has been postulated that low levels of moisture on a rail head can cause a lower co-efficient of traction when compared with a saturated rail head (during periods of prolonged rain for example), since the rain may wash away contaminants such as oil, oxides and wear debris (2,3). Another suggestion is that adhesion is at its lowest when the rail head is drying out (2).

Considering the possibility that adhesion is lowest as the rail head is drying out; then the train may perform multiple successful stops with dew present, but then attempt to stop as the rail head reaches a critical point in the drying out process, causing the train to experience low adhesion, seemingly inexplicably. If such a low adhesion event were to cause a station overrun or even a SPAD then the Mobile Operations Manager (MOM) would be called to inspect the rail head and would likely arrive to find a rail head that has dried since the incident.

In order to investigate the effect of atmospheric conditions, average environmental data (air temperature and precipitation) is plotted against daily station overruns for autumn 2012, in Figure 3 and Figure 4.

Figure 3 indicates a correlation between the average temperature over the country and the number of incidents which occur. Most noticeably, leading up to 27/10/12 the average temperature drops significantly and there is a large peak in incidents. This could be explained by the formation, and later drying out of dew on the rail head. Throughout the rest of the season incidents generally occur during, or immediately after, a drop in the average temperature, which may suggest clearer skies, high daily temperature ranges, and better conditions for dew formation.

[1] It should be highlighted that a relatively small sample size is collected (147 incidents in total) over this particular period – namely autumn 2012. It is suggested that the study is developed to incorporate a larger amount of data.

Unfortunately without humidity data it is difficult to conclusively prove whether or not these would result in dew formation.

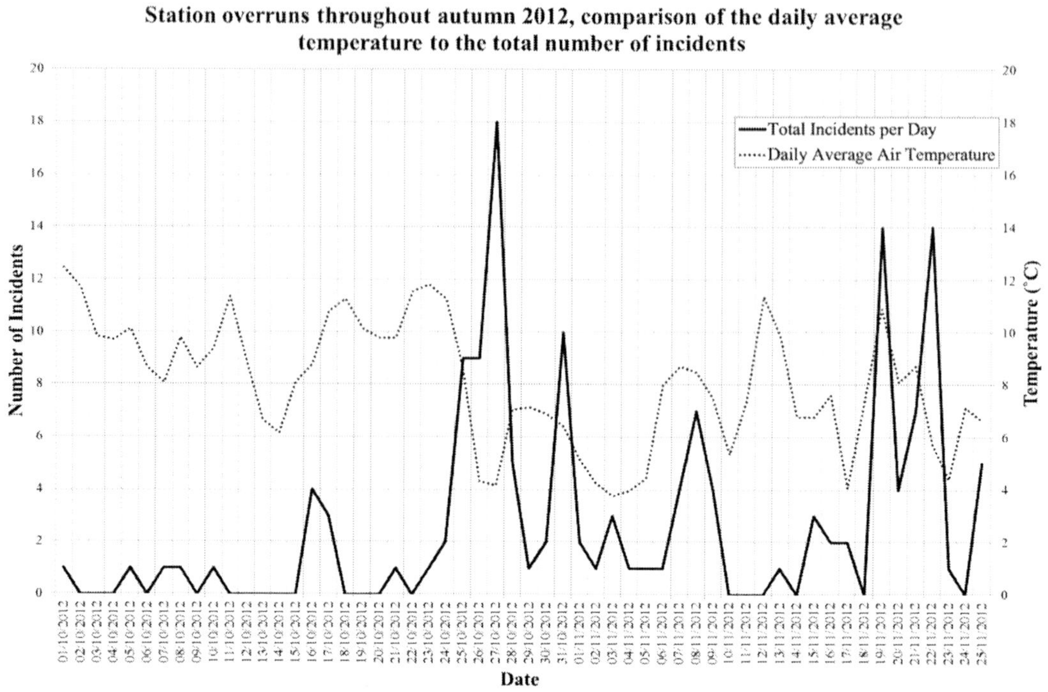


**Figure 3 - Autumn Incidents data-set 2012, comparison of the daily average air temperature to incidents in wet, dry or unrecorded weather conditions**

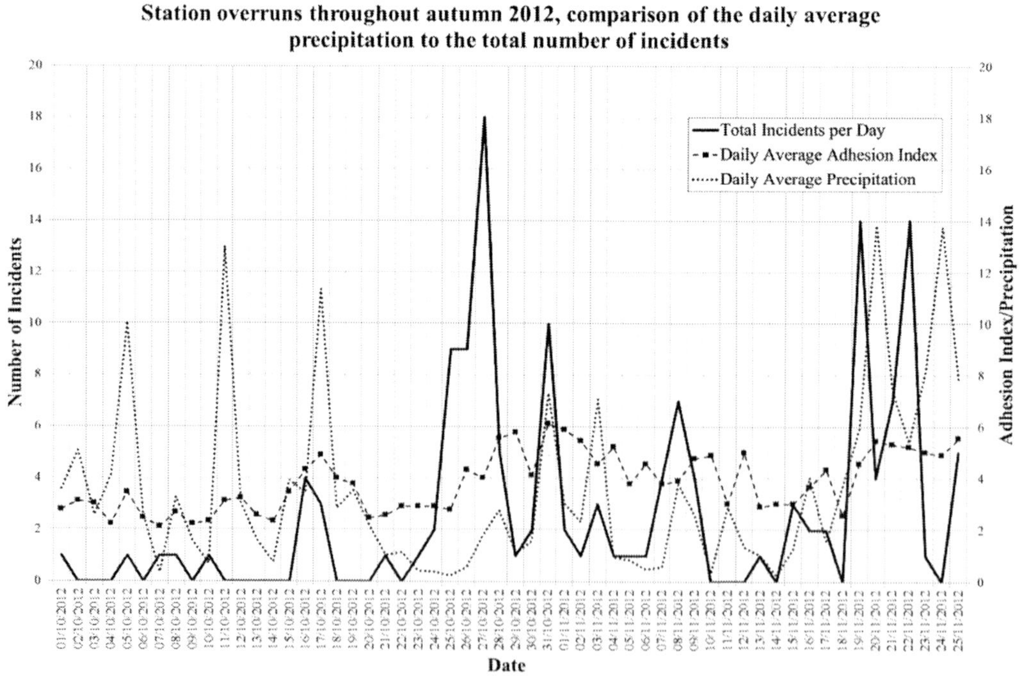


**Figure 4 - Autumn Incidents data-set 2012, comparison of the daily average precipitation and adhesion index to incidents in wet, dry or unrecorded weather conditions**

Furthermore, it is important to highlight that country-wide average daily temperature is likely to vary widely across the country. Hence, a direct correlation between temperature and low adhesion events would not be possible to derive from this information alone. This highlights the importance of good weather data gathered near and immediately during a low adhesion event, which is something recommended for future consideration.

Towards the end of the season there are two periods with high incident rates: 19/11/2012 and the 22/11/2012, the second of which occurs alongside a drop in air temperatures. Figure 4 indicated this might be due to precipitation.

Figure 4 shows that there was an increase in incidents at the end of the season when the adhesion index (an industry measure of leaf fall) was high, and a day before the precipitation increases. It has been proposed that both of these incident peaks are caused by leaf contamination and the air being humid due to a change in weather patterns before a major rainfall, rather than the morning dew interacting with oxides or other non-leaf contaminants. There is also a smaller peak at 31/10/2012 which might also be explained by a leaf-rainfall combination; whereas on 03/10/2012 there is significant rainfall but a lower adhesion index causing no significant increases in incidents. It is important to highlight the simplistic nature of this approach and to suggest more detailed weather and environmental observations are kept alongside low adhesion performance data.

It is also interesting to note that many of the incidents which occur on the dates of 19/11/2012 and 22/11/2012 occur later in the day when compared with those dates earlier in the season. They account for many of the 'wet/dry unrecorded' incidents occurring over the afternoon-evening in Figure 2 and Figure 3. The diurnal change of humidity for a typical day may account for this, showing an increase in the evening to levels typical of dawn for which low adhesion is more familiar (Figure 5). This would indicate that it is not just a drying rail which may suffer low adhesion but also one becoming gradually wet as dew begins to form.

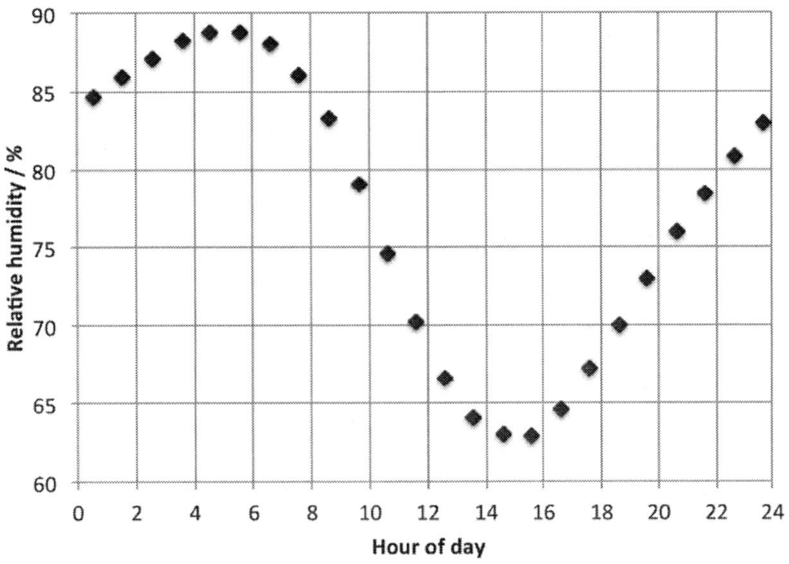


**Figure 5 - The diurnal cycle of relative humidity at Heathrow, average for thirty-year interval 1981–2010, extracted from (4)**

Evidence from autumn 2012 suggests that low adhesion events can be categorised into two main causes:

- A combination of high adhesion index (leaf fall) and precipitation or the environmental changes that occur on the lead up to precipitation.
- Morning dew post local air temperature drop interacting with contaminants such as oxides or leaves.

## 6. EVIDENCE BASED DEFINITION OF "WET RAIL SYNDROME"

Firstly, it is important to note from the Knowledge Map presented in Section 4 that formal evidence based literature specifically on the topic of wet rail phenomena is (relatively speaking) lacking in terms of quantity and quality of published papers. These are papers whose primary focus is low adhesion between rail and wheel under non-contaminated wet conditions. That is, however, not to say that those papers in other categories cannot help in the understanding and indeed definition of such phenomena.

Low adhesion events occur, or at least are recorded, most frequently during the autumn months (which are defined in NR/L2/OCS/021). Many of these events are attributed to the contamination of the rail head by crushed leaf debris. This study has further highlighted several facts about this particular phenomenon:

- That often, low adhesion events during this period are attributed to crushed leaf debris contamination with no real robust evidence to suggest this is definitely the case (this is evidenced by our own performance data analysis);
- Nevertheless that it certainly is a phenomenon which does occur and is repeatable in both track and laboratory testing environments (5,6);
- Furthermore, it is a phenomenon which is fairly well understood in terms of the mechanisms by which adhesion is reduced at the wheel / rail interface in this way (5–10);

To take the other extreme, it is also clear from academic understanding, laboratory data and indeed from an operational performance viewpoint that adhesion between a dry, clean wheel and a dry, uncontaminated rail head is high and that low adhesion events under these conditions are extremely rare (11).

If these two extremes are considered as boundary conditions, factors and events *between* these extremes in might help define some understanding of what has come to be termed "wet rail syndrome". Again therefore, this study has highlighted several facts about those *in-between* factors:

- That crushed leaf debris is not the only type of contaminant found on the rail head (12–14);
- That other types of contaminants will cause low adhesion between the wheel and rail when in solution with water according to a variety of fairly well understood mechanisms (12);
- That wheel and rail will indeed slide and slip with respect to one another under rolling contact when no contamination at all other than pure water itself is present as a "lubricant" (11,15–17);
- That, in terms of operational performance, low adhesion events (whether contamination is conclusively found or not) are more prevalent in "slightly wet" conditions, where the rail-head is just about to dry, or just about to become wet (this is evidenced in (2,3) and from our own performance data analysis);
- That the inverse is also true, namely that "very wet" conditions give rise to better adhesion performance when compared with those conditions best described as "slightly wet" (2);

These conclusions give rise to a sixth point, not taken directly from the literature but interpreted from an overall analysis:

- That this particular phenomenon (pertaining to "slightly wet" conditions) might also be repeatable in controlled environments (whether on the track or in the laboratory) and that to undertake study in this particular area might provide very useful insight in the field of wet rail adhesion;

It is important to carefully consider each of these facts and collate the conclusions, with reference to the pieces of evidence from which they are drawn, to determine a more encompassing and evidence based definition of "wet rail syndrome".

To take the issue of other third body contaminants: This study has revealed that there is much live and historical activity pertaining to the analysis and understanding of the effect of contaminants of the rail head. This data include the analysis of the way in which these contaminants are deposited and build up on the rail head (18), and indeed the exact mechanism by which adhesion between the wheel and rail is affected (12–14,18,19). Loosely speaking, contaminants other than those consisting primarily of leaf debris are categorised in the following ways:

- Oil, fuel and grease (etc.) from the train and that which is introduced by road vehicles at level crossings;
- Brake debris from brake pads and other parts of the braking system;
- Wear debris from the wheel or rail themselves;
- Oxidation products such as rust;
- Other foreign materials which are introduced to, or present at, the wheel / rail interface (such as coal dust on freight lines or kerosene, which has condensed onto the rail, on routes near airports for example)
- Any combination of the above (including leaf debris);

It is important to highlight these phenomena because it has been suggested in the past to consider "wet rail syndrome" as occurring under uncontaminated wet conditions only. In trying to define those low adhesion events (and the factors which cause them) outside of autumn, or indeed inside the autumn months but where no leaf debris is found, it must be considered that these other contaminants may be present and that they can have a significant effect on the level of adhesion between the wheel and rail.

Nevertheless, with only the possible exception of oil, fuel and grease, all the above highlighted contaminants only become a problem when water is also present (2), i.e. when they are in solution and form a paste. This is perhaps the reason for which "wet rail syndrome" has been conventionally so named, water is indeed a requirement even under contaminated conditions. Furthermore, and further reinforcing the view that water is the "overriding influence" (2), low adhesion between wheel and rail will occur in both track and laboratory environments when water is the only presence between the rail and wheel, i.e. when no other contaminants are present whatsoever. This is especially evidenced in the results of a number of laboratory testing programmes carried out under these specific conditions (11,16,20).

Finally, concerning the matter of the mechanisms by which water and suspended contaminates cause the reduction of adhesion at the wheel / rail interface. One curious clue as to this mechanism (or combination of mechanisms) is the fact that data pertaining to "slightly wet" rail (both in terms of controlled environments and in-service observations) is rife and conclusive; namely that adhesion is much lower when the rail head is on the point of drying, or on the point of becoming wet (2,3) (this is also evidenced in the current study in the performance data analysis).

Several reasons for this have been postulated. One popular comment is that when the rail is very wet the concentration of contaminant in solution is very low and the rail is washed clean. When the rail head is just about to dry (for example) the concentration of contaminant in solution is very high and therefore a paste forms which is trapped between the rail and wheel (2). These particular comments do not explain incidents where no observable contaminant is present or where, in controlled environments, it is known that clean water is the only "third body".

Under these conditions the phenomenon is perhaps less understood. Maybe when the rail head is just about to dry only a small amount of water is trapped between the rail and the passing wheel, whereas under wetter conditions the flow of water causes more favourable adhesion characteristics. This may go some way to explain the observation that "very wet" rails (under heavy rain conditions for example) give rise to better adhesion characteristics when compared with "slightly wet" rails, albeit worse when compared with fully dry conditions (11). Certainly, this is something to suggest for further research. In any case, the fact that low adhesion events occur under the influence of water alone is certainly not a myth but is evidenced on the track and in the laboratory.

It is also worth commenting on the difference between a rail head which is just about to dry and rail head which is just about to become wet. Clearly they comprise two different scenarios, and perhaps one transition gives rise to lower adhesion characteristics than the other. There is little information in the current body of knowledge to better understand this transient behaviour and is something to suggest for further work, in terms of laboratory and track testing.

Building towards a comprehensive evidence based definition of wet rail phenomenon, it is most relevant given the literature, data and scientific understanding available to incorporate the following themes:

- The phenomenon is a cause of low adhesion between rail and wheel, to the extent which causes problems of all potential levels of severity,
- Occurring as a result of water or moisture on the rail head,
- Often, but not exclusively involving other third body contaminants,
- That it is a particular concern (in that the likelihood and severity of low adhesion characteristics increases) when the rail head is on the point of becoming wet or on the point of becoming dry,
- That it is therefore defined as a transient phenomenon, likely a strong function of environmental factors such as air temperature, humidity and dew point,
- That mitigation will be inherently complex, in terms of how and when to deploy it,
- Nevertheless that with improved understanding it might be possible to suggest better ways to predict and mitigate against such low adhesion events.

**WET RAIL PHENOMENON**

**Poor adhesion conditions caused when low levels of moisture are present at the wheel / rail interface. These conditions are associated with dew on the rail head, very light rain, misty conditions and the transition between dry and wet rails at the onset of rain. They are not necessarily associated with the additional presence of other (non-water) rail head contaminants.**

**These conditions are not associated with continuous rain.**

## 7. CONCLUSIONS

Low adhesion on the railways occurs due to a vast array of extremely complex phenomena. The rail industry seeks to simplify and categorise these phenomena in order to understand and predict their effects, and to mitigate against them. While this simplification is necessary and practical, true understanding and effective mitigation is inherently complex due to the chaotic nature of the wheel / rail adhesion system. Nevertheless, some helpful conclusions can be drawn from this study.

From the extensive review on the current body of literature pertaining to railway adhesion, mitigation against low adhesion events and wet rail phenomena, the following conclusions can be drawn:

1) There is vast knowledge database containing good, conclusive and useful information on the phenomena of low adhesion during the autumn months due to crushed leaf debris on the rail head.
2) This information includes descriptions of the ways in which leaves are introduced into the wheel / rail contact and crushed on the rail head, the characteristics of the debris or "third body" which accumulates on the rail head, and on the precise mechanisms by which adhesion levels are reduced at the wheel / rail interface due to the presence of this particular contaminant.
3) There is also much live and historical activity to precisely investigate the characteristics of other so called "third body" contaminants such as oil, oxidation products, wear and brake debris, and others.
4) There is a good body of knowledge on contact theory including good understanding of friction coefficients between the wheel and rail as a function of environmental and operating conditions, including the presence of above mentioned contaminates.
5) There is some good data on adhesion between the wheel and rail in wet conditions, mostly in the form of laboratory small scale testing data focussing on the effect of surface roughness and contact pressure.
6) Nevertheless there are certainly gaps in the knowledge database in terms of detailed understanding of adhesion behaviour under such conditions.
7) Furthermore, while some sources indicate that adhesion phenomena under wet rail conditions are particularly sensitive to the transition from dry to wet or vice versa, very little formal academic or empirical investigation on this specific sensitivity is contained within the current body of knowledge.

Contained within the body of knowledge itself the following conclusions can also be drawn:

8) When visible contamination is present on the rail head it will dominate the adhesion characteristics between the wheel and the rail. The most effective way to mitigate against this is the use of properly functioning on-board sanders, and the treatment of the rail head by water jetting in conjunction with the application of adhesion modifiers.
9) Nevertheless adhesion characteristics at the wheel / rail interface can be significantly impacted by water alone, and that this impact is most significant when the levels of moisture are low.

From the review of the performance data made available to the delivery team during the 2010 and 2012 seasons the following conclusions can be drawn:

1) Colder average UK air temperatures appear to significantly influence adhesion performance, in that a drop in average air temperature is

associated with an increase in the number of low adhesion events in autumn 2012. This might also be explained by the fact that autumn frosts are associated with accelerated levels of leaf fall. However it is difficult to draw confident conclusions without more precise (geographically accurate) temperature and humidity data available.

2) Two main scenarios leading to low adhesion events have been suggested, these aren't exclusive and other factors may also contribute. These are:
    a. A combination of a leaf layer and rail head moisture, often from precipitation, or the environmental changes leading up to precipitation. These events may occur throughout the day, but especially during the afternoon.
    b. Morning dew interacting with contaminants such as oxides or leaves. These account for the high incident rate during peak morning service. Affected mornings tend to be during colder weather. There is a possibility that low adhesion events may be most likely to occur during the drying out of the rail head.

## 8. RECOMMENDATIONS

Recommendations for future research activity:

1) Controlled testing in small scale laboratory and / or full scale on-track testing environments focussed on understanding the effects of small amounts of moisture on the rail head
2) The study should be extended to determine the effect of wetting and drying of the rail head
3) Recommendations are also made to improve the quality of the environmental and adhesion data acquired and included in incident reports

Recommendations for improved operational performance:

1) Low adhesion incidents should be recorded with equal rigour and detail all year around (including outside of autumn) to aid investigations
2) Where possible, low adhesion events should be recorded along with local environmental conditions such as:
    a. Rail temperature
    b. Level of rainfall (if any), i.e. heavy, light rain, misty, dew, dry
    c. Rail head moisture levels
    d. Local humidity and dew points
3) Where SPADs are reported by MOMs, their arrival time should be noted
4) Reliable reporting of quantitative rail head moisture with more detailed moisture levels, and an indication of the moisture source (rain, dew, etc.) would aid investigations
5) A central data repository should be created for the storage and analysis of accurate and high quality KPI and environmental data

## ACKNOWLEDGMENTS

The collaborative project delivery team would like to formally thank RSSB for funding and guidance relating to the programme of research published in this paper, which is published with their kind permission.

The project steering group was also represented by Virgin Trains, Greater Anglia, Network Rail, and ATOC; all of whom are also acknowledged for their provision of data and technical support.

## REFERENCE LIST


(1) RSSB, (2014), Investigation into the Effect of Moisture on Rail Adhesion T1042 - June 2014. doi:10.1016/S2213-8587(14)70150-0.

(2) T.M. Beagley, C. Pritchard, (1975), Wheel/Rail Adhesion - The Overriding Influence of Water, Wear. 35 299–313.

(3) P. Dixon, (2012), Wessex Route - Wet Rails Research Update.

(4) Brimblecombe, P, Temporal humidity variations in the heritage climate of South East England, Heritage Science 2013, Vol 1, No 3, available from http://www.heritagesciencejournal.com/1/1/3

(5) U. Olofsson, K. Sundvall, (2004), Influence of Leaf, Humidity and Applied Lubrication on Friction in the Wheel–Rail Contact: Pin-on-Disc Experiments, Proc. Inst. Mech. Eng. Part F Rail Rapid Transit. 218 235–242.

(6) P.M. Cann, (2006), The "Leaves on the Line" Problem - A Study of Leaf Residue Film Formation and Lubricity Under Laboratory Test Conditions, Tribol. Lett. 24 151–158. doi:10.1007/s11249-006-9152-2.

(7) T.M. Beagley, I.J. Mcewen, C. Pritchard, (1975), Wheel/Rail Adhesion - The Influence of Railhead Debris, Wear. 33 141–152.

(8) W. Poole, (2007), Characteristics of Railhead Leaf Contamination.

(9) I.J. Mcewen, (1974), Adhesion and Leaves: Experiments with High Pressure Water Jets.

(10) C. Pritchard, M.A. Tanvir, (1973), A Basic Study of Wheel/Rail Adhesion 9. Further Observations on Leaves.

(11) K.-S. Baek, K. Kyogoku, T. Nakahara, (2007), An Experimental Investigation of Transient Traction Characteristics in Rolling–Sliding Wheel/Rail Contacts Under Dry–Wet Conditions, Wear. 263 169–179. doi:10.1016/j.wear.2007.01.067.

(12) Y. Zhu, (2011), Adhesion in the Wheel–Rail Contact Under Contaminated Conditions, Royal Institute of Technology, Stockholm.

(13) T.M. Beagley, (1975), Analytical Techniques for Examining Wear Debris.

(14) A.R. Savkoor, H. Ouwerkerk, (1995), Tribological Transitions due to Heat Dissipation during Braking on Contaminated Rails, Wear. 181-183 391–396.

(15) H. Chen, T. Ban, M. Ishida, T. Nakahara, (2002), Adhesion Between Rail/Wheel Under Water Lubricated Contact, Wear. 253 75–81.

(16) W.J. Wang, H. Wang, H.Y. Wang, J. Guo, Q.Y. Liu, M.H. Zhu, et al., (2013), Sub-Scale Simulation and Measurement of Railroad Wheel/Rail Adhesion Under Dry and Wet Conditions, Wear. 302 1461–1467. doi:10.1016/j.wear.2012.12.014.

(17) H. Chen, T. Ban, M. Ishida, T. Nakahara, (2006), Effect of Water Temperature on the Adhesion Between Rail and Wheel, Proc. Inst. Mech. Eng. Part J J. Eng. Tribol. 220 571–579. doi:10.1243/13506501JET75.

(18) S. Descartes, C. Desrayaud, E. Niccolini, Y. Berthier, (2005), Presence and Role of the Third Body in a Wheel–Rail Contact, Wear. 258 1081–1090. doi:10.1016/j.wear.2004.03.068.

(19) E. Niccolini, Y. Berthier, (2005), Wheel–Rail Adhesion: Laboratory Study of "Natural" Third Body Role on Locomotives Wheels and Rails, Wear. 258 1172–1178. doi:10.1016/j.wear.2004.03.028.

(20) H. Chen, T. Ban, M. Ishida, T. Nakahara, (2008), Experimental Investigation of Influential Factors on Adhesion Between Wheel and Rail Under Wet Conditions, Wear. 265 1504–1511. doi:10.1016/j.wear.2008.02.034.

# Development of a methodology for co-simulation of tilting trains in SIMPACK-SIMULINK

**N Kuka[(+)], C Ariaudo[(*)], R Verardi[(*)]**
[(*)] Railway Dynamics/Engineering, ALSTOM Ferroviaria S.p.A., Italy
[(+)] Railway Dynamics/Engineering, ALSTOM Transport, Italy

**ABSTRACT**

The aim of the paper is to present the latest developments and results of Alstom in the field of modelling and simulation of tilting vehicles. Starting from the existing model presented to IAVSD 2011, a new and more complex model has been developed for the closed-loop simulations of tilting vehicles in SIMPACK-SIMULINK environments, focusing on the Control and Command algorithms developed in SIMULINK and on the interface SIMPACK-SIMULINK.

A vehicle with electro-hydraulic tilting actuation and pneumatic Hold-off Device is studied, with the description of the main sub-systems and functional blocks.

Finally the functional and field validation is presented and results discussed.

**Keywords:** Railway dynamics, Multi-Body System (MBS), tilting vehicle, active system, validation

## 1. INTRODUCTION

In the last decades the railway operators have continuously challenged the railway industry to run faster on existing lines, as a very effective solution to increase the capacity of the lines, while maintaining or even improving the level of the passenger comfort and the running safety, and reducing the aggressiveness of the rolling stock on the track. Tilting trains are the right solution to meet such challenging requests, but a careful vehicle design and optimisation of the dynamic behaviour of the whole system is required.

Since the late sixties ALSTOM has developed different active tilting systems that allow a relevant speed increase in curves (up to 30% more), while keeping a very good level of ride comfort, thanks to the active rotation of the carbody around the longitudinal axis of the vehicle.

Alstom tilting technology has been installed on more than 2300 coaches (more than 309 tilting trains) and has covered more than 700 million km in service, considering all Pendolino trains currently operating worldwide on the principal railway networks (Italy, Germany, Switzerland, United Kingdom, Czech Republic, Finland, Russia, Spain, Portugal, Slovenia, etc.).

At this stage, a simplified model of the control and actuating system was still used, useful for early design phase and feasibility studies, while the full tilting system was modelled through compiled user-subroutine (Vampire environment), allowing good modelling of the system but not flexible enough to keep up with the continuous evolution of system architecture and features.

Thus a dedicated tool for the design and optimisation of the tilting vehicles, in closed-loop simulation, was necessary.

The paper presents the Alstom experience and latest methodology developments in the field of modelling and simulation of tilting vehicles.

The core of the work is the SIMPACK-SIMULINK Tilting co-simulation model, with a description of the mechanical model in SIMPACK, including the tilting kinematics, a description of the Simulink Model, made by different blocks representative of the physical system on the tilting train (real Control Unit and Actuating System), a description of the tool architecture with some details about the interfaces and functionality of the main modules. Finally an overview of the validation process and simulation results are presented and discussed.

## 2. TILTING TECHNOLOGY AND INCREASE OF SPEED

The "Tilting" railway vehicles are characterized by the possibility to incline the car-body inwards the curve, in order to allow higher speed in curve compared to conventional railway vehicles, increasing the not compensated lateral acceleration (n.c.a.) at track level, but maintaining the residual acceleration on the passenger at acceptable level.

The tilting vehicle can travel up to twice the value of maximum cant deficiency of a conventional vehicle. The consequence is that the lateral forces between wheel and rail are higher. In this scenario, the tilting device must have no impact (or, at least, minimum) on the repartition of wheel/rail forces: this is the case of tilting solutions designed by Alstom.

The Alstom tilting system considered in this study is reactive, that is the reference tilting signal/angle is calculated based on the information of the track geometry, vehicle speed, and accelerations measured by on-board sensors of the vehicle. Furthermore it takes into account a percentage of compensation of the n.c.a., optimized for the passenger's comfort. Several experiences have shown that, for comfort reason, it is not recommended to compensate 100% of the lateral acceleration on passengers.

The tilting command and control should be complemented by a system of monitoring and detection: given the complexity of the system it is necessary to monitor in real time the functionality of both the tilting as a whole, and each individual component, and detect any anomalies to intervene with appropriate corrective action.

For more details and a complete survey of the tilting trains in the world with their characteristics and performances, interested readers can refer to (1), (2), (3), (4), (5), (7) and (8).

## 3. ALSTOM PAST AND PRESENT IN THE SIMULATION OF TILTING VEHICLES

The design of tilting vehicles requires reliable simulation models, able to represent the topology and features of the vehicle and of the control and command logics, both in nominal conditions and in case of failures of the system or degradation of some of the components. From the 80s to the present day, the development of computational models in Alstom has progressed steadily.

In the 70s-80s, when commercial simulation codes in the railway dynamics field were not yet available, the first studies on tilting system were carried out developing in Fortran simple calculation models (see Fig. 1) with few Degrees of Freedom (DoF), to simulate just the carbody and bogie bolster rotation and to determine the actuation forces, basis for dimensioning of the electro-hydraulic tilting system.

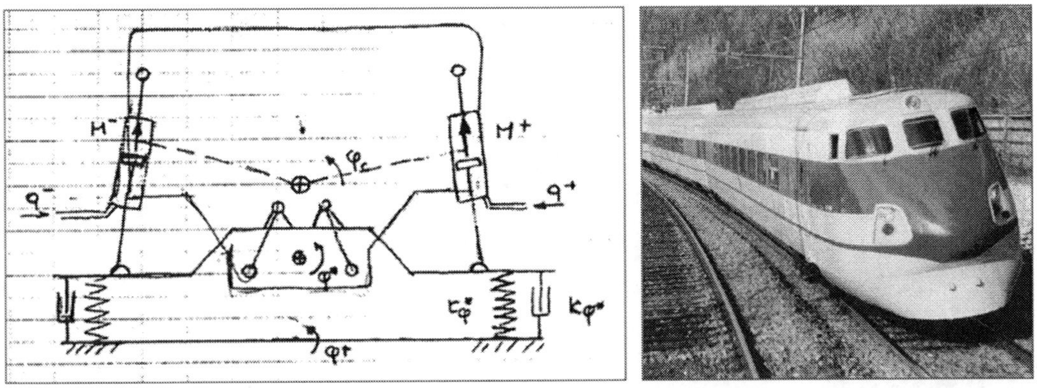

**Fig. 1 First homemade simulation model and 1st generation of Tilting trains**

The second generation of tilting trains was developed in the 90s in order to increase the performance of trains over the previous generation. Consequently, the simulation models were upgraded in order to simulate the whole system (mechanical model and tilting system), to achieve more comprehensive and reliable results: thus the Fortran model was enlarged with the bogie with its suspension, the actuating hydraulic circuit (complete model), the Control Unit (simplified model) and the Hold-Off Device (HoD).

Later, in the 2000s, with the large-scale deployment of commercial codes for multibody (MBS) simulation, the in-house simplified models have been replaced by MBS VAMPIRE and SIMPACK, also for study and development of the tilting trains. At this stage it was still used a simplified model of the Control and actuating system (Fig. 2).

In fact the tilting system is ideal, without any delays in actuation and therefore it is not realistic for some simulations scenarios. That's why this simplified model is only recommended for feasibility calculations and for sensitivity analysis on main parameters, in nominal operating condition.

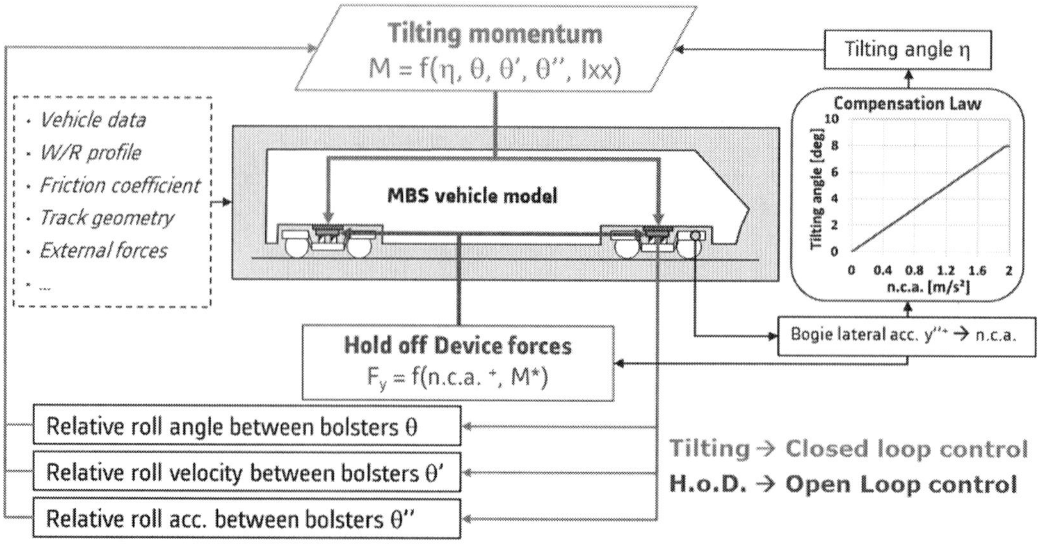


**Fig. 2 MBS and simplified tilting model**

The big evolution of computational models used for the study of tilting trains is represented by the closed-loop simulation: the model includes the conventional MBS model in-the-loop with the real Control Unit and Actuating System, and the diagnostics. Only in this way the mutual influence between all these elements of the train in all operating conditions can be studied. The Fig. 3 shows the main systems and subsystem considered during the modelling and the closed-loop simulation of the tilting vehicle.

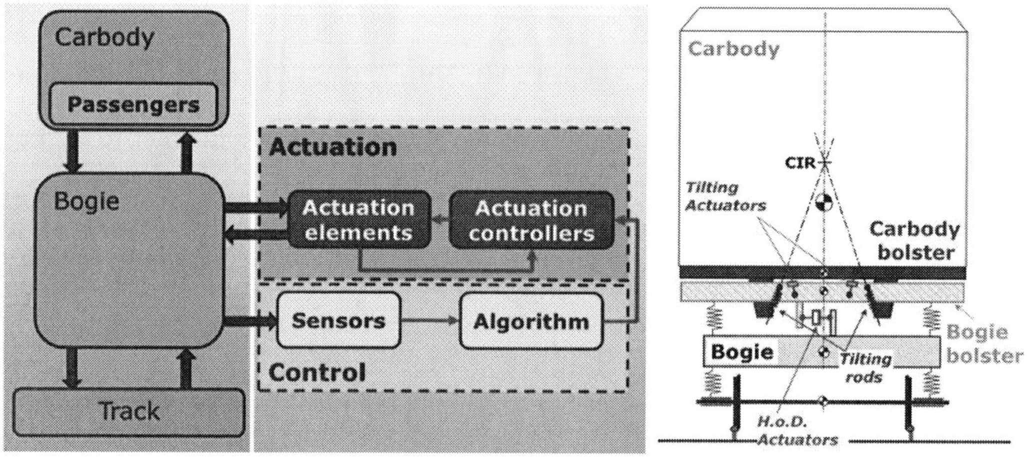


**Fig. 3 (left) Tilting vehicle closed-loop simulation; (right) mechanical model of the vehicle (Note: CIR = tilting centre)**

The integration of the tilting system within the MBS model can be done in different way. Usually one of the two following approaches is used:

- A compiled user-subroutine reproducing the Control Unit and actuating system in the same environment of the MBS code,
- A co-simulation between MBS (SIMPACK or VAMPIRE) and SIMULINK (a block diagram environment for multi-domain simulation of hydraulic, pneumatic, electric modules).

The first approach has been followed while working with VAMPIRE (Delta Rail Group Ltd) and its User Subroutine module, which allows the use of C or FORTRAN language to write the external code. The integrated model has been called

"VAMPIRE Tilting Simulator" and it has been fully described in papers (1), (2) and (6).

With this approach it is possible to reproduce quite well the real dynamic behaviour of the tilting vehicle in curves in normal operating condition, as demonstrated by the model validation process in (1), (2), (6). This model has been used to support the development of the latest generation of tilting algorithm, today installed on the fleets ICN (SBB), Class 390 (WCML in UK), ETR600/610 (Trenitalia and SBB) and Sm6 (Russia-Finland).

The limitation of this method lies in the fact that the tool, being a compiled user-subroutine, is not enough flexible to keep up with the continuous evolution of tilting system architecture and features.

This is the reason for the development and validation of the co-simulation model SIMPACK-SIMULINK, which will be treated extensively in the next chapter.

## 4. CO-SIMULATIONS IN SIMPACK-SIMULINK ENVIRONMENTS

Following the choice to adopt MBS commercial code SIMPACK (SIMPACK AG) as standard, a step forward has been made with the possibilities of holistic co-simulations with SIMPACK-Simulink, named hereafter "SIMPACK-Simulink Tilting Simulator".

Such a model allows the simulation of the behaviour of a complete tilting vehicle in normal modes, in degraded modes and in failure modes (including the transient phase of failure) for what concerns running safety (wheel/rail forces) and passenger ride comfort.

The co-simulation between SIMPACK and Simulink is managed by SIMAT interface (15): in this way both simulated subsystems (in Simulink and in SIMPACK) can take advantage of highly optimized integrators for their specific tasks in the original environments.

The Fig. 4 shows the overall architecture of the SIMPACK-Simulink Tilting Simulator, bringing out the different block-functions, functional dependencies, external input/output interfaces and communications points.

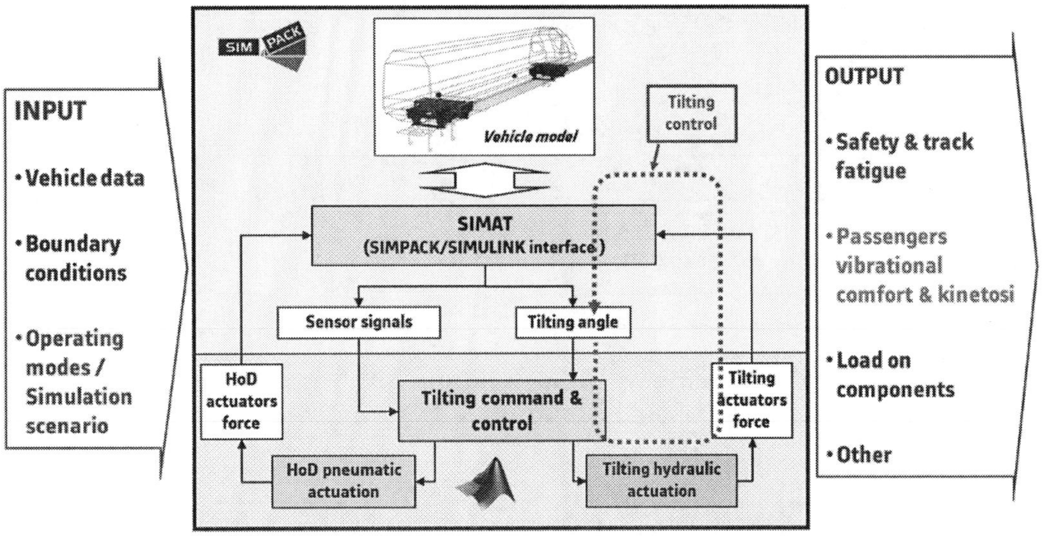


**Fig. 4 Overview of the SIMPACK-Simulink model**

This integrated model reproduces both the control systems of a generic tilting vehicle: the carbody tilting control, in closed loop, and the HoD control system, in open loop.

The SIMPACK-SIMULINK Tilting Simulator, being based on co-simulation and on a modular structure of the Simulink model, allows easy adaptation to different solutions and architectures and new modules can be easily added to complete or improve the representation of the command & control and actuation system.

The methodology has been developed to study the dynamic behaviour of the latest generation of tilting vehicles "ETR600/ETR610 Pendolino" (Fig. 5): the tilting system is located above the secondary suspension, the kinematics is made with rods/pendulum and the actuation is electro-hydraulic, with HoD.

**Fig. 5 ETR610 Tilting Bogie**

In Fig. 6 (left) the topology of the tilting bogie is shown, while Fig. 6 (right) clarifies the relative movements of carbody centre of gravity and of tilting centre during the tilting rotation.

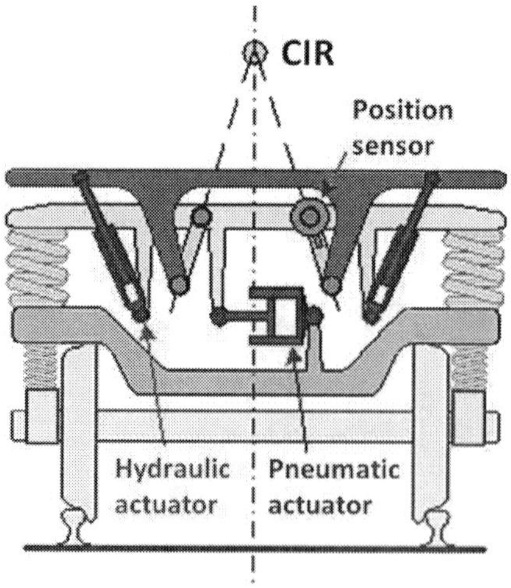


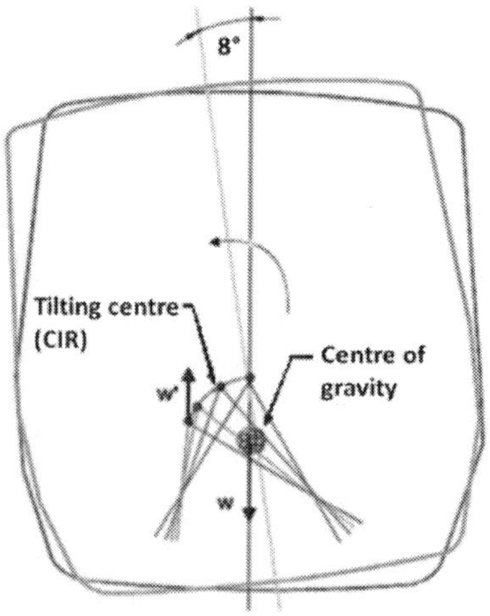


**Fig. 6 (left) Scheme of tilting system with rods/pendulum - (right) Kinematics of tilting**

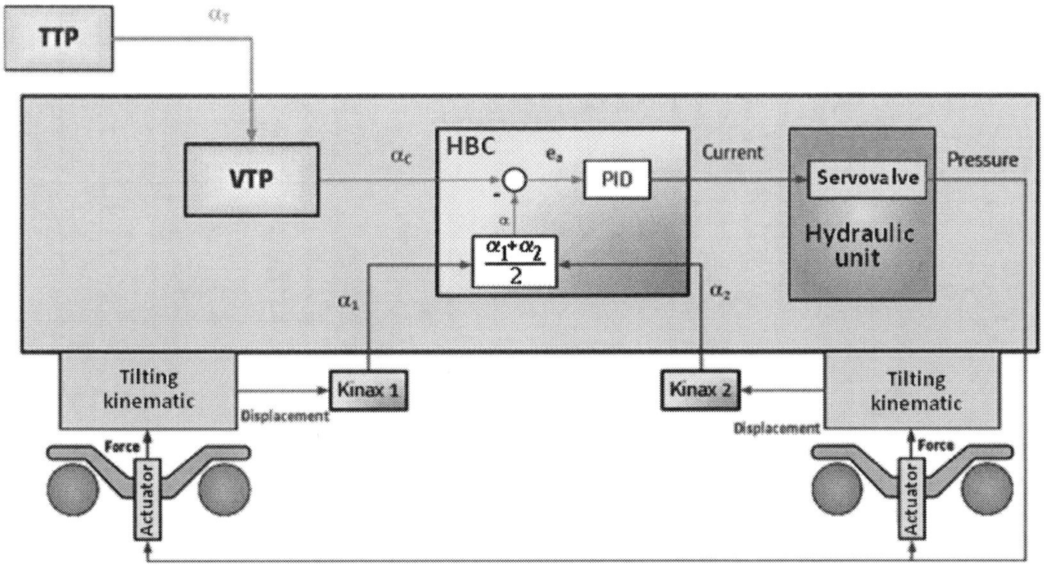


**Fig. 7 Block Scheme of Carbody Tilting Actuation (source Alstom)**

The Tilting System (Fig. 7), implemented in the ETR600/610 is made by:

- Sensors providing the information for the generation of tilting command (curves, cant, speed, accelerations)
- The electronic devices Train Tilting Processor (TTP) and Vehicle Tilting Processor (VTP) providing control and regulation
- The hydraulic system Hydrobox carbody (HBC) including power unit, pipes, actuators, etc.
- Pneumatic system (power unit, actuators, pipes, etc.)
- Sensors providing the information regarding the accelerations, forces, pressures, displacements (kinax) etc. and thus used for tilting actuations and control (feedback & diagnosis) in different operation conditions.
- Mechanical parts which assure the transmission of the forces/loads on the different parts of the system and the kinematics of the carbody during the rotation with respect to the bogie (tilting rods, etc.)

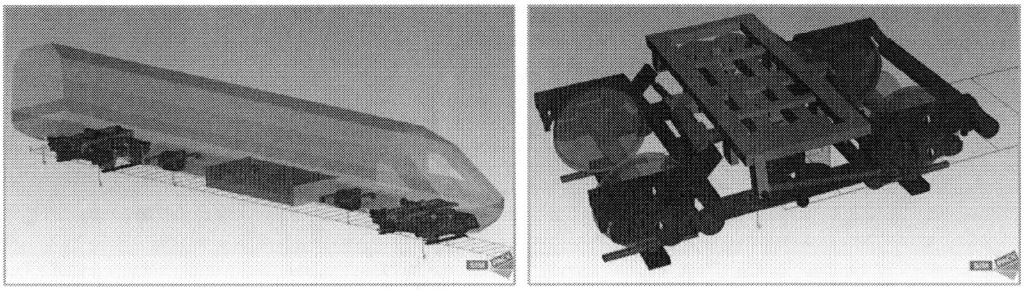

**Fig. 8 SIMPACK vehicle model**

The SIMPACK model (Fig. 8) consists of rigid bodies connected by elastic and viscous elements, including the body bolster and the bogie bolster, with rods.

The Simulink model of the tilting system is characterized by a modular structure (Fig. 9) that reproduces the main blocks of the system, i.e. the algorithm of calculation of the angle of carbody rotation (Tilting algorithm), the algorithm of calculation of the lateral displacement of the carbody (HoD algorithm), the hydraulic (Tilting) and pneumatic (HoD) actuations, diagnostics and supervision of the operation of the system.

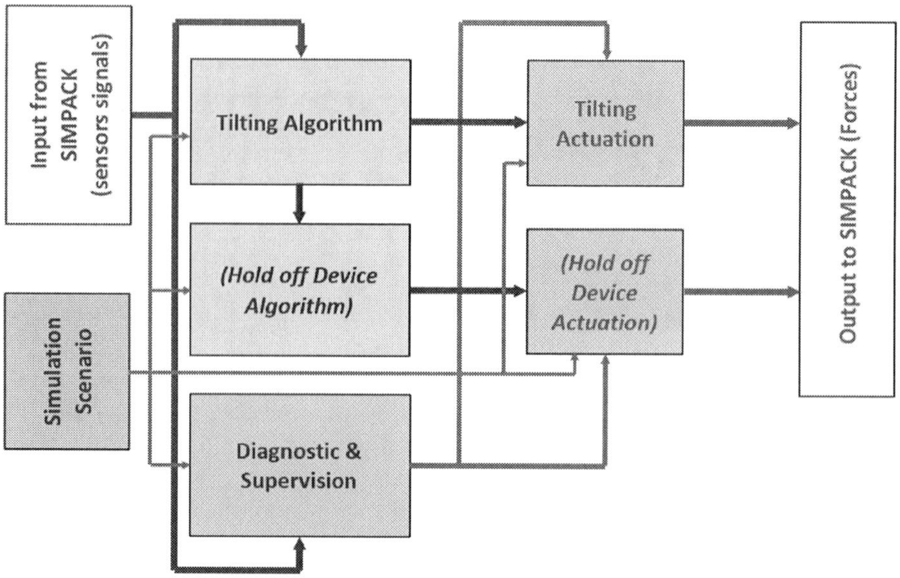


**Fig. 9 Simulink model of Tilting System: flow chart**

Below a brief description of the main modules is provided.

The Simulink model reproduces the functional behaviour of the real Train Tilting Processor, placed in the leading vehicle. The SIMULINK model of the tilting reactive algorithm (Fig. 10, taken from (3)), elaborates the command tilting angle and the HoD command, on the base of the sensors signals.

The lateral bogie acceleration, the roll and yaw angular speed of the bogie (X-Gyro, Z-Gyro) and train tachometers speed are the main input of this algorithm.

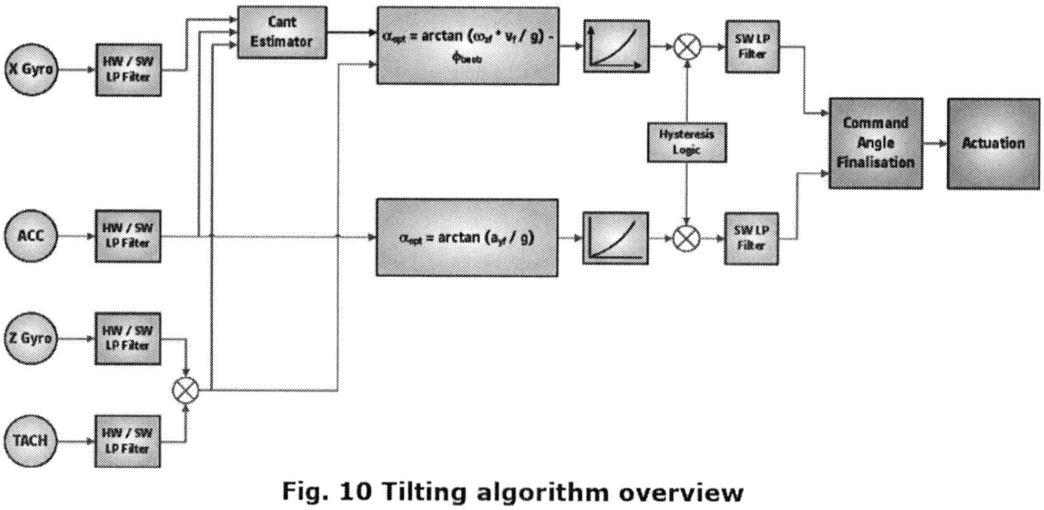


**Fig. 10 Tilting algorithm overview**

The Fig. 11 clarifies the scheme of the hydraulic tilting actuation, highlighting the isobaric connection between the actuators.

The modelling of the hydraulic actuation is based on the transfer function of the system, validated with the results of the bench tests and field tests of the tilting design solution. So it provides the correct behaviour of real hydraulic actuation in all the considered operation modes.

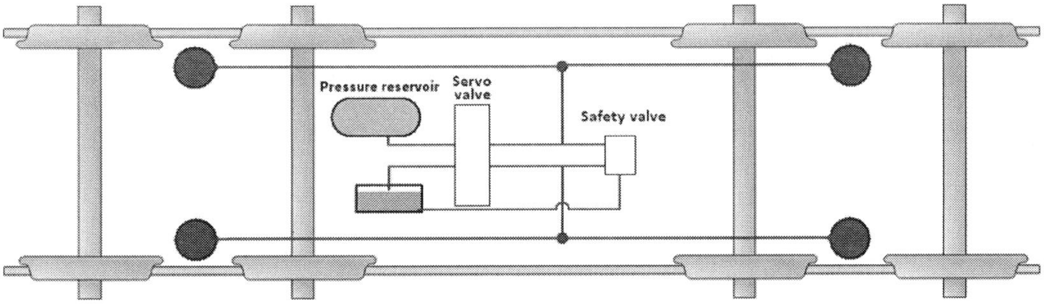


**Fig. 11 Carbody Tilt Actuation – Isobaric Connection (source Alstom)**

The Fig. 12 shows the functional diagram of the HoD pneumatic system modeled in Simulink. In this case all the relevant components of the real HoD system are reproduced in Simulink: compressor, air reservoir, pipes, electro-valves, actuators. This model uses the thermodynamics laws describing the behavior of fluids as a function of: gas constant and polytrophic law, the temperature, the length and the section of pipes, concentrated and distributed losses of hydraulic load and the real geometry of each components.

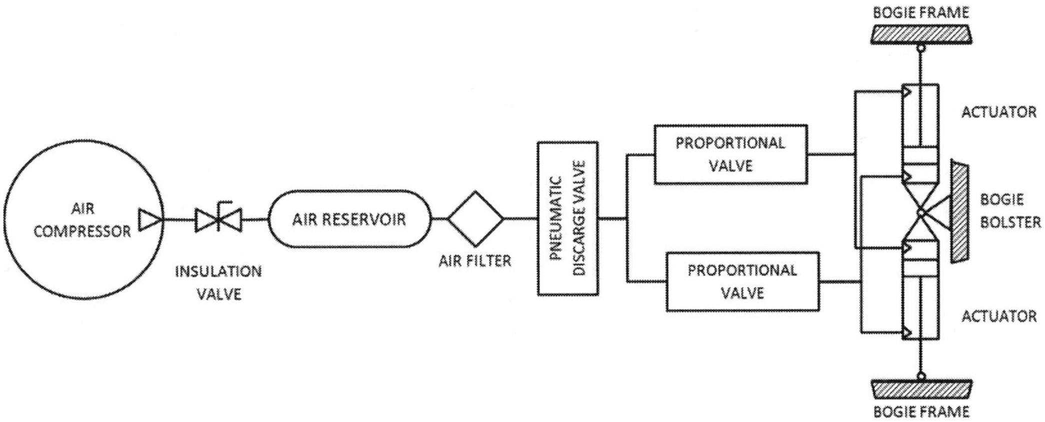


**Fig. 12 Functional scheme of the HoD pneumatic system**

In this way the behaviour of real hydraulic and pneumatic actuation in all the considered operation modes can be correctly simulated.

The Simulink model includes also some diagnostics and supervision control, such as the recognition of the rotation in the wrong direction or insufficient carbody rotation, which can have an impact on the dynamic behavior of the vehicle.

## 5. MODEL VALIDATION

Since SIMPACK-Simulink Tilting Simulator is based on a complex model, the validation process was divided in four main phases:

1. SIMPACK model validation
2. Functional validation of the Simulink Command & Control model
3. Functional validation of the SIMPACK-SIMULINK model
4. Field validation of the integrated model

The SIMPACK model validation went through the following steps:

- Model topology and characteristics based on best ALSTOM practice
- Model vehicle data based on available bench tests of some key design components (Anti yaw damper, bump-stops, etc.)
- Validation of the model based on static and quasi-static tests

For more details related to model description, simulations and validation of the MBS model through field tests the interested readers are referred to the works (1), (2), (6). In this work we focus on the modelling and validation of SIMULINK model of the tilting system and on the validation of whole SIMPACK-SIMULINK model.

The functional validation of the SIMULINK model is focused on tilting algorithm, both in normal operation conditions and in failure mode, using appropriate input data.

Test run on the line of the Gotthard Pass in Switzerland was simulated, which is in fact characterized by a high percentage of S-curves of small radius. The sensor signals (roll and yaw angular speed of the bogie and train speed tachometers) were provided as input to the algorithm of command and control and the consistency between the reference tilting angle in output and the expected response of the system has been verified.

The phase of the functional validation of the integrated model SIMPACK-SIMULINK has the objective of verifying that the co-simulation results match qualitatively the expected behaviour of the tilting vehicle.

As example of validation, some results of the SIMPACK-SIMULINK Tilting Simulator characterising the tilting system behaviour (tilting angles, forces in the tilting and HoD actuators), are shown here below, in comparison to corresponding values of the real system. All the comparisons show that the developed model can simulate normal operation modes, permanent failures condition of tilting system and transient failure modes of the tilting trains Pendolino ETR610 without significant restrictions.

All the results shown are related to a simulation on a right curve with radius R = 600 m, track cant h = 100 mm and with train speed = 144 kph (n.c.a. = 2.01 m/s²).

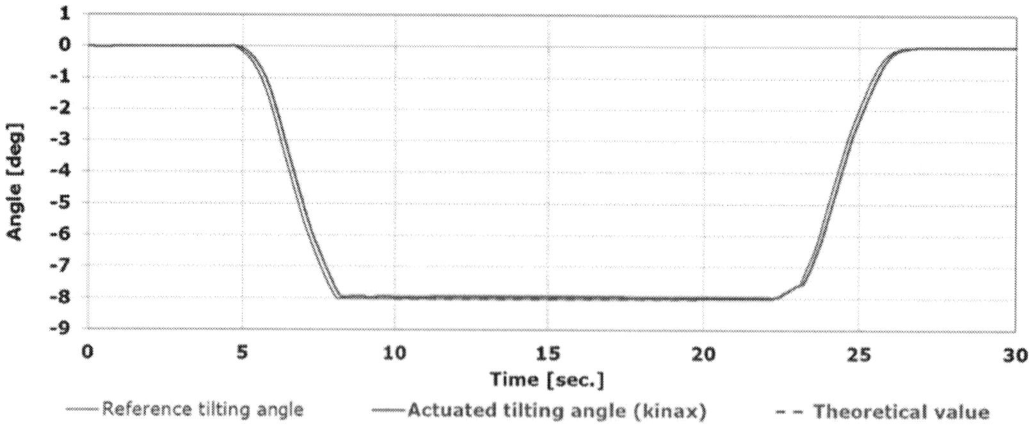


**Fig. 13 Reference and actuated tilting angles - Normal mode**

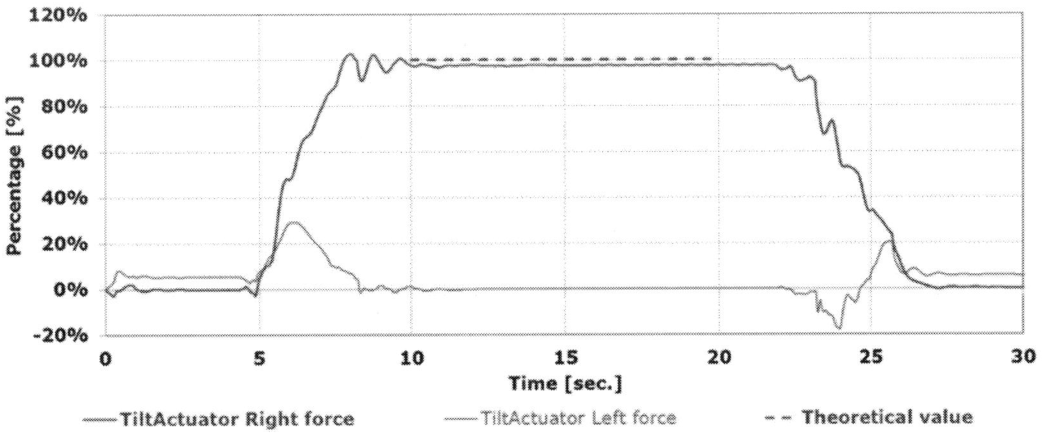


**Fig. 14 Tilting Actuators force - Normal mode**

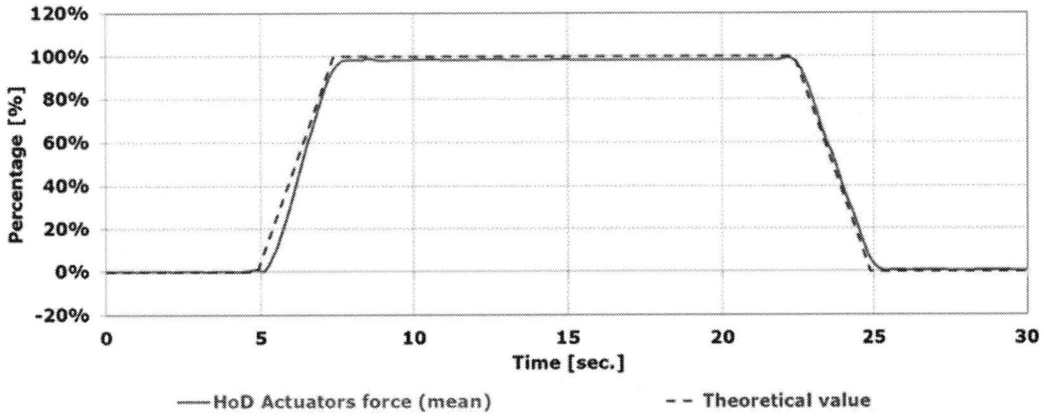


**Fig. 15 HoD Actuators force - Normal mode**

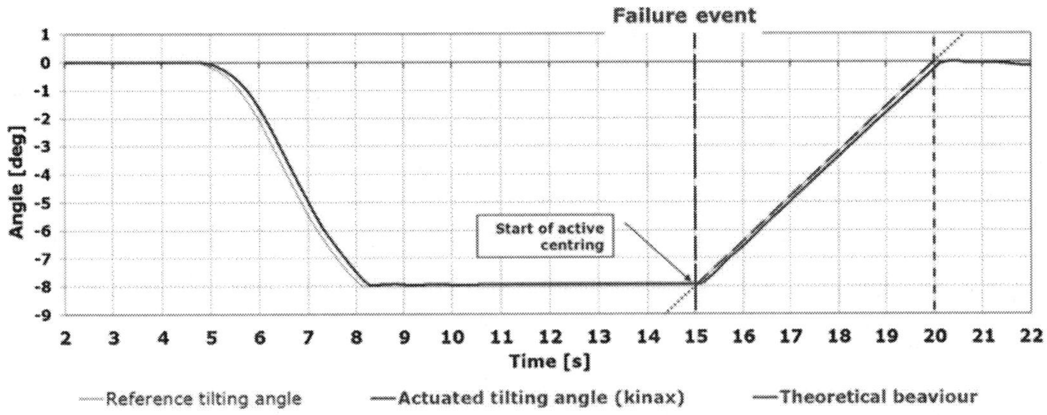


**Fig. 16 Reference and actuated tilting angles: Active centring in event of sudden failure**

The field validation is performed comparing test and simulation results obtained in similar boundary conditions.

The analysed quantities are representative of the static, quasi-static and dynamic running behaviour of the vehicle and of the tilting system operations:

- General kinematic quantities (train speed and n.c.a.)
- Tilting and HoD quantities (sensors signals, reference and actuated tilting angles, actuator forces, carbody/bogie lateral displacement)
- Wheel/Rail contact forces

The first step of the field validation consists in the reproduction of the results of the static in-door tests to verify the passive centring of the carbody and the accuracy of the tilting controller.

The passive centring test has the purpose of determining the residual angle of the carbody in the case of passive centring.

The second test has the purpose to demonstrate that the tilting control and actuation system can rotate the carbody reproducing the command signal with the expected accuracy.

The Fig. 17 shows that the model reproduces the real behaviour of the vehicle in terms of actuated tilting angles in response to an external command signal.

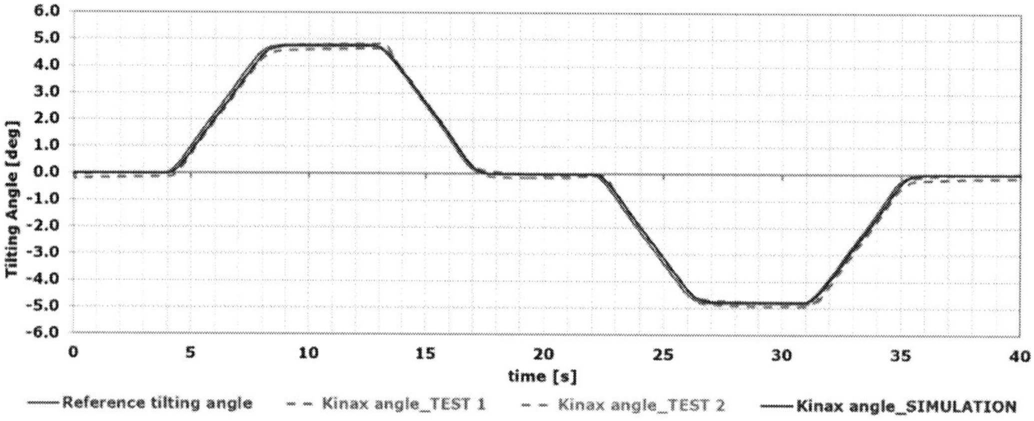


**Fig. 17 Comparison of static test and simulation results**

The validation of the SIMPACK-SIMULINK Tilting Simulator is finally completed with the investigation of the dynamic behaviour of the vehicle, both in nominal operating conditions of the tilting system, and both during the transient failures, in comparison with available on-line test results.

The validation results shown below refer to certification tests on Italian network.

Measurement of wheel/rail forces followed the normal method of EN14363 (11), with instrumented wheel set.

In the Fig. 18 to Fig. 25 are shown the results related to a right curve with R = 290 m of the Signa-Empoli line (line A) and to a left curve with R = 900 m of the Paola-Gioia Tauro line (line B) are shown.

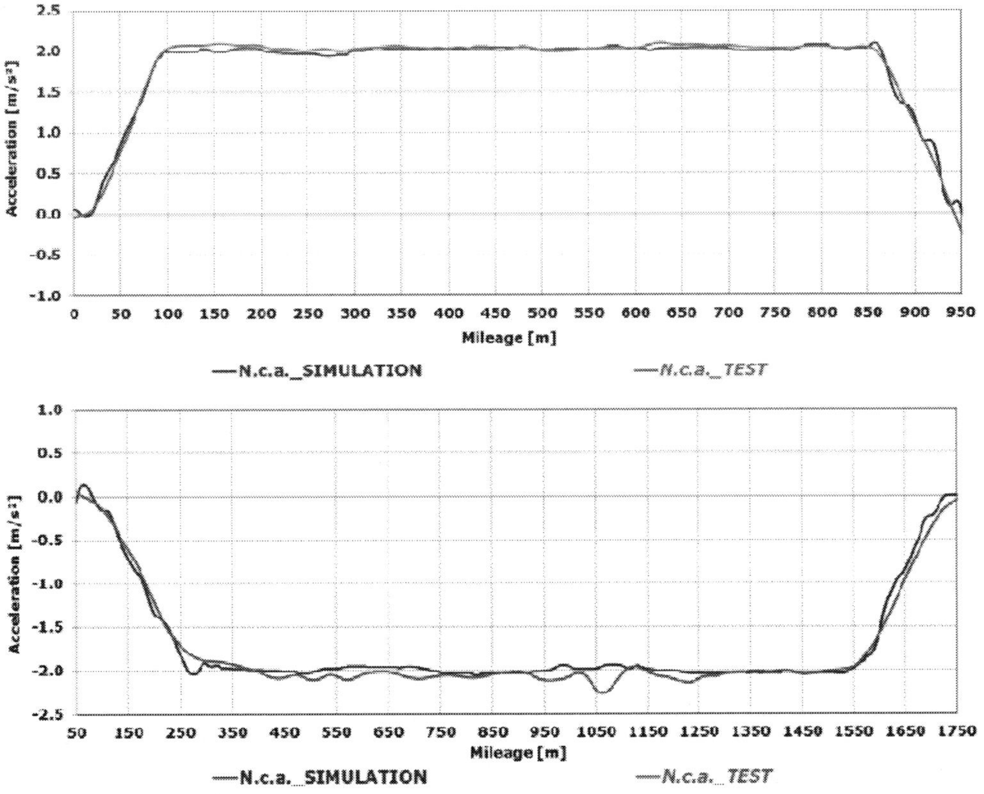


**Fig. 18 N.c.a. vs. mileage: (up) Line A, (down) Line B**

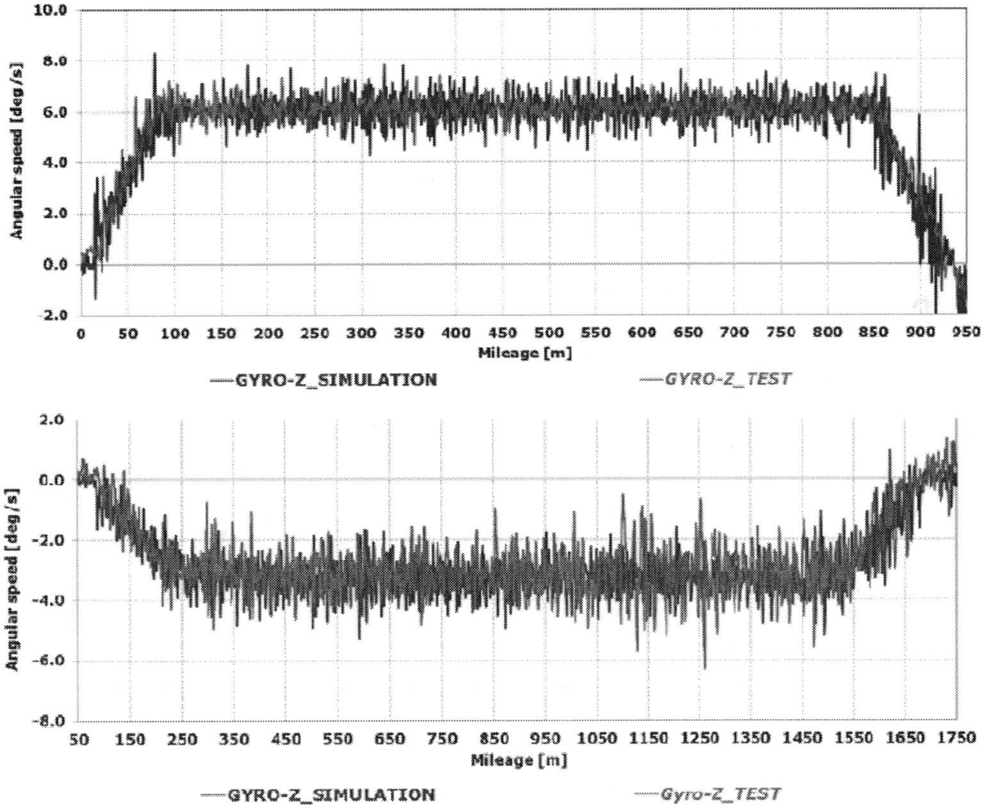


**Fig. 19 Yaw angular speed vs. mileage: (up) Line A, (down) Line B**

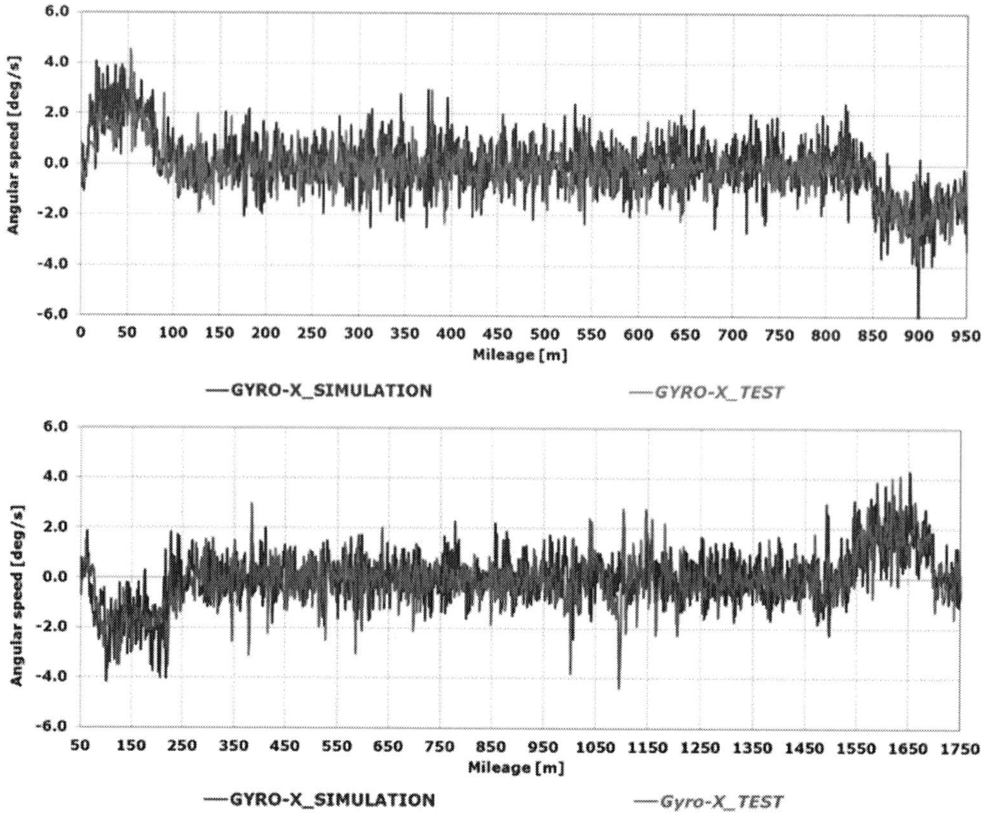


**Fig. 20 Roll angular speed vs. mileage: (up) Line A, (down) Line B**

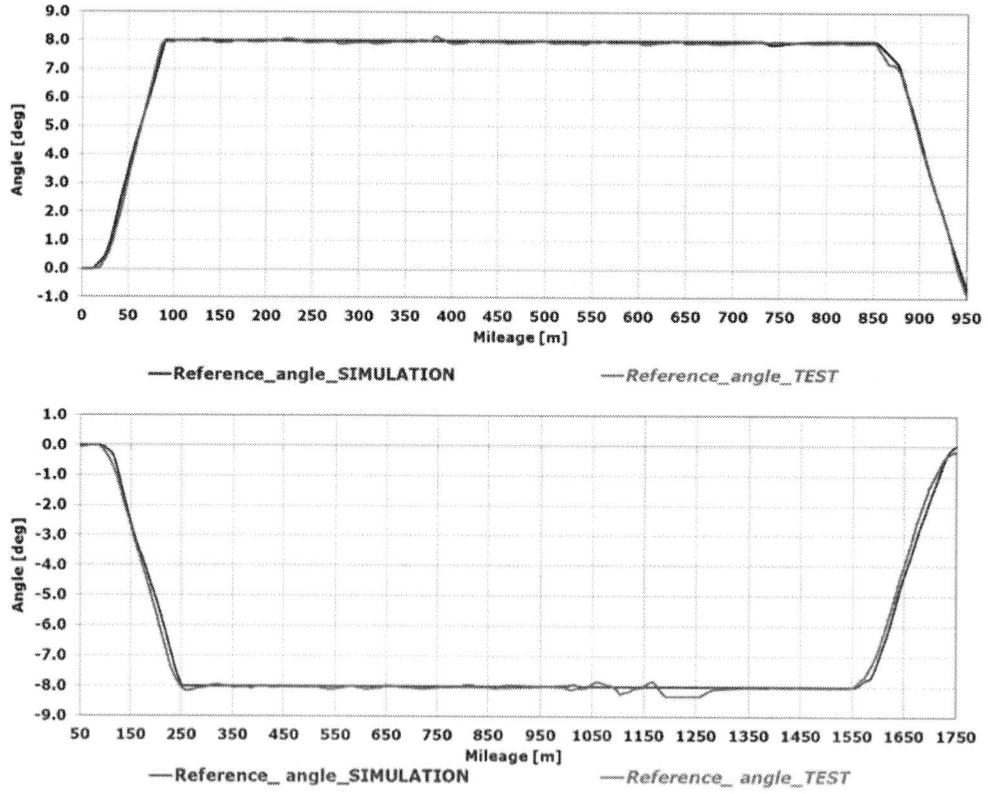


**Fig. 21 Reference tilting angles vs. mileage: (up) Line A, (down) Line B**

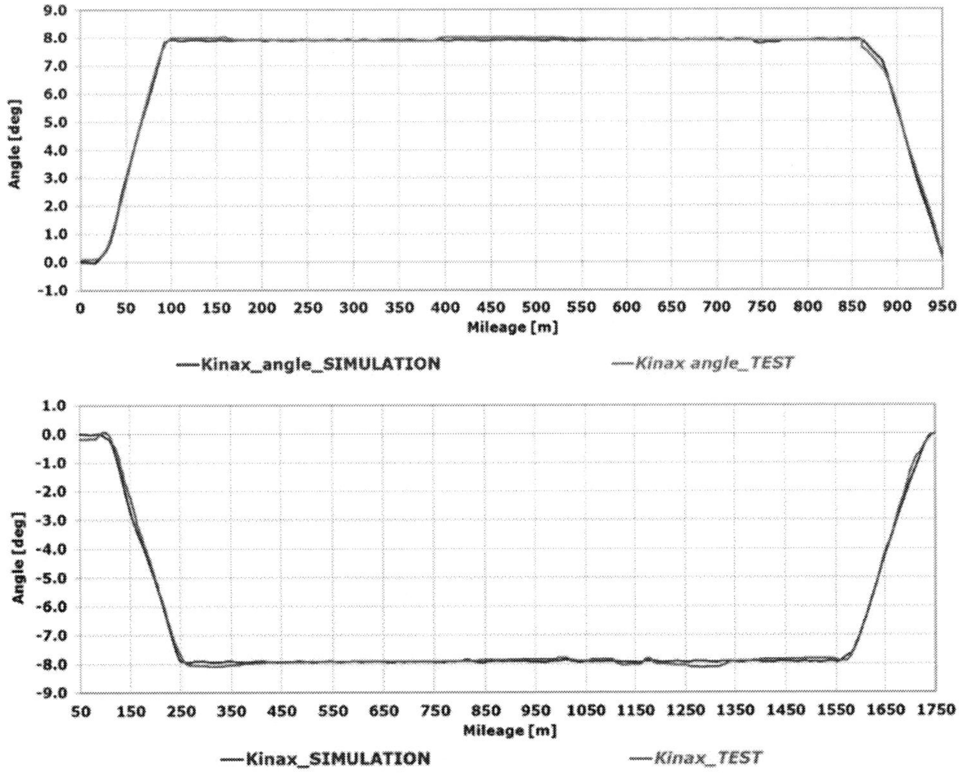


**Fig. 22 Actuated tilting angles (kinax) vs. mileage: (up) Line A, (down) Line B**

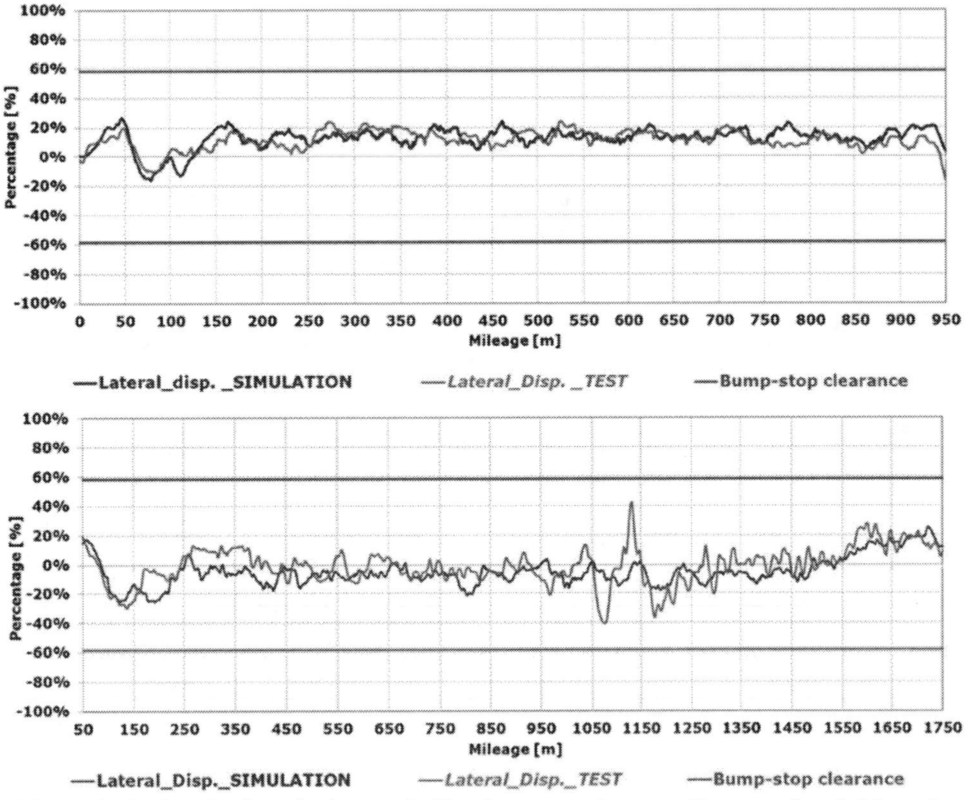


**Fig. 23 Carbody-bogie lateral displacement vs. mileage: (up) Line A, (down) Line B**

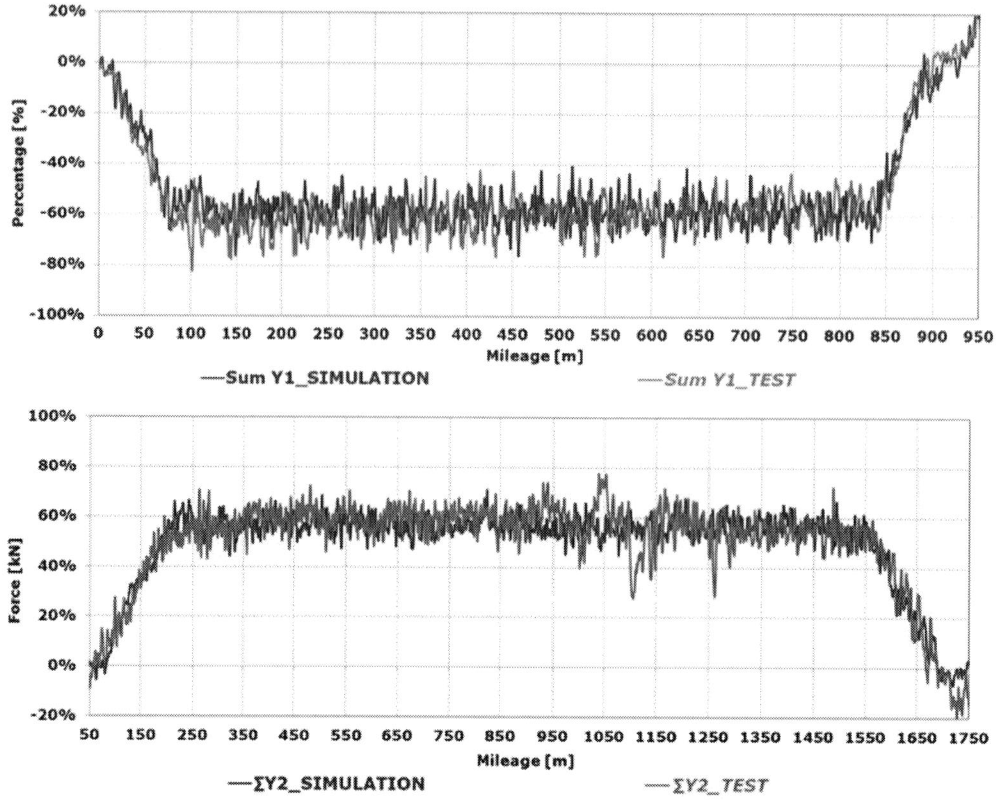


**Fig. 24 Track shift forces vs. mileage(up) Line A, (down) Line B**

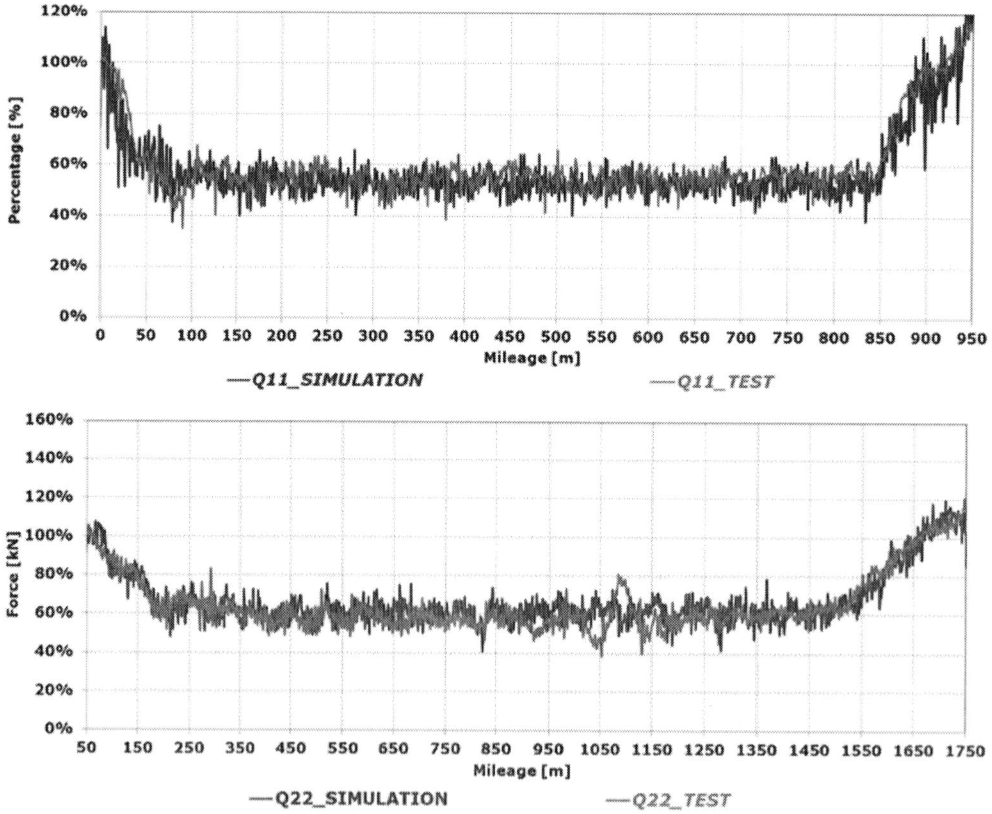


**Fig. 25 Vertical wheel forces vs. mileage(up) Line A, (down) Line B**

Furthermore the Fig. 26 and Fig. 27 show an example of comparison between test results and simulation in the case of sudden failure of the tilting system and of the HoD. The test was carried out on an S-curve of small radius of the Signa-Empoli line.

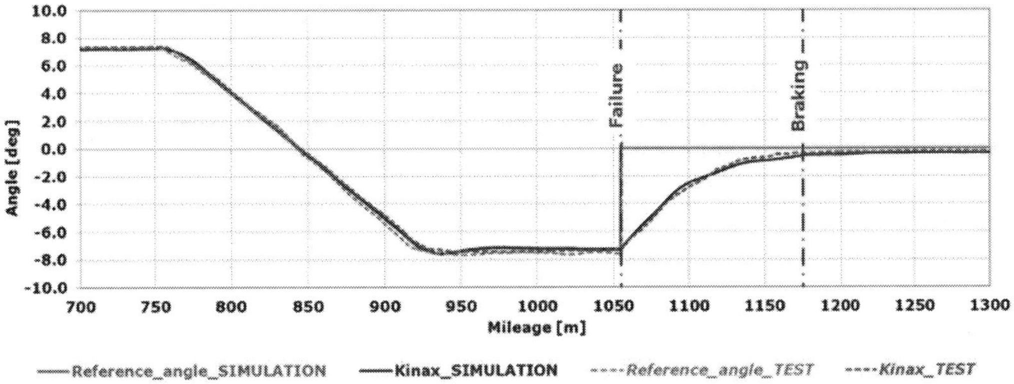


**Fig. 26 Sudden failure of tilting system in line A: Tilting angles vs. mileage**

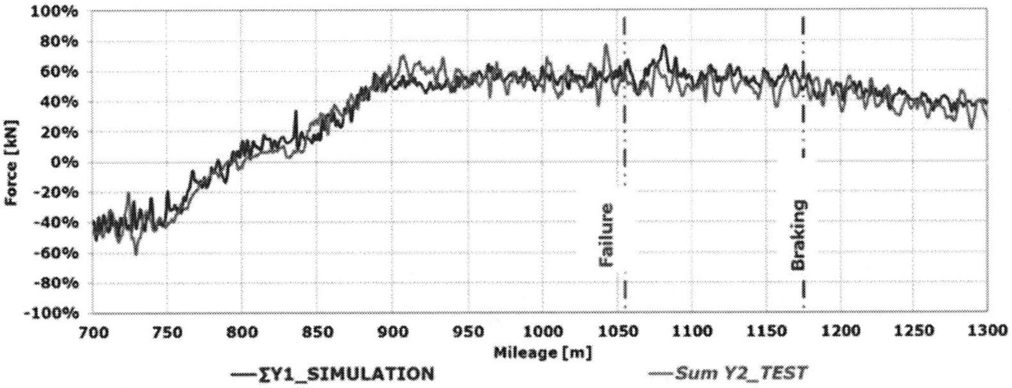


**Fig. 27 Sudden failure of tilting system in line A: track-shifting force vs. mileage**

It can be stated that while for the quasi-static values the comparison can be judged always very good, for the maximum values in the line B scenario, the comparison is not fully satisfying, principally due to the fact that track quality, rail profiles, friction coefficient used for simulations are different respect to the real track of the test.

This validated model may be used for the extension of homologation (like change of vehicle parameters or change of different boundary conditions) through verification via simulations of the assessment quantities, both static and dynamic, as defined by EN14363.

## 6. CONCLUSIONS AND FUTURE DEVELOPMENTS

The active tilting technology is continuously improving, to increase the performances in terms of speed, safety, and ride comfort, and to reduce the maintenance costs for vehicles and infrastructures.

This work has presented the development and validation of the "SIMPACK-SIMULINK Tilting Simulator" as a methodology dedicated to the design and optimisation of the tilting vehicles, in a co-simulation environment. We have highlighted the need to reliably simulate such a complex system, not only in normal operation conditions, but also in degraded or failure operation modes. Hence a particular attention is given to electric, hydraulic and pneumatic actuation systems modelling, command signal generation, control in the loop and diagnostics.

The main advantages of this methodology, in comparison to past approaches, are the flexibility in development phase and the accuracy in simulation of the dynamic behaviour in all operation modes of the tilting system.

Future comparison tests, carried out on a length of track on which full information about track geometry and wheel/rail contact condition can be collected, at the same time, could improve the validation of the model.

Anyhow, the validated model of this multi-physic system is applied to the design optimisation of the tilting vehicle and to the impact analysis of failure modes through simulations, thus reducing development costs and time to the market.

Some limitations of the model are related with the simplification of the hydraulic actuation system that could need tuning of some input parameters in case of relevant change of the hydraulic scheme. Further step will be therefore to build in Simulink the real hydraulic model of the tilting actuation, thus reducing the uncertainty related to the degree of experience of the designer in tuning the parameters and to the availability of test data.

## REFERENCES


(1) N. Kuka, C. Ariaudo, R. Verardi: Modelling and Simulation of Tilting Vehicles, The International Journal of Railway Technology, Volume 2, Issue 1, 2013, page 63-82

(2) N. Kuka, C. Ariaudo, R. Verardi: Modelling and simulation of tilting trains, 22nd International Symposium on Dynamics of Vehicles on Roads and Tracks, (IAVSD'11, Symposium proceedings) Manchester, UK, 14-19 August, 2011

(3) B. Dalla Chiara, G. Hauser and A. Elia: I treni ad assetto variabile: evoluzione, prestazioni e prospettive, Ingegneria Ferroviaria, 7-8 2008

(4) R. Persson: Tilting trains - Description and analysis of the present situation, Publ 0702, ISBN 978-91-7178-608-1, KTH Railway Group, Stockholm, Sweden, 2007

(5) G. Hauser: Alstom Tiltronix Anticipative Tilt Control, Le Rail, Sept. 2006

(6) N. Kuka, F. Balsi, and O. Maggi, VAMPIRE Tilting Simulator – An integrated vehicle model with the active elements control, VAMPIRE User's Conference in Derby, 20th May 2002

(7) G. Hauser: West Coast Main Lines Trains "Tilt Technology", IMechE Conference Birmingham, Nov. 2002

(8) G. Klaus Koenig, Gli elettrotreni ad assetto variabile – Storia del progetto Fiat Y1060: L'ETR400 e l'ETR450, Valerio Levi Editore, 1986

(9) N. Kuka, A. Böttcher, ALSTOM Experience with MBS Model Validation, DYNOTRAIN, Berlin 22-23 Mar. 2010

(10) S. Bruni, R. Roberti, Development of operations of tilting train on Italian network, World Congress on Railway Research 2001, Koln (Germany), 25-29 November 2001, 1-19

(11) EN14363 Railway applications – Testing for the acceptance of running characteristics of railway vehicles – Testing of running behaviour and stationary tests, 2005

(12) AEA Technology plc, "VAMPIRE User Manual - V 4.32", Derby, UK, 2004

(13) British-Rail, Research, "Introduction to VAMPIRE", AEA Technology plc., London, UK, 1997

(14) DeltaRail Group Ltd, "VAMPIRE Pro User Manual – V 5.02", Derby, UK, 2006

(15) SIMPACK AG "SIMPACK Reference Guide" SIMDOC 8.904 Build 58 Aug 31, 2010

# Apparently independently rotating wheelset – a possible solution for all needs?

**A Bracciali**
University of Florence, Italy
AB Consulting sas, Italy

**ABSTRACT**

The paper describes a patented innovative railway wheelset (*AIR Wheelset*) where the wheels are rotating supported by roller bearings and are torsionally connected via a shaft with proper stiffness. This arrangement can dramatically change maintenance operations while keeping optimal running stability at high speed. Better curving characteristics can be obtained by a version where the torque transmitted through the shaft is limited by torque limiters, possibly reducing rail corrugation and improving negotiation of tight curves.

## 1 INTRODUCTION

Independently rotating wheels (IRW) were developed in the last century for railway applications but, in practice, had some success only in trams. This is due to the lack of self-centring ability of a wheelset equipped with IRWs.

A new design (patented) of a wheelset is described in this paper, where the wheels are installed in the classical arrangement of IRWs but are connected via a shaft and possibly a clutch mechanism. The Apparently IRW wheelset (named *AIR Wheelset*) behaves as a conventional wheelset for large curve radii (higher speed) and as a dissipative limited-torque wheelset in tight curves (lower speed). It is expected that this arrangement leads to small modifications while running in straight track and may results in dramatic running dynamics improvements in tight curves, leading possibly to a large reduction of rail corrugation and wheels polygonization.

Another distinct advantage of the *AIR Wheelset* is that the axle is not rotating, leading to much higher safety coefficients, virtually eliminating any possibility of fracture in service. Wheelset maintenance is much easier, as wheels and bearings are the only parts subjected to maintenance. The *AIR Wheelset* is designed inspiring to the fail-safe philosophy, i.e. any failure in service does not jeopardize running safety. Its roots lie in the studies held in the '70s and in the '80s in the UK, Japan and Italy to reach a simple and effective solution. The optimal field of application of the *AIR Wheelset* is that of the full range of railway vehicles with inboard bearings, although the best performances can be obtained from the use in vehicle with axleload up to 25 t and a speed up to 200 km/h running in curvy lines.

# 2 HISTORICAL AND TECHNICAL BACKGROUND

## 2.1 Comparison of rail and road wheels

Road vehicles all have, with some marginal deviations, independently rotating wheels (IRW), i.e. wheels that although belonging to the same axle are not rotationally constrained. Driving wheels are normally connected through a differential gearbox, a gear mechanism that equally shares torque between the two wheels leaving their angular speed independent. Only in some niche applications, such as 4WD vehicle suitable for travelling on low-adhesion surfaces, differential gearboxes are supplemented by friction or viscous devices which link the wheels in some way, but their application makes the vehicle harder to steer and leads to higher tyre wear.

On the contrary, railway vehicles have always been equipped with a solid axle connecting the two wheels, realizing what is normally called a "railway wheelset".

The motion of a wheelset with perfectly coned wheels was studied in the 19$^{th}$ century by Klingel who derived his famous equation relating the length of a sinusoidal lateral movement to the conicity of the wheels, the wheel radius and the track gauge. As a result of an initial lateral displacement or lateral velocity, the wheelset tends to return towards the centred position.

The self-centring ability of a wheelset with tapered profiles is vital to the railway vehicle, in the sense that although railway safety against derailment is guaranteed by wheel flanges which interact with the rail gauge corner, this contact should intervene only where needed, i.e. in narrow curves or through S&C, but wheel flanges and railhead should not interact in tangent track to limit wear as much as possible.

While on a road vehicle the vehicle trajectory is left to the driver's judgment whose style may dramatically influence both comfort and tyre wear, no such freedom is left to railway vehicle which is fundamentally a laterally-guided, one-dimensional transportation mode, with bi-directional vehicles. Train drivers can only decide the motion along a predefined line (the track) acting on traction and braking forces. That's why self-centring ability is needed on trains composed of numerous vehicles (with rotationally constrained wheels) while it should be avoided on basically mono-directional vehicles as cars and lorries (with IRW).

IRWs have undisputable maintenance advantages. It is a common experience that changing a punctured wheel only requires to unscrew some nuts or bolts. On some vehicles (such as motorbikes) also the brake calliper may need to be removed in advance, but this is a rather fast and straightforward operation.

As railway wheels wear, they need to be changed from time to time. The standard maintenance operation requires to remove the wheelset from the vehicle, uncouple the suspensions, lower the wheelset (or lift the vehicle), then install a new or overhauled wheelset. After off-line axlebox and bearings removal, wheels have to be pressed off typically with the help of a high pressure oil injection in a dedicated groove.

## 2.2 Literature analysis

IRWs were initially used on railway vehicles to reduce longitudinal forces in tight curves, leading to solutions that are still in service with satisfaction in many trams. This solution intrinsically allows the use of low floor arrangements, where the wheels are connected by a "bridge" similarly to road vehicles. Architectures including IRWs are nowadays common in light rail and tram vehicles and are not further described here.

During high-speed running (approximately $v>200$ km/h), vehicles with conventional (coned) wheels may suffer lateral instability ("hunting") phenomena, i.e. self-excited lateral movements of the whole wheelset and of the bogie in the typical range 5÷8 Hz. Displacements are limited only by the shocks resulting from sudden interaction between wheel flanges and rail gauge corner. The appearance of hunting was a nightmare for railway engineers until the wheel-rail contact theory was developed in the '50s and the '60s of the last century, leading to the publication of the first railway vehicle dynamics software packages.

As IRWs have no self-steering effect and therefore no critical speed, their possible adoption in high speed vehicles originated many hopes especially in the '60s and the '70s, when commercial speeds of 300 km/h were looked for in Europe and in Japan. Nevertheless, they were never adopted for the reasons described in the following.

The review paper by Dukkipati et al. (1) summarizes all the experiences about IRWs until the end of the '80s. At that time it was already clear that most of the advantages claimed by the use of IRW in terms of running stability and curving were achievable but a major drawback could not be avoided.

In 1977, Hayden et al. (2) solved an 11-*dof* model of the dynamics of a bogie to analyze the influence of the bogie architecture on the critical speed, i.e. the speed at which an initial displacement does not damp out but increases. Although the model has all the limits intrinsic to the step solution of a system of linear differential equations, Figure 1 shows that the critical speed decreases when the torsional stiffness of the axle decreases. For the special case of $k_{AX}=0$, i.e. the case of IRWs, the critical speed is always zero as the stiffness matrix of the bogie is singular and an eigenvalue of the system is zero. This "*implies that there is no preferred equilibrium point and that the wheelset may wander randomly from side to side in response to track irregularity*". While severe hunting is prevented, "*there does exist the possibility of greater flange wear*".

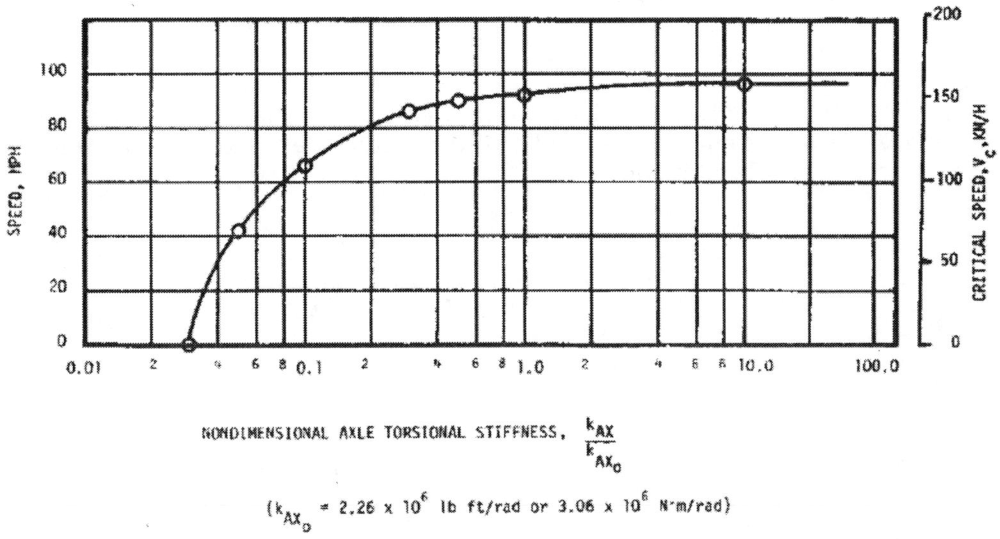


**Figure 1. Effect of axle torsional stiffness on critical speed on nominal 11-DOF model (2).**

Similar results were obtained by Doyle et al. (3) (Figure 2), confirming that bogies equipped with IRWs have no critical speed. Both these references show that reducing the torsional stiffness of the axle lowers the critical speed.

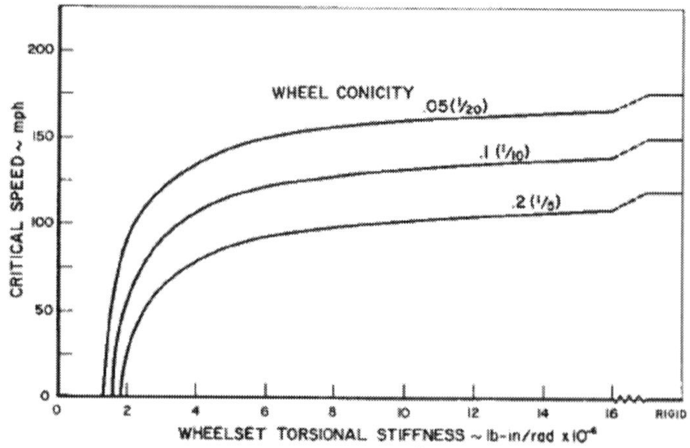


**Fig. 4** The effect of wheel conicity on critical speed with variable wheelset torsional stiffness. Conversion factors: (n-m/rad) = (.113) (lb-in./rad); (kph) = (1.61) (mph)

**Figure 2. Effect of axle torsional stiffness on critical speed with varying wheel conicity (3).**

Some researchers (4, 5, 6) found experimentally that wheelsets equipped with IRW may run displaced but without wheel flange providing that the wheelsets have a specially designed profile. During some tests on real vehicles, abnormal longitudinal vibrations were measured on the axleboxes. This demonstrates that the adoption of IRWs on existing vehicles is not trouble-free and that at least the stiffness of the primary longitudinal suspension has probably to be redesigned.

A search revealed that some inventors patented solutions (now expired) for motor wheels where torsion bars are used as transmission shafts. In patent (7) the wheels are rigidly connected via a hollow axle and the torque is transmitted with a long, flexible torsion bar, while in patent (8) wheels are connected through a differential gear mechanism and are therefore torsionally disconnected (they are subjected to the same torque when the motor is on but angular speed may be different). In the first case the wheelset can be considered conventional (torque is transmitted by the axle) while in the second case the wheels can be considered independent (when motor is off the torque is zero). So, none of the two inventions deals with problems arising from low torsional stiffness of the connection between the two wheels, at least when the vehicle is trailed.

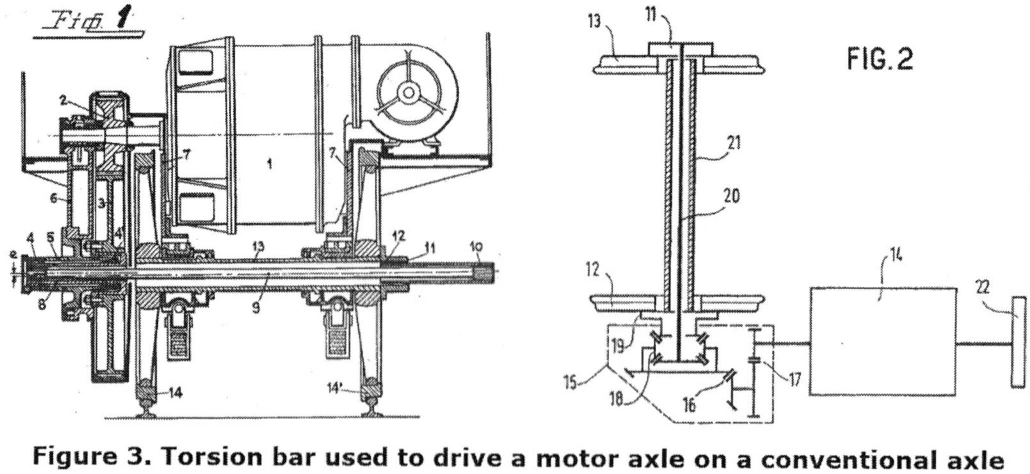


**Figure 3. Torsion bar used to drive a motor axle on a conventional axle (left, (7)). Torsion bar included in a differential gear mechanism to drive independently rotating wheels (right, (8)).**

The centring effect of wheel profiles on IRW wheelsets is limited to the so-called "gravitational stiffness", i.e. the fact that the resulting centring effect is not due to longitudinal creepage forces but to the different angle at the two wheel-rail contacts.

This conclusions gives rise to some concerns, as already in 1977 Doyle (2) noted that *"it has been determined that wheel conicity cannot be designed because the wheel contact geometry wears rapidly to a "worn wheel" profile"*. King (9) reports the Müller suggestion *"... designers should accept the wear forms of tyres and railheads as they are defined in the stable (worn) form..."*.

In more recent times, Wickens (10) cites Heumann saying that *"he argued that after reprofiling to a coned tread, tyre profiles tend to wear rapidly so that the running tread normally in contact with the rail head is worn to a uniform profile. This profile then tends to remain stable during further use, and is largely independent of the original profile and of the tyre steel"*.

Designing profiles is therefore an activity that is intrinsically fruitless and this is true also for the case of IRW wheelsets. Self centring due to gravitational effect has always demonstrated to be insufficient, such that already in 1970 Becker (11) closed the debate giving evidence that *"extensive experimental experience has shown that, indeed, the kinematic oscillation is absent but that one or other of the wheels tends to run in continuous flange contact"*. Resulting flange wear is therefore absolutely unacceptable.

## 2.3 Other types of IRW vehicles

Old literature amply justifies why no conventional railway vehicles equipped with bogies use IRWs today. The only effective and successful example of the use of IRWs is the family of vehicles produced by the Spanish company Talgo, but these vehicles have a particular architecture and don't use conventional bogies but portals with steered single wheels (known as *rodal* or *Talgo truck*). What follows therefore doesn't apply to these vehicles.

In more recent years only a few papers dealt with special applications of the IRW concept. This is attributable to the low interest that the technical community deserved to this concept for the reasons already discussed. Vehicle dynamics at high speed is nowadays effectively simulated through commercial software packages and running daily at 300 km/h is safe and comfortable. There is no need any more to resort to the IRW technique to avoid hunting.

The few references found concentrate on the possible applications of IRW to the so-called "mechatronic bogies", i.e. those bogies where axle orientation is managed by the proper combination of actuators and a control strategy, leading to "controlled steering" vehicles where the wheelsets have the best (radial) orientation and therefore minimize wheel and rail wear. Although interesting from a scientific point of view, to the author's knowledge the application of mechatronic bogies is not widespread yet; some applications of a similar concept, the so-called "self-steering bogies" techniques, were applied to bogies with conventional wheelset whose attitude is managed by linkages actuated by the rotational carbody - bogie frame angle. Neither mechatronic bogies nor passively steered wheelsets are further addressed in this paper, as they require the use of advanced technology (with the related reliability issues) and lie outside the scope of this paper.

In the case of low floor trams, already discussed, the absence of self-centring capabilities is such that tyres are often cylindrical and it is accepted that these vehicles run laterally displaced with some flanges continuously interacting with the grooved rail.

It can be summarized that the IRW technique is used only on low-speed vehicles with specific needs but is not used at all on conventional railway vehicles.

## 3 CURRENT MARKET PERSPECTIVES

Inboard bearings existed since the early railway era as steam locomotives often had external cylinders with rods that were not compatible with conventional axleboxes. When journal bearings were substituted by roller bearings, inboard bearings almost disappeared, with only few exceptions such as the Budd Silverliner EMU (USA, 1963), the Commonwealth Engineering 2000 Class railcar (Australia, 1979) and, more recently, the Leila Bogie (12). To the author's knowledge, none of these bogies was used on large scale.

More interesting for the scope of this paper is the story of the inboard bearings bogie B5000 developed by British Rail in the '80s. Thanks to its interesting dynamics characteristics in term of track friendliness, vehicles mounting it have the lowest track access charges to the UK railway network. The bogie is now built by Bombardier Transportation under the brand name *FLEXX Eco*. Class 220 Voyager UK (in service since 2001, 288 bogies), Class 222 Meridian UK (in service since 2003, 303 bogies) and Class 172 Turbostar UK (in service since 2011, 186 bogies), are good examples of the favour that that bogie received in its mother land.

Other manufacturers developed inboard bearings bogies for the UK market. Siemens is going to supply up to 1200 cars of the Desiro City Thameslink train which uses SF 7000 bogies (13, 14) and Hitachi has just rolled out the first units of Class 800/801 with specifically designed bogies (15).

This "all British" bogie family has quite recently started to spread across Europe and beyond. Bombardier Transportation was in fact subcontracted by Siemens to manufacture 1390 FLEXX Eco trailer bogies for the ICx trains for DB (16), designed for 19.5 t/axle and operating at 250 km/h with 860 mm diameter wheels. Other vehicles with FLEXX Eco bogies are Class 5 coach for Norway (in service since 2011, 122 bogies), Crossrail (in design, 1170 bogies), Riyadh Metro (in design, 188 bogies) and Stockholm C30 (in design, 768 bogies).

After 25 years from the roll out of the first B5000 bogie the market seems ready for the massive introduction of this kind of bogies.

## 4 DEVELOPMENT OF A NEW TYPE OF IRW WHEELSET

### 4.1 New possibilities for IRWs on inboard bearings bogies?

Technology advances open new fields for the application of IRWs on inboard bearings bogies that were not possible only a few years ago. The design that will be described in the next paragraph benefits in fact by some recent innovations:

- inboard bearings cannot be monitored as required by TSIs (Technical Specifications for Interoperability) by trackside HABD (Hot Axle Box Detectors). The recent standard (17), supporting the TSI Rolling Stock, allows the monitoring of bearing temperature by means of transducers integrated in the axlebox. This technology is nowadays rather simple to put in service thanks to the development of sensors, electronics and software, certainly much more easily than 30 or 40 years ago;
- inboard bearings "forced" the development of TBU (Tapered Roller Bearing Units) with a large internal diameter (in the order of 180-185 mm). This is crucial for the development of the new design. No bearings with these performances were available in the past;

- NDT with ultrasound probes designed to check the axle from a central bore machined in the axle is nowadays a standard. This technique is well proven and hollow axles are widely used while they almost didn't exist when the first IRWs solutions were attempted.

### 4.2 Concept of the *AIR Wheelset*

The *AIR Wheelset* (18) is mainly intended to improve maintenance operations, reducing timed and costs. With the *AIR Wheelset* it will be possible to change a wheel similarly to what happens on road vehicles, without the need to lower the bridge that is designed for infinite life. It is the perfect companion of internal frame bogies.

The goals that were defined during the design of the *AIR Wheelset* are:

- it had to reduce maintenance time and costs, needing simple tools, less labour expenses and less workshop space;
- it had to improve safety, with specific fail-safe features;
- it had to improve reliability, with continuous monitoring of running conditions;
- it had to give superior running characteristics in narrow curves, while keeping a good behaviour at higher speeds;
- it had be available in both motor and trailed versions.

The *AIR Wheelset* satisfies all these goals. As wheels with bearings are mounted on a hollow axle that is not rotating, an external observer could see the wheelset as a conventional IRW wheelset. In the developed arrangement the wheels are nevertheless *not independent* as they are connected through a stiff shaft which rotates in the bore of the hollow axle. For this reason the acronym *AIR Wheelset*, i.e. *Apparently Independently Rotating Wheels Wheelset*, was created.

One of the most specific features of the invention is that the trailed *AIR Wheelset* may have the wheels and the shaft connected by friction limiters set at a predefined value of maximum transmissible torque. This limits longitudinal forces in narrow curves possibly effectively reducing rail corrugation formation and growth.

A sketch of the *AIR Wheelset* possible layouts is shown in Figure 4. The case of the trailed wheelset with torque limiters is shown in Figure 5.

### 4.3 Wheel design

The introduction of higher load capacity bearings in the wheel seat area required a larger hub diameter and a longer wheel bore. The design started from a typical wheel with web-mounted brake disks and showed no criticalities in terms of both wheel manufacturing (hot rolling and turning operations) and stress resistance. The design of both the original and the modified wheel was successfully assessed according to current regulations on wheel design (19, 20).

Stresses in the hub area results much lower thanks to the absence of fit while the rest of the wheel is practically unaffected by the new design. The *AIR Wheelset* is therefore compatible with the majority of existing wheels with minor changes in the design and tooling processes.

| Maximum torque | Trailed *AIR Wheelset* | Motor *AIR Wheelset* |
|---|---|---|
| Zero (free) | 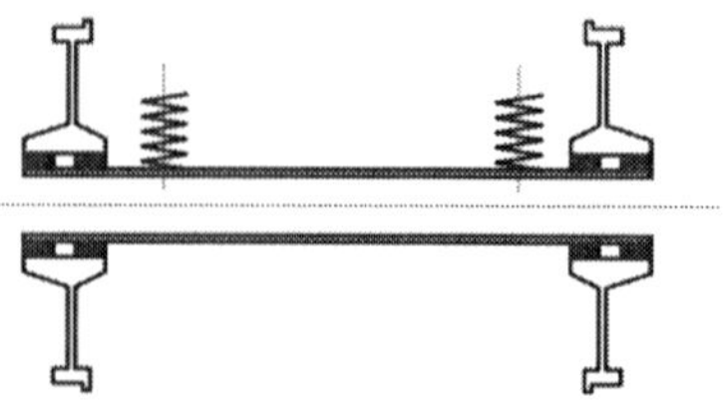 | Not applicable |
| Limited at 0.05 *PR* -0.2 *PR* *(torque limiter)* | 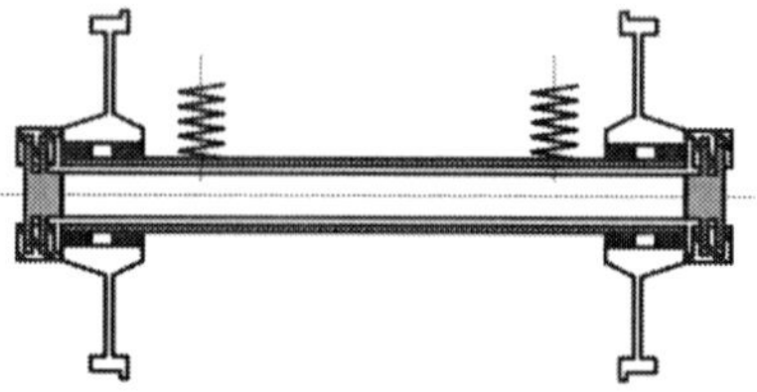 | Not applicable |
| Limited by wheel-rail adhesion limit *(rigid coupling)* | 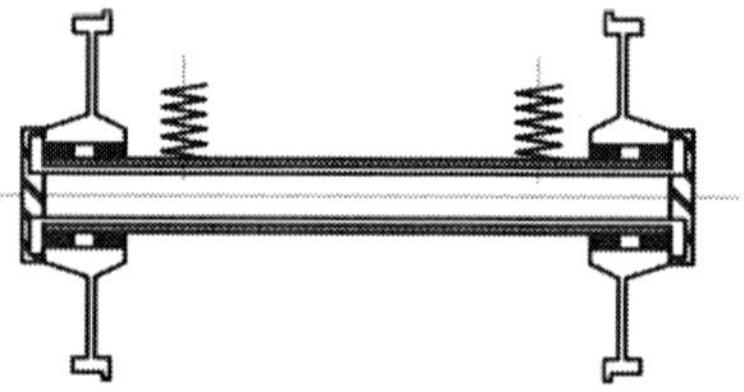 | 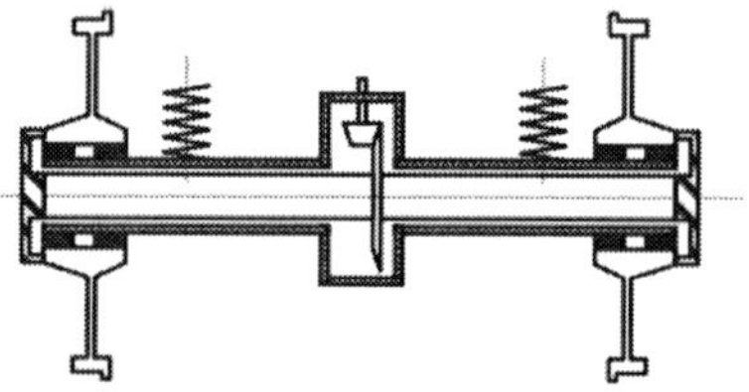 |

**Figure 4. Possible layouts of the *AIR Wheelset* (18). All solutions can be equipped with block brakes and/or up two discs per wheel (maximum four brake discs per wheelset).**

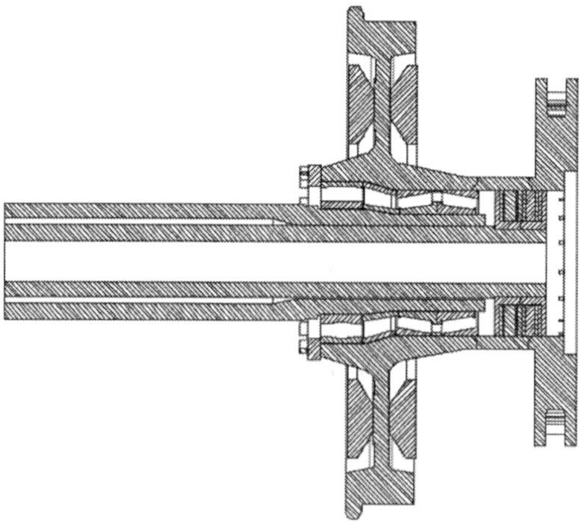

**Figure 5. Partial cross section of an *AIR Wheelset* equipped with friction torque limiter and two brake disks per wheel.**

### 4.4 Bearings design

European standards on the performance testing of railway bearings (21) considers a 1/2 factor in lateral load of bearings as the overall lateral force "applies to calculation of forces for one journal". This means that the lateral force is shared by the two axleboxes. Moreover, conventional wheelsets "equalize" the force on the two wheels through the axle. For example, if a force F is applied to the left wheel and a force –F is applied to the right wheel, bearings are not subjected to any lateral load. This clearly doesn't happen in *AIR Wheelset*, where the bearings of each wheel are substantially independent.

Units with two tapered bearings (TBU) mounted "back-to-back" (or in an "O" arrangement) are the current railway standard as they provide a reliable and economically advantageous way to support all the forces acting on an axlebox. The choice of the "O" arrangement is due to the fact that the axlebox must be firmly connected to the axle about torques acting along the vehicle axis.

*AIR Wheelset* bearings are loaded also by the torque generated by the lateral force acting on the wheel flange. Bearings had to be chosen looking for an increasing cross section of the bridge starting from the outer end, considering moreover both the speed limitations arising from the rotating outer ring of the bearings and the limited available space.

Two bearings are used on *AIR Wheelset*, one acting as a hinge (locating) and the other as a support (non-locating), a standard solution in machine design. This design satisfies all the requirements in terms of mounting sequence, tolerances, machinability, etc. of the wheel and of the bridge. It consists of:

- a tapered bearing unit mounted "face to face" (or in an "X" arrangement), with high lateral and vertical load capacity (locating);
- a CARB® bearing with very high radial capacity and a very high maximum speed (non-locating).

CARB® bearings were introduced by SKF in 1995 and allow high axial and angular displacements. All the other types of rolling bearings (ball, cylindrical rollers, needle, spherical, etc.) did not satisfy the requirements. These bearings are not commonly used in the railway sector and some development will be needed. With this solution for the first time a single wheel can withstand high vertical loads at high speeds. This is a noticeable improvement compared to classical IRW solutions (trams and light railway design for low speed and low axleload).

## 4.5 Bridge design

The *AIR Wheelset* design includes a bridge that is not subjected to rotating bending, so the classical approach to fatigue used in axles was not used. Loads on the bridge result from the superposition of the vertical static load on the axle and to lower forces in other directions (braking, steering, inertial). It can be assumed that bending in the vertical plane never reverses, i.e. the vertical load on a wheel never reaches zero (this conditions would lead to the uplift of the corresponding wheel).

Fatigue validation of the bridge design was therefore done considering it as a part of the bogie subjected to high inertia loads. While fatigue limits for solid axle made of EA1N and EA4T steels are respectively 166 and 180 MPa including a safety coefficient of 1.2 (22), the DVS 1612 standard (23) normally used in the design of bogie frames indicates in 240 MPa the fatigue limit for the S355 steel in practice in the whole $0<R=\sigma_{min}/\sigma_{max}\leq 1$ range. This carbon steel is largely used for bogie frame construction. Also the fatigue limit of castings made of G20Mn5 steel can be considered equivalent to that of S355, thereby allowing the bridge to be made as a combination of forged, welded and casted parts.

The bridge can be made with a geometry that better suits the needs of high strength and stiffness, leading to high safety coefficients. The current non-optimized design shows maximum stresses in the order of 100 MPa with an axleload of 20 t/axle for all load cases described in (19). This should avoid in the future catastrophes (24) due to axles failures under rotating bending fatigue. Being less critical, treatments developed to reduce the effect of impacts due to flying ballast (25) are not needed.

### 4.6 Shaft (torsion bar) design

As seen in par. 2.2, the *AIR Wheelset* needs to have the highest possible torsional stiffness to avoid an inadmissible reduction of the critical speed. A successful solution was found where the torsional stiffness of the shaft is approximately 40% of the stiffness of a conventional 160 mm diameter axle. Limited reduction in the critical speed can be efficiently tackled with minor suspension modifications.

European standard (22) indicates for trailed wheelsets a maximum torque of $0.2PR$ (where $P$ is the vertical wheel load and $R$ is the wheel radius), indirectly indicating that a maximum (differential) adhesion coefficient of 0.2 can be reached in practice. It is important to highlight that the single adhesion coefficient on each wheel can be higher, but that only the difference is able to produce a torque that stresses the shaft.

By convention, the torsional moment between running surfaces is selected at the value of $0.3P'R$ during braking, where $P'$ is the proportion of P braked with with the method of braking considered. It includes the torsional moment due to braking and the torsional moment due to curving and wheel geometry. The shaft equipping the *AIR Wheelset* can survive this torque without problems, as a shaft of much smaller size is used for example on 1.88 MW electric motors of the E652 3kV DC locomotive of Italian railways (B wheel arrangement). It is able to make two axles slipping simultaneously (26).

The shaft design is based on stiffness and therefore results not critical from the stress point of view. In case of failure of the shaft the *AIR Wheelset* continues to behave as a conventional IRW wheelset, without impact on safety (no hunting develops). It can be detected and fixed when the vehicle is maintained without further limitations.

### 4.7 Friction torque limiter (clutch) design

Although the torque that has to be transmitted by the torque limiters is rather high, the dissipated power is rather small, less than 2 kW per axle in a wide range of speed and curve radius. For an 80 t, 4 axle vehicle this corresponds to an equivalent extra slope of 0.375 ‰, which is negligible. In any case this resistance is lower than the one offered by conventional wheelsets.

Considering that the torque limiter is immersed in the airflow outside the vehicle, no heating problem are forecasted. Monodisc dry friction clutches can be used. The one specifically developed for the *AIR Wheelset* is currently patent pending and will be shown in a future paper.

### 4.8 Braking the *AIR Wheelset*

Braking the *AIR Wheelset* is not much different from a standard wheelset, with the obvious provisions that no axle-mounted discs can be used. Braking the wheel can be done by conventional tread braking, by using web-mounted brake discs or by using discs mounted on an extension of the axle/wheel (Figure 6).

Figure 5 shows a solution where both web-mounted and external disks are applied. This means that up to a total of 4 discs "per axle" can be used. Perfectly symmetric braking does not generate torsion on the shaft; any dissymmetry (or partial malfunctioning) in braking forces is equalized by the shaft as in conventional wheelsets.

**Figure 6. Two examples of brake discs mounted on the axle ends. The bogie to the right has internal bearings.**

## 5 RUNNING AND CURVING BEHAVIOUR OF THE *AIR WHEELSET*

As shown above the reduction in the torsional stiffness can lead to a reduction in the critical speed, although the results from old linear models should be used with care.

To prevent such limitations, a fully non-linear analysis of a vehicle equipped with *AIR Wheelset* was made with a specific tool developed for a commercial railway vehicle dynamics package (27). The results are rather interesting but cannot be discussed here for space reasons. The reader is referred to paper (28) for further details. In any case it can be anticipated that no sensible reduction in the critical speed are forecasted by the use of the *AIR Wheelset*.

The use of torque limiters promises great improvements in the curving behaviour of a vehicle equipped with this *AIR Wheelsets*. Eadie et al. (29) showed that the longitudinal traction coefficient, measured on a high speed train, exceeds 0.2 only at speeds lower than approximately 100 km/h (Figure 7), that is a quite common speed in tortuous lines. While the adhesion used in tangent track is rather low (see the "cloud" of points in the ±0.1 traction coefficient range) and would never lead to sliding in the torque limiter, it is a common experience that most of the curves with radius lower than 500 m are prone to rail corrugation.

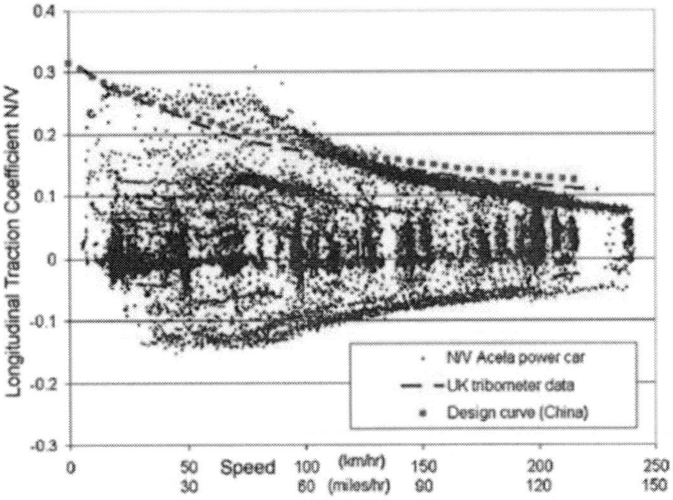


**Figure 7. Longitudinal traction coefficient for the leading axle of the power car (29).**

The possibility of a net relative rotation of the wheels along a curve leads to the non-necessity of the oscillatory stick-slip phenomenon that is intrinsic in a standard wheelset where the relative position of the wheels cannot change despite the fact that the distance run by the two wheels is different. More details on the complex wheel/rail interface issues arising from the adoption of the torque limiters on the *AIR Wheelset* will be published in (30) to which the reader is referred. It can be anticipated that with an accurate setting of the torque both rail and torque limiter low wear can be obtained, reducing maintenance costs.

Although extensive testing will be needed to verify the estimated performances, the *AIR Wheelset* promises to be a device that can effectively help to solve the never-ending conflict between guidance and stability. Stability is guaranteed as the *AIR Wheelset* is almost as torsionally stiff as a conventional wheelset, while curving attitude is dramatically improved in narrow curves thanks to the use of torque limiters. Friction damping supplied by torque limiters gives a new dimension to resonance phenomena typical in narrow curves with corrugated rails.

## CONCLUSIONS AND FURTHER DEVELOPMENTS

A new patented wheelset, named *AIR Wheelset* was introduced. It is the development and the combination of conventional and IRW wheelsets, with the aim of designing a wheelset that can dramatically improve rolling stock maintenance practice (31).

Summarizing, the *AIR Wheelset* is composed of the following elements:
- a non-rotating *bridge*, made of structural steel;
- two *wheels*, which are not press or shrink fitted on an axle but that simply house the bearings;
- *rolling bearings* that support vertical, lateral and braking forces and the resulting torques acting on each wheel;
- a stiff *shaft*, which torsionally connects the wheels ensuring stability at high speed;
- the *torque limiters* which releases the wheels in tight curves maintaining a predetermined torque while allowing relative finite rotations.

The next steps in the application of the *AIR Wheelset* necessarily pass through a validation of the concept on a roller rig or under a test vehicle.

## ACKNOWLEDGMENTS

The authors wishes to thank for their contribution Dr. Fabio Piccioli (Italcertifer, formerly University of Florence), Dr. Mauro Cavalletti (VI-Grade), Dr. Eddie Searancke (Bombardier),

## REFERENCE LIST

1. R. V. Dukkipati, S. Narayana Swamy, M.O.M. Osman, *Independently Rotating Wheel Systems for Railway Vehicles: A State of the Art Review*, Vehicle System Dynamics, 21(1), 297-330, 1992.
2. J.A. Hadden, E.H. Law, *Effects of Truck Design on Hunting Stability Of Railway Vehicles*, Journal of Engineering for Industry, Transaction of the ASME, 162-171, February 1977.

3. G.R. Doyle, Jr., R.H. Prause, *Hunting Stability of Rail Vehicles with Torsionally Flexible Wheelsets*, Journal of Engineering for Industry, Transaction of the ASME, 10-17, February 1977.
4. R. Panagin, *Carrello a ruote indipendenti*, Ingegneria Ferroviaria, 651-664, October 1982.
5. C. Casini, R. Panagin, G. Tacci, *La sperimentazione dei veicoli a ruote indipendenti presso le F.S. – parte 1*, La Tecnica Professionale, 266-272, maggio-giugno 1984.
6. C. Casini, R. Panagin, G. Tacci, *La sperimentazione dei veicoli a ruote indipendenti presso le F.S. – parte 2*, La Tecnica Professionale, 344-348, luglio-agosto 1984.
7. Patent US 2698526A (Josef Beier), 4 January 1955, *Shafting for power transmission in motor-driven vehicles.*
8. Patent EP 0428091A1 (Alsthom GEC), 22 May 1991, *Dispositif d'entraînement de roués indépendantes en rotation d'un véhicule ferroviaire è essieu moteur.*
9. B.L. King, *New Tyre Profiles for British Railways*, The Railway Gazette, 60-64, Jan. 19, 1968.
10. A. H. Wickens, *A History of Railway Vehicle Dynamics*, in Handbook of Railway Vehicle Dynamics, S. Iwnicki (ed.), Taylor and Francis, 2006.
11. P. Becker, *On the use of individual free rolling wheels on railway vehicles*, Eisenbahn Technische Rundschau, 19, 11, 1970.
12. Hecht M. and Keudel J.,LeiLa-DG. Bericht Nr. 30/2004, Technische Universität Berlin, 2004.
13. *Desiro City – Evolution in motion,* http://www.mobility.siemens.com, accessed on 4 October 2014.
14. *Thameslink bogies start rolling out*, Railway Gazette International, 44-15, Jan 2014.
15. N. Kingsley, M. Hughes, *IEP design awaits the acid test*, Railway Gazette International, December 2014.
16. E. Searancke, *FLEXX Eco – the leading lightweight passenger bogie design*, Bogies/Wheelsets Supplement, European Railway Review, Volume 20, Issue 2, 3-5, 2014.
17. European Standard EN 15437-2, *Railway applications - Axlebox condition monitoring – Interface and design requirements - Part 2: Performance and design requirements of on-board systems for temperature monitoring*, CEN, Brussels, September 2012.
18. AB Consulting sas di Andrea Bracciali & C., *Railway Wheelset With Partially Independent Wheels*", world patent PCT/IB2015/051855.
19. European Standard EN 13979-1:2003+A2 March 2011, *Railway applications - Wheelsets and bogies - Monobloc wheels - Technical approval procedure - Part 1: Forged and rolled wheels*. CEN, Brussels, March 2011.
20. UIC CODE 510–5 2nd edition, *Technical approval of monobloc wheels. Application document for standard EN 13979-1*. UIC, Paris, May 2007.
21. European Standard EN 12082:2007+A1, *Railway applications - Axleboxes - Performance testing*. CEN, Brussels, October 2010.
22. European Standard EN 13103:2009+A2, Railway applications - Wheelsets and bogies - Non powered axles - Design method, CEN, Brussels, July 2012.
23. Guideline DVS 1612, *Design and endurance strength evaluation for welded steel joints in railway vehicle construction, German association for welding and related procedures*, August 2009.
24. Viareggio train derailment, accessed on 22.09.2014, http://en.wikipedia.org/wiki/Viareggio_train_derailment
25. Lucchini RS, *Wheelset Protection*, retrieved from www.lucchinirs.com on 22.09.2014.
26. Locomotiva FS E.652, http://it.wikipedia.org/wiki/Locomotiva_FS_E.652, accessed on 22.09.2014.

27. A. Bracciali, G. Megna, *Dynamics of a railway vehicle with torsionally flexible axles*, 2015 VI-Grade Users Conference, Trieste, Italy, 15-16 April 2015.
28. A. Bracciali, G. Megna, *Running Dynamics Of Railway Vehicles Equipped With Partially Independently Rotating Wheelsets,* accepted for presentation to to the 24th International Symposium on Dynamics of Vehicles on Roads and Tracks (August 17-21, 2015, Graz, Austria).
29. E. Magel, A. Tajaddini, M. Trosino, J. Kalousek, *Traction, forces, wheel climb and damage in high-speed railway operations*, Wear 265 (2008) 1446–1451, doi:10.1016/j.wear.2008.01.036.
30. A. Bracciali, G. Megna, *Contact Mechanics Issues Of A Vehicle Equipped With Partially Independently Rotating Wheelsets*, accepted for presentation to CM2015 (10th International Conference on Contact Mechanics of Wheel / Rail Systems), Colorado Springs, Colorado, USA, 30 August 2015.
31. A. Bracciali, *Maintainability of wheelsets: a novel solution to save time and money*, abstract submitted to 11$^{th}$ World Congress on Railway Research, WCRR 2016, Milan 29.5-2.6.2016.

# Hunting stability and improvement measures for high speed vehicles

**J Zeng, L Wei, M Chi, J Wang**
State Key Laboratory of Traction Power, Southwest Jiaotong University, China

**ABSTRACT**

The hunting stability is an important issue for vehicle systems that determines the maximum operating speeds of the vehicles on tracks. In this paper, the research background for three main stability phenomena, including the "VIP Chair Shaking", "Bogie Alarm" and "Carbody Sway" problems, are introduced. Then, the wheel/rail contact relations are investigated and a modified wheel profile is recommended. In order to solve or relieve the stability problem, the conicity at the flange root for the standard new wheel is decreased and the conicity at the wheel profile end is increased. Furthermore, the numerical model for a high speed passenger car is established to verify the dynamic performance of the modified wheel profile. Finally, the dynamic test for the vehicle equipped with the modified wheels is carried out using full-scale roller test rig to validate the theoretical analysis. It can be seen from simulations and validation tests that the dynamic performance of the modified wheel LME17 matched with new rail maintains at the same level as the standard new wheel. The bogie lateral acceleration for the "LME17 worn wheel" matched with new rail is decreased compared with that of the standard new wheel. The lateral ride index for the modified wheel matched with "over-grinding rail" is smaller than that of the standard new wheel.

**Keywords:** Hunting stability; contact relations; modified wheel; roller rig test; bogie lateral acceleration; lateral ride index.

## 1. INTRODUCTION

China's high speed railways have been developed very rapidly recent years. Up to now the total high speed lines are more than 10,000km's long and every day there are more than 1000 EMU train sets operating on lines with commercial speed 300km/h. The hunting stability is an important issue for vehicle systems that determines the maximum operating speeds of the vehicles on tracks. For the Chinese high speed trains, it is occasionally found that some high speed vehicles appear hunting motions in the particular locations of the track after a certain operating mileages due to the wheel and rail profile wear which causes the lateral acceleration alarm of the bogie and abnormal vibrations of the carbody. But the hunting motions can vanish after re-profiling the wheel or grinding the rail head, which means that the stability behavior is improved by modifying the wheel/rail contact relations. Since the hunting stability will affect the comfort and safe operation of the train, high attentions have been paid by researchers and extensive studies on this important issue have been conducted theoretically and experimentally in recently years [1-5]. It is known that we can improve the vehicle hunting stability normally from two aspects. One is from the vehicle side, the suspension parameters of the bogie can be optimized or even using active suspensions to improve the hunting stability [6-9]. Another is from the side of wheel/rail contact relations, the stability can be improved by the better combination of wheel and rail profiles or even under worn conditions, and the equivalent conicity of wheel/rail contact is an important parameter that can be taken to evaluate the vehicle hunting stability [10-12].

The purpose of the paper is trying to understand the hunting instability mechanism caused by the wheel/rail interactions and to improve the vehicle stability from point of view of wheel/rail contact relations. Hence, a modified wheel profile with better wheel/rail interactions will be suggested according to the existing problems in the high speed train operation. Then the adaptability to the track for the modified wheel in new and worn conditions and the dynamic performance of the high speed vehicle are investigated and compared with results for the standard new wheel. Finally, the stability test for the vehicle equipped with the modified wheels is carried out using full-scale roller test rig to validate the theoretical analysis.

## 2. RESEARCH BACKGROUND

Usually the hunting motions of the vehicle system can be classified into two types, i.e., the bogie hunting that mainly behaves the vibrations of the bogie and carbody in the horizontal plane, and the carbody hunting that behaves the carbody vibrations in the horizontal plane. For the vehicle system, the bogie hunting normally appears at higher speed and the stability is impossible to recover once the system is unstable after the vehicle speed exceeds the critical speed. Normally the higher the wheelset equivalent conicity is, the lower the bogie hunting critical speed. While the carbody hunting usually arises at lower speed when the wheelset hunting frequency closes to the carbody yaw or roll frequencies, and the hunting instability only exists in a speed range. Fig.1 is the schematic diagram for the carbody hunting with respect to small and large wheelset equivalent conicity. It is seen that the hunting instability may emerge at ordinary running speed 200-350km/h for high speed train if the wheelset equivalent conicity is too small.

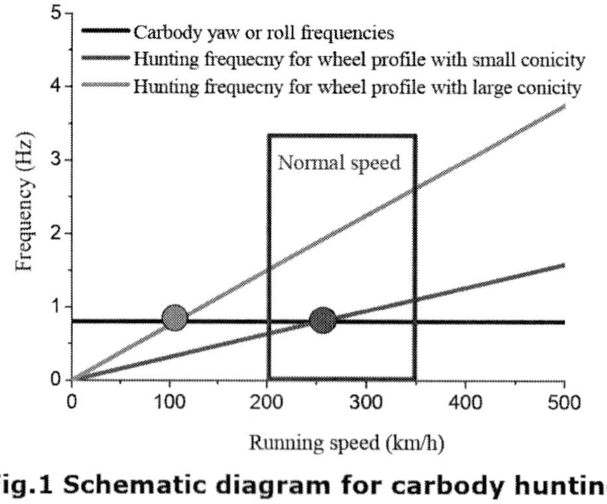


**Fig.1 Schematic diagram for carbody hunting**

The hunting motions occasionally appear in some particular locations of the track for the high speed EMUs in China, which causes lateral acceleration alarm of the bogie and abnormal vibrations of the carbody. The main reason is that the wheel-rail profile wear, especially the concentrated wheel tread wear leads to unreasonable wheel-rail matching relationship. There are the following three main phenomena happened in the operations of the high speed trains.

One is the "VIP Chair Shaking" problem with frequency around 9-12Hz which normally appears in the wheel wear later period when passing through the turnout section. Fig.2 shows the measured lateral accelerations of the carbody, bogie frame and axlebox in the turnout section in which it is clearly seen that there exists a common harmonic vibration with frequency 9.10Hz. It is found that the wheel flange root contact occurs due to the rail abnormal wear in the turnout section. Thus the resulted large wheel/rail contact equivalent conicity may cause high frequency hunting motion, and then excite the carbody local resonance and further cause the VIP chair shaking.

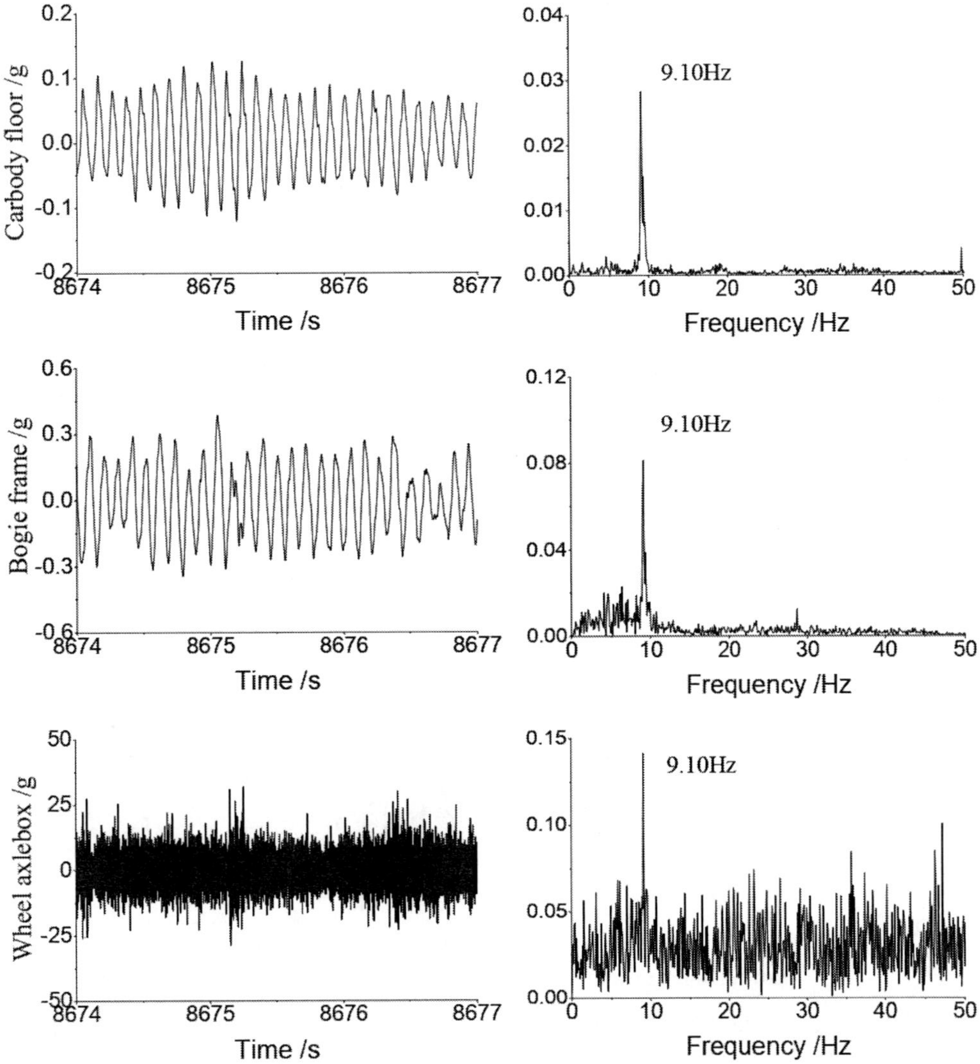


**Fig.2 Measured lateral accelerations of the carbody, bogie frame and axlebox**

The second is the "Bogie Alarm" problem with vibration frequency 7-8 Hz and the lateral acceleration of the bogie frame exceeds the limit value, which normally appears in the non-grinding rail section matched with the wheel concave wear. This alarm problem caused by hunting instability can be solved after grinding the rail head shown as Fig.3.

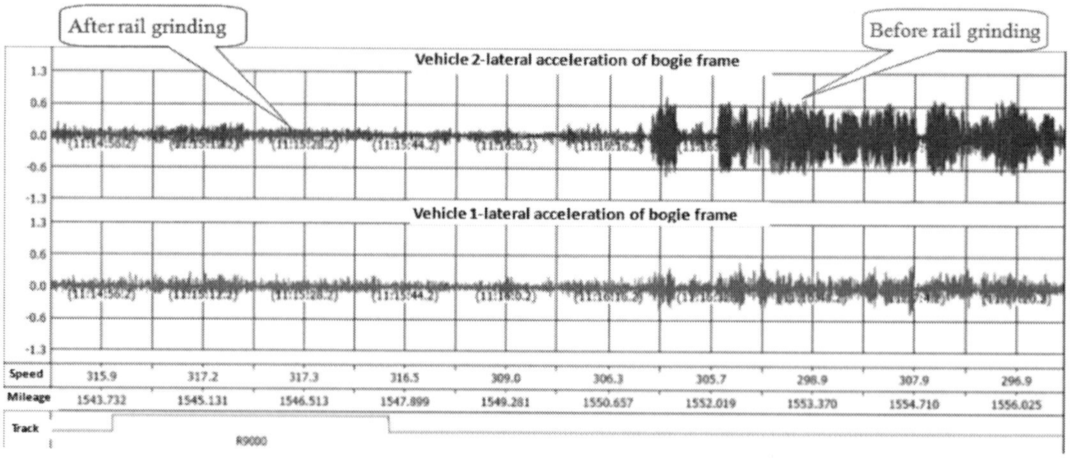


**Fig.3 Bogie frame lateral accelerations before and after rail grinding**

The third is the "Carbody Sway" problem with the vibration frequency lower than 3Hz which normally happens in the over-grinding rail section matched with the standard new wheel. In this case the over-grinding of rail gauge corner will cause small equivalent conicity and further result in carbody hunting. Fig.4 shows the comparison and equivalent conicity of the standard rail profile and the grinding rail profiles. In Fig.4, NW indicates the standard new wheel, NR and GR denote the standard new rail and grinding rail respectively, and L and R for the left and right rails. Fig.5 gives the actual measured caybody and bogie frame lateral accelerations which obviously display that a harmonic vibration exists. However, this phenomenon can gradually disappear due to the wheel profile wear after operating a certain mileage.

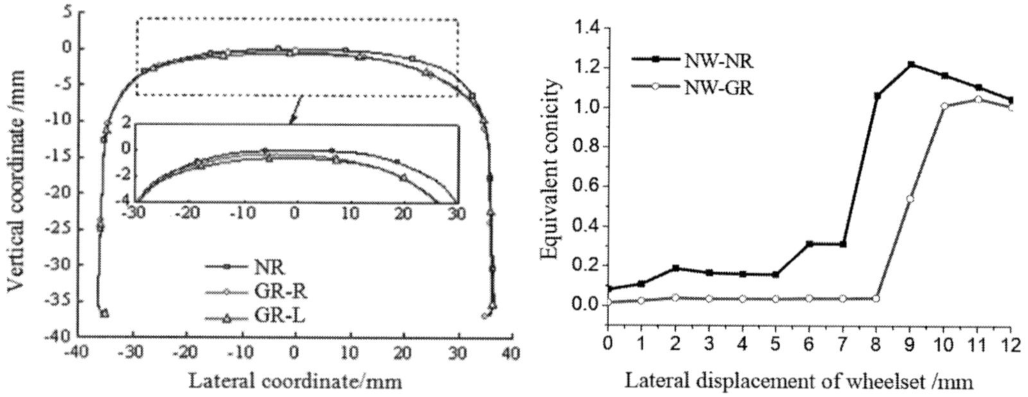


(a) Standard rail and grinding rail

(b) Equivalent conicity before and after grinding

**Fig.4 Rail over-grinding and equivalent conicities**

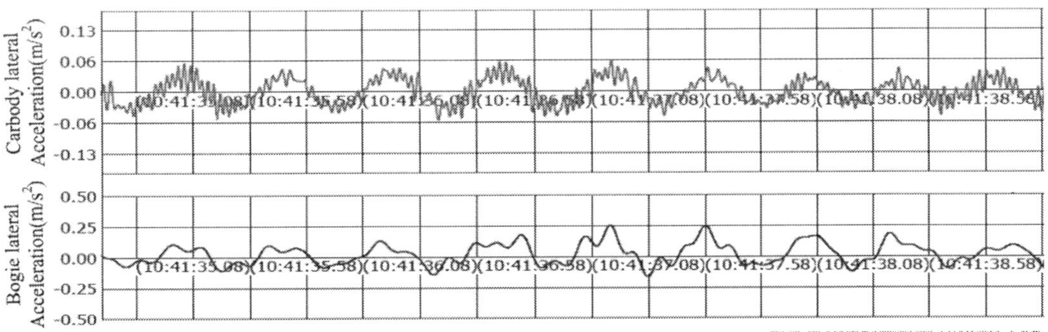


**Fig.5 Carbody and bogie frame lateral accelerations**

Therefore, it can be summarized that the resulted problems in the service period of the high speed trains are strongly related to the actual conditions of rail profiles. The wheel contact areas may be divided into three zones depicted in Fig.6, i.e., normal contact zone, high conicity zone and low conicity zone. The bogie hunting motion with high frequency may occur if the wheel/rail contact is located or close to the high conicity zone, and the carbody hunting with low frequency may appear if located in the low conicity zone. Thus, the wheel flange root contact during passing by the turnout lies in the high conicity zone that will cause the "VIP Chair Shaking" problem. The gauge corner contact due to not in time grinding of the rail matched with concave wear wheel will result in large equivalent conicity more than 0.35 that will cause the "Bogie Alarm" problem. And sometimes the over-grinding of the gauge corner may lead the wheel/rail contact to the outside of the wheel tread which makes the conicity smaller than 0.08 and results in "Carbody Sway" problem.

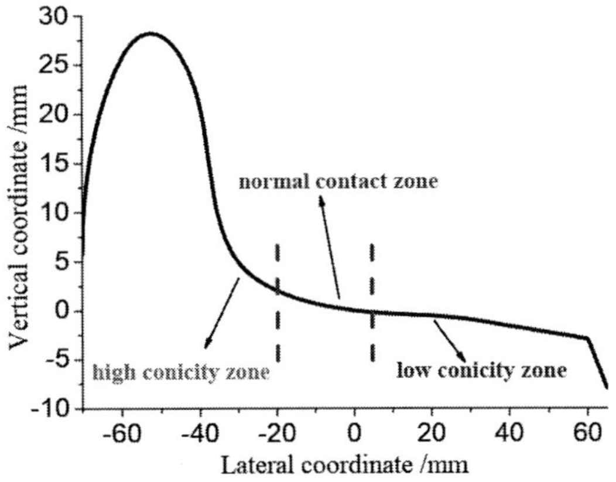


**Fig.6 Wheel contact zones**

## 3. MODIFICATION OF WHEEL/RAIL CONTACT RELATION

Fig.7 gives the comparison of the standard new wheel profile and the worn wheel profile (called WW) and Fig.8 illustrates the equivalent conicities varying with the wheelset lateral displacement for the new wheel matched with the new rail (NW-NR), worn wheel with worn rail (WW-WR) and new wheel with grinding rail (NW-GR). The contact point conditions for the three types of matched wheel/rail profiles are shown as Fig.9. It is seen that the wheel/rail contact points for new wheel matched with new rail (NW-NR) are mainly located around the rolling cycle and the rail top little towards inside, and the equivalent conicity is between 0.08-0.35 before flange contact. While the contacts for worn wheel matched with worn rail (WW-WR) are concentrated on the wheel flange root and inner gauge corner, which easily leads to the equivalent conicity exceeding 0.35 and causes bogie hunting. An efficient manner to deal with this issue is to increase the flange clearance by reducing the thickness of the flange so that the slope around the flange root can be reduced and further the equivalent conicity is decreased. But the contact points for new wheel matched with grinding rail (NW-GR) mainly lie in the positions of wheel tread end and rail top outside which will result in small equivalent conicity less than 0.08 and easily cause carbody hunting. The measure to solve this problem is to increase the slope of the wheel tread end in order to increase the equivalent conicity in this contact area.

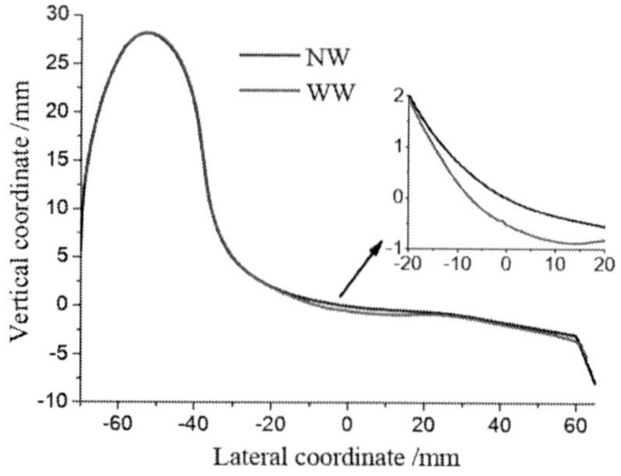


**Fig.7 Wheel profiles**

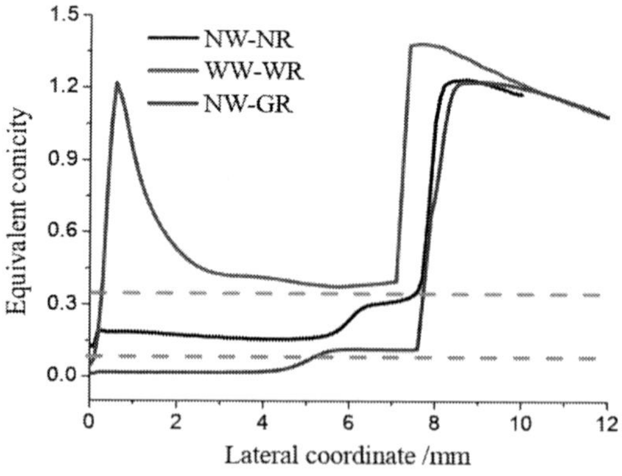


**Fig.8 Equivalent conicity**

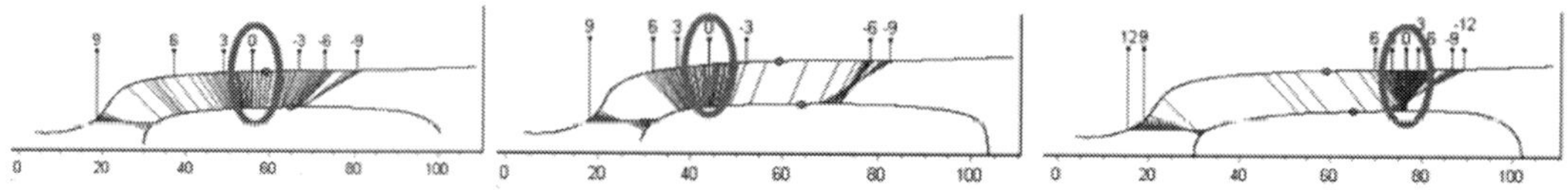

(a) New wheel/new rail (b) Worn wheel/worn rail (c) New wheel/grinding rail

**Fig.9 Wheel/rail contact**

Therefore, through the above analysis it is known that the vehicle hunting stability can be improved by modifying the wheel/rail contact relations. Thus, a modified new wheel profile named LME17 is designed shown as Fig.10, and the distribution of the wheel/rail contact points is illustrated in Fig.11. The comparison of the profiles and the equivalent conicity between modified wheel profile LME17 and the standard new wheel profile is depicted by Fig.12 and Fig.13. Firstly, it is seen that the profile and equivalent conicity of the normal working area for LME17 keep almost the same compared with NW which means that the advantages of the vehicle dynamic performance under normal operations are retained. Secondly, the flange thickness of LME17 is reduced 2mm so that the variation slope around the flange root cycle is decreased and then the equivalent conicity is reduced, thus the possibility of VIP chair shaking and bogie alarm can be reduced. Thirdly, the slope of the wheel tread end is increased and then the equivalent conicity is increased, thus the carbody sway phenomenon can be relieved.

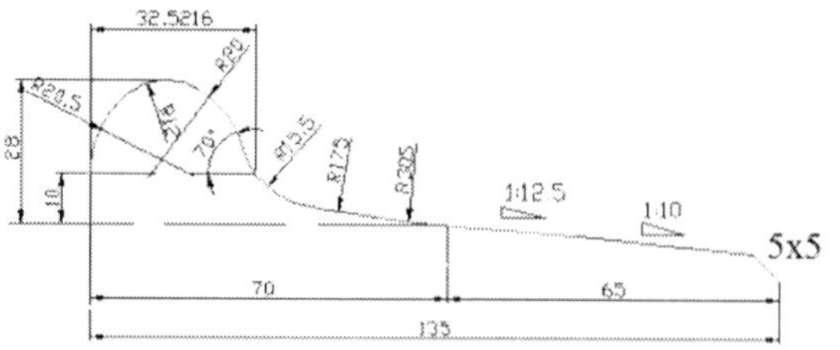


**Fig.10 Modified wheel profile LME17**

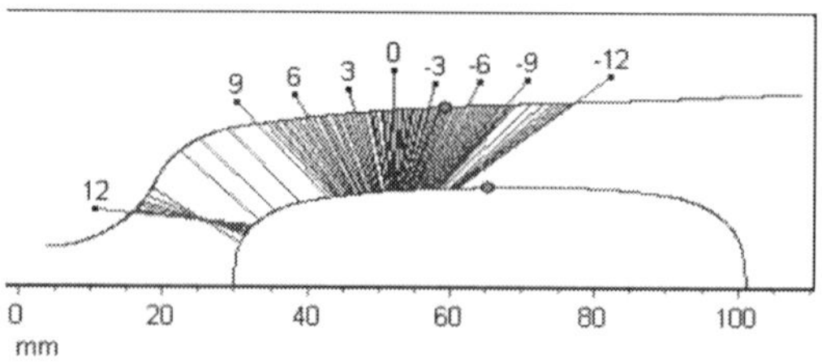


**Fig.11 Distribution of contact points for LME17**

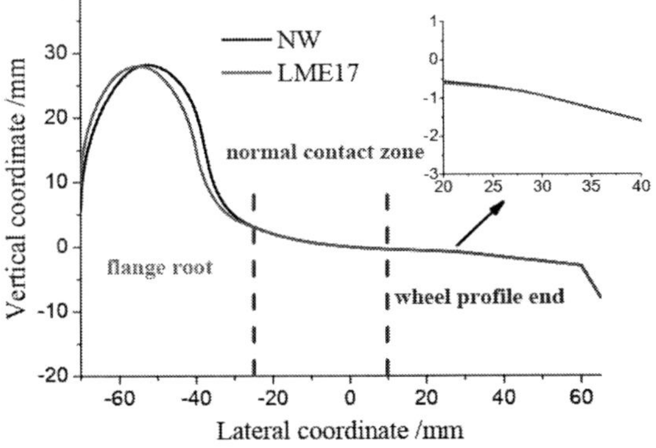


**Fig.12 Wheel profile comparison**

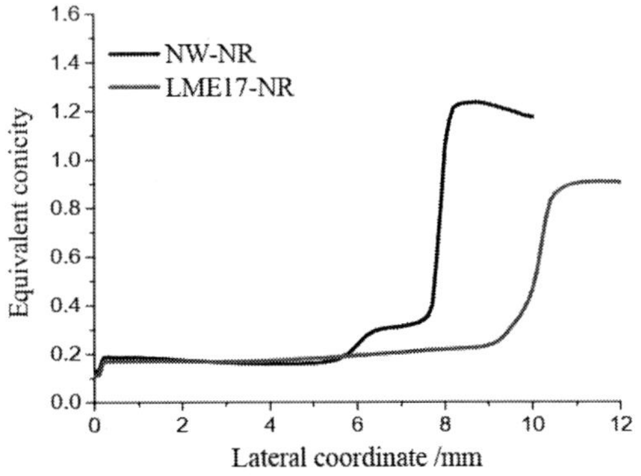


**Fig.13 Wheel/rail equivalent conicity comparison**

Fig.14 gives the wheel/rail contact stress comparison between LME17 and NW. It is known that the contact stresses for the two wheel profiles are almost the same, but the stress in the area of the flange root cycle for LME17 is smooth and much smaller compared with NW. Thus it is expected that the wheel wear of LME17 during operation will be less than the standard new wheel NW.

We know from Fig.15 that the equivalent conicity of standard new wheel matched with grinding rail may be less than 0.08 which can easily cause carbody hunting or sway, while the equivalent conicity of the worn wheel matched with the worn rail may be larger than 0.35 which can easily induce bogie alarm or bogie hunting. However for the modified wheel LME17 (see Fig.16), either its new wheel matched with the grinding rail or the worn wheel matched with the worn rail, the equivalent conicity is larger than 0.08 or less than 0.35 before flange contact, which indicates that the modified wheel LME17 has better adaptability to match with different rail profiles.

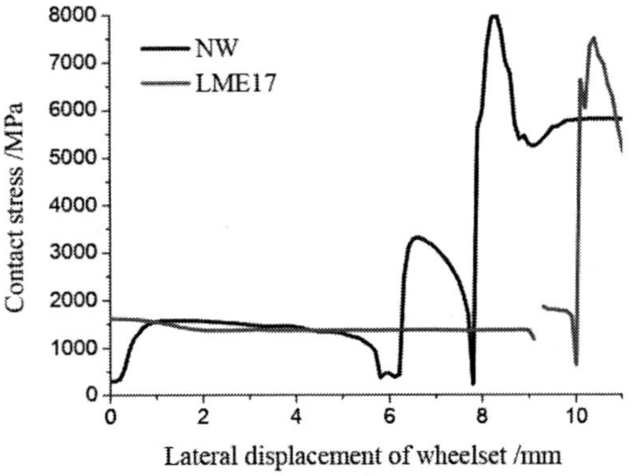


**Fig.14 Wheel/rail contact stress comparison**

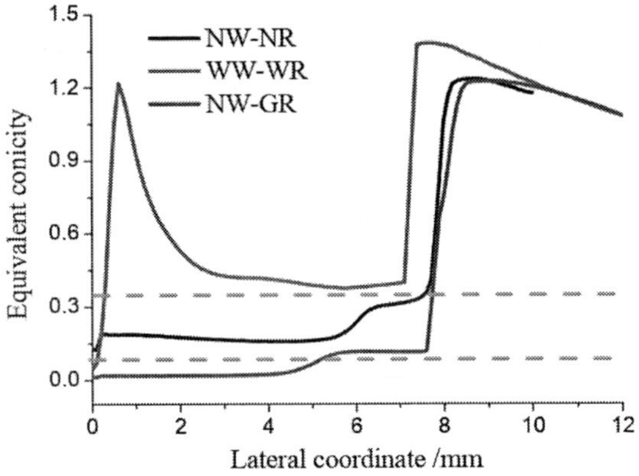


**Fig.15 Equivalent conicity for standard wheel**

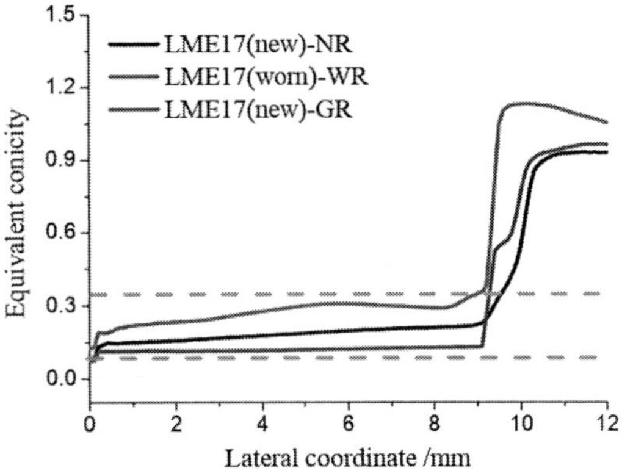


**Fig.16 Equivalent conicity for modified wheel**

Fig.17 displays the equivalent conicity of the standard new wheel and modified wheel LME17 varying with the track gauge. It is seen that the equivalent conicity of new wheel exceeds 0.35 when track gauge is reduced to 1430mm, and less than 0.08 when track gauge is 1445mm. However for the modified wheel LME17, the equivalent conicity is always between 0.08 and 0.35, which means that LME17 has better adaptability to the gauge variation.

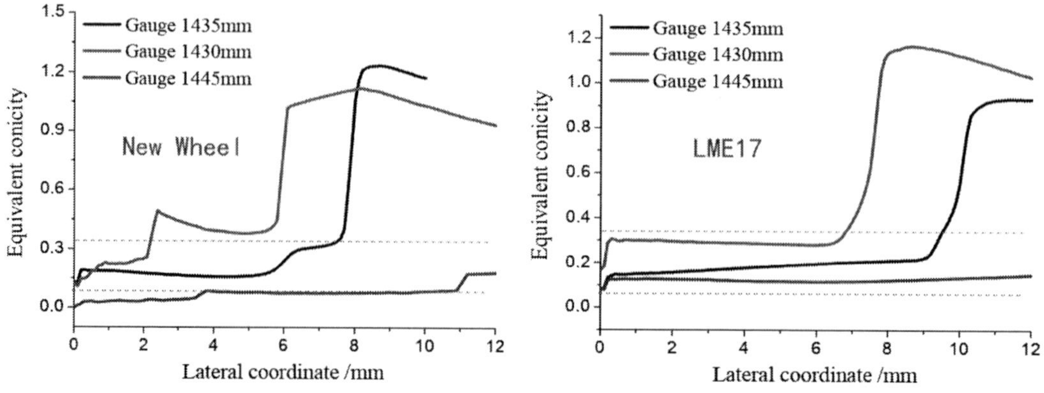


(a) Standard wheel (b) Modified wheel

**Fig.17 Equivalent conicity varying with gauge**

## 4. DYNAMIC PERFORMANCE ANALYSIS

In this section, the numerical model for a high speed EMU car is established to study the dynamic performance of the modified wheel profile LME17. The vehicle system is treated as multiple rigid bodies and the non-linear characteristics of suspension system are taken into consideration. The model consists of 44 independent degrees of freedom (DOFs), in which each wheelset has four independent DOFs, each axle box guiding arm has one independent DOF, each traction rod has one independent DOF, each bogie frame has six independent DOFs, and the carbody has six independent DOFs.

The comparisons of amplitude of limit cycle and lateral ride index for the new wheel matched with new rail are shown in Fig.18. It is seen that the critical speed for the modified wheel LME17 matched with new rail are slightly higher than that of the standard new wheel, and the lateral ride index for the wheel LME17 is decreased.

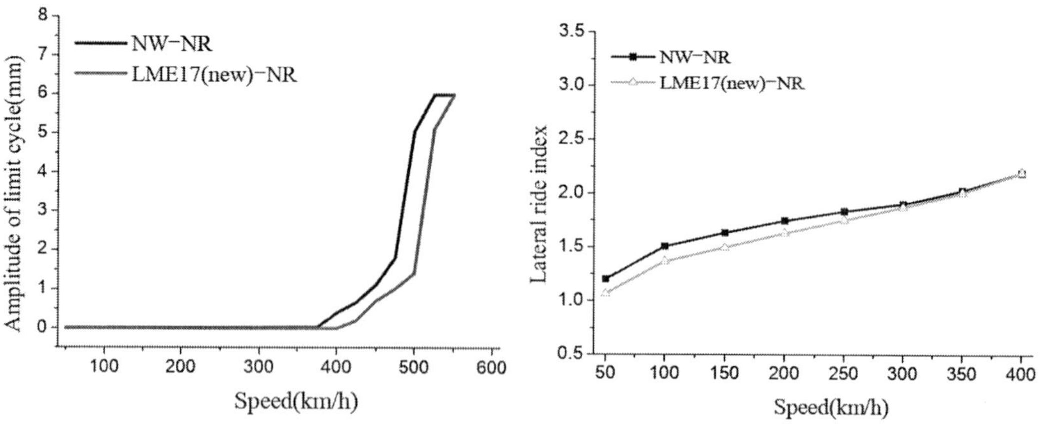


**Fig.18 Comparison of limit cycle and ride index for new wheel/new rail**

As the lack of actual worn profile for the LME17 wheel, the wheel concave wear is assumed in order to get the LME17 worn wheel and make further comparison with the worn profile of the standard new wheel. Fig.19 gives the comparisons of the wheelset bifurcation diagram and carbody ride index for the worn wheel matched with worn rail, which indicates that the critical speed for the LME17 wheel is significantly increased and the ride index is improved.

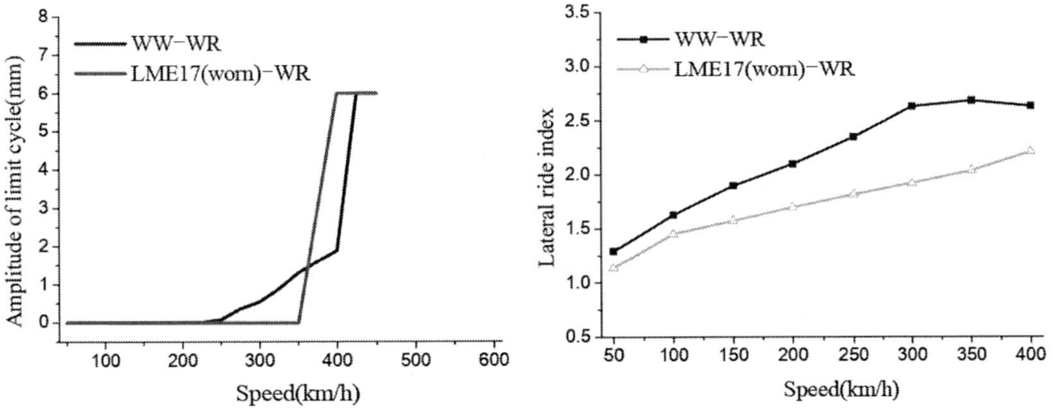


**Fig.19 Comparison of limit cycle and ride index for worn wheel/worn rail**

Fig.20 displays the comparisons of indices for the new wheel matched with over-grinding rail. It is shown that the carbody hunting problem in the speed range of 120-250 km/h can be solved by the new designed wheel LME17. The lateral ride for the new wheel matched with over-grinding rail is also improved.

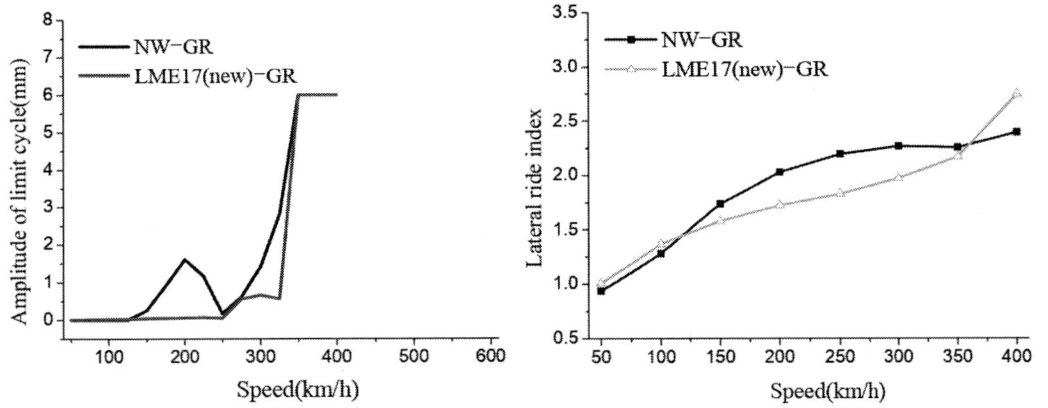


**Fig.20 Comparison of limit cycle and ride index for new wheel/grinding rail**

In normal contact situations, the dynamic performance of the modified wheel LME17 maintains at the same level compared with the standard new wheel. The dynamic performance for the worn wheel matched with worn rail can be improved by decreasing the conicity at the flange contact zone. Meanwhile, the low-frequency carbody hunting resonance problem can be solved by increasing the conicity at the wheel tread end.

## 5. TEST VALIDATION USING ROLLER RIG

The dynamic test for the vehicle equipped with the modified wheels was carried out using full-scale roller test rig to validate the theoretical analysis, see Fig.21. The dynamic tests on the roller rig include the stability test and ride comfort test. In order to reproduce the stability phenomenon mentioned above, adjustments to the roller gauge were conducted to simulate different wheel/rail contact conditions. In the condition of normal contact, the gauge of rollers kept at 1435mm with the conicity of 0.169, which aims at validate the new wheel matched with new rail. The gauge can be narrowed to 1429mm with the conicity of 0.319 so as to simulate the "worn wheel" matched with worn rail. The carbody hunting can be reproduced by widening the gauge to 1441mm with the conicity of 0.1 to simulate the new wheel combined with over-grinding rail.

**Fig.21 High speed vehicle on roller test rig**

Figure 22 shows the test results for the condition of standard and modified new wheels (NW and LME17) matched with new rail with the gauge of 1435mm. It is found that the bogie frame lateral acceleration for the modified wheel LME17 is close to that of the standard new wheel and the lateral ride index is slightly decreased.

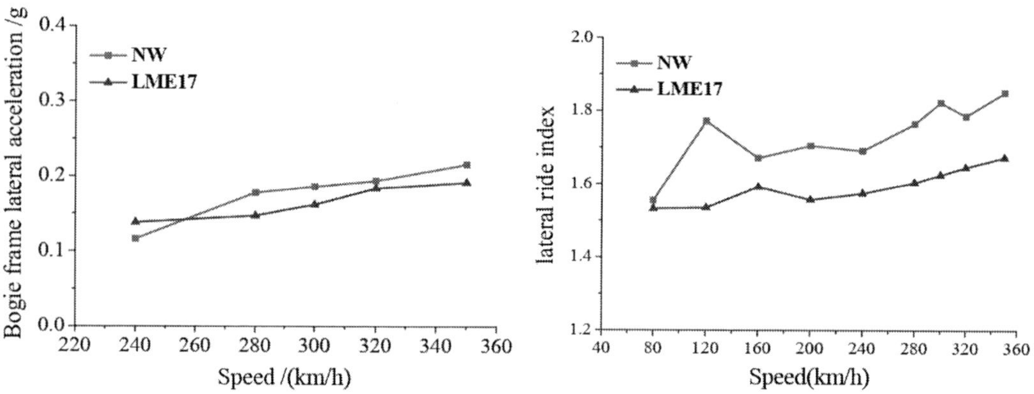


(a)Bogie lateral acceleration (b) lateral ride index

**Fig.22 Test results for new wheel/new rail at the gauge of 1435mm**

Figure 23 displays the comparison of test results for the "worn wheel" matched with "worn rail" with the gauge of 1429mm. It can be found that the bogie frame lateral acceleration for the modified wheel is decreased compared with that of the standard new wheel, which is expected to relieve the "Bogie Alarm" problem.

Figure 24 displays the comparison of test results for the new wheel matched with "over-grinding rail" with the gauge of 1441mm. According to the stability test, carbody hunting occurred at the speed of 240km/h for the standard new wheel. While for the modified wheel, the vehicle system kept at stable from 80km/h to the maximum test speed 350km/h, which indicates that the carbody hunting resonance can be solved by the new designed wheel. And also the lateral ride index for the modified wheel was much smaller than that of the standard new wheel.

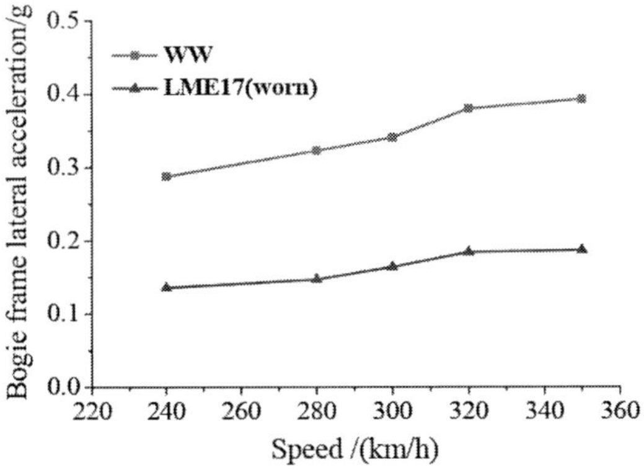


**Fig.23 Test results for "worn wheel/ worn rail" at the gauge of 1429mm**

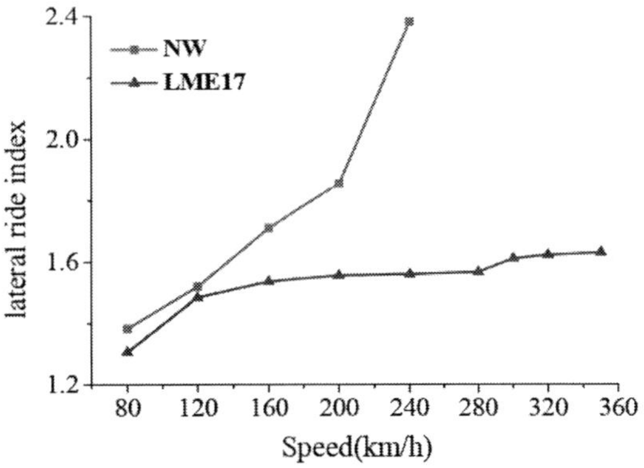


**Fig.24 Test results for "new wheel/ over-grinding rail" at the gauge of 1441mm**

## 6. CONCLUSIONS

The mechanism of hunting instability due to abnormal wheel/rail contact relations for high speed vehicles has been investigated and the following conclusions can be drawn:

(1) Three typical instability phenomena are introduced. One is the "VIP Chair Shaking" problem due to the wheel flange root contact for the worn wheel matched with rail abnormal wear in the turnout section. The second is the "Bogie Alarm" problem appearing in the non-grinding rail section matched with the wheel concave wear. The third is the "Carbody Sway" problem happening in the over-grinding rail section matched with the new wheel.

(2) A modified wheel profile named LME17 is designed in order to improve the vehicle hunting stability by modifying the wheel/rail contact relations. The profile and equivalent conicity of the normal working area for LME17 keep almost the same compared with the standard new wheel. The flange thickness of LME17 is reduced 2mm compared with the standard new wheel so that the variation slope around the flange root cycle is decreased and then the equivalent conicity is reduced. The slope of the wheel tread end is increased and then the equivalent conicity is increased.

(3) The verifications of dynamic performance for the modified wheel profile have been conducted using numerical simulations and roller rig test. It can be summarized that the dynamic performance of the modified wheel LME17 matched with new rail maintains at the same level as the standard new wheel. The bogie lateral acceleration for the "worn wheel" matched with new rail is decreased compared with that of the standard new wheel. The lateral ride index for the modified wheel matched with "over-grinding rail" is smaller than that of the standard new wheel.

(4) Through the study of this paper, it is known that the modified wheel LME17 can possess better stability compared with the standard new wheel. Therefore, it is necessary to conduct the field test for the modified wheel and further investigate its dynamic behavior in new and worn wheel conditions.

## ACKNOWLEDGEMENT

We would like to thank the National Natural Science Foundation of China under grand No.U1334206 for supporting this research.


## REFERENCES


[1] Gunter Schupp. Bifurcation Analysis of Railway Vehicles. Multibody System Dynamics (2006) 15: 25–50

[2] Krzysztof Zboinski, Miroslaw Dusza. Self-exciting vibrations and Hopf's bifurcation in non-linear stability analysis of rail vehicles in a curved track. European Journal of Mechanics A/Solids 29 (2010) 190–203

[3] H. Molatefi, M. Hecht and M. H. Kadivar. Critical speed and limit cycles in the empty Y25-freight wagon. Proc. IMechE, Part F: J. Rail and Rapid Transit, 2006, 220(4), 347–359

[4] O Polach. On non-linear methods of bogie stability assessment using computer simulations. Proc. IMechE, Part F: J. Rail and Rapid Transit, 2006, 220: 13-27

[5] P. D. Allen and S. D. Iwnicki. The critical speed of a railway vehicle on a roller rig. Proc. IMechE, Part F: J. Rail and Rapid Transit, 2001,215(2), 55–64

[6] H. Molatefi, M. Hecht and M. H. Kadivar. Effect of suspension system in the lateral stability of railway freight trucks. Proc. IMechE, Part F: J. Rail and Rapid Transit, 2007, 221(3), 399–407

[7] Pilkee Kim, Jeehyun Jung and Jongwon Seok. A Parametric Dynamic Study on Hunting Stability of Full Dual-Bogie Railway Vehicle. International Journal of Precision Engineering and Manufacturing Vol. 12, No. 3, pp. 505-519, 2011

[8] Karim H. Ali Abood and R. A. Khan. Improve hunting stability of railway carriage model using semi-active longitudinal primary stiffness suspension. International J. of Multidispl. Research & Advcs. in Engg.(IJMRAE). Vol. 3, No. I ( 2011), pp. 107-128

[9] Yung-Chang Cheng and Chin-Te Hsu. Hunting stability and derailment analysis of a car model of a railway vehicle system. Proc. IMechE, Part F: J. Rail and Rapid Transit, 2011 225: 1-16

[10] Kevin Sawley, Curtis Urban, Russell Walker. The effect of hollow-worn wheels on vehicle stability in straight track. Wear 258 (2005) 1100–1108

[11] O. Polach. Comparability of the non-linear and linearized stability assessment during railway vehicle design. Vehicle System Dynamics, 2006, 44 (Supp): 129-138

[12] O. Polach. Characteristic parameters of nonlinear wheel/rail contact geometry, Vehicle System Dynamics, 2010, 48: 19–36

# Failure analysis of railroad concrete crossties in center binding conditions using finite element method

**H Yu, B P Marquis, D Y Jeong**
Structures and Dynamics Division, Volpe National Transportation Systems Center, USA

## ABSTRACT

Finite element models are developed for railroad concrete crossties and employed to analyze tie responses to two center binding conditions, a center negative moment test condition and a hypothetical deteriorated ballast support condition, which can lead to center cracks and eventually sudden, catastrophic failure of concrete ties. Analysis results are compared with available test data, and cracking/failure patterns and force-displacement characteristics of center bound concrete ties are evaluated. The analyses predict critical wheel loads above which catastrophic tie failure is likely to occur under center binding conditions.

## 1. INTRODUCTION

Ties are an integral part of railroad tracks. The main functions of railroad ties include: anchoring rails against lateral and longitudinal movements, maintaining gauge and vertical position, and transmitting and distributing rail loads to ballast with reduced contact pressures (1). Because concrete can be engineered to meet specific service requirements and add overall stability and performance to railroad tracks, concrete ties have been employed in the U.S. rail industry since the 1960's (2). Concrete ties are also estimated to last twice as long as timber ties and therefore expected to reduce the life cycle cost despite incurring a higher initial cost. Over 30 million concrete ties are estimated to exist in track in North America.

Nevertheless, concrete ties have displayed multiple failure modes in service that have shortened their expected service lives and sometimes caused derailment accidents (3). One of the main failure modes of concern is center bound or center negative cracks. The combined effects of abraded tie bottom (from dynamically pumping against ballast) and ballast deterioration can lead to voids or gaps between ties and ballast. Uneven gaps toward the tie ends can give rise to center binding support conditions in which the center of a tie is supported while the ends behave like cantilever beams. Figure 1 shows examples of multiple center bound cracks on concrete ties. The concrete ties appear to be abraded at the bottom therefore losing significant portions of the cross sections, particularly at the tie ends.

The American Railway Engineering and Maintenance-of-Way Association (AREMA) manual on concrete ties specifies that monoblock concrete ties shall pass a center negative moment test, a diagram of which is reproduced in Figure 2 for a 102 in. (2,590.8 mm) tie (4). In this test, a concrete tie is placed upside down with its rail seats resting on two rubber supports, each being 30 in. (762 mm) away from the tie center. The tie bottom is then loaded by a force $P$ over a 6 in. (152.4 mm) length at the center until $P$ meets a specified pass/fail criterion. The concrete tie is deemed to pass the test if no structural cracking is observed at the completion of the test. Under the assumption that a single rail seat carries 50% of the wheel load, it is calculated that $P$ has the same magnitude as the wheel load. While the test does simulate a center binding support condition, it is unclear how faithful it resembles field conditions or how well a tie that passes this test would perform under field conditions.

This paper uses the finite element (FE) method to examine the behavior of concrete ties under two center binding conditions: the AREMA center negative moment test condition described above and a hypothetical deteriorated ballast support condition featuring an uneven gap between tie and ballast. Static analyses are conducted for the AREMA test, and the FE results are compared with and show reasonably good agreement with available test data, despite some simplifications adopted in modeling. Static and dynamic analyses are further conducted for the deteriorated ballast support condition. Cracking/failure patterns and force-displacement relations are then analyzed based on the simulation results.

**Figure 1 Multiple center bound or center negative cracks on concrete ties with abraded bottom.**

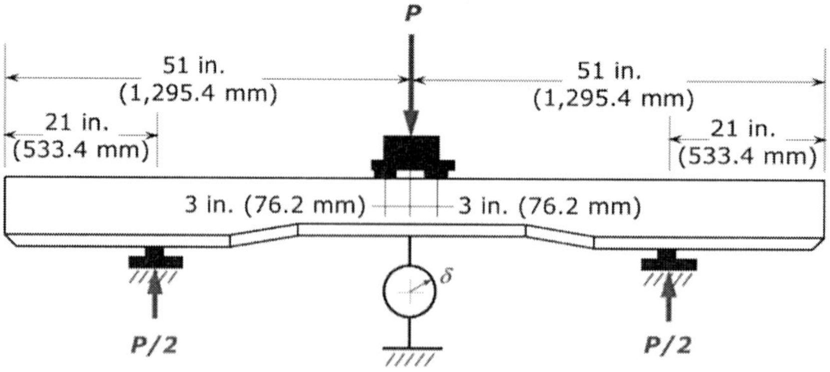


**Figure 2 AREMA center negative moment test specification for monoblock concrete ties (4), reproduced for a 102 in. (2,590.8 mm) tie.**

## 2. FINITE ELEMENT MODELING

The original, unused form of the concrete tie examined in this paper is 102 in. (2,590.8 mm) in length and 10 1/2 in. (266.7 mm) in base width. The heights of the tie at the center of rail seats are 8 in. (203.2 mm). The 36 in. (914.4 mm) long center section has a reduced height of 7 in. (177.8 mm). There are eight steel reinforcements in the form of indented seven-wire strands with a nominal diameter of 3/8 in. (9.525 mm). Each strand is pretensioned to 17.25 kip (76.73 kN). The concrete compressive strength is specified to be 4,000 psi (27.58 MPa) at prestress transfer and 7,500 psi (51.71 MPa) at 28 days. Concrete ties are modeled as a heterogeneous medium composed of steel reinforcements, concrete and their interfaces. Idealized quarter symmetry is assumed in all FE analyses.

### 2.1 Concrete modeling

A damaged plasticity model available in commercial FE software is applied in the material modeling of concrete. The model uses isotropic damaged elasticity in combination with isotropic tensile and compressive plasticity to represent the inelastic behavior of concrete (5). Tensile and compressive cracks/damages in concrete in terms of respective strength degradations are monitored by tensile and compressive damage variables defined in the model. The constitutive equations and material parameters needed to apply this model are described in detail in previous publications (6-7).

An unused concrete tie with the above geometry was subjected to the AREMA center negative moment test condition and loaded to failure under displacement control. Concrete cores were then drilled from the failed tie and subsequently underwent standard material testing. Table 1 shows the average concrete material parameters measured for the concrete tie. These basic material parameters are employed to define the constitutive relations for concrete in the FE analyses presented in this paper. Figure 3 shows the schematic of the full end section of the unused tie model.

**Table 1 Average concrete material parameters measured on concrete cores drilled from the unused tie.**

| | Density lb/ft$^3$ (kg/m$^3$) | Elastic modulus ksi (GPa) | Poisson's ratio | Split tensile strength psi (MPa) | Compressive Strength psi (MPa) |
|---|---|---|---|---|---|
| Unused tie | 140.5 (2250.6) | 4,941.0 (34.07) | 0.202 | 1,012.5 (6.98) | 10,138.5 (69.90) |

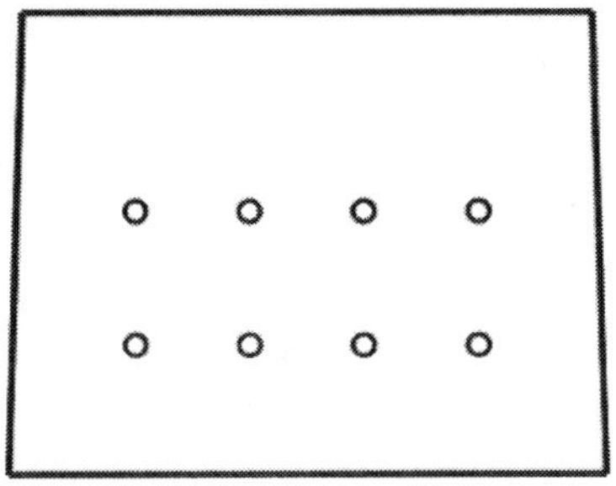

**Figure 3 Full end section of the unused tie model.**

## 2.2 Interface modeling

The strands have natural spiral surfaces that interlock with the matching inner surfaces of concrete. When the strands slip relative to the concrete as the pretension is released, there expected to be normal dilation along the strand-concrete interfaces that affects the overall bond behavior. A phenomenological bond model capturing such coupled tangential-normal behavior is currently under development. In such a modeling approach, the strand and concrete surfaces are homogenized, and cohesive elements of 0.005 in. (0.127 mm) thicknesses following predefined traction-displacement constitutive relations are employed between the strand and the concrete elements (5). In this paper, a user frictional model is adopted to simulate the interface behavior between the seven-wire strand and concrete. This is considered a simplification and expected to affect the concrete tie's flexural response to some extent. The user frictional model is included in the bond model development for seven-were strands (8), following the adhesive-frictional model development for the interface of a smooth wire (9). Here the interface of the 3/8 in. (9.525 mm) diameter strand is assigned a coefficient of friction of 0.45 and no initial adhesion.

## 2.3 Initial condition

As pretensioned concrete members, concrete ties come with initial stress states due to the pretension release/prestress transfer phase during manufacturing. The initial tension in the strands is assumed to be 156 ksi (1,075.6 MPa), and static analyses are conducted to simulate the pretension release process. The resulting stress state for the unused tie, whose longitudinal component contour is shown in Figure 4, serves as initial conditions in the subsequent FE analyses in which the tie model is further subjected to static/dynamic loading.

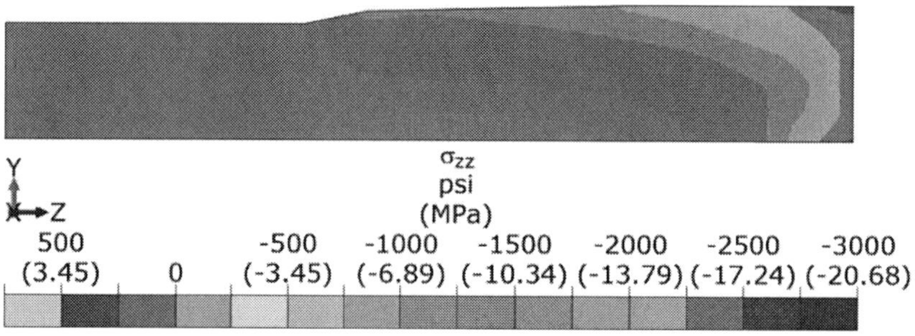


**Figure 4 Initial longitudinal stress contours in the unused tie model due to release of the initial tension in the strands.**

## 2.4 Simulation of deteriorated ballast

In addition to the AREMA center negative moment test, a hypothetical center binding scenario with deteriorated ballast support, as shown in Figure 5, is studied. In the latter case, a wedge shaped gap is assumed to exist between the concrete tie and the ballast. The gap is assumed to initiate at 1/3 of the half tie length from the tie center and grow linearly toward the tie end. The maximum gap is $\Delta h$ at the tie end. It is noted that both abraded tie bottom and ballast deterioration can contribute to tie-ballast gaps. The gap shown in Figure 5 apparently does not include the contribution from uniform tie abrasion or uniform ballast deterioration. A downward traction load is applied on the rail seat. Again the sum of the resultant forces on the rail seats of a full tie is calculated to be equivalent to the wheel load, provided that a single rail seat carries 50% of the wheel load. The Extended Drucker-Prager model is applied to the ballast material. Some heterogeneity is introduced in the ballast material as shown in Figure 5, in which "stiff ballast" is intended to simulate consolidated ballast, whereas "weak ballast" is intended to simulate pulverized ballast. The ballast is supported by a subgrade modeled as an elastic half space.

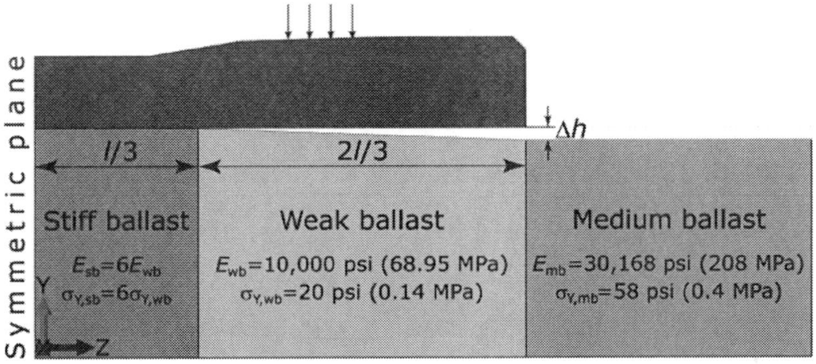


**Figure 5 A hypothetical center binding scenario due to deteriorated ballast support.**

### 2.5 Convergence in static analyses

Static analyses are conducted for both the AREMA center negative moment test condition and the hypothetical deteriorated ballast condition. The goals are to load the ties to complete failure in the analyses so that their ultimate center negative moment capacities can be evaluated. Because the ultimate failure of concrete ties is often sudden and catastrophic, it is usually very difficult for a static simulation to capture the post-peak behavior. It is therefore very important that the simulations go as far as possible in the pre-peak stages. This can be achieved by improving the convergence in the static analyses.

Two techniques are adopted to improve the convergence in the static analyses. First, as opposed to pure displacement or force methods, the Riks method that applies force conditions but uses the length of the equilibrium path as an additional variable in equation-solving iterations is employed (10). Second, artificial weak spots are introduced in the concrete elements to improve divergence cases in which adjacent elements are competing for entering the post-peak or softening stage of the constitutive curves. The Riks method alone or both techniques combined are effective in bringing all simulations to near the peak load without convergence issues.

### 2.6 Mesh sensitivity

The maximum mesh size for concrete elements is 1/2 in. (12.7 mm). Mesh sensitivity studies using maximum concrete element sizes of 2/3, 1/2 and 1/3 in. (16.9, 12.7 and 8.5 mm) show that the force-displacement curves are not very sensitive to the three different mesh sizes. The crack locations are not identical under each mesh when uniform concrete material properties are used. This is mainly due to the fact that concrete elements compete for entering the post-peak or cracking stage and that the locations of the cracking elements can be random. Assigning artificial weak spots in concrete can direct the cracks to form in known crack locations based on experimental or field observations. This is justified by the fact that actual concrete is heterogeneous with stronger and weaker spots distributed throughout the material.

## 3. VERIFICATION WITH TEST DATA

AREMA center negative moment tests were conducted on concrete ties until the ties failed catastrophically. Static simulations are conducted in the same manner until convergence issues stop the simulations. Force-displacement characteristics are calculated from both the test and the simulation data. As wheel loads are also supported by adjacent ties in track, the assumption here is that the directly loaded single tie under study carries 50% of a wheel load. Therefore, the force quantity *P* in Figure 2 corresponds to an equivalent wheel load. The rail seat displacement is calculated relative to the displacement of the center of the tie. The equivalent wheel load-relative rail seat displacement curves are plotted for the unused tie in Figure 6 and compared between the test and the FE analysis.

The simulation curve appears to generally track the test curve, though it stops at a lower relative rail seat displacement and a slightly lower peak load than the test curve. A few factors may have contributed to these discrepancies. While mechanical interlocking plays a prominent role in the behavior of actual strand-concrete interfaces, only friction is considered in the current interface modeling. Further, the concrete strengths in Table 1 are used to simulate the pretension release process, but the actual concrete strengths at prestress transfer during manufacturing can be considerably lower. The strength evolution is not considered in modeling because the FE software does not allow the change of material parameters during an analysis. Moreover, direct tensile strengths may be preferred over split tensile strengths in the constitutive modeling of concrete, but direct tensile strengths are more difficult to measure and unavailable in this case. Despite these limitations, the simulation curve agrees reasonably well with the test curve, thus lending credibility to the FE model developed for the concrete tie.

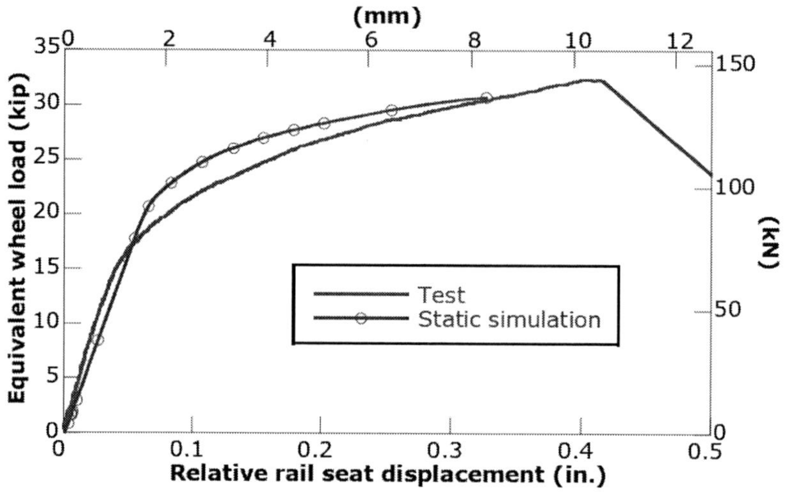


**Figure 6 Equivalent wheel load-relative rail seat displacement curves under AREMA center negative moment condition: test vs. simulation.**

## 4. ADDITIONAL SIMULATION RESULTS

Static and dynamic FE analyses are further conducted using the same concrete tie model but under the deteriorated ballast support condition shown in Figure 5. The cracking patterns and the force-displacement characteristics are then compared between the two center binding scenarios, under the AREMA center negative moment test condition and the deteriorated ballast support condition, respectively. The maximum gap $\Delta h$ for the deteriorated ballast support condition is set to be 2 in. (50.8 mm) in this paper.

Figures 7 and 8 show the tensile damage variable contours in the unused tie under the AREMA center negative moment test and the deteriorated ballast support conditions, respectively, obtained in static FE simulations. The tensile damage variable $d_t$ measures the degree of tensile strength degradation in concrete, with $d_t$=0 indicating undamaged concrete and $d_t$=1 indicating formation of cracks as the tensile strength is completely degraded. The contours show that the concrete tie experiences three similar cracking stages in both center binding scenarios. The cracking stages may be characterized in terms of crack depths, as the cracks appear on the top surface (A), reach the first row of strands (B) and then reach the second row of strands (C). It is observed that there are more center cracks that are also more spread out in the deteriorated ballast scenario than under the AREMA test condition.

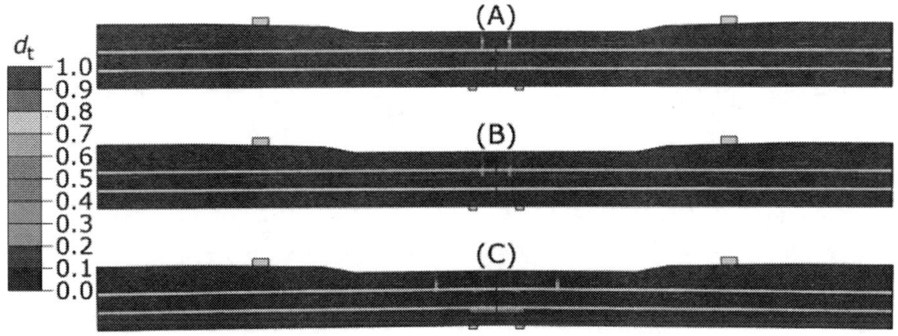


**Figure 7 Tensile damage variable contours showing three cracking stages in the static analysis of the unused tie under the AREMA center negative moment test condition: (A) cracks appearing on top surface, (B) cracks reaching first row of strands, and (C) cracks reaching second row of strands.**

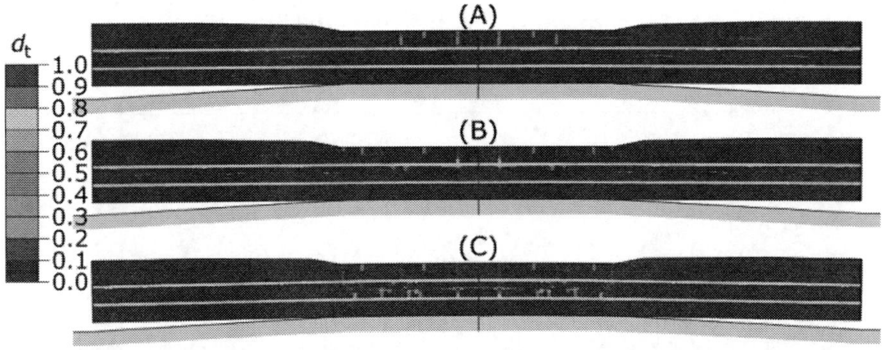


**Figure 8 Tensile damage variable contours showing three cracking stages in the static analysis of the unused tie under the deteriorated ballast support condition: (A) vertical cracks appearing on top surface, (B) vertical cracks reaching first row of strands, and (C) vertical/shear cracks reaching second row of strands.**

Figure 9 shows the equivalent wheel load-relative rail seat displacement curves highlighting the three cracking stages for the unused concrete tie in the static analyses of both center binding scenarios. Compared to the response under the AREMA test condition, the initial response is stiffer and the peak load is higher under the deteriorated ballast support condition.

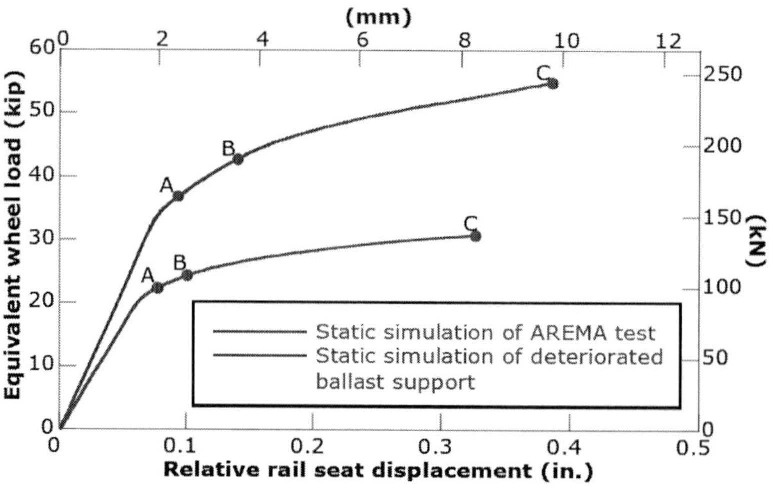


**Figure 9 Equivalent wheel load-relative rail seat displacement curves with indication of cracking stages for the unused concrete tie in static analyses of the two center binding scenarios.**

To further examine the post-peak behavior of the concrete ties, dynamic analyses are conducted in the deteriorated ballast support scenario. The equivalent wheel load evolution is assumed as shown in Figure 10 and applied on the unused tie model. The dynamic analysis predicts catastrophic failure of the concrete tie at around 60 kip (266.9 kN) equivalent wheel load. Figure 11 shows the $d_t$ contours at (D) initiation and (E) completion of the catastrophic tie failure, which is also accompanied by bond failure between the strands and the concrete.

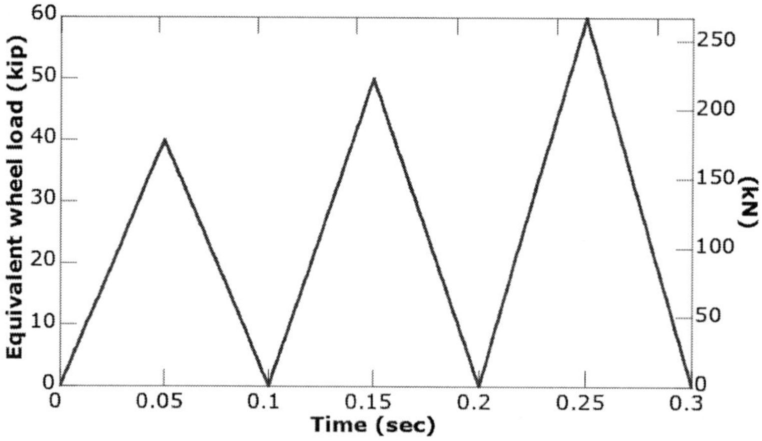


**Figure 10 Time history of dynamic equivalent wheel load.**

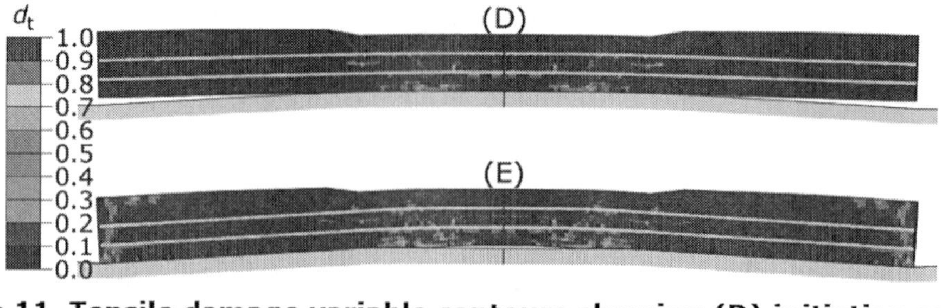


**Figure 11 Tensile damage variable contours showing (D) initiation and (E) completion of catastrophic failure observed in the dynamic analysis of the unused tie under the deteriorated ballast support condition.**

Figure 12 shows the complete set of equivalent wheel load-relative rail seat displacement curves obtained for the unused concrete tie from both the test and the static/dynamic analyses. The solid blue curve corresponds to the response to the 40-50-60 kip (177.9-222.4-266.9 kN) load cycles shown in Figure 10 with each cycle lasting 100 msec. The dotted blue curve corresponds to the response to one 60 kip (266.9 kN) load cycle that completes in 100 msec. The five cracking/failure stages are labeled on the curves. The 60 kip (266.9 kN) wheel load appears to be a critical load that causes the unused tie to fail catastrophically under the specific deteriorated ballast support condition.

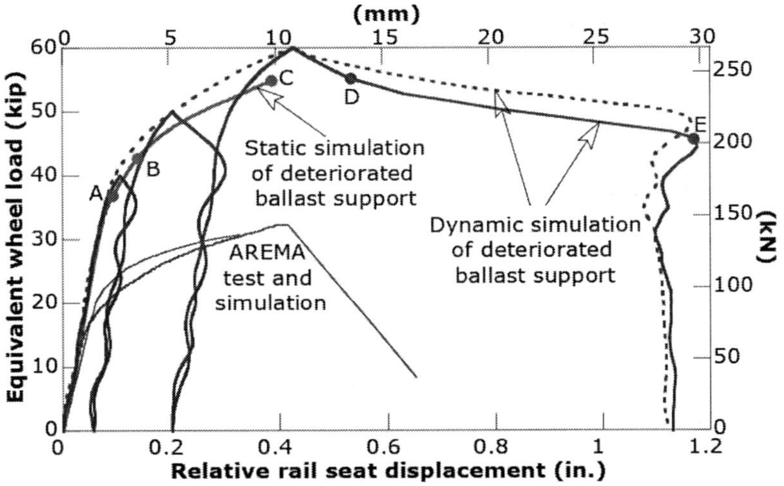


**Figure 12 Equivalent wheel load-relative rail seat displacement curves for the unused concrete tie obtained from test and static/dynamic analyses.**

## 5. CONCLUSIONS

An unused concrete tie model is developed and applied in FE analyses to simulate its response to two center binding conditions: the center negative moment test condition specified by AREMA for monoblock concrete ties, and a hypothetical deteriorated ballast support condition which is deemed more representative of the field condition. Under both conditions, the analyses show that the tie goes through similar cracking stages that can be characterized by crack depths. However, the AREMA test condition yields fewer center cracks than the deteriorated ballast support condition does. Moreover, the force-displacement relations obtained under the AREMA test condition are more compliant in general, and the peak load prior to failure is 44% lower, than under the deteriorated ballast condition.

The dynamic analyses indicate that the concrete tie can fail catastrophically when the wheel load reaches a critical magnitude. The critical wheel load for the unused concrete tie to fail is predicted to be around 60 kips (266.9 kN). Continued research is being conducted to evaluate the effect of tie abrasion and consequent loss of cross sections on the tie performance in center binding conditions.

## ACKNOWLEDGMENT

The work described in this paper was sponsored by the Office of Research, Development and Technology, Federal Railroad Administration, U.S. Department of Transportation. Directions provided by Messrs. Gary Carr, Cameron Stuart and Ali Tajaddini of the Track Research Division are gratefully acknowledged.

## REFERENCES


1. Kerr, A. D. Fundamentals of Railway Track Engineering. Simmons-Boardman Books, Inc., 2003.
2. Hanna, A. N. State-of-the-Art Report on Prestressed Concrete Ties for North American Railroads. *PCI Journal*, September-October, 1979, pp. 32-61.
3. Yu, H., D. Jeong, B. Marquis, and M. Coltman. Railroad Concrete Tie Failure Modes and Research Needs. *Transportation Research Board 94th Annual Meeting*, TRB15-0311, 2015.
4. *Manual for Railway Engineering, Chapter 30, Part 4: Concrete Ties*. American Railway Engineering and Maintenance-of-Way Association, 2010.
5. *Abaqus Analysis User's Manual*. Dassault Systèmes, 2012.
6. Yu, H., D. Y. Jeong, J. Choros, and T. Sussmann. Finite Element Modeling of Prestressed Concrete Crossties with Ballast and Subgrade Support. In *Proceedings of the ASME 2011 International Design Engineering Technical Conferences & Computers and Information in Engineering Conference*, DETC2011-47452, 2011.
7. Yu, H., and D. Y. Jeong. Railroad Tie Responses to Directly Applied Rail Seat Loading in Ballasted Tracks: A Computational Study. In *Proceedings of the ASME/ASCE/IEEE 2012 Joint Rail Conference*, Philadelphia, Pennsylvania, 2012.
8. Yu, H., and D. Y. Jeong. Finite Element Bond Models for Seven-Wire Prestressing Strands in Concrete Crossties. In *Proceedings of the 2015 Joint Rail Conference*, JRC2015-5758, San Jose, California, 2015.
9. Yu, H., and D. Y. Jeong. Bond between Smooth Prestressing Wires and Concrete: Finite Element Model and Transfer Length Analysis for Pretensioned Concrete Crossties. In *Proceedings of the 2014 ASCE Structures Congress*, Boston, Massachusetts, 2014.
10. Riks, E. An Incremental Approach to the Solution of Snapping and Buckling Problems. *International Journal of Solids and Structures*, 15(7), 1979, pp. 529-551.

# Insulated rail joint enhancement: testing and analysis

**A E Beagles, P Beaty, D I Fletcher, R Lewis**
Department of Mechanical Engineering, The University of Sheffield, UK
**B P Temple**
LB Foster Rail Technologies (UK) Ltd, UK

## ABSTRACT

Joints in railway track are vulnerable to failure, so most modern rail is welded. However, many signalling systems rely on sections of rail being electrically isolated from adjacent sections so insulated joints are necessary. At these joints several millimetres of insulation material, an "end post", separates the ends of the rails. To support the rail two pieces of steel (known as fish-plates) are bolted on either side of the rail web. All steel components are separated from each other by insulation material.

Conventional designs of rail joints are known to suffer from a range of failure modes: bolt failures, rail breaks, fish-plate failure, insulation degradation, and lipping (in which plastic flow in the rail ends bridges the gap between the rail ends). To assist in developing improved designs accelerated test procedures and sophisticated finite element models have been developed; this paper describes both. Results from physical testing and numerical modelling are presented and compared.

## 1 INTRODUCTION

Insulated rail joints (IRJ) are required in networks that verify the location of trains by monitoring the change in resistance between the rails. A typical insulated joint is shown in Figure 1. A gap between the ends of the running rails is filled with an insulating endpost. To support the rail, two steel fishplates are bolted to the rail through its web. Between the fishplates and the rail a layer of insulation is needed and the bolts are also insulated from the rail. For the most robust joints the fishplates are glued to the rails in a factory and pre-assembled units are welded into track.

The discontinuity in the stiffness of the running rail produces dynamic loading above that seen on continuous track, so rail joints experience more rapid degradation than plain track. This is a worldwide problem: Half of the European infrastructure managers contacted by Innotrack [1] reported that insulating rail joint failures were a major problem; in America "40% of the maintenance cost on rail is associated with rail joints" (this includes joints that are not insulating) [2] and failures, although rare, are often associated with derailment [3]; in Australia it has been found [4] that IRJ have lives of only 20% that of plain rail, and a large variability means that maintenance cannot readily be planned; in a recent two-year period, failures of IRJ cost Network Rail ten million pounds [5].

Failures can manifest as electrical or mechanical, but there is some linkage as many electrical failures result from mechanical damage. Electrical failures occur when the insulation is breached, for example when there is excessive plastic deformation (driven by mechanical loading of passing wheels) of the running rail (sufficient to extend across the gap filled by the endpost); these account for about 30% of IRJ failures on Network Rail infrastructure [6]. Mechanical failures can occur when bolts come loose, when a fishplate breaks, or a when a crack develops from a bolt hole in the rail; in glued joints such failures can be precipitated by debonding of the glue permitting increased relative movement.

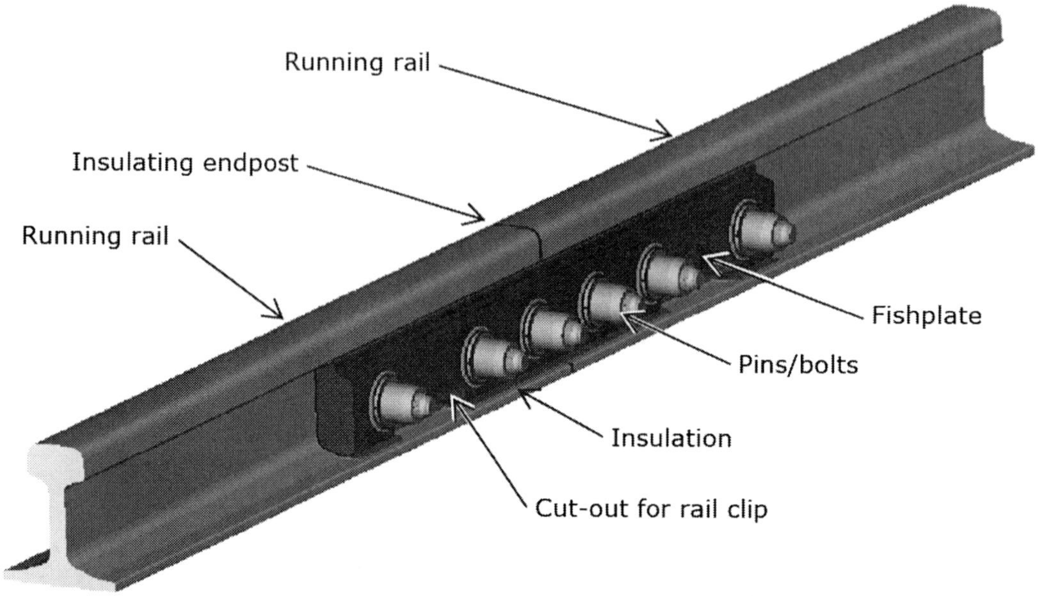


**Figure 1: A typical insulated rail joint**

The high cost of repair and consequences of failure of IRJ has made them the subject of numerous physical studies, e.g. [7-13], and numerical studies, e.g. [14-20]. This paper describes parallel research to develop both laboratory tests and supporting finite element analyses to enable enhancements to materials and geometries to overcome these failures.

## 2 TESTING

A range of different tests have been developed to enable changes to IRJ materials and geometries to be tested without the time and expense involved in on-track trials. These are described in the following subsections; more details are given in [21, 22].

### 2.1 Testing endpost and rail material for lipping

To investigate lipping (plastic flow of rail steel across the end-post – see Figure 3 for an example) slots were introduced into one of the discs in the Sheffield University Rolling and Sliding (SUROS) twin disc test machine [23] shown in Figure 2. This machine simulates rail-wheel contact by pressing disks of appropriate materials together, each rotated by an independent motor.

For these tests the disks were 47mm in diameter, had a contact width of 10mm, and were pressed together to produce a maximum contact pressure of 1.5GPa (a typical value for wheel-rail contact of heavily loaded vehicles). Tests were run at 400rpm (corresponding to about 2mph, appropriate for a joint near a loop) with a slip of 0.5% (appropriate for a driven wheel). A range of 3mm-deep slot-widths (0.5, 0.75 and 1mm) were considered and these were filled with either an epoxy-glass composite material, used mostly in glued IRJs, or a polyamide (PA6) used more commonly in non-glued IRJs.

The effect of different rail steels was also considered with tests using grades [24], R260, R350HT, and grade R260 laser-clad with a hard surface. The surface hardness (Vickers) of the discs was measured and was found to be about 350, 430 and 600HV respectively. The values for both the R260 and R350HT were significantly higher than the minima (which are equivalent to about 260 and 360HV respectively) given in the standard [24].

**Figure 2: Typical test pieces in SUROS machine and rail disc with endpost inserts**

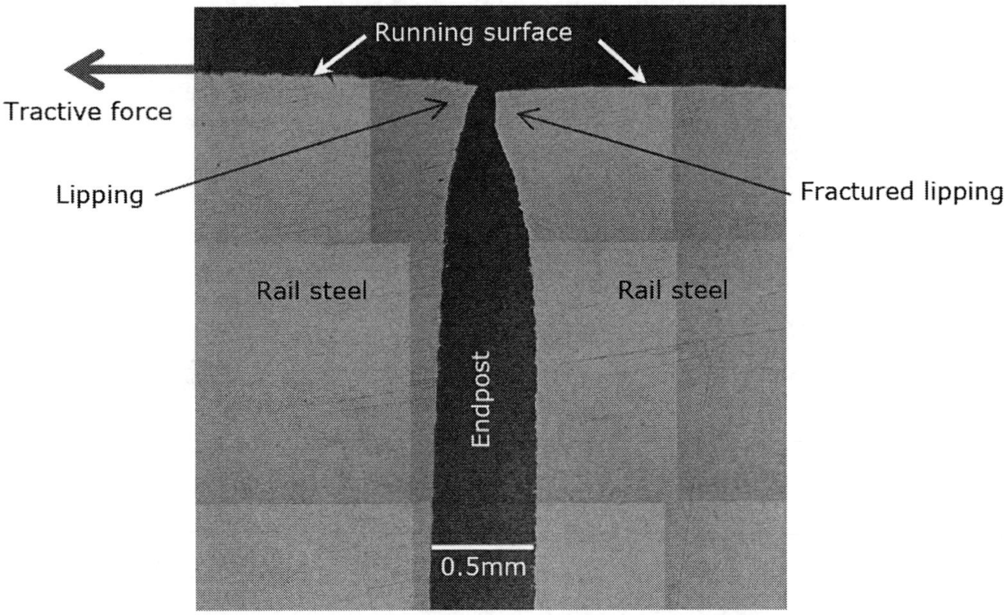


**Figure 3: Sectioned view of R260 grade steel disc with 0.5mm epoxy-glass endpost after 2000 load cycles**

A typical result is shown in Figure 3; a section through a 260 grade steel disc after 2000 cycles of loading. It can be seen that the test produced lipping across the slots, so was able to assess different combinations of rail steels and endpost materials. In these tests:

1. lipping was evident on both sides of the slot, evidence that traction is not required to cause this damage
2. width of slot had a small effect on the rate of lipping, so it is predicted that across larger gaps it would take longer for the steel to bridge the gap (time approximately proportional to width)[1]
3. softer polyamide provided less resistance to lipping than the harder epoxy (surface layers of PA6 wore away rapidly providing negligible benefit)
4. harder steel provided more resistance to lipping (rate was approximately halved for R350 compared to R260), the laser-clad surface having a rate of lipping slightly slower than that for R350

## 2.2 Glues and liners

Two types of tests were undertaken to assess the performance of different glues and liners used between the fishplates and the rail. They are discussed in the following two subsections.

### *2.2.1 Small-scale tests*

These tests used the test arrangement shown in Figure 4; this is based on European Standard BSEN 1465 [25] which provides all dimensions (100x25x1.6mm steel plates with 12.5mm overlap), but tests here had a thicker adhesive layer more representative of that in an IRJ. These tests are similar to the test detailed in ASTM D1002-05 [26] and discussed in [13]; it is concluded that due to the complex stress environment the results of this test should not be considered to define material properties but are primarily for comparison purposes between different adhesives. To simulate the manufacture of IRJ the steel plates were roughened with p60 abrasive paper and degreased with alcohol wipes prior to applying glue to one side of the joint. After this had fully cured the other steel piece was glued to it.

Four adhesive were assessed, identified here as A, B, C, and D. These are all two-part epoxy systems made from a resin and a hardener that were developed by the respective manufacturers for use in IRJ's. These were combined with either: a pultruded glass reinforced polyester resin liner (a stiff liner material), a glass fibre sheet (flexible woven mat), a Kevlar® sheet (a flexible woven mat made from the DuPont trademarked material), or no liner. These liners are used in various IRJ designs around the world; no installations without a liner have been identified. The samples (5 of each combination) were tested to identify peak force at failure on an Instron 8500 machine under displacement control at a rate of 0.01mm/sec.

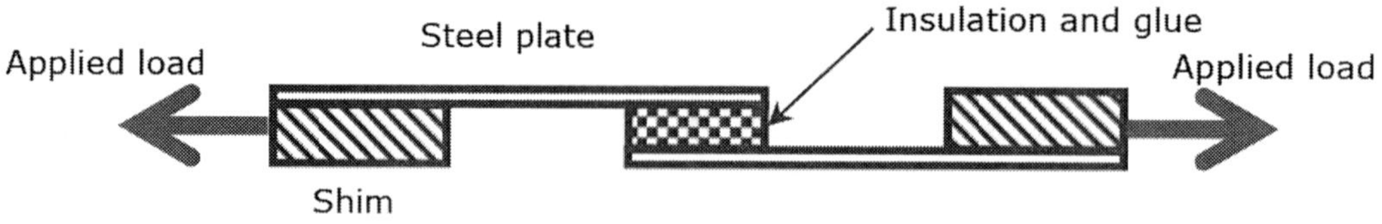


**Figure 4: Sketch (not to scale) of small-scale glue and liner shear tests**

---

[1] The dynamic response of the wheel disc to a gap in the rail disc will not be identical to that of a train wheel running across a gap in an IRJ due to different stiffness and damping in the two systems. However, if the dynamic loads in track are significantly larger then although the lipping may grow more rapidly it would also be more prone to fracture. The details of this phenomenon could be the object of further studies.

A typical set of results is shown in Figure 5. This is one of the tests of a Kevlar liner with adhesive D and shows rapid failure after the peak force is reached.

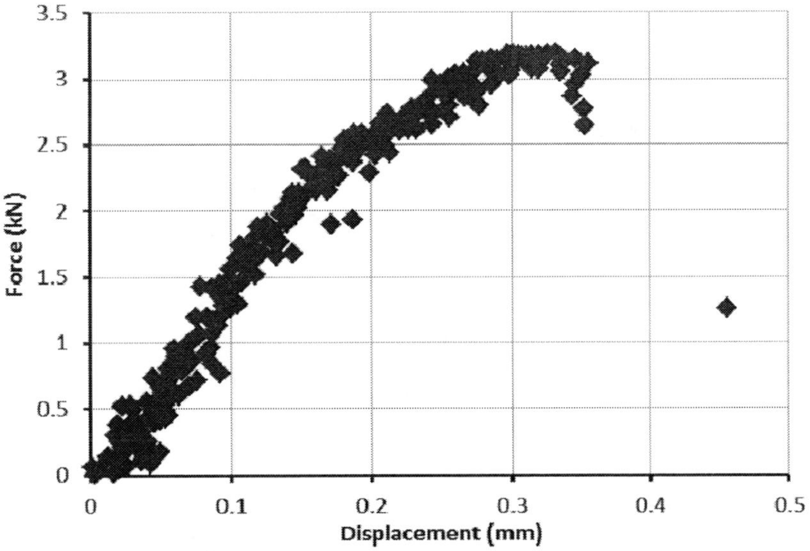


**Figure 5: Force/displacement plot for Kevlar with adhesive D**

The results for glue and liner combinations are shown in Figure 6 (the pultruded glass reinforced polyester resin liner was found to fail at a significantly lower stress with adhesive B and was not tested with the others).

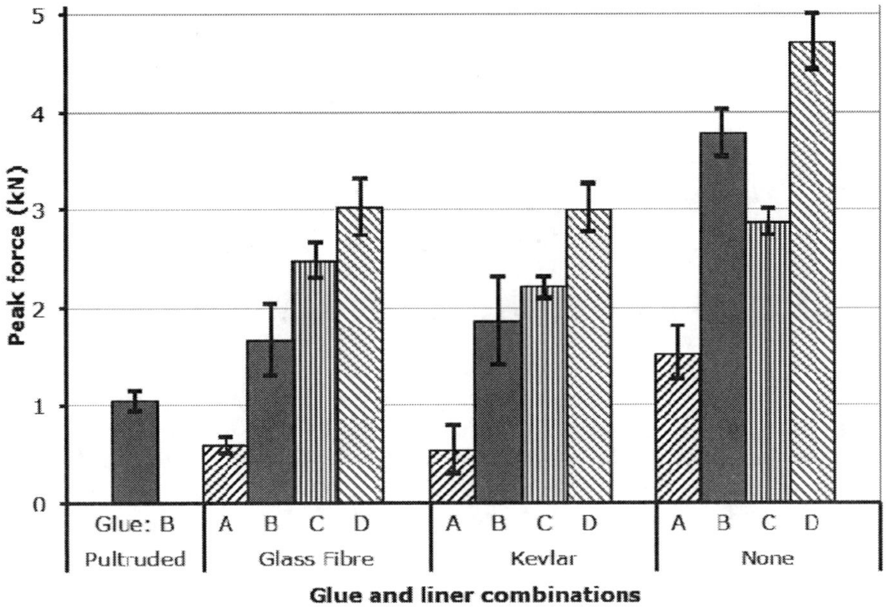


**Figure 6: Peak force at failure of small-scale tests, showing means and ranges of the five results for each case**

It can be seen from Figure 6 that:

1. for all adhesives, the presence of a liner reduced the shear strength of the joint, with adhesives A and B supporting at least twice as much load when there was no liner than when a liner was present
2. adhesive D supported higher shear stresses with all (or no) liners

3. the pultruded glass reinforced polyester resin liner supported no more than about two thirds of the applied force of the other liners (tested with the adhesive B only)

The first observation is consistent with the results in [13], where it was also found that dispersed fibres performed well. However, one function of the liner is to ensure that there is no danger that there will be a low resistance between the rail and the fishplate as would happen if the adhesive was squeezed from the joint during manufacture or cracked in service. Another observation from [13] is that different adhesives may deteriorate at different rates when exposed to the track environment; a factor that may influence the rankings of glues and liners.

### *2.2.2 Tests on glued IRJ sections*

The second tests of glues and liners used sections cut from joints that had been assembled using a controlled environment and procedures that are standard in the UK. These sections were tested in a hydraulic test machine using the rig illustrated in Figure 7. Load was applied at an increasing rate of 100kN/minute until the ultimate shear strength of the glue/liner combination was reached.

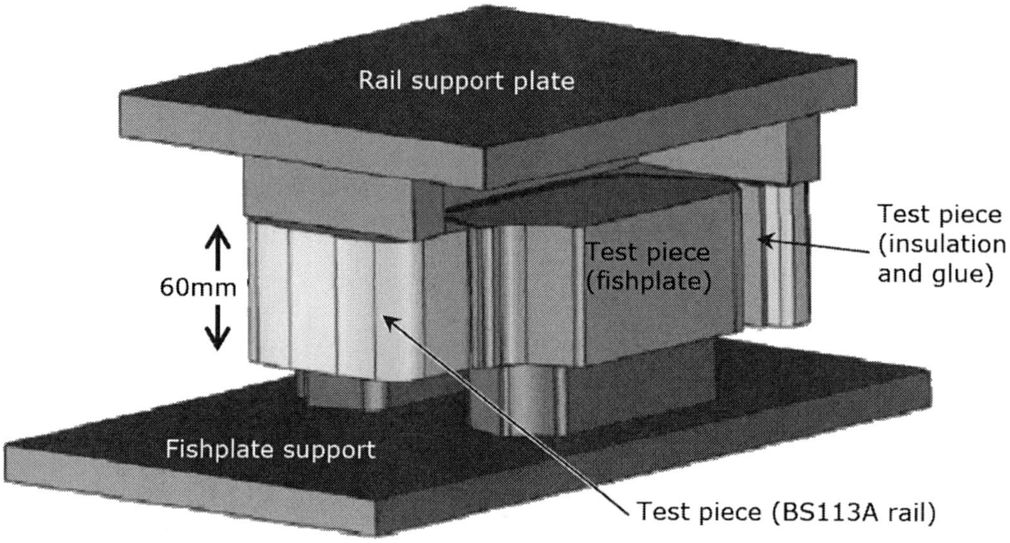


**Figure 7: Shear test of joint section**

Five different joint types were tested (four specimens of each). Two designs were used: standard (wedge-fit) fishplates not fitted to the web, and 'full-fit' joints, with a more uniform glue thickness. Three 'full-fit' joints with Kevlar liners had different fastening torques applied to assess the effect of different glue thickness (it had been expected that higher torques would produce thinner layers). The fourth 'full-fit' joint had a pultruded glass fibre liner while the standard joint had a glass fibre mat.

The maximum applied forces were converted to shear stress by dividing by the area of the joint (0.0166m²). The results are shown in Figure 8 and it can be seen that:

1. the fastening torque tests produced similar results (this was expected since the increased torque hadn't decreased the glue thickness as had been intended)
2. the full-fit glass-fibre liner survived a higher shear stress than the full-fit Kevlar ones
3. the full-fit glass-fibre joints generally survived higher shear stresses than the standard ones

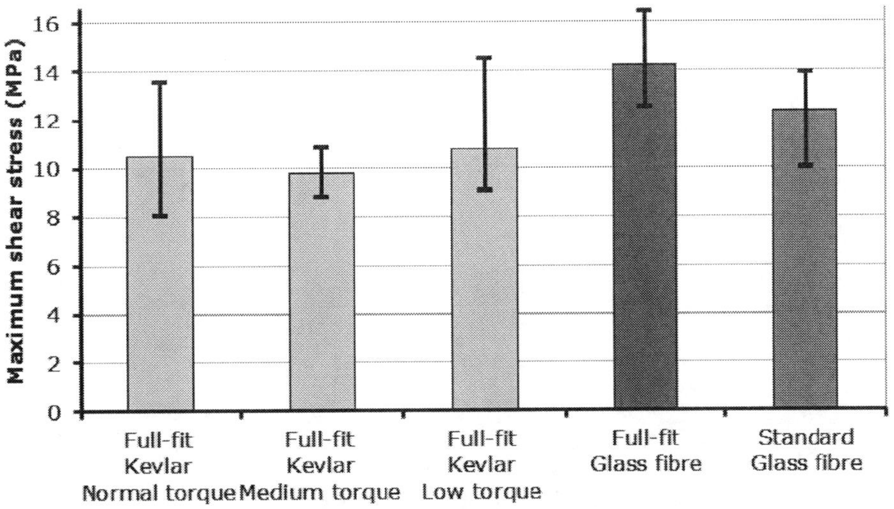


**Figure 8: Shear stress at failure for full-scale tests, showing means and ranges of four results for each case**

These tests are more representative than the small-scale tests in Section 2.2.1 of the loading seen in track, but significant edge effects and the influence of the compressive load provided by bolts are not included since these elements are cut away to form the test pieces.

## 2.3 Full scale bending tests

Two designs of IRJ, a full-fit and a (standard) wedge-fit as described in Section 2.2.2, were installed in 3m lengths of rail and tested in the four-point bending configuration shown in Figure 9. They both had a glass fibre liner glued to the fishplate with Temperange II adhesive. To accelerate failure four-hole joints were tested although these are rarely used in service (six-hole joints are more common). The joints, one specimen of each, were to be successively subjected to half a million cycles of loading from zero to the forces producing the following bending moments in the rail and joints between the actuators:

1. 40kNm: representing the maximum static load on UK infrastructure (25tonne axleload)
2. 67kNm: the maximum qualification test requirement, that from Australia [27]
3. 101kNm: representing extreme dynamic train loading on track (a dynamic factor of 2.5 multiplying the maximum static load)

However, both specimens failed before 250 thousand cycles of 101kNm. The failed specimens are shown in Figure 10 (wedge-fit) and Figure 11 (full-fit). Prior to failure it was observed that wear debris was accumulating on both joints, but more on the standard joint indicating that there was more relative movement resulting from debonding of the adhesive.

The standard design failed by fracture of the fishplates, whereas the full-fit design remained intact with failure occurring in the rail adjacent to the joint; subsequent investigation determined that the failure had initiated from a small defect in the foot of the rail. These results suggest that

1. the full-fit design is better than the standard one
2. it should be ensured that the foot of the rail adjacent to IRJ has no defects that could initiate fatigue cracks

3. there is scope to improve the design to remove the stress concentration at the end of the fishplate (see also finite element results in Figure 16)[2]

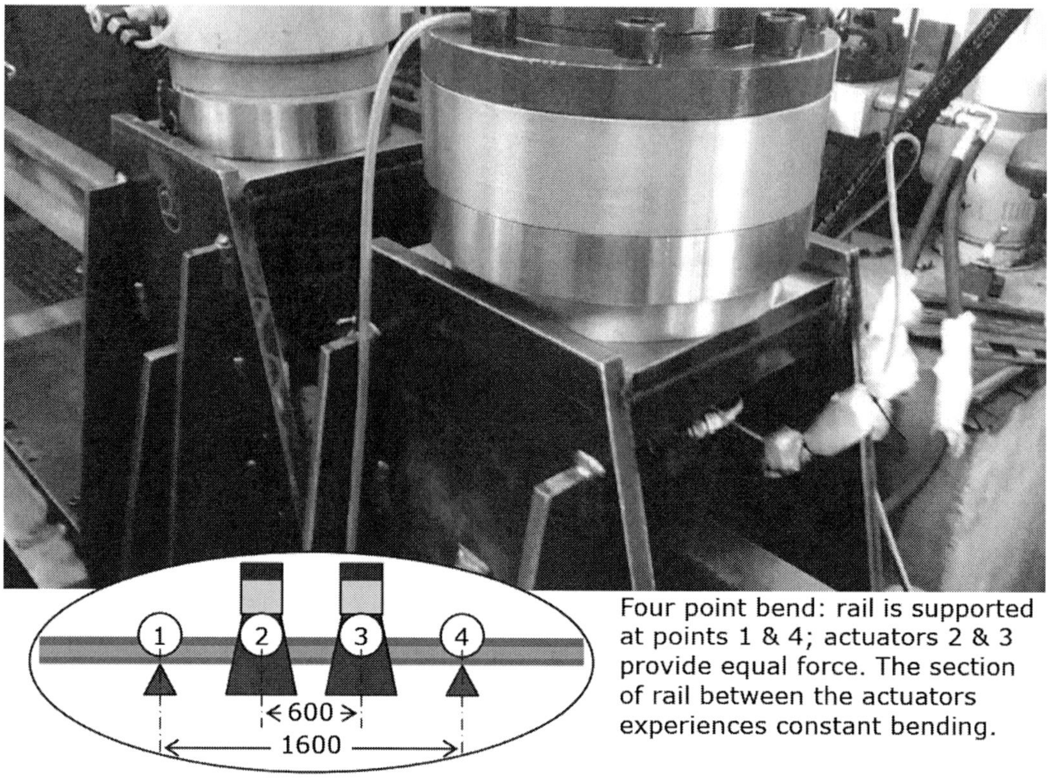


**Figure 9: Four-point bending rig**

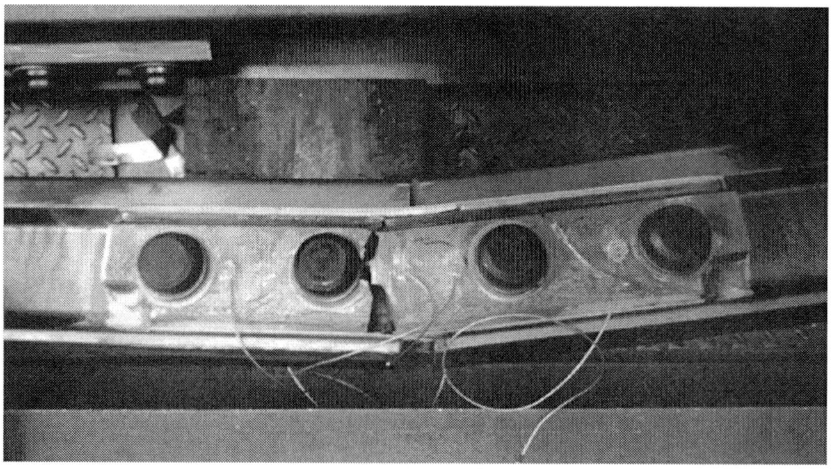

**Figure 10: Standard rail joint failure**

---

[2] A design of fishplate with rounded ends has been proposed [31]. The motivation was to strengthen the centre of the fishplate without increasing the total mass. The design may have had the added benefit of reducing bending stress concentration in the rails at the ends of the bars, but if so this is not mentioned.

**Figure 11: Full-fit rail joint failure**

## 3 FINITE ELEMENT MODELLING

Finite element models of the standard rail joints were established using geometry defined by Network Rail [28]. The model was created using the parametric design language in ANSYS® [29] and used mainly 8-noded hexahedral elements; a small volume around the spherical-ended cut-out for rail clips (see Figure 1) was meshed with tetrahedral elements, and the bolts, apart from the heads, were cylindrical beam elements. The loading was assumed to be vertical so that symmetry could be used meaning it was only necessary to model a quarter of the joint.

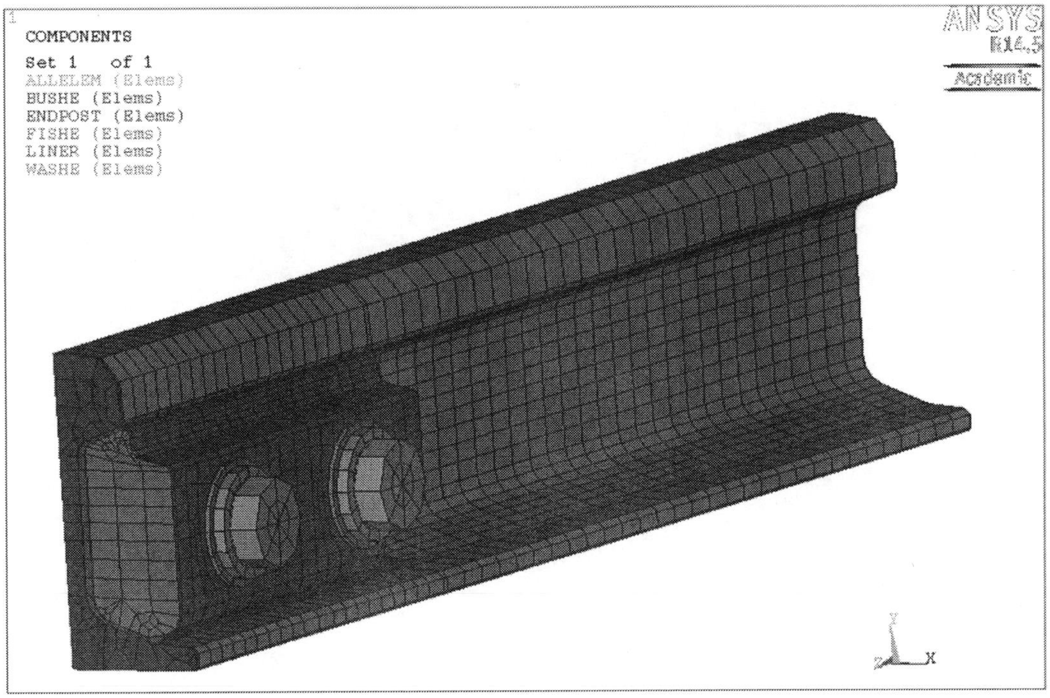


**Figure 12: Finite element mesh of rail and IRJ**

The model is shown in Figure 12 (note that the elements' mid-side nodes lie on relevant circular arcs, but this plot shows straight facets). The elements of the liner (LINER), bushes around bolts (BUSHE, these elements are not visible in Figure 12 as they are inside the joint), insulating washers (WASHE), and endpost (ENDPOST) were assigned elastic material properties of epoxy [17] (E=45GPa, v=0.19), with all other elements being steel (E=207GPa, v=0.3). Relevant symmetry conditions were applied to symmetry planes and the rail was supported (fixed vertical, y-direction, displacements) at the base of the remote end (800mm from the centre).

Two load steps were used: In the first the bolt load was applied as a tension of 200kN in the bolts (appropriate for a torque of 1020Nm [30] on an M29 bolt); in the second the bending moment was applied as a point force on the top of the rail at 300mm along the rail from the centre, matching the test described in Section 2.3.

Two situations were modelled: the glue fully supporting the interface (modelled by sticking the elements together) and glue having debonded from the surfaces (modelled using contact algorithms that calculate interface pressures and the amount of sliding).

The results for the principal stress in the fishplate for the two cases are shown in Figure 13 and Figure 14. It can be seen that the principal stress (and hence risk of fatigue failure) reduces marginally when the glue debonds.

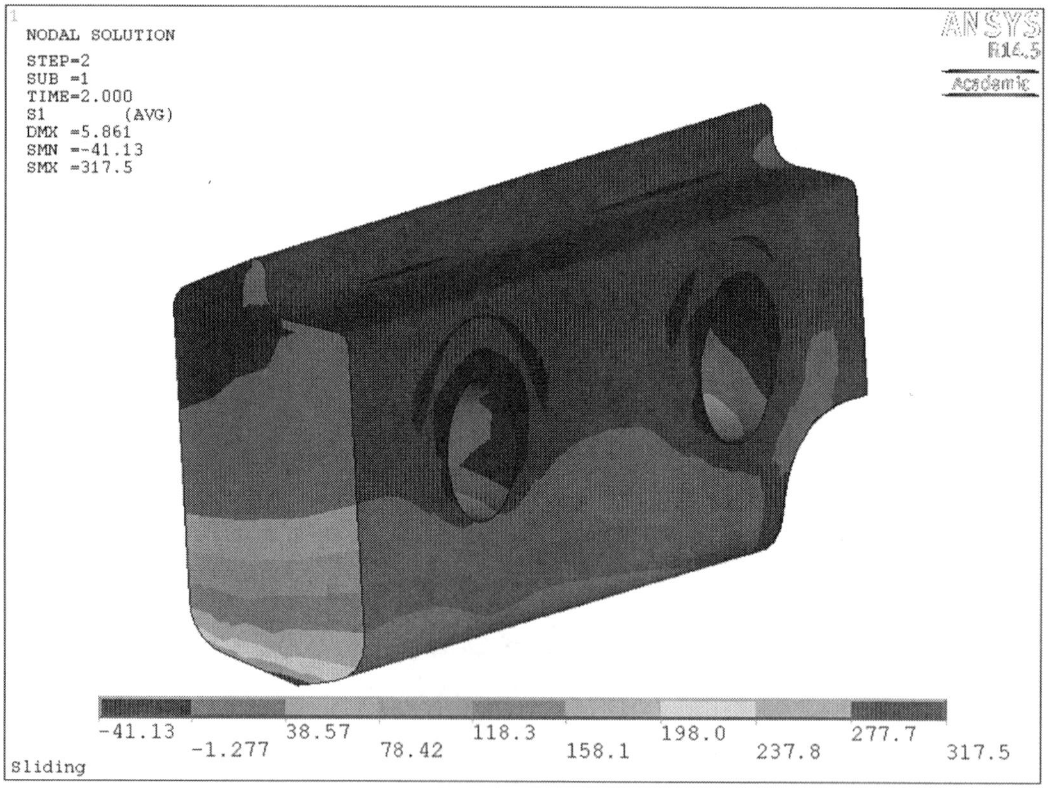


**Figure 13: Major principal stress in fishplate for debonded glue**

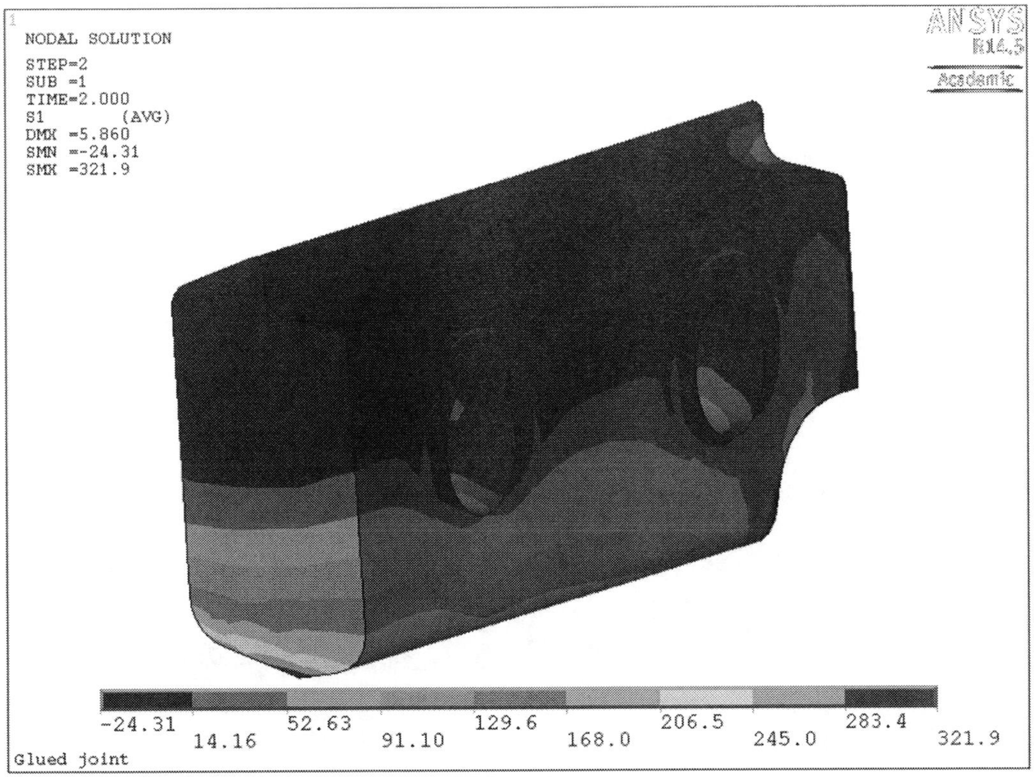


**Figure 14: Major principal stress in stuck fishplate**

The effect on the stress in the endpost is insignificant, as seen from results in Figure 15, the maximum tensile stress in the debonded (sliding) model drops marginally to 235MPa from 236MPa for the glued model.

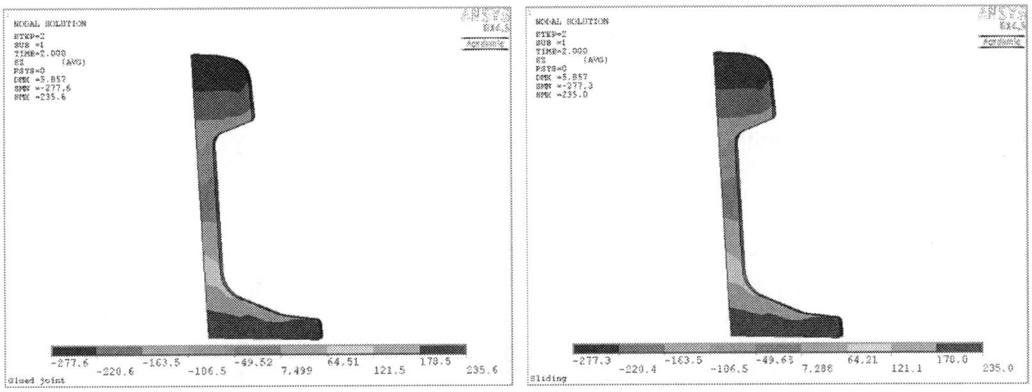

**Figure 15: Longitudinal (z) stress in endpost**

The stress in the rail is predicted to be at its maximum near the location that the crack initiated in the testing of the full-fit joint (at the foot of the rail near the end of the fishplate, compare Figure 11 and Figure 16). Although the modelling was of the wedge-fit joint the bending stress in the full-fit joint at this location is expected to be almost identical so modelling prediction and results of testing are consistent. Further testing and analyses are planned to provide more evidence.

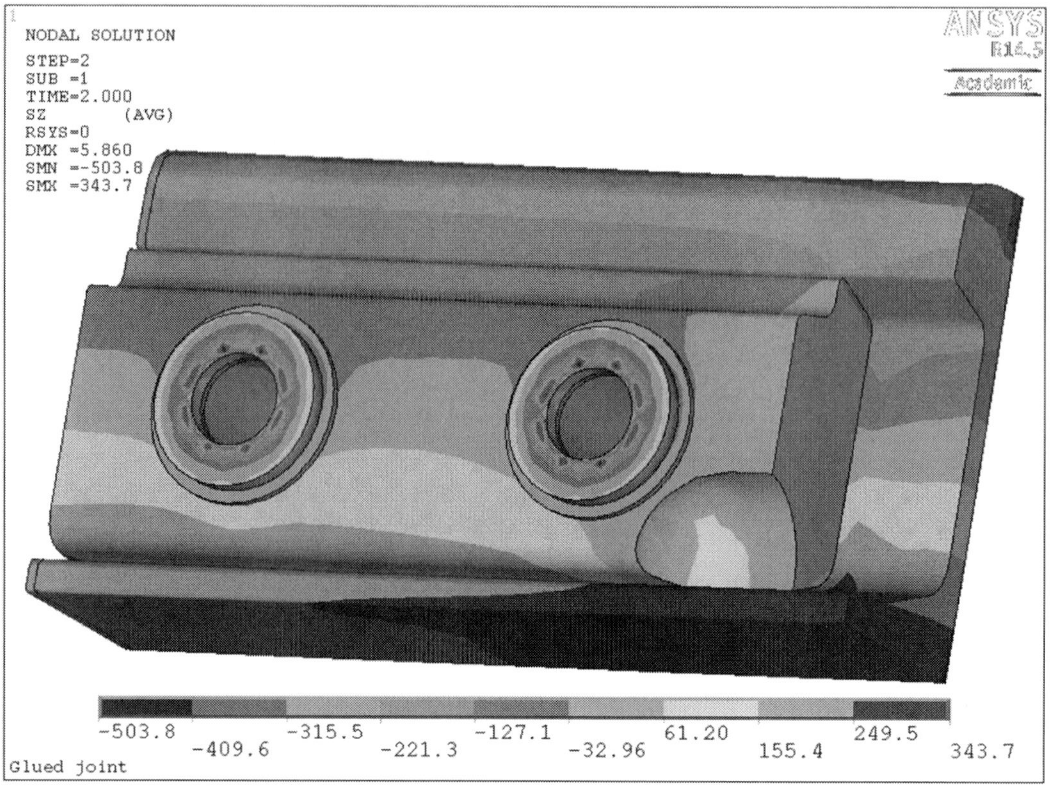


**Figure 16: Longitudinal stress in glued joint assembly**

## 4 CONCLUSIONS

A range of tests have been undertaken to investigate the effect of varying designs and material properties on the performance of insulated rail joints. Some novel tests have been undertaken and have the potential for further development to enable investigation of the behaviour of different materials and geometries for insulated joints. The physical fatigue testing of full joint assemblies and finite element modelling have been seen to be consistent.

Small-scale testing showed that

- the use of absorbent liners increases the shear capacity of the glued joint (having no liner, the largest shear capacity, is currently not feasible as it is not fail-safe)
- hard endpost materials are less prone to failure by lipping, and that (in these tests) the rate of lipping is not affected by the thickness of the endpost

Full-scale testing showed that a full-fit joint is more robust, although more tests should be undertaken to verify this.

A significant finding is that it is important to check the underside of rails adjacent to joints as high stresses here can lead to failure of the rail. A redesign of the ends of fishplates could reduce the stress concentration at these locations.

The finite element results suggest that the mechanical strength of joints is not adversely affected by debonding of glue in the joint. However, there may be adverse electrical effects if debonding reduces the resistance of the insulation, and even small relative motions can lead to fretting, as seen in the tests, and hence loss of material, reduction in bolt loads, and weakening of the joint. These factors were not assessed here but are the subject of on-going research.

## ACKNOWLEDGEMENTS

The testing work reported in Section 2 was carried out as part of a Knowledge Transfer Partnership (KTP) co-funded by LBFoster and the Technology Strategy Board and used some of their dedicated joint test equipment housed at The University of Sheffield.

This work was supported by the EC Framework 7 project SustRail: "The sustainable freight railway: Designing the freight vehicle – track system for higher delivered tonnage with improved availability at reduced cost", Grant 265740.

## REFERENCE LIST

[1] INNOTRACK Consortium. A report providing detailed analysis of the key railway infrastructure problems and recommendation as to how appropriate existing cost categories are for future data collection. European Commission 6th Framework Programme; 2009. D1.4.6.
[2] Zarembski AM. Forecasting of track component lives and its use in track maintenance planning. In: International Heavy Haul Railway Conference, Vancouver, Canada; 1991. p. 61–67.
[3] Jeong DY, Bruzek R, Tajaddini A. Engineering studies on joint bar integrity, part i: field surveys and observed failure modes. In: Proceedings of the 2014 Joint Rail Conference; 2014.
[4] Dhanasekar M, Bayissa W. Review of Insulated Rail Joints. CRC for Rail Innovation; 2009. R3.100.
[5] Network Rail. IRJ (Insulated Rail Joint) Head Repair from Selectequip; 2012. Http://youtu.be/vbpjPLKWJ24 (accessed 3 September 2014). You Tube.
[6] Network Rail. Identified failure modes of IBJs (based on Trust output); 2010. Sited in [21].
[7] INNOTRACK Consortium. Improved model for loading and subsequent deterioration of insulated joints. European Commission 6th Framework Programme; 2009. D4.2.3.
[8] Davis DD, Akhtar M, Kohake E, Horiszny K. Effects Of Heavy Axle Loads On Bonded Insulated Joint Performance. In: Proceedings of the AREMA 2005 Annual Conferences; 2005.
[9] Akhtar M, Davis D, O'Connor T. Revenue Service Evaluation of Advanced Design Insulated Joints. In: Proceedings of the AREMA 2008 Annual Conference; 2008.
[10] Li D, Clark D, McDaniel R, Allen L. Evaluation of Track Component Performance under Heavy Axle Load Revenue Service Conditions. In: Proceedings of the AREMA 2008 Annual Conference; 2008.
[11] Dhanasekar M, Bayissa W. Performance of square and inclined insulated rail joints based on field strain measurements. Proceedings of the Institution of Mechanical Engineers Part F-journal of Rail and Rapid Transit. 2012 March;226(F2):140–154.
[12] Peltier DC, Barkan CPL. Characterizing and Inspecting for Progressive Epoxy Debonding in Bonded Insulated Rail Joints. Transportation Research Record. 2009;2117(2117):85–92.
[13] Nicoli E, Dillard DA, Dillard JG, Campbell J, Davis DD, Akhtar M. Using standard adhesion tests to characterize performance of material system

options for insulated rail joints. Proceedings of the Institution of Mechanical Engineers Part F-journal of Rail and Rapid Transit. 2011 Sep;225(F5):509–521.

[14] Ding K, Dhanasekar M. Flexural behaviour of bonded-bolted butt joints due to bolt looseness. Advances in Engineering Software. 2007 August;38(8-9):598–606.

[15] Himebaugh AK, Plaut RH, Dillard DA. Finite element analysis of bonded insulated rail joints. International Journal of Adhesion and Adhesives. 2008 April;28(3):142–150.

[16] Igwemezie J, Nguyen AT. Anatomy of Joint Bar Failures: Taking a close look at the how and why of joint bar failures - and how to prevent the problem. Railway Track and Structures. 2009 July;7:31–37.

[17] Chen YC, Kuang JH. Contact stress variations near the insulated rail joints. Proceedings of the Institution of Mechanical Engineers Part F-journal of Rail and Rapid Transit. 2002;216(4):265–273.

[18] Wen Z, Jin X, Zhang W. Contact-impact stress analysis of rail joint region using the dynamic finite element method. Wear. 2005 March;258(7-8):1301–1309.

[19] Cai W, Wen Z, Jin X, Zhai W. Dynamic stress analysis of rail joint with height difference defect using finite element method. Engineering Failure Analysis. 2007 December;14(8):1488–1499.

[20] Carolan ME, Jeong DY, Perlman AB. Engineering Studies on Joint Bar Integrity: Part II - Finite Element Analyses. In: 2014 Joint Rail Conference. American Society of Mechanical Engineers; 2014. p. V001T01A002; 10 pages.

[21] Beaty P. Experimental Testing Procedures to Investigate and Improve Insulated Block Joint Design and Life Cycle [Master of Philosophy]. Department of Mechanical Engineering, The University of Sheffield; 2014.

[22] Beaty P, Lewis R, Temple BP, Marshall MB. Experimental modelling of rail end lipping in insulated joints. In: World Congress on Railway Research (WCRR) 2013, Sydney, Australia; 2013.

[23] Fletcher D, Beynon J. Development of a machine for closely controlled rolling contact fatigue and wear testing. Journal of testing and evaluation. 2000;28(4):267–275.

[24] CEN: European Committee for Standardization. Railway applications. Track. Rail. Vignole railway rails 46 kg/m and above. British Standards Institution; 2011. BS EN 13674-1.

[25] CEN: European Committee for Standardization. Adhesives. Determination of tensile lap-shear strength of bonded assemblies. British Standards Institution; 2009. BS EN 1465:2009.

[26] American Society for Testing and Materials. Standard test method for apparent shear strength of single-lap-joint adhesively bonded metal specimens by tension loading (metal-to-metal); 2005. D1002-05.

[27] Standards Australia Committee CE-002, Railway Track Materials. Railway Track Material, Part 12: Insulated Joint Assemblies. Standards Australia International; 2002. AS1085.12. Reconfirmed 2013.

[28] Network Rail. BS113A BR Mk III Glued Insulated Joint 4 & 6 Hole Type; 1987. Drawing RE/PW/268 D.

[29] ANSYS. ANSYS Parametric Design Language Guide; 2010. Release 13.0. Available from: http://www.ansys.com/.

[30] Network Rail. How to replace a fishbolt; 2003. Trackwork Instruction: 2G020.

[31] Ciloglu K, Frye PC, Almes S, Shue S. Advances in Bonded Insulated Rail Joints to Improve Product Performance. In: 2014 Joint Rail Conference, Colarado Springs. American Society of Mechanical Engineers; 2014.

# Real-time on-board condition monitoring of train axle bearings

**I Corni[1,2], N Symonds[1], R J K Wood[1], A Wasenczuk[2], D Vincent[2]**
[1] National Centre for Advanced Tribology at Southampton,
Faculty of Engineering and the Environment, University of Southampton, UK
[2] Perpetuum Ltd, UK

## ABSTRACT

Premature failure of rail axle bearings causes a significant increase in train operating costs and can impact train safety. A new on-board condition monitoring approach provided by Perpetuum Ltd is fitted on passenger trains to provide the operator with real-time information on bearing health. This new technology allows the detection of early bearing damage.

This paper reports an initial study to understand the source of vibrations. The final aim of the project is to connect the increasing vibration data to the bearing surface damage, measured with surface profilometry.

## 1 INTRODUCTION

Rolling element bearings are mechanical components widely employed in industrial applications, from rotating machinery to means of transport. Rolling element bearings consist of a complex structure (outer ring, inner ring, cage and rollers) and usually operate in harsh working environments (e.g. high temperature, rotational speed and loads) and therefore they are often at risk of failure.

Fatigue is the normal failure mode for bearings that are properly installed and maintained. Unfortunately, numerous bearings suffer early failures due to poor fitting, insufficient lubrication, overloading, overheating, contamination, unbalance and misalignment. Early failure can cause the breakdown of the machinery, decrease its efficiency and even reduce safety with risks to lives (1-3). For these reasons it is vital to monitor the bearing condition and lots of research has been recently undertaken in fault diagnosis using mainly vibrations (4-14), but also acoustic emissions (15-16) and temperature measurements. Changes in vibration signature have been employed for many years to detect degrading bearings before failure; but often the bearing vibrations cannot be directly measured because they are affected by vibrations from other parts of the apparatus (background noise) making condition monitoring more complex (3). In order to overcome this difficulty, numerous signal processing techniques (e.g. envelope spectrum analysis (10, 17), empirical mode decomposition (EMD) (2, 7, 8, 10, 18-19), Hilbert-Huang transform (HHT) (4, 7), wavelet transform (WT) (5, 12, 20) and minimum entropy deconvolution (MED) (17)) have been developed to identify fault characteristics precisely in the vibration data. The data used to develop these techniques were partly originated in the laboratory (1, 4, 6, 8, 10-14, 19), partly obtained from the Case Western Reserve University (CWRU) bearing data centre (6-7, 21-24) and only a few acquired from in-situ equipment (i.e. mining machines (18), servomotors running arbitrary motion profiles (9), helicopters (25-27) and traction fans (28)).

Train axle bearings are key elements in the integrity of railway wheel sets; ensuring their good condition is fundamental for train safety and this can be achieved using condition monitoring. The most common failure in axle bearings is due to rolling contact fatigue (RCF) of the outer race (29-30). Subsurface networks of cracks develop in the rolling direction by shear stresses and then deviate towards the surface causing macropits that represent a measurable surface damage (31-33).

This paper presents an initial study to correlate vibration data with bearing damage. Real-life vibration data are measured using wireless sensor nodes (WSNs) developed by Perpetuum Ltd and fitted to Southeastern operated rolling stock. This new technology monitors in real-time the vibrations of axle-bearings and successfully detects early bearing failures. The failed bearings are characterised by measuring the surface profile of the rolling area of the outer race.

## 2 CURRENT TECHNOLOGY

The degradation of axle bearings occurs with an increase in noise and vibration, followed by temperature (34). Nowadays, the increase in temperature is detected by trackside Hot Axle Box Detectors (HABDs); when the temperature increases the train has to be stopped immediately causing disruption to passengers and railway traffic. Therefore more research to detect precursors to run-away thermal failure using noise and vibrations has been carried out and it is still needed.

In the past 10 years technology employing noise has been developed; noise is generally detected using acoustic monitors that use the frequency of the sound to determine the damaged component (e.g. damaged bearing, noises from flanging, wheel flat, gearbox noises). Tsukahara *et al.* (34) compared the ability of two different Acoustic Bearing Detectors (ABD) (RailBAM and unknown) to detect eight faulty bearings installed on a French TGV; after 57 runs at speed between 15 and 130 km/h both ABDs located correctly 7 bearings out of 8. In the USA train axle bearings are monitored using TTCI Trackside Acoustic Detection System (TADS®) that identifies defects before overheating (35).

Due to the complexity of the train system not much research has been reported on condition monitoring of train axle bearings using vibrations. SKF Multilog online system consists of numerous modular sensors which monitor and transmit many bogie operating conditions (temperature and vibrations are measured for bearings). Laboratory experiments on 43 bearings with or without damage (previously operating on trains or artificially damaged) were carried out using train operational conditions; "good" bearings could be successfully distinguished from damaged bearings (36). Deutsche Bahn Systemtechnik has developed an on board bogie diagnostic system made of two units: the bogie-instrumentation with sensors (noise and vibration) and cables running to the diagnostic unit under the passenger's seat. This technology can successfully detect early bearing damage (37). Moreover, Bellaj *et al.* (38) compared results from track-side ABD and on-board vibration measurements for a high speed TGV fitted with eight faulty bearings. The on-board system showed increased vibration for damaged bearings and effectively detected faulty bearings for speed higher than 60 km/h. The track-side detector revealed 75% of failed bearings using multiple runs and speed higher than 60 km/h. Currently, none of these technologies show the ability to link the level of damage to a measurable parameter.

Perpetuum Ltd has developed a new on-board system for measuring local vibrations using wireless sensor nodes (WSNs). The system is fitted on several networks in the UK, but the current work concerns Southeastern trains. The on-board self-powered WSNs are bolted onto the bearing axle housing of the Southeastern Electrostar fleet (148 trains and 4,944 wheels) covering 540 miles of tracks in the south-east of England (Figure 1). All the wheels of all the trains are individually monitored during normal passenger service.

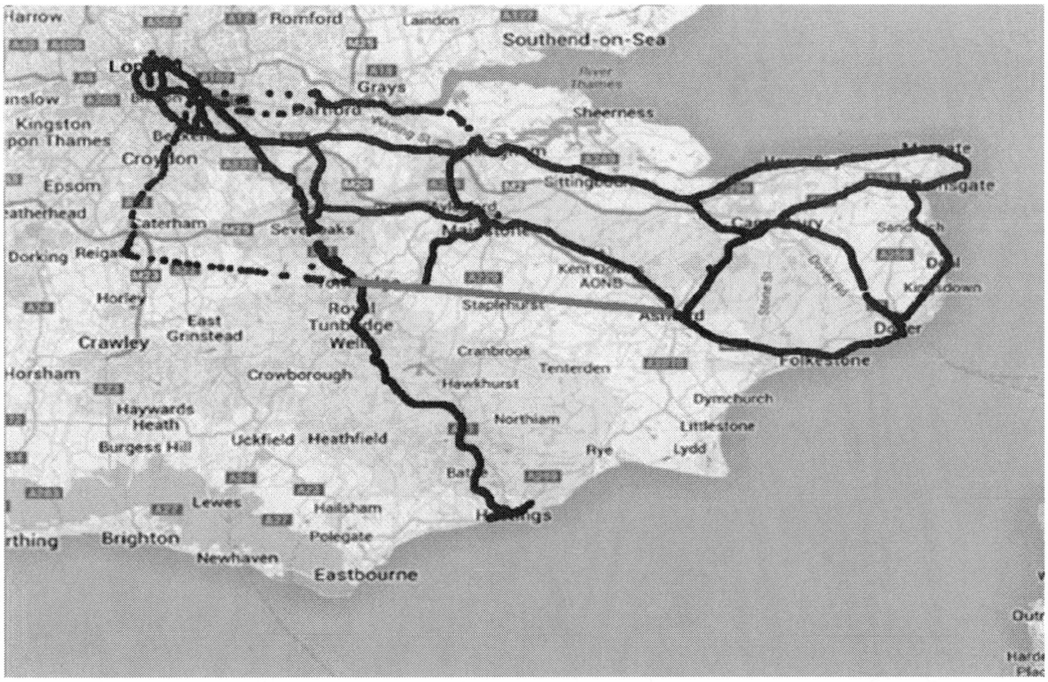


**Figure 1 – Map of the south east of England. Track covered by Southeastern trains is indicated in blue and orange; the track Tonbridge to Ashford International is in orange.**

The WSN measures vibrations using a tri-axial accelerometer; the acceleration data are recorded for 4 seconds every 3 minutes. The sensor records 16,000 points per second on each axis (x is the vertical direction, y is the direction of travel and z is the direction parallel to the axle); in 4 s of recording time 192,000 points are recorded. These data points (raw data) are used to calculate: *rms X*, *rms Y*, *rms Z*, *peak X*, *peak Y*, *peak Z* and vertical Fast Fourier Transforms (FFTs). These data combined with train number, wheel position, date and time, speed of the train, GPS location, direction of travel, temperature, bearing health index (BHI), wheel health index (WHI) are wirelessly sent to the Cloud and are available in real-time via a website. BHI and WHI values are calculated with complex algorithms. Depending on the daily mileage of the train an average of 300-400 data are received from each WSN.

The train operator has real-time access to BHI and WHI values and uses these to identify bearing and wheel issues, respectively.

## 3 METHODOLOGY

The vibration data are analysed with two aims:

a) Understand the sources of the vibrations;
b) Link the increasing vibrations to the condition of the removed bearings.

Part a) requires looking at large data sets from many "good" bearings travelling at different speeds and locations. Part b) is a larger task, beyond the scope of this paper, however two example bearings are shown to demonstrate the methods and techniques being used to look at the possible relationships between vibration data and the extent of bearing damage.

In the current commercial system, the faulty bearings are identified by increasing BHI values compared to what is considered normal. In this study, the data have been re-analysed comparing the vibrations of the failed bearing with the background vibrations of the other bearings on the same train covering the same track and subjected to the same speed profiles and loadings. This method clearly identifies the extra vibrations related to the damage.

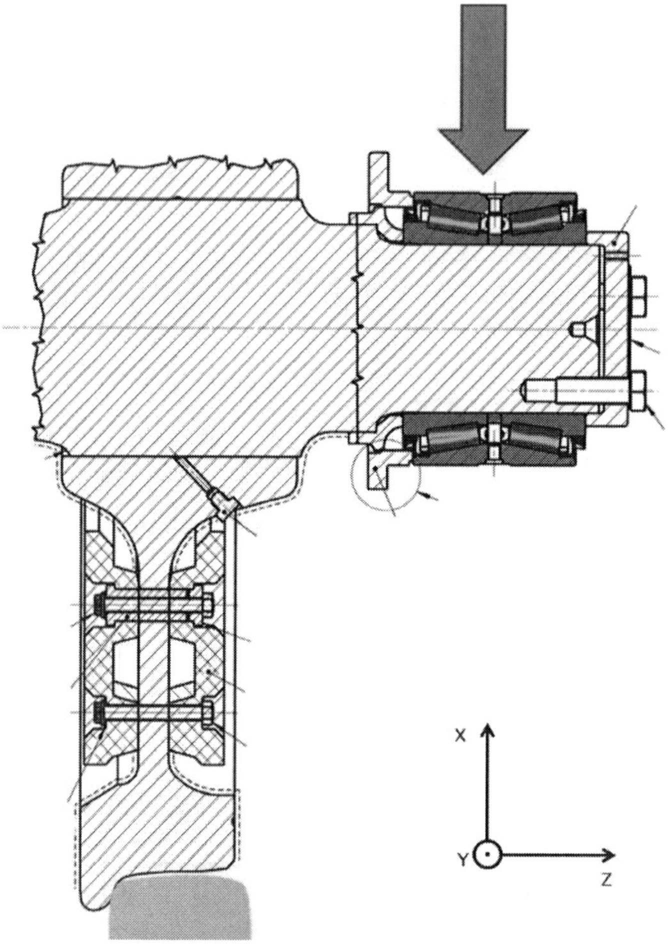


**Figure 2 – Schematic showing the bearing position with respect to wheel, axle and rail.**

For the purpose of this study, rms X is the parameter chosen to compare the vibrations. The vibration in X is solely affected by the loaded contact bearing-wheel-rail, as shown in Figure 2. Rms X is therefore influenced also by wheel damage. It has been confirmed that the bearings analysed are characterised only by high BHI values (and not by high WHI) confirming that the extra vibrations are only due to damage in the bearings.

Bearings with high BHI readings were removed by the operator and stripped at the Southeastern Ramsgate Maintenance depot. The bearings were then sent to the

University of Southampton for further analysis. All the bearings detected with high BHI values presented outer race damage.

The surface profile of the damaged outer race is measured at Taylor Hobson Limited (Leicester, UK) using a Talyrond 595. The bearing is positioned on a spindle top and five traces of the whole rolling area are measured with high data density (72,000 points over 360°). The traces are 10 mm apart as shown in Figure 3. The 0 mm trace is carried out outside the worn area and is used as a reference.

The volume missing on the outer race was measured using an Alicona Infinite Focus (Sevenoaks, UK), an optical non-contact profilometer. The scan was carried out using a horizontal resolution of 13.98 µm and a vertical resolution of 2.40 µm.

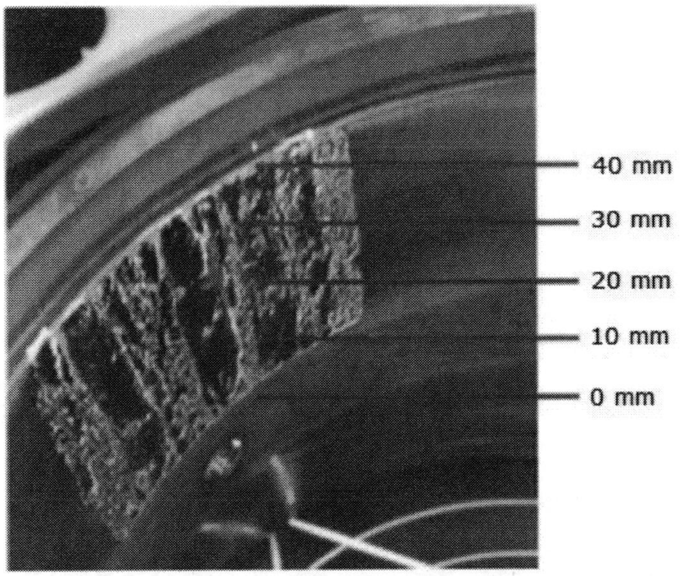


**Figure 3 – Bearing outer race with position of the 5 traces measured by Talyrond 595.**

## 4 RESULTS AND DISCUSSIONS

### 4.1 Factors affecting vibrations

By analysing data from numerous trains recorded during 10 months it emerged that a common vibration pattern for healthy wheels and bearings is the one shown in Figure 4. Figure 4 shows the vibrations in X, Y and Z for one wheel of one train travelling for one month all around the Southeastern train track (blue and orange track in Figure 1). This clearly demonstrates that the vibrations vary significantly in value from point to point. The vibration values change continuously but they are reproducible over time and therefore they could be used to map the railway line to check for changes in the infrastructure.

The changes in vibrations can be explained considering that the system bearing-wheel-axle-track (Figure 2) is a dynamically complex system with numerous variables affecting its behaviour. The vibrations are affected by:

- train track (e.g. fish plates, track points, junctions between tracks),
- wheel conditions,
- bearing conditions,
- loading,
- train speed and
- time (e.g. track conditions can change with time).

For this work, in order to be able to compare the vibrations of bearings from different trains; the data was filtered to eliminate most of these factors.

From the GPS data recorded, it is known that each train covers the whole Southeastern track (blue and orange in Figure 1) in a period between 2 and 6 weeks; therefore some of the factors affecting vibrations can be eliminated by using data from only one part of the track. The track *Tonbridge to Ashford International* (latitude: 51.1478 to 51.1927, longitude: 0.2639 to 0.8591 and heading: 86 to 116) (orange in Figure 1) has been chosen because: being a predominantly straight track (approximately 40 km long) there are no extra vibrations due to curves (extra load on wheels during turning) and the data are easily plotted using latitude or longitude. In a position detected by GPS there are at least 2 tracks (each track consists of two rails) going in opposite directions; only one direction (from Tonbridge to Ashford International) and only one rail (only one side of the train) are considered so that the contact rail-wheel is consistent and reproducible (this eliminates the effect of *train track*). The effect of time can be eliminated by comparing the faulty bearing with only the remaining "good" bearings on the same side of the train.

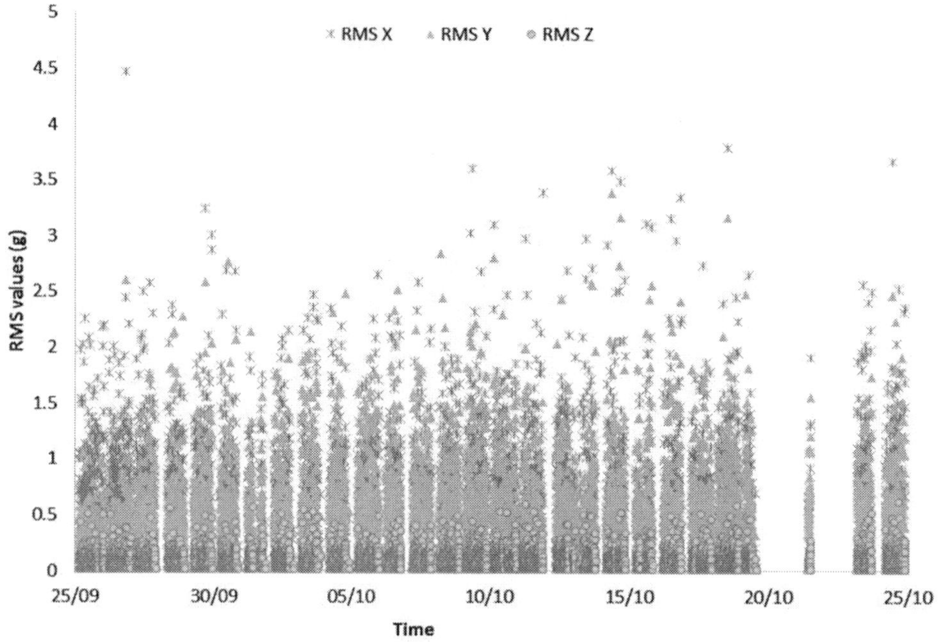


**Figure 4 – Rms vibrations in X, Y and Z for only a wheel going everywhere on the train track over a month (6,900 data in each direction).**

Comparing the variation of rms X (Figure 5a) and speed (Figure 5b) for the same part of the track, it is clear that the vibrations in X increase with train speed. Figure 5 shows data from 210 journeys from Tonbridge to Ashford International; the journeys are clearly repetitive with 5 stops (slow journeys) and the same speed variation. The few data from fast journeys have been eliminated. Therefore it could be concluded that generally trains run journeys with similar speed profiles on the same part of the track (e.g. same number of stops, same speed due to speed limits). Therefore it is possible to affirm that all the trains pass on a particular part of track always at the same speed (Figure 5b).

The *effect of speed* could be eliminated by using only one particular speed, but this has not been done because it would markedly decrease the number of points making comparison between different bearings very difficult.

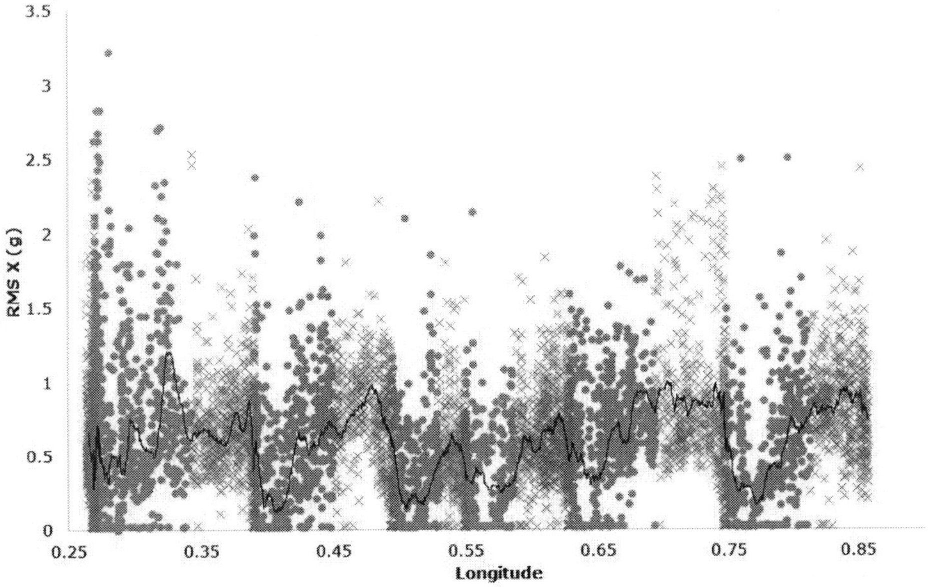


a)

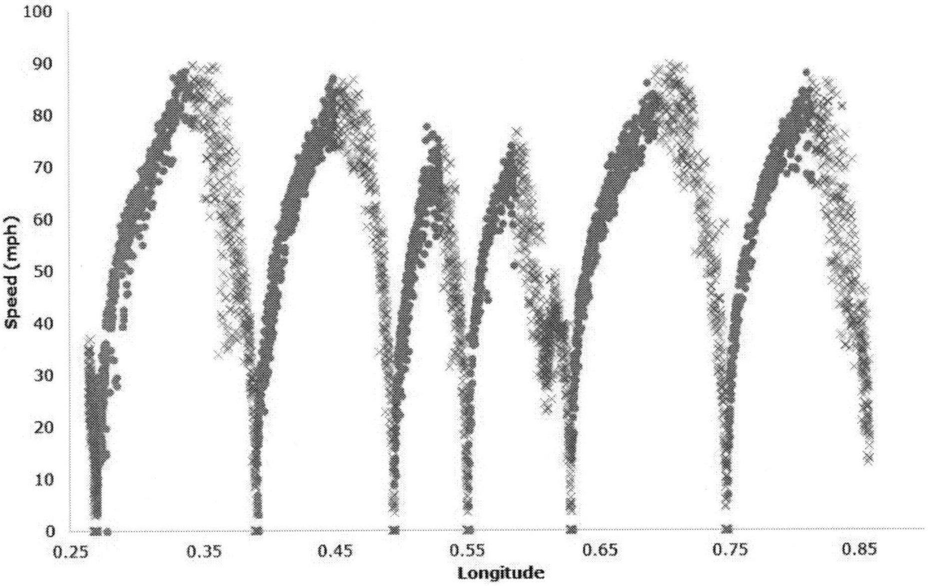


b)

**Figure 5 – a) vibration profile vs longitude (the black line shows the average vibration) and b) speed profile vs longitude for journeys of one train in a 3 month period going from Tonbridge to Ashford International (left to right). The data in orange represent accelerating time and those in blue decelerating and braking.**

### 4.2 Vibrations of failed bearings

By comparing data from four trains travelling from Tonbridge to Ashford International during the same 2-month period it has been demonstrated that all the 16 wheels from the same side of the train have the same baseline vibrations (when they are in good health) independently from the position along the train and from the train type (results not shown here). It is therefore possible to compare the vibrations of a damaged bearing with a *baseline vibration* characteristic for the same train, journey and period of time (data from 15 WSNs). Figure 6 shows two examples of this comparison.

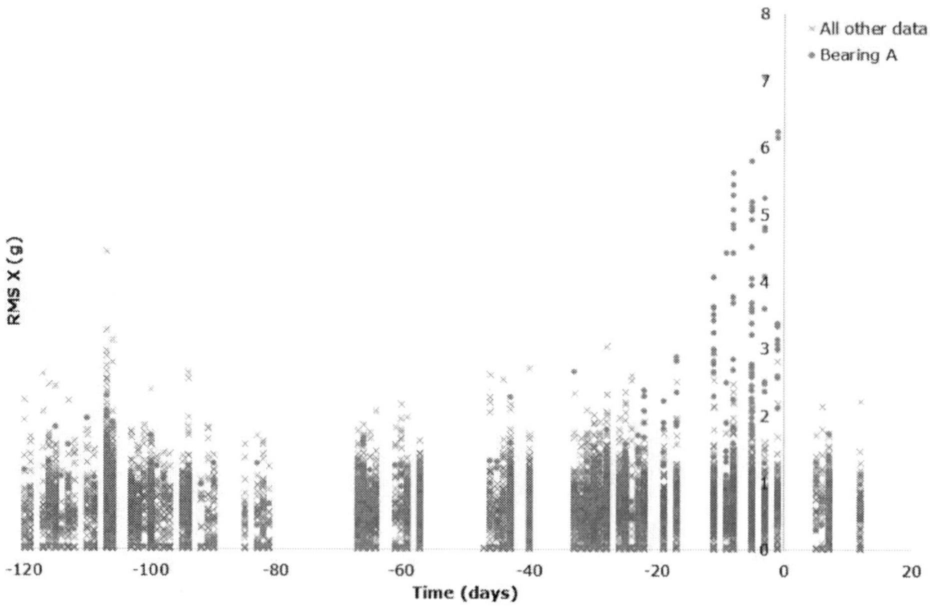

a)

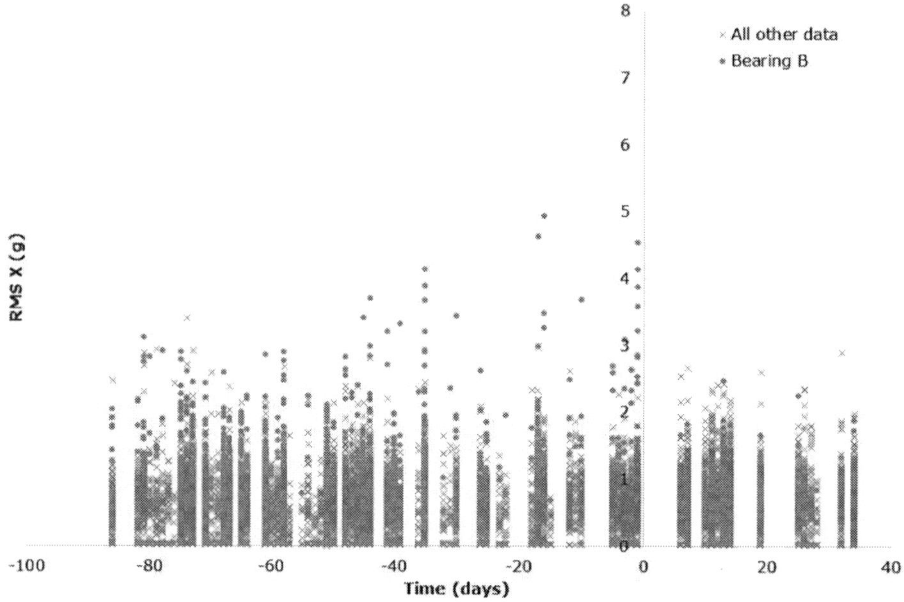

b)

**Figure 6 – Vibrations of one damaged bearing (in orange) compared with the baseline vibrations of the other bearings on the same side of the same train for the same journeys and the same period of time for a) bearing A and b) bearing B. Day 0 corresponds to bearing removal.**

For bearing A the vibration is very similar to the baseline vibrations until 11 days before bearing removal when the increase is very rapid. Rms X increases from 3 g to 7 g in 11 days. For bearing B the vibrations start to behave differently from the baseline 48 days before bearing removal and the increase is slow. The rms X value for this second bearing reaches only 5 g; rms X increases from 3 g to 5 g in 48 days. The vibrations of both WSNs decrease considerably after new bearings were fitted, further confirming that the excess vibrations were due only to the damage in the bearings.

Figure 7 compares the last 1,600 miles of both damaged bearings: the mean values of rms X are higher for bearing A than bearing B. During the last 400 miles bearing A displayed a mean rms X value 1.7 times greater than bearing B.

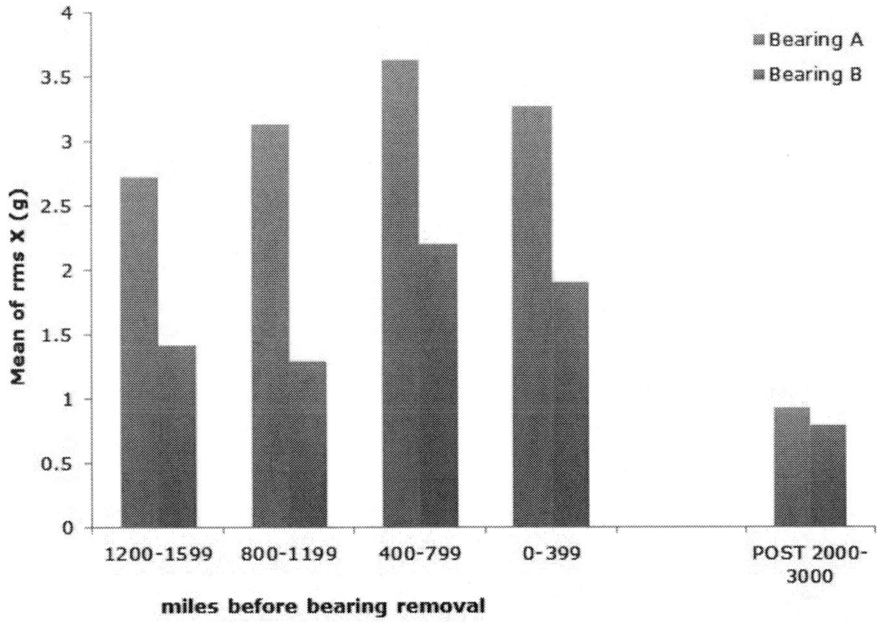


**Figure 7 – Mean rms X values recorded over the last 1,600 miles of use before removal of each bearing. The values from bearing A are significantly higher than those for bearing B.**

### 4.3 Outer race damage

Figure 8 shows the damaged outer races of the two bearings removed after the increase in vibrations shown in Figure 6. The damage observed has clear signs of rolling contact fatigue; more details are reported in (29). These two bearings will be used for this preliminary study to show the issues involved in linking vibration data to bearing damage. Bearing A shows 56 mm (36°) of uniform circumferential damage while bearing B shows a smaller damage over only 22 mm (14°). For every revolution of the train wheel 23 tapered rollers pass over the damage. The roller spacing of 15.65° suggests that only one roller was over the damage on bearing B, but two on bearing A. The damage on the outer race of bearing B is in two separate areas as if the two damages have been initiated separately. The damage observed in these bearings is similar to those reported in other studies (34, 36-37).

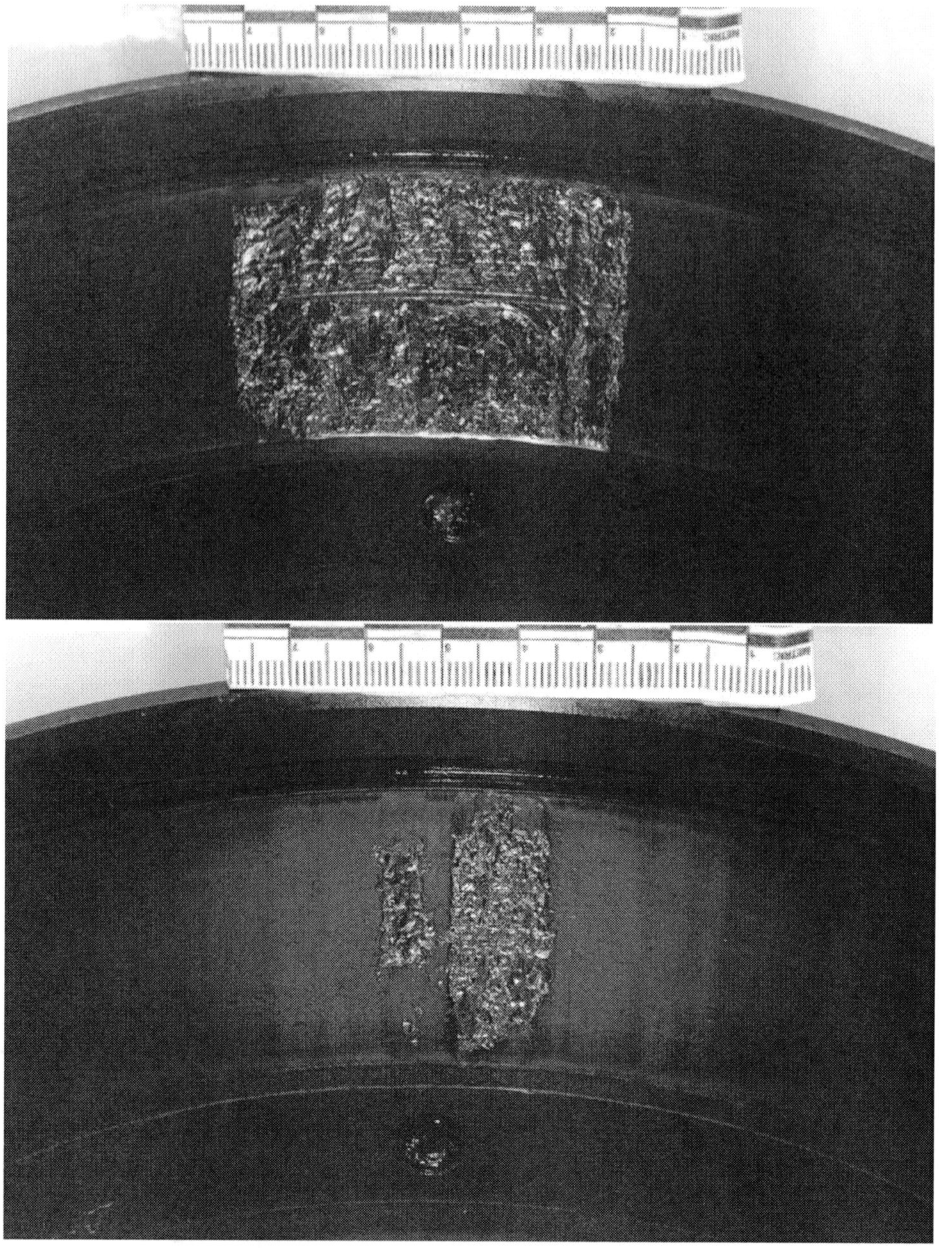

a)

b)

**Figure 8 – Examples of damage on the outer race for a) bearing A and b) bearing B.**

### 4.4 Surface profilometry of failed bearings

**Table 1 - Damage for bearings A and B.**

| Variable | Bearing A | Bearing B | Ratio A:B |
|---|---|---|---|
| **Damage arc (°)** | 35.82 | 14.39 | 2.5 : 1 |
| **Damage depth (average of 20 and 30 mm) (mm)** | 0.77 | 0.34 | 2.3 :1 |
| **Area loss (average of 20 and 30 mm) ($mm^2$)** | 23.69 | 3.43 | 7 : 1 |
| **Volume loss ($mm^3$)** | 900 | 75 | 12 : 1 |

The surface profiles of the rolling area of the outer race for bearing A and B are shown in Figure 9; for bearing A only one trace is reported because the damage is similar across the outer race. Table 1 clearly shows that the damage in bearing A is bigger than that in bearing B with the damage width and depth being more than double and the volume loss being almost 12 times larger.

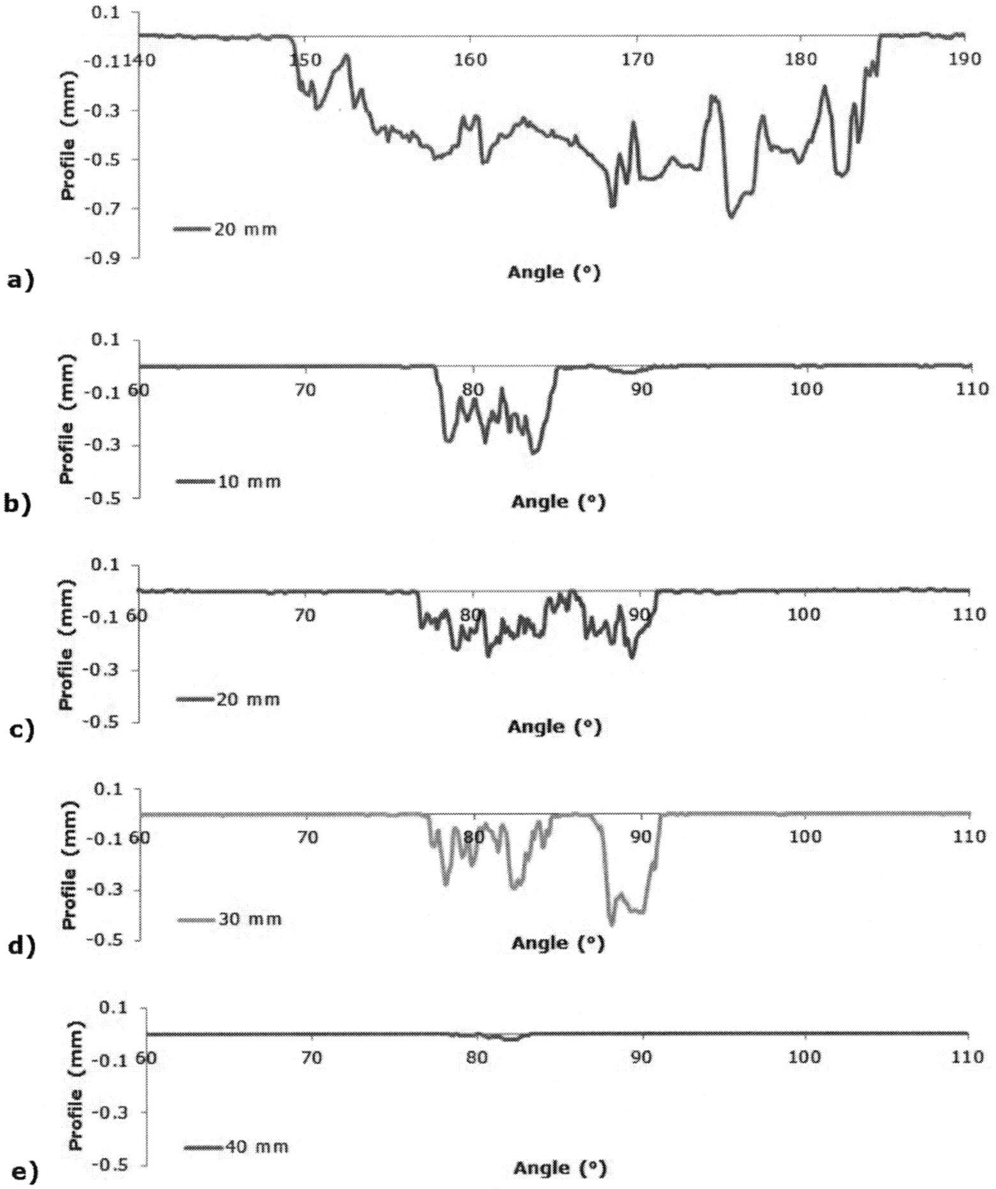


**Figure 9 – Surface profile of a) bearing A at 20 mm and bearing B at b) 10 mm, c) 20 mm, d) 30 mm and e) 40 mm.**

### 4.5 Correlation between damage and vibrations measured

From the two examples presented above, it seems that the extent of damage is comparable to the increase in vibrations. There is a loose proportionality that suggests that a bearing with twice the damage size will produce a doubling in the vibration amplitude. However, the bearing described in (29) (damage arc = 60°, damage depth = 0.8 mm, volume loss = 1,245 $mm^3$, rms X = 5) does not fit this pattern. More examples of bearings are needed to be able to understand this very complex system.

This two-bearing model is very simplistic, it has been observed that the measured vibrations grow with different slopes: 4 g in 11 days (bearing A) and 2 g in 48 days (bearing B) (Figure 6); indicating that other parameters are affecting the damage growth and in turn the vibrations measured. The loading profile is not considered to be a variable affecting these behavioural differences as both trains were proven to be subjected to the same conditions.

However, the following variables could influence the vibration/damage correlation:

- metallurgical differences (i.e. a bearing with a large subsurface flaw may present a unique damage/vibration pattern),
- isolated loading events that have not been captured or have been filtered out,
- history of the bearing before the WSN was fitted,
- physical growth of the damage (e.g. the shape, depth and number of RCF macropits could influence the vibration signature as could also the subsequent smoothing of the fracture surface).

## 5 CONCLUSIONS

- It has been shown that the vibrations present at a wheel axle bearing are very complex and influenced by many factors. Even a "good" bearing will produce a spread of vibration data in all three axes (Figure 4).

- Limiting this study to rms X and filtering the data to those collected while travelling on a length of straight track allow comparison of the vibrations of two damaged bearings (Figure 7).

- With further evaluation of more real-life train axle bearings, it is hoped that a relationship may be found linking bearing damage to vibrations, further improving the accuracy of condition monitoring and ultimately the safety of the train.


## ACKNOWLEDGEMENTS

The authors gratefully acknowledge the EPSRC IAA funding, Dr Liam Goodes for measuring the volume loss and the Southeastern staff for their support.


## REFERENCES


1. Gao, R. X., Jin, Y., Warrington, R. O. (1994) IEEE Transactions on instrumentation and measurement **43**, 216-219
2. Lei, Y., Lin, J., He, Z., Zuo, M. J. (2013) Mechanical Systems and Signal Processing **35**, 108–126 C
3. Lacey, S. J. (2008) Analysis, maintenance & asset management **23**, 32-42
4. Li, H., Zhang, Y., Zheng, H. (2010) Journal of Mechanical Science and Technology **23**, 291–301. doi:10.1007/s12206-008-1110-5
5. Li, H., Fu, L., Zheng, H. (2011) Journal of Mechanical Science and Technology **25**, 2731-2740
6. Ming, A. B., Qin, Z. Y., Zhang, W., Chu, F. L. (2013) Mechanical Systems and Signal Processing **41**, 141–154
7. Georgoulas, G., Loutas, T., Stylios, C. D., Kostopoulos, V. (2013) Mechanical Systems and Signal Processing **41**, 510–525
8. Feng, Z., Zuo, M. J., Hao, R., Chu, F., Lee, J. (2013) J. Vib. Acoust. **135**, 031013 1-21

9. Cocconcelli, M., Bassi, L., Secchi, C., Fantuzzi, C., Rubini, R. (2012) Mechanical Systems and Signal Processing **27**, 667–682
10. Tsao, W-C., Li, Y-F., Le, D. D., Pan, M-C. (2012) Measurement **45**, 1489–1498
11. Mohammadi, A., Safizadeh, M. S. (2013) Journal of Tribology **135**, 011102-1
12. Lei, Y., Lin, J., He, Z., Zi, Y. (2011) Mechanical Systems and Signal Processing **25**, 1738–1749
13. Wang, D., Tse, P. W., Tsui, K. L. (2013) Mechanical Systems and Signal Processing **35**, 176–199
14. Kar, C., Mohanty, A. R. (2004) Journal of Sound and Vibration **269**, 439–454
15. Mba, D., Rao, R. B. K. N. (2006) The shock and vibration digest **38**, 3-16
16. Elforjani, M., Mba, D. (2010) Engineering Fracture Mechanics **77**, 112–127
17. Jiang, R., Chen, J., Dong, G., Liu, T., Xiao, W. (2013) Proceedings of the Institution of Mechanical Engineers, Part C: Journal of Mechanical Engineering Science **227**, 1116-1129
18. Dybała, J., Zimroz, R. (2014) Applied Acoustics **77**, 195–203
19. Jiang, F., Zhu, Z., Li, W., Chen, G., Zhou, G. (2014) Meas. Sci. Technol. **25**, 025003
20. Yan, R., Gao, R. X., Chen, X. (2014) Signal Processing **96**, 1–15
21. Zhu, D., Gao, Q., Sun, D., Lu, Y., Peng, S. (2014) Signal Processing **96**, 80–89
22. Yu, G., Li, C., Zhang, J. (2013) Mechanical Systems and Signal Processing **41**, 155–175
23. Yang, W., Court, R. (2013) Measurement **46**, 2781–2791
24. Zhao, M., Jin, X., Zhang, Z., Li, B. (2014) Expert Systems with Applications **41**, 3391–3401
25. Symonds, N. (2004) Control and Reduction of Wear in Military Platforms NATO AVT-109, Williamsburg, US
26. Symonds, N. (2005) Insight - Non-Destructive Testing and Condition Monitoring **47**, 472-477
27. Zakrajsek, J. J., Dempsey, P. J., Huff, E. M., Augustin, M., Safa-Bakhsh, R., Duke, A., Ephraim, P., Grabil, P., Decker, H. J. (2006) Technical Report No. NASA/TM-2006- 214022
28. Trilla, A., Gratacòs, P. (2013) Chemical Engineering Transactions **33**, 733-738
29. Symonds, N., Corni, I., Wood, R. J. K., Wasenczuk, A., Vincent, D. ICEFA VI, manuscript in preparation
30. Zhao, J-S., Liu, W., Zhanga, Y., Feng, Z-J., Ye, J., Niu, Q-B. (2013) Mechanism and Machine Theory **69**, 185–199
31. Rosado, L., Forster, N. H., Thompson, K. L., Cooke, J. W. (2009) Tribology Transactions **53**, 29-41
32. Arakere, N. K., Branch, N., Levesque, G., Svendsen, V., Forster, N. H. (2009) Tribology Transactions **53**, 42-51
33. Forster, N. H., Rosado, L., Ogden, W. P., Trivedi, H. T. (2009) Tribology Transactions **53**, 52-59
34. Tsukahara, H., Aime, W., Burle, A., Chavance, D., Mouille, E. (2013) World Congress on Railway Research
35. Kalay, S., French, P., Tournay, H. (2011) World Congress of Railway Research
36. SKF, Railway technical handbook Volume 2 (2012) ISBN 978-91-978966-6-5
37. Müller, L., Sunder, R. (2013) World Congress on Railway Research
38. Bellaj, S., Pouzet, A., Mellet, C., Vionnet, R., Chavance, D. (2011) World Congress of Railway Research

# Low-power embedded system approach for axlebox bearing condition monitoring

**M Entezami*, L Saade*, E Stewart*, C Roberts*, P Weston*, R Lewis†**
* The University of Birmingham, UK
† TWI Ltd, UK

## ABSTRACT

Condition monitoring of axlebox bearings for in-service trains is becoming more of a priority for operators. Bearing faults can lead to unexpected downtimes of running lines and, in severe circumstances, can result in catastrophic failures such as derailments. To achieve higher operational reliability and to improve upon the efficacy of scheduled maintenance, it is beneficial to have a condition monitoring system that indicates early stage faults on running vehicles.

Over the past decade there have been many technological developments related to railway bearing analysis, particularly through the use of on-board vibration monitoring systems and wayside acoustic monitoring systems. Wide-scale adoption of these technologies by railway operators has yet to gain momentum due to current limitations in reliability, installation procedures and cost.

This paper presents the application of a prototype wireless embedded system for analysing vibration signals on axlebox bearings. This system has been designed to overcome one of the main limiting factors of train bearing monitoring which is the costly and technically prohibitive installation of monitoring system cables to carry power and data signals from the axlebox to the central monitoring unit.

**Keywords:** bearing monitoring, energy harvesting, embedded systems

## 1 INTRODUCTION

Preventative maintenance by monitoring the vibration of axlebox bearings is a scientific approach that has shown significant potential in maintenance management. Rolling element bearings, even newly manufactured ones, generate some level of vibration which, to a certain level, is acceptable. Increments in vibration levels are usually symptoms of abnormal machine performance. Current studies and laboratory experiments on bearing condition monitoring have demonstrated great potential in fault detection and isolation [1].

Vibration analysis is a well-known monitoring technique used to identify bearing faults [2]. This can be carried out on data collected from an accelerometer or other vibration sensor. Frequency analysis is most commonly used to characterise bearing faults, so sampling rates over 20 $kSs^{-1}$ are usually required, which results in large datasets.

Traditional bearing condition monitoring practices involves cabling from accelerometers on the bearing to a central point on the bogie where data is then routed into the body of the vehicle. The sensors are generally active devices that require power to be sourced from the vehicle body. In this paper, the focus is on the fault diagnosis part of a wireless bearing condition monitoring system about which more detail can be found in[3].

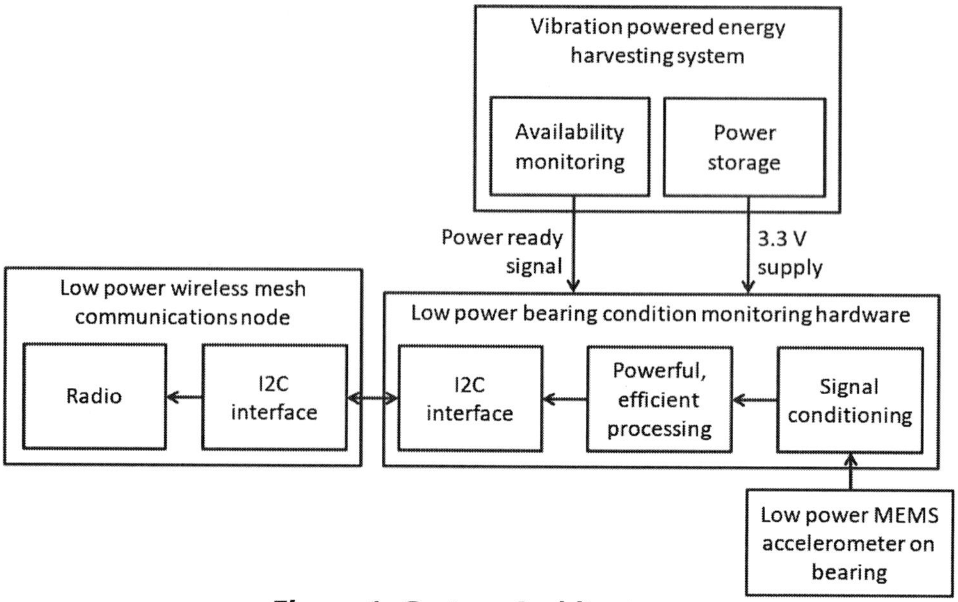


**Figure 1: System Architecture**

Data can be transmitted over a wireless link, however, power is still required at the sensors and for their end of the wireless data link. The system described in this paper uses an energy harvesting device to supply this power at 3.3 V, see Figure 1. Due to the limited amount of power that can be generated through vibration powered energy harvesting, the system must be frugal in its power usage.

A micro-electro-mechanical system (MEMS) accelerometer is attached to the axle box with a magnet. The raw vibration signal is then processed by a low power 32 bit microcontroller power embedded system resulting in short bearing condition messages. These are transmitted to the central monitoring unit over wireless node.

### 1.1 Processing

The processing methodology used in this project is a simplified version of a well-known technique called High Frequency Resonance Technique (HFRT) which is also known as envelope analysis. HFRT has been used on vibration and acoustic emission signals for bearing fault detection and diagnosis for decades [4].

The method extracts fault-related high frequency components of a signal, which are compared to known bearing fault characteristic frequencies. HFRT processing shifts low amplitude high frequency harmonics of bearing defect frequencies into a low frequency range. Envelope detection or amplitude demodulation is the technique of extracting the modulating signal from an amplitude-modulated signal. The result is the time history of the modulating signal. Envelope analysis is the FFT spectrum of the modulating signal.

The procedure for this method is as follows:

- Wait in sleep mode until enough energy is available from the energy harvesting system.
- Activate peripherals and sample data from the accelerometer then deactivate peripherals.
- Pass collected signal through a high-pass filter.
- Generate envelope using the Hilbert Transform [5] to identify individual impact events within the signal.
- Apply a low-pass filter to the enveloped signal keeping the low frequency content that is indicative of bearing faults.
- Convert the filtered signal into the frequency domain using a Fast Fourier Transform (FFT) to identify the characteristic frequencies of a potential fault.
- Report the highest peaks of the FFT along with a message indicating the health of the bearing over I2C to the wireless node and then return to sleep.

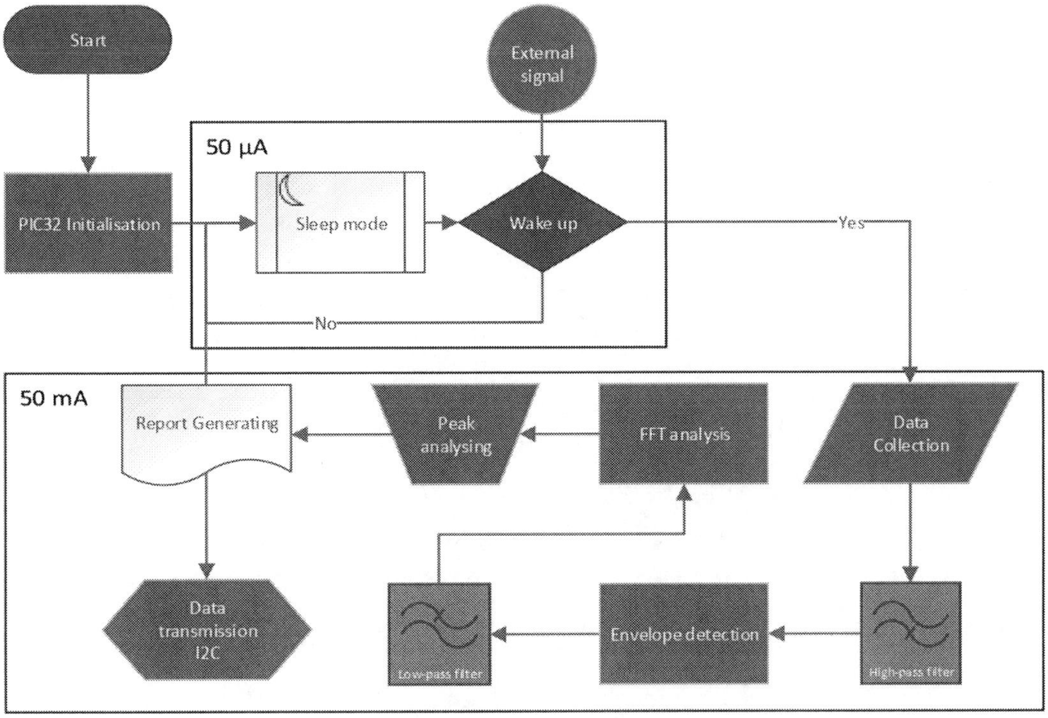


**Figure 2: Embedded code flow chart**

## 2 BEARING FAULTS

Different parts of bearings and different faults have different characteristic frequencies which are dependent on the bearing parameters such as number of rollers, roller diameter, pitch diameter, contact angle, and rotational speed.

Given the geometry of the bearing and the shaft speed, the frequencies of various faults types can be calculated. The following equations define the characteristic frequencies using well-known equations [6] with: $f$ the frequency of the item [Hz]; $\omega_r$ the shaft rotation rate (revolution/second) [Hz]; $RD$ the roller diameter [m]; $PD$ the mean roller race diameter [m]; $\beta$ the contact angle [rad]; and $n$ the number of rollers.

Fundamental Train Frequency (FTF) indicates a fault on the cage or mechanical looseness

$$f = \frac{\omega_r}{2} \left(1 - \frac{RD}{PD} \cos\beta\right) \qquad (1)$$

Roller Passing Frequency Outer Race (RPFO) indicates a local fault on the outer race

$$f = \frac{n}{2} \omega_r \left(1 + \frac{RD}{PD} \cos\beta\right) \qquad (2)$$

Roller Passing Frequency Inner Race (RPFI) indicates a local fault on the inner race

$$f = \frac{n}{2} \omega_r \left(1 - \frac{RD}{PD} \cos\beta\right) \qquad (3)$$

Roller Fault Frequency (RFF) is twice the Roller Spin Frequency (RSF) and indicates a local fault on the rolling element

$$f = \frac{PD}{2RD} \omega_r \left(1 - \left(\frac{RD}{PD} \cos\beta\right)^2\right) \qquad (4)$$

These equations and their corresponding frequencies can be used after a developing fault is detected to diagnose the cause of the incipient fault.

### 2.1 Test configuration and processing results

The automotive tapered roller bearing that is shown in Figure 3 was used to carry out the tests.

**Figure 3: Test bearing - type 801023AB supplied by FAG**

The geometry and specifications of the bearing are given in Table 1.

Figure 4 demonstrates an induced race fault on the outer race. This is minor damage to a small area on each outer race, consisting of surface roughening by means of an electrical discharge engraver. The damage length is about 3% of circumference.

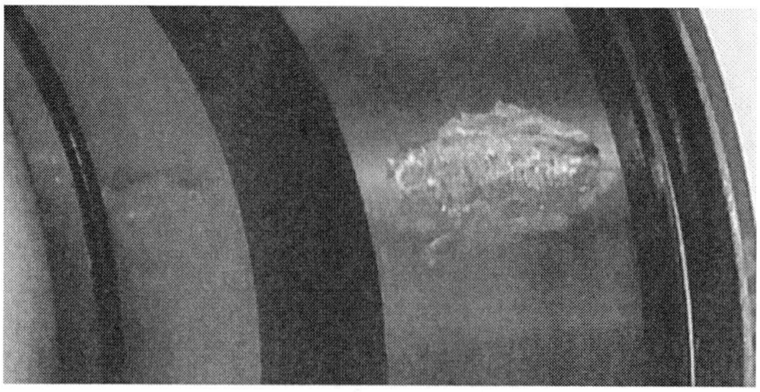

**Figure 4: Race fault**

**Table 1: Bearing 801023AB specifications**

| Item | Variable | Value | Units |
|---|---|---|---|
| Shaft diameter (in bearing) | d | 29 | mm |
| Casing diameter | D | 53 | mm |
| Casing length | C | 37 | mm |
| Roller diameter | | 5.3 | mm |
| Pitch diameter | | 42.5 | mm |
| Axle diameter | d2 | 35.9 | mm |
| Mass | m | 0.41 | kg |
| Vehicle weight/bearing, empty vehicle | | 295 | kg |
| Vehicle weight/bearing, all vehicle seats occupied | | 395 | kg |
| Roller angle | | 13.8 | ° |
| Number of rollers per cage | | 21 | |

Figure 5 illustrates raw signal from a bearing spinning at a constant speed of 1000 RPM with some scratches on the race.

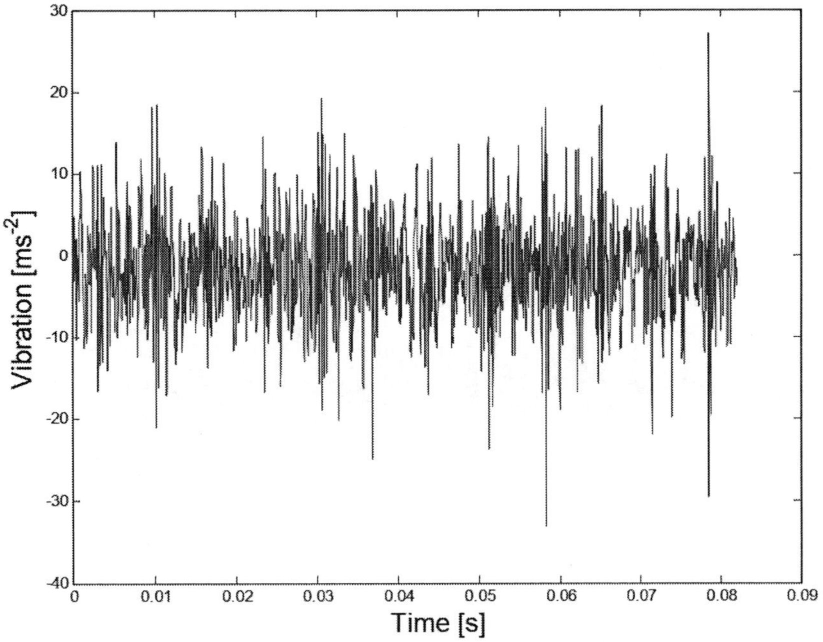


**Figure 5: Faulty race raw signal**

The signal was then passed through a 64$^{th}$ order 5 kHz FIR high-pass filter; the result of which is shown in Figure 6. The filter cut-off can be adjusted depending on the resonant frequency of the bearing.

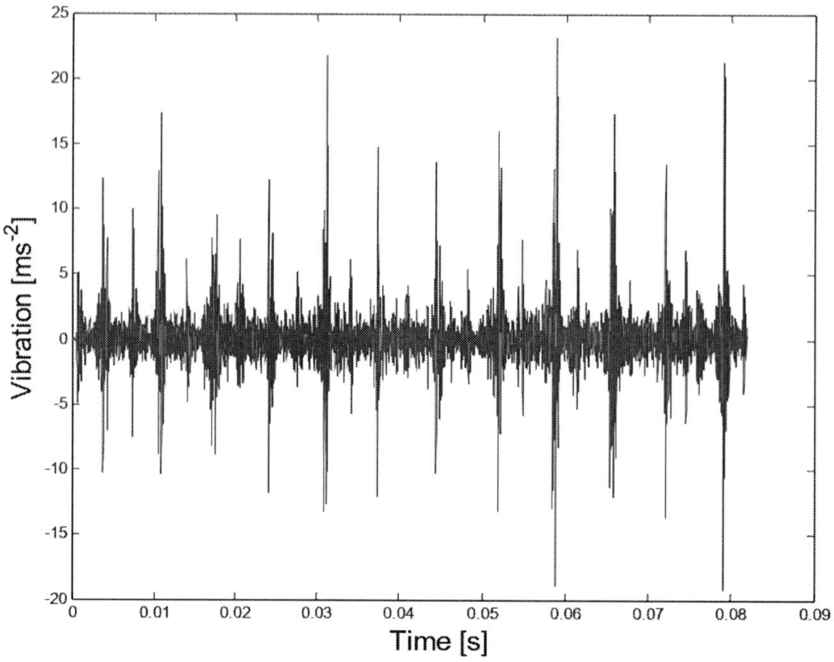


**Figure 6: Application of a high-pass filter on the raw signal**

The envelope of the filtered data was generated using the enveloping algorithm, demonstrated in Figure 7.

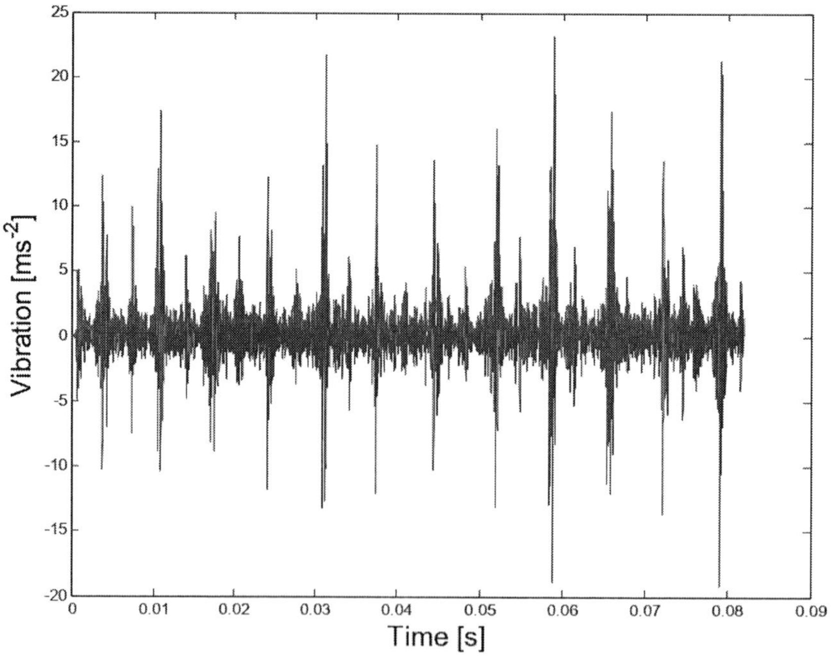


**Figure 7: Filtered signal and its envelope**

A 16th order 1500 Hz FIR low-pass filter was applied to the envelope to emphasise the low frequency contents.

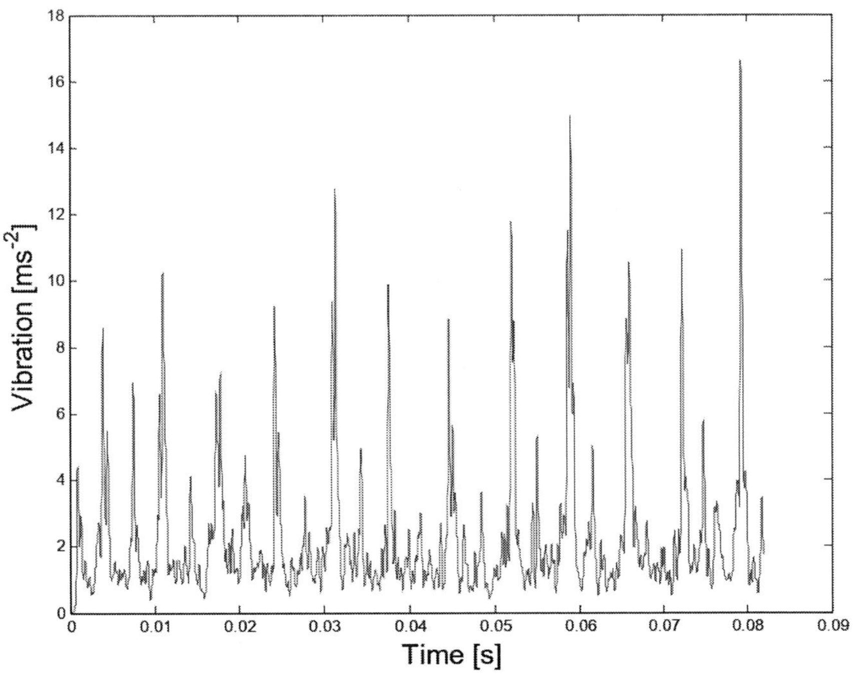


**Figure 8: Application of a low-pass filter on the envelope**

An FFT is then carried out on the filtered result of the envelope to allow the system to indentify the frequencies of interest. The FFT spectrum shown in Figure 9 clearly demonstrates the presence of FFT peaks at the race fault frequency and its harmonics.

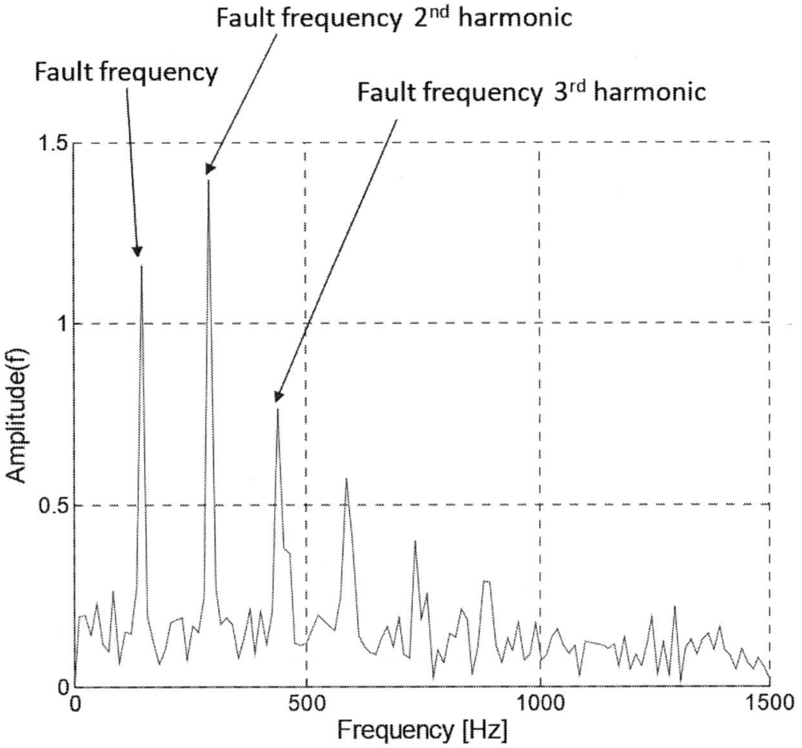


**Figure 9: FFT spectrum with a clear race fault**

### 2.2 Embedded system architecture

For this work, an industrial accelerometer capable of operating on less than 1 mA at 3.3 V and a low power 32 bit processor capable of running various digital signal processing algorithms are used to do the data acquisition and analysis. The processor is set to run at 40 MHz and utilises 50 mA at 3.3 V during the sampling and processing although the power consumption drops to 50 µA during sleep. The processing was optimised to work using integer arithmetic that allows it to generate suitable results in approximately 3 seconds. The data sampling rate was set to 22 kHz and FFT window has 4096 points, which provides ±5.37 Hz accuracy in the FFT spectrum.

### 2.3 Decision making

In order to make a decision using the FFT spectrum, the largest six peaks were identified and a crest factor (see eq. below) was calculated for each peak.

$$Crest\ Factor\ (CF) = \frac{Peak\ value}{RMS\ of\ FFT} \tag{5}$$

Depending on the severity of the fault, the CF usually increases and the fault appears on the spectrum along with a few related harmonics. A threshold based method was applied to report the signals with high CFs and existing harmonics as faulty.

Figure 10 shows the peaks with the highest CF values which also happen to be harmonics related to the race fault.

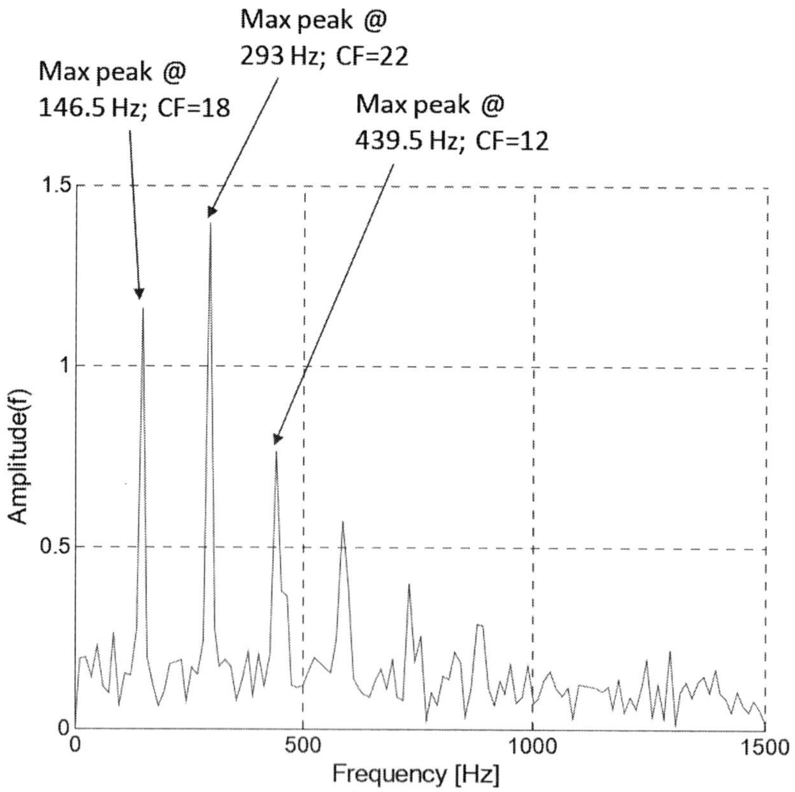


**Figure 10: FFT spectrum showing peaks of interest**

The system reports these CF and frequency values along with a flag to identify if the collected data represents a healthy or faulty condition.

## 2.4 Results

It is estimated that trains would need to run at no less than 60 mph (96 $kmh^{-1}$) to generate enough energy with the current energy harvesting system. Tests were carried out on a faulty and healthy roller bearing to evaluate the low power monitoring system. The shaft speed was 1200 RPM, as expected for a 100 mph train (depending on wheel diameter). Figure 11 shows minor damage to a small area of one roller in each cage. The fault length is about 10% of the roller circumference.

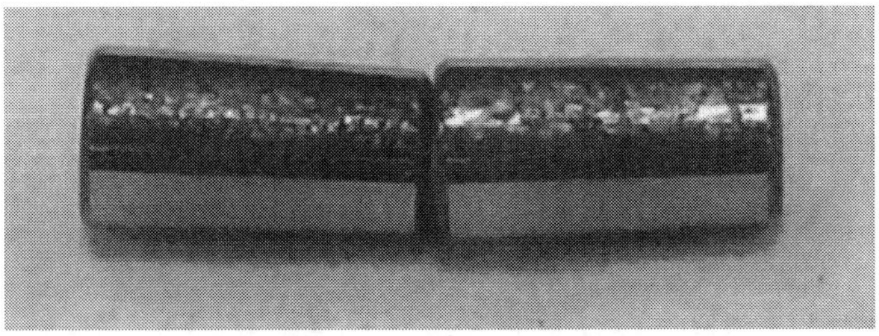

**Figure 11: Faulty roller**

The results were continually sent from the monitoring system to a PC over a serial link for demonstration purposes. A screenshot of the outputs for the faulty bearing can be seen in Figure 12. The system reports the number of harmonics and the level of fault, which is the sum of the crest factors for the 6 largest peaks.

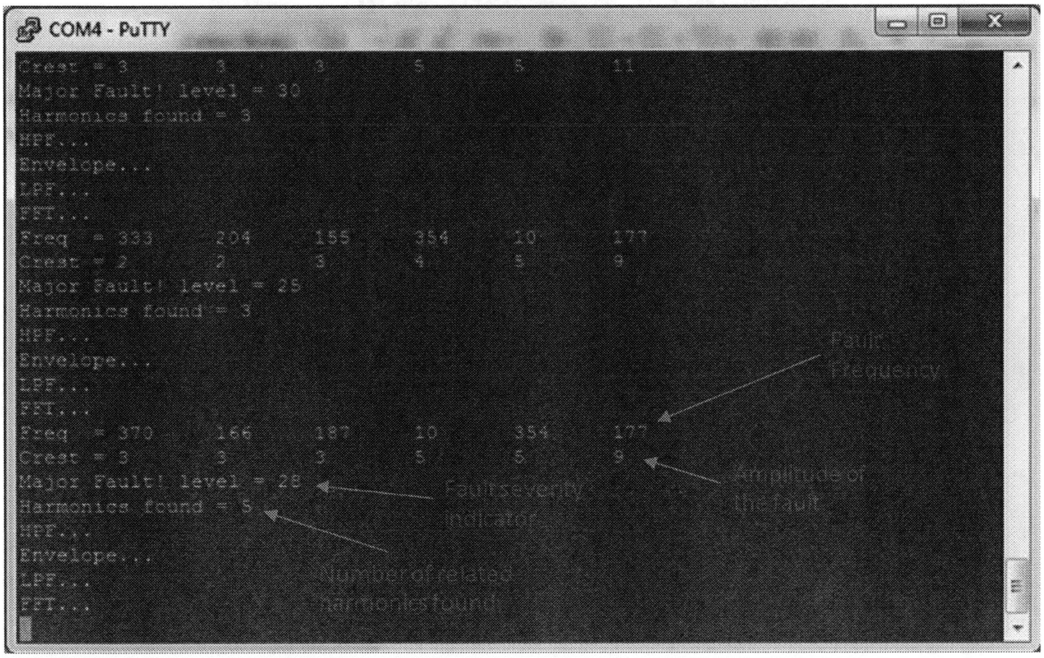


**Figure 12: Faulty bearing screenshot**

Healthy bearings would not generate any significant peaks with related harmonics. Two examples for the healthy bearing test are shown in Figure 13. Most crest factor values are 2 or lower indicating an insignificant level of fault severity.

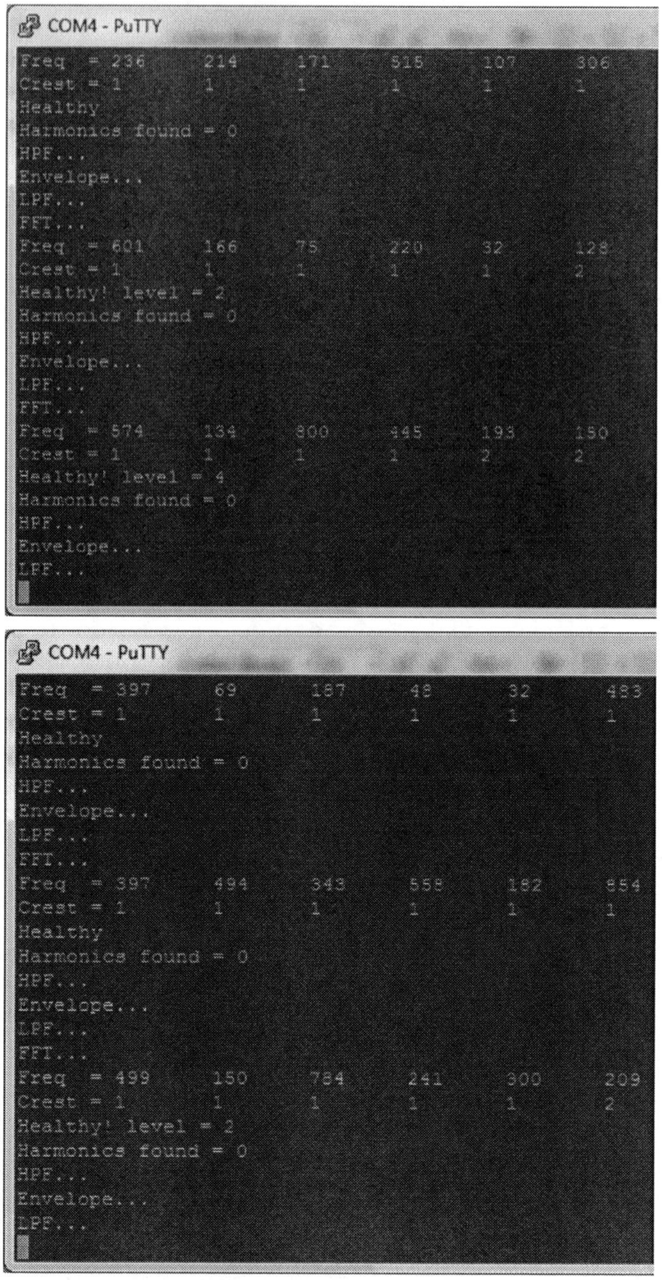


**Figure 13: Healthy bearing screenshots**

## 3 CONCLUSIONS

An embedded system using a low power microcontroller was designed to acquire data from an industrial accelerometer and carry out bearing fault detection algorithms. The algorithms were evaluated on a test rig and a method for automatic fault detection was presented.

The system showed great potential in its ability to take measurements and report faults in a short period of time whilst operating on low power. The results from tests conducted on a bearing rig demonstrate a consistent and reliable technique that successfully differentiates between faulty and healthy bearings.

## 4 ACKNOWLEDGEMENTS

The WiRailCom project is funded by the European Union's Seventh Framework Programme for research, technological development and demonstration under grant agreement no. 313941. The project partners are: TWI Ltd (TWI Technology Centre Wales, UK), Airtren S.L (Spain), Reuschling (Germany), BK Telematics Ltd (Greece), The University of Birmingham (UK), MerMec, (Italy), Senceive Ltd (UK).

## 5 REFERENCES

[1] M. Kashiwagi, C. d. Costa and M. H. Mathias, "Development of Diagnosis System for Rolling Bearings Faults on Real Time Based on FPGA," in *International Conference on Renewable Energies and Power*, Santiago de Compostela, Spain, 2012.

[2] J. A. Stamm and Y. S. Shin, *Machinery diagnostics via mechanical vibration analysis using spectral analysis techniques,* Monterey, California: Naval Postgraduate School, 1988.

[3] R. W. Lewis, S. Maddison and E. J. C. Stewart, "An extensible framework architecture for wireless condition monitoring applications for railway rolling stock," in *IET Railway Condition Monitoring (RCM)*, Birmingham, 2014.

[4] M. S. Darlow, R. H. Badgley and G. W. Hogg, "Application of High-Frequency Resonance Techniques for Bearing Diagnostics in Helicopter Gearboxes," Mechanical Technology Inc Latham N Y, Technical Report, US Army Air Mobility Research and Development Laboratory, 1974.

[5] MathWorks, "Envelope Detection - MATLAB & Simulink Examples," [Online]. Available: http://www.mathworks.co.uk/help/dsp/examples/envelope-detection.html. [Accessed 14 April 2014].

[6] C. Scheffer and P. Girdhar, "Machinery fault diagnosis using vibration analysis," in *Pratical machinery vibration analysis & predictive maintenance*, Amsterdam, Elsevier, 2004, pp. 110-115.

# Modern innovations on the application of the rotating ultrasonic probe to in-service inspection of freight solid axles: experiments and MAPOD approach

**M Carboni*, S Cantini****
* Dept. Mechanical Engineering, Politecnico di Milano, Italy
** Lucchini RS SpA, Italy

**ABSTRACT**

The present scenario of freight applications in Europe is characterized by a trend towards rolling stock renewal and new approaches to periodical inspection and maintenance. A specific ultrasonic technique, based on more than 40 years of service experience on passenger wheel-sets inspection, is here proposed for periodical NDT on freight wheel-sets.

In the present research, a Model-Assisted Probability of Detection approach is considered. In particular, the derivation of the Probability of Detection curves of the rotating probe and an optimization of its calibration block are carried out considering both experimental tests and numerical simulations.

## 1 INTRODUCTION

The in-service maintenance of axles for freight applications is nowadays one of the main topic for railway safety working groups at European level. The present scenario shows a trend toward new procedures, especially looking for a higher level of safety and, at the same time, a reduction of the total life cycle cost of wheel-sets. Due to the typical in-service damages, a "Damage Tolerant" design (1) (2) requires periodic NDT inspections and typically applies ultrasonic (UT) and magnetic particles (MT) testing. In particular, it is worth mentioning that the purpose of such in-service inspections is to detect any damage, and subsequent propagation of cracks up to failure, due to characteristic phenomena, such as corrosion-fatigue or impact of the ballast. These phenomena have not the same probability of occurrence along the entire axle but they are related to the naked surface and to the most stressed zones, so it is important to ensure a good sensitivity for the entire inspected axles with particular attention to the most critical areas. From this point of view, freight applications usually adopt solid axles and the UT technique is mainly based on manual or semi-automatic inspection from axle surface with angled probes.

An alternative technique is based on more than 40 years of Italian service experience on passenger wheel-sets inspection; the axle is inspected by both ends by angled probes fit on a rotating probe-holder to be magnetically fastened at the ends of the mounted axle. This technique allows the inspection of axle critical sections with different angled probes (Figure 1) each being dedicated to a specific surface portion. The inspecting procedure considered in the present study (3) prescribes a UT contact inspection from the two end-faces of the solid axle. The probe is applied on either end to fully inspect the axle, diminishing the effects of attenuation. The transducers are sequentially activated by the operator thank to a selector switch mounted on the probe and the inspection is performed by a complete rotation of the probe on his axis (Figure 1). Since solid axles have three threaded blind holes, located on the contact surface of the probe, needed to mount the axle-box cap (Figure 2), the possibility to perform a complete inspection of the surface is partially impeded. Therefore, for every converging transducer a diverging one is also available to supersede the holes.

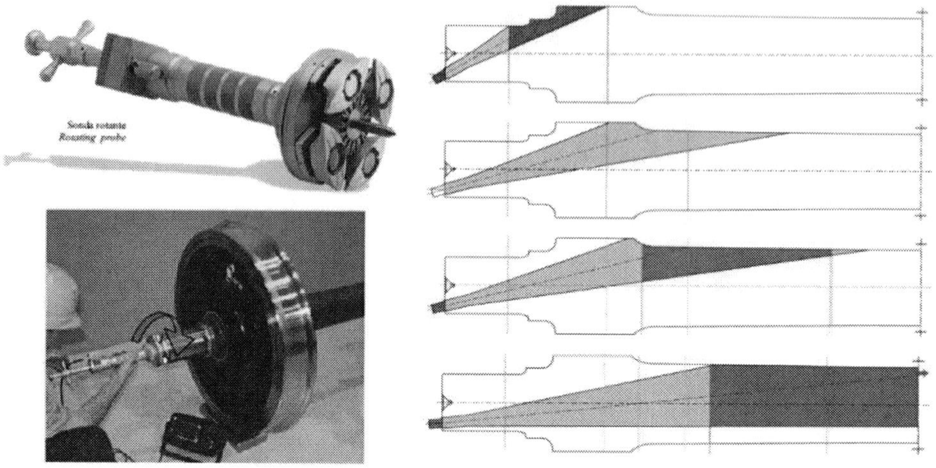

**Figure 1 – Rotating probe for ultrasonic inspection of solid axles.**

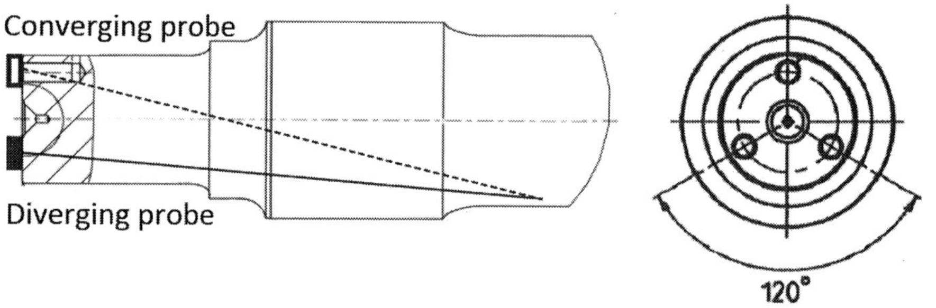


**Figure 2 – Threaded holes preventing an efficient inspection of the solid axle.**

The structural integrity of the wheel-set is then based, together with other factors, on the reliable determination of the Probability of Detection (POD) curve (4) (5). An accurate definition of the POD curve is a prerequisite for the reliable application of a NDT procedure. Actually, the derivation procedure is quite expensive because a large amount of experimental evidence is required, especially considering the derivation of the confidence band. Furthermore, the experimentally derived POD curve is strictly related to the considered application and cannot be transferred to other ones, even if similar. For these reasons, the Model-Assisted POD (MAPOD) approach was developed (6) allowing the partial replacement of experimental evidence with numerical results.

The purpose of this work is then the derivation of the Probability of Detection curves of the rotating probe (7) and an optimization of its calibration block. Adopting the inspecting procedure developed by Lucchini RS (3), the optimization began with experimental tests, carried out on a novel sample block for the present rotating probe, and their comparison to numerical simulations by the CIVA[nde] 11 dedicated software package (8). The validated numerical model was then adopted for the optimization by a MAPOD approach.

## 2 POD CURVES OF THE PRESENT ROTANTING PROBE

To evaluate the performance of the rotating probe applied to solid axles made of EA1N steel, three of them were prepared by introducing, in different sections (Figure 3), a series of artificial defects obtained by EDM. They were characterized by both concave (representative the fretting-fatigue cracks observed at press-fit seats) and convex (representing the classical fatigue cracks observed on the body and along geometrical transitions) shapes. In particular, each section was characterized by three defects of the same shape, but different reflecting area, angularly spaced by 120°.

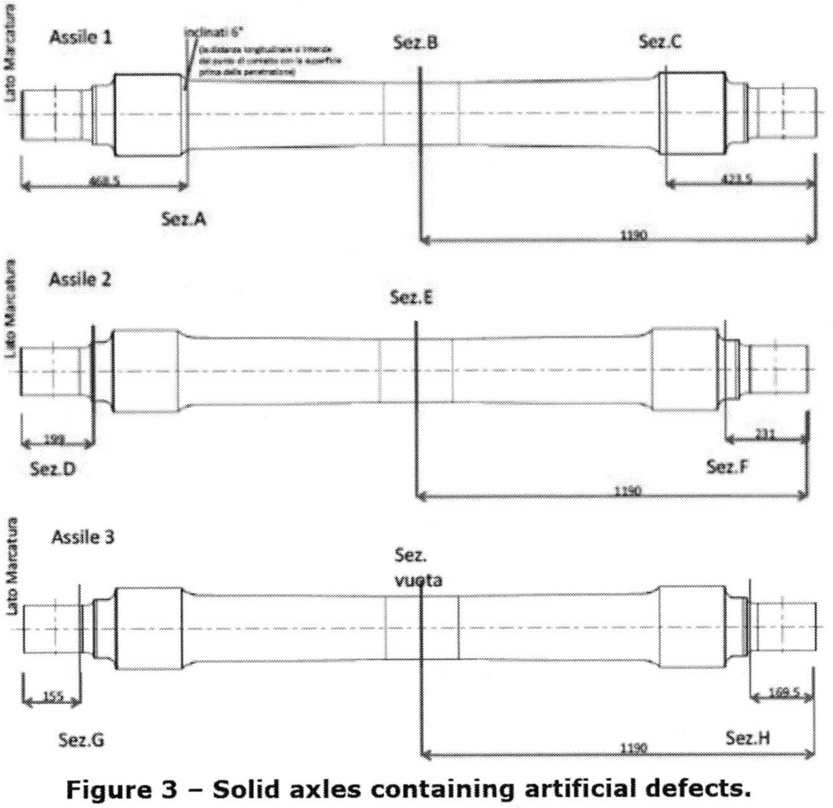


**Figure 3 – Solid axles containing artificial defects.**

All defects were inspected by the present rotating probe, connected to a digital Gilardoni RDG500 UT flaw detector, equipped with piezoelectric crystals having diameter of 20 mm and a nominal frequency equal to 4 MHz. Longitudinal waves were employed and their speed was set to 5900 m/s. The refraction angles chosen for the inspection of the different sections containing artificial defects were:

- convergent longitudinal waves at 6°: Sec. B, Sec. E
- convergent longitudinal waves at 16°: Sec. A, Sec. C
- convergent longitudinal waves at 30°: Sec. D, Sec. F
- divergent longitudinal waves at 6°: Sec. G, Sec. H

For each defect, the echo amplitude was recorded as a function of its reflecting area.

Since very few experimental data can be acquired, for each section, from the sample blocks shown in Figure 3, a MAPOD approach could be actually effective. In the present study, all the numerical simulations were carried out by the dedicated software CIVA$^{nde}$ 11 (8). The model was previously validated by simulating all the available experimental cases (7). The "complete" MAPOD approach (6) was then applied. In particular, the applied numerical procedure is essentially based on the Monte Carlo statistical method: before each simulation, a value of the source (or sources) of variability is randomly extracted from the appropriate statistical distribution used to define the numerical set-up. In the end of all the programmed simulations, the obtained results should be reasonably representative of the effects of the selected source of variability on the results themselves.

It is now necessary to point out some peculiarities of the rotating probe. It is designed so that each transducer can inspect an area of the axle different from the others and, therefore, every critical zone can be observed by only one of the available transducers. It can be concluded that the characterization of the reliability of the rotating probe consists in the collection of POD curves related to each transducer mounted in it. For simplicity, it will be analyzed here, without losing generality in the description of the method, the performance of just two sections (B + E and G). Figure 4 shows the obtained results. In both cases, for brevity, only one source of variability was considered: the location of the defect along the longitudinal coordinate of the axle. Moreover, for both scenarios, such a position was characterized by a uniform statistical distribution. This was possible due to the following considerations:

- in the case of section B+E (Figure 4a), the initiation of fatigue cracks occurs in correspondence of in-service damages such as corrosion pits and impacts from ballast. It is not possible to predict the position of occurrence of such damages and, as a consequence, all the active sections of the axle should be described by the same probability of occurrence;
- in the case of section G (Figure 4b), the damage occurs in terms of fretting-fatigue at the press-fit seat at a maximum distance of 20 mm from the edge of the press-fit. It is again not possible to provide a preferential position, but all those within the 20 mm will have the same chance of seeing the initiation of a crack.

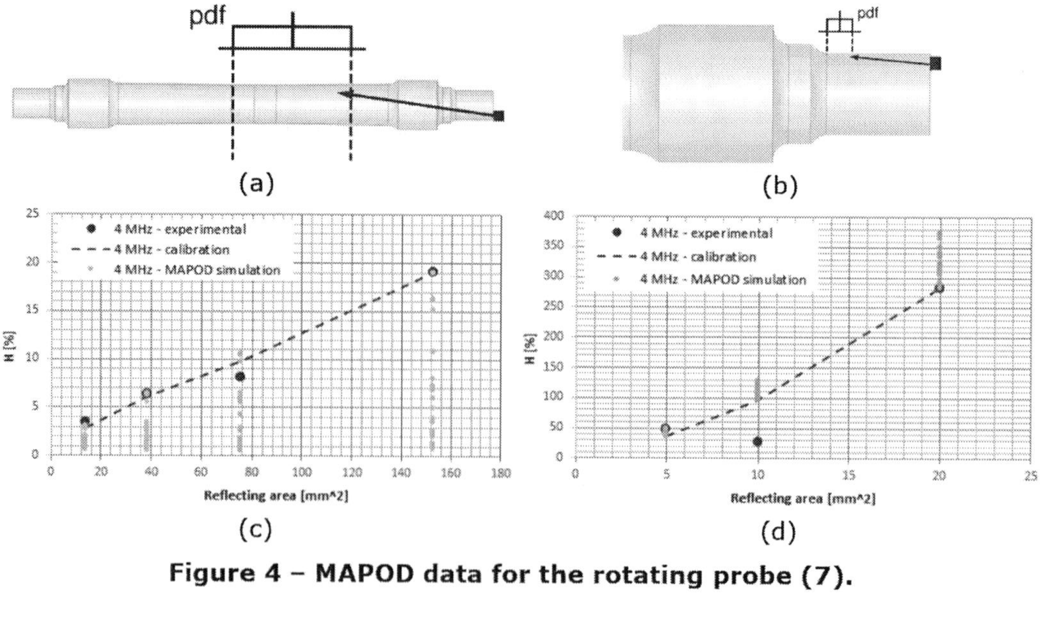


**Figure 4 – MAPOD data for the rotating probe (7).**

For each dimension of experimental artificial defect, thirty random values of the source of variability were extracted and as many simulations were run, for a total of 240 simulations for section B+E (Figure 4c) and 180 for section G (Figure 4d). The choice of thirty extractions for each scenario derives from the statistical theory that requires a minimum of 29 findings for the definition of the 95% confidence band.

The "signal response" experimental data shown in Figure 4 were statistically elaborated by means of Berens' approach (9). A defect is considered as detected if its response exceeds some predefined decision threshold corresponding to the response of the defect to be detected with 50% probability. The choice of the decision threshold is therefore one of the most critical points in the definition of POD curves. In this work, it was assumed equal to the value of the UT response obtained, by CIVA$^{nde}$ simulations, from the sample defect defined in Lucchini RS's procedure (3) for calibrating the rotating probe. Figure 5 shows the experimental POD curves obtained for sections B+E and G. No confidence bands are given due to the very small available experimental sample size.

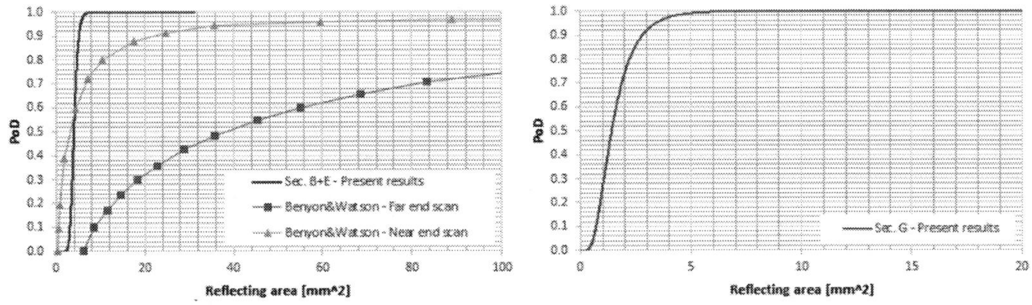


**Figure 5 – POD curves for the present rotating probe.**

Figure 5 shows also a comparison of the POD curve obtained at section B+E with the traditional POD curves derived by Benyon and Watson for solid axles (10). It is necessary to add that several details behind the derivation of the experimental curves by Benyon & Watson are not publicly known, but it is possible to conclude that the here-obtained POD curves are systematically better than the "far end scan" inspection and comparable with the "near end scan" one.

Numerical results allowed then to draw some conclusions about the calibration block adopted for the present rotating probe. As can be seen in Figure 4c and Figure 4d, the experimental UT responses seem to lie on the upper bound of the numerical ones for section B+E and on the lower one for section G, so resulting non conservative for the former and too conservative for the latter. This observation prompted the need for a further optimization of the sample block and the rotating probe itself, as described in the following.

## 3 OPTIMIZED POD CURVES OF THE PRESENT ROTANTING PROBE

A novel sample block specifically designed and manufactured by an EA1N freight axle (Y25 type boogie) characterized by a helical arrangement of planar flaws machined by EDM (Figure 6), was experimentally used to optimize the rotating probe technique. All the flaws are 2 mm deep and 10 mm long (convex shape), they are longitudinally spaced of 50 mm along the axle with 45° relative angular position. This positioning of the flaws guarantees there are no mutual interactions between the incident sound waves. For the sake of simplicity, the block has no threaded holes at its ends.

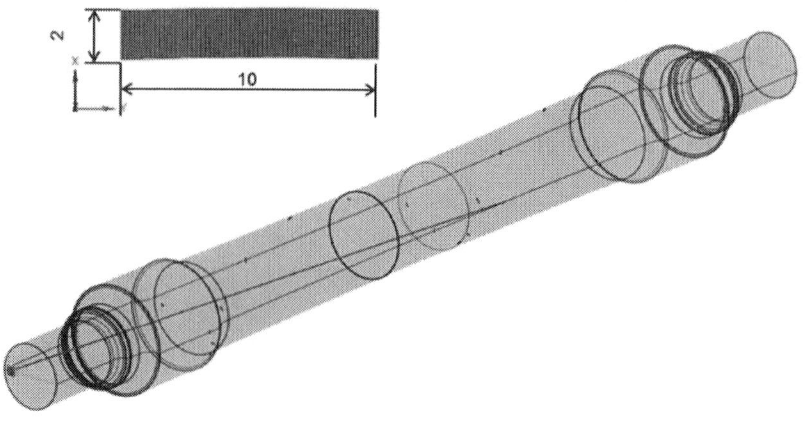


**Figure 6 – Novel calibration block for the rotating probe.**

### 3.1 Characterization of the structural attenuation in EA1N

Before carrying out the experimental tests on the novel sample block, the structural attenuation of EA1N was characterized by the methodology described in (11). In particular, multiple backwall reflections were acquired from a cylindrical specimen made of EA1N steel (Figure 7) using a conventional longitudinal wave transducer (GE B4S, D = 24 mm, 4 MHz). The frequency of the testing probe was chosen so to be the same of the rotating probe's transducers. The law describing the dependence of the structural attenuation on the frequency is (11):

$$\alpha(f) = \alpha_0 \cdot \left(\frac{f}{f_0}\right)^{\eta} \qquad (1)$$

In order to estimate the empirical parameters $\eta$ and $\alpha_0$, the test was also repeated using a probe with a different central frequency (GE B5S, D = 24 mm, 5 MHz).

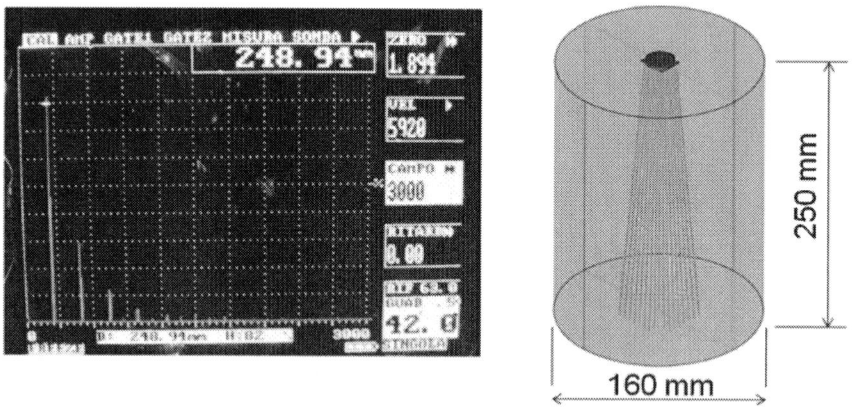


**Figure 7 – Characterization of structural attenuation.**

Assuming the EA1N steel as an isotropic and homogeneous medium, a constant decrease of the response $\Delta dB_{i,structural}$ between two adjacent backwall echoes is expected due to structural attenuation. The decay law, however, is the sum of a term related to the effect of structural attenuation and of another one related to the geometrical divergence of the sound beam, which is well described in the literature (11) and by the DGS diagrams provided by the manufacturers of the probes.

Consequently, $\Delta dB_{i,structural}$ can be determined subtracting the geometrical term $\Delta dB_{i,geometrical}$ from the total measured decrease $\Delta dB_{i,TOT}$:

$$\Delta dB_{i,structural} = \Delta dB_{i,TOT} - \Delta dB_{i,geometrical} + \gamma \quad (2)$$

where the term $\gamma$ stands for the measurement error and represents the variability of the measurement parameters governed by the operator, the error of invasiveness linked to the presence of the probe on the surface of reflection and the partial reflectivity of the bottom surface for the effect of surface roughness. This error term is here neglected due to the averaging operation applied to the measurements. In particular, six consecutive backwall echoes were acquired allowing the determination of five attenuation values (Figure 8). The value of the attenuation parameter $\alpha$ was then obtained as their average and resulted to be equal to 13.97 dB/m. Such a value is to be considered a conservative estimate, i.e. a "worst case", because keeps into account for the old axles produced by obsolete processes and is higher than the nowadays-typical value obtained by Lucchini RS during the qualification of axles during production, which is equal to 9 dB/m. Anyway, the here-derived value will be introduced in numerical simulations to build a correct description of the physical model.

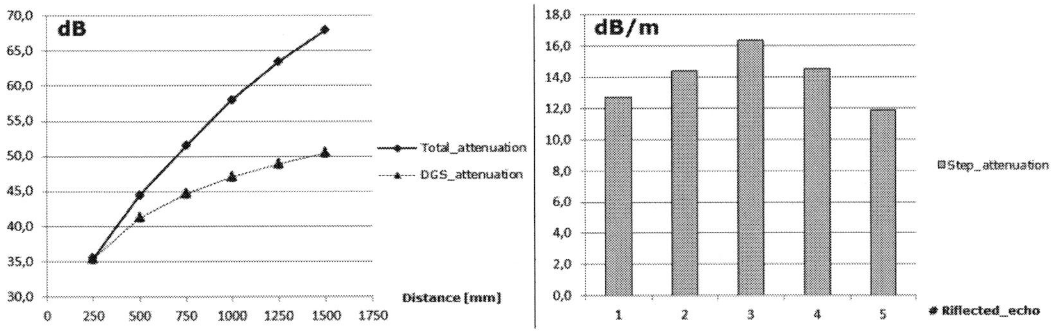


**Figure 8 – Structural attenuation in EA1N steel grade.**

### 3.2 Experimental tests on the novel sample block

The rotating probe was first calibrated according to (3) and using a calibration block with four different flaws (Figure 9). Every transducer of the probe requires an independent calibration and uses just one of the flaws. The dimension and the position of each flaw are specifically designed to set the correct sensitivity level for the corresponding region of the axle. Figure 10 then shows the implementation of the experimental tests on the novel sample block.

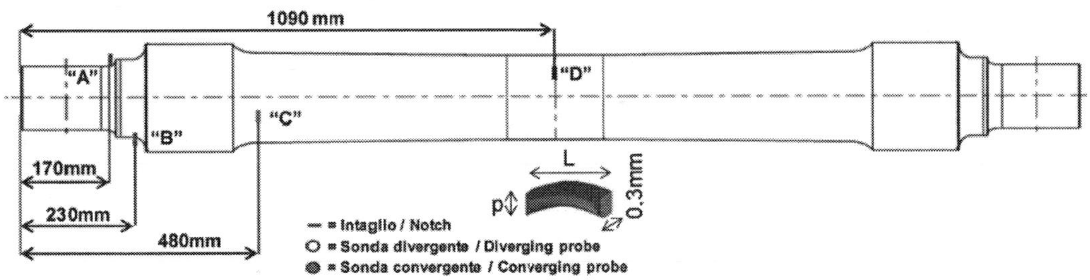


**Figure 9 – Calibration block for the rotating probe.**

**Figure 10 – Experimental tests on the novel sample block.**

Figure 11 shows the experimental evidence of those transducers of the rotating probe able to inspect the central body of the sample block. For the sake of brevity, in the following only these transducers will be considered, keeping in mind that all the others were succesfully optimized, as well. As can be seen, the trend of the responses is very similar between the different transducers, as it is the gain needed to obtain an echo amplitude at the 80% of the screen.

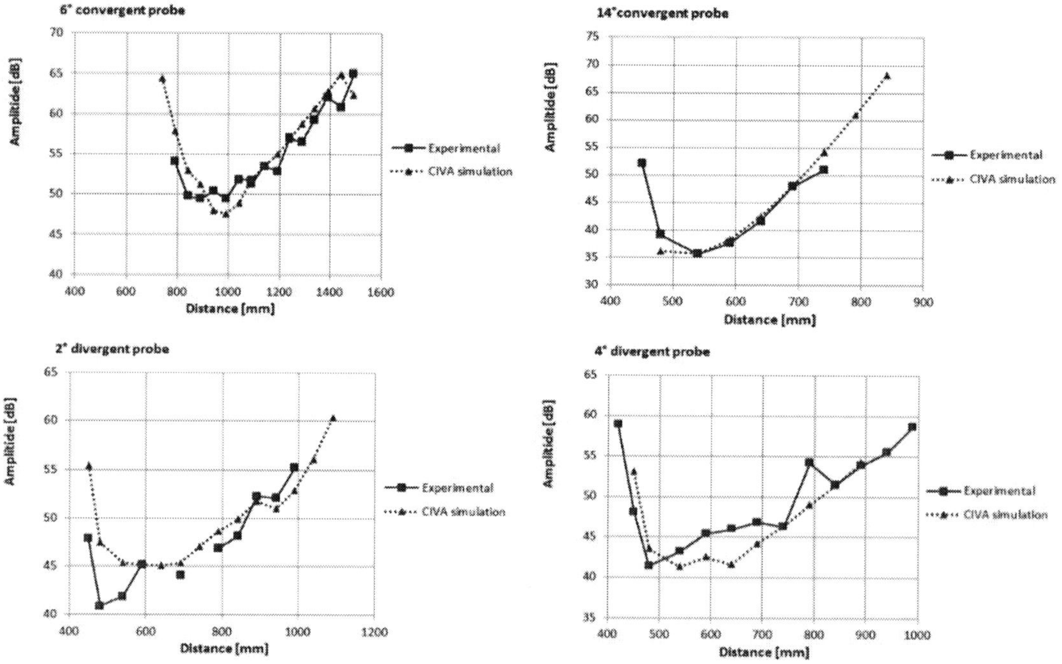


**Figure 11 – Experimental results and comparison with numerical simulations.**

### 3.3 Numerical simulations of the novel sample block

Numerical simulations were carried out by CIVA$^{nde}$ 11 dedicated software package (8). Both the sample block (Figure 6) and the rotating probe were modelled considering all the needed factors, included the structural attenuation of the material determined in Section 3.1.

The comparison between experimental and numerical results is shown in Figure 11 and highlights a very good correlation. Moreover, Figure 12 shows the comparison of experimental and numerical scatters moving the defects along the longitudinal direction and within the active region of the considered probes. As can be seen, also in this case the correlation is very good and suggests the successful validation of the numerical model, which is adopted in the following in order to optimize the rotating probe.

### 3.4 Optimization of the rotating probe by means of MAPOD curves

By means of CIVA$^{nde}$ simulation software, a numerical MAPOD curve can be derived by setting a suitable statistical model of the considered inspection and adopting a Monte Carlo methodology on one or more stochastic parameters, as already described in (7). Here, the chosen characteristic parameter is the reflecting area of the defect, while the stochastic one is its longitudinal position within the active region of the 6° convergent probe, which is taken as an example of optimization in the following. A typical set of numerical results obtained by such a simulation is shown in Figure 12.

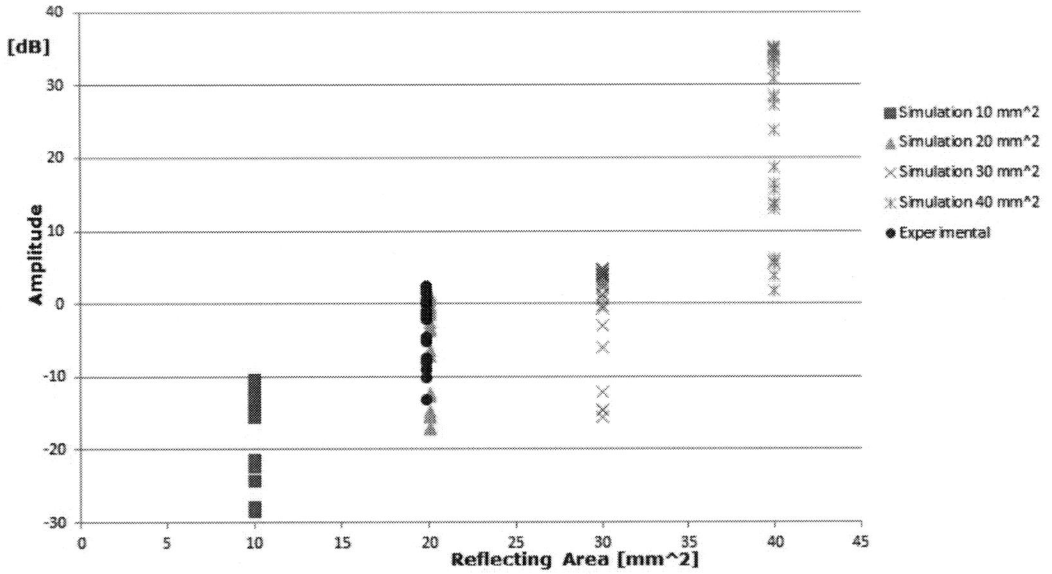


**Figure 12 – Experimental vs. numerical scatter.**

The optimization of the 6° convergent probe was then carried out by deriving the MAPOD curves for three different angles of the probe (6°, 7° and 8°) and looking for the best performance. It is worth mentioning that, for each selected angle, 30 stochastic extractions were simulated, so to be able to derive the 95% confidence band on a robust statistical sample. Figure 13 shows the results and it can be concluded that a better performance for the rotating probe would be achieved adopting a 8° convergent transducer instead of the 6° one.

Figure 14 shows the present experimental response, provided by the four examined transducers, to the flaws of the novel sample block. The aim of the optimization process is now to increase further the response of the defects so to flatten to the lowest possible energy level the pattern of responses for each individual transducer. This will be carried out in the next steps of the research thank to the MAPOD methodology described in the present paper.

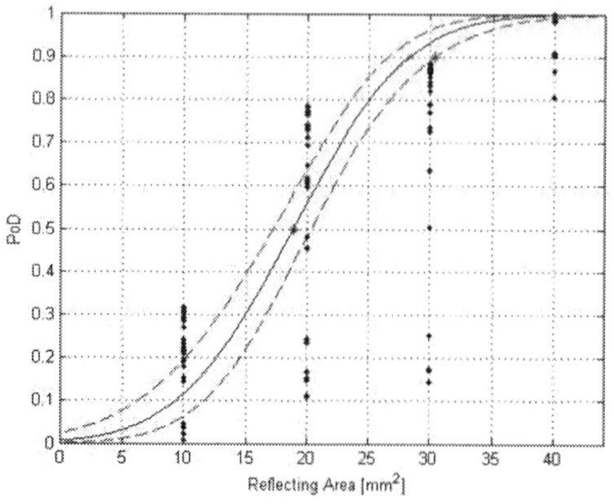


**6° convergent**

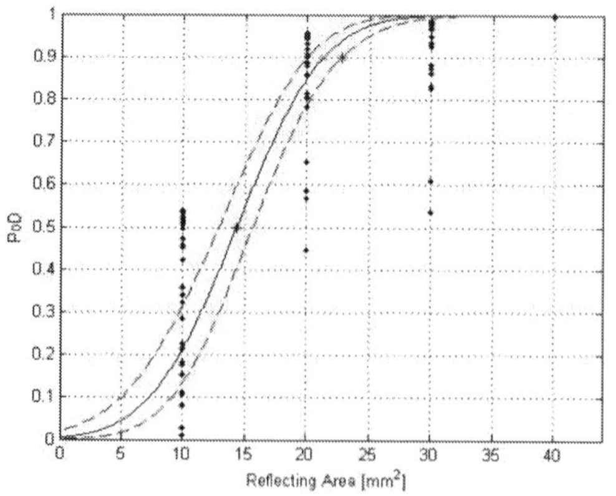


**7° convergent**

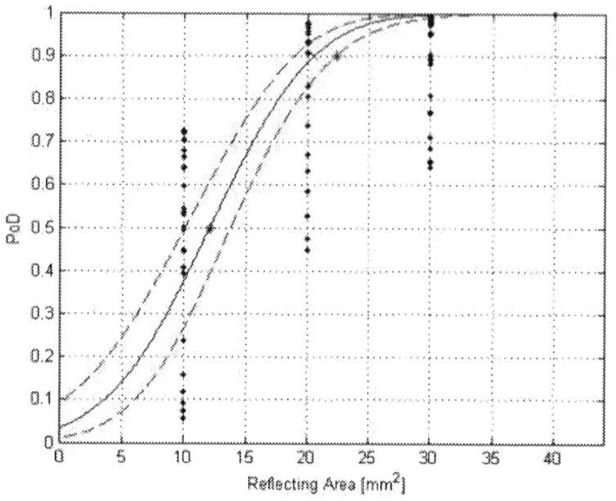


**8° convergent**

| Probe | $AREA_{90/95}$ $[mm^2]$ |
|---|---|
| 6° | 30.42 |
| 7° | 22.80 |
| 8° | 22.28 |

**Figure 13 – MAPOD curves for the optimization of the 6° convergent probe.**

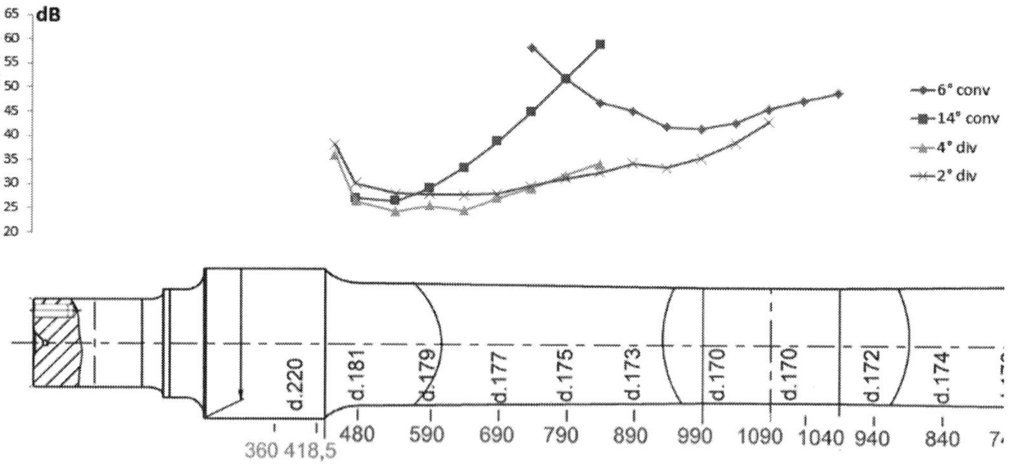


**Figure 14 – Present experimental response to artificial flaws.**

## 4 CONCLUDING REMARKS

A Model-Assisted Probability of Detection approach is here considered in order to optimize the present rotating probe for inspecting solid freight axles. An optimization of the calibration block for the present rotating probe was successfully carried out considering both experimental tests and numerical simulations.

**ACKNOWLEDGEMENTS**

The authors are sincerely grateful to Mr. L. Girlanda for the help given to the present research.

**REFERENCE LIST**

(1) Cantini, S. & Beretta, S. (Editors) (2011) LRS-Techno Series 4.
(2) Grandt, A.F. Jr. (2003) John Wiley & Sons Inc., Hoboken, US.
(3) Benzoni, F., Cantini, S. & Tonelli, L. (2013) Technical Instruction QUA IT 142 Rev.1, Lucchini RS S.p.A., Lovere (BG).
(4) Georgiou, G.A. (2006) Research Report 454, HSE Books, Health and Safety, Executive, UK.
(5) ASM (1997) ASM handbook - Vol. 17: Non-destructive evaluation and quality control.
(6) Thompson, R.B., Brasche, L., Knopp, J. & Malas, J. (2006) Proc. Aging Aircraft Conference.
(7) Carboni, M., Cantini, S. & Gilardoni, C. (2013) $5^{th}$ European-American Workshop on Reliability of NDE.
(8) $CIVA^{nde}$ 11 (2013) User's manual.
(9) ASM. ASM handbook - Vol. 17: Non-destructive evaluation and quality control. 1997.
(10) Benyon J.A., Watson A.S., The use of Monte Carlo Analysis to increase inspection intervals, Proc. International Wheelset Congress, Rome, Italy, 2001.
(11) Krautkrämer, J. & Krautkrämer, H. (1990) Ultrasonic Testing of Materials, Springer.

# Robust multi-train trajectory optimisation for real world conditions using a 'noisy' genetic algorithm

**J Goodwin**[a], **D I Fletcher**[a], **R F Harrison**[b]
[a] Department of Mechanical Engineering, University of Sheffield, UK
[b] Department of Automatic Control & Systems Engineering, University of Sheffield, UK

In planning and managing rail network traffic there are aims such as minimising travel times and, increasingly, the consideration of how driving style can influence traction energy consumption. The optimisation of trajectories by defining the control points at which traction, coasting and braking should begin has been well studied for a single train. This research is already feeding into Driver Advisory Systems, but only recently has the problem has been extended to the multi-train case, where network interactions between different trains must be considered.

Optimisations are currently capable of predicting trajectories that will achieve their time or energy consumption aims if followed closely. However, in reality, railway networks are subject to many routine uncertainties (e.g. small differences in the traction characteristics of nominally identical trains, varying dwell times in stations, varying train loading) and also a limited accuracy of control application (i.e. differences in exact application point for traction or coasting). Suggested driving patterns that appear to be optimal can be vulnerable, producing severely sub-optimal outcomes under real world conditions, that is, they are not robust.

In this paper robustness of planned trajectories are investigated by introducing uncertainties, through probabilistic variation of the control point distances and the dwell times in stations, within a genetic algorithm (GA) based multi-train optimisation process. Trade-offs are explored between highly optimised trajectories that are unlikely to be achieved, and slightly less optimal trajectories that are robust to real world disturbances. The newly developed procedure is shown to find control strategies of significantly increased robustness.

## 1 INTRODUCTION

Due to increased traffic demands, the railway system in Great Britain has become increasingly busy and in some areas the capabilities of the system are restricting the number of people and trains that can be moved. This means trains are operating at increased frequency and as a consequence the likelihood of perturbations propagating across the network is increased. To avoid the reductions in service quality this could cause, a number of measures have been implemented across the network. In some areas physical network capacity has be increased through infrastructure improvement such as to track, stations, or signalling systems. More generally, improvements in operation have been incentivised through the setting of ambitious targets by regulatory organisations, and brought to bear through a system of financially penalties and incentives. However, another consideration becoming increasingly important is the desire to minimise the energy consumption of rail operations. This is driven by both environmental concerns linked to the combustion of fossil fuels and the rising cost of the energy itself, which is an increasingly significant incentive.

### 1.1 Optimisation in uncertain systems

Most trajectory planning work (for both single and multiple trains) is focussed on finding the optimal solution for a predefined schedule. Usually control seeks to maintain the scheduled traverse time for a length of track, whilst minimising the energy consumption. [1] The problem is constrained by the dynamic performance limits of trains and restrictions on velocity and headways (imposed to ensure safe operation). The traditional formulation of the muti-train trajectory planning problem can be formalised as:

$$f(\underline{\boldsymbol{X}}) \rightarrow \min \tag{1}$$

where $\underline{\boldsymbol{X}}$ is a control strategy for all trains on the network, and $f(\underline{\boldsymbol{X}})$ is a cost function usually based on the traverse times and energy consumptions of all train journeys

It is assumed that if the 'optimal' control strategy is identified, then it can be implemented resulting in optimal performance of the system. However, the 'optimum' is likely to lie on the limit of the feasibility boundary so will be very sensitive to noise. [2] In reality there are many small uncertainties that most current models do not consider. These mean that if most optimised control strategies (for a noiseless system) were applied to real operation then it is likely they would not perform as well as expected, and may in-fact result in severely sub-optimal outcomes. Most of these uncertainties fall into two different groups classified by Chen et. al. [3] as:

Type 1 – variations in performance caused by variations in uncontrolled parameters.

Type 2 – variations in performance caused by variations in the design (control) variables.

When controlling a train it is just not possible for all drivers to consistently follow a control strategy perfectly accurately. Automatic train control systems can achieve a very repeatable application of control, but modern systems deliberately make small variations in control to avoid problems with localised track wear in the positions where control actions are applied. Either way, there will be variation in the position of control point application equivalent to Type 2 uncertainty. Including this uncertainty the problem becomes:

$$f(\underline{\boldsymbol{X}} + \delta) \rightarrow \min \tag{2}$$

where $\delta$ is the uncertainty in the in control application

Likewise, in the context of train control, Type 1 uncertainties may be caused by variation in the properties of the vehicles, track being traversed or movement authority given. For example, even when composed of the same types of rolling stock, the mass, resistance, traction and braking characteristics of nominally identical trains will vary slightly. Also, some characteristics of the line, for example, the level of adhesion or overhead line voltage, are unlikely to be constant and will be affected by external factors. The combination of these uncertainties in train and track means that the expected rate of acceleration or deceleration of a train will not be known precisely in advance. Other external factors can also influence the definition of the problem being solved. Variation in station dwell times may extend or reduce the target traverse time for a journey, while equipment failures may cause large delays and may even stop services altogether. Including both Type 1 and 2 variations the problem becomes:

$$f(\underline{\boldsymbol{X}} + \delta, a) \rightarrow \min \quad \textbf{(3)}$$

where $a$ represents the changing operating conditions of the system

The function $f(\underline{\boldsymbol{X}} + \delta, a)$ has an expectancy distribution, $F(\underline{\boldsymbol{X}})$, found by integration over $\delta$ and $a$. Since the optimum of $f(\underline{\boldsymbol{X}})$ is not necessarily the same as the 'robust optimum' of $F(\underline{\boldsymbol{X}})$, there is often a trade-off between the quality and the robustness of solutions. [4]

### 1.2 Techniques used in 'noisy' genetic algorithms

It has been widely noted that GAs can still effectively optimise systems in the presence of noise. Arnold [5] found that they performed particularly well on problems with both high dimensionality and high noise, when compared to a number of other direct optimisation strategies. This ability to find robust solutions in the presence of uncertainty has been demonstrated in many papers (see [4] for a comprehensive review), which consider both Type 1 and 2 uncertainties. In many practical situations the analytical form of the expectancy distribution, $F(\underline{\boldsymbol{X}})$, is not known but numerical approximations can be used to estimate its statistical properties. Each time the fitness function in Eq (3) is evaluated its value will vary stochastically, but the mean expected value $\hat{F}(\underline{\boldsymbol{X}})$ can be estimated by averaging a number of random samples.

$$\hat{F}(\underline{\boldsymbol{X}}) = \frac{1}{N}\sum_{i=1}^{N} f(\underline{\boldsymbol{X}} + \delta, a) \quad \textbf{(4)}$$

where $N$ is the number of times a control strategy is resampled

This is referred to as *explicit resampling* as each candidate solution is evaluated multiple times. The larger N is used, the more likely that $\hat{F}(\underline{\boldsymbol{X}})$ is close to the actual mean but since evaluation of candidate solutions is usually the most computationally expensive part of an optimisation there is a balance between getting a good enough estimate of the mean expected fitness against the effort spent achieving this. However, GAs don't just operate on a single candidate solution, but simultaneously on a population of candidate solutions. Clustering of candidate solutions tends to occur in promising areas of the search space making it likely that a number of similar candidate solutions are evaluated during each iteration of the algorithm, thus reducing the influence of noise. This is known as *implicit resampling*. In the extreme case (of infinite population) Tsutsui showed that noise does not affect GAs that use a fitness proportional selection scheme. [6] So, the effect of noise may be reduced either by explicit resampling (more evaluations of each candidate solution) or implicit resampling (increasing the population size).

## 2 METHODS

A previous multi-train trajectory planning model described in [7] used a Genetic Algorithm (GA) to seek a near optimal control strategy for multiple trains operating on a simple railway network. This optimisation was extended in [8] and shown to consistently find good solutions for the system studied. However, these studies and similar work to-date by other authors did not consider Type 1 or 2 uncertainties.

### 2.1 Uncertain control point application

Adapting the model described in [8] to introduce Type 2 uncertainty into the system was achieved by temporarily adding a zero mean normally distributed random distance to the position of each control point (CP) for the duration of each simulation. Instead of the next control action being applied as soon as the train passed the distance specified in the control sequence the control action was applied with the specified level of uncertainty (Figure 1).

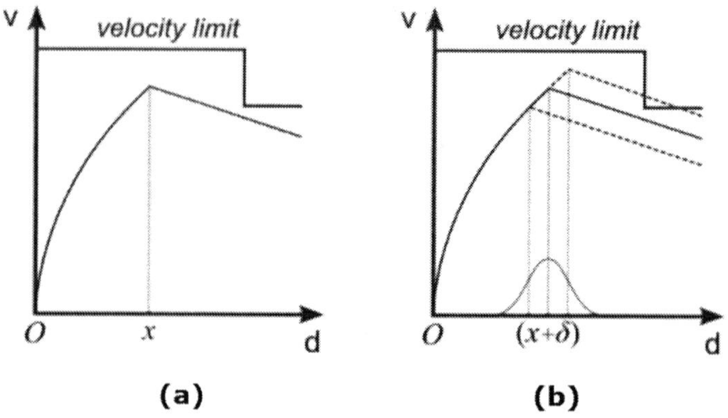


**Figure 1 Effect of uncertainty in control point application on train velocity trajectories: (a) When applied without any noise, a control strategy may consistently pass close to a feasibility boundary without the possibility of violations arising. (b) When implemented with noise the same control strategy may break the safety constraints, which will result in it being evaluated as 'invalid' for that simulation.**

However, this modification led to two closely related situations becoming possible, both of which required alterations to the algorithms discussed in [8].

Firstly, each time a candidate control strategy was simulated there was a chance that the trajectories taken did not maintain safe operation (either by exceeding the speed limits or headway constraints). To discriminate against invalid candidates and maintain the same population of valid candidate solutions in the modified optimisation, any candidate solution evaluated as invalid was discarded and a new candidate probabilistically selected to replace it. This policy of re-selection (and subsequent re-simulation) is equivalent to using explicit resampling to find the expected probability of a candidate being valid, biased towards greater resampling of high performing candidate solutions. Instead of the probability of a candidate solution passing to the next generation being proportional to its expected fitness rank, the total probability is now proportional to its expected fitness rank multiplied by the probability that it is valid.

Secondly, it is possible that the 'best' control strategy found by the optimisation in fact relies on very specific deviations on control point application. These deviations would be unlikely to reliably occur in reality which could easily result invalid operation if the control strategy was implemented without these deviations present. To screen out these cases the final population was re-simulated without any noise and any candidates found to be invalid were discarded.

**2.2 Uncertainty of dwell times at stations**

Although the running times of trains are themselves subject to some variation this is within the control of the drivers or dispatchers in most cases, whereas station dwell times are inherently unpredictable and harder to control. [9] This is largely due to the boarding and alighting of passengers, the speed of which is affected by many uncontrollable variables, such as the number and configuration of train doors, and the configuration of the platform to reduce congestion between alighting and joining passengers. In addition to the fleet characteristics there can also be specific local issues such as the familiarity of travellers with the journey, the rate of passenger arrival at the station, and the volume of luggage being carried. For example, an airport station having mostly occasional travellers carrying large items of luggage would be expected to have different dwell time characteristics to a station with mainly daily commuters without luggage (cf. Figure 2(a)).

Variation in station dwell times was introduced into the model as an example of Type 1 uncertainty. Similar to the introduction of CP noise, uncertainty in the dwell time was implemented by replacing deterministic dwell times with stochastic ones. Modification of the existing algorithms in [8] enabled the optimisation to continue functioning effectively. The actual departure time of a train is now given by Eq (5).

$$\text{departure time} = \max\begin{pmatrix}\text{arrival time + stochastic dwell time,}\\ \text{arrival time + minimum dwell,}\\ \text{scheduled departure time,}\\ \text{headway conflicts resolved}\end{pmatrix} \tag{5}$$

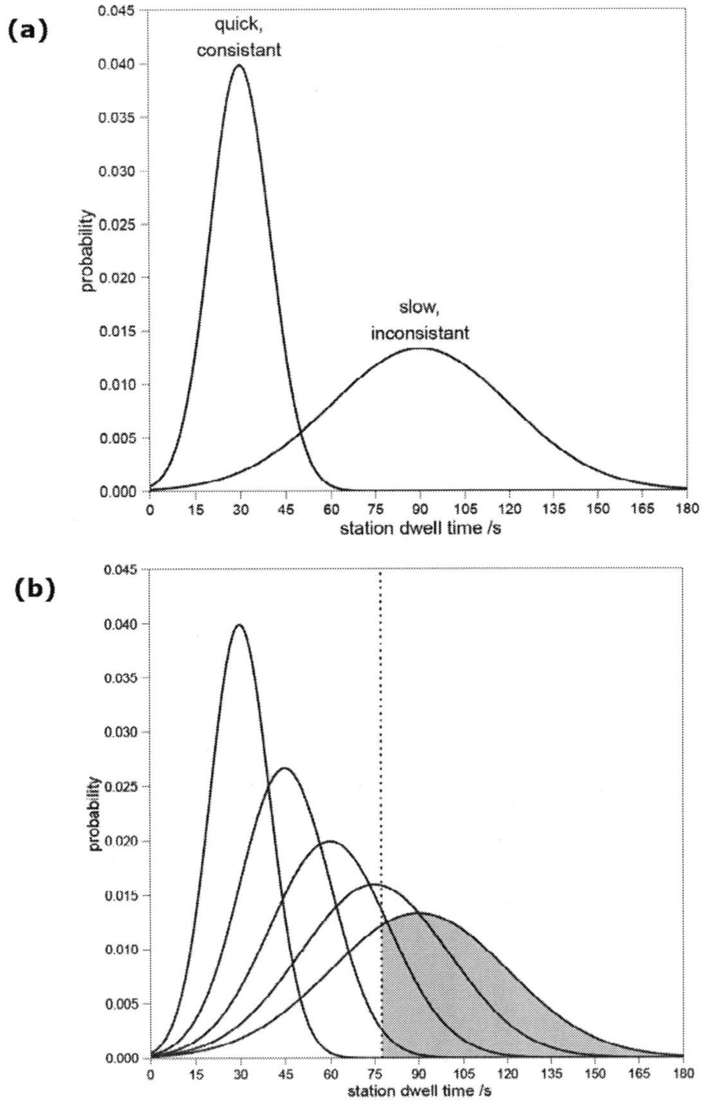


**Figure 2 Distribution in dwell times. (a) Illustrates the two distribution extremes. (b) The probability distributions used in this system when introducing noise in the dwell time. The situation dependant minimum dwell time (dashed vertical line) is dependent on the minimum possible dwell time, the scheduled departure time, and train dispatching. Eq (5) is used to combine these constraints with a dwell time probability distribution (in this case, mean = 90 s, sd = 30 s) to give the distribution of possible dwell times (shaded area) possible in that specific situation.**

Different studies suggest different ways to represent station dwell times, including: normally distributed [10], log-normally distributed [9], Weibull distributed [11]. For simplicity a normal distribution is assumed in this study, and implemented by stochastically varying the station dwell time. To enable the uncertainty in dwell time to be varied smoothly between the two extreme examples shown in Figure 2(a), the standard deviation was chosen to always be one third of the mean, with distributions having mean = [30, 45, 60, 75, 90] s and standard deviation = [10, 15, 20, 25, 30] s respectively. This resulted in the series of distributions shown in Figure 2(b), although in real application of the model these could be tuned to the characteristics of a specific station, line, fleet or known passenger behaviour. It is worth noting that after combining this with operational constraints (introduced in Eq (5)) the actual distribution of dwell times will end up being more positively skewed (see shaded area in Figure 2(b)), and so agree better with the majority of these other studies.

Current operational practice is more accurately reflected by defining a schedule, rather than the target traverse times used in [8]. This meant it was necessary to adopt a different cost function, show in Eq (6).

$$\text{cost function} = c_T\left(\sum_{i=1}^{M} \min(0, \text{arrival time} - \text{scheduled arrival})\right) + c_E\left(\sum_{i=1}^{M} \text{energy consumed on journey}\right) \quad \textbf{(6)}$$

where $c_T$ is the financial cost of arriving late, $c_E$ is the cost of energy, and $M$ is the total number of journeys taken on the network

The financial cost of arriving late ($c_T$) is similar to the current UK concept of 'delay minutes', but without any attribution to the causer of the delay. When optimising train movement of the system as a whole, attributing delay is no-longer meaningful as it is the total cost for the system which is being minimised even if this may disadvantage some particular services. This is an implementation issue beyond the scope of the work presented here, but it is expected that better estimates of optimal system performance will help quantify the costs attributed to different parties in the event of disruption.

One further modification necessary to make the GA procedure function well in a noisy environment was the method of identifying the best candidate solution found. Without noise, the best candidate solution was simply the control strategy that was evaluated as having the lowest cost function score. This 'all-time top scoring' control strategy could be found at any time during the optimisation progress. However, in a noisy system the best scores no longer identify the candidates with the best expected score, but instead were often outliers from candidates with fairly mediocre expected scores. For example, in a system where extended dwell times are introducing large delays, there is still a small chance that all dwells times will be small, so the solution evaluated under these improbable conditions will have an artificially high score. To overcome this, the expected mean cost and probability of being valid was estimated explicitly (using Eq (4) with N = 50) for all the individual candidate solutions in the final population only. These measures were then combined, and the control strategy with the lowest (expected cost / expected probability of being valid) chosen as the output of the optimisation. This method of identifying the best control strategy is similar to the effective selection pressure on the population used during to optimisation process. Explicit averaging over the final population required an additional (N*population_size) evaluations to be carried out, but was found to greatly improve the quality of the final candidate solution chosen.

## 2.3 Model parameters

The underlying model, optimisation parameters, and system being modelled, are identical to those used in [8] (optimisation G4, network N1), except for the modifications explained above. The same number of train routes and stops were retained but, given the use of a new cost function in Eq (6), a target schedule was defined based on a scheduled dwell time of one minute (Table 1).

**Table 1 Target schedule for the system modelled.**

| | Scheduled time /s | | | | |
|---|---|---|---|---|---|
| | **First stop** | | **Second stop** | | **Third stop** |
| **Train** | (arrive) | depart | arrive | depart | arrive |
| 1 | (-60) | 0 | 720 | 780 | 1500 |
| 2 | (-60) | 0 | 720 | 780 | 1500 |
| 3 | (-60) | 0 | 660 | 720 | 1380 |

The GA parameters used in this investigation are shown in Table 2. As is often the case with GAs, no systematic optimisation of parameters has been conducted. However, the GA performance achieved is more than sufficient to demonstrate the validity of the proposed technique for increased robustness. Eiben observes that parameters may have different optimum values throughout the optimisation process, and suggests that a suboptimal choice of parameter function (varying with the number of generations) can often lead to better results than a suboptimal choice of a 'rigid' parameter. [12] Illustrative of this, most GA parameters were kept constant as the optimisation progressed but it was found that a linear decrease in the size of mutation, with each generation, yielded improved performance.

**Table 2 Model and GA parameters used in this investigation.**

| **Parameter** | **Value** |
|---|---|
| M (initial size of mutation) /m | start = 200<br>end = 0 |
| $\alpha$ (roulette wheel selection) | 0.05 |
| population_size | 100 |
| number of generations | 200 |
| $c_T$ (cost of delay minutes) /£/min | 100 |
| $c_E$ (cost of energy) /£/kWh | 0.15 |
| N (re-samplings when explicitly averaging) | 50 |
| minimum_dwell /s | 30 |

Although only nominal estimates of dwell times, noise levels, $c_T$ and $c_E$ are used the overall patterns of results observed is expected to remain the same if different, more system specific, values were used instead.

Since optimisations by GAs are not deterministic, 10 repeats were carried out for each combination of training noise being investigated. After completion of each optimisation, the performance of the chosen control strategy was then explicitly estimated, using Eq (4) (N = 50), at each different level of utilisation noise. It is important to draw a distinction between *training noise* (the level of uncertainty during the optimisation process) and *utilisation noise* (the level of uncertainty in the system where the optimised control strategy is applied). The system was first optimised without any training noise, with results given in section 3.1. Increased levels of training noise in control point (CP) accuracy and dwell time were introduced separately in sections 3.2 and 3.3 respectively, to investigate their effect on cost and probability of validity over a range of utilisation noise level. Finally, the combined performance with a high training noise in both control point accuracy and dwell time simultaneously is investigated in section 3.4.

## 3 RESULTS AND DISCUSSION

### 3.1 No training noise

The first series of optimisations were carried out without any training noise during the optimisation process, the results from which are shown in Figure 3. This is equivalent to a conventional optimisation developed for application in 'perfect' conditions. With that in mind, it would be expected that it's performance when utilised with no control point or dwell time noise would be good, but would deteriorate markedly under real conditions.

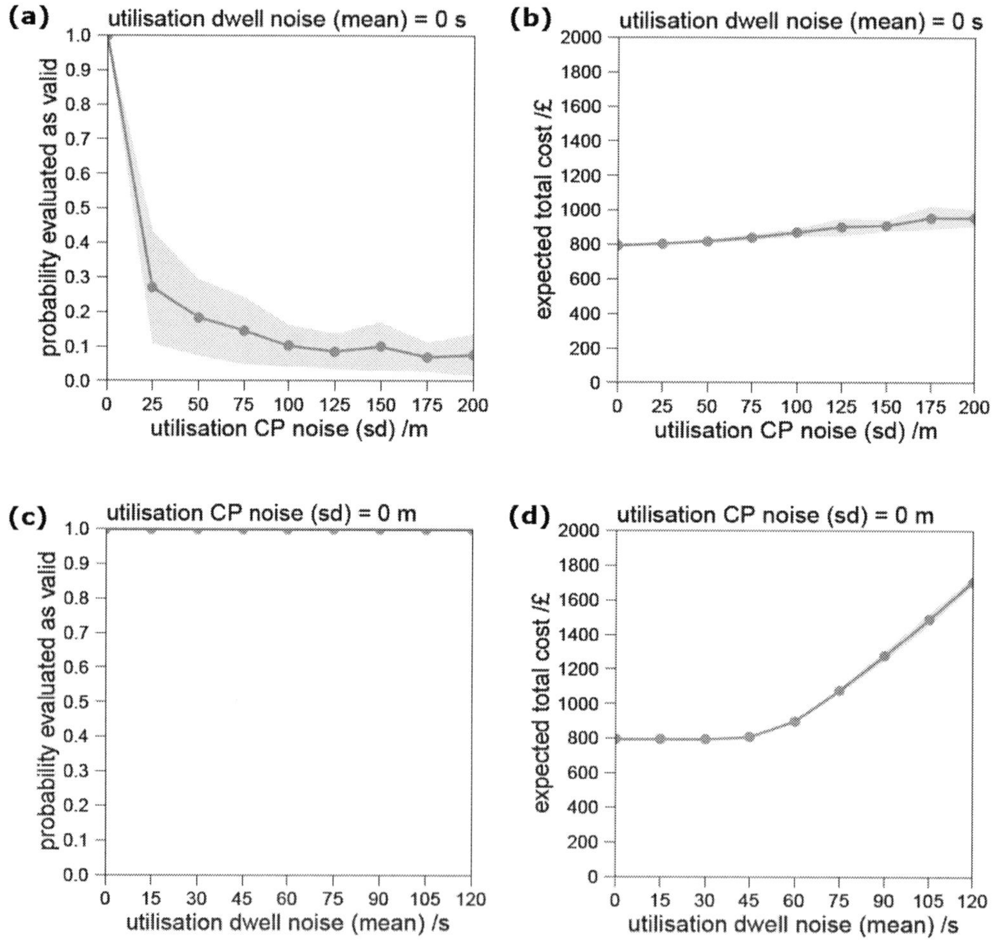


**Figure 3 The average robustness of control strategies found under conditions of zero training noise is investigated by simulating there utilisation with different noise levels. Shaded areas show the standard deviations. Variation in control point (CP) utilisation noise on: (a) the probability the control strategy is evaluated as valid, (b) the expected cost of operations. Variation in utilisation dwell time noise on: (c) the probability the control strategy is evaluated as valid, (d) the expected cost of operations.**

From Figure 3(a) it can be seen that the probability of an optimised control strategy being valid drops quickly as the variation in CP application is increased, and stays at a low level. Given that there are sections where the line speed drops below the holding speed, and therefor restricts the optimisation, we would expect the optimum trajectory to pass as close to these speed limits as possible. Even when a small amount of CP noise is introduced during the utilisation of a control strategy the probability of speed limit violations occurring (through a situation similar to the one illustrated in Figure 1) becomes large. Far from being surprising, this lack of robustness is exactly the behaviour we would expect from near optimal solutions to the noiseless problem. Similarly, above a certain threshold, increases in utilisation dwell time mean and spread have an almost linear effect on the expected cost of a control strategy, i.e. increasing poor performance. Again, this behaviour can be rationalised for the system being studied. The scheduled dwell times are 60 s, so the stochastic dwell time has to have a significant chance of being greater than this in order for any delay (and therefore additional cost) to be incurred.

In the system studied, the utilisation dwell time distribution appears to have no effect on the validity of control strategies, Figure 3(c), and the variation in utilisation CP noise has only a small influence on the expected cost of a control strategy, Figure 3(d). These relations did not significantly change for different combinations of training noise so are not discussed further.

### 3.2 Control point application training noise

The second series of optimisations were carried out with different levels of control point application training noise during the optimisation process, the results from which are shown in Figure 4.

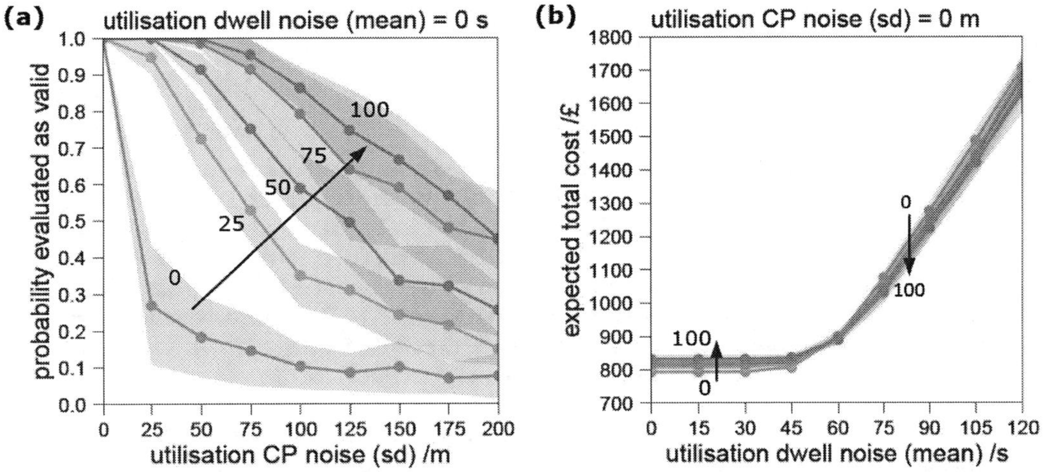


**Figure 4 Control strategies were optimised using different levels of control point (CP) training noise (labelled on the figure), but without any dwell training noise, during the optimisation process. Average robustness is investigated by simulating their utilisation at different noise levels. Shaded areas show the standard deviations. (a) Variation in the probability that control strategies is valid with control point (CP) noise level during utilisation. (b) Variation in the expected cost of operations with dwell time noise level during utilisation.**

Figure 4(a) shows that increasing the CP training noise leads to the optimisation finding control strategies that are significantly more robust to variation in CP application. However, from Figure 4(b) it can be seen that when utilised at a zero CP and dwell noise there is a very small increase in cost of having this increase in robustness. Even in the worst case of this, seen at a utilisation CP noise of 100 m, increasing the CP training noise from 0 to 100 m causes the probability of the control strategy being evaluated as valid to increase from 0.10 to 0.86 (860%), but the expected cost (utilisation CP noise = 0 m) to increase by only £38 (4.8%).

It can also be seen from Figure 4(b) that increased CP training noise also causes the secondary benefit of a slight increase in the robustness of the system to variation in station dwell times. As will be discussed in section 3.3, this is likely to be due to increased recovery time being introduced into the journeys. Introducing one type of noise can lead to the system becoming more robust to other types of noise that were not considered during optimisation.

### 3.3 Station dwell time training noise

The next series of optimisations were carried out with different levels of dwell time training noise during the optimisation process, the results from which are shown in Figure 5.

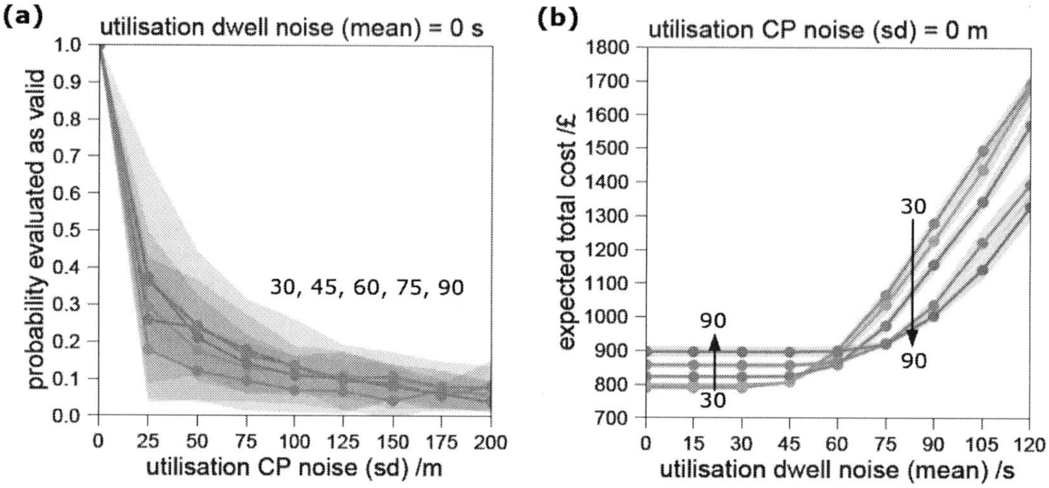


**Figure 5 Control strategies were optimised using different levels of dwell training noise (labelled on the figure), but without any control point (CP) training noise, during the optimisation process. Average robustness is investigated by simulating their utilisation at different noise levels. Shaded areas show the standard deviations. (a) Variation in the probability that control strategies is valid with control point (CP) noise level during utilisation. (b) Variation in the expected cost of operations with dwell time noise level during utilisation.**

It can be seen from Figure 5(b) that increasing the dwell time training noise leads to solutions that have a lower expected total cost when utilised at high dwell noise (ie. are more robust). However, this increase in robustness is accompanied by increased expected total cost when utilised at low dwell noise. This suggests that the optimisation is working effectively, because the solutions being found are well suited to the environment they were trained for, but that care should be taken to make sure that the training noise matches the noise level at which the solution with be utilised. This is highlighted in Table 3 where the performance when the utilisation noise matches the training noise is found to be much better than when the utilisation noise is fixed but the training noise varied.

**Table 3 The expected benefit of from using different levels of training noise varies depending on how well they match the noise level at which the optimised control strategy is utilised.**

| | | Training dwell noise (mean) /s | | | | |
|---|---|---|---|---|---|---|
| | | **30** | **45** | **60** | **75** | **90** |
| Expected % cost saving (relative to training dwell noise = 0) | **utilisation noise = training noise** | 0 | 0 | 4 | 14 | 21 |
| | **utilisation noise = 0** | 0 | -1 | -4 | -8 | -13 |

The reason for this trade-off, between the cost of solutions (utilised at dwell noise = 0) and their increased robustness, can be understood by looking at the traverse times of the trains and the corresponding energy consumptions. For convenience the notation (CP_dwell) will be used to describe the training noise levels used during optimisation. The average scheduled journey time in the system modelled is 700 s and it can be seen from Table 4 that the average journey time when both training CP and dwell noise are zero gives less than one second recovery time. This results in the lowest average speed and therefore the lowest energy consumption. If the dwell training noise is increases to 90 s (0_90) than the average recovery time increases to about 33 s per journey.

**Table 4 Average properties of the control strategies resulting from different combinations of training noise (all utilised at, CP noise = dwell noise = 0).**

| | Training noise (CP_dwell) | | | |
|---|---|---|---|---|
| | **0_0** | **100_0** | **0_90** | **100_90** |
| Average speeds /m/s | 43.0 | 43.3 | 45.1 | 45.2 |
| Average journey times /s | 699 | 695 | 667 | 665 |
| Average energy /kWh | 881 | 916 | 994 | 1027 |

This allows punctual operation in systems where there is a significant probability that dwell times will be longer than scheduled, at the cost of running faster and using more energy. Interestingly, fast running (in order to build up recovery time) is similar to typical driver behaviour reported in [13] but in this case has been found by a direct optimisation, which has no 'understanding' of the system it is optimising or prior knowledge of existing operational concepts. The recovery times found here are different for each journey and may be hard to estimate them in advance as they depend, not just on the scheduled times, but also on the different speed limits and headway restrictions experienced by each train as it moves through the network.

### 3.4 Both control point application and station dwell time training noise

In sections 3.2 and 3.3 the proposed procedure has shown success in finding increasingly robustness solutions to variation in CP application and dwell time, respectively. However, in real systems there may be uncertainty in many parameters *simultaneously*, so it important to investigate the performance of the proposed procedure in this situation. The performance when high levels of CP and dwell training noise are used simultaneously is shown in Figure 6.

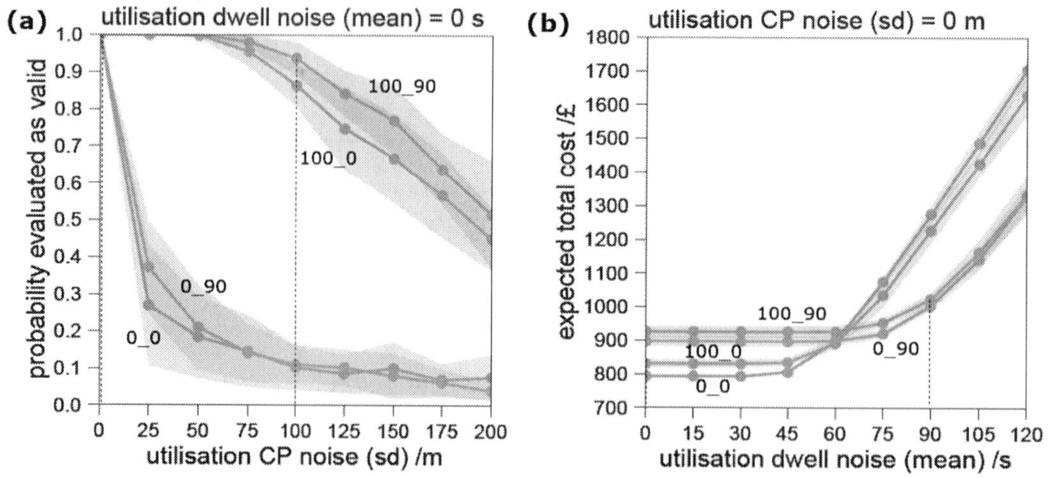


**Figure 6 Comparison of the performance of optimised solutions found with no training noise (0_0), the highest levels of CP and dwell training noise applied separately (100_0 and 0_90 respectively) and also *simultaneously* (100_90). The dotted lines emphasise the level of the training noise used.**

It can be seen from Figure 6(a) that the robustness of control strategies to noise in CP application and dwell times is predominantly influenced by the respective training noise level of CP and dwell used during optimisation. In terms of maximising robustness to CP variation the performance of the 100_90 optimisation is very close to the performance of the 100_0 optimisation, and actually seems to be slightly better. Similarly, Figure 6(b) shows the performance of the 100_90 optimisation is close in performance to the 0_90 solution, for maximising the robustness to dwell time variations. In this second case including both training noises simultaneously seems to have led to a marginally worse performance than a large dwell training noise alone. There is clearly room for more complete optimisation of the use of training noise to increase robustness, however, most importantly the results in Figure 6 show that the optimisation is capable of finding solutions that are robust to two different types of uncertainty simultaneously. The levels of robustness produced are similar to the case in which the uncertainties were considered independently.

## 4 CONCLUSION

When planning train trajectories it is important to consider the robustness of control strategies if they are to be implemented in real systems. Real systems contain many uncertainties, such as the accuracy of control point application or variations in station dwell times. If these are not considered during the optimisation process then it is unlikely that the control strategies found will be robust enough to perform as predicted in real operation.

A new, genetic algorithm based, optimisation procedure has been described which seeks to find robust solutions to the multi-train trajectory planning problem. The procedure is easily generalizable to include many different uncertainties in the system, but here it was shown to be effective in finding robust control strategies in the presence of two different types of uncertainty: the accuracy of control point application, and variation in stations dwell times. These uncertainties were first considered separately, before it was shown that they could be considered simultaneously in the optimisation and still achieve similar levels of robustness. For both types of uncertainty investigated a trade-off between robustness and the cost of the solution was found, reminding us that robustness doesn't come 'for free'.

This means that for best results the training noise level used during the optimisation progress should reflect, as accurately as possible, the noise level that will be experienced when the optimised control strategy is utilised.

## 5 REFERENCES


[1] M. McClanachan and C. Cole, "Current train control optimization methods with a view for application in heavy haul railways," *Proceedings of the Institution of Mechanical Engineers, Part F: Journal of Rail and Rapid Transit,* vol. 226, no. F1, pp. 36-47, 2012.

[2] H.-G. Beyer and B. Sendhoff, "Robust optimization – A comprehensive survey," *Comput. Methods Appl. Mech. Engrg.,* vol. 196, p. 3190–3218, 2007.

[3] W. Chen, J. K. Allen, K.-L. Tsui and F. Mistree, "A Procedure for Robust Design: Minimizing Variations Caused by Noise Factors and Control Factors," *Transactions of the ASME,* vol. 118, pp. 478-485, 1996.

[4] Y. Jin and J. Branke, "Evolutionary Optimization in Uncertain Environments - A Survey," *IEEE Transactions on Evolutionary Computation,* vol. 9, no. 3, pp. 303-317, 2005.

[5] D. V. Arnold and H.-G. Beyer, "A Comparison of Evolution Strategies with Other Direct Search Methods in the Presence of Noise," *Computational Optimization and Applications,* vol. 24, pp. 135-159, 2003.

[6] S. Tsutsui and A. Ghosh, "Genetic Algorithms with a Robust Solution Searching Scheme," *IEEE Transactions on Evolutionary Computation,* vol. 1, no. 3, pp. 201-208, 1997.

[7] L. Yang, K. Li, Z. Gao and X. Li, "Optimizing trains movement on a railway network," *Omega,* pp. 619-633, 2012.

[8] J. C. J. Goodwin, D. I. Fletcher and R. F. Harrison, "Multi-train trajectory optimisation to maximise rail network energy efficiency under travel-time constraints," *Proc IMechE Part F: J Rail and Rapid Transit,* 2014, UNDER REVIEW.

[9] I. Martínez, B. Vitoriano, A. Fernández and A. P. Cucala, "Statistical dwell time model for metro lines," *WIT Transactions on The Built Environment,* vol. 96, pp. 223-232, 2007.

[10] W. H. K. Lam, C. Y. Cheung and Y. F. Poon, "A Study of Train Dwelling Time at the Hong Kong Mass Transit Railway System," *Journal ofAdvanced Transportation,* vol. 32, no. 3, pp. 285-296, 1998.

[11] J. Yuan, Stochastic Modelling of Train Delays and Delay Propagation in Stations, Delft: Eburon Academic Publisher, 2006.

[12] A. Eiben, R. Hinterding and Z. Michalewicz, "Parameter Control in Evolutionary Algorithms," *IEEE Transactions on Evolutionary Computation,* vol. 3, no. 2, pp. 124-141, 1999.

[13] RSSB, "Driver advisory information for energy management and regulation - Stage 1 Report (T724)," 2009. [Online]. Available: http://www.sparkrail.org/_layouts/Rssb.Spark/Attachments.ashx?Id=75NEMTS3ZVHP-8-2899

# SafeCap: the train advisory system for real-time traffic management

**A Iliasov, A Romanovsky**
Newcastle University, School of Computing Science, UK

## ABSTRACT

SafeCap is an advisory system which operates as a layer above the fixed-block signalling in order to improve track capacity and the operational safety of junctions and stations. It does this through real-time control which projects the current state of the train configuration along with its set of commands into the future. Control software allows the traffic to be quickly stabilised should there be a need to accommodate additional trains, recover from train delays, and, generally, sustain a high rate of train flow when there are limited opportunities for implementing stability margins in a schedule.

## 1 INTRODUCTION

The on-going series of the SafeCap projects (sponsored by EPSRC and RSSB) focuses on improving capacity of railway nodes (junctions and stations) while ensuring their safety [1]. The projects have now developed a formal approach to safety verification and an integrated tooling environment that supports a domain-specific interface to help signalling engineers, formal safety verification and simulation used to calculate node capacity for various traffic scenarios. This work serves as a sound foundation for the development of the SafeCap train advisory system.

Static optimisation of junctions and stations critically depends on knowledge about train schedules and traffic mix. There is a danger of over-optimising to such a degree that stability is heavily compromised. To counter this, static optimisation should be somewhat defensive and leave some slack in train spacing. Real-time control can make the best use of this slack by assessing, in real time, the current state of trains around a station, acting to control signalling and, if applicable, ATO (Automated Train Operation) programmes.

Automated Train Operation-capable trains and moving-block signalling present the best opportunities for real-time control. The main function of the latter is an accurate projection of the current train configuration and control commands into the future. By projecting a set of alternative control scenarios, one can explore future configurations and choose the locally optimal control. Control software could alter acceleration/ braking curves, set individual train speed limits, adjust dwelling times, and choose platforms. It should allow the traffic to be quickly stabilised should there be a need to accommodate an extra train, recover from train delays, and, generally, sustain a high rate of train flow when there are limited opportunities for implementing stability margins in a schedule.

Albeit a simple idea, technologically real-time control is extremely challenging. To provide meaningful control logic for each particular station, and to be able to predict the time of arrival and the speed of approaching trains, one needs to consider a large context area. Furthermore, finding an optimum in such a setting is bound to rely on probabilistic and genetic algorithms that require a fairly large test base, in the order of tens of millions of control scenarios. Such scenarios would have to be generated and processed within a tiny time budget: one to ten seconds. Fortunately, the problem is amenable to massive parallelisation, and there are computer science techniques and specialised hardware to achieve this today.

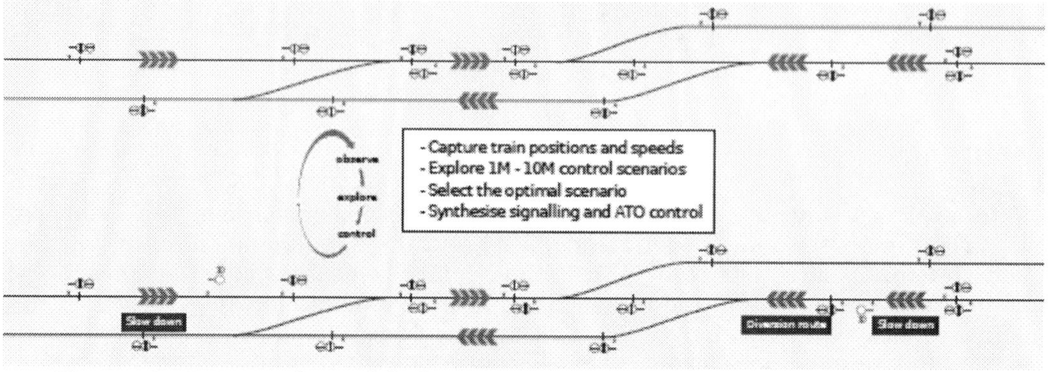


**Figure 1. A CUDA-based [2] simulation can explore many millions of control scenarios in just a few seconds, to predict and pre-empt delays and disruptions.**

Figure 1 shows a schematic representation of a railway model, where each train behaviour is pre-computed depending on the state of the route inter-locking. Essentially, before commencing movement each train computes a number of scenarios in its near horizon, and selects to act upon the one that renders an optimal track capacity. Since at every time slice such computation is carried out by every train present in the simulation, the horizon might become increasingly more difficult to compute, with the complexity determined by amount of interference between trains. Hence, a parallel implementation is a good candidate for an optimisation which has potential for drastically increasing overall efficiency.

## 2 OVERVIEW

Our aim is to improve capacity and operational stability of a railway network. By capacity we understand the ability of a network to accommodate specific traffic schedules. The stability property ensures that railway operation is able to recover (i.e., run to schedule) promptly after a disruption caused by a late or unscheduled train.

One common technique for assessing capacity is the UIC 406 "Capacity Calculation Method" [3] that measures how well a given layout and signalling are able to accommodate some specific schedule and whether there is enough capacity left to introduce extra traffic. Stability is achieved through a degree of redundancy in the spatial domain: time gaps between successive trains are used to mitigate knock-on effect of delayed trains and isolated 'service' gaps help to schedule unplanned traffic.

The measures advocated by UIC 406 to increase stability achieve this at the expense of capacity: larger time gaps between trains mean fewer trains per unit of time. Time gaps may be smaller if the trains are slower (on average) but this again severely impacts capacity. Thus, for the critical sections of railway infrastructure, traditional schedule-based approach to capacity and stability is too restricting.

At a finer level of granularity, the available capacity of a network is distributed quite unevenly. A straight section of track offers more capacity than, for instance, a curved one (assuming a speed limit is enforced on a curved section). Not much can be done to rectify this other constructing new tracks or using tilting trains. However, the major sources of capacity restriction are areas where tracks intersect and trains stop, i.e., junctions and stations. And, unlike curved tracks, we believe there is an opportunity for capacity improvement in junctions and stations by adding a train advisory layer on top of the existing signalling and interlocking. The guiding idea is to redistribute capacity consumption from areas of restricted capacity (junctions, stations) to areas where there may be an excess capacity (e.g., straight line). Our thesis is that this leads the better overall capacity utilisation. Another important step is realising that assessing in static context, with a layout and schedule alone, necessarily leads to conservative solutions that try to minimize occurrence of significant service degradation at the cost of somewhat lower overall performance. In contrast, a reactive control system, operating in real-time, has far more opportunities for delivering higher capacity as extreme events may be ruled out with a high degree of confidence for the time window of control.

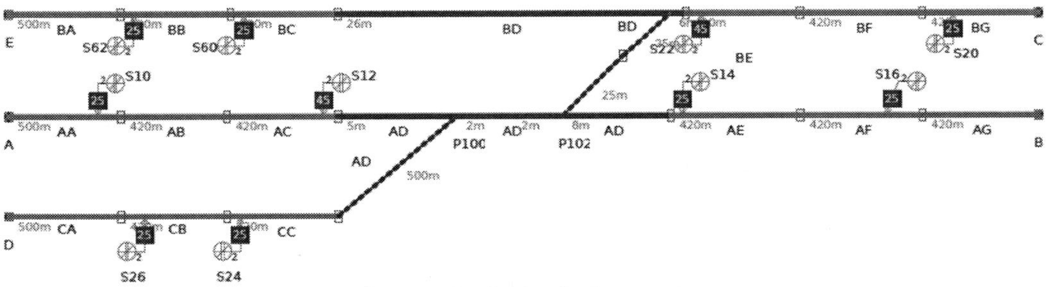


**Figure 2. Sample Layout.**

As an illustration, consider the layout in Figure 2. Assume that running trains over a path entering the layout at A and leaving at B delivers capacity cap(A_B) when there is no traffic on other paths; let us denote the same metric for paths from C to D and from C to E as cap(C_D) and cap(C E). Also let cap($p_1$, ..., $p_n$) be the overall capacity of simultaneous traffic on paths $p_1$, ..., $p_n$. For our example we have that:

$cap(A\_B) + cap(C\_E) = cap(A\_B, C\_E)$
$cap(A\_B) + cap(A\_B) \geq cap(A\_B, C\_D)$

In other words, when some paths share an area of track the resulting interference may prevent attaining the full capacity of each path considered in isolation. Whereas paths do not interfere, like A_B and C_E in our example, their combined capacity is a simple summation of individual capacities. If we account for the interference between some two path pi and $p_j$ with the term $int(p_i, p_j) \geq 0$ then the overall capacity is sum of isolated path capacities minus the overall path interference:

$$cap(p_1, \dots, p_n) = \sum_i cap(p_i) - \sum_{i \neq j} int(p_i, p_j) \quad (1)$$

Hence, to maximize capacity one needs to minimize term $\sum_{i \neq j} int(p_i, p_j)$. The inference of paths $p_i$ and $p_j$ is a consequence of interference of individual trains on these paths, i.e., the delays introduced on either path due traffic on another path. The ordering of trains on a path may be fixed by a route reservation policy, commonly defined in a train schedule. Our approach, however, requires a degree of flexibility in route reservation where the priority of reservation of routes with shared track elements is dynamically computed by the proposed advisory system.

## 3 CONTROL AND OPTIMISATION

Interference *int(*$p_i$*,* $p_j$*)* may be determined from the observation of train runs (using, for instance. a simulator tool) and comparing the measured capacity *cap(*$p_i$*,* $p_j$*)* to *cap(*$p_i$*)* + *cap(*$p_j$ *)*. Each of these terms is completely defined by a record of individual train runs - record of train position over time. A detailed train run is a costly to compute and manipulate for a railway junction of any significant size. This limitation becomes prohibitive in the context of real-time control. In lieu of a suitable explicit form of term *int(*$p_i$*,* $p_j$*)* that gives an accurate quantitative measure of interference, it is necessary to devise a simpler, proxy metric that only correlates with actual interference value to provide a qualitative measure of interference. As such metric we propose to employ a record of times when circuits containing points are locked and freed by trains. Such a record *T*, defines, for each point, a sequence which elements are triples of path, circuit entry time and circuit occupation time:

$$T = \left\{ \begin{matrix} R_j : \ldots, (p_i^j, \tau_i^j, \Delta_i^j), (p_{i+1}^j, \tau_{i+1}^j, \Delta_{i+1}^j), \ldots, \left(p_{i_k}^j, \tau_{i_k}^j, \Delta_{i_k}^j\right), \ldots \\ \ldots \end{matrix} \right.$$

In the above, $R_j$ is the name of a train detection circuit (e.g., AD in Figure 2) which contains point(s) or diamond crossing(s); $p_i^j$ are path identifiers; $\tau_i^j, \Delta_i^j$ are positive real values that define the moment and extent of time when circuit $R_j$ was occupied by a train from the respective path.

The example in Figure 3 shows simulated run over the layout from Figure 2 obtained with the SafeCap modelling platform [4]. Six trains, divided evenly between paths AB and CD are scheduled to enter the layout in 25 second intervals and trains on AB have priority in route reservation. The time vs. distance plot in 3 shows the first train on CD being delaying at signal S22. Other CD trains are also slightly delayed by traffic on AB. Slight wobble of the AB distance plot is the effect of sub-optimal two aspect signalling.

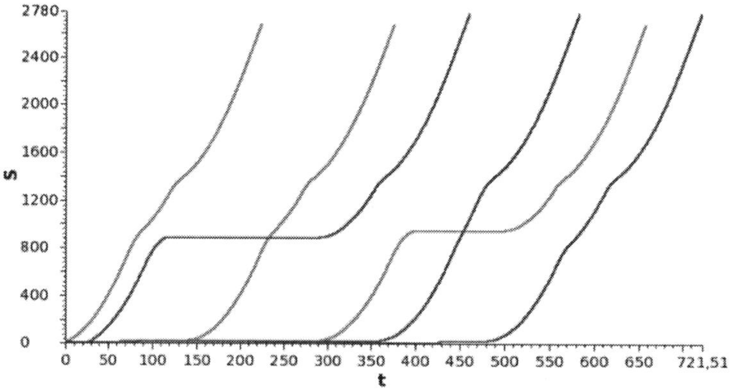


$$AD \; : \; (AB, 126.9, 8.9), (AB, 278.9, 9.1), (CD, 353.6, 11.6), (CD, 476.7, 11.7)$$
$$BD \; : \; (CD, 351.7, 5.6), (CD, 474.35, 4.94), (CD, 403.6, 5.1)$$

**Figure 3. Train run plot and shared circuit traversal histories. In green (light grey) are identical trains on paths between nodes A, B; in red (dark grey) are trains between nodes C, D.**

Figure 3 also gives an excerpt of shared circuit traversal histories. One point of interest is that the second train on line CD occupies AD for a somewhat shorter duration of time, i.e., it enters circuit boundary with a higher speed. By changing train schedule and introducing extra delays (i.e., an extra stop or longer dwelling time), one may observe a differing record *T'*. It is useful to identify the ways some two records *T* and *T'* may differ. One possibility is that a train arrives at the boundary of a circuit slightly later or earlier and occupies it for a different duration of time. We record such a change with a transformation rule called shift:

$$shift(T,j,i,t_o,t_1) = \begin{cases} R_j : \dots, (p_i^j, \tau_i^j + t_0, \Delta_i^j + t_1), \dots \\ \dots \end{cases}$$

One may also observe, due to a differing schedule or a change in route reservation policy, a different order of paths in the history record of a circuit. Rule swap defines a transformation where two previously adjacent paths $p_j$ and $p_{j\ is+1}$ are swapped:

$$swap(T,j,i) = \begin{cases} R_j : \dots, (p_{i+1}^j, \tau_{i+1}^j, \Delta_{i+1}^j), (p_i^j, \tau_i^j, \Delta_i^j) \dots \\ \dots \end{cases}$$

If the traffic pattern remains invariant (i.e., the same number trains travel over same paths), any possible record *T′* may be obtained from any other record T by applying a sequence S of transformation rules shift and swap: *T′ = S(T)*. The essence of our technique is based on the following conjecture: small changes to a record history T corresponds to a small variation in interference value int($p_i$,...,$p_k$).

### 3.1 Real-time control

The SafeCap [5] advisory system implements a real-time control loop for the supervision of train movement at the sub-signalling level with the aim of increasing overall system capacity or meeting specific operation goals. It is assumed that the system has a complete view of a controlled area including position and velocity of each train and the state of movable track equipment. This view cumulatively defines the sensors of the control system. Its sole actuator - the way it influences train behaviour - is the command interface to an automated train operation subsystem (ATO), assumed to be already in place. The control system issues commands to ATO to slow down or speed up specific trains. The core of the control is an optimisation procedure based on the notion of shared circuit traversal history, presented above. An optimal control is identified by finding a sequence of transformation rules s leading to a significant and positive change in capacity.

The diagram in Figure 4 presents the overall architecture of the developed control system. The first step, state snapshot, polls railway infrastructure and sensing equipment to acquire train and equipment state. A simulator tool is instantiated with the captured state and its simulation run provides the initial record history T.

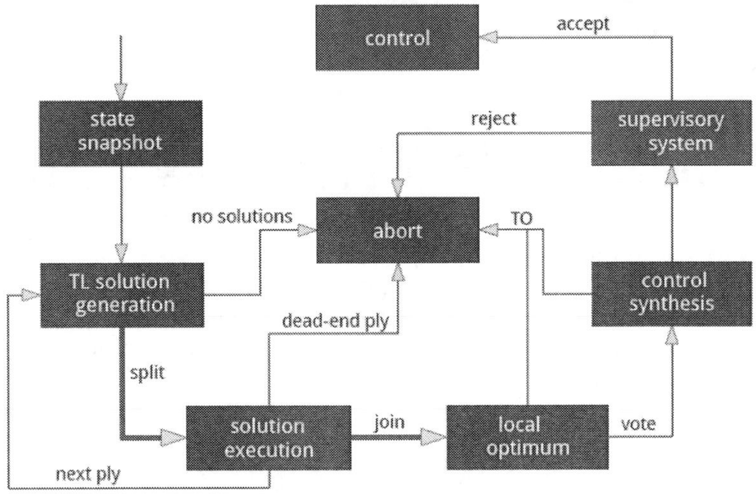


**Figure 4. SafeCap real-time control system.**

The *solution generation* step employs a probabilistic procedure to generate a set of transformation sequences $S^{(1)}$ such that, at this stage, each sequence has just one transformation rule.

In general, we are looking for solutions that are likely to yield significantly different performance. Many of these solutions will offer a lower capacity than the original situation; some proportion may also be not realizable. Realizability of a transformation rule requires that every train, starting from the captured state, is able, in a simulator tool, to meet the constraints of a transformed record. To remove solutions which unrealisable or offer lower performance, we employ an event-based train simulator that is able quickly confirm the capacity utilisation and the feasibility of the imposed constraints. The capacity utilisation metric may be a general purpose one (i.e., track utilisation) or specific to an area and situation (e.g., station call punctuality).

The result of the first step is a set $S^{(1)}$ of viable one-step transformation sequences. We call such set the first-ply solution and, more generally, refer to $S^{(n)}$ as n-th ply solution. The ply index describes how far, in the terms of shared train circuits, the optimisation procedure has advanced. The first ply solution considers only the initial points and crossings on each paths, the second ply adds all the second ones, and so on. Each ply adds a new dimension to the solution space: each transformation sequence s from ply i is expanded to a set of sequences q such that s is a prefix for every sequence in q.

The set of first-ply solutions is fed back into the solution generation procedure to obtain the candidate set $S^{(2)}$ of second-ply solutions. These are again filtered by realizability and capacity estimates and provide the basis for the next ply. The depth of the solution generation procedure is limited by the number of circuits with movable elements that occur on a path. The largest such number (i.e., a path with three points, all on separate circuits) defines the index of the last ply. Once all solutions are generated and filtered, those remaining are ranked to identify one optimal solution. The ranking criterion is specific to the current operational goals and the area controlled. The control loop executed every 5 to 10 seconds. This time window is small enough to maintain a useful degree of feedback in the control loop and large enough to be able to consider a large number of potential solutions.

We apply certain heuristics to efficiently explore a set of candidate solutions.

### 3.2 The swap heuristic

The aim of the swap heuristic is to generate a relatively small number of promising candidate solutions by changing order trains over shared circuits.

The simplest possibility is a random search; it does not deliver, in our experiments, any solutions within reasonable (10-15) seconds time. A more sophisticated option is do probabilistic gradient search using a technique such as simulated annealing. With this technique the likelihood of swapping some two paths and $p_{i+1}^j$ and $p_i^j$ decreases with the time distance $|\tau_{i+1}^j - \tau_i^j|$ between these instances of traversal. Thus, if some two trains travel over a point one right after another, the heuristic is likely to generate a candidate solution where the order of the trains is swapped for this point. If the trains are separated by a considerable time gap, the swap becomes less likely.

Figure 5 demonstrates an application of simulated annealing to the example from Figure 2. An optimal solution here is the alternating order of trains. This heuristic determined 6 or 7 (depending on random seed) candidate solutions out of 720 possible combinations. The best of the solutions (Figure 5, left) delivers 21%

increase in track utilisation having considered. The worst of these solutions delivers 17% capacity decrease.

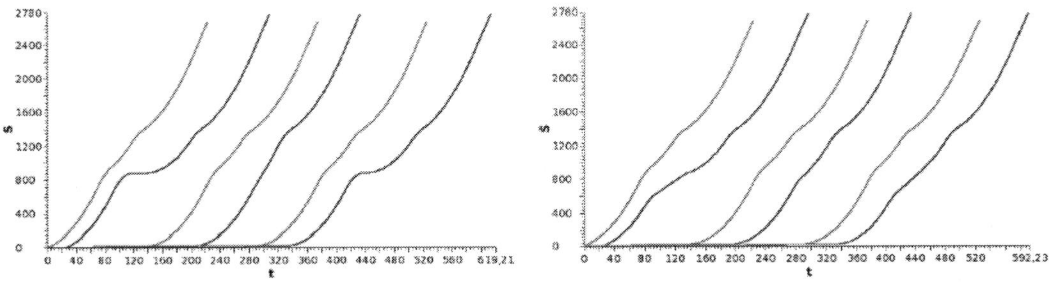


**Figure 5. Train runs of optimal solution; the left plot is an optimal train ordering by the swap heuristic; the right plot adds the speed control of shift heuristic.**

Unfortunately, the performance of simulated annealing is too unpredictable and generally, it is to slow for the purpose of real time control. The vast majority of candidate solutions it generates are infeasible (incompatible with signalling). To improve performance of simulated annealing we have introduced a stage that efficiently (that is, without running a full simulation) filters out all the infeasible solutions.

This stage is a form of strategic game played over a route graph of railway topology. Each 'player' controls the trains originating from some source node; the goal is to get each train as quickly as possible to an exit node.

Each player controls tokens (trains) starting at some node. The moves are constrained by signalling and interlocking rules and trains may be moved only when a valid route reservation is made. Since signalling rules are in discrete logic rules, the solution space is countable and may be, in principle, explored to deliver an optimum according to some reward function. Although the explicit state space is extremely large (grows as a factorial of a number of points in the layout), symmetry reduction made possible complete state exploration within milliseconds even for large layouts (> 100 points). For state exploration we have initially employed our own simple model checking tool; a new version is based on a translation into the ProB model checker that offers both higher performance and degree of confidence.

### 3.3 The shift heuristic

The aim of this heuristics is to adjust the way a train approaches the boundary of a route containing a shared train circuit. With conventional signalling, a train slowing down to a red signal has no knowledge of the cause of the signal being red nor how soon the cause will cease to be and let the signal show a proceed aspect.

We are going to override this behaviour and guide the train to approach a red signal at speed. Knowing the position and capabilities of each train, it is possible to predict exactly the point in time when the route containing shared circuits becomes available to any given train. Thus, we can command a train to arrive at the route boundary right at the moment the route is available for the train.

No matter how long a train had to wait for a point and its containing route to be allocated to the train, the moment this happens the train should be entering the route with the maximum feasible speed. Given this rule is maintained for all the trains, the occupation of the critical sections of railways (those containing points and switching crossing) is minimized (within the limitation of a current solution).

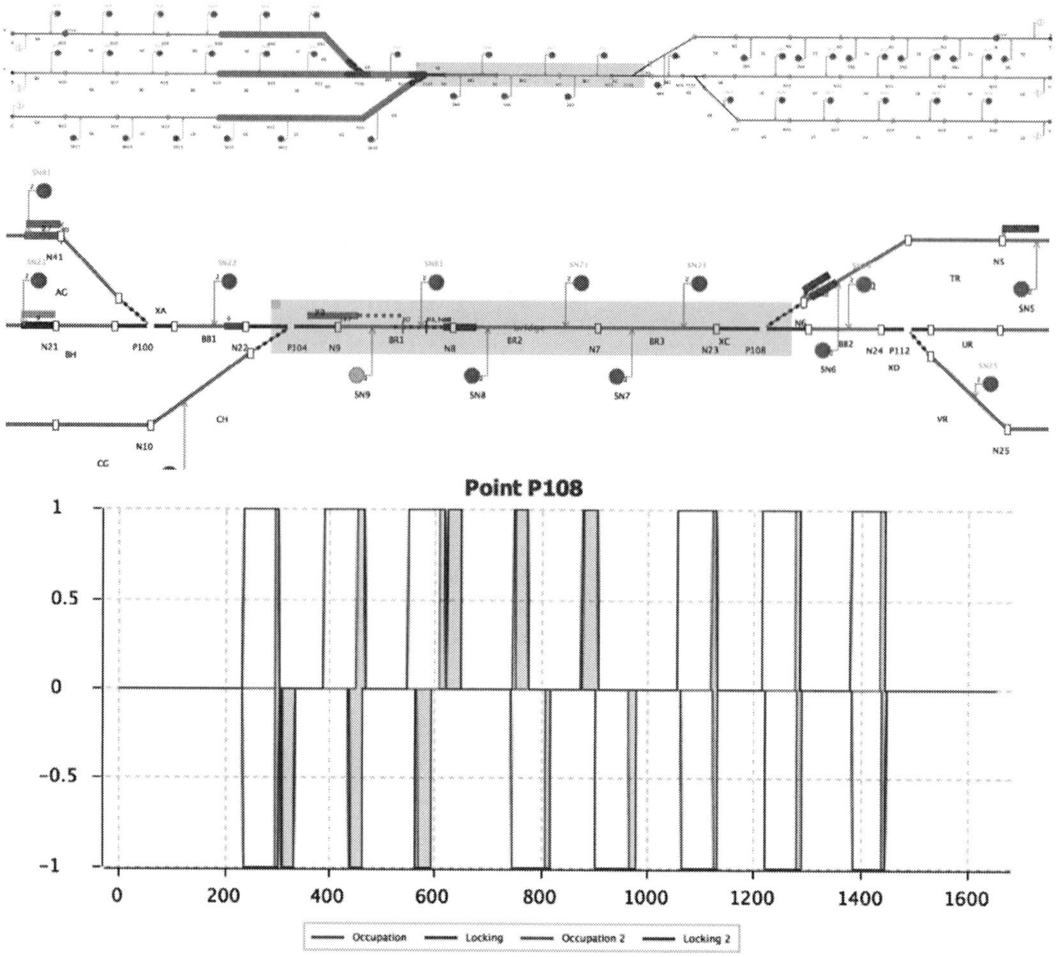


**Figure 6. Shift heuristics performance for a layout with a single-track middle section. The first diagram shows the area of single track; the second diagram overlays un-optimized and optimized solutions (rectangle depict trains and dotted lines show their immediate braking distance; the raised pictograms are the trains of an optimized solutions). The third diagram shows the old (above 0) and new (below zero) traversal history for point P108. There is a small but significant gain in the locking and occupation times.**

Although such behaviour does not violate route-based signalling principles (train should never pass a red signal), the risk of passing a red signal increases as a train approaches it at speed. The two principal safety violation scenarios are as follows: a detection circuit detects another train on the route, and there is a fault in movable track equipment. The former case may be largely excluded with a small (e.g., 5 sec) control window: the advisory system would be able to detect stationary or unusually slow train and direct other trains accordingly. The latter case when, for instance, a point lock is not acquired, can be managed by ascertaining that at all times there is a sufficient braking distance for ATP to stop a train in front of a faulty point.

The right-hand side plot in Figure 6 shows a combination of the optimal solution found by swap and approach speed control computed by the shift heuristic. As evident from the plots, the heuristic prevents two of the trains from stopping and thus making them traverse the shared area quicker. This delivers approx. 8% increase in capacity.

### 3.4 Synthesising ATO control

The selected optimal solution (in the form of transformation sequence $s_j$) must be transformed into a set of train control rules. The first step consists in overriding of the existing route reservation and point locking schedule in order to realise swap rules of the optimal solution $s_j$ . Since from T' we order the relative order of traversal of every shared circuit (and thus point and diamond crossing), it is not straightforward to compute such new route reservation schedule. The second kind of rules, the shift rule, speeds up or slows down a train. The effect of these rules is synthesised using a dedicated simulation procedure where train time of traversal of shared circuits is reflected in waypoints on a train path.

A waypoint W = (d; t) is defined by its offset d within the train path and the arrival time t. If every train respects the constraints set by waypoints, one would observe history record T' = $s_j$(T) (record T is a project of state snapshot based on signalling and scheduling rules alone). The set of waypoints is handed over to a third-party ATO subsystem. This subsystem is assumed to analyse waypoints for feasibility and safety and generate corresponding driving profiles. It is the obligation of the ATO subsystem to reject waypoints that may violate signalling constraints. If any single waypoint is rejected, all waypoints are cleared and the system falls back into it original state, until the next control window.

## 4 CONCLUSION

We have briefly presented our findings on real-time advisory control system. The system attempts to stabilise performance in presence of disturbances by issuing ATO control commands computed by a relatively simple optimisation engine. We are working on applying the technique to a large-scale, real world railway operation spanning busy stations. The system illustrated in the present paper is aimed at computing a myriad of scenarios for each entity in order to determine the optimal solution in terms of track capacity at any given time-step and route region. By taking advantage of the multi-core graphics processor, the simulation could compute scenarios into the future for each train entity in parallel, simultaneously, and return the optimal scenario, therefore increasing the computation speed that the CPU would have to carry. In addition to that, a very large number of train entities could be simulated in parallel using the appropriate hardware and memory management paradigms.

## ACKNOWLEDGMENTS

This work is supported by the EPSRC SafeCap Impact Acceleration Account and by the RSSB SafeCap for FuTRO project. We are grateful to Simon Chadwick and David Gill, our colleagues in Siemens Rail Automation, for the useful discussions.

## REFERENCES

[1] Iliasov, A., Lopatkin, I., & Romanovsky, A. SafeCap: advanced computer science techniques for railways of tomorrow. EURAILmag. Issue 28. 2013.
[2] Nickolls, J., Buck, I., Garland, M. & Skadron, K., Scalable parallel programming with CUDA. Queue, 6(2), 2008.
[3] UIC, Kodex 406, 2004.
[4] Iliasov, A., Lopatkin, I. & Romanovsky, A., The SafeCap platform for modelling railway safety and capacity. Proc. SAFECOMP'13, Springer, 2013.
[5] Iliasov, A. Lopatkin, I. Mihut, A. & Romanovsky, A. Real-time ATO reconfiguration for operational stability. Computers in Railways XIV: Railway Engineering Design and Optimization 135, WIT Press. 2014.

# Evaluation of bogie rotational resistance

**T Tanaka, K Iida, M Suzuki, T Iida, N Watanabe, T Miyamoto**
Vehicle Mechanics Laboratory, Railway Dynamics Division,
Railway Technical Research Institute, Japan

To prevent derailment such as flange climbing derailment, precise understanding of the vehicle dynamics is important. The authors have developed bogie rotational resistance test machine which has many innovative capability to reveal the characteristics of the railway bogies. The test machine can measure the rotation resistance of a bogie directly by rotating it set to a carbody. By using this, some distinctive characteristics of the rotation resistance have become clear. A simulation model based on the test result was established, which enabled more realistic estimation of lateral force during the curve passage of the bogie.

## 1. INTRODUCTION

To ensure running safety of railway vehicles, a systematized evaluation method of describing the probability of the derailment is necessary. In Japan, a regulation for the speed up of the conventional railway has been established and a variety of the indicators for preventing from the derailment are described there. Running tests are carried out in order to confirm the running safety of individual type of railway vehicles. However, the flange climbing derailment caused by the train, which runs through a tight curve at low speed has not been eradicated so far, and the precise assessment of the safety condition in terms of the matching between the vehicle and the track is needed.

One parameter to be reconsidered in order to achieve more precise assessment is the rotational resistance of railway bogies. A stationary test to measure the rotational resistance using a test rig is required in the European Standards [1]. We developed a bogie rotational resistance test machine which can measure not only the rotational resistance as mentioned in European Standards but also other bogie's behaviors under many types of the rotating condition. This machine can measure the rotational resistance of the bogie directly by rotating it set to a carbody. This rotation test makes it possible to estimate the rotational resistance precisely. Using results of this test, the precise method to analyze the characteristics of the bogie rotation has been established and the details of them have been revealed. The modeling of these characteristics was carried out to be applied to a simple equation called the Lateral-force/Wheel-load estimation equation which is used to estimate the running safety in the curve passage.

In this paper, the overview of the evaluation method of the running safety in the curve passage are described focusing on the detail of the bogie rotational resistance test, the Lateral-force/Wheel-load estimation equation, and the future plan to improve the evaluation of the running safety are described.

## 2. EVALUATION OF RUNNING SAFETY DURING TIGHT CURVE PASSAGE

There are many factors which may cause the flange climbing derailment. The main factors due to the vehicle are the followings: the static wheel load and its unbalance, the stiffness of the suspensions, the height of the center of gravity of the carbody, and so on. Those due to the track are the followings: the curve radius, the cant, the track irregularity, and so on.

Generally, the derailment coefficient calculated from the ratio of the lateral force to the wheel load of the outside wheel is used to estimate the risk of the derailment. In case of the flange climbing derailment, the critical value of the derailment coefficient is given by Nadal's equation as:

$$\left(\frac{\mathrm{Q}}{\mathrm{P}}\right)_{\mathrm{cri}} = \frac{\tan\alpha - \mu_e}{1 + \mu_e \tan\alpha} \tag{1}$$

where, $\left(\frac{Q}{P}\right)_{cri}$ is the critical derailment coefficient, $\alpha$ is the flange angle of the wheel, and $\mu_e$ is the equivalent coefficient of the friction between flange and rail. One method of estimating eq. 1 is the precise time-historical simulation. However, this method is less flexible in simulating various combinations of the vehicle and track condition. To solve this, a Lateral-force/Wheel-load estimation equation has been developed for the purpose of calculating the lateral force and the wheel load in the curve passage using approximative simple equations [2]. This equation is a combination of the generation mechanism of the wheel load and lateral force and the analytical solution based on the experimental result.

In this paper, we focus on the equation for estimating the lateral force. The detail of the whole equation appears in Ref.[2]. In this equation, the lateral force of the outside wheel is written as

$$\mathrm{Q_0} = \overline{\mathrm{Q_I}} + \Delta\mathrm{Q_{AS}} + \Delta\mathrm{Q_{AD}} + \Delta\mathrm{Q_{unsp}} \ , \tag{2}$$

where $\overline{Q_I}$ is the steady lateral force of the inner side wheel, $\Delta Q_{AS}$ is the steady lateral force of the wheelset, $\Delta Q_{AD}$ is the fluctuation component coming from the rail irregularity, and $\Delta Q_{unsp}$ is the shocking component of the lateral force at the rail joint.

During the tight curve passage, the flange of the outside wheel is in contact with the outside rail pushing to its inner side. To resist against this force, friction force acts between the inner side wheel and the rail pushing. This friction force behaves as reaction force and makes the influence on the lateral force of the outside wheel. This friction force corresponds $\overline{Q_I}$ in eq.2 and written as

$$\overline{\mathrm{Q_I}} = \kappa P_i \ , \tag{3}$$

where $P_i$ is the wheel load of the inner side. The coefficient $\kappa$ depends on the shape of the wheel surface, the curvature radius, and the friction coefficient between wheel and rail. The value $\kappa$ under the various conditions is calculated by the time-historical simulation and the simulation results are compiled in a database. We refer this database as a wheel/rail condition which we want to calculate.

When considering a bolsterless bogie, the steady lateral force of the wheelset $\Delta Q_{AS}$ is consist of the terms related to the centrifugal force and the air-spring distortion as

$$\Delta\mathrm{Q_{AS}} = W_0\left(\frac{v^2}{gR} - \frac{C}{G}\right) + \beta \times \frac{2kb_2^2c}{aR} \ , \tag{4}$$

where $W_0$ is the static axle load, $v$ is running speed, $G$ is gauge, $C$ is cant, $R$ is curve radius, $g$ is the acceleration of gravity, $2b_2$ is the width between right and left air springs as a secondary suspension, $2a$ is the wheel base of the bogie, $2c$ is distance between two bogies of the vehicle, $k$ is stiffness of the secondary spring, and $\beta$ is correction coefficient for the air-spring distortion. In Eq.4, the first term and second term in the right-hand side describe the centrifugal force and the force derived from the air-spring distortion, respectively. The coefficient $\beta$ is introduced in order to conform the lateral force of the wheel set caused by the air-spring distortion calculated by this equation to that estimated by the vehicle dynamics simulation.

Track irregularities, especially alignment irregularities cause vehicle vibrations. The inertia force according to the vehicle vibrations makes some influence on the lateral force of the wheelset. The shocking variations occurring at rail joints also affect to the lateral force. Those effects are given as

$$\Delta \mathrm{Q_{AD}} = 3W_0 k_Q \sigma_z V \ , \tag{5}$$

$$\Delta \mathrm{Q_{unsp}} = \left(\frac{\varepsilon}{100}\right) \times \left(\frac{2500}{R} + 4\right) \frac{V}{100} \ , \tag{6}$$

where $k_Q$ is the variation coefficient of track shifting force given by the running test analysis, $\sigma_z$ is the standard deviation of alignment irregularities, $\varepsilon$ is the effective ratio of the variation of the lateral force.

The calculation result using the lateral-force estimation equation is generally consistent with the result obtained from the running test [2]. On the other hand, some differences between them were found in some situations. One of the possible sources of this deviation was considered as inadequate estimation accuracy of the force caused by the air-spring distortion because the strength of this force was not confirmed experimentally.

## 3. EVALUATION OF THE BOGIE ROTATIONAL RESISTANCE

### 3.1 Bogie Rotational Resistance Test Machine

The bogie rotational resistance test machine was developed in order to evaluate accurately the rotational resistance of the bogie. An outside view of this machine is shown in Fig.1. The schematic view and the specifications are shown in Fig.2 and Tab.1, respectively. The main part of this machine is a rotation table. Arc-like guide rails are equipped under the table, which can rotate around the axis of rotation smoothly. One of the bogies of a carbody is put on the rotation table, and the other one is fixed on the rail located outside of this table. This rotation table is driven by two electric actuators. A load cell is set at the joint between the actuator and the rotation table in order to measure the rotation moment. Load cells for measuring the wheel load and the lateral force are put at each contact point of wheel and rail in order to measure fluctuation of the wheel load

**Fig.1 A view of the bogie rotation resistance test machine with test bogie**

and the lateral force when the bogie is rotated. The layout of the rails on the rotation table can be changed flexibly so that the rotation resistance of several types of bogies as shown in Tab.1 can be tested.

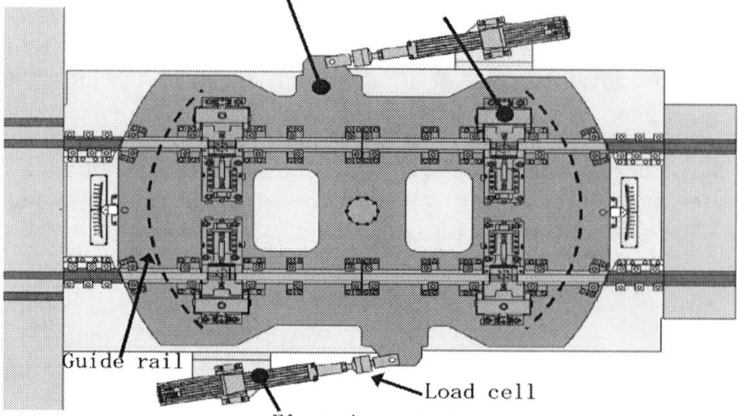


(a) Top view

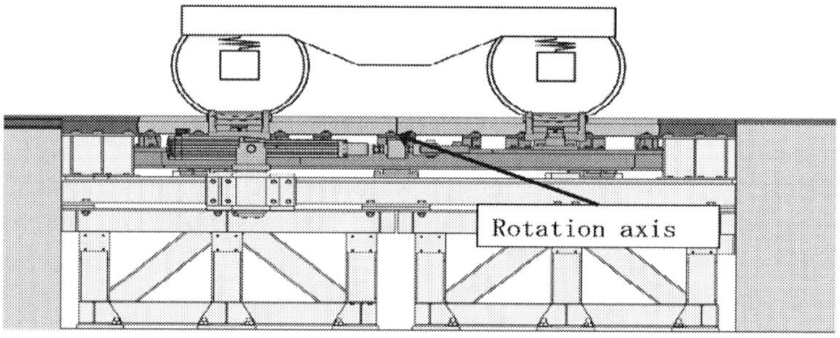


(b) Front view

**Fig.2 The schematic view of the bogie rotational resistance machine**

**Tab.1 The specification of the bogie rotation resistance test machine**

| | |
|---|---|
| Bogie | Gauge : 1067, 1435mm<br>Wheel base : 2100, 2150, 2500mm<br>Maximal axle load :125kN |
| Motion | Maximal rotation angle : 6deg<br>Maximal rotation speed : 3deg/s |

There are two types of the rotation modes in this machine: constant-speed rotation mode and sinusoidal variation mode. In the former mode, the table rotates at a constant speed until an arbitrary angle. The static characteristics of the bogie rotation are measured by this mode. The basic test of bogies prescribed in EN14363 also can be carried out using this mode. In the latter mode, table rotates with an arbitrary frequency and magnitude of the rotation around the initial position. The dynamic characteristic of the bogie rotation can be measured by this mode. In addition, the moment of inertia of the bogie in the yaw motion can be measured.

## 3.2 Methods of analyzing bogie rotational characteristics

### *3.2.1 Analysis method of the rotation resistance*

When a bolsterless bogie is put on the bogie rotational resistance test machine and rotated within a certain angle by reciprocatory motion, the measured rotation moment is expected as shown in Fig.3. The rotation moment is considered to be separated into three components as: restoring, viscous damping and friction component. The restoring component corresponds to *k* in Fig.3. The viscous damping and friction component appear as a mixed quantity *D* in Fig.3.

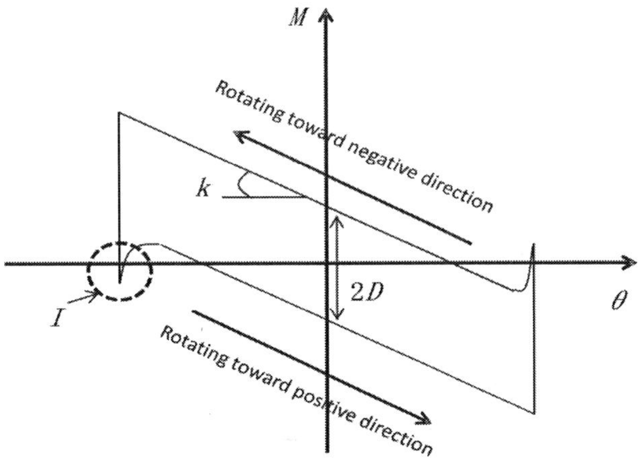


**Fig.3 Typical rotation moment**

$\theta$ appeared in Fig.3 is the rotation angle of the bogie. Here, we define $M_-(\theta)$ and $M_+(\theta)$ as the rotation moment measured when the rotation table moves backward and forward, respectively. Each component can be extracted by comparing $M_-(\theta)$ and $M_+(\theta)$ at the same rotation angle. By ignoring the inertia component described as $I$ in Fig.3, the restoring component can be written as

$$\frac{1}{2}\left(M_-(\theta)+M_+(\theta)\right)=\sum_{s} k_s L_s \theta \ , \qquad (7)$$

where, $\theta$ is the rotation angle, k is the spring constant of bogie components, suffix s shows the number of the source elements generating restoring force e.g. the air spring, and $L_s$ is the effort distance of each element of the restoring force.

Also, the viscous damping and friction component can be written as

$$\frac{1}{2}\left(M_-(\theta)-M_+(\theta)\right)=\left|\sum_{d} c_d L_d \dot{\theta}\right|+\left|\sum_{b} L_b h_b\right|+\left|L_t \cdot F_t\right| \ , \qquad (8)$$

where, c is the coefficient of damping of bogie components, suffix d shows the number of source elements generating damping force e.g. dampers, h is the coefficient of friction of bogie components, suffix b shows the number of source elements generating friction force, $F_t$ is the friction force coming from the rotation table itself, $L_d$, $L_b$, $L_d$ is the effort distance of each element. As regards the right-hand side of Eq.8, the first to third terms correspond to damping force, friction force of the bogie, and the friction force of the rotation table, respectively. Using Eq.8, the sum of the viscous damping and friction component can be extracted since the friction force of the rotation table is measured by the characteristic test of the rotation table itself. The viscous damping and friction component can be separated using the test results obtained at various rotation speeds because only the damping component depends on the rotation speed $\dot{\theta}$.

### *3.2.2 Analysis method of the inertia radius*

Assuming the bogie is set to the rotation table and rotated by a moment $M_{\mathrm{I}}$ with sinusoidal variation mode, the motion equation around the center of the rotation is written as

$$\left(m_{\mathrm{B}} \cdot i_{\mathrm{B}}^2 + I_{\mathrm{t}}\right) \cdot \ddot{\theta} = -M_{\mathrm{I}} - \left|L_{\mathrm{t}} \cdot F_{\mathrm{t}}(\dot{\theta},\ m_a)\right| \qquad (\dot{\theta} \geq 0)\ , \tag{9}$$

where, $m_{\mathrm{B}}$ is the mass of the bogie, $i_{\mathrm{B}}$ is the inertia radius of the bogie, $I_{\mathrm{t}}$ is the moment of inertia of the rotation table. $I_{\mathrm{t}}$ and $F_{\mathrm{t}}$ can be obtained by the characteristic test of the rotation table. Taking into account the sinusoidal motion of the rotation table, Eq. 9 can be rewritten as,

$$M_{\mathrm{I}} = \left(m_{\mathrm{B}} \cdot i_{\mathrm{B}}^2 + I_{\mathrm{t}}\right) \cdot (2\pi f)^2 \cdot \theta - \left|L_{\mathrm{t}} \cdot F_{\mathrm{t}}(\dot{\theta},\ m_a)\right| \tag{10}$$

where $f$ is the rotation frequency. Therefore, from the test of a sinusoidal variation mode in which the rotation frequency is constant, the inertia radius of the bogie can be estimated by investigating the relationship between the rotation angle and the rotation moment.

## 3.3 Examples of the analysis result

### *3.3.1 Analysis result of the rotation resistance*

The test result of a typical commuter type bogie rotated by a constant-speed rotation mode is described in this section. Here the rotation speed of the table was 0.5 deg/s and the maximal rotation angle was 4.5deg. The estimated bogie angle in which this bogie passes through a tight curve of a radius of 100m is 4deg. Thus, this test can reproduce the behavior of the bogie during a tight curve passage. The bogie was rotated under some conditions. From the rotation tests under several conditions, we have clarified the following facts: the characteristic hysteresis loop and the non-linearity of the moment at a large rotation angle under the normal condition of the bogie, the large rotation moment under the deflated condition of air springs, and the damping effect of yaw dampers. Also, we carried out this test by alternately attaching and detaching other components of the bogie such as lateral damper and link. The effects of those components to the rotation moment are vanishingly small at this type of bogie.

In this paper, the overview of the detailed analysis of the bogie under the normal condition is described as an example. The analysis result about the restoring component and viscous damping/friction components are shown in Fig.4 and Fig.5, respectively. Those results are obtained using Eq.7 and Eq.8. The test results are substituted into the left hand side of each equation. The friction of the rotation table appeared in Eq.8 is subtracted in Fig.5.

In Fig.4, a line which approximates the analysis result by a margin of not more than 1.6deg is also shown. The slope of this line is -5.0kNm/deg. The stiffness in the bogie rotating direction of the air springs estimated from the specification is (-4.7 $\pm$ 0.47) kNm/deg. Therefore, the stiffness of the air springs is considered as a dominant element of the rotation moment in the case of this bogie. On the other hand, the absolute value of the measured moment is lower than the approximate line when the bogie rotates beyond plus or minus 1.6 deg. At a small rotation angle, the direction of the restoring force generated by the air springs is along to the rotating direction. When the rotation angle becomes large, the direction of the restoring force is diffused into various directions due to the large deformation of the air spring. This is why the non linearity at a large rotation angle appears.

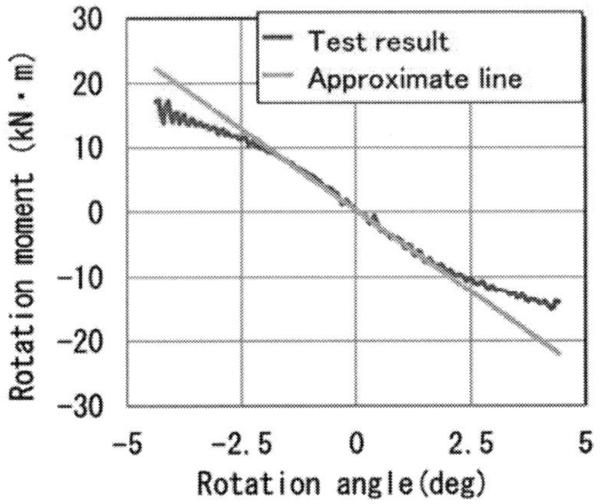


**Fig.4 The restoring component of the bogie rotation**

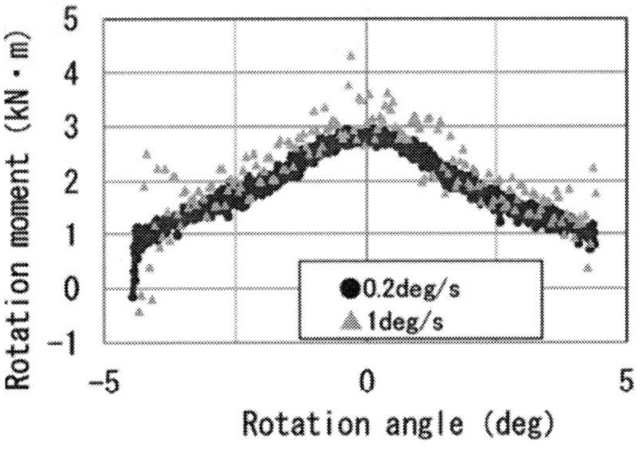


**Fig.5 The viscous damping/friction component of the bogie rotation**

In Fig.5 the test results of different rotation speeds are shown. There are not large difference between the result of 0.2deg/s and 1 deg/s of the rotation as shown in Fig.5. Thus, the viscous damping component is considered to be relatively small and the friction component is dominant in the moment shown in Fig.5. The possible element of this friction is the inner friction of the air springs. This friction component increases with the rotation from the stationary position, which is considered to be the characteristics of this friction component. Therefore, the moment at 0 deg is larger than that at large rotation angle in Fig.5.

### *3.3.2 Analysis result of the inertia radius*

To measure the inertia radius of the yaw motion of the bogie, a test using the sinusoidal variation mode was carried out. A bogie without a carbody was put on the table in this test. Measured actuator force on the condition that the frequency is 0.2 Hz and the amplitude of the rotation is 4 deg is shown in Fig.6. As expected from Eq.10, the slope of the approximate line indicated in Fig.6 is determined by the rotation frequency, the inertia and friction of the rotation table, the mass of the bogie, and the inertia radius of the bogie. Since those values except for the inertia radius of the bogie are given, the inertia radius can be obtained by this test result. The estimated inertia radius from this test is from 1.17 to 1.19m. On the other hand, the designed value of that is 1.13m. From this result, it is confirmed that the precise measurement of the inertia radius of bogies can be achieved using this method.

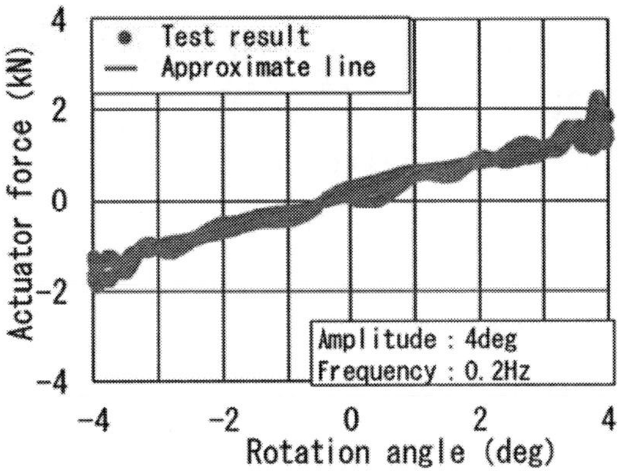


**Fig.6 The actuator force measured in the sinusoidal variation mode**

## 3.4 Adopting the result to the evaluation of running safety

### *3.4.1 The bogie rotation model*

To adopt the result measured by the bogie rotation test to the evaluation of the running safety, an improved model for describing the precise characteristics of the rotation moment is proposed. As described in a previous section, the rotation resistance of the test bogie depends mainly on the characteristics of the air springs. Therefore, the air spring model is modified so that the rotating test result and the simulation result is well matched. In the present model, the restoring force is dealt with as being proportional to the bogie angle. In the proposed model, the non linearity at the large rotation angle is considered. In addition, two types of friction force are introduced in order to represent the characteristic form of the friction of the air springs. The parameters to define the proposed model were tuned so that the proposed model represents the test result well. The comparison between the result using the proposed model and the test result is shown in Fig.7.

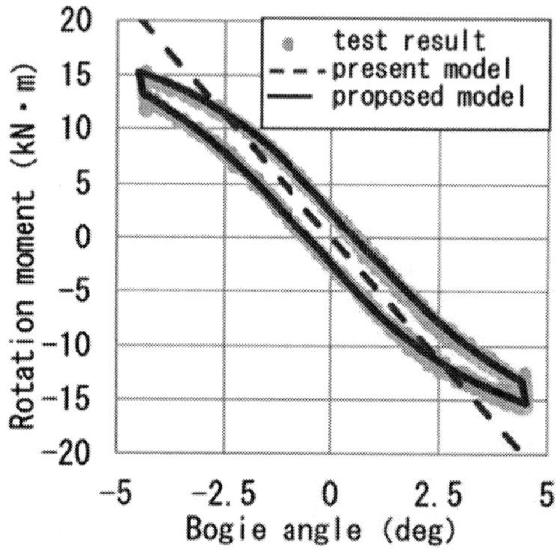


**Fig.7 The comparison between proposed model and test result**

### *3.4.2 Application of the model to the vehicle dynamics simulation*

The proposed model is applied to the vehicle dynamics simulation of the curve passage of vehicles. We simulate the vehicle running through two curves tuning to the same direction. The first curve radius is set at 160m and the second one is set at 100m. The comparison between the simulation result using the present model and the proposed model is shown in Fig.8. Here, the lateral force of the outside wheel and the air spring force in the rotating direction are shown. In the proposed model, smaller lateral force and air spring force are observed in the steady curve region. This is caused by the decrease of the air spring stiffness at the large bogie angle. When the bogie angle is suddenly changed e.g. the vehicle enters in the transition curve after the steady curve, air spring force is also quickly changed in the proposed model. This is one of the effects coming from the friction force.

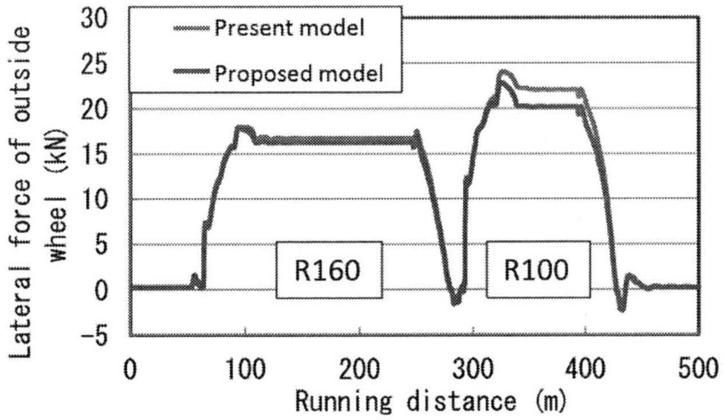


(a)Lateral force of outside wheel

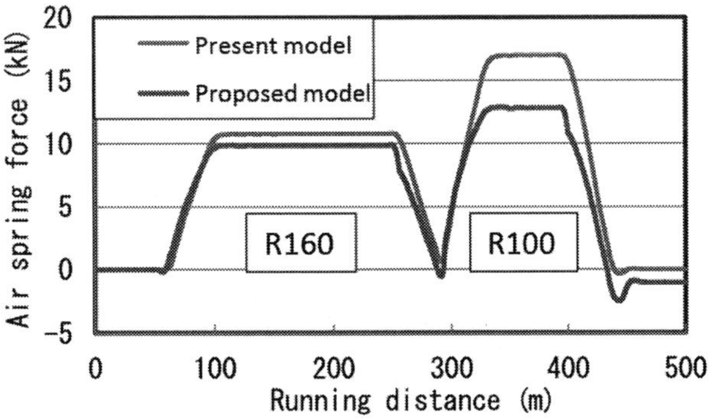


(b) Air spring force

**Fig.8 The simulation result of curve passage using present model and proposed model**

### *3.4.3 Application of the model to the Lateral-force estimation equation*

When the proposed model is applied to the Lateral-force estimation equation, the second term of the right-hand side of Eq. 10 can be turned to

$$\mathrm{F}_1 = \frac{\beta b_2}{a} \times \left( \frac{2k' b_2 \, c}{R} \pm \frac{D}{2} \right), \tag{11}$$

where, $\mathrm{F}_1$ is the lateral force derived from the air-spring distortion and $k'$ is the restoring force with the nonlinearity at the large rotation angle. To represent the hysteresis effect of the bogie rotation moment, the experimentally-modeled friction force $D$ is introduced in the proposed model. The comparison of lateral force induced by the air-spring distortion between the present calculation and the proposed calculation is shown in Fig.9. In the proposed estimation, the lateral force is estimated as a region surrounded by the upper and lower value in Eq.11.

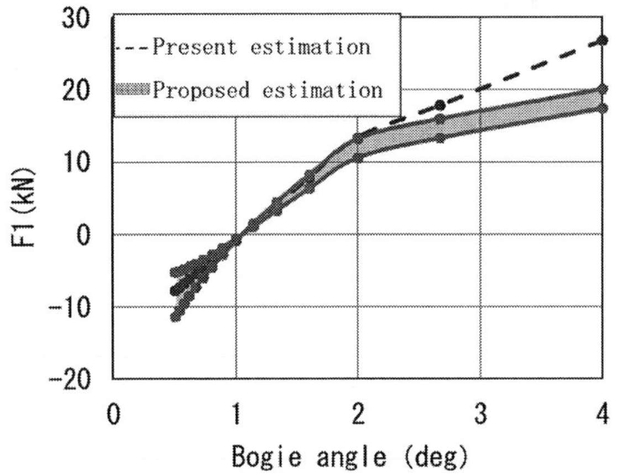


**Fig.9 The comparison of the lateral force derived from the air-spring distortion (F1) between the present and proposed estimation**

## 4. CURRENT PROJECTS FOR IMPROVING THE EVALUATION OF RUNNING SAFETY

In order to eradicate the flange climbing derailment, we are studying from various points of view such as introducing new indicators which are useful to assess the running safety, measuring and analyzing method of such indicators, and so on. Through those studies, we are challenging to ensure the analytical calculation related to the evaluation of the running safety. Eventually, we hope that the developed evaluation method improves the safety of both vehicles and tracks.

Focusing on the improvement of the Lateral-force/Wheel-load estimation equation, we are trying to take into account the dynamic motion of the bogie and track as a next step. For example, we are studying about the effect of the motion of the yaw dampers of the bogie during the curve passage. We are also investigating the bogie's rotation moment more precisely. For an example, we are trying to take into account the effect of the tangential force generated at the four wheels in a bogie on the bogie moment. We are also trying to estimate the large lateral force which is produced by the vehicle's passage through a turnout or a track with the short wavelength irregularity.

Running tests which lead to the flange climbing will be also carried out. Through this test, we are going to survey the flange climbing phenomena more precisely and try to develop an evaluation to secure safety operation. In parallel, a new bogie structure with a low incidence of flange climbing has been developed and the characteristic test of the bogie will be carried out.

## 5. CONCLUSION

The lateral force/wheel load estimation equations are the one of the useful tools to estimate the risk of the flange climbing derailment. To improve these equations or vehicle dynamics simulations, we developed the bogie rotational resistance test machine for railway vehicles. Through the test result by using this machine, the more accurate estimation of the lateral force or the air spring force can be achieved. This new method can contribute to the development of the assessment of running safety.

The derailments are caused by a variety of factors. The individual surveys of the possible factors of the derailment and the accumulation of the surveys are important in establishing a systematize evaluation method against the derailment. We are going to keep studying the derailment phenomena in order to eradicate the derailment accidents.

## REFERENCE LIST


[1] EN 14363:2005, "Railway applications - Testing for the acceptance of running characteristics of railway vehicles - Testing of running behaviour and stationary tests".

[2] Takai, H., Uchida, M., Muramatsu, H., Ishida, H., (2002) Quarterly Report of RTRI Vol. 43, No. 3, 119-124.

# Investigation of the application of adjustable air suspension under traction control for heavy haul locomotives

**M Spiryagin[1], V Spiryagin[2], P Wolfs[3], C Cole[1], Y Q Sun[1], I Persson[4]**
[1] Central Queensland University, Centre for Railway Engineering, Australia
[2] Rostov Electric Locomotive Repair Plant, Design Office, Russia
[3] Central Queensland University, Power and Energy Centre, Australia
[4] AB DEsolver, Sweden

**ABSTRACT**

During operation in traction or braking modes, a locomotive's body weight distributes between bogies in different proportions depending on many factors. Each bogie and wheelset thus realises a different traction coefficient. The adhesion coefficient is not only dependent on the tractive or braking effort, but also on vertical wheel loading. Designs of a Co-Co locomotive equipped with adjustable secondary air suspension to optimise adhesion are studied in this work. The proposed solution is verified during numerical experiments using rail vehicle multibody modelling software. Results show that it is possible to improve weight utilisation compared with a standard locomotive equipped with rubber elements.

## 1. INTRODUCTION

A recent study (1) shows that different types of traction control do not provide a perfect solution for the realisation of maximum possible tractive efforts for heavy haul locomotives equipped with three axle bogies. However, it has been demonstrated that a good traction control strategy can definitely lead to the improvement of locomotive performance (2,3). It should also be emphasised that the traction behaviour is also dependent on the parameters of the primary and secondary suspension systems (4-8). Mechanical optimisation of the suspension is not a very attractive option because, in such cases, it can only be optimised for some specific tracks. In order to get more universal results, it seems very reasonable to work on the characteristics of the secondary suspension and to use an active spring suspension instead of a conventional design. As discussed in (9), for vertical suspension this can be achieved by means of an "air pump" controlled air spring. From literature reviews for designs of rail vehicles (9-11), it is possible to see that major efforts are focused on lateral dynamics (stability and guidance) and ride comfort. Traction studies in this field have not been performed at all, although some comments on this issue for heavy haul locomotives were made in experimental and theoretical work published in 1983 (12). However, no further development on this air suspension design has been carried out.

This paper presents a proposal for the implementation of an active secondary suspension system equipped with air springs. In order to show the impact of such designs, the study performed in this paper covers three different designs of the secondary suspension, these being conventional springs, plus uncontrolled and fully controlled air springs. Taking into account that the studied locomotive is equipped with a bogie traction control system, a common design approach for typical heavy haul locomotives, this paper aims to systemise the combination of aspects for both systems and to provide recommendations for further development by means of analysing results obtained from numerical studies performed in Gensys multibody software (13).

## 2. SECONDARY SUSPENSION DESIGN AND ITS MODELLING APPROACH

The basic design of the proposed heavy haul locomotive is shown in Figure 1. Eight air springs are installed on the locomotive, i.e. four springs are mounted on each bogie as secondary suspension system elements. They substitute for the rubber elements used in the conventional design. Each air spring has its own individual air supply and control system. The symbols as presented in Figure 1 are: k – air spring stiffness; c – damping coefficient; F – controllable force.

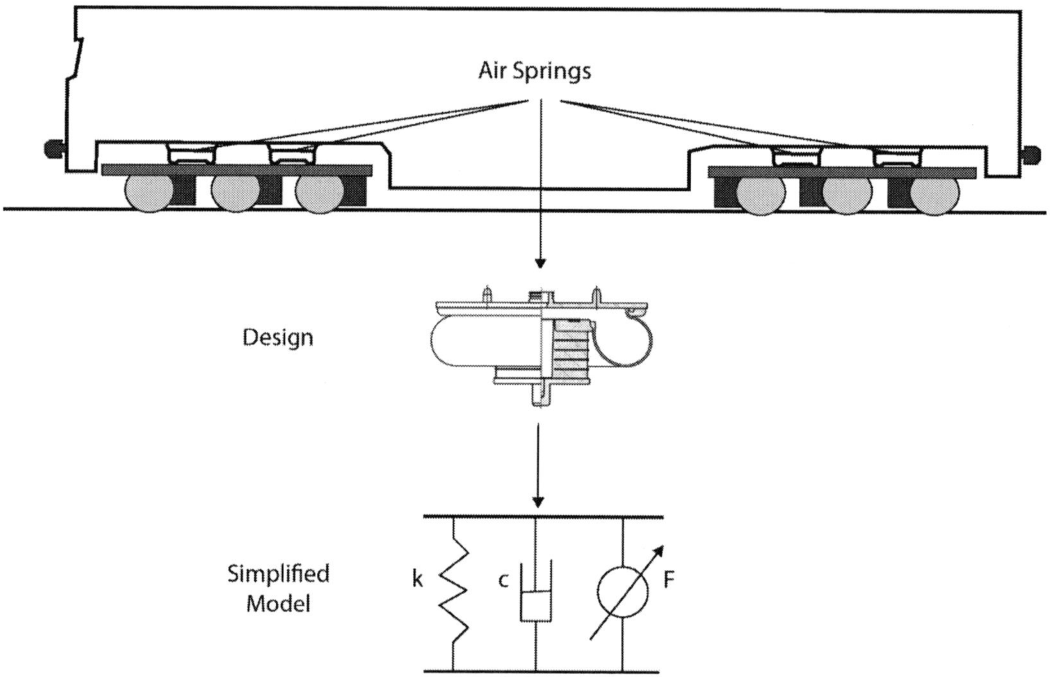


**Figure 1. Heavy haul locomotive equipped with the secondary air spring suspension and its simplified modelling approach**

A lot of work has been undertaken on the modelling of air springs for vehicle dynamics (14,15). Most of the models try to establish very accurate relationships with data obtained from laboratory experiments. The main disadvantage of such an approach is that it requires a great number of numerical adjustments, which commonly are not required for the development of a concept design. Taking into account that the traction coefficient is highly dependent on weight distribution, it looks reasonable to focus on the air spring modelling in the vertical direction. The simplified model, which has been verified in experiments (16), is also shown in Figure 1. Determination of stiffness and damper parameters are given detailed discussion in (17,18).

However, in this paper the stiffness, $k$, in N/m can be obtained as (19):

$$k = \frac{n A_{eff}^2 p_o}{V_o} \quad (1)$$

where $p_o$ is the pressure in the air spring, N/m$^2$; $A_{eff}$ is the air spring effective area, m$^2$; $V_o$ is the air spring nominal volume, m$^3$; $n$ is the polytropic index, which equals 1 for slow static deformation of the air spring and 1.3-1.4 for dynamic deformation (19). In this paper, the assumption has been made that the polytropic index is equal to 1.4 for the simulation of dynamic processes in air springs.

The work (12) states that the air spring area changes by approximately 5 percent when the height is changed under testing, therefore, the assumption of constant effective area has been made for this study. The main characteristics of the air spring, presented in Table 1, have been chosen based on the air spring design for a 138 tonne locomotive with three axle bogies described in (12).

**Table 1. The main characteristics of the air spring (12)**

| Parameter | Value |
|---|---|
| Height | 0.21 m |
| Effective area | 0.212 m$^2$ |
| Nominal volume | 0.027 m$^3$ |
| Working pressure | 550000 N/m$^2$ |

One more assumption needs to be made on the maximum allowable pressure. The work (12) shows that the dynamic pressure change under lateral dynamics studies for curves can add approximately 20 percent to the working pressure in air springs. Therefore, it looks reasonable to assume that the maximum pressure can be limited to $10^6$ N/m$^2$ as done in (11).

## 3. MODELLING IN GENSYS

Three simulation cases are investigated to show traction behaviour of locomotives equipped with conventional, non-controlled and controlled secondary suspension systems. Only two multibody models are required for these three cases.

### 3.1. Locomotive in Gensys

For this study, the simulation methodology developed in the previous studies (3,20,21) has been used for the investigation of locomotive movement under traction. Two mechanical models of a heavy haul locomotive with a semi-steering bogie design have been implemented in Gensys with the characteristics presented in Table 2. These models have different elements in their secondary suspension system as shown in Figure 2. One model has characteristics as installed on the standard locomotive (see Figure 2(a)) and the second one (see Figure 2(b)), is for the locomotive shown in Figure 1.

The model consists of a locomotive car body, two bogie frames, twelve axle boxes and six wheelsets. All bodies are modelled as rigid masses with 6 degrees of freedom. Constraints between each wheelset and its axle boxes are used in the model. In the lateral direction, a linear spring element is used between an axle box and a wheelset in order to model a possible clearance between these two bodies. In wheel/rail contact modelling, the rails are modelled as massless bodies. There are three springs normal to the wheel/rail contact surface, so that three different contact surfaces can be in contact simultaneously. The normal contact forces are solved by these three springs. The rails are connected to the track via springs and dampers in the lateral and vertical directions. The calculations of creep forces are made based on the method developed by Polach (22).

**Table 2. Parameters for the multibody model**

| | |
|---|---|
| ***Vehicle body*** | |
| Centre of gravity, vertical distance above top of rail | 1.930 m |
| Mass | 90000 kg |
| Moment of inertia, roll/ pitch/ yaw | 174002/3727195/3706796 kg·m$^2$ |
| ***Bogie frame*** | |
| Centre of gravity, vertical distance above top of rail | 0.733 m |
| Mass | 12121 kg |
| Moment of inertia, roll/ pitch/ yaw | 5318/37007/41029 kg·m$^2$ |
| ***Axle box*** | |
| Centre of gravity, vertical distance above top of rail | 0.5335 m |
| Mass | 200 kg |
| Moment of inertia, roll/ pitch/ yaw | 50/50/50 kg·m$^2$ |
| ***Axles*** (with the traction motor mass shared between bogie and axles) | |
| Centre of gravity, vertical distance above top of rail | 0.5335 m |
| Mass | 2893 kg |
| Moment of inertia, roll/ pitch/ yaw | 2067/1387/2067 kg·m$^2$ |
| ***Secondary suspension*** | |
| Vertical stiffness per spring *(conventional design)* | 7870000 N/m |
| Vertical stiffness per spring *(design with air springs)* | 1281000 N/m |
| Vertical damper per spring *(design with air springs)* | 1000 Ns/m |
| ***Other dimensions*** | |
| Axle Spacing | 1.9 m |
| Bogie Spacing | 13.70 m |

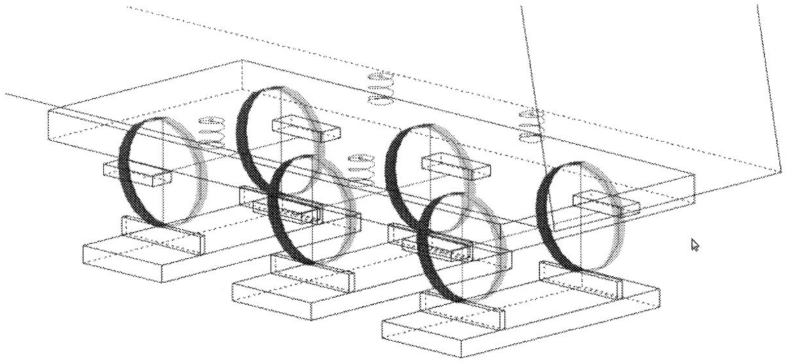

**(a)**

**(b)**

**Figure 2. Locomotive secondary suspension model**
**(a) conventional secondary suspension**
**(b) air spring secondary suspension**

### 3.2. Control systems

Two traction control systems are needed for this study. The traction control system that acts for each bogie which is used both for the case of a standard locomotive and for a locomotive with uncontrolled secondary suspension system is shown in Figure 3(a). The advanced traction control system, which also includes a control of air spring secondary suspension, is shown in Figure 3 (b). Both traction control systems have been created as subroutines in Gensys.

The variables shown in Figure 3 are $T_{ref}$ = reference torque; $T_{ref^*}$ = reference torque generated by the control system; $T_{in}$ = input motor torque; $T_{wheels}$ = traction torque applied to the axles; $\Delta T$ = torque reduction; $\omega$ = angular velocity of a reference axle; $\omega_1$, $\omega_2$, $\omega_3$ = angular velocity of front, middle and rear axles respectively; $s_{est}$ = estimated longitudinal slip and $s_{opt}$ = optimal longitudinal slip, $h_1,..,h_8$ - heights for each air spring; $F_{a1},..,F_{a8}$ - actuator forces acting in the vertical direction and dependent on the supply pressure to each air spring. The dynamics of the inverter and traction motor(s) has been modelled by means of a low-pass filter. In this study, we assume that similar sensors as described in (11) can be used for a proposed suspension control subsystem, having an individual algorithm for each air spring based on simple fuzzy logic rules that tries to equalise heights of the air springs for each bogie, i.e. the suspension control of a front bogie is independent from the suspension control of a rear bogie.

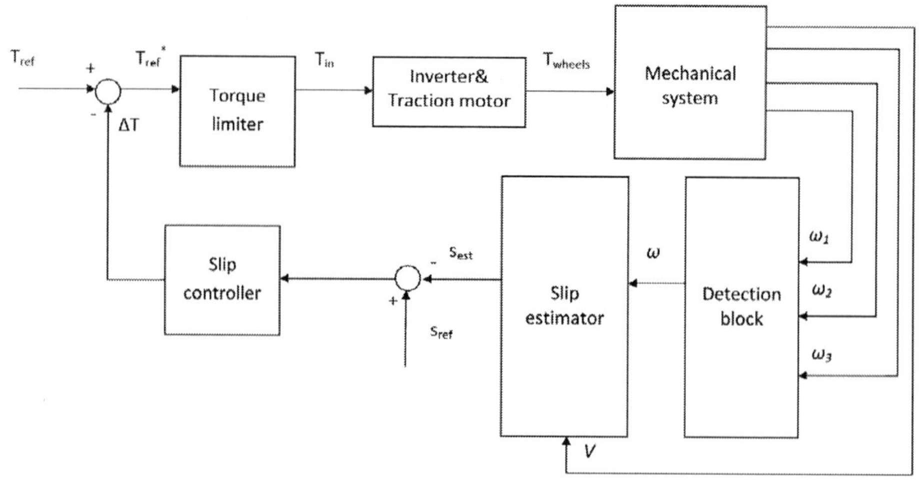


(a)

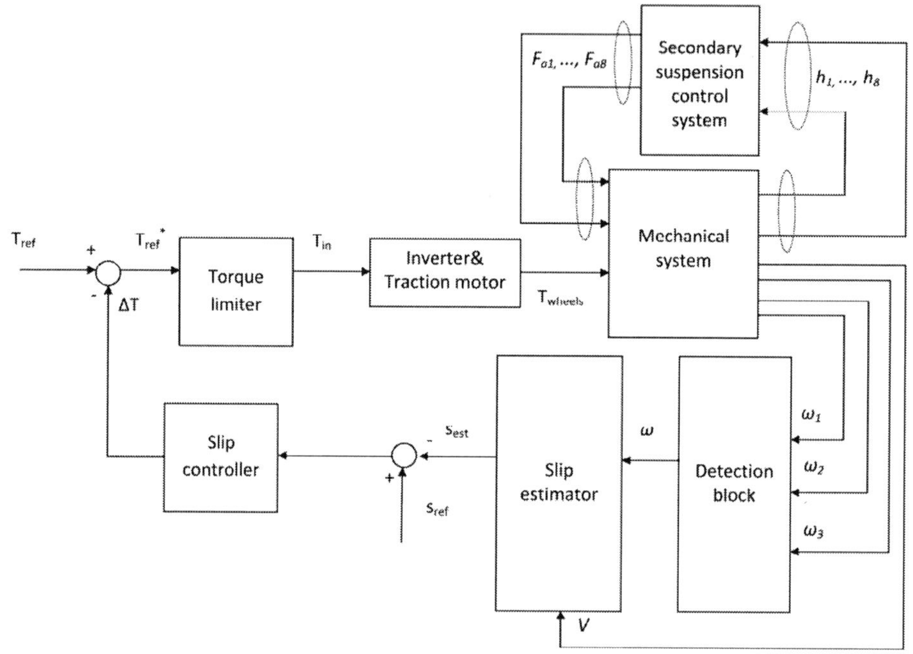


(b)

**Figure 3. Traction control algorithms**
**(a) standard approach, (b) an advanced approach**

## 4. SIMULATION AND RESULTS

The locomotive runs on straight track with ideal track conditions. All-weather adhesion of 35% is used as the limit because it can be assumed that this level can be realised with a 97% probability on dry rail.

The adjustable longitudinal coupler force has been attached to the locomotive model in order to have a constant speed of 22 km/h, which has been chosen for this study because it is the continuous mode speed, i.e., at which the locomotive experiences maximum creep forces.

As mentioned previously, three simulation cases have been performed in order to compare three secondary suspension design variants. The comparison of results for axle load distributions, traction coefficients and tractive efforts for each design variant are shown in Figures 4-6. The maximum tractive efforts have been applied to the wheelsets in order to perform this study. In theory, the maximum tractive effort of 420 kN assumed for this speed can be achieved with the optimal slip value set to 0.08.

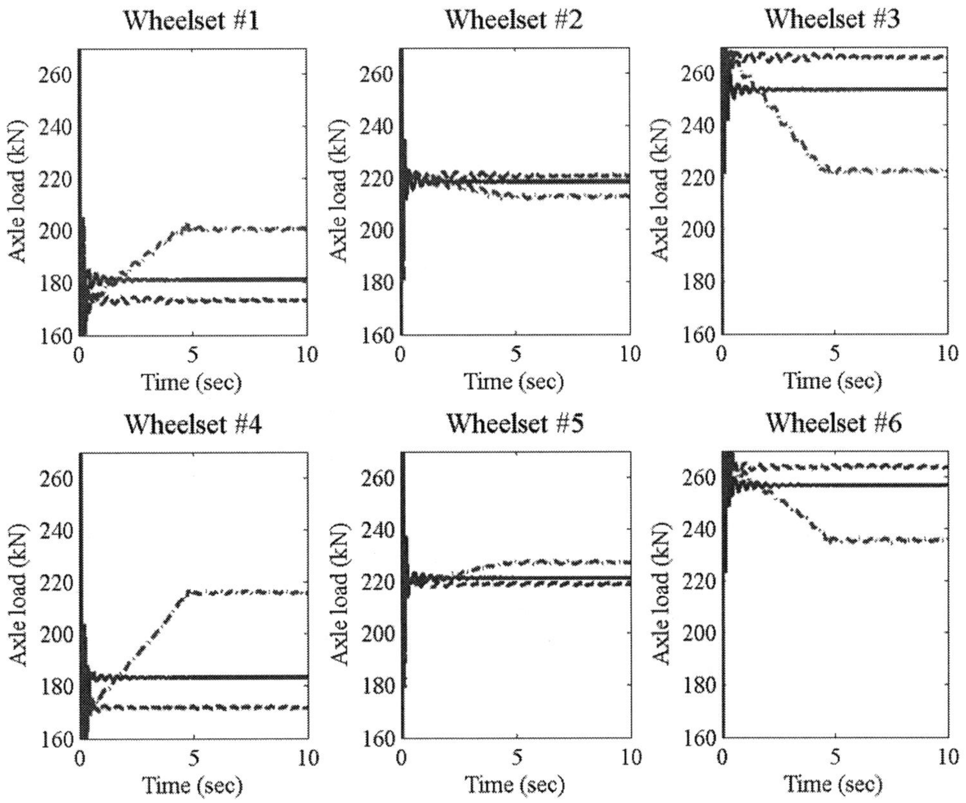


**Figure 4. Axle load results in the time domain**
**_____ (blue line) conventional secondary suspension**
**- - - (green line) uncontrolled air spring secondary suspension**
**-.- (red line) controlled air spring secondary suspension**

The axle load results presented in Figure 4 show that the axle loads are not perfectly distributed in the cases of conventional and uncontrolled secondary suspension design. These results show only a slight difference between these two designs. However, the weight transfer for the controlled secondary suspension design almost allows equalising weight loads between axles, which is a good sign and would lead to the reduction of vertical load from a wheel to a rail under traction. Figure 5 also confirms more benefits for the controlled secondary suspension design; the middle and trailing axles are very close to reach a required traction coefficient of 0.35. Meanwhile, the uncontrolled secondary suspension design leads to a significant reduction of traction coefficient even in comparison with the conventional design. All these design variations result in the lowest tractive efforts being achieved by the uncontrolled design as shown in Figure 6. However, the design with controlled secondary suspension seems a good solution because it shows a 10 percent increase in drawbar pull.

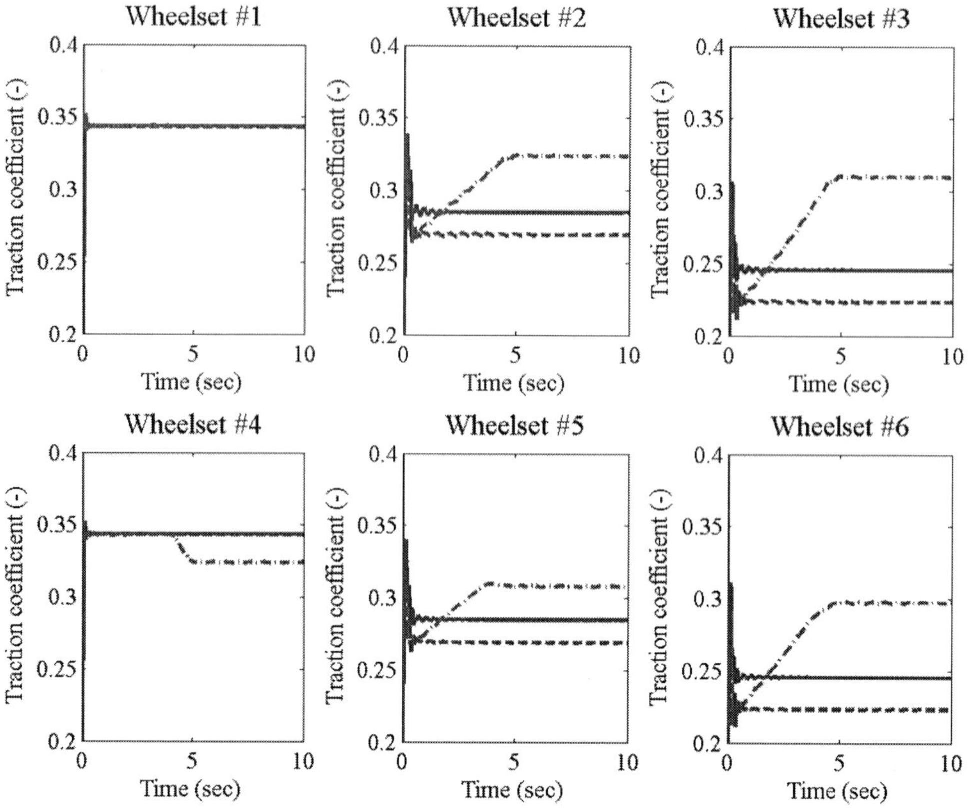


**Figure 5. Axle traction coefficient results in the time domain**
**_____ (blue line) conventional secondary suspension**
**- - - (green line) uncontrolled air spring secondary suspension**
**-.- (red line) controlled air spring secondary suspension**

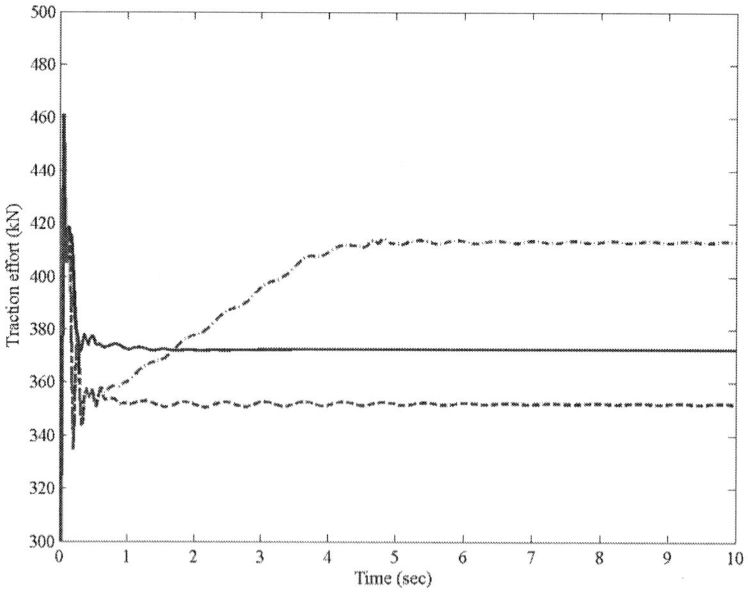


**Figure 6. Locomotive tractive effort results in the time domain**
**_____ (blue line) conventional secondary suspension**
**- - - (green line) uncontrolled air spring secondary suspension**
**-.- (red line) controlled air spring secondary suspension**

Summarising the above, the uncontrolled air spring suspension gives worse results for all output parameters due to the soft stiffness characteristics and there is no reason to use it on heavy haul locomotives when high traction is required. Similar assumptions can be found in (12). The conventional secondary suspension shows better results than uncontrolled suspension equipped with air springs, but it still cannot reach the performance of a locomotive equipped with a controlled air suspension system.

## 5. CONCLUSIONS

The traction behaviour of heavy haul locomotives equipped with different secondary suspension systems under a bogie traction control system has been investigated using the Gensys multibody software package. The results confirm that an advanced traction control system shows a better traction performance in comparison with a conventional locomotive. Furthermore, there is no reason to install an uncontrolled air spring suspension on a heavy haul locomotive because a worse axle load distribution is generated by this design in the traction mode.

This paper has some limitations due to the usage of a simplified air spring model. Therefore, further research in this direction is strongly recommended, and that should be focused on the more detailed modelling of air spring behaviour and its implementation in multibody dynamics codes. A more detailed study into secondary suspension control is also required because it should also take into account lateral dynamics, especially in curved track.

Finally, it is necessary to indicate that this paper does not provide a final solution, but it proves a design concept and it opens more topics which might be explored in the future.

## REFERENCES


1. M. Spiryagin, C. Cole, Y.Q. Sun, T. McSweeney, V. Spiryagin, N. Gorbunov, A. Golubenko, Heavy haul locomotive suspension behaviour under bogie and individual wheelset traction control strategies, 23rd Symposium on Dynamics of Vehicles on Roads and Tracks (IAVSD2013), Qingdao, China, 19-23 August, 2013.
2. M. Yamashita, T. Soeda, Development of re-adhesion control method considering axle-weight transfer of electric locomotive, Quarterly Review of Railway Technical Research Institute, 52(1) (2011), 7-12.
3. M. Spiryagin, Y. Q. Sun, C. Cole, S. Simson, I. Persson, Development of traction control for hauling locomotives, Journal of System Design and Dynamics, 5(6) (2011), 1214-1225.
4. H.A. Marta, K.D. Mels, G.S. Itami, Design features and performance characteristics of the high traction, three- axle truck, ASME/IEEE Joint Railroad Conference, Jacksonville, Florida, USA, March 14-15, 1972.
5. N. Gorbunov, Theory and practical realization of the system approach in the construction of the carriage of locomotive, DSc thesis, East Ukrainian National University, Lugansk, Ukraine, 2006. (in Russian)
6. V. Spiryagin, Improvement of dynamic interaction between the locomotive and railway track, PhD thesis, East Ukrainian National University, Lugansk, Ukraine, 2004. (in Russian)
7. S. Simson, Three axle locomotive bogie steering, simulation of powered curving performance. Passive and active steering bogies, PhD thesis, Central Queensland University, Rockhampton, Queensland, Australia, 2009.

8. M. Spiryagin, H. H. Yoo, K. S. Lee, V. Spiryagin, M. Gorbunov, Investigation of influence of constraints with radius links on locomotive axle load distribution and wheelset steering ability, Journal of Mechanical Science and Technology, 27(7), (2013), 629-640.
9. R.M. Goodall, R.A. Williams, A. Lawton, P.R. Harborough, Railway vehicle suspensions in theory and practice, Vehicle System Dynamics: International Journal of Vehicle Mechanics and Mobility, 10(2-3) (1981), 210-215.
10. A. Orvnas, Active secondary suspension in trains – A literature survey of concepts and previous work, Publ. 0803, ISBN 978-91-7415-144-2, Division of Rail Vehicles, Department of Aeronautical and Vehicle Engineering, Royal Institute of Technology, Stockholm, Sweden, 2008.
11. N. Docquier, Multiphysics modelling of multibody systems: Application to railway pneumatic suspensions, PhD thesis, Université catholique de Louvain, 2010.
12. N.A. Lobachev, Pneumatic support and connection system between the locomotive body and bogies, PhD thesis, All-Union Scientific Research Diesel Locomotive Institute, Kolomna, USSR, 1983. (in Russian)
13. AB DEsolver, GENSYS.1410 Reference Manual. Available at: http://gensys.se/ref_man.html.
14. S. Bruni, J. Vinolas, M. Berg, O. Polach, S. Stichel, Modelling of suspension components in a rail vehicle dynamics context, Vehicle System Dynamics: International Journal of Vehicle Mechanics and Mobility, 49(7) (2011), 1021-1072.
15. S. Alfi, S. Bruni, G. Diana, A. Facchinetti, L. Mazzola, Active control of airspring secondary suspension to improve ride quality and safety against crosswind, Journal of Rail and Rapid Transit, 225(1) (2011), 84-98.
16. J. Wang, Nonlinear modelling and h-infinity model reference control of pneumatic suspension system, PhD thesis, Iowa State University, Ames, Iowa, USA, 2012.
17. M. Presthus, Derivation of air spring model parameters for train simulation, MSc thesis, Division of Fluid Mechanics, Department of Applied Physics and Mechanical Engineering, Lulea University of Technology, Lulea, Sweden, 2002.
18. M. Berg, An airspring model for dynamic analysis of rail vehicles, Division of Railway Technology, TRITA-FKT Report 1999:32, ISRN KTH/FKT/R-99/32-SE, Department of Vehicle Engineering, Royal Institute of Technology, Stockholm, Sweden, 1999.
19. B.M. Ibraev, Parameters of pneumatic suspension of a rail bus for Kazakh railways, PhD thesis, Moscow State Railway University (MIIT), Moscow, Russia, 2009. (in Russian)
20. M. Spiryagin, A. George, Y. Q. Sun, C. Cole, T. McSweeney, S. Simson, Investigation on the locomotive multibody modelling issues and results assessment based on the locomotive model acceptance procedure, Journal of Rail and Rapid Transit, 227(5), (2013), 453-468.
21. M. Spiryagin, C. Cole, Y. Q. Sun, M. McClanachan, V. Spiryagin, T. McSweeney, Design and simulation of rail vehicles, Ground Vehicle Engineering Series, CRC Press, 2014.
22. O. Polach, Creep forces in simulations of traction vehicles running on adhesion limit, Wear, 258 (2005), 992-1000.

# Wireless mesh sensors are transforming asset monitoring in rail

**S Maddison, B Smith**
Senceive Ltd, UK

## ABSTRACT

Wireless sensor network systems as supplied by Senceive are now totally accepted as a robust, cost effective, precise and reliable tool for the condition monitoring of rail, geo-technical and structural assets. They are providing for remote monitoring of assets through the use of low cost, battery powered, high precision, self-configuring wireless sensor systems which can be readily installed with a minimum of effort. They also, of course, avoid the need for wiring or indeed, in the case of optical systems, the need for frequent cleaning and extensive set-up and configuration. Extensive development over the past 2-3 years has led to a highly significant step change in functionality by Senceive in wireless RCM, including for example: battery life extended to over 15 years and the provision of industrial levels of network robustness and data stability. Additionally, and as a direct reflection of this success over recent years, Senceive's wireless systems have also been approved and extended into two new important and challenging asset classes: Rail Tunnels and Track Bed.

## 1 INTRODUCTION

The use of wireless for remote condition monitoring for railway infrastructure has long been a subject of trial and discussion e.g. (1) (2) and much more recently (3) (4), but reports of commercial deployments are much more limited e.g. (5) (6) reflecting the substantial gestation that the technology has required in order to become accepted as a robust mechanism for Remote Condition Monitoring.

After a long period of iterative development, trialling and product acceptance, Senceive's wireless sensor networks are now becoming, or indeed have become, the established means of wireless Railway Condition Monitoring (6), with considerable benefits over the alternatives in terms of ease of deployment, speed of installation, and the ability to readily reconfigure and redeploy. Considerable technical advances have been made that are for example extending battery life to 15 years whilst reporting every 10-15 minutes, and system sizes to greater than 100 nodes with integrated stable high precision sensors.

Prior to more recent success in tunnels and on track beds, Senceive's deployments were typically on remote and difficult to reach assets, such as bridges, earthworks, retaining walls and viaducts, as shown for example in Figure 1. Monitoring requirements were generally associated with linear movement sensors for measuring displacement e.g. for cracks, and tilt meters for measuring angular movement in a structure or on an earthwork.

Now with a range of patented fixings, deployments can typically be achieved in a single shift or less, and data can be accessed remotely prior even to leaving site. This minimises track possessions and staff working time in hazardous locations. Furthermore the system's ability to operate in remote locations enables the monitoring of assets where otherwise manual inspection would be the only alternative.

**Figure 1: Installation of wireless nodes in difficult locations**

As previously stated, significant advances have been made in the use of wireless for the successful monitoring of two new additional and important classes of asset: railway tunnels and railway track. In tunnels, wireless has proven itself through its speed and ease of installation, and also flexibility during intense engineering works, and offering a monitoring solution in confined spaces where there was no viable alternative. On track, for the monitoring of track tilt (cant) and associated twist parameters, wireless is now proving itself as a superior alternative to optical monitoring through its high stability, high resolution measurements and lack of cleaning, maintenance or configuration requirements.

This paper describes these applications in greater detail, drawing on the experience of a number of successful deployments over the past 2-3 years.

## 2 WIRELESS STRUCTURAL MONITORING SYSTEM

The acceptability of wireless has been and still is, in many instances, held back because of some wireless solutions being unreliable, difficult to install, insufficiently robust or scalable, having short battery life and even the need for mains power for parts of the system.

Senceive has solved these problems through the experiences learned from some 10 years of challenging geo-technical deployments, the lessons of which have been incorporated in a wide ranging and step change development programme, resulting in the launch of its highly efficient next generation FlatMesh system.

Whilst the system has always been a "true wireless mesh", recent improvements impact all aspects of the whole system. The FlatMesh firmware, installed on each wireless node, enables a network of such nodes to communicate efficiently with each other effecting network-wide data transmission, which ultimately provides an integrated monitoring system. Mesh networks have a "flat structure", whereby every node communicates with its neighbours on an equal status basis, without hierarchy and without the need for any specialised router or concentrator nodes, see Figure 2. This enables simple and quick installation, through its self-organising operating mechanisms. This also provides a very robust system, as with redundant wireless paths it is tolerant of individual node or path failures, as well as being readily reconfigurable, accommodating the addition or removal of nodes.

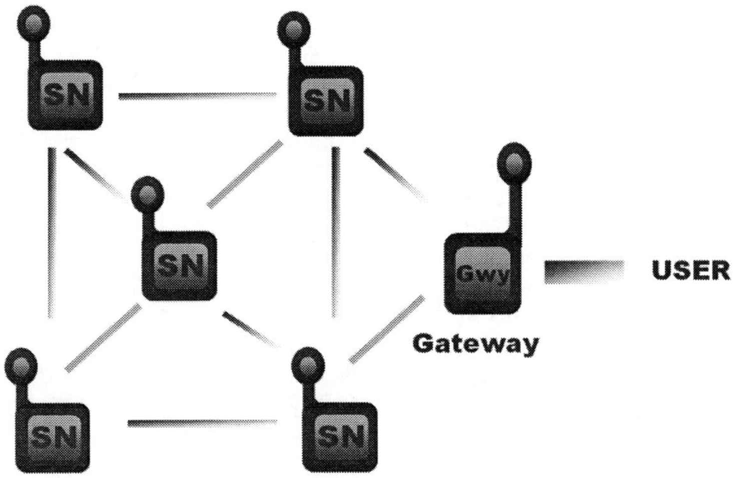


**Figure 2: Key elements of FlatMesh system showing Sensor Nodes (SN)**

Extensive development of proprietary algorithms and firmware within the node enables low power consumption, yielding an unprecedented 10-15 year's operational life with 10-15 minute reporting, powering both high precision sensors and wireless communications from a single 'D' cell battery. In the event of a node failing or radio paths being blocked, data is automatically re-routed to ensure a seamless flow of information back to the user with greater than 99.9% data transmission reliability.

Different sensor classes each present their own distinct challenges in terms of interface, operation, power management and stability. The design challenges to integrate a wide range of different sensor types and to operate them with extreme energy efficiency have been successfully surmounted, and system improvements continue to be made. A single node can now be fitted with three sensor interface modules, and each of these modules can support distinct sensor types and applications, and one or more sensor channels depending on the class. Simpler single function sensors are significantly reduced in size and cost, for specific larger scale applications, Figure 3.

**Figure 3: Single function reduced size sensor node**

## 3 TUNNEL MONITORING DURING ENGINEERING AND CONSTRUCTION WORKS

Three projects on London Underground (LU) demonstrate the effectiveness of wireless monitoring in Tunnels: Bond Street grouting works, BSBS tunnel relining, and major demolition and re-development above the Northern Line. The first of these was a short term remedial back-grouting project, to stabilise the tunnel running through a geology with particular challenges. The second is a long term project replacing concrete linings with modern cast iron segments. In both of these, monitoring is required for understanding the immediate impact of works during engineering activities, and to evaluate ongoing stability during running hours. In both cases the system needs to be quickly installed, readily redeployed by technicians as works progress, and provide data accessible to the engineering teams in situ - as well as remotely during running hours.

The third example is a long term site re-development with fixed deployment of sensors over 5 years on both track and tunnel lining, and where optical systems could not be deployed due to lack of space and other considerations.

### 3.1 Remedial grouting project

Typical challenges for such works in underground rail tunnels are that there should be little or no planned interruption to the train running schedules. All works normally have to be carried out during the night-time 'engineering hours' typically between 01:10 and 04:30. The critical question in this situation was: How to monitor dynamically in real time whilst grouting works were in progress and to ensure the tunnel remained stable?

Conventional means of monitoring would have been impractical. A system of fixed wired sensors would take months to install, requiring several thousand sensor points at a prohibitive expense. This would also involve a spaghetti of wiring that would potentially interfere with access to the grouting valves in the tunnel lining, quite apart from having to be installed right through the area from the start, rather than just at a particular area of shift work. Optical systems would not work as there would not be the required lines of sight with the grouting crew in situ, with all their equipment, pipes, access scaffolding, and the same issues applied to manual methods, see Figure 4.

**Figure 4: Wireless sensors installed on tunnel rings around grouting works**

The solution was to use Senceive's battery powered wireless tilt meters, positioned around the cast iron segments of the tunnel lining, to measure any rotational movement in each segment. This data was then used to derive linear measurements of lining movement, and hence changes in the two key parameters of squat (i.e. tunnel height) and horizontal diameter.

Data samples are transmitted from the originating sensor node, via its neighbours, which pass the data back to a network gateway, in this case connected to a laptop for use in the tunnel. The big attractions of using wireless tilt sensors were the absence of wires, the ability to provide measurements with all the crew and equipment in action, the speed with which they could be deployed, and the ability to move them as required to the locations of the particular shift's work as work progressed.

Shortly into the project the value of continuing to monitor during normal operating hours was realised. Clearly it was not possible to leave a laptop running close to the running tunnels, and a suitable location was identified along a cross passage, up a shaft and into a passage well away from the track. It was a matter of ten minutes to place a sequence of 'repeater' nodes (i.e. nodes without sensors) to link the monitoring system back to where the laptop was left at the end of shift.

In order to mount the 100 sensors quickly and robustly to the tunnel lining, so that they could be deployed, baselined and fully functioning all within 10 minutes from the start of shift, a patented system of magnetic brackets was designed. These brackets were designed collaboratively with Halcrow/CH2M Hill, Figure 5.

**Figure 5: Wireless sensors with patented magnetic mountings**

Fast data reporting was required, and the data was to be presented in an accessible and friendly way for the monitoring engineers; this was so that they could interact in real time with the lead shift grouting engineer and meant a lot of data from a large number of sensors with minute by minute reporting. Senceive's 'Web Monitor' interface software (Figure 6) was deployed on a laptop to present the data in a number of simple, but suitably accessible visuals, with neat user friendly graphical displays. With the proprietary tunnel algorithms incorporated into the Web Monitor, the necessary computations of the data produced the composite tunnel distortion measurements required.

All necessary approvals were achieved with London Underground, going through a rigorous process of evaluation and testing of the system, both for EMC compatibility and, also very importantly, for mechanical stability and safety.

In practice, it only took a couple of shifts to train up the operatives to place the nodes with confidence. Once located and with the first data coming back, the nodes could be 'baselined' and thus made ready for monitoring. Operatives quickly became aware of the sensors, the need to avoid accidentally moving them, and to notify the monitoring engineer if this did occur. When this did occasionally happen it was straightforward to re-baseline the node, and carry on.

To provide validation, and also as part of the handback process, manual measurements were taken at the end of shift – diametrically at different angles across the tunnel. These confirmed measurements from the wireless system, and were also cross checked where permanent long term sensors were already installed, showing a high degree of correlation and the success of the design.

The grouting project took four months to complete, and the use of Senceive's wireless monitoring system saved at least £1m to the overall project cost. That is without any consideration of the other benefits from maximising grouting time and real time decision making, as well of course in minimising issues of installation or deployment "hassle". The project was submitted for the prestigious International Tunnelling Awards, and won the Product Innovation Category in the autumn of 2012, in conjunction with Costain/Laing O'Rourke and Halcrow/Atkins.

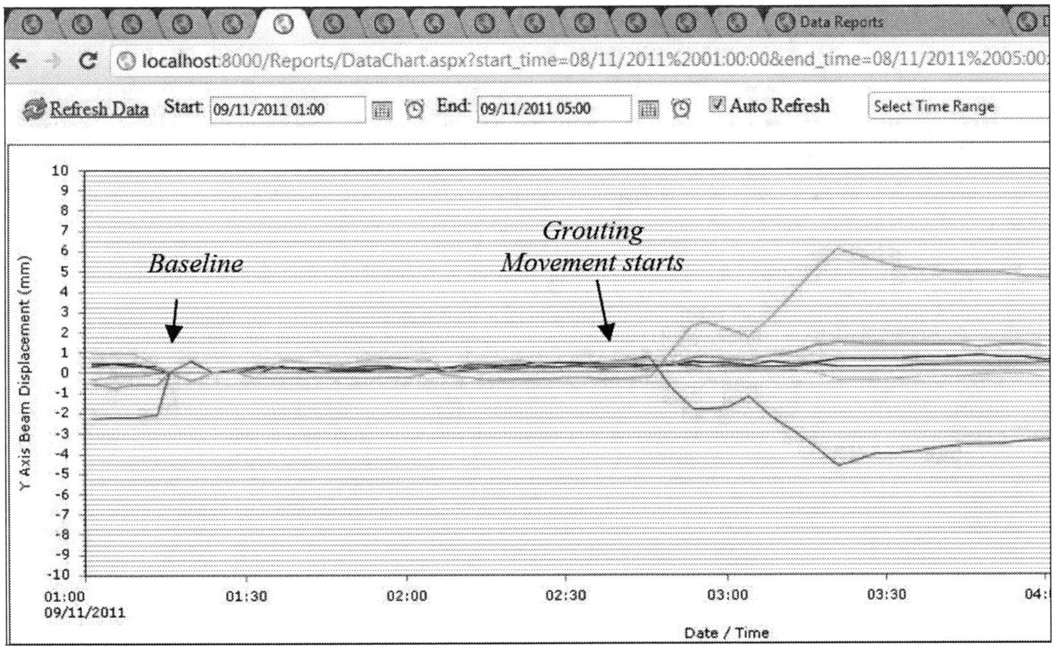


**Figure 6: Graphical display showing segment movement during grouting**

### 3.2 Tunnel lining replacement

Similarly challenging is a 3 year project for tunnel lining replacement that London Underground / Tubelines is carrying out on another part of the network. A 200m length of tunnel lining comprising concrete segments is being progressively replaced with cast iron segments, shift by shift in engineering hours, using two specially designed trains. Senceive's wireless tilt sensors are deployed ahead of the works, leapfrogging ring by ring as the works progress; further rings of sensors are deployed behind, some following the work as it moves along the tunnel, and some being left behind long term, to check on the tunnel stability post-works (Figure 7).

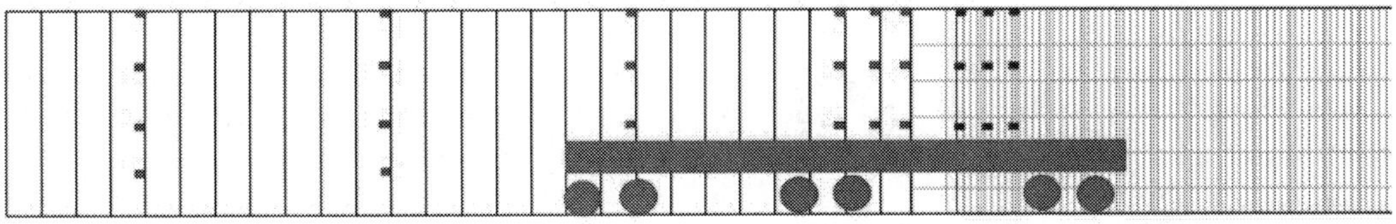

**Figure 7: Schematic of sensors on both sides of the works in the tunnel**

This is an exciting and highly demanding environment where sensors are installed in the immediate proximity of heavy engineering activity (Figure 8), and the sensors are deployed and redeployed by the engineering team operatives. The system has been further developed subsequent to the initial trials, enabling the data to be accessed not only from the nearest platform at any time, as was originally required, but also by the engineers on the train. All access is performed wirelessly and in real time.

The system is demonstrably easy to use, and the sensors are showing extremely good long term stability, in terms of the measurements taken and the data generated.

**Figure 8: Wireless sensors in the immediate proximity of the tunnel lining replacement machine**

### 3.3 Long term deformation monitoring from construction works

Through 2013, systems have been installed in other tunnels running underneath a swathe of London, including those associated with massive redevelopment above the Northern Line, as well as LU and BT tunnels to monitor the effects of boring for London Power Tunnels.

For long term stability monitoring of LU tunnels in close proximity to the major demolition and construction works above the Northern Line, nodes are deployed around the tunnel rings, in order to determine distortion, as in the cases above. Additionally they are fitted along the middle of the track and on the tunnel crown in order to determine tunnel and track heave and settlement longitudinally. In this particular case, the long term stability of the wireless sensors, in an easy deployable package, is absolutely essential to determine behaviour of the tunnel and ensure that it is within the predicted bounds of movement, see Figure 9.

In this project, the wireless system has displaced the use of optical methods, based on automated total stations (ATS), as the very limited space available in the tunnel envelope meant that this method would simply not have been possible. Periodic manual surveying, precise levelling as well as using optical reflectors mounted on the nodes, provides an independent validation of the stability and precision of the system. Although backed up by this routine manual surveying, the latter is not only highly labour intensive, but of course only provides snapshots in comparison to the much more frequent measurements carried out automatically by the wireless system. It turns out that the Senceive system has other advantages over optical methods, in terms of stability, resolution and accuracy, as described in the next section below.

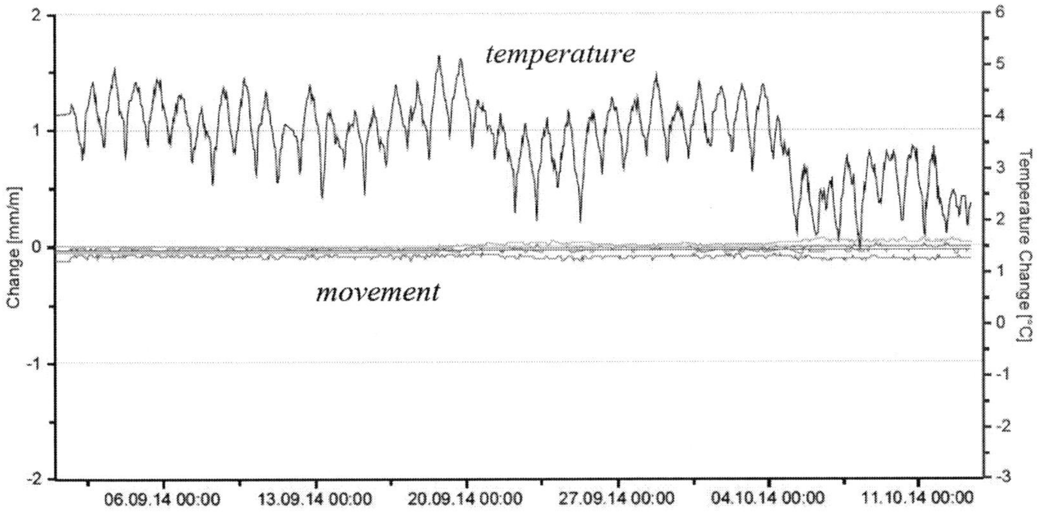


**Figure 9: Tunnel measurements in Northern Line**

# 4 RAILWAY TRACK MONITORING

## 4.1 Cant and twist

The Docklands Light Railway (DLR) station at Pudding Mill Lane in East London was to be demolished and both the track and station to be moved to make way for the Crossrail tunnel that surfaces next to it. Morgan Sindall are responsible for the works, and are required to monitor the multiple Network Rail tracks that run alongside as well as the DLR tracks, to ensure that there is no adverse movement whilst the works are in progress, for a period of some 2 years.

Following earlier issues of stability and maintenance with an optical ATS system, they sought a new solution for measuring track tilt and twist. A pilot deployment was made involving Senceive's high precision wireless tilt sensors, which were attached directly to the sleepers on the DLR track, with a view to proving this as a viable alternative to optical methods. The 6 month trial successfully showed outstanding stability and resolution of track cant change to considerably less than 0.1mm. This led to the design of a hybrid system, with Senceive's wireless tilt sensors mounted on the sleepers at intervals along the track, to measure tilt, twist and longitudinal settlement, cross correlated with a reduced system of optical measurements to detect possible slew.

This has resulted in the installation of a complete monitoring system of ca. 700 sensors extended over several hundred metres of parallel tracks (Figure 10), with a number of solar powered (and therefore totally mains power free) GSM/GPRS enabled gateways. These gather the data and forward it back to a secure web server. The speed of installation has also proved to be a significant benefit, with the installation being effected quickly and efficiently during limited track possessions. It has also been necessary to move the gateways at different phases of the construction work, with the construction and relocation of protection barriers. Moving the gateways was quickly and readily undertaken without any interruption of service or monitoring.

**Figure 10: Wireless sensors deployed on track sleepers to measure cant**

The cant change measurements are presented directly on a graphical representation. A sample showing diurnal movements is shown in Figure 11. These measurements are additionally processed to generate both relative and absolute twist measurements based on measured and design characteristics of the track. A series of trigger levels are configured specific to different track sections which generate colour coded visible and email / SMS alerts. The data is routinely validated by precise levelling surveys which confirm the outstanding accuracy of the system.

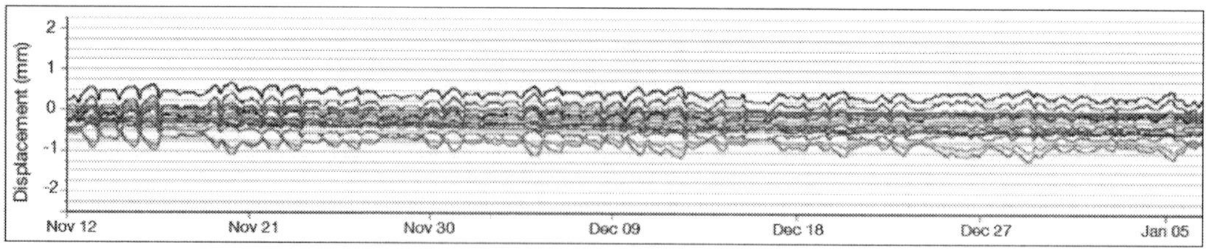


**Figure 11: Long term cant measurements**

## 4.2 Track heave monitoring

As part of the Victoria Station Upgrade Project an underpass is to be tunnelled directly beneath the existing sub-surface District and Circle line brick tunnel. The ground beneath the track is permeable, and first needs to be treated using TAM grouting during engineering hours, prior to tunnelling the underpass. The risk was the potential to heave the tracks, so the client, TWBN, required a real time very high precision monitoring solution that could be quickly deployed on the rails for each shift. Lines of sight could not be guaranteed, precluding the use of an optical solution, and a wired system would be too clumsy.

During grouting, movement could occur in any part of the 30 metre zone across two separate tracks. Senceive's patented magnetic mountings were specially upgraded to affix the 20 wireless sensors directly onto the surface of the rail, Figure 12. These were configured to alert for any movement in excess of 0.009 degrees of angular tilt, measured at one minute reporting intervals. All this was carried out in a totally wire and mains power free environment.

**Figure 12: Track heave monitoring in real time**

The sensors communicate real time to a gateway attached to a local PC which has a graphic display and specially developed highly audible alerts built in, Figure 12, to ensure someone does not have to be permanently stationed by the laptop during the short engineering shift. This allows the engineer to use his time more effectively away from the monitor and be able to communicate to the grouting team, and only carry out manual and precise levelling if and when alarms are sounded, in order to validate the movement. The data was incredibly stable, only measuring real movement, with no false alarms, see Figure 13. A full precise levelling survey was carried out at the end of each shift, providing a comprehensive validation of the measurements.

This system was installed on site within 3 weeks of the initial proposal. Senceive worked alongside the client's team for the first two night shifts, to ensure all was well and the system was fully understood, and since then all works and deployments have been carried out exclusively by TWBN engineers. The system has been a resounding success.

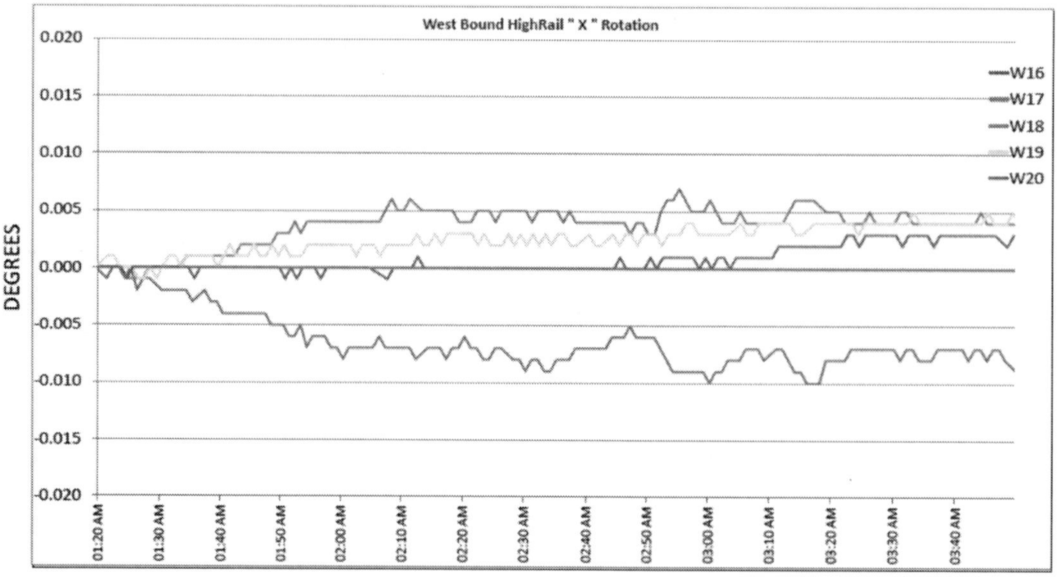


**Figure 13: Track heave movements during shift**

## 5 CONCLUSION

After a long gestation period and following extensive funded development Senceive has now shown that its unique and proprietary wireless sensing is able to perform reliably and with high precision and stability.

This has taken almost a decade of experience with demanding clients, and challenging geo-technical and structural deployments, incorporating many lessons into the development of this wireless sensor platform. It has almost uniquely come to be accepted in the UK as a primary means of wirelessly monitoring structural railway assets in a wide variety of different applications.

Step change improvements in terms of reliability, battery life, robustness and stability of sensor readings, along with the benefits of speed of deployment, flexibility and the ability to easily reconfigure and redeploy make it uniquely valuable for certain problem areas such as tunnel engineering works, locations where there is extremely limited space, and where there really is no viable monitoring alternative, including labour intense manual systems. Moving forward, wireless is complementing or indeed displacing other wired or optical solutions where they might previously have been seen as the first and only choice.

The prime benefit of wireless is of course that there are no wires. In the railway environment in particular this minimises personnel hazards, track access and possessions, and allows for the monitoring of difficult to access assets, where it might otherwise not even be cost effective to attempt automated monitoring. Indeed in the Rail sector wireless has potential for rolling stock as well, and Senceive has been providing the wireless backbone for a collaborative project for detecting wheel bearing degradation on wagon bogies (7).

Senceive's Wireless Sensor Networks are changing the face of Remote Condition Monitoring of Railway assets.

## ACKNOWLEDGEMENTS

Senceive would in particular like to thank the following for their support: London Underground/ Tubelines; Network Rail; Costain; Morgan Sindall; Halcrow C2HM; Arup; Getec/Keller; Amey; Taylor Woodrow/Bam Nuttall; TPS Consult/Carillion; SAL; Technology Strategy Board.

## REFERENCE LIST

(1) E. Aw, J. Germaine, A. Whittle. "Monitoring System to Diagnose Rail Track Settlement due to Subgrade Problems", *Soil and Rock America 2003, 12th Panamerican Conference on Soil Mechanics and Geotechnical Engineering*, Vol 2, pp 2715-2720 (2003)

(2) E. Aw, I. Stoianov, A. Whittle, J. Germaine. "A Wireless Remote Monitoring System: Application in the Northeast Corridor Railtrack", *Seventh International Symposium on Field Measurements in Geomechanics*, (2007)

(3) A. Bilbao, D. Hoover, J. Rice, J. Champman. "Ultra-low power wireless sensing for long-term structural health monitoring", *Sensors and Smart Structures Technologies for Civil, Mechanical, and Aerospace Systems 2011*, Proc. of SPIE Vol. 7981, (2011)

(4) D-D. Ho, K-D. Nguyen, P-Y. Lee et al. "Wireless structural health monitoring of cable-stayed bridge using Imote2-platformed smart sensors", *Sensors and Smart Structures Technologies for Civil, Mechanical, and Aerospace Systems 2012*, Proc. of SPIE Vol. 8345, (2012)

(5) B. Back. "The use of wireless technology for reliable remote condition monitoring in the rail industry", IET Seminar Digest Vol.2008 (12216), (2008)

(6) M. Britton, M.S. Maddison. "Towards the reality of Intelligent Infrastructure with Wireless Meshed Sensors", *4th IET International Conference on Railway Condition Monitoring*, (2008).

(7) R.Lewis, S.Maddison, E.Stewart. "An extensible framework architecture for wireless condition monitoring applications for railway rolling stock", *6th IET International Conference on Railway Condition Monitoring*, (2014)

# Deploying a low energy power model within data loggers and sensors for the remote monitoring of assets

**R J Preece**[*+], **T M Hanif**[*], **R J Amos**[*], **E J C Stewart**[+]
[*] Arrowvale Electronics, UK
[+] University of Birmingham, UK

## ABSTRACT

Arrowvale Electronics, in partnership with the University of Birmingham have developed an energy management utility for usage in the deployment of remote energy harvesting data loggers, this utility allows for application specific energy management configurations. This model uses information about the types of harvested energy available, hardware configuration and functional capabilities to firstly provide an estimated feasibility of deployment. This information is then used to develop an initial configuration file for the hardware, the utility can then use data collected by physical hardware to fine-tune the configuration to its operating environment. This allows each piece of hardware to adapt its functionality to its surroundings. Detailed are two case studies of where the utility has been used.

This utility has so far been integrated into a trackside 'Energy Harvesting Data Logger' designed for applications such as embankment slippage monitoring, water levels or orientation of temporary speed restriction board. In this instance, data is fed back into Network Rail's intelligent infrastructure system. This system is designed to allow for rapid deployment and remote configuration of the loggers.

Additionally the utility has been incorporated into a vehicle based system, forming the backbone of Arrowvale Electronics wireless remote condition monitoring product range. These energy harvested sensors enable to remote monitoring of train borne systems from bearings to door operation and from sanding systems to status lamp monitoring. Each application has been developed and tuned using the utility to determine which energy harvesting source is most applicable and how best to operate wireless connectivity.

## BACKGROUND

Due to rapid progress in the development of microelectronics in recent decades it is now possible to achieve significantly more from increasingly smaller packages. This smaller component size has often resulted in a reduction in the power requirement of the components (1).

A wider result of such an explosion in the availability of such low powered devices is that the floodgates have been opened to a range of handheld (or smaller) devices that are either connected directly to the internet or interconnected between devices; this is known as the Internet of Things (2). Such interconnection has thus far been used in the commercial sector to add additional functionality to household appliances, entertainment systems and heating. In most cases, this involves adding a web connected microprocessor into a new design of an existing product and using the additional functionality to report the status of the equipment and allow the user to remotely control the devices settings, an example of such as system is British Gas' Hive system which offers the user increased control over the central heating system within the home. Within industry, deployment of 'connected devices' has been focused in the field of Remote Condition Monitoring.

Remote Condition Monitoring or RCM is the monitoring of the health or operation of a piece of equipment whilst it is in the field. In some cases the data can be stored locally and downloaded when required, but the ideal system is one where data is streamed near real-time to a remote server for analysis. Over the previous decade such monitoring has tended to be post initial installation and in direct response to a problem where more detailed information is required to diagnose a recurring fault. This process has begun to shift where remote condition monitoring has become a standard installation on all new equipment.

Within the rail industry, remote condition monitoring has been used for a variety of purposes, from the analysis of driver performance (3) to the monitoring of system components (4). These two installations show the two opposite sides of Railway Remote Condition Monitoring. The first example is done using existing data logging capability of the train (the On Train Monitoring Recorder (OTMR) Black Box recorder); in this case, data which would otherwise be stored locally until accessed by depot staff is automatically downloaded from the OTMR and relayed over the internet to a server. This data is then analysed to assess driver performance and improve operational efficiency. The second example above is a 'retro-fitted' example, where an additional system is added to the existing vehicle to monitor the condition of the axles, bearings and wheels. Such a system allows the fleet engineer to remotely monitor the condition of the fleet, and arrange maintenance prior to an in-service failure, reducing unexpected downtime and any delay minutes.

A major barrier to a full roll out of any retrofitted condition monitoring system is cost, it is difficult to justify the initial installation and setup cost of any system without a business case to support it. On vehicle installations can often be hindered by the sheer cost of installing new wiring or equipment to the vehicle. As a result of this some systems, such as the Perpetuum system (4), has shunned wiring the system together and instead used wireless technology to interconnect devices within the system, allowing for full train coverage at a local level without the need to install cabling throughout the vehicle.

Off-vehicle, such condition monitoring system can be just as difficult to install, with safety cases, power connectivity and site access it can be almost impossible to justify the deployment of a remote condition monitoring system, such systems are currently installed on points monitoring equipment and signalling power supplies as an early warning system, however for more wide-spread coverage a more flexible system is required allowing for rapid deployment and multiple functionality.

For both of the above examples it has not been considered how such system would be powered. One method which is increasing in popularity is that of Energy Harvesting. Off Vehicle, such technology is already in use for powering trackside equipment such as temporary lighting for night working. Network Rail has also recent upgraded Blackfriars Bridge to include a 1.1 MW solar array; this installation generates 900,000 kWh per year.

To enable the deployment of energy harvested products, data must first be collected about the hardware which is to be deployed. This can often be as simple as recording the maximum power requirement for each component in the system and summing them together. This however is a very inefficient method, modern microelectronics is often better viewed as a 'duty cycled' system, where there are peaks and troughs in the power requirement of the system (5). For example, if a data logger is only periodically monitoring an input then the power requirement when the system is not monitoring can be significantly reduced. Depending on the required time resolution, or 'how often the data needs to be recorded', the 'on-time' of a system can be reduced to less than 1 %.

This paper looks at the implementation of a component level power model to calculate the power requirements of an Energy Harvesting Data Logger in various operating modes. The paper then looks at the development of the supporting software and how the system can be tuned using real world data, finally the paper looks at how, once verified and iteratively improved, the model is then applied to other hardware platforms and applications.

## ENERGY HARVESTING DATA LOGGER

The Energy Harvesting Data Logger was designed as a remote data logger aimed at solving the problem posed by monitoring of assets in locations where conventional power sources are not available. This opened the door to the usage of energy harvesting power sources. To facilitate the use of energy harvested power sources, a low power electronics approach is required. For the hardware development of the Energy Harvesting Data Logger, each required circuit was re-designed from the ground up to use low powered components.

Further to this, a new architecture for a low power data logger was devised, this architecture focussed on isolating power from parts of the logger when not required, by doing so, it is possible to reduce the 'sleep' current requirement of the logger to a minimal level, with a very low power microprocessor at the core of the logger the unit can operate with a sleep current measured in µA (micro Amps).

The energy supplied by any harvested source will be irregular, for example solar panels harvest less to no energy during the night, and wind turbines do not harvest when there is not wind. By integrating a power storage capability (in the form of either a battery of a super-capacitor – shown in Figure 1) the logger is able to continue to function without harvested energy, or through time windows with low levels of harvested energy.

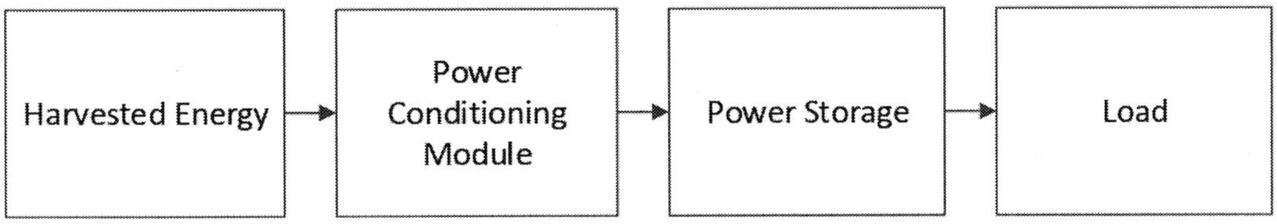


**Figure 1 - Energy harvesting power supply block diagram**

By both reducing the sleep current requirements of the hardware and using a large reservoir of stored energy it is possible to operate the Energy Harvesting Data Logger persistently without a requirement to change batteries, reducing track access time and expanding the locations in which monitoring can occur.

The Energy Harvesting Data Logger incorporates a 3G modem alongside its local data storage to allow the offload of data to a remote server. This removes a further common requirement for staff to access the equipment to retrieve the data and opens the door for real time data transfer where necessary. The 3G modem is the highest single power requirement of the Energy Harvesting Data Logger; this limits the frequency of use if the unit is to remain a feasible deployment (on a power front).

Accompanying the Energy Harvesting Data Logger hardware is a combination of embedded software and a PC based configuration utility, these two applications work together as part of an Energy Management Utility to assist with the configuration and practicality of deployment of the Energy Harvesting Data Logger. The Energy Management Utility has three separate sections: a functional configuration utility, a power configuration utility and an iterative improvement utility.

The configuration utility is similar to those used by conventional data loggers; it allows the user to configure the various input channels, setup simple filtering or thresholds and implement a very simple rules engine. For the Energy Harvesting Data Logger, this section of the utility also allows the user to set up the 3G connectivity options for data offload.

The power configuration utility works alongside the configuration utility to assist the user in an initial estimate for deployment feasibility, this essentially works by analysing the setup chosen in the configuration utility and combining it with further information such as battery configuration, energy harvesting source and geolocation to give a yes or no answer as to the feasibility of the deployment. The most obvious causes for a no answer would commonly be due to scenarios such as 'too high frequency of server update' or 'too low level of harvested energy' in such situations the utility can suggest alternative configurations to assist the user in achieving a workable solution.

The final section of the energy management utility works as an iterative improvement utility. This utility works in tandem with the on-board firmware to develop a 'history' of the systems power level as well as monitor the power requirement of the various operating functions within the unit. These readings aim to give an accurate picture of the device in operation enabling the utility to iteratively improve the deployment. Coupled with this each set of data received in this way allows the utility to feed back to improve the accuracy of future developments. At a simple level, for each reading a delta is made against the predicted reading, this delta is then added to a moving average weighted towards the existing value to remove any anomalous results. If the predicted value is significantly and persistently less or more than the recorded value, future predicted values will be amended accordingly.

## ADDITION OF A WIRELESS MODULE

A primary expansion target for the Energy Harvesting Data Logger was the integration of a sub 1 GHz wireless connection to surrounding sensors, various wireless protocols and methods of integration were studied before deciding on an OEM (Original equipment manufacturer) module driven by an additional external processor, this was done for two primary reasons, firstly the greater amount of control offered over the power usage of the module – i.e. the capability to completely isolate power from the wireless module, and secondly the ability to have full control over the primary processor without concerns of external processor overheads.

Before the module was added to the unit, a simulation of the module was added to the power model, from this it was possible to see the effect that installing the module would have on the feasibility of the module. For ease of implementation the wireless module was added as an additional 'expansion card' within the model, this allowed the continued use of the original model whilst development took place.

The major issue encountered with the addition of the sub 1 GHz wireless link was the operating method of the wireless device. For the wireless network to operate effectively the 'Master' node must remain powered at all times as a data concentrator. The enables the 'slave' nodes within the system to power on when required and transmit data. Whilst this effectively reduces the power requirement of the sensor nodes, it greatly increases the requirement for the master. For example, on one wireless device the power requirements for the various operating modes are as shown in Table 1.

**Table 1 - Example wireless module power requirements**

| Mode | Current (Power when operating at 3.3 VDC) |
|---|---|
| Transmit | 48 mA (158 mW) |
| Receive | 27 mA (89 mW) |
| Sleep | 1.7 μA (5.6 μW) |

Whilst receive mode consumes less power than transmit mode, the requirement is still $10^4$ times greater than the ideal requirement of sleep mode. Depending on the initialisation time and start-up mode of module used and the required duty cycle of the application, with such an efficient sleep mode it is possible that keeping the module in sleep mode may be more effective than completely turning it off.

It was proposed that the master node could operate in a duty cycled mode similar to the slave node. This was however avoided as an option due to the stress this places on keeping an accurate clock on each slave module. When a microcontroller is in a sleep mode it is common practice to fall back onto an RC (Resistor Capacitor) oscillator due to their low power requirement relative (50 %) to operating a crystal oscillator. RC oscillators are less accurate than their crystal counterparts, to such a degree that temperature and voltage variations can result in up to 5 % variation in output frequency (in the worst case), in a standard application this can manifest as a drift of multiple seconds per day. Whenever the radios are connected the clocks can be brought back into sync, but this will not help when the radios wake up to communicate and have moved so far out of sync that the operating windows do not overlap. Addressing this functional failure is deemed a higher priority than the energy consumption concerns.

To initially counter this issue it was decided to model the system (using the Energy Harvesting Data Logger as the master) without local data collection. I.e. the Energy Harvesting Data Logger operates in one of two modes. Mode 1 is operating with the local data acquisition channels and the radio link disabled and mode 2 is has the local channels disabled, whilst operating the radio link. From modelling of the system, the power consumption was calculated as roughly equivalent.

## DEVELOPMENT OF A NEW PRODUCT

Adding a wireless module to a product displays the capability of the power model to be modified to demonstrate how changes to the configuration and hardware will impact the functionality. The following section focusses on using the model developed in the previous sections to design a completely new product.

To compensate for the removal of the sensing capability from the primary Energy Harvesting Data Logger a supporting Energy Harvesting Sensor is to be developed. The design process will work in the following stages.

Firstly a proposed top-level block diagram for the sensor will be devised; this block diagram will focusses on the major components of the system as shown in Figure 2.

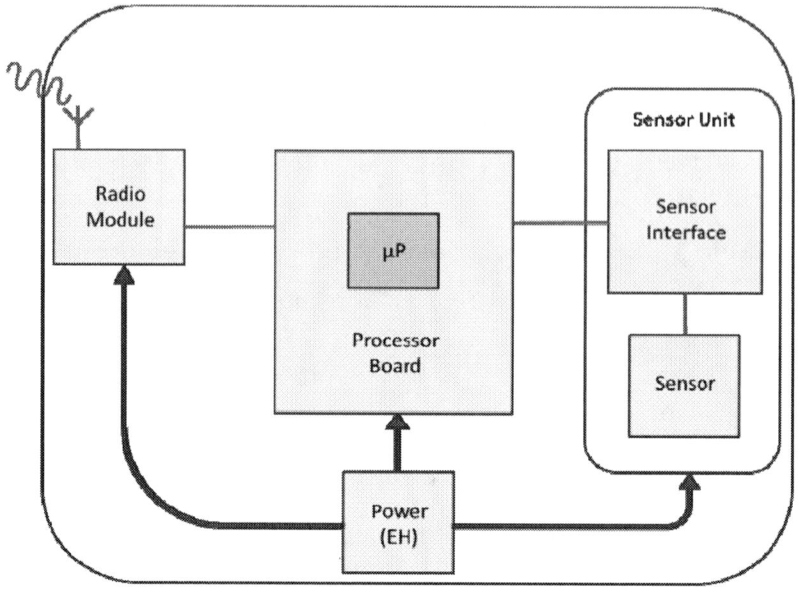


**Figure 2 - Energy harvesting sensor block diagram**

From this block diagram it is possible to focus on the technical detail of each section. For the radio module, the module chosen for the Energy Harvesting Data Logger will be duplicated so that any difficulty in pairing of the devices will be reduced. The processor will be a smaller version of the one used in the main Energy Harvesting Data Logger, this smaller version has most of the functionality of the larger processor, but with smaller amounts of connectivity and less general purpose ports, often used for inputs or outputs. This is not an issue as it is proposed for the Energy Harvesting Sensor to only have one or two inputs or sensors.

The sensor part of the unit depends on the application; current target applications are similar to those on the main Energy Harvesting Data Logger, i.e. Digital inputs for sensors such as tilt, orientation and basic detection and analogue inputs for sensors such as moisture and temperature.

The proposed power source is a small solar panel such as those used for recharging 'solar lamps' as used in garden decoration, whilst the amount of energy harvested is minimal relative to the amount collected by larger industrial solar panels, unlike such installations, the solar cell can be fitted to or integrated into the enclosure of the unit itself, this enables the unit to be rapidly deployed in a wider range of locations.

The main advantage of the Energy Harvesting Sensor is that in contrast to the original Energy Harvesting Data Logger deployment do not need to be wired to the sensors. This reduces the cost of the installation, increases the ease of installation and increases the potential area of an installation. This last advantage is key, as it enables an installation to cover both sides of the running rail without the need for wiring between sides.

Drilling deeper into each section begins to put together a component list, when selecting components and designing circuits it is important to consider the parasitic power draw of certain components, for example, if a 'pull-up' resistors, used to ensure that digital logic settles to an expected level when not being 'driven', are selected with incorrect values or used inappropriately the overall power requirement of the system will increase, this will reduce the efficiency of the system, and reduce the expected lifetime when operating on stored power.

To utilise the power model for this design, it requires a calculation of the hardware requirement for the various software modes of the sensor. Within the power model there are three standard power modes, each covering a separate operation. The first operating mode is a data acquisition mode, in this mode sensors are powered up and the inputs to the system are recorded. The second mode is an update mode, when updating the system connects to the master Energy Harvesting Data Logger and relays any recorded data. Finally the system has a sleep mode, in this mode power usage is reduced to a minimum to preserve stored power. Depending on the required application, the system will often spend up to 99 % of its time in sleep mode. In larger systems, such as the main Energy Harvesting Data Logger, there can be multiple instances of each phase, i.e. there are multiple input cards which each require reading separately.

From this information it is a matter of inputting the calculated values into the power model to calculate detailed information about the operation of the new hardware before the design is even complete. The most useful information which can be extracted at this stage is the 'power requirement' of the Energy Harvesting Sensor, this is often the first question asked when designing an energy harvesting solution.

As the hardware development continues, real-world values can be fed back into the power model to increase its accuracy and give better predictive capabilities for future deployments.

## SUMMARY AND CONCLUSION

This paper has shown a method of how a power model can be developed for a specific piece of electronic hardware, in this case an Energy Harvesting Data Logger. By accurately calculating the operating power of the hardware in various software modes it has been shown how an estimate of average power usage and peak power requirement can be determined, and that sometimes this can differ from the naturally assumed values.

Secondly this paper demonstrates how the usage of an accurate and trusted power model can be used to aid with product development, by modelling the effects on the overall system of adding or removing hardware, changing software variables and of changing the system configuration it is possible to demonstrate the required functionality and feasibility of each setup without performing hardware or software changes for the initial stages of prototyping, this can reduce cost by reducing the amount of development iterations.

Additionally, the transferability and versatility of the power model is shown through the demonstration of how the power model can be easily translated to different hardware and be involved in the hardware design even before prototype design has begun. The process for setting up the power model is shown through a breakdown of the required stages; through isolation of the high-level functions of the power model from the base layer, hardware details of power requirements, the hardware configuration or in fact the entire hardware, can be changed beneath the overlying higher-level functionality of the power model without affecting its performance.

The transfer of the power model to the Energy Harvesting Sensor, has led to the possibility of other such transfers. The next stage of development is to characterise additional power supplies, at present the model supports 1 W, 10 W and 30 W solar panels and a wind turbine, these energy sources are not abundantly available when considering on-vehicle remote condition monitoring. For on vehicle solutions, energy can often be harvested from either vibration (the closer to the rail the better) or through the Seebeck effect, where voltage is developed due to a temperature gradient. Future work will study the feasibility and practicality of using such sources to power a sensor similar to the Energy Harvesting Sensor discussed above.


## ACKNOWLEDGEMENTS

Parts of this project have been funded by a Knowledge Transfer Partnership, co-ordinated by the Technology Strategy Board. This Knowledge Transfer Partnership was between Zerpetz Ltd (T/A Arrowvale Electronics) and The University of Birmingham.


## REFERENCES


(1) Kish, L. B., (2004) Circuits, Devices and Systems, IEE Proceedings. **151:2** 190-194. Doi: 10.1049/ip-cds:20040434

(2) Holler, J., et al. (2014) From Machine-to-Machine to the Internet of Things: Introduction to a New Age of Intelligence. Academic Press

(3) Green, S. R., (2011) Railway Condition Monitoring and Non-Destructive Testing (RCM 2011), 5th IET Conference on. 1-5. Doi: 10.1049/cp.2011.0582

(4) Reynolds, L., Wasenczuk, A., (2006) Railway Condition Monitoring, 2006. The Institution of Engineering and Technology International Conference on. 194-197

(5) Hsu, J. et al. (2006) Low Power Electronics and Design, 2006. ISLPED'06. Proceedings of the 2006 International Symposium on. 180-185 Doi: 10.1109/LPE.2006.4271832

# Application of FRA new vehicle-track system qualification requirements

**B P Marquis**
Structures and Dynamics Division, Volpe National Transportation Systems Center, USA

**A Tajaddini**
Office of Research and Development, Federal Railroad Administration, USA

**H-P Kotz**
Siemens – Infrastructure and Cities, Rail Systems Division Locomotives and Components, BG EN SDS VD, Austria

**W Breuer**
Siemens – Infrastructure and Cities, Rail Systems Division Locomotives and Components, EN SC, Germany

**M Trosino**
National Railroad Passenger Corporation, USA

## ABSTRACT

Vehicle/Track Interaction (VTI) Safety Standards aim to reduce the risk of derailments and other accidents attributable to the dynamic interaction between moving vehicles and the track over which they operate. On March 13, 2013, the Federal Railroad Administration (FRA) published a final rule titled "Vehicle/Track Interaction Safety Standards; High-Speed and High Cant Deficiency Operations" [1] which amended the Track Safety Standards (49 CFR Part213) and the Passenger Equipment Safety Standards (49 CFR Part 238) in order to promote VTI safety under a variety of conditions at speeds up to 220 mph. Among its main accomplishments, the final rule revises standards for track geometry and enhances qualification procedures for demonstrating vehicle trackworthiness by taking advantage of the results of computer modeling and simulation.

The Track Safety Standards provide safety limits for maximum allowable track geometry variations for all nine FRA Track Classes — i.e., safety "minimums." These limits serve to identify conditions that require immediate attention because they may pose or create a potential safety hazard for North American service proven equipment. While these conditions occur infrequently, they define the worst-case track conditions that can exist before a vehicle is required to reduce its speed or other corrective action must be taken.

To promote the safe interaction of rail vehicles with the track over which they operate (i.e. wheels stay on track, the track structure or vehicle itself is not overloaded, or the dynamic environment does not cause injury to passengers), these worst-case track conditions must be considered in the design of suspension systems. In particular, rail vehicle suspensions must be designed to control the dynamic response to these more demanding track geometry conditions such that wheel/rail forces and vehicle accelerations remain within prescribed thresholds (VTI safety limits), at all allowable speeds associated with a particular track class. In order to evaluate the suitability of a particular equipment design for the existing track conditions found in the U.S. (i.e. evaluate vehicle trackworthiness), the combination of on-track testing and simulation is used to check vehicle performance, derailment tendency, and stability. On-track testing is conducted on a track "representative" of the operating route while computer simulations are done using an analytically defined minimally compliant track (MCAT). The requirements depend upon the anticipated maximum operating speed and maximum cant deficiency. This paper provides an overview of these requirements and an application of these standards to Amtrak's new electric locomotives designed and built by Siemens. The process used for selection of a "representative" track and validation of computer simulation models is also discussed.

## 1. INTRODUCTION

Qualification of a rail vehicle in accordance with FRA regulations mandates testing to assess safety, vehicle performance, and vehicle/infrastructure compatibility. Such tests evaluate a vehicle's response to track conditions before commencement of a new operating service. While testing has long been recognized as a critical element of the qualification process [1, 2], testing is often regarded as a costly, logistically cumbersome, disruptive, and time demanding process. While testing does raise confidence in the likelihood of safe rail operation, the results are often limited because it is unfeasible to perform exhaustive testing. A balance must be struck between the right amount of testing (not enough testing adds risk to a program, while testing too much can be very costly and may add unnecessary burden) and the use of simulations to improve the comprehensiveness of a qualification by extending its evaluation to conditions beyond those assessed in a given test and likely to be encountered in service.

As specified in the new FRA final rule, simulations using computer models are now required during the vehicle qualification process as an important tool for the assessment of vehicle performance. As a result, the final rule institutes more comprehensive and cost-effective equipment qualification requirements. These simulations are intended not only to augment on-track, instrumented performance assessments but also to provide a means for identifying vehicle dynamic performance issues prior to service to validate the suitability of a vehicle design for operation over its intended route. In order to evaluate safety performance, as part of the vehicle qualification process, simulations are required using both a measured track geometry segment representative of the full route, and an analytically defined track segment containing geometry perturbations representative of minimally compliant track conditions for the respective track class.

## 2. OVERVIEW OF FRA QUALIFICATION

Figure 1 below provides a general framework for the different elements that make up the overall FRA vehicle/track system qualification process. There are two major branches of the qualification process. The left hand branch involves simulations using a model representative of the vehicle design (black circle) and a virtual track (orange circle) to assess vehicle performance for safety over maximum allowable track geometry perturbations. The right hand branch involves on-track testing using an instrumented prototype vehicle (black circle) and a representative track (orange circle) to assess vehicle performance for safety over conditions unique to the route. Together these branches provide a comprehensive assessment of safety and compatibility between a particular vehicle design and the infrastructure over which it may operate. For this qualification process to work, it is important to validate the model that is used in the left hand branch. A methodology for model validation is outlined in the bottom branch (tan square). In this methodology there are two validation steps: a "static" validation involving three checks (three purple circles) and a "dynamic" validation involving comparisons between modeling and test results over an excitation track (orange circle). The objective of the static part is to prove the vehicle parameters while the dynamic part should evaluate the vehicle-track interaction.

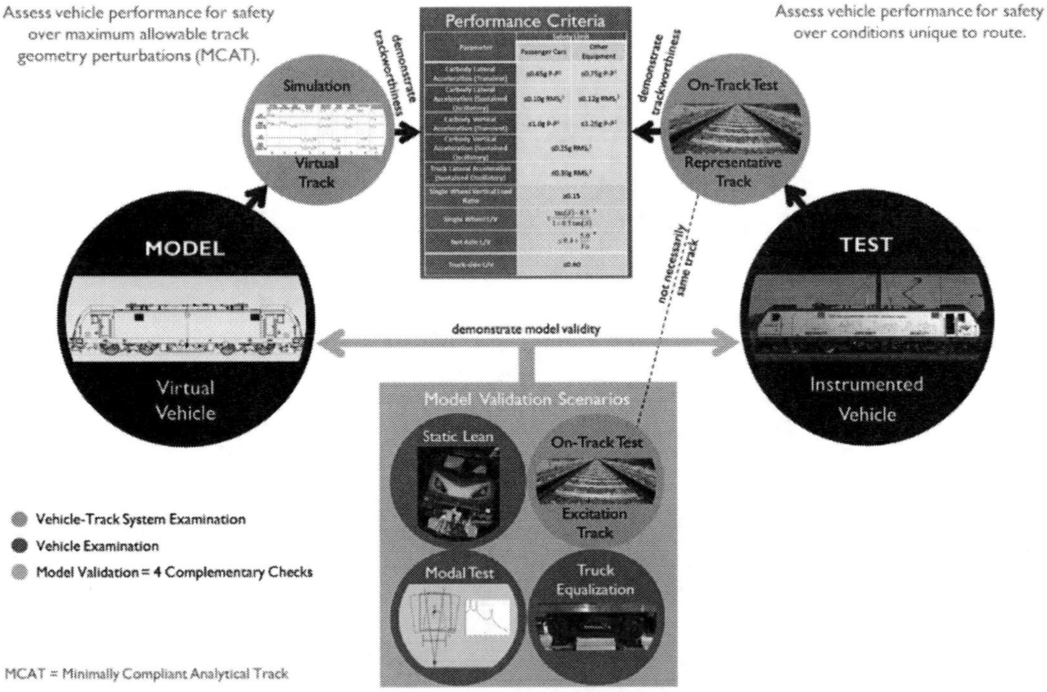


**Figure 1: Compliance with 49 CFR 213.345, Vehicle-Track System Qualification.**

Each of the elements above will be described in greater detail in the sections below. In addition, Figure 2 shows a summary of the test and simulation requirements for "new vehicle types" and "qualified vehicle types". The requirements depend upon the anticipated maximum operating speed and maximum cant deficiency. In general, the requirements for demonstrating safe performance increase as speed and cant deficiency increase. Since, under these conditions vehicle response can be larger and the consequence of a derailment can be more severe, instrumented wheelsets are required to measure wheel-rail forces on a track representative of the entire route planned for the intended operation. For "qualified vehicle types" (those vehicles whose dynamic characteristics are similar to equipment operating in an existing similar service), the amount of required simulations and testing is significantly reduced in light of the more comprehensive initial assessment. Previously, the rule did not permit such a reduction and required complete testing over the entire route.

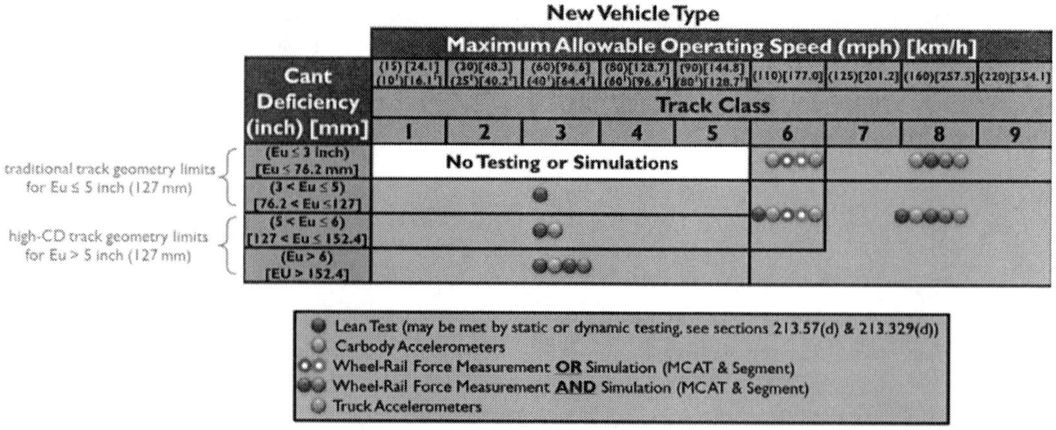


[1]Max allowable operating speed for freight trains

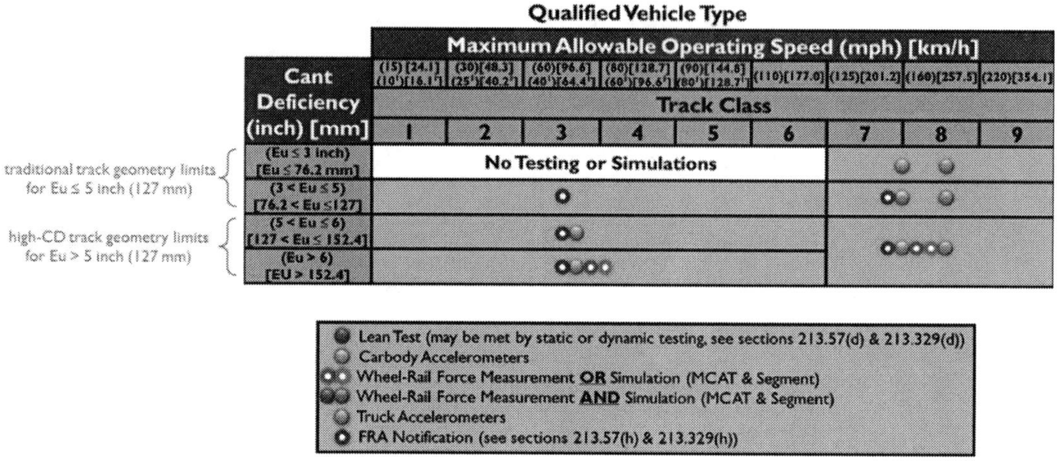


[1]Max allowable operating speed for freight trains

**Figure 2: Testing/Simulation Requirements in 49 CFR 213.345, Vehicle-Track System Qualification.**

## 2.1 Vehicle Track Interaction Safety Limits

Table 1 summarizes the FRA Vehicle-Track Interaction Safety Criteria found in 49 CFR 213.333. These limits on wheel/rail forces and vehicle accelerations are intended to promote the safe interaction of rail vehicles with the track over which they operate. In particular they ensure that wheels stay on track, and vehicle dynamics do not overload either the track or vehicle, or cause injury to passengers. These limits define safety from a performance standpoint and are based on operational and vehicle qualification experience (data), the results of simulation studies (modeling), current research results, and consideration of international practices. They serve as the foundation for the qualification process shown in Figure 1 above.

**Table 1: FRA Vehicle-Track Interaction Safety Criteria (49 CFR 213.333).**

| | Safety Concern | Parameter | Safety Limit: Passenger Cars | Safety Limit: Other Equipment | Filter/Window |
|---|---|---|---|---|---|
| Accelerations | Ride Safety/Ride Quality | Carbody Lateral Acceleration (Transient) | $\leq$0.65g P-P[1] | $\leq$0.75g P-P[1] | 1 sec window |
| | | Carbody Lateral Acceleration (Sustained Oscillatory) | $\leq$0.10g $RMS_t$[2] | $\leq$0.12g $RMS_t$[2] | 4 sec window 4 sec sustained |
| | | Carbody Vertical Acceleration (Transient) | $\leq$1.0g P-P[1] | $\leq$1.25g P-P[1] | 1 sec window |
| | | Carbody Vertical Acceleration (Sustained Oscillatory) | $\leq$0.25g $RMS_t$[2] | | 4 sec window 4 sec sustained |
| | Stability/Hunting | Truck Lateral Acceleration (Sustained Oscillatory) | $\leq$0.30g $RMS_t$[2] | | 2 sec window 2 sec sustained |
| Wheel Forces | Wheel Unloading | Single Wheel Vertical Load Ratio | $\geq$0.15 | | 5 foot window |
| | Wheel-climb Derailment | Single Wheel L/V | $\leq \frac{\tan(\delta)-0.5}{1+0.5\tan(\delta)}$[3] | | 5 foot window |
| | Track Shift | Net Axle L/V | $\leq 0.4+\frac{5.0}{Va}$[4] | | 5 foot window |
| | Rail Rollover | Truck-side L/V | $\leq$0.60 | | 5 foot window |

[1] Peak-to-peak value
[2] Root mean squared with linear trend removed
[3] $\delta$ – Flange angle in radians
[4] Va – Vertical axle load in kips

## 2.2 Representative track testing

Because simulations are now required to evaluate the performance of various vehicle and train designs subject to different minimally compliant track irregularities (discussed in greater detail below), it is important that on-track testing also examine other aspects important to vehicle behavior. It is particularly important that such testing evaluate vehicle responses to steady state and time varying changes in conditions or characteristics that are unique to a particular route or operation such as rail profile, curvature, gage, and rail cant. Given the importance of curving performance in overall vehicle behavior, selection of a "representative track" for on-track testing is generally based primarily on examining the distribution of curvature for the intended operating route.

For illustration purposes, an example "representative track" for Amtrak's Northeast Corridor (NEC) is presented. Figure 3 shows a cumulative distribution function of curvature[1] for three track segments; (1) the entire NEC territory which extends approximately 500 miles from Boston to Washington D.C., (2) the segment from Boston to New York ("North End") which extends approximately 250 miles, and (3) the zone from New York to Washington D.C. ("South End") which extends approximately 250 miles. The plot shows that the "North End" of the railroad has higher curvature than the "South End". The differences in curvature characteristics for the different segments can affect the performance of the vehicle. As a result, a separate representative track was defined for each segment.

---

[1] The degree of curvature is a measure of the sharpness of a curve defined in the United States as the central angle D subtended at the center of curvature by a chord of 100 feet. The relation between the degree of curvature D and curve radius R is R = 5729.58/D where D is expressed in degrees and R in feet [3].

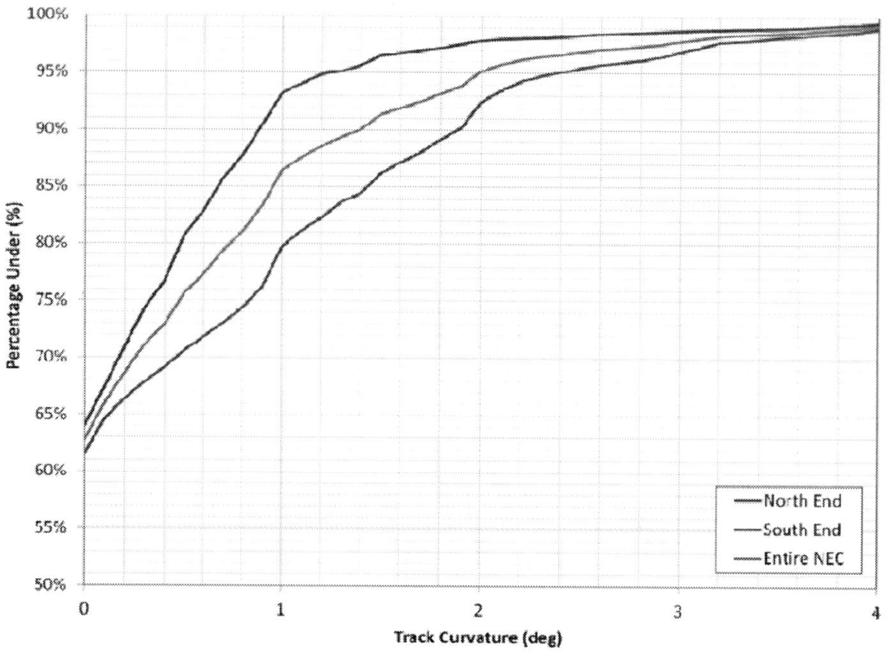


**Figure 3: Cumulative distribution function of curvature on NEC.**

Candidate test zones were identified based on a comparison of the curvature distribution computed using a 50 mile moving window to the curvature distribution computed using each of the segments. A level of correlation between the candidate zones and the entire segment was developed based on the RMS of the difference between the 50 mile window and the entire segment, Figure 4.

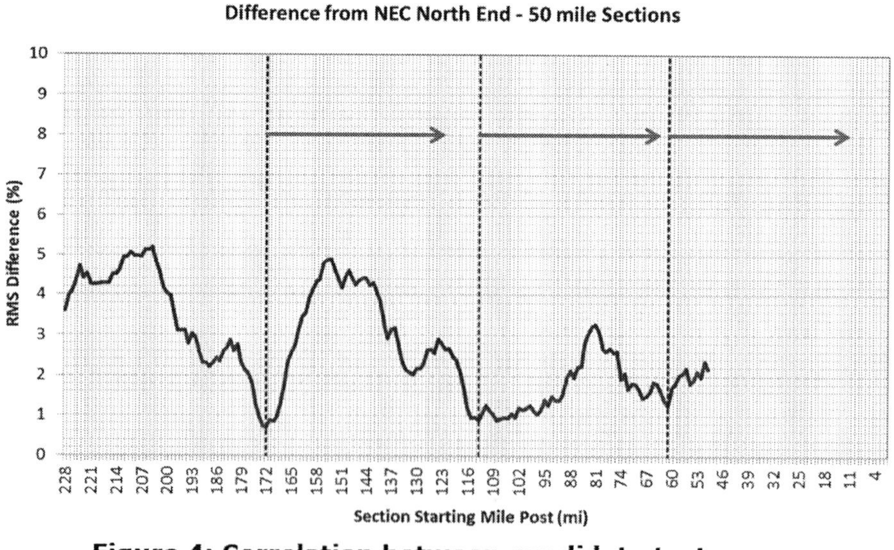


**Figure 4: Correlation between candidate test zones (defined by running 50 mile window) and entire track segment.**

From this analysis, three candidate test zones were chosen for the "North End" segment. The candidate test zones were considered representative of the route since the RMS of the difference between the 50 mile window and the entire segment was less than 3%. An overlay of the cumulative distribution function for these three candidate zones with the entire segment is shown in Figure 5. Since the curvature distribution in each test zone closely matches that of the route, and the target test speeds and cant deficiencies are achievable in these zones, the dynamic response of the representative segments is expected to match the response of the entire route. The actual test zone, MP 185 to MP 140, was selected based on operational considerations as there are convenient turn-around points with pockets to lay up the train and review data without infringing on regular train operation. Figure 6 highlights the representative test zones for both the "North End" and "South End".

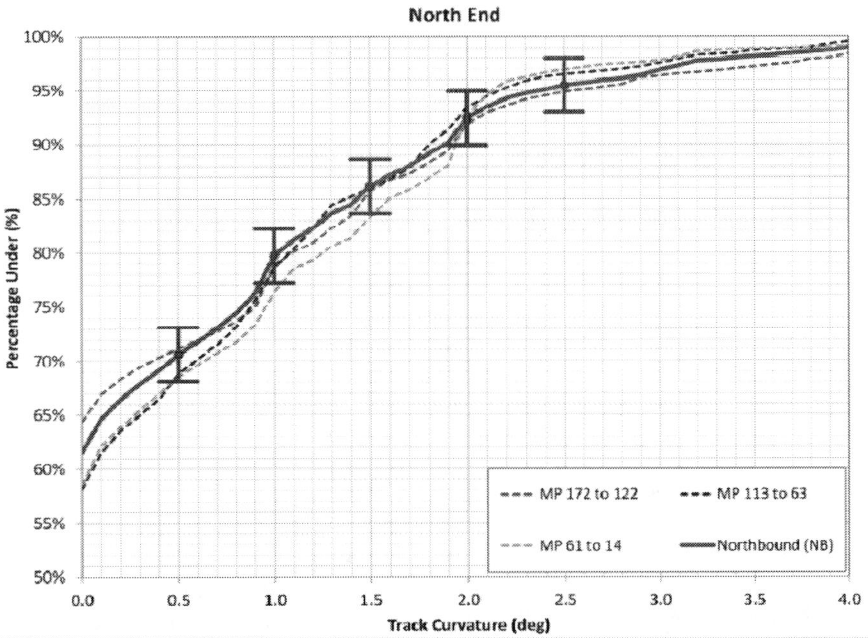


**Figure 5: Cumulative distribution function of curvature in candidate test zones and "North End" segment.**

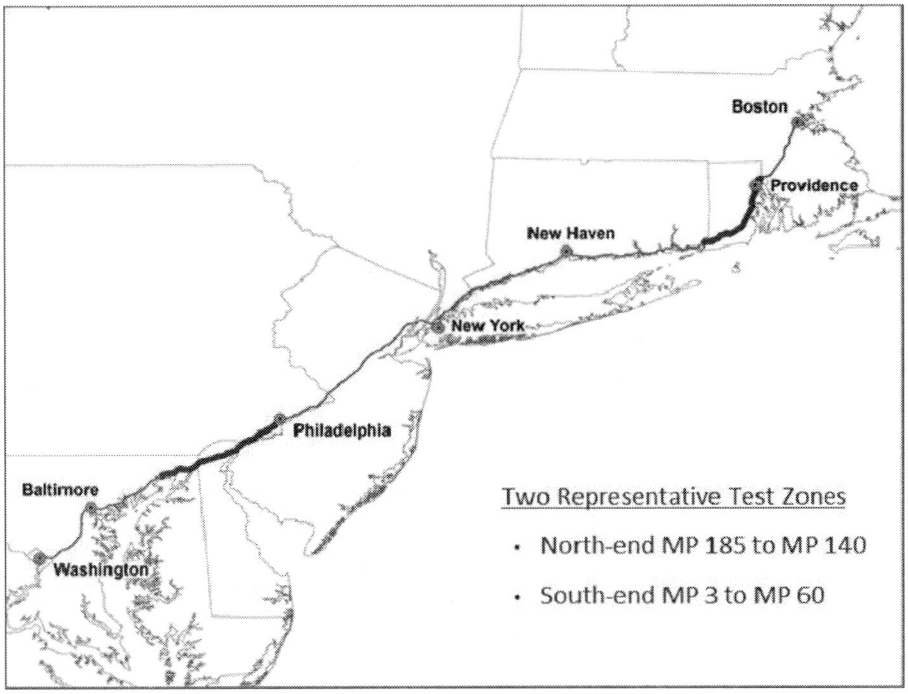


**Figure 6: NEC Representative Test Zones.**

## 2.3 Simulations with Virtual track

Computational models have become practical and reliable tools for understanding the dynamic interaction of vehicles and track, as a result of advancements made over the last few decades. Such models are capable of assessing the response of vehicle designs to a wide range of track conditions corresponding to the limiting conditions allowed for each class of track. As indicated in Figure 2 above, in some of these cases simulations are required as part of the assessment of vehicle performance (derailment tendency, stability, etc.) using both a measured track geometry segment representative of the full route, and an analytically defined track segment containing geometry perturbations representative of minimally compliant track conditions for the respective track class— Minimally Compliant Analytical Track (or MCAT). Appendix D of the final rule is a new appendix containing the requirements for the use of computer simulations to demonstrate compliance with the vehicle/track system qualification testing requirements specified in subpart G of the rule.

MCAT is intended to be used to qualify both new vehicles for operation and vehicles previously qualified (on other routes) for operation over new routes. MCAT consists of nine sections, Figure 7; each section is designed to test a vehicle's performance in response to a specific type of perturbation (hunting perturbation, gage narrowing, gage widening, repeated and single surface perturbations, repeated and single alinement perturbations, short warp, and combined down-and-out perturbations). Appendix D contains tables specifying the amplitudes $a_1$ through $a_{13}$ and wavelengths $\lambda$ to be used in the simulations. Typical simulation parameters (that vary) include: speed, cant deficiency, gage, and wheel profile.

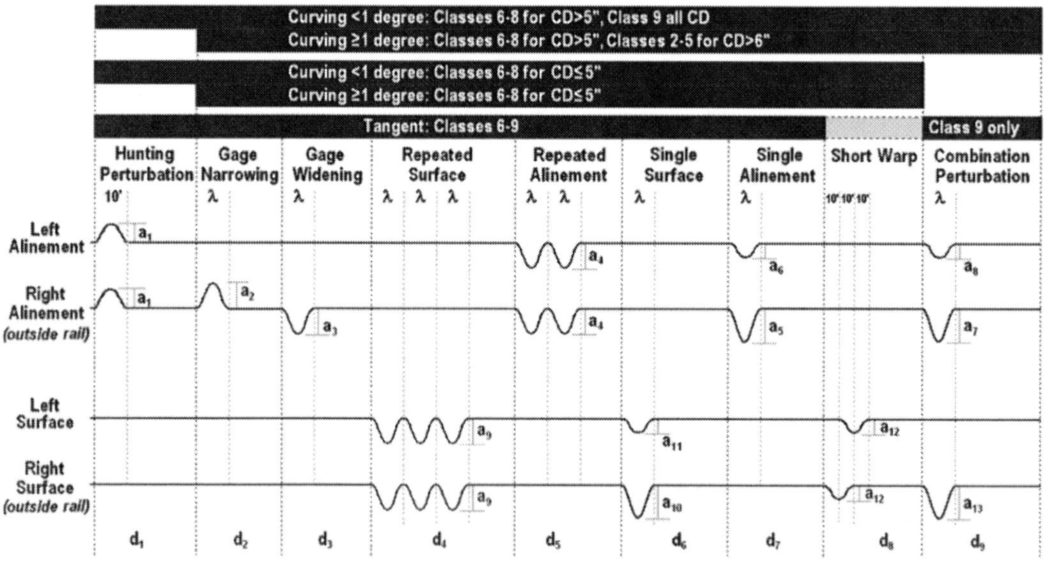


**Figure 7: Minimally Compliant Analytical Track (MCAT).**

In general, MCAT is used to model a wide range of conditions not tested (there are an infinite number of possible track geometry variations that can occur in track structures and on-track testing assesses performance for a snapshot of conditions at a single point in time). The FRA-maximum allowable amplitudes discussed previously, which are generally not present on actual track, are used as inputs to the simulation. Such an analysis provides data for assessment of equipment performance when the track conditions become poorer than those present during the on-track testing, i.e. deteriorated track geometry exists. While its primary purpose is to ensure compatibility of the equipment with the existing infrastructure, data from such an analysis may also be used to identify potential safety concerns that may need to be addressed with other appropriate mitigating action(s). Example mitigating actions include ensuring that the track over which the equipment is operating is maintained to tighter standards appropriate for the specific equipment type, placing operational restrictions on the equipment, or both [4]. In addition to providing a more comprehensive assessment of safety and compatibility between a particular vehicle design and the infrastructure over which it may operate, other significant benefits of such an MCAT analysis include better vehicle designs (vehicles designed to account for limiting track conditions generally perform better on typical quality tracks) and a reduction in the amount of required testing when a piece of equipment is moved to another corridor.

### 2.4 Model validation

As simulations are increasingly used worldwide as part of vehicle qualification, there is a great deal of interest in developing procedures for validating models used for evaluation of vehicle performance, derailment tendency, and stability [5]. To include simulation into its regulations, FRA has developed and applied a validation methodology. This methodology for validating the model is outlined in the bottom branch of Figure 1 (tan square). In this methodology there are two validation steps: a "static" validation involving three checks (three purple circles) and a "dynamic" validation involving comparisons between modeling and test results over an excitation track (orange circle). The objective of the static part is to prove the vehicle parameters and the dynamic part should evaluate the vehicle-track interaction.

The three "static" tests (purple circles) characterize behavior of the vehicle in isolation. Comparisons between model and test results for static tests provide confirmation of different portions of the vehicle model. The static lean test confirms that the model reasonably represents the center of gravity of the vehicle; the wheel load equalization test confirms that the model reasonably represents the primary suspension of the vehicle; and the modal test confirms that the model reasonably represents the mass and suspension properties of the vehicle (mass, inertia, suspension stiffnesses, and damping characteristics). Together, these three "static" checks represent a part of the model validation process.

The final part involves comparison between on-track testing and simulation using measured track geometry inputs. The track segment for this part, referred to as the "excitation" track, should contain irregularities that are large enough to excite significant, nonlinear motions of the car. Results measured in the tests are then compared with results predicted by simulations in order to determine that the vehicle model is able to replicate all key responses observed during the tests.

## 3. APPLICATION TO AMTRAK'S NEW SIEMENS BUILT LOCOMOTIVES

The requirements described above have been applied during the recent procurement of Amtrak's new ACS64 electric locomotives designed and built by Siemens. Amtrak qualified the ACS64 Locomotive for operation with a maximum operating speed of 125 mph and 5 inches of cant deficiency on the NEC between Boston and Washington D.C.

As per the new 49 CFR 213.345, Amtrak conducted both testing and simulations to evaluate the dynamic response of the vehicle. The combination of tests and simulations provided a comprehensive assessment of the effect of track conditions (track geometry, track curvature, etc.) and vehicle speed on the dynamic response of a vehicle. Simulations using MCAT track segments were performed to assess the vehicle dynamic response to maximum allowable track geometry variations provided in the FRA Track Safety Standards that may be present anywhere on the route. These simulations were run and provided as part of its report to the FRA. In addition, testing was conducted over two track segments that were representative of the NEC, one test zone on the "North-end" between Boston and New York (described above) and one test zone on the "South-end" between New York and Washington D.C.

The steps described above were followed to validate the model used for conducting the MCAT simulations. Figure 8 shows the measured and simulated values for the car body roll angle during the static lean test. Figure 9 shows the measured and simulated values for wheel unloading during the static lean test. In both cases the simulation results agree well with the test results, the difference of the test results and the simulation results is below 5%. Based on these results the location of the center of gravity can be calculated. Figure 10 shows the resulting path of the center of gravity during the static lean test. The results give a height of 57.3 inches for the center of gravity at zero lean. The projected height of the center of gravity is 52.7 inches, a relative error of 8.7%.

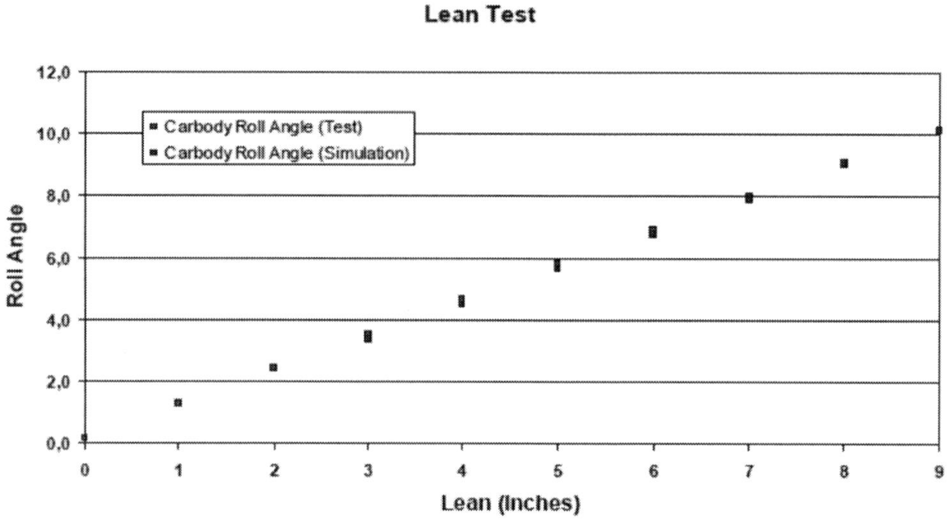


**Figure 8: Carbody roll angle – simulation vs test.**

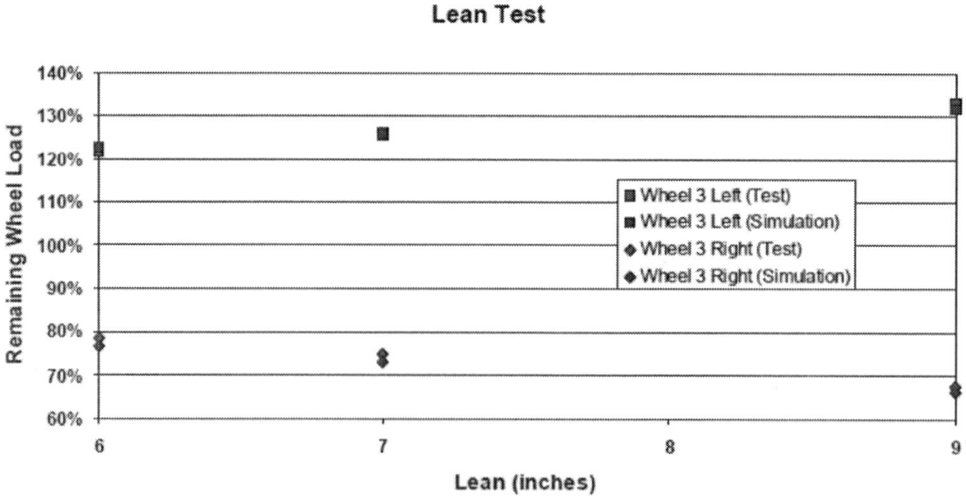


**Figure 9: Wheel unloading WS 3 – simulation vs test.**

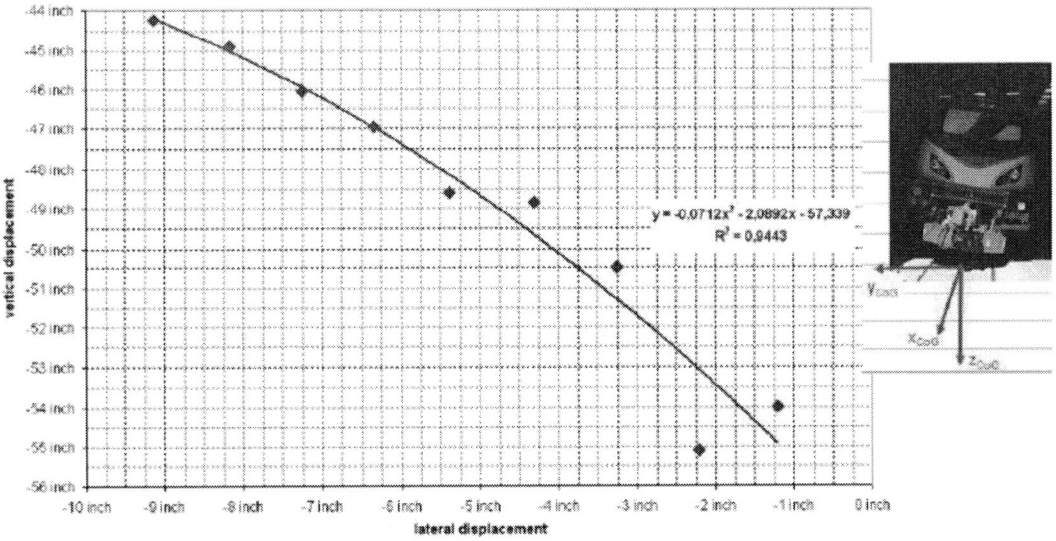


**Figure 10: Location of the center of gravity during the static lean test relative to the track in the global coordinate system.**

Figure 11 shows a comparison of the measured and simulated wheel loads for lowering wheel L3 3 inches. The absolute difference between the test and simulation is below 5% in terms of wheel unloading.

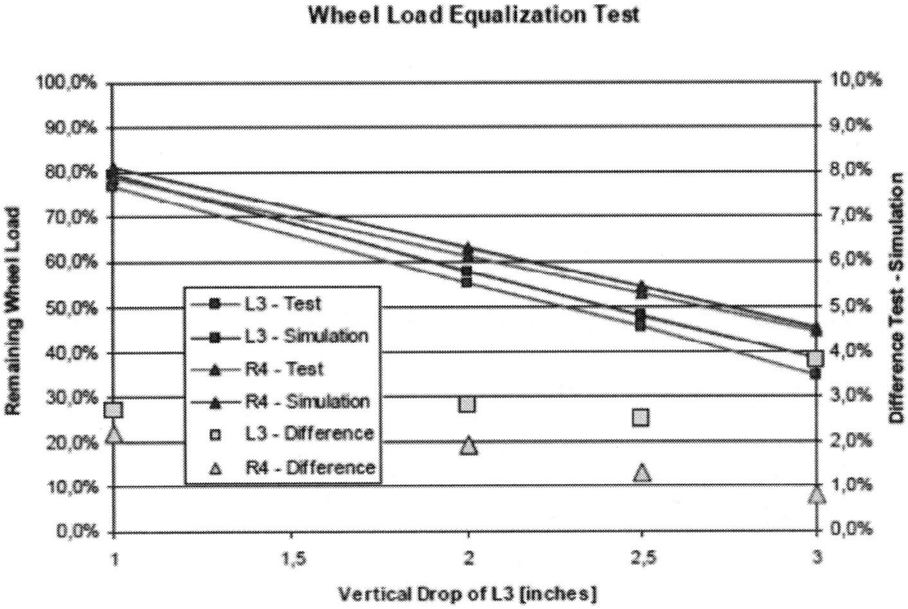


**Figure 11: Wheel load equalization – L3 dropped.**

Figure 12 shows a comparison of the measured and simulated frequencies for the locomotive. The results show an overall accuracy better than 10%. Hence, the simulation model can be considered as validated with respect to the eigen frequencies. Therefore the wedge test validates the correct modeling of the inertia and mass parameters of the carbody, truck and proves the overall feasibility of the simulation model, together with the results of the static lean and the truck equalization results.

| Mode | Carbody | | | | Truck | | |
|---|---|---|---|---|---|---|---|
| | sway/roll about x | | pitch about y | bounce in z | sway/roll about x | pitch about y | bounce in z |
| | lower | upper | | | | | |
| Wedge Test | 0.65 | 1.33 | 1.33 | 1.25 | 7.5 | 4.97 | 4.74 |
| Simulation Model | 0.56 | 1.3 | 1.44 | 1.22 | 7.23 | 4.72 | 4.67 |

**Figure 12: Comparison of the eigen frequencies of the simulation model with the dominant modes excited by wedge test runs.**

Figure 13 show an example of several comparisons made between on-track measurements and simulations using measured track geometry data. The results agree reasonably well. Certain conditions or input parameters (such as wheel-rail friction, track stiffness, exact speed-profile, and traction forces) can explain the differences. Comparisons were also made with other channels and other locations to demonstrate model validity.

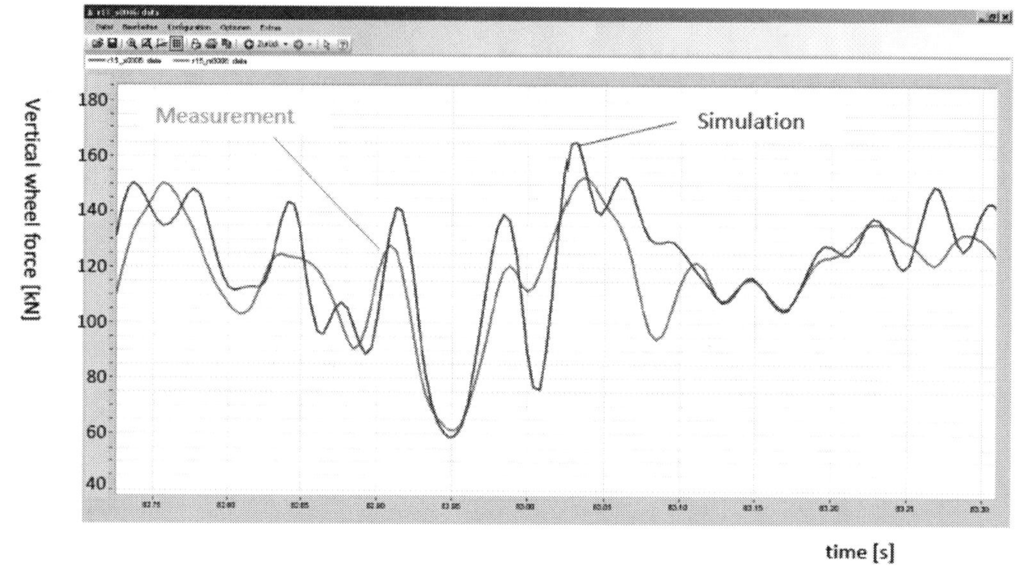


**Figure 13: Vertical wheel load – simulation vs. on-track test.**

## 4. CONCLUSIONS

As per the new 49 CFR 213.345, operators are required to conduct both testing and simulations to evaluate the dynamic response of the vehicle. The combination of tests and simulations are intended to provide a comprehensive assessment of the effect of track conditions (track geometry, track curvature, etc.) and vehicle speed on the dynamic response of a vehicle. Simulations using MCAT track segments are used to assess the vehicle dynamic response to maximum allowable track geometry variations provided in the FRA Track Safety Standards that may be present anywhere on the route. Testing will assess the response of the vehicle to the actual curvatures and other conditions present on the route and help to ensure safe operation.

## ACKNOWLEDGMENT

The work described in this paper was sponsored by the Office of Research and Development, Federal Railroad Administration, U.S. Department of Transportation.

## REFERENCES


1. U.S. Department of Transportation, Federal Railroad Administration, 49 CFR Parts 213 and 238, Vehicle/Track Interaction Safety Standards; High-Speed and High Cant Deficiency Operations; Final Rule March 13, 2013.
2. EN 14363:2005, "Railway applications-Testing for the acceptance of running characteristics of railway vehicles-Testing of running behaviour and stationary tests", CEN, Brussels, 2005.
3. "Practical Guide to Railway Engineering", American Railway Engineering and Maintenance-of-Way Association, 2003.
4. B. Marquis, J. LeBlanc, and A. Tajaddini. "Vehicle Track Interaction Safety Standards." In Proceedings of the ASME/ASCE/IEEE 2014 Joint Rail Conference, Colorado Springs, CO, 2014.
5. O. Polach and J. Evans. "Simulations of Running Dynamics for Vehicle Acceptance: Application and Validation." The 2nd International Conference on Railway Technology: Research, Development and Maintenance (RAILWAYS 2014) in Ajaccio. Corsica. 8-11 April 2014.

# The development and application of a lean future state design methodology to the railway possession and isolation authorisation process

**S Stiles**
University of Nottingham, SHE Solutions Ltd, UK

**W Garner**
Coventry University, UK

## ABSTRACT

The application of lean techniques to remove waste within processes is widespread within the Rail Industry and provides a useful means of continuous improvement and refinement. However to get a notable improvement in efficiency and effectiveness within the service part of the industry; a new methodology is required to facilitate a different way of thinking which results in a step change, similar to that experienced in lean manufacturing by Just in Time (JIT) manufacture. This paper combines lean techniques, human factors analysis and best practice communication technology from the fire service to analyse the Rail Possession and Isolation Authorisation Process. It creates a lean future state vision with benefits in quicker response time, improved process traceability and enhanced safety integrity. It adapts the manufacturing lean methodology and recommends a 5 step guide for the key services within the industry.

## 1. THE 24/7 RAILWAY AND ENGINEERING WORK

As the UK Rail Industry moves towards operating a 24/7 railway, the time taken out to undertake maintenance and project works is at a premium. It is therefore essential that any waste or delays incurred whilst taking possession and isolation of the line are eliminated, thus maximising the time for the engineering activity to be completed or releasing time for the safe operation of trains. However, maintaining protection for the safety of the workforce remains paramount.

It is recognised that the current standards for the authorisation and control of possession and isolation works result in unnecessary lengthy delays between trains stopping running, electricity being isolated and engineering works being able to commence (RSSB 2004). Research into the effectiveness of current safety arrangements, has identified that the quality of safety critical communication is variable, with errors commonplace often contributing to safety incidents, delays and possession over-runs.

Lean systems design has been successfully implemented across a number of industry and service sectors to improve productivity through simplifying the process to make an effective contribution towards the end to end process time. Improving process quality can also be aligned with reducing opportunities for error, which is essential for such a safety critical work activity. A detailed current state analysis of the existing arrangements has identified opportunities for both error and waste in the form of duplication, rework, delays, over processing and motion (Garner and Stiles 2013). Existing processes could benefit from lean attributes such as; reduced response time, standardisation, visualisation, transparency, empowerment, 'takt' and information flow. By applying a lean approach to the existing safety critical processes, opportunities for process simplification and a reduction of human error can be identified, whilst ensuring that workforce and system safety are not compromised.

A previous study (Garner and Stiles 2013) adapted the lean techniques to critically appraise the current state within the Railway Possession and Isolation Authorisation Process (RPAIP), this paper continues the research by developing a lean future state methodology to facilitate the value added flow redesign within the existing RPIAP.

## 2. ACADEMIC REVIEW

Developing a future state design should focus on making value flow and this is normally achieved at two levels; detailed low level operations and a higher planning level . The detailed operations level is already common in the Rail Industry where simple tools such as improved housekeeping and standard operations are employed to remove the waste and opportunity for delay, thus increasing the productivity of the workers at a local level. However to make value truly flow, a second high level planning redesign is also required. In manufacturing this is often achieved through Just in Time and cellular production, however in a service industry, there are no similar panaceas; as it is generally information that flows through a process rather than a product. A high level redesign should therefore look to employ techniques that enable information to flow in a controlled manner to the respective parties, without delay and interruption. Converting series flow to parallel flow (refer to figure 1) is a best practice that is common place in the emergency services' command and control environment (Robinson, 2013). Broadcasting data in parallel dramatically reduces the timeline to get information to the point of use.

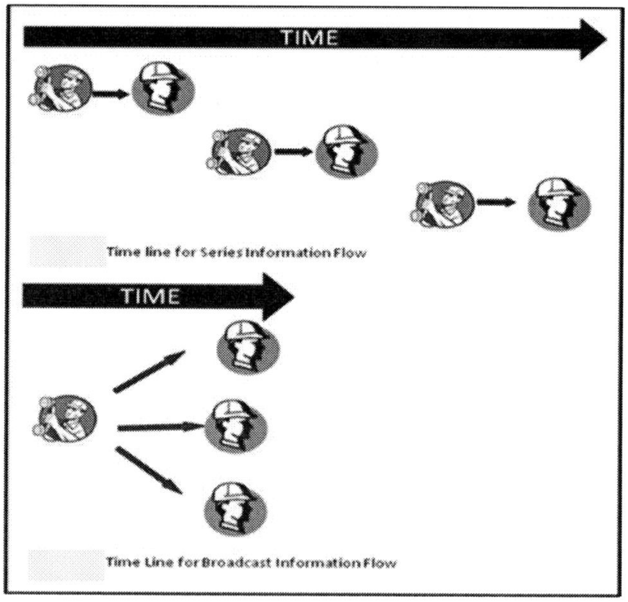

**Figure 1 comparison of time for information flow**

In manufacturing, transparency is a key element that facilitates flow at a high level. This can be achieved within the RPIAP future state by incorporating new communication technology; which will also improve the quality of the information transmitted and remove the opportunities for error.

Future state mapping is used to show the flow at a high level within manufacturing, linking JIT and single piece flow cells to the control systems. This technique cannot be easily applied to a Service Industry as it is typically information not products that flow. Data Flow Diagrams (DFD) however are used within the Software

Industry to demonstrate information flow and can be adapted to represent the future state within the RPIAP. According to Boucher (2006) DFD focus on how data is used and transformed. This DFD technique can be adapted to map the data flow for the RPIAP and enables the process to be visualised from a different perspective; identifying the gateway outcomes, key stakeholders and the interfaces for different levels of access and control.

Safety of the operational railway and the workforce is of prime concern to the UK Rail Industry. According to several RAIB investigation reports there have been a number of communication-based errors which have contributed to safety-related events. Furthermore a study by Gregory Harland Ltd (2006) identified 35.7% of trackwork incidents involved ineffective communication, with 88% of miscommunication attributed to human error, specifically failure to communicate the correct information. Therefore any RPAIP future state design should enhance existing safety critical communication arrangements.

Safety barrier analysis is a tool to understand what barriers were in place that should have prevented an accident from occurring (Hammerl and Vanderhaegen 2012). Sklet (2006) stated that safety barriers were important to prevent, identify and manage human errors that occur. Therefore, it is essential to understand the significance of any safety barriers within the current processes and where these have to be maintained in the RPAIP future state.

The RPIAP demands that individuals have complete and up to date accurate understanding of the environment which is essential to undertake work activities safely, a construct referred to as situational awareness (Golightly et al 2012). An individual's understanding of the environment they are working in will affect how they make decisions and their subsequent behaviours. This is relevant for the RPIAP where the process is heavily reliant on information flow (Endsley 1999); verbal communications detailing locations of sites of work, protection limits, exact location on the rail infrastructure etc. The very nature of the RPIAP has a high dependency on attention and working memory, and where these requirements are subjected to over-stress. This predisposes the process to human error. Therefore the future state design should address the importance of situational awareness, with individuals having accurate, up to date and complete information to hand.

When technology is being introduced it is important to consider the human interface of the end user (Constantine and Lockwood 1999). User-centred design is often applied to the design of any computer-based system. This process focuses on the needs, wants and limitations of end users, taking these factors into account in the final product design. There is an International Standard detailing 6 key principles of user-centred design; ISO 9241-210 (2010) Human-centred design for interactive systems. There are three principles particularly relevant. The RPIAP should:

- make simple, common tasks easy, communicating clearly and simply in the user's own language.
- make all needed options and materials for a given task visible without distracting the user with extraneous or redundant information.
- keep users informed of actions or interpretations, changes of state or condition, and errors or exceptions that are relevant and of interest to the user.

A thorough understanding of the end-users and their work environment is critical; and a balance between prompt communication and disruptions during safety critical activities should also be considered.

These factors are important for ensuring the effectiveness of the future state design, maintaining safety and removing opportunities for error.

## 3. PROPOSED METHODOLOGY

The RPIAP is fundamentally a communication based permit to work process between many stakeholders. The objective of the process is to achieve a safe work environment for the completion of engineering works whilst there are no trains running, and to hand back the railway for the safe passage of trains afterwards. This is therefore different to manufacturing processes which are based on delivering a physical product, and requires an adapted application of lean principles.

Lean thinking has been applied in Manufacturing Industry in the UK since the 1990's, based on the Japanese working practices developed by Toyota. The 5 steps of lean (Womack and Jones 2003) are adapted for the RPIAP (refer to Table 1).

**Table 1 lean approaches**

| Traditional Lean Approach | Service Industry Approach |
|---|---|
| 1. Identify value,<br>2. Map the current state (identifying waste)<br>3. Make value flow<br>4. Create a pull type system<br>5. Continuous improvement | 1. Identify value<br>2. Critically appraise the current state at a detailed level<br>3. Carry out high level gateways analysis<br>4. Simplify the process; make communication flow<br>5. Provide transparent and live information to all |

When applying the 5 steps of lean in a service industry it is easy to apply the detailed low level waste analysis and standardisation described by steps 1 and 2. Typically a swim lane variant of a current state maps is used, to visualise the flow of data between key stakeholders.

However steps 3 and 4 require further adaptation into the service industry. The typical method of making value flow (step 3) is not applicable for the communication flow predominant in the Rail Industry. Other Service Industries i.e. Health, apply four rules (Jimmerson 2007); i) Clearly specified activities, ii) Connections by direct communications between people, iii) Simple pathways to deliver service, iv) Continuous improvement. This methodology encourages analysis to move from a detailed waste removal process to a much higher level, challenging whether there is a more effective overall process. Information technology has an obvious role to contribute at this stage as it enables both parallel flow of data and high visibility.

Step 4 explains how control systems are shared visually, enabling all staff to check if they are 'on time', empowering staff to monitor their own progress and respond/request assistance accordingly. This relates closely to the situational awareness proposed within human factors research and could be adapted into the RPIAP future state design.

Step 5 requires continuous improvement; performance measures and transparent information enabling management to review the effectiveness of the service process and also empowers workers to identify opportunities for improvement.

## 4. CASE STUDY – RAILWAY POSSESSION AND ISOLATION AUTHORISATION PROCESS FUTURE STATE DEVELOPMENT

The following case study describes how the adapted 5 steps of lean have been applied to make value flow through the RPIAP.

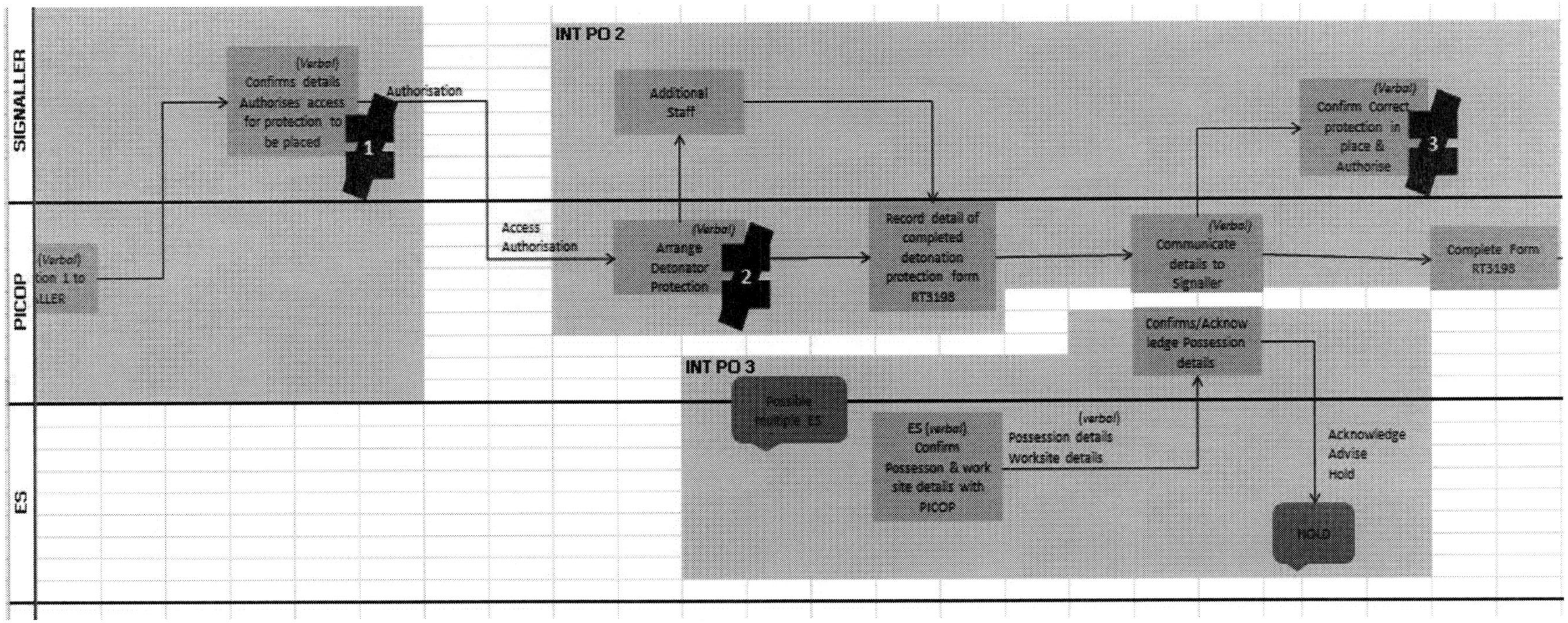


**Figure 2 extract of an adapted swim lane current state map**

### 4.1 Step 1 identify value and step 2 critical appraisal

An area of critical importance within the RPIAP and the 'value add' of the process, is the provision of a safe working environment. The current state swim lane map identifies the safety barriers (value add) as shown in Figure 2.

The seven wastes of lean were adapted for the RPIAP to represent 8 categories: Error (verbal); Error (positional), Waiting (staff), Excessive Communication (manual), Excessive Communication (verbal), Over Processing (duplication, over authorisation), Over Production and Capacity Lost (repeated tasks in series, lack of synchronisation), Motion (walking). The current process depicted in the swimlane was critically appraised in a previous study (Garner and Stiles 2013) and demonstrated that a high level of waste existed in the process.

### 4.2 Step 3 identify high level gateways

Lean manufacturing suggests that removing the waste at this stage will make value flow. However in the service industry, simply removing the wastes, would just facilitate doing the same process quicker. A different way of thinking is required to create a step change. Gateways analysis was used to view the process differently from a higher level, identifying the process inputs/ outputs for each stage gateway, see an extract in Figure 3.

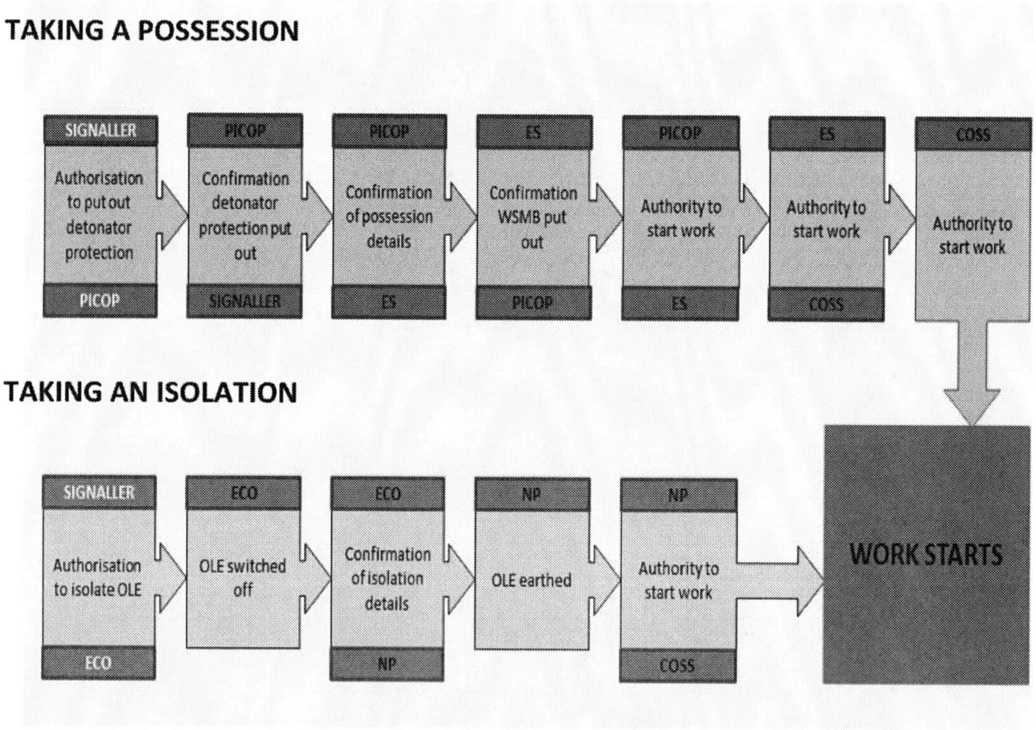


**Figure 3 extract of high level gateways of RPIAP**

The opportunities for the eight wastes were collated at each gateway section and SIPOC analysis applied to identify potential areas for improvement. Table 2 summarises the wastes within each gateway. Due to the safety critical nature of the existing RPIAP process the identification of safety barriers within each of the gateway stages, was also undertaken, ensuring that the safety of the workforce in the future state design.

## Table 2 summary of waste within each gateway

| Result of the waste | | | | | | | | | |
|---|---|---|---|---|---|---|---|---|---|
| | | INT PO 1 | INT PO 2 | INT PO 3 | INT PO 4 | INT PO 5 | INT PO 6 | INT PO 7 | START ENGINEERING WORK |
| | POSSESSION | Authorisation to put out protection | Confirmation detonator protection put out | Confirmation of possession details | Confirmation WSMB put out | Authority to start work | Authority to start work | Authority to start work & safety briefing | |
| Rework | Error - verbal communication | | | | | | | | |
| | Error – positioning | | | | | | | | |
| Waiting | Delays for staff | | | | | | | | |
| Motion | Excessive communication (manual) | | | | | | | | |
| | Excessive communication (verbal) | | | | | | | | |
| Over production | Lost capacity due to series communication | | | | | | | | |
| Over processing | Too much clarification/ authorisation/ duplication | ? | ??? | ? | ??? ?? | ?? | ??? | ? | |
| Safety Barriers | | | | | | | | | |

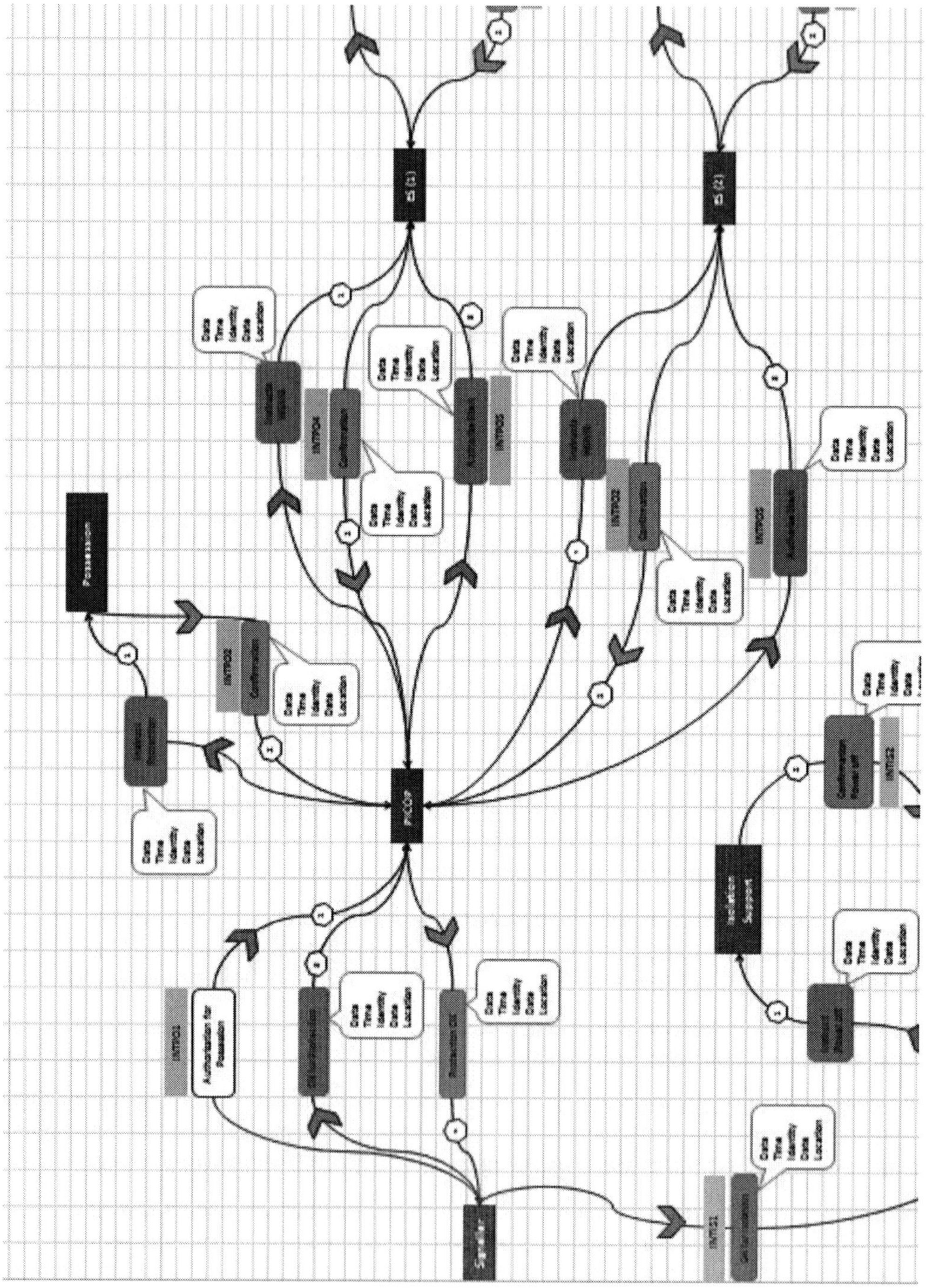

**Figure 4 data flow diagram of RPAIP**

## 4.3 Step 4 simplify the process: make communication flow

Step 4 is adapted from the lean manufacturing methodology to use data flow diagram (DFD) to map communication flow extract is given in figure 4.

The DFD highlights the transfer of outcome information for each stakeholder, which in turn shows the information centres. The study found that RPIAP would benefit from broadcasting data in parallel to several recipients, rather than the current method of telephoning each recipient in series, and recording data manually. The adoption of broadcasting data will facilitate the effective design of the new RPAIP system as it shortens the overall time taken for communication flow (refer to Figure 1).

A method of broadcasting relies on variable levels of access to information amongst stakeholders within the RPIAP, which are stored and tracked centrally through a databank. The aim of the databank is to give duty holders transparency of the authorisation process; flowing towards the start of work at the work site, giving the opportunity to prepare in advance and improve situational awareness (V). However, in order to maintain validity and confidence in the process, action to change the status on the databank was only given to those duty holders with authority (A). An example of databank information exchanges are identified in table 3.

**Table 3 extract from databank**

| Gateway | Signaller | PICOP | Possession Support (1) | Possession Support (2) | ES (1) | ES (2) |
|---|---|---|---|---|---|---|
| INTPO1 | A | A | V | V | V | V |
| INTPO2 | V | A | A | V | V | V |
| INTPO3 | | A | | | A | A |
| INTPO4 | | A | | | A | A |
| INTPO5 | | V | | | A | A |

## 4.4 Step 5 provide transparent and live information to all

Continuous improvement and enhanced safety (step 5) within the RPIAP requires data transparency and live information. By analogy, best practice from the fire service can be developed for the RPIAP and the new broadcasting technology would seek to include:

- Secure encrypted communications on different bearers
- Deployable on hosted or private infrastructure
- Two step process with a passcode to prevent accidental triggers
- User identification based on login in and authority codes
- Access only to possessions that are assigned to each individual/role
- Visibility of progress of the whole possession, with restricted authority to take action
- Each role in the RPIAP has actions configured to support the safe method of work
- Traceability of communication interfaces for ongoing monitoring

A significant factor in the success of the new technology is the availability and suitability of a delivery network with deterministic data communication facility throughout the country. Based on the systems already working within the fire service, figure 5 depicts a mock screen for secure access to the RPIAP control system and figure 6 shows the improvements to transparency and audit trail that can be achieved through new communication technology via the mock screen shots.

**Figure 5 mock screen for access to RPAIP app**

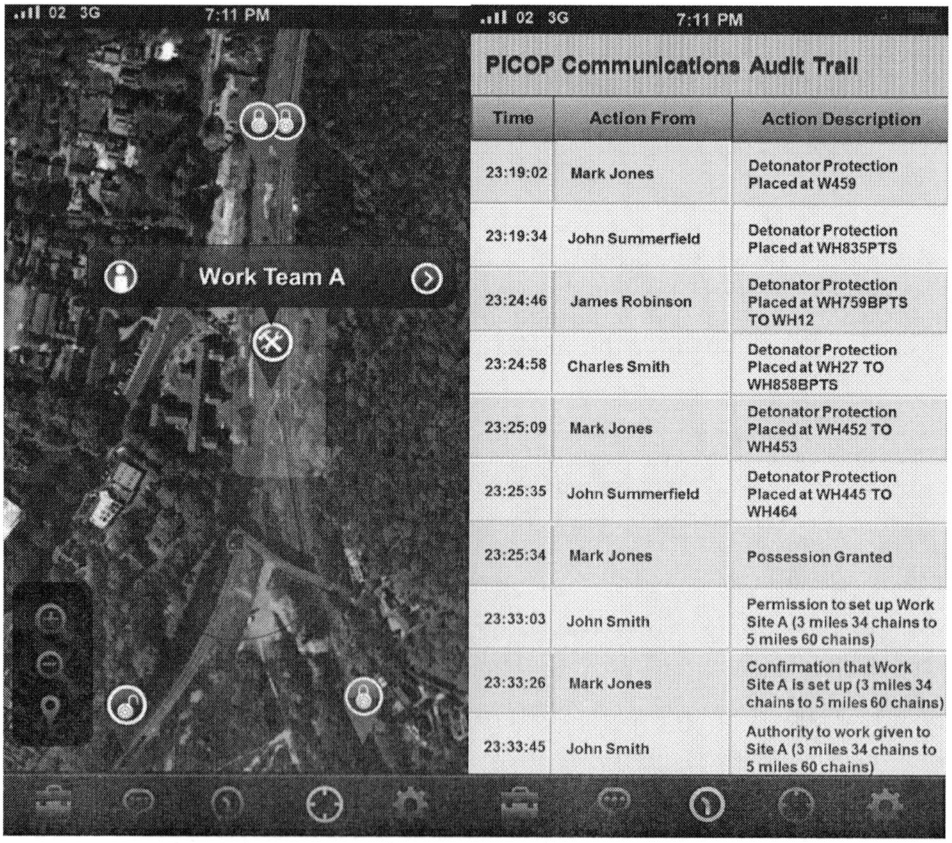


**Figure 6 mock possession screen and audit trail**

## 5. BENEFITS

Reviewing the future state design, it can be seen that many benefits would be derived. The methodology developed in this study does not simply speed up the current process but evaluates the best means to deliver the required outcomes (safety, efficiency, delivery) at each stage to all of the stakeholders in the process through implementation of broadcasting technology which will:

- Reduce: delay, error, complexity, paperwork
- Improve: traceability, situational awareness, transparency (readiness to start work), management visibility, performance measurement, real time data and monitoring
- Empower continuous improvement
- Maintain/enhance safety barriers.
- Increase time to complete engineering works or availability to run trains.

### 5.1 Potential capital savings

In order to quantify the potential benefit of the proposed use of broadcasting technology, a calculation based on possession data (note this excludes isolations) can be estimated with the following assumptions (refer to table 4).

**Table 4 assumptions**

| | |
|---|---|
| 1. | A typical possession consists of the following personnel:<br>- 1 PICOP<br>- 2 ES<br>- 3 COSS reporting to each ES<br>- 4 workers reporting to each COSS<br>- 33 personnel in total |
| 2. | A typical communication is estimated at 5 minutes to include; dialling number, convey information, read back information and manually log information. |

If a 5 minute call made in series could be eliminated through the broadcasting of data, during the gaining of possession and then again when returning the possession, a reduction in the current timeline can be seen as shown in table 5.

**Table 5 time line for 5 min in series phone call saving (assumes 33 personnel)**

| | For every 5 min phone call in SERIES Waiting Time for Possession | | | | | | For every 5 min phone call Time lost to run trains | | | | |
|---|---|---|---|---|---|---|---|---|---|---|---|
| Time (mins) | 5 | 10 | 15 | 20 | 25 | | 5 | 10 | 15 | 20 | 25 |
| Communication | PICOP - ES 1 | PICOP - ES2 | | | | | | | | ES2 - PICOP | ES1 - PICOP |
| Communication | | ES1 - COSS1 | ES1 - COSS2 | ES1 COSS 3 | | | | ES1 - COSS 3 | ES1 - COSS 2 | ES1 - COSS 1 | |
| Communication | | | ES2 -COSS 4 | ES2- COSS 5 | ES2 - COSS 6 | | ES2 - COSS 6 | ES2 - COSS 5 | ES2 - COSS 4 | | |

There are approximately 93,500 possessions in a year undertaken in the UK. Considering the above assumptions, the estimated saving of 2.5 million hours is achieved by eliminating one 5 minute phone call made in series to a team of 2 ES with 3 COSS each, representing a work team of 4. The current state map has shown that there are several 'in series' phone calls made during the RPAIP, so these savings can be multiplied further, by the number of calls eliminated.

Although these figures are based on the assumptions listed, they do demonstrate the scale of the savings that can be made through use of broadcasting technology.

## 6. DISCUSSION

The application of tried and tested lean techniques, have been adapted for the UK Rail Industry providing opportunity for continuous improvement. A previous critical appraisal of the current state (Garner and Stiles 2013) reveals the complexity and interdependency of the RPIAP, with many iterative loops and cascading of repeated communications throughout the chain of command. The high level of opportunity for waste, delays and errors provides a catalyst for a lean approach to be applied to the RPIAP focusing on improved efficiency, effectiveness and safety.

This study has identified that it is necessary to combine a lean approach with due cognisance of human factors techniques, such as safety barrier analysis and situational awareness to ensure that the safety critical nature of the RPIAP is maintained within any future state design. This adapted lean approach seeks to remove waste, improve efficiency, and give greater data flow with real time accurate information through new technology and data broadcasting.

Broadcasting data also improves transparency, potentially enhances safety and traceability, and facilitates continuous improvement. The introduction of broadcasting technology introduces further considerations with regards to both the delivery network and the deployment of the handset. Consideration should be given to best practice from both the emergency services, where their control and command software is used in these safety critical conditions.

If broadcasting technology is to be used for RPIAP, then both direct Network Rail employees and contracting staff will require access to the handsets. The handsets will need to be developed with specific usability requirements addressed and for solutions to be robustly tested for human interfaces. Human-centred design will support the overall effectiveness of the RPIAP future state (RSSB 2008). It is important to consider the product from a usability perspective; how the product will be operated safely when working trackside, the attention needed and time taken to perform the necessary functions, ease of use when wearing gloves, use of product at night, during periods of sunshine and glare, and in inclement weather.

The RPIAP future state design develops a proposal for feasibility; up to a technology readiness level 3 (Mankins 1995). The next stage would be technology development where the usability requirements would be a prominent feature; understanding user aims, characteristics and values.

The Railway and Other Guided Transport Systems (Safety) Regulations 2006 place specific statutory duties on organisations within the industry. In response to these duties the Rail Industry has developed specific arrangements for the introduction of new technology through a product acceptance process and ongoing governance detailed within rail industry standards. The future state design has potential to impact on both of these aspects.

The prevailing culture within the industry is to be resistant to change, therefore even the minor changes associated with the RPIAP future state design would be a challenge to implement. The RSSB (2008) recognise that failing to consider the human factor is likely to cause problem when introducing new technology. The use of socio-technical design is recommended to avoid such problems, where both the technical and social aspects of the technology are developed (Mumford 1983). The Rail Industry Product Acceptance Process ensures that the interfaces with the wider Rail Industry are also addressed as part of any product development, design and approval.

The RPIAP future state design has been reviewed against the existing Rail Industry standards; GE/RT8000 HB9, 11, 12, 16, NR/L2/OHS/019 and NR/SP/ELP/29987. The impact of the recommendations on the existing standards has been identified as minimal. Whilst there would be some minor revision needed to the standards, the actual process, accountabilities, responsibilities and duties detailed within the standards listed above would not change.

Although the impact would be minimal it is not underestimated that even such minor change would be a challenge to the industry in terms of managing the change effectively; this is even more significant with reference to the prevailing infrastructure maintenance culture.

The UK Rail Industry is governed by numbers of standards, defining either processes to follow or requirements to be met. The industry approach to standards will bear on innovations and changes that may arise from the 24/7 drive for improvement, and will need to be considered in translating the themes that have been identified within this study into reality. This study aims to support the 24/7 railway through the application of lean techniques to identify opportunities for improvement, whilst maintaining the safety of the engineering workforce.

## 7. CONCLUSION

A methodology has been developed that enables a lean solution to be applied to the RPIAP which does not simply speed up the current process but evaluates the best means to deliver the required outcomes (safety, efficiency, delivery) at each stage to all of the stakeholders in the process.

A future state design has been developed; including a future state map, accompanying data flow diagram, broadcasting databank and user specification. The introduction of broadcasting data reduces waste and inefficiency arising from lost time, improves transparency, enhances safety and traceability and facilitates continuous improvement aligned with the lean approach.

The proposed changes, both technological and cultural are already demonstrated as best practice within the emergency services and should therefore pose limited risk within the rail industry, with the potential benefit acknowledged as high impact.

## 8. ACKNOWLEDGEMENT

Robinson, A, 3TC software, http://www.3tcsoftware.com/

## 9. REFERENCES

Boucher, T, 2006, Design of Industrial Systems, Academic Press

Constantine L., and Lockwood, L. (1999) *Software for Use: A Practical Guide to the Essential Models and Methods of Usage-Centered Design.* Reading, MA: Addison Wesley

Endsley, M.R. 1999 *'Situation awareness and human error: designing to support human performance'* Proceedings of the High Consequence Systems Surety Conference Albuquerque, NM

Garner, W and Stiles S, 2013 applying lean techniques to analyse railway possessions and isolations to identify opportunities for more effective procedures, Proceedings of the Rail Human Factors Conference, London

Golightly, D. Balfe, N. Sharples, S. And Loew E. 2012 'Measuring situation awareness in rail signalling' in *Rail Human Factors around the World by Wilson,* J.R. Mills, A. Clarke, T. Rajan, J. And Dadashi, N. 2012 CRC Press

Gregory Harland Ltd 2006 *The Collection of Railway Safety Critical Communication Error Data* RSSB

Hammerl, M. And Vanderhaegen, F. 2012 'Human factors in the railway system safety analysis process' in *Rail Human Factors around the World by Wilson*, J.R. Mills, A. Clarke, T. Rajan, J. And Dadashi, N. 2012 CRC Press pages

ISO 9241-210 (2010) Human-centred design for interactive systems

Jimmerson C. 2007 *A3 problem solving for Healthcare*, NY Healthcare Press

Mankins, J.C. 2005 *Technology Readiness Levels - A White Paper* Advanced Concepts Office, Office of Space Access and Technology NASA

Mumford E. (1983) Desig*ning Human Systems* Manchester Business School cited in RSSB 2008 *Understanding Human Factors – a guide for the railway industry* RSSB

Robinson, A. (2013) Personal Communication, 3TC Software

RSSB 2004 *Assessing the balance between trackworker safety and overall system safety* RSSB

RSSB 2008 *Understanding Human Factors – a guide for the railway industry* RSSB

Sklet, S. 2006 'Safety barriers; definition, classification and performance' *Journal of Loss Prevention in the Process Industries* 19 pages 494 – 506

Womack, Jones 2003 *Lean Thinking,* Free Press, USA

# The variability of generalised cost weighting parameters

**S Pani[1], N G Harris[2], F Schmid[1]**
[1] Birmingham Centre for Railway Research and Education, School of Civil Engineering, University of Birmingham, UK
[2] Railway Consultancy Ltd, UK

**ABSTRACT**

Although generalised cost theory underpins most modern transport planning, it is based on a number of assumptions, not least in terms of the weightings applied to different journey elements. Within a study or a geographical area, these are often assumed to be the same and constant, largely because it is difficult to capture and apply more accurate and directly relevant values.

The authors of the paper suggest that variations in three of the weighting parameters of the generalised cost applicable to rail journeys affect the accuracy of modelling, and may lead to errors in the demand calculations if the changes are not considered carefully: First is the use of electronic media and communication equipment that appears to be reducing the factor associated with in-vehicle time. Secondly, average feelings about values of time appear to be bi-modally distributed, which means that quoted averages may not accurately reflect the true behaviour of anyone. Thirdly, those familiar with the travel environment appear to adopt weightings lower than those adopted by passengers unfamiliar with the particular mode, an issue which may affect management decision-taking.

## 1 INTRODUCTION

Generalised cost theory has been the accepted theoretical framework for modern transport planning for about 50 years. The Generalised Cost Function (GCF) underpins both the large-scale multimodal models of major conurbations, as operated by city governments and national transport authorities, e.g., Transport for London's Railplan model (TfL, 2010) and Australia's National Guidelines for Transport System Management (Australian Transport Council, 2006). The approach is also used in academic studies of traveller behaviour and has been written-up extensively (e.g. Harris & Godward, 1992, chapter 1).

However, as noted by Grey (1978), its application in project appraisals and behavioural understanding is reliant on the validity of a number of assumptions. Even if one were to accept that human beings acting rationally may not take quantified decisions for each and every travel choice, transport planners have to assume that passengers have innate personal valuations for the different elements that make up the journey time and that they apply these instinctively when choosing modes or combinations of modes. To enable consistent decision-making, the parameters are normally set out in a detailed protocol, such as the UK Governments' webtag procedure, and may include a differentiation by market segment, e.g., business vs leisure Values of Time (VoT).

There have been detailed studies by others into the variability of parameter valuations, for instance by those seeking to undertake discrete choice-type experiments, where it is important to understand whether respondents are genuinely trading off one attribute against another. Börjesson et al (2012) note that methods of analysis which exclude non-traders can lead to the adoption of incorrect mean values. Peer et al (2012) note that scheme evaluation needs to understand both the expected distributions of attributes (such as journey time) and also the distribution of their valuations. They suggest that, whilst simulations can help estimate the former, less is known about the latter and they also make the useful point that variability is assessed by the user either by comparison to a rough estimate (e.g. averaged across a year) or a finer estimate (which may be able to take into account prevailing conditions on the day).

Conventionally, the values for weighting access, egress and waiting times are taken as twice those of in-vehicle time and are set in the context of exogenous, as well as endogenous factors driving demand (Wardman, 2006). Allowance is sometimes made, e.g., in studies of business travel, for the adjustment of actual travel time by the time spent usefully (Fickling et al, 2009). However, some authors are beginning to see weaknesses in the general approach that arise from the use of standard parameters for all scheme beneficiaries. For instance, Hultkrantz (2013) has noted that there is an impact on the valuation of time saved in high-speed rail projects, which depends on the previous mode used by the beneficiary. In summary, therefore, there is a chance of making a significant mis-investment in a railway project because the assumed values of responses do not reflect the true mean.

In this paper, the authors report on work carried out as part of MSc research at the University of Birmingham. They provide three insights into reasons for the variability of generalised cost weighting parameters as applied to rail journeys, which allow us to understand and thus avoid errors in the estimation of transport demand that might be introduced otherwise. After description of the research methodology in Section 2, the insights are covered in the three sub-sections of Section 3 reflecting respectively the impact of electronic media and communications, the bi-modality of results and the impact of familiarity.

## 2 RESEARCH METHODOLOGY

The study began with the design, distribution and analysis of a pilot survey that became the template for the main study. As part of a demonstration of the principles of generalised cost theory, the survey described to respondents three possible rail journeys that they might undertake and indicated to them the generalised cost weightings which would be applied conventionally. The researchers then asked the respondents to suggest their own parameter weightings. The pilot provided the survey format and preliminary results that assisted in the development of the final survey. The pilot was completed by post-graduate students of the 2012-2013 Railway Systems Engineering and Integration MSc programme at the University of Birmingham. The three pilot scenarios were as follows: (i) commute to job, (ii) travel to a job interview and (iii) travel to visit friends in London. The last column was left as an open question, allowing respondents to state how they spent their time in each segment of the journey, for example, listening to music). Figure 1 presents one scenario as an example.

The value of this survey towards the research into traveller behaviour was its contribution to the development of the survey. The pilot survey showed that a Stated Preference type survey, in which different travel scenarios were broken up into segments, could provide a means for capturing information on traveller behaviour that would be sufficient to update the Generalised Cost Function and better understand travellers' perception and behaviour during rail journeys. However, it also proved that the survey had to be redeveloped to ensure that the complex topic of generalised cost and the questions were presented in a way that would be easy to understand by the general public.

| | Activity | Time (min) | Standard Weight | Result of Calculation | 'My own Weight' | Result of Calculation | 'Activities' |
|---|---|---|---|---|---|---|---|
| Travel to a Job Interview | Car to Local Station | 10 | 1 | | | | |
| | Wait for quarter-hourly Train Service | | 2 | | | | |
| | Train In-Vehicle Time | 30 | 1 | | | | |
| | Change to Hourly Train Service | | 10 | | | | |
| | Wait for Hourly Train Service | | 2 | | | | |
| | Train In-Vehicle Time | 120 | 1 | | | | |
| | Walk to Interview Hotel | 10 | 1 | | | | |
| | Fare | £60 | £20/h | | | | |
| | Generalised Cost | | | | | | |

What would be the impact of a 5 min delay?
What would be the impact of a 10 min delay?

**Figure 1 Format for pilot survey generalised cost scenario**

The first step in redeveloping the survey was to determine the target audience. The main constraint that limited the scope of the study was the available time and resources. The most effective option was to limit the study to the United Kingdom (UK). Limiting the population sample allowed for travel scenarios to be focused on those that may be encountered while travelling in the Greater London region.

In addition, limited time and resources also meant that the best option for distributing the survey would be through an online survey. Other options, such as distributing questionnaires on-board trains or conducting interviews with individual passengers would require lengthy planning, cooperation, and coordination with various organisations, most importantly the Train Operating Companies (TOCs). Using an online survey allowed quick and easy distribution, although its use limited the population sample to those with access to the internet.

The online survey was constructed in such a way that responders were presented with one of three travel scenarios or journeys. These were based on the scenarios of the pilot study and the experience gained thereby. The three scenarios were: a commute to work (Scenario 1), travelling to a job interview (Scenario 2) and travelling into London to visit friends (Scenario 3). The commute scenario is a travel journey where the passenger is most likely to be familiar with the route and has a routine at every stage of the journey; this familiarity helps determine the traveller's behaviour. This is in contrast to the job interview scenario, where the traveller might not be familiar with the end destination or routes to get there, whilst there also may be nervousness or stress about the interview. The journey into London to meet friends is the type of scenario where the passenger may be more relaxed, may or may not be familiar with the route, but may have more leeway on the time of arrival.

In addition to the familiarity of the journeys, survey also considerered the different lengths of journeys. The commute scenario was developed as the shortest of the three, while the job interview was the longest. In summary, although the scenarios were developed to try and cover different types of standard trips that a traveller might make, it was also developed in part to determine whether the above considerations are true by measuring the *annoyance* and stress of travellers during the various segments of the journey.

The travel scenarios were constrained in scope by limiting the options of modes of transport. Typically when calculating the generalised cost of a specific journey, it is necessary to consider all available travel modes, which includes the use of several different modes for each segments of the journey. Since the project focuses on 'travel behaviour of rail passengers' the major journey segments were restricted to rail travel only.

Figure 2 is used to summarise the three travel scenarios by segment and shows the time needed to complete each of the segments. Waiting times for the initial train journey were not provided - instead, a question was incorporated asking how early the respondents would arrive at the station before their train departure. This information would allow for a comparison of situations where passengers know the train frequency compared to when they do not know it.

In addition to the travel scenarios, participants were asked to provide general information on their demographics. To encourage a higher rate of response for these questions the surveys were made anonymous and the responder had the option not to provide this information. The exceptions to this were the questions that asked whether respondents had experience in the public transport industry and railway industry. This was due to the expectation that there could be a higher rate of bias of respondents with experience of the industries.

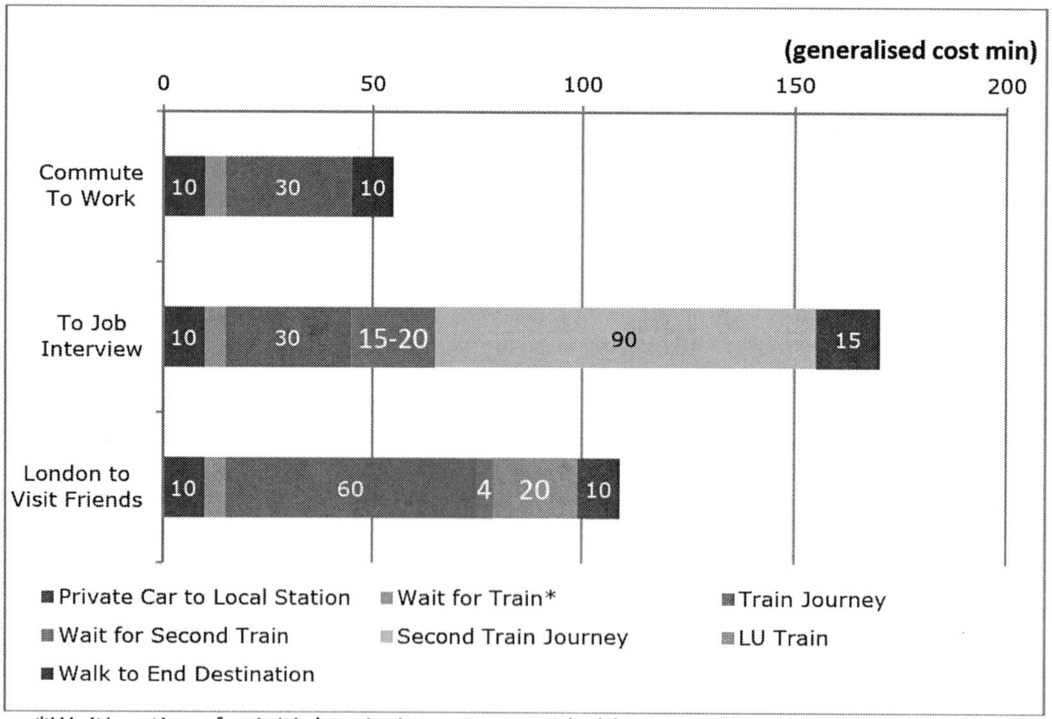


*Waiting time for initial train journey provided by survey responders.

**Figure 2 Travel scenarios**

The major sets of data collected included the activities performed during each segment of each journey (see Figure 3 for the presentation of the question), the annoyance ranking of completing each segment (see Figure 4), annoyance due to delays (see Figure 5), annoyance due to an interchange during peak and non-peak hours (see Figure 6) and, finally, experienced stress that might be incurred due to delays (see Figure 7) and due to performing interchanges.

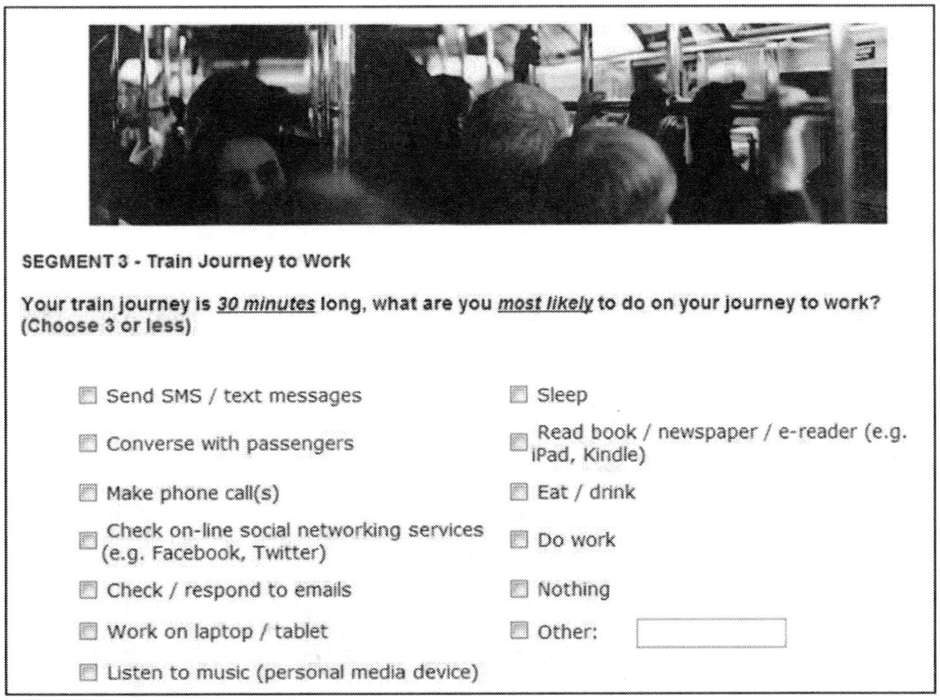


**Figure 3 Example question for most likely activities while on board train**

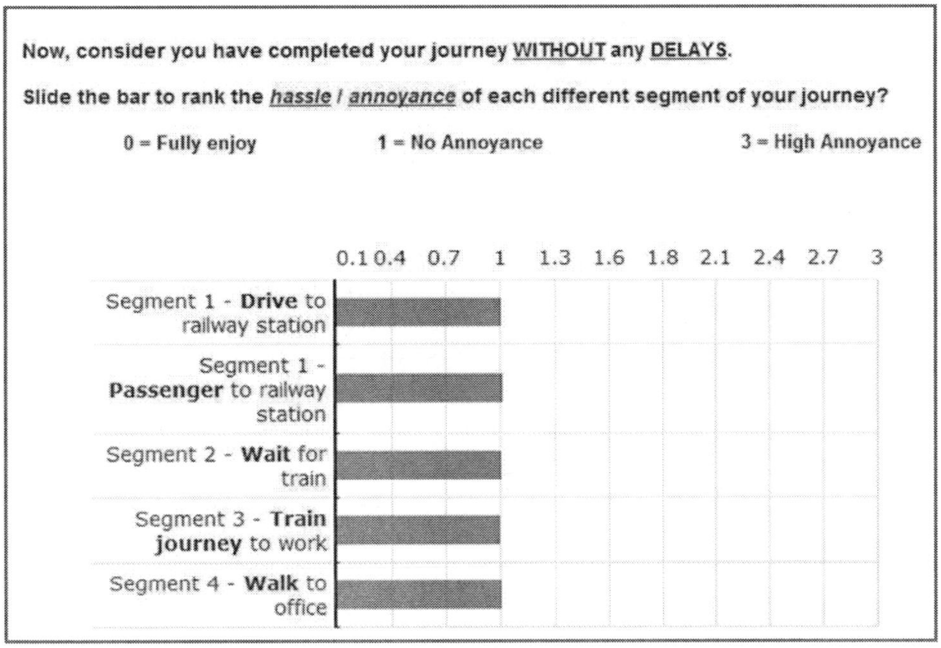


**Figure 4 Example question for rating *annoyance* levels in scenario 1 – commute to work**

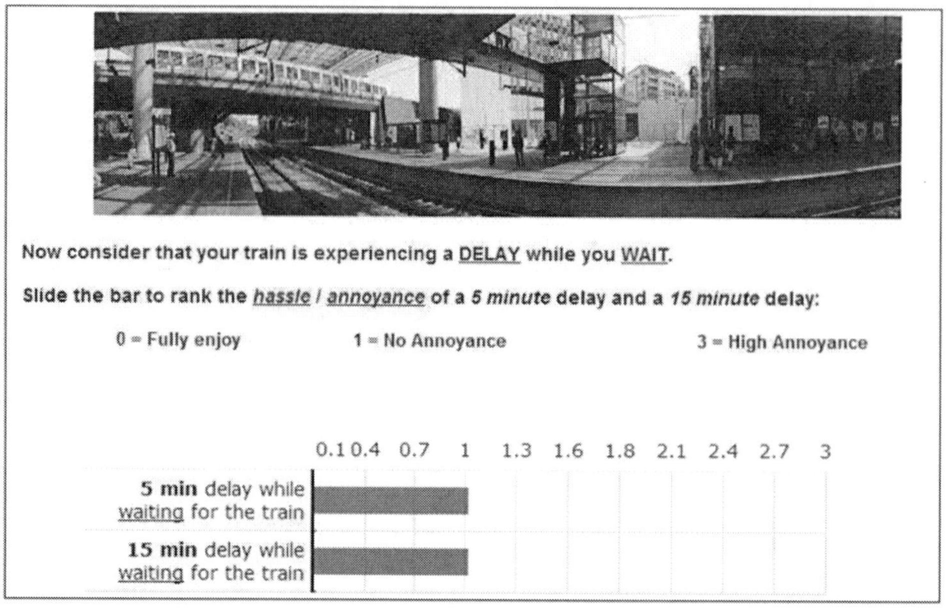


**Figure 5 Example question for rating *annoyance* due to delay**

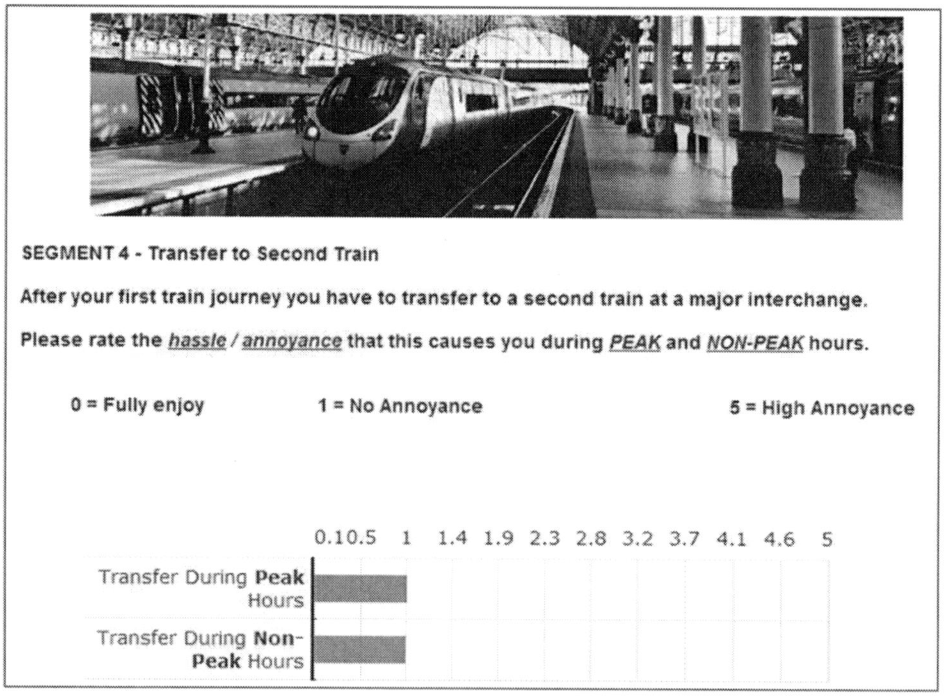


**Figure 6 Example question for rating *annoyance* due to interchange**

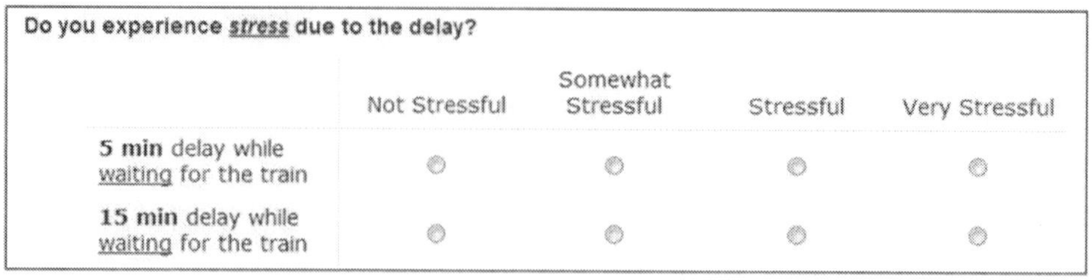


**Figure 7 Example question for rating stress experienced due to delay**

To test the survey and to ensure that there were no errors or flaws in the flow of the survey and that it was easy to follow, a pilot survey was performed before fully distributing the survey. The results of the pilot survey proved that no major changes were required in the presentation of the travel scenarios. Nevertheless, there were some small adjustments that were made, including adding additional options in multiple-choice questions, adjusting the scale on the *annoyance* rating bar and adding questions in the demographics section of the survey.

The main survey distribution was executed in a similar fashion to that of the pilot survey. Email and online social networks were used to distribute the survey. 600 individuals were invited to participate in the survey; however, it is estimated that the total sample size was closer to 650. An exact figure is not known since responders were encouraged to share the survey link in an attempt to increase the total response rate.

The survey was activated on 16 November 2012 and allowed to run for four weeks, until 14 December 2012, at which point the link to the survey was deactivated and no more surveys could be submitted. A total of 106 responses were received. Twenty (20) of the submitted responses completed only the demographics section or less of the survey and were discarded as no useful information could be gained for the analysis. Two responses completed the majority of the questions for one scenario. Since the final sample size and the number of incompletesurveys was low, it was decided that the responses for the answered questions could still be used for the final analysis. Finally, 86 responses were taken into account for the analysis and this provides a final response rate of approximately 13.2%.

The survey was formatted so as to encourage responses to more than one scenario. This meant that it was possible to gather more data than it would be possible if individuals were only required to provide answers to one scenario. Of the total of 86 completed surveys, 73 per cent completed one scenario, 6 per cent completed two scenarios and 21 per cent completed all three scenarios. This presentation format was highly successful and allowed for the capture of 48 per cent more total responses per travel scenario. The final count of completed responses per travel scenario was:

- 40 responses – Commute to work (Scenario 1)
- 41 responses – Journey to job interview (Scenario 2)
- 46 responses – Journey to London to visit friends (Scenario 3)

Overall this is still a relatively small sample size, especially considering that the demographics data is still representative of only 86 individuals. Any type of correlation or extrapolation to the population is not possible, for this reason. However, the study still indicates interesting trends among the sample population, which merit further study and analysis.

# 3 RESULTS OF SURVEY

## 3.1 Effects of Electronic Media and Communication Systems

Overall 'Listen to music (personal media device)' was among the top 3 choices for activities during every segment of the journey, with the exception of the segments by private car, where the radio ranked higher. On average, 35 per cent of passengers chose it as an activity they were most likely to perform during any of the given scenarios. This activity requires some kind of mobile technology that may or may not offer wireless connectivity. Looking at a correlation of activities with annoyance ratings was not possible, due to a combination of a small sample size and a large range of provided activities, namely, a total of 14 options, including an option to provide a custom response. The high number of variables led the correlation tests to fail.

Had the ratings been grouped further, had a greater number of responses been available or the number of variables been reduced it would have allowed the data to be correlated better. Still, the impact of wireless devices used during travel can be observed: 'Checking social online media (i.e., Facebook)' and 'Send SMS / text messages' both averaged approximately 19 per cent as a chosen activity, making it one of the three most likely activities to be performed during the journeys. The categories of 'Check /respond to emails' averaged 12 per cent and 'Make phone calls' 9 per cent, both rating as popular activities. The chart shown in Figure 8 summarizes the activities involving electronic media by percentage of each activity chosen for each journey segment. The percentages given in Figure 8 are cumulative, as respondents were allowed to choose up to 3 activities per segment, see Figure 3 for example question.

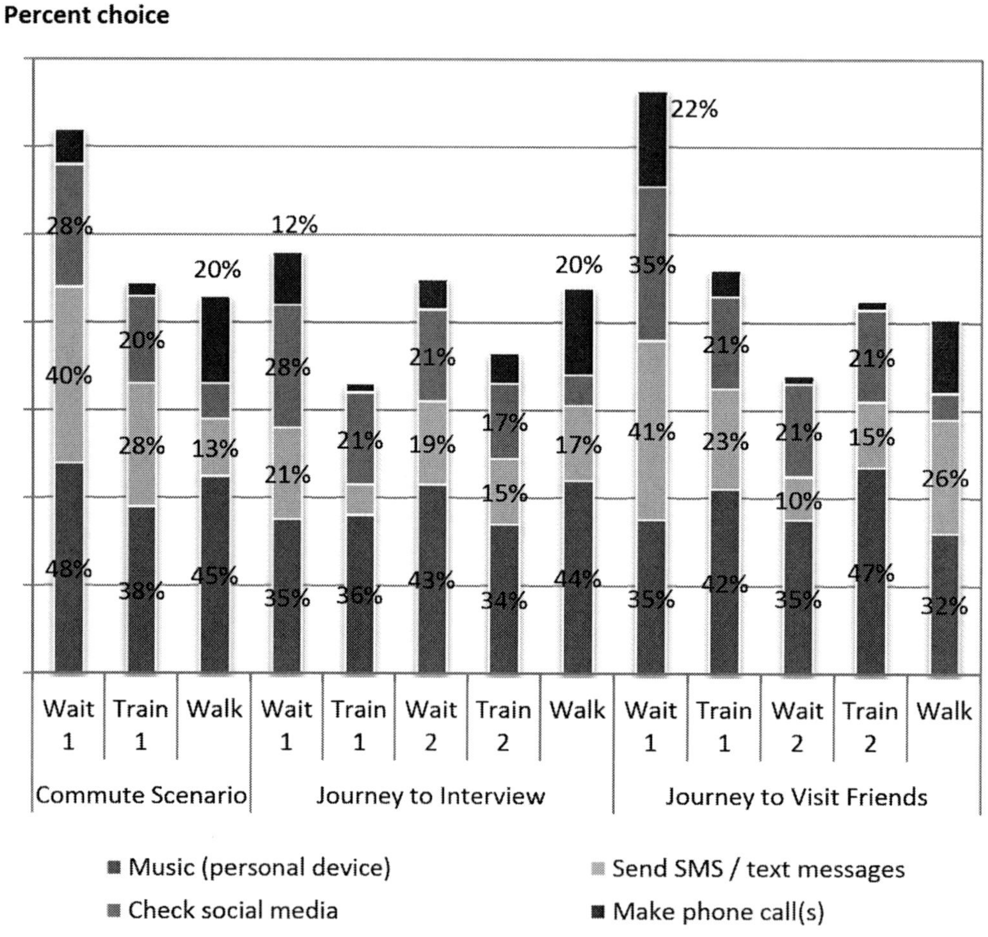


**Figure 8 Electronic media and communications usage by travel scenario**

### 3.2 Bi-modality of Weightings

The results collected on the *annoyance* of performing each of the journeys was the data that required the most careful analysis. In order to determine whether the data was statistically significant a t-test was required; however, to ensure the validity of the t-test the data must be tested for normality. The k-sample Anderson-Darling test for normality was used on each of the sets of annoyance data, as per D'Agostino (1986). The Anderson-Darling test is used by setting the null hypothesis to: "the data are from a normal distribution". All the *annoyance* weighting results failed this normality test and hence it was not possible to confirm that the gathered data followed a normal distribution. This can be attributed to the small sample size.

Large standard deviations and very similar means did not provide enough information to understand what the results represented, so histograms were created. The graphical representation of the results presented a clearer understanding about the travellers' *annoyance* with aspects of the travel for each scenario and made comparison more relevant than what would have been achieved with a simple comparison of means.

The histograms were created using 'bins' which were calculated using the average and the standard deviation for each segment. Eight bins were used to create the histograms. The x-axis on the histograms represents the *annoyance* rating (the values correspond to the weighting factors as used in the GCF), while the y-axis represents the frequency of the chosen rating. The scale given to passengers to rank *annoyance* of performing each of the journeys was from 0.1 to 3.

The results of the *annoyance* ranking of completing a journey over all scenarios were lower than the standard weighting factors that are used for the calculation of the GC. However, no significant conclusion on the weightings could be gathered through this study. The values range from 0.74 to 1.13.

The histograms throughout all scenarios were bimodal and skewed towards the left, as can be seen for example in Figure 9. The bi-modality of the histograms showed that there are two cohorts within the population - those who rank each segment significantly lower and those who average a more 'neutral' ranking. The results also followed the industry standard of in-vehicle weightings being used as neutral, and having waiting factored slightly higher. The survey used a similar scale to that used in the transport industry and the values provided in the results were lower than neutral suggesting a possible trend towards lower than standard values.

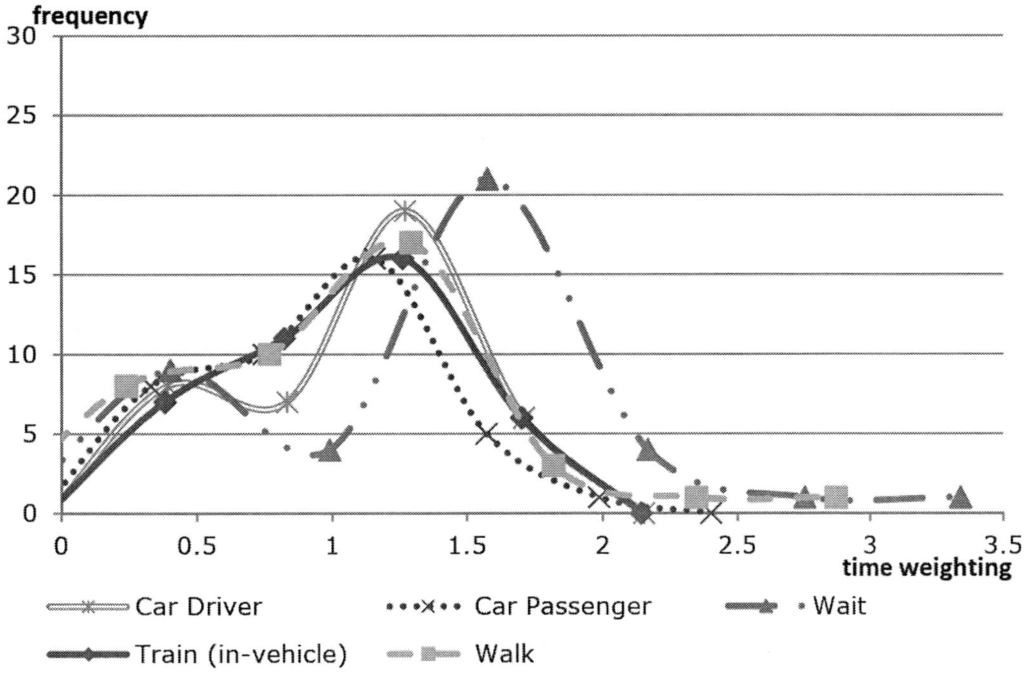


**Figure 9 Scenario 1: annoyance rating histograms**

The scale for rating *annoyance* due to delays was the same as that used for overall annoyance from 0.1 to 3.0. Two different delay options were given: a 5 minute delay and a 15 minute delay. Delays could arise in both the waiting segments and in the in-vehicle segments of each journey.

Figures 10 and 11 present histograms for the rating results for Scenario 2 of the study. It is clear that the 5-minute delays skew to the right, owing to the lower *annoyance* rating, while the 15-minute delays skew to the left. This shows the clear difference of *annoyance* due to the length of the delay, while the location where the actual delay occurs has less of an impact on the results.

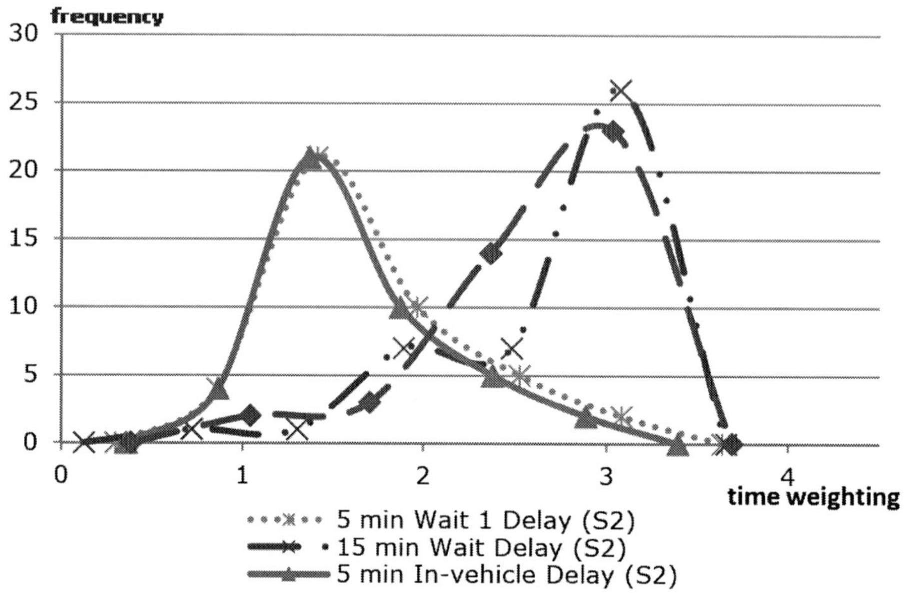


**Figure 10 Scenario 2 delay annoyance histograms (chart 1)**

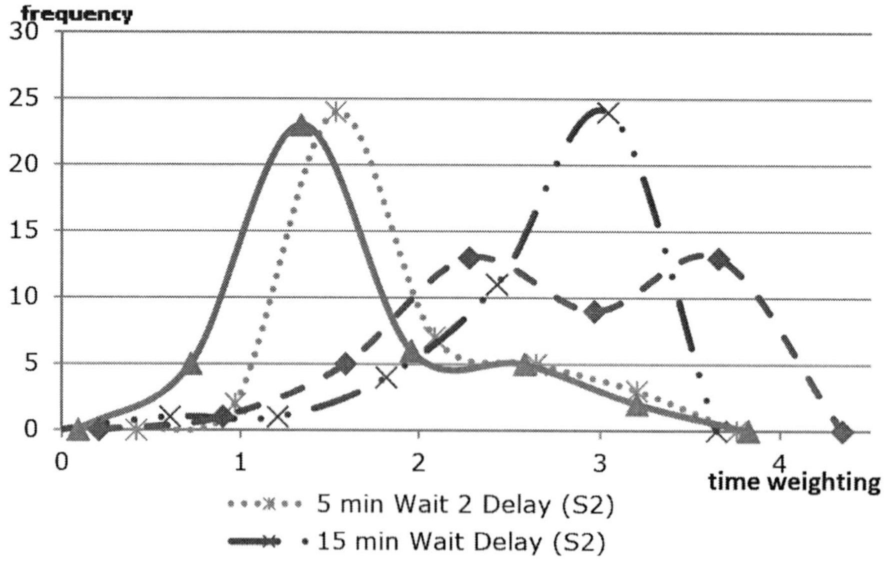


**Figure 11 Scenario 2 delay annoyance histograms (chart 2)**

It is important to note the bi-modal shape of the histogram in Figure 11, for a 15-minute in-vehicle delay during the second leg of the journey in Scenario 2. The bi-modality of the histogram could be attributed to the length of the train journey. The first segment in Scenario 2 is a 30 min journey where the impact of a 15-minute delay would be greater hence the clear left skew. However, although on a 90-minute journey the *annoyance* ratings are still skewed, meaning still greater annoyance due to the longer delay, the rating is not as consistent. It leads to the assumption that the length of a journey also plays an important role in the *annoyance* due to a delay.

The study's focus on interchange was realised for a train-to-train interchange. Therefore, only Scenario 2 and 3 had questions relating to interchange *annoyance*. The impact of an interchange on Generalised Cost is calculated differently in the GCF: the number of interchanges during the journey is multiplied by an 'interchange penalty time', the perceived time it takes to complete the interchange. It does not feature an explicit *annoyance* weighting factor. To understand better the impact of an interchange it was decided to follow the same format that was used in previous survey questions with the exception being that the maximum available rating was increased from 3.0 to 5.0 (see Figure 12).

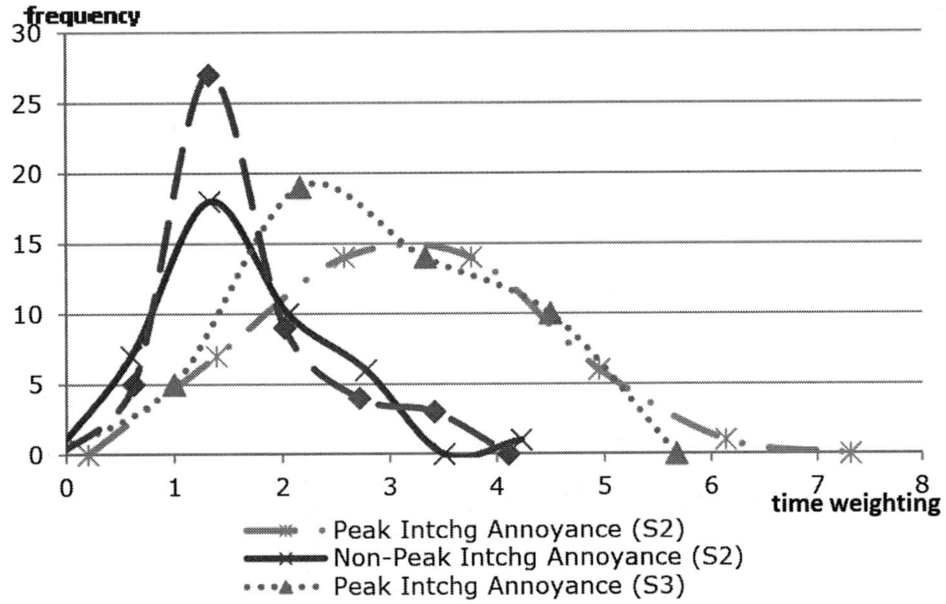


**Figure 12 Interchange annoyance rating histograms**

Experienced *stress* was an important factor to measure during delays as it is an additional component to the experienced *annoyance*. Travellers were asked to estimate the level of stress they were likely to experience during delays and while performing interchanges. It is important in these types of questions to differentiate *stress* from annoyance. Stress ratings followed patterms similar to those from the annoyance results: the length of the delay had the most effect. A more significant difference was observed in the effects on *stress* by location: the results show that there is a trend of higher *stress* due to delays while waiting than delays incurred while in-vehicle, whilst the effects on stress are also greater than those on *annoyance*. Nonetheless, the factor that created a larger effect was the length of the delay. Figures 13 and 14 present the results of *stress* ratings for Scenario 2.

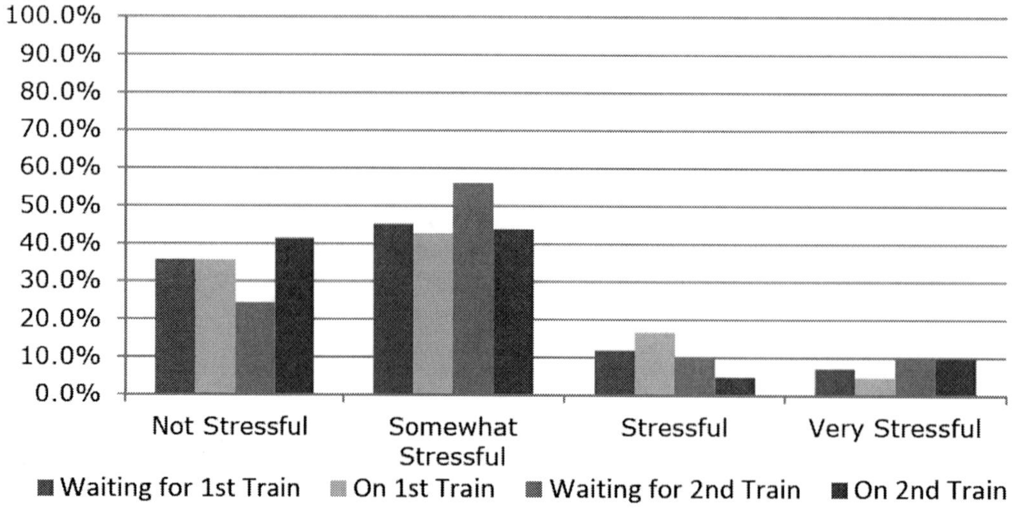


**Figure 13 Percentage of travellers experiencing stress due to a 5 min delay in Scenario 2**

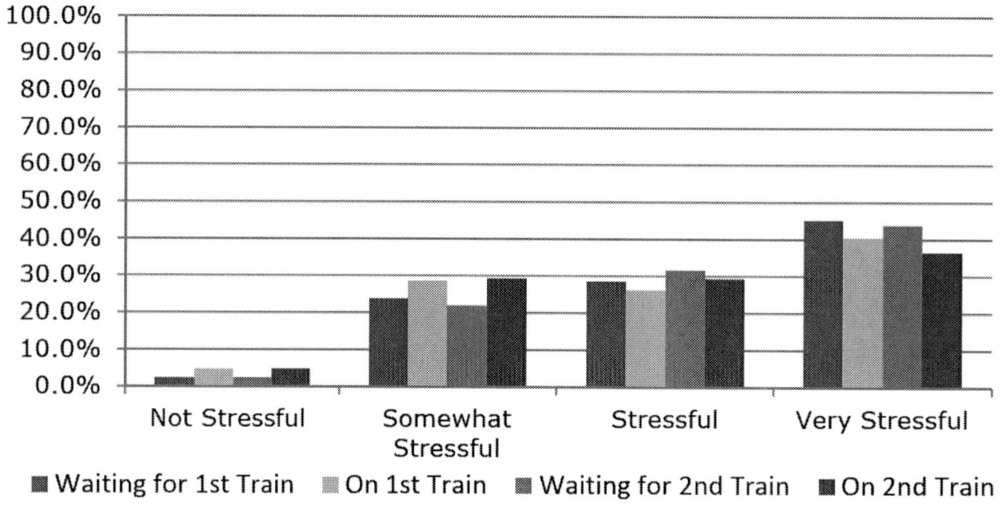


**Figure 14 Percentage of travellers experiencing stress due to a 15 min delay in Scenario 3**

### 3.3 Familiarity with the Environment

A number of the respondents were associated with the transport industry – indeed, some worked for railway companies. However, this raised the possibility that their parameter weightings for stress and annoyance could be different from those of the wider population: railway staff may have access to greater information about service disruption, or be requesting help from colleagues (as opposed to strangers). On the other hand, rail staff could be travelling on staff passes, so the fares element of their personal GCF could be minimal, distorting any Value of Time trade-offs. These possible biases are important, as they could colour management's judgment about the relative value of interventions in the service, potentially leading to investments which do not focus on the needs of fare paying passengers.

In fact, results show that those familiar with the railway travelling environment, those who have work experience in the transport industry and travel by train at least once per week, have time valuations only 3.5% lower than the average, so this appears not to be a significant issue. Further study is warranted considering the small sample size of this study.

## 4 DISCUSSION

The findings reported above have a variety of impacts. The lower weighting for in-vehicle time apparently realised by those using electronic media raises the same sorts of issues as highlighted by some commentators on the proposed High-Speed 2 rail line in Britain, namely, that in-vehicle journey time reductions may not be as important as previously-thought. However, rather than being limited to business travel, which is not the majority journey purpose, even on InterCity routes, this potentially extends this feeling to the majority of passengers and suggests that attention to other elements of the journey (e.g., access, waiting or fares) may be more important in increasing the demand for rail services and rail's market share.

The bi-modality of results shows that great care must be taken in predicting passenger behaviour. This might be seen more clearly with an equivalent example concerning a possible average behaviour prescribed on the basis of a mix of business and leisure passengers: the former are likely to be significantly time-limited but monetarily less-constrained, whereas the latter are likely to be money-constrained. The average response may not reflect accurately either market segment. The results suggest, therefore, that we need to segment our analyses more carefully in order to avoid such spurious results; fortunately, modern IT capabilities (in terms of data collection and storage) make this rather less of a problem than in previous generations.

A particular example of bi-modality might accrue from variations in familiarity, which we have also demonstrated to be significant. Unfamiliar passengers may welcome extra announcements, whilst frequent passengers may positively dislike them, thereby masking any average response. However, we should also point out that many railway planning decisions are taken (unsurprisingly) by railway managers and consultants who are familiar with their own environment. Our analysis here should act as a reminder for them to consider the railway environment from the perspective of the infrequent or non-user, who generally represent significant markets yet to be captured.

## 5 CONCLUSIONS AND RECOMMENDATIONS

The valuation of rail travel time is a key element in assessing the benefit of rail schemes, and has conventionally used standard parameters within a generalised cost framework. However, the research reported here provides more detailed information in three areas, helping us to understand more about the behaviours underpinning any average valuations and also the distribution of those valuations amongst the sampled populations.

This research has shown that passengers using a range of electronic media and communications systems, such as laptops, mobile phones and tablets may perceive in-vehicle time on trains as being about 15% less than the traditionally-assumed value of 1 (minute per actual minute). This may, to some extent, shift the balance of potential investment in railway systems away from speed improvements per se to those investments designed to reduce the impact of access time to, and waiting time for, trains.

Although the average value of weightings for access and waiting may be close to standard rail industry values, our results highlight the range of real responses, with significant numbers of respondents recording values around 50% lower or higher than the mean. This has implications for the application of standard results through the scheme development process, since the average expected response is unlikely to be achieved, if the 'average' passenger does not exist. Improved segmentation is the obvious recommendation here, as the smaller the market segment, the smaller the expected range.

Those familiar with the railway travelling environment have been shown to have time valuations lower than the average – but only by 3.5%. Although railway industry staff may be something of a special group, they are instrumental in making decisions which take the industry forward, so could be biasing investments into areas which are of different relevance to other users and non-users. Ideally, schemes might instead be focussed on those which were likely to attract the greatest net increase in rail demand – and these could be schemes which have different characteristics from those schemes actually chosen.

## REFERENCES


Australian Transport Council (2006), 'National Guidelines for Transport System Management in Australia', Vol. 4, Urban Transport, Bureau of Transport and Regional Economics, Canberra, Australia, ISBN: 0-9802880-3-7.

Börjesson, M, Fosgerau, N & Algers, F (2012), 'Catching the Tail: Empirical Identification of the Distribution of the Value of Travel Time', *Trans. Res. A* 46 pp. 378-391.

D'Agostino, Ralph B. (1986) 'Tests for the normal distribution. *Goodness-of-fit techniques'* (1986), pp. 367-419.

Fickling, R et al (2009), 'Productive Use of Rail Travel Time and the Valuation of Rail Travel Time Savings for Business Travellers', report for the UK's Department for Transport.

Grey, A (1978), 'The Generalised Cost Dilemma', *Transpn.* 7. pp. 261-280.

Harris, N G & Godward, E W (eds) (1992), 'Planning Passenger Railways', TPC, Glossop (256pp).

Hultkrantz, L (2013), 'A Note on High-Speed Rail Investments and Travelers' Value of Time', *Jnl. Rail Trans. & Man.* 3 pp. 14-21.

Peer, S, Koopmans, C C & Verhoef, E T (2012), 'Prediction of Travel Time Variability for Cost-Benefit Analysis', *Trans. Res. A* 46 pp. 79-90.

Transport for London (2010), www.tfl.gov.uk/assets/downloads/Item07-ECPP-Oct-10-Modelling.pdf (accessed 16/06/13)

Wardman, M (2006), 'The Demand for Rail Travel and the Effects of External Factors', *Trans. Res. E* 42 pp. 129-148.

# Influence of varying input parameters on model validation results

**G Götz**[1], **O Polach**[2]
[1] Technische Universität Berlin, Institute of Rail Vehicles, Germany
[2] Bombardier Transportation (Switzerland) AG, Switzerland

**ABSTRACT**

This study contains an evaluation of the model validation method developed in the European research project DynoTRAIN. Varying input parameters of the method are investigated, namely, the section length, number, and selection of sections as well as varying parameters of the simulation model. The overall validation results (the fulfilment of the validation criteria) are very rarely influenced either by the variation of sections selected for validation, the use of a higher number of sections than the minimum of 12 or by the use of longer sections than according to EN 14363. But the investigations also showed that sections with single large deviations between simulation and measurement can influence the validation results. The validation methodology is able to recognize errors in model parameters, if they have a relevant influence on the running dynamics behaviour of the evaluated vehicle.

## 1 INTRODUCTION

The application of multi-body simulations to analyse rail vehicle running dynamics during vehicle design and development is nowadays very common (1). It is increasingly recognised to use the results of multi-body simulations for rail vehicle acceptance as a possible means to reduce the amount of on-track tests and costs. The application of simulations instead of physical testing, however, requires a reliable validation of the simulation model. The definition of such validation process, criteria and quantitative limits was one of the objectives of the EU funded research project DynoTRAIN (2).

The proposed methodology for the model validation was evaluated by using validation investigations of different vehicle types and vehicle models in different multi-body simulation tools. However, the influence on validation results of varying input parameters of the simulation model as well as test section compilation, number and length used for the evaluation remains open. Due to lack of time and resources, this was not investigated in the DynoTRAIN project.

In this study, the results of two different rail vehicles, namely, the Bim passenger coach 547.5 and the Sgns691 freight wagon with Y25 bogies (empty, loaded), are used to investigate the sensitivity of input parameters on model validation results. The aim with these investigations is to assess the robustness of the validation method and its usefulness in order to recognise an error in model parameters and an incorrect modelling.

# 2 DEVELOPMENT OF MODEL VALIDATION METHODOLOGY

## 2.1 Measurements and validation exercises

Validation studies carried out in the framework of DynoTRAIN project represented a unique activity of testing, simulations, comparisons with measurements and evaluations. The on-track measurements included an electric locomotive, a passenger coach, empty and loaded freight wagons with Y25 bogies and a freight wagon unit consisting of two two-axle wagons (one empty, one loaded) interconnected by short coupler. The test runs were conducted in four European countries (Germany, France, Italy, and Switzerland) with speeds up to 200 kph without freight vehicles and up to 120 kph using the complete test train including freight wagons. The test train was equipped with ten force measuring wheelsets and a simultaneous on-line recording of track irregularities as well as rail profiles.

The simulations were carried out using several multi-body vehicle models, built in different simulation tools by different project partners. The comparisons between simulation and measurement covered over 1000 simulations using a set of the same test sections for all investigated vehicles.

The results were assessed by three different validation approaches:

- Assessment based on measured quantities, filtered and processed in analogy with EN 14363:2005 (3);
- Subjective engineering assessment using a simple "Yes/No" method by project partners as well as during a workshop with invited experts, where a set of selected plots were assessed by 26 workshop attendees;
- Validation metrics, i.e. quantitative measures comparing simulation and measurement in the time histories with the aim to maintain agreement with engineering judgement.

As a result of these validation investigations, the comparisons of simulation and measurement using a set of quantities based on EN 14363 were proposed as a suitable methodology for model validation in the context of vehicle acceptance (4,5,6). A statistical approach has been selected to assess an overall agreement calculating mean value and standard deviation of differences between simulation and measurement for 12 agreed quantities for a minimum of 12 test sections representing the conditions for vehicle acceptance.

The proposed validation method allows identification of model weaknesses and a model improvement by adjusting uncertain model parameters (4). Furthermore, the effect of the knowledge of track irregularity data and the measured wheel and rail profile data was evaluated, see (5). Moreover, the effect of using stationary tests for model adjustments to improve the agreement between the measurement and simulation of on-track tests was evaluated, see (6).

## 2.2 Methodology to validate multi-body simulation models

The model validation criteria and limits proposed as the result of DynoTRAIN project are based on 12 quantities evaluated in analogy to EN 14363, see Table 1.

A minimum of two different measuring signals per quantity have to be evaluated. To fulfil this requirement for the sum of guiding forces, two force measuring wheelsets are necessary. Therefore, four (instead of two) available quasistatic and dynamic wheel/rail force value pairs (Y, Q and Y/Q) can be used for the model validation. The specified quantities have to be evaluated using a minimum of three sections from each test zone according to EN 14363, thus at least 12 test sections. Considering these minimum requirements, the validation assessment consists of a total of 432 compared pairs (6).

**Table 1: Quantities for model validation**

| | | | | |
|---|---|---|---|---|
| Quasi-static values of wheel-rail quantities | Guiding force $Y_{qst}$ | Wheel load $Q_{qst}$ | Ratio $Y/Q_{qst}$ | Sum of guiding forces $\Sigma Y_{qst}$ |
| Dynamic values of wheel-rail quantities | Guiding force $Y_{max}$ | Wheel load $Q_{max}$ | Ratio $Y/Q_{max}$ | Sum of guiding forces $\Sigma Y_{max}$ |
| Vehicle body accelerations | Vertical | | Lateral | |
| | Rms-value | Maximum value | Rms-value | Maximum value |

The fulfilment of validation criteria is assessed using the mean value and the standard deviation for a set of minimum 24 values representing differences between simulated and measured values of each quantity. The vehicle model is considered as validated if the mean as well as the standard deviation of differences simulation – measurement of all quantities do not exceed their respective validation limits; see (4,5,6) for details.

This validation methodology was developed in the framework of the project DynoTRAIN using a set of 17 selected sections from the measurements conducted in that project. Test zone 1 (straight track and very large radius curves) was represented by four sections; two sections were from large radius curves (test zone 2, R > 600 m); five sections represented small radius curves (test zone 3, 400 m ≤ R ≤ 600 m) and six were from very small radius curves (test zone 4, 250 m ≤ R < 400 m). Because of their speed limitation, the freight wagons were tested only in one of the selected four test sections in test zone 1. Thus, the DynoTRAIN validation exercises consisted of 14 test sections for freight wagons or 17 test sections for other vehicles, respectively.

The evaluated sections were limited to the part of track with constant curvature (full curve part), however, they were longer than the sections length specified according to EN 14363. The lengths of the sections evaluated in DynoTRAIN were:

- 1760 m – 3200 m in test zone 1, compared to 250 m according to EN 14363 (for speed up to 200 kph);
- 190 m – 360 m in test zone 2, compared to 100 m according to EN 14363;
- 250 m – 680 m in test zone 3, compared to 100 m according to EN 14363;
- 110 m – 771 m in test zone 4, compared to 70 m according to EN 14363.

The minimum of the test sections specified for the model validation according to the methodology proposed by DynoTRAIN was agreed as three sections from each test zone according to EN 14363, thus at least 12 sections. The length of the test section should be in minimum at least the section length of the particular test zone according to EN 14363. However, the sections used for validation can be longer; any limitation regarding the length of the sections used for validation is not specified. It cannot be excluded that the number, the length and the selection of the sections used in the validation process as well as varying model parameters may influence the results of this validation process.

## 3 INVESTIGATED VEHICLES

The results of two different types of rail vehicles (tested in the DynoTRAIN framework) are used to carry out the presented investigations:

- Bim passenger coach 547.5 with Minden Deutz 36 bogies;
- Empty and loaded freight wagon Sgns691 with Y25 bogies.

All following simulation results base on simulations carried out with measured track irregularities, measured track layout and measured wheel/rail profiles. Both vehicles are modelled with SIMPACK and all bodies are modelled as rigid bodies.

The Bim passenger coach 547.5 with Minden Deutz 36 bogies (MD 36, see Figure 1) is part of the standard UIC-X express coach of the former "Deutsche Bundesbahn" (state railway of the former Federal Republic of Germany). These vehicles are in operation since 1952, typically in D-trains or inter-city trains. The coach has a maximum speed of 200 kph, a wheelset base distance of 2.5 m, a bogie base distance of 19.0 m and a wheel diameter of 0.95 m. The simulation model of the passenger coach (normal operation mode, unloaded) consists of one car body and two bogies series MD 36. The main components are conventional wheelsets, axle boxes, axle guides for wheelset guidance, coil springs for primary and secondary suspension and dampers in different spatial directions, bogie frame and bogie bolster. Friction forces occur between bogie bolster and car body and also in the secondary spring support. (7)

The Sgns691 freight wagon with Y25 bogies (see Figure 1) is a four axle flat wagon for containers and swap bodies. The Y25 bogie is the most common bogie in Europe and in operation in different variants since 1965. The wagon has a maximum speed of 120 kph; a wheelset base distance of 1.8 m, a bogie base distance of 14.2 m and a wheel diameter of 0.92 m. The freight wagon is modelled in empty and in loaded vehicle condition with a maximum axle load of 22.5 t. Each simulation model consists of one car body and two bogies series Y25. The bogie consists of conventional wheelsets, axle boxes, primary suspension consisting of two spring packs for each axle box to connect the axle box with the bogie frame, one Lenoir suspension element per axle box acting as a friction damper, a secondary layer between bogie frame and car body consisting of a centre of pivot in the bogie centre and two side bearers (above the left and right longitudinal girder of the bogie frame).

**Figure 1: Y25 bogie of freight wagon (left) and MD 36 bogie of passenger coach (right)**

## 4 METHOD TO INVESTIGATE VARYING INPUT PARAMETERS ON MODEL VALIDATION RESULTS

The investigation of varying input parameters on model validation results are conducted as follows: The sections used for validation exercises in the DynoTRAIN project are split in sections with length according to EN 14363 to investigate varying input parameters on model validation results, see Figure 2. This results in 89 sections for the passenger coach and 55 sections for the freight wagons (freight wagons were excluded from test runs with speed above 120 kph). The calculated results based on DynoTRAIN sections and EN 14363 sections can differ, see Table 2. The model validations are then repeated with different compilations of sections, different length of sections and different number of applied sections.

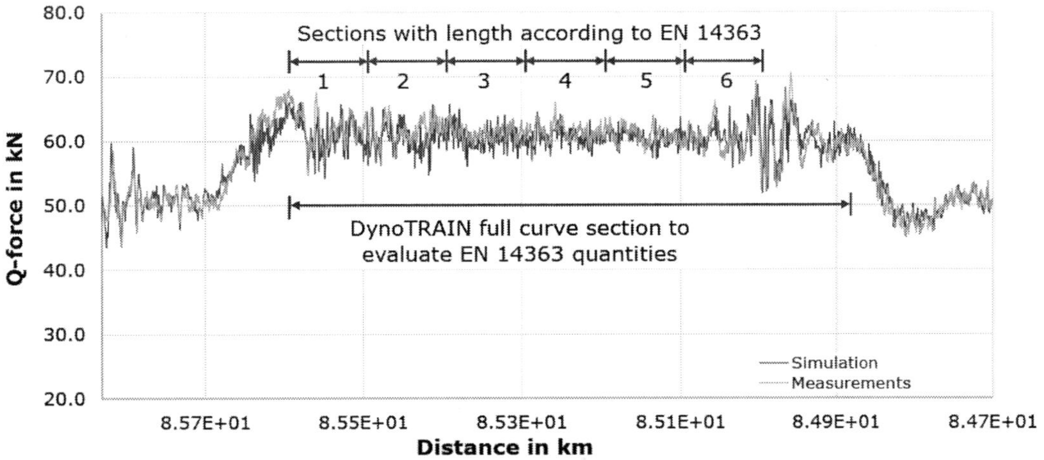


**Figure 2: Difference between the full curve length as used in DynoTRAIN and the sections according to EN 14363 (curve radius R = 581 m, test zone 3, curve length 680 m)**

For the investigations of varying input parameters, a scatter of the selected model parameter is defined. The parameter is increased successive by 80% and reduced successive by 40% from the initial value. For each parameter variation, the DynoTRAIN sections are simulated and the results are used for model validation.

**Table 2: Influence of the separation of full curve in to sections with length according to EN 14363 (Q-force, max value (99.85 percentile), curve radius R = 581 m, test zone 3, full curve length 680 m)**

| EN 14363 section | 1 | 2 | 3 | 4 | 5 | 6 | Full curve length |
|---|---|---|---|---|---|---|---|
| Simulation in kN | 65.7 | 65.7 | 64.5 | 64.1 | 63.7 | 68.9 | 68.3 |
| Measurements in kN | 67.6 | 65.3 | 63.8 | 66.0 | 64.0 | 69.3 | 69.6 |
| Difference in kN | -1.8 | 0.5 | 0.7 | -1.8 | -0.3 | -0.5 | -1.2 |

The fulfilment of validation criteria was assessed using the mean value and the standard deviation of differences between simulation and measurement (5). These differences are calculated for each value and quantity from simulations and measurements. Dependent on the sign of the measured value, this difference is transformed as follows: If the magnitude of the simulation value is higher than that of the measurement (simulation overestimating the measurement), the difference is positive, and vice versa, see (5) and (6). The model validation methodology specifies for each quantity a validation limit for the standard deviation of differences between simulation and measurement. For the mean values of differences simulation - measurement, the limit values are proposed to be equal to two-third of the limit for the standard deviation. Then, the values for the mean and the standard deviation are normalised by the proposed validation limits to achieve a better comparability of the validation results. Table 3 shows the mean values and the standard deviations with the correspondent limit and normalised values at the example of the section length variation for the passenger coach (chapter 5.1, initial section length, see Figure 5).

**Table 3: Example of mean values and standard deviations of differences between simulation and measurement together with correspondent limit values and normalised values**

| Value | Unit | Mean | Limit value | Normalised value | Std. dev. | Limit value | Normalised value |
|---|---|---|---|---|---|---|---|
| $Y_{qst}$ | kN | -0.75 | 3.33 | -0.225 | 2.30 | 5.00 | 0.460 |
| $Q_{qst}$ | kN | 0.35 | 3.98 | 0.088 | 1.99 | 5.98 | 0.333 |
| $(Y/Q)_{qst}$ | - | -0.015 | 0.047 | -0.325 | 0.048 | 0.070 | 0.683 |
| $\Sigma Y_{qst}$ | kN | -1.28 | 4.00 | -0.319 | 3.12 | 6.00 | 0.521 |
| $Y_{max}$ | kN | 0.49 | 6.00 | 0.082 | 3.39 | 9.00 | 0.376 |
| $Q_{max}$ | kN | 1.83 | 5.98 | 0.307 | 2.99 | 8.94 | 0.333 |
| $(Y/Q)_{max}$ | - | 0.010 | 0.067 | 0.143 | 0.053 | 0.100 | 0.525 |
| $\Sigma Y_{max}$ | kN | 1.65 | 6.00 | 0.275 | 3.40 | 9.00 | 0.378 |
| $\ddot{y}^*_{rms}$ | $m/s^2$ | 0.029 | 0.10 | 0.292 | 0.057 | 0.15 | 0.381 |
| $\ddot{z}^*_{rms}$ | $m/s^2$ | 0.009 | 0.10 | 0.088 | 0.043 | 0.15 | 0.283 |
| $\ddot{y}^*_{max}$ | $m/s^2$ | 0.040 | 0.267 | 0.152 | 0.136 | 0.40 | 0.340 |
| $\ddot{z}^*_{max}$ | $m/s^2$ | 0.014 | 0.267 | 0.052 | 0.105 | 0.40 | 0.262 |

### 4.1 Particularities while comparing results of simulation with measurement

The comparisons of simulation with measurement results of the passenger coach showed in the framework of DynoTRAIN project only few sections with high deviations; a high level of agreement was rather typical. The results of the freight wagons were contrary: Deviations, partly high, occurred more often. The term "deviation" can be understood in two different ways:

a) A constant low or high deviation between simulation and measurement signals in the full curve part (typical for wheel/rail forces)
b) A single large deviation between amplitudes of simulation and measurement signals.

Figure 3 shows two examples of measured and simulated Q-forces. In the left diagram, the deviations between both signals remain in the full curve part almost at a constantly high level. In comparison, the right figure shows a constantly good agreement in the full curve part; the absolute deviation changes only slightly and remains at a low level.

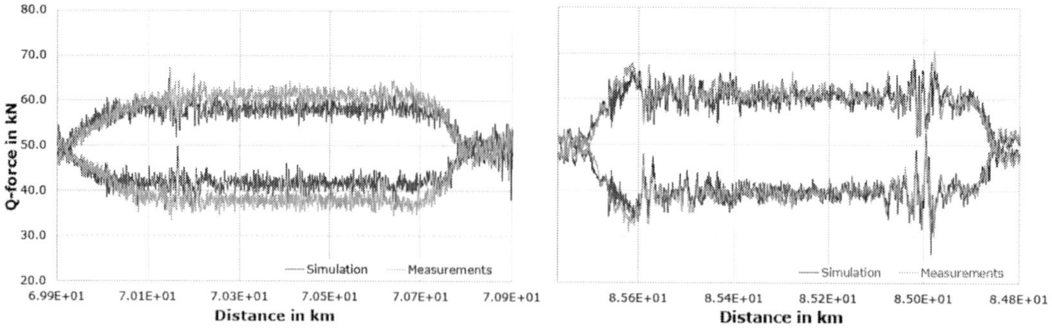


**Figure 3: Exemplary sections with an almost constant high deviation (left) and a very small deviation (right) between simulation and measurement in full curve part (Q-force, passenger coach)**

Figure 4 shows an exemplary section with a single large deviation in the acceleration signal. The overall agreement is quite satisfying; deviations between simulation and measurements are very low. But between km 76.0 and km 75.9, a single large deviation occurred.

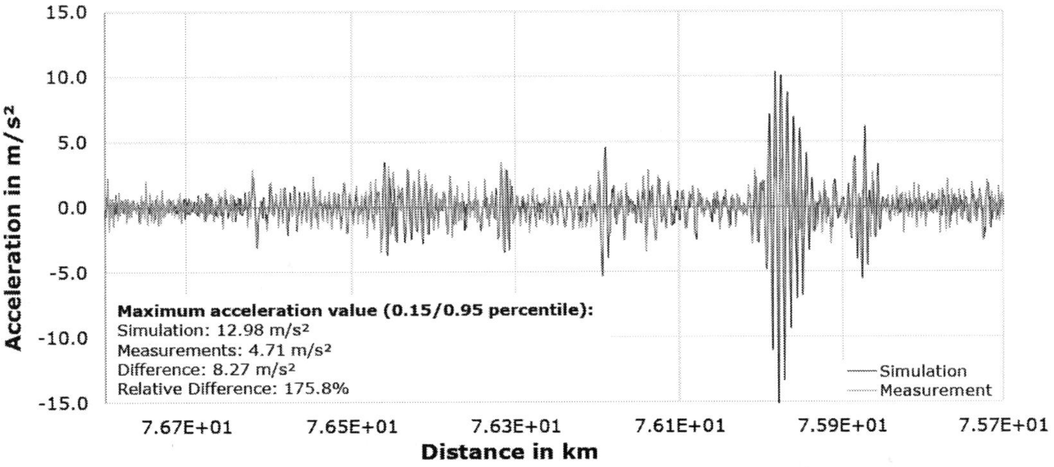


**Figure 4: Exemplary section with single large deviation between simulation and measurement (freight wagon, empty, car body vertical acceleration above bogie 1)**

## 5 INFLUENCE OF VARYING INPUT PARAMETERS ON MODEL VALIDATION RESULTS

### 5.1 Variation of section length

The influence of section length on model validation results is investigated using a set of 12 sections (3 per test zone) with a length according to EN 14363 that are successive increased by a percentage value in which 100% means a doubling and 200% a triplication of the length. The results of the passenger coach for this influence analysis are shown in Figure 5. Although the selected section length is increased by 200%, the validation results of the quasistatic wheel/rail forces are only slightly influenced; the values remain nearly at the same level. As shown in Figure 3, it is typical to have a constant high or low deviation between simulation and measurement signals of wheel/rail forces in the full curve part, which can explain this low influence of the section length on the validation results. In comparison, the validation results of the rms values of the vertical car body accelerations are more influenced; probabilities are higher that deviations can occur and change in the full curve part more frequently; new parts of sections with higher deviations are accordingly caught. But for the passenger coach, longer sections do not change the overall model validation result; this vehicle model is in each investigated case assessed to be validated.

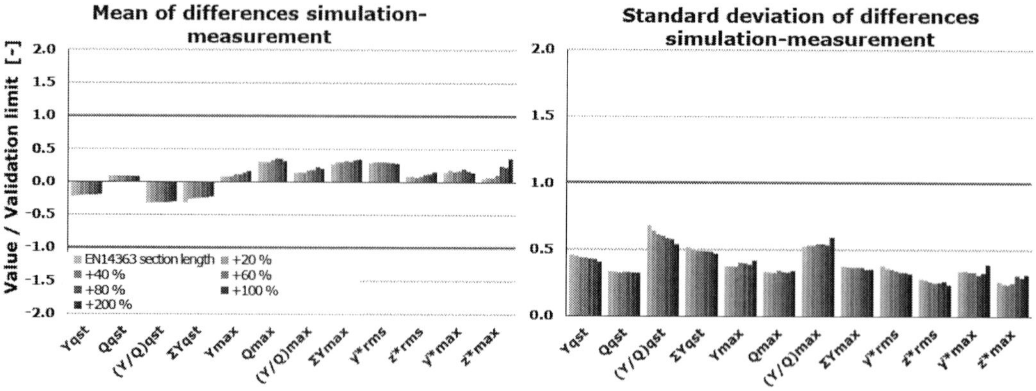


**Figure 5: Normalised values of mean and standard deviation of differences simulation - measurement for varying section length (passenger coach)**

Sections with quasistatic and/or single large deviations between simulation and measurement are more typical for the empty and loaded freight wagon. Contrary to the expectation, the length increase of the selected sections do not influence the validation results strongly, see Figure 6 and Figure 7. The results remain for both loading conditions nearly at the same level and only a slight scatter is partly visible. As with the results of the passenger coach (see Figure 3), also the freight wagon results show usually a constant high or low deviation of wheel/rail forces in the full curve part, thus a low influence on validation results. The scatter of the validation results of the acceleration values are also relatively low which means that sections with (single) large deviations are not caught, or that the rms acceleration values in the selected sections do not change strongly in function of the length increase. In summary, the model validation results remain unchanged; both models of the freight wagon are still assessed to be not validated.

Based on the results of this influence analysis, a rather low effect on the validation results of wheel/rail forces can be expected when increasing the section length. The deviation between simulation and measurement in the full curve part can be either high or low, but remains almost constant. Only the maximum values of the wheel/rail forces may have a higher influence on validation results, if single large deviations occur. A higher influence on the validation results of the acceleration values is observed. This is because the deviations can occur at different positions almost randomly due to their stochastic character.

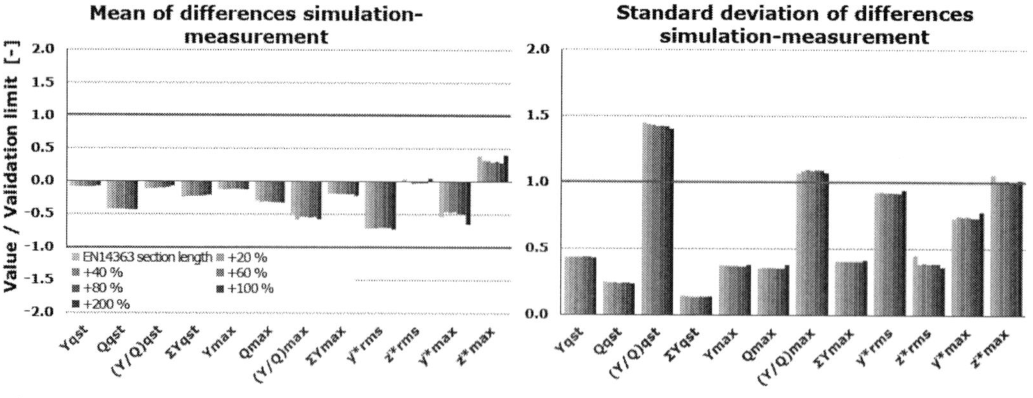


**Figure 6: Normalised values of mean and standard deviation of differences between simulation and measurement for varying section length (freight wagon; empty)**

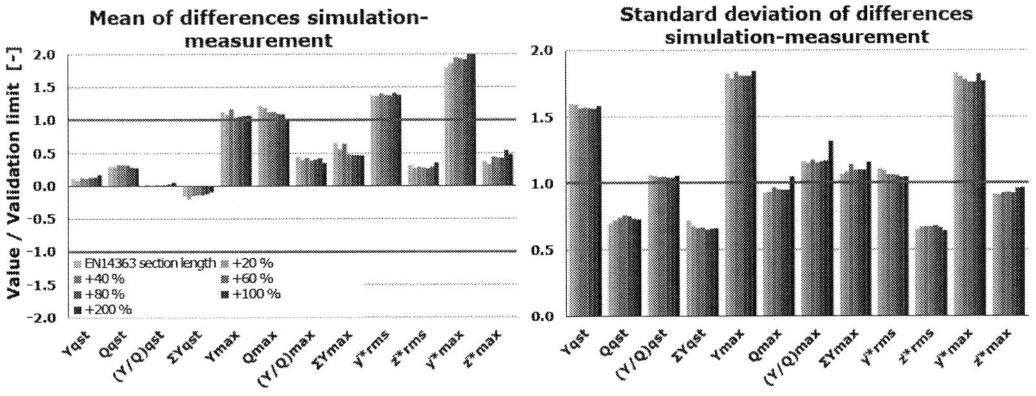


**Figure 7: Normalised values of mean and standard deviation of differences between simulation and measurement for varying section length (freight wagon; loaded)**

## 5.2 Variation of section number

The influence of section number on the validation results is investigated using a set of 12 sections (3 per test zone) which are increased in steps by 8 (freight wagon) or 12 (passenger coach) until the maximum available sections of 55 (for freight wagon) or 89 (for passenger coach), respectively. No section is considered twice. The sections are chosen randomly.

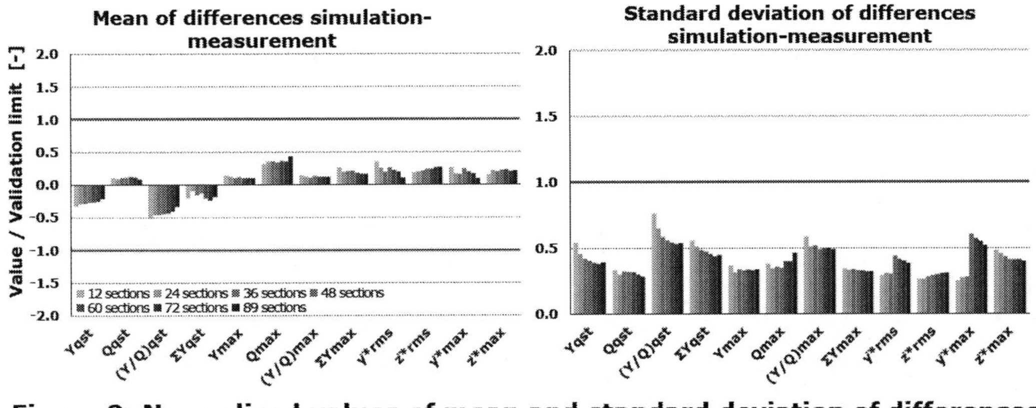


**Figure 8: Normalised values of mean and standard deviation of differences between simulation and measurement for increasing section number (passenger coach)**

The results of this influence analysis for the passenger coach are shown in Figure 8. The values of the wheel/rail forces are only slightly influenced. Some values such as quasistatic Q-forces are almost not affected; the quasistatic Y-force shows a slight trend of improvement. Since for this vehicle the overall level of agreement of the wheel/rail forces and car body accelerations is basically good, the scatter remains rather small and more sections have only a slight influence on the validation results. For every set of section number, the passenger coach model is always assessed to be validated. In comparison to the passenger coach, the influence analysis for the freight wagon results in a higher scatter of the validation results, see Figure 9. The first 12 randomly selected sections (loaded vehicle condition) provide a validation result of the quasistatic Y-force that changes the sign of the difference while increasing the section number. In the loaded vehicle condition, the selection of the first 12 sections leads also to an exceedance of the threshold value (maximum Y-force) but the validation value decreases the more sections are considered. Higher deviations, especially in the guiding forces of the freight wagons, occur much more often than for the passenger coach. Their influence on the validation results, if randomly selected, can be reduced the more sections are considered. The scatter of the acceleration values is significantly higher; comparisons of simulation with measurement show more often deviations. As with the values for the wheel/rail forces, the more sections are considered for the validation, the more their influence can be reduced. This could lead to the argumentation that even for models with lots of deviations, a sufficient amount of sections can lead to a change of the model validation result. But even if all 55 available sections are considered, there are still a couple of values remaining above the threshold value. For every section number, the loaded as well as empty freight wagon remain to be assessed as not validated.

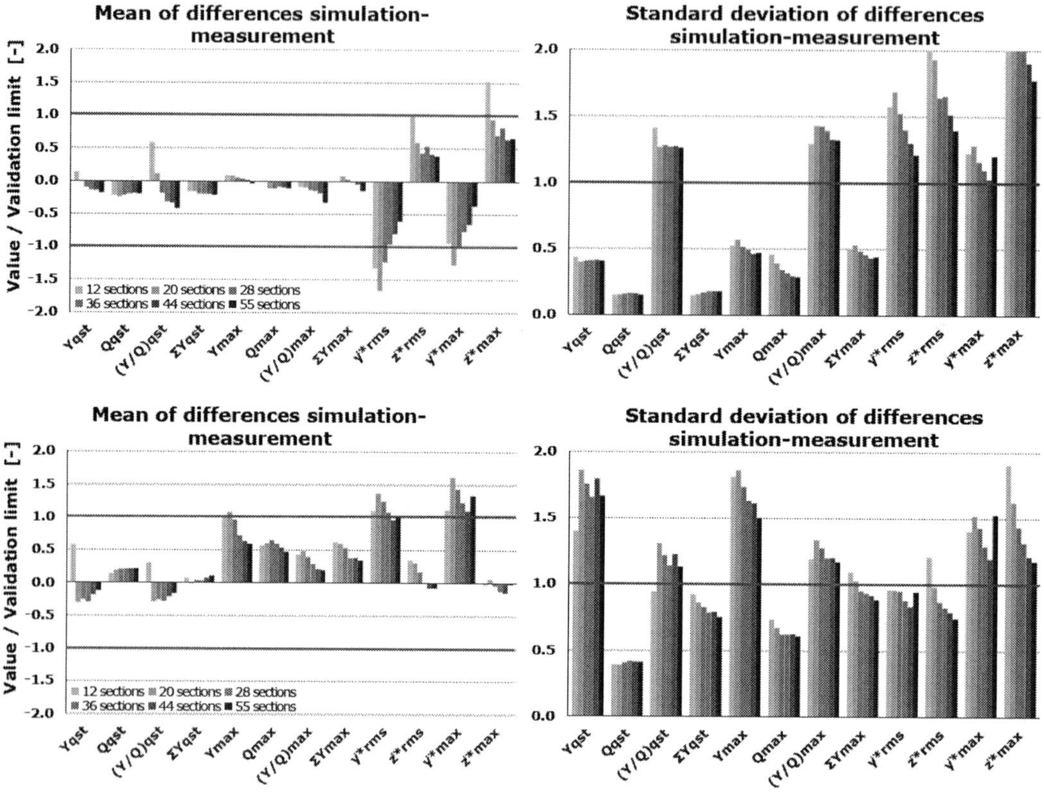


**Figure 9: Normalised values of mean and standard deviation of differences between simulation and measurement for increasing section number (freight wagon; top: empty; bottom: loaded)**

Based on the results of this influence analysis, it can be concluded that the influence of the section number on the validation results of wheel/rail forces is rather low if the overall level of agreement between simulation and measurement is high. If deviations occur more often and at different positions, the use of more sections can improve or worsen the validation results dependent on the selected sections. The results of this investigation refute that a sufficiently high number of sections may lead to an influence on the model validation result. Even if all available sections are considered for the validation, both freight wagon models remain to be not validated.

## 5.3 Variation of section compilation

The influence of varying section compilation is investigated using six different compilations of 12 sections (three sections per test zone). The sections are chosen rather randomly but to investigate their influence also some sections with higher deviations are purposefully considered. The validation results of each case are compared with the validation results of 14 (freight wagon) and 17 (passenger coach) sections which were used in DynoTRAIN project. Figure 10 shows the validation results of the passenger coach for different section compilations. The result of the compilation five has to be highlighted. The model validation result is affected and the assessment of the simulation model changes from validated to be not validated. The standard deviation of differences between simulation and measurement for the maximum lateral car body acceleration exceeds the threshold value. In that compilation only sections with higher deviations are purposefully selected. For other section compilations, the model remains to be assessed as validated.

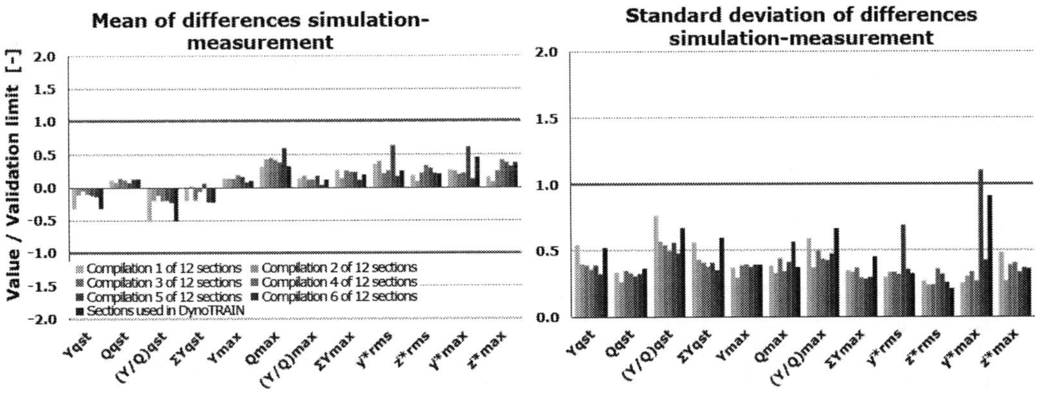


**Figure 10: Normalised values of mean and standard dev. of diff. simulation - measurement for varying section compilation (passenger coach)**

In comparison, the section selection for the freight wagons lead to a higher scatter of the validation results, see Figure 11 and Figure 12. Much more values vary above or below threshold value and are similarly higher or lower compared to the corresponding DynoTRAIN values. Due to this scatter, doubts could raise up, the model validation result can change from not validated to be validated by selecting only sections with small deviations between simulation and measurement. However, the presented evaluation demonstrates that this is rather impossible; the investigated selection of sections did not lead to such a result. The empty and loaded freight wagon model remains to be not validated in all investigated cases.

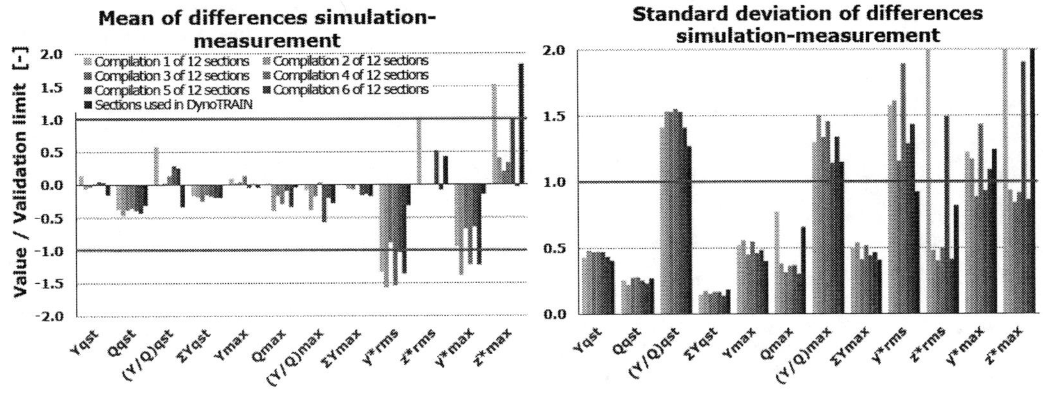


**Figure 11: Normalised values of mean and standard deviation of differences between simulation and measurement for varying section compilation (freight wagon; empty)**

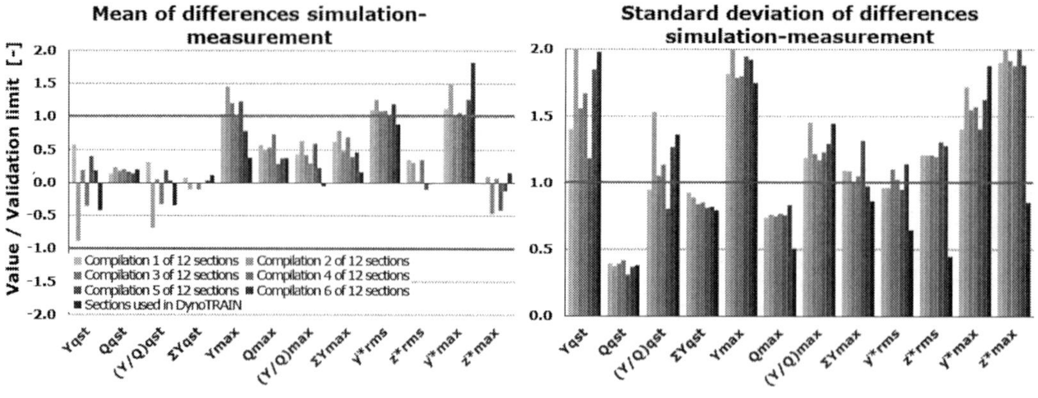


**Figure 12: Normalised values of mean and standard deviation of differences between simulation and measurement for varying section compilation (freight wagon; loaded)**

Based on the results of this influence analysis, it can be concluded that the selection of the sections can lead to a high scatter of the validation results, but the model assessment remains not changed in the majority of investigations. Only for one compilation of sections (passenger coach), the model validation result changed from validated to not validated. The validation results of the freight wagons as models with more often occuring deviations between simulation and measurement, are more influenced by the selection of sections. Nevertheless, a change of the model validation from not validated to validated did not occur within the conducted investigations.

## 5.4 Variation of simulation model parameters

Running dynamics experts use multi-body simulation tool to create complex vehicle models with the aim to analyse their running dynamics. Despite the greatest diligence, possible sources of error by adjusting the parameters exist, e.g. faulty model data. These effects are investigated for the passenger coach at the example of varying longitudinal primary spring stiffness and vertical secondary spring stiffness to analyse their influence on the model validation results. For each of the two variations, the initial stiffness value is reduced successive to minus 40% and increased successive to plus 80%. For each parameter variation, DynoTRAIN sections are used for the simulation.

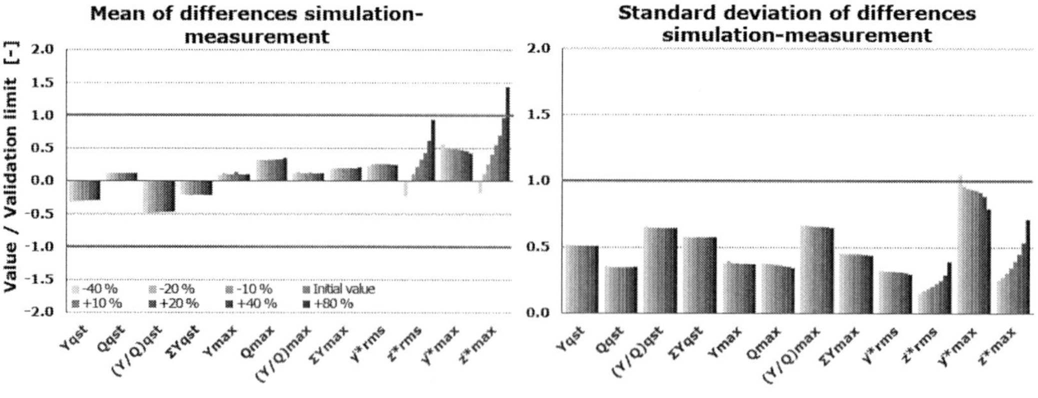


**Figure 13: Normalised values of mean and standard deviation of differences between simulation and measurement for varying vertical secondary spring stiffness (passenger coach)**

Figure 13 shows the validation results while varying vertical secondary spring stiffness. The spring stiffness of the secondary coil spring has almost no influence on the validation values of the wheel/rail forces. A higher influence can be observed in the validation values of the vertical car body accelerations. Stiffer springs lead to an increase of the validation results by a degree in which the model is assessed to be not validated. A stiffness increase by 80% describes a value outside the normal scatter of this stiffness value and in that case the error is recognised by the model validation methodology.

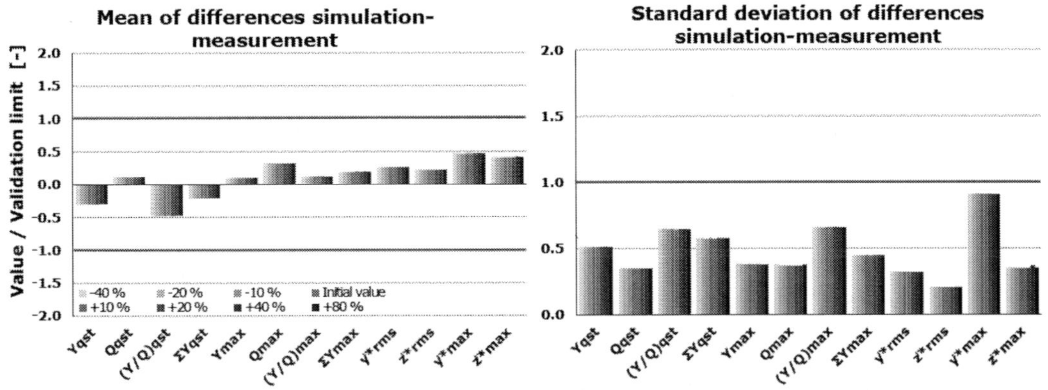


**Figure 14: Normalised values of mean and standard deviation of differences between simulation and measurement for varying longitudinal primary spring stiffness (passenger coach)**

Figure 14 shows the validation results while varying the longitudinal primary spring stiffness. The axle guidance of the passenger coach consists of steel leafs and is therefore very stiff. The variation of the spring stiffness shows almost no influence on the model validation results that means neither an increased stiffness value by 80% nor a reduced stiffness value by 40% lead to varying validation results. The stiff axle guidance limits the movement of the axle strongly so that a stiffer or softer stiffness values have almost no effect on forces of the wheel/rail contact and accelerations of the car body. The running dynamic of the passenger coach is not influenced by such an insensitive parameter. In every case, the simulation model is assessed to be validated.

Based on this two results it can be concluded that the proposed model validation methodology is able to recognize errors in model parameters if the parameter are sensible to the vehicle running dynamics.

## 6 CONCLUSIONS

In this study, the influence of varying input parameters on the model validation results is investigated, namely, section length, number, and selection of sections as well as varying parameters of the simulation model. The investigations are carried out at the example of two rail vehicles: Bim passenger coach 547.5 (empty) and Sgns691 freight wagon in empty and loaded condition. The results demonstrate that single sections with single large deviations can have a strong influence on the model validation results. Nevertheless, the overall validation results (the fulfilment of validation criteria) is very rarely influenced either by the variation of the sections selected for the validation, the use of a higher number of sections than the minimum of 12 or the use of longer sections than specified in EN 14363. This confirms the robustness of this model validation methodology. The model validation result is rather insensitive to the selected number and length of the sections used for comparisons between simulation and measurements; the model remains to be validated or not validated. The validation result changed only in one case within the presented investigations, but it should be highlighted that in that case sections with high deviations have been selected purposefully. Sensible model parameters influencing running dynamics are discovered by this validation methodology if they have a relevant influence on the running dynamics behaviour of the evaluated railway vehicle.

## FUNDING

This investigation used data from the research project DynoTRAIN supported by the European 7th Framework Programme, contract number: 234079. The research presented in this paper received no specific grant from any funding agency.

## REFERENCE LIST

(1) Polach, O., Evans, J. (2013) Simulations of running dynamics for vehicle acceptance: Application and validation. International Journal of Railway Technology, Vol. 2, pp. 59-84

(2) Web: www.triotrain.eu, assessed on 7th of October 2014 at 6:15 pm

(3) EN 14363:2005 Railway applications—Testing for the acceptance of running characteristics of railway vehicles—Testing of running behaviour and stationary tests, CEN, Brussels, 2005

(4) Polach, O., Böttcher A., Vannucci, D., Sima, J., Schelle, H., Chollet. H., Götz, G., Garcia Prada, M., Nicklisch, D., Mazzola, L., Berg, M. and Osman, M. (2014) Validation of multi-body models for simulations in authorisation of rail vehicles. In: Zobory, I. (ed.), Proceedings of the 9th International Conference on Railway Bogies and Running Gears, Budapest, 9 - 12 September, 2013. BME, Budapest, pp. 187-196

(5) Polach, O. and Böttcher, A. (2014) A new approach to define criteria for rail vehicle model validation, Vehicle System Dynamics 52:sup1: pp. 125-141

(6) Polach O, Böttcher, A., Vannucci, D., Sima, J., Schelle, H., Chollet, H., Götz, G., Garcia Prada, M., Nicklisch, D., Mazzola, L., Berg, M. and Osman, M. (2014) Validation of simulation models in context of railway vehicle acceptance. Proceedings of the Institution of Mechanical Engineers, Part F: Journal of Rail and Rapid Transit, OnlineFirst, Web: http://pif.sagepub.com

(7) Götz, G. (2014) Comparison of simulation and measurements from on-track tests for model validation. In: SIMPACK News (03/2014), pp. 23-25

# A comparison of full scale and model scale measurements of train aerodynamic characteristics

**C Baker, A Quinn, H Hemida, M Sterling, M Gallagher, J Morden**
School of Civil Engineering, Birmingham Centre for Railway Research and Education, University of Birmingham, UK

## ABSTRACT

This paper describes the ongoing project "The measurement of train aerodynamic phenomena in operational conditions", which aims to compare experimental and numerical measurements of train aerodynamic phenomena with full scale on-train and trackside measurements. This paper sets out the various experimental measurements and numerical calculations that are being carried out. Two sets of results are presented – a comparison between full scale trackside pressure and slipstream measurements with model measurements and CFD calculations; and full scale measurements made of the pressures caused by trains passing or running through tunnels. Further work on the project is outlined.

## 1 INTRODUCTION

In 2011 work began on a major railway aerodynamic project funded by the Engineering and Physical Sciences Research Council entitled "The measurement of train aerodynamic phenomena in operational conditions". This project addresses a wide range of railway aerodynamic issues and involves on-track and on-train measurements, model scale measurements and CFD calculations. The range of aerodynamic issues that need to be considered in train design is wide [1], [2] and relies on a mixture of full scale and model scale experiments and numerical calculations. The fundamental hypothesis of the project is that the techniques that have been used in the past to predict aerodynamic effects, and are embedded within codes of practice, are likely to predict magnitudes of pressures, velocities, forces etc. that are higher than those observed in practice, where other effects – such as track roughness, variability in meteorological conditions etc. are likely to usually obscure aerodynamic effects to some extent, and because of this some of the current design methodologies are unnecessarily restrictive and conservative. This leads to the following major project aim.

*To investigate and measure a range of aerodynamic phenomena observed in real train operation, both relative to the train and relative to a fixed point at the trackside, and to compare how such effects match model scale measurements and various types of CFD calculation, and thus to test the validity, or otherwise, of the above hypothesis.*

The specific objectives of the project are as follows.

- To carry out a range of conventional physical model tests and computational trials to measure the velocity and pressure field around a 200kph Class 43 train, commonly referred to as the HST.

- To instrument the Network Rail High Speed High Speed Measuring Train (which is based on Class 43 vehicles) to measure aerodynamic phenomena during its operation on all main line routes in the UK. This includes measurement of cross wind forces, train passing pressures and pressure transients in a wide variety of tunnels.
- To carry out a series of measurements at a trackside location on the Western Main Line (a route where the Class 43 is widely used), such that the pressure and velocity fields close to such trains can be measured under operational conditions.
- To compare the results of the model scale and computational trials to the on-train measurements to determine the adequacy of the predictions of established test methods to give the cross wind forces, moments, displacements and on-train pressure transients found in operational conditions.
- To compare the results of the model scale and computational trials to the on-track measurements to determine the adequacy of predicted slipstream velocities and trackside pressure fields with those measured in operational conditions.
- To thus investigate the fundamental hypothesis outlined above and to develop possible methods in which testing and codification procedures could be changed to make them more adequately reflect operational conditions, and thus to make the design process for new trains less conservative and restrictive.

In section 2 of this paper we outline the trackside and train based measurements that have been carried out, together with the physical model tests and the numerical calculations. Section 3 then presents a comparison between trackside and physical model measurements and CFD calculations of pressure transients due to a passing train. Section 4 presents a similar set of data for slipstream velocity measurements and also describes the effect of ballast shoulder height on the measured velocities. Section 5 presents data measured on board a test train of the pressure transients caused by passing trains and passage through tunnels. Finally some concluding remarks are made in section 6, together with a brief description of ongoing and future work on the project.

## 2 THE MEASUREMENTS AND CALCULATIONS

### 2.1 Trackside measurements

The full scale test campaign is planned to be carried out at Uffington on the Western Main Line (figure 1). The main series of tests will take place in the spring of 2015. However, a preliminary series of tests were carried out in the summer of 2012. At this point there is a double track, 200kph railway which is used by around 6 Class 43 trains an hour, in either two power car plus seven coach formation (2+7), or two power car plus eight coach formation (2+8). In the preliminary tests measurements were made of the slipstream velocities at various distances from the side of the train using ultrasonic anemometers and pressure transients at the side of the train using static pressure probes. As far as possible the CEN / TSI methodology was followed [3], [4]. The main series of experiments will thus give a thorough description of the flow field around the train in operational conditions and will, of themselves, give useful information concerning the pressure transients around the trains (which need to be known for the calculation of train induced forces on trackside structures), and the train slipstream velocities (of importance in assessing the safety of passengers or trackside workers). In the preliminary tests, slipstream measurements were made at 1.1m above the cess (0.7m above the top of rail) and 3.05m from the track centre line and pressure measurements at 1.0m above the top of the rail and 2.0m from the track centre line.

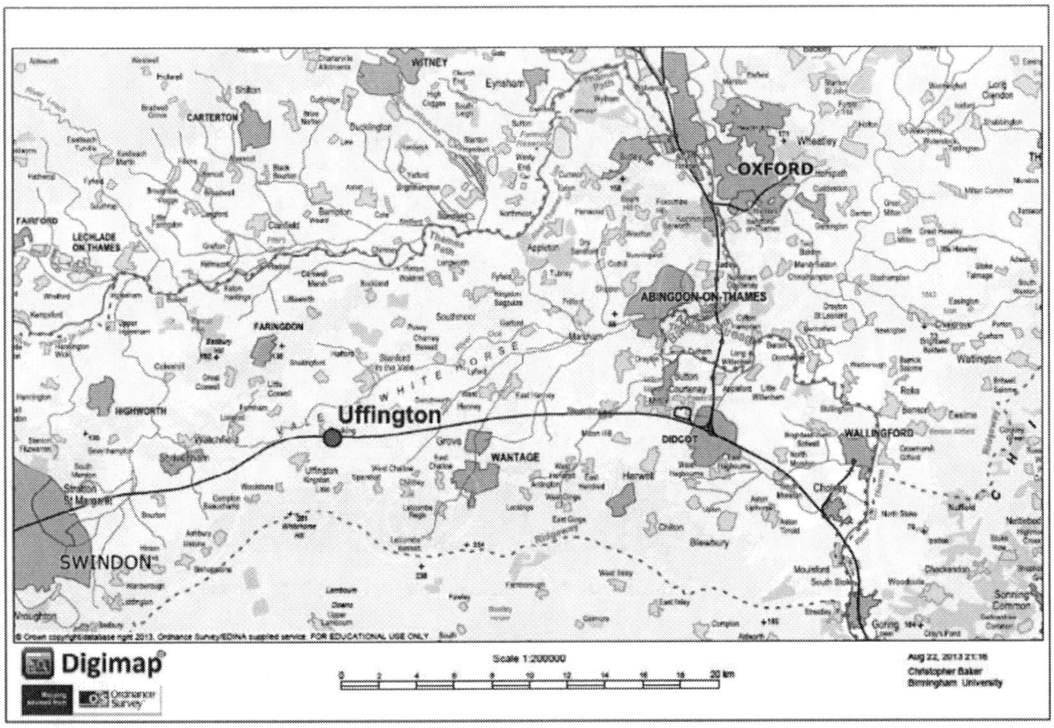


**Figure 1 The trackside measurements site at Uffington on the Western Main Line**

## 2.2 Class 43 measurements

Measurements have been made on the Network Rail "New Measurement Train" (NMT), a Class 43 train that is used to measure track geometry and condition in the UK (indicative positions shown in figure 2). The train runs over every UK main line at operational speeds (up to 200kph) on a regular (fortnightly) pattern. Two sets of measurements have been carried out. The first measured the pressure across the train nose of the power car 43062, which, together with the output from a nose-mounted pitot tube, enables the instantaneous yaw angle and aerodynamic speed of the flow to be obtained. The second set of measurements was made on a loop of surface pressure tapping points around the rear of the same power car (indicative positions shown in figure 3). These measurements are recorded in real-time alongside GPS position and speed information. This information has been used to compare with the measurements of cross wind pressures from wind tunnel tests and the TRAIN Rig measurements (see below), but this aspect of the work will not be reported here. In addition these pressure tappings measure the transient pressures associated with passage through tunnels and will thus give an extremely useful database of the pressure transients in all main line UK tunnels. A schematic of the experimental set up is given in figure 4. Essentially the pressures are measured by pressure transducers mounted just behind the tappings, with a backing pressure provided by an on board reservoir, and all the data recorded by data acquisition hardware and software in the rear of the power car. This is supplemented by other data from the NMT – specifically vehicle speed and GPS location data (figure 5).

**Figure 2 The Class 43 New Measurement Train**

**Figure 3 NMT pressure measurement locations**

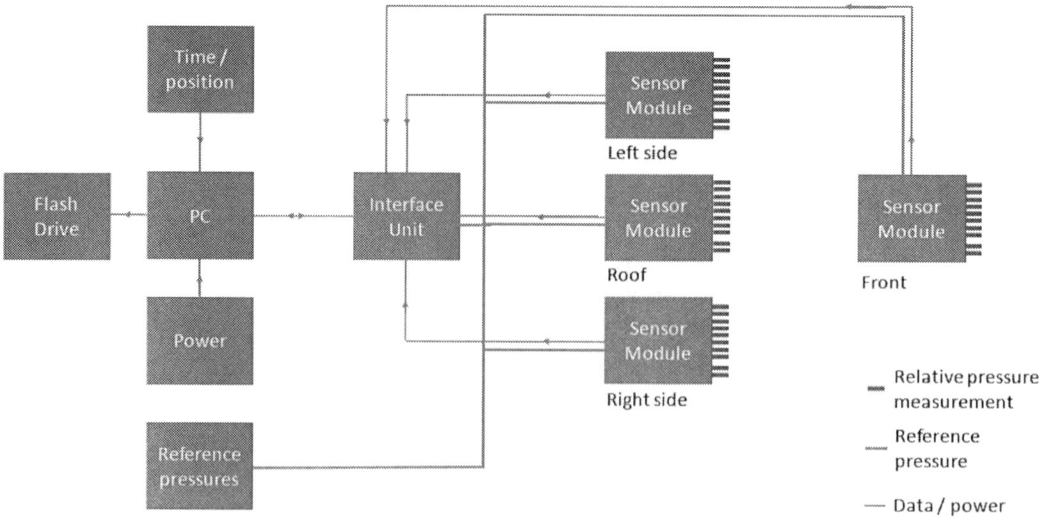


**Figure 4 NMT data acquisition system**

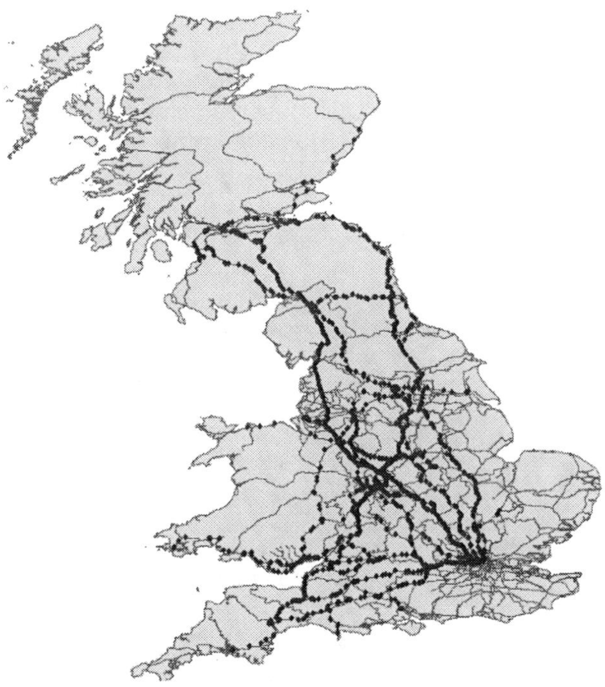

**Figure 5 GPS data indicating train position**

### 2.3 TRAIN Rig measurements

The TRAIN (Transient Railway Aerodynamics INvestigation) Rig has been developed specifically for conducting experimental investigations into railway aerodynamics but is also suitable for studies into road and ground vehicle aerodynamics and other aerodynamic applications (figure 6). The overall length of the facility is 150m with a 50m long test section. The normal model scale that is used is 1/25th, and speeds of up to 80m/s can be achieved. The propulsion system consists of an anchored elastic bungee rope that is tensioned and released, which pulls along a firing carriage connected to the train via a rope and pulleys. A cross flow facility is provided adjacent to the test section to permit tests to be carried out on the effects of winds on vehicles and the influence of winds on vehicle slipstream flows. The specific advantages of moving model experiments over conventional full scale and wind tunnel experiments are that its use avoids the uncontrolled ambient conditions and expense associated with full scale tests, and model test geometry can be varied relatively quickly and cheaply; it can be used to investigate transient phenomena, such as train passing pressures and slipstreams; and that the relative motion between the train and the track is properly modelled.

A number of different types of test are being carried out on the rig using a 1/25th 4 car (2+2 consist) scale model of the Class 43 (figure 7). The first set of tests aimed to replicate the full scale measurement positions on the Great Western Mainline. The train speeds were around 40 m/s to meet CEN compliance of a minimum Re number of 250 000. Experiments were also carried out at the same height as in the full scale measurements (0.7m above the rail), and ensemble averages of the measured pressures and velocities were formed from around 20 runs of the rig. The second set of tests aimed to quantify the effect of ballast shoulder height on the slipstream velocity at (and near to) the TSI position of 2 m from track centreline and 0.2 m above TOR. Three ground configurations were selected - flat ground, a typical UK height ballast shoulder (0.3m) and a TSI compliant ballast shoulder representative of European conditions (0.7m). The slope of both ballast shoulders was 2:3. The Class 43 was fired at about 36 m/s and ensemble averages were formed from 20 runs within (+/-) 3% of this target speed.

The train velocity was recorded with two sets of photoelectric sensors spaced equally between the test section. Air velocity (in u, v and w components) and static pressures were recorded using specialist turbulent flow measurement probes (Cobra probes). The TSI characteristic gust velocities were then calculated - the mean plus two standard deviations of the ensemble of maximum one second (full scale) gusts from around 20 runs.

**Figure 6 The TRAIN Rig**

**Figure 7 HST Class 43 model on the TRAIN rig**

### 2.4 CFD calculations

Steady Reynolds Average Navier Stokes (RANS) and unsteady Delaayed Detached Eddy Simulation (DDES) CFD simulations of the flow around a $1/25^{th}$ scale four car Class 43 have been carried out using an OpenFOAM generated mesh (figure 8) with a total of 44.1 million cells. Air speed was set at 40 m/s which results in a Reynolds number of around 300,000 based on the train height. The RANS approach uses the k-ω SST turbulence model, The DDES approach uses the Spalart-Allmaras turbulence model for the RANS regions. Results from the DDES approach are time averaged to enable comparison to the RANS and TRAIN rig results. Both sets of simulations were conducted using OpenFOAM 2.1 on the University BlueBEAR system. One RANS simulation required 32 processors for 48 hours, whilst one DDES simulation used 80 processors for 288 hours.

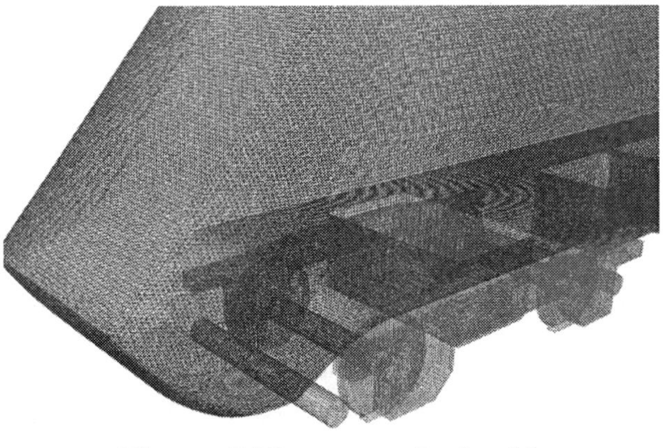

**Figure 8 The numerical grid**

## 3 PRESSURE TRANSIENT COMPARISONS

Figure 9 shows a comparison between the pressure coefficient transients measured in the full scale experiments for both 2+7 and 2+8 HST sets, those measured on the TRAIN Rig and those calculated using the two types of CFD methodology. The full scale measurements were made at 1.0m above the top of rail at a distance of 2.0 from the track centre line, which were chosen for convenience rather than to correspond with the TSI measurement positions. The model scale and CFD measurement positions were at equivalent positions. Note that some of the results have been "split" two coaches before the end of the train, to allow an appropriate comparison between the 4 coach experiments and calculations, the 8 vehicle train and the 9 vehicle train. It can be seen that in general the agreement is good, with similar forms for all the measurements and calculations. There is some difference in the magnitudes of the nose pressure coefficient peaks as can be seen in table 1 below. The negative peak of the TRAIN Rig measurements is particularly high, almost certainly due to the underbody blockage caused by the rig firing mechanism.

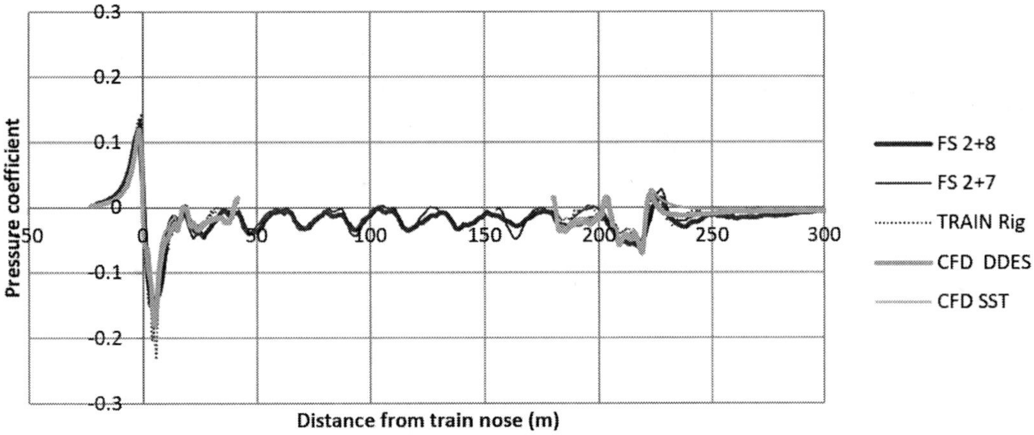


**Figure 9 Pressure transient measurements (1m above top of rail, 2 m from track centre line)**

**Table 1 Maximum and minimum values of nose pressure peaks**

| | Full scale 2+7 | Full scale 2+8 | TRAIN Rig | CFD DDES | CFD SST |
|---|---|---|---|---|---|
| Maximum | 0.135 | 0.118 | 0.144 | 0.119 | 0.120 |
| Minimum | -0.145 | -0.147 | -0.232 | -0.181 | -0.163 |

## 4 SLIPSTREAM COMPARISONS

Figure 10 shows a comparison of the ensemble averages of slipstream velocity for the full scale measurements at 3.0m from the track centre line, 0.7m above the top of rail for the full scale experiments for the 2+7 and 2+8 units, the TRAIN Rig measurements and two types of CFD simulation. For the runs that were used in the calculation of the full scale ensembles, only a minority had ambient wind conditions of less than 2m/s as required by the TSI / CEN methodology. [3], [4]. This can be clearly seen by the non-zero velocities upstream of the train nose in the figure. Again the 2+7, TRAIN Rig and the CFD calculations have been split two coaches before the end of the train. It can be seen that there is a considerable difference between the full scale data and the TRAIN Rig / CFD data. The former shows a build up of the flow speed along the train side and a peak in the ensemble around the latter half of the train. The latter show a sharp peak in the wake of the train. The full scale results are perhaps somewhat surprising as most full scale data for other passenger trains show peaks in the train wakes rather than along the train side [5], [6] – indeed, such as is found in the experimental results and the CFD. Also one should be aware of the nature of the full scale data. Figure 11 shows the individual runs that go to make up the HST 2+8 ensemble. There can be seen to be a very great deal of variation, with a range of peak values, and some runs showing no peak at all. This effect is presumably due to slight variations in wind direction that carries the wake to or from the measurement position. There is clearly more work to be done here and in particular further full scale data is required to check if the present dataset is anomalous in any way.

Experiments were also carried out using model ballast shoulders of different heights. The TSI methodology allows a shoulder height of between 0.5m and 1.0m. The height of the ballast shoulder at Uffington was lower – of the order of 0.3m. Measurements were made of the slipstream velocities on the TRAIN Rig over 20 runs of the rig for flat ground and ballast shoulder heights of 0.3 and 0.7m, and the TSI characteristic velocity then determined for each ballast shoulder height and measurement position. This is calculated by calculating the maximum 1 second moving average gust in each run, and then finding the mean plus two standard deviations of the ensemble of gust measurements. It can be seen from table 2 that the TSI characteristic velocity normalised by the train velocity falls significantly as ballast shoulder height increases, presumably because the less intense part of the wake impinges on the measuring probe. This raises questions concerning the robustness of the TSI methodology – in effect "better", or lower, values can be obtained by testing at sites with a high ballast shoulder.

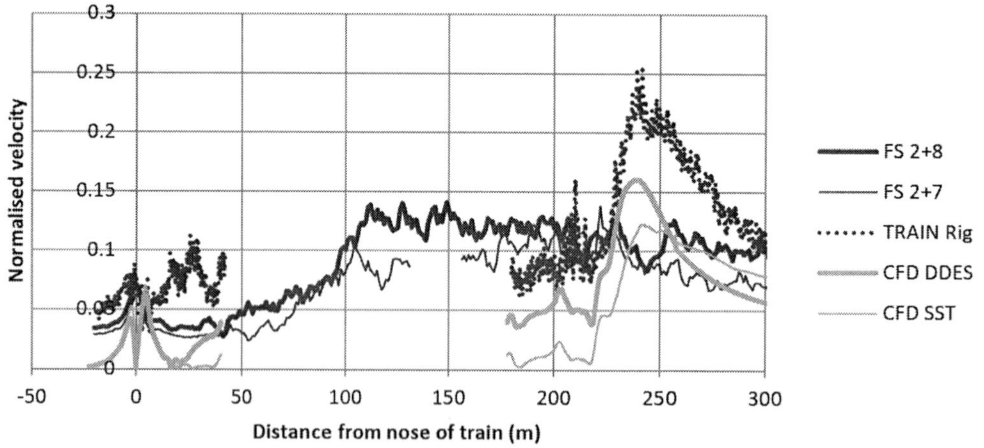


**Figure 10 Normalised velocity ensembles for the different sets of experiments and calculations**

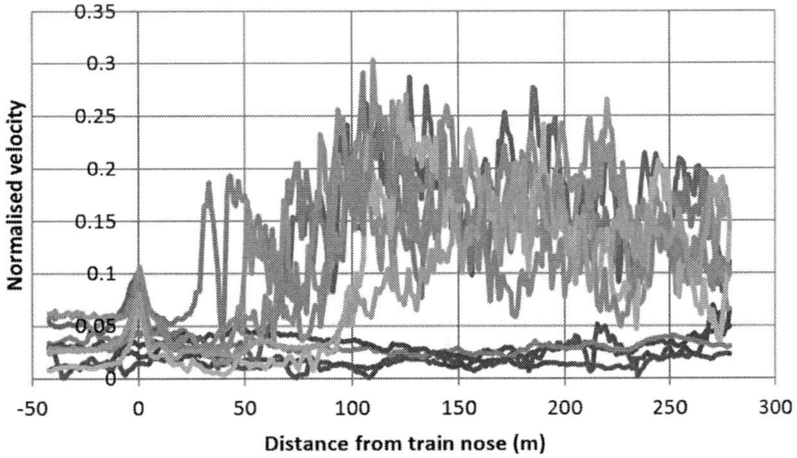


**Figure 11 Individual normailised velocity measurements for the HST 2+8 ensemble**

**Table 2 TRAIN Rig measurements of normalised slipstream velocities at TSI position for different ballast shoulder heights**

| | TSI position 0.2m ATOR | | | Uffington position 0.7m ATOR | | |
|---|---|---|---|---|---|---|
| Ballast shoulder configuration | Mean velocity | Standard deviation | TSI velocity | Mean velocity | Standard deviation | TSI velocity |
| Flat ground | 0.236 | 0.027 | 0.292 | 0.200 | 0.035 | 0.270 |
| Uffington height (0.3 m) | 0.213 | 0.026 | 0.266 | 0.187 | 0.032 | 0.251 |
| European height (0.7 m) | 0.188 | 0.028 | 0.245 | 0.160 | 0.042 | 0.244 |

## 5 TUNNEL AND TRAIN PASSING MEASUREMENTS

In this section we briefly consider the transient pressures measured on the side of the Class 43 NMT caused by passage through tunnels and by passing trains. Figures 12 and 13 show the pressures measured as the train passed through Ampthill "fast" tunnel at 182 kph. This is a 645m long double track tunnel on the Midland Main Line. Figure 12 shows the pressures measured at approximately mid height on the leading car and figure 13 the pressures measured on a trailing car. The effects of the airshafts within the tunnel can be seen, as a series of peaks and troughs in the pressure trace, as can the pressure waves reflected from the end of the tunnel. The utility of these measurements is that they provide a reference set of pressure characteristics for all tunnels on major lines in the UK that can be used to calibrate tunnel pressure transient models which can then be used to investigate tunnel modifications and improvements of different sorts.

Finally figure 14 shows the pressure transients measured on the leading coach of the train as it is passed in the other direction by a Class 390 Pendolino. The NMT speed at this point was 18 $ms^{-1}$ and the passing train speed is estimated to be 56 $ms^{-1}$. The large pressure transients on the side of the passing train caused by nose and tail pressure peaks are very apparent. In this case the magnitude is around 1100Pa, well below the 1440Pa value in the current Railway Group Standards.

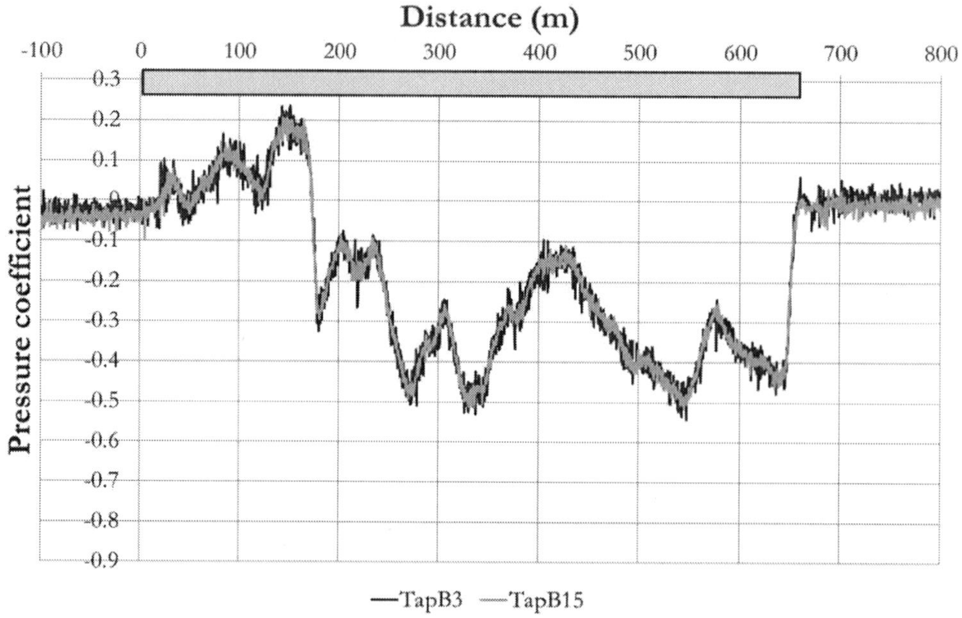


**Figure 12 Pressure transient measurements on the class 43 train side of the leading vehicle when entering Ampthill tunnel (taps on different sides of the train)**

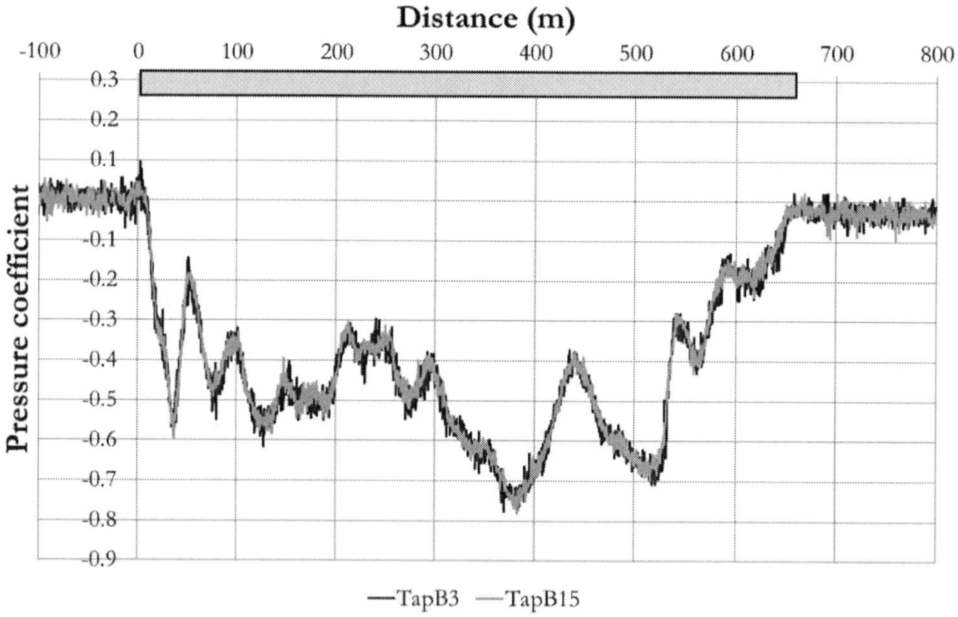


**Figure 13 Pressure transient measurements on the class 43 train side of the rear vehicle when entering Ampthill tunnel (taps on different sides of the train)**

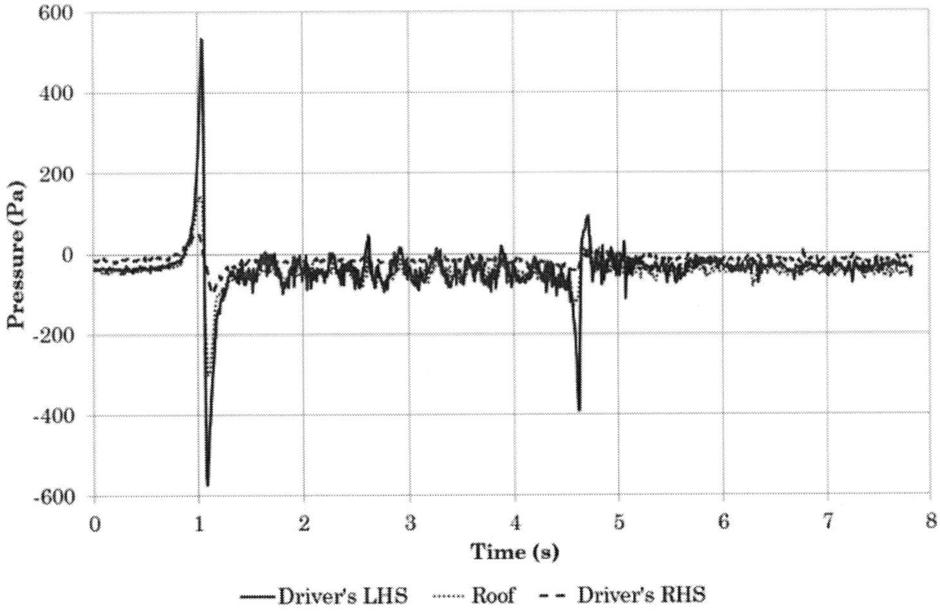


**Figure 14 Pressure transient caused by passing train (Class 390)**

## 6 CONCLUDING REMARKS

From what has been presented in earlier sections, the following conclusions can be drawn.

- The measurements carried out on the NMT have produced large quantities of high quality data for the pressure transients due to passage through tunnels and due to passing trains (and, although not presented in this paper, for the effect of cross winds). The tunnel transient data in particular will have a long term utility in that it characterises the behaviour of every tunnel on the UK main line network.
- The preliminary full scale measurements made at the Uffington site have enabled the pressure transients and slipstreams caused by passing Class 43 trains to be measured, although the ambient wind speeds during the measurement period were rather higher than would have been ideal.
- It has been shown that moving model experiments on the TRAIN Rig and the CFD calculations can adequately predict nose pressure transients caused by a passing train, but there is significant disagreement with the full scale data for ensemble averaged slipstream velocities. The latter may be partly due to the seemingly anomalous nature of the full scale measurements.

Whilst the work on the overall project is proceeding well, more work remains to be carried out, in particular further trackside measurements are required and a full comparison of the NMT data with wind tunnel, TRAIN Rig and CFD results. There is also a need for more trackside slipstream data for conditions where the wind speeds are lower and within the TSI limits.

**REFERENCES**


[1] C J Baker "Train aerodynamics – a review; Part 1 Fundamentals" Aeronautical Journal 118, 1201, 2014

[2] C J Baker C J "Train aerodynamics - a review; Part 2 Applications" Aeronautical Journal 118, 1202, 2014

[3] CEN, Railway applications — Aerodynamics — Part 4: Requirements and test procedures for aerodynamics on open track, CEN EN 14067-4:2005+A1, 2009

[4] TSI, "EU Technical Specification for Interoperability Relating to the 'Rolling Stock' Sub-System of the Trans-European High-Speed Rail System" HS RST TSI, 2008/232/EC, 2008

[5] C J Baker, A Quinn, M Sima, L Hoefener, R Licciardello, "Full-scale measurement and analysis of train slipstreams and wakes: Part 1 Ensemble averages." Proceedings of the Institution of Mechanical Engineers, Part F: Journal of Rail and Rapid Transit, doi:10.1177/0954409713485944, 2013

[6] C J Baker, A Quinn, M Sima, L Hoefener, R Licciardello, "Full-scale measurement and analysis of train slipstreams and wakes. Part 2 Gust analysis", Proceedings of the Institution of Mechanical Engineers, Part F: Journal of Rail and Rapid Transit, doi:10.1177/0954409713488098, 2013

# Preliminary results of monitoring audio frequency track circuits

**L Saade, P Weston, C Roberts**
University of Birmingham, UK

## ABSTRACT

Track circuits form a critical part of the UK rail signalling infrastructure. Condition monitoring of track circuits has the potential to improve the availability of track circuits, hence improving service levels. Audio-frequency track circuit condition monitoring is relatively new and little work has been done on the subject. This paper gives some preliminary results obtained from instrumentation installed on a cascade of track circuits. Both the transmitter and receiver voltages were monitored, making it possible to identify whether a fault is located at the receiver or transmitter end of a track circuit. Shunted and un-shunted behaviour can be examined using the data collected, but this paper concentrates on the long-term variations in transmitter and un-shunted receiver levels. In addition, feed-through from other local transmitters is shown.

## ABBREVIATIONS

| | |
|---|---|
| AC | Alternating current |
| AF | Audio frequency |
| DC | Direct current |
| FFT | Fast Fourier transform |
| HVI | High voltage impulse |
| IRJ | Insulated rail joint |
| NR | Network Rail |
| RMS | Root mean square |
| RX | Receiver |
| TC | Track circuit |
| TU | Tuning unit |
| TX | Transmitter |

## 1 INTRODUCTION

Track circuits are used in the railway industry to identify when a section of track is occupied. Track circuits work by injecting some form of electrical signal along the track, to be detected at a distant location. A train present in the track circuit area diverts the transmitted signal away from the detector; the absence of a received signal signifies an occupied track section.

The simplest track circuits are direct current (DC) track circuits such as the one shown in Figure 1. The track is modelled with inductance and capacitance that is more relevant to alternating current (AC) track circuits. More complex designs [1, 2] include the use of AC track circuits, both at 50 to 175 Hz and at audio frequencies (1.5 to 8.4 kHz), as well as high voltage impulse (HVI) track circuits.

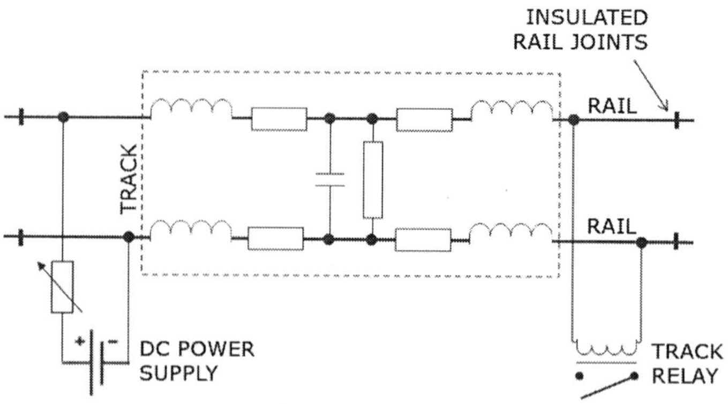


**Figure 1: A simplified model of a track circuit including inductance, conductance and capacitance**

Track circuits are one of the main forms of train detection on the UK rail infrastructure and audio frequency (AF) track circuits are popular because they do not require insulated rail joints (IRJs) to define the extent of the track circuit. Instead, AF track circuits have transmitter (TX) units that generate a signal around a specific frequency that is only detected by the corresponding receiver (RX) unit. The frequencies alternate along the track with track circuit boundaries defined by two track tuning units (TUs) connected to the rails that form a tuned zone (Figure 2). Each TU is tuned to contain each particular frequency within the track circuit [3].

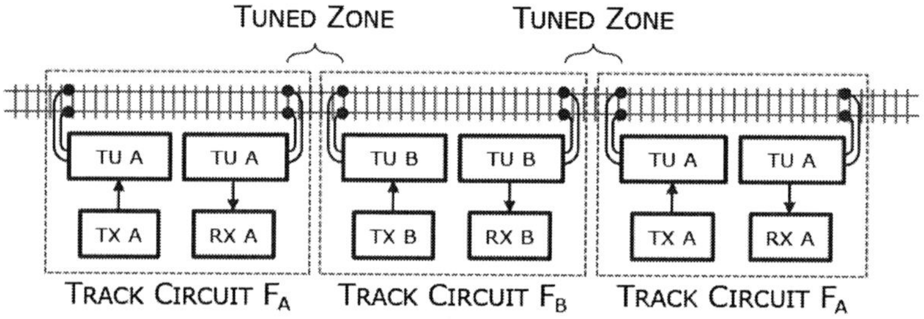


**Figure 2: AF track circuit setup with alternating frequencies, $F_A$ and $F_B$**

The functionality of track circuits is affected by the condition of the infrastructure on which they operate. Some of the biggest causes of track circuit failures are poor connections and contamination of the wheel/rail interface. Failures can be categorised as either right-side failures (where the signalling indicates occupied even if there is no train present) or wrong-side failures (where the signalling indicates clear even though a train is in the section) [2]. Right-side failures can result in significant delays whereas a wrong-side failure can be catastrophic. One of the examples shown in the sequel shows a right side failure.

A program of condition monitoring of railway assets (known as Intelligent Infrastructure) is underway in the UK in an attempt to reduce the number of delay minutes caused by infrastructure failures. Condition monitoring of track circuits typically involves monitoring the voltage at the detector relay and flagging up anomalous values. Monitoring systems have been developed for Intelligent Infrastructure that can monitor DC or power frequency (50 Hz) AC track circuits [4] but they have not yet incorporated monitoring of AF track circuits as there is no signal that can be directly monitored by a typical railway infrastructure condition monitoring system.

Monitoring AF track circuits requires intelligent sensors that can turn complex time-domain waveforms into useful data that can be monitored for changes that indicate incipient failures [5]. Work is continuing to develop this ability to support the Intelligent Infrastructure vision at Network Rail.

## 2 TRACK CIRCUIT MONITORING SYSTEM

A modular track circuit monitoring system has been designed to monitor a multitude of track circuits, including DC, AC (50 Hz and similar) and AF track circuits (such as TI21, Aster, SF15 and FS2500, FS2550) in significantly more detail than simply monitoring the track relay state [6].

The monitoring system includes a logger that is used to keep a record of all the collected data that is later processed into useful information. The logger can use global positioning system (GPS) signals to maintain the accuracy of an on-board real-time clock calendar. This provides accurate time stamping of data and events so that time-aligned comparison of results from geographically separated loggers, for example at the transmitter and receiver ends of a track circuit when they are located in different line-side cabinets. The logger can record data onto a Secure Digital (SD) card continuously for up to 2 years.

As part of this system, intelligent sensor nodes have been developed to allow data from track circuits at audio frequencies (up to 10 kHz) to be processed on-board. These nodes feature 2 isolated channels capable of measuring AC and DC signals with a third order analogue low-pass filter on every channel to reject high-frequency interference and act as an anti-aliasing filter. The connection into the sensor node can be from a passive clamp-on current sensor, or from a custom made voltage interface module that allows safe connection to the signalling system. Sensor nodes can be designed to interface with any type of track circuit signal.

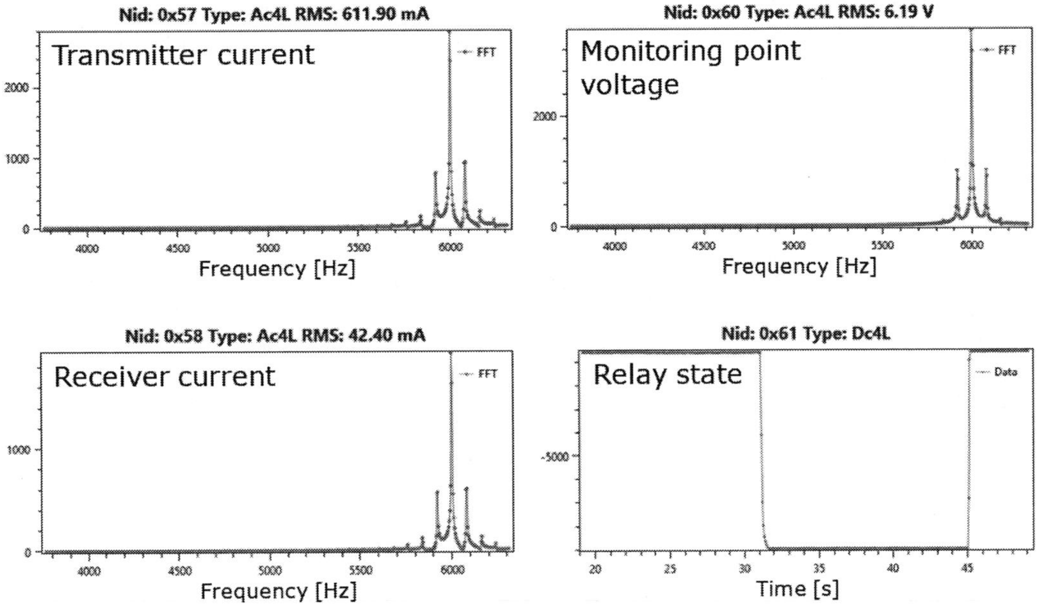


**Figure 3: Sample data streaming from a 6000 Hz FS2550 track circuit with ±80 Hz FSK modulation**

The AF track circuit sensor nodes have an on-board processor that computes a 4096 point Fast Fourier Transform (FFT) four times a second on the input signals from which important information such as the RMS levels in a given frequency range, sideband RMS levels (where the carrier is modulated), as well as the frequencies of the carriers and sidebands, can be computed. Figure 3 shows the type of data that is made available from performing an FFT on a signal with frequency-shift-keyed (FSK) modulation of 80 Hz applied to the carrier.

Short time synchronisation messages are broadcast by the logger board on an RS422 bus on a backplane to ensure that all the sensor nodes in the system collect data synchronously. The processed data from the sensor nodes is then output onto the RS422 bus to be logged onto the logger's SD card. The accurate timing (to within about 1 ms) of logged data enables track circuits with geographically separated transmitters and receivers to be time-aligned, and also allows for the possibility of monitoring crosstalk or feed-through problems.

Currently, the logger and sensor nodes can monitor up to 4 AF transmitters/ receivers (logging 4 FFTs per second) or up to 16 DC quantities, or a combination of both. The modular format allows for the necessary nodes to be installed depending on the monitored track circuit type. The current implementation only logs to SD card however the modular format with the RS422 backplane would allow for an additional module that monitors the data and detects events that can be transmitted over some communication channel (internet, 3G, or other radio channel) and also upload selected detailed data to a central server.

This modular system was developed and put into practice on a live track in Wales, as described in the next section.

## 3 TEST INSTALLATION

A 10 km section of track in Wales was selected for testing the instrumentation. This length of track is divided into 14 sequential track circuits set up in a cascade so that when one detects a train, it cuts the power to the previous subsection, causing detection to be propagated through the cascade until it reaches the signalling relay (Figure 4). This cascaded arrangement makes it difficult to identify which (sub) track circuit has a fault when there is a fault in the track circuit as a whole.

Twelve of the sub track circuits are AF Aster or SF15 units (which are interchangeable) but there are also 2 DC track circuits, one at the start of the whole track circuit, and one near the end. More details of Aster track circuits can be found in [7].

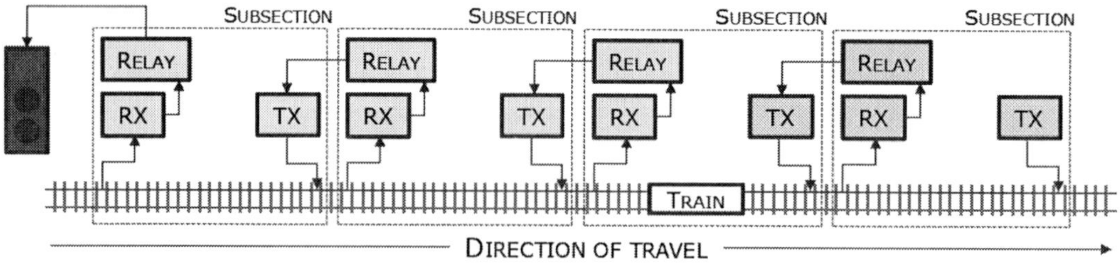


**Figure 4: Cascaded track circuits**

Twelve track circuit monitoring units were built specifically to monitor the receiver and transmitter ends of Aster/SF15 track circuits and two were built to monitor the DC track circuits. These units were installed along the track in the location cabinets that service the cascaded track circuit.

The units save four FFTs every second from the receiver input voltage, which comes from a track transformer [7]. This signal also happens to contain the information from the adjacent transmitter, unlike more modern TI21 track circuits where the receiver and transmitter voltages are not available together.

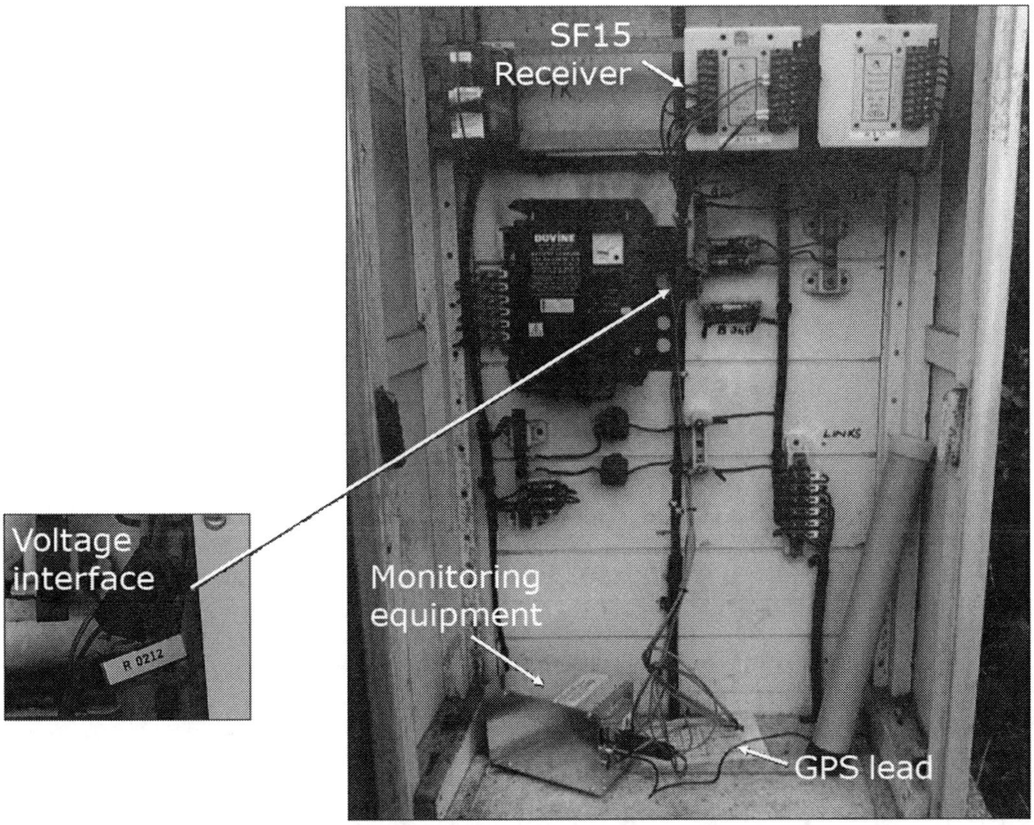


**Figure 5: Track circuit monitoring unit customised for monitoring Aster/SF15**

Figure 5 shows the monitoring equipment in a location cabinet connected to the receiver voltage and the relay drive voltage. The voltage connections are made through a 'voltage interface' (shown in the figure) that allows safe connection to the signalling system. The transmitter voltage is detected at the input to the receiver, a useful feature of Aster/SF 15 track circuit implementation.

Two and a half months after the installation, data were collected and processed to see if anything of interest had occurred. The next section shows some samples of the collected data.

## 4 EXAMPLE RESULTS

This section shows some examples of the results from the data collected from the cascaded track circuit. The variation in received and transmitted RMS voltage are shown over two 24 hour periods and then feed-through is demonstrated in a three minute window.

### 4.1 Receiver and transmitter RMS voltages over 24 hours

One of the things that has been observed is the receiver levels changing over the course of a day. Figure 6 shows the receiver levels of all the track circuits in the cascade (the DC levels TC/1 and TC/13 are reduced by a factor of 10). When a train goes past, the receiver levels drop nearly to zero. Monitoring the behaviour of the shunted track circuits is very important, but in these examples the longer-term unshunted levels are examined.

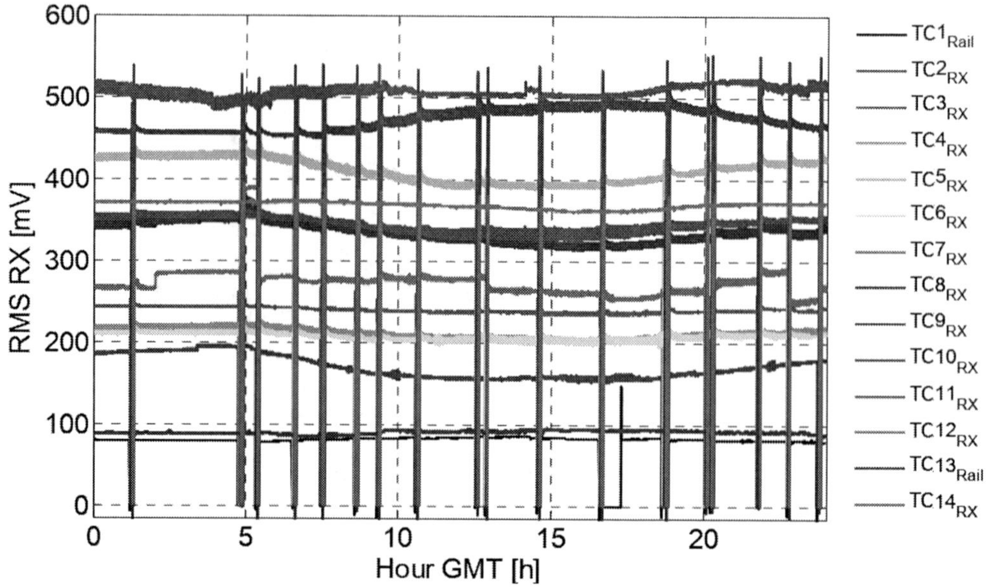


**Figure 6: Receiver voltages (30 June 2014)**

On the 30 June 2014, the receiver level for TC/3 is almost constant all day while TC/4, TC/6 and TC/14 remain relatively constant, dipping down slightly during the day. Most of the other track circuits show some drifting, generally downwards during the day as the temperature increases and the back up again as it cools down. However, TC/9 shows the opposite trend.

TC/10 and TC/11 show a significant fuzziness in their lines that is caused by a small amount of feed-through from nearby track circuits with the same frequency. However, note that TC/14 is quite clean - see §4.3.

There is an instance of a right side failure in this example. The TC/2 level drifts to a low level around 17:00. A train goes past, shunting the TC/2 track circuit and when the train has exited the cascade, the TC/1 track circuit does not show clear but continues to show occupied for some time after the train has gone. This is because the TC/2 receiver level has dropped so low that the receiver does not pick up once the train has passed. This right side failure was identified by NR and resulted in a maintenance team visiting and doing some shunt tests. If the data shown here had been available remotely, this failure and the identity of the affected track circuit within the cascade would have been instantly recognised.

The second fault that can be identified on this day is that TC/7 is stepping up and down, generally when a train passes. This variability in the received level does not affect the track circuit operation, but it suggests that there is a poor connection somewhere. This could develop into a failure in the future.

Figure 7 shows the transmitter levels (as seen via the track transformer) on the same day. Most transmitter levels are fairly constant during the day. However, TC/10, for example, has more variability than the others. TC/7 shows step changes, generally when a train has passed. This indicates a possible poor connection somewhere within the TC/7 track circuit.

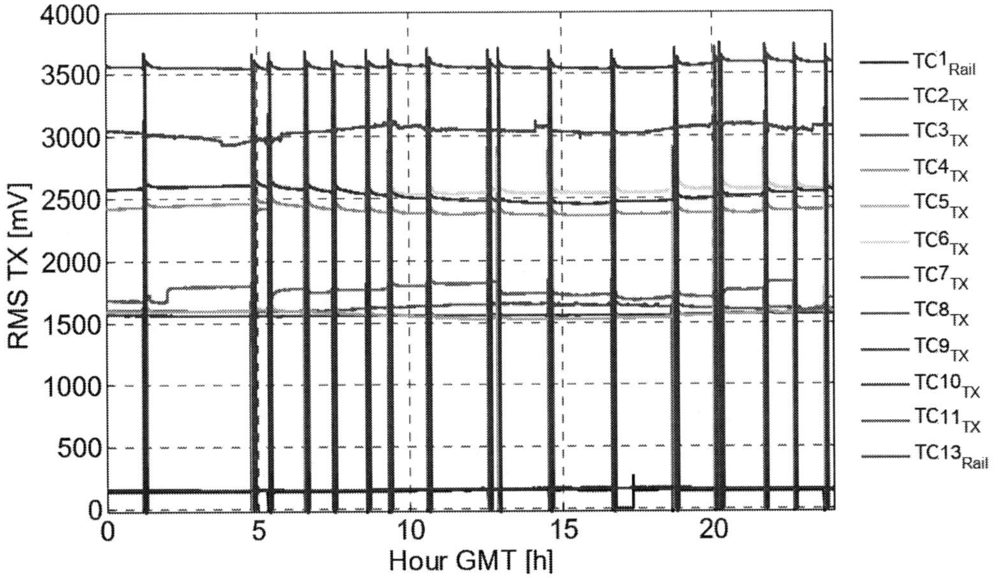


**Figure 7: Transmitter voltage, 30 June 2014**

Since both the transmitter and receiver signals for each track circuit are available, they can be considered jointly. The TC/7 receiver level shows step changes that are the same as the steps in the TC/7 transmitter level, so the problem with the track circuit is located at the transmitter end. If there was a poor connection at the receiver end, then the transmitter would not show step changes, although the TC/6 transmitter might show small step changes instead.

Note that the TC/2 transmitter level is almost constant but the TC/2 receiver level varies. Next, the same data are examined on a day two weeks later.

## 4.2 Receiver and transmitter RMS voltages over another 24 hours

This subsection shows the same signals on another day. Figure 8 shows the receiver voltages and Figure 9 the transmitter voltages (via the track transformer) on the 15 July 2014.

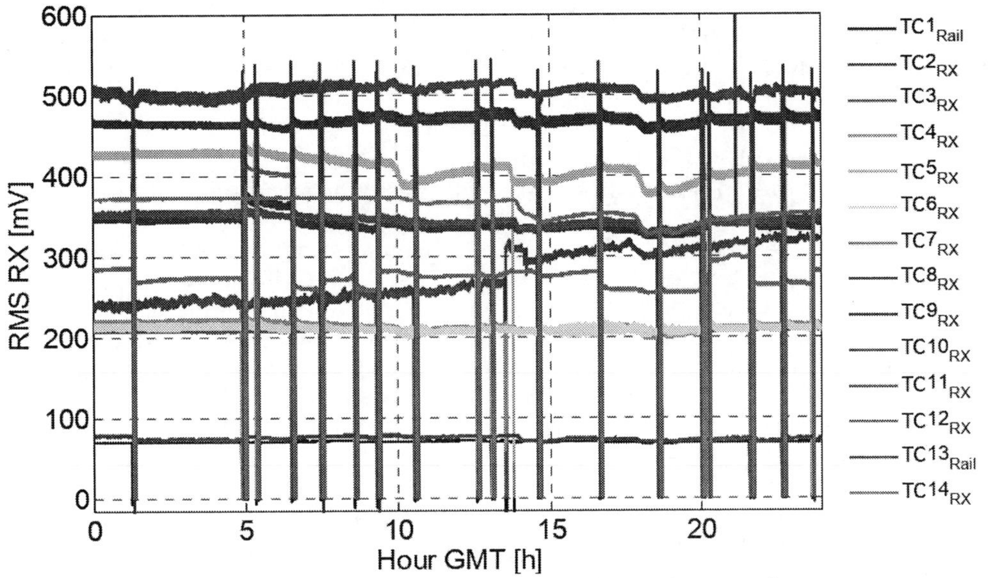


**Figure 8: Receiver voltages (15 July 2014)**

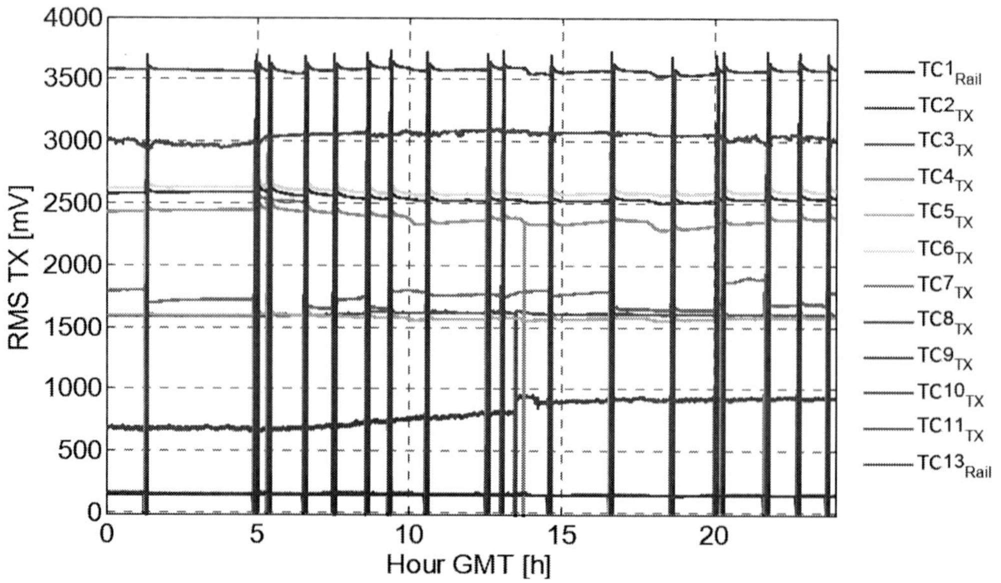


**Figure 9: Transmitter voltages (15 July 2014)**

On this day, the behaviour of the TC/2 track circuit has changed compared to its behaviour in the previous example (from two weeks earlier). The receiver level is now significantly higher than it was before, reducing the likelihood of a right side failure. However, the transmitter level is lower than before and is now showing significant drift. Previously, the TC/2 transmitter level was nearly constant. The drift is visible in both the transmitter and receiver, so it is most likely associated with some fault at the transmitter end. Given the step change in the middle, this could indicate a poor connection at the transmitter end.

Some track circuits show a reduction in their receiver level, and sometimes in the transmitter level too, when there is rainfall. Examples of this can be seen around 10:00, 14:00 and 18:00 most prominently in the receiver level of TC/5, and also in the transmitter level of TC/4. This could be to do with changes in the ballast resistance when it rains, or perhaps because of the effect of rainfall on a tuned zone (which is shared between TC/5 RX and TC/4 TX), but could also be caused by a drop in temperature during rainfall.

The examples given so far are seen on the timescale of a whole day. Other effects can be seen over shorter timescales.

### 4.3 Feed-through

Figure 10 shows the un-shunted receiver voltages over a three minute interval.

The TC/9 and TC/4 receiver levels show a sinusoidal variation that is believed to be caused by 'beating' between the expected signal and the signal from a nearby transmitter (or transmitters) with the same frequency. The beating is caused by the slight difference in the transmitter frequencies. The receiver RMS level varies as the interfering signals alternately augment and cancel part of the expected receiver voltage. The feed-through can therefore be monitored by observing the amplitude of the sinusoidal variation. If this increases beyond a pre-defined level, then there is a feed-through problem.

The TC/14 receiver consistently shows no discernible feed-through as it is surrounded by insulated rail joints.

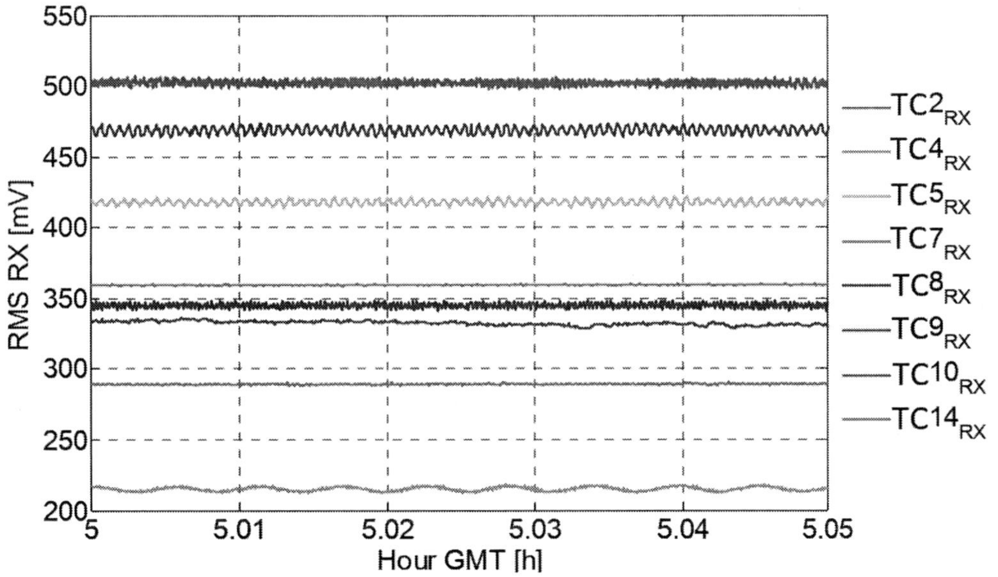


**Figure 10: Example of feed-through**

## 5 CONCLUSIONS AND FURTHER WORK

A modular track circuit monitoring system, capable of logging and processing data from a wide range of track circuits has been developed. Twelve SF15 and Aster U type audio frequency track circuits are currently being monitored in a site in Wales. The data collected from the first two and a half months of operation have shown that the long-term behaviour of the track circuits has the potential to be used to inform the maintenance process. The transmitter and receiver ends are being monitored. This provides a clue as to which end of a track circuit has developed a fault. Feed-through has been seen, but not at a level that would cause concern, and one right side failure has been observed.

One of the key obstacles to implementation is the need for a direct connection to the track circuit signals. The use of non-contact current transducers as opposed to voltage sensors has yielded positive results and reduces the risk of injecting unwanted signals into the track circuit.

A future goal is to develop a low-cost, non-contact system with the addition of event detection based on simulation and field tests. These events can then be transmitted to a central server over a 3G link, for example.

## 6 ACKNOWLEDGEMENTS

This research has been made possible through the support and funding of the Railway Safety and Standards Board project T579-04, Network Rail and London Underground.

## 7 REFERENCES

[1] R. T. Sewell, "Evolution of electronics in railway signalling from the track circuit standpoint," in *IEEE Vehicular Technology Conference*, 1985.

[2] Railtrack, "GK/RC0752 General Information on Track Circuits," Safety and Standards Directorate, London, 1998.
[3] Railtrack, "GK/RC0761 TI.21 Audio Frequency Track Circuits," Safety and Standards Directorate, London, 1996.
[4] C. R. Bell, "Remote condition monitoring of phase sensitive ac track circuits," in *IEE Seminar on Railway condition monitoring*, 2005.
[5] P. Weston, J. Chen, E. Stewart and C. Roberts, "Condition monitoring of audio frequency track circuits," in *International Conference on Railway Engineering - Challenges for Railway Transportation in Information Age, 2008. ICRE 2008.*, Hong Kong, 2008.
[6] L. Saade, P. Weston and C. Roberts, "Deep Monitoring of Audio Frequency Track Circuits," in *IET Railway Condition Monitoring (RCM)*, Birmingham, 2014.
[7] British Railways Board, "The Aster type 'U' jointless track circuits for non-electrified lines, BR13448/2," 1976.

# A 10 Gbit/s Ethernet infrastructure for future-proofed railway communications

**G Koczian, S Walker**
CSEE, University of Essex, UK
**G Howell, B Simpkin**
LPA Connection Systems, UK

## ABSTRACT

The UK and European Rail industry is enjoying renewed business and public custom because end-to-end journey times are now competitive with short-haul flights. There is now the further expectation that train wireless connectivity and "infotainment" experience will vie with that of aircraft. With 600 passengers, a 10 Mbit/s bandwidth expectation per customer equates to 6 Gbit/s aggregate throughput; a serious bottleneck.

This paper describes an upgrade of on-train communication capacity to 10 Gbit/s or 10GBASE-T Ethernet using rail-approved, copper-media cabling. In essence, data-centre technical advances have been deployed in the hostile train environment whilst acquiring all the requisite rail approvals along the way.

## 1 INTRODUCTION

Travellers will increasingly require connectivity that supports entertainment, social media engagement and business activity. For example, entertainment on longer journeys might involve Internet surfing, email and social networking. In particular, live web updates on news, weather, sports are becoming a "must have" in many parts of the globe. Train operators are doubly challenged to provide the growing demands for real-time information on board for trains. Firstly, train-to-shore connectivity provision is a non-trivial exercise at Gbit/s data rates. Secondly, bottleneck-free distribution of the data through the train is a further pressing concern. However, the potential financial rewards are significant. For example, a survey showed that 72% of business travellers are more likely to use trains than airplanes or cars if there is a good Wi-Fi connection on board (1).

Although, there is much current discussion around the train-to-shore connectivity issue, there are also many unresolved issues concerned with the distribution of data within the train; this latter being the focus of this paper.

In section 2, we will note that provision of the 10 Gbit/s Ethernet capacity for on-train communication can only be achieved by careful attention to each layer of the infrastructure. At the physical connection level or "PHY Layer", as it is usually termed in the data manipulation context, the discussion will be restricted to copper media only, as it is widely accepted as reliable in the Railway Industry. Both cable type and acceptable connector configurations must be considered quite apart from the electrical transmission issues of 10GBASE-T connectivity. These problems have been overcome in a fully rail-approved solution. Section 3 will deal with higher-level issues, such as the management and distribution of the data throughout the length of the train. In many respects, the problem is similar to that of a data centre so advantage may be taken of commercial off-the-shelf-solutions (COTS); the remaining concern being electrical and mechanical ruggedization for the train environment. Section 4 will briefly consider options for the future including 25 GbE, which will facilitate further backbone enhancements. Some conclusions are then drawn.

## 2 BASIC PRINCIPLES OF RAIL-APPROVED HIGH-SPEED CONNECTORS

An electromagnetic (EM) analysis and a careful testing regime of both connectors and cables have resulted in a certified 10GBASE-T backbone for trains. With full Ethernet switching and signal regeneration, length extension on a practically unlimited basis is possible. Figure 1 below shows examples of rail-approved "4S-10G ™" connector builds (2). There is no EM shielding between the four pins in the connector. However, the screen of the four-pair cable (known as Cat6 or Cat7) is connected to the metal shell, satisfying the requirements for 10GBASE-T.

**Figure 1. Examples of rail-approved "4S-10G ™" connector builds.**

This means, of course, that two sets of "4S-10G ™" connectors are required and up to four sets can be accommodated as shown in the centre of figure 1. Although cabling is short on trains (55m) relative to the TIA spec, the number of connectors will usually exceed the two covered by the specifications. Hence, the performance of connectors and cabling certified to Ethernet standards cannot be guaranteed in a train installation. Industry standard 8-wire connectors and cables are not used in the inter-car gap. The use of larger conductor diameters and increased stranding assure long-life in a high flex environment.

Whilst the foregoing arguments satisfy all the theoretical constraints, the real-world electrical performance must be verified. This was carried out using a calibrated and certified test instrument (FLUKE DTX-1800). As shown in figure 2 below, the two parts of the test instrument were simply connected to all four pairs of ~100m data cable including a connector (not shown on instrument test report) and a complete characterization, including a pass/fail decision, carried out.

The results are discussed briefly as they form an important part of the PHY layer characterization. For convenience, the boxes are numbered in sequence.

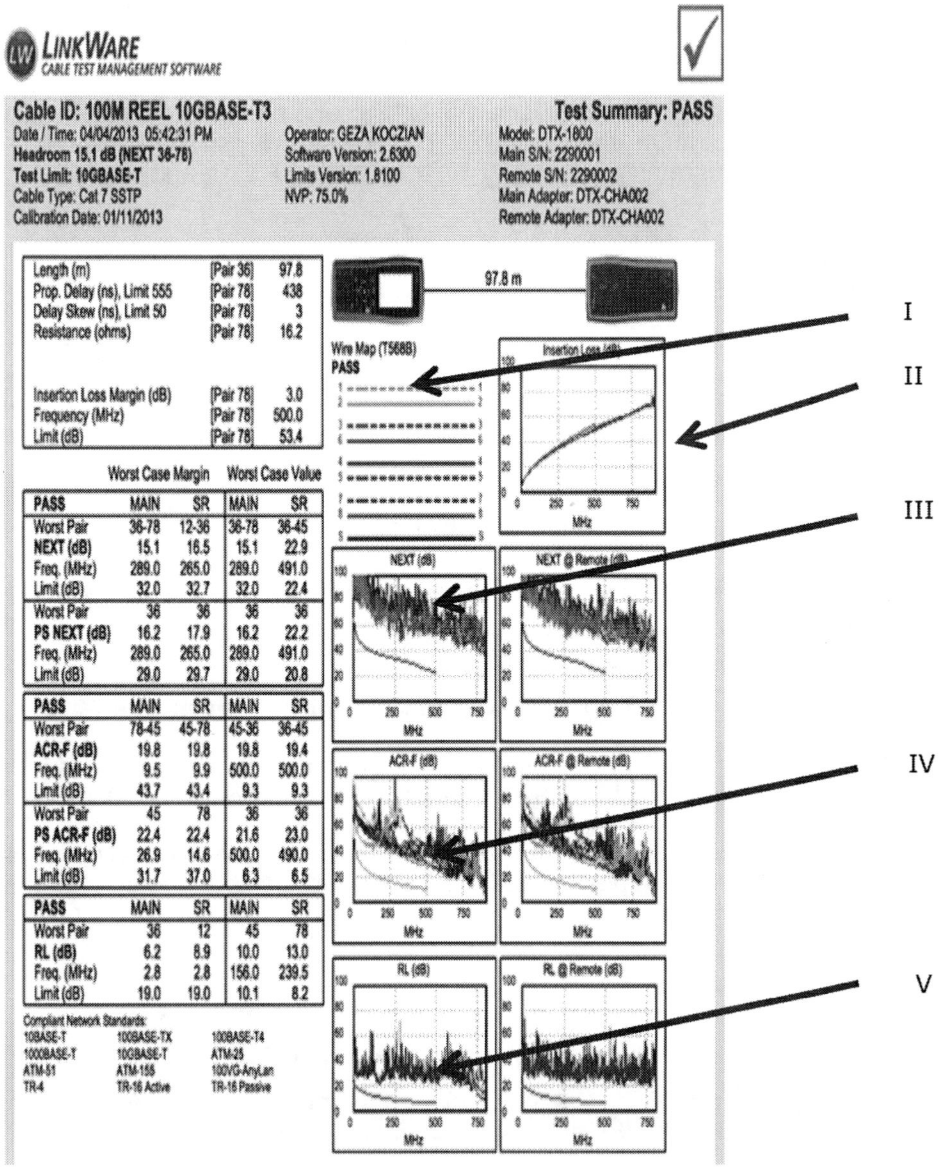


**Figure 2. Example test report of the cable/connector analyser.**

Test I shows that the correct pair designation has been made. In test II, the link loss is compared to that expected for 10GBASE-T. It can be seen that it rises to ~50dB at the maximum 500 MHz frequency used. In dB terms, at 3dB margin, this is the closest to a fail in the test series. In III, Near-End Crosstalk or NEXT showing how much signal transmitted on one pair is returned on an adjacent pair, with the worst case highlighted. Attenuation Crosstalk Ratio Far-end (ACR-F) as in IV is a derived figure produced by subtracting the insertion loss of the disturbing pair from the Far-End Crosstalk (FEXT) this pair induces in an adjacent pair. This normalizes the results for length. FEXT is the proportion of a disturbing signal, which appears on the far end from a transmitter and on an adjacent pair. We may note that Power Sum Attenuation to Crosstalk Ratio–Far End (PSACR-F) is the power sum of multiple cable pairs. Measurement V or return loss show how good match the cable/connector is to the desired 100 ohm. Finally, a summary pass/fail decision made as shown in the top right corner of the report. These tests may carried out on different lengths of cable with multiple sets of connectors to assess the performance in a representative real-life train environment where car length is typically up to 25m with a consist of up to 11 cars (3).

Overall, these tests form the foundation for a complete train infrastructure as the ~100m reach is more than enough for a typical car and connector combination. The next step is to provide high-end 10GBASE-T servers and test the complete set up for RF and impulse immunity according to the EN50121 series standards e.g. (4) and RIA 12 (5).

## 3 HIGHER-LEVEL CONNECTIVITY ISSUES

Having laid the foundations for correct PHY layer operation with 10GBASE-T, the electronic data management issue in trains must be considered. Figure 3 below shows a template model for conventional data centre wiring. It is fortunate that it bears some basic similarity to cable disposition on a train as will be seen.

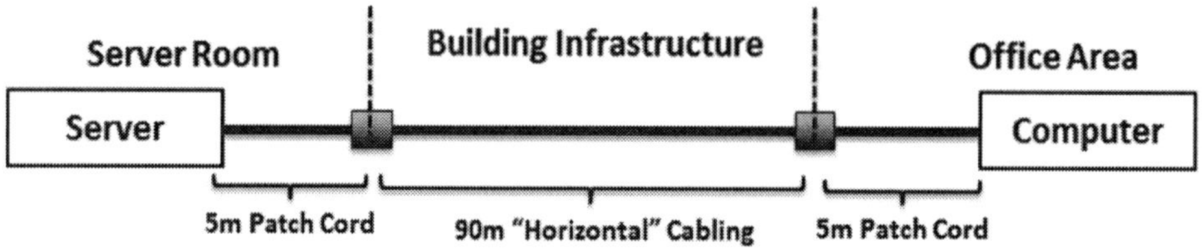


**Figure 3. Standard data centre cable deployment model (6).**

This stems from the fact that the 5m patch cords correspond to inter-car jumpers, whilst the 90m horizontal cable is more than enough for a typical 25 m car length.

Figure 4 is a schematic of a worst-case train distribution where the Ethernet switches are located at opposite ends of adjacent coaches, which makes the correspondence quite clear.

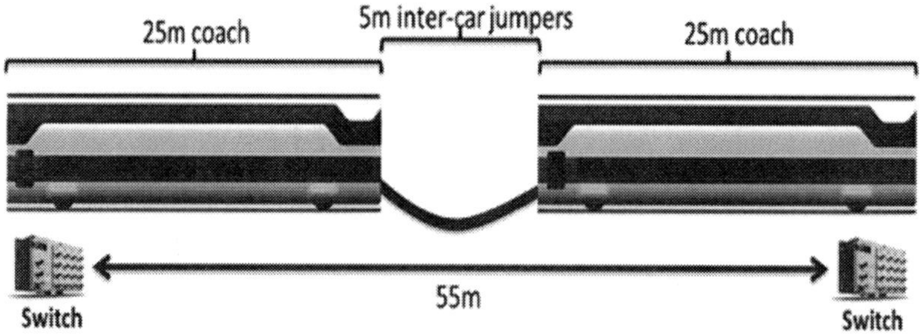


**Figure 4. Ethernet cable deployment on trains.**

The 55m length is also much less than the 90m data centre specification. As the cable is now terminated in Ethernet switches, it is possible to carry out data throughput and integrity checks.

One of the subtleties involved in obtaining the highest data throughputs, is to change the standard payload of up to 1500 bytes for one of up to 9014 bytes; the so-called "Jumbo" frame in figure 5 below.

| Preamble | Des. Add | Sour. Add | Type | Data | FCS |
|---|---|---|---|---|---|
| **8 Bytes** | **6 Bytes** | **6 Bytes** | **2 Bytes** | **46 - 1500 Bytes** | **4 Bytes** |

- Preamble: For synchronization
- Dest. Add: Destination address
- Sour. Add: Source address
- Data: Actual Payload (video, email, picture...etc)
- FCS: Frame Check Sequence

**Jumbo frames:**

- Ethernet frames with 4088bytes or 9014bytes of data

**Figure 5. Standard and "Jumbo" Ethernet frames.**

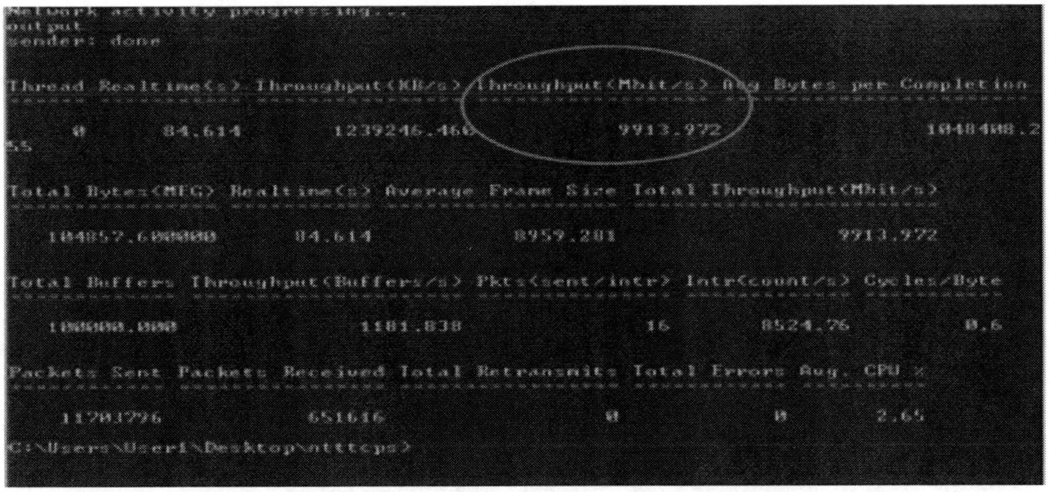

**Figure 6. Screen shot of live throughput testing.**

Figure 6 above shows real-life test results after 100s of GBytes of video data had been transferred. For example, 470 GBytes can be transferred in just over 6 minutes; underlining the degree of future-proofing offered. The throughput as circled is 9.91 Gbit/s; not exactly 10 Gbit/s as there are various overheads to be transmitted. It is also noteworthy that no retransmissions or errors were detected during the EM tests mentioned previously.

It is now possible to carry out certified throughput and packet loss testing both in electrically "clean" and "noisy" environments. This latter situation was provided by EN50121 series and RIA 12 certified test-houses. Damaged packets are simply re-transmitted so loss of a packet is not a frequent occurrence. Nevertheless, re-transmission of existing packets will leave less capacity for new ones leading to ever-increasing capacity loss. A guideline figure for acceptable loss is ~1% capacity reduction. Beyond this point, capacity reduces in an exponential fashion as progressively more time is spent sending repeat packets. Furthermore, as 10GBASE-T Ethernet has very powerful digital data error correction, it is difficult to cause complete cessation of traffic; almost to the point of cable damage or connector disengagement.

Once the data has passed through a switch, it is effectively regenerated back to clean electrical signals again. This offers the prospect of numerous added value services such, passenger information, passenger entertainment, high resolution CCTV, 4k forward facing/ pantograph camera, train-to-shore connectivity or, in fact, any IP-based peripherals. Although 10 Gbit/s wireless requires further investigation, the backward compatibility of Ethernet ensures that 1 Gbit/s or 100 Mbit/s throughputs can be auto-negotiated. This data is then passed to a distributed management system using the 10GBASE-T connectivity described. One of the most bandwidth greedy services is video entertainment. Fortunately, the original data is highly compressed such that a data rate of just 5 Mbit/s will suffice for a high-definition video. Consequently, ~2000 video channels can be delivered. This is more than enough for every seat on a 590 passenger, 11 car consist to have an individual video channel. Furthermore, this data can be conveniently stored on-train.

## 4 OPTIONS FOR THE FUTURE AND CONCLUSIONS

The backbone and connector techniques described here are not restricted to 10GBASE-T. Already, data centres are calling for 25 and 40GBASE-T copper-media cabling and this new technology may be carried over to future trains superseding the case of 10GBASE-T as presented here. Cable length may be restricted to 30m but this is still enough for a 25m car (7). Train-to-shore capacity is also likely to develop in the future and will be greatly assisted in its distribution by the high-capacity backbone described here.

In conclusion, we have shown how data centre technology can be applied successfully in the railway environment. EM interference issues have been overcome leaving implementations to be limited by the imagination on the one hand and financial constraints on the other.

## 5 ACKNOWLEDGEMENTS

The authors gratefully acknowledge the support of Knowledge Transfer Partnership (KTP) award 008968. The aims of each KTP programme are to facilitate the transfer of knowledge and technology and the spread of technical and business skills to the company, stimulate and enhance business-relevant research and training undertaken by the knowledge base, and enhance the business and specialist skills of a recently qualified graduate. The KTP associate from the University of Essex will be presenting at the conference.

## 6 REFERENCES

1. Wi-Fi may tempt train travellers [Online]. Available: http://news.bbc.co.uk/1/hi/technology/3729583.stm
2. LPA Niphan Systems launches 10 Gbit/sec connector [Online]. Available: http://www.lpa-group.com/News-65-133-NewEventDetail.aspx
3. British Rail Class 390 [Online]. Available: http://en.wikipedia.org/wiki/British_Rail_Class_390
4. Railway Standards EN 50121-1-X (2006) Series [Online]. Available: http://www.yorkemc.co.uk/consultancy/railway-consultancy/en-50121-railway-emc-standards/
5. RIA Technical Specifications [Online]. Available: http://www.riagb.org.uk/ria-technical-specifications/
6. ISO/IEC/IEEE 8802-3:2014 Standard for Ethernet [Online]. Available: http://www.iso.org/iso/home/store/catalogue_tc/catalogue_detail.htm?csnumber=64882
7. 10GBASE-T Today or 40GBASE-T Tomorrow [Online]. Available: http://www.belden.com/blog/datacenters/10GBASE-T-Today-or-40GBASE-T-Tomorrow.cfm

# Revising US passenger railcar occupant volume integrity requirements

**J Gordon**
Federal Railroad Administration, Office of Research and Development,
U.S. Department of Transportation, USA

## ABSTRACT

The Federal Railroad Administration (FRA) is developing new regulations addressing the occupant volume integrity (compressive end strength) of passenger rail cars. The new rules are being adopted to accommodate the introduction of rail equipment designed to alternate standards that will provide a level of safety equivalent to that of conventionally-designed vehicles. The fundamental change in the regulations involves applying the proof load on the collision load path rather than on the line of draft, as has been longstanding U.S. practice. Alternatively-designed passenger equipment must be shown to comply with one of the following loading scenarios:

- 3,560 kN (800,000 lbf) with no permanent deformation
- 4,450 kN (1,000,000 lbf) with limited plastic deformation
- 5,340 kN (1,200,000 lbf) without exceeding the crippling strength of the car.

Full-scale tests have been performed to determine whether these scenarios adequately represent the compressive end strength of conventionally-designed passenger equipment. This paper includes a description of and selected results from the full-scale crippling load test program and illustrates that the proposed load levels and performance requirements are reasonably reflective of the strength of conventional equipment. Alternatively-designed equipment compliant with the new requirements will achieve the safety-equivalence goal.

## 1 BACKGROUND

The FRA Office of Research and Development conducts research to inform rulemaking to improve crashworthiness and occupant protection for passenger railroad equipment operated in the U.S. Acknowledging that equipment designed to more modern, performance-based standards is highly desired by U.S. passenger railroads, FRA is focusing on implementing crashworthiness regulations that are compatible with emerging safety technologies. In a passenger train collision or derailment, the principal crashworthiness risks include the loss of survival space inside the passenger compartment due to crushing of the carbody structure and, as the train decelerates, the risk of secondary impacts of passengers with interior surfaces. Resistance to loss of survival space is referred to as occupant volume integrity (OVI). Since May 2003, FRA, with the assistance of the Volpe Center, has conducted substantial research on rail equipment crashworthiness to establish the technical basis for more performance-based regulations to respond to the needs of the industry. Ensuring that equipment designed to alternate standards possesses

OVI equivalent to that of conventionally-designed passenger equipment is the cornerstone of the revised regulatory philosophy.

Current FRA regulations require that Tier I passenger equipment (with maximum operating speed of 200 km/h [125 mph]) sustain a static compressive end-load of 3,560 kN (800,000 lbf) applied longitudinally on the line of draft without permanent deformation (1). This traditional approach to demonstrating OVI is relatively simple to accomplish. Since it is intended to be non-destructive to the test article, the required load is applied and released, and measurements and visual observations confirm or refute the absence of permanent deformation. Passenger rail equipment built to alternative design standards generally incorporates crash energy management (CEM) principles and employs crush zones and deformable structures outboard of the occupant volume which collapse during impact and absorb collision energy while preserving passenger survival space. In order to provide maximum effectiveness, CEM-equipped vehicles must possess a strong occupant volume to ensure that vehicle crushing is restricted to areas outboard of the occupant compartment. As such, this equipment will generally not pass the traditional line of draft buff strength test since permanent deformation cannot be avoided.

FRA convened the Engineering Task Force (ETF) of the Passenger Safety Working Group of the Railroad Safety Advisory Committee (a government-labor-industry body formed to develop consensus rulemaking) in 2009 to conceive new standards for assessing crashworthiness generally, and OVI in particular, of alternatively-designed equipment not built according to traditional U.S. practice (2). The primary goal of the new standards is to ensure OVI equivalence of new equipment with that of existing rolling stock. This new standard differs from the conventional approach primarily in that loads are introduced into the occupant volume through the collision load path, whereas the conventional requirement places loads along the line of draft. For passenger vehicles equipped with crash energy management (CEM) components, collision loads are applied at the interface between the occupant volume and the CEM components. Originally conceived as guidance for railroads seeking a waiver of the conventional FRA regulations in order to deploy alternatively designed passenger equipment, the ETF criteria and procedures describe the analysis and testing which must be performed and submitted to FRA to demonstrate compliance with the new requirements. The ETF criteria and procedures are now in the process of being codified into regulation.

The ETF OVI requirements for alternatively-designed passenger equipment require compliance with one of the following loading scenarios:

- 3,560 kN (800,000 lbf) with no permanent deformation
- 4,450 kN (1,000,000 lbf) with limited plastic deformation
- 5,340 kN (1,200,000 lbf) without exceeding the crippling strength of the car.

In this context, the crippling strength is the maximum load (applied on the collision load path) that can be sustained by the occupant volume. The ETF crippling strength requirement is the extreme means by which OVI equivalence between conventionally- and alternatively-designed equipment is determined, and is founded on previously-performed testing and analysis of typical US passenger equipment (3).

## 2 MOTIVATION and APPROACH

FRA conducted a test program culminating in a destructive test of a conventionally-designed passenger car in order to:

- provide additional validation of the ultimate (crippling) strength of the occupant volume of conventional equipment,
- provide the opportunity for two independent teams to perform the analysis that would be expected to be provided to FRA to demonstrate compliance with the ETF criteria and procedures related to OVI, and
- produce documentation which could serve as a model for railroads seeking approval from FRA to acquire and place into service equipment not compliant with the existing regulations.

Confirmation of the crippling strength of conventionally-designed equipment is important since the ETF OVI requirements are based in part on this metric. Previous testing was performed on a pair of cars of identical construction. In this test, a different design was used to provide additional validation of the ETF criterion.

A manufacturer must supply information to its customer demonstrating that the equipment it is providing is compliant with applicable regulations; this information is subsequently provided to FRA for the same purpose. Regulatory compliance demonstration is increasingly reliant on numerical analysis and simulation, which is validated by elastic structural tests. The validated model is used to show compliance with requirements for which tests were not performed. Although carbody destructive tests, such as the one described here, are rarely performed by manufacturers when developing data to demonstrate compliance with structural requirements, FRA sought to use this opportunity to develop information, which can be publicly shared, illustrating how the ETF criteria and procedures could be applied in practice. Two modelling teams were organized to perform analyses, informed by the calibration test data, of the crippling test and produce mock documentation representative of that which would be presented to FRA to demonstrate compliance with the regulations.

The testing was conducted in two parts:

- an elastic test in which 3,560 kN (800,000 lbf) was applied to the carbody to ensure that it was sufficiently structurally sound for the crippling test and to develop calibration data for the modelling teams, and
- a crippling (destructive) test which involved loading the occupant volume to the point at which it could no longer sustain load to establish its OVI.

The general approach is shown in Figure 1 and generally follows the ETF recommendations. The final crippling load test has been added to this program in order to provide the additional information on OVI described above.

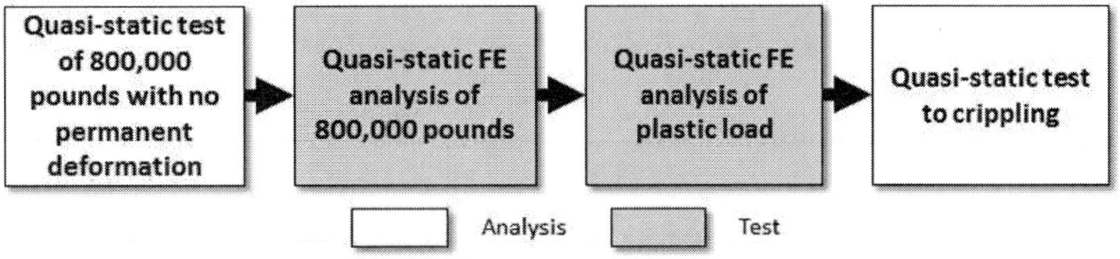


**Figure 1. ETF approach to OVI assurance with additional crippling test component added as part of this research program.**

## 3 TEST PROGRAM

### 3.1 Overview

Transportation Technology Center, Inc. (TTCI) performed two full-scale tests on a Budd M1 Car 9614 at the Transportation Technology Center in Pueblo, CO. This particular car had been previously modified to include CEM elements on both ends. In order to evaluate OVI, the CEM elements were removed to allow loads to be placed where they would be introduced into the occupant volume in the event of a collision.

In the first test, conducted on March 13, 2013, a 3,560 kN (800,000 lbf) load was applied through the two floor-level CEM pockets. This test was performed to assess the integrity of the test article (since it had been damaged during a previous train-to-train collision test) and to develop data which would be provided to the analysis teams to calibrate their respective models in advance of the crippling load test. The crippling load test was performed on July 17, 2013 during which loads were introduced through the floor and roof level CEM pockets. The test setup is shown in Figure 2 and is identical to that used in prior similar tests (3).

**Figure 2. Test car in loading fixture. Arrows indicate approximate location at which loads are applied for crippling load test.**

Before testing, all significant damage observed on the car was repaired. A patch was applied to one location on the right side sill, and to maintain structural symmetry, an identical patch was applied in the same location on the opposite side of the car. Material characterization tests were performed on selected structural members using specimens excised from a sister car. These data were provided to the analysis teams for incorporation into their respective finite element models. The squareness of the car in the test fixture was surveyed prior to commencing testing. The survey indicated that the car was not precisely aligned within the fixture, resulting in slightly non-perpendicular alignment between loading rams and the end structure of the car. This information was provided to the modelers to allow for appropriate corrections.

### 3.2 Instrumentation

Arrays of strain gauges and displacement transducers (string potentiometers) were applied to the test car. The instrumentation arrangement was the same for both tests. Strain gauges (81) were installed at nine longitudinal stations along the length of the car. A typical cross-sectional strain gauge distribution is shown in Figure 3. At certain locations, individual strain gauges were shifted slightly from their desired positions to avoid structural details. The spatial position of each gauge was recorded and provided to the analysts to permit extraction of model results at the relevant locations.

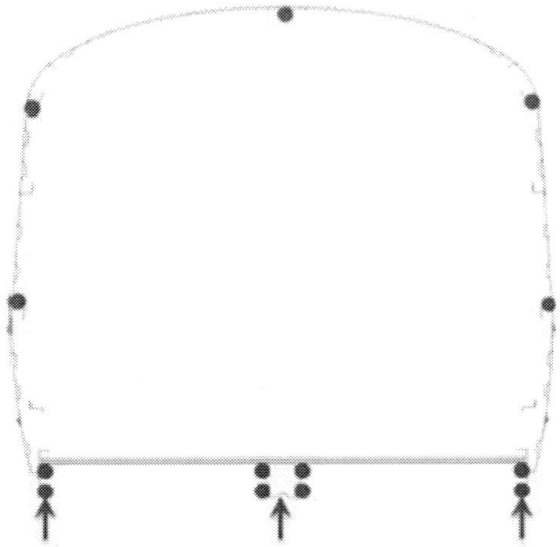

**Figure 3. Cross-sectional distribution of strain gauge instrumentation at typical longitudinal station.**

String potentiometers (63) were arranged in triads to permit measurement of the three-dimensional displacement and were installed along the center and side sills at each cross section, with the exceptions of cross-sections 2 and 8 since these sections coincided with the bolster locations. The displacement transducer arrangement is shown in Figure 4. The spatial position of each triad was recorded and provided to the analysts to permit extraction of model results at the relevant locations.

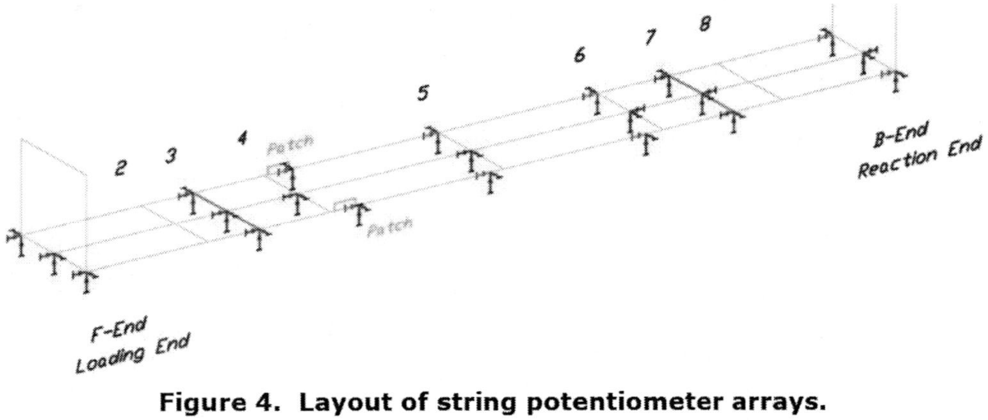


**Figure 4. Layout of string potentiometer arrays.**

The four load actuators were equipped with pressure and displacement transducers to monitor the stroke and applied load at each corner on the F-end of the car to which loads were applied, as shown in Figure 2.

### 3.3 Proof test

The 3,560 kN (800,000 lbf) test was performed first in order to provide calibration data for the modelers. The load was applied to the floor-level CEM pockets only in gradually-increasing increments and reduced to 90 kN (20,000 lbf) after each load increment. This strategy allows for verification that the measurement equipment is functioning properly. Since the instrumentation is monitored in real-time, cycling the load allows comparisons to be made at the same load level multiple times throughout the test. Substantially different measurement observations at the same load would indicate problems with the test setup or instrumentation which might warrant suspending the test. Prior to application of the final (maximum) load, the load was reduced to 9 kN (2,000 lbf) as required by the American Public Transportation Association (APTA) Standard PR-CS-S-034-99, Rev 2 (4). Reduction of the load to nearly zero allows for confirmation that permanent deformation has not occurred before application of the maximum load. The loading rate was approximately 1.8 cm/min (0.7 in/min). Load, strain and displacement sensor readings were collected. Review of the data following the test suggested locations at which buckling could be expected during the crippling test.

Following the proof test, slight buckling of the left and right side sills was observed at station 6 as shown in Figure 5. However, upon removal of the load a mere 0.25 cm (0.1 in) permanent change in length of the car was measured, providing evidence (despite the pre-existing damage) that the car was suitable for use in the crippling test.

**Figure 5. Slight buckle on lower flange of left side sill following proof test. Similar damage observed on right side sill at same location.**

### 3.4 Crippling test

The loading procedure for the crippling test generally followed the same protocol employed for the proof test with one exception. Following application of the 2,670 kN (600,000 lbf) load increment, the load was reduced to 9 kN (2,000 lbf). The remainder of the test was performed under stroke control. The actuators were advanced in 0.64 cm (0.25 in) increments, with a dwell after each increment, until crippling of the carbody occurred.

Buckling first occurred in the roof at a total load of approximately 3,115 kN (700,000 lbf) followed by a corresponding buckle at 4,272 kN (1,000,000 lbf) at approximately the same position on the roof at the opposite end of the car as shown in Figure 6 and Figure 7. Roof buckling did not occur during the 3,560 kN (800,000 lbf) proof test because in that test the loads were applied to the floor-level CEM pockets only which resulted in upward bending of the carbody, putting the roof structure in tension.

At a total load of approximately 4,700 kN (1,100,000 lbf), complete crippling of the center sill and side sills occurred. Previous results obtained from similarly-performed tests of a Budd Pioneer cab car indicated crippling strength of this conventionally-designed passenger car to be between 5,120 - 5,296 kN (1,150,000 - 1,190,000 lbf) (2). Recall that M1 car 9614 had sustained damage during a previous test program. The major damage was repaired prior to commencing testing; however some local areas of damage remained, which initiated critical buckling of the underframe and lowered the apparent crippling strength of the carbody.

The test results provide additional validation of the extreme carbody strength requirement proposed by the ETF. Additionally, based upon the results of the three crippling tests performed to date, the crippling strength of conventional equipment operated in the U.S. can be conservatively estimated to be 5,340 kN (1,200,000 lbf).

**Figure 6. First roof buckle at 3,115 kN (~700,000 lbf).**

**Figure 7. Second roof buckle at 4,272 kN (~1,000,000 lbf).**

## 4 ANALYSIS

Since demonstration of the OVI of modern passenger rail equipment designed to alternative standards (especially those incorporating CEM technology) cannot be practically achieved using the conventional proof test approach, the ETF criteria and procedures acknowledge that extensive computer modelling supported by limited physical testing is necessary to accomplish this. The modelling activity associated with this test series was conceived to provide publishable results that could be used as guidance for manufacturers in developing information to present to FRA to show compliance with FRA structural requirements.

Generally, such testing and analysis would be performed by the equipment manufacturer and submitted to FRA for review and acceptance, with that information treated confidentially. Destructive tests, such as the one described here, are typically not performed by manufacturers. Instead, elastic tests are performed on instrumented equipment and the results are validated with a numerical (i.e., finite element) model. Once validated against the test data to within a reasonable tolerance (ETF recommendations are ±10% for displacements and ±20% for strains), the model can be used to demonstrate vehicle response to loading conditions not tested. Demonstration of compliance with regulations is achieved through submission to FRA of the test plan, test data, a demonstration of model validation, and model results that illustrate compliance with the relevant regulations.

Two modeling teams were assembled to develop prototype analyses and documentation of the crippling test according to the ETF procedures. The teams operated independently, and were provided the same information related to the car design, and the proof and material test results. The teams included Arup (in collaboration with TTCI) and the Volpe Center. This parallel modeling effort was devised to demonstrate that two different modelers, utilizing different approaches (e.g. different modeling techniques, different software packages, etc.) could each produce results comparable to the test measurements.

The complete details of each team's approach, methodology and results will be published in the final report describing the entire activity. Selected results are presented here. The finite element (FE) models prepared by each team are shown in Figure 8.

| Solver: LS-DYNA<br>Number of elements: 700,000<br>(a) | Solver: ABAQUS/Explicit<br>Number of elements: 760,000<br>(b) |
|---|---|

**Figure 8. Finite element models developed by Arup (a) and Volpe (b).**

The first test performed on car 9614 was a 3,560 kN (800,000 lbf) load applied to the floor-level energy absorbers. The data from this test were used by both Arup and Volpe to calibrate their respective FE models. Once calibrated, these models could be used to predict the expected response of the car during the crippling load test. Figure 9 shows the results for vertical deflection under the 3,560 kN (800,000 lbf) axial proof load as provided by each team. It should be noted that each team has chosen to present the results differently. For example, Arup has adjusted the test data in its comparison to remove rigid body motion of the carbody observed during the test. In Figure 9(b) the Volpe results are presented in raw form. For each plot, the test data have also been plotted alongside the FE results.

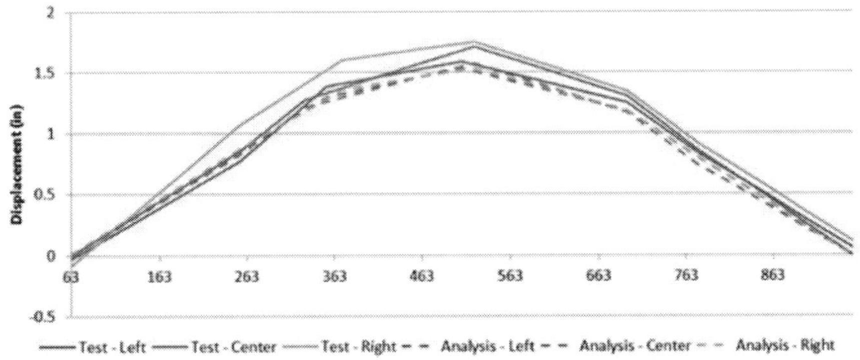


(a) Arup

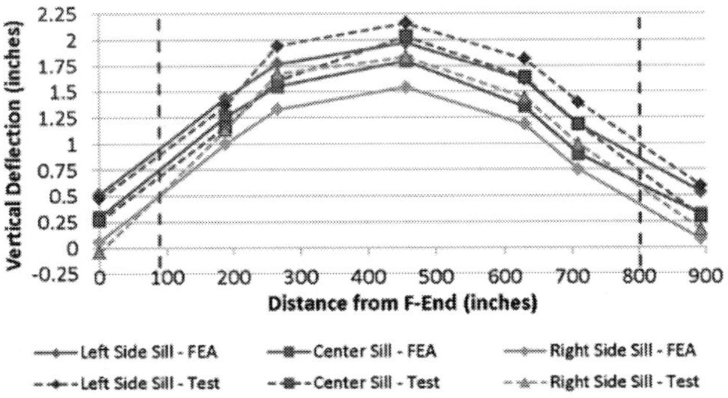


(b) Volpe

**Figure 9. Comparison of Arup-Volpe vertical displacement results from 3,560 kN (800,000 lbf) center, left and right side sills (1 in = 2.54 cm).**

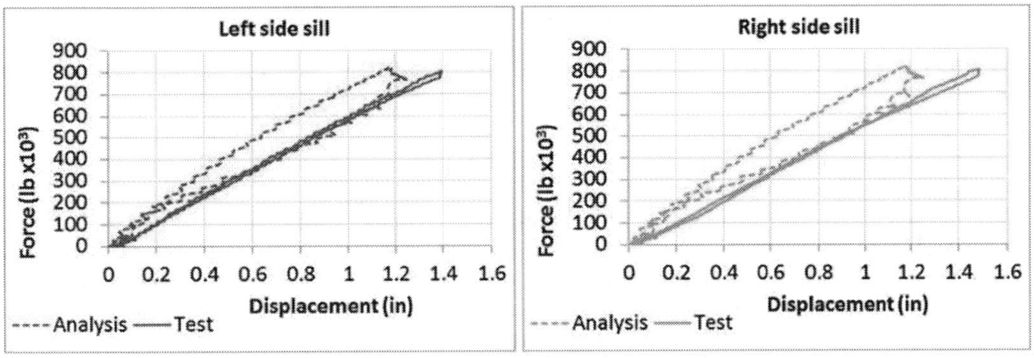


(a) Arup

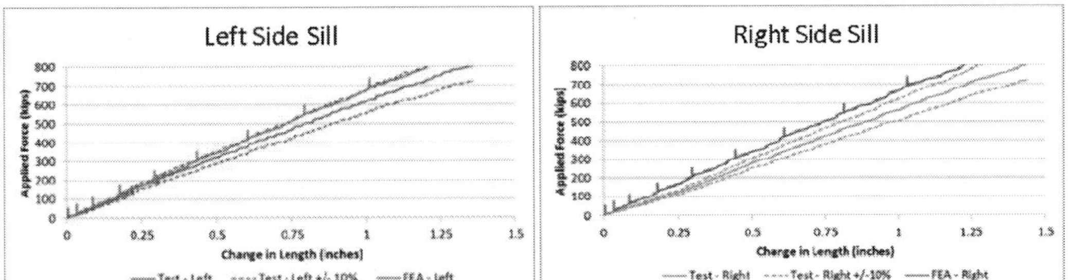


(b) Volpe

**Figure 10. Comparison of Arup-Volpe force-displacement results in 800,000 lbf test for left and right side sills (1 kip = 4.448 kN; 1 in = 2.54 cm).**

Figure 10 contains force-versus-displacement plots for the left and right side sills from both sets of finite element results as well as the 3,560 kN (800,000 lbf) validation test. Arup's results are on the top of this Figure, and Volpe's results are on the bottom. For both set of results, the applied force is plotted against the change-in-length of the car obtained from both the left side sill and the right side sill.

Once the FE models were calibrated using data from the first test, the models were used to simulate the crippling response of the car.

Table 1 displays several key results from both numerical models and the crippling test. This Table includes the roof buckling load, which is the maximum load introduced into the car through the two roof-level loading points. The crippling load is the maximum load the occupant volume sustained during the test, and is equal to the sum of the loads applied at the four loading locations. These data suggest similar outcomes from the analysts, whereas the test values are lower for the reasons described in Section 3.4.

**Table 1. Crippling strength estimates.**

| Parameter | Arup | Volpe | Test |
|---|---|---|---|
| Roof buckling load, kN, (klbf) | ~1,335 (300) | 1,312 (295) | 890 (200) |
| Crippling load, kN (klbf) | 5,921 (1,331) | 6,072 (1,365) | 4,272 (1,100) |

Figure 11 contains two sets of force-versus-displacement plots from the crippling test and analyses. The top portion of this figure contains Arup's force-versus-displacement results and the bottom portion contains Volpe's results. Three curves have been plotted for each set of data. The total floor load is the sum of the loads applied by the two floor-level actuators, the total roof load is the sum of the loads applied by the two roof-level actuators, and the total load is the sum of all four actuators.

It is important to note that a crippling test would not likely be performed by a manufacturer as part of its compliance demonstration. Elastic analyses require very simple material characterization and the companion validation tests are simple (and inexpensive) to perform. The material model in the crippling test requires much more sophistication to properly establish both the elastic and plastic behaviors of the materials of construction. If the material properties are not appropriately modeled, the simulation may predict crippling loads in excess of what the physical carbody can sustain (a non-conservative estimate). This is especially relevant when carbodies are constructed of alternative materials. The carshells tested in this program were of stainless steel construction.

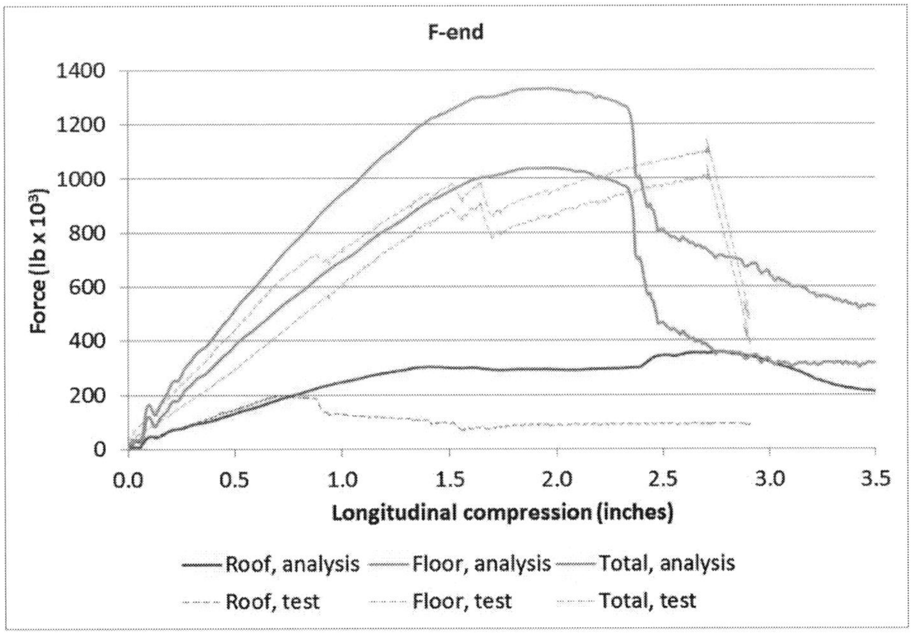


(a) Arup

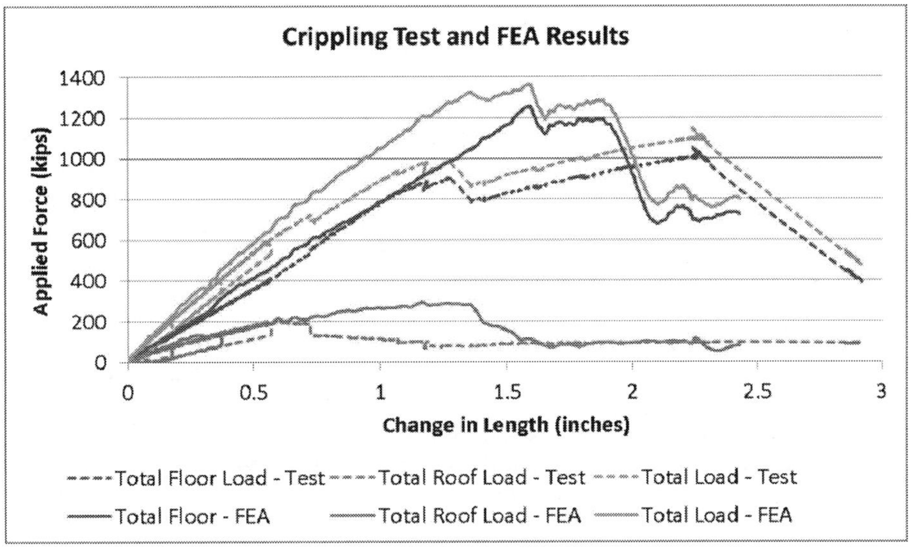


(b) Volpe

**Figure 11. Comparison of Arup-Volpe force-displacement results in crippling test for floor, roof, and total loads (1 kip = 4.448 kN; 1 in = 2.54 cm).**

Figure 12 shows the force versus displacement results for three crippling tests performed by FRA. The crippling load is denoted by the peak force. These data suggest that the ETF criterion for equivalent occupant volume integrity (5,340 kN or 1,200,000 lbf) is a conservative estimate of the structural capacity of conventionally-designed passenger rolling stock. Applying this criterion to an alternative design will, at the occupant volume level, result in equivalent safety.

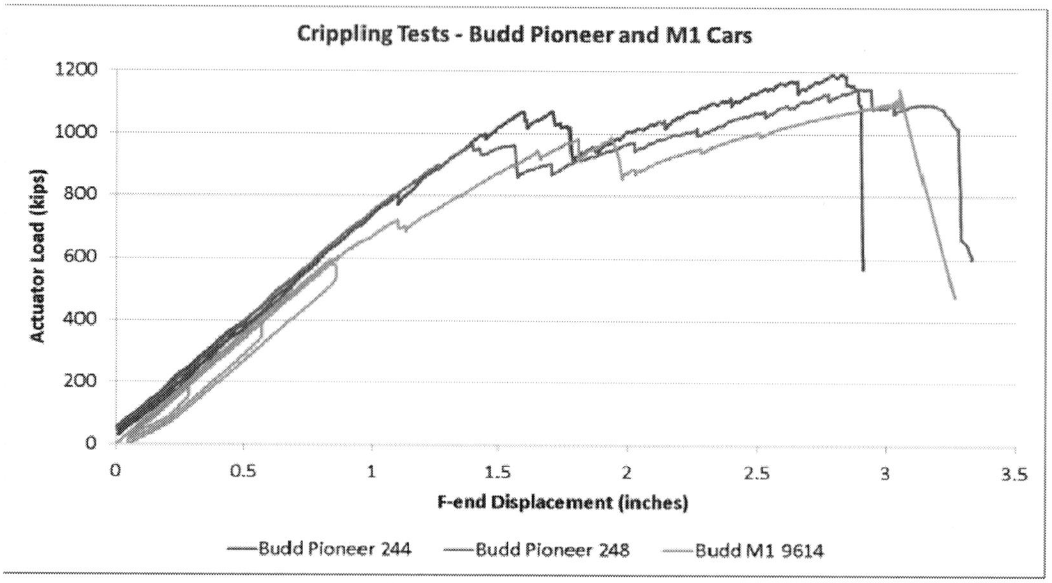


**Figure 12. Force-displacement results for three crippling tests performed by FRA (1 kip = 4.448 kN; 1 in = 2.54 cm).**

## 5 CONCLUSIONS

This occupant volume integrity test program generally accomplished the goals it sought to achieve. The destructive crippling test produced additional information on the strength of the occupant volume of conventional US passenger equipment. Although the occupant volume strength observed in this test was slightly lower than that derived from prior testing, it is believed that the difference is due to pre-existing damage sustained during previous high-energy impact tests which was not repaired prior to the crippling test.

From an analytical perspective, the two teams present high-level results which are in reasonable qualitative agreement. Each was able to produce estimates of the deflected shape of the carbody under the 3,560 kN (800 klb) elastic load which were is good agreement with the test results. Each team, however, overestimated the crippling strength of the car. This was likely due to the fact that the analytical models do not account for preexisting damage, the presence of which resulted in reduced crippling strength.

As the final report will reveal, the details provided by each analyst team are substantially different. Each took different approaches in how the finite element model was constructed and degree of structural detail accounted for (weldments, roof corrugations, etc.) as well as how the material data derived from coupon tests were treated. Thus, particular results differ between the two analysts. As a practical matter, compliance documentation will be prepared by the equipment manufacturer according to internal methodologies and best practices.

It should also be noted that the purpose of this exercise was not to compare the model results directly, since in practice FRA would never be in the position of reviewing compliance documentation submitted by two entities for the same piece of equipment. Rather, the activity was intended to show how the ETF criteria and procedures could be applied in practice in a form that could be made publically available.

FRA intends to use the "mock" submittals developed during this research program to inform planned activities to develop a "suggested practices" document which will serve as industry guidance in the preparation of compliance documentation for FRA review and acceptance.

## REFERENCE LIST


1. Code of Federal Regulations, Title 49, Transportation, Volume 2, Subtitle B (Parts 200-299) Federal Railroad Administration, Department of Transportation.
2. Carolan, Michael et al. "Technical Criteria and Procedures for Evaluating the Crashworthiness and Occupant Protection Performance of Alternatively Designed Passenger Rail Equipment for Use in Tier I Service," DOT/FRA/ORD-11/22, October 2011. http://www.fra.dot.gov/eLib/details/L01292
3. Carolan, M., Perlman, B., and Tyrell, D., "Alternative Occupied Volume Integrity (OVI) Tests and Analyses," U.S. Department of Transportation, DOT/FRA/ORD-13/46, October 2013. http://www.fra.dot.gov/eLib/details/L04847
4. APTA PR-CS-S-034-99, Rev 2, Standard for the Design and Construction of Passenger Railroad Rolling Stock, American Public Transportation Association, Washington, DC.

# Evaluating options for a fault tolerant railway system

**C Fecarotti, J Andrews, R Remenyte-Prescott**
Department of Civil Engineering, The University of Nottingham, UK

## ABSTRACT

The greater utilisation of the existing infrastructure exacerbates the degradation processes and reduces the time slots for maintenance, thus increasing system vulnerability to disruptions. Fault tolerance is the key property of complex systems to achieve high levels of service reliability. However, the railway system is neither designed nor managed to be fault tolerant. A Petri net simulation model is presented in this paper aims at evaluating opportunities in system structure, operation and maintenance in order to ensure acceptable levels of service reliability following system failures. The Petri net is combined with a fault tree module for the creation of failure scenarios.

**Keywords**: fault tolerance, Petri nets, simulation.

## 1. INTRODUCTION

The increasing demand on the UK railway network for both passengers and freight traffic is challenging the existing capacity. A potential solution to meet these requirements is to increase the number of operational hours with a stated objective of the Railway Technical Strategy to be a 24/7 operation. However, greater utilisation of the existing infrastructure will accelerate the degradation processes and reduce the time slots available for maintenance, thus making the system more vulnerable to disruptions. In order to assure the required level of service reliability, the railway system must be maintained to reduce system failures as well as designed and operated so as to improve its resilience to disruptions.

Fault tolerance is the key property required of complex transport systems to achieve a high level of service reliability. However, the railway system is currently neither designed nor managed to be fault tolerant. The railway network is currently structured as a series system, and as such even when all of its components achieve high reliability, it is in itself unreliable. Network redundancy has long been implemented by complex engineering systems in order to achieve the required levels of system reliability. Many sections of the railway network already have some potentialities to introduce network redundancy, which have not been exploited yet. For instance long sections of network consist of four parallel tracks for fast and slow trains. The current signalling system usually provide each line with the safety requirements for one-way operation, therefore all regular movements have to be made with the direction of traffic and "wrong movements" have to be authorised by special instructions. The underlying idea of this research is that parallel lines maintained to run fast trains and equipped with a bi-directional signalling system, would become "interchangeable". Then the optimal location of crossovers would provide the flexibility to achieve the required level of performance.

This paper aims at evaluating potential opportunities in system structure, its operation and maintenance in order to ensure that an acceptable service can be attained following failures on the system. A structured methodology implementing simulation, failure analysis and optimization techniques is suggested to analyse options concerning the design, operation and maintenance strategies which may contribute to improve the reliability and availability of the network. A railway modelling tool based on timed coloured Petri nets has been developed to investigate the effectiveness of the selected options judged against the performance indicator of total service delay time. Railway operations are simulated according to a given timetable and the disruptions to the service resulting from technical failures of the system are predicted.

## 2. BACKGROUND

The concept of railway service reliability is usually associated with timetable robustness. Robust timetables are designed in order to handle small disturbances during operation. However, even robust timetables cannot cope with larger disturbances due to technical failures. The effects of timetable characteristics on service reliability is investigated in (1), where the authors also attempt to create more homogeneous timetable by reducing interdependencies between trains. Comparison of the alternative timetables is performed by means of simulation. Simulation tools are widely used to evaluate the effects of primary delays and timetable robustness on trains operation. NEMO and SIMONE are simulation software developed by ProRail to analyse complex and large scale train networks. In (2) the authors describe the use of SIMONE and in (3) they apply the tool to evaluate the effects of infrastructure and operational changes in some critical sections of the Dutch railway network, on the timetable and network stability. OpenTrack® and RailSys® are microscopic models for a more detailed description of the infrastructure and they are usually used to analyse smaller networks. OpenTrack® is applied in (4) to analyse the relation among capacity, block section length, timetable and operational program in high density lines for both normal and perturbed conditions. The authors in (5) also use OpenTrack® for the evaluation of the impact of a new real traffic management system and train control on the system stability and capacity. In (6) the authors use RailSys® to support the improvement of timetable stability for a section of the German railway network. Applications of railway simulation for timetable construction and timetable robustness assessment are also presented in (7) and (8) respectively. However, since this literature is mainly focused on evaluating and improving timetables characteristics, only small disturbances are considered by generating small primary delays at stations.

By contrast, the focus of this research is on the capability of the network to cope with large disturbances that cannot be handled by a robust timetable only. Fault tolerance, implemented by means of redundancy, is a fundamental property that allows complex systems to achieve high levels of reliability. One of the contributions of this research is the application of fault tolerance and redundancy concepts for improving reliability of railway service. Fault tolerance of the railway network is intended here as the capability of the network to provide more diversionary and alternative routes to trains when sections of track are unavailable due to technical failures or maintenance operations.

## 3. A SYSTEM-WIDE ANALYSIS: THE PROPOSD METHODOLOGY

Design, operation and maintenance strategies all contribute to the overall system performance. The decisional process supporting the choice of a particular asset management strategy is very complex and must take into consideration both operational and economic constraints. Therefore, a structured methodology is needed for a systematic evaluation and optimisation of possible solutions. Figure 1 summarizes the main aspects of the proposed methodology which implement simulation, optimization and failure modelling techniques to assess the effectiveness of different solutions.

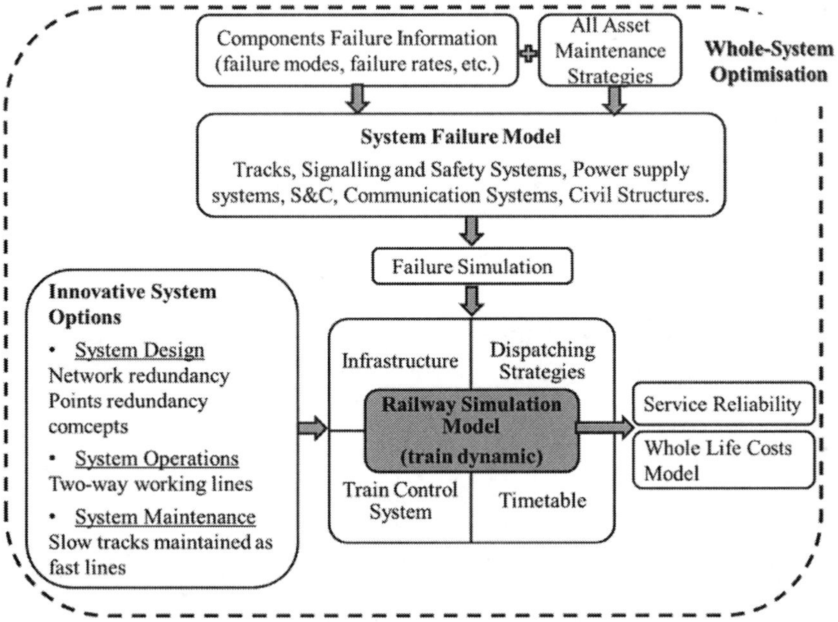


**Figure 1. The methodology.**

Options concerning with the design, operation and maintenance strategies which may contribute to improve the reliability and availability of the network are identified, and their impact on system performance is evaluated by means of simulation. Failure analysis is applied to model disruptions in the network and optimization is performed in order to improve current solutions subject to operational and resource constraints.

If parallel lines were maintained in order to allow the use of fast trains and equipped with a bi-directional signalling system, then the optimal location of crossovers to move trains between parallel tracks would provide the flexibility needed to achieve the required level of performance. Although the options listed above are suggested as possible solutions to provide path redundancy on open lines, they can also contribute to provide diversionary routes at junctions, depending on the junction layout. Switching and crossings (S&C) play a central role in connecting the railway network providing flexibility to railway operation by guiding trains from one track to another and enabling lines to cross paths. However, S&C, particularly at junctions, are considered critical elements since their failures can lead to delays and journey cancellations. Nonetheless, S&Cs are the only elements allowing to provide a multi-routed network. This paper suggests the introduction of additional crossovers on open lines for moving trains between parallel tracks when sections of tracks are unavailable, rather than new junctions. If the switches cannot be properly set to the required position because of a failure, the previous crossover can be used to re-route approaching trains. Besides, S&C

failures are not the only contribution to system unreliability. Signalling, telecommunication, electrifications, track faults and all the other assets contribute to system unreliability and crossovers give the possibility to provide diversionary routes when sections of tracks are unavailable. Furthermore, the optimisation process will exclude solutions which are not cost effective and will select the number and location of crossovers where it is required.

# 4. THE RAILWAY SIMULATION MODEL

## 4.1 Introduction to Coloured Petri nets

A Petri net is a directed, weighted bi-partite graph consisting of two separated sets of nodes called places and transitions connected by direct arcs. Places represent physical resources, conditions or the state of the system. Places contain tokens, whose number and distribution over the network indicates the state of the system at a given time and is defined as "marking" of the Petri net. Transitions represent events that make the state of the system change. Transitions govern the evolution of the marking over time, by removing and adding tokens from and to the input and output places respectively, thus determining the dynamic of the system. At the same time the marking of the net along with the multiplicity of the arcs determine the enabling conditions for each transition. Inhibitor arcs stop transitions from firing under given circumstances. In coloured Petri net (9), information are associated with tokens and enabling conditions may include functions associated with arcs involving tokens' colours. Figure 2 shows a small Petri net before and after the transition fires.

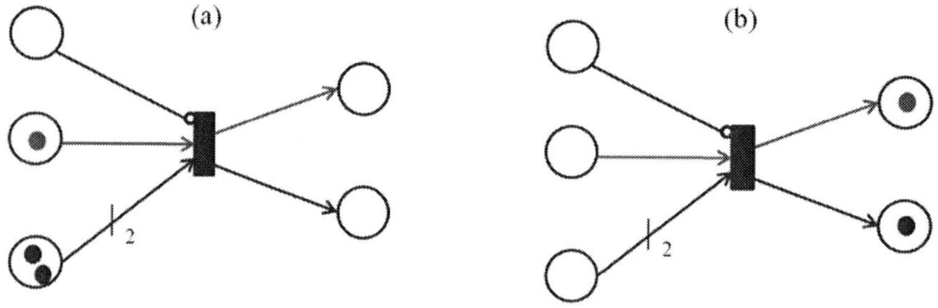


**Figure 2. Coloured Petri net before (a) and after (b) firing.**

Places and transitions are represented by circles and rectangle respectively; arrows represent input and output arcs while inhibitor arcs have circular head rather than arrow head. A timestamp can be associated with tokens, indicating when the token can be removed from its current position. Similarly, a firing time can be attributed to transition. The firing time can be deterministic or be sampled from distributions if the transition represents a stochastic event.

The Petri net method provides a mathematical and graphical representation of complex systems and is particularly suited for describing discrete event systems. Conditions and causal relationships between events that characterise the system behaviour can be modelled. The mathematical framework provided by Petri nets allows for a rigorous description of railway signalling, controlling and dispatching rules that govern the safe movement of trains through the network. As will be shown later in this section, the elements of the railway systems have an equivalent in the Petri net model. Coloured tokens represent trains, places represent block sections while transitions represent the event of a train entering or exiting a block section or a switch. Inhibitor edges represent semaphores forbidding or allowing train movements according to state of the section of track ahead. Furthermore, the marking of the net provides information on the state of the overall network in terms of trains location over the network at any time.

## 4.2 The railway network model

In this paper a network performance evaluation tool is used to assess the impact of different infrastructure configuration on railway operation. In the model, developed in a C++ language programming, events are processed in a chronological order according to a synchronous processing technique. The Petri Net method is used to model the railway system and its dynamics, including the safety and control rules for train movements. The network consists of three main modules: block sections, stations and switches. The modules are connected according to the infrastructure layout. Tokens represent trains and contain information including train length, speed, acceleration and braking rates, and the scheduled route. The timestamps of tokens indicate the running time calculated upon the train characteristics, the path length and allowed speed. The marking of the network represents the distributions of trains over the network at any time. Disruptions are simulated during operation implying one or more block sections to be unavailable or speed restrictions to be imposed for a given time interval. In such circumstances assigned paths may not be available and conflicts may arise between trains leading to delays and journey cancellations. Conflicting trains are dispatched on a first-come first-served basis, but priority can be considered as well for determining the order of trains. When some of the originally assigned routes are no longer feasible, alternative paths are selected according to the distributions of crossovers over the network. The location of crossovers clearly affects the selection of alternative paths thus contributing to the overall system performance. The total delay time D, given by equation 1 below, is used to measure system performance and therefore compare different infrastructure configurations.

$$D = \sum_{\substack{i=1 \\ s=1}}^{I,S} \left(t_{is} - t_{is}^{d}\right), \tag{1}$$

where $t_{is}$ is the scheduled arrival time of train i at station s, and $t_{is}^{d}$ is the actual arrival time of the considered train at the same station. I and S are the number of trains and the number of stations respectively.

### *4.2.1 The block section module*

The model implements a fixed block system. Tracks are divided into block sections, defined as sections of track between two consecutive signals. Each block section is equipped with track clear detection devices that detect whether the block section is occupied or clear, and communicate the state of the section to signals. A single block section can be occupied by only one train at a time. A slow train requires only the approaching block section to be clear in order to advance while fast trains require the availability of two block sections ahead. The Petri net for the block section is represented in Figure 3.

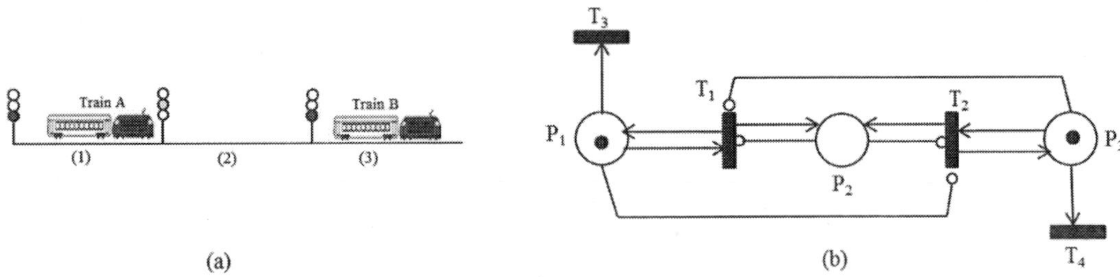


**Figure 3. Petri net track model.**

Places P1, P2 and P3 represent three adjacent block sections. The transitions along with the input, output and inhibitor arcs model the rules for train movement. The inhibitor arcs in particular forbid the firing of transitions when the corresponding place is already marked, acting as semaphores stopping a train when the requested resource is not available. Different colours are associated with tokens in order to distinguish slow and fast trains, and the direction of travel as detailed in table 1. Two colours are used to specify the head and the tail of the train in order to simulate train movements through the network with the exact track occupation time.

**Table 1. Tokens colours and corresponding meanings.**

| Token-Colour id | Meaning |
|---|---|
| *a* | Direction of travel: up (left to right in the pictures) |
| *b* | Direction of travel: down (right to left in the pictures) |
| *c* | Slow train |
| *d* | Fast train |
| *e* | Head of train |
| *f* | Tail of train |

According to the colour of the tokens contained in the input places, transitions T1 and T2 represent the movement of trains going up-ward (left to right) and down-word (right to left). Firing of transition T3 and T4 represent the exit of a train from block section 1 and 3 respectively. The enabling conditions of each transition, as well as the new marking after firing, depend on the colours of the initial marking. Figure 4 and 5 show how the system state changes after firing of transitions enabled by different initial markings.

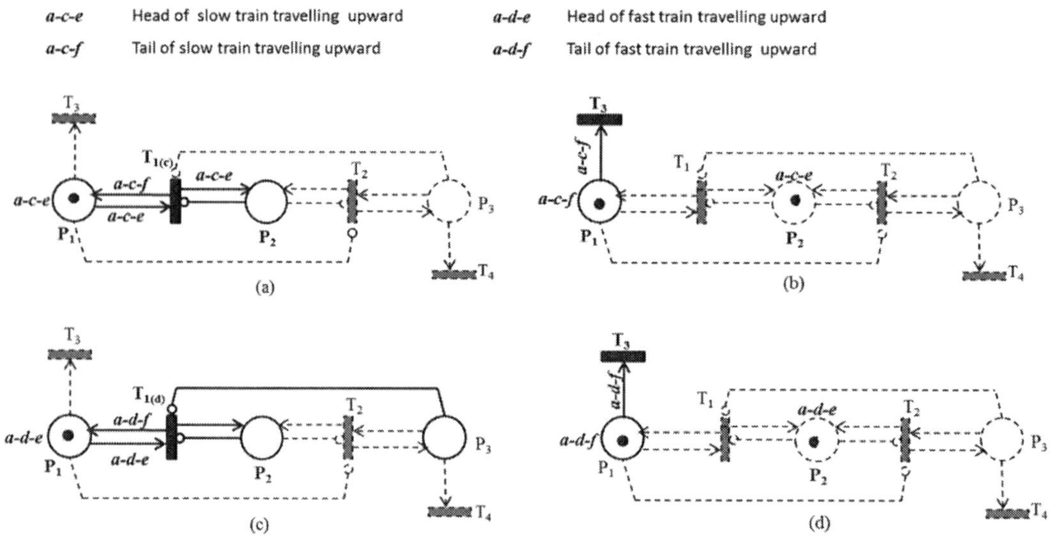


**Figure 4. Occurrence colours enabled by a train-token travelling upward, of slow type (4.1 and 4.2) and fast type (4.3 and 4.4).**

Given the initial marking, the dotted lines represent the elements of the Petri net which do not contribute to the enabling conditions. Therefore, a token with colours a-c-e in place P1 (Figure 4.1) representing the head of a slow train travelling up-ward, will enable the firing of transition T1 when place P2 is empty. Firing of transition T1 given this initial marking will involve the following steps: (1)removing the token from place P1, (2) add the same token to place P2 and (3) add a token with colours a-c-f will be add to place P1, representing the tail of the same slow train travelling upward in block section 1. Figure 6 shows the Petri net obtained from the connection of consecutive block sections.

*b-c-e* Head of slow train travelling downward

*b-c-f* Tail of slow train travelling downward

*b-d-e* Head of fast train travelling downward

*b-d-f* Tail of fast train travelling downward

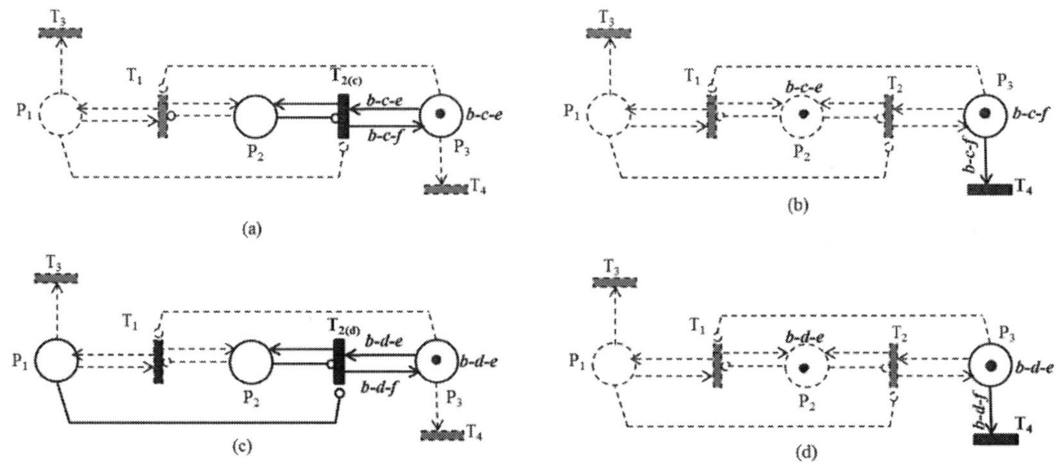


**Figure 5. Occurrence colours enabled by a train-token travelling downward, of slow type (5.1 and 5.2) and fast type (5.3 and 5.4).**

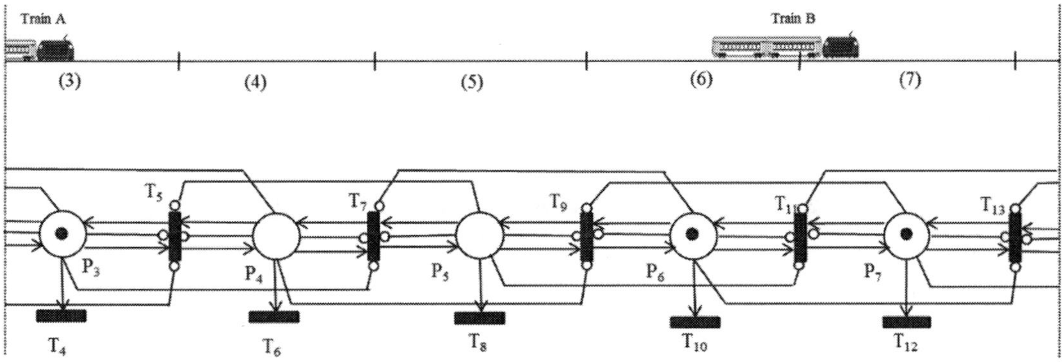


**Figure 6. Petri net for a section of track accounting for five consecutive block sections.**

### *4.2.2 The station module*

Stations are characterised by the number of platforms that determine the maximum number of trains that can be host in the station. The Petri net model for a station with four platforms is depicted in Figure 7.

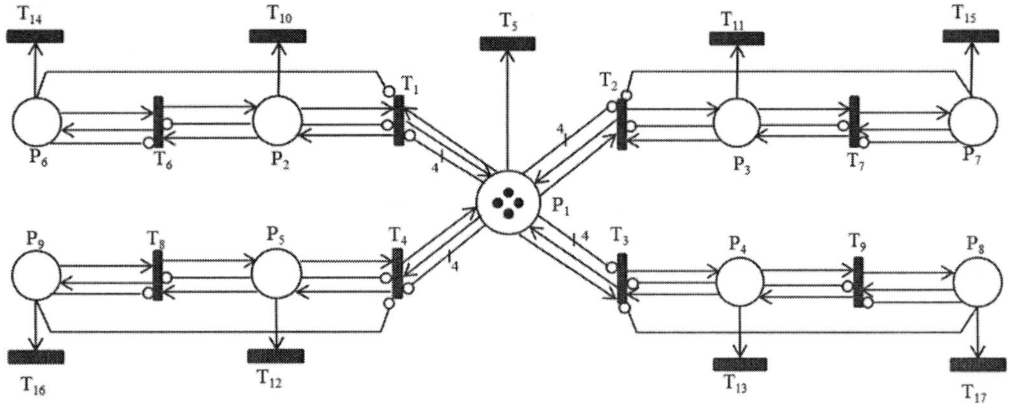


**Figure 7. Petri net model for a railway station with four platforms.**

Place P1 representing the station, can contain up to four tokens. This capacity constraint is ensured by the inhibitor arcs with multiplicity four, that allow firing of transitions T1, T2, T3 and T4 to add a token in place P1(a train entering the station) only if P1 contains less than 4 tokens. The station module is connected to the adjacent block section modules either directly as shown in Figure 7, or by means of a switch module, depending on the station layout. Tokens contained in place P1 enable one of transitions T1 to T4 depending on their colour. Then timestamp of tokens will determine the firing time. The colours defined for this Petri net are the same detailed in Table 1.

### 4.2.3 *The switch module*

Railroad switches provide connections between tracks. They can be found at junctions where two lines converge or diverge, as well as on open lines, to form a crossover connecting parallel tracks. Point blades can be positioned into two end positions, "normal" and "reverse" position, the former usually corresponding to the straight direction. The Petri net model for the diverging junction in Figure 8, is depicted in Figure 9.

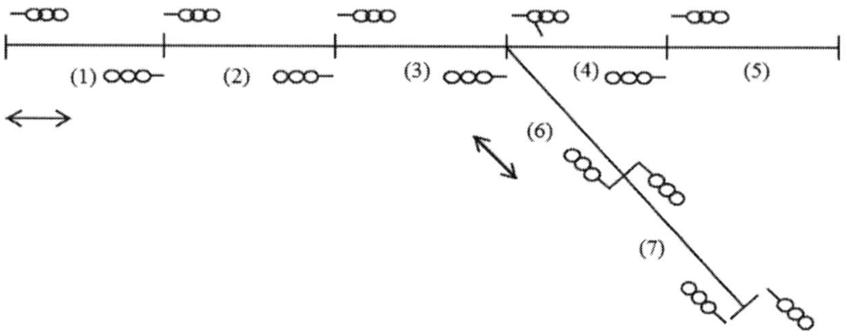


**Figure 8. Schematic representation of a junction area.**

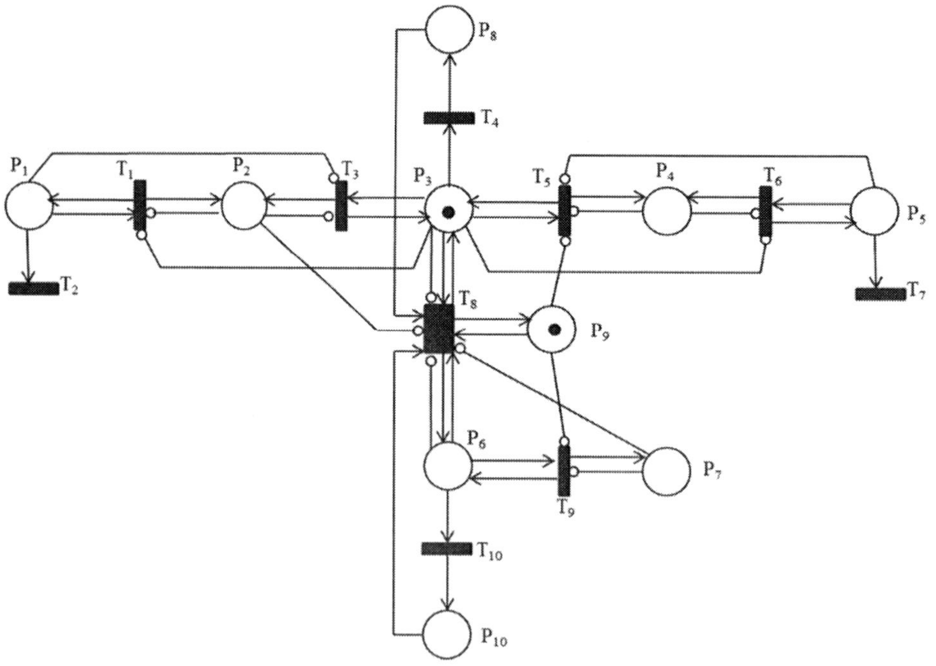


**Figure 9. Petri net for the junction area in Figure 8.**

Place P9 represents the position of the switch point, while the other places represent the block sections in the junction area. P9 contains tokens of only two different colours indicating one of the possible point positions. Firing of transition T8 is responsible for "changing" the token contained in P9 according to the current marking of the net. Places P8 and P10 do not represent any physical resource but are used to represent the conditions for transition T8 to fire. If P8 is marked block section 3 is clear and therefore a train approaching from block section 6 can cross the switch, given that all the other enabling conditions on the surrounding block sections are met. Analogously, when P10 is marked, a train from block section 3 is enabled to cross the switch and enter block section 6.

## 5. FAILURE SCENARIOS AND ALTERNATIVE ROUTE SELECTION

In real situations, disruptions may result from the failure of any subsystem or component, and may occur anywhere in the network at any time. The effect of failures on the system does not depend only on the failure itself but on the possibility to isolate the obstructed area in order to divert traffic and minimize the impact on service. This flexibility on open lines is provided by crossovers that allow to divide the network in clusters. A cluster here is defined as a section of track between two consecutive crossovers. According to the position of the switches (normal or diverging), trains may traverse or overcome a cluster. For instance, if any block section between two consecutive crossovers becomes unavailable, the entire cluster between the two crossovers will be closed in order to reroute trains. However if the nature of the failure is such to affect a bigger section of track, the selection of the alternative path may involve more consecutive clusters. Fault trees (10) are built for each cluster in order to predict the combinations of events resulting in the unavailability of the cluster. Basic events can either be failures affecting a single block section within the cluster, an example being a signal or a track circuit failure. However, more consecutive block sections can be simultaneously unavailable due to a common failure such as a fault in the electrification system. The events in the fault tree are linked by means of "gate symbols" whose meaning is detailed in table 2.

**Table 2. Gate symbols.**

| Symbol | Name | Causal Relation |
|---|---|---|
|  | AND | Output event occurs if all of the input events occur. |
|  | OR | Output event occurs if at least one of the input events occurs. |
|  | NOT | Output event occurs if none of the input events occurs. |

The OR gate implies that the output event will occur if any of the input event happens. The AND gate implies that the occurrence of the output event requires all the input events to occur simultaneously. Finally the output to a NOT gate occur as long as none of the input events occur. For instance, unavailability of cluster C2 only (Figure 10) requires that at least one of the block sections in cluster C2 must be unavailable due to a failure, and simultaneously all the block sections in cluster C7 must be available along with switches S2 and S3, and clusters C1 and C3.

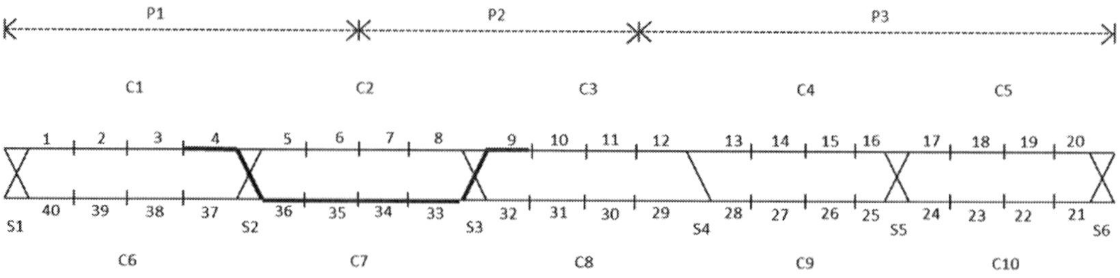


**Figure 10. Example of cluster identification.**

The fault tree representing the combination of events leading to the unavailability of cluster C2 only, is depicted in Figure 11.

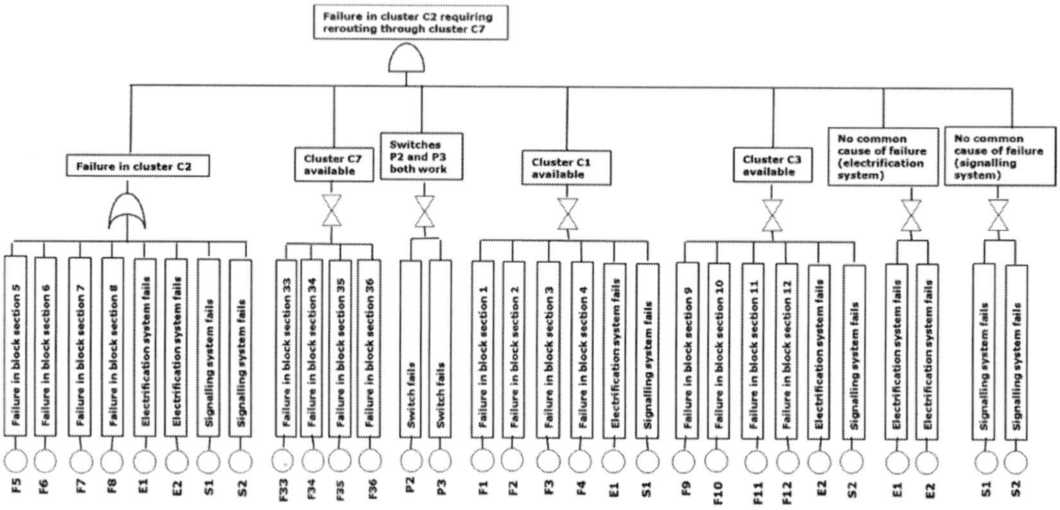


**Figure 11. Fault tree for failure in cluster C2.**

If a Boolean variable is defined for each basic event which is TRUE if the event occurs and FALSE if it does not, then the Boolean expression for the top event is given by:

$$C7 = (F5 + F6 + F7 + F8 + P1 + P2).(\overline{F33}.\overline{F34}.\overline{F35}.\overline{F36}.\overline{P2}.\overline{P3}.\overline{E1}.\overline{E2}.\overline{S1}.\overline{S2}) \qquad (2)$$

where $Fi$ indicates a failure in block section $i$, $Pj$ indicates a failure of point $j$, $E_k$ indicates a failure in the electrification section $k$ and $S_m$ indicates a failure in the signalling section $m$.

## 6. NUMERICAL EXAMPLE

The network analysed in this paper is depicted in Figure 12. The total length of track is 30 miles, of which 13 miles consist of four parallel tracks for fast and slow services running in both directions of travel, while the remaining 17 miles consist of a double track. One junction is located at the 8th mile and allows traffic to flow to and from station C. The basic network includes a unidirectional signalling system for all lines and one crossing between stations D and E, to merge the four-track section into the double track section. A fixed block signalling system is considered, with all block sections having the same length (1 mile). Trains run according to a cyclic timetable along the lines A-C and A-F. The railway is operated for a period of 20 hours with trains running every 15 or 30 minutes depending on the type of service. The electrification and signalling system is sectioned as shown in Figure 12, where three sections (E1, E2, and E3) are identified for the electrification system, and two (S1 and S2) for the signalling system. For the network in Figure 12, the model has been initially applied to three different distributions of crossovers. The final aim is to compare the total delay resulting from disruptions, for different numbers of crossovers. Results from Monte Carlo simulations show that for a given timetable, the total delay decreases with the number of crossovers available to reroute trains around the unavailable sections of track (Table 3).

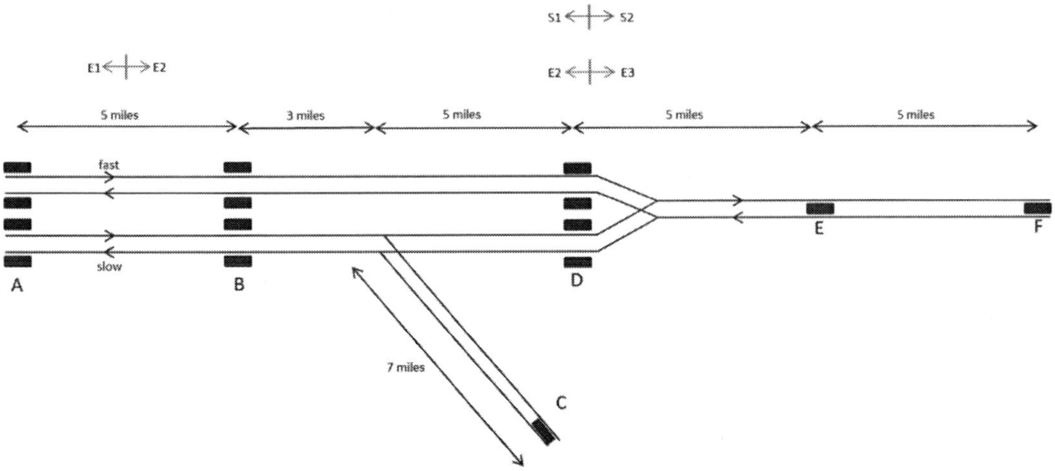


**Figure 12. Network layout.**

**Table 3. Simulation results.**

| N°of crossovers | Avarage delay per train (mins) |
|---|---|
| 5 | 9.5 |
| 4 | 15.5 |
| 3 | 24.3 |

## 7. CONCLUSIONS

Due to the increasing utilisation of the ageing network, the railway industry will have to adopt new solutions in order to provide a railway infrastructure capable of meeting the increasing demand for improved performance. The trend towards 24/7 railway operation will require a flexible and highly connected network in order to ensure the continuity of service under both planned and unplanned network unavailability. In this research, the application of fault tolerance and redundancy concepts for improving reliability of railway service is suggested. A methodology for the systematic evaluation of possible solutions to provide alternative routes to trains following system failures is discussed. The methodology implements simulation, optimization and failure analysis to assess the effectiveness of different solutions. A railway modelling tool based on the Petri net method is used to assess the performance of different network layouts in terms of total delay. Fault tree analysis is applied for the systematic evaluation of failure scenarios and identification of alternative paths. Simulations show that an increased number of crossovers distributed over the network would reduce the total delay in case of large disturbances. The model, once included into an optimisation routine, will be used to determine the most effective location of switches and crossings, as well as the sections of the network that would benefit from a bi-directional signalling, while solutions which are not cost effective will be automatically excluded. This would support prioritisation of investment plans to provide the most robust network that the railway industry can afford. With the in-built fault trees generations model, the simulation model can be also used to create perturbation plans if the behaviour of the system has to be analysed under a given disruption scenario, for example in order to identify optimal compensating strategies.

## REFERENCES


(1) Vromans M.J.C.M., Dekker R. and Kroon L.G., Reliability and heterogeneity of railway services. European Journal of Operation Research, 172, 647-665, (2006).

(2) Middelkoop D. and Bouwman M., SIMONE: Large Scale Train Network Simulation. In Proceeding of the 2001 Winter Simulation Conference, ed. B. A. Peters, J. S. Smith, D. J. Medeiros and M. W. Rohrer, (2001).

(3) Middelkoop D. and Bouwman M., Testing the stability of the rail network. In Computer in Railways VIII, ed. J. Allan, R. J. Hill, C. A. Brebbia, G. Sciutto and S. Sone. Southampton, United Kingdom: WIT Press, (2002).

(4) Dicembre A. and Ricci S., Railway traffic on high urban corridors: capacity, signalling and timetable. Journal of Rail Transportation Planning and Management, 1, 59-68. (2011).

(5) Luethi M., Weidmann U.A., Laube F. B., Medeossi G., Rescheduling and Train Control: A New Framework for Railroad Traffic Control in Heavily Used Networks, Proceedings of the Transportation Research Board 86th Annual Meeting, Washington DC, January 21-25, (2007).

(6) Demitz J., Hubschen C. and Albrecht C., Timetable stability – using simulation to ensure quality in a regular interval timetable. In Computer in Railways IX, ed. J. Allan, R. J. Hill, C. A. Brebbia, G. Sciutto and S. Sone. Southampton, United Kingdom: WIT Press (2004).

(7) Carey, M. and Carville, S., Testing schedule performance and reliability for train stations, Journal of the Operational Research Society, Vol. 51, pp. 666-682, (2000).

(8) Salido M. A., Barber F. and Ingolotti L., Robustness for a single railway line: analytical and simulation methods. Expert System with Applications, 39, 13305-13327, (2012).

(9) Jensen, K., and Kristensen, L. M., Colored Petri nets. Modelling and validation of concurrent systems. Springer-Verlag Berlin Heidelberg. (2009).

(10) Andrews J.D. and Moss T.R., Reliability and Risk Assessment, 2nd ed., Professional Engineering Publishing (2002).

# The paper that eliminates paper

**M Davies**
Software Solutions, Interfleet, UK

## 1. INTRODUCTION

The 'Internet of Things', mobile devices and their apps are a ubiquitous part of modern life. The UK rail industry is using mobile devices and industry focussed apps to help drive innovation and to put information at the fingertips of key staff throughout the industry.

This paper focuses on mobile apps that have been developed by Interfleet to add value to railway staff, such as 'Timetable Advisory System' (TAS) and 'Rail Companion'. TAS provides the train driver with information on how the train they are driving is progressing relative to timetable. Rail Companion is a powerful new industry information portal delivering a suite of information apps directly to the working environment of rail staff involved in the industry.

The paper will explore the apps, where they are being used in the UK, their benefits as well as the practical software and engineering challenges surrounding the implementation of mobile devices in the train environment. The paper will also look at future trends associated with rail orientated "big data", such as energy metering and condition monitoring, how and where they are being implemented and what apps may exist to support railway staff in the future.

## 2. BACKGROUND

The third generation wireless telephone technology (3G) was launched in 2001 which helped to start the revolution of collecting and delivering remote data real time. The standard suggests minimum data rate of 2 Mbit/s when stationary and 384 kbit/s whilst moving.

The fourth generation (4G) has two different flavours, Mobile WiMAX and Long Term Evolution (LTE). The first releases of both flavours were not fully compliant with the International Telecommunications Union-Radio communications sector (ITU-R) standards but backwards compatible compliant versions of the above two flavours were standardised during 2011. 4G has been widely marketed as providing mobile broadband Internet access. The 4G standard sets peak speed requirements at 1Gbit/s whilst stationary and 100Mbit/s whilst moving. With speeds that can match traditional broadband landline speeds, the continued upgrading of cellular infrastructure and data charges starting to fall, users are quickly discovering the opportunities for mobile working without the need to find Wi-Fi hotspots. People are now looking for devices that can support their mobile working.

Back in 1996 the Nokia Communicator series was launched which many recognise as the first smartphone. A smartphone is defined as a mobile phone with additional computing and connectivity capabilities. In 1999 Research In Motion (RIM) BlackBerry launched its first two-way pager model and users had to wait until 2003 for models which also contained an integrated phone.

Smartphones evolved from phones with PDA/media player capability to devices with cameras, media players, GPS navigation, web browsing and more recently touch screens and third party applications. The original Apple iPhone running the iOS operating system was released in June 2007 and was widely considered to be the first of the modern smartphones providing a combination of inbuilt functionality and a platform to develop, deliver and use third party applications. Google released version 1.0 of its Android operating system in September 2008 and the first smartphone to use this operating system was launched by HTC in October 2008 as a competitor to the Apple iPhone. Currently approximately 80% of the mobile phones sales worldwide are smartphones using the Google Android or Apple iOS operating system.

In April 2010 Apple released their first tablet computer (iPad) using the same iOS operating system and this was quickly followed by the Samsung Galaxy Tab in September 2010 using the Android operating system. Although not the first tablet computers, these devices quickly caught the imagination of consumers, especially with the similarities to the smartphones they had become accustomed to and reliant upon. Tablet computers combined the computing power of a laptop, the built in connectivity of a smart phone with a large touch screen.

In October 2012 Microsoft released Windows 8.0, which has helped to introduce a viable alternative to the Android and iOS based smartphones and tablet.

## 3. TABLETS AND SMART PHONES WITHIN THE UK RAIL INDUSTRY

The UK Rail Industry has been slow to react and adopt smartphone and tablet technology, especially within operational staff.

There were several barriers to overcome for the introduction of smartphone and tablet computers which led to the UK Rail Industry being slow to embrace this new technology.

Senior management within companies initially perceived tablets and smart phones as luxury personal devices that were delicate and would not be suitable for use in the harsh cab environment. IT teams within organisations also had concerns regarding the security and management of devices. IT teams were experienced in implementing restrictive rule sets in traditional Microsoft and Blackberry environments. In 2011 organisations within the rail industry took inspiration from other sectors and began to empower users by managing devices through IT policies rather through device restrictions. iPhones and iPads were issued to front line staff before senior management and users were enabled to install business and personal apps and content onto the device. Organisations found that users engaged with the device including keeping the device charged, with them at all times and often bought their own cases to protect the devices. Several projects found that the return rate due to damage of personally issued iPhones and iPads was substantially lower than ruggedized mobile devices which were used purely for business functions.

Developments in Mobile Device Management (MDM) platforms have also helped to reassure IT teams that smartphones and tablets can be easily and reliably managed in a similar manner to their previous Blackberry estate.

The Rail Industry has been struggling with the issue of mobile phones causing a distraction within the train cab environment for some time. Other industries have suffered from similar issues, including road haulage, however due to the legislation within the UK rail industry and high profile train accidents including the derailment near Santiago de Compostela, Spain in 2013, companies perceive this as a critical risk to their business and requiring the use of such a device in the cab environment could promote misuse. Many tablets and smartphones have been designed to prevent apps from enabling and disabling core functionality and interacting with other apps as this is perceived as a security risk within the device. However recent developments in MDM platforms and the tablet operating systems have enabled developers to be able to implement measures within the apps to put the devices into a controlled mode. This functionality has been considered essential by most safety case holders on the UK Railway and is helping to remove a barrier preventing such technology being used in the cab environment.

Initially companies allowed their procurement teams to lead tablet implementation with very loose requirements for their devices e.g. tablet >= 7". This unfortunately led to tablets being purchased and rolled out which do not deliver the full benefit to the organisations and users. By allowing operational teams to lead projects, supported by the IT and procurement teams, devices could be selected upon their ability to deliver the right benefit and engagement with the user.

## 4. TIMETABLE ADVISORY SYSTEM

In 2011 Interfleet identified a gap within the UK Rail market for a cheaper and similar alternative to the more complex Driver Advisory Systems (DAS) which were being implemented by freight and passenger operators. Driver Advisory Systems are bespoke systems which provide precise advice to train drivers to minimise energy usage whilst maintaining timetable. Most systems are retrofitted to existing rolling stock and provide detailed speed information to the driver to allow them to maintain an optimal glide path. The introduction of DAS was complex from an engineering perspective, with complex installation into existing trains requiring installation of a screen within the cab environment, GPS receivers, communication modules and in some instances connections into existing on board systems. Data flow had to be very carefully managed to ensure that accurate safety information including speed restrictions was reliablely and consistently delivered to the DAS units to avoid unsafe advice being given. The introduction of DAS also caused issues with users as it was perceived by many as telling a highly skilled driver how to drive a train and could be used as a spy in the cab.

Interfleet's approach was to develop a simpler solution which could be delivered via a tablet computer. Interfleet's Timetable Advisory System (TAS) product was developed as a revolutionary tablet app that monitors train performance against its schedule. Interfleet sought guidance and support from Network Rail, freight and passenger operating companies to develop an iPad app and supporting management and reporting portal. A collaborative and transparent approach to development enabled a rapid move through conception, production, development and trial.

TAS uses GPS to plot the exact location of a train and provides the train driver with a range of real-time information on how the train they are driving is progressing relative to the timetable. TAS provides real-time feedback to the driver on the train's performance against schedule and can therefore be used to develop efficient driving styles, maintain right-time running, enhance passenger comfort, reduce energy consumption and prevent driver-related delays. The introduction of TAS has greatly increased the quality of information around Sectional Running Times (SRTs) and delays and has contributed to improving railway operations, ensuring trains run safely, efficiently and punctually.

The benefits from the TAS app can be very quickly realised and can be quantified easily. By running a small trial, operators were quickly able to identify electric and diesel savings. This information helped to build business cases for the introduction of TAS and issuing tablet computers. Additional benefits which were openly acknowledged as substantial were more difficult to quantify to support business cases but were realised very quickly through the introduction of TAS including identifying and correcting section running times.

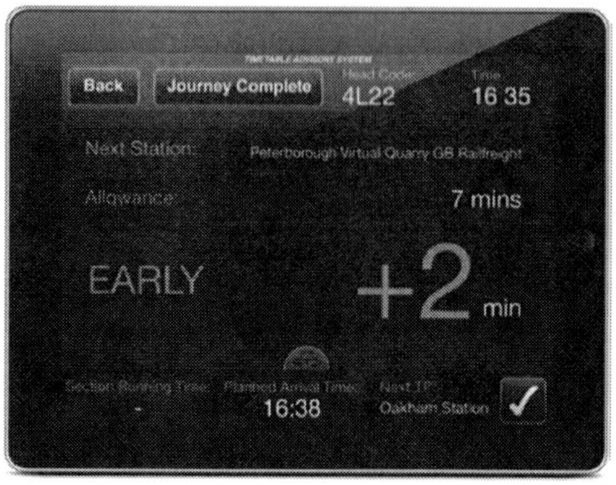


**Figure 1 - TAS advisory screen**

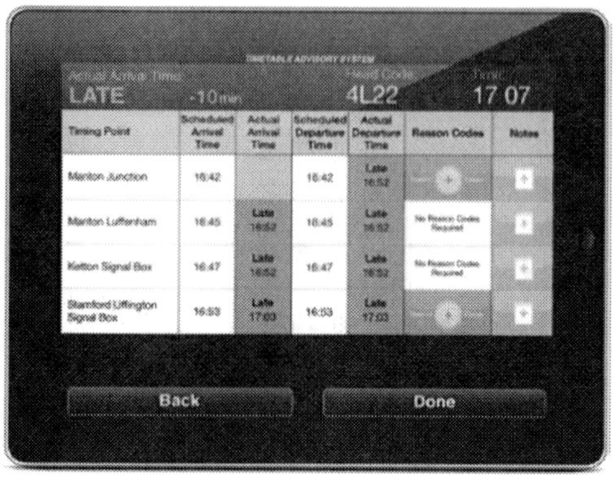


**Figure 2 - TAS delay capture**

TAS was originally designed to operate on iPad tablet devices in a driver's cab (allowing for a simple cab installation, as no separate antennas are required for GPS and comms traffic). The mobile tablet device allows the driver to transport their 'information hub' from cab to cab - providing greater flexibility than a fixed in-cab device - and lost or broken devices can be replaced quickly with a minimum loss of service. Using standard equipment and protocols made the approvals process much simpler for the introduction of tablet computers in the cab environment as for example 3G, 4G, Wi-Fi and Bluetooth do not need to be specifically considered for electromagnetic interference.

During the development of the TAS app, Interfleet developed a proof of concept to ensure that the GPS receiver within the iPad was sufficient. Interfleet found that the location services capability within the iPad was actually superior to GPS receivers used within other systems on the train. The iPad uses a combination of cellular, Wi-Fi, and GPS to determine the location of the device and if there is not a clear line of sight to GPS satellites the iPad can use details of cellular phone masts and crowd-sourced Wi-Fi information to determine an approximate location. Interfleet also found that the cellular communication was more than sufficient in the cab environment unlike some saloons which are affected by reflective windows.

Interfleet carefully designed and implemented the solution to engage with the driver, requiring continued use of their skill and knowledge. Interfleet also promoted the personal issue of devices rather than devices secured into the cab environment. By issuing a premier product to users and allowing them to use the device for both work and personal purposes, users quickly became engaged with the device. Without specific instructions from the company, users ensured that the devices were always charged and were put into cases and well looked after. Quite often the cases were purchased by the user and not charged back to the company.

Key considerations during the development of the TAS iPad app were User Experience (UX) and User interface(UI). Carefully designing a simple and intuitive app, reduced the time and effort of train users but also by sticking to the guidelines and paradigms within the tablet operating system enabled the user to use the TAS app in a similar manner to any other app which they use personally or for work.

The TAS app is distributed through the Apple App Store infrastructure which allows automatic updates to be distributed to the user community. By ensuring that all new features are configurable remotely and are seamlessly integrated with the TAS user interface allows updates to be pushed out quickly and easily to the disparate user community.

There were some challenges with the introduction of TAS and tablet computers. As described above the rail industry has particular concerns regarding mobile devices within the cab environment and the distraction they can cause. Initially operators were managing this though guidelines but with the recent developments in MDM platforms operators have started to put physical measured in place on the device. These physical measures have been well received by the industry and has helped to reassure operators that tablet computers can be introduced into the cab safely. Interfleet has supported a number of operators and has provided support to demonstrate the benefits of TAS and tablet computers and to reassure them that there was not a spy in the cab.

## 5. RAIL COMPANION

Following the success of Interfleet's TAS app, in 2012 working with operators and drivers to develop a number of innovative ideas, Interfleet developed a concept called Driver's Companion to help support train drivers (and operators) via the use of tablet devices and replace paper based documentation and form processes. It was intended that app suite would replace the majority of the driver's bag and the functions of a drivers' depot. This concept was successfully submitted to the 2013 RIA/RSSB innovation competition and received funding from RSSB to complement the internal development funding provided. The original concept was to develop the following functionality for an iPad device:

- Personalised dashboard for each driver
- Driver's manuals and information including immediate access to Rule Book, Sectional Appendix, PONs & WONs, Temporary Speed Restrictions, Company/Safety Notices, Vehicle Technical Information and Driver's Fault Guides
- Online booking on, monitoring of driving hours, rest day management and interfacing with existing systems
- Forms library - ability to download regularly used forms and complete them online, such as defect book, incident form etc
- Simple driver reporting tool allowing quick 'one-touch' reporting of common problems such as poor adhesion, trespassers, low voltage, rough riding or vegetation overgrowth and including automatic email of location, headcode, driver ID and the nature of the problem
- Ride hot spot detection using the in-built accelerometers of the tablet device to monitor and record vibration

From the original requirements the suite grew in functionality and was renamed to Rail Companion to illustrate its capability to all rail personnel. Rail Companion is a powerful new industry information portal delivering a suite of information apps directly to the working environment of rail staff, including train drivers, crew, operations managers, maintenance staff and a range of other personnel involved in the industry.

**Figure 3 - Rail Companion**

Accessed via a tablet device such as an iPad, Rail Companion provides live updates and enables real-time reporting of a variety of events and information.

The primary benefits of Rail Companion are to effectively collect and deliver the right information, right on time to rail industry staff in a controlled and streamlined way, via a single tablet device. Rail Companion brings benefits for staff across many different areas of the rail industry. Some benefits include:

- Quick and efficient capture of consistent and reliable data
- Quick and efficient distribution of time critical information
- Fully auditable processes of data capture and distribution
- Wealth of reports and analysis to support and/or replace manual processes
- Quick to adapt to changing requirements
- Helps to close information flow loops of existing process
- Reduce printing, distribution and associated administration costs
- Dramatically reduces what a driver has to carry
- Single sign on which gives users access to all Rail Companion Mobile Suite apps
- Quick and easy integration to a variety of industry information sources

The personal hub gives the user access to all the different Rail Companion mini-apps using just a single login and without the need to re-enter their security credentials. The personal hub draws key actions and information from each of the mini-apps to provide a central dashboard of information and alerts personal to the user. Each task and update will automatically direct users to the corresponding information within the suite.

The Personal Hub also allows the user to view and manage the information about them. The profile page allows users to see all the company information used within Rail Companion to control the information that they can access. The user can also edit information about them which is not managed centrally by the company.

The user can also view the distribution lists they are currently subscribed to, either automatically through their profile information or through manual subscription. The user can also subscribe and unsubscribe from distribution lists which are not automatically or centrally controlled.

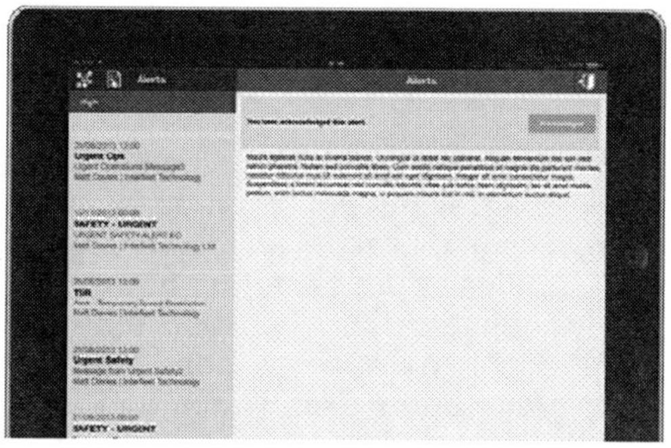

**Figure 4 - Rail Companion alerts**

The Alert system allows information to be quickly and easily distributed to a user base. The features include:

- Distribution list, driven by manual subscription or from criteria on profile information
- Alerts can be prioritised and categorised according to their importance and subject
- The app will automatically poll the management server to check for updates but when offline all information is stored locally and changes cached until connectivity is restored
- Acknowledgement of alerts can be requested and tracked
- Full audit trail of alert – information includes when and where each user receives, reads and acknowledges alert

- High priority alerts can be distributed which will be displayed within the Personal Hub on top of all other content and must be read and acknowledged
- Alerts are time critical and can be distributed ahead of time but only displayed after a specific time and can be expired and retracted from each user's device

The Forms system is designed to allow information to be quickly and consistently captured and collated using customisable forms which make full use of inbuilt functionality of the device including location services, photo and video attachments. The features include:

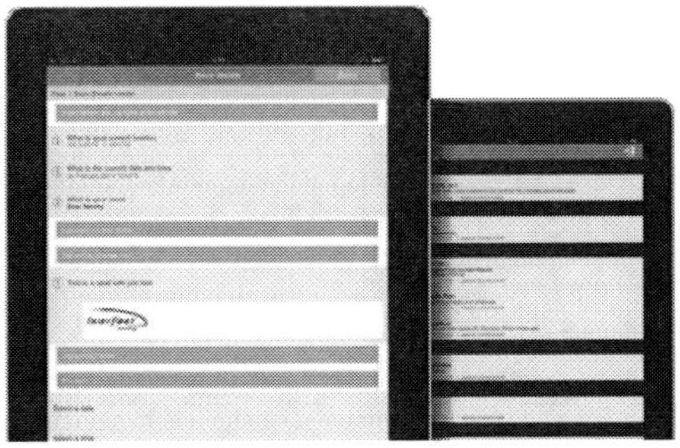

**Figure 5 - Rail Companion forms**

- Distribution lists, driven by manual subscription or from criteria on profile information
- Task list of actions
- Ability to view in portrait or landscape device orientation
- Form template version control
- Customisable folder structure
- Download form templates for viewing and completing offline
- Saving partially completed forms
- Immediate transmission of completed and partially completed forms to the management portal
- Full audit trail of forms – information includes when and where each user receives form template, initiates and completes a form
- The app will automatically poll the management server to check for updates but when offline all information is stored locally and changes cached until connectivity is restored
- Management of time critical form templates which can be distributed ahead of time but only displayed after a specific time and can be expired and retracted from each user's device
- Form designer to quickly and easily create and edit form templates - Rail Companion Management Portal
- Automatic fields which are populated from personal profile details (e.g. name, job title and depot) and by device information including location and current date time
- Supports a range of form controls:-
    - Text (single and multi-line)
    - Pick lists (single and multiple)
    - Slider
    - Location
    - Signature
    - Drawing
    - Location
    - Media attachments
    - Supports required fields and default values

Special 'One-Touch' Forms can be raised within any Rail Companion app to allow users to instantly create a form when time is short, automatically capturing key information and transmitting it back to the management application. The full form can then be completed at a later, more convenient time.

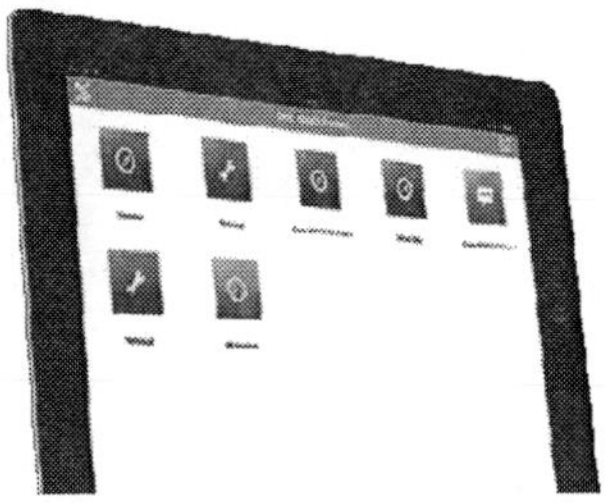

**Figure 6 - Rail Companion one-touch forms**

The aim of Documents functionality was to achieve everything that was possible with current paper based processes and everything possible with static electronic documents but better.

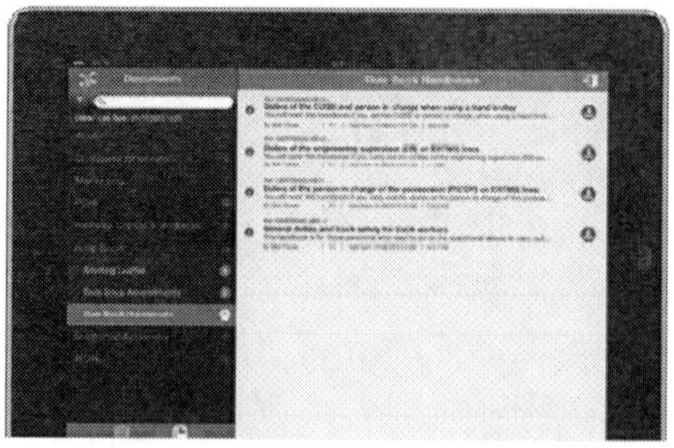

Figure 7 - Rail Companion documents

The Documents system includes the following features:

- Distribution lists, driven by manual subscription or from criteria on profile information
- Task list of actions
- Ability to view in portrait or landscape device orientation
- Document version control
- Customisable folder structure
- Customisable document properties
- Download documents for viewing offline
- Ability to search within a document and its meta data
- Zoom functionality
- Bookmark functionality
- Ability to highlight and annotate a user's own copies of documents
- Full audit trail of documents – information includes when and where each user receives document meta data, downloads, reads and acknowledges document
- Ability to send document to Wi-Fi/Bluetooth printer
- The app will automatically poll the management server to check for updates but when offline all information is stored locally and changes cached until connectivity is restored
- Management of time critical documents which can be distributed ahead of time but only displayed after a specific time and can be expired and retracted from each user's device
- Ability to retrieve document statistics

The Rough Riding system contains algorithms designed to utilise inbuilt functionality of the iPad device to monitor and capture areas of poor ride quality.

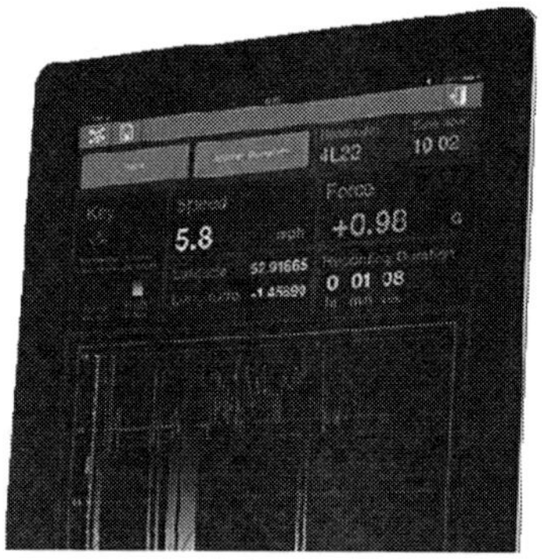

Figure 8 - Rail Companion rough riding

The Rough Riding app includes the following features:

- Automatically captures information including rough riding details, journey, date/time and location
- Automatically transfers information back to the management portal for reporting and analysis
- Can be run as a standalone app or integrated with Interfleet's Timetable Advisory System

From within the Rail Companion Personal Hub the user is able to book on and off shift without the requirement to use a manual or telephone based booking on/off system.

The Booking On/Off functionality includes the following features:

- Automatic capture of user's details, date time and location information

- Ability to capture comments and variations
- Automatically transferred back to management portal for transfer into existing systems

The Rail Companion Time Management system is designed to integrate with the Booking On/Off functionality allowing a user to see their rostered work.

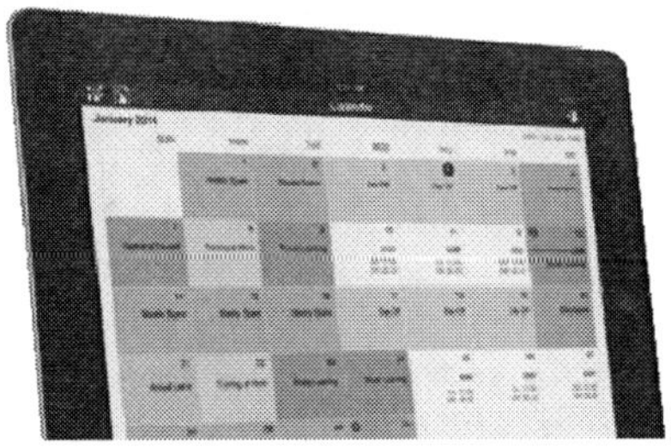

**Figure 9 - Rail Companion time management**

The Time Management app includes the following features:

- View calendar
- View diagram information
- View and accept any changes to rostered work
- Integration with booking on/off functionality
- Fully audit trail

A corresponding web based central management and reporting portal hosted on clyx.net allows for the full configuration and reporting on the information managed through Rail Companion.

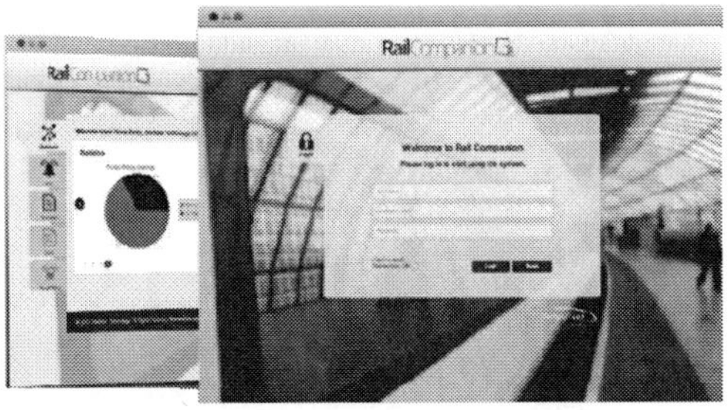

**Figure 10 - Rail Companion management and reporting portal**

## 6. DATA, DATA EVERYWHERE

The core of Rail Companion is distributing information, but the suite is also designed to capture a wealth of associated data. For example when a document is uploaded, the system will automatically capture when and who uploaded the document, when it was distributed and when, where and who was notified, downloaded, read and acknowledged the document. But in addition to capturing and reporting on this activity, Rail Companion can also record information on the document use, including which pages were read, how often and for how long. This will give a wealth of data which can then be interrogated and used to identify which documents are written badly, because people keep having to re-read the same information or which documents should be restructured as people don't read pages 10-20. This sounds fantastic in theory but in practice the volume of data being collected could be vast and the data is not worth anything until you can glean information from this data.

The more we monitor the more data we capture and have to store, manage and analyse. Many make the assumption that if you don't want to report or analyse the data you capture then it has no value. However with data storage and analysis becoming increasingly easy, many organisations are exploring the opportunities and associated challenges by storing and analysing large data sets. These large data sets can then be combined and analysed at a later date to give additional insight.

Big data is a term that is used to encompass any collection of data sets that are so large and complex that it becomes difficult to process using traditional data processing. Challenges include being able to store, manage, analyse, search, visualise and protect the volume of data. However the power of combining large data sets is vast. Combining data sets, which in isolation may seem unconnected, can often lead to correlations being found. For example combining weather information to engine failures can help to spot trends and issues which monitoring just the brake temperature might not and combining passenger numbers with social media feeds can give insight into passenger behaviour. There are a range of different analysis tools on the market allowing a range of different data sources to be combined and analysed.

For some years, there have been initiatives within the UK Rail Industry to connect systems together and centralise data sources, which has started to create consistent shared data sets. However the majority of these data sets are not openly available within and outside of the industry. Network Rail have made inroads through their Open Rail Data feeds, however this is only an extremely small percentage of the amount of data being captured and held. By making these feeds available Network Rail and the industry have already found that people are using this data for a variety of purposes, including some that are already bringing benefit to the industry. There are a lot of concerns about releasing data to the public, especially commercially sensitive data however the more datasets that are available the more they can be combined and analysed, which would give more information and insight to the industry.

## 7. SUMMARY

Although the Rail Industry was slow to respond to the introduction of smartphones and tablet computers, it is now realising the benefits of introducing the latest consumer technology to all areas of the rail industry.

When deploying such technology the user has to be the centre of attention, they need to be engaged with the device and technology. Personally issuing a device to users and allowing them freedom to use the device to its maximum, encourages the user to take ownership and familiarise themselves with using the device and apps. Often a user will find clever and innovative ways to use the device and apps to support them in their work life.

When developing an app, especially one which is to be used by a wide user base, with the minimum of disruption and training, it is essential to design the user interface to maximise the user experience of any solution. An app should be as easy to use as the most common apps, for example social media. So by transferring common interactions and paradigms users will automatically feel more comfortable.

The capability to collect a wealth of different parameters should be built into every solution and the data stored and made widely available for analysis to get the best return in information. For example by capturing as many channels as feasible through a remote condition monitoring installation the data can then be combined with remote condition monitoring from the infrastructure, operational data and passenger data. This could then be used to identify trends and patterns in the channels that the user was interested in but it might also identify trends and patters in other areas. This data could then be combined with other apparently unconnected datasets for example social media and weather to give even more perspective.

As more data is captured via consumer technology, for example through fitness tracking wristbands, made available more quickly and readily through smart watches and analysed and presented in new mediums such as augmented reality, it is important for the rail industry to keep pace, engage with consumer technology and adopt as quickly as possible to support companies and rail staff.

# Theoretical and experimental analyses of the HXN3 locomotive cab vibration isolation system

**C-S Li[a], T-W Qu[b], S-H Luo[a], W-H Ma[a]**
[a] State Key Laboratory of Traction Power, Southwest Jiaotong University, P.R. China
[b] Dalian Locomotive and Rolling Stock Co., Ltd., CNR Croup, P.R. China

**ABSTRACT**

This paper focuses on the analyses of the HXN3 diesel locomotive cab isolation system. The multi-dimension frequency response function (FRF) concept is applied to assess the isolation performance of the HXN3 cab isolation system. The transmissibility (longitudinal, lateral and vertical) of the cab isolation system is derived. In order to assess the isolation performance under the high frequency vibration, the rigid-flexible coupling model of the HXN3 cab isolation system is established and the transmissibility with and without consideration of cab deformation is compared. The theoretical analysis results indicate that the HXN3 cab isolation system cannot suppress the lateral excited vibration effectively and the lateral structural vibration will be caused by the high frequency lateral vibration coming from the underframe; the vertical resonant vibration of the HXN3 cab cannot be avoided, while in the contrast of the lateral vibration, the vertical structural vibration of the cab is not significant. The HXN3 vibration experiment was conducted to validate the theoretical analyses. The experimental results illustrated a good agreement with that of the theoretical analysis.

**Keywords:** HXN3 diesel locomotive; cab; vibration isolation; frequency response function; rigid-flexibility coupling

## 1. INTRODUCTION

Noise, vibration, and harshness (NVH) in railway vehicles have received increased attention in recent years. Although vibration isolation methods maintain a sustainable development in railway vehicle industry, the diesel locomotives are lagging behind the high-speed trains in this respect by several years.

Additional vibration isolation systems are required by diesel locomotives to separate cabs from the vibration excited by engines and generators. Different from the domestic DF series diesel locomotives [1, 2], pivot-mount isolation systems are applied to HXN3 diesel locomotives for separating cabs from underframes. While as a matter of fact, the vibrations of the HXN3 cabs are not isolated effectively, especially the lateral and the vertical vibrations.

According to the vibration isolation theory, the vibration attenuation can only be obtained when input frequencies are greater than $\sqrt{2}$ times the natural frequency of the isolation system. For input frequencies which are below $\sqrt{2}$ times of the natural frequency, especially those which are close to the natural frequency, the vibration level of the isolated equipment is actually increased compared with that of the base. The stiffness of the isolator dominantly determines the resonance frequency of the system which is equal to the natural frequency, and the viscous damping is often introduced to reduced vibration amplitude at resonance. However, if the effect of the introduced damping is basically linear, the isolation system can only offer a good effectiveness for the high excitation frequencies, and if the effect of the introduced viscous damping is linear, as the damping level is increased to reduce the transmissibility in the resonant region, the transmissibility is increased in the region where isolation is required, as a result, the range of frequencies where vibration isolation can be achieved is reduced.

In order to improve the property of the linear isolation system, the authors have recently proposed the introduction of nonlinear stiffness and nonlinear viscous damping and have conducted a series of studies.

In Ref. [3-5], a negative stiffness structure (NSS) was introduced to improve vibration isolation performance of the vehicle seat under low excitation frequencies. Theoretical analysis of the proposed system was conducted via establishing a single degree of freedom (SDOF) model and calculating the transmissibility of the proposed system at the steady state for each of the excitation frequency. Simulation results reveal that this type of nonlinear stiffness isolation system has larger frequency region of isolation than that of the linear system. Experiments were also conducted, and experiment results confirmed that the isolation property of the system with NSS was greatly improved at low excited frequency region.

In Ref. [6-8], a quasi-zero-stiffness (QZS) isolator was analyzed which comprises a vertical spring and two oblique springs. It was proved that the QZS isolation system outperformed the linear system by simulation and experiment.

In Ref. [9-10], it was proved that a new design strategy for an anti-vibration mount could increase isolation by lowering the natural frequency of the mount, whilst maintaining the same static load bearing capacity which was based on the concept of the High Static Low Dynamic Stiffness (HSLDS).

There is a well-known dilemma associated with linear viscous damping systems. That is when the linear damping coefficient is increased, the force transmissibility under both excitation and force excitation when excitation frequency $\Omega < \sqrt{2}\omega_0$ is further reduced, but the performance when excitation frequency $\Omega > \sqrt{2}\omega_0$ is contrarily deteriorated.

In order to get a better vibration isolation performance, isolators with nonlinear damping have been studied by authors to explore the potential nonlinear benefits in vibration control [11-13]. It is proved that the cubic order nonlinear damping can produce much better isolation performance, i.e., obvious peak suppression at resonant frequency and very close transmissibility to system linear damping over non-resonant frequencies under force and base displacement excitations.

In theoretical studies, single degree of freedom (SDOF) models are widely used in isolation property analyses. Based on the concept of frequency response function (FRF), the transmissibility of isolation systems is calculated to measure the vibration transmission at different frequencies.

In this paper, a novel theoretical analysis is introduced to assess the isolation performance of the HXN3 cab isolation system. Based on the concept of frequency response function (FRF), the study first establishes the governing equations of a 6-D (6-dimension) HXN3 cab isolation system, and model the frequency response function (FRF) matrix. Secondly, the longitudinal, lateral, and vertical isolation performance of the system are assessed via calculating the longitudinal, lateral, and vertical transmissibility. Then a MD (Multi-dimension) model of the cab isolation system is established with the consideration of the cab flexibility, and the transmissibility is compared with that of the cab isolation system with neglecting the cab flexibility to assess the isolation property under high-frequency excited vibrations. Finally, an experiment is conducted and the vibration accelerations of the cab and the underframe are obtained at the diesel-working rotating speed. It is proved that the HXN3 cab isolation system cannot isolate the vibration coming from the underframe which is excited by the diesel and the generator by numerical simulation and experiment.

## 2. VIBRATION ISOLATION ANALYSIS WITHOUT CONSIDERING CAB FLEXIBILITY

### 2.1 Model description of the HXN3 cab isolation system

As shown in Fig.1, the HXN3 cab pivot-mount isolation system consists of two rubber bushings at the back corner of the cab to control the frequency and magnitude of the vertical, lateral, and longitudinal natural frequencies, two vertical coil springs at the front corner of the cab to control a low cab pitch natural frequency, two vertical dampers next to coil springs to reduce or eliminate the magnification of track, and two lateral links at the front of the cab to not only keep the vertical and longitudinal motion, but also restrict lateral motion, thereby, control yaw natural frequency of the cab.

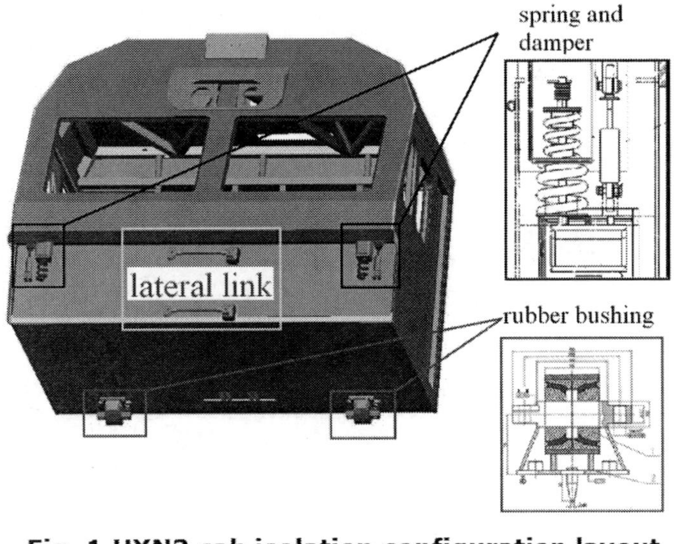


**Fig. 1 HXN3 cab isolation configuration layout**

To assess the performance of HXN3 locomotive cab isolation system, a mathematical model is developed, which shown in Fig.2, with considering the system as a 6-dimension structure with dampers, springs and rubber bushings located at the underframe. The values of the parameters of the system are as shown in Table 1.

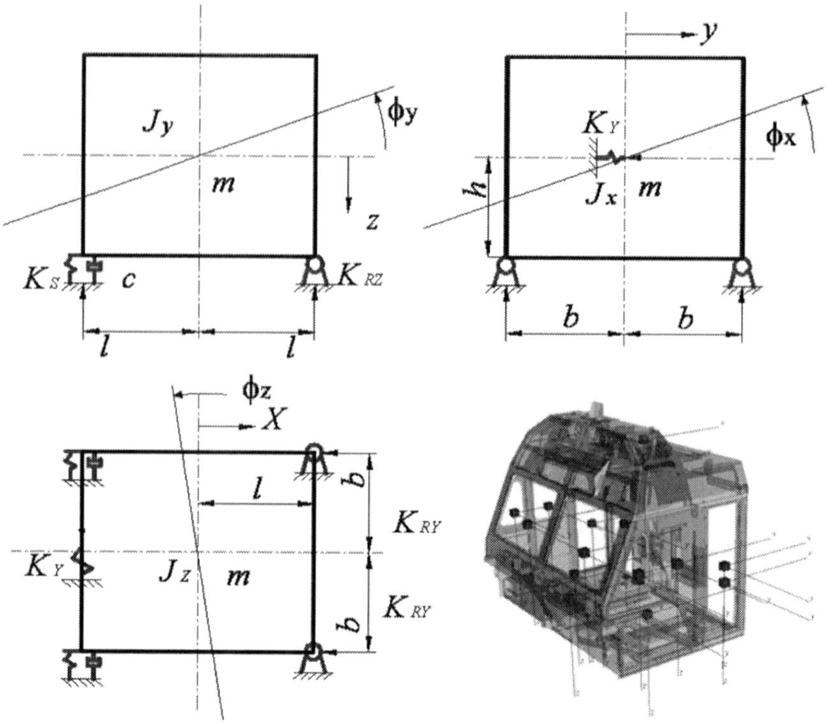


**Fig. 2 6-dimension multi-body dynamic model of the HXN3 cab isolation system**

**Table 1 Suspension parameters of the cab isolation system**

| Symbol | Description | Units |
|---|---|---|
| $m$ | Cab mass | $3.27\times10^{3}$ *kg* |
| $J_XJ_YJ_Z$ | Inertia tensor | |
| $K_S$ | Spring stiffness | $7.2\times10^{5}$ *N/m* |
| $K_{RXZ}$ | Rubber bushing radial stiffness | $1.6\times10^{7}$ *N/m* |
| $K_{RY}$ | Rubber bushing axial stiffness | $3.3\times10^{7}$ *N/m* |
| $C$ | Suspension damping | $9.0\times10^{4}$ *Ns/m* |
| $K_Y$ | Lateral link stiffness | $3.3\times10^{7}$ *N/m* |

The governing equations of the 6-D (6-dimension) HXN3 cab isolation system with underframe excitation are derived. Each dimension equation of motion for the system is:

Longitudinal vibration: $$m\ddot{x}+K_{RXZ}(x+\phi_z)+K_{RXZ}(x-\phi_z)=F_X \quad (1)$$

Lateral vibration: $$m\ddot{y}+K_{RY}(y-\phi_x)+K_{RY}(y+\phi_x)+K_Y y=F_Y \quad (2)$$

Vertical vibration: $$m\ddot{z}+2K_S\left(z+l\phi_y\right)+2K_{RXZ}(z-l\phi_y)+2c(\dot{z}+l\dot{\phi}_y)=F_Z \quad (3)$$

Roll vibration:

$$J_x\ddot{\phi}_x + (K_S + K_{RXZ})\left(z + l\phi_x\right)b + (K_S + K_{RXZ})(z - l\phi_x)b + c(\dot{z} - l\dot{\phi}_x)b + c(\dot{z} + l\dot{\phi}_x)b = F_\alpha \tag{4}$$

Pitch vibration:

$$J_y\ddot{\phi}_y + 2K_S\left(z + l\phi_y\right)l + 2K_{RXZ}(z - l\phi_y)l + 2c(\dot{z} + l\dot{\phi}_y)l = F_\beta \tag{5}$$

Yaw vibration:

$$J_z\ddot{\phi}_z + K_{RY}(x + l\phi_z)l + K_{RY}(x - l\phi_z)l + 2K_Y\phi_z l = F_\gamma \tag{6}$$

The governing equations of this 6-dimension structure can be written in following matrix form.

$$M\ddot{z} + C\dot{z} + Kz = F_e \tag{7}$$

Where $z = \begin{bmatrix} x & y & z & \phi_x & \phi_y & \phi_z \end{bmatrix}^T$ is the displacement vector of the cab isolation system, $F_e$ is the excited force coming from the underframe and $\omega$ is the corresponding frequency. *M, C* and *K* is the mass, damping and stiffness matrix of the cab isolation system, respectively.

$$M = \begin{pmatrix} m & 0 & 0 & 0 & 0 & 0 \\ 0 & m & 0 & 0 & 0 & 0 \\ 0 & 0 & m & 0 & 0 & 0 \\ 0 & 0 & 0 & J_x & 0 & 0 \\ 0 & 0 & 0 & 0 & J_y & 0 \\ 0 & 0 & 0 & 0 & 0 & J_Z \end{pmatrix}, C = \begin{pmatrix} 0 & 0 & 0 & 0 & 0 & 0 \\ 0 & 0 & 0 & 0 & 0 & 0 \\ 0 & c & 0 & 0 & 2lc & 0 \\ 0 & 0 & 2bc & 0 & 0 & 0 \\ 0 & 0 & 2l_C c & 0 & 2l^2 c & 0 \\ 0 & 0 & 0 & 0 & 0 & 0 \end{pmatrix},$$

$$K = \begin{pmatrix} 2\xi_{XZ}K_{RXZ} & 0 & 0 & 0 & 0 & 0 \\ 0 & 2(\xi_{RY}K_{RY} + \xi_Y K_Y) & 0 & 0 & 0 & 0 \\ 0 & 0 & 2(\xi_S K_S + \xi_{XZ}K_{RXZ}) 0 & & 0 & 2l(\xi_S K_S - \xi_{XZ}K_{RXZ}) \\ 0 & 0 & 2b(\xi_S K_S + \xi_{XZ}K_{RXZ}) & 0 & 0 & 0 \\ 0 & 0 & 2l(\xi_S K_S + \xi_{XZ}K_{RXZ}) & 0 & 2l^2(\xi_S K_S + \xi_{XZ}K_{RXZ}) & 0 \\ 2\xi_{XZ}K_{RXZ}l & 0 & 0 & 0 & 0 & 2l\xi_Y K_Y \end{pmatrix}$$

Eq. (7) can be Fourier-transformed as

$$(-\omega^2 M + \omega C + K)X(\omega) = F(\omega) \tag{8}$$

The impedance matrix of the cab isolation system is shown as

$$Z(\omega) = \frac{F(\omega)}{X(\omega)} \tag{9}$$

The frequency response function matrix of the cab isolation system is shown as

$$H(\omega) = Z^{-1}(\omega) = \frac{X(\omega)}{F(\omega)} = (-\omega^2 M + \omega C + K)^{-1} \tag{10}$$

The cab isolation system is a 6-input/6-output system (the cab isolation system is a 6-dimension structure and the input forces comes from 6 different directions). The frequency response function (FRF) matrix $H(\omega)$ of this system is a 6×6 matrix which is shown below.

$$H(\omega)=\begin{pmatrix} H_{C_xE_x}(\omega) & H_{C_yE_x}(\omega) & H_{C_zE_x}(\omega) & H_{C_\alpha E_x}(\omega) & H_{C_\beta E_x}(\omega) & H_{C_\gamma E_x}(\omega) \\ H_{C_xE_y}(\omega) & H_{C_yE_y}(\omega) & H_{C_zE_y}(\omega) & H_{C_\alpha E_y}(\omega) & H_{C_\beta E_y}(\omega) & H_{C_\lambda E_y}(\omega) \\ H_{C_xE_z}(\omega) & H_{C_yE_z}(\omega) & H_{C_zE_z}(\omega) & H_{C_\alpha E_z}(\omega) & H_{C_\beta E_z}(\omega) & H_{C_\lambda E_z}(\omega) \\ H_{C_xE_\alpha}(\omega) & H_{C_yE_\alpha}(\omega) & H_{C_zE_\alpha}(\omega) & H_{C_\alpha E_\alpha}(\omega) & H_{C_\beta E_\alpha}(\omega) & H_{C_\lambda E_\alpha}(\omega) \\ H_{C_xE_\beta}(\omega) & H_{C_yE_\beta}(\omega) & H_{C_zE_\beta}(\omega) & H_{C_\alpha E_\beta}(\omega) & H_{C_\beta E_\beta}(\omega) & H_{C_\lambda E_\beta}(\omega) \\ H_{C_xE_\lambda}(\omega) & H_{C_yE_\lambda}(\omega) & H_{C_zE_\lambda}(\omega) & H_{C_\alpha E_\lambda}(\omega) & H_{C_\beta E_\lambda}(\omega) & H_{C_\lambda E_\lambda}(\omega) \end{pmatrix}$$

Where $H_{C_xE_x}(\omega)$ is the longitudinal ($C_x$) transmissibility of the cab isolation system in response to the longitudinal ($E_x$) excited force coming from the underframe and so on.

## 2.2 Numerical simulation result and analysis

It is found that the HXN3 cab isolation system cannot isolate the lateral and the vertical excited vibration effectively. Therefore, the theoretical analysis of the HXN3 cab isolation system is focused on the lateral and the vertical. The working rotating speed ($n$) of the diesel engine is 1000rpm. The frequency ( $\omega$ ) of the excited vibration, generated by the diesel engine and the generator, is 16.7Hz by approximate calculation using the formula below.

$$\omega = n / 60 \tag{11}$$

For this 6-dimension structure, there are 6 normal modes which are associated with resonant frequencies can be determined by using mode analysis technique, and the results are listed in Table 2.

**Table 2 Natural frequency of the 6-dimension cab isolation system**

| Mode | Damping ratio | Frequency/Hz |
|---|---|---|
| 1 | 0.1506 | 3.8794 |
| 2 | 0.0284 | 13.2062 |
| 3 | 0.0207 | 15.9716 |
| 4 | 0.0009 | 31.0331 |
| 5 | 0.0160 | 34.3780 |
| 6 | 0.0111 | 43.4328 |

In order to assess the performance of the HXN3 cab isolation system, numerical simulation studies are conducted. In the numerical simulations, the longitudinal, lateral and vertical transmissibilities of the system are calculated, as are shown in Fig.3~Fig.5, which are represented by $H_{C_xE_x}(\omega)$ , $H_{C_yE_y}(\omega)$ , $H_{C_zE_z}(\omega)$ in the frequency response function (FRF) matrix $H(\omega)$ respectively. In the following simulations studies, the excited frequency $\omega$ will be varied from 0.1Hz to 200Hz.

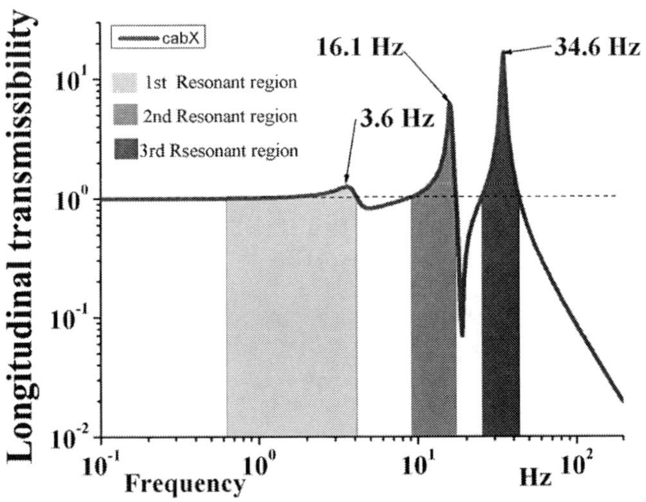


**Fig. 3 The longitudinal transmissibility of the cab isolation system**

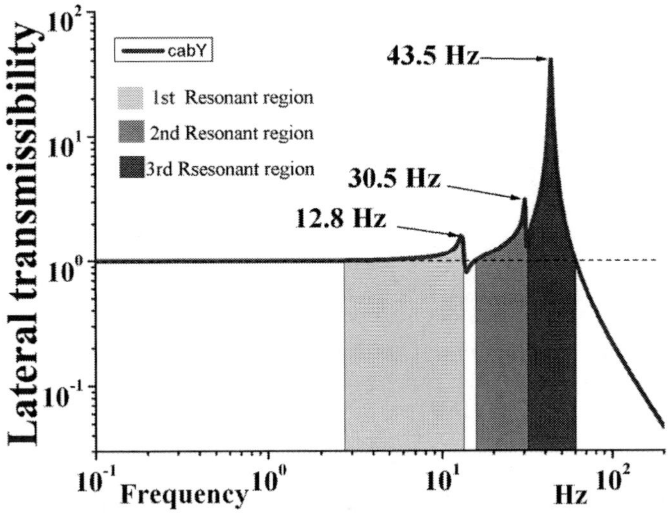


**Fig. 4 The lateral transmissibility of the cab isolation system**

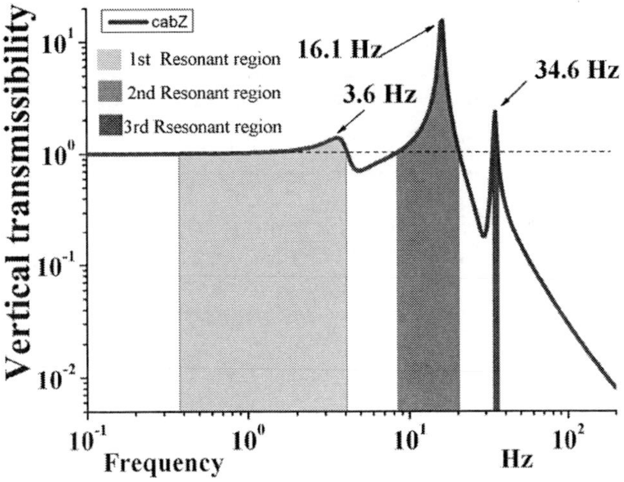


**Fig. 5 The vertical transmissibility of the cab isolation system**

As shown in Fig.3~Fig.5, there are 3 resonant regions in the transmissibility curve of each direction, and the resonant frequencies are equal to the natural frequencies which are shown in Table 2 because the longitudinal vibration and the vertical vibration are coupled. There is a coupling relationship between the longitudinal vibration and the vertical vibration. Therefore, the longitudinal resonant frequencies are equal to the vertical.

Fig.4 shows the lateral transmissibility of the HXN3 cab isolation system. The three resonant regions are close to each other. The lateral transmissibility will not be underneath 1 until the excited frequency is over 60Hz. So the lateral vibration isolation will work effectively with the excited frequency ( $\omega$ ) exceeding 60Hz. The transmissibility at the excited frequency (16.7Hz) derived by approximate calculation, which is 1.3, is greater than 1. Therefore, the cab isolation system cannot suppress the lateral excited vibration coming from the diesel engine.

Fig.5 shows the vertical transmissibility of the HXN3 cab isolation system. Different from the lateral, there are 2 isolation regions between the three vertical resonant regions, so the cab isolation system could suppress the low-frequency excited vibration. But the $2^{nd}$ resonant frequency (16.1Hz) is very close to the excited frequency (16.7Hz). Therefore, the resonant vibration will happen likely. And that is the reason why the vertical vibration of the HXN3 cab is large relatively.

The numerical results are very close to the actual situation. The lateral excited vibration cannot be suppressed by the HXN3 cab isolation system, and the vertical resonant vibration of the cab will occur. Further analysis is conducted to investigate the HXN3 cab structure vibration via establishing MBS (Multi-Body System) model with considering the cab flexibility.

## 3. VIBRATION ISOLATION ANALYSIS CONSIDERING CAB FLEXIBILITY

The excited frequency of the diesel engine was derived by approximate calculating. The frequency doubling of the excited frequency will cause the cab structural vibration. In this part, a rigid-flexible coupling model is established, and the transmissibilities at different direction are calculated with considering the cab structural flexibility.

### 3.1 Model description of the HXN3 cab isolation system

The HXN3 cab consists of the BIW (Body-In-White) and the vehicle equipments, such as the driver seats, the windows, the doors, the control table, and the air conditioner. The mass of the BIW and equipments is shown in Table 3.

**Table 3 The mass of the BIW and the cab equipment**

| | BIW | Driver seat | Window | Door | Control Tab | Air-conditioner |
|---|---|---|---|---|---|---|
| Mass/*Kg* | $1.69e^{3}\times 1$ | $100\times 2$ | $20\times 6$ | $150\times 4$ | $150\times 1$ | $150\times 1$ |

As shown in Fig.6, the FE (Finite-Element) model of the HXN3 cab BIW (Body-In-White) is imported into the HXN3 cab model. In this rigid-flexible coupling system, 85 nodes of the FE model are selected to analyze the deformation of the cab structure under the excited vibration, and the equipments are all considered as rigid bodies.

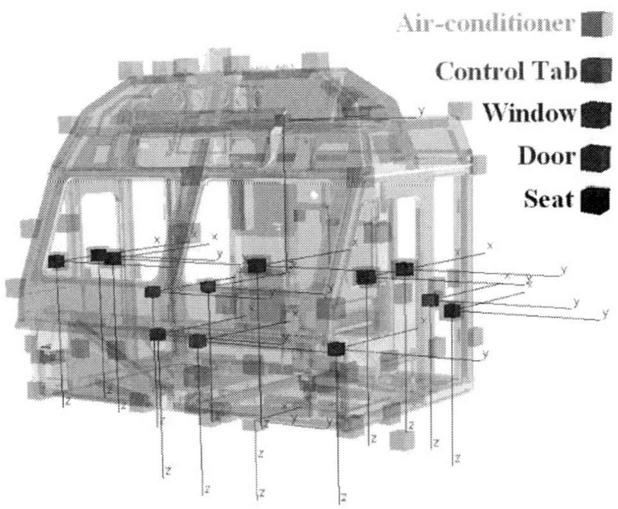


**Fig. 6 The rigid-coupling model of the HXN3 cab isolation system**

### 3.2 Numerical simulation result and analysis

For the rigid-flexible coupling model, there are 14 normal modes, and the damping ratio and frequency are listed in Table 4.

**Table 4 Natural frequency of the rigid-flexible coupling model**

| Mode | Damping ratio | Frequency/Hz | Mode | Damping ratio | Frequency/Hz |
|---|---|---|---|---|---|
| 1 | 0.1142 | 2.9510 | 8 | 0.0127 | 42.0629 |
| 2 | 0.0250 | 9.4495 | 9 | 0.0212 | 46.7976 |
| 3 | 0.0047 | 13.5645 | 10 | 0.0274 | 52.5318 |
| 4 | 0.0221 | 14.8578 | 11 | 0.0196 | 69.0739 |
| 5 | 0.0027 | 25.2260 | 12 | 0.0360 | 77.6145 |
| 6 | 0.0040 | 27.7565 | 13 | 0.0912 | 95.1143 |
| 7 | 0.0162 | 34.7260 | 14 | 0.0420 | 106.3476 |

Compared with the HXN3 cab 6-dimension structure, the normal modes increases to 14 and the highest natural frequency is larger after considering the cab deformation. The last 8 principal models are rigid-flexible coupled.

The transmissibilities of the HXN3 cab isolation system at different directions with considering the cab deformation, which are shown in Fig.7-Fig.9, are calculated by the dynamic software SIMPACK. In Fig.7-Fig.9, the dashed line is the transmissibility without considering cab deformation and the red bold line is the transmissibility with considering cab deformation.

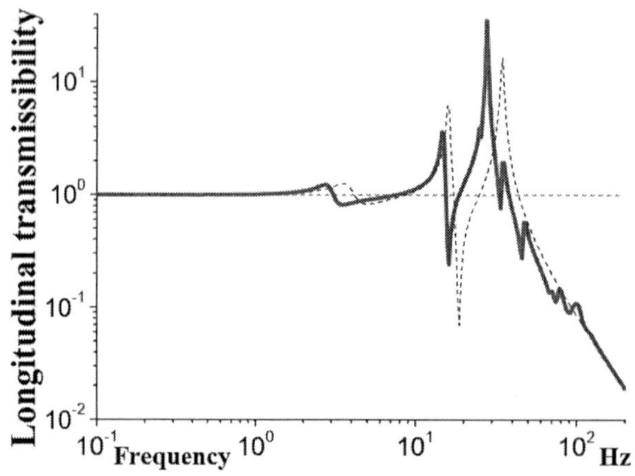


**Fig. 7 The longitudinal transmissibility comparison of the cab isolation system**

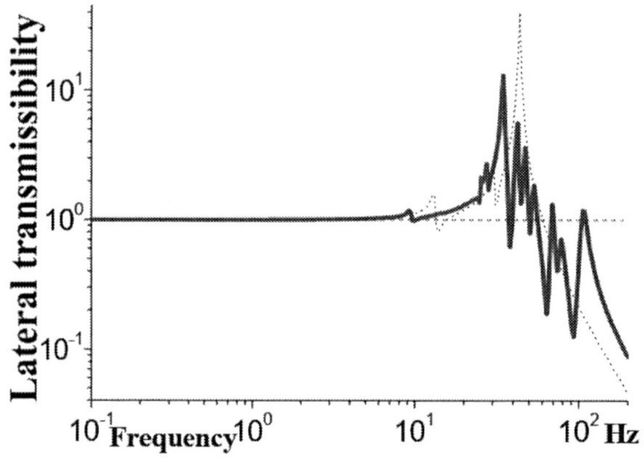


**Fig. 8 The lateral transmissibility comparison of the cab isolation system**

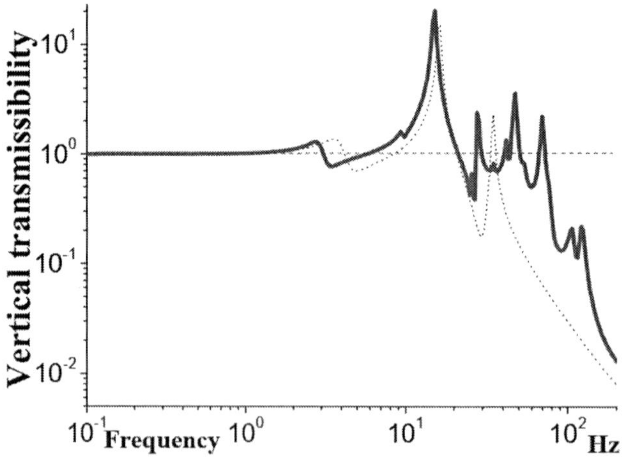


**Fig. 9 The vertical transmissibility comparison of the cab isolation system**

As shown in Fig.7-Fig.9, there are more resonant peaks appearing in the higher frequency region, after considering the cab deformation, especially the vertical transmissibility.

Fig.7 shows that the longitudinal transmissibility curve of the HXN3 cab isolation system is basically the same before and after the consideration of the cab deformation. Due to the cab deformation, the 3 longitudinal resonant frequencies decrease slightly, and there is barely not any new resonant peak appearing. Therefore, the cab deformation has a little impact on the longitudinal vibration isolation property of the system.

As shown in Fig.8-Fig.9, there are more new resonant peaks appearing in the lateral and vertical transmissibility in higher frequency region (30Hz-200Hz), compared with the longitudinal. Due to the cab deformation, the lateral and vertical transmissibility increase, and the resonant frequency region expands in higher frequency region. The cab deformation has a great impact on the lateral and vertical vibration isolation system property of the system. Therefore, the structural resonances will occur due to the doubling-frequency vibration.

## 4. EXPERIMENT APPARATUS AND RESULTS

### 4.1 Experiment setup

An experiment is conducted to assess the vibration of the HXN3 diesel locomotive. Acceleration sensors are located on the floor of the cab, the front part of the underframe, and the mounting positions of the engine and generator, so that the accelerations of different positions at different engine-working speeds are obtained. The experimental set-up is shown in Fig.10.

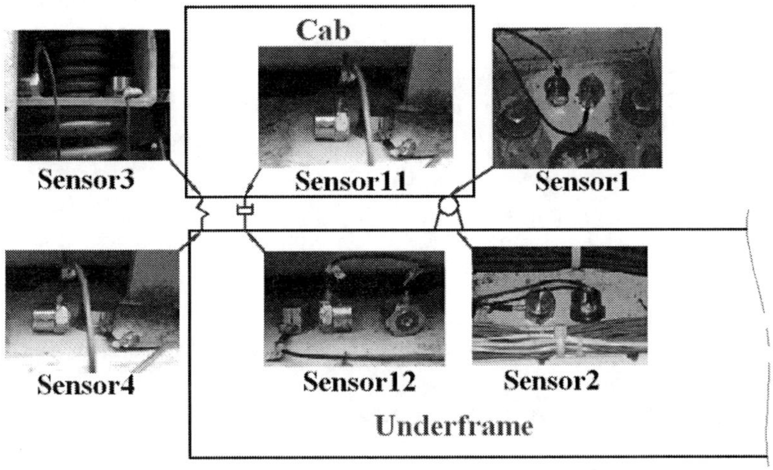


**Fig. 10 Experimental set-up for HXN3 diesel locomotive vibration experiment**

### 4.2 Experiment results

This section presents the isolation performance observed from the experiment. The vibration acceleration PSD (Power Spectral Density) of different sensors is derived, shown in Fig.11 and Fig.12.

Fig.11 shows the lateral and vertical vibration acceleration PSD at the cab mounting position of the underframe. The main frequencies of the lateral and vertical excited vibration acceleration are about 17Hz, which is very close to the result of the approximate calculation. It is evident that the high frequency component of the lateral and vertical excited vibration is significant. Therefore, the HXN3 cab isolation system should isolate the excited frequency (17Hz) effectively, and meanwhile, suppress the cab structural vibration effectively.

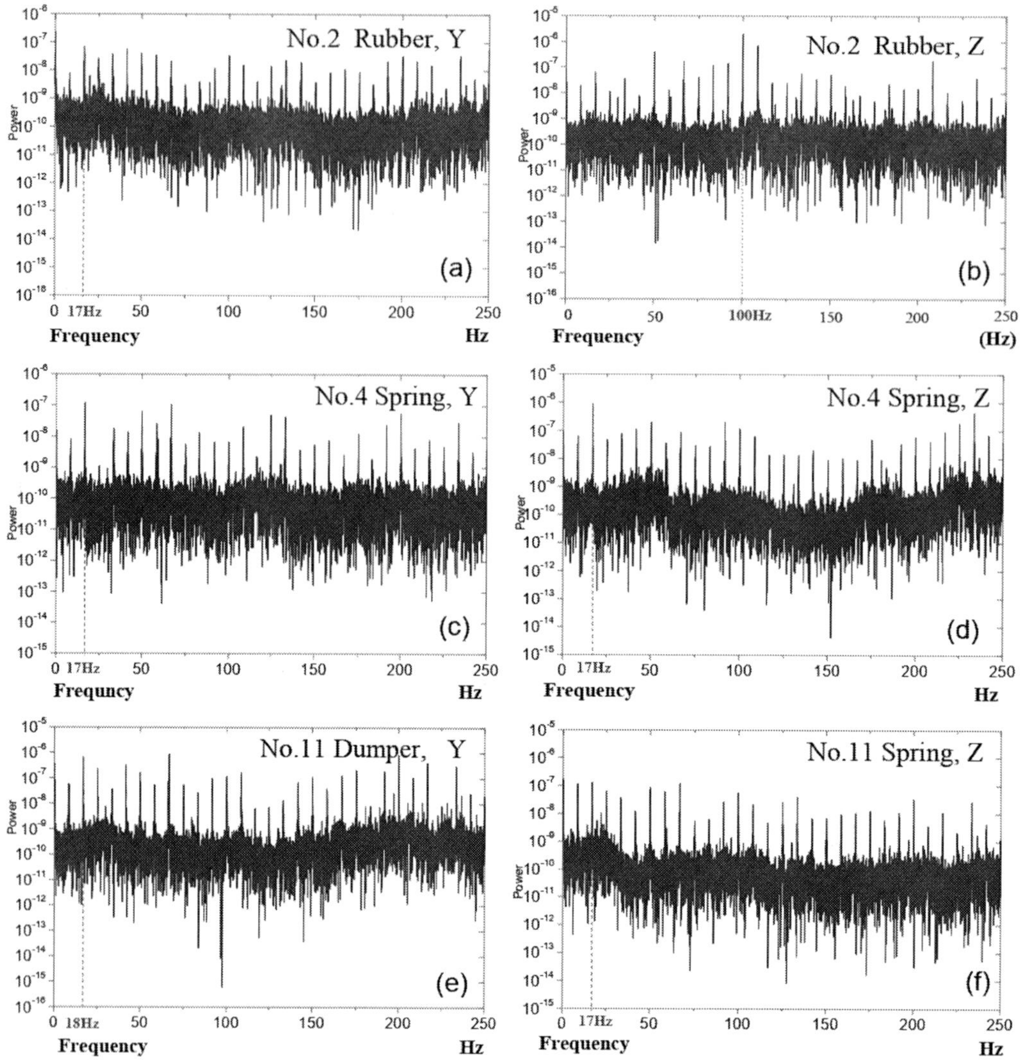


**Fig. 11 The lateral and vertical acceleration PSD at the front of underframe**

Fig.12 shows the lateral and vertical vibration isolation PSD at the floor of the HXN3 cab. The main frequencies of the cab lateral and vertical vibration acceleration are all about 17Hz, which are equal to the excited vibration frequency. The high frequency component of the lateral cab vibration is significant, however, the high frequency component of the cab vertical vibration is not.

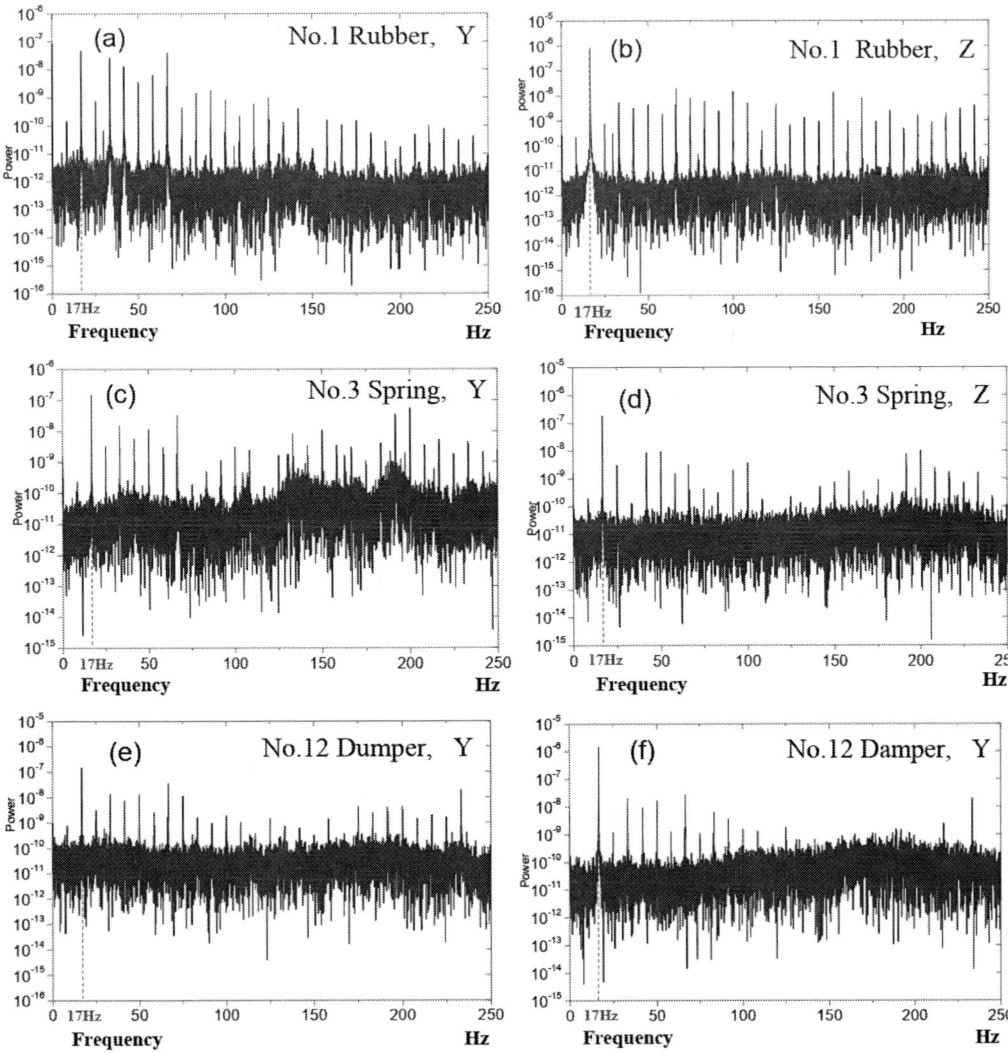


**Fig. 12 The lateral and vertical acceleration PSD at the floor of the cab**

## 4.3 Comparison of the experimental and theoretical results

The experimental results indicate that the cab isolation system could not suppress the 17Hz lateral excited vibration coming from the underframe, and the lateral structural vibration will occur under the high frequency excited vibration, while the vertical cab resonant vibration will occur but the vertical structural vibration is not significant in the contrast of the lateral. The experimental results agree with the theoretical analysis.

## 5. CONCLUSIONS

Vibration isolation is a very important problem for a wide range of vehicle engineering practice. In this study, the isolation performance of the HXN3 cab isolation system was theoretically analyzed by using the concept of multi-dimension frequency response function. The transmissibilities of the cab at different direction were compared before and after considering the cab deformation, so that the isolation performance under both the lower frequency region and the higher frequency region was derived. A verification experiment was conducted and the experiment results agreed with the theoretical analysis.

The following conclusions were drawn:

1) The dominant frequency of the excited vibration was derived by an experiment, which was equal to the approximate calculation result. The experiment results revealed that the high frequency component of the lateral and vertical excited vibration was significant.
2) The HXN3 cab isolation system could not suppress the lateral excited vibration effectively and the lateral structural vibration would be caused by the high frequency lateral excited vibration coming from the underframe generated by the diesel engine and the generator.
3) The vertical resonant vibration of the HXN3 cab could not be avoided. While in the contrast of the lateral vibration, the vertical structural vibration of the cab is not significant.
4) Some analysis procedure can be widely used to assess the isolation performance of other isolation systems.

Future work will focus on the optimization of the HXN3 cab isolation system in order to improve the isolation performance.

## REFERENCES


[1] Zhang Shu tang,WANG Shuang qiang. Research on the issue of the resonance of DF4 locomotive's cab. Diesel Locomotives. 2002(05)

[2] Wang Xiu–ping. Analysis & Treatment of Vibration of DF5 Locomotive. Internal Combustion Engine & Powerplant. 2010(02)

[3] Thanh Danh Le, Kyoung Kwan Ahn. A vibration isolation system in low frequency excitation region using negative stiffness structure for vehicle seat. Journal of sound and vibration. 330(2011) 6311-6335

[4] Zeqi Lu, Michael J. Brennan, Tiejun Yang, Xinhui Li, ZhigangLiu.Dynamics and power flow behaviour of a nonlinear vibration isolation system with a negative stiffness mechanism. Journal of sound and vibration. 332(2013) 167-183

[5] C.-M. Lee, A.H. Bogatchenkov, V.N. Goverdovskiy,Y.V. Shynkarenko, A.I. Temnikov. Position control of seat suspension with minimum stiffness. Journal of sound and vibration. 292(2006) 435-442

[6] A. Carrella, M.J.Brennan, I.Kovacic, T.P.Waters. On the force transmissibility of a vibration isolator with quasi-zero-stiffness. Journal of sound and vibration. 322(2009) 707-717

[7] Daolin Xu, QipingYu, Jiaxi Zhou, S.R.Bishop. Theoretical and experimental analyses of a nonlinear magnetic vibration isolator with quasi-zero-stiffness characteristic. Journal of sound and vibration. 332(2013) 3377-3389

[8] A. Carrella, M.J. Brennan, T.P. Waters. Static analysis of a passive vibration isolator with quasi-zero-stiffness characteristic. Journal of sound and vibration. 301(2007) 678-689

[9] A.D. Shaw, S.A. Neild, D.J. Wagg. Dynamic analysis of high static low dynamic stiffness vibration isolation mounts. Journal of sound and vibration. 332(2013) 1437-1455

[10] N. Zhou, K. Liu. A tunable high-static–low-dynamic stiffness vibration isolator. Journal of sound and vibration. 329(2009) 1454-1273

[11] Z.Q. Lang, X.J.Jing, S.A.Billings, G.R.Tomlinson, Z.K.Peng. Theoretical study of the effects of nonlinear viscous damping on vibration isolation of sdof systems. Journal of sound and vibration. 323(2009) 352-365

[12] Zhenlong Xiao, XingjianJing, LiCheng. The transmissibility of vibration isolators with cubic nonlinear damping under both force and base excitations. Journal of sound and vibration. 332(2013) 1335-1354

[13] Bin Tang, M.J. Brennan. A comparison of two nonlinear damping mechanisms in a vibration isolator[J]. Journal of sound and vibration. 332(2013) 510-520

[14] P.Thureau,D.Lecter, An Introduction to the Principle of Vibrations of Linear Systems, Stanley Thornes (Publishers) Ltd., UK, 1981
[15] Hariss, C.H.: Shock and Vibration Handbook. McGraw Hill, New York, 1996
[16] Ashrafiuon, H. and Nataraj, C.: Dynamic Analysis of Engine Mount Systems. J. Vib. Acoustic 114 (1992), pp. 79–83
[17] M. Roseau, Vibrations in Mechanical Systems, Springer, Berlin, 1987
[18] R.S. Sharp, D.A. Crolla, Road vehicle suspension system design: a review, Vehicle System Dynamics 16 (1987) 167-192

# Vibration isolation design of under chassis equipments for high-speed EMU trains

**D Gong, J Zhou, W Sun**
Institute of Railway & Urban Mass Transit Research, Tongji University, China

## ABSTRACT

To investigate the effects of the under chassis suspended equipment on carbody eigenfrequencies, a numerical method for calculating the eigenfrequencies of a high-speed EMU vehicle carbody with suspended equipment is proposed, and the design of the equipment suspension parameter according to the principles of mode matching, transmissibility matching and consider the optimal vibration of both carbody and equipment as the objective is presented. Results show that by tuning the equipment suspension stiffness, the first vertical bending frequency of carbody will have a 'frequency hopping' phenomenon; The first vertical bending frequency of carbody could be enhanced by means of equipment with optimized suspension parameter to suppress the vibrations of both the carbody and the equipment.

**Keywords:** High-speed EMU trains, under chassis equipment, vibration isolation design, mode matching.

## NOTATION

| | |
|---|---|
| $m_c$ | carbody mass |
| $I_c$ | carbody pitch inertia |
| $EI$ | carbody bending stiffness |
| $\mu$ | carbody internal damping coefficient |
| $m_t$ | bogie mass |
| $I_t$ | bogie pitch inertia |
| $k_s$ | secondary spring stiffness per bogie |
| $c_s$ | secondary damper coefficient per bogie |
| $k_p$ | primary spring stiffness per axle |
| $c_p$ | primary damper coefficient per axle |
| $L$ | carbody length |
| $l_b$ | half of bogie centres |
| $l_w$ | half of bogie wheelbase |
| $l_i$ | location of suspended equipment |
| $\tau_i$ | time delay between wheelsets |
| $\bar{W}_w$ | vertical movement of the wheelset |
| $z(x,t)$ | carbody vertical vibration displacement |
| $\delta(x)$ | the Dirac function |
| $\omega$ | the circular frequency |
| $G(x, \xi)$ | Green's function |
| $f_{eq}$ | natural frequency of independent suspended equipment |
| $k_{eq}$ | stiffness of rubber spring |
| $m_{eq}$ | equipment mass |
| $f_{Fixed}$ | the first vertical bending frequency of carbody with fixed equipment |
| $f_i$ | the $i$-th order eigenfrequency of carbody with suspended equipment |
| $f_{1st}$ | the first vertical bending frequency of carbody with suspended equipment |
| $\zeta$ | the damping ratio of the suspended equipment |
| $\gamma$ | the ratio of the excitation frequency to the suspended equipment natural frequency |

## 1 INTRODUCTION

Distributed power technology is widely utilized in high-speed EMU trains, subway and urban rail vehicles. Comparing to the trains with centralized power, distributed power trains can achieve flexible marshalling, high-power traction, full using of wheel-rail adhesion, and has lower wheel/rail forces, etc [1]. Distributed power means there will be more equipments mounted under the chassis, and many of the equipments have large mass and even vibration sources. Usually, the equipment can be divided into two types according to the mounting methods, i.e. fixed equipment and suspended equipment. Fixed equipment means the equipment is welded or fixed under the vehicle's chassis directly, while suspended equipment is the equipment that hanged elastically under the chassis by using vibration isolation components, such as rubber spring whose material is compact, and which can provide three-dimensional stiffness and certain damping. Equipment suspended elastically is more welcome to the high-speed EMU trains, not only it has great effects on the vibration reduction especially at higher frequencies, but also the elastic components can automatically balance the preload force discrepancy on each suspended bases of the equipment. However, the selection of equipment hanging scheme should consider its influences on the carbody modal frequencies and the carbody and the equipment vibrations. Gong studied the influences of under chassis equipment suspension stiffness on carbody bending frequency by utilizing a simplified 2 DOFs model [1]. He found that the carbody bending frequency could be enhanced by equipment suspended. Clas investigated the advantages with an elastically suspended equipment on decreasing the amplitudes of the carbody's oscillations of deflection [2]. Sun et al. researched the effect of under chassis equipment parameters on vehicle ride quality by simulation [3]. Shi [4] and Gong [5] proposed to utilize dynamic vibration absorber theory in vibration suppression carbody vibration with under chassis equipment.

In this paper a vertical theory model of carbody with under chassis equipments is established first, and the numerical method for calculating the eigenfrequencies of the coupled system is derived. Then, the "frequency hopping" phenomenon of carbody first vertical bending mode is presented and investigated by frequency response, eigenfrequency calculation and testing. Finally, a new optimum design method of equipment suspension parameter based on mode matching, transmissibility matching and consider the optimal vibration of both carbody and equipment as the objective is described and validated by an in-situ test.

## 2 THEORY

Consider a vertical railway carbody model with $n$ under chassis equipments (Fig. 1). The carbody is modelled as an Euler-Bernoulli beam supported on two secondary suspensions [6, 7], only vertical vibrations of the carbody, i.e. bounce, pitch and flexible bending modes are considered. Each frame has two degrees of freedom i.e. bounce and pitch. It is assumed that the no wheel jump happens when the vehicle runs on track, which means the wheelsets vertical movements are exactly the same as the track irregularity. According to the coordinates definition in Fig.1, the carbody partial differential equation and the suspended equipment motion equation can be obtained

$$EI\frac{\partial^4 z(x,t)}{\partial x^4}+\mu I\frac{\partial^5 z(x,t)}{\partial t\partial x^4}+\rho\frac{\partial^2 z(x,t)}{\partial t^2}=\sum_{i=1}^{2}k_s[z_{t1}-z(l_i,t)]\delta(x-l_i)+\sum_{i=1}^{2}c_s[\dot{z}_{t1}-\dot{z}(l_i,t)]\delta(x-l_i)+\dots$$

$$\dots+\sum_{j=3}^{n}k_j[z_j-z(l_j,t)]\delta(x-l_j)+\sum_{j=3}^{n}c_j[\dot{z}_j-\dot{z}(l_j,t)]\delta(x-l_j) \quad (1)$$

$$\begin{aligned} m_t\ddot{z}_{ti} &= k_s[z(l_i,t) - z_{ti}] + c_s[\dot{z}(l_i,t) - \dot{z}_{ti}] - k_p(2z_{ti} - z_{w(2i-1)} - z_{w(2i)}) + \ldots \\ &\ldots - c_p(2\dot{z}_{ti} - \dot{z}_{w(2i-1)} - \dot{z}_{w(2i)}) \qquad i = 1,2 \end{aligned} \tag{2}$$

$$m_j\ddot{z}_j = -k_j[z_j(t) - z(l_j,t))] - c_j[\dot{z}_j(t) - \dot{z}(l_j,t)], \quad j = 3,\ldots,n \tag{3}$$

where $z(x,t)$ denotes the vertical vibration displacement of the carbody at point $x$ and time $t$, $\delta(x)$ is the Dirac function. When the system experiences free vibration, the solutions of the above equations have the following form

$$z(x,t) = \bar{W}(x)e^{\mathrm{i}\omega t}, \quad z_{ti} = \bar{W}_{ti}e^{\mathrm{i}\omega t}, \quad z_j = \bar{W}_j e^{\mathrm{i}\omega t} \tag{4}$$

where $\omega$ is the circular frequency. This allows Eqs.(1) ~ (3) to be reduced to the following simplified form

$$\bar{W}'''' - \kappa^4\bar{W} = \frac{1}{EI + \mathrm{i}\omega\mu I}\left\{\sum_{i=1}^{2}\Lambda_s\left[\bar{W}_i - \bar{W}(l_i)\right]\delta(x - l_i) + \sum_{j=3}^{n}\Lambda_j\left[\bar{W}_j - \bar{W}(l_j)\right]\delta(x - l_j)\right\} \tag{5}$$

$$\bar{W}_{ti} = \hat{K}_s\bar{W}(l_i) + \hat{K}_p(\bar{W}_{w(2i-1)} + \bar{W}_{w(2i)}) \tag{6}$$

$$\bar{W}_j = \hat{K}_j\bar{W}(l_j) \tag{7}$$

where

$$\kappa^4 = \frac{\omega^2\rho}{EI + \mathrm{i}\omega\mu I}, \quad \Lambda_s = k_s + \mathrm{i}\omega c_s, \quad \Lambda_j = k_j + \mathrm{i}\omega c_j$$

$$\hat{K}_p = \frac{k_p + \mathrm{i}\omega c_p}{k_s + \mathrm{i}\omega c_s + 2(k_p + \mathrm{i}\omega c_p) - \omega^2 m_t}, \quad \hat{K}_s = \frac{k_s + \mathrm{i}\omega c_s}{k_s + \mathrm{i}\omega c_s + 2(k_p + \mathrm{i}\omega c_p) - \omega^2 m_t}, \quad \hat{K}_j = \frac{k_j + \mathrm{i}\omega c_j}{-\omega^2 m_j + k_j + \mathrm{i}\omega c_j} \tag{8}$$

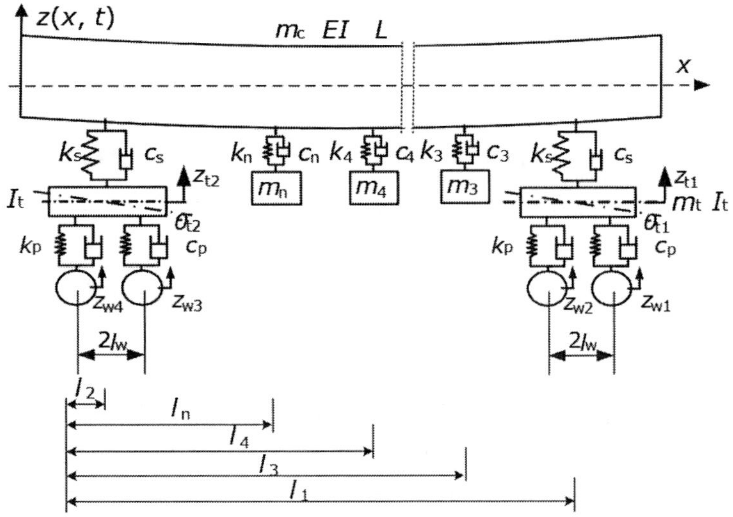


**Fig.1 A vertical model of carbody with under chassis equipments**

Assuming that Green's function $G(x, \xi)$ is the solution of the equation below

$$G''''(x,\alpha) - \kappa^4 G(x,\alpha) = \frac{\delta(x - \alpha)}{EI + \mathrm{i}\omega\mu I} \tag{9}$$

Green's function $G(x, \xi)$ is the response at position $x$ with unit concentration force at an arbitrary position $\xi$ of carbody [8]. The Green's function $G(x, \xi)$ can be derived from equation (9) by utilizing Laplace transformation [9]. Based on the superposition principle, the solution of equation (9) can be given as

$$\bar{W}(x) = \int_0^L f(\xi)G(x,\xi)\mathrm{d}\xi \tag{10}$$

The functional form of $f(\xi)$ is given by Eq.(5). Utilizing the properties of the Dirac function and performing the integration one obtains

$$\bar{W}(x) = \sum_{i=1}^{2} \Lambda_s \left[\bar{W}_{ti} - \bar{W}(l_i)\right] G(x,l_i) + \sum_{j=3}^{n} \Lambda_j \left[(\hat{K}_j - 1)\bar{W}(l_j)\right] G(x,l_j) \tag{11}$$

One solves the $\bar{W}(x)$ of equation (11) at the all points of springs and equipment attachments, i.e. at $x = l_i$ and $l_j$. This gives $n$ equations with $n$ unknowns $\bar{W}(l_1)$, $\bar{W}(l_2)$, $\bar{W}(l_3),\ldots,\bar{W}(l_n)$, the equations can be written in the matrix form as

$$\mathbf{B}\tilde{\mathbf{Z}} = \mathbf{c} \tag{12}$$

Where

$$\tilde{\mathbf{Z}} = [\bar{W}(l_1),\ \ \bar{W}(l_2),\ \ \bar{W}(l_3),\ \ \cdots,\ \ \bar{W}(l_n)]^{\mathrm{T}} \tag{13}$$

$$\mathbf{B}=\begin{bmatrix} 1+\Lambda_s(1-\hat{K}_s)G(l_1,l_1) & \Lambda_s(1-\hat{K}_s)G(l_1,l_2) & \Lambda_3(1-\hat{K}_3)G(l_1,l_3) & \Lambda_4(1-\hat{K}_4)G(l_1,l_4) & \cdots & \Lambda_n(1-\hat{K}_n)G(l_1,l_n) \\ \Lambda_s(1-\hat{K}_s)G(l_2,l_1) & 1+\Lambda_s(1-\hat{K}_s)G(l_2,l_2) & \Lambda_3(1-\hat{K}_3)G(l_2,l_3) & \Lambda_4(1-\hat{K}_4)G(l_2,l_4) & \cdots & \Lambda_n(1-\hat{K}_n)G(l_2,l_n) \\ \Lambda_s(1-\hat{K}_s)G(l_3,l_1) & \Lambda_s(1-\hat{K}_s)G(l_3,l_2) & 1+\Lambda_3(1-\hat{K}_3)G(l_3,l_3) & \Lambda_4(1-\hat{K}_4)G(l_3,l_4) & \cdots & \Lambda_n(1-\hat{K}_n)G(l_3,l_n) \\ \Lambda_s(1-\hat{K}_s)G(l_4,l_1) & \Lambda_s(1-\hat{K}_s)G(l_4,l_2) & \Lambda_3(1-\hat{K}_3)G(l_4,l_3) & 1+\Lambda_4(1-\hat{K}_4)G(l_4,l_4) & \cdots & \Lambda_n(1-\hat{K}_n)G(l_4,l_n) \\ \vdots & \vdots & \vdots & \vdots & \ddots & \\ \Lambda_s(1-\hat{K}_s)G(l_n,l_1) & \Lambda_s(1-\hat{K}_s)G(l_n,l_2) & \Lambda_3(1-\hat{K}_3)G(l_n,l_3) & \Lambda_4(1-\hat{K}_4)G(l_n,l_4) & \cdots & 1+\Lambda_n(1-\hat{K}_n)G(l_n,l_n) \end{bmatrix} \tag{14}$$

$$\mathbf{c} = \Lambda_s \hat{K}_p \bar{W}_w \begin{bmatrix} (1+e^{-i\omega\tau_2})G(l_1,l_1) + (e^{-i\omega\tau_3} + e^{-i\omega\tau_4})G(l_1,l_2) \\ (1+e^{-i\omega\tau_2})G(l_2,l_1) + (e^{-i\omega\tau_3} + e^{-i\omega\tau_4})G(l_2,l_2) \\ (1+e^{-i\omega\tau_2})G(l_3,l_1) + (e^{-i\omega\tau_3} + e^{-i\omega\tau_4})G(l_3,l_2) \\ (1+e^{-i\omega\tau_2})G(l_4,l_1) + (e^{-i\omega\tau_3} + e^{-i\omega\tau_4})G(l_4,l_2) \\ \vdots \\ (1+e^{-i\omega\tau_2})G(l_n,l_1) + (e^{-i\omega\tau_3} + e^{-i\omega\tau_4})G(l_n,l_2) \end{bmatrix} \tag{15}$$

in which, $\bar{W}_w$ is the vertical movement of the wheelset and also the track irregularity according to the assumption, $\tau_i$ is the time delay of the $2^{nd} \sim 4^{th}$ wheelsets respect to the $1^{st}$ one, gives

$$\tau_2 = 2l_w / v,\quad \tau_3 = 2l_b / v,\quad \tau_4 = 2(l_w + l_b) / v \tag{16}$$

One can obtain the eigenfrequencies of the coupled system by setting the determinant of the matrix **B** equals to zero, yielding the frequency function by appropriate techniques.

$$\det \mathbf{B}(\omega) = 0 \tag{17}$$

## 3 CASE STUDIES AND DISCUSSION

In this section, the influences of the suspended under chassis equipment on carbody frequency will be further investigated. Due to the requirement of the balanced distribution of wheel loads, heavy equipments will be deployed near the carbody centre, while the lighter ones which have less effect on the carbody vibration will be deployed uniformly at both sides of the carbody. In the case study, an equipment (a transformer) which mounted under the carbody centre by rubber springs is considered, its mass is 6559 kg. The stiffness of rubber spring $k_{eq}$ can be determined by Eq.(18),

$$f_{eq} = \frac{1}{2\pi}\sqrt{\frac{k_{eq}}{m_{eq}}} \tag{18}$$

where, $f_{eq}$ is the natural frequency of independent suspended equipment, which is one of the most important factors of rubber spring designing.

The damping ratio of rubber spring is usually in the range of 0.025~0.20 but can not be too high, otherwise, the rubber is easy to heat and its process of aging and creep would be accelerated [10], in this paper, the damping ratio of rubber spring is set to 0.06. The vehicle parameters of the studied high-speed EMU train in Table 1 are considered.

**Table 1 The vehicle parameters of a typical high-speed EMU train**

| | | | |
|---|---|---|---|
| carbody mass | 29080 kg | secondary spring stiffness per bogie | 260 kN/m |
| carbody pitch inertia | $1.42\times10^{6}$ kgm$^2$ | secondary damper coefficient per bogie | 20 kNs/m |
| carbody bending stiffness | $3.64\times10^{9}$ Nm$^2$ | primary spring stiffness per axle | 2000 kN/m |
| carbody internal damping coefficient | 0.015 | primary damper coefficient per axle | 20 kNs/m |
| bogie mass | 2200 kg | carbody length | 24.5 m |
| bogie pitch inertia | 1200 kgm$^2$ | half of bogie centres | 8.75 m |
| equipment mass | 6559 kg | half of bogie wheelbase | 1.25 m |

### 3.1 Frequency response

The vehicle is firstly excited by harmonic excitations on four wheelsets at the same time to analyze the frequency response. Fig.2 shows the relations between the natural frequency of independent suspended equipment $f_{eq}$ and the displacement transmissibility from wheelset to the carbody centre. It can be revealed that comparing to the carbody of fixed equipment, the carbody and the suspended equipment make a new coupled system, the original first vertical bending frequency of carbody $f_{Fixed}$ is separated into two new ones, one lower $f_1$ and one higher $f_2$. The mode shape of the lower eigenfrequency $f_1$ is that carbody bends in phase with equipment vertical movement, and for the higher eigenfrequency $f_2$ is that they vibrate opposite in phase [1]. One can also find that, with the increasing of $f_{eq}$, the vibration response of carbody at $f_1$ enhances, while that at $f_2$ decreases sharply. When $f_{eq}$ rises to higher than $f_{Fixed}$, the carbody will vibrate more at $f_1$ which is negative for the ride quality of carbody. The stiffer the suspended rubber spring, the more the coupled system looks like the carbody with fixed equipment.

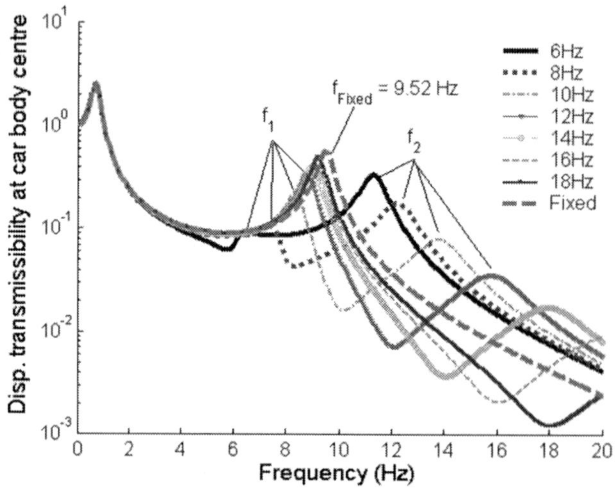


**Fig.2 Displacement Transmissibility at carbody centre**

Fig.3 depicts the displacement transmissibility difference between carbody and equipment $T_c$-$T_{eq}$. It is easy to find that when $f_{eq}$ is lower than $f_{Fixed}$, the equipment will be oscillating most at the lower $f_1$ and the carbody at the higher $f_2$, in this case, the major mode at the system's lower eigenfrequency is equipment motion, while the higher eigenfrequency is carbody bending motion. In other words, $f_1$ and $f_2$ may be defined as the nominal frequencies of equipment motion and carbody first vertical bending modes of the coupled system, respectively. When $f_{eq}$ is higher than $f_{Fixed}$, the difference value of displacement transmissibility at $f_2$ almost equals to zero, while that at $f_1$ begins to decline until it equals zero when $k_{eq}$ grows to infinite, meanwhile, $f_1$ will converge to $f_{Fixed}$, on this occasion, $f_1$ corresponds to the nominal frequency of carbody first vertical bending mode of the coupled system.

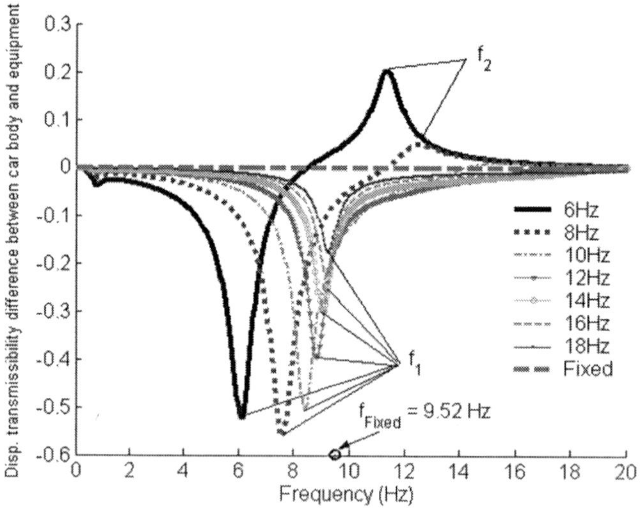


**Fig.3 Disp. transmissibility difference between carbody and equipment**

When $f_{eq}$=8 Hz, the equipment mass is set to be 1500 kg, 3000 kg and 6559 kg, respectively, the effects of different weights of the suspended equipment mass on the displacement transmissibility of carbody centre is depicted in Fig.4. From the results presented one concludes that when the suspended mass becomes heavier, the vibration response of carbody at $f_2$ will decline, meanwhile $f_2$ will move to the right. That is to say, with an appropriate suspension stiffness, the equipment seems to be removed from the carbody, which results in the carbody first vertical bending frequency being enhanced comparing to that of carbody with fixed equipment. Furthermore, due to the movement of carbody bending and suspended equipment in inverse phase, it will help attenuate the carbody vibration to some extent. The heavier the equipment weight with reasonable suspension, the better.

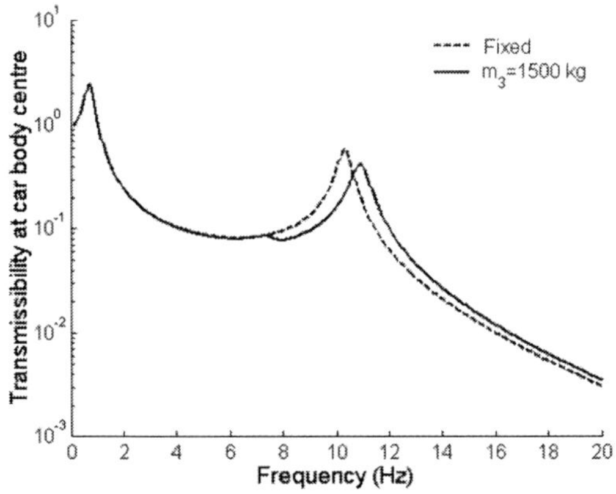


**a) $m_{eq}$=1500 kg**

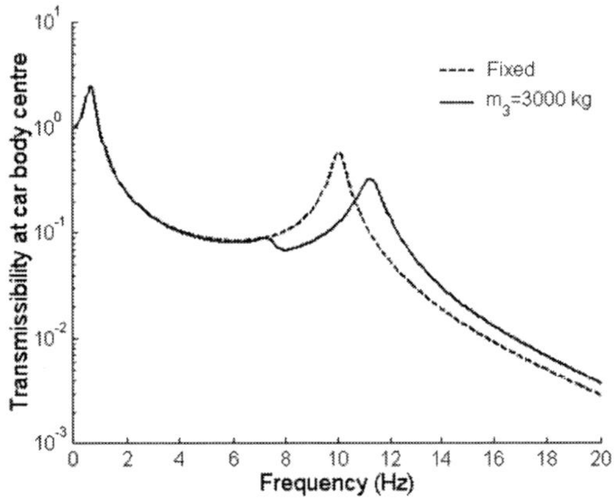


**b) $m_{eq}$=3000 kg**

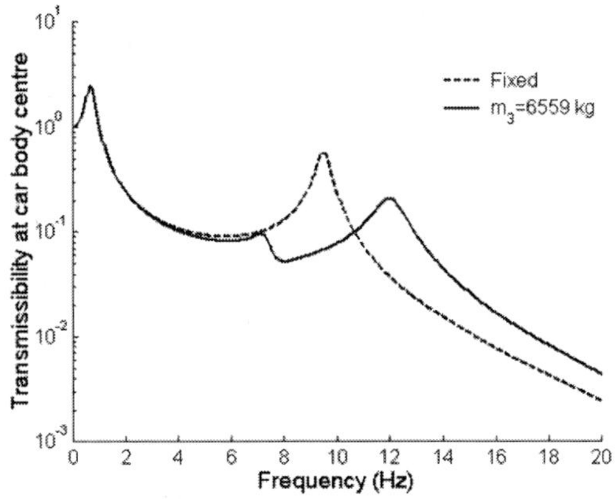


**c) $m_{eq}$=6559 kg**

**Fig.4 Comparison of different weights of the suspended masses**

### 3.2 Eigenfrequency calculation and testing

To further study the effects of suspended stiffness on carbody eigenfrequencies, considering the parameters shown in Table 1, the first four eigenfrequencies of the coupled system are calculated by means of Eq.(17), and the results are listed in Table 2 for different values of $f_{eq}$. In addition, the first vertical bending frequency of carbody is calculated to validate the result of Eq.(17) by using the finite element model (FEM) of the EMU train, which is listed in the last column of Table 2. The last two rows list the first vertical bending frequencies of carbody with fixed equipment and without equipment, respectively. Table 2 illustrates that, according to the analysis of Section 3.1, the calculated carbody first vertical bending frequencies for different values of $f_{eq}$ by Eq.(17) are identified and shown in italics, and they are in good agreement with the results of FEM. Increasing $f_{eq}$ leads to a 'frequency hopping' phenomenon of the first vertical bending frequency of carbody. The 'frequency hopping' happens when $f_{eq}$ bigger than $f_{Fixed}$ as shown in Fig.5. Furthermore, this changing of $f_{eq}$ almost has no effect on the second and third order eigenfrequencies of the system.

**Table 2 The eigenfrequency calculation results by using Eq.(17) and FEM**

| $f_{eq}$(Hz) | Eq.(17) | | | | FEM |
|---|---|---|---|---|---|
| | $f_1$ | $f_2$ | $f_3$ | $f_4$ | $f_{1st}$ |
| 6 | 6.15 | *11.34* | 24.03 | 57.25 | 11.58 |
| 7 | 6.93 | *11.72* | 24.03 | 57.30 | 12.00 |
| 8 | 7.56 | *12.26* | 24.03 | 57.35 | 12.51 |
| 9 | 8.04 | *12.95* | 24.03 | 57.41 | 13.25 |
| 10 | *8.39* | 13.77 | 24.03 | 57.48 | 8.50 |
| 11 | *8.63* | 14.68 | 24.03 | 57.56 | 8.77 |
| 12 | *8.81* | 15.67 | 24.03 | 57.65 | 8.97 |
| 13 | *8.94* | 16.70 | 24.03 | 57.75 | 9.12 |
| 14 | *9.03* | 17.75 | 24.03 | 57.85 | 9.19 |
| 15 | *9.10* | 18.82 | 24.03 | 57.97 | 9.28 |
| 16 | *9.16* | 19.89 | 24.03 | 58.10 | 9.30 |
| 17 | *9.21* | 20.97 | 24.03 | 58.25 | 9.35 |
| 18 | *9.24* | 22.04 | 24.03 | 58.40 | 9.42 |
| Fixed | -- | 9.52 | 24.03 | 50.40 | 9.73 |
| No eq. | -- | 11.55 | 24.08 | 57.13 | 11.55 |

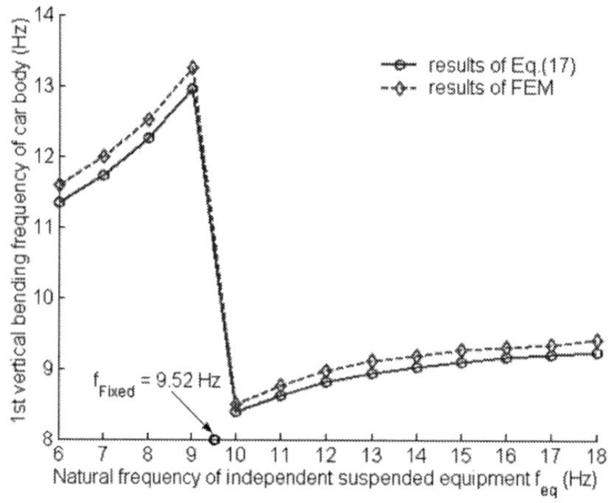


**Fig.5 'Frequency hopping' phenomenon of carbody**

A modal test is carried out to validate the carbody 'frequency hopping' phenomenon as shown in Fig.6. In the test, seven equidistant cross sections of the studied carbody along its length and the four mounting bases of the transformer are arranged to deploy acceleration sensors, the natural frequency $f_{eq}$ of the transformer is tuned to be 6 Hz, 7 Hz and 10 Hz, respectively. The identified eigenfrequency of the test are shown in Fig.7. As can be seen, by tuning the $f_{eq}$ to be respective 6 Hz and 7 Hz, the carbody first vertical bending frequency is respective 11.79 Hz and 12.04 Hz with the mode shape of carbody bends out of phase with equipment motion. When tuning the $f_{eq}$ to be 10 Hz, the first vertical bending frequency will be 8.85 Hz with the mode shape of carbody bends in phase with equipment motion. The test results agree with the above calculation results well. From the results presented one concludes that, if the frequency of the studied carbody first vertical bending mode is recommended over 10 Hz which is positive for the carbody vibration [6], the value of $f_{eq}$ should be lower than 9.52 Hz.

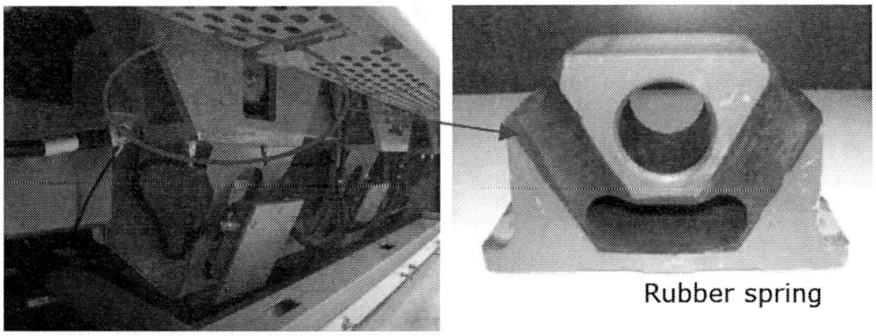


**Fig.6 The tested vehicle, equipment and rubber spring**

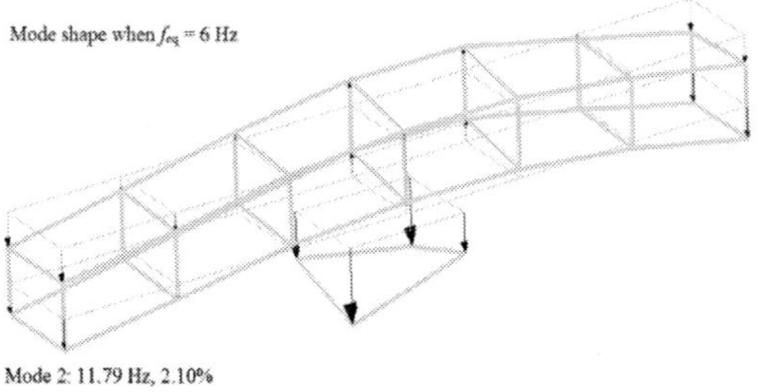


**(a) $f_{eq}$ = 6 Hz (out of phase)**

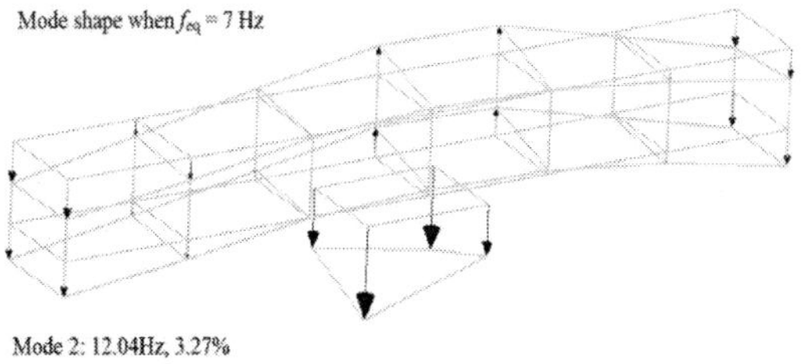


**(b) $f_{eq}$ = 7 Hz (out of phase)**

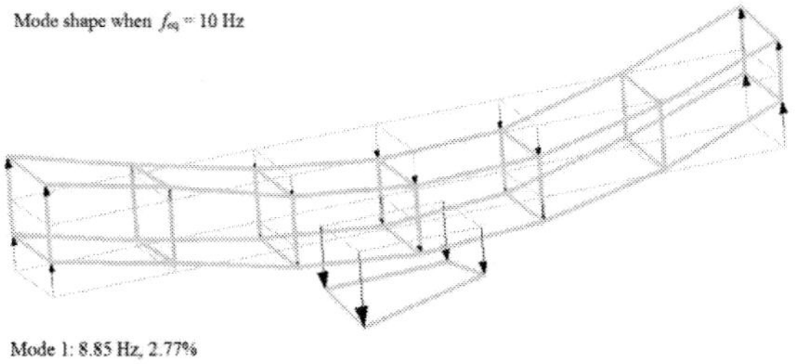


**(c) $f_{eq}$ = 10 Hz (in phase)**

**Fig.7 Mode shapes**

## 4 DESIGN OF EQUIPMENT SUSPENSION PARAMETERS BASED ON MODE MATCHING

Obviously, although the under chassis equipment suspended elastically can enhance the carbody first bending frequency to some extent, if the suspended parameters of equipment are not appropriate, it will probably lead to the equipment abnormal vibration and carbody localized resonance, or even affect the vehicle operation reliability and safety. To avoid these problems happen, an optimum design method of equipment suspension parameter on account of mode matching, transmissibility matching and taking the optimal vibration of both carbody and equipment as the objective are proposed in this section.

The mode matching adopted between carbody and suspended equipment is based on the theory of vibration isolation because of its good effect on mechanical vibration reduction. Generally, there are two types of vibration isolation, one is negative vibration isolation which aims to reduce the vibration transmitted from the supporting base to the equipment and the other is positive vibration isolation which aims to reduce the vibration transmitted from the equipment to the surrounding objects. According to the vibration characteristics of railway carbody and suspended equipment, the vibration transmitted from carbody to suspended equipment is suggested to be designed as negative vibration isolation, while the opposite vibration transmitted from suspended equipment to carbody is suggested to be designed as positive vibration isolation. The dimensionless transmissibility of negative vibration isolation is given by [11],

$$|T|=\sqrt{\frac{1+4\zeta^2\gamma^2}{(1-\gamma^2)^2+4\zeta^2\gamma^2}} \tag{19}$$

where $\zeta=\frac{c_{eq}}{2\sqrt{k_{eq}m_{eq}}}$ is the damping ratio of the suspended equipment, $\gamma=f_{ex}/f_{eq}$ is the ratio of the excitation frequency to the suspended equipment natural frequency. According to Eq. (19), no matter what the value of $\zeta$ is, to achieve good performance of vibration isolation, i.e. $|T|<1$, $\gamma$ should be greater than $\sqrt{2}$, i.e.,

$$\text{Condition 1,}\ \frac{f_{ex1}}{f_{eq}}>\sqrt{2} \tag{20}$$

If $\frac{f_{ex1}}{f_{eq}}$ could only be in the range of less than $\sqrt{2}$ for other reasons, $\gamma$ should be no more than 0.4 [12], one has,

$$\text{Condition 2,}\ \frac{f_{ex2}}{f_{eq}}\leq 0.4 \tag{21}$$

Generally, the vertical excitation sources of carbody are mainly composed of the vibration of carbody rigid modes (no more than 2Hz) and elastic modes. On account of the aforementioned conditions and the vibration contribution of carbody modes [13, 14], in this case, $f_{ex1}$ will be the first vertical bending frequency of the carbody without equipment, i.e. 11.55 Hz attained from Table 2 and $f_{ex2}$ equals 2 Hz.

Also the $f_{eq}$ needs to follow the transmissibility matching in Section 3, i.e.,

$$\text{Condition 3,}\ f_{eq}<9.52\ \text{Hz} \tag{22}$$

Consequently, one obtains the range of $f_{eq}$, i.e. 5.0 Hz to 8.2 Hz. Since the excited frequencies of some equipments such as motor or air compressor, etc., are always higher than this frequency rage, it also satisfies with the requirements of positive vibration isolation design. The optimal value of $f_{eq}$ can be further obtained by considering the optimal vibration of both carbody and equipment as the objective.

A 3D rigid-flexible coupled dynamic model of the studied EMU train in Table 1 is established by utilizing the multibody dynamic software SIMPACK, varying the running speed, the relations between $f_{eq}$ and the vertical vibrations of carbody and the equipment are calculated and shown in Fig.8 and Fig.9. Without loss of generality, the new wheel and worn wheel profiles of LMA are considered in the calculation, respectively, and the frequency range of $f_{eq}$ is extended to be 5 Hz to 12 Hz. Synthesizing the results of the different running speeds, the vibrations of carbody and the equipment are relatively low when $f_{eq}$ is in the range of 7 Hz to 8 Hz, which comforms to the condition of mode matching design.

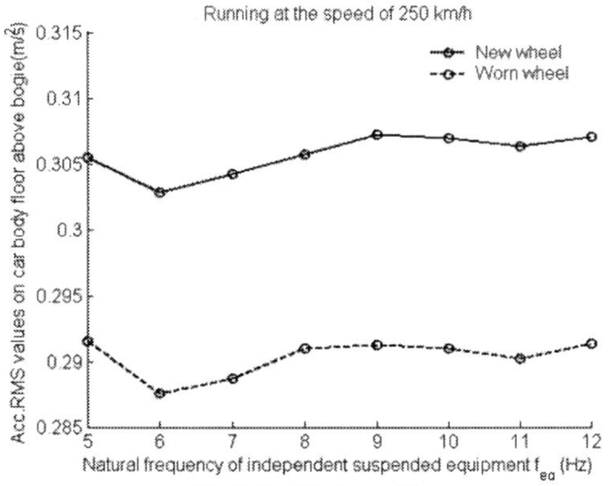


**(a) 250 km/h**

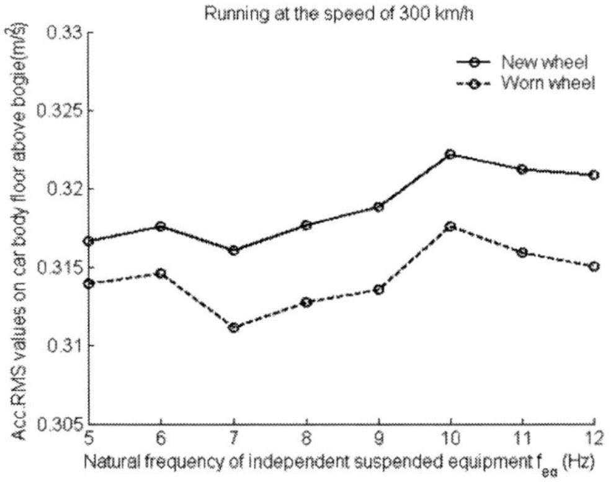


**(b) 300 km/h**

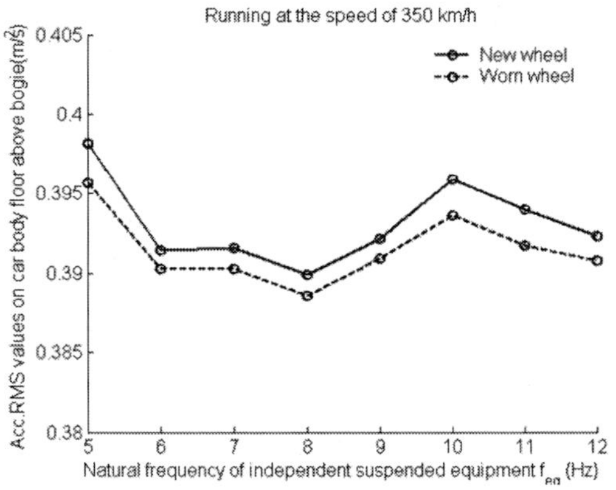


**(c) 350 km/h**

**Fig.8 Acc. RMS values on carbody floor above bogie**

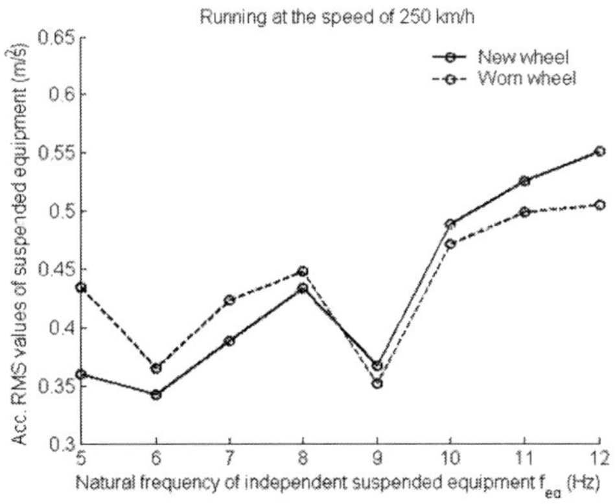


**(a) 250 km/h**

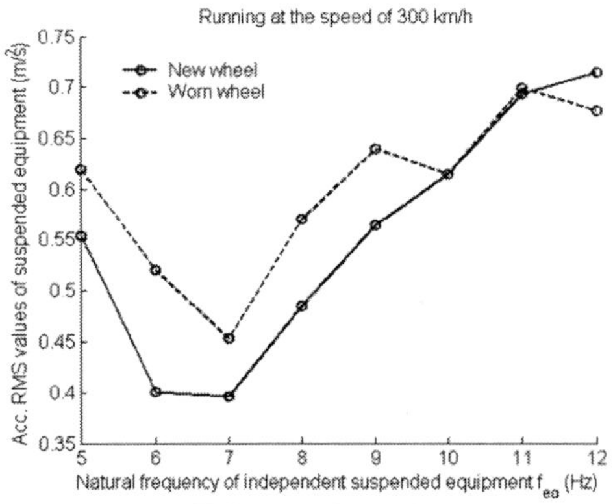


**(b) 300 km/h**

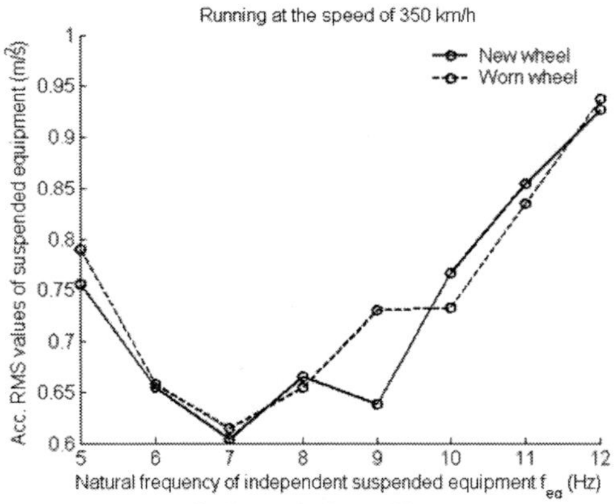


**(c) 350 km/h**

**Fig.9 Acc. RMS values of suspended equipment**

Ride index is a necessary tool to estimate the subjective comfort feelings of majority passengers with the objective physical variables of running vehicles. Sperling index is one of the ride index criteria widely used by the railway industry [15]. Fig.10 shows the calculation results of Sperling index on carbody floor above bogie. Due to the distinction of weighting value at different frequency of Sperling index, the tendency of the curves shown in Fig.10 is not exactly the same as that in Fig.8. However, for the studied vehicles service speed at 300~350km/h, it is more reasonable when $f_{eq}$ equals 7 Hz. It is noted that the stiffness of rubber spring will probably vary with the environmental temperature about 0.7 to 1.3 times than its original, which means when $f_{eq}$= 7 Hz, its actual value might be extented to in the range of 5.9 Hz to 8.0 Hz.

Nonetheless, it is still smaller than the hopping frequency as shown in Fig.5 and satifies with the requirements of mode matching design.

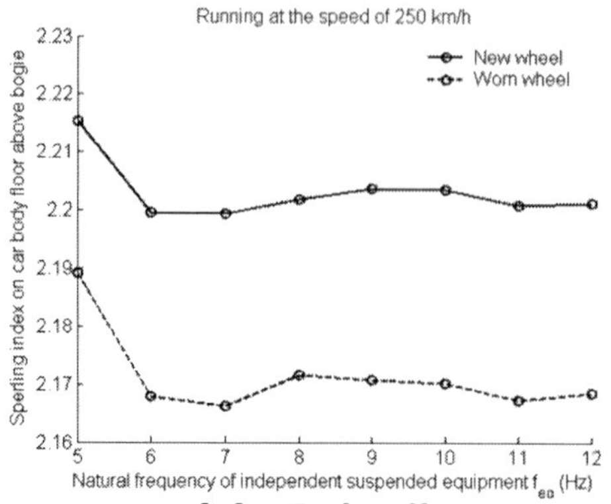


**(a) 250 km/h**

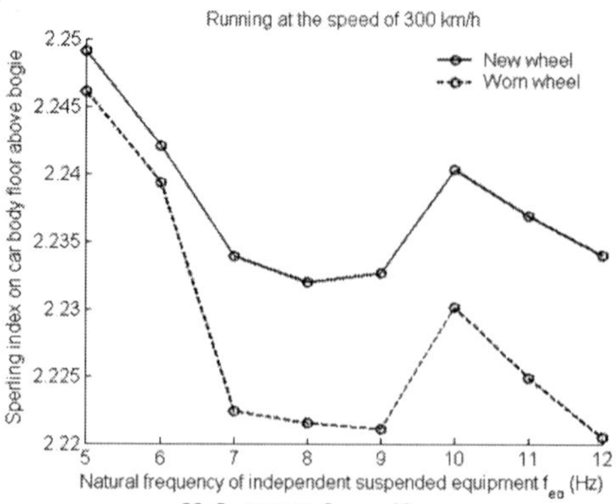


**(b) 300 km/h**

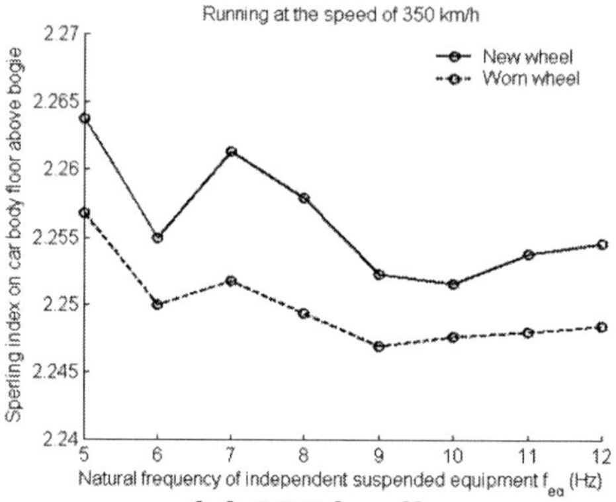


**(c) 350 km/h**

**Fig.10 Sperling Index on carbody floor above bogie**

When $f_{eq}$=7 Hz, the vibration transmission from carbody to suspended equipment in the frequency range of 0 Hz to 2 Hz and above 7 Hz are illustrated in Fig.11. The vibration transmission from suspended equipment to carbody based on positive vibration isolation has the same results as those in Fig.11. One can find that in the range of 0~2 Hz, the vibration transmission from carbody to equipment is magnified. The vibration starts to be attenuated at 9 Hz, when the frequency greater than 12Hz, 43% of the vibration can be reduced. At 100Hz, the vibration reduction efficiency rises to 99%, which has great effects on suppressing the vibration and noise transmission between the carbody and suspended equipments.

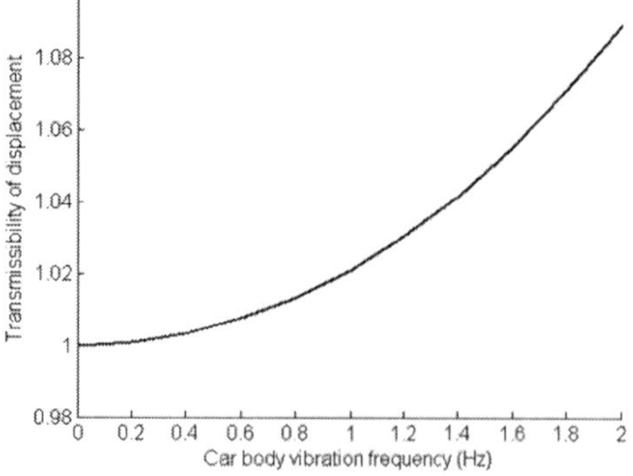


**(a) 0~2Hz**

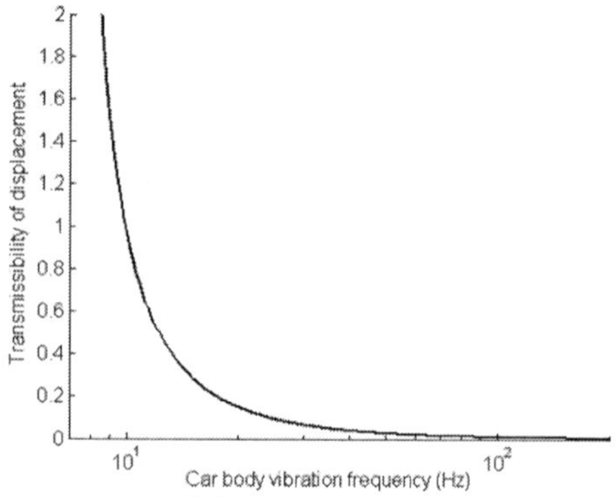


**(b) above 7Hz**

**Fig.11 Vibration transmission analysis**

## 5 IN-SITU TESTS

With the proposed method, the rubber spring of the suspended equipment for the studied high-speed EMU trains was designed and verified by a in-situ test. The test results of the vertical frequency response of the equipment and the acceleration RMS values of the equipment and the carbody structure above it are shown in Fig.12. It can be seen that the frequency design of the equipment is reasonable. The vibration frequencies of the equipment converge to less than 2 Hz and about 7 Hz, which respectively are the frequencies of carbody rigid modes and the equipment motion, meanwhile, the carbody vibration is not severe according to EN12299 [16] and no resonant vibration happens in the overall running speed range. In spite of there exists a little discrepancy between the suspended equipment frequency of the test and that of the design due to coupling with the carbody, it has little effect on the vibration isolation performance.

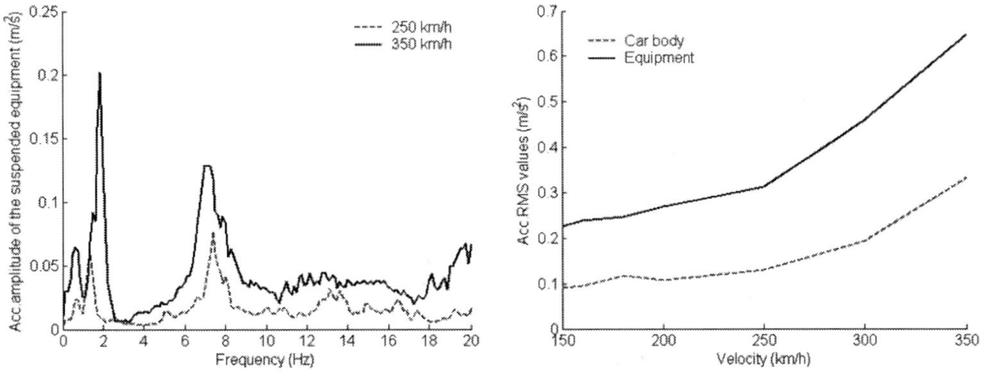


**(a) Frequency response of the equipment (b) RMS values of acceleration**

**Fig.12 Test results of equipment and carbody**

## 6 CONCLUSIONS

The influences of the under chassis equipment on the carbody eigenfrequencies are studied by means of a numerical method and verified by FE method and a modal test. The rubber spring parameter design of the suspended equipment on the basis of mode matching, transmissibility matching and taking the optimal vibration of both carbody and equipment as the objective are presented and verified by an in-situ test. Results show that increasing the natural frequency of independent suspended equipment $f_{eq}$ leads to a 'frequency hopping' phenomenon of the first vertical bending frequency of carbody. The 'frequency hopping' happens when $f_{eq}$ passes through the first vertical bending frequency of carbody with fixed equipment. The first vertical bending frequency of carbody could be enhanced by means of equipment with optimized suspension parameter. When the natural frequency of independent suspended equipment is designed according to the proposed principles, the vehicle can obtain a good ride quality and the equipment vibration is not severe in the overall running speed range.

## REFERENCES


1 Gong, D., Zhou, J. S., Sun, W. J. and Shen, G. *Analysis of flexible resonant vibration of railway carbody and design of equipment suspension parameters.* Proceeding of the International Symposium on Dynamics of Vehicles on Roads and Tracks (23rd IAVSD), 19-23 August, 2013, Qingdao, China.

2 Clas, H. *Active Damping of Carbody Vibrations.* Master of Science thesis, Stockholm, Department of Mechanics of the Royal Institute of Technology, 1998.

3 Sun W. J., Gong D. and Zhou J. S. *Influences of Suspended Equipment under Carbody on High-speed Train Ride Quality*. Procedia Eng, 2011, 16: 812-817.
4 Shi H. L., Luo R., Wu P. B., Zeng J., and Guo J. Y. *Application of DVA theory in vibration reduction of carbody with suspended equipment for high-speed EMU*. Science China Technological Sciences, 2014, 57(7): 1425-1438.
5 Gong D., Zhou, J. S. and Sun, W. J. *On the resonant vibration of a flexible railway carbody and its suppression with a dynamic vibration absorber*. J. Vibration and Control, 2013, 19(5): 649-657.
6 Zhou, J. S., Goodall, R., Ren, L. and Zhang, H. *Influences of Carbody Vertical Flexibility on Ride Quality of* Passenger *Railway Vehicles*. Proc. IMechE Part F: J. Rail and Rapid Transit, 2009, Vol.223, pp.461-471.
7 Suzuki, Y. and Akutsu, K. *Theoretical analysis of flexural vibration of carbody*. Quarterly Report of RTRI, 1990, 31(1): 42-48.
8 Foda, M. A. and Albassam, B. A. *Vibration confinement in a general beam structure during harmonic excitations. J.* Sound Vibr., 2006, 295: 491-517.
9 Abu-Hilal, M. *Forced vibration of Euler-Bernoulli beams by means of dynamic Green functions*. J. Sound Vibr., 2003, 267: 191-207.
10 Yan, J. *Mechanical Vibration Isolation Technology*. Shanghai, Shanghai Scientific and Technical Literature Publishing House, (in Chinese) 1985.
11 Clarence, W.de, S. *Vibration Damping, Control and Design*. CRC Press Inc. 2007.
12 Gong, D. *Study on Flexible Vibration and Control of Carbody for a High-Speed Railway Train*. Doctoral dissertation, Shanghai, Tongji University, (in Chinese) 2012.
13 Carlbom, P. *Combining MBS with FEM for Rail Vehicle Dynamics Analysis*. Multibody System Dynamics, 2001, Vol.6, pp.291-300.
14 Schandl, G., Lungner, P., Benatzky, C., Kozek, M. and Stribersky, A. *Comfort enhancement by an active vibration reduction system for a flexible railway carbody*. Vehicle System Dynamics, 2007, 45(9): 835-847.
15 Kim, Y.-G., Kwon, H.-B., Kim, S.-W., Park, C.-K., and Kim, T.-W. *Correlation of ride comfort evaluation methods for railway vehicles*. Proc. IMechE, Part F: J. Rail and Rapid Transit, 2003, 217(F2), 73–87.
16 EN12299. *Railway applications - Ride comfort for passengers – Measurement and evaluation*, 2009.

# Flexible carbody vertical vibration characteristic of high-speed train

**H Cao** [1,2], **W Zhang** [1], **B Miao** [1]
[1] Traction Power State Key Laboratory, Southwest Jiaotong University, P.R. China
[2] School of Manufacturing Science and Engineering, Southwest University of Science & Technology, P.R. China

## ABSTRACT

For high-speed train carbody elastic vibration affecting suspension components safety and ride comfort, a vertical dynamics model for high-speed train was established. Vibration characteristic and transfer relationship under flexible effect was studied in frequency domain. Analysis results show that symmetrical modes response is zero and antisymmetric modes response is maximum in some specific wavelengths, antisymmetric modes response is zero and symmetrical modes response is maximum in others. The resonance speed and resonance wavelength of first vertical bending vibration have significant impact on train operation. The higher speed, the higher first bending frequency is required.

**Keywords:** High-speed train; Flexible carbody; Vibration mechanism; Modal; Ride index

## 1 INTRODUCTION

Fast and comfort are the core competiveness for high-speed train with other transport. With the increasing train speed, track excitation frequency is increased and the carbody is increasingly lighter in weight, which leads to more flexibility, lower modal frequency. This could cause a serious vibration, such vibration often occurs in a frequency range to which passengers aboard train are very sensitive, thus affects ride comfort. Meanwhile, the vibration causes the carbody fatigue, affecting the dynamic performance and service life [1-3]. In order to improve the performance of the vehicle vertical dynamics, several methods have already proposed. The dynamic vibration absorber and flexible damper are installed under carbody and the control effects are analyzed by absorber parameters and installation style [4-6]. The appropriate constrained damping layers is designed by Zeng et al. to obtain good damping effect[7]. A novel method for vibration reduction is the application of an active control system directly to the flexible structure of the railway carbody in order to attenuate the elastic vibrations. Both passive and active control schemes have already been proposed [8]. Above research focus on the factors and measures of improvement ride quality under the elastic body structure, few studies on the vibration mechanism and transfer relationship under flexible effect. The structure of this paper is outlined below.

Firstly, a vertical dynamics model for high-speed train was established based on rigid-flexible coupled dynamics, vertical vibration characteristic under flexible effect is analyzed by analytical method. Resonance effect and filtering effect exist in elastic vibration. The occurrence conditions are derived. Finally, the vibration influence factors and the ride comfort improvement methods are analyzed, which could instruct the structure design and running parameters selection.

## 2 VERTICAL RIGID-FLEXIBLE COUPLED DYNAMICS MODEL

The vertical rigid-flexible coupled dynamics model for high-speed train is as shown in Figure 1. It includes carbody, bogie frames, wheelsets and suspensions. The carbody is modeled as a simple Euler-Bernoulli beam in consideration of only vertical vibration, i.e., bounce, pitching, and bending modes, while the wheelsets and bogie frames are considered as rigid bodies. As a result, the total number of independent degrees of freedom of the rigid vehicle system is six, namely bogie frames bounce $z_{b1}$ and $z_{b2}$, bogie frames pitching $\theta_{b1}$ and $\theta_{b2}$, carbody bounce $z_c$, and carbody pitching $\theta_c$. When *N* flexible modes are considered, the *N* modal coordinate can deem the flexible vibration freedom. So, the total degrees of freedom is 6+*N*. The parameters of a typical high-speed passenger vehicle are shown in Table 1.

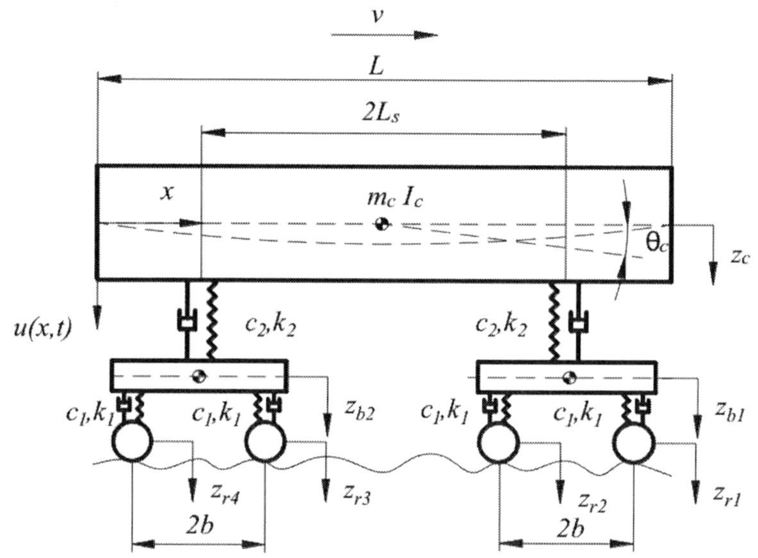


**Figure 1 Dynamic model of vehicle system**

When the vertical displacement of carbody being expressed as u(x, t), the appropriate partial differential equation of the carbody is[9]

$$EI\frac{\partial^4 u(x,t)}{\partial x^4}+\rho A\frac{\partial^2 u(x,t)}{\partial t^2}+\phi I\frac{\partial^5 u(x,t)}{\partial x^4 \partial t}=\sum_{i=1}^{2}F_i\delta(x-x_i) \quad (1)$$

Where $F_i$ is the forces acted by second suspensions and $x_i$ is the second suspension displace, $i=1$ means front bogie and $i=2$ means rear one. $\delta(x)$ is the delta function. $F_i$ is

$$F_i=c_2[\dot{u}(x_i,t)-\dot{z}_{bi}]+k_2[u(x_i,t)-z_{bi}] \quad (2)$$

**Table 1 Parameters symbol and value**

| Parameter | Value | Meaning |
|---|---|---|
| $m_c$ /t | 26 | Mass of carbody |
| $I_c / (t \cdot m^2)$ | 1300 | Carbody pitch inertia |
| $\rho A / (t \cdot m^{-1})$ | 1.07 | Mass per meter |
| $EI / (kN \cdot m^2)$ | $3.58 \times 10^6$ | Bending rigidity |
| $m_b$ /t | 2.44 | Bogie frame mass |
| $I_b /(t \cdot m^2)$ | 1.4 | Bogie frame pitch inertia |
| $c_1$ / (kN·s/m) | 30 | Primary damping coefficient per axle |
| $c_2$ / (kN·s/m) | 50 | Secondary damping coefficient per bogie |
| $k_1$ / (kN/m) | 2400 | Primary spring coefficient per axle |
| $k_2$ / (kN/m) | 380 | Secondary spring coefficient per bogie |
| $L$ /m | 24.5 | Length of carbody |
| $L_S$ /m | 8.75 | Half of bogie centres |
| $f_1$ | 10.02 | First bending mode frequency |
| $\xi_1$ | 1.5% | First bending mode damping ratio |
| $f_2$ | 27.83 | Second bending mode frequency |
| $\xi_2$ | 1.5% | Second bending mode damping ratio |

It is assumed that the shape function and modal coordinate of ith mode are $Y_i(x)$ and $q_i(t)$, respectively. When the rigid modes are included with the flexible modes in $z(x,t)$, the first mode of the carbody is chosen as the bounce of rigid mode and its shape function is taken as $Y_1(x) = 1$. The second mode is the pitch and its shape function is $Y_1(x) = L/2 - x$ accordingly under the coordinate definition as in Figure 1. When N modes are considered, the vertical displacement of the carbody can be written as

$$u(x,t) = z_c(t) + (x - L/2)\theta_c(t) + \sum_{i=3}^{N} Y_i(x) q_i(t) \tag{3}$$

the shape functions are taken as

$$Y_i(x) = ch\beta_i x + \cos\beta_i x - \frac{ch\beta_i L + \cos\beta_i L}{Sh\beta_i L + \sin\beta_i L}(sh\beta_i x + \sin\beta_i x) \tag{4}$$

Where $\beta_i = \dfrac{(2i+1)\pi}{2L}$. By substituting equation (3) into equation (1) and integrating along the length of carbody, considering the orthogonality of shape functions, the carbody and bogies vibration equations are:

$$
\left.\begin{aligned}
&\ddot{q}_i(t) + 2\xi_i\omega_i\dot{q}_i + \omega_i^2 q_i = \frac{Y_i(x_1)}{m_c}F_1 + \frac{Y_i(x_2)}{m_c}F_2 \\
&m_c\ddot{z}(t) = \sum_{i=1}^{2} F_i \\
&I_c\ddot{\theta}(t) = \sum_{i=1}^{2} F_i(x_i - L/2) \\
&m_b\ddot{z}_{b1} + c_2\left[\dot{z}_{b1} - \dot{u}(x_1,t)\right] + k_2\left[z_{b1} - u(x_1,t)\right] + c_1(2\dot{z}_{b1} - \dot{z}_{r1} - \dot{z}_{r2}) + k_1(2z_{b1} - z_{r1} - z_{r2}) = 0 \\
&m_b\ddot{z}_{b2} + c_2\left[\dot{z}_{b2} - \dot{u}(x_2,t)\right] + k_2\left[z_{b1} - u(x_2,t)\right] + c_1(2\dot{z}_{b2} - \dot{z}_{r3} - \dot{z}_{r4}) + k_1(2z_{b2} - z_{r3} - z_{r4}) = 0 \\
&I_b\ddot{\theta}_{b1} + c_1 b\left[2b\dot{\theta}_{b1} - \dot{z}_{r1} + \dot{z}_{r2}\right] + k_1 b[2b\theta_{b1} - z_{r1} + z_{r2}] = 0 \\
&I_b\ddot{\theta}_{b2} + c_1 b\left[2b\dot{\theta}_{b2} - \dot{z}_{r3} + \dot{z}_{r4}\right] + k_1 b[2b\theta_{b2} - z_{r3} + z_{r4}] = 0
\end{aligned}\right\} \tag{5}
$$

Where $\omega_i^2 = \frac{EI\beta_i^4}{\rho}, 2\xi_i\omega_i = \frac{\phi I\beta_i^4}{\rho}$

If $\{\mathbf{z}\} = \{z_c\ \theta_c\ z_{b1}\ z_{b2}\ \theta_{b1}\ \theta_{b2}\ q_3 \cdots q_N\}^T$ (6)

Writing equations (5) in matrix form gives

$$
[\mathbf{M}]\{\ddot{\mathbf{z}}\} + [\mathbf{C}]\{\dot{\mathbf{z}}\} + [\mathbf{K}]\{\mathbf{z}\} = [\mathbf{C_r}]\{\dot{\mathbf{z}}_\mathbf{r}\} + \{\mathbf{K_r}\}\{\mathbf{z_r}\} \tag{7}
$$

In which $[\mathbf{M}],[\mathbf{K}],[\mathbf{C}]$ are the inertia, stiffness, and damping matrices, respectively, and $[\mathbf{C_r}],[\mathbf{K_r}]$ are the excitation damping and stiffness matrices, $\mathbf{z_r}$ is the track irregularities.

## 3 THE FREQUENCY RESPONSE FUNCTION (FRF)

According to Equation (7), the FRF matrix $H(\omega), H_a(\omega)$ of the carbody displacement and acceleration can be obtained by

$$
\left.\begin{aligned}
&\mathbf{H}(\omega) = [-\omega^2\mathbf{M} + j\omega\mathbf{C} + \mathbf{K}]^{-1}[-j\omega\mathbf{C_r} + \mathbf{K_r}] \\
&\mathbf{H}_a(\omega) = -\omega^2\mathbf{H}(\omega)
\end{aligned}\right\} \tag{8}
$$

Matrices obtained by Equations (8) are the multiple input multiple output (MIMO) displacement and acceleration FRF matrices, whose elements are the single input single output (SISO) FRFs from the single wheelset to the carbody response, which do not include the time delay effects between wheelsets. For the vehicle model in Figure 1, the track inputs vector is

$$
\{\mathbf{z}_r(t)\} = \begin{Bmatrix} z_{r1}(t) \\ z_{r2}(t) \\ z_{r3}(t) \\ z_{r4}(t) \end{Bmatrix} = \begin{Bmatrix} z_{r1}(t) \\ z_{r1}(t-\tau_1) \\ z_{r1}(t-\tau_2) \\ z_{r1}(t-\tau_3) \end{Bmatrix} \tag{9}
$$

Where $\tau_1 = \frac{2b}{V}$, $\tau_2 = \frac{2L_s}{V}$, $\tau_3 = \frac{2(b+L_s)}{V}$. By Fourier transforming the above equation one has

$$\{\mathbf{Z}_r(\omega)\} = \begin{Bmatrix} 1 \\ e^{-j\omega\tau_1} \\ e^{-j\omega\tau_2} \\ e^{-j\omega\tau_3} \end{Bmatrix} Z_{r1}(\omega) \tag{10}$$

Therefore, the system displacement responses in the frequency domain are

$$\{\mathbf{Z}(\omega)\} = \mathbf{H}(\omega) \begin{Bmatrix} 1 \\ e^{-j\omega\tau_1} \\ e^{-j\omega\tau_2} \\ e^{-j\omega\tau_3} \end{Bmatrix} Z_{r1}(\omega) \tag{11}$$

By defining

$$\{\mathbf{H}_c(\omega)\} = \mathbf{H}(\omega) \begin{Bmatrix} 1 \\ e^{-j\omega\tau_1} \\ e^{-j\omega\tau_2} \\ e^{-j\omega\tau_3} \end{Bmatrix}, \{\mathbf{H}_{ac}(\omega)\} = \mathbf{H}_a(\omega) \begin{Bmatrix} 1 \\ e^{-j\omega\tau_1} \\ e^{-j\omega\tau_2} \\ e^{-j\omega\tau_3} \end{Bmatrix} \tag{12}$$

$H_c(\omega)$ and $H_{ac}(\omega)$ become the displacement and acceleration single input multiple output (SIMO) FRFs from track input to vehicle responses. As these SIMO FRFs include time delay effects between wheelsets, they are called correlated FRF matrices [10]. The PSDs of vehicle displacement and acceleration responses can be obtained by

$$\left.\begin{aligned} \mathbf{S}_z(\omega) &= \mathbf{H}_c^*(\omega)\mathbf{S}_\omega(\omega,V)\mathbf{H}_c^T(\omega) \\ \mathbf{S}_{\ddot{z}}(\omega) &= \mathbf{H}_{ac}^*(\omega)\mathbf{S}_\omega(\omega,V)\mathbf{H}_{ac}^T(\omega) \end{aligned}\right\} \tag{13}$$

In which $\mathbf{H}_c^*(\omega), \mathbf{H}_c^T(\omega)$ are the conjugation and simple transpose of the FRF matrix $\mathbf{H}_c(\omega)$ $\mathbf{S}_\omega(\omega,V)$ is the displacement response PSD matrices at running speed V.

## 4 FLEXIBLE CARBODY VIBRATION ANALYSIS

In carbody rigid and flexible vibration, the shape functions can be divided into symmetric modes and antisymmetric modes according shape function changing with the position. The bounce, first bending, and third bending functions are symmetric modes. The pitch, second bending, and fourth bending functions are antisymmetric modes. The displace amplitude of FRF of symmetric modes and antisymmetric modes can be obtained according to formula (12):

$$\left.\begin{aligned} \left|H_c^S(\omega)\right| &= 2\left|H^S(\omega)\right|\sqrt{(1+\cos\omega\tau_1)(1+\cos\omega\tau_2)} \\ \left|H_c^{AS}(\omega)\right| &= 2\left|H^{AS}(\omega)\right|\sqrt{(1+\cos\omega\tau_1)(1-\cos\omega\tau_2)} \end{aligned}\right\} \tag{14}$$

For symmetric modes, when the wavelength of track irregularity input

$$\lambda = \frac{4b}{2n-1} or \lambda = \frac{4L_s}{2n-1}, n = 1,2,3... \tag{15}$$

carbody response is zero and the corresponding frequency are

$$f = \frac{V(2n-1)}{4b} \text{ or } f = \frac{V(2n-1)}{4L_s} \tag{16}$$

Similarly, for antisymmetric modes, the wavelength and frequency of none response of carbody are

$$\left.\begin{aligned} \lambda &= \frac{4b}{2n-1} or \lambda = \frac{2L_s}{n}, n = 1,2,3... \\ f &= \frac{V(2n-1)}{4b} orf = \frac{Vn}{2L_s}, n = 1,2,3... \end{aligned}\right\} \tag{17}$$

According to equations (16) and (17), there will be no response when $f = \frac{V(2n-1)}{4b}$. Therefore, this phenomenon is called as the filtering phenomenon. For symmetric modes, when the wavelengths of track irregularity input

$$\lambda = \frac{2b}{n} or \lambda = \frac{2L_s}{n}, n = 1,2,3... \tag{18}$$

There will be maximum carbody responses. This wavelengths are called symmetric resonance wavelengths. In fact, the wheel base of bogie and resonance wavelengths are small, the corresponding input energy is also small. Therefore, the resonant frequencies are

$$f = \frac{nV}{2L_s}, n = 1,2,3... \tag{19}$$

Similarly, the antisymmetric modes resonance wavelengths and the resonance frequencies are

$$\lambda = \frac{4L_s}{2n-1}, f = \frac{(2n-1)V}{4L_s}, n = 1,2,3... \tag{20}$$

The resonance frequencies are related to the length of bogie center and speed of the vehicle. When natural frequencies are equal to the resonance frequencies, the carbody will set up resonance, the corresponding speed are called resonance speed [11]. According to Table 1, the carbody resonance speeds are shown in Figure 2. As Figure2 is shown, the rigid vibration resonance speeds are generally less than the vehicle running speed. The first vertical bending resonance speed is nearby vehicle running speed. It is a significant impact on the body vibration. When vehicle parameters is settled, the vehicle running speed should avoid to conflict the first bending resonance speed. The corresponding wavelengths of second and higher resonance speed are small and have less impaction on vehicle.

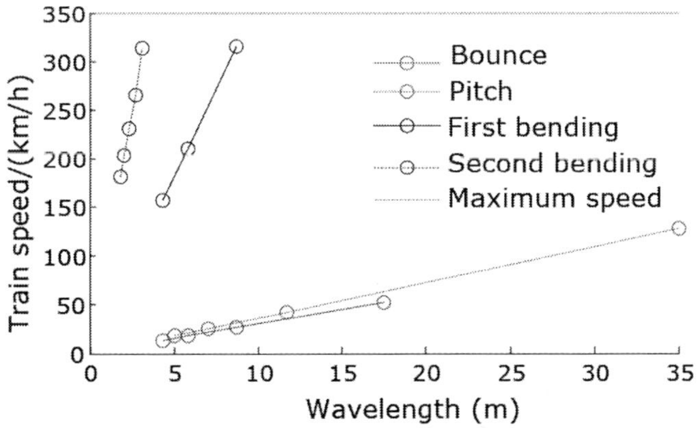


**Figure 2 Carbody resonance speed**

It is considered that first bending resonance vibration will be excited when bounce vibration transmissibility is maximum and second vertical bending resonance vibration will be excited when pitch vibration transmissibility is maximum. The above analysis shows that the carbody will produce the modal resonances as long as the natural frequency is consistent with its resonant frequency, regardless of the symmetric modes and antisymmetric modes. According to equations (15)-(20), when the symmetric modes response is maximum antisymmetric mode response is zero and vice versa.

Because acceleration power spectrum density (PSD) is related to the acceleration transmissibility, it is necessary to analyze the modal acceleration transmissibility. According to table 1 parameters, the modal acceleration transmissibility shows in Figure 3. These show that the elastic modal vibration acceleration transmissibility is larger than that of rigid, the firs vertical bending vibration acceleration transmissibility is maximum. Therefore, the carbody vibration acceleration increase rapidly and ride comfort deteriorate when first vertical bending resonance happen.

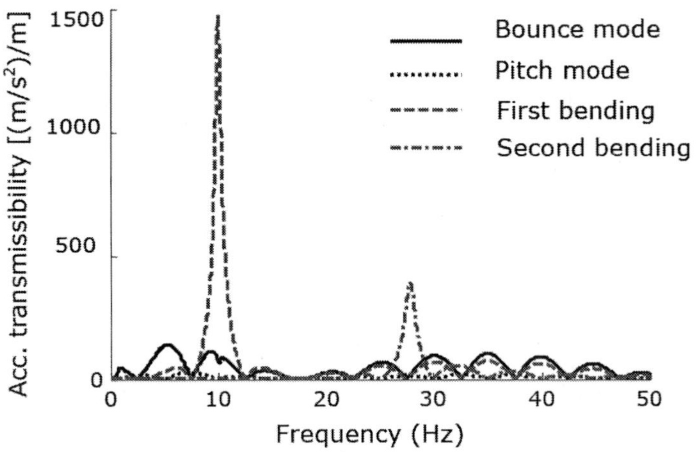


**Figure 3 Modal acceleration transmissibility**

The vibration acceleration is related to track input except for acceleration transmissibility. The high and low excitation track unevenness PSDs are illustrated in Figure 4. It can be seen that the track excitation power decreases for the frequencies grow up. So, it is important to avoid the low frequencies resonance for carbody vibration.

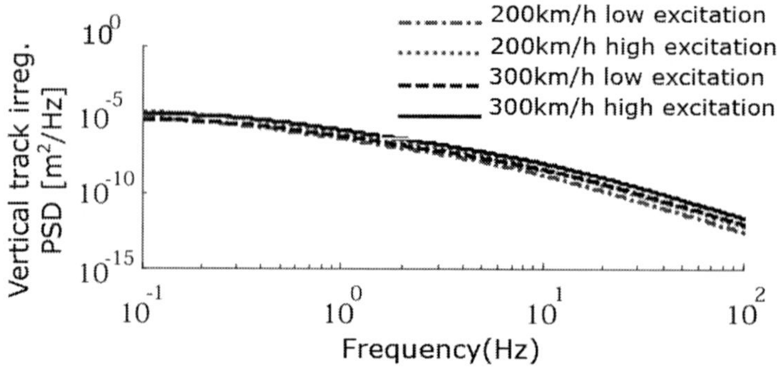


**Figure 4 High-speed Track irregularity PSD**

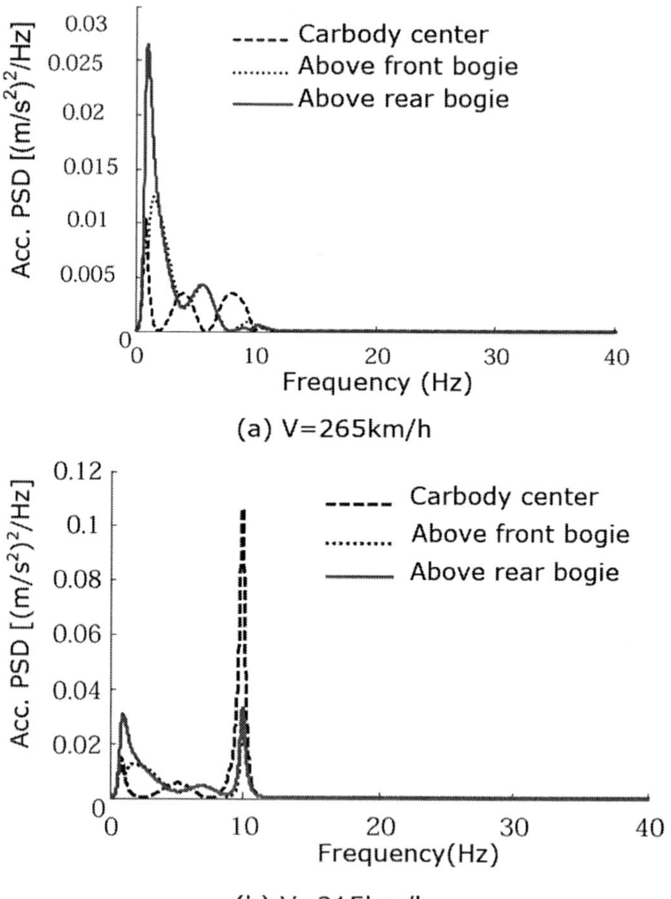


(b) V=315km/h

**Figure 5 Vertical vibration acceleration PSD of flexible carbody**

When only considering the first four carbody modes, using original values for the typical high-speed vehicle parameters shown in Table 1 and vehicle running speed of 265 km/h(antisymmetric modes resonance speed) and 315km/h(symmetric modes resonance speed), acceleration PSDs at the centre of carbody and above the centre of bogie are shown in Figure 5, from which one can find that the carbody vibration is changed with frequences and positions, each mode has different energy proportion in vibration. In 265km/h, the acceleration PSDs are gradually declined with increasing frequency. So the elastic vibrations have less power than rigid. Because of antisymmetric effect, the vibration above rear bogie is more violence than the center and the front bogie. In 315km/h, it has the same vibration energy above the front and rear bogie because it is the node of first bending vibration shape function. The center is the peak of first bending vibration shape function, the flexible vibration energy is about seven times of rigid at the center. The carbody appears strong elastic resonance.

The relationship between the vibration and ride quality can be measured by Sperling indexes. Figure 6 shows that the ride quality indexes will increase with the vehicle speed. The ride indexes at the center are changed bigger than that above the bogie because of the resonance effect and filtering effect of first vertical bending vibration.

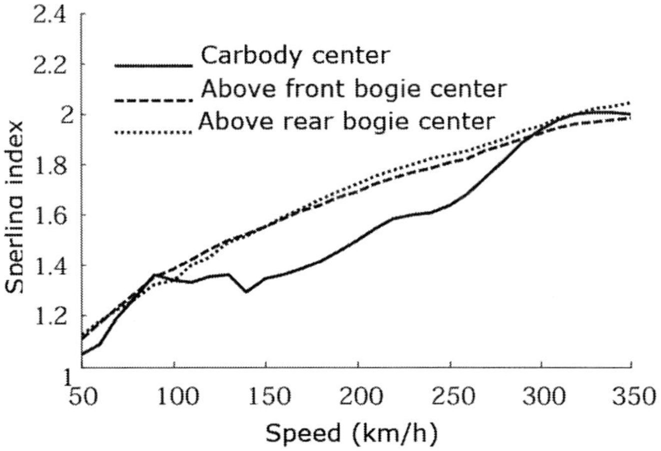


**Figure 6 Ride index vs. train speed**

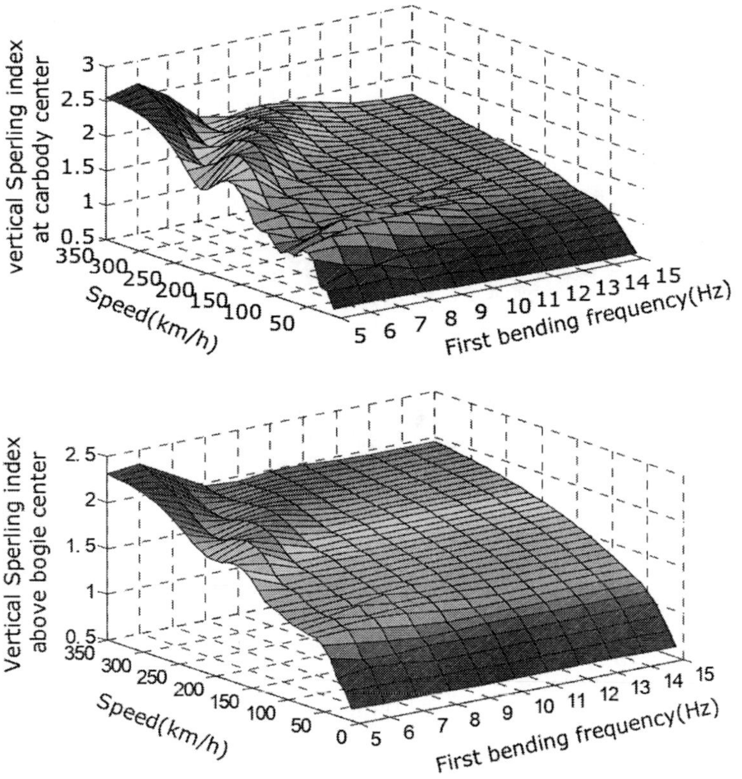


**Figure 7 Ride index vs. train flexibility**

As mentioned before, the first vertical bending mode has the greatest influence on the vibration of carbody. The relations between the ride quality and the first vertical bending frequencies are illustrated in figure7. Due to the filtering effect and resonance effect, ride indexes have a drastic change with the first vertical bending frequency, it is the peak at resonance frequency and the trough at filtering frequency. The higher vehicle speed the more dramatic ride quality is changed. When bending frequencies get to certain values the ride indexes reach to constant values. The value is greater when the speed is higher. This means that the higher vehicle speed the greater stiffness of the carbody is required. Increasing bending frequency means that the mass of carbody and energy consumption are increased. For an existing vehicle, it is important to avoid the confliction of running speed and resonance speed. Because second suspension points are the node of first bending mode, it has a smaller change of ride indexes above bogie than the center.

When the first bending frequency is less than 8Hz and vehicle has a high speed, the Sperling index has reached about 2.5. It is necessary to take measures to control ride quality. Usually, it can be achieved by changing the damping ratio of the structure and optimization of vehicle suspension parameters. The relationship between Sperling indexes and damping ratio are illustrated and evaluated in Figure 8 when the vehicle speed is 300km/h. Results show that when damping ratio increases from 0.015 to 0.15 ride index can be decreased by about 13 percent. Increasing the damping ratio can significantly reduce the vibration of carbody. The structure surface damping layers processing, passive or active damping control measures can be taken to improve the carbody damping [12-14].

Vertical suspension parameters have an important influence on ride quality. The vertical suspension stiffness component is limited by the vehicle structure. So damping coefficients change can be taken to improve ride quality. Figure 9 illustrates the ride quality changes with the damping coefficients and speed, which shows that higher primary damping reduces the rigid and flexible vibration. Higher primary damper coefficient may deteriorate the maximum dynamic wheel-rail forces. So the appropriate damping is helpful to improve ride comfort. Secondary damping coefficient has an optimal value, the higher vehicle speed, the smaller damping is needed. In Figure 9, the optimum value of 200km/h is 20kN•s/m, the optimum value of 300km/h is 15kN•s/m. The reason is that secondary vertical damping should decrease appropriately to dissipate vibration with the increase of speed and excitation frequency.

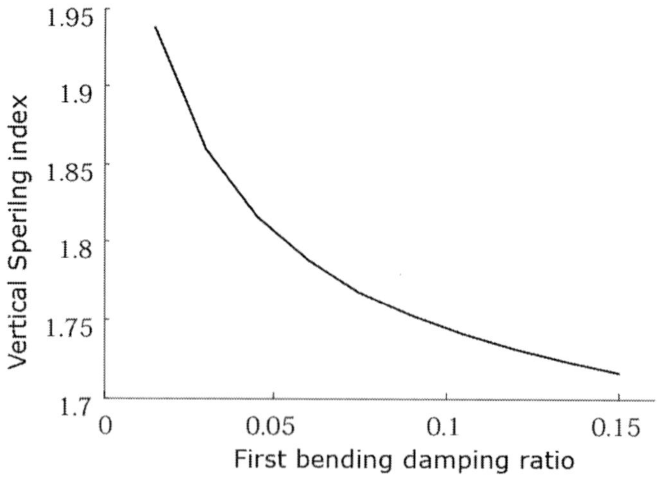


**Figure 8 Effect of carbody structural damping on ride index**

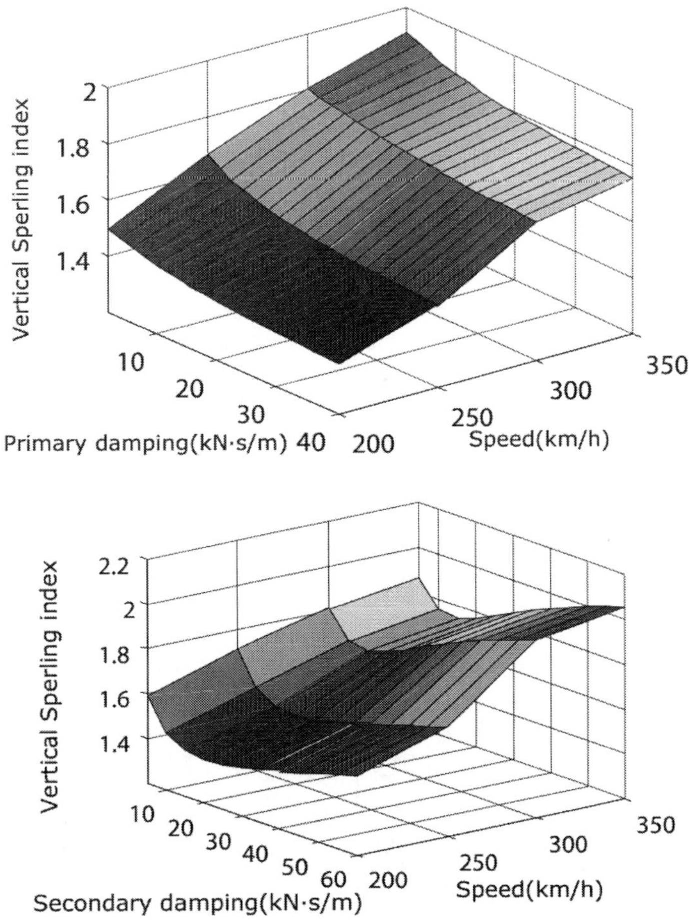


**Figure 9 Effect of vertical damping on ride index**

## 5 CONCLUSIONS

(1) The modal vibration has the resonance effect and filtering effect, in particular track wavelength, symmetric modes vibration is zero and antisymmetric modes vibration is strongest, and vice versa.

(2) Each mode has its corresponding resonance frequency, the first bending mode acceleration transmissibility is maximum and has the largest contribution to vibration.

(3) The first bending resonance speed has a dramatic impact on ride indexes. This speed is equal to the vehicle speed, the elastic resonance happen. So it is necessary to avoid this situation. The rigid resonance speed is lower than running speed and has a little impact on ride quality. A lower bending frequency can deteriorate the vehicle ride quality. So when running speed of a vehicle is higher the larger bending frequency is required.

(4) Improve the primary damping and the structure damping, appropriate reducing the secondary damping can improve the vehicle ride quality. When damping ratio increases from 0.015 to 0.15 ride index can be decreased by about 13 percent. The optimum secondary damping of 200km/h is 20kN s/m and the value of 300km/h is 15kN s/m.

## REFERENCES


[1] Suzuki, H. (1998). Research trends on riding comfort evaluation in Japan. *Proceedings of the Institution of Mechanical Engineers, Part F: Journal of Rail and Rapid Transit*, *212*(1), 61-72.

[2] Carlbom, P. F. (2001). Combining MBS with FEM for rail vehicle dynamics analysis. *Multibody System Dynamics*, *6*(3), 291-300.

[3] Diana, G., Cheli, F., Collina, A., Corradi, R., & Melzi, S. (2002). The development of a numerical model for railway vehicles comfort assessment through comparison with experimental measurements. *Vehicle System Dynamics*, *38*(3), 165-183.

[4] Gong, D., Zhou, J. S., & Sun, W. J. (2012). On the resonant vibration of a flexible railway car body and its suppression with a dynamic vibration absorber. *Journal of Vibration and Control*, 1077546312437435.

[5] ZHOU, J. S., Gong, D., SUN, W. J., & REN, L. H. (2009). Influence of Vertical Elasticity of Carbody of Railway Passenger Vehicles on Ride Quality. *Journal of the China Railway Society*, 31(2),*32-37*.

[6] Zhou, J., Goodall, R., Ren, L., & Zhang, H. (2009). Influences of car body vertical flexibility on ride quality of passenger railway vehicles. Proceedings of the Institution of Mechanical Engineers, Part F: Journal of Rail and Rapid Transit, 223(5), 461-471.

[7] Zeng, J., Luo, R. (2007). Vibration analysis of railway passenger car systems by considering flexible carbody effect. *Journal of the China Railway Society*, *29*(6), 19-25.

[8] HUANG, C. H., ZENG, J., & WU, P. B. (2010). Study on carbody flexible vibration reduction for railway passenger carriage. *Engineering Mechanics*, *27*(12), 250-256.

[9] Suzuki, Y., & Akutsu, K. (1990). Theoretical analysis of flexural vibration of car body. *Railway Technical Research Institute, Quarterly Reports*, *31*(1).

[10] Cossalter, V., Doria, A., Garbin, S., & Lot, R. (2006). Frequency-domain method for evaluating the ride comfort of a motorcycle. *Vehicle System Dynamics*, *44*(4), 339-355.

[11] Cheli, F., & Corradi, R. (2011). On rail vehicle vibrations induced by track unevenness: Analysis of the excitation mechanism. *Journal of Sound and Vibration*, *330*(15), 3744-3765.

[12] Huang, C. H., & Zeng, J. (2010). Flexural vibration suppression of car body for high-speed passenger car based on constrained damping layers. *Journal of Traffic and Transportation Engineering*, 10(1): 36-42.

[13] Kozek, M., Benatzky, C., Schirrer, A., & Stribersky, A. (2011). Vibration damping of a flexible car body structure using piezo-stack actuators. *Control Engineering Practice*, *19*(3), 298-310.

[14] Bruni, S., Goodall, R., Mei, T. X., & Tsunashima, H. (2007). Control and monitoring for railway vehicle dynamics. *Vehicle System Dynamics*, *45*(7-8), 743-779.

# ICE-S
# Vehicle reaction measurement and track geometry measurement on the same measuring train Results of the comparison of the two different track inspection methods

**T Kolbe, R Kratochwille**
Deutsche Bahn AG, DB Systemtechnik GmbH,
Department for Vehicle Dynamic Tests, Germany

**ABSTRACT**

The Deutsche Bahn AG periodically has to carry out track inspection measurements as the basic input information for track maintenance work. The track inspection measurements for high speed lines are not only geometry measurements with track recording cars. The track inspection measurements are in addition vehicle reaction measurements with the reference train called ICE-S. Actually the Deutsche Bahn AG faces to the new situation, where data for track geometry and vehicle reaction should be collected at the same time by the ICE-S for all high speed lines. The comparison of the results of the two different track inspection methods will be the main part of the paper.

## 1 INTRODUCTION

The Deutsche Bahn AG regularly has to carry out track inspection measurements as the basic input information for track maintenance work. The track inspection measurements for high speed lines are not only geometry measurements with track recording cars (track geometry measurement coaches, for Deutsche Bahn AG called RaiLab). The track inspection measurements are in addition vehicle reaction measurements with the reference trains called ICE-S.

The ICE-S is equipped with accelerometers and instrumented wheelsets. The vehicle reactions measurement controls the interaction between vehicle and track with regard to running safety, track loading and ride characteristics for the local high speed of the track. Defects in the track which produce an unacceptable vehicle reaction have been detected. If defects in the track are identified, they will be rectified by the maintenance team immediately.

Until 2013 the ICE-S was only equipped with the vehicle dynamic measuring technique.

Since January 2014 a second measuring coach is included in the ICE-S following the strategy of multi-measuring trains for more time and cost efficient track inspection. This measuring coach 02 carries the technology for the track geometry measurement which is a duplicate of the RaiLab-technology.

Today there is a complete new situation for Deutsche Bahn AG. Track geometry and vehicle reaction are measured periodically at the same time on the same test train on every high speed line.

This paper describes the principles of the track geometry measurement, of the vehicle reaction measurement and its measuring techniques. In addition examples of the results are shown.

The comparison of the results of the two different track inspection methods is the main part of the paper. It will be shown which geometry measuring results lead to high vehicle reaction. It will also be presented which track geometry situation will be found if a high vehicle reaction is measured (influence of the wavelength etc.). Typical measuring results for high speed lines will be presented depending on track design, track elements, temperature and season as well as the temporal development of track defects. The presentation will highlight a better understanding of vehicle/track-interaction.

A summary will close the paper.

## 2 TRACK GEOMETRY MEASUREMENTS

The measurement of the track geometry has been done by so called track recording cars and is defined as safety relevant. The way and the quality of track geometry measurement is described and defined in the Euro-Norm EN 13848 [6]. The relevant standard for track geometry inspection for the network of the Deutsche Bahn AG is the Ril 821.2001 [5].

The track recording cars of the DB Netz AG (infrastructure company of Deutsche Bahn AG) are the well-known RaiLab-coaches [8]. The measuring system bases on an inertial measuring platform and optical laser-technology.

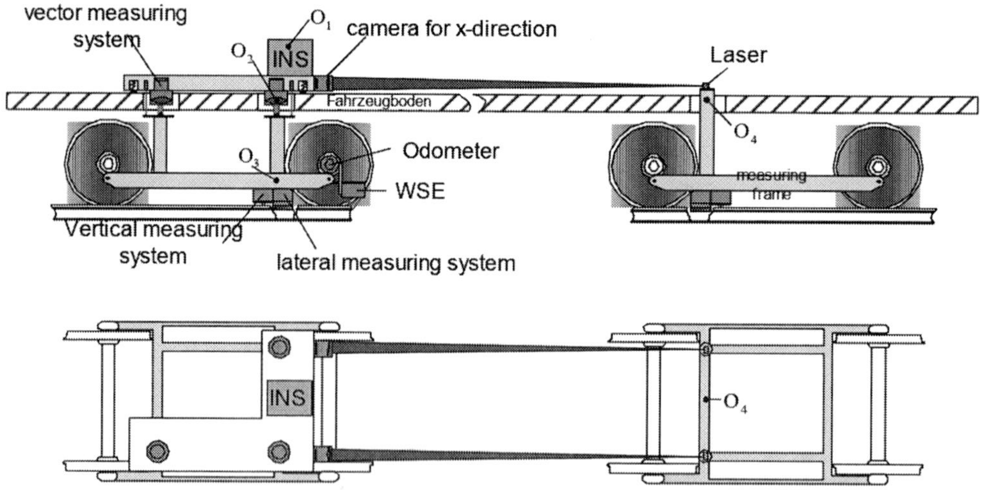


**Picture 01: principle of RaiLab measuring [8]**

The picture 01 gives an overview of the principle of the measuring technique of With the RaiLab. There are sensors for the horizontal (lateral measuring system) and vertical (vertical measuring system) measurement of the distance between rail and measuring frame. The optical distance measurement exists two times for each reassuring frame. The measuring frames are fixed at the bearings of the wheelsets. The heart of the system is the inertial measuring platform (INS) in the coach. This system provides the absolute position relative to the North Pole. The distance between the measuring frames and the INS is quantified with the help of laser based measuring systems. The twist of the coach is identified with the help of two extra cameras. All optical sensors work on the principle of optical triangulation without mechanically moved elements [8].

This technology is a further development of the former cord measuring system.

The following parameters identify the track geometry:

- Gauge,
- Cant,
- Curvature,
- y- and z- profile of the rail (level and alignment),
- The twist and cant deficiency are calculated.

These values describe the deviation between the actual position of the rail and the designed position of the rail. Maximum values for each parameter are defined in [5, 6]. An example of the output of the track geometry measurement can be seen in picture 02.

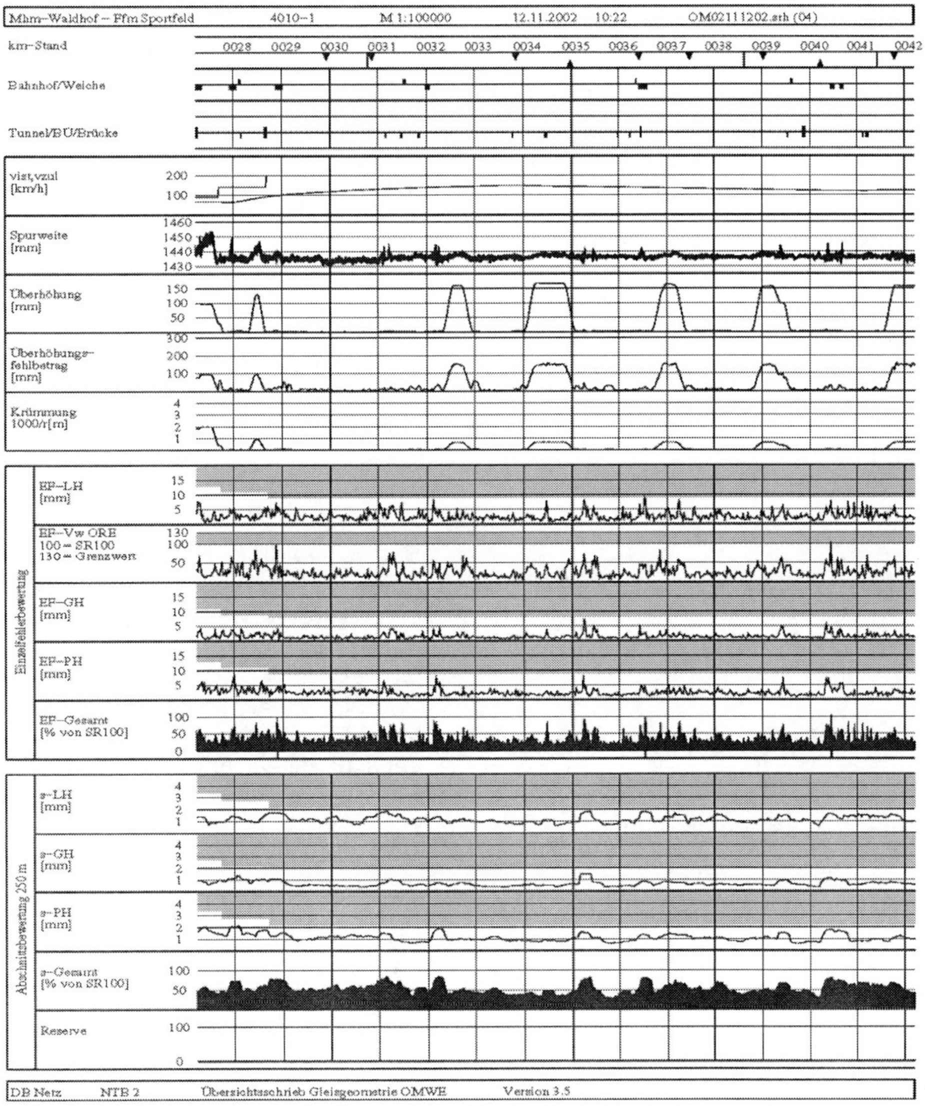


**Picture 02: output of track geometry measurement [7]**

## 3 VEHICLE REACTION MEASUREMENTS

The high speed lines and tilting train lines of Deutsche Bahn AG repeatedly have been inspected by special test trains of Deutsche Bahn AG. These test trains are equipped with measuring technique especially with accelerometers and instrumented wheelsets. The sensors measure the vehicle reaction on each track. The vehicle reaction measurement is part of the regular track inspection of Deutsche Bahn and is described in the Deutsche Bahn AG standards [1, 2]. The vehicle reaction measurement is defined as safety relevant.

The vehicle reaction measurement controls the interaction between vehicle and track with regard to running safety, track loading and ride characteristics for the local high speed of the track. With the vehicle reaction measurement is proven that the track resists to the higher impact resulting from high speed trains or tilting trains. Defects in the track which cause an unacceptable vehicle reaction are detected. If defects in the track have been detected, maintenance work follows immediately.

The inspection trains are equipped with measuring wheelsets (instrumented wheelsets) and accelerometers on every suspension level. The whole measuring principle is closely related to the homologation principle according to EN 14363 [3] and UIC 518 [4]. At the moment the ICE-S performs test runs for track inspection on the high speed network of Deutsche Bahn AG as well as on the network of Österreichische Bundesbahn (ÖBB) and Schweizerische Bundesbahnen (SBB).

The ICE-S is the reference train for all the high speed vehicles of Deutsche Bahn AG. More detailed the ICE-S includes two different reference vehicles: the power head TK1 and the passenger coach MW1.

The power head TK1 is the reference vehicle for all ICE1/2 power heads and all fast locomotives (BR120, BR101, BR182).

The passenger coach MW1 is the reference vehicle for all ICE1/2 coaches, ICE3 (BR403, BR 406, BR 407), ICE-T (BR 411, BR415) and all conventional fast passenger coaches.

The following values are measured for the track inspection:

- $\Sigma Y_1$ , $\Sigma Y_2$ — Sum of guiding force wheelset 1 and 2,
- $Y_{11}$ , $Y_{12}$ — guiding force wheelset 1,
- $Q_{11}$ , $Q_{12}$ — vertical force wheelset 1 and 2,
- $(Y/Q)_{11}$, $(Y/Q)_{12}$, $(Y/Q)_{21}$, $(Y/Q)_{22}$ — quotient (Y/Q),
- aq — lateral acceleration wheelset level,
- $y_{pp}{}^{+}$ — lateral acceleration bogie level,
- $y_{ppF}{}^{*}$ — lateral acceleration car body level,
- $z_{ppF}{}^{*}$ — vertical acceleration car body level.

Maximum values and limits for all of these parameters are defined in [2].

The measuring plan of the inspection train ICE-S can be seen in picture 03.

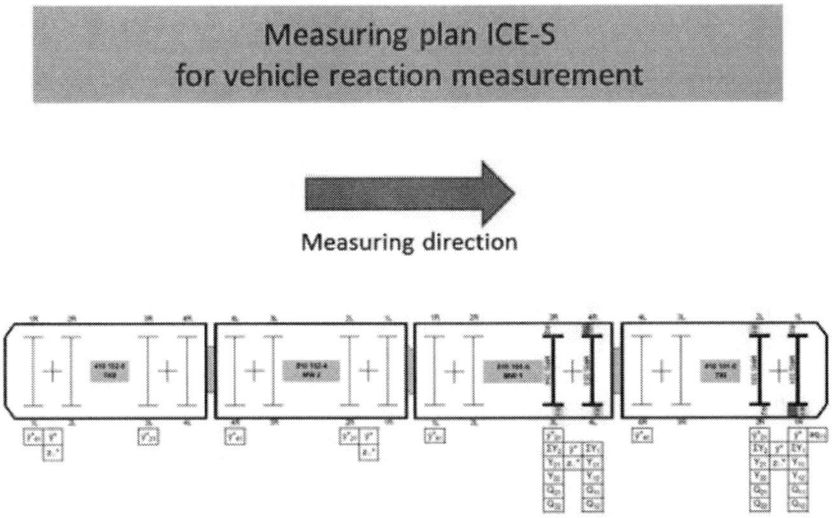


**Picture 03: measuring plan ICE-S [7]**

An example of the results of the vehicle reaction measurement for the power head TK1 can be seen in picture 04.

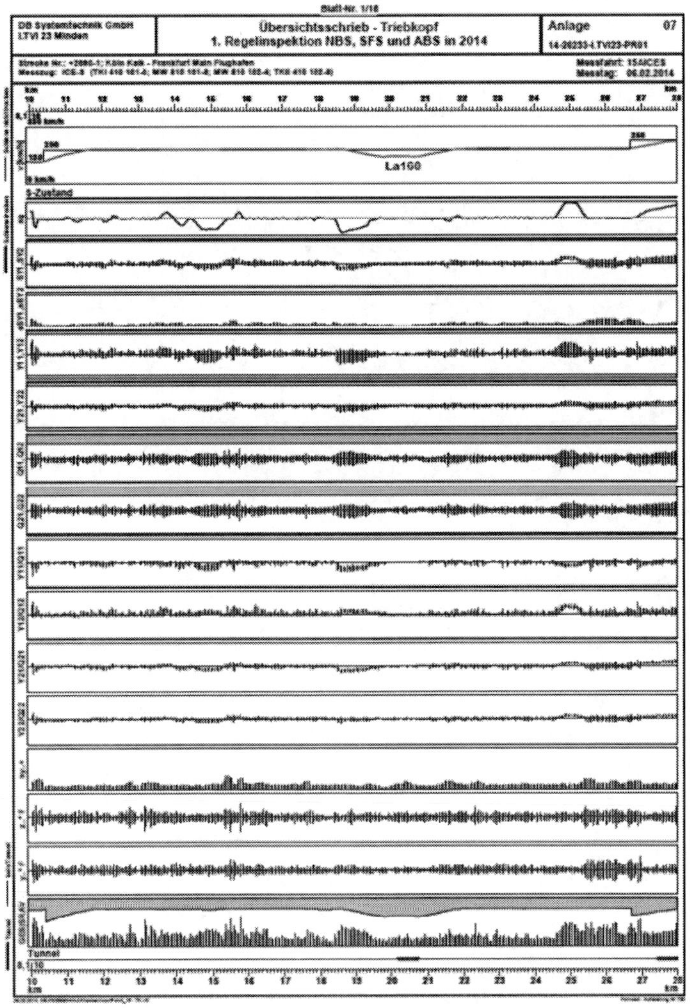


**Picture 04: result vehicle reaction measurement power head TK1 [7]**

## 4 ICES TODAY

The ICE-S was only equipped with the vehicle reaction measuring technique in the past (until 2013).

Since January 2014 a second measuring coach has been integrated into the ICE-S for the optimization of multi-measuring trains strategy. This situation increases the operation time and makes costs of track inspection efficient. This measuring coach 2 carries the technology for the track geometry measurement which is a duplicate of the RaiLab-technology following Deutsche Bahn AG standard 821.2001 [5] and EN13848 [6].

Actually the Deutsche Bahn AG faces to the new situation, where data for track geometry and vehicle reaction should be collected at the same time by the ICE-S for all high speed lines ($v_{max}$ > 160 km/h) twice a year (about 22,400 km track). The picture 05 shows the actual train configuration of the ICE-S.

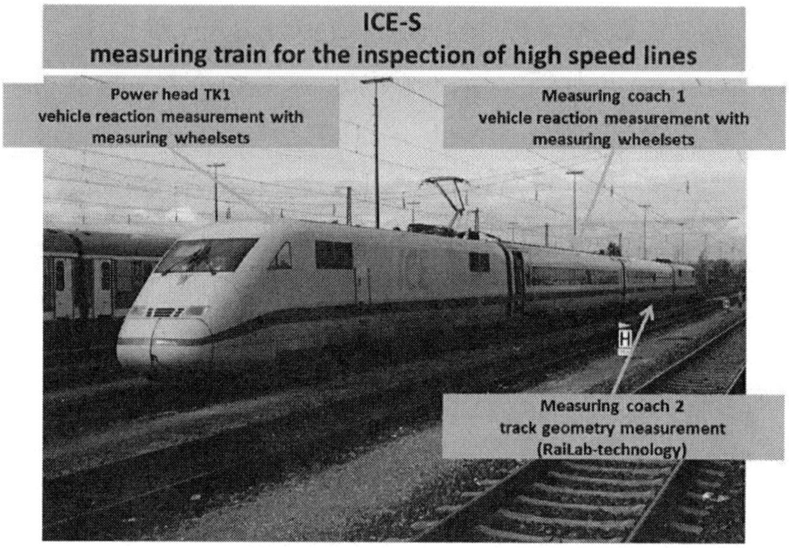


**Picture 05: actual train configuration of ICE-S**

## 5 BASIC PROBLEMS

There are today two different track inspection methods on board the ICE-S:

- Vehicle reaction measurement,
- Track geometry measurement.

Testing and approval of railway vehicles from the point of view of their dynamic behaviour (UIC 518 [4], EN 14363 [5]) are the origin of the limits for the vehicle reaction measurement. The limit values for the wheel/rail-forces are specific for each vehicle.

The track geometry measurement limits have known an empirical development in the last 40 years. They cover the safety for the general fleet of railways vehicles on all lines of the network. The limits are more challenging with the increase of the line speed [5, 8].

It is not astonishing that the two inspection methods are not harmonised quite well for all cases. This can be seen in picture 06:

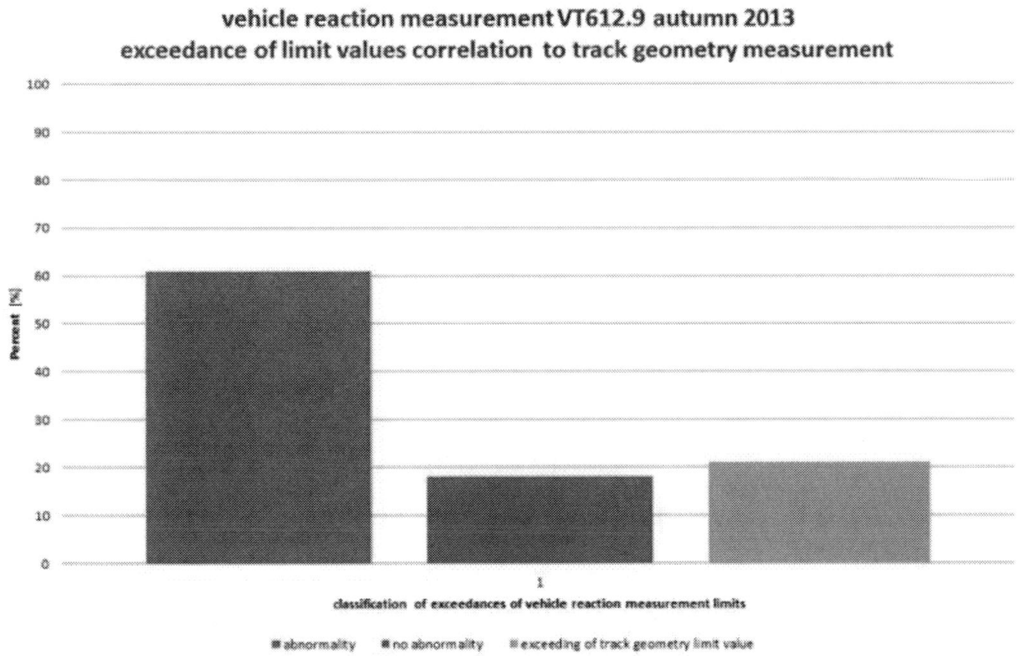


**Picture 06: exceeding of vehicle reaction measurement limits**

Picture 06 gives a classification of exceedings of the vehicle reaction measurement limits and its correlation to track geometry measurement. The results of the vehicle reaction measurement and the track geometry measurement have been compared for the period of autumn 2013. For 61 % of the exceedances of vehicle reaction measurement limits an abnormality in the geometry of this track section can be found. For 18 % of the exceedances of vehicle reaction measurement limits no abnormality in this track section can be found. Everything is well maintained from the point of view of track geometry. For only 21 % of the cases an exceedance of vehicle reaction measurement limits and an exceedances for track geometry measurement could be detected.

## 6 OVERVIEW OF EXCEEDANCES OF VEHICLE REACTION MEASUREMENT LIMITS

For the vehicle reaction measurement wheel/rail-forces and its derived values (Sum Y, $(Y/Q)_a$) are measured. Accelerations on the level of the wheelset, the bogie frame and in the car body are measured as well.

The exceeding of limits for the wheel/rail-forces is very rare (less 2 %). The exceeding of the limits for the uncompensated acceleration on the level of the wheelset is a problem of settling of the cant. This is not found on slab track. It is found on lines with conventional ballasted track and today it is also very rare on the network of Deutsche Bahn AG (less 3 %). The acceleration measured on the bogie frame cover the problem of instability in track sections with high equivalent conicity caused by wear of the rail heads. Today this problem is well controlled due to huge grinding programs of the rail heads on the high speed lines. Over 90 % of the exceedings of limits of vehicle reaction measurement for the high speed lines are accelerations in the car body. Based on that, it makes sense to focus on these cases.

The following picture 07 gives an overview over the exceedings of the car body accelerations of the ICE-S.

**Overview exceedings accelerations ypp* und zpp***

| | TK1 | | | | MW1 | | | | MW2 | | | | TK2 | | | |
|---|---|---|---|---|---|---|---|---|---|---|---|---|---|---|---|---|
| | ypp* | | zpp* | | ypp* | | zpp* | | ypp* | | zpp* | | ypp* | | zpp* | |
| | > 2,5 m/s² | > 3,25 m/s² | > 2,5 m/s² | > 3,25 m/s² | > 2,5 m/s² | > 3,25 m/s² | > 2,5 m/s² | > 3,25 m/s² | > 2,5 m/s² | > 3,25 m/s² | > 2,5 m/s² | > 3,25 m/s² | > 2,5 m/s² | > 3,25 m/s² | > 2,5 m/s² | > 3,25 m/s² |
| **Spring 2013** | 5 | 0 | 76 | 4 | 11 | 0 | 1 | 0 | 1 | 0 | 8 | 1 | 9 | 0 | 140 | 22 |
| **Autumn 2013** | 16 | 2 | 84 | 3 | 0 | 0 | 22 | 3 | - | - | - | - | 13 | 0 | 182 | 20 |
| **Spring 2014** | 4 | 0 | 29 | 1 | 0 | 0 | 16 | 0 | 0 | 0 | 9 | 0 | 15 | 0 | 121 | 11 |

**Picture 07: exceedings of car body accelerations**

Picture 07 shows the results for the car body exceedings for the campaigns in spring 2013, autumn 2013 and spring 2014 for the ICE-S. The exceedings are split for the four vehicles of the train: power head TK1 and TK2 and the coaches MW1 und MW2. The table differentiates between horizontal (ypp*) and vertical (zpp*) accelerations. Most of the exceedings are vertical accelerations. The two power heards show the most exceedings. The leading power head (TK1) has got less exceedings than the trailing power head TK2. The campaigns in spring are performed in February and beginning of March. In this time you will find strong frost in Germany. The ballasted tracks are more fixed and so track irregularities are not so often. Because of that the number of exceedings in autumn is much higher than in the spring campaign.

## 7 EXAMPLE FOR A HIGH SPEED LINE

The picture 08 shows the results of the accelerations for the vehicle reaction measurement on the high speed line (number 1733_1) from Hannover Hbf to Würzburg Hbf. It is the oldest high speed line in Germany and has been released for the traffic in 1991. The chart includes all exceedings for the track 1 which is the direction to the south from Hannover Hbf to Würzburg Hbf for all 327.4 km of the line. The line speed is 280 km/h which is performed along the whole line with the exception of the stations Göttingen, Kassel-Wilhelmshöhe and Fulda. The line is constructed with ballasted track. The minimum radius of the line is 5100 m. The maximum inclination is equal to 12.5 ‰. The line is used for the IC and ICE service over the day and for freight trains in the night.

Picture 08 shows the results for the car body exceedings for the campaigns in spring 2013, autumn 2013 and spring 2014 for the ICE-S. Generally the values are higher in spring (cold, frosty) than in autumn. It is eye-catching that most of the exceedances are measured regularly. Some exceedings of the limits occur once-time.

It can be demonstrated that most of the exceedings are located at special track elements. The running over switches or most of the time over viaducts and bridges have provoked the high acceleration values. Sometimes the high accelerations appear at the beginning or end of a tunnel. Some values can't be related to a track element.

This characteristic is very typical for a high-speed line with ballasted track. The characteristic of the newer high-speed lines with slab track is clearly different. The number of exceedings is very low. But the few numbers of exceedings can also be related to special track elements like viaducts.

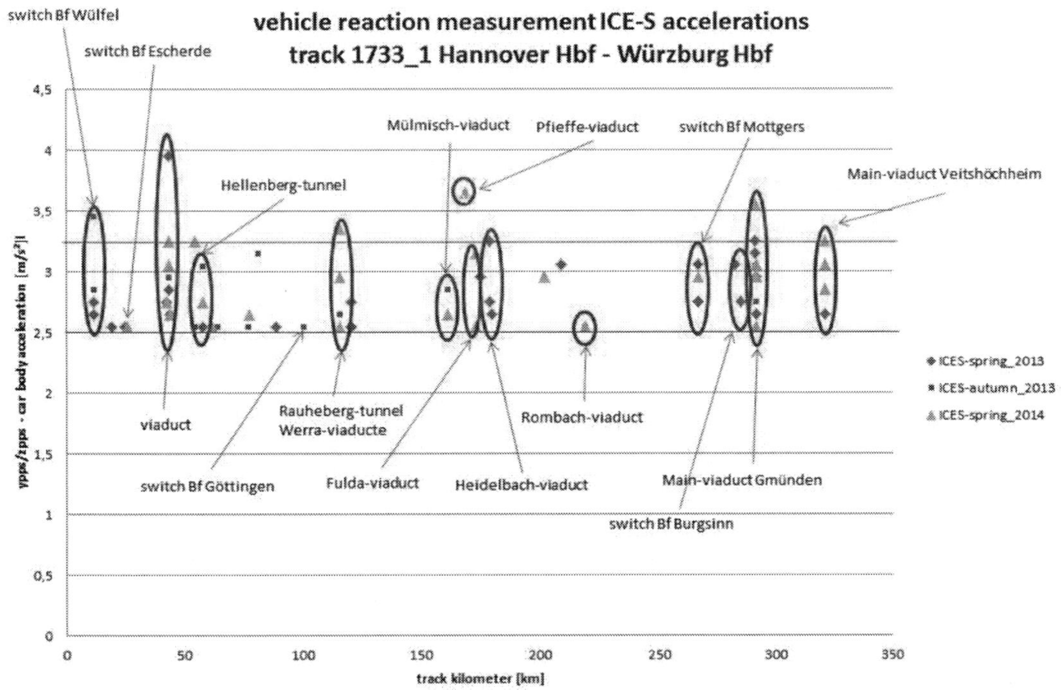


**Picture 08: results high speed line Hannover Hbf – Würzburg Hbf**

## 8 CORRELATION OF Z-ACCELERATION AND VERTICAL ALIGNMENT

A vertical movement and acceleration of the car body is caused by vertical deviation of the track. The vertical acceleration zpp* is measured with the technique of the ICE-S for the vehicle reaction measurement. The vertical alignment (LH) is the difference between the desired position and the actual position of the rail. This is measured by the track recording car and the result is a height in millimetres (mm).

The analyses of the z-profiles of the track geometry measurement have to be done for three different wavelength ranges λ according EN 13848 [6]:

- D1: 3 m < λ ≤ 25 m,
- D2: 25 m < λ ≤ 70 m,
- D3: 70 m < λ ≤ 150 m.

The wavelength D1 is the standard analyse for the track geometry cars and has done almost realtime. The wavelengths D2 and D3 can only be analysed for the signals in a post process.

The picture 09 shows the correlation of the vertical accelerations zpp* greater than 3,25 m/s² and the vertical alignment LH.

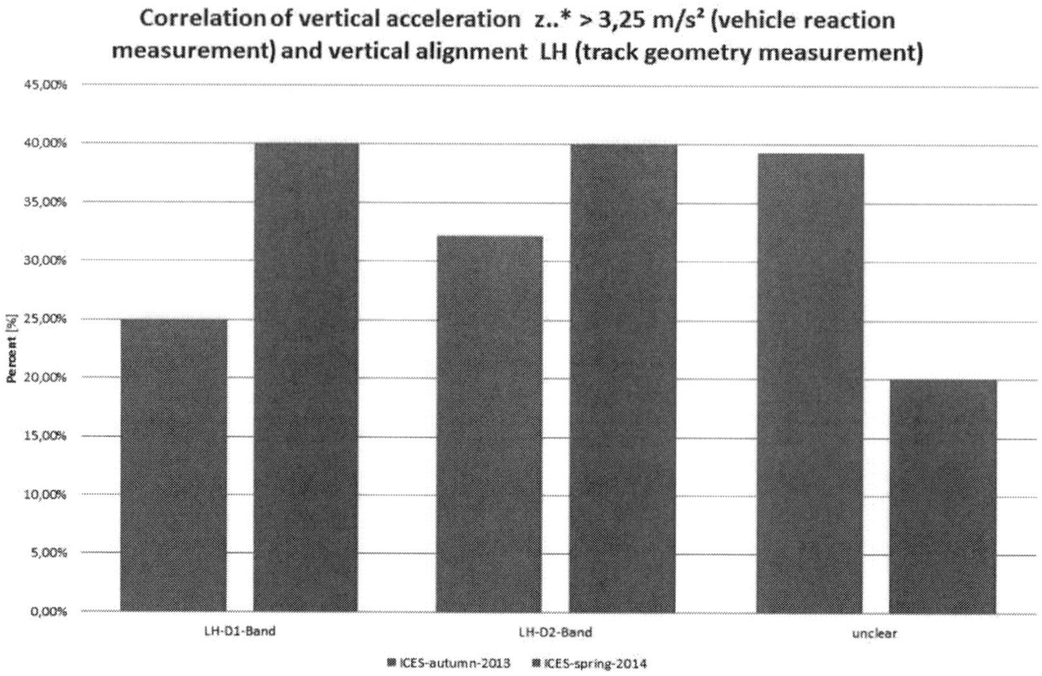


**Picture 09: correlation of vertical accelerations zpp* and vertical alignment LH**

The results have been analysed for the ICE-S campaign in autumn 2013 and in spring 2014. The picture shows the following for the campaign in autumn 2013: for 25 % of the exceedings of the vertical acceleration there is also an exceeding in the vertical alignment in the wavelength range D1 ($3\ m < \lambda \leq 25\ m$). For about 34 % of the exceedings of the vertical acceleration a correlation to exceedings in the vertical alignment in the wavelength range D2 ($25\ m < \lambda \leq 70\ m$) could be found. And for about 40 % of the exceedings there is no correlation and the situation is unclear. For the campaign in spring 2014 the situation is similar. A better correlation of the D1 (40 %) and D2 (40 %) range could be found. Summarising there is a number between 20 to 40 % of the exceedings of the vertical acceleration where no correlation to the measuring results of the track geometry car could be found.

The picture 10 shows a more detailed analysis of the situation for an example. The speed [km/h] of the ICE-S is written on the x-axis. The acceleration of the car body [$m/s^2$] in z-direction can be seen on the y-axis of the sheet. Only accelerations higher than 3.25 $m/s^2$ have been taken into account.

The values for the acceleration are marked for the different heights of vertical alignment (wavelength D1) with different coloured symbols.

It can be seen that the amplitude of the car body acceleration is not dependent to the speed. Also it has to be remarked that the amplitude of the car body acceleration is independent to the vertical alignment of the track. The highest values of car body acceleration have been found in combination with track deviations with longer wavelength (D2) or in situations where no higher deviation in vertical direction of the rail could be measured.

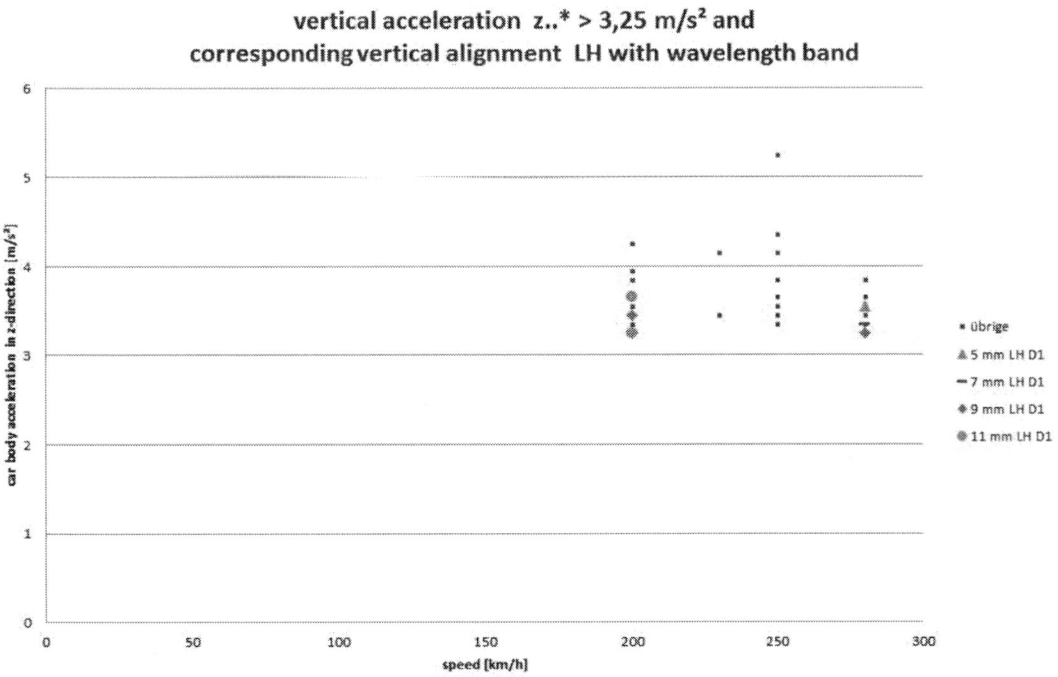


**Picture 10: correlation of vertical acceleration and vertical alignment**

The table in picture 11 shows the criteria for vertical alignment in millimetres according the Deutsche Bahn AG standard Ril 821.2001 [5]. SRA, SR100 and SRlim are the three different levels of these criteria.

| | Criteria for vertical alignment | | |
|---|---|---|---|
| Speed range | SRA | SR100 | SRlim |
| 160 km/h < v ≤ 230 km/h | 6 mm | 9 mm | 11 mm |
| v > 230 km/h | 5 mm | 7 mm | 9 mm |

**Picture 11: criteria for vertical alignment [5]**

The comparison of the information of picture 10 and 11 underlines the statement that sometimes the two different measuring principles produce contradictory information. For example the track quality seems to be acceptable (SRA level) from the point of geometry measurement but the accelerations in the car body are not acceptable from the point of vehicle reaction measurement (> 3.5 m/s²). But it also has to be realised that for some cases there is a good correlation of the two measuring principles. It will be the research work of the future to identify a better understanding of vehicle/track-interaction in these cases. This will lead to a new approach of track inspection and different inspection limit values as a next step.

## 9 SUMMARY

The Deutsche Bahn AG periodically has to carry out track inspection measurements as the basic input information for track maintenance work. The two main principles of track inspection for the high speed lines in Germany have been presented in this paper. Both measuring principles are today together on board the high speed measuring train ICE-S which is used for the periodical inspection of the high speed lines.

The technique of the track geometry measurement has been described and examples showed the output of the system.

The vehicle reaction measurement was explained in combination with an example of the measuring results.

It could be demonstrated generally that the two inspection methods are not harmonised quite well for all cases.

The character of the exceedings of the vehicle reaction measurement has been specified. The influence of the season, track design, viaducts, bridges and special track elements could be shown.

The correlation of the results of the two different track inspection methods was the main part of the paper. The influence of the wavelength could be explained. The comparison of the results underlines the statement that sometimes the two different measuring principles produce contradictory information. But it also has to be realised that for some cases there is a good correlation of the two measuring principles. It will be the research work of the future to identify the relationship of the vehicle/track-interaction in these cases. This will lead to a new approach of track inspection and different inspection limit values.

## REFERENCE LIST


[1] Ril 820.0120, Bautechnik, Leit-, Signal- und Telekommunikationstechnik – Grundlagen des Oberbaus – Richtlinie "Oberbautechnische Freigabe" der DB AG. Stand 01.01.2007.

[2] Ril 821.2002, Bautechnik, Leit-, Signal- und Telekommunikationstechnik – Oberbau inspizieren – Richtlinie "Fahrtechnische Inspektion" der DB AG. gültig ab 04.09.2012.

[3] DIN EN 14363:2005, Bahnanwendung – Fahrtechnische Prüfung für die fahrtechnische Zulassung von Eisenbahnfahrzeugen – Prüfung des Fahrverhaltens und stationäre Versuche. Berlin, DIN, Beuth-Verlag, Oktober 2005.

[4] UIC-KODEX 518, Fahrtechnische Prüfung und Zulassung von Eisenbahnfahrzeugen – Fahrsicherheit, Fahrwegbeanspruchung und Fahrverhalten. UIC, Paris, Oktober 2009, 4. Ausgabe.

[5] Ril 821.2001. Bautechnik, Leit-, Signal- und Telekommunikationstechnik – Oberbau inspizieren – Richtlinie "Prüfung der Gleisgeometrie mit Gleismessfahrzeugen" der DB AG. gültig ab 01.01.2010.

[6] EN 13848:2008, Railway applications – Track – Track geometry quality, part 1 – 5, Bahnanwendungen – Oberbau – Gleislagequalität, Teil 1 – 5. Berlin, DIN, Beuth-Verlag, November 2008.

[7] Rainer Kratochwille, Thomas Kolbe: Fahrtechnische Prüfungen zur Oberbauinspektion. 12. Internationale Schienenfahrzeugtagung, Dresden, 2012.

[8] Herbert Zück: Neues Messverfahren zur Erfassung der Gleisgeometrie. Deine Bahn, 2012, No.4, page 15 – 19.

# Detection of vertical track faults using an instrumented in-service vehicle: Considering efficient real-time fault detection

**G J Yeo, P F Weston, C Roberts**
Birmingham Centre for Railway Research and Education, University of Birmingham, UK

**ABSTRACT**

An in-service vehicle instrumented with a low cost Inertial Measurement Unit (IMU) comprising of three accelerometers and three gyroscopes can provide information about the condition of the track traversed by the train. Data from the pitch-rate gyroscope alone can be used to identify short dips in the track caused by defects such as wet spots, ballast settlement and hanging sleepers. This paper considers how an in-service inertial logging system can be augmented to process data in real-time and, ultimately, allow certain faults to be detected on-board the vehicle. This would enable fault reports to be generated and sent using a cellular network connection, allowing for prompt assessment of faults and for maintenance to be scheduled before faults increase in severity.

## 1 INTRODUCTION

An IMU mounted on a bogie of an in-service vehicle provides an inexpensive method of track condition monitoring, compared to a dedicated Track Recording Vehicle (TRV). An IMU system logs data from three accelerometers, three gyroscopes (gyros), a tachometer (tacho) and a Global Positioning System (GPS) device during the day-to-day operation of an in-service vehicle. This produces a large quantity of data that can be used to assess the quality of the track on which the vehicle operates. Generally, measurements of a section of track occur much more frequently using an in-service vehicle than using a TRV, providing several measurements per day rather one measurement every few months.

One problem arising from the use of this system is that a large quantity of data is produced, as the vehicle may be operating around 18 hours per day. Whilst numerous processing methods are available (1–3) for detecting vertical geometry faults, these are generally computationally complex and consequently unsuitable for large quantities of data. Future work will also require faults to be detected at the time that measurement occurs (or shortly thereafter). Efficient processing in a real-time analysis situation is an important consideration. Another problem arising from the large data quantity is that it is impossible to transmit this data using a 3G or GSM-R connection.

In this paper, the use of a versine calculation is considered, which is a fast operation. By calculating versines from measured vertical curvature, potential faults can be identified. Further processing can then be applied to the immediate area of the detected fault in order to verify it and provide a detailed evaluation if necessary. Consideration is also given to how short fault reports may be generated containing relevant data from around the identified fault site, suitable for transmission using a cellular network connection.

## 2 EXISTING IN-SERVICE TRACK MEASUREMENT SYSTEM

The University of Birmingham has installed an IMU onto a bogie of a Southern Trains Class 377 Electrostar EMU (Figure 1). The system collects inertial data from routes on the UK rail network south of London. The IMU contains a 3-axis accelerometer and three gyroscopes, all of high quality. The system accepts an input from a tacho on one wheelset of the bogie on which the IMU is mounted, and also records positional data from an on-board GPS receiver. The GPS antenna is mounted in the roof of the train, directly above the IMU.

**Figure 1: IMU mounted on vehicle bogie**

The system collects data from all of these sensors, as well as data from some other monitoring systems. The bogie-mounted IMU uses six inertial sensors, allowing many elements of track geometry to be calculated. However, in this paper, consideration is given to how a subset of the inertial sensors could be used for the purpose of detecting dips in the loaded vertical profile of the track which could indicate the presence of voids beneath the sleepers, or other problems with the track structure.

Currently, data from the IMU system is collected onto a portable hard disk drive (HDD) connected to an on-board logging computer. This disk is periodically exchanged so that data can be retrieved and processed. Data is not transmitted in any way owing to the high data-rate produced by the system. A forthcoming Future Rail project for void detection demands wireless reports of the faults as they occur.

## 3 SHIFT TO ON-BOARD PROCESSING

The current measurement system (Figure 2) performs data acquisition and storage only on-board the train. All of the processing is done remotely.

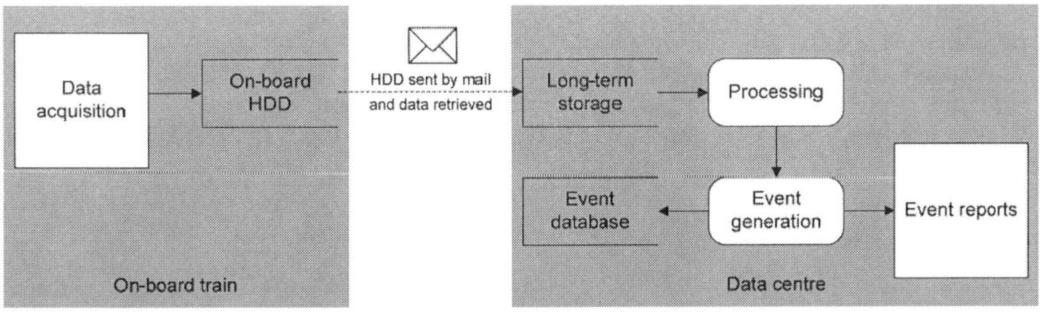


**Figure 2: Flow of data in existing system**

The proposed system (Figure 3) would perform a primary processing stage on-board the train. This processing stage would contain only limited processing; enough for event detection to occur. Event packets would then be generated containing data acquired around the time of the event. Packets may then be transmitted using a cellular connection to the data centre or other remote location. Upon reception, secondary processing is performed to analyse faults in greater detail, after which event reports can be produced.

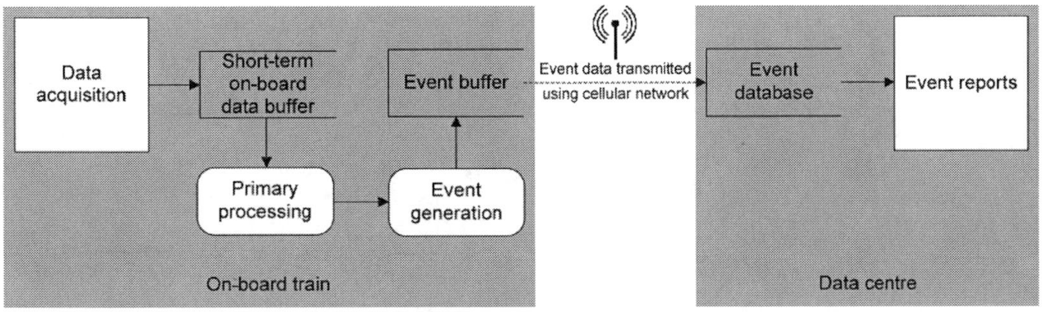


**Figure 3: Flow of data in proposed system**

This system distributes the processing which has the advantage of being able to report defects as soon as they are measured. By performing further processing remotely from the train, the analysis of the data can be much more in-depth, as it is not limited by the real-time performance of the on-board PC.

## 4 DATA PROCESSING

Data processing methods for deriving the vertical displacement of the bogie from accelerometer and gyro measurements have been well established in previous work (1,3). Mid-wavelength vertical defects such as those introduced by voids and wet spots, can be identified by applying versine calculations to the vertical curvature without the need to derive the vertical displacement. Vertical displacement is, however, a useful diagnostic when visualising a fault. It is therefore a good example of the kind of processing that could be done remotely.

### 4.1. Domain conversion

Conversion of data from the time domain to the spatial domain is a necessary step in the on-board processing, as versine calculation must be done in the spatial domain. In (1, 3) a resampling algorithm, which uses the speed vector $V_x$, is used to convert data from samples in the time domain to samples in the spatial domain. In the Southern Trains system, IMU channels are each logged at a rate of 250 $Ss^{-1}$.

This can be converted on-board to 8 $Sm^{-1}$ which is adequate for the analyses performed here.

Current processing calculates curvature in the time domain, which is then converted to the spatial domain. However, for real-time processing, it is more efficient to convert the incoming IMU data (vertical accelerometer/pitch-rate gyro, with tachometer) to the spatial domain before performing any further processing.

### 4.2. Vertical curvature

Curvature can be found using the vertical sensing accelerometer (Equation 1) or by using the pitch-rate gyro (Equation 2).

$$K_z = \frac{\ddot{Z}}{V_x^2} \quad (1)$$

$$K_\phi = \frac{\dot{\phi}}{V_x} \quad (2)$$

Where $\ddot{Z}$ and $\dot{\phi}$ are vectors containing the vertical acceleration and pitch rate, respectively. $V_x$ is a vector containing the train's speed in the $x$ (longitudinal) direction. Because the measurement data is now in the spatial domain, equations 3 and 4 can be used instead of 1 and 2, resulting in reduced on-board processing.

$$K_z = \frac{\ddot{Z}}{ds^2}(\Delta T)^2 \quad (3)$$

$$K_\phi = \frac{\dot{\phi}}{ds^2}\Delta T \quad (4)$$

Either of these methods can be used to estimate the vertical curvature of the track from the respective sensor measurements. In fact both sensors have advantages and disadvantages to their use. At low vehicle speeds, insufficient vertical acceleration is generated for a vertically-sensing accelerometer to produce a signal above its noise level. In this situation, a pitch-rate gyro produces better results, as a slow bogie pitch is still detectable. The pitch-rate gyro can produce acceptable results at speeds as low as 1 $ms^{-1}$, which is useful on an in-service vehicle that slows and stops regularly. At vehicle speeds of 10 $ms^{-1}$ or higher, the vertically-sensing accelerometer tends to produce better results.

Each sensor also responds differently to differing wavelengths of vertical track geometry. Figure 4 shows how the motion of the bogie on the track produces a filtering effect on the output produced by each sensor.

Vertical geometry features with a wavelength around double the wheelbase of the bogie cause little or no vertical acceleration at the center of the bogie, but the pitch can be detected. Conversely, geometry features with a wavelength the same or near to the bogie wavelength will cause little or no bogie pitch, but the vertical acceleration will be detectable.

In selection of which sensor is to be used, the length of the defects being measured must be taken into account. In this work defects around 4–16 m in length are to be detected, which makes the gyroscope the best choice. Its operation at low speeds also makes it suitable for a regularly stopping in-service vehicle.

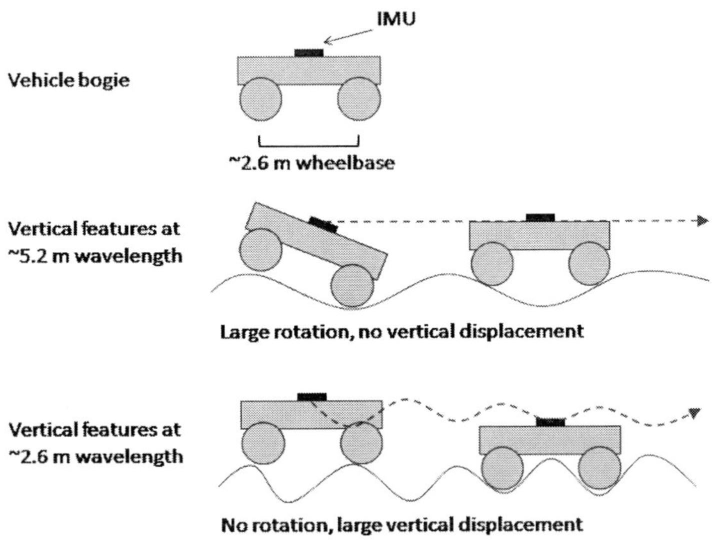


**Figure 4: The filtering effect of the vehicle bogie**

### 4.3. Using versines for dip detection

Versine calculation provides an efficient way to indicate potential faults in the loaded vertical track profile. A versine is the vertical distance to the track at the mid-point of a straight line between two end points located on the track (Figure 5). Weston *et al.* show how versines have been used to detect faults from logged inertial data (1). Versines can be calculated using a convolution of the vertical geometry with a suitable impulse response with 1 in the middle, -0.5 at the outer extents and zeros elsewhere. The double integration required to convert curvature to vertical displacement can be combined with the versine filter resulting in a triangular-shaped impulse response. Versines may then be calculated by convolving this impulse response with the curvature calculated using equation 1 or 2, followed by a high-pass filter.

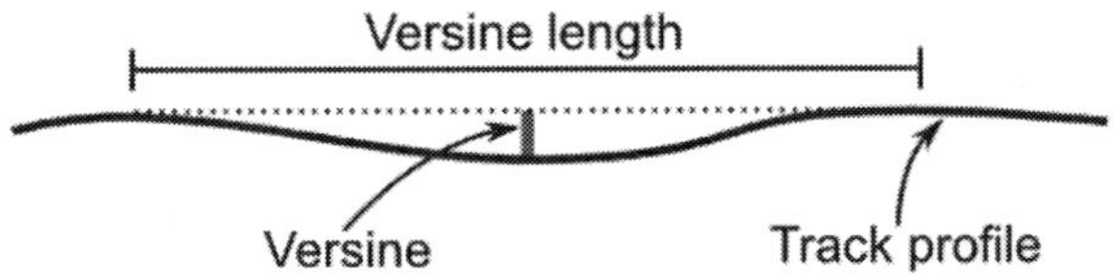


**Figure 5: An example of a versine measurement**

The advantage of the versine method is that vertical geometry features with lengths around the versine length will be emphasised, allowing for easier detection. Mid-wavelength faults such as a wet-spots or hanging sleepers are typically in the range 4–16 m.

### 4.4. Processed data

Figures 6 and 7 show two versines of lengths 16 and 4 m respectively, calculated for the same section of track, before and after tamping had occurred. The gyro-derived pitch-rate and speed were used to compute the vertical curvature. The first figure shows large positive and negative peak excursions at ~480 m, however, these are not evident in the second figure. Using this information, it can be inferred that there are no short geometry faults such as hanging sleepers present, but there is a long geometry fault. In fact, the section of track studied here is the approach to a level crossing (around 500 m) where the track stiffness is likely to be highly variable. It can also be seen that the maintenance performed did improve the track geometry.

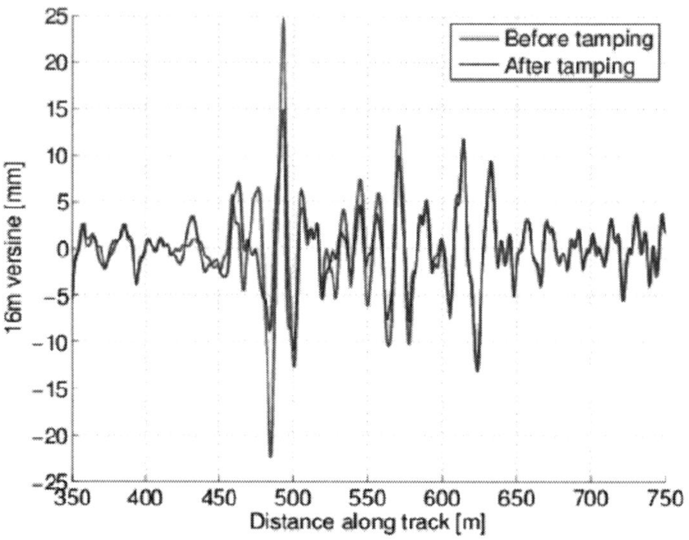


**Figure 6: 16 m versines before and after track maintenance**

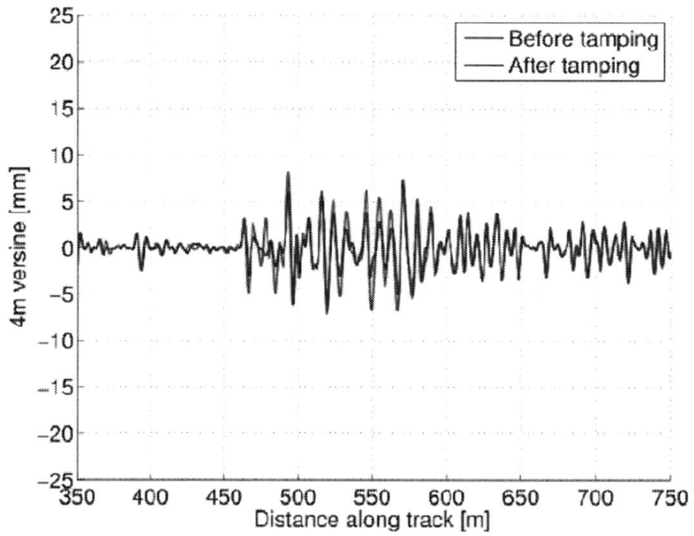


**Figure 7: 4 m versines before and after track maintenance**

By setting a threshold on the versine data, faults such as the one found in Figure 6, could be detected automatically and used to trigger the production of a fault report. By using multiple threshold levels, the event can be graded with a basic measure of severity.

## 5 ENHANCED POSITIONAL PROCESSING

Acquisition of suitable positional data is an area which must be given reasonable consideration when detecting faults in real time. Detected faults must be reported including the location where the fault occurred. In some scenarios such as single track, or double track where each track operates in a single direction, GPS positional data may be adequate to provide the position of the train at the time of measurement. However, in many areas such as those with multiple tracks in each direction or areas with poor or no GPS coverage, the error level of the GPS is often too high to discern which track the vehicle is travelling on. In such a scenario an alternative approach is required.

The inertial sensors can provide information about changes in position and orientation of the train, and the tacho can measure the distance travelled by the train. Using a detailed electronic map of the rail network being covered, various methods (3–6) make it possible to supplement available GPS data with tacho and inertial data. This can generate a position estimate with reasonable accuracy and certainty, however a better estimate can be achieved, particularly at junctions, by using reverse processing passes. At a junction, once the end location of the vehicle is known (e.g. which direction the train took at a switch) previous position estimates can be corrected accordingly. In a real-time processing situation, reverse passes are more difficult to implement as the data is being collected in forwards order.

## 6 TRANSMITTING EVENT DATA

The PC used on the Southern Trains system includes an on-board 3G module. This will be used to transmit reports from the logging system. Each event packet will be uploaded via the cellular network connection to a server where it is stored and further processing can be performed. Careful consideration must be made about the size of the event reports and their ordering in a transmission queue. The bandwidth of the cellular network connection and the likely up-time of the system must also be taken into account; a connection may not be available in areas of poor signal reception.

### 6.1. Event packet contents

Efficient transmission can be achieved by minimising the quantity of data sent in an event packet. Upon first consideration it may seem most practical to send the versine data which has already been calculated. However, the information that this conveys about the identified fault is limited. If the raw (unprocessed) data were sent instead, the versine data could be calculated again at the receiving end.

It is necessary to include data measured from the track either side of the event in order to be able to see the fault data in context. How much data can be sent is dependent upon the bandwidth and up-time of the connection, but from inspection of other faults (e.g. Figure 6) it can be concluded that around 100–200 m of data either side of the event is acceptable.

Sensor data (vertically-sensing accelerometer and/or pitch-rate gyro) and tachometer data would need to be transmitted as vectors. Time can be transmitted as a single start value, as the sample rate is constant. It is also necessary to transmit GPS positional data so that the fault can be located. For a short span of data (200–400 m of track), the positional data could simply comprise of a start and end position, as tacho data is available allowing intermediate positions to be identified. The end position allows the direction of travel to be determined. It would also be prudent to include the position datum at the exact time of the fault for verification of its location.

### 6.2. Event packet priorities

During the course of a vehicle's journey, there may be times when a cellular network connection is unavailable. At these times, event packets must be buffered so that they can be transmitted when the connection becomes available again. Some journeys may have very limited connection opportunities. In this case many event reports may need to be transmitted in a short period of time. In this case it is preferential to transmit events with the greatest severity first in case the connection is lost again.

## 7 REMOTE PROCESSING

Once an event packet is received at the server, it can be stored and the data it contains can be processed. Domain conversion is performed again, and curvature and versines can be calculated.

One of the most useful tools for inspecting a vertical track fault is the vertical displacement data. This can be calculated from the curvature data, $K_z$ or $K_\phi$ as shown in Equation 5.

$$\hat{z} = \iint K \, ds \, ds \tag{5}$$

Figure 8 shows the vertical displacement data for the same section of track inspected in Figure 6.

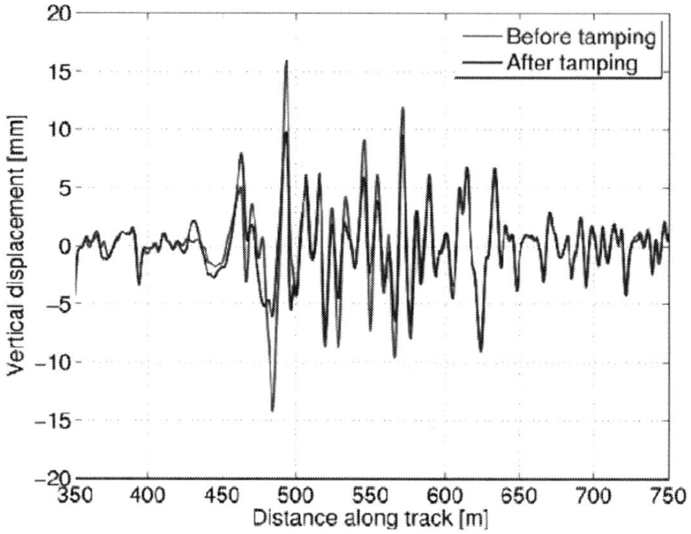


**Figure 8: Vertical displacement before and after tamping**

This method of deriving vertical displacement has the drawback that it only uses curvature from either the vertically-sensing accelerometer, or the pitch-rate gyro. A method exists (3) which allows measurements from both sensors to be combined to give the optimal result. This uses a modified Bryson-Frazier (mBF) filter to combine curvatures calculated from each sensor, and results in vertical geometry features at wavelengths detected by either sensor to be shown in a single vector.

Using information produced at the post-processing stage, the potential fault identified by the on-board system can be verified. A fault report will then be generated by the system, containing a location, the fault severity level, and vertical displacement graphs.

## 8 CONCLUSIONS

A real-time vertical track fault detection system can be created by considering each component of the system against the requirements. The Southern Trains inertial logging system discussed in Section 2 can be augmented with real-time processing by developing processing software to be run on its on-board computer which would compute versines at 4, 8 and 16 m and apply a number of thresholds to detect faults. The threshold levels can be adjusted to set the severity levels of potential faults which will be transmitted. When a potential fault, or a group of potential faults close together are detected, an event packet can be generated, queued according to severity, and transmitted as soon as a 3G connection is available.

Event packets will be uploaded to a server, where further processing will take place to verify the faults and generate detailed reports accordingly. For development purposes, the Southern Trains inertial logging system will continue to log raw (unprocessed) data onto a portable HDD to be exchanged periodically so that data can be retrieved. This way, data can be post-processed so that any faults undetected by the on-board system can be identified, and the on-board software modified accordingly.

Future systems could be made more economical by using only a pitch-rate gyro for vertical geometry measurement, and a yaw-rate gyro for navigational aid. The system would also require a GPS receiver and a tacho input for positioning purposes.

## ACKNOWLEDGEMENTS

The authors would like to thank Track 21 for an EPSRC grant (EP/H044949/1) making this research possible. Thanks also to Southern Railways for agreeing to allow our logging equipment to be installed on one of their vehicles.

## REFERENCE LIST

(1) P. F. Weston, C. S. Ling, C. Roberts, C. J. Goodman, P. Li, and R. M. Goodall, *Monitoring vertical track irregularity from in-service railway vehicles*, Proceedings of the Institution of Mechanical Engineers Part F - Journal of Rail and Rapid Transit, vol. 221, pp. 75-88, Mar. 2007.

(2) S. Alfi and S. Bruni, *Estimation of long wavelength track irregularities from on board measurement*, in Proceedings of the 4th IET International Conference on Railway Condition Monitoring, pp. 1-6, June 2008.

(3) G. Yeo, P. F. Weston, and C. Roberts, *The utility of continual monitoring of track geometry from an in-service vehicle*, in Proceedings of the 6th IET Railway Condition Monitoring Conference, Sept. 2014.

(4) O. Heirich, P. Robertson, A. C. Garca, and T. Strang, *Bayesian train localization method extended by 3D geometric railway track observations from inertial sensors*, 2012.

(5) S. Saab, *A map matching approach for train positioning. I. development and analysis*, IEEE Transactions on Vehicular Technology, vol. 49, pp. 467-475, Mar. 2000.

(6) S. Saab, *A map matching approach for train positioning. II. application and experimentation*, IEEE Transactions on Vehicular Technology, vol. 49, pp. 476-484, Mar. 2000.

# Determination of vertical alignment of track using accelerometer readings

**E J OBrien[1], C Bowe[2], P Quirke[1,2]**
[1] School of Civil, Structural and Environmental Engineering, University College Dublin, Ireland
[2] Iarnród Éireann Irish Rail, Technical Department, Engineering & New Works, Ireland

## ABSTRACT

Railway track vertical alignment is an important indicator of serviceability condition. Through comparisons with past history, track alignment also informs maintenance planning. The vertical alignment of a railway track excites a dynamic response in a train which can potentially be used to determine that alignment. A method is proposed in this paper for the detection of the alignment through an analysis of vehicle accelerations resulting from the train/track dynamic interaction. The Cross Entropy optimisation technique is applied to determine the railway track profile heights that best fit the measured accelerations at and above a railway carriage bogie. Such an approach, using relatively low-cost accelerometers fixed to trains in regular service, would provide inexpensive daily 'drive-by' track monitoring to complement and compare data collected by the Track Recording Vehicle (TRV). The use of a TRV is the current preferred method used to determine railway track profiles using laser based methods. Numerical validation of the concept is achieved by using a 2-dimensional quarter-car dynamic model for the railway carriage and bogie to infer the track profiles in the longitudinal direction. The interaction model is implemented in MATLAB. The track is modelled as an infinitely stiff beam featuring various grades of rail irregularity which excite the vehicle inducing a dynamic response. Ten vertical elevations are found at a time which give a least squares fit of theoretical to measured accelerations. In each time step, half of these elevations are retained and a new optimisation is used to determine the next ten elevations along the length of the track. The optimised displacements are collated to determine the overall rail track profile over a finite length of railway track. This paper reports the results of the numerical simulations and the plans that are underway to further develop the model and test the concept in field trials.

**Keywords**: Railway; Track; Vertical Alignment; Drive-by; Vehicle Rail Interaction; Cross Entropy.

## 1 INTRODUCTION

Increased demand on railway networks is reducing the time available to carry out inspections necessary to determine track condition. As a result, the collection of acceleration and other dynamic parameters from sensors mounted on in-service vehicles is becoming more desirable as a tool for monitoring the condition of railway track. Dynamic measurements are preferable to optical measurements which have a tendency to perform poorly in dirty railway environments. The drive-by nature of this continuous track monitoring (CTM) system has the potential to provide 'real time' feedback to railway infrastructure managers on the condition of their network. This makes possible the forecast of track defect development, verification of quality of repairs and the improvement of maintenance management [1]. A CTM system has been used by Deutsche Bahn since 2004. The system, using accelerometers fixed to axle boxes on their ICE 2 high-speed trains provides a periodic assessment of track geometry.

Traditionally, railway infrastructure managers assess the condition of their network using a Track Recording Vehicle (TRV): a specialised, instrumented train which periodically collects geometric data of the railway track including track gauge, longitudinal profile, alignment, superelevation irregularity (cross level or cant) and twist. European Standard EN13848 [2] is a series of standards defining the approach for evaluating railway track using TRVs for European Union member states. However, these vehicles are expensive to run and may disrupt regular services during their operation.

Other authors have investigated the use of vehicle acceleration measurements to determine the roughness of road profiles [3, 4]. Harris *et al.* [4] used the Cross Entropy combinatorial optimisation technique to characterise vehicle model parameters and road surface profiles using measured vehicle acceleration responses. Recently, the possibility of using inertial methods to estimate rail profiles using acceleration data measured on a vehicle has gained considerable interest. Real *et al.* [5] use a Fourier transform to solve the vehicle equations and find the transfer function that relates the input function and the output function in the frequency domain. The solution is then reverted into the time domain by applying an inverse Fourier transform to estimate the track profile. A mixed acceleration data filtering approach is used by Lee *et al.* [6] for stable displacement estimation and waveband classification of the irregularities in the measured acceleration. Investigations into the use of accelerometers and rate gyroscopes to estimate track geometry is presented by Weston *et al.* [7].

In this paper the Cross Entropy combinatorial optimisation method, as described by de Boer *et al.* [8], is adapted to determine rail profiles and the difficulties in making the transition from estimating road profiles to estimating rail profiles are discussed. To the authors' knowledge, the Cross Entropy combinatorial optimisation method has not previously been used to determine rail profiles.

This paper is divided into the following sections. First a brief overview of existing profile estimation methods using vehicle accelerations is presented. The next section explains the numerical model used for the simulations. Following this, the Cross Entropy Method is described and the optimisation process used in this paper is discussed. This is followed by the results of the optimisation that validates the methodology for a range of profiles. Finally, an overview of the results and suggestions for further research are presented.

## 2 RAIL PROFILE AND VEHICLE MODEL

In order to assess the application of the combinatorial optimisation technique for estimating the rail profile for given accelerations, numerical simulation in MATLAB is used to generate the dynamic response of an instrumented vehicle. The generation of the rail profile and the procedure of calculating the vehicle dynamic responses are described in this section.

A series of rigid rail profiles are included in the simulations. It is the passage of the vehicle across the irregularities on these profiles that excite it and thereby invoke the dynamic response. A perfectly smooth profile featuring a bell-shaped normal distribution 'pothole' is first used to demonstrate the capabilities of the algorithm. Three random rail profiles of varying degrees of roughness are also generated and used in numerical tests.

Power Spectral Density (PSD) is used in several countries (including the US, Germany, China and France) to classify track quality according to its irregularity spectrum [9]. For this paper, three track profiles with random vertical irregularity are generated using theUS Federal Railroad Administration (FRA) PSD function $S(\Omega)$, Eqn.1[10]. The FRA function is chosen due to its common use in the literature [11, 12].

$$S(\Omega) = \frac{A_v \Omega_c^2}{(\Omega^2 + \Omega_r^2)(\Omega^2 + \Omega_c^2)} \tag{1}$$

where $\Omega$ is the spatial frequency, and coefficients $A_v$, $\Omega_r$, $\Omega_c$, are related to the grade of the track and are given in the Table 1.

**Table 1. FRA American Railway Standard PSD Coefficients**

| Line Grade | Quality | $A_v$ | $\Omega_r$ [rad/s] | $\Omega_c$ [rad/s] |
|---|---|---|---|---|
| Class 4 | Very Poor | $2.39\times10^{-5}$ | $2.06\times10^{-2}$ | 0.825 |
| Class 5 | Poor | $9.35\times10^{-6}$ | $2.06\times10^{-2}$ | 0.825 |
| Class 6 | Moderate | $1.5\times10^{-6}$ | $2.06\times10^{-2}$ | 0.825 |

The model of vehicle-rail interaction, which is used in this study, is shown in Fig. 1. The vehicle consists of two masses $m_1$ representing the quarter carriage mass and $m_2$ representing the suspension (bogie and wheelset) half mass. Each mass has a single degree of freedom (DOF), and are connected by an elastic spring $k_1$ and damping system $c_1$ representing the secondary suspension of the vehicle. The stiffness and damping of the bogie system are represented by $k_2$ and $c_2$ respectively and these parameters characterize the primary suspension of the vehicle. The sprung vehicle model is connected to the rail profile through its primary suspension system. The rail is modelled as an infinitely stiff beam of length $L$, with irregular vertical profile $r(x)$. This model is similar to vehicle descriptions used in other studies [13].

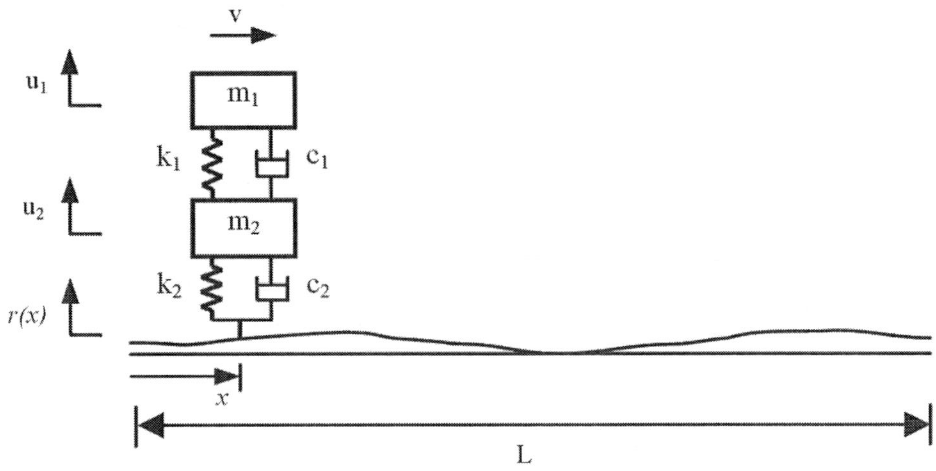


**Fig. 1. 2DOF Vehicle Rail Model**

The vehicle properties are listed in Table 2. The vibrations of the main vehicle body mass are dependent on the primary and secondary suspension systems which are approximated by the adopted model. Vehicle properties have been taken from a recently published paper using a similar model [14]. The vehicle travels over the profiles at a constant velocity, *v*, of 108 km/hr (30 m/s). The authors acknowledge that the quarter car vehicle properties represent the upper half of the train vehicle, thereby only approximating the effect of the train wheels which exhibit high stiffness and negligible damping effects. The motive behind this decision is explained further in Section 4. This simplified vehicle is chosen to demonstrate the capabilities of the algorithm.

**Table 2. Properties of the vehicle**

| Property | Unit | Symbol | Value |
|---|---|---|---|
| Mass of main body | kg | $m_1$ | 7900 |
| Mass of bogie and wheelset | kg | $m_2$ | 512.5 |
| Damping of Secondary Suspension | Ns/m | $c_1$ | $15\times10^3$ |
| Damping of Primary Suspension | Ns/m | $c_2$ | $5\times10^3$ |
| Stiffness of Secondary Suspension | N/m | $k_1$ | $7.3\times10^5$ |
| Stiffness of Primary Suspension | N/m | $k_2$ | $5\times10^5$ |
| Vehicle velocity | m/s | $v$ | 30 |

A train/track interaction model is used to represent the vehicle travelling over the rail profile. The dynamic interaction between the vehicle and the rail is executed in MATLAB. The equations of motion of the vehicle model expressed in the time domain are as follows:

$$m_1\ddot{u}_1 + c_1(\dot{u}_1 - \dot{u}_2) + k_1(u_1 - u_2) = 0 \tag{2}$$

$$m_2\ddot{u}_2 - c_1\dot{u}_1 + (c_1 + c_2)\dot{u}_2 - k_1u_1 + (k_1 + k_2)u_2 = c_2\dot{r} + k_2r \tag{3}$$

where $r$ is the rail profile and $\dot{r}$ is the first derivative of the rail profile with respect to time. These equations can be represented in matrix form as follows:

$$\begin{bmatrix} m_1 & 0 \\ 0 & m_2 \end{bmatrix}\begin{Bmatrix} \ddot{u}_1 \\ \ddot{u}_2 \end{Bmatrix} + \begin{bmatrix} c_1 & -c_1 \\ -c_1 & c_1 + c_2 \end{bmatrix}\begin{Bmatrix} \dot{u}_1 \\ \dot{u}_2 \end{Bmatrix} + \begin{bmatrix} k_1 & -k_1 \\ -k_1 & k_1 + k_2 \end{bmatrix}\begin{Bmatrix} u_1 \\ u_2 \end{Bmatrix} = \begin{Bmatrix} 0 \\ c_2\dot{r} + k_2r \end{Bmatrix} \tag{4}$$

i.e.,

$$\mathbf{M_v\ddot{u}_v} + \mathbf{C_v\dot{u}_v} + \mathbf{K_vu_v} = \mathbf{f_v} \tag{5}$$

where $\mathbf{M_v}$, $\mathbf{C_v}$ and $\mathbf{K_v}$ are the mass, damping and stiffness matrices of the vehicle respectively. The vectors, $\mathbf{\ddot{u}_v}$, $\mathbf{\dot{u}_v}$ and $\mathbf{u_v}$ are the vehicle accelerations, velocities and displacements respectively. The displacement vector of the vehicle is, $\mathbf{u_v} = \{u_1u_2\}^T$. The vector $\mathbf{f_v}$ contains the time varying dynamic forces applied by the vehicle to the rail: $\mathbf{f_v} = \{0\ F\}^T$ where F is the dynamic force,

$$F = c_2\dot{r} + k_2r \tag{6}$$

The dynamic equations of motion of the system are solved using the Wilson-Theta integration method [15,16]. The value of the parameter, $\theta$ =1.420815, is used for unconditional stability in the integration scheme. This represents an optimal value as determined by [17].

## 3 CROSS-ENTROPY METHOD

The Cross-Entropy (CE) method is a combinatorial optimisation technique used in this paper to infer a series of rail profile elevations from measurements of vehicle response to them. Simply stated, the CE method is an iterative procedure which firstly generates a *population* of trial solutions according to a specified random mechanism and then updates the parameters of the random mechanism in order to produce an improved population of solutions in the next generation of estimates [4]. The second step involves minimising an objective function which is used to determine the fit of the data generated to the reference data being used.

For this study we wish to generate a population of rail profiles and determine which profile estimates generate vehicle responses most similar to the actual measured response, referred to from here as the reference acceleration signal. The reference acceleration signal is taken from the response of the vehicle described in Section 2, traversing a rail profile, as described in Section 4.

At the beginning of the algorithm, an initial normal distribution (defined by a mean and standard deviation) of values for each of the unknown rail profile elevations is assumed. Using the predefined mean and standard deviation, a population of rail profile estimates is pseudo-randomly generated. Following this, the vehicle track interaction is simulated for each profile in the population, returning an acceleration signal for the DOF being analysed. An objective function is then minimised to determine the best matches to the reference acceleration signal. For this paper the objective function (Eqn.7) is defined as the sum of the squares of the differences between accelerations calculated for each trail profile, and the reference acceleration signal.

$$O(t) = \sum_{t=1}^{T} \left(\ddot{u}_{est,t} - \ddot{u}_{meas,t}\right)^2 \tag{7}$$

where O(t) is the objective function to be minimised, $\ddot{u}_{est,t}$ is the acceleration signal generated by the vehicle traversing the estimated profile, $\ddot{u}_{meas,t}$ is the reference acceleration signal, *t* is the timestep, and *T* is the total number of timesteps in the acceleration signal. The value of the objective function for each profile estimation in the population is ranked and, in this case, the best 10% of trial profiles are retained. The mean and standard deviation of the best profiles are calculated and used in a Monte Carlo simulation to generate a new population of profile estimates and the vehicle track interaction is simulated for the new population. This process is repeated until convergence of $\ddot{u}_{est,t}$ and $\ddot{u}_{meas,t}$ is reached. To increase computational efficiency, this iterative process is limited to 25 iterations during which it is observed that the desired convergence generally occurs.

Depending on the sampling interval and length of the profile, there may be a large number of unknowns in the problem. This means that a very large population size would be required and there is a risk that the algorithm may converge prematurely to a false solution. To overcome this problem the optimisation is split into a number of phases. Using this method, the phase location is represented by a 'window' of the profile. A number of unknowns, *n,* are determined within the window before moving to the next phase. At the end of each phase, the first $n/2$ best estimates of the profile heights are saved as the estimated profile, and the remaining $n/2$ profile heights are used as the first $n/2$ means for the next phase of $n$ unknowns. To increase the efficiency of the algorithm, the remaining $n/2$ means values for the next phase are taken as the $n^{th}$ mean from the previous phase. Standard deviation is also reset to account for the relative uncertainty in profile heights further along the phase window being analysed. This is achieved by increasing the standard deviation in an array from 0.1 in increments of $1/n$ to 1. Fig. 2 provides a schematic of the phasing procedure used to determine long profiles.

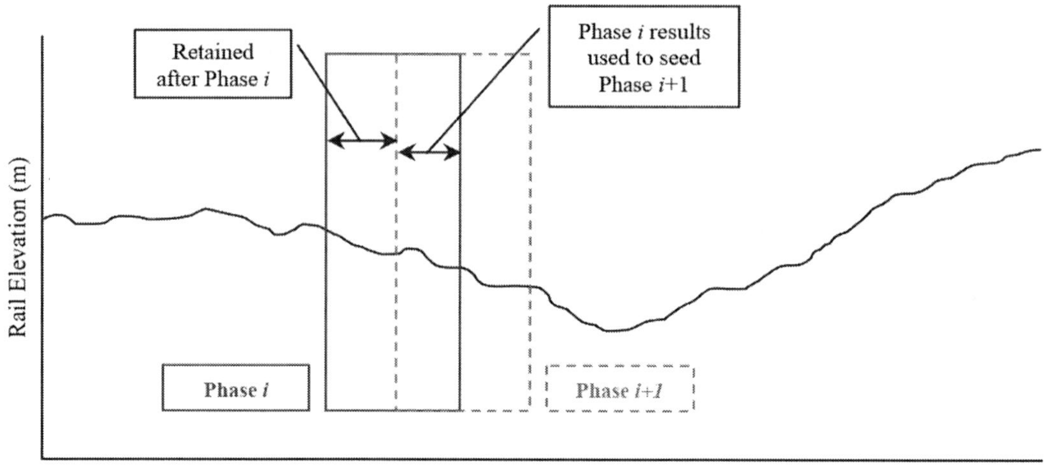


**Fig. 2. 'Windowing' of Rail Profile in Phases**

## 4 NUMERICAL ASSESSMENT

### Test Profile

Numerical assessment of the Cross Entropy method is carried out in MATLAB. Firstly, a relatively smooth test rail profile, 20m in length, featuring a normal distribution 'bell' curve to represent an irregularity is used to demonstrate the capabilities of the algorithm. This 'pothole' irregularity, located at 5.0m, has a depth of 0.002m and a 'standard deviation' pothole width of 0.5. The quarter car vehicle is run across the profile and acceleration responses are generated. Following this, the Cross Entropy method, as described in Section 3, is executed in MATLAB with the parameters presented in Table 3. The acceleration signal from the bogie degree of freedom, $m_2$, is used as the reference acceleration.

**Table 3. Cross Entropy Method Parameters**

| Property | Value |
|---|---|
| Length of Profile Inferred per Phase | 0.25m |
| Number of Profile Estimates per Phase | 10 |
| Initial Mean | 0 |
| Initial Standard Deviation | 0.5 |
| Size of Population of Estimates | 150 |
| Percentage of Population Transferred to Next Iteration | 10% |
| Iterations per Phase | 25 |

The computational time required to infer a profile 20m in length was about 2h using a 2.67 GHz processor and 6.0 GB RAM running on MATLAB. It is anticipated that improvements in algorithm efficiency and employment of more powerful parallel processors will improve on this computational time. These efficiencies are required to allow the inference of longer track profile lengths so that track geometry parameters within the wavelength ranges stipulated in EN 13848, [2] are determined and comparable track geometry results are generated from the algorithm.

The result of the rail profile estimation for the first test profile is shown in Fig. 3. An excellent estimate is generated with small errors: in the region of ~0.018mm.

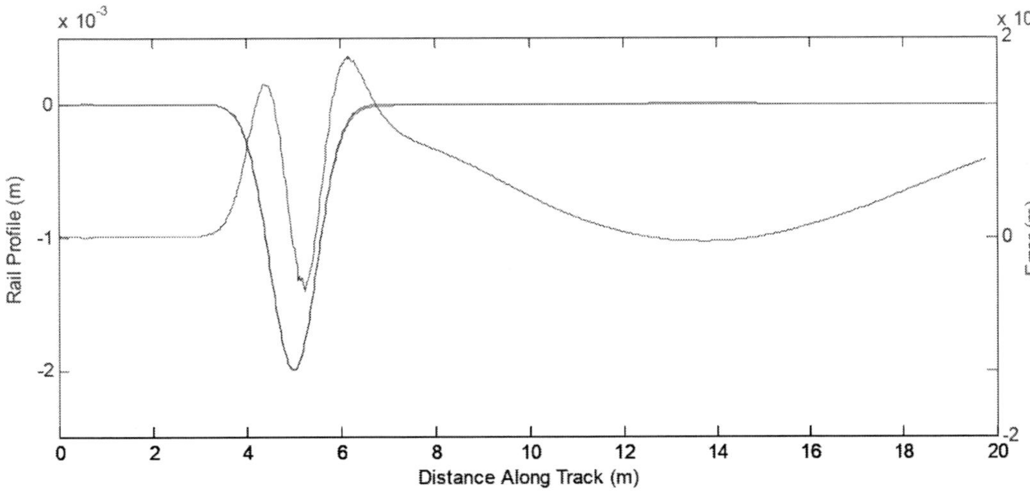


**Fig. 3. Estimated (red) and actual (blue) rail profile verses distance, and errors in estimation (green) for Test Profile**

**FRA Rail Profiles**

The FRA PSD function presented in Eqn. 1 is used to generate 3 rail profiles of varying roughness. The profiles range from Class 4 (roughest) to Class 6 (smoothest for non high speed rail applications). The PSDs for the profiles are presented in Fig. 4.

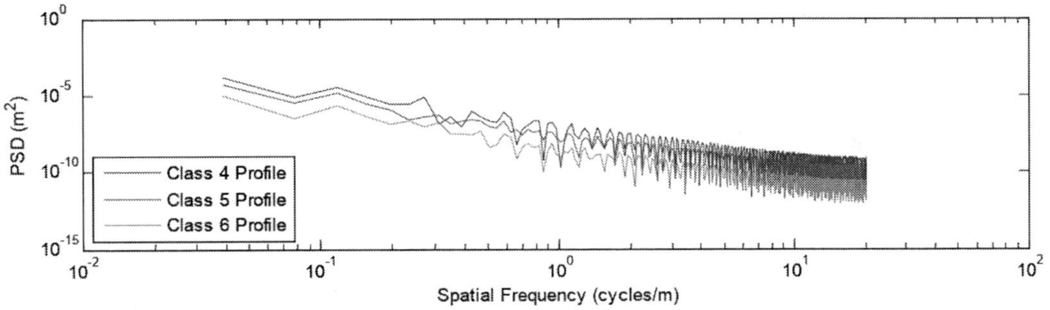


**Fig. 4. Power Spectral Density for 3 FRA Rail Profiles used in analysis**

Again, the quarter car vehicle is run across the profiles and reference acceleration responses are generated. The increase in the number of irregularities and rate of elevation changes exhibited in the FRA profiles excite the vehicle in a more random fashion resulting in 'spikier'acceleration data. An alternative quarter-car vehicle model representing the train wheel and half-bogie sub-system was tested using the algorithm; however, the high stiffness of the train wheel (ignoring the influence of the non-linear wheel-rail contact phenomenon) resulted in very spiky acceleration signals. It was found that the objective function used in this algorithm did not perform satisfactorily when processing the spiky data leading to the adoption of a simplified model ignoring axle box accelerations and using bogie sprung mass acceleration signals to infer the rail profile with more satisfactory results. Other authors [4] have employed the Cross Entropy method to determine road profiles using vehicle quarter cars exhibiting a relatively low tyre stiffness compared to the high wheel stiffness observed in rail vehicles, achieving good results.

The results of the rail profile estimation for the FRA profiles are shown in Fig. 5. Excellent estimates for all three profiles are generated by the algorithm, again with small errors.

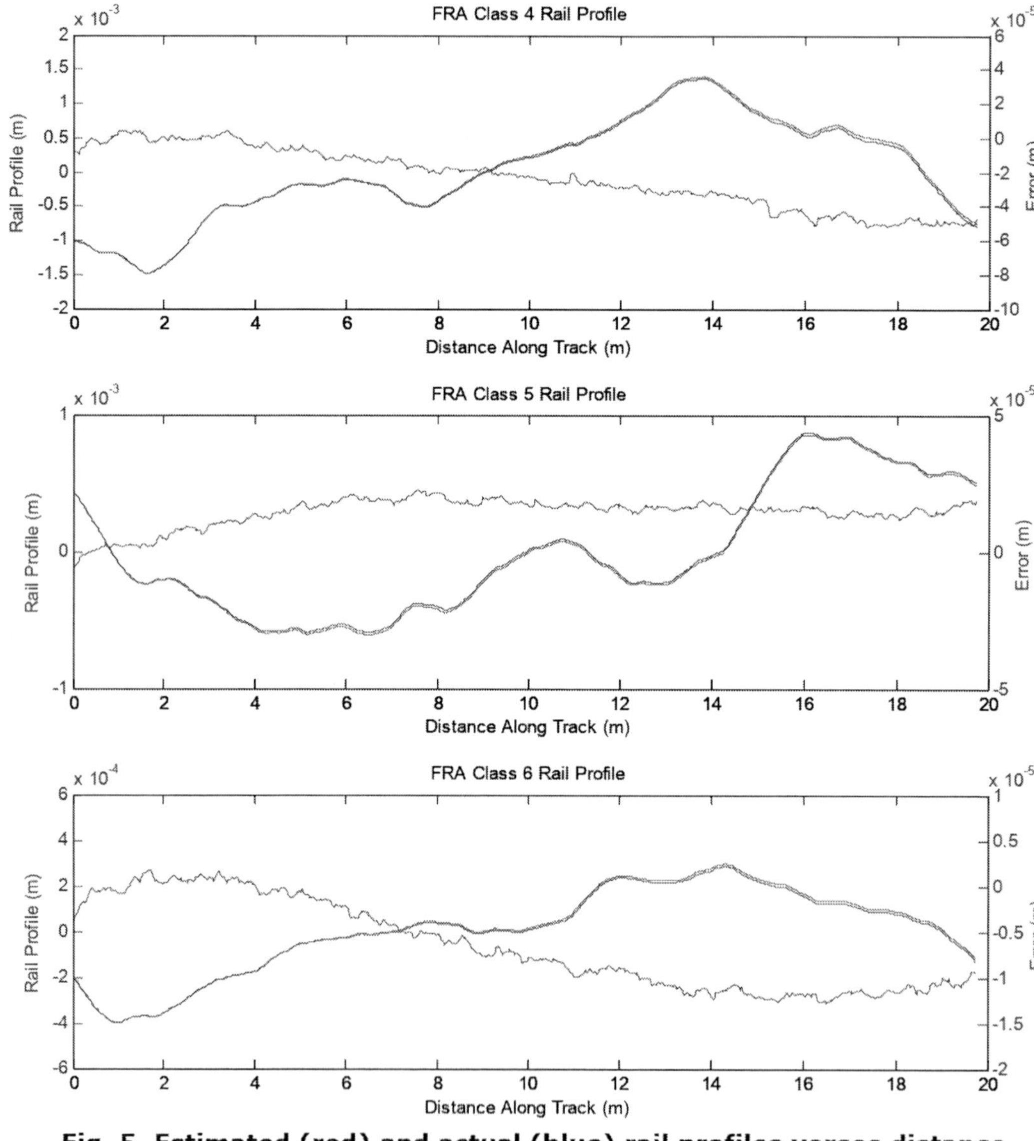


**Fig. 5. Estimated (red) and actual (blue) rail profiles verses distance, and errors in estimation (green) for FRA Rail Profiles**

It is observed that there is a gradual drift in the estimated rail profile which increases with distance from the origin. Harris *et al.* [4] attributed this drift to unavoidable errors in the estimation of target accelerations and any differences between the reference and observed accelerations will be apparent in the estimated profile as they are double-integrated with respect to time. It is not clear from the PSD comparison for the Class 4 profile (Fig. 6) whether this drift is due to low frequency or high frequency errors as the PSD profile for the actual profile and the estimate are well matched.

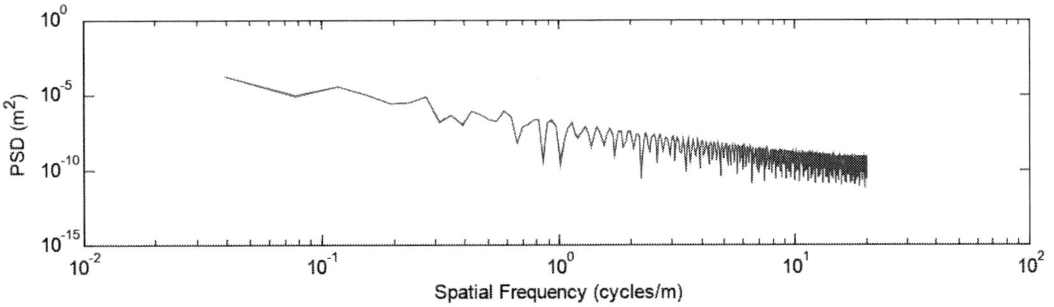


**Fig. 6. Estimated (red) and actual (blue) rail profile power spectral densities for FRA Class 4 Rail Profile**

## 5 SUMMARY AND DISCUSSION

A method for the estimation of the vertical alignment of various railway track profiles using the Cross Entropy combinatorial optimisation method has been presented. It is found that the estimated rail profiles produced by the method provide a very good fit to the actual profiles.

The analysis was carried out using a simplified railway vehicle quarter-car model using acceleration signals generated by the bogie to infer the rail profile with satisfactory results. It was found that spiky acceleration signals generated by the high stiffness properties of a train wheel in alternative vehicle arrangements were difficult to process in the algorithm. The authors acknowledge that the computational efficiency of the algorithm will need to be greatly improved prior to implementation.

Further work planned includes the transformation of the axle box acceleration signals into a form more suited to the optimisation process while retaining time-domain information. Methods such as the continuous wavelet transform (CWT) and Hilbert Transform are being investigated. The Cross Entropy method may then be used to determine the rail profile by changing the objective function to minimising the sum of squares of the differences between the intensity of the frequencies observed in the acceleration data at each time-step.

It is planned to increase the complexity of the vehicle model used in the algorithm, possibly to a more realistic configuration featuring four wheels and two bogie sets. The addition of a track model featuring rails (with profile), sleepers and ballast would allow for the deflection of track under the vehicle to be considered. The authors acknowledge that these improvements, and allowing for varying vehicle speed, must be considered prior to implementation.

The authors are in the process of acquiring accelerometers to attach to an Irish Rail train vehicle so that acceleration measurements can be obtained from a section of track. It is planned to use this data to estimate the vertical profile of the track and compare to values recorded on the TRV.

From the results shown in this paper, it can be concluded that the Cross Entropy method has the potential to be used to estimate and classify rail profiles using vehicle acceleration data. Accurate estimation of railway track vertical alignment using sensors mounted on in-service vehicles has the potential to provide a valuable tool to inform maintenance planning.

## ACKNOWLEDGMENTS

The research presented in this paper was carried out as part of the Marie Curie Initial Training Network (ITN) action FP7-PEOPLE-2013-ITN. The project has received funding from the European Union's Seventh Framework Programme for research, technological development and demonstration under grant agreement number 607524.

## REFERENCES

[1] Nielsen, J., Berggren, E., Lölgen, T., Müller, R., Stallaert, B., Pesqueux, L., (2013), 'Overview of methods for measurements of track irregularities important for ground-borne vibration' *Railway Induced Vibration Abatement Solutions (RIVAS)* Deliverable D2.5. (www.rivas-project.eu).

[2] Railway Applications - Track - Track Geometry Quality - IS EN 13848 - CEN European Committee for Standardization.

[3] Gonzalez, A., OBrien, E.J., and McGetrick, P.J. (2012),'Identification of damping in a bridge using a moving instrumented vehicle', *Journal of Sound and Vibration*, 331 (18), 4115-4131.

[4] Harris, N.K., Gonzalez, A., OBrien, E.J., McGetrick, P., (2010), 'Characterisation of pavement profile heights using accelerometer readings and a combinatorial optimisation technique', *Journal of Sound and Vibration*, 329, 497-508.

[5] Real, J. Salvador, P, Montalbán, L. and Bueno, M. (2011), 'Determination of rail vertical profile through inertial methods', *Proceedings of the Institution of Mechanical Engineers, Part F: Journal of Rail and Rapid Transit 2011,* 225, 14-23.

[6] Lee, J. S., Choi, S., Kim, S.S., Park, C., and Kim, Y.G. (2012), 'A mixed filtering approach for track condition monitoring using accelerometers on the axle box and bogie', *IEEE Transactions on Instrumentation and Measurement*, Vol.61, No. 3, March 2012.

[7] Weston, P.F., Ling, C.S., Roberts, C., Goodman, C.J., Li, P. and Goodall, R.M. (2007), 'Monitoring vertical track irregularity from in-service railway vehicles' *Proceedings of the Institution of Mechanical Engineers, Part F: Journal of Rail and Rapid Transit 2007*, 221, 75-88.

[8] de Boer, P.T., Kroese, D., Mannor, S. and Rubinstein, R., (2005), 'A tutorial on the Cross-Entropy method'. *Annals of Operations Research*, 134(1): 19-67.

[9] Berawi, A., (2013), 'Improving Railway Track Maintenance Using Power Spectral Density'. A Dissertation Submitted in Partial Fulfilment of the Requirements for the Degree of 'Doctor of Philosophy in Transport Systems - Universidade Do Porto.

[10] White, R.C., Limbert, D.A., Hederick, J.K. and Cooperrider, N.K., (1978), 'Guideway-suspension tradeoffs in rail vehicle systems, Report ERC-R-78035', Arizona State University Engineering Research Centre, Arizona, USA.

[11] Lei, X., and Noda, N.-A.. (2002), 'Analyses of Dynamic Response of Vehicle and Track Coupling System with Random Irregularity of Track Vertical Profile', *Journal of Sound and Vibration*, 258(1), 147-165.

[12] Yang, Y.N., Yau, J.D. and Wu, Y.S., (2004), *Vehicle-Bridge Interaction Dynamics,* World Scientific Publishing Co., Singapore.

[13] Frýba, L., (1996), *Dynamics of Railway Bridges.* Thomas Telford: London.
[14] Cantero, D. and Basu, B., (2014), 'Railway infrastructure damage detection using wavelet transformed acceleration response of traversing vehicle'. *Structural Control and Health Monitoring. 2014.* Published online in Wiley Online Library (wileyonlinelibrary.com). DOI: 10.1002/stc.1660.
[15] Bathe, K.J. and Wilson, E.L., (1976), *Numerical methods in finite element analysis,* Prentice Hall, Englewood Cliffs, UK.
[16] J.W. Tedesco, W.G. McDougal and C.A. Ross, (1999), *Structural Dynamics, Theory and Applications*, California, USA: Addison Wesley Longman.
[17] Weaver, W. and Johnston, P.R., (1987), *Structural dynamics by finite elements*, Prentice-Hall, Indiana, USA.

# Crashworthiness research on diesel multiple unit fuel tanks

**K M Jacobsen, M E Carolan**
Volpe National Transportation Systems Center, US Department of Transportation, USA

## ABSTRACT

Throughout the United States (US), where highways are at capacity and commuters are looking for more economical and eco-friendly means of transportation, many cities are expanding existing or planning new commuter rail lines. The growing demand for these unique small-scale rail lines has created a niche market for self-propelled vehicles that operate in short consists over short routes. In the US the demand is for Diesel Multiple Units (DMU). European and Asian manufacturers are the primary suppliers of DMU equipment to the US market.

Currently, US fuel tank standards apply to both conventional locomotives and alternative locomotives such as DMUs. Code of Federal Regulation (CFR) requires that Tier I (operations at speeds of 125 mph and less) locomotive fuel tanks have minimum structural properties adequate to sustain a prescribed set of static load conditions. In this paper, a comparison will be made of the existing fuel tank standards and regulations that DMU fuel tanks are designed to meet both domestically and abroad.

The Federal Railroad Administration's Office of Research and Development is conducting research into passenger locomotive fuel tank crashworthiness. An objective of the research it to determine how well existing regulations for conventional fuel tanks apply to alternative fuel tanks, such as DMU tanks. Current research is intended to increase understanding of the impact response of fuel tanks under dynamic impact conditions. Analyses and preliminary testing, including full-scale dynamic impact tests of tanks, are underway. This paper reports the preliminary findings of test results, describes future plans for testing and discusses development of fuel tank standards for DMUs.

## 1 BACKGROUND

The growing demand for capacity on the US rail network for passenger train service must be balanced by the need to share existing right-of-way. Currently, the US railroad system is dominated by freight operations. As such, passenger trains time-share track usage with freight trains. Short lines dedicated to passenger service are a small percentage of US railroad operations and are focused primarily on serving urban areas and the surrounding suburbs. The focus on freight operations is likely to continue as the US transportation layout of existing roadways and existing railroad lines leaves few opportunities for building entirely new large-scale railroad lines. In 2009, the US Congress passed the American Recovery and Reinvestment Act (ARRA) which committed $8 billion for intercity passenger rail projects and rail congestion grants, with priority for high-speed rail (1). The availability of funding from the Federal government has encouraged proposals in metropolitan areas for rail infrastructure upgrades and expansions to existing rail lines. Between the economic needs caused by overcrowded highways around most major cities and the available funding from the government, many metropolitan areas developed proposals for rail infrastructure upgrade, rail line expansions, and new equipment. This amplified a trend of expanding existing commuter rail lines such as Sonoma-Marin Area Rail Transit (SMART) in the San Francisco Bay Area, the Denver Regional Transportation District (RTD), and the Denton County Transportation Authority (DCTA) outside of Dallas, Texas. Services such as these are looking to operate consists of 2-4 railcars over distances of 16 kilometers to 85 kilometers, with a desire to have flexibility in the consist length.

In the last twenty years the US passenger railcar manufacturers have slowly gone out of business. There is currently only one passenger railcar manufacturer, US Railcar, a reincarnation of Colorado Railcar, in the US. It produces a series of DMUs, used for short commuter rail operations and scenic railroad lines. The market for passenger rail equipment is dominated by European and Asian manufacturers. Many of these manufacturers have begun to open North American subsidiaries, including Bombardier Transportation, Stadler USA, and Sumitomo Corporation of America.

The need to share track with freight operations is a challenge for vehicles designed to operate on dedicated track, such as DMUs. This issue is exhibited by European manufacturers that win railcar procurements in the US. The equipment designed and manufactured by such entities must demonstrate compliance with US standards and regulations. The fuel tank has shown to be a particularly challenging issue in demonstrating compliance.

## 2 EXISTING FUEL TANK REGULATIONS

Fuel tank integrity standards should set a minimum level of protection against an impact to the fuel tank that might puncture the tank and cause the release of diesel fuel. As seen in numerous accidents in the US including Membane, North Carolina in 2010, Goodwell, Oklahoma in 2012, Hoxie, Arkansas in 2014, and Concord, MA in 2010, breach of fuel tanks in train collisions and derailments increases the potential of injury and fatality. The outbreak of a fire can inhibit the emergency egress of passengers and crew in both passenger trains and freight operations. With trains powered by conventional locomotives, fuel tanks are located on equipment occupied by crew only. For alternative locomotives, such as DMUs, the passengers are in closer proximity to the fuel tank. The risk profile of such an arrangement is enhanced in the event of puncture of the fuel tank with subsequent fire.

Figure 1 shows the fuel tank from an exemplar incident in Concord, MA. A passenger locomotive struck a track component (joint bar) that was within the right-of-way, resulting in puncture of the fuel tank. The inset image shows an exemplar joint bar fitting into the puncture on the bottom surface of the fuel tank. While this incident did not result in a fire, a large quantity of diesel fuel was spilled along the right-of-way. Additionally, the passenger train became stranded, as the locomotive was unable to provide power once it had lost its fuel supply.

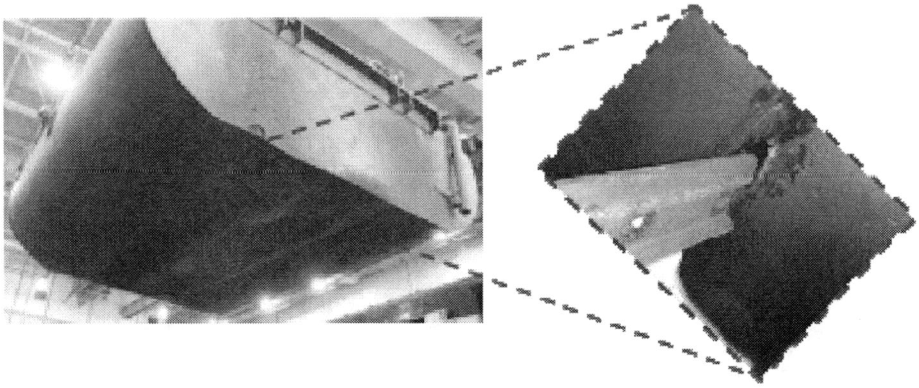

**Figure 1. Locomotive Fuel Tank Punctures from Concord, MA**

Existing fuel tank standards are based upon freight train operations. A typical fuel tank capacity on a freight locomotive ranges from 7,570 L to 30,280 L (2,000 gallons to 8,000 gallons). A typical fuel tank on a passenger locomotive ranges from 3,785 L to 7,570 L (1,000 gallons to 2,000 gallons). A DMU fuel tank may range from 1,515 L to 3,030 L (400 gallons to 800 gallons). Despite this range of fuel tank sizes, all passenger equipment is required to meet the same regulations. Per FRA regulatory definition of a locomotive, DMUs are treated as an alternative type of locomotive.

There are three sets of industry standards that may apply to fuel tanks used on US railroads, depending on the intended service. The CFR (2) mandates that freight locomotive fuel tanks must meet standards set by the Association of American Railroads (AAR), an industry group. The CFR contains a separate set of requirements for locomotives used in passenger service (3). Passenger locomotive fuel tanks may also be subject to standards developed by the American Public Transportation Association (APTA), an industry group. A railroad authority may include additional requirements, beyond the applicable standards and regulations, within its equipment specification to meet its own unique needs or preferences.

The passenger locomotive requirements contained within the CFR comprise static loads that are meant to represent a set of possible derailment or impact scenarios encountered in train operation. There are three load cases which must be evaluated: a jack-knifed locomotive, a minor derailment of the locomotive, and a side impact to the fuel tank. There is an additional design requirement intended to provide a minimum level of penetration resistance through a combination of material thickness and strength. Additionally there is a requirement for the valves and fittingz\\s to be flush with the tank. Protection of the fill controls is encouraged to be considered but no specifics are defined.

**Table 1. Summary of CFR requirements**

| Load Case | Condition to be Met | Pass/Fail Criterion |
|---|---|---|
| 1 - Minor Derailment | One-half locomotive weight at 2g loaded on end plate within +/- 8" of rail position. | Without exceeding the ultimate strength of the material. |
| 2 - Jackknifed Locomotive | One-half locomotive weight at 2g loaded transversely at the center of tank. Load distributed in a 2" band between the longitudinal centerline of the tank and the edge of the tank bottom. | Without exceeding the ultimate strength of the material. |
| 3 - Side Impact | A 200,000 pound load at the longitudinal center of the tank distributed over an area of six inches by forty-eight inches at a height of thirty inches above the rail. | Without exceeding the ultimate strength of the material. |
| 4 - Penetration Resistance | • The minimum thickness of the sides, bottom sheet and end plates of the fuel tank shall be equivalent to a 5/16-inch steel plate with a 25,000 pounds-per-square-inch yield strength (where the thickness varies inversely with the square root of yield strength).<br>• The lower one third of the end plates shall have the equivalent penetration resistance by the above method of a 3/4-inch steel plate with a 25,000 pounds-per-square-inch yield strength. | |

The APTA standards are similar to the AAR/FRA requirements except they cite more details for evaluating each load case and also list exceptions for load cases if the manufacturer can show other evidence indicating that that case will not occur. In some standards the manufacturer is allowed to take into account the surrounding structure and the placement of the tank.

There is no equivalent set of standards in Europe or Asia that addresses fuel tank integrity. The only applicable requirement for fuel tanks is an attachment strength requirement in EN12663(4). Market research of foreign manufacturers reveal that fuel tanks have been designed to meet strength standards for transit equipment such as trucks or buses. This is explained by the fact that many manufacturers of DMU equipment also specialize in Light Rail Vehicle (LRV) equipment, designed to operate on roadways.

The lack of foreign standards for fuel tank design and the difference in risk environment between conventional locomotive fuel tanks and fuel tanks on DMUs are two motivating factors for FRA's current research program into the puncture resistance of fuel tanks.

## 3 FUEL TANK RESEARCH PROGRAM

### 3.1 Goals

The objective of the fuel tank research is to provide the technical basis needed to develop fuel tank integrity requirements appropriate for alternative fuel tanks, such as those on DMUs. The research methodology incorporates a process of developing dynamic impact conditions from the accident history of related equipment, idealizing conditions for testing, conducting tests to measure the deformation behavior of sample fuel tanks, and applying those results to create a standardized impact scenario for DMU fuel tanks.

In accidents, fuel tanks are subjected to dynamic loading, often including a blunt or raking impact from various components of the rolling stock or trackbed. An accident survey, conducted as part of the FRA research program, describes accidents and derailments that occurred in the US on freight and passenger trains involving fuel tank punctures over the last two decades (5). The survey shows that fuel tank punctures can occur on any exposed location of the tank by a variety of impacting objects, including front or rear impact by an adjacent component as it detaches in an event, a side impact or bottom impact by a rigid part of another rail vehicle in the event of a rollover, a side-swiping impact with another train or railcar, or a direct impact by a piece of debris. Each scenario is categorized by its resultant loading type and in general there are two loading conditions leading to punctures: blunt impacts and raking impacts.

### 3.2 Testing Plans

Current research is intended to increase understanding of the impact response of fuel tanks under dynamic loading. Two test setups will be used to evaluate blunt and raking collision loads on fuel tank specimens. The blunt impact scenario was developed and tested on a series of conventional passenger locomotive fuel tanks. The blunt impact scenario will be used to test DMU fuel tanks. Additional test plans include designing a test setup to evaluate a raking impact against the fuel tank. Analyses, including finite element modeling of the fuel tank, are developed based upon known information, prior to each test, to plan the details of the test.

## 4 PRELIMINARY TESTS AND ANALYSES

### 4.1 Test setup

FRA has conducted a series of preliminary impact tests and analyses to characterize the behavior of locomotive fuel tanks when subjected to a blunt impact on their bottom surfaces. These tests are referred to as preliminary because they were primarily conducted to evaluate the test setup (e.g. support conditions, instrumentation requirements, etc.) in anticipation of performing additional impact tests on modern DMU fuel tanks. These preliminary tests utilized fuel tanks from retired passenger locomotives that had been donated to FRA as test articles. Since these locomotives are no longer capable of providing power, the fuel tanks were removed and used in this impact testing program. Figure 2 shows one of the locomotives whose fuel tank was used in this program. The bottom half of this image shows the fuel tank's position on the underframe of the locomotive.

Three fuel tanks were available from existing FRA test locomotives. These fuel tanks were referred to by the numbers assigned to their parent locomotives. The three fuel tanks were referred to as tank 202, tank 232, and tank 234. All three fuel tanks had approximately a 6,810 L (1,800 gallon) capacity. While all three fuel tanks were removed from similar F-40 type passenger locomotives, each of the three fuel tanks was of a unique design. Tanks 232 and 234 had approximately the same outer shell dimensions and plate thicknesses, but each featured a unique arrangement of internal baffles. Tank 202 was of an entirely different design, featuring squared-off sides and a unique baffle arrangement of its own. All three tanks are shown below, in Figure 3. The left column of this figure shows a photograph of the actual tank. The center column shows an inverted view of the tank's exterior, taken from the finite element (FE) model. The right column of this figure shows the baffle arrangement within the FE model of each tank.

**Figure 2. Typical fuel tank position on passenger locomotive**

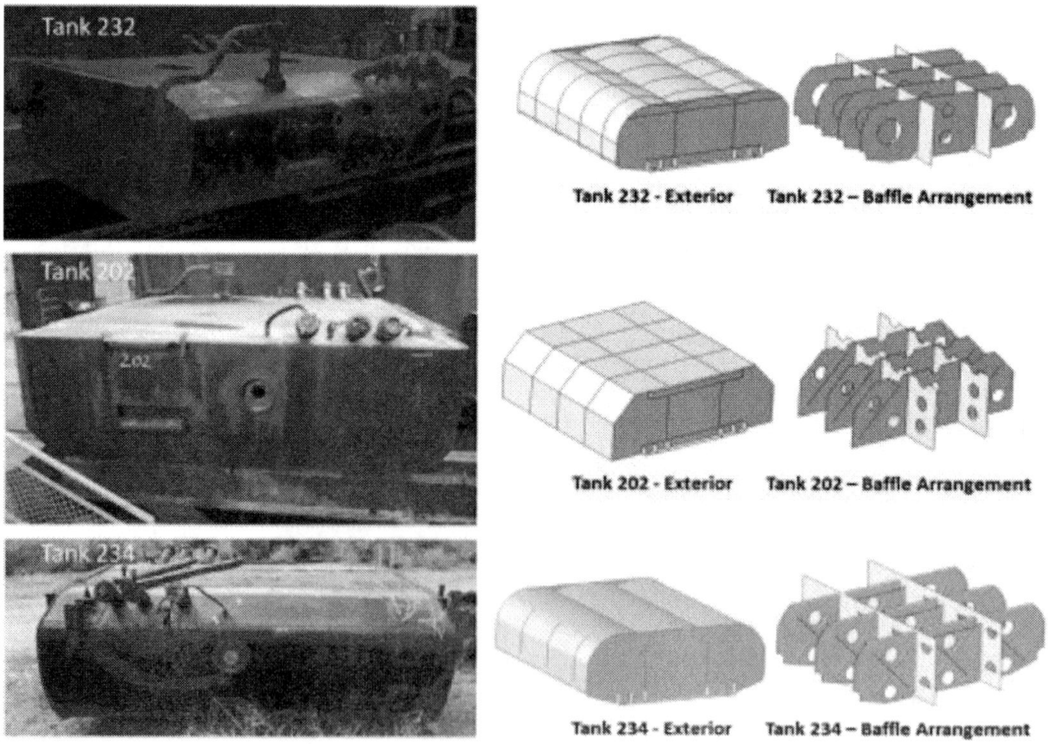


**Figure 3. Tanks 232, 202, and 234 used in this program**

In the testing program, the fuel tanks were removed from the locomotives and mounted to a rigid, vertical wall. The tanks were attached to the wall through the bolt holes that would normally be used to attach the fuel tanks to the underside of the locomotive during service. This mounting scheme approximates an impact condition that could occur if the locomotive were to have derailed and come to rest on its side, exposing the fuel tank to secondary impacts. Figure 4 shows the locomotive fuel tank mounted to the impact wall. The inset portion of this image shows the bolted connections at two of the fuel tank mounting locations.

**Figure 4. Locomotive fuel tank mounted to impact wall**

The fuel tank was struck by a dynamic impact vehicle equipped with a 30.5 cm x 30.5 cm (12 inch x 12 inch) ram head. The impact vehicle was formerly used as a piece of maintenance-of-way equipment and has been modified to accommodate ram heads of different shapes and sizes. With the ram head installed, the impact cart weighed approximately 6,350 kg (14,000 pounds) for this test. Figure 5 shows the impact cart with the ram head used in the fuel tank impact tests.

**Figure 5. Impact cart with ram head used in tests**

The primary results from the impact tests included the deceleration-time history of the impact cart and the deformation mode of the tank, including whether puncture occurred. Tri-axial accelerometers (i.e. vertical, lateral, and longitudinal) were installed on the impact cart at five locations. These locations included the leading-end cross member, the left and right side sills, the center cross-member, and the trailing-end cross member. A pair of speed sensors was mounted on the left and right side sills to measure the speed of the cart just prior to impact. Each test was documented using high-speed and conventional-speed video cameras on the ground, and conventional-speed video onboard the impact vehicle. A summary of the instrumentation used in each fuel tank impact test is given in Table 2.

**Table 2. Instrumentation summary**

| Instrumentation | Location | Quantity |
|---|---|---|
| Speed Sensors | Mounted on impact vehicle | 2 |
| Tri-axial Accelerometers | Mounted on impact vehicle | 5 |
| High-speed video | Adjacent to track | 2 |
| Conventional-speed video | Adjacent to track | 4 |
| Conventional-speed video | Mounted on impact vehicle | 1 |

The three blunt impact tests performed in this preliminary phase of the research program are summarized in Table 3.

**Table 3. Summary of Impact Tests**

| Tank | Test Date | Impact Speed | Impact Energy | Result |
|---|---|---|---|---|
| 202 | October 9, 2013 | 10 km/h (6.2 mph) | 24 kJ (17,980 ft-lbs) | 3.8 cm (1.5 inch) residual dent No puncture |
| 232 | October 8, 2013 | 7.25 km/h (4.5 mph) | 13 kJ (9,470 ft-lbs) | 12.7 cm (5 inch) residual dent No puncture |
| 234 | August 20, 2014 | 18 km/h (11.2 mph) | 79.5 kJ (58,660 ft-lbs) | 21.6 cm (8.5 inch) residual dent No puncture |

### 4.2 FE analyses

As a part of this research program, FE analysis (FEA) has been used in conjunction with the testing to investigate the behavior of the fuel tanks under blunt impact conditions. FEA was used to guide instrumentation placement, establish the constraints to be placed on the test article, and determine the desired impact speeds. Testing and analysis were performed in two distinct phases during this program. During phase 1, non-puncture tests were conducted on tanks 202 and 232. Following these tests, FE models were updated based upon post-test measurements of tank geometry and materials. These models served as the starting point for the puncture modeling performed in phase 2 of this research. Phase 2 consisted of modeling tank 234 up to its point of puncture. Tank 234 was then subjected to a higher-speed impact test to help estimate its puncture behavior.

Because all three fuel tanks were salvaged from retired locomotives, technical drawings and material specifications were not available to generate the precise details for the FE models of the fuel tanks. This information needed to be developed as a part of the program. Prior to beginning the test series, external measurements were made of all tanks. Using a small camera (boroscope) passed through one of the existing fuel gage holes in the tank's exterior, a rough approximation of the baffle arrangement within each tank was obtained. Because the integrity of the tanks could not be compromised prior to the tests, material coupons for metallurgical testing could not be taken from any tanks until after the tests. Rather, hardness measurements were made on the exterior surfaces of the tanks to provide some data on the materials of construction. The flowchart in Figure 6 provides an overview of the testing and analysis program.

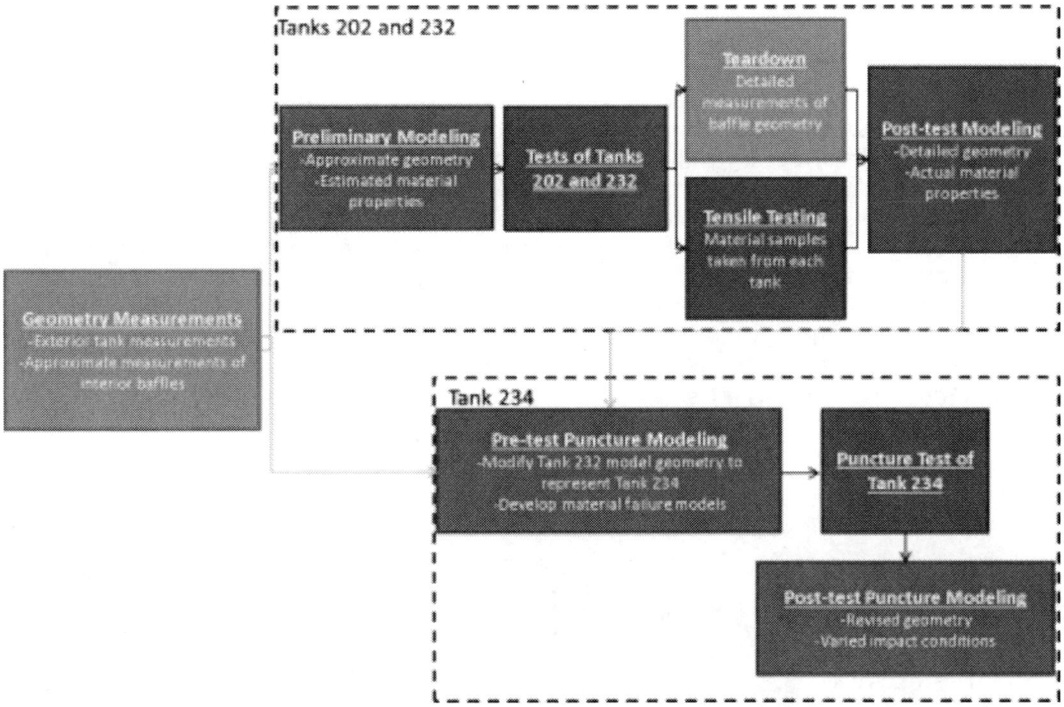


**Figure 6. Flowchart of tests and analyses**

Following the tests of tanks 202 and 232, a teardown was conducted where sections of the exterior of the tank were removed, providing access to the interior of the tank. This teardown permitted engineers to inspect the post-test damage to the interior baffles, as well as to obtain detailed measurements on the baffles and their connections to the outer surfaces of the tank. During this teardown, material coupons were cut from different tank surfaces (e.g. longitudinal baffles, lateral baffles, bottom sheet, etc.) and subjected to tensile testing. This tensile testing provided stress-strain data for the actual materials of construction making up tanks 202 and 232. The detailed geometry and actual stress-strain relationships were used to update the preliminary FE models to more accurately represent tanks 202 and 232. These models were then evaluated at the actual test speeds and compared with the test results.

The impact test of tank 234 was planned as a puncture test, to provide an estimate of the amount of energy necessary to cause puncture of a tank of this type. Based on the geometry measurements that were made to tanks 232 and 234, it was apparent that the tanks had the same exterior measurements. Based on this assertion and the fact that the exterior sheets were both the same thickness, it was assumed that the materials of construction of these two tanks were the same. The stress-strain measurements made on tank 232's materials were used as inputs to the model of tank 234, since the exterior sheets of the two tanks were of the same geometry and thickness. Because the test of tank 234 was intended as a puncture test, additional material parameters required definition within the FE model so that material failure could be modeled.

Based on boroscope images, it was determined that tank 234 had a different baffle design and arrangement from either other tanks. The FE model required updating to account for the actual geometry of the baffles within tank 234. A post-test teardown of tank 234 revealed some differences between the baffle arrangement seen using the boroscope and the baffle arrangement in the impacted region of the tank. This resulted in a need to make post-test changes to the FE model to reflect the actual baffle arrangement.

The commercial FE software Abaqus/CAE was used to construct the models, and Abaqus/Explicit was used to execute the simulations (6). An explicit FE solver was necessary to capture the dynamic impact event, as well as the highly non-linear material and deformation behaviors experienced by each fuel tank during impact. The model included a deformable tank, deformable ram, and deformable impact cart. While the intent of the test was to impart all of the ram's initial kinetic energy to the tank, it was apparent from review of the test measurements that the cart itself experienced some deformation, therefore dissipating some impact energy. The cart was included in the post-test models to examine the effects of the deformable cart and estimate energy not imparted to the tank itself. The FE model is shown in Figure 7. In this figure, the mesh is shown on the left side, and the mesh hidden is on the right side.

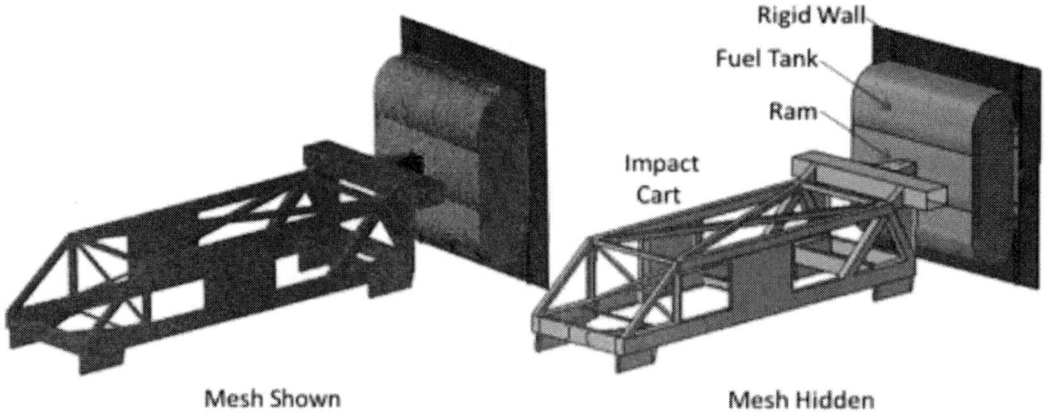


**Figure 7. FE model showing mesh (left) and with mesh hidden (right)**

Several modeling techniques were common to all three FE models (i.e. tank 202, 232, and 234). In all three models, the tank, the ram, and the impact cart were modeled as deformable parts. A summary of the parts in the three FE models is given in Table 4.

**Table 4. Summary of parts in FE models**

| Part Name | Deformable | Weight N (lbf) | Mass kg (lbf-s$^2$/inch) | Number of Elements |
|---|---|---|---|---|
| *Impact Cart* | Deformable | 62,275 (14,000) | 6,350 (36.23) | 66,265 |
| *Ram* | Deformable | 4,003 (900) | 408 (2.33) | 3,240 |
| *Tank 202* | Deformable | 19,950 (4,485) | 2,034 (11.61) | 266,586 |
| *Tank 232* | Deformable | 16,592 (3,730) | 1,692 (9.65) | 91,978 |
| *Tank 234* | Deformable | 15,346 (3,450) | 1,565 (4.05) | 253,300 |
| *Rigid Bolt* | Rigid | -- | -- | 80 |
| *Rigid Wall* | Rigid | -- | -- | 8,197 |

In all models, the deformable cart and ram were given an initial velocity in the longitudinal direction. In all three models, the tank was constrained to the wall via contact with rigid "bolts". These "bolts" were based on the dimensions of the actual bolts and washers used to attach the tank to the rigid blocks on the wall. In the model, the "bolts" were initially in contact with the mounting bracket on the tank. The "bolts" were rigid bodies and were constrained in all 6 displacement degrees-of-freedom. Figure 8 shows the initial and boundary conditions on the model. The inset image shows a close-up view of one attachment "bolt".

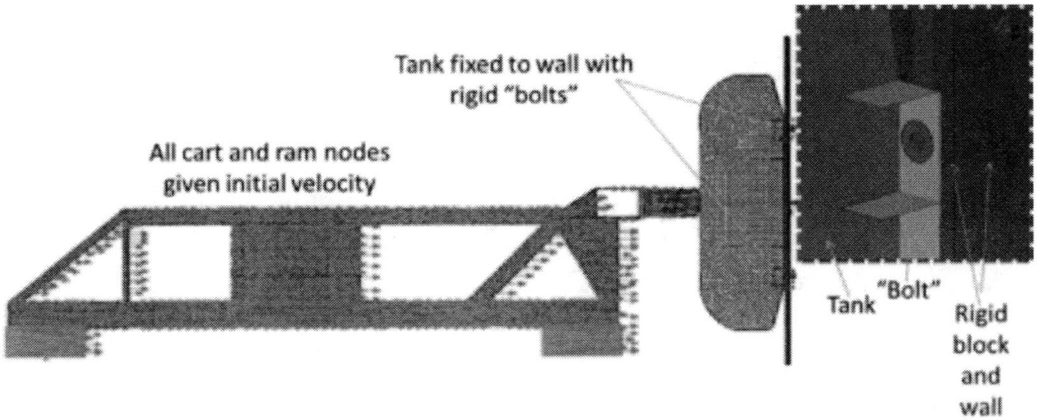


**Figure 8. Boundary conditions in FE model**

The primary results of the FE models included the force-versus-time response of the impact cart as well as the overall mode of deformation. In each test or analysis, the longitudinal acceleration measured or calculated on the left and right sides of the impact cart were multiplied by the total mass of the cart to obtain the impact force. Figure 9 contains a plot of the force-versus-time response of tank

202 on the left and tank 232 on the right. These two impact responses are different from each other, primarily due to the effects of the internal baffle arrangements. Tank 202's baffles were attached directly to the bottom sheet of the tank. Thus, the initial response is stiff, until the baffle buckles. Following this buckling, the force level continues to increase as the impact event continues. Tank 232 featured a gap between the baffles and the bottom sheet of the tank. The response is initially soft, until this gap closes and the bottom sheet contacts the baffles.

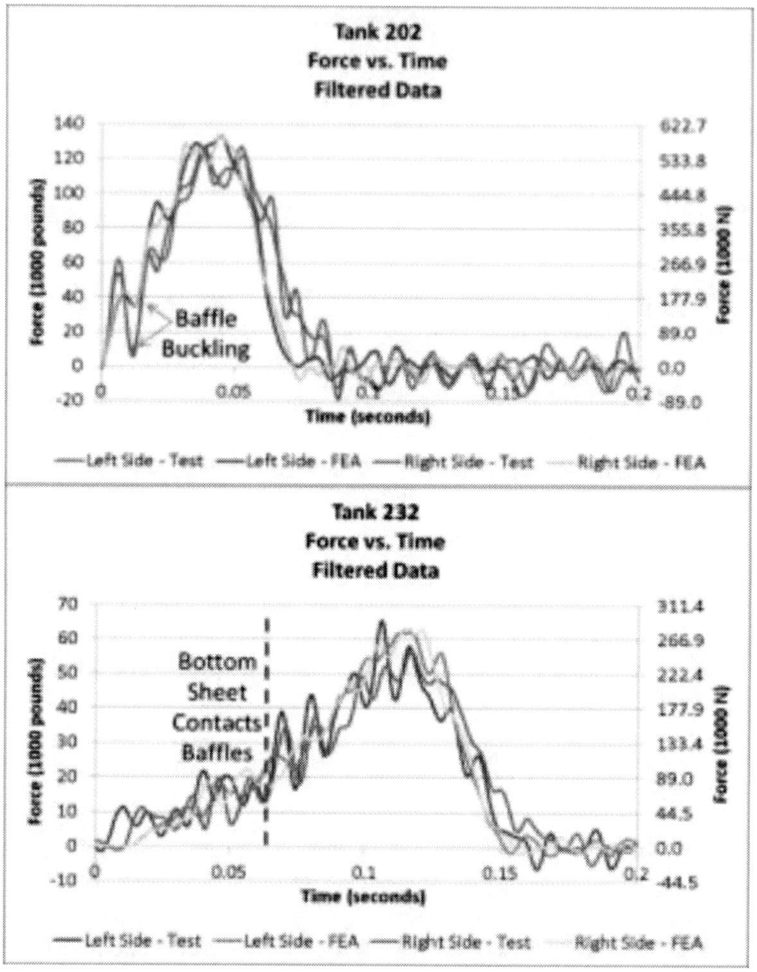


**Figure 9. Test and FEA force versus time results from tank 202 (top) and tank 232 (bottom)**

Tank 234 was tested subsequent to tanks 202 and 232. The impact test of tank 234 occurred at a higher speed than the previous tests, 11.2 mph (18 km/h). Because of this higher speed, the damage to this tank was more widespread. However, the tank did not puncture under this impact condition. The force versus time response for tank 234 is shown in Figure 10. Because this tank featured a similar structure to that of tank 232, the initial impact response is also similar. Initially, the tank offers fairly little resisting force to the impacting object. The force increases more quickly once the bottom sheet of the tank has contacted the baffles within the tank. The large drop-off in force is associated with buckling of the internal baffles. Following this drop, the force continues to climb as the buckled baffles begin to contact themselves. The tank crushes at a fairly constant force until the impactor comes to a stop, whence the force decreases.

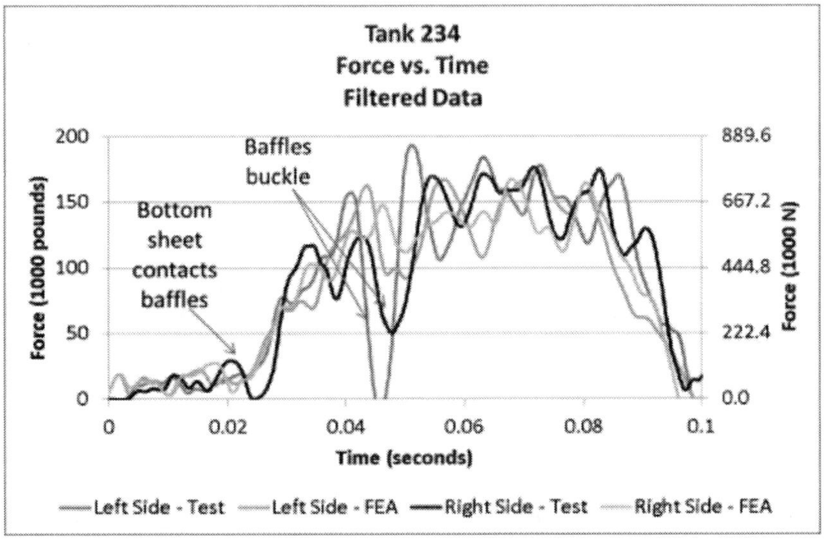


**Figure 10. Test and FEA force versus time results from tank 234**

## 5 DISCUSSION

The results of the three blunt impacts to used passenger locomotive fuel tanks have provided valuable information on the impact response of the fuel tanks, as well as on the details of construction that have an effect on this response. While the CFR requirements for puncture resistance require a combination of bottom sheet thickness and strength, these tests have demonstrated that the arrangement of internal baffles also plays a role in determining the puncture behavior of the fuel tank. While the baffles are typically intended to prevent fuel from sloshing within the tank, it may be important to also consider the structural influences of baffles to a tank's puncture resistance when discussing the topic of fuel tank integrity.

The recent research on dynamic impacts of conventional fuel tanks provides some key results that can be transferred to inform the design of fuel tanks. Strategic construction of tanks can likely improve fuel tank puncture resistance. This research indicates that design planning should include careful consideration of material properties beyond the yield strength consideration required in the regulations (e.g. higher-ductility materials may offer greater puncture resistance). Additional considerations that have come to the attention of the researchers through this program include the possible effects of baffle arrangement on puncture resistance. While it has not yet been tested in this program, external shielding of the fuel tank may also offer improvements to the puncture resistance of fuel tanks. Additionally, surveys of accidents show that strategic placement of the tank can aid in protecting the tank from impacts (e.g. placing carbody structural members between the side of the DMU and the tank).

FRA's research program is focused on establishing the technical basis needed to develop performance-based fuel tank standards, applicable to a range of designs. A more detailed description of this research is provided in a series of FRA papers (10) (11).

There are two key reasons prompting the review of standards for DMU fuel tanks. The first stems from the history of fuel tank regulations in the US, which were created for freight equipment and adopted for passenger equipment. With the size of DMU fuel tanks nearly a tenth of the capacity of freight fuel tanks, applying existing requirements to DMU tanks is challenging and may not be appropriate. The second impetus for looking at DMU standards is the growing demand for DMU equipment in the US for small-scale passenger line expansions or start-ups desire flexible consist arrangements. Finally, with US passenger equipment manufacturers near-extinct, most new equipment is provided by European or Asian manufacturers. Equipment from foreign manufacturers must be made to demonstrate compliance with the CFR or a waiver must be submitted to FRA for consideration as "alternatively compliant".

Collaboration between the government-led researchers, the passenger rail industry and the passenger rail equipment manufacturers is needed to develop a set of requirements that ensure fuel tank integrity for alternative fuel tanks. It is suggested that a set of performance-based requirements be developed based upon an understanding on dynamic impact conditions known to puncture fuel tanks.

## REFERENCE LIST


(1) "Recovery and Reinvestment Act of 2009", Public Law 111-5, US Congress, February 17, 2009.
(2) 49 CFR 229.217 "Fuel tank."
(3) 49 CFR "Appendix D Part 238 – Requirements for External Fuel Tanks on Tier I Locomotives."
(4) EN12663. "Railway applications - Structural requirements of railway vehicle Bodies." European Committee For Standardization. January 14, 2000.
(5) Jacobsen, K., "Fuel Tank Crashworthiness: Loading Scenarios," American Society of Mechanical Engineers, Paper No. JRC2011-56077, March 2011.
(6) Abaqus 6.13-1. Dassault Systèmes, Providence, RI, 2013.
(7) US Department of Transportation, Federal Railroad Administration, Code of Federal Regulations, Title 49, Part 238, Appendix D— "Requirements for External Fuel Tanks on Tier I Locomotives".
(8) American Association of Railroads, S-5506 "Performance Requirements for Diesel Electric Locomotive Fuel Tanks".
(9) APTA SS-C&S-007-98 REV.1 "Standard for Fuel Tank Integrity on Non-Passenger Carrying Passenger Locomotives."
(10) Jacobsen, K., Llana, P., Carolan, M., Llana, P., "Test Requirements of Locomotive Fuel Tank Blunt Impact Tests," Proceedings of the 2013 ASME Rail Transportation Division Fall Technical Conference, RTDF2013-4701, October 2013.
(11) Jacobsen, K., Carolan, M., Perlman, A.B., "Conventional Fuel Tank Blunt Impact Tests: Test and Analysis Results," Proceedings of the 2014 ASME Joint Rail Conference, JRC2014-3786, April 2014.

# Locomotive crashworthiness research

**D C Tyrell, P Llana**
Volpe National Transportation Systems Center
U.S. Department of Transportation, USA

**ABSTRACT**

The Federal Railroad Administration (FRA) of the U.S. Department of Transportation conducts research on locomotive crashworthiness. The research approach includes four phases:

1. Accident investigations to assemble sequences of events leading to injury and fatality.
2. Locomotive performance is analyzed, and potential improvements are explored.
3. Specifications are developed, using the research results.
4. Locomotives are designed to the evolved standard and introduced into service.

As technological advances show promise for improved performance, the phases proceed in an evolutionary fashion, with continuous research leading to continuously improving standards and safer designs.

Recent accidents' consequences suggest that locomotives built to the requirements of U.S. Code of Federal Regulations Title 49 Part 229 Appendix E (Appendix E) are often more crashworthy than would be expected of older locomotives. In the 1990s, FRA conducted research in response to the Rail Safety Enforcement and Review Act of 1992. In 1997, at the request of the FRA, the Locomotive Crashworthiness Working Group was formed. Members of the working group included railroads, suppliers, and labor organizations. Further research was conducted and Appendix E was drafted and recommended to FRA. After the Working Group disbanded, FRA continued research on locomotive crashworthiness. Recently component tests were conducted of energy-absorbing anti-climbers and couplers. Full-scale impact tests are being planned, in part to show crashworthiness compatibility of modified locomotives with a range of freight and passenger equipment. Results of this research are currently being used by the Liquefied Natural Gas (LNG) Tender Technical Advisory Group of the Association of American Railroads (AAR) to develop crashworthiness requirements for LNG tenders and locomotives.

## 1 INTRODUCTION

The Federal Railroad Administration (FRA) Office of Research, Development, and Technology conducts research on the full range of railroad safety topics. This research includes advancing and enhancing technologies for rail equipment safety and facilitating their implementation in the railroad industry. Work is done to identify safety concerns, such as assessing the likelihood and damage resulting from accidents and the development of accident scenarios of concern. Safety strategies for mitigating the consequences of the scenarios, including development of technologies for improved occupied volume protection, injury prevention, fuel containment, and glazing impact resistance. This research produces the information needed to understand and apply the technology to the rail industry. The information may then be used to engineer equipment and verify its performance, inform policy decisions, and support standards development.

A timeline of selected accidents, noteworthy documents, and locomotive crashworthiness research activities is shown in Figure 1. The continued occurrence of train accidents has motivated the government and industry to strive to provide adequate locomotive crashworthiness. In the 1970s and 1980s, the National Transportation Safety Board (NTSB), an agency of Congress, investigated a number of train accidents which resulted in recommendations for increased locomotive crashworthiness (1, 2, 3, 4). These recommendations led the Association of American Railroads (AAR), an industry organization, to issue Locomotive Crashworthiness Standard S-580 (5) and Congress to request FRA's assessment of locomotive crashworthiness (6). FRA then went onto develop crashworthiness regulations, from recommendations made by an FRA-led government-industry committee, the Railroad Safety Advisory Committee (RSAC) (7). RSAC was formed in 1996 to develop recommendations for improving railroad safety, taking advantage of the Federal Advisory Committee Act of 1972. In parallel, the AAR revised S-580 based on recommendations of RSAC (8). In 2012, Amtrak petitioned FRA to accept an alternative crashworthiness standard for the Siemens-built ACS-64 (9). This alternative standard was developed from criteria and procedures for passenger trainsets with crash-energy management features (10) and dynamic load requirements for passenger cab car end frames (11). Currently, AAR is developing standards for Liquified Natural Gas (LNG) fuel tenders.

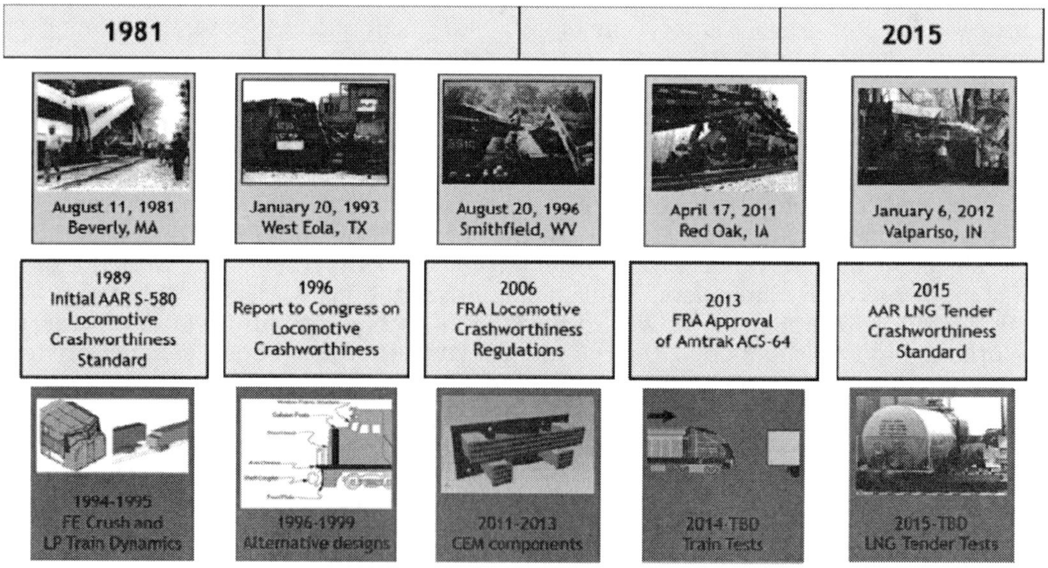


**Figure 1. Timeline of selected accidents, noteworthy documents, and locomotive crashworthiness research activities**

FRA has been continuously conducting research on locomotive crashworthiness since the mid-eighties. To support FRA's response to the U.S. Congress, research on locomotive crashworthiness was conducted. This research led to the development of finite-element analysis techniques for evaluating the crush behavior of locomotive structures and lumped-parameter analysis techniques for evaluating the collision-trajectories of the locomotives and cars that make up trains (12).

Follow-on research was conducted to help support the development of the current FRA regulations and industry standards (13, 14, 15, 16). This research focused on strategies for improving locomotive crashworthiness. Subsequently, additional research was conducted to further refine locomotive analysis techniques (17, 18) and to explore further improvement strategies (19, 20, 21, 22). FRA research on passenger equipment crashworthiness was used by Amtrak in submitting its petition for an alternative crashworthiness standard for the Siemens-built ACS-64 (9). FRA research results are being made available to AAR, to help support development of LNG locomotive and fuel tender crashworthiness standards.

## 2 FRA CRASHWORTHINESS RESEARCH AND INDUSTRY PARTICIPATION

The approach taken in locomotive crashworthiness research is to identify problematic areas from accident investigations, to develop alternative strategies for addressing the problems, to compare the effectiveness of the alternatives and traditional strategies, and to support implementation of effective crashworthiness strategies. When the NTSB identified deficiencies in locomotive crashworthiness, the effectiveness of modified collision posts, short hoods, anti-climbers, and window structures were subsequently evaluated. As the research was conducted, the results were shared with RSAC Locomotive Crashworthiness Working Group and documented in technical papers and reports. The working group used this information to inform development of industry standards and recommendations for FRA regulations. The influences of the new standards were reflected in the next generation of rail equipment. This cycle of research and application of results is illustrated in Figure 2. This cycle has four facets:

1. Accidents are investigated to estimate the sequence of events leading to injury and fatality.
2. The performance of conventional equipment is analyzed, and potential improvements are analytically explored. Conventional and improved designs are tested, and the results compared to determine the potential degree of improvement.

**Figure 2. Engineering research approach to locomotive crashworthiness**

3. The results of the accident investigations, analyses, and testing are used to help develop specifications and standards
4. Improved equipment is designed according to the evolving standards, and introduced into service.

These facets constitute an evolutionary cycle, in which continuous research leads to continuously improving standards and safer designs.

# 3 CURRENT PRACTICES

The generation of locomotives compliant with the current FRA crashworthiness regulations has been in service sufficiently long that the cycle is repeating. While there seems to be general agreement among government, industry, and labor that locomotives compliant with current standards are more crashworthy than older locomotives, there appears to be room for improvement. As part of the results of its investigation of the Red Oak accident, the NTSB has recently issued recommendations for further improvement of locomotive crashworthiness (23). In a 23 mph collision with maintenance-of-way equipment, the two locomotive cab occupants were fatally injured.

## 3.1 Regulations and standards

Locomotive crashworthiness in the U.S. is governed by FRA regulations and AAR standards. The FRA regulations in Appendix E (24) are performance-based, while the AAR standards in S-580 (8) are design-based. Performance-based requirements include impact scenarios that are intended to bound the range of potential accidents and incidents. Compliance can typically be demonstrated with computer simulations, non-destructive car tests, and destructive component tests. Performance standards permit a wide range of design approaches. Design-based requirements include static loads that are intended to result in particular design features. These features are expected to be effective in the range of expected accident and incident conditions. Compliance can typically be demonstrated with classical manual engineering calculations and non-destructive tests. Design standards often implicitly assume a particular design features, such as buff stops. These and other loads, described in detail in S-580, are intended to provide the same level of crashworthiness for a traditionally-designed North American locomotive as the performance-based federal regulations (7).

There are two scenario prescribed in Appendix E (24). These scenarios are shown schematically in Figure 3. In both scenarios, the locomotive is impacted by an idealized proxy object at a prescribed speed. In the first scenario, the proxy object impacts the full width of the locomotive. In the second scenario, the proxy object impacts the corner of the locomotive. The outcome of both scenarios must not result in more than 305 mm of intrusion into the cab, the occupied space of the locomotive.

There are many load cases prescribed in AAR S-580. These include load cases for the underframe and collision posts. Briefly, the underframe is required to support a load of 4.45 MN when loaded on the buff stops. (The buff stops are a design feature necessary to support the draft gear and coupler system typically used on U.S. railroads.) The underframe must support this load without any permanent deformation. Compliance can be shown with a non-destructive quasi-static test, first loading the underframe and then carefully inspecting it. Collision posts must be able to sustain loads of 3.34 MN at the attachment to the underframe and 2.23 MN at a height 762 mm above the underframe. The short hood must be able to support a load of 1.78 MN applied to the upper corner of the short hood over an area that is 305 mm wide and 762 mm tall.

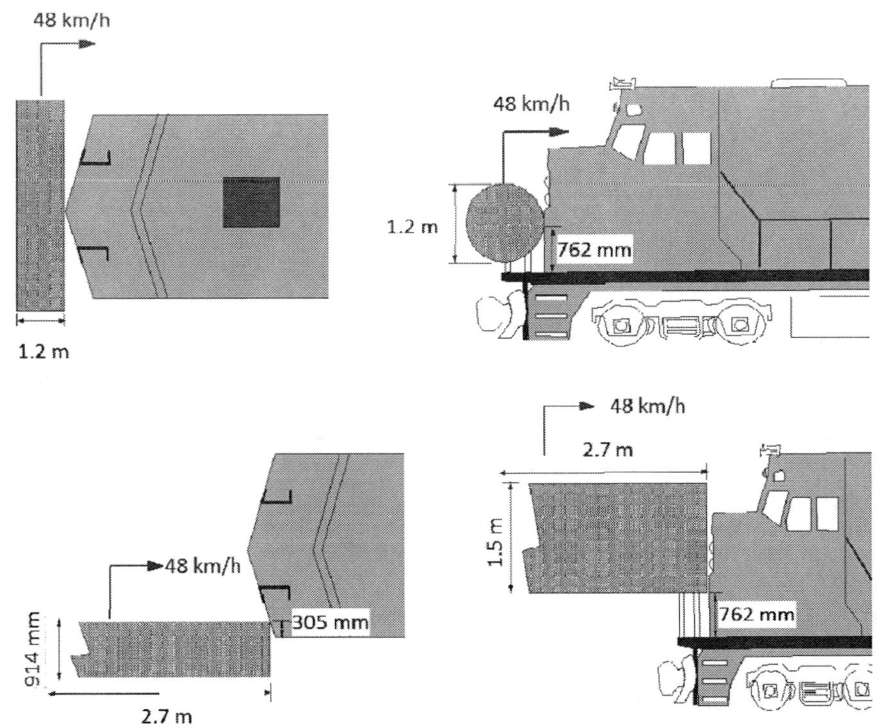


**Figure 3. Schematics adapted from Appendix E locomotive impact scenarios**

In addition to FRA Appendix E and AAR S-580, FRA regulations allow petition for approval of alternative locomotive crashworthiness designs. Amtrak petitioned FRA for such approval of the ACS-64 electric locomotive (9). This locomotive was designed and manufactured by Siemens, based on its Vectron locomotive platform design. The crashworthiness aspects of this locomotive were designed to scenarios adapted from the grade-crossing collision scenarios described in U.S. Code of Federal Regulations Title 49 Part 238 Appendix F (11) and the train-to-train collision scenario described in a criteria and procedures report (10). This petition was accepted by FRA after careful review.

### 3.2 Selected accidents showing strengths and room for improvement of current practice

On January 6, 2012, there was a train accident involving three freight trains in Valparaiso, Indiana (25). None of the crew members of the three trains were fatally injured. All three trains were initially travelling in the same direction, westbound. While travelling at approximately 43 mph, the lead locomotive of the second train collided with the trailing tank car of the stopped train. During the collision, the lead locomotives of the second train derailed onto the adjacent track. Shortly after coming to rest, the lead locomotive of the third train collided with the derailed locomotives of the second train. The third train was travelling approximately 50 mph when it collided with the derailed locomotives of the second train. A photograph of the accident is shown in Figure 4, with the locomotives from the first train on the left and near the middle, and a locomotive from the first train toward the right. The crashworthiness of the locomotives, all of which were designed to AAR S-580 and to Appendix E, is believed to have helped the crew members survive.

**Figure 4. Photograph of Valparaiso accident (24)**

On April 17, 2011, a freight train travelling at 23 mph collided with the rear of a standing maintenance-of-way equipment (MoW) train (23). Both crew members on the freight train were fatally injured. The crew members of the thirty-five car MoW train were on the locomotive at the head end, and were not fatally injured. As seen in Figure 5, the accident resulted in several maintenance-of-way equipment cars overriding the impacting locomotive. The photograph shows the impacting locomotive's modular crew cab was detached and partially crushed as a result of being overridden, resulting in two fatalities. While deformed, the anti-climber and collisions posts remained essentially intact. The initial conditions of this accident appear to be within the range intended for current locomotive crashworthiness requirements.

**Figure 5. Photograph of Red Oak accident**

## 4 ONGOING RESEARCH

### 4.1 Crash Energy Management

As shown by the Red Oak accident, locomotives are susceptible to override when they collide with another vehicle, and the consequences can be catastrophic. Research has shown that the addition of structural features to the forward end of a locomotive can greatly reduce the propensity for override (20). These features include:

1. Push-back couplers, and
2. Deformable anti-climbers.

Push-back couplers allow the ends of the vehicles to engage prior to the build-up of large forces and moments that might lead to lateral or vertical buckling of the vehicles with respect to one another. Deformable anti-climbers manage the load path between the vehicles, deforming gracefully and inhibiting the formation of a ramp.

Push-back coupler and deformable anti-climber proof-of-concept designs have been developed and are being tested (20, 21, 22). These designs were developed for a conventional platform-style locomotive. Computer simulations were used extensively in their development. Component tests have been conducted, and full-scale service and impact tests are planned.

The design for the push-back coupler employs a conventional AAR H-type coupler head with a modified shank attached to a push-back yoke and deformation tube (20). Figure 6 shows a view from below of the push-back coupler inside the draft gear pocket of a locomotive. The draft gear pocket has been redesigned to provide more than 305 mm of additional stroke after the pushback coupler is exhausted. The push-back coupler is attached to the draft gear pocket by the coupler support assembly (shown in yellow in Figure 6) with 12 shear bolts, six on each side. The six shear bolts on the right side are shown in Figure 6 in green. These bolts are designed to fail at a total load of approximately 4.45 MN once the energy-absorbing stroke of the push-back coupler has been exhausted. After the shear bolts fail, the entire coupler support assembly, or 'sliding lug', slides back, so that the load through the coupler is effectively zero. Because the coupler can no longer support a longitudinal load, the load path is shifted to the deformable anti-climber.

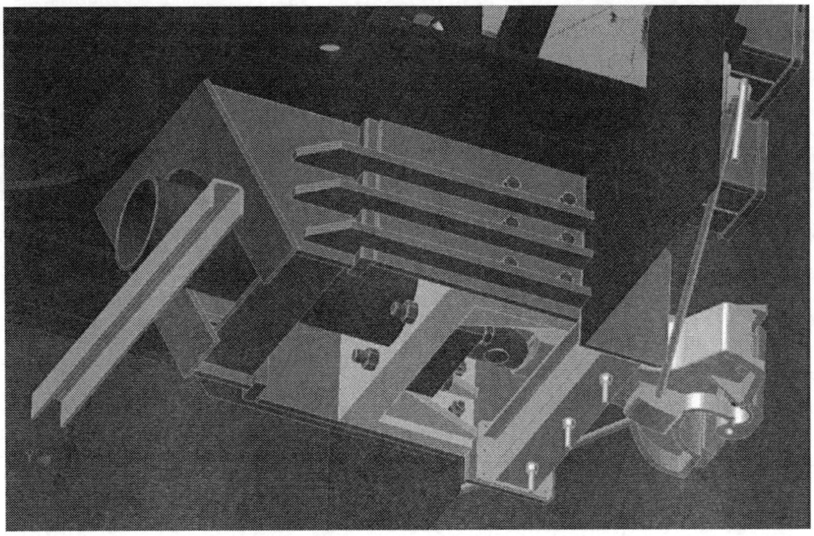

**Figure 6. View of the shear bolts (green) that attach the push-back coupler to the sides of the draft gear pocket**

The locomotive deformable anti-climber was designed to be effective with a wide range of equipment, including conventional locomotives, conventional passenger-carrying cab cars, and freight cars (20). The design for the deformable anti-climber employs four progressive buckling tubes (crush tubes) welded onto the front plate of the locomotive: two tubes located at the base of the short hood, and two tubes located beneath them. This difference in elevation helps the anti-climber to be effective in impacts with equipment which has a range of underframe heights.

Figure 7 shows a detailed view of the deformable anti-climber/pushback coupler system. In a collision with a conventional locomotive, the upper crush tubes are designed to interact with heavy gussets that are welded to the front plate of the locomotive as part of the conventional anti-climber. The upper crush tubes are connected laterally by a ribbed front plate. In the event of a collision with a cab car, this plate also interacts with the collision posts at the end of the cab car, allowing the upper crush tubes to absorb energy. The lower set of crush tubes is designed to interact with the buffer beam of a colliding cab car. In a collision with a conventional locomotive, these lower tubes do not participate in the early stages of the collision. In a collision with a center beam flat car-type freight car, neither set of tubes participates in the early stages of the collision. Eventually, they interact with the bulkhead of the freight car. Not shown in Figure 7 is support structure which has been added to the locomotive for the purpose of transferring impact loads into the underframe.

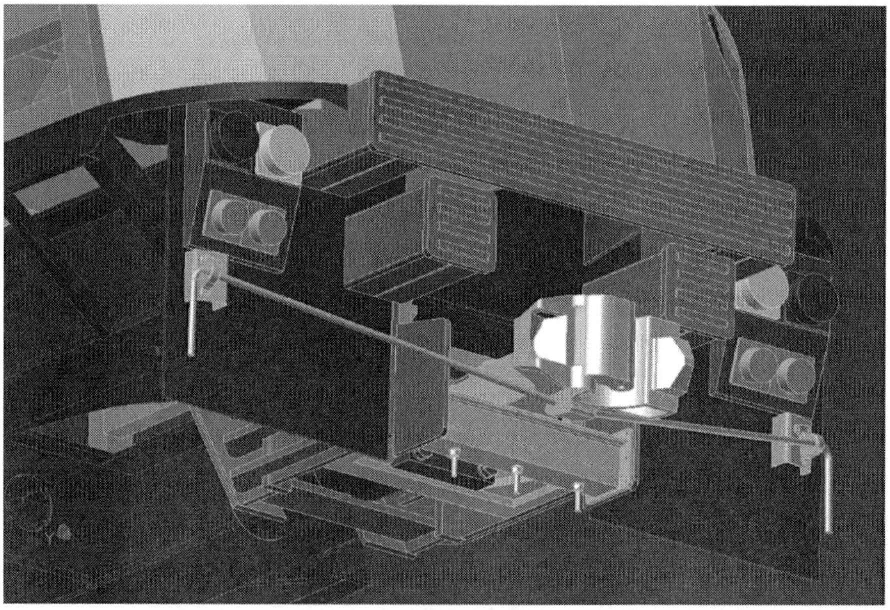

**Figure 7. Detailed view of the deformable anti-climber/push-back coupler system**

Finite element (FE) models were constructed for the conventional and modified locomotives, a cab car, and a freight locomotive (21). These models were appropriately combined to simulate CEM locomotive collisions with conventional locomotive, a conventional cab car, and a freight car. The simulations showed that the pushback coupler and deformable anti-climber designs met their deformation mode and energy absorption requirements. There was no ramp formation or uncontrolled deformation in the CEM locomotive or conventional vehicles, and, of particular importance, there was no override of one vehicle onto another in any of the collision scenario cases.

Pushback coupler and deformable anti-climber test articles were analyzed, constructed, and tested (22). Finite element models of the test articles were developed and exercised to e predict of the performance of the test articles. Dynamic testing was conducted at TÜV SÜD Rail GmbH[1], in Görlitz, Germany. Figure 8 shows pre- and post-test photographs of the pushback coupler test article and Figure 9 shows pre- and post-test photographs of the deformable anti-climber test article. For the pushback coupler test, the test article was mounted to the initially moving freight car, which impacted a fixed wall. For the deformable anti-climber test, the test article was mounted to the fixed wall and impacted by the freight car. The tests were successful in demonstrating the effectiveness of the two design concepts. Test results were consistent with finite element model predictions in terms of energy absorption capability, force-displacement behavior and modes of deformation.

**Figure 8. A photograph of the push-back coupler test article mounted on the impact car and the push-back coupler deformation tube after impact**

**Figure 9. Photographs of the deformable anti-climber test article before and after impact**

Work is ongoing to retrofit pushback couplers and deformable anti-climbers onto conventional locomotives and conducting full-scale dynamic impact tests of colliding cars and colliding trains. Service tests to measure the impact speed at which the pushback coupler just triggers are planned, as well as higher-speed impact tests to measure effectiveness of the components with conventional locomotives, cab cars,

[1] TÜV SÜD Group is a commercial consulting and testing company which owns and operates a railroad test facility in Görlitz, Germany.

and freight cars. These tests are planned for the summer of 2015 at the Transportation Technology Center (TTC) in Pueblo, Colorado[2].

The overridden locomotive involved in the Red Oak accident was compliant with the latest FRA regulations and AAR standards. A push-back coupler and deformable anti-climber on the locomotive might have mitigated the severity of the collision. When the current regulations and standards were adopted, the technology for push-back couplers and deformable anti-climbers was not sufficiently mature to allow their introduction into service. This research effort has endeavored to develop this technology further and provide the technical basis for including push-back couplers and deformable anti-climbers in future specifications.

**4.2 LNG**

Currently, there is much interest in the U.S. in using (LNG) as a locomotive fuel. The relative cost of LNG is about 1/6 the cost of diesel, for the same amount of latent energy. LNG is a pressurized cryogenic fluid, and for the same locomotive travel distance about twice the volume of LNG is needed as diesel fuel. There is not sufficient space available on a locomotive platform to double the capacity of the fuel tank. In order to maintain the same travel range as a diesel-fueled locomotive, a fuel tender is needed for an LNG-fueled locomotive. There are many safety considerations in implementing LNG as a railroad industry-wide locomotive fuel. One consideration is the crashworthiness of the locomotive and tender.

LNG fuel tenders are being developed in phases. Legacy fuel tenders, from research conducted by Union Pacific and Burlington Northern Railroads in the 1990s, are to be used to develop reliable engine and tender fuel handling components. These legacy fuel tender designs are based on traditional pressurized tank car designs. A single tender is intended to supply two locomotives in a locomotive-tender-locomotive configuration. Simultaneously, the Association of American Railroads is developing standards for production LNG fuel tenders.

FRA is sharing research results with the industry working group developing crashworthiness specifications for production tenders, the AAR Locomotive Committee/LNG Technical Advisory Group (LNG TAG). FRA is working with the industry to define the collision scenarios of concern, evaluate the crashworthiness performance of legacy LNG fuel tenders, evaluate the crashworthiness performance of alternative-design LNG fuel tenders, and compare legacy and alternative-design LNG tender crashworthiness performance. The LNG tender crashworthiness standards are expected to incorporate cost-effective and practical research results and also reflect industry experience.

Five collision scenarios have been accepted by the LNG TAG, which are listed in Table 1. These scenarios include a train-to-train collision, a grade-crossing impact, rollover, shell impact, and head impact. The train-to-train scenario results in a high compressive load being applied to the tender. The grade-crossing and rollover scenarios result in high loads being applied to the valves and plumbing. The shell and head scenarios results in concentrated loads being applied to the side of the tank and to the end of the tank. These scenarios were accepted by the LNG Tag as bounding the range of impacts that the LNG tender may experience in service.

[2] TTC is a U.S. DOT owned facility that is managed by FRA and operated by Transportation Technology Center Incorporated (TTCI), a subsidiary of AAR.

**Table 1. Candidate Collision Scenarios for LNG Fuel Tender Crashworthiness**

| # | Scenario | Mode | Threat to Tender |
|---|---|---|---|
| 1 | Head-on collision between LNG locomotive-led freight train and diesel locomotive-led freight train | Compressive impact load on tender underframe | LNG tender underframe cripples |
| 2 | Collision of highway vehicle into LNG tender at grade-crossing | Blunt impact of LNG tender side valves and fittings | Valve and fitting failure |
| 3 | Derailment and rollover of LNG tender | Blunt impact of LNG tender roof valves and fittings | Valve and fitting failure |
| 4 | Derailment and rail car impact into LNG tender tank shell | Blunt impact to LNG tender tank shell | Loss of LNG tank integrity |
| 5 | Derailment and rail car impact into LNG tender tank head | Blunt impact to LNG tender tank head | Loss of LNG tank integrity |

Preliminary analyses of the legacy tenders have been conducted for FRA and shared with the LNG TAG. Underframe strength and interaction of the tender with the locomotives has been identified as a particular concern. Work is ongoing to develop and evaluate alternative crashworthiness strategies for LNG tenders and locomotives. Results from the locomotive CEM research are being considered for adaptation to LNG fuel tenders, as are results from hazmat tank car integrity research (26).

## 5 SUMMARY

The continued occurrence of train accidents has motivated the government and industry to strive to improve locomotive crashworthiness. FRA has been conducting research on locomotive crashworthiness continuously since the early 1990s. This research is coordinated with research on passenger train crashworthiness and with research on hazmat tank car integrity. The research results are applied to help develop federal regulations and industry standards, which result in equipment with improved-crashworthiness being put into service. The accidents, research, standards, and incrementally improved equipment constitute steps in an evolutionary cycle, in which continuous research supports continuously improving standards and safer designs.

Current practice for locomotive crashworthiness includes federal regulations and industry standards. The federal regulations for locomotive crashworthiness are performance-based and include impact scenarios that are intended to bound the range of potential accidents and incidents. Compliance can typically be demonstrated with computer simulations, non-destructive car-level tests, and destructive component tests. The industry standards are design-based and include static load requirements that are intended to result in particular design features that are effective over the range of expected accident and incident conditions. Compliance can typically be demonstrated with classical manual engineering calculations and non-destructive tests. The design-based industry standards are intended to provide the same level of crashworthiness for a traditionally-designed North American locomotive as the performance-based federal regulations.

Recent accidents show the strengths and weaknesses of current locomotive crashworthiness design practice in the U.S. On January 6, 2012 in Valparaiso, Indiana, one train rear travelling at 43 mph rear ended a second train. A third train, travelling at 50 mph, collided with the derailed locomotives of the first train. Although injured, all train crew members survived. On April 17, 2011 a freight train travelling at 23 mph collided with standing maintenance of way equipment. All of the locomotive crew members were fatally injured when the locomotive was overridden by the equipment.

Current research on locomotive crashworthiness has focused on inhibiting override and managing the load path between the locomotive in the event of collision with other rail equipment. Crash energy management features, including pushback couplers and deformable anti-climbers, have been developed which provide crashworthiness compatibility with passenger-carrying cab cars, conventional locomotives, and freight cars. Deformable anti-climbers and pushback couplers have been successfully component tested. Fullscale car-to-car and train-to-train tests are planned of modified locomotives.

The latest in crashworthiness technology is being applied to assure that LNG fuel tenders and LNG-fueled locomotives are at least as crashworthy as diesel-fueled locomotives. FRA is working with the industry to define the collision scenarios of concern, evaluate the crashworthiness performance of legacy LNG fuel tenders, evaluate the crashworthiness performance of alternative-design LNG fuel tenders, and compare legacy and alternative-design LNG tender crashworthiness performance. The LNG tender crashworthiness standards are expected to incorporate cost-effective and practical research results and also reflect industry experience.

## REFERENCE LIST


(1) National Transportation Safety Board, 1982, "Head On Collision of Boston & Maine Corp Extra 1731 East & MBTA Train No. 570 on Former Boston & Maine Corp. Tracks, Beverly, Massachusetts, August 11, 1981," Railroad Accident Report, NTSB/RAR-82-01.

(2) National Transportation Safety Board, 1983, "Rear End Collision of Two Burlington Northern Railroad Company Freight Trains Pacific Junction, Iowa, April 13, 1983," Railroad Accident Report, NTSB/RAR-83/09.

(3) National Transportation Safety Board, 1987, "Rear End Collision and Derailment of Two Union Pacific Freight Trains Near North Platte, Nebraska on July 10, 1986," Railroad Accident Report, NTSB/RAR-87/03.

(4) National Transportation Safety Board, 1989, "Head-on collision between Iowa Interstate Railroad Extra 406 East with release of Hazardous Materials near Altoona, Iowa, July 30, 1988", Railroad Accident Report, NTSB/RAR-89/04.

(5) Association of American Railroads Technical Services Division, Mechanical Section - Manual of Standards and Recommended Practices, "Locomotive Crashworthiness Requirements, Standard S-580," Adopted: 1989.

(6) Federal Railroad Administration, "Locomotive Crashworthiness and Cab Working Conditions Report to Congress", Office of Safety Assurance and Compliance, 1996.

(7) Cothen, G., Schulte, C., Horn, J., Tyrell, D., "Consensus Rulemaking at the Federal Railroad Administration," TR News, Transportation Research Board, Number 236, January-February 2005.

(8) Association of American Railroads Technical Services Division, Mechanical Section - Manual of Standards and Recommended Practices, "Locomotive Crashworthiness Requirements, Standard S-580," Adopted: 1989, Revised, 1994.

(9) Department of Transportation, Federal Railroad Administration, (Docket Number FRA–2012–0036), "Petition for Alternative Locomotive Crashworthiness Design," Federal Register, Vol. 77, No. 98, May 21, 2012.
(10) Carolan, M., Jacobsen, K., Llana, P., Severson, K., Perlman, B., and Tyrell, D. (eds). "Technical Criteria and Procedures for Evaluating the Crashworthiness and Occupant Protection Performance of Alternatively Designed Passenger Rail Equipment for Use in Tier I Service." U.S. Department of Transportation, DOT/FRA/ORD-11/22. October, 2011.
(11) U.S. Department of Transportation, Federal Railroad Administration, "49 CFR Part 238 Passenger Equipment Safety Standards; Front End Strength of Cab Cars and Multiple-Unit Locomotives; Final Rule," Federal Register, Vol. 75, No. 5, January 8, 2012.
(12) Mayville, R., Stringfellow, R., Rancatore, R. and Hosmer, T., 1995, "Locomotive Crashworthiness Research – Volumes 1 to 5," Federal Railroad Administration/U. S. Department of Transportation, FRA/ORD-95/08.1 through 08.5.
(13) Tyrell, D., Severson, K., Marquis, B., Martinez, E., Mayville, R., Rancatore, R., Stringfellow, R., Hammond, R., Perlman, A. B., 1999(a), "Locomotive Crashworthiness Design Modifications Study," Proceedings of the 1999 IEEE/ASME Joint Railroad Conference, April 13-15, 1999, IEEE Catalog Number 99CH36340, ASME RTD Volume 16.
(14) Tyrell, D., Severson, K., Marquis, B., Perlman, A. B., 1999(b), "Simulation of an Oblique Collision of a Locomotive with an Intermodal Container", The Proceedings of the 8th ASME Symposium on Crashworthiness, Occupant Protection, and Biomechanics in Transportation, November 1999, ASME.
(15) Tyrell, D., Martinez, E., Wierzbicki, T., 1999, "Crashworthiness Studies of Locomotive Wide Nose Short Hood Designs," draft paper, accepted for publication at the 8th ASME Symposium on Crashworthiness, Occupant Protection and Biomechanics in Transportation, Nashville, TN.
(16) Martinez, E., Tyrell, D., "Alternative Analyses of Locomotive Structural Designs for Crashworthiness," Rail Transportation, American Society of Mechanical Engineers, RTD-Vol. 19, 2000.
(17) Kokkins, S., Kong, W., and Kasturi, K., 'Locomotive Crashworthiness Research: Modeling, Simulation, and Validation," DOT/FRA/ORD-01/23, Final Report, July 2001.
(18) Samavedam, G., and Kasturi, K., "Full-Scale Locomotive Dynamic Crash Testing and Correlations: Locomotive Consist Colliding with Steel Coil Truck at Grade Crossing (Test 3)," DOT/FRA/ORD-11/16.I, Final Report, September 2011.
(19) Xin, X., Parida, B., Zaouk, A., Dana, N., and Punwani, J., "Impact Analysis of Innovative Shock Energy Absorber and Its Application in Improving Railroad Safety," 12$^{th}$ International LS_DYNA Users Conference, 2012.
(20) Llana, P., Stringfellow, R., "Preliminary Development of Locomotive Crashworthy Components," American Society of Mechanical Engineers, Paper No. JRC2011-56104, March 2011.
(21) Llana, P., Stringfellow, R., "Preliminary Finite Element Analysis of Locomotive Crashworthy Components," American Society of Mechanical Engineers, Paper No, RTD2011-67006, September 2011.
(22) Llana, P., Stringfellow, R., Mayville, R., "Finite Element Analysis and Full-Scale Testing of Locomotive Crashworthy Components," American Society of Mechanical Engineers, Paper No. JRC2013-2546, April 2013.
(23) National Transportation Safety Board, 1993, "Collision of BNSF Coal Train With the Rear End of Standing BNSF Maintenance-of-Way Equipment Train Red Oak, Iowa April 17, 2011," Railroad Accident Report, NTSB/RAR-12/02.
(24) U.S. Department of Transportation, Federal Railroad Administration, "49 CFR Parts 229 and 238 Locomotive Crashworthiness; Final Rule," Federal Register, Vol. 71, No. 124, June 28, 2006.

(25) National Transportation Safety Board, Railroad Accident Brief, Accident No. DCA-12-FR-002, NTSB/RAB-13/03, August 20, 2013.

(26) Carolan, M.E. Jeong, D.Y., Perlman, A.B., Murty, Y.V., Namboordri, S., Kurtz, R., Elzey, R.K., Ananikitpaiboon, S., Tunna, L., Fries, R., "Application of Welded Steel Sandwich Panels for Tank Car Shell Impact Protection," Final Report DOT/FRA/ORD-13/19, April 2013.

# Radial expansion of metal tubes: static testing and simulation correlation

**C Moreno**[a], **R Beaumont**[a], **D J Hughes**[a], **T Williams**[b], **R Dashwood**[a]
[a] WMG, University of Warwick, UK
[b] OLEO International, UK

## ABSTRACT

Railway passive safety regulation states that numerical models of areas of large deformation, such as crumple zones, must be validated by appropriate physical testing. The radial expansion of tubes is used as an energy absorption mechanism in the event of crashes in railway vehicles. Here we present the physical test and finite element analysis correlation activity of scaled radial expanding tubes.

Dimensionally scaled tubes were tested under quasi-static conditions. Four different dies, with expansion ratio (the ratio of die outer diameter to tube inner diameter) R = 1.10 and R = 1.25 and die conical semi-angle A = 15° and A = 30°, were used to crush the expanding tubes. Numerical simulation models were constructed to predict the behaviour of the tubes. Correlation between simulation and static test was achieved by matching the deformed shape and energy absorption of the specimens. The friction coefficient was inferred by adjusting its value in the numerical model until the simulation matched the test results.

The friction coefficient ($\mu$) was predicted to be on the range 0.07-0.11, depending on the expansion ratio and semi-angle of the crushing dies. These values are slightly larger than the value frequently quoted in the literature of $\mu$ = 0.05.

## 1 INTRODUCTION

Railway passive safety regulation (1) states that numerical simulation is accepted to predict the crashworthiness behaviour of areas of large deformation, such as crumple zones. In addition, the numerical models used for this purpose must be validated by appropriate physical testing.

Radial expansion tubes, or deformation tubes, are widely used energy absorbers in the railway industry. Owing to their non-collapsible nature, energy absorbers based on tube radial expansion are particularly well suited to resist any oblique loading while dissipating energy. Several rail energy absorption providers manufacture deformation tubes. Example of these is OLEO International deformation tubes, as shown in Figure 1.

The radial expansion of tubes has attracted academic interests for the last ten years but had been used in industry for longer than that time. As an energy absorber, the expansion tube presents two sought after features: a remarkable constant load and good anti-climbing behaviour. Its main drawback is that the stroke efficiency is limited to up to 50%. The energy dissipation is achieved by means of plastic deformation and friction.

**Figure 1: OLEO deformation tube: none-deformed (left) and deformed (right) state**

(2) provided an analytical model of the tube expansion process. Friction was not accounted for in this model. The model was assessed against an FEA model which, in turn, was verified by an experimental test. The FEA result coincided with results from test when the friction coefficient was set to $\mu = 0.2$. The theoretically predicted results were found to be in good agreement with the simulation and test results, despite the lack of inclusion of friction effects on the former.

(3) investigated the conditions under which the expansion of tubes are expected to be successful. It was found that, for large punch semi-angles, the process was limited by local buckling, whereas for smaller semi-angles, the process was limited by ductile fracture. The FEA model suggested a coefficient of friction $\mu = 0.02$ and $\mu = 0.2$ for lubricated and dry conditions respectively.

(4) performed an investigation of expansion tubes as energy absorbers. The coefficient of frictions were assumed to be $\mu = 0.25$, $\mu = 0.10$ and $\mu = 0.05$ respectively. An analytical model was developed with the following assumptions: no strain hardening and no wall thickness variation. The model was assessed with the help of a FEA axisymmetric model. Good agreement was found between the test, the analytical model and numerical model.

(5) used a numerical model to simulate different geometries of expansion tubes assemblies in order to study the onset of local buckling. From the numerical simulation, a coefficient of friction $\mu = 0.05$ was inferred.

(6) presented an analytical model to predict the mean force and energy absorption of expansion tubes. A numerical model was created to assess the accuracy of the analytical model. A coefficient of friction in the range of 0.05 to 0.35 was included in the simulation. The discrepancy between the numerical and analytical model increased from 1% to 16%, as the ratio wall thickness to inner diameter increased. It was also found that the analytical model was more accurate for small values of friction.

(7) studied the expansion of tubes and the effect of tube wall thickness and die semi-angle. A numerical model was employed, which included a coefficient of friction $\mu = 0.05$. The simulation reproduced accurately the tests characteristics, except for the initial load peak.

Table 1 shows a summary of the coefficients of friction quoted in the literature. Most authors agree that the coefficient of friction for lubricates tubes is small, of the order of $\mu = 0.05$.

**Table 1: Friction coefficients quoted in the literature**

| | (2) | (3) | (4) | (5) | (6) | (7) |
|---|---|---|---|---|---|---|
| μ | 0.20 | 0.02 (lubricated) 0.20 (dry) | 0.05 (coated) 0.10 (dry) 0.25 (blasted) | 0.05 | 0.05-0.35 | 0.05 |

## 2 EXPERIMENTAL QUASI-STATIC TESTING OF EXPANSION TUBES

Quasi-static expansion tests on cylindrical specimens were performed using a conical die to study the expansion mechanism. Four different dimensions of conical dies were used in the experiments, featuring combinations of two expansion rates and two cone semi-angle.

Cold drawn tubes, made of mild steel E355 as per BS EN 10305-4:2011 (8), were used as specimens for the expansion test. The specimens were cut to a length of 150 mm. The end faces were machined. The tubes were tested in their as-received condition. The dimensions of the tubes were 30 mm outer diameter and 2 mm wall thickness. The ratio of the outer diameter to thickness was 15. The nominal mass of the tubes was 204 g.

The conical dies were made of tool steel EN24. The dies were hardened by oil quenching. A diamond pyramid hardness (DPH) test indentation test was performed on the dies. The specimens had an average hardness Hv = 385 kgf/mm2. According to the formula YS = -90.7 + 2.876 * Hv, specified by (9), the corresponding yield strengths was $\sigma_y$ = 1016 MPa. There were two dies with expansion ratio R = 1.10 and two dies with expansion ratio R = 1.25, with cone semi-angle A = 15° and A = 30° in each case. The dimensions of the dies are shown in Figure 2.

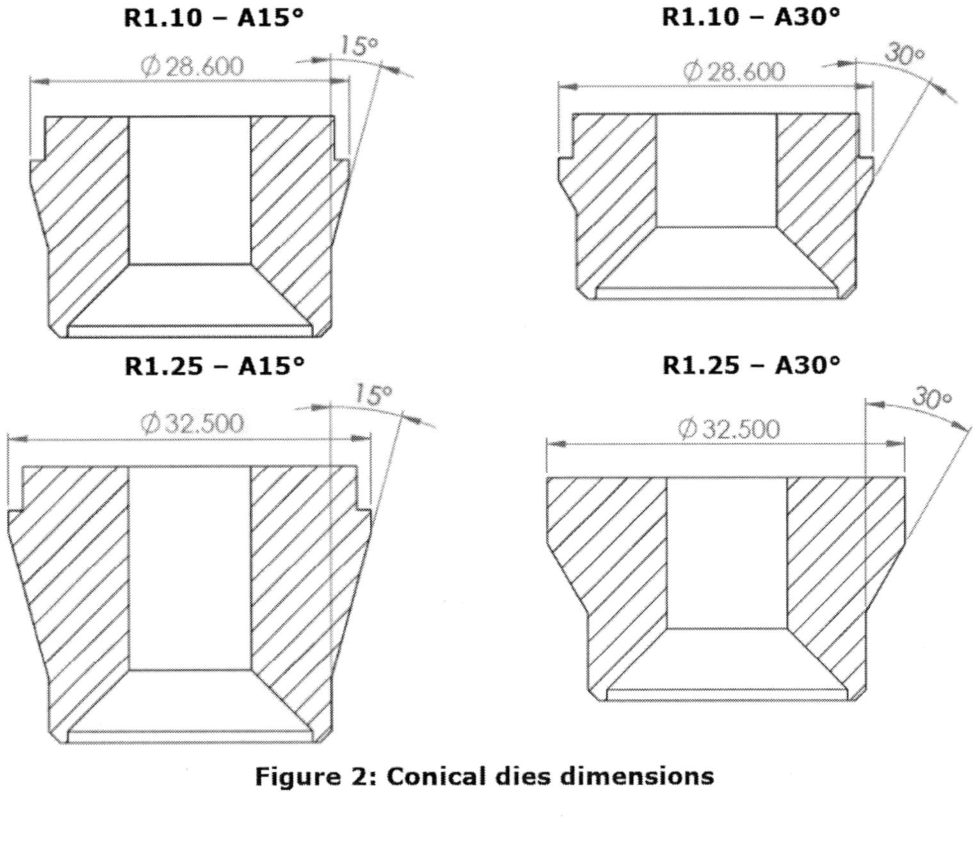


**Figure 2: Conical dies dimensions**

The material properties of the E355 tubes were determined through tensile test. The test pieces were cut from the tube by waterjet cutting process. The test pieces dimensions and test procedure were as per BS EN 6892-1:2009 (10). Figure 3 shows the stress-strain curve of one of the coupons cut from the E355 steel tubes. The material yield strength was 383 MPa.

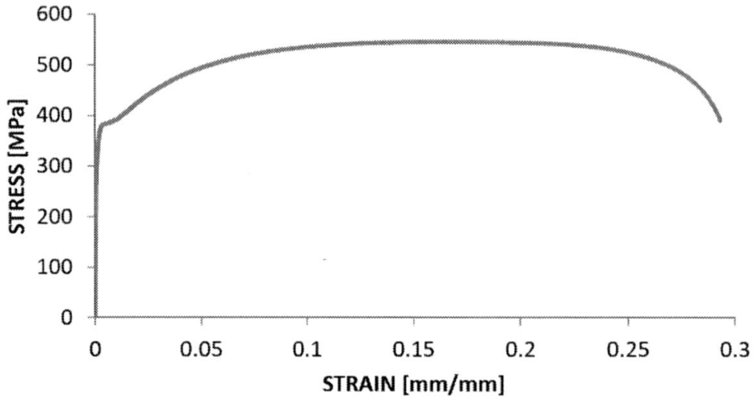


**Figure 3: Engineering stress-strain curve of E355 mild steel obtained from tensile test**

## 2.1 Test procedure

A screw-driven Instron 5800R 100kN test machine was used to crush the specimens. Figure 4 shows the test assembly arrangement. Table 2 shows the mass of the dies and impactor assembly. Copper grease was applied on the die radius surface and around the tube bore. The die was fit onto the tube. The machine crosshead was located under the assembly support. It compressed the tube against the die and platen at a constant speed rate of 16 mm/min. The compressive load was applied until a stroke of 115 mm was achieved. The applied load and the crosshead displacement were recorded by the test equipment. Three specimens were crushed for each type of specimen.

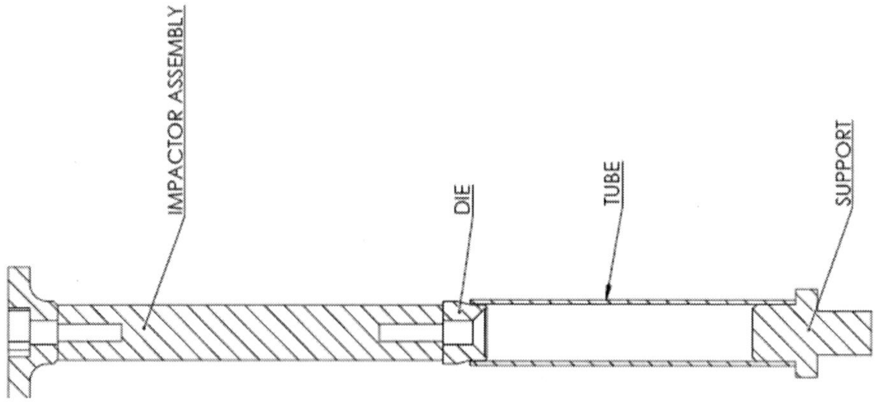


**Figure 4: Schematic representation of the expansion tube assembly**

**Table 2: Mass of the dies and impactor assembly**

| IMPACTOR ASSEMBLY | R110A15 DIE | R110A30 DIE | R125A15 DIE | R125A15 DIE |
|---|---|---|---|---|
| 1097 g | 67 g | 55 g | 104 g | 82 g |

## 2.2 Static test results

The tubes were crushed between the die and the machine crosshead. Figure 5 shows a sequence of the crushing of a R110A15 specimen. The specimens appeared to deform as expected, although several specimens exhibited a slight tilt or bending of the deformed section after crushed.

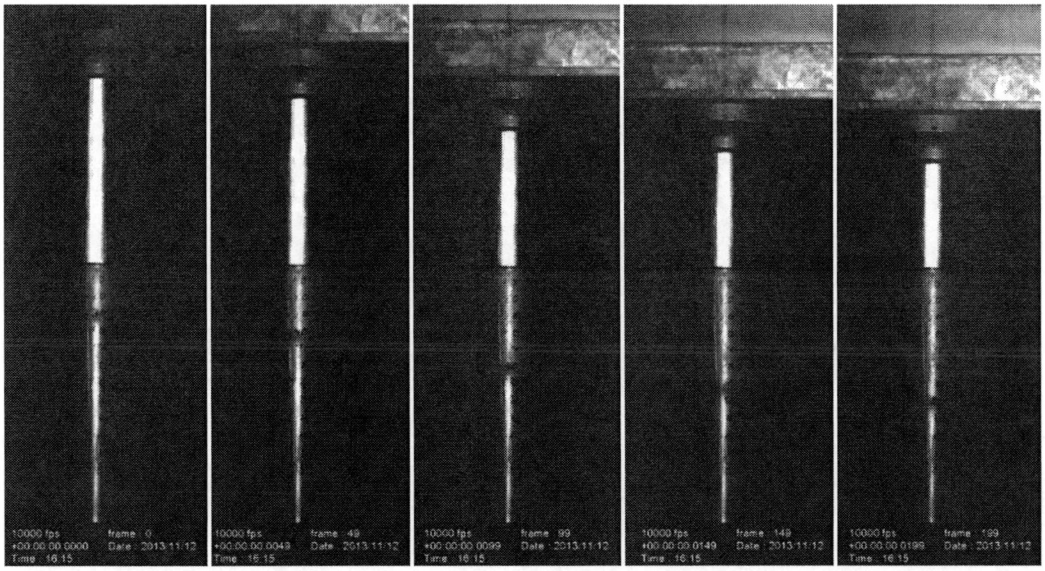

**Figure 5: Sequence of the crushing of a R110A15 specimen**

Three test specimens were crushed for each type of die. Figure 6 shows the load-displacement curves of all specimens tested. It can be observed that the experiment were very repeatable. Figure 7 shows each type of the deformed specimens.

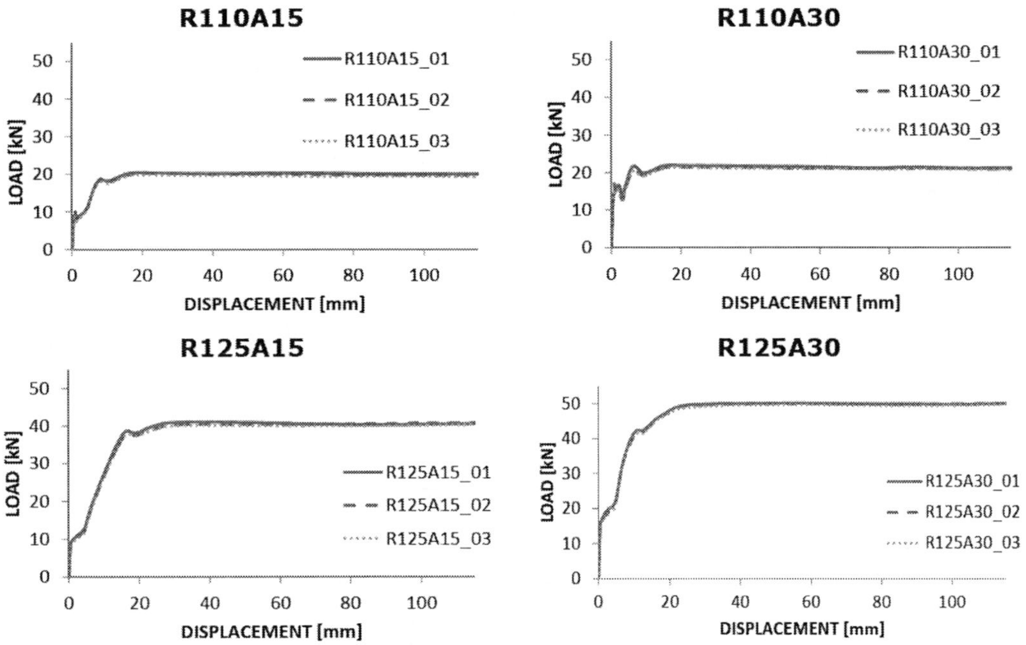


**Figure 6: Load-displacement characteristics of tubes crushed by the four different dies**

**Figure 7: Deformed specimens, R110A15, R110A30, R125A15 and R125A30, from left to right (specimen 125A15 was cut in half to retrieve the die)**

Figure 8 presents the characteristic load-displacement curves of tubes crushed by each type of die.

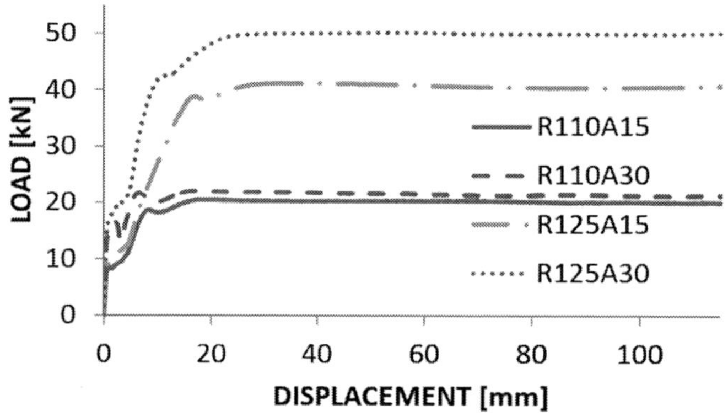


**Figure 8: Comparative load-displacement properties of tubes crushed by each type of die: R110A15, R110A30, R125A15 and R125A30**

The energy absorption characteristics of the specimen crushed are summarised in Table 3. The specific energy absorption was calculated twice: including the impactor mass and rejecting it.

The increase of energy absorption between specimens with expansion ration R = 1.10 and R = 1.25 is 49%, calculated from specimens R110A15 and R125A15. The increase of energy absorption between specimens with conical semi-angle A = 15° and A = 30° is 20%, calculated from specimens R125A15 and R125A30. Unfortunately, the energy absorption characteristics of die R110A30 cannot be used for comparative purposes, because the die deformed plastically under the compressive load, as shown in Figure 9. As a result, the cone semi-angle has most certainly been reduced.

In addition, those two dies with expansion ratio R = 1.10, R110A15 and R110A30, present a polished conical surface which does not correspond to the dies' original appearance and evidence of possible rounding of the edge between the conical/ cylindrical transition of the die.

**Table 3: Energy absorption characteristics of the specimens**

| TEST SAMPLE | PEAK LOAD (kN) | MEAN LOAD (kN) | ENERGY (kJ) | SPECIFIC ENERGY ABSORPTION -WITH DIE MASS- (J/g) | SPECIFIC ENERGY ABSORPTION -NO DIE MASS- (J/g) | FORCE EFFICIENCY (%) |
|---|---|---|---|---|---|---|
| R110A15_01 | 20.5 | 19.5 | 2.247 | 1.6 | 11.0 | 95 |
| R110A15_02 | 20.2 | 19.3 | 2.219 | 1.6 | 10.9 | 96 |
| R110A15_03 | 19.8 | 18.8 | 2.166 | 1.6 | 10.6 | 95 |
| R110A30_01 | 22.0 | 21.2 | 2.437 | 1.8 | 11.9 | 96 |
| R110A30_02 | 21.6 | 20.9 | 2.401 | 1.8 | 11.8 | 97 |
| R110A30_03 | 21.5 | 20.7 | 2.383 | 1.8 | 11.7 | 97 |
| R125A15_01 | 41.2 | 38.1 | 4.376 | 3.1 | 21.5 | 92 |
| R125A15_02 | 40.9 | 37.7 | 4.340 | 3.1 | 21.3 | 92 |
| R125A15_03 | 40.4 | 37.4 | 4.302 | 3.1 | 21.1 | 93 |
| R125A30_01 | 50.2 | 47.5 | 5.462 | 3.9 | 26.8 | 95 |
| R125A30_02 | 50.2 | 47.3 | 5.444 | 3.9 | 26.7 | 94 |
| R125A30_03 | 49.9 | 47.0 | 5.405 | 3.9 | 26.5 | 94 |

**Figure 9: R110A15, R110A30, R125A15 and R125A30 conical dies after test**

## 3 FINITE ELEMENT MODELLING AND SIMULATION OF THE QUASI-STATIC TESTING

A numerical model of the expansion tube has been developed to predict the performance of the tube. The model comprises three parts: the die, the deformable tube and the rigid tube support, as shown in Figure 10. The die has been modelled following two approaches: as a rigid part and as an elasto-plastic part.

The commercial software LS-DYNA was used for the simulation. The elements were fully integrated hexahedral solids. Fully integrated elements were selected to avoid hourglassing (11). An automatic penalty-based contact formulation was selected for the two contact areas: between the die and the deformable tube (*CONTACT_AUTOMATIC_SURFACE_TO_SURFACE) and tube self-contact between the curled strips and the un-deformed section of the tube (*CONTACT_AUTOMATIC_SINGLE_SURFACE). The choice of the friction coefficient will be discussed in the next section. The LS-DYNA material model, *MAT_PIECEWISE_LINEAR_PLASTICITY, models plasticity by introducing a monotonically increasing curve. The curve in Figure 3 was converted into true stress-true strain curve and imported into the FEA material card. The stress-strain properties of the die material, EN24, were not known and were approximated from the yield strength of the material, specified above. The die part was given a constant velocity motion (*BOUNDARY_PRESCRIBED_MOTION) alongside the axis of the deformable tube.

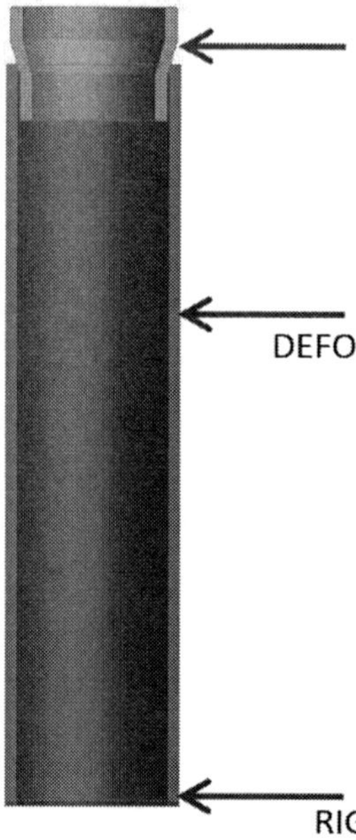


**Figure 10: Section of the expansion tube showing its three components: die [1], deformable tube [2] and rigid support [3]**

The expanded tubes, due to the axial plastic compression, were shorter than the original tubes. The results of the numerical models showed that the final length of the tubes, for a given expansion ration, depended exclusively on the friction coefficient between the tube and the die. This effect could be explained because the resultant of the friction load is purely axial, thus, it only causes compression to the expansion tube. Table 4 shows the final length of the deformed tube and the energy absorbed, and compares it against the FEA predicted values. The values of friction coefficients which match the test results are also included in the table.

**Table 4: Deformed length of the specimens, measured from test specimens and from the simulation, including the coefficient of friction which matches the test results**

| | LENGTH (mm) | | |
|---|---|---|---|
| **DIE** | **TEST** | **SIMULATION** | **FRICTION μ** |
| **R110A15** | 143.5 | 143.5 | 0.07 |
| **R125A15** | 134.4 | 134.5 | 0.11 |
| **R125A30** | 129.0 | 129.1 | 0.09 |

Both approaches, rigid and elasto-plastic dies, are modelled and run. Table 5 shows the energy absorption comparison between the test tubes and the simulation (with rigid dies and elasto-plastic dies). It can be noted that the models with the elasto-plastic die are marginally more accurate than the models which comprise the rigid die.

It can also be noted from Table 5 that the simulation results for specimen R110A15 deviated significantly from the test results. This confirms that the R110A15 deformed plastically during the event.

It is believed that, once the simulation is capable of correlating the tubes deformed dimensions, the predicted coefficients of friction are accurate. The inferred values of the coefficients of friction are consistent with the values quoted in the literature, shown in Table 1, albeit slightly larger.

**Table 5: Energy absorption comparison between the test tubes and the simulation (with rigid dies and elasto-plastic dies)**

| | | ENERGY (kJ) | | | | |
|---|---|---|---|---|---|---|
| SPECIMEN | μ | TEST | RIGID DIE | Δ | ELASTO-PLASTIC DIE | Δ |
| **R110A15** | 0.07 | 2.166 | 2.302 | 6.28% | 2.294 | 5.91% |
| **R125A15** | 0.11 | 4.340 | 4.391 | 1.18% | 4.320 | -0.46% |
| **R125A30** | 0.09 | 5.444 | 5.529 | 1.56% | 5.484 | 0.73% |

## 4 CONCLUSION

The simulation has shown that the deformed tube length depends mainly on the friction coefficient. The effect of friction is reflected both on the tube deformed length and on the energy absorption. The friction coefficient values which match the energy absorption are 0.07, 0.07, 0.11 and 0.09 for the dies R110A15, R110A30, R125A15 and R125A30 respectively. The predicted deformed length of the tubes also matches the observations during test. This range of values of friction is slightly larger than the value $\mu = 0.05$ often quoted in the literature, as shown in Table 1.

## REFERENCE LIST


1. British Standards, *Railway applications — Crashworthiness requirements for railway vehicle bodies*, in *BS EN 15227:2008+A1:2010*2010.
2. Lu, Y.-H., *Study of tube flaring ratio and strain rate in the tube flaring process.* Finite Elements in Analysis and Design, 2004. **40**(3): p. 305-318.
3. Almeida, B.P.P., M.L. Alves, P.A.R. Rosa, A.G. Brito, and P.A.F. Martins, *Expansion and reduction of thin-walled tubes using a die: Experimental and theoretical investigation.* International Journal of Machine Tools and Manufacture, 2006. **46**(12–13): p. 1643-1652.
4. Shakeri, M., S. Salehghaffari, and R. Mirzaeifar, *Expansion of circular tubes by rigid tubes as impact energy absorbers: experimental and theoretical investigation.* International Journal of Crashworthiness, 2007. **12**(5): p. 493-501.
5. Ahn, K., J.S. Kim, and H. Huh. *The effects of local buckling on the crash energy absorption of thin-walled expansion tubes.* in *Numisheet 2008*. 2008. Interlaken, Switzerland.
6. Karrech, A. and A. Seibi, *Analytical model for the expansion of tubes under tension.* Journal of Materials Processing Technology, 2010. **210**(2): p. 356-362.
7. Yang, J., M. Luo, Y. Hua, and G. Lu, *Energy absorption of expansion tubes using a conical–cylindrical die: Experiments and numerical simulation.* International Journal of Mechanical Sciences, 2010. **52**(5): p. 716-725.
8. British Standards, *Steel tubes for precision applications - Technical delivery conditions - Part 4: Seamless cold drawn tubes for hydraulic and pneumatic power systems*, in *BS EN 10305-4:2011*2011.
9. Pavlina, E.J. and C.J. Van Tyne, *Correlation of Yield Strength and Tensile Strength with Hardness for Steels.* Journal of Materials Engineering and Performance, 2008. **17**(6): p. 888-893.
10. British Standards, *Metallic materials - Tensile testing - Part 1: Method of test at room temperature (ISO 6892-1:2009)*, in *BS EN 6892-1:2009* 2009.
11. Hallquist, J.O., *LS-DYNA Theory Manual*, 2006, Livermore Software Technology Corporation: Livermore, California.

# Lean principles and sustainable operation of high-speed railways

**D Hasegawa**[*1]**, G Nicholson**[*1]**, C Roberts**[*1]**, F Schmid**[*1]**, V Novak**[*2]
[*1]Birmingham Centre for Railway Research and Education,
The University of Birmingham, UK
[*2]Virginia Polytechnic Institute and State University, USA

**ABSTRACT**

In various businesses, waste is minimised and value increased by applying 'lean' principles. Adopting a lean perspective, we have analysed the operation of a planned new high-speed railway to identify parameters that have the potential to influence positively the resource requirements during construction and the long term sustainable operation. We have found that, by optimising rather than minimising running times, combined with better management of turnarounds at termini, we can reduce both the first and the life-cycle cost of high-speed railways. Despite reducing land-take in city centres and lowering energy consumption, it is still possible to ensure reliable operations.

## 1. INTRODUCTION

In the climate of rising concern about environmental and economic problems, transport should be discussed in the context of sustainability. High-speed railways are often regarded as the most "green" mode of transport for long distance passenger travel. However, ill-considered operations planning can decrease the system efficiency of high-speed railways in terms of resource usage during construction and operation. Stakeholders of high-speed railway have increased the maximum speed of operation from 210 km/h of the first Japanese Shinkansen line to over 300 km/h of LGV Est. Indeed, the new British high-speed railway project High Speed Two (HS2) is designed for a top speed of 400 km/h and is planned to operate at 330 km/h in 2026 [1]. Higher speed generally increases the energy consumption during operation and makes the requirements for infrastructure and vehicles more severe, which in turn requires greater input of resources. Thus, operations should be planned carefully to avoid any unnecessary use of resources.

Timeabling is the core process of operational planning and determines the technical specification of infrastructure and rolling stock. A good timetable can reduce resource usage. For example, Swiss Federal Railways successfully optimized their timetable and reduced the infrastructure and rolling stock requirements in 6$^{th}$ June 1972; optimized arrival times at termini assisted the efficient rostering of rolling stock and optimised timing of passing trains on single track reduced the requirement for double track [2]. In a similar way, high-speed railways can reduce its resource usage and enhance its sustainability. The so-called 'lean' philosophy, adopted by manufacturing industry since the 1960s [3], provides a set of principles that can be applicable in the planning of high-speed railways. These principles are applied in various businesses and in the public sector to eliminate waste and increase value in the production process of both tangible and intangible products. There have been a few previous studies about the application of lean principles on the railway, however, specific applications have been presented rarely. In this paper, lean principles are adopted for optimization of a timetable to reduce resource usage and increase value for the customers and the business.

## 2. MODELLING OF TIMETABLE AND REQUIREMENTS FOR FACILITIES

The objectives of timetabling are maximizing customer satisfaction and lowering the cost while maintaining quality of service. Customers value absolute safety, short travel times, stable and frequent services, comfortable accommodation and low ticket price. Infrastructure managers and train operating companies seek to operate with absolute safety, minumum infrastructure and rolling stock and low operational cost. There are trade-offs between each objective. In this paper, cyclic timetabling is studied, where the service pattern repeats each cycle time. A cyclic timetable is common in high-speed rail for marketing reasons and because it results in low variation in resource use.

### 2.1. Cyclic timetable modelling

Cyclic timetable reapeats the same timeble every cycle time and intervals between trains are generally same. For a simple double track high-speed railway network which has two termini (*n*, *m*) and service frequency *F* in cycle time *T* at both termini, the time interval between two consecutive departures from terminus *n* is given by

$$t_d^{i+1}(\mathrm{n}) - t_d^i(\mathrm{n}) = {}^T\!/_F \quad ( i \leq F \in \mathbb{N} ), \tag{1}$$

where $t_d^i(\mathrm{n})$ is the departure time of the *i* th train service at terminus *n* in cycle time *T*, $t_d^{i+1}(\mathrm{n})$ is the departure time of next train at terminus *n*. In the same way, the time interval between two consecutive arrival trains at terminus *n* is given by;

$$t_a^{j+1}(\mathrm{n}) - t_a^j(\mathrm{n}) = {}^T\!/_F \quad ( j \leq F \in \mathbb{N} ). \tag{2}$$

where $t_a^j(n)$ is the arrival time of the *j* th train service at terminus *n* in cycle time *T*, $t_a^{j+1}(\mathrm{n})$ is the arrival time of the next train service at terminus *n*. Cycle time *T* is generally 60 minutes for high-speed railway and departure time $t_d^1(\mathrm{n})$ is fixed at 0 minutes. The running time of train service *j* from terminus *m* to terminus *n* ($RT_{nm}$) is given by

$$RT_{nm} = t_a^j(\mathrm{n}) - t_d^j(\mathrm{m}) \in [l_{nm}\,,\ u_{nm}], \tag{3}$$

where, $l_{nm}$ and $u_{nm}$ are the lower and upper limits of the running time from terminus *m* to *n*. This time consists of the minimum running time, dwell time at intermediate stations and a running time supplement for robustness. For a cyclic timetable, the running time for the opposite direction ($RT_{mn}$) is fixed as same:

$$RT_{nm} = RT_{mn} = t_a^i(m) - t_d^i(n) \in [l_{mn}\,,\ u_{mn}]. \tag{4}$$

Assume that the *j* th train service arrives at terminus *n*, turns around there and departs in the opposite direction to *i* th train service, the turnaround time at terminus *n* ($TT_n$) is given by

$$TT_n = t_d^i(n) - t_a^j(n) \in [l_n\,,\ u_n], \tag{5}$$

where, $l_n$ and $u_n$ are the lower and upper limits of the turnaround time at terminus *n* [4]. This time includes time for alighting, handover, cleaning, restoking, checks for departure, boarding and a time supplement for robustness. The handover and checks for the next departure are conducted after passengers have alighted, and handover and checks for departure are short enough to be excluded from the consideration in comparison with the time for restoking and cleaning. Boarding is normally allowed conducted only once the cleaning is finished.

### 2.2. Robustness

Robustness is the characteristic of a railway infrasracture and timetable combination that reduces overall delays when real process times deviate slightly from the design values due to changing passenger loads, different driving styles of each driver and varying weather conditions [5]. To create robust timetables, time supplements and buffer times are added. A time supplement is added to try running times to allow trains to recover by themselves. It is also added to turnaround times at termini to cope with the delayed arrival of trains so that the next train departure in the opposite direction can be on time. Buffer time is added between two consecutive trains that share the same track to prevent delay propagation, where an initial delay affects other trains and produces additional delays. In a cyclic timetable, if the time supplements are already included in the running time and turnaround time, the robustness depends on the spacing of consecutive trains sharing the same track [4]. Greater time on space separation between trains increases the amount of buffer time, resulting in higher robustness. However, the required service frequency in the cycle time does not allow the train separation to be arbitrary. In this context, a lower service frequency timetable has better robustness and, for timetables with the same service frequency, the timetable with identical intervals should be the most robust.

### 2.3. Requirement for infrastructure and rolling stock

The precise calculation of the required number of platforms and sets of rolling stock necessitates a detailed timetable. However, at the planning stage, it is possible to estimate these numbers roughly by using the running time, turnaround time, service frequency and cycle time. If all the platforms are fully utilized, the number of platforms required at the terminus *n* is given by

$$NP_n = ceiling\,(\frac{TT_n+HT_n}{T/F})\,, \tag{6}$$

where $HT_n$ is the required headway time between arrival and departure of trains at the terminus *n* due to the layout of the tracks and the signalling set-up. If the sum of turnaround time and technical headway time is less than or equal to *T/F*, the number of platforms is only one. The number of rolling stock required is given by

$$NR = ceiling(\frac{2RT_{nm}+TT_n+TT_m}{T/F})\,, \tag{7}$$

where the numerator of the right hand side of equation (7) is the time required for one train to become available for departure from the origin terminus *n*, after finishing its round trip. If the time is less than or equal to the train's departure time plus *T/F*, the number of sets of rolling stock required is only one [4]. In the following section, we shall show how a 'lean' approach can be used to improve performance and reduce resource use.

## 3. LEAN

Firstly, a brief explanation of 'lean' principles is given and then an example of the application of 'lean' to timetabling is offered.

### 3.1. Lean principles

The term lean was introduced following research into the higher production efficiency of Toyota Motor Corporation compared with other automobile manufacturing companies [3]. As a result of further research, the principles of Toyota's production methods were generalized as the lean principles [6]: defining the concept of value, using the principles to identify and eliminate non value-added activities and to develop mistake proofing procedures, in a continuing process of improvement. The application of lean principles to aeroplane travel was discussed as one of the examples in a later book [7]. A significant amount of literature about specific applications of lean in service industry such as health, education, banks, airlines and hotels is listed [8]. However, in railway operations, the application of lean principles has rarely been studied.

### 3.2. Value stream mapping and analysis

"What is the value for customers?" We have to answer this question before conducting value stream mapping (VSM), which helps to diagnose and differentiate the value contributions of different activities [9]. Customers use trains to reach a particular destination in safety, on time and quickly. From this, it could be said that trains create value for customers when they are moving with passengers. In contrast, when a train rests at a station or in a depot, it does not create any value. Running times on services between stations are value added (VA) steps, dwell times at stations for boarding and alighting are VA and turnaround time at termini, other than for alighting and boarding are non value added (NVA) steps. An example of a VSM for one train traveling between two termini is shown in Figure 1.

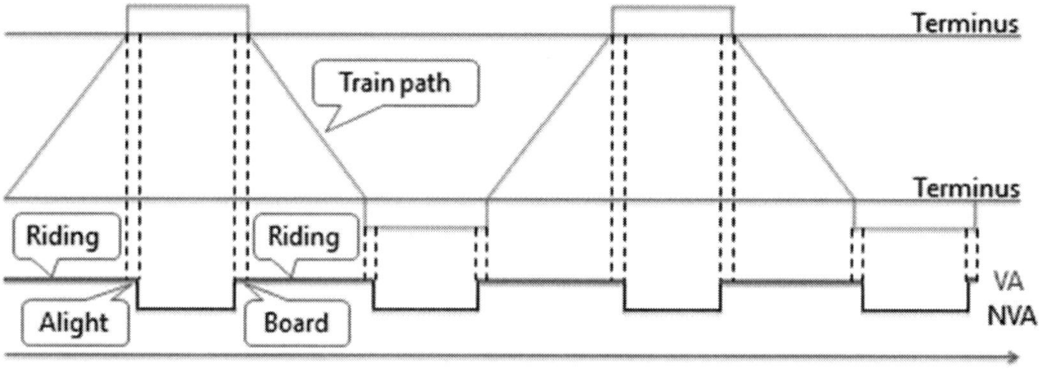


**Figure 1: Value stream map for one train with two termini**

Turnaround time at a terminus consists of time for alighting, handover, restocking of supplies, cleaning, checks for departure and boarding. Within that, time for handover, restocking, cleaning and checks for departure are NVA for customers. According to lean principles, NVA steps are always considered as waste. In Figure 2, 3 NVA steps by three trains are shown to take place at the same time and may increase the number of platforms required at the terminus. Currently high-speed railway rolling stock is typically between 200 m and 400 m long [10]. Longer trains require longer platform lengths, which directly increases the cost of construction. Also, termini of high-speed railways are generally located in the centre of large cities such as London, Paris and Tokyo, which means that the cost of land take and construction is extremely high. Conversely, if high-speed railway operation is possible with fewer platforms at termini by reducing the NVA steps during the turnaround time for requirements other than alighting and boarding, the initial cost of high-speed railway can be reduced and the non necessary usage of additional massive quantities of resources can be avoided.

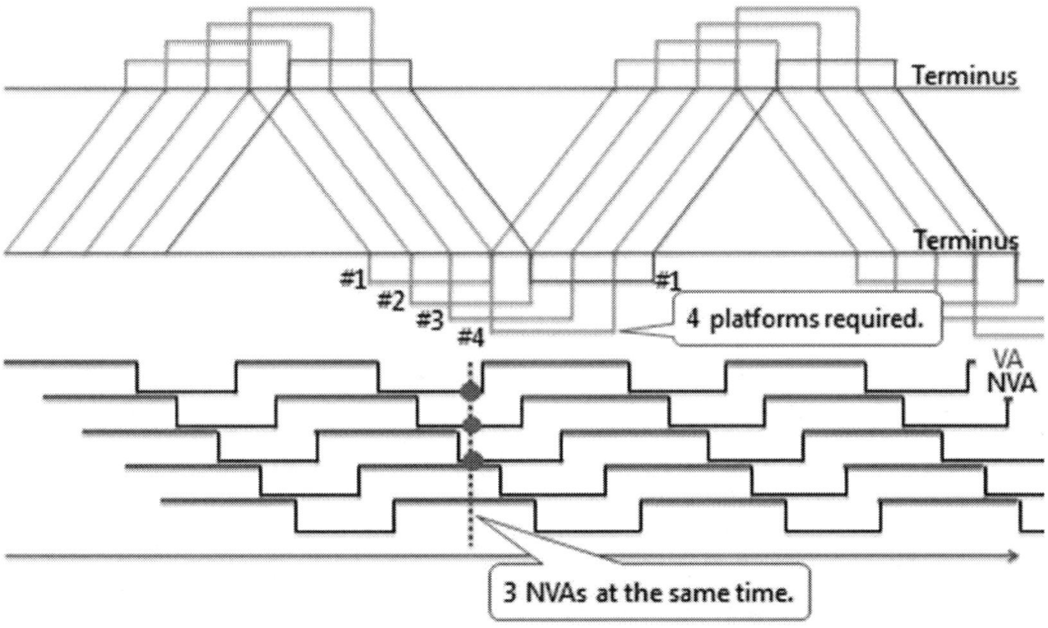


**Figure 2: Waste (platforms) associated with NVA (turnaround time used for activities other than alighting and boarding)**

### 3.3. Just In Time and running time optimization

Just In Time (JIT) is an important lean concept for eliminating waste [11], [12]. For example, in a manufacturing industry context, if the distribution of products is scheduled to follow immeditately after the completion of production, with no delay, the need for storage is eliminated. This analogy is applicable for turnarounds at termini. Turnaround time is time waiting for the next departure, and the platforms for the waiting trains are the warehouse to store the products. Departure times are fixed for commercial reasons in a cyclic timetable, so shorter turnaround should be realised by shifting the arrival time later in equation (5). Also, to maintain the robustness of the timetable, the amount of time supplement for running and turnaround should be kept at the original level. Thus, the running time should be extended to realise the target faster turnaround time (see Figure 3). Clearly, the passengers experience a longer period on board the train but the overall journey time can be held by reducing the NVA access time, e.g., by providing multiple escalators along the platforms.

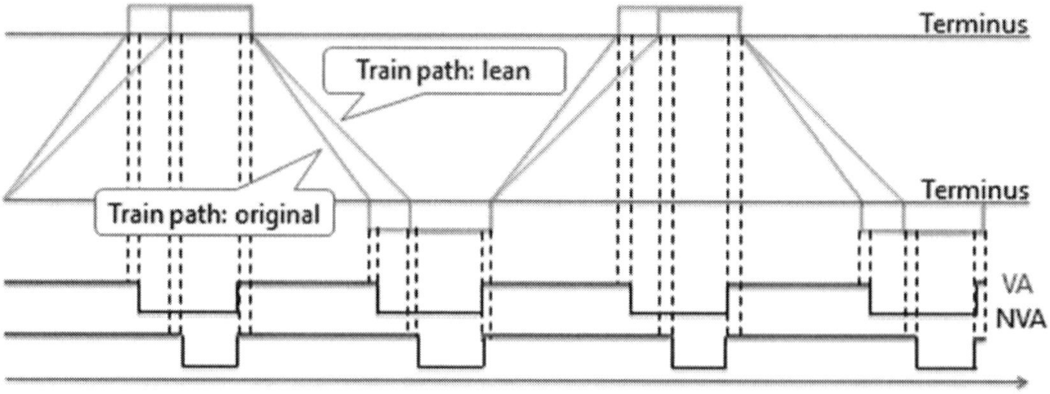


**Figure 3: Comparison between original and lean case with faster turnaround and longer running time**

An additional type of benefit of running time optimization is saving of propulsion energy, because it decreases the operational speed. It can be regarded as the elimination of unnecessary very high speed. The time supplement for turnaround is for absorbing arrival delays and thus keeping the next departure of the train on time. If the punctuality of arrivals is improved through an additional running time supplement, decreased supplemental time for turnaround should not have a negative effect on the robustness of the timetable. This lean operation can thus reduce the number of platforms required without decreasing the level of robustness (Figure 4), however, from equation (7), the number of trains required is the same as that for the original timetable because the sum of running time and turnaround is the same as in the original case.

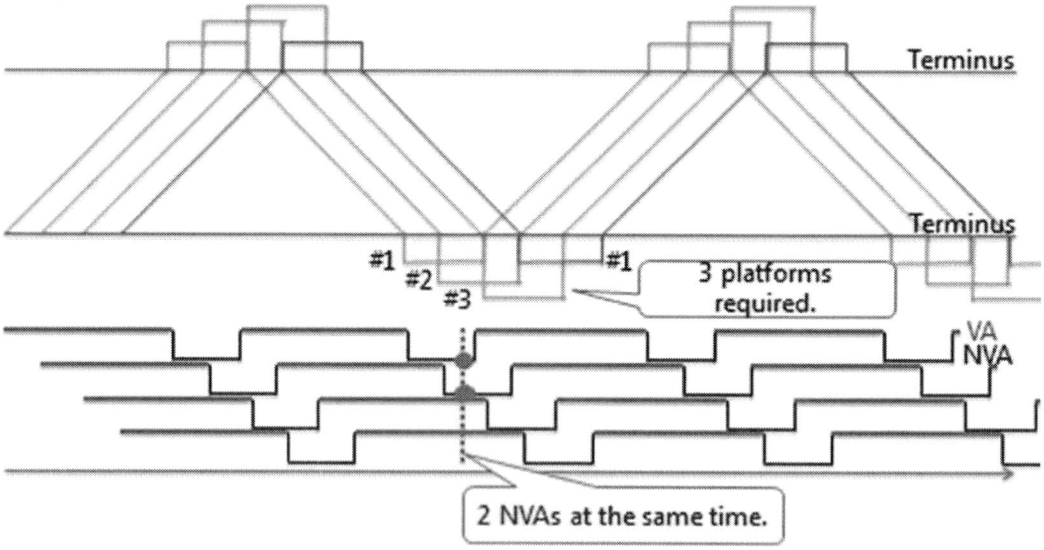


**Figure 4: Associated benefit of faster turnarounds (fewer platforms) and running time extension**

## 4. IMPLICATIONS OF SHORTER JOURNEY TIME

How does an extended running time affect customer safisfaction? A widely accepted premise of transportation planning is that journey time is wasted time and, hence, that reducing journey time increases demand [13]. However, the benefit of reducing journey time has recently become a topic. Lyons and Urry [14] question the traditional approach of travel demand analysis, which is based on the assumption that journey time is completely unproductive and they shows several ways of productive use of the time in order to challenge the assumption. In a subsequent paper, they collected data on the use of the journey time in railway travel by age for Great Britain and showed that, even in the age range from 16 to 25, only 30% of passengers thought the journey time was wasted time [15]. Berry and Hamilton [16] showed the data of mobile phone use by age on train and on platforms in Melbourne and found that less than 30% of people in the age range from 41 to 60 and less than 20% of people older than 60 had never used a mobile phone during a jurney. Other than voice calls, there are several ways to use a mobile phone on the train or platform. Gripsrud and Hjorthol [17] also criticized the conventional assumption of unproductive journey time by collecting data of journey time use on trains in Norway and showed that only 10% of passengers did not use the time for any productive purpose at all. In the context of high-speed railway, Givoni and Banister [18] point out that the development of high-speed railway in Japan and France did not aim to reduce journey time but increase capacity and question that the Britatin will set a new standard for higher speed in order to construct a new high-speed railway. They also suggest an optimal speed for high-speed railways in terms of holistic planning of transport and less carbon emission that recommends the average speed should be about 200 km/h and route distance should be less than 500 km [19]. From these studies, it could be said that the benefit of shorter journey time is overestimated. Conversely, extended running time may have less negative effect on travel demand than we assume currently. However, further research on this topic should be counducted for a detailed quantitative discussion.

## 5. CASE STUDY

A case study of this lean approach to railway improvement is presented in this section. High Speed Two (HS2) is chosen as a case study.

### 5.1. High Speed Two

HS2 will start its operation with a maximum speed of 330 km/h from London to Birmingham in 2026 and wiil be extend to Manchester and Leeds in 2033. It aims to operate 18 trains per hour from London Euston station [20]. This terminus is planned to have 11 platforms for HS2 [21]. From these figures and equation (6), if the required headway time between arrival and departure at Euston were to be 3 minutes, the allowable turnaround time at the terminus can be estimated as 33 min.

### 5.2. Turnaround and platforms

There are faster turnaround examples of on both existing high-speed railways and conventional inter city services: the turnaround for Tohoku, Joetsu and Hokuriku Shinkansen service at Tokyo station takes 12 minutes, while the Tokaido Shinkansen service takes 16 minutes at Tokyo station and the Pendolinos at London Euston station takes 25 minutes. Based on these figures, the number of platforms required for different service frequencies and turnaround times for a cyclic timetable are estimated for the range of turnaround times from 12 minutes to 33 minutes by equation (6) (Table 1). For the planned service frequency of 18 trains per hour, turnaround times of 30, 27, 23, 20, 17, 13 minutes are thresholds for fewer platforms. For example, by reducing the turnaround time by 3 minutes, 10 platforms become sufficient for the operation of 18 trains per hour. Conversely, in spite of reducing the number of platforms, 11 platforms and 30 minutes of turnaround time could allow the operation of 20 trains per hour.

**Table 1: The number of platforms required for different service frequencies (F: trains/hour) and turnaround time (TT: min) with the required headway time between arrival and departure of 3 minutes.**

| TT \ F | 15 | 16 | 17 | 18 | 19 | 20 |
|---|---|---|---|---|---|---|
| 12 | 4 | 4 | 5 | 5 | 5 | 5 |
| 13 | 4 | 5 | 5 | 5 | 6 | 6 |
| 14 | 5 | 5 | 5 | 6 | 6 | 6 |
| 15 | 5 | 5 | 6 | 6 | 6 | 6 |
| 16 | 5 | 6 | 6 | 6 | 7 | 7 |
| 17 | 5 | 6 | 6 | 6 | 7 | 7 |
| 18 | 6 | 6 | 6 | 7 | 7 | 7 |
| 19 | 6 | 6 | 7 | 7 | 7 | 8 |
| 20 | 6 | 7 | 7 | 7 | 8 | 8 |
| 21 | 6 | 7 | 7 | 8 | 8 | 8 |
| 22 | 7 | 7 | 8 | 8 | 8 | 9 |
| 23 | 7 | 7 | 8 | 8 | 9 | 9 |
| 24 | 7 | 8 | 8 | 9 | 9 | 9 |
| 25 | 7 | 8 | 8 | 9 | 9 | 10 |
| 26 | 8 | 8 | 9 | 9 | 10 | 10 |
| 27 | 8 | 8 | 9 | 9 | 10 | 10 |
| 28 | 8 | 9 | 9 | 10 | 10 | 11 |
| 29 | 8 | 9 | 10 | 10 | 11 | 11 |
| 30 | 9 | 9 | 10 | 10 | 11 | 11 |
| 31 | 9 | 10 | 10 | 11 | 11 | 12 |
| 32 | 9 | 10 | 10 | 11 | 12 | 12 |
| 33 | 9 | 10 | 11 | 11 | 12 | 12 |

### 5.3. Optimization of the running time, energy use and service robustness

Based on the discussion in section 3.3, the running time must be optimized if faster turnarounds are possible. The extension of running time can be achieved by decreasing the maximum speed of operation. An example of the effect of lower speeds on running time and energy consumption of HS2 is shown in Figure 5. For instance, a 3 minutes faster turnaround with 10 platforms requires a 3 minute extension of the running time, which decreases the maximum speed by 30 km/h and saves energy use for traction and auxiliary needs by 11.4%. However, the level of robustness and the number of sets of rolling stock required is the same as for the original timetable.

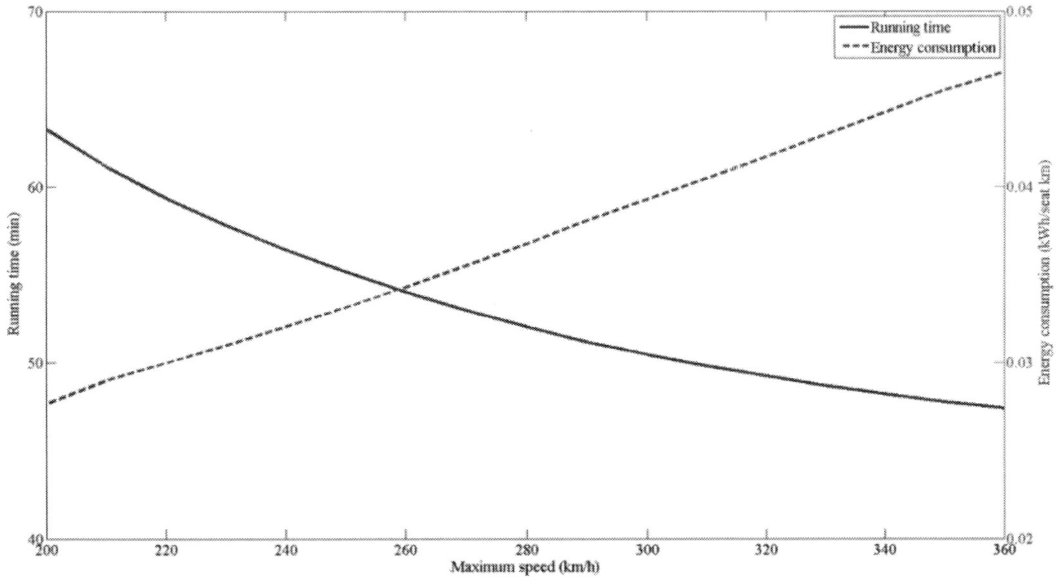


**Figure 5: Running time and energy consumption for different maximum speed of operation for the route of HS2 from London Euston station to Birmingham Curzon street station** [22]

## 6. CONCLUSIONS

By adopting lean principles, the turnaround time other than what is needed for alighting and boarding is identified as NVA and the related waste, additional platform usage, is highlighted. Faster turnaround is studied as a possible solution to reduce the waste, with the potential benefit of fewer platforms and, thus reduced capital investment and initial input of resources. It is suggested that the optimization of running time is preferable to using the minimum running time, in order to support a just-in-time arrival of the train in the station and maintain the established departure schedule. The combination of an extended running time along with the faster turnaround will maintain the same level of robustness as the original timetable. Furthermore, extending the running time can be regarded as elimination of a wasted speed enhancement on top of the optimal speed and gives an additional benefit of energy saving, which contributes to long term savings of operational cost and resource usage. Further study is recommended on the quantitative discussion of the negative effect on the extension of running time on travel demand and detailed investigation on robustness by multi train simulation with different allocation patterns of time supplements.

## REFERENCES


[1] Department for Transport, "The strategic case for HS2," 2013.

[2] T. Graffagnino, "The Swiss one-hour-model for improved service and reduced cost," 2013.

[3] D. Jones, D. Roos, and J. Womack, *Machine that Changed the World*. 1990.

[4] L. Peeters, "Cyclic Railway Timetable Optimization," 2003.

[5] R. M. P. Goverde and I. A. Hansen, "Performance indicators for railway timetables," *2013 IEEE Int. Conf. Intell. Rail Transp. Proc.*, pp. 301–306, Aug. 2013.

[6] J. Womack and D. Jones, *Lean thinking: banish waste and create wealth in your corporation*. Simon and Schuster, 1996.

[7] J. Womack and D. Jones, *Lean solutions: how companies and customers can create value and wealth together*. Simon and Schuster, 2006.
[8] M. F. Suarez-Barraza, T. Smith, and S. M. Dahlgaard-Park, "Lean Service : A literature analysis and classification," *Total Qual. Manag.*, vol. 23, no. 4, pp. 359–380, 2012.
[9] M. Rother and J. Shook, *Learning to see: value stream mapping to add value and eliminate muda*. Lean Enterprise Institute, 2003.
[10] UIC, "World High Speed Rolling Stock," 2013.
[11] T. Ohno, *Toyota Production System: Beyond Large-Scale Production*. Productivity press, 1988.
[12] Y. Monden, *Toyota Production System: An Integrated Approach to Just-In-Time, 4th Edition*. CRC Press, 2011.
[13] S. R. Jara-Diaz, "Allocation and valuation of travel time savings," in *Handbooks in Transport 1*, 2000, pp. 303–319.
[14] G. Lyons and J. Urry, "Travel time use in the information age," *Transp. Res. Part A Policy Pract.*, vol. 39, no. 2–3, pp. 257–276, Feb. 2005.
[15] G. Lyons, J. Jain, and D. Holley, "The use of travel time by rail passengers in Great Britain," *Transp. Res. Part A Policy Pract.*, vol. 41, no. 1, pp. 107–120, Jan. 2007.
[16] M. Berry and M. Hamilton, "Changing Urban Spaces: Mobile Phones on Trains," *Mobilities*, vol. 5, no. 1, pp. 111–129, Feb. 2010.
[17] M. Gripsrud and R. Hjorthol, "Working on the train: from 'dead time' to productive and vital time," *Transportation (Amst).*, vol. 39, no. 5, pp. 941–956, Mar. 2012.
[18] M. Givoni and D. Banister, "Speed: the less important element of the High-Speed Train," *J. Transp. Geogr.*, vol. 22, pp. 306–307, May 2012.
[19] D. Banister and M. Givoni, "High-Speed Rail in the EU27: Trends, Time, Accessibility and Principles," *Built Environ.*, vol. 39, no. 3, pp. 324–338, 2013.
[20] HS2 Ltd, "HS2 Project Specification," 2012.
[21] HS2 Ltd, "LONDON- WEST MIDLANDS ENVIRONMENTAL STATEMENT," 2013.
[22] D. Hasegawa, G. L. Nicholson, C. Roberts, and F. Schmid, "The impact of different maximum speeds on journey times, energy use, headway times and the number of trains required for Phase One of Britain's High Speed Two line," *WIT Trans. Built Environ.*, vol. 135, pp. 485–496, Jun. 2014.

# Consideration of the trade-offs between operational energy consumption and emissions and embedded energy consumption and emissions in rail infrastructure

**J A Pritchard**
Transportation Research Group, University of Southampton, UK

**ABSTRACT**

Railway infrastructure can have a big effect on energy consumption and emissions. Operationally, aspects of the infrastructure such as gradients and tunnels can impact the energy consumption of a train running along a route, whilst some components, such as bridges and tunnels can have relatively high embedded energy and emissions. This can lead to trade-offs - for example, earthworks and engineering structures to minimise the gradient (and hence operational energy consumption) may result in greater embedded energy and emissions. This paper outlines some of the findings from initial analysis of operational and embedded energy and emissions.

## 1 INTRODUCTION

In recent years, reductions in both greenhouse gas (GHG) emissions and energy consumption have become increasingly desirable. This has been partially driven by concerns about climate change, and stringent targets for GHG emissions have been set as a result. For example, the UK Climate Change Act 2008 established a legally binding target to reduce (GHG) emissions relative to 1990 by at least 80% by 2050 (HM Government, 2011). There are other good reasons to reduce energy consumption, including cost and concerns about resource depletion, but there is also a direct link between energy consumption and GHG emissions. The internal combustion engine (ICE) produces GHG emissions at the point of use, whilst the generation processes for over 70% of the electricity generated in the UK between April 2011 and March 2012 directly resulted in $CO_2$ emissions (Department of Energy & Climate Change, 2012).

Previous literature reviews, including one conducted by Arup (2009), have noted that much of the published data in the past were very much concerned with operational energy use and emissions, to the neglect of other areas. More recently, however, it has become apparent that other aspects of the life-cycle of a transport system are significant enough to warrant further consideration. For example, Chester & Horvath (2009), suggest that for the railway systems they considered, active operation accounted for less than 40% of the overall energy consumption.

Chester & Horvath's analysis shows that the energy expended in infrastructure construction and maintenance (the embedded energy of the infrastructure) is a significant component of the overall energy consumption of a railway (Chester &

Horvath, 2009, Figure 1). This paper begins with a literature review, looking at the embedded energy in railway infrastructure and notes that the design of a railway line, including the amount of bridges, tunnels and other civil engineering structures can significantly impact the estimated embedded energy.

The paper then goes on to consider the potential impacts of infrastructure design on operational energy consumption and emissions. Empirical energy consumption data have been obtained and analysed, and the results are discussed here. Theoretical data for the energy consumption of a train running in a tunnel (as opposed to open track) are also presented.

A brief discussion follows, noting the importance of choice of metric, the potential for uncertainty, and the need to consider other aspects of sustainability.

## 2 EMBEDDED ENERGY & EMISSIONS IN RAIL INFRASTRUCTURE

Embedded energy and emissions in infrastructure typically include three main sources. Firstly, there is the embedded energy and emissions of the materials, which takes in to account raw material extraction and manufacturing processes. Secondly, there is the energy expended and emissions produced during the infrastructure construction. Finally, the materials and processes associated with infrastructure maintenance should also be considered. The transportation of materials to the construction site is also an important consideration, and will in this case be considered as part of the construction and maintenance processes.

### 2.1 Choosing a metric for the embedded energy and emissions of rail infrastructure

When considering only the construction of rail infrastructure, excluding any subsequent maintenance activities, a single figure for each of the embedded energy and emissions can be calculated, representing the total energy expended and GHGs emitted from the materials and by the construction processes. Although this can be valuable, disadvantages of presenting the data in this format include the fact that it may not be easy to make meaningful comparisons with other infrastructure projects or with other aspects of the railway. This is because total figures do not take into account the lifespan of the infrastructure or the usage of the system. For the same reason, it can also be difficult to include the embedded energy and emissions of maintenance activities.

Chester & Horvath (2009) present their findings in terms of passenger-km, which enables direct comparisons to be made with operational aspects of the systems considered. In order to present the data on this basis, however, the total number of passenger-km travelled over the infrastructure during its lifetime must be estimated. In other literature reviewed, embedded carbon and energy are typically given per distance per year. This is harder to compare directly with the energy and emissions of the operational aspects of a railway, but doesn't require any knowledge of expected passenger usage. In both cases, however, the expected lifespan of the infrastructure must be known.

### 2.2 Infrastructure lifespan and the problem of accounting for existing infrastructure

The infrastructure lifespan over which embedded energy and emissions should be allocated can be determined by a number of things. For example, some components, such as ballast or running rails, wear out over time, and would likely be in need of replacement after a certain number of years. For other components, wear and tear is less of a limiting factor, but the lifespan may instead be determined by the length of time over which it would be reasonable to assume the railway remains in active use. For example, the carbon footprinting undertaken by Ademe, SNCF, & RFF (2009) assumes that the high-speed line in question has an "operating and maintenance phase" of 30 years, even though it may be fairer to assume that bridges, tunnels and buildings typically have a life span of 100 years (Baron, Martinetti, & Pepion, 2011).

The selection of a lifespan over which embedded energy and emissions can be allocated can have a big impact on the data which are ultimately presented. Baron et al. compared calculations for the construction of particular lines where the lifespan of tunnels and bridges is 100 years with calculations made for the same lines on the assumption that bridges and tunnels only have a 60 year lifespan. The relative size of the embedded energy and emissions in these components means that the effect on the life-cycle calculations for the whole project was significant in each case.

There are also questions about how to account for existing infrastructure. It could be argued that when assessing the provision of new transport services over existing infrastructure, the embedded energy and emissions of the infrastructure have already been accounted for and have no bearing on the new services. On the other hand, if existing infrastructure is to be included in calculations, it should not necessarily be assessed by modern standards; Baron et al. note that construction methods and processes have changed such that using modern standards to assess a tunnel built 20 or 30 years ago could serve to underestimate the impact of the embedded energy and carbon.

### 2.3 Embedded energy & emissions in the materials themselves

Data are available for the embedded carbon and energy in construction materials, for example in a database compiled at the University of Bath (Hammond & Jones, 2011). This means that it is theoretically possible to estimate the embedded carbon and energy in the materials of a particular infrastructure project, although detailed knowledge of the quantities and of the types of material is required. In the case of concrete, for example, the embedded energy and carbon can vary significantly, and Hammond & Jones strongly advise against simply using the "general" value they provide.

Some aspects of rail infrastructure, such as the track itself in the case of the railway, are already well documented; for example, details of the most popular track designs are given in an Arup report (Ceney, n.d.) and can – for example in the case of the Rheda 2000 Slab Track System – be supplemented by details from the manufacturer themselves (Rail.One GmbH, 2011). Other aspects of the infrastructure are currently more difficult to quantify – for example, tunnels and bridges tend to be more bespoke – although estimations are available in current literature (for example, Baron et al., 2011). It is also worth noting that much of the available data for materials are subject to various assumptions. Stimpson (2011) outlines some of the pitfalls associated with carbon footprint data and cites a recent project which concludes that the University of Bath data for embedded carbon in concrete is inaccurate. In this case, the estimates used by the University of Bath were likely to be overstating reality, which is arguably better than underestimating embedded carbon but is still not ideal.

### 2.4 Energy & emissions associated with construction processes

The construction of rail infrastructure can be difficult to quantify, especially since some rail projects rely on bespoke machinery (for example, a Tunnel Boring Machine may be designed and built for a single project). Some details of construction techniques, plant and labour hours have been documented, and could be expected to vary little from project to project; for example, some details for the different types of track are given in a report by Dunne & Ceney (2005), and some work has been done to gather data for some standard types of machinery. On the other hand, some aspects of a project will be more variable, such as the landscaping required at a particular location and the transport distances of materials to and waste from the construction site.

### 2.5 Energy & emissions associated with maintenance

Infrastructure maintenance is not always considered in detail in the literature, with Network Rail (2009) citing a lack of information. Baron et al. note that to some extent maintenance can is accounted for by considering the reduced life span of some of the key components (such as rails or ballast) compared with others – assuming that the major maintenance of these components entails replacement when they are life-expired, this can be absorbed in to the cost of construction. However, Ademe, SNCF, & RFF (2009) suggest that maintenance costs should include the transport needed to conduct track inspections (typically helicopter or car), the energy used by ballasting, tamping, grinding and weeding machinery, and the embedded energy of the various materials (including not just the components of the track which need replacement but other sundries such as weed killer).

### 2.6 Sample embedded emissions data

Baron et al. (2011) estimated the embedded carbon-dioxide ($CO_2$) emissions in high-speed rail infrastructure. A summary of their estimates for the different aspects of constructing such a railway is given in Table 1.

**Table 1 - A summary of the estimated embedded $CO_2$ emissions in a high-speed railway line (Data source: Baron et al).**

| Aspect | Estimated $CO_2$ emissions (t/km/year) | Notes |
|---|---|---|
| **Conception Phase** | 0.45 | Includes office works for planning a high speed line prior to construction. Based on data for the LGV Mediterranee line |
| **Earthworks** | 5 to 22 | Estimates based on different TGV lines |
| **Track** | 22.8 (ballasted track)<br>31.6 (slab track) | The biggest source of emissions is the steel for the rails |
| **Bridges/Viaducts** | 68 (small bridges) to 183 (large and high viaducts over valleys) | |
| **Tunnels** | 172 to 243 | |
| **Railway Equipment** | 3.5 | |
| **Stations** | 33 to 82 | |

It is clear that some components have particularly high embedded emissions compared with others. For this reason, the make-up of a railway line can have a big effect on the overall embedded emissions. For example, the high-speed line between Taipei and Kaohsiung has an extremely high proportion of tunnels and viaducts, with only 9% of the line being on "normal" track; as a result, the embedded emissions (176.5t $CO_2$ per km per year) are significantly higher than those estimated for the LGV Mediterranee high-speed line in France, given as 68t $CO_2$ per km per year (Baron et al., 2011).

Baron et al. have not provided explicit data for infrastructure maintenance. However, estimates for the maintenance of a high-speed line are given by Ademe, SNCF, & RFF (2009). They suggest that the emissions from the maintenance of 140 route-km (double track) of the European Eastern LGV line are estimated to amount to 19,900 t$CO_2$e over a 30 year operating period. This amounts to 4.7 t$CO_2$e per km per year. It is noted that this figure is in terms of carbon-dioxide equivalent ($CO_2$e) not just $CO_2$ as used by Baron et al, and that it may include replacement materials already attributed to construction in Table 1. Nonetheless, it may help to gauge the relative significance of maintenance when considering embedded energy and emissions.

### 2.7 A note about scope and boundaries

Although the data presented in Section 2.6 are potentially useful in getting a sense of scale for embedded emissions, particularly with regard to the relative contributions of the different components, it should be noted, however, that all of the calculations are based on specific projects and may not be easily generalised. Furthermore, when assessing embedded energy and carbon, it can be difficult to define the scope and system boundaries. For example, when considering some tunnels on the Channel Tunnel Rail Link (CTRL) in the UK, Workman & Soga (2004) explicitly excluded the embedded energy of the factory which produced materials for the tunnel lining. It could be argued that if the factory were built specifically for the construction of the line then it should have been included in the calculations – the key point is that a line has to be drawn somewhere, and that if different projects are to be compared, an understanding of the boundaries used is necessary to ensure consistency.

A key component of embedded energy and emissions appears to be transport, both of materials and of construction and maintenance personnel. Indeed, Ademe et al. (2009, p.15) suggest that "much of the carbon production during maintenance operations arises from personnel travelling by road." Assumptions made about transport, and distance travelled, can have a big impact on overall figures.

## 3 VARIATION OF OPERATIONAL ENERGY CONSUMPTION & EMISSIONS BETWEEN ROUTES

### 3.1 Analysis of empirical data

In recent years, some train operators in the UK have fitted meters to their electric train fleets, in order to better monitor electricity consumption. Two such operators have made data available for analysis. By matching the raw data with information about the allocation of particular trains to services and details of the UK rail network, it was possible to produce data for the net energy consumption of a range of suburban and intercity trains for a number of different routes and services. The metric chosen for energy consumption was kWh per train-km, and associated emissions were estimated from emissions data for UK electricity consumption; in 2013 DEFRA estimated that the greenhouse gas emissions per kWh of electricity consumed (all scopes, including losses associated with transmission) were in the region of 0.56kg $CO_2$e (Department for Environment Food and Rural Affairs, 2013).

### 3.2 Observed variation between different routes

Figure 1 and Figure 2 show the route profiles for two intercity routes for which energy data were analysed. The gradient profile, line speed limit, and the position of the station stops on the services considered are shown. It was found that for "Intercity Route 1" the mean energy consumption on the outbound services was 7% higher than the overall mean energy consumption for the train-type, whilst the mean energy consumption on the inbound (return) services was 7% lower than the overall mean. Similarly, for "Intercity Route 2", the mean energy consumption was

2% above the overall mean for outbound services and 2% below the overall mean for inbound services.

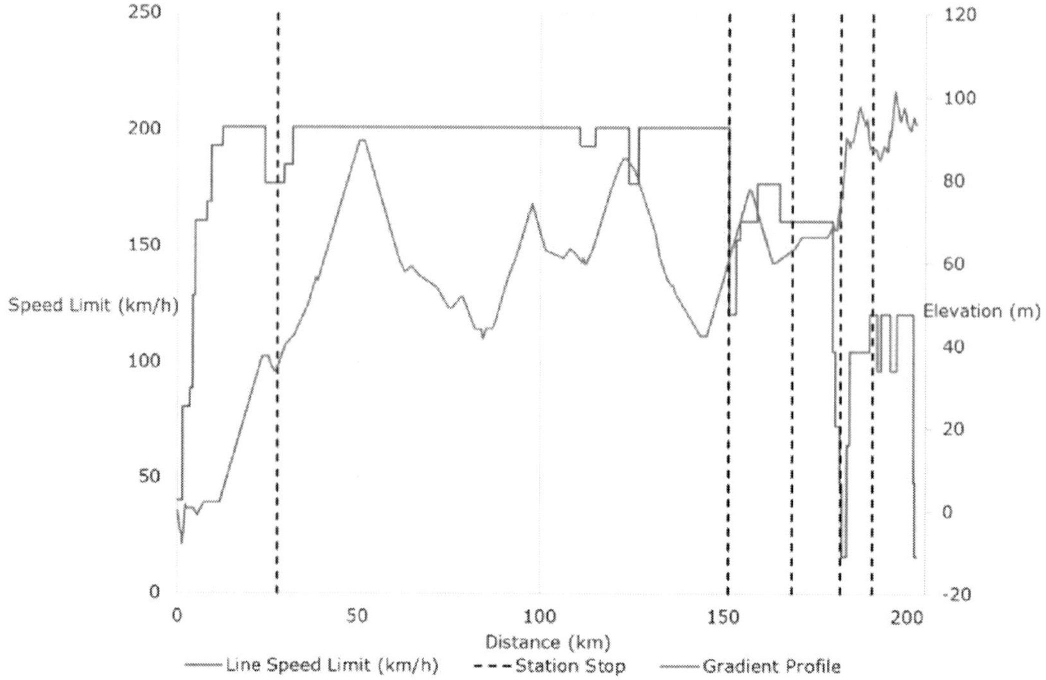


**Figure 1 - The route profile for "Intercity Route 1"**

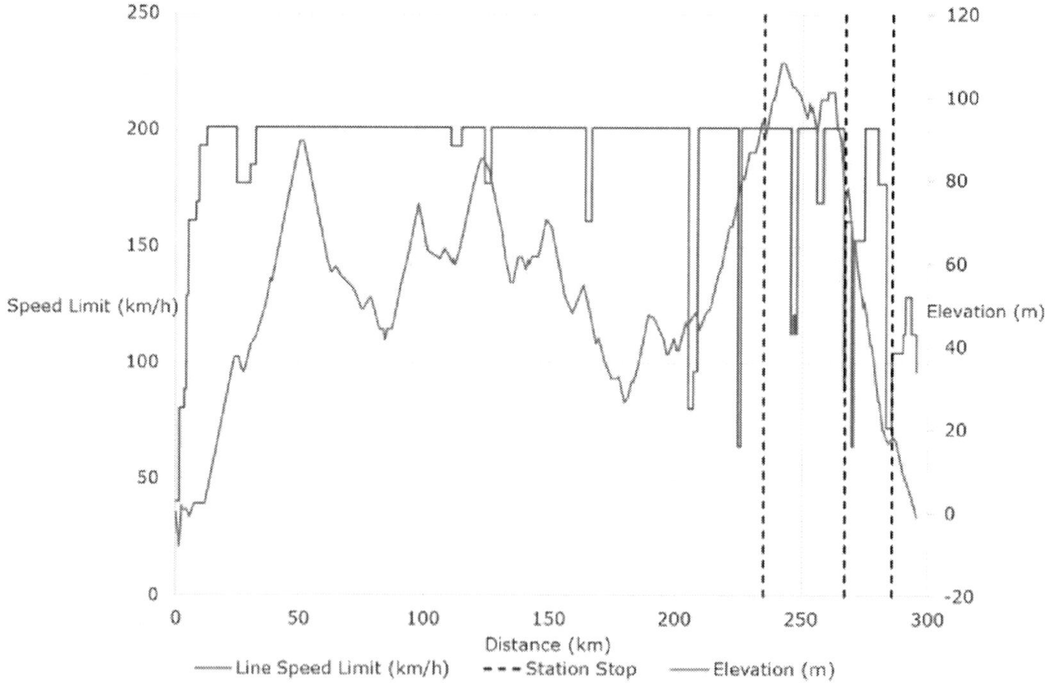


**Figure 2 - The route profile for "Intercity Route 2"**

In both cases, the final section of the return route involves a comparatively long downhill stretch, which is thought to be a key reason why the observed energy consumption is below average for the type of train, and the outbound (with an initial uphill stretch) is above average. "Intercity Route 1" is predominantly uphill (outbound) and downhill (inbound), whilst "Intercity Route 2" has less overall change in elevation. Furthermore, the stops on "Intercity Route 1" are mainly on an upward gradient on the outbound journey, whereas those on "Intercity Route 2" are sited on a mixture of uphill and downhill sections in each direction. This is thought to be why the variation from the average is greater for "Intercity Route 1" than it is for "Intercity Route 2."

Figure 3 shows graphically how the observed energy consumption was found to vary, relative to the overall mean observed for the train type, for the two intercity routes considered here, and three suburban services.

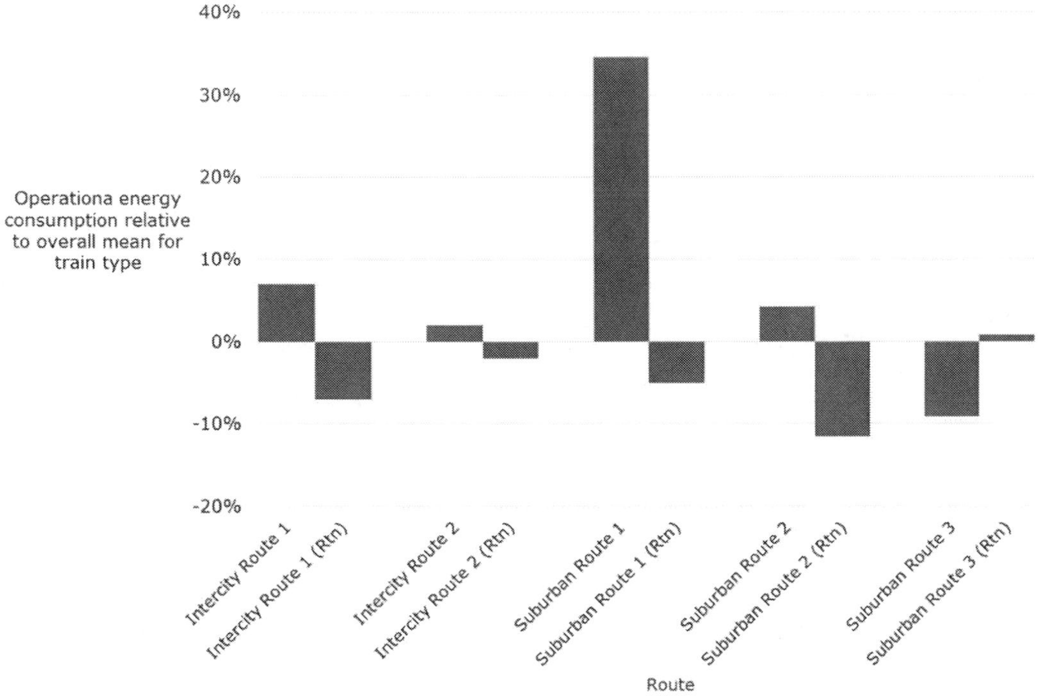


**Figure 3 - The observed variation in operational energy consumption for different routes**

The intercity services show a symmetry between the outbound and return routes, such that the overall impact for a return train journey is negligible. Even if it wasn't, some of the factors on a macro scale are hard to address – for example, the destination of Intercity Route 1 is about 100m further above sea-level than the origin. Of more interest are local variations; one reason for the greater variation on suburban services is that the routes are much shorter, and so a particular change in gradient will have a much larger overall impact. This is compounded by the fact that at on a more local scale, station positioning and speed limit profiles can be important – a station stop on a downhill stretch will limit opportunities for coasting or cruising at reduced power.

## 4 CASE STUDY: TUNNELS

It was shown in Section 2.6 that railway tunnels can have particularly high values of embedded energy and emissions relative to open track. They can also increase operational energy consumption, due to increased air-resistance. Table 2 shows the results of some simulations conducted by HS2 Ltd (2009) for a high-speed train (AGV) running at a constant 320 km/h over 10km.

**Table 2 - Simulated increase in energy consumption of a high-speed train in a tunnel relative to one on open track (Data Source: HS2 Ltd, 2009)**

| Tunnel Diameter (m) | Increase in work done by train relative to that in the open air (%) |
|---|---|
| 8.5 | 94 |
| 9.8 | 64 |
| 12 | 39 |

The dependence on the diameter of the tunnel leads to a trade-off; the quantity of materials for the tunnel lining (and the embedded energy and emissions there-in) is proportional to the surface area inside the tunnel.

Workman & Soga (2004) investigated twin-bore tunnels with diameter 7.15m and length 7km. They found that the total embedded energy was 899.1TJ, of which 65% was attributable to the materials. On this basis, Table 3 shows how an increase in material quantities might be expected to increase the overall embedded energy. Other costs associated with a larger bore would be expected to lead to an additional increase in embedded energy, particularly if the excavated material needed to be transported off site (in the case of the tunnels studied by Workman & Soga, the material was deposited locally). Some aspects, such as the manufacture of the tunnel boring machine, would not be expected to change significantly.

**Table 3 - Estimated increase in embedded energy due to increased material quantities**

| Tunnel Diameter (m) | Increase in internal surface area (%) | Estimated increase in embedded energy (%) due to increase in material quantity |
|---|---|---|
| 7.15 | - | - |
| 8.5 | 19 | 12 |
| 9.8 | 37 | 24 |
| 12 | 68 | 44 |

Using data from HS2 Ltd (2009) and Workman & Soga (2004), the variation in life-time energy consumption with tunnel diameter was estimated for three different scenarios given in Table 4. The results are shown in Figure 4.

**Table 4 - Operating scenarios considered**

| | Scenario 1 | Scenario 2 | Scenario 3 |
|---|---|---|---|
| **Operational Lifespan (years)** | 30 | 30 | 60 |
| **Operating hours per day** | 18 | 18 | 18 |
| **Trains per hour** | 4 | 10 | 4 |

It can be seen from Figure 4 that there is an optimum diameter that minimises the total embedded and operational energy consumption, which depends on the assumptions made about the lifespan of the tunnel and the intensity of operations. There are some caveats. Firstly, the change in embedded energy shown only considers the increase in material quantities (Table 3), whilst the embedded energy of the track itself, or any other necessary components (Table 1) are not included at all. Secondly, the operational energy is for a high-speed train (AGV) at 320km/h. This would be expected to vary for different trains, and for different speeds (at slower speeds, which may be more typical, the operational energy consumption would be lower). Lastly, the assumption is made that energy costs do not vary with time – operational energy costs can be discounted over time and increased use of renewable resources in the future could also be considered. However, Figure 4 serves to illustrate the potential trade-offs involved with railway tunnels.

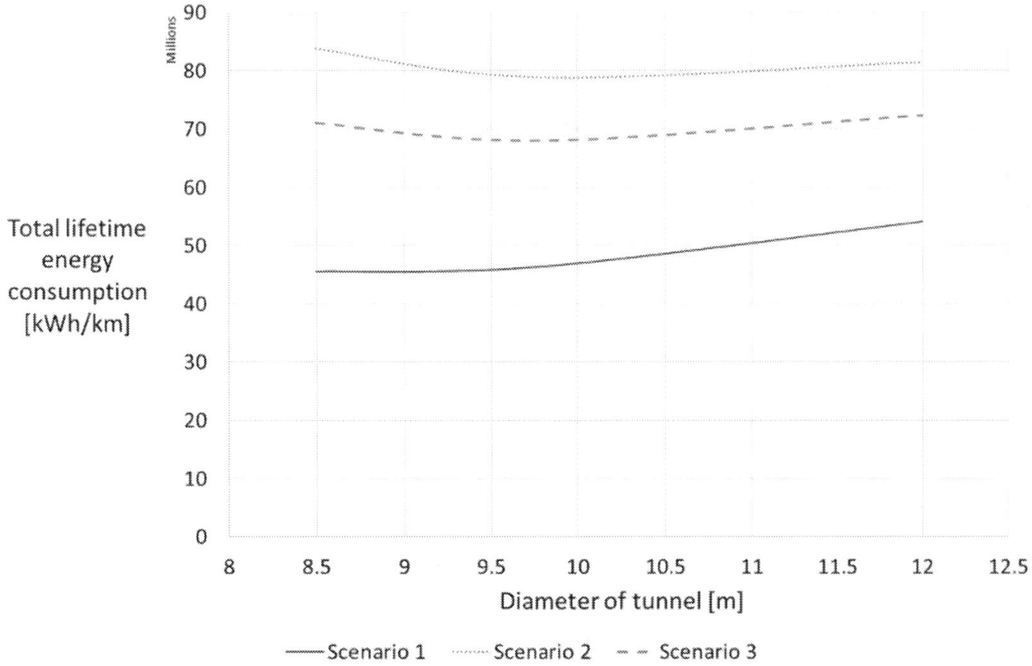


**Figure 4 - Variation in lifetime energy consumption with tunnel diameter**

## 5 DISCUSSION

### 5.1 Trade-offs between operational energy and emissions and embedded energy and emissions

It is clear from Section 2 that the construction and maintenance of railway infrastructure can be directly linked with significant energy consumption and greenhouse gas emissions. The embedded energy and emissions are also dependent on the design of the infrastructure, with some components such as tunnels and bridges being much more energy and carbon intensive than simple open track. It is also clear from Section 3 that the design of the infrastructure – including gradients and tunnels – can impact the operational energy consumption. This gives rise to the notion of trade-offs in rail infrastructure design when it comes to energy and emissions, and the notion that there may be a balance to be struck between minimising operational energy consumption and emissions and minimising embedded energy and emissions.

It may be desirable, for example, to use earthworks and civil engineering structures like viaducts to build a more level railway. Any reduction this might bring in operational energy and emissions will be offset to some degree by an increase in embedded energy emissions. Where the balance lies, however, is dependent on a number of factors, including line speed limits, station placement and intensity of usage of the railway, all of which can influence the relative proportions of operational and embedded energy and emissions over the lifetime of the system.

When it comes to emissions, these relative proportions are also dependent on how the trains themselves are powered, and on how clean the source of electricity is; if the trains are powered by electricity from renewable sources, the embedded emissions are likely to be much more significant. The construction industry is heavily reliant on diesel machinery, and some industrial processes involve the release of greenhouse gases. The embedded emissions from earthworks given in Table 1 are particularly variable, because they are very much dependent on the transport of materials and whether or not the soil needs drying out using quicklime.

It has been seen that railway tunnels increase both the embedded and the operational energy and emissions, and that the tunnel diameter is an important factor when considering the balance between the two. Where that balance lies will depend on the intensity of usage of the line.

For this reason, one of the most important considerations when ascertaining where the balance lies is the choice of metric (Section 2.1). The same metric needs to be used for both embedded and operational aspects so that direct comparisons can be made. The definition of system boundaries (Section 2.7) also needs to be clear, and the resulting potential for uncertainty and variation depending on what is included and what is not included acknowledged. It is suggested that pursuit of a single figure for each of the embedded and operational energy consumption (or emissions) risks relying too heavily on assumptions which are impossible to accurately quantify, and that future analysis should instead be undertaken in terms of estimated ranges.

### 5.2 Other aspects of sustainability

It is important not to lose sight of other aspects of sustainability when considering infrastructure design. Cuttings, for example, may be dug out not just to provide a good alignment for the operation of the trains, but to screen the railway from its surroundings in order to minimise noise and visual intrusion. Similarly, although the location of station stops was briefly referred to in Section 3.2, the placement of stations and the level of train service provided have important social and economic consequences which should not be ignored. Tunnels are another example of railway infrastructure which may contribute to overall sustainability goals, in spite of a negative impact on energy consumption and emissions. As well as reducing the noise and visual impacts of the railway, they may be the only way of providing a rail link without other serious economic and social consequences – for example, an alternative railway could require the destruction of farmland, the partial demolition of an urban area and may serve to sever communities.

## 6 CONCLUSIONS

It has been shown that infrastructure needs to be included when considering the energy and greenhouse gas emissions of a railway system, both because the embedded energy and emissions can be significant and because infrastructure design can impact operations. It may be necessary to find a balance between a design which minimises the embedded energy and emissions, and one which reduces the operational impact. At a macro level, some penalties, such as overcoming an overall change in elevation between two places, may be unavoidable, but it is thought that at a more local level, use of earthworks and civil engineering structures should be properly considered.

Various factors will influence the balance between embedded energy and emissions and operations, including the intensity of usage, the assumed lifespan of the project, the source of electricity and the distance over which construction materials need to be transported. It is important to use a metric which allows embedded and operational aspects to be compared on the same basis, and there needs to be clarity about the underlying assumptions and boundaries. Work is ongoing to develop tools to estimate the interaction between embedded and operational energy and carbon, and a level of uncertainty is unavoidable. This work needs to be seen in light of other aspects of sustainability.

## 7 REFERENCES


Ademe, SNCF, & RFF. (2009). *1st Bilan Carbone - global railway carbon footprint. The Rhine-Rhone LGV serving Sustainable Europe*.

Arup. (2009). *Initial Report Literature Search and Knowledge Review*.

Baron, T., Martinetti, G., & Pepion, D. (2011). Carbon footprint of high speed rail. Retrieved from http://trid.trb.org/view.aspx?id=1134978

Ceney, H. (n.d.). *Selection of track form*. Solihull.

Chester, M., & Horvath, A. (2009). Environmental assessment of passenger transportation should include infrastructure and supply chains. *Environmental Research Letters*, *4*(2), 024008. doi:10.1088/1748-9326/4/2/024008

Department for Environment Food and Rural Affairs. (2013). 2013 GHG Conversion Factors (Complete Offline DataSet). Retrieved from http://www.ukconversion factorscarbonsmart.co.uk/

Department of Energy & Climate Change. (2012). *Fuel Mix Disclosure Table. Energy Statistics*. Retrieved October 03, 2012, from http://www.decc.gov.uk/en/content/cms/statistics/energy_stats/fuel_mix/fuel_mix.aspx

Dunne, J., & Ceney, H. (2005). *Britpave Ballast and Slab Track Cost Comparison Report*. Solihull.

Hammond, G., & Jones, C. (2011). *Inventory of Energy and Carbon (ICE)*. Retrieved from http://www.bath.ac.uk/mech-eng/sert/embodied/

HM Government. (2011). *The Carbon Plan: Delivering our low carbon future*. Retrieved from https://www.gov.uk/government/uploads/system/uploads/attachment_data/file/47613/3702-the-carbon-plan-delivering-our-low-carbon-future.pdf

HS2 Ltd. (2009). *HS2 Traction Energy Modelling*. Retrieved from http://hs2.xplodelite.com/publications/HS2-Traction-Energy-Modelling-49802

Network Rail. (2009). *Comparing environmental impact of conventional and high speed rail*. Retrieved from http://www.networkrail.co.uk/documents/About%20us/New%20Lines%20Programme/5878_Comparing%20environmental%20impact%20of%20conventional%20and%20high%20speed%20rail.pdf

Rail.One GmbH. (2011). Rheda 2000 Ballastless Track System. Retrieved from http://www.railone.com/fileadmin/dateien/03_Broschueren/EN/Rheda2000_EN_2011_ebook.pdf

Stimpson, J. (2011). Carbon footprints: lies, damned lies, and statistics? *New Civil Engineer*. Retrieved from http://www.nce.co.uk/opinion/jo-stimpson/carbon-footprints-lies-damned-lies-and-statistics/8618424.blog

Workman, R., & Soga, K. (2004). *The Embodied Energy of Tunnel Construction*. Cambridge.

# How accident investigation can drive research

**W Rasaiah**
Rail Accident Investigation Branch, UK

## 1. INTRODUCTION

Throughout railway history, lessons learned from accident investigation have driven many aspects of railway design and technology, often through research. It is therefore useful to review how safety lessons from accident investigations, which may themselves involve research, are currently disseminated to the UK rail industry and academia and ask if there is a better way.

This presentation is based on the experience of the Rail Accident Investigation Branch (RAIB) which began is operations in October 2005. The views expressed here are therefore also based on UK experience and do not necessarily reflect the situation in other countries.

## 2. ACCIDENT INVESTIGATION TO RESEARCH – IS THERE A BETTER WAY?

Generally, railway research is undertaken to achieve improvements in safety, performance or efficiency. The role of an independent accident investigation body is to identify safety lessons and make recommendations by means of its public investigation reports[1], to prevent or minimise recurrence or minimise consequence. Recommendations are addressed to duty holders through the national safety authority, which is responsible for ensuring that each recommendation is duly considered in the context of applicable legal health and safety requirements.

The majority of safety recommendations made by the RAIB involve improvements to standards, procedures, training, or design, and cover railway operations, infrastructure and rolling stock. They do not usually involve research but in some notable cases, which are covered in the case studies, the causal chain of the accident and/or the potential risk associated with a recurrence may require that some form of research is recommended.

Generally, there are four potential routes from accident investigation to research:

a) Research during the course of an investigation, led by the accident investigation body and supported by industry. The RAIB will take the lead in carrying out the required research to fulfil its duty to establish the causal chain of an accident to a sufficient degree that it can make evidence based recommendations. By its nature, the scope of such research will be limited to meet the requirements of the investigation (case study b).

---

[1] RAIB investigation reports can be found at:
www.raib.gov.uk/publications/investigation_reports.cfm

b) Research undertaken by the industry as part of its implementation of a safety recommendation from the RAIB (or potentially from an industry formal investigation) to carry out specific research. In such cases the RAIB will have made the recommendation because it considers it is necessary to bring about a required safety improvement. The research could be undertaken in house by the duty holder or by a consultancy or academia; depending on resources and skills. This is the most common route from accident investigation to research (case study a).

c) Research areas linked to safety lessons or recommendations from the RAIB which have been considered by industry but not implemented on the grounds that it considers the costs are likely to be disproportionately greater than the benefits. As a consequence of the non-implementation, significant areas of understanding and knowledge may not be developed because those ideas are not sufficiently aired for others (e.g academia) to consider (case studies c and d).

d) Research which the circumstances of a single accident may not justify, and therefore the RAIB may decide it is not appropriate to make a formal recommendation for the research. In such cases the RAIB may highlight the potential for safety learning, in its report in a separate section from the recommendations.

Currently, there are well proven mechanisms for routes (a) and (b). However, there remains a need for developing routes (c) and (d) so that academia is kept informed of potential research ideas, which it may consider fit better with its own goals than those of industry. For example, a research idea may be suitable for undergraduate or post graduate projects to take forward. If early research finds that some of these projects have significant merit, it could spawn more significant programmes of research and innovation thereafter, which could filter back into industry. The key to developing routes (c) and (d) is good communication of research ideas between the accident investigation body, industry and academia and the provision of technical assistance where it is appropriate to do so.

## 3. THINKING WIDER THAN THE ACCIDENT

A pre-requisite for developing routes (c) and (d) is the willingness of the rail industry and researcher to think wider than the actual accident that has been investigated and which is the source of a particular research recommendation or safety learning. Historically, the industry would not have achieved current levels of active and passive safety for example, if the safety authority and industry had not thought wider than the watershed accidents which drove forward those key changes. Two examples of this are:

- Train protection and warning system (TPWS) and other SPAD reduction measures were introduced in the UK following Lord Cullen's Inquiry into the Ladbroke Grove rail accident of 5 October 1999, in which 31 people died and hundreds suffered injury. Since the introduction of TPWS in 2002-03, there have not been any significant collisions and no loss of life as a result of a train collision. By comparison, during the 17 year period 1981-97, there were 709 collisions comprising end-on, side-on and buffer stop collisions[1] resulting in over 100 weighted equivalent fatalities[2].

[2] 1 equivalent fatality = 10 serious = 200 minor injuries

- The introduction of passive safety measures in UK rolling stock following the Clapham accident of 12 December 1988, in which 35 people died, 69 and 415 people suffered serious and minor injuries respectively [1]. Safety improvements to rolling stock structures and interiors were introduced following a significant programme of crashworthiness research and development. Although those measures were designed principally to mitigate the consequences of end-on collisions, they have shown significant benefits in a high speed derailment (case study d).

In both cases a significant level of 'safety vision' beyond the circumstances of the actual accidents themselves, was required to provide the necessary impetus for change. Those changes were arguably greater than could be readily justified by standard industry cost-benefit analyses.

## 4. CASE STUDIES

The presentation will cover four examples of research ideas which have resulted from accident investigations by the RAIB; these are summarised briefly below.

(a) *Understanding derailment margins and the effect of offset loads*: following the derailment of an intermodal freight train on 15 October 2013 near the former Primrose Hill station near London [2], the RAIB made a recommendation for the rail industry to work together to assess the risk of derailment from asymmetric loading of wagons and to identify and adopt reasonably practicable control measures to mitigate that risk.

(b) *Understanding the effect of inadequate fits on critical bearings of final drive gearboxes:* during the investigation of a high speed passenger train derailment on 20 February 2010 at East Langton, which resulted from an axle failure [3], the RAIB and industry carried out tests to understand how one of the final drive output bearings had failed. The investigation found a general lack of understanding about the effect of loose fits on bearing behaviour which led to the tests and a better understanding of the criticality of getting the interference right in the first place.

(c) *Lateral retention of bogies in derailments:* following the derailment of a passenger train on 1 February 2008 after it collided with debris from a footbridge that had been knocked down onto the track at Barrow upon Soar [4], the RAIB made a recommendation for research into the practicability of design elements on the bogie that could limit the degree of deviation from the track following derailments.

(d) *Effect of multi-axial acceleration pulses on passengers in derailments:* On 23 February 2007, a high speed passenger train became derailed at a set of faulty facing points near Grayrigg in Cumbria, while travelling at 95 mph (153 km/h) [5]. All nine vehicles derailed and five vehicles rolled over onto their sides. The leading vehicle became detached from the rest of the train and rotated through 190 degrees. Among the four crew and 105 passengers, one passenger was fatally injured; 28 passengers, the train driver and one other crew member received serious injuries and 58 passengers received minor injuries. The RAIB made five recommendations relating to rolling stock; one of these related to research to study the effect of multi-axial pulses on passengers and interior safety.

## 5. REFERENCES


1. W G Rasaiah, 'Watershed UK accidents – lessons learned, Proc Instn Mech Engrs Vol 216 Part F: J Rail and Rapid Transit, April 2002.
2. 'Derailment at Primrose Hill / Camden Road West Junction, 15 October 2013', RAIB report 21/2014, October 2014.
3. 'Passenger train derailment near East Langton, Leicestershire', 20 February 2010, RAIB report 01/2012, January 2012.
4. 'Collision of a train with a demolished footbridge, Barrow upon Soar, 1 February 2008', RAIB report 18/2008, September 2008.
5. 'Derailment at Grayrigg, 23 February 2007', RAIB report 20/2008, October 2008.

# ON-TIME: Improving customer satisfaction through increased capacity and decreased delays

**M Dasigi**
Control Command and Signalling, STE, Network Rail, UK

## ABSTRACT

ON-TIME, a EU funded project led by Network Rail has developed new methods and processes to increase customer satisfaction by increasing the available capacity on the European railway network and decreasing overall delays. In the project, specific emphasis was placed on approaches for alleviating congestion at bottlenecks. Case studies considered included passenger and freight services along European corridors and on long distance main-line networks and urban commuter railways.

A key emphasis throughout this project was that the innovations developed should be implementable to solve real-life problems. This case study presents the six innovations of the project delivered by a multi-disciplinary team of 19 partners(1) consisting of Railway Infrastructure Managers, Industry/Supply Chain Companies, SMEs and Academia working together for three years.

## 1 INTRODUCTION

The overall aim of the EU funded ON-TIME project was to improve railway customer satisfaction through increased capacity and decreased delays for passenger and freight trains. The ON-TIME was delivered by a multi-disciplinary team of 19 partners consisting of Railway Infrastructure Managers, Industry / Supply Chain Companies, SMEs and Academia led by Network Rail.

The ON-TIME project has developed new methods and processes to help maximise the available capacity on the European railway network and to decrease overall delays in order to both increase customer satisfaction and ensure that the railway network can continue to provide a dependable, resilient and green alternative to other modes of transport. In the project, specific emphasis is being placed on approaches for alleviating congestion at bottlenecks. Case studies considered included passenger and freight services along European corridors and on long distance main-line networks and urban commuter railways. Current best practices have been identified and developed by examining national projects previously carried out by railway undertakings.

The focus of this case study is the innovations and test cases that were used in proving the concepts. The example of the East Coast Mainline in 2018 is discussed. The ON-TIME website can be found at http://www.ontime-project.eu and contains the public deliverable of the project.

## 2 OBJECTIVES

The objectives of the ON-TIME project were:

1. To improve management of the flow of traffic through bottlenecks to minimise track occupancy times.
2. To reduce overall delays through improved planning techniques and improve service dependability through improved traffic management techniques.
3. To improve the traffic flow by providing effective, real-time information to traffic controllers and drivers, thus enhancing system performance.
4. To provide customers of passenger and freight services with reliable and accurate information, particularly in the event of disruptions.
5. To improve and move towards the standardisation of the information provided to drivers to allow improved real-time train management on international corridors and system interoperability.
6. To provide a means of updating and notifying actors of changes to the timetable to enable them to use the information effectively.

The key outputs of the project are six innovations in the area of railway planning and operations management. These are:

Innovation 1: The development of standardised definitions and methods.

Innovation 2: The development of improved methods for timetable construction.

Innovation 3: The development of algorithms to either automatically provide control or provide decision support to controllers.

Innovation 4: The development of methods, processes and algorithms that provide decision support when events occur that require changes to the disposition of assets and resources, potentially across multiple networks, undertakings, operators and/or countries.

Innovation 5: The development of standardised, interoperable approaches for the communication and presentation of information to drivers and controllers.

Innovation 6: The development of an information architecture to support the communication of standardised and contextualised train control data.

The take up of the ON-TIME innovations would result in a step-change in railway capacity by reducing delays and improving traffic fluidity. These improvements will be delivered through innovations in principles, processes and tools in the following areas:

- Integrated Traffic Management, Optimised Timetables with Decision Support Tools and Automation: The ON-TIME project has developed improved methods of timetable construction that can be used for long term planning, for real time operating, and for replanning during disruptions (Innovations 2, 3 &4). This is done using the same processes and algorithms but applying different constraints (e.g. cost functions, operational constraints).
- Centrally Guided Train Operation (CGTO), Advanced DAS: CGTO uses closed-loop control to communicate control centre decisions, which consider the operational situation, to the train and is the next step in driver advisory systems.
- Extended RailML standards, CAD based signalling plans electronically imported into railway operations simulator and Open source data architecture: The extended RailML standards developed by the ON-TIME project have been proposed to the UIC. The event-based architecture has been designed with open interfaces for use with the HERMES railway operations simulator and the traffic control systems from Ansaldo STS.

The ON-TIME innovations have been evaluated and demonstrated using a variety of complex problems present across the European rail network:

- Sweden: Iron Ore Line
- United Kingdom: East Coast Main Line
- The Netherlands: corridors through s'Hertogenbosch
- Italy: Bologna Node

## 3 RESEARCH OUTLINES AND OUTCOMES

This section presents the innovations of the project individually. With each of these, the state-of-the-art as it was at the start of the project and the developments within the project are presented.

### 3.1 Innovation 1: Standardised definitions and methods to create interoperable processes

#### *3.1.1 State of the art*

Before the start of ON-TIME project, there was a lack of shared knowledge and documentation about underlying railway operations management processes for strategic planning, tactical planning, operational traffic control and train driving. There was also a lack of methods to connect the timetable planning and operational traffic control processes.

The process for timetable planning and operation is harmonised in Europe by Rail Net Europe, European legislation, Network statements and European corridors for freight traffic. The process for timetable planning is divided in one year timetable and ad hoc adjustments of the timetable. The ad hoc timetable process meets the operational process. There is a need to close the loop and better connect timetable planning and operational methods. There is also a need for better evaluation and feedback.

To estimate available capacity, number of possible train paths and robustness of the system, the IMs have common methods. The UIC 406 method outlines how track occupancy can be calculated. For traffic simulations there are commercial tools used by the IMs.to help the planners to provide quality timetables. There is also a need to simulate the operational process. In ON-TIME a simulator system, HERMES, has been used and a demonstrator system has been developed by Ansaldo.

There is a growing need is to have common procedures for IMs, RUs and Industry to share and exchange data. In ON-TIME, data has been transferred from IMs traffic simulation systems to the simulator HERMES and the demonstrator developed by Ansaldo and NTT Data. RailML is a standard for transferring data.

#### *3.1.2 Research aims and objectives*

The main objective is to give a process framework for timetable planning and operation, to demonstrate how innovations can be implemented into practice and how we can measure the benefits of innovations. The improved methods in timetable planning and decision support in traffic control and train driving will:

- decrease track occupancy in bottlenecks
- improve punctuality
- improve energy efficiency

A second objective is to outline a framework for future research in timetable planning, methods for handling minor perturbations, major disruptions and train driver advisory systems.

### *3.1.3 Key Research outputs*

- Defined measure called Quality of Service which includes factors such as punctuality and delays and can be used to evaluate various approaches to regulation and optimisation algorithms.
- State-of-the-art studies and review of present technology readiness level (TRL) for traffic planning and timetabling (WP3) level 3, traffic control under minor perturbations (WP4) level 3, Operational management in the event of large disruptions (WP5) level 3, Driving Advisor system (WP6) level5.
- Specification of locations and scenarios for ON-TIME simulator system and demonstrator system. Sweden – Norway: Iron Ore Line Kiruna – Narvik, single track line and border crossing, United Kingdom: East Coast Main Line, double and multiple track line, Italy: Bologna station, node with many merging lines, Netherlands: Corridors through s' Hertengenboersch, network, complex node.
- Simulations were undertaken for evaluating the developed systems for solving perturbations and disruptions in the different scenarios.
- Specification of a strategy for putting methods into practice, including guidelines for a deployment process.

## 3.2 Innovation 2: Improved methods for timetable construction

### *3.2.1 State of the art*

The scientific literature on railway timetabling mainly considers macroscopic optimisation models without concern for how to get accurate input parameters to set up the macroscopic model. On the other hand, the railway operations literature considers microscopic methods for calculating running times and blocking times given any infrastructure and signalling configuration, as well as microscopic methods for conflict detection and computing capacity consumption using timetable compression. The timetabling practice shows a similar separation, with either macroscopic models to compute network timetables using normative input, or microscopic blocking-time based tools for detailed planning on corridors and at stations, but without support for network optimisation. Timetable evaluation on feasibility, stability or robustness is typically applied - if at all - after the timetable construction using simulation tools with unclear procedures as to how the results are used to improve the timetable design.

### *3.2.2 Research aims and objectives*

The key objectives of the ON-TIME project in this area of robust and resilient timetables were to develop methods to reduce overall delays through the use of improved planning techniques that are able to provide better timetables capable of coping with normal statistical variations in operations, as well as minor perturbations. The research has specifically focused on the development of robust and resilient timetables, i.e. timetables that are able to cope with variations that occur every day and are designed to be easily recoverable in the event of incidents or disturbances.

### *3.2.3 Research outputs*

The key outputs related to Innovation 2 have been:

The development of common railway timetabling and capacity estimation methods appropriate for use by all EU member states that reflect customers' satisfaction and enable interoperability, more efficient use of capacity, higher punctuality and less energy consumption;

Further development of methods for robust cross-border timetables and integration of timetables between different regional and national networks improving interoperability and efficient corridor management including standardised approaches for exchanging timetable information between stakeholders; improved timetable quality, stability, robustness, reliability and effectiveness; validated development methods, through benchmarking, using a number of standard, real-world case studies.

ON-TIME developed a classification of Timetabling Design Levels depending on the explicit incorporation of performance measures in the timetable design process with increasing performance with respect to dealing with delays and disturbances.

The Timetabling Design Levels (TDL) are as follows:
- TDL 0 (Low quality). No conflict detection nor stability analysis is incorporated in the design process leading to poor timetables unless traffic is very low;
- TDL 1 (Stable). Stability analysis is an integrated part in the timetabling process so that the time allowances are proven to have good delay absorption behaviour for minor perturbations;
- TDL 2 (Feasible). Both conflict detection and stability analysis are incorporated in the design process resulting in proven conflict free and stable timetables so that trains will run undisturbed in normal traffic;
- TDL 3 (Robust). Robustness analysis is incorporated in the design process on top of conflict detection and stability analysis, resulting in proved robust timetables that can cope with normal statistical variations in operations;
- TDL 4 (Resilient). Rescheduling measures are taken into account in the timetabling design process resulting in flexible timetables that enable efficient rescheduling in a late (ad-hoc) stage of the timetabling process or in real-time traffic management to deal with larger perturbations.

An innovative three-level timetabling design approach has been developed as a prototype path to the highest Timetabling Design Levels 3 or 4. This performance-based timetabling process approach takes timetabling KPIs explicitly into account, including feasibility, infrastructure occupation, stability, robustness, resilience, travel times and energy efficiency. The approach contains three levels:
- Microscopic for highly detailed local computations;
- Macroscopic for aggregated network optimisation; and
- Fine-tuning for corridors.

Feasibility and stability are guaranteed by microscopic blocking time models, which feed a macroscopic model that optimises the timetable at the network level incorporating a Monte Carlo stochastic simulation model for robustness evaluation. These two levels iteratively compute a robust conflict-free and stable timetable. The fine tuning level computes energy-efficient speed profiles and optimises the timetable of the local trains on the corridors between main stations using stochastic optimisation with respect to dwell time distributions and energy consumption. This represents a sustainability dimension on top of the performance with respect to delays and disruptions.

This timetabling framework can be used to find optimal cyclic and non-cyclic timetables. Resilience is taken into account with respect to scheduling ad-hoc freight paths. The timetabling algorithms allow inserting additional (freight) train paths whilst sufficient residual capacity must be reserved to guarantee that a stable conflict-free timetable can be found. The freight paths are specified in a multilayer freight path catalogue with various maximum speed paths (e.g. 80/100/120 km/h) on specified corridors. Depending on the maximum speed of a requested freight path, the passenger timetable might have to be adjusted slightly to maintain a conflict-free timetable. This procedure allows a multilayer timetable with a basic passenger timetable and additional freight paths of different speeds to be selected from a catalogue on a first-come-first served basis.

The key academic work that has been undertaken is:

- Development of micro-macro network transformations;
- Microscopic conflict detection and capacity consumption;
- Macroscopic network timetable optimization including stochastic robustness evaluation;
- Computation of energy-efficient speed profiles;
- Stochastic optimization of optimal energy-efficient timetables on corridors between main nodes.

Moreover, the various timetabling processes have been implemented and integrated into an overall architecture, including microscopic running time calculations and conflict detection, macroscopic timetable optimisation, energy-efficient speed profile computations and timetable performance evaluation. The algorithms work with standard RailML input and deliver a RailML timetable at microscopic level, including energy-efficient speed profiles.

## 3.3 Innovation 3: Real-time traffic control algorithms

### *3.3.1 State of the art*

For many years, algorithms for real-time conflict detection and resolution have been described in the scientific literature. Only recently have these algorithms been able to solve problems of practical relevance in real-time. They are, however, still not applied in daily operation in large railway networks for two main reasons:

1. The benefit of the algorithms is difficult to predict.
2. Operational traffic control systems (TCSs) currently in operation by nationally-acting railway infrastructure managers are not easy to extend.

In order to change this, the project defined and tested a flexible system design for railway traffic management based on extensible interfaces. This will lead to a situation where hardware equipment installed on the track-side remains usable for long periods of time, but software optimisation components and hardware used for non-safety-critical calculations, such as traffic management, can easily be exchanged and extended depending on the current and future state of the art.

### *3.3.2 Research aims and objectives*

The objective of this part of the project was to develop a framework for a modular traffic management system (Perturbation Management Module), where independent modules collaborate.

Therefore, methods and tools for real-time traffic state monitoring and prediction, conflict detection and conflict resolution, including train speed optimization, had to be developed or expanded to fulfil the requirements of this modular architecture.

### *3.3.3 Research outputs*

The Perturbation Management Module (PMM) can be divided in four main sub-modules that allow traffic to be effectively managed in real-time when perturbations are observed in the network.

The first sub-module is the so-called Traffic State Monitoring (TSM) module, which is responsible for monitoring current traffic conditions by collecting, via track-side and train-side sensors, all the information relating to both the traffic and the infrastructure.

The second sub-module is called Conflict Detection and Resolution (CDR) and represents the most important part of the PMM. It is triggered cyclically in normal operating conditions and firstly involves a call to the Traffic State Prediction function, which forecasts the state evolution of traffic (positions, speeds of trains) within a certain time period ahead called the "prediction horizon". If conflicts exist, the Track Conflict Resolution function is executed, which computes a new Real-Time Traffic Plan (RTTP). This RTTP is used to derive route setting commands (Automatic Execution of the Real-Time Traffic Plan).

The third sub-module is the Train Path Envelope Computation (TPEC) that aims to identify the time allowances available in real operation that can be exploited by a train to adopt an energy-saving driving strategy without running late with respect to the timetable.

A fourth sub-module is the Human-Machine Interface (HMI), which is focused on giving real-time information to the operators (dispatchers, traffic controllers) by a screen visualization of the current traffic state, e.g. through a schematic infrastructure view, as well as the predicted traffic state from the RTTP, e.g. through the so-called train graph (time-distance diagram).

### *3.3.4 Implementation tasks*

The separation of tasks into different modules has proven to be feasible. The obtained solution has been designed to work on general cases. However, it turned out during experiments that the update processes, in particular of the real-time traffic plan as core object of real-time traffic management, need to be defined more specifically in order to become more robust. To allow better predictions of traffic and more information about train behaviour, interlocking and ATP rules need to be available; this would require further extensions in the data modelling. For continuing development, simulation engines need to be developed with an open architecture similar to the one proposed in the project, in order to enable testing of smaller modules. The most critical success factor of the algorithms in WP4 is a correct supply of static and dynamic data.

For a real-world implementation, comprehensive processes for data management first need to be established by the railway infrastructure managers (e.g. concerning data changes, construction work, timetable changes). These update processes must then be considered for real-time traffic management.

### *3.3.5 Case study*

The East Coast Main Line (ECML) in the UK is a complex and busy corridor linking London with the North East . There is mixed traffic, consisting of commuter routes, two long distance intercity routes and key freight paths. The ECML has been the test location for GB Rail investigations into the application of Traffic Management (TM) and ERTMS. This route has been used as one of four test cases for the management of small perturbations.

One algorithm applied for real time traffic management on the ECML is the Differential Evolution Junction Rescheduling Model (DEJRM) developed by The University of Birmingham. It optimally reschedules train timings at junctions in the event of minor perturbations.

The capability of the DEJRM is demonstrated by its application to ECML delay scenario 4, described below.

- Time period: From 7am to 10am on a weekday
- Description: Signal SIG2764 fails at 7:16:00 until 7:23:00, causing a passenger train to be delayed, which results in a conflict with a freight train at approximately 7:25:00am at the junction to the ECML. A number of trains are also affected at Finsbury Park station.

## 3.4 Innovation 4: Improved decision support – Handling major perturbations

### *3.4.1 State of the art*

Recovering from a disrupted situation to a feasible state in the network requires railway operators to perform changes in the timetable such as cancelling, rerouting or re-timing trains, changing the order of departure at stations, maintaining or dropping connections between trains, and also to perform reallocation of rolling stock and changes in crew schedules.

Various forms of objective functions are considered, which focus mainly on minimizing customer dissatisfaction by minimizing deviations from the intended timetable or minimizing expected delays.

Further objectives include minimizing deviations from the original rolling stock allocation plan, as well as costs related to the rescheduling and cancellation of crew tasks. This recovery problem is very complex and needs to be solved in real-time; it is therefore often heuristically solved manually by the railway operators or by using fast combinatorial optimization algorithms. Furthermore, the problem is usually split up into three main phases that may be defined as timetable rescheduling, rolling stock rescheduling and crew rescheduling.

The timetable rescheduling problem is solved with a list of emergency scenarios. However, there is no emergency scenario available when several disruptions occur at the same time. A combination of contingency plans has to be used in such cases. This is often done in a non-automatic way, by using the experience of the practitioners, especially when large disruptions occur, such as the unpredictable unavailability of some tracks or train failure or line fault leading to a complete closure of the line.

Most current solutions deal with a single rescheduling phase. There are just a few approaches that integrate two phases, namely either timetable and rolling stock rescheduling, or timetable and crew rescheduling. One research goal of ON-TIME was to work on further integration of the three main rescheduling phases (timetable, rolling stock and crew rescheduling).

### *3.4.2 Research aims and objectives*

The main objective of Innovation 4 was to focus on traffic changes and resource management strategies to deal with large scale disruptions . These objectives were:

1. To design and validate effective intelligent decision support strategies and tools for the optimal human supervisory control of the recovery processes in case of a large disruption;
2. To evaluate the state-of-the-art in optimisation algorithms strategies and stakeholder processes and information flow for managing large scale disruptions;
3. To specify the integration of the real-time traffic and asset management procedures, optimisation models and tools;
4. To develop algorithms for resource management in the case of a large disruption;
5. To design and validate effective intelligent decision support strategies and tools for the optimal human supervisory control of the recovery processes in case of a large disruption.

### 3.4.3 Research outputs

The key outputs related to Innovation 4 have been:

1. State-of-the-art of recovery algorithms for real-time railway optimization.
2. Questionnaire on best practices for modes of human-automation co-operation and monitoring state of resources put in use in case of railway disruptions, from the point of view of the infrastructure managers involved in ON-TIME.
3. A human factors study with structured interview method to determine a set of representative incidents and to analyze the processes associated with the incidents, and the decisions and strategies required to mitigate their impact.
4. Elicitation of a generic workflow of the recovery process with SysML activity diagrams.
5. Design of a tool with a framework that consists of a closed loop in which each rescheduling phase (i.e. timetable, rolling stock and crew rescheduling) is solved by an efficient algorithm to find a good feasible solution and gets feedback from the other phases in order to obtain a good feasible solution for the whole system.
6. Benchmarking of the tool developed within a distributed architecture in the case of a disruption scenario in the Dutch network.

### 3.4.4 Case study

The case study area is an important part of the Dutch railway network. The network is bounded by Utrecht Centraal in the northwest, Tilburg in the southwest, Eindhoven in the southeast and Arnhem in the east. On this network, about 90% of the trains are domestic passenger trains operated by Nederlandse Spoorwegen (NS). The case study only considered the timetable, rolling stock and crew schedule of these trains.

Disruption scenarios considered were:

1. Accident with a person near Rosmalen in the morning peak. As a consequence no train traffic is possible for 3 hours between Den Bosch and Oss. Passengers from s'Hertogenbosch to Nijmegen can use the bus or travel via Utrecht and Arnhem.
2. Signalling problems near Culemborg. As a consequence no train traffic is possible for 2 hours between Utrecht and Geldermalsen. Passengers and/or trains moving from north to south (Utrecht–s'Hertogenbosch–Eindhoven) can travel through Arnhem.

Since these disruptions have an impact on the whole Dutch railway network, the algorithms for timetabling (at a macro level), rolling stock and crew rescheduling also need data for the rest of the schedule of domestic trains. The main assumption is that in the remainder of the country, all trains run on time. Our experimental results show that the disruption management module computes feasible resource schedules in a couple of minutes. Such solution times are acceptable in practice. This shows that our iterative algorithm can be applied in a practical setting for the disruption management process.

## 3.5 Innovation 5: Centrally Guided Train Operation (CGTO)

### 3.5.1 State of the art

Prior to the start of the ON-TIME project some Infrastructure Managers and Train Operators had already performed significant research activities and tests on communication of control centre decisions to train drivers. However, only one control-centre-connected system in an operationally specific area had been applied to real operations.

On the other hand the use of driving advisory systems (DAS) disregarding current operational conditions has increased in recent years. Notwithstanding the great progress in DAS, the consideration of current operational conditions has remained an open issue.

The high interest of Infrastructure Managers, Train Operators and the Railway Industry in developing driving advisory systems with control centre connection went together with the need to agree on common standards of communication to ensure interoperability and prevent different developments leading to a wide range of incompatible systems within the European railway network.

### 3.5.2 Research aims and objectives

The main objective is to prove the concept of communicating driving advice in a DAS based on control centre decisions, leading to smoother traffic flow in order to:

- decrease track occupancy in bottlenecks;
- increase energy efficiency.

To allow the interoperable use of DAS throughout Europe the second objective is to propose a standardised data format for communication of operational decisions (e.g. speed advice) between control centres and trains.

### 3.5.3 Research outputs

The key outputs related to Innovation 5 have been:

1. State of the art analysis of 22 DAS (most of them not operational), of which only 8 are control-centre-connected, leading to the identification of key functions . Based on existing experiences, three design alternatives distributing these functions differently between control centre and on-board components have been analysed and described as system architectures:
    - DAS-C: mainly central intelligence
    - DAS-I: distributed intelligence
    - DAS-O: mainly on-board intelligence
2. Specification and implementation of an XML-interface data format supporting all three system architectures mentioned above and enabling bidirectional communication between central and on-board components.
3. Enhancement of existing algorithms for optimising train speed profiles (trajectory computation) to match the time constraints set by control centre decisions.
4. Evaluation of different methods to present advice to drivers (e.g. speed vs. time targets) based on simulator studies and expert interviews, examination of useful contextual advice (e.g. icons displaying the reason for advice) and experimental implementation of the proposed HMI design.
5. Implementation of a demonstrator for all components, calculating and displaying driving advice based on control centre decisions.

### 3.5.4 Implementation tasks

In order to apply Centrally Guided Train Operation in real operations these steps should be followed after the end of the ON-TIME project:

- Infrastructure managers have to upgrade their Traffic Management Systems to support driving advice (e.g. to collect the operational data required and generate target points) and set up a communication server with the specified interface.
- Train operators and industry have to include the specified interface in existing or newly developed DAS and enhance internal algorithms to consider targets set by control centre decisions in their calculation. Optionally, enabling the communication of current train data back to the control centre could further increase the quality of control centre decisions.
- Standardisation of the specified interface should be taken forward.

## 3.6 Innovation 6 Standardised ICT architecture supporting interoperability of operational data between industry stakeholders

### *3.6.1 State of the art*

Prior to the start of the ON-TIME project some Infrastructure Managers and Train Operators had already performed significant research activities and tests on communication of control centre decisions to train drivers. However, only one control-centre-connected system in an operationally specific area had been applied to real operations.

On the other hand the use of driving advisory systems (DAS) disregarding current operational conditions has increased in recent years. Notwithstanding the great progress in DAS, the consideration of current operational conditions has remained an open issue.

The high interest of Infrastructure Managers, Train Operators and the Railway Industry in developing driving advisory systems with control centre connection went together with the need to agree on common standards of communication to ensure interoperability and prevent different developments leading to a wide range of incompatible systems within the European railway network.

### *3.6.2 Research aims and objectives*

The main objective is to prove the concept of communicating driving advice in a DAS based on control centre decisions, leading to smoother traffic flow in order to:
decrease track occupancy in bottlenecks;
increase energy efficiency.

To allow the interoperable use of DAS throughout Europe the second objective is to propose a standardised data format for communication of operational decisions (e.g. speed advice) between control centres and trains.

### *3.6.3 Research outputs*

The ON-TIME data dictionary was developed to help the project team identify the types and content of messages that may be required in the context of railway operations. It contains over 300 concepts relevant to the domain, along with definitions and mappings to the railML data model as appropriate. In order to facilitate easy multi-site access to the resource the dictionary was implemented as a wiki, this had the added benefit of enabling cross-references between entries to be embedded as hyperlinks, helping users to easily navigate between related entries.

The ON-TIME architecture consists of a collection of loosely coupled data providers and data processing modules that communicate via the passing of event messages. The architecture utilises a publish-subscribe communication pattern, allowing modules to join or leave the system at any time without interfering with the communications to other elements of the framework. The well-known open source RabbitMQ messaging platform has been used to provide this functionality. RabbitMQ runs on all major operation systems and clients can be written using a wide range of programming languages, including Java, .NET, Ruby, Python, Erlang, PHP and C/C++. This flexibility is vital to the wide-scale adoption of the ON-TIME architecture.

Real-time data is passed as events in near-real-time from the live rail network / rail network simulator to the traffic management modules; when disruptions occur re-planning takes place and new events informing stakeholders of updated traffic management plans / crew rostas etc. are raised within the system.

Non-real-time data used by ON-TIME (for example infrastructure data and the current timetable) are cached by the platform and provided via an on-demand, read-only public interface allowing it to be accessed as required by any system. By providing a single-source-of-truth for these key, largely static data resources, the ON-TIME architecture reduces the risk of inconsistencies arising between stakeholder systems.

ON-TIME module integration specification allows generic processing modules to be added to the ON-TIME framework through the implementation of a standard set of interfaces. In particular the specification covers subscription to event queues, the production/consumption of event messages, and access to static data resources.

For the representation of messages to be exchanged within the system, the ON-TIME team needed to make use of an open, extensible set of data models that aligned well with the terminology set identified in the data dictionary. A mapping was created between the dictionary terms and the railML format that is rapidly gaining traction within the industry, and the degree of overlap between the two established. The coverage of the ON-TIME data dictionary by the railML model was good, with only a few gaps found. These included information on interlockings, ETCS-type train control concepts, disruption information, crew duty assignments and resource conflicts. Where gaps existed project-specific data model extensions were created, and this work has included contributions back to the railML community in the form of the candidate railML interlocking model.

## 4 NEXT STEPS

The ON-TIME project has connected timetable planning and decision support for the operational process:
Infrastructure managers and system suppliers need to develop methods to export and import data with de facto standards as railML.
There is a need for continuing algorithm development and integration to provide decision support tools for Railway Operations.
There is a need for further development of interactive solutions, i.e. systems where human traffic controllers can interact with the decision support tools developed in ON-TIME.
There is a need of further research in timetable planning, methods for handling minor perturbations, major disruptions and train driver advisory systems. Further research will be done in EU projects i.e. Capacity4Rail 201310 – 201709.

(1) ON-TIME Partners: D'Appolonia SPA (Project coordinator), Network Rail Infrastructure Ltd., Rete Ferroviaria Italiana, Deutsche Bahn AG, Société Nationale Des Chemins de Fer Francais, Trafikverket, Ansaldo STS, NTT Data Italia S.P.A, University of Birmingham, École Polytechnique Federale de Lausanne, Erasmus Universiteit Rotterdam, Institut Français des Sciences et Technologies des Transports, de l'Amenagement et des Reseaux, The University of Nottingham, Technische Universitaet Dresden, Technische Universiteit Delft, Universita di Bologna, Uppsala Universitet, Graffica Ltd., Transrail Sweden Aktiebolag.

# AUTHOR INDEX